STATISTICS/STATIS

Structural Statistics for Industry and Services

Volume 1

CORE DATA

Statistiques de structures de l'industrie et des services

Volume 1

DONNÉES DE BASE

2000

ORGANISATION FOR ECONOMIC CO-OPERATION AND DEVELOPMENT

Pursuant to Article 1 of the Convention signed in Paris on 14th December 1960, and which came into force on 30th September 1961, the Organisation for Economic Co-operation and Development (OECD) shall promote policies designed:

- to achieve the highest sustainable economic growth and employment and a rising standard of living in Member countries, while maintaining financial stability, and thus to contribute to the development of the world economy;
- to contribute to sound economic expansion in Member as well as non-member countries in the process of economic development; and
- to contribute to the expansion of world trade on a multilateral, non-discriminatory basis in accordance with international obligations.

The original Member countries of the OECD are Austria, Belgium, Canada, Denmark, France, Germany, Greece, Iceland, Ireland, Italy, Luxembourg, the Netherlands, Norway, Portugal, Spain, Sweden, Switzerland, Turkey, the United Kingdom and the United States. The following countries became Members subsequently through accession at the dates indicated hereafter: Japan (28th April 1964), Finland (28th January 1969), Australia (7th June 1971), New Zealand (29th May 1973), Mexico (18th May 1994), the Czech Republic (21st December 1995), Hungary (7th May 1996), Poland (22nd November 1996), Korea (12th December 1996) and the Slovak Republic (14th December 2000). The Commission of the European Communities takes part in the work of the OECD (Article 13 of the OECD Convention).

ORGANISATION DE COOPÉRATION ET DE DÉVELOPPEMENT ÉCONOMIQUES

En vertu de l'article 1er de la Convention signée le 14 décembre 1960, à Paris, et entrée en vigueur le 30 septembre 1961, l'Organisation de Coopération et de Développement Économiques (OCDE) a pour objectif de promouvoir des politiques visant :

- à réaliser la plus forte expansion de l'économie et de l'emploi et une progression du niveau de vie dans les pays Membres, tout en maintenant la stabilité financière, et à contribuer ainsi au développement de l'économie mondiale ;
- à contribuer à une saine expansion économique dans les pays Membres, ainsi que les pays non membres, en voie de développement économique ;
- à contribuer à l'expansion du commerce mondial sur une base multilatérale et non discriminatoire conformément aux obligations internationales.

Les pays Membres originaires de l'OCDE sont : l'Allemagne, l'Autriche, la Belgique, le Canada, le Danemark, l'Espagne, les États-Unis, la France, la Grèce, l'Irlande, l'Islande, l'Italie, le Luxembourg, la Norvège, les Pays-Bas, le Portugal, le Royaume-Uni, la Suède, la Suisse et la Turquie. Les pays suivants sont ultérieurement devenus Membres par adhésion aux dates indiquées ci-après : le Japon (28 avril 1964), la Finlande (28 janvier 1969), l'Australie (7 juin 1971), la Nouvelle-Zélande (29 mai 1973), le Mexique (18 mai 1994), la République tchèque (21 décembre 1995), la Hongrie (7 mai 1996), la Pologne (22 novembre 1996), la Corée (12 décembre 1996) et la République slovaque (14 décembre 2000). La Commission des Communautés européennes participe aux travaux de l'OCDE (article 13 de la Convention de l'OCDE).

FOREWORD

This *Structural Statistics for Industry and Services* (SSIS) publication is the continuation of the *Industrial Structure Statistics* (ISS) and constitutes the seventeenth edition of the series. As in previous editions, it comes in two volumes. Volume 1 - Core Data - provides detailed annual statistics on a number of structural variables for industrial and service sectors. However, following the recommendations of the Statistical Working Party of the Industry Committee (SWIC), disaggregated national accounts have been suppressed from the publication. Data presented in this publication facilitates an assessment of the evolution of the economic structure of OECD Member countries. The publication was prepared by the Statistics Directorate in the OECD Secretariat under the auspices of the SWIC. The Secretariat wishes to thank the representatives of all the national authorities who have assisted in the preparation of this publication.

All series published in Volume 1, as well as data for earlier years, are held in the SSIS database, (previously called *Information System on Industrial Structures* - ISIS), and are available on CD-ROM and on line at *www.sourceoecd.org*. Moreover, the electronic version contains more detailed statistics relating to variables and to industrial and service sectors which are not presented in the paper publication. These statistics are published under the responsibility of the Secretary-General of the OECD.

Volume 2 presents annual energy consumption data in manufacturing sectors. It was prepared by the Energy Statistics Division of the International Energy Agency (IEA). The energy data have also been collected under the auspices of the Statistical Working Party of the Industry Committee. These data are also available on CD-ROM and on line at *www.sourceoecd.org*.

AVANT-PROPOS

Cette publication *Statistiques des Structures de l'Industrie et des Services* (SSIS) est la continuation des *Statistiques des structures industrielles* (SSI) et constitue la dix-septième parution de la série. Comme lors des éditions précédentes, elle comprend deux volumes. Le Volume 1 - Données de base - présente des données annuelles détaillées pour les secteurs de l'industrie et des services. Cependant, selon les recommandations du Groupe de travail statistique du Comité de l'industrie (SWIC), les comptes nationaux désagrégés par industrie ont été supprimés de la publication papier. Les données présentées permettent d'évaluer l'évolution de la structure économique des pays Membres de l'OCDE. Cette publication a été préparée par la Direction Statistique du Secrétariat de l'OCDE sous l'égide du SWIC. Le Secrétariat tient à remercier les représentants de toutes les administrations nationales qui ont apporté leur concours à la préparation de cette publication.

L'ensemble des statistiques publiées dans le Volume 1 ainsi que celles portant sur des années antérieures sont contenues dans la base de données SSIS (précédemment appelée *Système d'Information sur les Structures Industrielles* - SISI) et sont disponibles sur CD-ROM et en ligne via *www.sourceoecd.org*. En outre, la version électronique contient davantage de statistiques détaillées relatives à des variables et à des secteurs de l'industrie et des services qui ne sont pas présentés dans la publication papier. Ces statistiques sont publiées sous la responsabilité du Secrétaire général de l'OCDE.

Le Volume 2 présente des données annuelles de consommation d'énergie dans l'industrie manufacturière. Il a été préparé par la division des statistiques de l'énergie de l'Agence Internationale de l'Energie (AIE). Les données sur l'énergie ont elles aussi été collectées sous l'égide du Groupe de travail statistique du Comité de l'industrie. Ces données sont également disponibles sur CD-ROM et en ligne via *www.sourceoecd.org*.

Table of contents - Table des matières

INTRODUCTION

OECD statistics on industry and services

The OECD Secretariat collects a wide range of statistics on industry and services, mostly under the auspices of the Statistical Working Party of the Industry Committee. The quarterly publication *Indicators of Industrial Activity (IIA)*, renamed *Indicators of Activities for Industry and Services (IAIS)* from 2001 onwards, provides an overall view of the most recent short-term economic development, and presents monthly indices of output, turnover, orders, producer prices and employment as well as qualitative information from business tendency surveys. More detailed annual data derived from business surveys and foreign trade statistics, covering a number of structural variables are published in *Structural Statistics for Industry and Services (SSIS)*. The *OECD Database for Structural Analysis (STAN)*, based on *SSIS*, was developed to ensure international comparability of survey-based national industrial statistics as well as their compatibility with national accounts. The publication contains a limited list of variables for most OECD countries. Two more specific publications provide annual data relating to service activities: *Services: Statistics on Value Added and Employment* and *Services: Statistics on International Transactions*.

Trends and structural shifts in industry, presented in numerous detailed statistical tables and graphs, are examined in two other OECD publications, issued alternatively every two years, *Science, Technology and Industry Outlook* and *Industry and Technology: Scoreboard of Indicators*, for use by policy makers and analysts.

Output from the computerised statistical databases is available both as printed publications, and in computer-readable form such as microcomputer diskettes. Further information on these databases is available on the World Wide Web at http://www.oecd.org/std or at http://www.oecd/org/dsti/eas/.

Process of Data Collection

The United Nations Statistical Commission, at its twenty-seventh session, recommended that OECD, in collaboration with the United Nations Industrial Development Organisation (UNIDO), take responsibility for the collection and dissemination of world-wide industrial statistics. The collection of industrial statistics through a joint OECD/UNIDO questionnaire is now co-ordinated with UNIDO as part of an on-going joint effort to collect world-wide data for this sector.

The questionnaire used to collect statistics for this year's publication again includes a segment on energy consumption by industry as part of another joint project with the International Energy Agency. The aim of such co-ordinated collection is to minimise the reporting burden for national agencies and to harmonise internationally comparable industrial structure data.

Furthermore, there has been considerable discussion by analysts on the need for comparable data for emerging areas of economic activity, in particular, for the services sector. This discussion led the Statistical Working Party of the Industry Committee to endorse the collection of statistics for services, according to the International Standard Industry Classification (ISIC) Revision 3. A number of countries has been able to provide statistics on services which are presented in Part C of this publication.

Methodological Notes

The previous edition, named *Industrial Structure Statistics* included data classified according to both versions of the International Standard Industrial Classification, ISIC Revision 2 (adopted in 1968 by the United Nations Organisation, UNO - *Statistical Papers, Series M, No. 4, Rev. 2, Add. 1*, New York, 1971) and ISIC Revision 3 (adopted in February 1989 by the UNO - *Statistical Papers, Series M, No. 4, Rev. 3*, New York, 1990). Furthermore, it included business survey data as well as disaggregated national accounts by sector.

In the current paper edition, named *Structural Statistics for Industry and Services*, data are still classified according to both versions of ISIC. Most tables with ISIC Revision 2 data have however been suppressed where data have not been updated for more than two years. By way of compensation, a number of new ISIC Revision 3 tables have been added and are presented for the first time in the paper edition. In particular, information related to services is published for a greater number of countries and variables. The electronic edition of this publication contains all available data for Revisions 2 and 3 of the ISIC classification.

Structural Statistics for Industry and Services are divided in four parts: the first contains statistics according to ISIC Revision 2; the second contains statistics according to ISIC Revision 3 for industrial sectors; the third contains statistics according to ISIC Revision 3 for service sectors; the fourth provides available methodological information. However, contrary to earlier editions, statistics in the three first parts relate to data derived directly from business survey and foreign trade data only ; data from national accounts disaggregated by activity are not any more presented in the paper version. Nevertheless, data previously published remain in the electronic edition.

Business survey figures comprise the results of sample surveys conducted by national statistical organisations in all sectors of the economy. Since the sampling procedures vary, the percentage of the economy covered will vary between countries. Some, but not all, countries "gross up" the data to account for the unsampled portion.

Foreign trade statistics are normally derived from customs figures and are product-based. Survey results, whether the base unit is the establishment or the enterprise, allocate an entire unit's transactions to one industry according to its principal activity. An adjustment to reallocate the unit's secondary activities to a more appropriate industry may or may not have been made, depending on the country.

All qualitative information provided to the Secretariat by Member countries are published in Section I of Part D. This Section includes definitions of variables as provided by countries in reply to the OECD methodological survey. These definitions may present various conceptual differences between countries. For business survey concepts and definitions, readers may refer to the *International Recommendations for Industrial Statistics* [Statistical Papers Series M, No. 4/Rev. 2 (United Nations publication, Sales No. E.83.XVII.8) and to Eurostat's *Legal Texts relating to the Structural Business Statistics*, August 1999.

INTRODUCTION

Statistiques de l'OCDE sur l'industrie et les services

Le Secrétariat de l'OCDE collecte une large gamme de statistiques sur l'industrie et les services, le plus souvent sous l'égide du Groupe de Travail Statistique du Comité de l'industrie. La publication trimestrielle *Indicateurs des activités industrielles (IAI)*, renommée *Indicateurs des activités de l'industrie et des services (IAIS)* à partir de 2001, donne un aperçu général de l'évolution économique à court terme la plus récente et présente des indices mensuels de production, chiffre d'affaires, commandes, prix à la production et emploi ainsi que des données tirées d'enquêtes de conjoncture. Des données annuelles plus détaillées résultant d'enquêtes industrielles et de statistiques du commerce extérieur et couvrant un certain nombre de variables structurelles sont présentées dans les *Statistiques des structures de l'industrie et des services (SSIS)*. La *Base de données STAN de l'OCDE pour l'Analyse de l'Industrie* fondée sur *SSIS* et développée dans le but d'assurer la comparabilité internationale des statistiques d'enquêtes d'entreprises effectuées dans les pays Membres ainsi que leur compatibilité avec les comptes nationaux comprend un nombre limité de variables pour la plupart des pays de l'OCDE. Deux autres publications plus spécifiques fournissent des données annuelles sur les services. Il s'agit de *Services : Statistiques sur la valeur ajoutée et l'emploi* et de *Services : Statistiques sur les échanges internationaux.*

Les tendances et les mutations structurelles de l'industrie, présentées dans de nombreux tableaux et graphiques détaillés, sont examinées dans deux autres publications de l'OCDE, diffusées alternativement tous les deux ans, *Les perspectives de la science, de la technologie et de l'industrie* et *Industrie et technologie : tableau de bord d'indicateurs,* à l'usage des décideurs politiques et analystes.

Les résultats provenant de ces banques de données sont non seulement disponibles sous forme de publications imprimées, mais aussi sur supports lisibles par ordinateur tels que disquettes de micro-ordinateur. De plus amples informations sur ces bases de données sont disponibles sur le site Internet aux adresses suivantes http://www.oecd.org/std ou http://www.oecd/org/dsti/eas/.

Processus pour la collecte des données

La Commission Statistique des Nations Unies, lors de sa vingt-septième réunion, a recommandé que l'OCDE, en collaboration avec l'Organisation des Nations Unies pour le Développement Industriel (ONUDI), prenne la responsabilité de la collecte et de la diffusion des statistiques industrielles au niveau mondial. La coordination de la collecte des statistiques des structures industrielles, au moyen d'un questionnaire commun à l'OCDE et à l'ONUDI, relève d'un effort conjoint permanent pour rassembler des données au niveau mondial dans ce domaine.

Le questionnaire utilisé pour la collecte de statisticque de la présente publication continue également d'inclure une demande d'informations sur la consommation d'énergie par industrie dans le cadre d'un autre projet commun avec l'Agence Internationale de l'Energie. L'objectif d'une telle collecte coordonnée est de minimiser les charges de transmission de données incombant aux instituts nationaux et d'harmoniser les données sur les structures industrielles de telle sorte qu'elles soient comparables au plan international.

En outre, en 1998, d'importantes discussions ont eu lieu sur la nécessité de disposer de données comparables relatives à des secteurs émergents de l'activité économique, en particulier pour le secteur des services. Elles ont conduit le Groupe de Travail Statistique du Comité de l'Industrie à approuver la collecte de statistiques sur les services, selon la Classification Internationale Type par Industrie (CITI) révision 3. Un certain nombre de pays a été en mesure de transmettre des statistiques sur les services qui sont présentés Partie C de cette publication.

Notes méthodologiques

L'édition précédente des *Statistiques des structures industrielles* contenait des données selon les deux versions de la Classification Internationale Type par Industrie, la CITI Révision 2 (adoptée en 1968 par l'Organisation des Nations Unies, ONU - *Etudes Statistiques, Série M, n° 4, Rév. 2, Add. 1*, New York, 1975) et la CITI Révision 3 (adoptée en février 1989 par l'ONU - *Etudes Statistiques, Série M, n° 4, Rév. 3,* New York, 1990). En outre, elle comprenait des données d'enquêtes d'entreprises aussi bien que des comptes nationaux désagrégés par secteur.

Dans la présente publication papier renommée *Statistiques des structures de l'industrie et des services*, les données sont encore publiées selon les deux versions de la CITI. La plupart des tableaux de données en CITI Révision 2 ont été supprimés parce qu'ils n'avaient pas été mis à jour depuis plus de deux ans. En revanche, de nouveaux tableaux présentant des données en CITI Révision 3 ont été ajoutés et sont publiés pour la première fois dans cette édition papier. En particulier, des informations relatives aux services sont publiées pour un plus grand nombre de pays et de variables. L'édition électronique de cette publication contient toutes les données disponibles selon les Révisions 2 et 3 de la classification CITI.

Les *Statistiques des structures de l' industrie et des services* sont divisées en quatre parties. Dans la première partie sont présentées les statistiques en CITI Révision 2 ; dans la deuxième partie les statistiques en CITI Révision 3 sur les activités industrielles ; dans la troisième partie les statistiques en CITI Révision 3 sur les activités de services ; la quatrième partie contient les informations méthodologiques disponibles. Cependant, contrairement aux éditions précédentes, les statistiques publiées dans chacune des trois premières parties correspondent seulement aux données provenant directement des enquêtes statistiques auprès des entreprises et aux données sur le commerce extérieur ; les données des comptes nationaux désagrégées par activité ne sont plus présentées dans la version papier. Néanmoins, les données préalablement publiées, mais non mises à jour, restent dans la version électronique.

Les résultats d'enquêtes auprès des entreprises proviennent d'enquêtes par échantillon menées par les instituts nationaux de statistiques dans tous les secteurs de l'économie. Etant donné la variété des méthodes d'échantillonnage, le pourcentage des industries couvertes varie d'un pays à l'autre. Certains pays «gonflent» les données pour prendre en compte la portion non couverte par l'enquête.

Les statistiques du commerce extérieur sont issues des sources douanières et sont fondées sur une classification par produit. Pour les résultats d'enquêtes, quelle que soit l'unité de base -- établissement ou entreprise --, l'ensemble des transactions d'une unité est alloué à une industrie suivant l'activité principale. Quelques pays procèdent à un ajustement pour allouer les activités secondaires d'une unité à une classe industrielle plus appropriée.

Toutes les informations qualitatives fournies au Secrétariat par les pays sont détaillées dans la section I de la partie D. Cette section inclut les définitions des variables telles qu'elles ont été fournies par les pays en réponse au questionnaire méthodologique de l'OCDE. Ces définitions peuvent présenter un certain nombre de différences de concepts entre pays. Pour les concepts et définitions des enquêtes d'entreprises, le lecteur peut se référer aux *Recommandations internationales concernant les statistiques industrielles* [Etudes statistiques série M, No. 48 (publication des Nations Unies, numéro de vente F.68.XVII.10)] et aux *Textes législatifs relatifs aux Statistiques structurelles sur les entreprises* d'Eurostat, d'août 1999.

CONVENTIONAL SIGNS AND ABBREVIATIONS
SIGNES ET ABREVIATIONS CONVENTIONNELS

Units	Unités
Units	Unités
Thousands	Milliers
Millions	Millions
Billions	Milliards

Other signs		Autres signes	
. .	Not available	. .	Non disponible
0	Nil or negligible	0	Nul ou négligeable
.	Decimal point	.	Point décimal

Currency		Monnaie
Australia	A$	Australie
Austria	Sch	Autriche
Belgium	BF	Belgique
Canada	C$	Canada
Czech Republic	CK	République Tchèque
Denmark	DKr	Danemark
Finland	Mk	Finlande
France	FF	France
Germany	DM	Allemagne
Greece	Dr	Grèce
Hungary	Ft	Hongrie
Iceland	IKr	Islande
Ireland	Ir£	Irlande
Italy	L	Italie
Japan	Y	Japon
Korea	Won	Corée
Luxembourg	LF	Luxembourg
Mexico	NM$	Mexique
Netherlands	Gld	Pays-Bas
New Zealand	NZ$	Nouvelle-Zélande
Norway	NKr	Norvège
Poland	PLN	Pologne
Portugal	Esc	Portugal
Spain	Ptas	Espagne
Sweden	SKr	Suède
Switzerland	SF	Suisse
Turkey	TL	Turquie
United Kingdom	£	Royaume-Uni
United States	$	États-Unis

VARIABLES

Data availability is limited to what member countries can provide. The paper version of this publication is restricted by size, please note that additional countries, variables and sectors are available on the diskette product.

La disponibilité des données reste limitée par la possibilité qu'ont les pays de transmettre des informations. La version papier de la publication étant limitée par le nombre de pages, veuillez noter que davantage de pays, de variables et de secteurs sont disponibles dans la version électronique.

English		**Française**
Production	--	Production
Production at factor costs	--	Production au coût des facteurs
Value added	--	Valeur ajoutée
Value added at factor costs	--	Valeur ajoutée au coût des facteurs
Investment	--	Investissement
Investment in machinery and equipment	--	Dépenses en machines et équipement
Capital stocks	--	Stocks de capital
Employment, total	--	Emploi, total
Employment, employees	--	Emploi, salariés
Employment, operatives	--	Emploi, ouvriers
Employment, females	--	Emploi, femmes
Establishments	--	Etablissements
Enterprises	--	Entreprises
Compensation of labour	--	Salaires et charges sociales
Compensation of labour, employees	--	Salaires et charges sociales, salariés
Compensation of labour, operatives	--	Salaires et charges sociales, ouvriers
Employers' social costs	--	Charges sociales des employeurs
Total wages and salaries	--	Salaires et traitements
Wages and salaries, employees	--	Salaires et traitements, salariés
Wages and salaries, operatives	--	Salaires et traitements, ouvriers
Hours worked by operatives	--	Heures ouvrées, salariés
Hours worked by employees	--	Heures ouvrées, ouvriers
Exports	--	Exportations
Imports	--	Importations

PART A

PARTIE A

ISIC REVISION 2

CITI REVISION 2

Structural business statistics and foreign trade statistics

Statistiques structurelles sur les entreprises et statistiques du commerce extérieur

List of countries
Liste des pays

Note :

In most instances, as countries convert to the ISIC Revision 3 industry classification, they cease to provide ISIC Revision 2 data. Only where countries have either updated ISIC Revision 2 data in the last 2 years, or alternatively, have not yet fully converted to ISIC Revision 3, will data be supplied in the paper ISS publication. Historical data series are available (and are more detailed) on the diskette version of this publication. Due to size restrictions a limited number of variables are presented in this paper version. Additional variables are available on the electronic version of the publication.

Dans la plupart des cas, dès que les pays fournissent leurs données en CITI révision 3, ils cessent de transmettre des données en CITI révision 2. C'est seulement dans le cas où les pays ont ou bien mis à jour leurs données en CITI révision 2 au cours des deux dernières années, ou bien ne sont pas encore passés en CITI révision 3, que des données en CITI révision 2 continuent à être publiées sur la version papier de la publication SSI. Les séries de données historiques sont disponibles (et de manière plus détaillée) sur la version disquette de cette publication. Pour des raisons de limite en matière de taille de publication, un nombre réduit de variables est présenté sur cette version papier. D'autres variables sont disponibles dans la version électronique de cette publication.

ICELAND

Table IC.1 PRODUCTION - PRODUCTION

Millions of IKr (Current Prices)

ISIC revision 2		1991	1992	1993	1994	1995	1996	1997	1998
31	**Food, beverages and tobacco**	99 801	97 911	97 219	107 173	109 325	117 900	116 676	..
311.2	Food	95 519	94 054	93 500	103 044	105 237	113 533	112 001	..
313	Beverages	4 282	3 857	3 719	4 129	4 088	4 367	4 676	..
314	Tobacco	..	..	..	..	..	..	..	..
32	**Textiles, apparel and leather**	6 331	6 290	5 892	6 549	7 465	8 498	8 516	..
321	Textiles	3 346	3 156	3 013	3 145	3 852	4 524	4 603	..
3213	Knitting mills	416	405	478	459	501	612	645	..
322	Wearing apparel, except footwear	1 837	1 963	1 940	2 087	2 114	2 234	2 179	..
323	Leather and fur products	1 042	1 129	926	1 298	1 477	1 716	1 721	..
324	Footwear, except rubber or plastics	105	43	12	19	22	24	14	..
33	**Wood products and furniture**	6 640	6 387	5 581	5 710	5 764	6 626	7 373	..
331	Wood products, except furniture	224	218	227	250	286	341	391	..
332	Furniture, fixtures, except metal	6 416	6 169	5 354	5 460	5 478	6 285	6 982	..
34	**Paper, paper products and printing**	12 119	12 088	12 108	12 512	13 263	14 634	15 308	..
341	Paper and paper products	1 480	1 416	1 494	1 801	1 906	2 252	2 217	..
3411	Pulp, paper and paperboard articles	1 480	1 416	1 494	1 801	1 906	2 252	2 217	..
342	Printing and publishing	10 639	10 672	10 614	10 711	11 357	12 381	13 091	..
35	**Chemical products**	9 867	9 328	9 127	9 699	10 293	11 200	12 711	..
351	Industrial chemicals	2 981	2 565	2 188	2 318	2 364	2 478	2 495	..
3511	Basic industrial chemicals except fertilizers	2 981	2 565	2 188	2 318	2 364	2 478	2 495	..
3512	Fertilizers and pesticides	..	..	..	..	..	..	..	..
3513	Synthetic resins and plastic materials	..	..	..	..	..	..	..	..
352	Other chemicals	2 864	2 713	2 854	3 013	3 248	3 623	4 600	..
3521	Paints, varnishes and lacquers	1 310	1 323	1 394	1 460	1 525	1 601	1 582	..
3522	Drugs and medicines	..	..	..	..	..	..	..	..
3523	Soap, cleaning preparations, perfumes, cosmetics	1 555	1 391	1 459	1 553	1 723	2 022	3 019	..
3529	Other chemical products	..	..	..	..	..	..	..	..
353	Petroleum refineries	..	..	..	..	..	..	..	..
354	Miscellaneous petroleum and coal products	60	74	56	32	51	64	417	..
355	Rubber products	..	..	..	..	..	..	..	..
356	Plastics products	3 961	3 976	4 030	4 336	4 631	5035	5198	..
36	**Non-metallic mineral products**	6 218	5 785	5 540	5 920	5 984	6373	7417	..
361	Pottery, china, earthenware	109	107	88	87	95	83	89	..
362	Glass and glass products	639	605	549	545	550	555	602	..
369	Non-metallic mineral products nec.	5 470	5 073	4 903	5 287	5 338	5734	6726	..
37	**Basic metal industries**	10 134	9 921	11 154	13 796	15 106	15961	18894	..
371	Iron and steel	1 854	1 923	2 733	2 959	3 431	4134	3942	..
372	Non-ferrous metals	8 279	7 998	8 421	10 837	11 674	11827	14952	..
38	**Fabricated metal prod.; machinery & equip.**	13 154	12 572	12 267	13 111	14 973	20364	22838	..
381	Fabricated metal products	10 896	10 045	9 717	11 086	12 698	17501	19709	..
3813	Structural metal products	..	..	..	..	..	..	..	..
382	Non-electrical machinery	..	..	..	..	..	..	..	..
3821	Engines and turbines	..	..	..	..	..	..	..	..
3822	Agricultural machinery and equipment	..	..	..	..	..	..	..	..
3823	Metal and wood working machinery	..	..	..	..	..	..	..	..
3824	Special industrial machinery	..	..	..	..	..	..	..	..
3825	Office, computing and accounting machinery	..	..	..	..	..	..	..	..
3829	Other non-electrical machinery and equipment	..	..	..	..	..	..	..	..
383	Electrical machinery	..	..	..	..	..	..	..	..
3831	Electrical industrial machinery	..	..	..	..	..	..	..	..
3832	Radio, TV and communication equipment	..	..	..	..	..	..	..	..
3833	Electrical appliances and housewares	..	..	..	..	..	..	..	..
3839	Other electrical apparatus and supplies	..	..	..	..	..	..	..	..
384	Transport equipment	2 258	2 528	2 550	2 025	2 275	2 863	3 129	..
3841	Shipbuilding	2 258	2 528	2 550	2 025	2 275	2 863	3 129	..
3842	Railroad equipment	..	..	..	..	..	..	..	..
3843	Motor vehicles	..	..	..	..	..	..	..	..
3844	Motorcycles and bicycles	..	..	..	..	..	..	..	..
3845	Aircraft	..	..	..	..	..	..	..	..
3849	Other transport equipment	..	..	..	..	..	..	..	..
385	Professional and scientific equipment	..	..	..	..	..	..	..	..
3851	Professional and scientific equipment, nec	..	..	..	..	..	..	..	..
3852	Photographic and optical goods	..	..	..	..	..	..	..	..
3853	Watches and clocks	..	..	..	..	..	..	..	..
39	**Other manufacturing industries**	5 277	5 366	5 575	6 076	7 437	8 848	9 552	..
3901	Jewellery and related articles	260	245	229	239	255	275	288	..
3	**Total manufacturing**	169 541	165 648	164 461	180 545	189 608	210 403	219 285	..
1	**Agriculture, hunting, forestry and fishing**	67838	65341	68071	70412	72656	77914	76611	..
2	**Mining and quarrying**	..	..	..	..	..	..	..	..
21	Coal mining	..	..	..	..	..	..	..	..
22	Crude petroleum and natural gas	..	..	..	..	..	..	..	..
23.29	Other mining	..	..	..	..	..	..	..	..
4	**Electricity, gas and water**	22497	24185	25308	26700	27986	29012	31325	..
5	**Construction**	85372	80002	82329	82530	80510	88458	109841	
6.9	**Total services**	398 830	408 304	414 727	442 434	461 938	514 389	555 467	..
6	**Wholesale & retail trade; restaurants & hotels**	70512	72133	68493	72973	77006	85577	90496	..
7	**Transport, storage and communication**	56292	56100	60863	67048	70550	84902	92488	..
8	**Finance, insurance, real estate, business serv.**	88591	91368	91937	96023	99287	110942	127211	..
9	**Community, social and personal services**	183435	188703	193434	206390	215095	232968	245272	..

Note: ISIC 3511 includes 3512.

Table IC.2 VALEUR AJOUTÉE - VALUE ADDED

Millions de IKr (Prix courants)

1991	1992	1993	1994	1995	1996	1997	1998		CITI révision 2
28 079	25 872	28 159	31 744	30 227	30 306	34 423	..	**Alimentation, boissons et tabacs**	**31**
26 696	24 511	27 013	30 560	28 881	28 896	32 760	..	Alimentation	311.2
1 383	1 361	1 145	1 184	1 346	1 409	1 663	..	Boissons	313
..	..	..	..	..	..	..	..	Tabacs	314
2 349	2 483	2 289	2 630	2 953	3 399	3 208	..	**Textiles, habillement et cuir**	**32**
1 012	1 138	1 028	1 126	1 334	1 593	1 620	..	Textiles	321
157	193	214	163	219	247	287	..	Bonneterie	3213
789	846	850	933	943	1 029	990	..	Habillement, à l'exclusion des chaussures	322
527	497	407	569	671	771	595	..	Cuir et articles en cuir	323
21	2	3	2	5	5	3	..	Chaussures, sauf en caoutchouc et plastique	324
2 820	2 964	2 434	2 279	2 373	2 514	2 640	..	**Bois et meubles**	**33**
72	91	103	94	104	136	120	..	Ouvrages en bois	331
2 748	2 872	2 331	2 185	2 269	2 379	2 520	..	Meubles et accessoires	332
6 363	5 977	5 735	6 443	6 644	7 653	7 656	..	**Papier, impression et édition**	**34**
601	539	613	705	698	739	766	..	Papier et articles en papier	341
601	539	613	705	698	739	766	..	Pâte à papier, papier et carton	3411
5 762	5 438	5 122	5 738	5 945	6 914	6 889	..	Imprimerie et édition	342
3 741	4 077	3 950	3 766	3 853	4 123	4 321	..	**Produits chimiques**	**35**
1 067	1 080	854	742	667	770	739	..	Industrie chimique	351
1 067	1 080	854	742	667	770	739	..	Chimie de base, à l'exception des engrais	3511
..	..	..	..	..	..	..	..	Engrais et pesticides	3512
..	..	..	..	..	..	..	..	Résines synthétiques et matières plastiques	3513
1 019	1 226	1 311	1 203	1 352	1 333	1 566	..	Autres produits chimiques	352
524	558	588	592	534	614	531	..	Peintures, vernis et laques	3521
..	..	..	..	..	..	..	..	Produits pharmaceutiques	3522
495	668	723	611	818	719	1 034	..	Produits de nettoyage	3523
..	..	..	..	..	..	..	..	Autres produits chimiques	3529
..	..	..	..	..	..	..	..	Raffineries de pétrole	353
21	35	25	14	24	30	194	..	Derives du pétrole et du charbon	354
..	..	..	..	..	..	..	..	Industrie du caoutchouc	355
1 634	1 737	1 761	1 806	1 810	1 991	1 822	..	Ouvrages en matière plastique	356
2 637	2 684	2 471	2 784	2 656	2 715	3 119	..	**Produits minéraux non métalliques**	**36**
26	1	56	41	49	43	48	..	Grès, porcelaines et faïences	361
252	283	244	241	242	242	275	..	Industrie du verre	362
2 359	2 402	2 171	2 502	2 365	2 430	3 796	..	Produits minéraux non métalliques, nca	369
1 356	1 854	2 208	4 277	4 265	4 492	5 793	..	**Industries métallurgiques de base**	**37**
340	380	957	1 032	1 380	1 560	1 248	..	Sidérurgie	371
1 016	1 474	1 252	3 245	2 884	2 932	4 545	..	Métaux non ferreux	372
6 563	5 695	5 456	5 317	6 093	8 107	10 223	..	**Ouvrages en métaux, machines et matériel**	**38**
5 245	4 215	4 105	4 223	4 864	6 540	8 470	..	Ouvrages en métaux	381
..	..	..	..	..	..	..	..	Eléments de construction en métal	3813
..	..	..	..	..	..	..	..	Machines non électriques	382
..	..	..	..	..	..	..	..	Moteurs et turbines	3821
..	..	..	..	..	..	..	..	Machines agricoles	3822
..	..	..	..	..	..	..	..	Machines pour le travail du bois et du métal	3823
..	..	..	..	..	..	..	..	Machines et matériel spéciaux	3824
..	..	..	..	..	..	..	..	Machines de bureau et à calculer	3825
..	..	..	..	..	..	..	..	Machines et matériel, nca	3829
..	..	..	..	..	..	..	..	Machines électriques	383
..	..	..	..	..	..	..	..	Machines électriques industrielles	3831
..	..	..	..	..	..	..	..	Appareils de radio, télévision et télécom.	3832
..	..	..	..	..	..	..	..	Appareils électroménagers	3833
..	..	..	..	..	..	..	..	Appareils électriques, nca	3839
1 318	1 480	1 351	1 094	1 229	1 568	1 754	..	Matériel de transport	384
1 318	1 480	1 351	1 094	1 229	1 568	1 754	..	Construction navale	3841
..	..	..	..	..	..	..	..	Matériel ferroviaire	3842
..	..	..	..	..	..	..	..	Véhicules automobiles	3843
..	..	..	..	..	..	..	..	Motocycles et cycles	3844
..	..	..	..	..	..	..	..	Construction aéronautique	3845
..	..	..	..	..	..	..	..	Matériel de transport, nca	3849
..	..	..	..	..	..	..	..	Matériel professionnel et scientifique	385
..	..	..	..	..	..	..	..	Equipement professionnel	3851
..	..	..	..	..	..	..	..	Matériel photo et optique	3852
..	..	..	..	..	..	..	..	Montres et horloges	3853
1 726	1 815	2 061	2 023	2 465	2 982	3 249	..	**Autres industries manufacturières**	**39**
128	125	111	110	115	120	122	..	Joaillerie	3901
55 633	53 420	54 763	61 263	61 528	66 291	74 632	..	**Total industries manufacturières**	**3**
41 597	40 475	41 502	41 884	44 549	46 629	42 084	..	**Agriculture, chasse, sylviculture et pêche**	**1**
..	..	..	..	..	..	..	..	**Industries extractives**	**2**
..	..	..	..	..	..	..	..	Extraction du charbon	21
..	..	..	..	..	..	..	..	Production du pétrole brut et gaz naturel	22
..	..	..	..	..	..	..	..	Extraction d'autres minéraux	23.29
12 714	13 530	14 345	14 654	16 025	16 392	18 263	..	**Électricité, gaz et eau**	**4**
28 409	28 802	28 178	27 146	27 977	28 153	33 700	..	**Construction**	**5**
249 569	254 236	258 128	274 623	284 852	312 383	340 135	..	**Total services**	**6.9**
45 369	45 499	44 687	48 109	49 821	54 727	55 635	..	**Commerce de gros & détail ; rest. & hotels**	**6**
25 326	25 585	26 481	32 931	33 556	34 912	34 992	..	**Transport, entrepôts et communications**	**7**
62 286	64 937	65 675	67 122	67 265	73 149	89 366	..	**Banques, assur., aff. imm., serv. aux entrepr.**	**8**
116 588	118 215	121 285	126 461	134 210	149 595	160 142	..	**Services a la collect, sociaux et personnels**	**9**

Note: ISIC 3511 includes 3512.

ICELAND

Table IC.3 EMPLOYMENT - EMPLOI

Thousands

ISIC revision 2		1991	1992	1993	1994	1995	1996	1997	1998
31	**Food, beverages and tobacco**	11 440.0	10 524.0	10 463.0	10 898.0	11 005.0	11 376.0	11481.0	..
311.2	Food	11 105.0	10 180.0	10 169.0	10 592.0	10 726.0	11 070.0	11162.0	..
313	Beverages	325.0	334.0	294.0	306.0	273.0	306.0	318.0	..
314	Tobacco	10.0	10.0	..	..	6.0	..	..	..
32	**Textiles, apparel and leather**	1 450.0	1 293.0	1 190.0	1 205.0	1 266.0	1 357.0	1393.0	..
321	Textiles	601.0	459.0	455.0	487.0	524.0	591.0	576.0	..
3213	Knitting mills	60.0	72.0	67.0	72.0	85.0	84.0	79.0	..
322	Wearing apparel, except footwear	525.0	533.0	488.0	518.0	527.0	501.0	527.0	..
323	Leather and fur products	283.0	285.0	235.0	188.0	207.0	258.0	286.0	..
324	Footwear, except rubber or plastics	41.0	16.0	12.0	11.0	9.0	7.0	4.0	..
33	**Wood products and furniture**	1 354.0	1 239.0	1 110.0	1 092.0	1 090.0	1 143.0	1185.0	..
331	Wood products, except furniture	42.0	36.0	44.0	37.0	43.0	46.0	56.0	..
332	Furniture, fixtures, except metal	1 312.0	1 203.0	1 066.0	1 055.0	1 047.0	1 097.0	1130.0	..
34	**Paper, paper products and printing**	2 395.0	2 276.0	2 153.0	2 080.0	2 210.0	2 288.0	2255.0	..
341	Paper and paper products	233.0	236.0	235.0	228.0	237.0	271.0	285.0	..
3411	Pulp, paper and paperboard articles	233.0	236.0	235.0	228.0	237.0	271.0	285.0	..
342	Printing and publishing	2 162.0	2 040.0	1 918.0	1 852.0	1 973.0	2 017.0	1970.0	..
35	**Chemical products**	1 196.0	1 128.0	1 067.0	1 058.0	1 088.0	1 110.0	1191.0	..
351	Industrial chemicals	250.0	248.0	236.0	217.0	197.0	185.0	186.0	..
3511	Basic industrial chemicals except fertilizers	250.0	248.0	236.0	217.0	197.0	185.0	186.0	..
3512	Fertilizers and pesticides	..	..	..	..	..	..	..	..
3513	Synthetic resins and plastic materials	..	..	..	..	..	..	..	..
352	Other chemicals	416.0	369.0	381.0	355.0	391.0	398.0	424.0	..
3521	Paints, varnishes and lacquers	181.0	161.0	169.0	172.0	169.0	167.0	167.0	..
3522	Drugs and medicines	..	..	..	..	..	..	..	..
3523	Soap, cleaning preparations, perfumes, cosmetics	235.0	208.0	212.0	183.0	222.0	231.0	256.0	..
3529	Other chemical products	..	..	..	..	..	..	..	..
353	Petroleum refineries	..	..	..	..	..	..	..	..
354	Miscellaneous petroleum and coal products	10.0	10.0	9.0	..	1.0	1.0	37.0	..
355	Rubber products	..	..	..	..	..	..	..	..
356	Plastics products	520.0	501.0	441.0	486.0	499.0	526.0	545.0	..
36	**Non-metallic mineral products**	975.0	928.0	852.0	790.0	817.0	745.0	762.0	..
361	Pottery, china, earthenware	18.0	33.0	29.0	18.0	31.0	25.0	19.0	..
362	Glass and glass products	122.0	112.0	108.0	104.0	97.0	72.0	87.0	..
369	Non-metallic mineral products nec.	835.0	783.0	715.0	667.0	689.0	648.0	656.0	..
37	**Basic metal industries**	812.0	816.0	753.0	691.0	673.0	689.0	722.0	..
371	Iron and steel	190.0	193.0	161.0	157.0	162.0	169.0	173.0	..
372	Non-ferrous metals	622.0	623.0	592.0	534.0	511.0	520.0	549.0	..
38	**Fabricated metal prod.; machinery & equip.**	2 826.0	2 672.0	2 540.0	2 356.0	2 513.0	2 787.0	3031.0	..
381	Fabricated metal products	2 196.0	2 097.0	1 978.0	1 974.0	2 095.0	2 264.0	2474.0	..
3813	Structural metal products	..	..	..	..	..	..	..	..
382	Non-electrical machinery	..	..	..	..	..	..	..	..
3821	Engines and turbines	..	..	..	..	..	..	..	..
3822	Agricultural machinery and equipment	..	..	..	..	..	..	..	..
3823	Metal and wood working machinery	..	..	..	..	..	..	..	..
3824	Special industrial machinery	..	..	..	..	..	..	..	..
3825	Office, computing and accounting machinery	..	..	..	..	..	..	..	..
3829	Other non-electrical machinery and equipment	..	..	..	..	..	..	..	..
383	Electrical machinery	..	..	..	..	..	..	..	..
3831	Electrical industrial machinery	..	..	..	..	..	..	..	..
3832	Radio, TV and communication equipment	..	..	..	..	..	..	..	..
3833	Electrical appliances and housewares	..	..	..	..	..	..	..	..
3839	Other electrical apparatus and supplies	..	..	..	..	..	..	..	..
384	Transport equipment	630.0	575.0	562.0	382.0	418.0	523.0	557.0	..
3841	Shipbuilding	630.0	575.0	562.0	382.0	418.0	523.0	557.0	..
3842	Railroad equipment	..	..	..	..	..	..	..	..
3843	Motor vehicles	..	..	..	..	..	..	..	..
3844	Motorcycles and bicycles	..	..	..	..	..	..	..	..
3845	Aircraft	..	..	..	..	..	..	..	..
3849	Other transport equipment	..	..	..	..	..	..	..	..
385	Professional and scientific equipment	..	..	..	..	..	..	..	..
3851	Professional and scientific equipment, nec	..	..	..	..	..	..	..	..
3852	Photographic and optical goods	..	..	..	..	..	..	..	..
3853	Watches and clocks	..	..	..	..	..	..	..	..
39	**Other manufacturing industries**	598.0	570.0	556.0	624.0	699.0	767.0	860.0	..
3901	Jewellery and related articles	60.0	62.0	58.0	63.0	65.0	56.0	53.0	..
3	**Total manufacturing**	23 043.0	21 444.0	20 681.0	20 794.0	21 360.0	22 263.0	22880.0	..
1	**Agriculture, hunting, forestry and fishing**	13 554.0	12 868.0	12 771.0	12 177.0	11 992.0	11 811.0	11322.0	..
2	**Mining and quarrying**	..	..	..	..	..	..	..	..
21	Coal mining	..	..	..	..	..	..	..	..
22	Crude petroleum and natural gas	..	..	..	..	..	..	..	..
23.29	Other mining	..	..	..	..	..	..	..	..
4	**Electricity, gas and water**	1 269.0	1 259.0	1 243.0	1 297.0	1 364.0	1 415.0	1393.0	..
5	**Construction**	10 972.0	10 743.0	10 570.0	10 364.0	9 783.0	9 810.0	10246.0	..
6.9	**Total services**	76 002.0	76 730.0	76 791.0	78 029.0	79 217.0	81 289.0	82992.0	..
6	**Wholesale & retail trade; restaurants & hotels**	18 128.0	18 093.0	17 762.0	17 917.0	18 351.0	19 042.0	20118.0	..
7	**Transport, storage and communication**	8 622.0	8 445.0	8 105.0	8 185.0	8 428.0	8 745.0	8817.0	..
8	**Finance, insurance, real estate, business serv.**	10 484.0	10 655.0	10 675.0	10 533.0	10 604.0	11 045.0	11537.0	..
9	**Community, social and personal services**	38 768.0	39 537.0	40 249.0	41 394.0	41 834.0	42 457.0	42520.0	..

Note: ISIC 3511 includes 3512.

Table IC.4 EMPLOI, SALARIÉS - EMPLOYMENT, EMPLOYEES

Milliers

1991	1992	1993	1994	1995	1996	1997	1998	CITI révision 2	
11 226.0	10 326.0	10 276.0	10 704.0	10 811.0	11 193.0	11 304.0	..	**Alimentation, boissons et tabacs**	**31**
10 891.0	9 982.0	9 982.0	10 399.0	10 532.0	10 887.0	10 986.0	..	Alimentation	311.2
325.0	334.0	294.0	306.0	273.0	306.0	318.0	..	Boissons	313
10.0	10.0	..	..	6.0	..	..	..	Tabacs	314
1 384.0	1 233.0	1 139.0	1 158.0	1 217.0	1 310.0	1 351.0	..	**Textiles, habillement et cuir**	**32**
579.0	440.0	436.0	468.0	506.0	572.0	559.0	..	Textiles	321
57.0	70.0	65.0	70.0	83.0	81.0	76.0	..	Bonneterie	3213
493.0	502.0	464.0	495.0	501.0	477.0	509.0	..	Habillement, à l'exclusion des chaussures	322
272.0	275.0	227.0	184.0	201.0	254.0	279.0	..	Cuir et articles en cuir	323
40.0	16.0	12.0	11.0	9.0	7.0	3.0	..	Chaussures, sauf en caoutchouc et plastique	324
1 229.0	1 066.0	943.0	925.0	925.0	983.0	1 033.0	..	**Bois et meubles**	**33**
24.0	17.0	28.0	21.0	27.0	30.0	37.0	..	Ouvrages en bois	331
1 205.0	1 049.0	915.0	904.0	897.0	953.0	996.0	..	Meubles et accessoires	332
2 287.0	2 159.0	2 043.0	1 968.0	2 101.0	2 182.0	2 166.0	..	**Papier, impression et édition**	**34**
231.0	233.0	231.0	224.0	234.0	268.0	283.0	..	Papier et articles en papier	341
231.0	233.0	231.0	224.0	234.0	268.0	283.0	..	Pâte à papier, papier et carton	3411
2 056.0	1 926.0	1 812.0	1 745.0	1 867.0	1 914.0	1 883.0	..	Imprimerie et édition	342
1 176.0	1 107.0	1 050.0	1 044.0	1 069.0	1 094.0	1 175.0	..	**Produits chimiques**	**35**
248.0	246.0	234.0	217.0	197.0	185.0	186.0	..	Industrie chimique	351
248.0	246.0	234.0	217.0	197.0	185.0	186.0	..	Chimie de base, à l'exception des engrais	3511
..	..	..	..	..	..	..	..	Engrais et pesticides	3512
..	..	..	..	..	..	..	..	Résines synthétiques et matières plastiques	3513
412.0	366.0	377.0	353.0	389.0	397.0	422.0	..	Autres produits chimiques	352
181.0	161.0	169.0	172.0	169.0	167.0	167.0	..	Peintures, vernis et laques	3521
..	..	..	..	..	..	..	..	Produits pharmaceutiques	3522
231.0	205.0	208.0	181.0	219.0	230.0	255.0	..	Produits de nettoyage	3523
..	..	..	..	..	..	..	..	Autres produits chimiques	3529
..	..	..	..	..	..	..	..	Raffineries de pétrole	353
10.0	10.0	9.0	..	1.0	1.0	37.0	..	Derives du pétrole et du charbon	354
..	..	..	..	..	..	..	..	Industrie du caoutchouc	355
506.0	485.0	429.0	474.0	482.0	511.0	531.0	..	Ouvrages en matière plastique	356
936.0	892.0	820.0	749.0	776.0	706.0	731.0	..	**Produits minéraux non métalliques**	**36**
8.0	23.0	22.0	8.0	21.0	14.0	10.0	..	Grès, porcelaines et faïences	361
113.0	102.0	98.0	91.0	82.0	60.0	77.0	..	Industrie du verre	362
815.0	767.0	700.0	650.0	674.0	632.0	644.0	..	Produits minéraux non métalliques, nca	369
810.0	813.0	753.0	691.0	673.0	689.0	722.0	..	**Industries métallurgiques de base**	**37**
188.0	190.0	161.0	157.0	162.0	169.0	173.0	..	Sidérurgie	371
622.0	623.0	592.0	534.0	511.0	520.0	549.0	..	Métaux non ferreux	372
2 559.0	2 403.0	2 278.0	2 090.0	2 228.0	2 506.0	2 733.0	..	**Ouvrages en métaux, machines et matériel**	**38**
1 960.0	1 857.0	1 740.0	1 730.0	1 834.0	2 009.0	2 206.0	..	Ouvrages en métaux	381
..	..	..	..	..	..	..	..	Eléments de construction en métal	3813
..	..	..	..	..	..	..	..	Machines non électriques	382
..	..	..	..	..	..	..	..	Moteurs et turbines	3821
..	..	..	..	..	..	..	..	Machines agricoles	3822
..	..	..	..	..	..	..	..	Machines pour le travail du bois et du métal	3823
..	..	..	..	..	..	..	..	Machines et matériel spéciaux	3824
..	..	..	..	..	..	..	..	Machines de bureau et à calculer	3825
..	..	..	..	..	..	..	..	Machines et matériel, nca	3829
..	..	..	..	..	..	..	..	Machines électriques	383
..	..	..	..	..	..	..	..	Machines électriques industrielles	3831
..	..	..	..	..	..	..	..	Appareils de radio, télévision et télécom.	3832
..	..	..	..	..	..	..	..	Appareils électroménagers	3833
..	..	..	..	..	..	..	..	Appareils électriques, nca	3839
599.0	546.0	538.0	360.0	394.0	497.0	527.0	..	Matériel de transport	384
599.0	546.0	538.0	360.0	394.0	497.0	527.0	..	Construction navale	3841
..	..	..	..	..	..	..	..	Matériel ferroviaire	3842
..	..	..	..	..	..	..	..	Véhicules automobiles	3843
..	..	..	..	..	..	..	..	Motocycles et cycles	3844
..	..	..	..	..	..	..	..	Construction aéronautique	3845
..	..	..	..	..	..	..	..	Matériel de transport, nca	3849
..	..	..	..	..	..	..	..	Matériel professionnel et scientifique	385
..	..	..	..	..	..	..	..	Equipement professionnel	3851
..	..	..	..	..	..	..	..	Matériel photo et optique	3852
..	..	..	..	..	..	..	..	Montres et horloges	3853
528.0	495.0	486.0	551.0	629.0	701.0	796.0	..	**Autres industries manufacturières**	**39**
36.0	41.0	38.0	39.0	42.0	36.0	34.0	..	Joaillerie	3901
22 132.0	20 493.0	19 784.0	19 880.0	20 430.0	21 365.0	22 011.0	..	**Total industries manufacturières**	**3**
7 314.0	6 832.0	7 438.0	7 015.0	6 962.0	7 050.0	6 768.0	..	**Agriculture, chasse, sylviculture et pêche**	**1**
..	..	..	..	..	..	..	..	**Industries extractives**	**2**
..	..	..	..	..	..	..	..	Extraction du charbon	21
..	..	..	..	..	..	..	..	Production du pétrole brut et gaz naturel	22
..	..	..	..	..	..	..	..	Extraction d'autres minéraux	23.29
1 269.0	1 259.0	1 243.0	1 297.0	1 364.0	1 414.0	1 393.0	..	**Électricité, gaz et eau**	**4**
8 248.0	8 089.0	8 040.0	7 907.0	7 331.0	7 402.0	7 794.0	..	**Construction**	**5**
68 276.0	68 986.0	69 354.0	70 483.0	71 628.0	73 853.0	75 845.0	..	**Total services**	**6.9**
16 706.0	16 717.0	16 483.0	16 664.0	17 138.0	17 908.0	19 086.0	..	**Commerce de gros & détail ; rest. & hotels**	**6**
6 652.0	6 589.0	6 392.0	6 532.0	6 830.0	7 202.0	7 332.0	..	**Transport, entrepôts et communications**	**7**
8 907.0	9 069.0	9 129.0	8 879.0	8 905.0	9 386.0	9 946.0	..	**Banques, assur., aff. imm., serv. aux entrepr.**	**8**
36 011.0	36 611.0	37 350.0	38 408.0	38 756.0	39 357.0	39 481.0	..	**Services a la collect, sociaux et personnels**	**9**

Note: ISIC 3511 includes 3512.

ICELAND

Table IC.5 SALAIRES ET CHARGES SOCIALES, SALARIÉS - COMPENSATION OF LABOUR, EMPLOYEES

Millions de IKr (Prix courants)

ISIC revision 2		1991	1992	1993	1994	1995	1996	1997	1998
31	**Food, beverages and tobacco**	17 725	17 534	18 028	19 274	20 251	21 803	23 265	..
311.2	Food	16 921	16 777	17 293	18 559	19 462	20 946	22 319	..
313	Beverages	803	757	735	715	788	857	946	..
314	Tobacco	..	..	..	..	..	..	..	..
32	**Textiles, apparel and leather**	1 904	1 919	1 671	1 805	1 957	2 298	2 488	..
321	Textiles	802	711	690	693	813	1 004	1 110	..
3213	Knitting mills	102	77	77	79	101	104	103	..
322	Wearing apparel, except footwear	661	776	644	710	739	816	839	..
323	Leather and fur products	411	419	325	390	394	467	533	..
324	Footwear, except rubber or plastics	30	13	11	12	11	11	6	..
33	**Wood products and furniture**	2 281	2 216	1 694	1 544	1 827	1 626	2 051	..
331	Wood products, except furniture	37	41	47	44	55	61	90	..
332	Furniture, fixtures, except metal	2 244	2 175	1 647	1 500	1 772	1 565	1 961	..
34	**Paper, paper products and printing**	4 436	4 104	4 203	4 689	4 924	5 523	5 953	..
341	Paper and paper products	441	409	384	436	461	552	611	..
3411	Pulp, paper and paperboard articles	441	409	384	436	461	552	611	..
342	Printing and publishing	3 995	3 695	3 819	4 253	4 463	4 971	5 342	..
35	**Chemical products**	2 212	2 327	2 341	2 213	2 280	2 576	2 974	..
351	Industrial chemicals	623	686	609	484	483	495	509	..
3511	Basic industrial chemicals except fertilizers	623	686	609	484	483	495	509	..
3512	Fertilizers and pesticides	..	..	..	..	..	..	..	..
3513	Synthetic resins and plastic materials	..	..	..	..	..	..	..	..
352	Other chemicals	537	530	638	644	658	808	992	..
3521	Paints, varnishes and lacquers	251	285	332	338	332	406	350	..
3522	Drugs and medicines	..	..	..	..	..	..	..	..
3523	Soap, cleaning preparations, perfumes, cosmetics	286	245	306	306	327	402	642	..
3529	Other chemical products	..	..	..	..	..	..	..	..
353	Petroleum refineries	..	..	..	..	..	..	..	..
354	Miscellaneous petroleum and coal products	16	19	13	8	13	16	104	..
355	Rubber products	..	..	..	..	..	..	..	..
356	Plastics products	1 037	1 093	1 081	1 077	1 126	1 256	1 368	..
36	**Non-metallic mineral products**	1 881	1 813	1 478	1 489	1 515	1 571	1 818	..
361	Pottery, china, earthenware	70	71	32	2	19	33	20	..
362	Glass and glass products	190	189	162	143	141	151	168	..
369	Non-metallic mineral products nec.	1 621	1 553	1 284	1 344	1 356	1 386	1 629	..
37	**Basic metal industries**	1 948	1 879	1 696	1 708	1 729	1 879	2 175	..
371	Iron and steel	424	471	385	419	478	549	610	..
372	Non-ferrous metals	1 523	1 408	1 311	1 289	1 251	1 330	1 565	..
38	**Fabricated metal prod.; machinery & equip.**	4 961	4 543	4 028	3 720	4 495	6 105	7 702	..
381	Fabricated metal products	3 829	3 253	2 839	2 843	3 477	4 837	6 253	..
3813	Structural metal products	..	..	..	..	..	..	..	..
382	Non-electrical machinery	..	..	..	..	..	..	..	..
3821	Engines and turbines	..	..	..	..	..	..	..	..
3822	Agricultural machinery and equipment	..	..	..	..	..	..	..	..
3823	Metal and wood working machinery	..	..	..	..	..	..	..	..
3824	Special industrial machinery	..	..	..	..	..	..	..	..
3825	Office, computing and accounting machinery	..	..	..	..	..	..	..	..
3829	Other non-electrical machinery and equipment	..	..	..	..	..	..	..	..
383	Electrical machinery	..	..	..	..	..	..	..	..
3831	Electrical industrial machinery	..	..	..	..	..	..	..	..
3832	Radio, TV and communication equipment	..	..	..	..	..	..	..	..
3833	Electrical appliances and housewares	..	..	..	..	..	..	..	..
3839	Other electrical apparatus and supplies	..	..	..	..	..	..	..	..
384	Transport equipment	1 132	1 290	1 188	877	1 019	1 269	1 449	..
3841	Shipbuilding	1 132	1 290	1 188	877	1 019	1 269	1 449	..
3842	Railroad equipment	..	..	..	..	..	..	..	..
3843	Motor vehicles	..	..	..	..	..	..	..	..
3844	Motorcycles and bicycles	..	..	..	..	..	..	..	..
3845	Aircraft	..	..	..	..	..	..	..	..
3849	Other transport equipment	..	..	..	..	..	..	..	..
385	Professional and scientific equipment	..	..	..	..	..	..	..	..
3851	Professional and scientific equipment, nec	..	..	..	..	..	..	..	..
3852	Photographic and optical goods	..	..	..	..	..	..	..	..
3853	Watches and clocks	..	..	..	..	..	..	..	..
39	**Other manufacturing industries**	1 157	1 212	1 305	1 446	1 707	2 148	2 523	..
3901	Jewellery and related articles	58	67	55	44	56	59	64	..
3	**Total manufacturing**	38 504	37 546	36 443	37 887	40 685	45 529	50 948	..
1	**Agriculture, hunting, forestry and fishing**	22 507	22 524	23 375	23 777	24 701	26 592	24 799	..
2	**Mining and quarrying**	..	..	..	..	..	..	..	..
21	Coal mining	..	..	..	..	..	..	..	..
22	Crude petroleum and natural gas	..	..	..	..	..	..	..	..
23.29	Other mining	..	..	..	..	..	..	..	..
4	**Electricity, gas and water**	3 440	3 643	3 706	3 965	4 247	4 514	5 147	..
5	**Construction**	16 723	15 642	14 716	14 205	14 181	15 259	18 053	..
6.9	**Total services**	120 982	125 377	125 598	132 032	141 317	156 819	166 826	..
6	**Wholesale & retail trade; restaurants & hotels**	29 970	31 119	30 182	31 583	33 102	36 703	37 966	..
7	**Transport, storage and communication**	14 113	14 083	13 884	15 269	17 526	20 007	22 065	..
8	**Finance, insurance, real estate, business serv.**	19 426	20 066	19 512	19 719	21 130	23 839	25 277	..
9	**Community, social and personal services**	57 556	60 185	62 090	65 534	69 492	76 270	81 518	..

Note: ISIC 3511 includes 3512.

ISLANDE

Table IC.6 **ETABLISSEMENTS - ESTABLISHMENTS**

Units

1991	1992	1993	1994	1995	1996	1997	1998		CITI révision 2
711	686	722	763	747	746	..	..	**Alimentation, boissons et tabacs**	**31**
703	676	715	756	740	738	..	..	Alimentation	311.2
7	9	7	7	6	7	..	..	Boissons	313
1	1	..	..	1	1	..	..	Tabacs	314
193	184	171	190	190	211	..	..	**Textiles, habillement et cuir**	**32**
67	65	64	72	72	82	..	..	Textiles	321
21	21	19	21	20	21	..	..	Bonneterie	3213
102	98	85	96	95	105	..	..	Habillement, à l'exclusion des chaussures	322
21	19	20	19	20	21	..	..	Cuir et articles en cuir	323
3	2	2	3	3	3	..	..	Chaussures, sauf en caoutchouc et plastique	324
313	372	382	388	383	380	..	..	**Bois et meubles**	**33**
36	37	38	41	39	42	..	..	Ouvrages en bois	331
277	335	344	347	344	338	..	..	Meubles et accessoires	332
342	353	369	390	414	434	..	..	**Papier, impression et édition**	**34**
10	9	10	11	14	14	..	..	Papier et articles en papier	341
10	9	10	11	14	14	..	..	Pâte à papier, papier et carton	3411
332	344	359	379	400	420	..	..	Imprimerie et édition	342
93	91	83	85	97	92	..	..	**Produits chimiques**	**35**
6	6	7	7	7	6	..	..	Industrie chimique	351
6	6	7	7	7	6	..	..	Chimie de base, à l'exception des engrais	3511
..	..	..	..	..	..	..	..	Engrais et pesticides	3512
..	..	..	..	..	..	..	..	Résines synthétiques et matières plastiques	3513
26	25	22	25	25	22	..	..	Autres produits chimiques	352
7	7	5	5	5	4	..	..	Peintures, vernis et laques	3521
..	..	..	..	..	..	..	..	Produits pharmaceutiques	3522
19	18	17	20	20	18	..	..	Produits de nettoyage	3523
..	..	..	..	..	..	..	..	Autres produits chimiques	3529
..	..	..	..	..	..	..	..	Raffineries de pétrole	353
2	2	2	..	2	1	..	..	Derives du pétrole et du charbon	354
..	..	..	..	..	..	..	..	Industrie du caoutchouc	355
59	58	52	53	63	63	..	..	Ouvrages en matière plastique	356
121	112	116	126	127	120	..	..	**Produits minéraux non métalliques**	**36**
24	24	21	24	24	27	..	..	Grès, porcelaines et faïences	361
23	22	24	24	26	24	..	..	Industrie du verre	362
74	66	71	78	77	69	..	..	Produits minéraux non métalliques, nca	369
2	2	2	2	2	3	..	..	**Industries métallurgiques de base**	**37**
1	1	1	1	1	1	..	..	Sidérurgie	371
1	1	1	1	1	2	..	..	Métaux non ferreux	372
558	574	576	602	623	651	..	..	**Ouvrages en métaux, machines et matériel**	**38**
488	504	511	538	566	587	..	..	Ouvrages en métaux	381
..	..	..	..	..	..	..	..	Eléments de construction en métal	3813
..	..	..	..	..	..	..	..	Machines non électriques	382
..	..	..	..	..	..	..	..	Moteurs et turbines	3821
..	..	..	..	..	..	..	..	Machines agricoles	3822
..	..	..	..	..	..	..	..	Machines pour le travail du bois et du métal	3823
..	..	..	..	..	..	..	..	Machines et matériel spéciaux	3824
..	..	..	..	..	..	..	..	Machines de bureau et à calculer	3825
..	..	..	..	..	..	..	..	Machines et matériel, nca	3829
..	..	..	..	..	..	..	..	Machines électriques	383
..	..	..	..	..	..	..	..	Machines électriques industrielles	3831
..	..	..	..	..	..	..	..	Appareils de radio, télévision et télécom.	3832
..	..	..	..	..	..	..	..	Appareils électroménagers	3833
..	..	..	..	..	..	..	..	Appareils électriques, nca	3839
70	70	65	64	57	64	..	..	Matériel de transport	384
70	70	65	64	57	64	..	..	Construction navale	3841
..	..	..	..	..	..	..	..	Matériel ferroviaire	3842
..	..	..	..	..	..	..	..	Véhicules automobiles	3843
..	..	..	..	..	..	..	..	Motocycles et cycles	3844
..	..	..	..	..	..	..	..	Construction aéronautique	3845
..	..	..	..	..	..	..	..	Matériel de transport, nca	3849
..	..	..	..	..	..	..	..	Matériel professionnel et scientifique	385
..	..	..	..	..	..	..	..	Equipement professionnel	3851
..	..	..	..	..	..	..	..	Matériel photo et optique	3852
..	..	..	..	..	..	..	..	Montres et horloges	3853
144	158	157	174	153	160	..	..	**Autres industries manufacturières**	**39**
33	34	33	39	35	36	..	..	Joaillerie	3901
2 477	2 532	2 578	2 720	2 736	2 797	..	..	**Total industries manufacturières**	**3**
7 275	7 055	6 423	6 344	6 177	5 892	..	..	**Agriculture, chasse, sylviculture et pêche**	**1**
..	..	..	..	..	..	..	..	**Industries extractives**	**2**
..	..	..	..	..	..	..	..	Extraction du charbon	21
..	..	..	..	..	..	..	..	Production du pétrole brut et gaz naturel	22
..	..	..	..	..	..	..	..	Extraction d'autres minéraux	23.29
44	44	43	41	42	41	..	..	**Électricité, gaz et eau**	**4**
4 229	4 168	4 252	4 201	4 270	4 260	..	..	**Construction**	**5**
17 831	18 598	19 005	20 368	21 083	..	..	..	**Total services**	**6.9**
3 678	3 657	3 674	3 785	3 831	3 922	..	..	**Commerce de gros & détail ; rest. & hotels**	**6**
2 573	2 444	2 358	2 301	2 234	2 229	..	..	**Transport, entrepôts et communications**	**7**
4 073	4 174	4 184	4 536	4 742	4 863	..	..	**Banques, assur., aff. imm., serv. aux entrepr.**	**8**
7 507	8 323	8 789	9 746	10 276	10 976	..	..	**Services a la collect, sociaux et personnels**	**9**

Note: ISIC 3511 includes 3512.

ICELAND

Table IC.7 ENTERPRISES - ENTREPRISES

Units

ISIC revision 2		1991	1992	1993	1994	1995	1996	1997	1998
31	**Food, beverages and tobacco**	571	565	604	643	643	649	640	..
311.2	Food	564	556	598	637	637	642	633	..
313	Beverages	7	9	6	6	6	7	7	..
314	Tobacco	..	..	..	..	..	..	..	..
32	**Textiles, apparel and leather**	186	175	163	184	181	201	194	..
321	Textiles	64	60	58	68	65	73	74	..
3213	Knitting mills	20	20	18	20	19	20	19	..
322	Wearing apparel, except footwear	99	94	83	94	93	104	98	..
323	Leather and fur products	20	19	20	19	20	21	20	..
324	Footwear, except rubber or plastics	3	2	2	3	3	3	2	..
33	**Wood products and furniture**	303	367	377	383	377	373	387	..
331	Wood products, except furniture	34	37	37	40	38	41	50	..
332	Furniture, fixtures, except metal	269	330	340	343	339	332	337	..
34	**Paper, paper products and printing**	336	351	366	388	413	432	424	..
341	Paper and paper products	10	9	10	11	14	14	15	..
3411	Pulp, paper and paperboard articles	10	9	10	11	14	14	15	..
342	Printing and publishing	326	342	356	377	399	418	409	..
35	**Chemical products**	86	85	78	82	94	90	100	..
351	Industrial chemicals	6	6	7	7	7	6	8	..
3511	Basic industrial chemicals except fertilizers	6	6	7	7	7	6	8	..
3512	Fertilizers and pesticides	..	..	..	..	..	..	..	..
3513	Synthetic resins and plastic materials	..	..	..	..	..	..	..	..
352	Other chemicals	21	20	19	23	24	21	24	..
3521	Paints, varnishes and lacquers	6	5	5	5	5	4	4	..
3522	Drugs and medicines	..	..	..	..	..	..	..	..
3523	Soap, cleaning preparations, perfumes, cosmetics	15	15	14	18	19	17	20	..
3529	Other chemical products	..	..	..	..	..	..	..	..
353	Petroleum refineries	..	..	..	..	..	..	..	..
354	Miscellaneous petroleum and coal products	2	2	2	..	2	1	2	..
355	Rubber products	..	..	..	..	..	..	..	..
356	Plastics products	57	57	50	52	61	62	66	..
36	**Non-metallic mineral products**	116	108	112	122	124	118	114	..
361	Pottery, china, earthenware	22	23	20	23	23	26	23	..
362	Glass and glass products	23	22	24	24	26	24	23	..
369	Non-metallic mineral products nec.	71	63	68	75	75	68	68	..
37	**Basic metal industries**	2	2	2	2	2	3	4	..
371	Iron and steel	1	1	1	1	1	1	1	..
372	Non-ferrous metals	1	1	1	1	1	2	3	..
38	**Fabricated metal prod.; machinery & equip.**	547	557	563	593	614	639	681	..
381	Fabricated metal products	477	487	498	530	557	575	616	..
3813	Structural metal products	..	..	..	..	..	..	..	..
382	Non-electrical machinery	..	..	..	..	..	..	..	..
3821	Engines and turbines	..	..	..	..	..	..	..	..
3822	Agricultural machinery and equipment	..	..	..	..	..	..	..	..
3823	Metal and wood working machinery	..	..	..	..	..	..	..	..
3824	Special industrial machinery	..	..	..	..	..	..	..	..
3825	Office, computing and accounting machinery	..	..	..	..	..	..	..	..
3829	Other non-electrical machinery and equipment	..	..	..	..	..	..	..	..
383	Electrical machinery	..	..	..	..	..	..	..	..
3831	Electrical industrial machinery	..	..	..	..	..	..	..	..
3832	Radio, TV and communication equipment	..	..	..	..	..	..	..	..
3833	Electrical appliances and housewares	..	..	..	..	..	..	..	..
3839	Other electrical apparatus and supplies	..	..	..	..	..	..	..	..
384	Transport equipment	70	70	65	64	57	64	65	..
3841	Shipbuilding	70	70	65	64	57	64	65	..
3842	Railroad equipment	..	..	..	..	..	..	..	..
3843	Motor vehicles	..	..	..	..	..	..	..	..
3844	Motorcycles and bicycles	..	..	..	..	..	..	..	..
3845	Aircraft	..	..	..	..	..	..	..	..
3849	Other transport equipment	..	..	..	..	..	..	..	..
385	Professional and scientific equipment	..	..	..	..	..	..	..	..
3851	Professional and scientific equipment, nec	..	..	..	..	..	..	..	..
3852	Photographic and optical goods	..	..	..	..	..	..	..	..
3853	Watches and clocks	..	..	..	..	..	..	..	..
39	**Other manufacturing industries**	137	155	154	170	149	156	160	..
3901	Jewellery and related articles	32	34	33	38	35	36	35	..
3	**Total manufacturing**	2284	2365	2419	2567	2597	2661	2704	..
1	**Agriculture, hunting, forestry and fishing**	7209	7005	6363	6285	6120	5837	5588	..
2	**Mining and quarrying**	..	..	..	..	..	..	..	..
21	Coal mining	..	..	..	..	..	..	..	..
22	Crude petroleum and natural gas	..	..	..	..	..	..	..	..
23.29	Other mining	..	..	..	..	..	..	..	..
4	**Electricity, gas and water**	30	31	29	26	25	25	26	..
5	**Construction**	4075	4023	4073	4049	4128	4125	4269	..
6.9	**Total services**	17 315	18 182	18 528	19 845	20 571	21 482	21 362	..
6	**Wholesale & retail trade; restaurants & hotels**	3594	3598	3637	3749	3788	3875	3879	..
7	**Transport, storage and communication**	2531	2417	2309	2254	2190	2187	2153	..
8	**Finance, insurance, real estate, business serv.**	4059	4165	4175	4529	4731	4849	4837	..
9	**Community, social and personal services**	7131	8002	8407	9313	9862	10571	10493	..

Note: ISIC 3511 includes 3512.

Table TU.1 **PRODUCTION - PRODUCTION**

Milliards de TL (Prix courants)

1991	1992	1993	1994	1995	1996	1997	1998	CITI révision 2	
60 478	112 069	198 092	393 526	793 325	1 392 564	2 764 204	..	**Alimentation, boissons et tabacs**	**31**
41 860	82 395	144 151	302 081	614 801	1 069 066	2 220 251	..	Alimentation	311.2
7 691	12 552	23 053	42 489	83 587	141 407	269 588	..	Boissons	313
10 928	17 121	30 888	48 956	94 937	182 090	274 365	..	Tabacs	314
47 367	100 935	172 218	441 247	881 785	1 465 714	3 201 428	..	**Textiles, habillement et cuir**	**32**
32 427	67 059	114 251	300 162	572 620	951 345	2 160 695	..	Textiles	321
3 824	11 151	22 183	58 619	101 404	131 871	294 118	..	Bonneterie	3213
13 219	30 177	51 695	125 698	277 092	459 790	888 712	..	Habillement, à l'exclusion des chaussures	322
710	1 812	2 941	9 384	19 260	35 983	94 744	..	Cuir et articles en cuir	323
1 011	1 887	3 332	6 002	12 813	18 596	57 277	..	Chaussures, sauf en caoutchouc et plastique	324
3 197	7 847	14 832	23 485	52 192	103 184	201 388	..	**Bois et meubles**	**33**
2 440	5 751	10 525	15 881	34 988	60 892	117 288	..	Ouvrages en bois	331
756	2 096	4 307	7 604	17 204	42 292	84 101	..	Meubles et accessoires	332
9 311	18 306	40 216	82 657	169 151	272 802	537 039	..	**Papier, impression et édition**	**34**
5 343	9 591	17 910	43 178	109 840	116 519	216 477	..	Papier et articles en papier	341
3 072	6 282	10 780	24 604	58 362	55 321	103 035	..	Pâte à papier, papier et carton	3411
3 969	8 715	22 306	39 479	59 310	156 283	320 562	..	Imprimerie et édition	342
80 873	141 110	231 119	510 104	1 052 901	1 817 282	3 666 026	..	**Produits chimiques**	**35**
14 694	26 064	41 866	96 925	205 556	296 078	545 988	..	Industrie chimique	351
1 652	11 390	5 510	8 470	18 452	31 796	65 891	..	Chimie de base, à l'exception des engrais	3511
6 082	8 508	18 929	31 104	67 864	118 728	194 391	..	Engrais et pesticides	3512
6 960	6 165	17 428	57 351	119 240	145 554	285 705	..	Résines synthétiques et matières plastiques	3513
14 649	26 374	50 008	111 185	234 797	386 142	816 324	..	Autres produits chimiques	352
2 426	3 848	7 641	18 826	43 735	59 021	135 127	..	Peintures, vernis et laques	3521
7 161	11 783	23 001	47 358	93 774	160 888	328 084	..	Produits pharmaceutiques	3522
3 397	7 330	12 917	28 123	58 307	98 656	218 523	..	Produits de nettoyage	3523
1 665	3 413	6 449	16 877	38 981	67 578	134 590	..	Autres produits chimiques	3529
34 743	59 344	95 285	204 529	389 053	764 442	1 552 063	..	Raffineries de pétrole	353
8 350	12 170	13 536	27 571	60 842	94 323	197 426	..	Derives du pétrole et du charbon	354
4 378	8 777	14 827	33 005	70 305	110 702	231 280	..	Industrie du caoutchouc	355
4 058	8 380	15 597	36 889	92 348	165 595	322 944	..	Ouvrages en matière plastique	356
16 344	31 023	57 527	119 651	221 064	384 280	809 925	..	**Produits minéraux non métalliques**	**36**
2 953	4 404	8 585	22 340	35 699	54 070	119 975	..	Grès, porcelaines et faïences	361
3 304	6 448	10 882	22 348	49 047	76 802	173 351	..	Industrie du verre	362
10 087	20 171	38 059	74 963	136 319	253 407	516 599	..	Produits minéraux non métalliques, nca	369
29 303	50 503	101 097	239 195	448 896	618 982	1 365 916	..	**Industries métallurgiques de base**	**37**
24 593	42 234	86 661	207 373	376 551	526 775	1 125 466	..	Sidérurgie	371
4 711	8 269	14 435	31 822	72 345	92 208	240 450	..	Métaux non ferreux	372
65 047	126 293	243 832	399 133	885 015	1 557 350	3 428 461	..	**Ouvrages en métaux, machines et matériel**	**38**
8 874	18 637	32 451	61 398	139 886	250 000	537 321	..	Ouvrages en métaux	381
1 672	3 238	7 060	13 221	28 602	56 574	115 485	..	Eléments de construction en métal	3813
15 083	27 583	50 908	95 428	202 261	341 634	699 839	..	Machines non électriques	382
229	499	937	1 524	3 876	6 869	12 436	..	Moteurs et turbines	3821
1 979	3 405	7 103	10 670	33 345	62 644	126 337	..	Machines agricoles	3822
395	979	1 300	2 851	4 594	11 415	25 168	..	Machines pour le travail du bois et du métal	3823
1 146	2 133	4 311	8 902	21 115	40 969	91 175	..	Machines et matériel spéciaux	3824
100	287	659	1 054	3 610	5 730	23 412	..	Machines de bureau et à calculer	3825
11 234	20 281	36 596	70 428	135 720	214 006	421 311	..	Machines et matériel, nca	3829
17 743	31 067	54 548	97 196	199 314	351 081	717 361	..	Machines électriques	383
3 327	5 049	9 744	17 645	39 998	68 185	124 277	..	Machines électriques industrielles	3831
9 429	16 718	28 926	44 372	76 447	156 385	349 636	..	Appareils de radio, télévision et télécom.	3832
950	2 417	4 206	7 050	17 688	36 325	60 763	..	Appareils électroménagers	3833
4 037	6 884	11 672	28 129	65 182	90 185	182 686	..	Appareils électriques, nca	3839
22 562	46 729	101 718	137 079	321 940	581 852	1 397 437	..	Matériel de transport	384
643	1 101	1 754	3 442	7 110	12 663	29 460	..	Construction navale	3841
..	1 187	..	..	5 478	9 267	25 352	..	Matériel ferroviaire	3842
20 413	42 884	95 039	124 839	297 252	535 645	1 290 868	..	Véhicules automobiles	3843
328	808	1 481	2 387	6 249	13 691	25 959	..	Motocycles et cycles	3844
..	743	..	..	..	..	25 047	..	Construction aéronautique	3845
..	4	18	39	..	..	751	..	Matériel de transport, nca	3849
785	2 276	4 207	8 032	21 614	32 784	76 503	..	Matériel professionnel et scientifique	385
431	1 083	3 782	7 538	20 352	30 174	70 772	..	Equipement professionnel	3851
195	179	320	416	1 083	..	..	..	Matériel photo et optique	3852
..	8	105	78	178	..	..	..	Montres et horloges	3853
658	1 296	2 286	4 943	9 628	20 212	55 122	..	**Autres industries manufacturières**	**39**
66	197	347	1 076	1 726	5 626	..	..	Joaillerie	3901
312 579	589 382	1 061 219	2 213 940	4 513 957	7 632 371	16 029 510	..	**Total industries manufacturières**	**3**
..	..	..	..	..	..	..	..	**Agriculture, chasse, sylviculture et pêche**	**1**
..	..	36 603	70 324	131 821	261 172	..	..	**Industries extractives**	**2**
..	..	15 981	30 005	54 664	101 714	..	..	Extraction du charbon	21
..	..	5 550	12 393	20 912	50 011	..	..	Production du pétrole brut et gaz naturel	22
..	..	10 732	20 364	39 445	86 049	..	..	Extraction d'autres minéraux	23.29
..	..	110 467	204 649	363 386	784 045	..	..	**Électricité, gaz et eau**	**4**
27 305	..	112 999	193 801	350 543	725 847	1 609 148	..	**Construction**	**5**
..	..	..	..	..	..	..	..	**Total services**	**6.9**
..	..	..	..	..	..	..	..	**Commerce de gros & détail ; rest. & hotels**	**6**
..	..	..	..	..	..	..	..	**Transport, entrepôts et communications**	**7**
..	..	..	..	..	..	..	..	**Banques, assur., aff. imm., serv. aux entrepr.**	**8**
..	..	..	..	..	..	..	..	**Services a la collect, sociaux et personnels**	**9**

Note: Break in series between 1991 and 1992. Some data are not given for reasons of confidentiality. ISIC 385 includes a figure for other professional goods (Turkish SIC 3854).
Data for 1997 are provisional.

TURKEY

Table TU.2 VALEUR AJOUTÉE - VALUE ADDED

Billions of TL (Current Prices)

ISIC revision 2		1991	1992	1993	1994	1995	1996	1997	1998
31	**Food, beverages and tobacco**	24 741	43 300	76 584	136 209	272 988	435 431	764 449	..
311.2	Food	13 716	26 741	47 638	91 679	187 931	307 603	569 589	..
313	Beverages	4 698	7 441	13 361	23 857	38 674	53 830	110 923	..
314	Tobacco	6 327	9 119	15 585	20 672	46 383	73 998	83 937	..
32	**Textiles, apparel and leather**	18 741	38 583	66 261	159 673	292 589	497 306	1 087 701	..
321	Textiles	13 809	27 571	46 584	115 248	203 241	339 444	746 185	..
3213	Knitting mills	1 363	3 828	8 648	17 876	29 070	41 123	100 663	..
322	Wearing apparel, except footwear	4 313	9 734	17 395	39 359	78 850	142 583	290 916	..
323	Leather and fur products	222	513	991	2 811	5 467	9 467	28 893	..
324	Footwear, except rubber or plastics	397	766	1 291	2 255	5 032	5 811	21 707	..
33	**Wood products and furniture**	1 128	2 913	5 547	7 934	19 253	34 874	77 856	..
331	Wood products, except furniture	755	1 998	3 666	4 422	12 170	17 973	38 023	..
332	Furniture, fixtures, except metal	373	915	1 882	3 512	7 083	16 901	39 833	..
34	**Paper, paper products and printing**	4 126	7 383	16 795	35 888	57 195	102 625	188 642	..
341	Paper and paper products	2 322	3 126	6 675	17 753	41 022	46 614	78 304	..
3411	Pulp, paper and paperboard articles	1 467	1 934	4 355	10 228	23 433	23 521	37 038	..
342	Printing and publishing	1 804	4 257	10 120	18 135	16 173	56 011	110 337	..
35	**Chemical products**	36 509	69 174	116 421	246 865	497 032	821 472	1 777 269	..
351	Industrial chemicals	5 516	10 768	16 288	46 108	86 857	118 596	232 530	..
3511	Basic industrial chemicals except fertilizers	812	4 336	2 771	4 635	9 555	15 499	33 096	..
3512	Fertilizers and pesticides	2 387	3 037	6 419	9 797	17 560	39 967	67 794	..
3513	Synthetic resins and plastic materials	2 317	3 395	7 098	31 676	59 741	63 131	131 640	..
352	Other chemicals	6 734	12 830	26 399	54 855	107 662	181 802	382 372	..
3521	Paints, varnishes and lacquers	759	1 527	3 257	8 500	16 167	21 892	65 959	..
3522	Drugs and medicines	3 760	6 333	13 298	26 915	45 569	78 983	151 417	..
3523	Soap, cleaning preparations, perfumes, cosmetics	1 451	3 476	6 947	12 706	31 053	55 757	112 211	..
3529	Other chemical products	763	1 494	2 896	6 734	14 874	25 168	52 785	..
353	Petroleum refineries	19 048	35 054	55 211	105 305	210 979	380 387	874 319	..
354	Miscellaneous petroleum and coal products	1 519	2 915	4 688	9 704	23 756	33 813	59 859	..
355	Rubber products	2 209	4 692	7 874	17 341	36 820	48 497	119 499	..
356	Plastics products	1 484	2 915	5 960	13 551	30 958	58 377	108 690	..
36	**Non-metallic mineral products**	9 127	17 827	34 754	70 335	119 422	199 852	424 281	..
361	Pottery, china, earthenware	1 962	3 030	5 680	15 363	21 918	30 047	64 769	..
362	Glass and glass products	1 780	3 850	6 740	13 396	27 950	44 772	88 522	..
369	Non-metallic mineral products nec.	5 386	10 947	22 334	41 576	69 554	125 033	270 990	..
37	**Basic metal industries**	9 237	15 520	33 413	85 386	116 225	176 668	411 090	..
371	Iron and steel	7 478	12 569	28 696	74 945	90 105	149 709	349 889	..
372	Non-ferrous metals	1 759	2 951	4 716	10 441	26 121	26 958	61 202	..
38	**Fabricated metal prod.; machinery & equip.**	26 732	52 568	101 518	176 589	345 806	597 975	1 366 116	..
381	Fabricated metal products	4 031	8 022	14 085	27 036	56 068	96 939	221 265	..
3813	Structural metal products	713	1 320	2 697	4 913	10 663	18 413	40 917	..
382	Non-electrical machinery	5 792	11 704	20 020	43 959	78 606	128 329	273 443	..
3821	Engines and turbines	117	244	366	567	1 380	2 332	4 390	..
3822	Agricultural machinery and equipment	578	1 181	1 902	3 220	14 057	14 656	36 565	..
3823	Metal and wood working machinery	220	491	617	1 449	2 021	5 640	11 698	..
3824	Special industrial machinery	531	927	1 853	4 361	8 056	15 586	34 989	..
3825	Office, computing and accounting machinery	42	98	220	225	797	1 681	4 410	..
3829	Other non-electrical machinery and equipment	4 304	8 763	15 063	34 137	52 294	88 435	181 392	..
383	Electrical machinery	7 648	13 833	25 412	45 686	87 043	149 039	291 084	..
3831	Electrical industrial machinery	1 671	2 491	4 679	8 007	17 401	30 279	52 387	..
3832	Radio, TV and communication equipment	4 236	7 730	14 346	23 834	38 964	71 873	151 382	..
3833	Electrical appliances and housewares	297	909	1 529	2 682	7 155	14 641	22 657	..
3839	Other electrical apparatus and supplies	1 443	2 704	4 858	11 163	23 523	32 247	64 657	..
384	Transport equipment	8 870	17 826	39 948	56 339	114 223	210 190	543 379	..
3841	Shipbuilding	402	819	1 343	2 515	4 846	7 735	22 046	..
3842	Railroad equipment	..	895	..	..	3 579	4 917	18 305	..
3843	Motor vehicles	7 632	15 219	35 497	48 434	98 810	183 759	473 115	..
3844	Motorcycles and bicycles	168	379	659	919	2 223	4 933	8 276	..
3845	Aircraft	..	514	..	..	..	..	21 243	..
3849	Other transport equipment	..	1	8	16	..	..	394	..
385	Professional and scientific equipment	392	1 182	2 053	3 570	9 866	13 479	36 945	..
3851	Professional and scientific equipment, nec	235	541	1 861	3 294	9 230	12 331	34 818	..
3852	Photographic and optical goods	95	87	160	261	588	..	..	..
3853	Watches and clocks	..	2	32	15	47	..	..	..
39	**Other manufacturing industries**	314	664	1 020	2 397	4 384	9 706	26 167	..
3901	Jewellery and related articles	32	102	182	509	857	2 798	..	..
3	**Total manufacturing**	130 654	247 932	452 313	921 275	1 724 894	2 875 908	6 123 571	..
1	**Agriculture, hunting, forestry and fishing**	..	..	..	..	..	..	..	..
2	**Mining and quarrying**	8 930	..	28 243	59 320	112 133	218 353	..	..
21	Coal mining	3 695	..	11 210	25 218	46 379	84 709	..	..
22	Crude petroleum and natural gas	3 062	..	4 930	11 522	19 539	47 264	..	..
23.29	Other mining	..	..	8 590	16 876	32 632	69 429	..	..
4	**Electricity, gas and water**	3 382	..	63 837	96 445	157 841	423 984	..	..
5	**Construction**	..	..	47 383	76 199	140 802	292 605	660 395	..
6.9	**Total services**	..	..	..	..	..	..	..	..
6	**Wholesale & retail trade; restaurants & hotels**	..	..	..	..	..	..	..	..
7	**Transport, storage and communication**	..	..	..	..	..	..	..	..
8	**Finance, insurance, real estate, business serv.**	..	..	..	..	..	..	..	..
9	**Community, social and personal services**	..	..	..	..	..	..	..	..

Note: Break in series between 1991 and 1992. Some data are not given for reasons of confidentiality. ISIC 385 includes a figure for other professional goods (Turkish SIC 3854).
Data for 1997 are provisional.

Table TU.3 INVESTISSEMENT - INVESTMENT

Milliards de TL (Prix courants)

1991	1992	1993	1994	1995	1996	1997	1998		CITI révision 2
1 268	2 592	7 035	10 236	21 243	49 240	92 500	..	**Alimentation, boissons et tabacs**	**31**
1 052	2 027	4 502	6 826	16 783	37 852	69 964	..	Alimentation	311.2
155	442	1 766	1 331	2 499	10 450	13 899	..	Boissons	313
61	123	767	2 078	1 961	937	8 638	..	Tabacs	314
2 290	7 001	7 839	20 990	54 732	141 668	249 460	..	**Textiles, habillement et cuir**	**32**
1 917	6 221	6 657	16 370	45 203	116 294	209 065	..	Textiles	321
188	3 205	995	4 526	6 233	15 782	24 322	..	Bonneterie	3213
339	695	1 051	4 165	8 250	22 336	33 972	..	Habillement, à l'exclusion des chaussures	322
17	45	71	250	612	2 282	3 999	..	Cuir et articles en cuir	323
16	41	60	206	667	755	2 423	..	Chaussures, sauf en caoutchouc et plastique	324
90	371	670	1 436	3 485	6 996	11 099	..	**Bois et meubles**	**33**
62	329	518	1 253	2 707	4 979	6 700	..	Ouvrages en bois	331
29	42	152	184	778	2 017	4 400	..	Meubles et accessoires	332
877	764	2 835	5 263	10 181	15 542	31 898	..	**Papier, impression et édition**	**34**
407	46	688	1 446	6 825	8 583	20 908	..	Papier et articles en papier	341
358	230	402	738	2 627	3 668	8 532	..	Pâte à papier, papier et carton	3411
471	717	2 147	3 817	3 357	6 960	10 990	..	Imprimerie et édition	342
2 612	4 514	7 154	16 897	41 751	82 373	151 810	..	**Produits chimiques**	**35**
517	1 228	1 926	3 978	5 688	10 701	36 604	..	Industrie chimique	351
102	545	305	413	1 010	1 769	8 177	..	Chimie de base, à l'exception des engrais	3511
220	289	240	522	1 756	4 751	462	..	Engrais et pesticides	3512
194	394	1 381	3 042	2 922	4 181	27 965	..	Résines synthétiques et matières plastiques	3513
805	865	2 148	3 134	7 202	14 808	35 524	..	Autres produits chimiques	352
78	93	251	277	981	2 664	7 817	..	Peintures, vernis et laques	3521
473	427	1 361	1 177	3 160	7 205	14 590	..	Produits pharmaceutiques	3522
179	100	189	866	949	3 131	9 892	..	Produits de nettoyage	3523
74	246	347	814	2 111	1 808	3 225	..	Autres produits chimiques	3529
321	1 335	580	115	11 933	24 607	13 112	..	Raffineries de pétrole	353
320	243	473	1 391	2 825	5 876	10 941	..	Derives du pétrole et du charbon	354
375	425	815	2 339	3 444	6 381	20 799	..	Industrie du caoutchouc	355
275	417	1 211	5 940	10 659	20 001	34 831	..	Ouvrages en matière plastique	356
2 212	2 872	4 357	12 160	19 387	50 357	99 569	..	**Produits minéraux non métalliques**	**36**
394	477	710	2 429	4 499	10 219	16 024	..	Grès, porcelaines et faïences	361
928	305	1 045	1 639	5 926	13 801	37 603	..	Industrie du verre	362
890	2 090	2 602	8 092	8 962	26 337	45 942	..	Produits minéraux non métalliques, nca	369
883	2 726	5 340	9 206	23 231	77 958	56 673	..	**Industries métallurgiques de base**	**37**
696	2 451	4 927	8 184	21 646	75 047	48 813	..	Sidérurgie	371
187	275	413	1 022	1 586	2 911	7 860	..	Métaux non ferreux	372
3 674	8 721	12 183	28 193	39 191	92 251	201 964	..	**Ouvrages en métaux, machines et matériel**	**38**
348	3 874	2 153	5 191	6 318	15 264	26 297	..	Ouvrages en métaux	381
50	- 1	412	2 335	992	4 334	7 230	..	Eléments de construction en métal	3813
602	1 015	2 270	3 284	7 398	23 298	34 727	..	Machines non électriques	382
10	59	12	15	142	226	460	..	Moteurs et turbines	3821
28	90	307	224	1 192	3 386	3 637	..	Machines agricoles	3822
26	243	37	15	173	620	1 418	..	Machines pour le travail du bois et du métal	3823
- 4	89	624	193	936	2 145	4 726	..	Machines et matériel spéciaux	3824
1	12	13	10	287	47	100	..	Machines de bureau et à calculer	3825
540	520	1 277	2 828	4 668	16 874	24 386	..	Machines et matériel, nca	3829
867	1 225	2 358	4 896	7 732	21 801	32 876	..	Machines électriques	383
174	270	673	1 300	2 014	3 596	7 474	..	Machines électriques industrielles	3831
525	439	844	1 662	2 813	8 981	12 544	..	Appareils de radio, télévision et télécom.	3832
45	179	297	152	491	2 072	2 631	..	Appareils électroménagers	3833
124	338	545	1 781	2 414	7 151	10 226	..	Appareils électriques, nca	3839
1 836	2 454	5 235	14 260	16 053	29 300	101 443	..	Matériel de transport	384
17	19	85	414	507	609	1 101	..	Construction navale	3841
..	28	..	..	32	48	1 108	..	Matériel ferroviaire	3842
1 770	2 310	4 899	13 788	15 174	26 520	96 572	..	Véhicules automobiles	3843
5	22	127	77	70	1 476	1 362	..	Motocycles et cycles	3844
..	75	..	..	..	..	1 294	..	Construction aéronautique	3845
..	..	1	6	..	..	5	..	Matériel de transport, nca	3849
21	154	166	561	1 690	2 588	6 621	..	Matériel professionnel et scientifique	385
15	74	144	548	1 650	2 513	6 555	..	Equipement professionnel	3851
2	6	15	12	38	..	..	..	Matériel photo et optique	3852
..	..	7	2	1	..	..	..	Montres et horloges	3853
21	53	894	428	589	2 269	7 003	..	**Autres industries manufacturières**	**39**
2	9	29	54	184	636	..	..	Joaillerie	3901
13 927	29 614	48 307	104 809	213 791	518 654	901 975	..	**Total industries manufacturières**	**3**
..	..	..	..	..	..	..	..	**Agriculture, chasse, sylviculture et pêche**	**1**
..	..	2 780	5 376	7 446	18 868	..	..	**Industries extractives**	**2**
..	..	1 802	3 616	4 537	6 026	..	..	Extraction du charbon	21
..	..	72	512	24	164	..	..	Production du pétrole brut et gaz naturel	22
..	..	799	952	1 620	10 770	..	..	Extraction d'autres minéraux	23.29
..	..	17 700	14 869	27 901	47 518	..	..	**Électricité, gaz et eau**	**4**
..	..	1 920	4 300	10 972	23 998	55 879	..	**Construction**	**5**
..	..	..	..	..	..	..	..	**Total services**	**6.9**
..	..	..	..	..	..	..	..	**Commerce de gros & détail ; rest. & hotels**	**6**
..	..	..	..	..	..	..	..	**Transport, entrepôts et communications**	**7**
..	..	..	..	..	..	..	..	**Banques, assur., aff. imm., serv. aux entrepr.**	**8**
..	..	..	..	..	..	..	..	**Services a la collect, sociaux et personnels**	**9**

Note: Break in series between 1991 and 1992. Some data are not given for reasons of confidentiality. ISIC 385 includes a figure for other professional goods (Turkish SIC 3854).
Data for 1997 are provisional.

TURKEY

Table TU.4 DÉPENSES EN MACHINES ET ÉQUIPEMENT - INVESTMENT IN MACHINERY AND EQUIPMENT

Billions of TL (Current Prices)

ISIC revision 2		1991	1992	1993	1994	1995	1996	1997	1998
31	**Food, beverages and tobacco**	818	1 677	3 341	7 688	13 858	30 801	57 763	..
311.2	Food	696	1 294	1 869	5 043	10 788	24 108	42 023	..
313	Beverages	94	326	1 019	965	1 360	6 276	10 155	..
314	Tobacco	29	57	453	1 680	1 710	417	5 585	..
32	**Textiles, apparel and leather**	1 848	5 920	7 013	16 339	43 606	109 954	197 224	..
321	Textiles	1 640	5 499	6 357	13 462	37 942	96 062	172 404	..
3213	Knitting mills	158	3 117	1 029	4 000	4 706	13 044	18 906	..
322	Wearing apparel, except footwear	188	364	566	2 592	4 708	12 027	20 175	..
323	Leather and fur products	9	25	45	126	414	1 172	2 995	..
324	Footwear, except rubber or plastics	12	33	45	160	542	693	1 651	..
33	**Wood products and furniture**	64	314	560	1 144	2 448	5 310	8 238	..
331	Wood products, except furniture	52	289	430	1 034	1 876	4 161	5 629	..
332	Furniture, fixtures, except metal	12	25	131	110	572	1 149	2 609	..
34	**Paper, paper products and printing**	719	378	1 909	4 103	8 049	11 839	24 525	..
341	Paper and paper products	381	23	525	1 205	5 931	7 629	16 210	..
3411	Pulp, paper and paperboard articles	338	202	290	613	1 885	3 274	6 824	..
342	Printing and publishing	339	355	1 383	2 898	2 118	4 210	8 315	..
35	**Chemical products**	1 965	2 864	5 283	11 128	32 600	69 635	113 957	..
351	Industrial chemicals	437	1 070	1 767	4 192	4 596	9 242	31 812	..
3511	Basic industrial chemicals except fertilizers	92	518	253	316	773	1 253	7 001	..
3512	Fertilizers and pesticides	187	178	192	491	1 176	3 824	246	..
3513	Synthetic resins and plastic materials	159	374	1 322	3 384	2 646	4 165	24 564	..
352	Other chemicals	553	607	1 339	1 747	4 104	9 901	17 550	..
3521	Paints, varnishes and lacquers	39	39	124	120	510	1 582	4 591	..
3522	Drugs and medicines	307	295	805	596	1 749	4 684	6 954	..
3523	Soap, cleaning preparations, perfumes, cosmetics	156	70	127	458	524	2 440	4 682	..
3529	Other chemical products	51	202	283	574	1 321	1 195	1 322	..
353	Petroleum refineries	289	345	284	100	11 853	24 262	12 221	..
354	Miscellaneous petroleum and coal products	137	153	205	905	1 413	3 504	7 847	..
355	Rubber products	349	367	681	2 206	2 345	5 085	15 017	..
356	Plastics products	199	322	1 008	1 978	8 288	17 642	29 511	..
36	**Non-metallic mineral products**	1 704	2 434	3 509	8 540	15 779	39 988	76 900	..
361	Pottery, china, earthenware	328	398	567	2 070	3 573	9 033	13 840	..
362	Glass and glass products	694	279	557	1 363	5 506	12 306	33 129	..
369	Non-metallic mineral products nec.	682	1 756	2 385	5 107	6 701	18 649	29 931	..
37	**Basic metal industries**	613	2 383	3 650	7 727	19 079	71 066	39 259	..
371	Iron and steel	474	2 156	3 295	6 942	17 757	68 741	33 780	..
372	Non-ferrous metals	138	227	355	784	1 322	2 325	5 479	..
38	**Fabricated metal prod.; machinery & equip.**	3 019	6 743	9 301	22 586	26 608	64 543	147 786	..
381	Fabricated metal products	259	3 003	1 272	2 922	3 721	11 244	17 385	..
3813	Structural metal products	37	45	91	602	754	2 822	4 228	..
382	Non-electrical machinery	403	602	1 866	2 766	5 197	17 266	25 767	..
3821	Engines and turbines	7	58	10	6	121	207	444	..
3822	Agricultural machinery and equipment	23	43	254	136	953	1 761	1 814	..
3823	Metal and wood working machinery	17	35	25	- 5	70	465	1 048	..
3824	Special industrial machinery	8	54	542	95	527	1 086	2 928	..
3825	Office, computing and accounting machinery		4	1	2	186	17	17	..
3829	Other non-electrical machinery and equipment	348	409	1 034	2 532	3 341	13 730	19 516	..
383	Electrical machinery	696	962	1 902	4 224	5 539	13 860	22 387	..
3831	Electrical industrial machinery	133	227	574	1 101	1 331	2 838	6 145	..
3832	Radio, TV and communication equipment	447	326	584	1 393	1 972	5 038	6 240	..
3833	Electrical appliances and housewares	27	142	293	118	356	1 169	2 138	..
3839	Other electrical apparatus and supplies	89	267	451	1 613	1 880	4 816	7 865	..
384	Transport equipment	1 646	2 036	4 123	12 273	10 823	19 539	76 584	..
3841	Shipbuilding	10	8	43	375	345	188	265	..
3842	Railroad equipment	..	23	..	..	31	12	172	..
3843	Motor vehicles	1 602	1 957	3 939	11 688	10 402	18 622	74 179	..
3844	Motorcycles and bicycles	3	19	55	10	- 135	348	1 185	..
3845	Aircraft	..	28	..	..	..	..	780	..
3849	Other transport equipment	..	..	1	4	..	..	2	..
385	Professional and scientific equipment	15	140	138	400	1 329	2 633	5 664	..
3851	Professional and scientific equipment, nec	10	71	118	394	1 298	2 591	5 603	..
3852	Photographic and optical goods	2	5	14	5	30	..	..	..
3853	Watches and clocks			6	1	1	..	..	..
39	**Other manufacturing industries**	17	37	236	383	417	1 627	3 798	..
3901	Jewellery and related articles	..	8	25	31	85	444	..	..
3	**Total manufacturing**	10 767	22 750	34 801	79 637	162 444	404 763	669 450	..
1	**Agriculture, hunting, forestry and fishing**	..	..	..	..	..	..	..	..
2	**Mining and quarrying**	..	..	1 250	1 788	2 562	7 090	..	..
21	Coal mining	..	..	402	871	1 413	1 182	..	..
22	Crude petroleum and natural gas	..	..	11	329	23	130	..	..
23.29	Other mining	..	..	814	543	837	5 067	..	..
4	**Electricity, gas and water**	..	..	11 088	13 107	20 023	31 455	..	..
5	**Construction**	..	..	953	2 091	5 071	9 431	23 337	..
6.9	**Total services**	..	..	..	..	..	..	..	..
6	**Wholesale & retail trade; restaurants & hotels**	..	..	..	..	..	..	..	..
7	**Transport, storage and communication**	..	..	..	..	..	..	..	..
8	**Finance, insurance, real estate, business serv.**	..	..	..	..	..	..	..	..
9	**Community, social and personal services**	..	..	..	..	..	..	..	..

Note: Break in series between 1991 and 1992. Some data are not given for reasons of confidentiality. ISIC 385 includes a figure for other professional goods (Turkish SIC 3854).
Data for 1997 are provisional.

Table TU.5 **EMPLOI - EMPLOYMENT**

Milliers

1991	1992	1993	1994	1995	1996	1997	1998		CITI révision 2
172	185	178	169	169	172	177	..	**Alimentation, boissons et tabacs**	**31**
124	141	135	131	134	138	143	..	Alimentation	311.2
15	14	14	12	11	11	11	..	Boissons	313
32	30	29	26	24	23	23	..	Tabacs	314
251	289	292	289	319	355	400	..	**Textiles, habillement et cuir**	**32**
172	188	190	187	199	224	258	..	Textiles	321
19	27	34	31	32	30	37	..	Bonneterie	3213
69	87	89	90	106	118	123	..	Habillement, à l'exclusion des chaussures	322
4	6	6	7	7	7	11	..	Cuir et articles en cuir	323
6	7	8	6	7	6	9	..	Chaussures, sauf en caoutchouc et plastique	324
16	23	23	21	20	23	26	..	**Bois et meubles**	**33**
12	15	14	14	12	12	14	..	Ouvrages en bois	331
4	7	8	7	8	11	13	..	Meubles et accessoires	332
33	35	34	34	35	36	33	..	**Papier, impression et édition**	**34**
21	20	20	20	21	19	20	..	Papier et articles en papier	341
13	14	13	13	13	11	11	..	Pâte à papier, papier et carton	3411
12	15	15	14	14	17	13	..	Imprimerie et édition	342
90	95	94	90	93	98	108	..	**Produits chimiques**	**35**
29	28	26	25	23	22	22	..	Industrie chimique	351
6	16	5	4	4	4	4	..	Chimie de base, à l'exception des engrais	3511
9	5	8	7	7	7	7	..	Engrais et pesticides	3512
13	6	13	13	12	11	12	..	Résines synthétiques et matières plastiques	3513
26	27	27	26	29	31	35	..	Autres produits chimiques	352
4	4	4	4	4	5	5	..	Peintures, vernis et laques	3521
12	12	13	12	13	14	15	..	Produits pharmaceutiques	3522
6	6	6	5	6	6	8	..	Produits de nettoyage	3523
5	5	5	5	6	7	7	..	Autres produits chimiques	3529
5	5	5	5	5	5	5	..	Raffineries de pétrole	353
5	4	4	5	4	4	5	..	Derives du pétrole et du charbon	354
11	13	14	13	12	12	15	..	Industrie du caoutchouc	355
14	18	18	17	19	23	27	..	Ouvrages en matière plastique	356
66	70	66	65	66	68	74	..	**Produits minéraux non métalliques**	**36**
11	10	10	11	12	11	12	..	Grès, porcelaines et faïences	361
13	12	11	10	10	11	13	..	Industrie du verre	362
42	48	45	43	44	46	49	..	Produits minéraux non métalliques, nca	369
74	71	69	63	64	58	57	..	**Industries métallurgiques de base**	**37**
58	56	55	51	52	47	44	..	Sidérurgie	371
16	15	14	12	12	11	13	..	Métaux non ferreux	372
194	212	218	200	202	218	249	..	**Ouvrages en métaux, machines et matériel**	**38**
37	43	44	41	42	48	57	..	Ouvrages en métaux	381
7	10	11	10	9	11	13	..	Eléments de construction en métal	3813
45	50	51	48	47	49	55	..	Machines non électriques	382
2	2	1	1	1	1	1	..	Moteurs et turbines	3821
9	9	8	8	8	9	9	..	Machines agricoles	3822
3	5	3	3	3	3	4	..	Machines pour le travail du bois et du métal	3823
7	8	8	8	9	10	11	..	Machines et matériel spéciaux	3824
	1	1	1	1	1	1	..	Machines de bureau et à calculer	3825
25	26	29	27	25	26	29	..	Machines et matériel, nca	3829
44	46	45	40	41	45	51	..	Machines électriques	383
10	9	11	10	10	12	13	..	Machines électriques industrielles	3831
18	20	18	15	14	15	17	..	Appareils de radio, télévision et télécom.	3832
3	5	4	4	5	6	7	..	Appareils électroménagers	3833
12	12	11	11	13	13	15	..	Appareils électriques, nca	3839
64	68	73	66	65	70	78	..	Matériel de transport	384
5	5	5	4	4	4	4	..	Construction navale	3841
..	8	..	..	6	6	6	..	Matériel ferroviaire	3842
47	51	57	51	50	55	62	..	Véhicules automobiles	3843
1	2	2	2	2	3	2	..	Motocycles et cycles	3844
..	3	..	..	..	..	3	..	Construction aéronautique	3845
..				..	..		..	Matériel de transport, nca	3849
5	5	5	5	6	6	7	..	Matériel professionnel et scientifique	385
3	3	5	4	5	5	6	..	Equipement professionnel	3851
1	1	1	1	1	..	..	..	Matériel photo et optique	3852
..					..	..	..	Montres et horloges	3853
4	5	6	6	6	7	8	..	**Autres industries manufacturières**	**39**
1	1	1	2	2	2	..	..	Joaillerie	3901
901	985	979	937	974	1037	1 133	..	**Total industries manufacturières**	**3**
..	..	..	..	..	..	..	..	**Agriculture, chasse, sylviculture et pêche**	**1**
89	..	92	83	82	83	..	..	**Industries extractives**	**2**
66	..	64	57	56	53	..	..	Extraction du charbon	21
1	..	1	1	1	1	..	..	Production du pétrole brut et gaz naturel	22
..	..	18	17	17	21	..	..	Extraction d'autres minéraux	23.29
37	..	95	96	90	88	..	..	**Électricité, gaz et eau**	**4**
..	..	156	181	190	202	222	..	**Construction**	**5**
..	..	..	..	..	..	..	..	**Total services**	**6.9**
..	..	..	..	..	..	..	..	**Commerce de gros & détail ; rest. & hotels**	**6**
..	..	..	..	..	..	..	..	**Transport, entrepôts et communications**	**7**
..	..	..	..	..	..	..	..	**Banques, assur., aff. imm., serv. aux entrepr.**	**8**
..	..	..	..	..	..	..	..	**Services a la collect, sociaux et personnels**	**9**

Note: Break in series between 1991 and 1992. Some data are not given for reasons of confidentiality. ISIC 385 includes a figure for other professional goods (Turkish SIC 3854).
Data for 1997 are provisional.

Table TU.6 **EMPLOI, SALARIÉS - EMPLOYMENT, EMPLOYEES**

Thousands

ISIC revision 2		1991	1992	1993	1994	1995	1996	1997	1998
31	**Food, beverages and tobacco**	171	183	176	168	168	171	177	..
311.2	Food	124	139	133	130	133	138	143	..
313	Beverages	15	14	14	11	11	11	11	..
314	Tobacco	32	30	29	26	24	23	23	..
32	**Textiles, apparel and leather**	251	287	291	288	318	355	400	..
321	Textiles	172	188	189	186	199	224	257	..
3213	Knitting mills	19	27	34	31	32	30	37	..
322	Wearing apparel, except footwear	69	87	89	90	106	118	123	..
323	Leather and fur products	4	6	6	7	7	7	11	..
324	Footwear, except rubber or plastics	6	7	7	6	7	6	9	..
33	**Wood products and furniture**	16	22	22	21	20	23	26	..
331	Wood products, except furniture	12	15	14	13	12	12	13	..
332	Furniture, fixtures, except metal	4	7	8	7	8	11	13	..
34	**Paper, paper products and printing**	33	35	34	34	35	36	33	..
341	Paper and paper products	21	20	20	20	21	19	20	..
3411	Pulp, paper and paperboard articles	13	14	13	13	13	11	11	..
342	Printing and publishing	12	15	15	14	14	17	13	..
35	**Chemical products**	90	95	93	90	92	98	108	..
351	Industrial chemicals	29	28	26	25	23	22	22	..
3511	Basic industrial chemicals except fertilizers	6	16	5	4	4	4	4	..
3512	Fertilizers and pesticides	9	5	8	7	7	7	7	..
3513	Synthetic resins and plastic materials	13	6	13	13	12	11	12	..
352	Other chemicals	26	27	27	26	29	31	35	..
3521	Paints, varnishes and lacquers	4	4	4	4	4	5	5	..
3522	Drugs and medicines	12	12	13	12	13	14	15	..
3523	Soap, cleaning preparations, perfumes, cosmetics	6	6	6	5	6	6	8	..
3529	Other chemical products	5	5	5	5	5	6	7	7	..
353	Petroleum refineries	5	5	5	5	5	5	5	..
354	Miscellaneous petroleum and coal products	5	4	4	5	4	4	5	..
355	Rubber products	11	13	13	13	12	12	15	..
356	Plastics products	14	17	18	17	19	23	27	..
36	**Non-metallic mineral products**	66	70	66	65	66	68	73	..
361	Pottery, china, earthenware	11	10	10	11	12	11	12	..
362	Glass and glass products	13	12	11	10	10	11	13	..
369	Non-metallic mineral products nec.	42	48	44	43	44	46	48	..
37	**Basic metal industries**	74	71	69	63	64	58	57	..
371	Iron and steel	58	56	55	51	52	47	44	..
372	Non-ferrous metals	16	15	14	12	12	11	13	..
38	**Fabricated metal prod.; machinery & equip.**	194	211	217	199	201	218	248	..
381	Fabricated metal products	36	43	44	41	42	48	56	..
3813	Structural metal products	7	9	10	10	9	11	13	..
382	Non-electrical machinery	45	50	50	48	47	49	55	..
3821	Engines and turbines	2	2	1	1	1	1	1	..
3822	Agricultural machinery and equipment	9	9	8	8	8	8	9	..
3823	Metal and wood working machinery	3	5	3	3	3	3	4	..
3824	Special industrial machinery	7	8	8	8	9	10	11	..
3825	Office, computing and accounting machinery	..	..	1	1	1	1	1	..
3829	Other non-electrical machinery and equipment	25	26	29	27	25	26	29	..
383	Electrical machinery	44	46	44	40	41	45	51	..
3831	Electrical industrial machinery	10	9	11	10	10	12	12	..
3832	Radio, TV and communication equipment	18	20	18	15	14	15	17	..
3833	Electrical appliances and housewares	3	5	4	4	5	6	7	..
3839	Other electrical apparatus and supplies	12	12	11	11	13	13	15	..
384	Transport equipment	64	68	73	66	65	70	78	..
3841	Shipbuilding	5	5	5	4	4	4	4	..
3842	Railroad equipment	8	8	..	..	6	6	6	..
3843	Motor vehicles	47	51	57	50	50	55	62	..
3844	Motorcycles and bicycles	1	2	2	2	2	3	2	..
3845	Aircraft	2	3	..	..	..	..	3	..
3849	Other transport equipment	..	..	..	..	..	..		..
385	Professional and scientific equipment	5	5	5	5	6	6	7	..
3851	Professional and scientific equipment, nec	3	3	5	4	5	5	6	..
3852	Photographic and optical goods	1	1	1		1	..	..	..
3853	Watches and clocks	..	..	..	..	..	..	..	..
39	**Other manufacturing industries**	4	5	6	6	6	7	8	..
3901	Jewellery and related articles	1	1	1	2	2	2	..	..
3	**Total manufacturing**	900	979	975	933	971	1 034.1	1 130.9	..
1	**Agriculture, hunting, forestry and fishing**	..	..	..	..	..	..	..	..
2	**Mining and quarrying**	..	..	91	82	82	82	..	..
21	Coal mining	..	..	64	57	56	53	..	..
22	Crude petroleum and natural gas	..	..	1	1	1	1	..	..
23.29	Other mining	..	..	17	16	16	20	..	..
4	**Electricity, gas and water**	..	..	95	96	90	88	..	..
5	**Construction**	..	..	148	174	183	195	216	..
6.9	**Total services**	..	..	..	..	..	..	..	..
6	**Wholesale & retail trade; restaurants & hotels**	..	..	..	..	..	..	..	..
7	**Transport, storage and communication**	..	..	..	..	..	..	..	..
8	**Finance, insurance, real estate, business serv.**	..	..	..	..	..	..	..	..
9	**Community, social and personal services**	..	..	..	..	..	..	..	..

Note: Break in series between 1991 and 1992. Some data are not given for reasons of confidentiality. ISIC 385 includes a figure for other professional goods (Turkish SIC 3854).
Data for 1997 are provisional.

Table TU.7 **SALAIRES ET CHARGES SOCIALES , SALARIÉS - COMPENSATION OF LABOUR, EMPLOYEES**

Milliards de TL (Prix courants)

1991	1992	1993	1994	1995	1996	1997	1998	CITI révision 2	
6 299	14 373	22 657	34 324	54 057	92 095	187 298	..	**Alimentation, boissons et tabacs**	**31**
4 380	10 784	16 260	23 746	39 091	70 155	143 208	..	Alimentation	311.2
681	1 160	2 144	3 069	5 361	8 061	14 891	..	Boissons	313
1 238	2 428	4 253	7 509	9 606	13 879	29 199	..	Tabacs	314
7 145	12 114	22 739	39 987	64 811	136 149	300 358	..	**Textiles, habillement et cuir**	**32**
5 661	9 288	17 552	30 311	46 021	98 563	210 433	..	Textiles	321
396	852	2 099	8 553	5 301	9 827	24 122	..	Bonneterie	3213
1 233	2 346	4 314	8 191	15 795	33 403	76 158	..	Habillement, à l'exclusion des chaussures	322
64	153	261	827	1 873	2 343	7 661	..	Cuir et articles en cuir	323
186	327	613	658	1 122	1 840	6 106	..	Chaussures, sauf en caoutchouc et plastique	324
495	1 061	2 427	2 810	4 015	8 873	17 251	..	**Bois et meubles**	**33**
393	813	1 502	2 086	2 687	4 923	9 699	..	Ouvrages en bois	331
101	247	925	724	1 327	3 950	7 551	..	Meubles et accessoires	332
1 477	2 722	4 415	7 187	12 237	22 530	42 720	..	**Papier, impression et édition**	**34**
947	1 647	2 786	4 521	8 514	13 059	27 822	..	Papier et articles en papier	341
672	1 298	2 148	3 368	6 005	8 265	17 552	..	Pâte à papier, papier et carton	3411
530	1 075	1 628	2 666	3 723	9 471	14 898	..	Imprimerie et édition	342
5 405	9 840	17 667	29 774	48 690	85 759	186 327	..	**Produits chimiques**	**35**
1 785	3 286	5 287	8 475	13 273	22 326	43 747	..	Industrie chimique	351
313	1 934	817	1 146	2 092	3 368	6 023	..	Chimie de base, à l'exception des engrais	3511
535	641	1 624	2 615	4 023	6 864	13 158	..	Engrais et pesticides	3512
937	710	2 846	4 715	7 158	12 095	24 566	..	Résines synthétiques et matières plastiques	3513
1 678	3 005	6 167	8 554	16 133	31 668	68 408	..	Autres produits chimiques	352
327	346	1 375	1 032	2 000	3 983	8 695	..	Peintures, vernis et laques	3521
808	1 556	2 779	4 424	8 342	16 333	35 679	..	Produits pharmaceutiques	3522
278	640	1 188	1 741	3 061	5 799	12 347	..	Produits de nettoyage	3523
264	463	825	1 357	2 730	5 552	11 687	..	Autres produits chimiques	3529
463	866	1 536	2 743	4 053	5 365	14 522	..	Raffineries de pétrole	353
349	546	928	1 773	2 726	4 459	9 744	..	Derives du pétrole et du charbon	354
698	1 342	2 341	4 489	7 814	11 506	26 195	..	Industrie du caoutchouc	355
433	796	1 407	3 739	4 690	10 435	23 711	..	Ouvrages en matière plastique	356
2 733	5 461	9 090	15 182	24 004	43 458	89 000	..	**Produits minéraux non métalliques**	**36**
408	733	1 336	2 240	4 039	6 707	14 760	..	Grès, porcelaines et faïences	361
793	1 606	2 022	3 457	7 355	13 092	28 170	..	Industrie du verre	362
1 531	3 122	5 731	9 485	12 610	23 659	46 070	..	Produits minéraux non métalliques, nca	369
5 008	9 178	14 049	21 955	30 086	51 535	85 815	..	**Industries métallurgiques de base**	**37**
4 182	7 770	11 714	18 923	25 332	44 304	66 453	..	Sidérurgie	371
826	1 408	2 335	3 033	4 754	7 231	19 362	..	Métaux non ferreux	372
8 780	16 333	28 921	51 945	74 837	141 754	315 475	..	**Ouvrages en métaux, machines et matériel**	**38**
1 278	2 333	4 220	11 309	10 710	21 996	53 390	..	Ouvrages en métaux	381
255	468	922	1 327	2 096	4 468	10 264	..	Eléments de construction en métal	3813
2 018	3 759	6 413	12 168	16 727	30 444	68 379	..	Machines non électriques	382
86	143	287	264	429	755	1 621	..	Moteurs et turbines	3821
355	737	1 092	1 881	2 895	5 405	11 526	..	Machines agricoles	3822
128	308	410	516	587	1 231	3 322	..	Machines pour le travail du bois et du métal	3823
257	471	899	1 357	2 822	4 759	10 949	..	Machines et matériel spéciaux	3824
14	23	44	75	141	213	984	..	Machines de bureau et à calculer	3825
1 177	2 077	3 681	8 075	9 854	18 081	39 977	..	Machines et matériel, nca	3829
2 110	3 888	6 634	9 194	16 693	33 101	73 465	..	Machines électriques	383
477	731	1 502	2 253	4 046	7 975	17 313	..	Machines électriques industrielles	3831
1 017	2 059	3 458	4 470	7 774	15 484	33 813	..	Appareils de radio, télévision et télécom.	3832
75	181	291	435	932	2 224	4 646	..	Appareils électroménagers	3833
540	917	1 383	2 036	3 942	7 418	17 694	..	Appareils électriques, nca	3839
3 202	6 032	11 105	18 576	29 185	53 115	113 897	..	Matériel de transport	384
286	498	799	1 318	1 836	3 052	7 030	..	Construction navale	3841
379	693	..	..	2 993	4 718	9 360	..	Matériel ferroviaire	3842
2 322	4 394	8 140	13 671	21 904	40 254	88 028	..	Véhicules automobiles	3843
40	113	223	271	581	1 375	2 629	..	Motocycles et cycles	3844
174	334	..	..	..	..	6 749	..	Construction aéronautique	3845
..	1	3	5	..	..	101	..	Matériel de transport, nca	3849
172	321	549	698	1 522	3 099	6 342	..	Matériel professionnel et scientifique	385
133	228	511	653	1 393	2 804	5 803	..	Equipement professionnel	3851
24	18	29	40	115	..	..	..	Matériel photo et optique	3852
..	1	8	4	13	..	..	..	Montres et horloges	3853
105	223	387	554	996	2 312	6 019	..	**Autres industries manufacturières**	**39**
9	49	76	130	253	588	..	..	Joaillerie	3901
37 447	71 305	122 350	203 719	313 732	584 466	1 230 262	..	**Total industries manufacturières**	**3**
..	..	..	..	..	..	..	..	**Agriculture, chasse, sylviculture et pêche**	**1**
..	..	16 299	23 333	34 153	50 280	..	..	**Industries extractives**	**2**
..	..	13 084	18 212	26 345	35 850	..	..	Extraction du charbon	21
..	..	357	666	827	1 078	..	..	Production du pétrole brut et gaz naturel	22
..	..	1 475	2 484	3 756	8 908	..	..	Extraction d'autres minéraux	23.29
..	..	14 441	23 510	36 670	62 649	..	..	**Électricité, gaz et eau**	**4**
..	..	8 144	14 485	26 416	49 568	126 005	..	**Construction**	**5**
..	..	..	..	..	..	..	..	**Total services**	**6.9**
..	..	..	..	..	..	..	..	**Commerce de gros & détail ; rest. & hotels**	**6**
..	..	..	..	..	..	..	..	**Transport, entrepôts et communications**	**7**
..	..	..	..	..	..	..	..	**Banques, assur., aff. imm., serv. aux entrepr.**	**8**
..	..	..	..	..	..	..	..	**Services a la collect, sociaux et personnels**	**9**

Note: Break in series between 1991 and 1992. Some data are not given for reasons of confidentiality. ISIC 385 includes a figure for other professional goods (Turkish SIC 3854).
Data for 1997 are provisional.

TURKEY

Table TU.8 WAGES AND SALARIES, EMPLOYEES - SALAIRES ET TRAITEMENTS, SALARIÉS

Billions of TL (Current Prices)

ISIC revision 2		1991	1992	1993	1994	1995	1996	1997	1998
31	**Food, beverages and tobacco**	5 515	10 273	16 820	27 591	45 370	77 372	158 167	..
311.2	Food	3 812	7 249	11 736	18 947	32 618	58 749	120 094	..
313	Beverages	601	968	1 676	2 569	2 884	5 518	10 346	..
314	Tobacco	1 102	2 056	3 408	6 074	368	575	1 542	..
32	**Textiles, apparel and leather**	6 145	9 658	16 905	26 663	53 792	111 925	242 900	..
321	Textiles	4 888	7 344	12 844	19 700	38 412	81 258	169 678	..
3213	Knitting mills	338	683	1 610	2 305	4 402	8 186	19 587	..
322	Wearing apparel, except footwear	1 040	1 917	3 361	5 752	12 878	27 286	61 810	..
323	Leather and fur products	53	122	205	665	1 563	1 914	6 328	..
324	Footwear, except rubber or plastics	163	275	495	546	939	1 467	5 084	..
33	**Wood products and furniture**	421	846	1 520	2 234	3 334	7 121	13 993	..
331	Wood products, except furniture	336	649	1 097	1 653	2 239	3 893	7 875	..
332	Furniture, fixtures, except metal	86	197	423	581	1 094	3 228	6 118	..
34	**Paper, paper products and printing**	1 282	2 190	3 531	5 837	10 525	19 156	36 470	..
341	Paper and paper products	824	1 320	2 240	3 795	7 407	11 150	23 938	..
3411	Pulp, paper and paperboard articles	588	1 036	1 731	2 858	5 273	7 146	15 238	..
342	Printing and publishing	458	869	1 291	2 042	3 118	8 006	12 532	..
35	**Chemical products**	4 762	8 209	13 607	23 438	43 675	75 651	160 544	..
351	Industrial chemicals	1 570	2 664	4 184	7 267	11 985	19 778	38 534	..
3511	Basic industrial chemicals except fertilizers	274	1 517	657	960	1 815	3 042	5 364	..
3512	Fertilizers and pesticides	472	548	1 332	2 231	3 634	6 064	11 309	..
3513	Synthetic resins and plastic materials	824	599	2 195	4 076	6 536	10 672	21 861	..
352	Other chemicals	1 480	2 533	4 374	7 097	14 522	28 619	59 584	..
3521	Paints, varnishes and lacquers	300	297	451	820	1 819	3 544	7 306	..
3522	Drugs and medicines	704	1 304	2 300	3 663	7 686	14 994	31 338	..
3523	Soap, cleaning preparations, perfumes, cosmetics	244	541	940	1 464	2 528	5 100	10 662	..
3529	Other chemical products	233	391	683	1 150	2 489	4 982	10 278	..
353	Petroleum refineries	418	742	1 266	2 324	3 675	4 275	11 658	..
354	Miscellaneous petroleum and coal products	307	462	722	1 290	2 385	3 957	8 320	..
355	Rubber products	613	1 167	1 963	3 731	7 231	10 367	22 781	..
356	Plastics products	373	641	1 098	1 730	3 878	8 655	19 667	..
36	**Non-metallic mineral products**	2 398	4 261	6 566	10 436	19 867	36 347	74 225	..
361	Pottery, china, earthenware	355	569	1 060	1 777	3 327	5 678	12 360	..
362	Glass and glass products	710	1 216	1 652	2 650	5 853	11 192	24 072	..
369	Non-metallic mineral products nec.	1 333	2 476	3 854	6 009	10 687	19 477	37 792	..
37	**Basic metal industries**	4 473	6 950	10 969	17 184	25 165	44 659	71 710	..
371	Iron and steel	3 756	5 883	9 189	14 704	21 103	38 411	55 801	..
372	Non-ferrous metals	717	1 067	1 780	2 479	4 061	6 248	15 910	..
38	**Fabricated metal prod.; machinery & equip.**	7 660	13 187	23 504	34 644	65 932	124 289	273 387	..
381	Fabricated metal products	1 111	1 847	3 257	5 257	9 229	18 733	45 177	..
3813	Structural metal products	222	364	705	1 074	1 771	3 739	8 552	..
382	Non-electrical machinery	1 761	2 983	5 023	7 739	14 301	26 527	59 028	..
3821	Engines and turbines	76	111	180	212	387	673	1 435	..
3822	Agricultural machinery and equipment	308	524	912	1 470	2 426	4 528	9 606	..
3823	Metal and wood working machinery	114	254	318	420	492	1 039	2 858	..
3824	Special industrial machinery	225	391	708	1 109	2 413	4 064	9 287	..
3825	Office, computing and accounting machinery	12	19	35	56	113	170	838	..
3829	Other non-electrical machinery and equipment	1 025	1 683	2 870	4 472	8 470	16 054	35 004	..
383	Electrical machinery	1 845	3 203	5 574	7 639	15 010	29 635	64 589	..
3831	Electrical industrial machinery	412	600	1 253	1 879	3 679	7 099	15 019	..
3832	Radio, TV and communication equipment	901	1 719	2 978	3 770	7 048	14 205	30 726	..
3833	Electrical appliances and housewares	65	148	237	323	797	1 894	3 840	..
3839	Other electrical apparatus and supplies	466	736	1 107	1 667	3 486	6 437	15 003	..
384	Transport equipment	2 791	4 892	9 213	13 438	26 076	46 703	99 241	..
3841	Shipbuilding	255	413	668	1 134	1 641	2 646	6 074	..
3842	Railroad equipment	..	567	1 104	1 889	2 774	4 229	8 169	..
3843	Motor vehicles	2 012	3 521	6 735	9 171	19 370	35 147	76 359	..
3844	Motorcycles and bicycles	35	95	178	218	518	1 177	2 323	..
3845	Aircraft	..	296	527	1 022	..	..	6 234	..
3849	Other transport equipment	..	1	2	4	..	..	81	..
385	Professional and scientific equipment	152	261	437	571	1 316	2 690	5 352	..
3851	Professional and scientific equipment, nec	118	184	297	351	1 206	2 445	4 918	..
3852	Photographic and optical goods	20	15	23	33	98	..	..	..
3853	Watches and clocks	..		7	3	12	..	..	..
39	**Other manufacturing industries**	90	170	293	440	829	1 863	4 896	..
3901	Jewellery and related articles	7	39	56	103	209	477	..	..
3	**Total manufacturing**	32 746	55 744	93 713	148 466	268 488	498 382	1 036 292	..
1	**Agriculture, hunting, forestry and fishing**	..	..	..	..	..	..	..	..
2	**Mining and quarrying**	..	..	..	..	30 760	42 747	..	..
21	Coal mining	..	..	..	..	23 755	30 809	..	..
22	Crude petroleum and natural gas	..	..	..	..	754	985	..	..
23.29	Other mining	..	..	..	..	6 251	10 953	..	..
4	**Electricity, gas and water**	..	..	..	..	34 425	57 097	..	..
5	**Construction**	..	..	..	..	20 771	32 692	101 597	..
6.9	**Total services**	..	..	..	..	..	..	..	..
6	**Wholesale & retail trade; restaurants & hotels**	..	..	..	..	..	..	..	..
7	**Transport, storage and communication**	..	..	..	..	..	..	..	..
8	**Finance, insurance, real estate, business serv.**	..	..	..	..	..	..	..	..
9	**Community, social and personal services**	..	..	..	..	..	..	..	..

Note: Break in series between 1991 and 1992. Some data are not given for reasons of confidentiality. ISIC 385 includes a figure for other professional goods (Turkish SIC 3854).
Data for 1997 are provisional.

Table TU.9 CHARGES SOCIALES DES EMPLOYEURS - EMPLOYERS' SOCIAL COSTS

Milliards de TL (Prix courants)

1991	1992	1993	1994	1995	1996	1997	1998	CITI révision 2	
784	3 520	5 837	6 733	8 687	14 723	29 131	..	**Alimentation, boissons et tabacs**	**31**
568	3 112	4 524	4 799	6 473	11 406	23 114	..	Alimentation	311.2
80	132	468	500	851	1 100	2 105	..	Boissons	313
136	276	845	1 434	1 363	2 217	3 912	..	Tabacs	314
1 000	1 694	5 834	13 324	11 018	24 224	57 457	..	**Textiles, habillement et cuir**	**32**
774	1 293	4 708	10 611	7 608	17 305	40 755	..	Textiles	321
59	129	489	6 248	899	1 641	4 534	..	Bonneterie	3213
193	343	952	2 439	2 917	6 117	14 348	..	Habillement, à l'exclusion des chaussures	322
11	20	55	162	309	429	1 333	..	Cuir et articles en cuir	323
23	37	118	112	183	373	1 022	..	Chaussures, sauf en caoutchouc et plastique	324
74	146	907	576	681	1 752	3 258	..	**Bois et meubles**	**33**
58	108	405	434	448	1 030	1 825	..	Ouvrages en bois	331
16	38	502	143	233	722	1 433	..	Meubles et accessoires	332
195	378	884	1 350	1 712	3 375	6 250	..	**Papier, impression et édition**	**34**
123	242	547	725	1 107	1 910	3 884	..	Papier et articles en papier	341
84	194	417	510	732	1 119	2 315	..	Pâte à papier, papier et carton	3411
72	136	337	625	605	1 465	2 366	..	Imprimerie et édition	342
643	1 231	4 060	6 336	5 014	10 108	25 783	..	**Produits chimiques**	**35**
215	480	1 103	1 209	1 288	2 549	5 213	..	Industrie chimique	351
39	322	160	186	277	326	659	..	Chimie de base, à l'exception des engrais	3511
63	77	291	384	388	800	1 849	..	Engrais et pesticides	3512
113	81	651	639	622	1 422	2 705	..	Résines synthétiques et matières plastiques	3513
197	347	1 794	1 457	1 611	3 048	8 824	..	Autres produits chimiques	352
27	40	924	213	181	439	1 389	..	Peintures, vernis et laques	3521
104	175	479	761	656	1 340	4 341	..	Produits pharmaceutiques	3522
35	78	248	276	534	699	1 685	..	Produits de nettoyage	3523
31	54	142	207	241	569	1 409	..	Autres produits chimiques	3529
45	94	270	435	378	1 090	2 864	..	Raffineries de pétrole	353
42	70	206	469	341	502	1 424	..	Derives du pétrole et du charbon	354
85	141	378	758	583	1 139	3 414	..	Industrie du caoutchouc	355
60	100	309	2 009	813	1 780	4 044	..	Ouvrages en matière plastique	356
335	725	2 524	4 746	4 138	7 111	14 775	..	**Produits minéraux non métalliques**	**36**
53	105	276	463	712	1 028	2 400	..	Grès, porcelaines et faïences	361
83	186	371	807	1 502	1 900	4 098	..	Industrie du verre	362
198	434	1 877	3 476	1 923	4 183	8 278	..	Produits minéraux non métalliques, nca	369
535	1 575	3 080	4 772	4 921	6 876	14 104	..	**Industries métallurgiques de base**	**37**
426	1 351	2 525	4 218	4 228	5 893	10 652	..	Sidérurgie	371
109	224	555	553	693	983	3 452	..	Métaux non ferreux	372
1 120	2 012	5 417	17 301	8 905	17 466	42 088	..	**Ouvrages en métaux, machines et matériel**	**38**
167	294	963	6 052	1 481	3 263	8 214	..	Ouvrages en métaux	381
34	59	217	252	325	728	1 712	..	Eléments de construction en métal	3813
257	467	1 390	4 429	2 426	3 917	9 352	..	Machines non électriques	382
10	20	107	52	42	82	186	..	Moteurs et turbines	3821
48	78	180	411	469	877	1 920	..	Machines agricoles	3822
15	33	91	95	94	192	464	..	Machines pour le travail du bois et du métal	3823
32	53	192	248	409	695	1 661	..	Machines et matériel spéciaux	3824
2	3	9	19	27	43	146	..	Machines de bureau et à calculer	3825
152	279	811	3 603	1 384	2 027	4 974	..	Machines et matériel, nca	3829
265	476	1 059	1 555	1 684	3 466	8 877	..	Machines électriques	383
65	92	249	374	367	875	2 294	..	Machines électriques industrielles	3831
116	252	480	699	726	1 279	3 086	..	Appareils de radio, télévision et télécom.	3832
10	27	54	113	135	330	806	..	Appareils électroménagers	3833
74	106	276	369	456	981	2 690	..	Appareils électriques, nca	3839
411	739	1 892	5 138	3 109	6 412	14 656	..	Matériel de transport	384
32	55	131	184	195	406	956	..	Construction navale	3841
..	..	..	..	219	488	1 191	..	Matériel ferroviaire	3842
310	545	1 405	4 499	2 534	5 107	11 669	..	Véhicules automobiles	3843
5	14	46	53	63	198	305	..	Motocycles et cycles	3844
..	..	..	..	..	..	515	..	Construction aéronautique	3845
..	..	1	1	..	..	20	..	Matériel de transport, nca	3849
20	37	113	126	206	409	990	..	Matériel professionnel et scientifique	385
14	26	105	118	187	359	885	..	Equipement professionnel	3851
3	2	6	8	17	..	..	..	Matériel photo et optique	3852
		2	1	2	..	..	..	Montres et horloges	3853
15	31	94	114	167	449	1 123	..	**Autres industries manufacturières**	**39**
..	..	20	27	44	110	..	..	Joaillerie	3901
4 700	11 310	28 637	55 253	45 244	86 084	193 971	..	**Total industries manufacturières**	**3**
..	..	..	..	..	..	..	..	**Agriculture, chasse, sylviculture et pêche**	**1**
..	..	1 580	1 640	3 393	7 533	..	..	**Industries extractives**	**2**
..	..	1 233	1 205	2 590	5 041	..	..	Extraction du charbon	21
..	..	30	24	73	93	..	..	Production du pétrole brut et gaz naturel	22
..	..	91	285	393	1 747	..	..	Extraction d'autres minéraux	23.29
..	..	986	1 518	3 306	8 098	..	..	**Électricité, gaz et eau**	**4**
..	..	1 692	3 060	5 645	9 875	24 408	..	**Construction**	**5**
..	..	..	..	..	..	..	..	**Total services**	**6.9**
..	..	..	..	..	..	..	..	**Commerce de gros & détail ; rest. & hotels**	**6**
..	..	..	..	..	..	..	..	**Transport, entrepôts et communications**	**7**
..	..	..	..	..	..	..	..	**Banques, assur., aff. imm., serv. aux entrepr.**	**8**
..	..	..	..	..	..	..	..	**Services a la collect, sociaux et personnels**	**9**

Note: Break in series between 1991 and 1992. Some data are not given for reasons of confidentiality. ISIC 385 includes a figure for other professional goods (Turkish SIC 3854). Data for 1997 are provisional.

TURKEY

Table TU.10 HOURS WORKED - HEURES OUVRÉES

Thousands

ISIC revision 2		1991	1992	1993	1994	1995	1996	1997	1998
31	**Food, beverages and tobacco**	268 000	293 000	283 000	276 000	274 000	278 000	282 000	..
311.2	Food	189 000	218 000	212 000	213 000	217 000	226 000	231 000	..
313	Beverages	20 000	19 000	19 000	16 000	14 000	15 000	16 000	..
314	Tobacco	60 000	57 000	52 000	48 000	42 000	37 000	35 000	..
32	**Textiles, apparel and leather**	505 000	578 000	588 000	573 000	643 000	723 000	811 000	..
321	Textiles	348 000	330 000	385 000	369 000	408 000	465 000	527 000	..
3213	Knitting mills	39 000	55 000	71 000	63 000	67 000	60 000	76 000	..
322	Wearing apparel, except footwear	137 000	171 000	177 000	179 000	208 000	232 000	243 000	..
323	Leather and fur products	9 000	12 000	11 000	13 000	15 000	14 000	21 000	..
324	Footwear, except rubber or plastics	12 000	14 000	15 000	12 000	13 000	13 000	20 000	..
33	**Wood products and furniture**	29 000	42 000	43 000	38 000	37 000	45 000	52 000	..
331	Wood products, except furniture	21 000	28 000	27 000	25 000	22 000	24 000	26 000	..
332	Furniture, fixtures, except metal	7 000	14 000	15 000	13 000	15 000	21 000	26 000	..
34	**Paper, paper products and printing**	56 000	61 000	53 000	50 000	55 000	54 000	54 000	..
341	Paper and paper products	39 000	38 000	31 000	30 000	35 000	32 000	32 000	..
3411	Pulp, paper and paperboard articles	25 000	26 000	19 000	18 000	21 000	16 000	15 000	..
342	Printing and publishing	18 000	23 000	22 000	20 000	20 000	22 000	22 000	..
35	**Chemical products**	145 000	153 000	149 000	138 000	142 000	153 000	166 000	..
351	Industrial chemicals	50 000	48 000	43 000	40 000	38 000	36 000	35 000	..
3511	Basic industrial chemicals except fertilizers	12 000	27 000	10 000	8 000	8 000	7 000	7 000	..
3512	Fertilizers and pesticides	12 000	9 000	10 000	9 000	9 000	8 000	8 000	..
3513	Synthetic resins and plastic materials	25 000	13 000	23 000	23 000	21 000	21 000	21 000	..
352	Other chemicals	33 000	33 000	33 000	29 000	33 000	37 000	39 000	..
3521	Paints, varnishes and lacquers	5 000	5 000	5 000	5 000	5 000	6 000	6 000	..
3522	Drugs and medicines	13 000	13 000	13 000	11 000	11 000	12 000	13 000	..
3523	Soap, cleaning preparations, perfumes, cosmetics	8 000	7 000	7 000	6 000	8 000	9 000	10 000	..
3529	Other chemical products	7 000	8 000	8 000	7 000	8 000	10 000	10 000	..
353	Petroleum refineries	9 000	9 000	9 000	8 000	8 000	8 000	8 000	..
354	Miscellaneous petroleum and coal products	7 000	6 000	6 000	7 000	5 000	6 000	6 000	..
355	Rubber products	20 000	23 000	25 000	23 000	22 000	23 000	27 000	..
356	Plastics products	26 000	32 000	33 000	31 000	35 000	43 000	51 000	..
36	**Non-metallic mineral products**	124 000	133 000	126 000	121 000	123 000	129 000	142 000	..
361	Pottery, china, earthenware	23 000	21 000	23 000	23 000	25 000	22 000	25 000	..
362	Glass and glass products	26 000	25 000	23 000	21 000	20 000	22 000	27 000	..
369	Non-metallic mineral products nec.	75 000	86 000	80 000	76 000	79 000	84 000	90 000	..
37	**Basic metal industries**	142 000	134 000	129 000	126 000	124 000	110 000	107 000	..
371	Iron and steel	116 000	108 000	105 000	105 000	103 000	90 000	83 000	..
372	Non-ferrous metals	26 000	26 000	24 000	21 000	21 000	20 000	24 000	..
38	**Fabricated metal prod.; machinery & equip.**	332 000	371 000	380 000	335 000	345 000	376 000	445 000	..
381	Fabricated metal products	68 000	80 000	81 000	75 000	79 000	90 000	107 000	..
3813	Structural metal products	13 000	18 000	20 000	18 000	18 000	21 000	24 000	..
382	Non-electrical machinery	74 000	86 000	85 000	80 000	77 000	82 000	97 000	..
3821	Engines and turbines	2 000	2 000	2 000	1 000	1 000	1 000	1 000	..
3822	Agricultural machinery and equipment	14 000	15 000	13 000	12 000	12 000	13 000	16 000	..
3823	Metal and wood working machinery	5 000	3 000	5 000	5 000	5 000	5 000	6 000	..
3824	Special industrial machinery	11 000	15 000	16 000	15 000	16 000	18 000	21 000	..
3825	Office, computing and accounting machinery		1 000	1 000	1 000	1 000	1 000	1 000	..
3829	Other non-electrical machinery and equipment	42 000	46 000	48 000	45 000	42 000	43 000	51 000	..
383	Electrical machinery	72 000	78 000	75 000	64 000	67 000	76 000	87 000	..
3831	Electrical industrial machinery	17 000	15 000	19 000	16 000	15 000	20 000	21 000	..
3832	Radio, TV and communication equipment	28 000	31 000	27 000	22 000	20 000	21 000	26 000	..
3833	Electrical appliances and housewares	6 000	9 000	9 000	7 000	9 000	11 000	13 000	..
3839	Other electrical apparatus and supplies	20 000	22 000	20 000	19 000	23 000	23 000	28 000	..
384	Transport equipment	110 000	120 000	130 000	107 000	110 000	119 000	142 000	..
3841	Shipbuilding	7 000	7 000	6 000	5 000	6 000	6 000	7 000	..
3842	Railroad equipment	..	12 000	..	..	10 000	9 000	11 000	..
3843	Motor vehicles	85 000	95 000	107 000	86 000	87 000	94 000	115 000	..
3844	Motorcycles and bicycles	2 000	3 000	4 000	3 000	4 000	5 000	5 000	..
3845	Aircraft	..	3 000	..	..	..	..	4 000	..
3849	Other transport equipment	..				..	..		..
385	Professional and scientific equipment	8 000	8 000	9 000	8 000	11 000	10 000	13 000	..
3851	Professional and scientific equipment, nec	5 000	5 000	8 000	7 000	9 000	9 000	11 000	..
3852	Photographic and optical goods	1 000	1 000	1 000	1 000	1 000	..	..	..
3853	Watches and clocks	..					..	..	..
39	**Other manufacturing industries**	8 000	10 000	11 000	11 000	11 000	12 000	15 000	..
3901	Jewellery and related articles	1 000	2 000	3 000	3 000	4 000	4 000	..	..
3	**Total manufacturing**	1 609 000	1 775 000	1 761 000	1 668 000	1 753 000	1 882 000	2 073 000	..
1	**Agriculture, hunting, forestry and fishing**	..	..	..	..	..	..	..	..
2	**Mining and quarrying**	172 000	..	160 000	154 000	147 000	151 000	..	..
21	Coal mining	129 000	..	106 000	103 000	96 000	92 000	..	..
22	Crude petroleum and natural gas	2 000	..	2 000	2 000	1 000	1 000	..	..
23.29	Other mining	..	..	36 000	34 000	34 000	42 000	..	..
4	**Electricity, gas and water**	69 000	..	154 000	170 000	139 000	141 000	..	..
5	**Construction**	..	..	..	..	..	..	..	..
6.9	**Total services**	..	..	..	..	..	..	..	..
6	**Wholesale & retail trade; restaurants & hotels**	..	..	..	..	..	..	..	..
7	**Transport, storage and communication**	..	..	..	..	..	..	..	..
8	**Finance, insurance, real estate, business serv.**	..	..	..	..	..	..	..	..
9	**Community, social and personal services**	..	..	..	..	..	..	..	..

Note: Break in series between 1991 and 1992. Some data are not given for reasons of confidentiality. ISIC 385 includes a figure for other professional goods (Turkish SIC 3854).
Data for 1997 are provisional.

Table TU.11 ETABLISSEMENTS - ESTABLISHMENTS

Units

1991	1992	1993	1994	1995	1996	1997	1998		CITI révision 2
945	2 163	1 966	1 860	1 792	1 821	1 861	..	**Alimentation, boissons et tabacs**	**31**
822	2 020	1 824	1 728	1 664	1 690	1 736	..	Alimentation	311.2
75	96	96	91	90	92	88	..	Boissons	313
48	47	46	41	38	39	37	..	Tabacs	314
1 536	3 316	3 133	2 976	3 148	3 329	3 629	..	**Textiles, habillement et cuir**	**32**
734	1 433	1 344	1 290	1 388	1 594	1 887	..	Textiles	321
180	380	358	328	325	338	418	..	Bonneterie	3213
699	1 591	1 533	1 446	1 496	1 466	1 389	..	Habillement, à l'exclusion des chaussures	322
57	151	136	128	148	150	205	..	Cuir et articles en cuir	323
46	141	120	112	116	119	148	..	Chaussures, sauf en caoutchouc et plastique	324
163	477	443	418	413	419	493	..	**Bois et meubles**	**33**
115	285	262	246	237	239	280	..	Ouvrages en bois	331
48	192	181	172	176	180	213	..	Meubles et accessoires	332
205	395	365	355	354	370	388	..	**Papier, impression et édition**	**34**
92	150	140	143	155	154	170	..	Papier et articles en papier	341
23	60	56	45	48	44	47	..	Pâte à papier, papier et carton	3411
113	245	225	212	199	216	218	..	Imprimerie et édition	342
525	947	918	905	902	932	1 041	..	**Produits chimiques**	**35**
67	92	86	89	78	78	92	..	Industrie chimique	351
34	50	42	39	38	35	40	..	Chimie de base, à l'exception des engrais	3511
18	16	19	18	17	18	20	..	Engrais et pesticides	3512
15	26	25	31	23	25	32	..	Résines synthétiques et matières plastiques	3513
170	265	265	255	279	288	304	..	Autres produits chimiques	352
35	53	52	54	62	63	64	..	Peintures, vernis et laques	3521
53	65	67	61	63	66	81	..	Produits pharmaceutiques	3522
35	65	64	58	60	61	62	..	Produits de nettoyage	3523
47	82	82	82	94	98	97	..	Autres produits chimiques	3529
5	5	5	5	6	6	6	..	Raffineries de pétrole	353
27	32	32	33	31	36	41	..	Derives du pétrole et du charbon	354
86	156	154	143	137	131	161	..	Industrie du caoutchouc	355
170	397	376	381	371	393	437	..	Ouvrages en matière plastique	356
480	860	827	832	840	842	874	..	**Produits minéraux non métalliques**	**36**
25	46	46	55	54	63	64	..	Grès, porcelaines et faïences	361
39	75	71	71	75	81	91	..	Industrie du verre	362
416	739	710	706	711	698	719	..	Produits minéraux non métalliques, nca	369
259	421	376	345	403	374	413	..	**Industries métallurgiques de base**	**37**
176	297	262	239	284	263	295	..	Sidérurgie	371
83	124	114	106	119	111	118	..	Métaux non ferreux	372
1 162	2 516	2 433	2 324	2 277	2 390	2 546	..	**Ouvrages en métaux, machines et matériel**	**38**
366	828	812	767	742	767	845	..	Ouvrages en métaux	381
84	235	218	202	189	206	211	..	Eléments de construction en métal	3813
299	770	716	714	687	713	745	..	Machines non électriques	382
5	7	6	5	5	4	5	..	Moteurs et turbines	3821
44	115	106	109	104	113	122	..	Machines agricoles	3822
34	98	85	89	80	69	71	..	Machines pour le travail du bois et du métal	3823
66	180	177	177	188	205	215	..	Machines et matériel spéciaux	3824
6	18	19	22	20	17	20	..	Machines de bureau et à calculer	3825
144	352	323	312	290	305	312	..	Machines et matériel, nca	3829
238	429	402	359	377	416	441	..	Machines électriques	383
65	114	115	107	104	117	119	..	Machines électriques industrielles	3831
53	105	101	79	79	85	92	..	Appareils de radio, télévision et télécom.	3832
35	53	50	39	46	49	57	..	Appareils électroménagers	3833
85	157	136	134	148	165	173	..	Appareils électriques, nca	3839
215	414	418	395	379	405	425	..	Matériel de transport	384
19	37	35	36	35	39	42	..	Construction navale	3841
4	4	3	4	5	5	5	..	Matériel ferroviaire	3842
184	352	349	327	314	331	347	..	Véhicules automobiles	3843
6	17	24	20	20	23	23	..	Motocycles et cycles	3844
2	2	2	2	2	2	3	..	Construction aéronautique	3845
..	2	5	6	3	5	5	..	Matériel de transport, nca	3849
44	75	85	89	92	89	90	..	Matériel professionnel et scientifique	385
19	42	63	72	73	69	70	..	Equipement professionnel	3851
13	16	17	14	16	18	18	..	Matériel photo et optique	3852
..	1	5	3	3	2	2	..	Montres et horloges	3853
55	106	106	112	101	106	116	..	**Autres industries manufacturières**	**39**
8	24	25	28	22	26	31	..	Joaillerie	3901
5 330	11 201	10 567	10 127	10 230	10 583	11 361	..	**Total industries manufacturières**	**3**
..	..	..	..	..	..	..	..	**Agriculture, chasse, sylviculture et pêche**	**1**
856	..	1 665	1 616	1 656	1 944	..	..	**Industries extractives**	**2**
231	..	217	221	235	215	..	..	Extraction du charbon	21
37	..	53	54	59	60	..	..	Production du pétrole brut et gaz naturel	22
..	..	1 281	1 247	1 242	1 562	..	..	Extraction d'autres minéraux	23.29
1 872	..	2 629	2 698	2 855	3 001	..	..	**Électricité, gaz et eau**	**4**
..	..	11 771	11 962	12 382	12 940	13 129	..	**Construction**	**5**
..	..	..	..	..	..	..	..	**Total services**	**6.9**
..	..	..	..	..	..	..	..	**Commerce de gros & détail ; rest. & hotels**	**6**
..	..	..	..	..	..	..	..	**Transport, entrepôts et communications**	**7**
..	..	..	..	..	..	..	..	**Banques, assur., aff. imm., serv. aux entrepr.**	**8**
..	..	..	..	..	..	..	..	**Services a la collect, sociaux et personnels**	**9**

Note: Break in series between 1991 and 1992. Some data are not given for reasons of confidentiality. ISIC 385 includes a figure for other professional goods (Turkish SIC 3854).
Data for 1997 are provisional.

PART B

PARTIE B

ISIC REVISION 3, INDUSTRY

CITI REVISION 3, INDUSTRIE

Note :

More variables and sectors are available on the electronic version of the publication.

D'autres variables et d'autres secteurs sont disponibles dans la version électronique de cette publication.

Structural business statistics and foreign trade statistics
Statistiques structurelles sur les entreprises et statistiques du commerce extérieur

List of countries - Liste des pays

AUSTRALIA

Table AS.1 PRODUCTION - PRODUCTION

Millions of A$ (Current Prices)

ISIC revision 3				1990-91	1991-92	1992-93	1993-94	1994-95	1995-96	1996-97	1997-98
C	10_14		Mining and quarrying	..	..	32 206	33 306	33 658	37 315	42 492	..
	10		Mining of coal and lignite; extraction of peat	..	..	8 917	9 185	9 140	10 078	10 722	..
	11		Crude petroleum & natural gas; service activities	..	..	7 724	7 382	6 843	7 654	12 465	..
	13		Mining of metal ores	..	..	..	..	..	..	15 199	..
	14		Other mining and quarrying	..	..	4 308	4 682	5 338	5 224	5 106	..
D	15_37		Total manufacturing	..	..	173 638	183 855	191 372	202 162	208 279	..
	15		Food products and beverages	..	..	36 922	39 528	40 829	42 408	44 004	..
		151	Production of meat, fish, fruit, vegetables, oils & fats	..	..	13 357	13 757	13 882	13 941	14 135	..
		152	Dairy products	..	..	5 411	6 311	6 144	6 591	7 024	..
		153	Grain mill and starch products; animal feeds	..	..	3 222	3 356	3 547	3 793	3 840	..
		154	Other food products	..	..	9 231	10 181	11 004	11 686	10 058	..
		155	Beverages	..	..	5 701	5 924	6 250	6 396	6 955	..
	16		Tobacco products	..	..	848	899	935	879	963	..
	17		Textiles	..	..	4 614	4 832	4 775	4 794	4 780	..
		171	Spinning, weaving and finishing of textiles	..	..	2 106	2 110	2 179	2 201	2 146	..
		172	Other textiles	..	..	1 572	1 665	1 591	1 628	1 677	..
		173	Knitted and crocheted fabrics and articles	..	..	936	1 057	1 005	965	957	..
	18		Wearing apparel and fur	..	..	4 036	4 114	4 297	4 420	4 500	..
		181	Wearing apparel, except fur apparel	..	..	3 414	3 453	3 475	3 635	3 673	..
		182	Dressing and dyeing of fur; articles of fur	..	..	622	661	822	785	827	..
	19		Leather and articles of leather; footwear	..	..	701	722	685	695	727	..
		191	Leather and articles of leather	..	..	79	79	82	112	103	..
		192	Footwear	..	..	622	643	603	583	624	..
	20		Wood & products of wood & cork, exc. furniture	..	..	5 556	6 099	6 633	6 030	6 061	..
		201	Sawmilling and planing of wood	..	..	4 166	4 356	4 684	4 268	..	..
		202	Wood products; articles of cork, straw, etc.	..	..	1 390	1 742	1 949	1 762	..	..
	21		Paper and paper products	..	..	..	..	..	..	..	..
	22		Printing, publishing and recorded media	..	..	16 721	17 442	18 836	19 381	20 359	..
		221	Publishing	..	..	..	..	..	..	..	..
		222	Printing and related activities	..	..	..	..	..	..	..	..
		223	Reproduction of recorded media	..	..	..	..	..	..	..	..
	23		Coke and petroleum products; nuclear fuel	..	..	8 236	8 094	7 346	6 998	7 990	..
		231	Coke oven products	..	..	219	305	402	377	345	..
		232	Refined petroleum products	..	..	8 017	7 790	6 943	6 622	7 645	..
		233	Processing of nuclear fuel	..	..	..	..	..	..	..	..
	24		Chemicals and chemical products	..	..	11 481	12 315	13 820	14 329	17 151	..
		241	Basic chemicals	..	..	6 451	6 837	7 805	7 800	10 394	..
		242	Other chemical products	..	..	5 030	5 478	6 015	6 529	6 757	..
		243	Man-made fibres	..	..	..	..	..	..	..	..
	25		Rubber and plastic products	..	..	33 840	35 257	36 091	37 781	39 014	..
		251	Rubber products	..	..	1 408	1 432	1 466	1 527	1 559	..
		252	Plastic products	..	..	5 205	5 641	5 970	6 325	6 157	..
	26		Other non-metallic mineral products	..	..	8 322	8 663	8 869	8 384	8 624	..
		261	Glass and glass products	..	..	1 066	1 177	1 155	1 160	1 066	..
		269	Non-metallic mineral products, nec	..	..	7 256	7 486	7 714	7 224	7 558	..
	27		Basic metals	..	..	..	..	..	..	..	..
		271	Basic iron and steel	..	..	..	..	..	..	..	..
		272	Basic precious and non-ferrous metals	..	..	..	..	..	..	..	..
		273	Casting of metals	..	..	..	..	..	..	..	..
	28		Fabricated metal products, exc. machinery	..	..	..	..	..	..	..	..
		281	Structural metal prod., tanks, reservoirs, etc.	..	..	..	..	..	..	..	..
		289	Other fabricated metal products	..	..	..	..	..	..	..	..
	29		Machinery and equipment, nec	..	..	41 130	43 064	44 822	48 296	48 400	..
		291	General purpose machinery	..	..	..	..	..	..	..	..
		292	Special purpose machinery	..	..	..	..	..	..	..	..
		293	Domestic appliances, nec	..	..	2 014	2 236	2 105	2 273	2 091	..
	30		Office, accounting & computing machinery	..	..	913	1 013	1 345	1 434	1 371	..
	31		Electrical machinery & apparatus, nec	..	..	3 826	4 230	4 150	4 415	4 465	..
		311	Electric motors, generators and transformers	..	..	2 007	2 226	1 939	2 198	2 431	..
		312	Electricity distribution & control apparatus	..	..	..	..	..	..	..	..
		313	Insulated wire and cable	..	..	1 128	1 217	1 344	1 455	1 378	..
		314	Accumulators, primary cells & primary batteries	..	..	247	226	329	286	144	..
		315	Electric lamps and lighting equipment	..	..	444	561	538	476	512	..
		319	Other electrical equipment, nec	..	..	..	..	..	..	..	..
	32		Radio, TV, communication equipment	..	..	2 999	2 764	3 503	3 762	4 102	..
		321	Electronic valves, tubes; other electronic components	..	..	1 398	914	1 498	1 638	1 892	..
		322	TV/radio transmitters; line teleph./telegr. apparatus	..	..	1 601	1 850	2 006	2 124	2 210	..
		323	TV/radio receivers; sound and video app.; assoc. gds.	..	..	..	..	..	..	..	..
	33		Medical, precision & optical instrum.; watch & clocks	..	..	1 393	1 649	1 663	1 600	1 599	..
		331	Medical appl.; instruments & appl. for measuring, etc.	..	..	..	..	..	..	..	..
		332	Optical instruments and photographic equipment	..	..	..	..	..	..	..	..
		333	Watches and clocks	..	..	..	..	..	..	..	..
	34		Motor vehicles, trailers and semi-trailers	..	..	10 796	12 972	14 240	14 545	15 303	..
		341	Motor vehicles	..	..	7 118	8 998	9 658	9 462	10 174	..
		342	Bodies for motor vehicles, trailers & semi-trailers	..	..	697	733	943	970	1 012	..
		343	Parts & accessories for motor vehicles & their engines	..	..	2 981	3 241	3 639	4 113	4 117	..
	35		Other transport equipment	..	..	4 181	4 202	4 043	4 274	4 387	..
		351	Building and repairing of ships and boats	..	..	2 081	1 808	1 721	2 097	1 984	..
		352	Railway and tramway locomotives and rolling stock	..	..	553	957	936	738	875	..
		353	Aircraft and spacecraft	..	..	1 496	1 397	1 323	1 374	1 459	..
		359	Transport equipment, nec	..	..	52	41	63	65	69	..
	36		Furniture; manufacturing, nec	..	..	4 984	5 381	5 456	5 388	5 882	..
		361	Furniture	..	..	3 665	4 082	4 178	4 130	4 424	..
		369	Manufacturing, nec	..	..	1 319	1 298	1 277	1 258	1 459	..
	37		Recycling	..	..	..	..	..	..	..	..
E	40_41		Electricity, gas and water supply	..	..	24 707	27 154	28 605	29 007	30 222	..
	40		Electricity, gas, steam and hot water supply	..	..	20 298	22 667	23 139	22 803	23 976	..
	41		Collection, purification and distribution of water	..	..	4 409	4 487	5 465	6 204	6 246	..
F	45		Construction	..	..	45 934	47 465	51 255	55 041	59 549	..

Note: Data cover the fiscal year ending 30 June. ISIC 182 includes part of 191; ISIC 201 includes part cf 202.
ISIC 311 includes 312 and 319;ISIC 321 includes 323. Part of ISIC 11, 403, 41, are not covered by the inquiry..

AUSTRALIE

Table AS.2 **VALEUR AJOUTÉE - VALUE ADDED**

Millions de A$ (Prix courants)

1990-91	1991-92	1992-93	1993-94	1994-95	1995-96	1996-97	1997-98			CITI révision 3
..	..	27 448	..	..	..	16 800	16 973	Activites extractives	C	10_14
..	..	5 118	..	..	..	3 722	4 026	Extraction de charbon, de lignite et de tourbe		10
..	..	7 762	..	..	..	6 409	6 075	Pétrole brut et gaz naturel ; activités annexes		11
..	..	2 244	..	..	..	6 683	..	Extraction de minerais métalliques		13
..	..	1 597	..	..	..	1 376	1 325	Autres activités extractives		14
..	..	67 965	..	..	..	63 810	..	Total activites manufacturieres	D	15_37
..	..	13 065	..	..	11 177	11 112	..	Produits alimentaires et boissons		15
..	..	4 102	..	..	3 167	3 205	..	Viande, poisson, fruits, légumes, huiles et graisses		151
..	..	1 689	..	..	1 564	1 380	..	Produits laitiers		152
..	..	1 411	..	..	780	709	..	Travail des grains; prod. amyl. ; alim. pour animaux		153
..	..	..	..	..	3 784	3 081	..	Autres produits alimentaires		154
..	..	2 474	..	..	1 883	2 134	..	Boissons		155
..	..	486	..	..	215	369	..	Produits à base de tabac		16
..	..	1 081	..	..	1 547	1 599	..	Textiles		17
..	..	869	..	..	701	739	..	Filature, tissage et achèvement des textiles		171
..	..	654	..	..	565	550	..	Autres articles textiles		172
..	..	427	..	..	281	310	..	Etoffes et articles de bonneterie		173
..	..	1 650	..	..	1 426	1 453	..	Articles d'habillement et fourrure		18
..	..	1 494	..	..	1 259	1 289	..	Articles d'habillement autres qu'en fourrure		181
..	..	156	..	..	167	164	..	Prép. et teinture des fourrures ; articles en fourrure		182
..	..	..	..	..	219	248	..	Cuir et articles en cuir ; chaussures		19
..	..	..	..	..	24	31	..	Cuir et articles en cuir sauf chaussures		191
..	..	308	..	..	195	217	..	Chaussures		192
..	..	2 161	..	..	2 176	2 072	..	Bois, articles en bois, sauf meubles		20
..	..	1 687	..	..	1 553	..	..	Sciage et rabotage du bois		201
..	..	475	..	..	622	..	..	Articles en bois, liège, vannerie, sparterie		202
..	..	..	..	..	..	..	..	Papier, articles en papier et carton		21
..	..	..	..	..	8 325	8 654	..	Edition, imprimerie et reproduction		22
..	..	..	..	..	..	..	..	Edition		221
..	..	..	..	..	..	..	..	Imprimerie et activités annexes		222
..	..	..	..	..	..	..	..	Reproduction de supports enregistrés		223
..	..	1 721	..	..	1 072	1 318	..	Coke. et prod. pétroliers ; comb. nucléaires		23
..	..	70	..	..	104	73	..	Cokéfaction		231
..	..	1 652	..	..	968	1 245	..	Produits pétroliers raffinés		232
..	..		..	..	..	..	..	Traitement de combustibles nucléaires		233
..	..	..	..	..	4 325	5 067	..	Produits chimiques		24
..	..	..	..	..	2 394	3 016	..	Produits chimiques de base		241
..	..	..	..	..	1 931	2 051	..	Autres produits chimiques		242
..	..	..	..	..	..	..	..	Fibres synthétiques ou artificielles		243
..	..	2 951	..	..	11 260	11	..	Art. en caoutchouc et matières plastiques		25
..	..	606	..	..	560	579	..	Articles en caoutchouc		251
..	..	2 345	..	..	2 210	2 127	..	Articles en matières plastiques		252
..	..	3 581	..	..	2 902	2 657	..	Autres produits minéraux non métalliques		26
..	..	587	..	..	504	384	..	Verre et articles en verre		261
..	..	2 993	..	..	2 398	2 273	..	Produits minéraux non métalliques, nca		269
..	..	..	..	..	..	..	..	Produits métallurgiques de base		27
..	..	6 668	..	..	..	..	..	Sidérurgie et première transformation de l'acier		271
..	..	..	..	..	..	..	..	Première transf. métaux précieux et non ferreux		272
..	..	..	..	..	..	..	..	Fonderie		273
..	..	..	..	..	..	..	..	Ouvrages en métaux, sauf machines		28
..	..	..	..	..	..	..	..	Constr. et menuis. métal.; réservoirs; générateurs		281
..	..	..	..	..	..	..	..	Autres ouvrages en métaux		289
..	..	..	..	..	15 196	14 766	..	Machines et matériels, nca		29
..	..	..	..	..	..	..	..	Machines d'usage général		291
..	..	..	..	..	..	..	..	Machines d'usage spécifique		292
..	..	845	..	..	659	636	..	Appareils domestiques, nca		293
..	..	385	..	..	263	243	..	Mach. de bureau, comptables, information		30
..	..	1 696	..	..	1 397	1 446	..	Machines et appareils électriques, nca		31
..	..	894	..	..	788	855	..	Moteurs, génératrices, transformateurs électriques		311
..	..	..	..	..	..	..	..	Matériel électrique de distribution et de commande		312
..	..	488	..	..	379	362	..	Fils et câbles électriques isolés		313
..	..	118	..	..	69	55	..	Accumulateurs et piles électriques		314
..	..	196	..	..	161	173	..	Lampes électriques et appareils d'éclairage		315
..	..	..	..	..	..	..	..	Autres matériels électriques, nca.		319
..	..	1 310	..	..	1 312	1 620	..	Equipements et appareils de radio, TV, etc.		32
..	..	557	..	..	609	682	..	Tubes et valves électroniques;autres composants		321
..	..	753	..	..	704	938	..	Emetteurs radio/TV; app. téléphonie et télégraphie		322
..	..	..	..	..	..	..	..	Récepteurs radio/TV, son et image; art. associés		323
..	..	783	..	..	650	638	..	Instr. médicaux, de précision ; optique ; horlog.		33
..	..	..	..	..	..	..	..	Instruments médicaux et de précision, etc.		331
..	..	..	..	..	..	..	..	Instruments d'optique ; matériel photographique		332
..	..	..	..	..	..	..	..	Horlogerie		333
..	..	3 206	..	..	3 882	4 208	..	Véhicules automobiles; remorques, etc.		34
..	..	1 603	..	..	2 057	2 387	..	Véhicules automobiles		341
..	..	290	..	..	365	319	..	Carross. automobiles, remorques, semi-remorques		342
..	..	1 313	..	..	1 461	1 502	..	Pièces et accessoires pour automobiles et moteurs		343
..	..	1 707	..	..	1 473	1 595	..	Autres matériels de transport		35
..	..	631	..	..	607	612	..	Construction et réparation de navires		351
..	..	226	..	..	285	298	..	Matériel ferroviaire roulant		352
..	..	827	..	..	555	662	..	Construction aéronautique et spatiale		353
..	..	24	..	..	27	23	..	Autres équipements de transport		359
..	..	2 087	..	..	2 020	2 069	..	Meubles ; activités de fabrication, nca		36
..	..	1 497	..	..	1 540	1 571	..	Meubles		361
..	..	590	..	..	481	498	..	Activités de fabrication, nca		369
..	..	..	..	..	..	..	..	Récupération		37
..	..	..	..	..	..	14 331	14 467	Electricité, gaz et eau	E	40_41
..	..	..	..	..	..	10 710	10 843	Prod. et dist. d'électricité gaz, vapeur d'eau chaude		40
..	..	..	..	..	..	3 621	3 624	Captage, épuration et distribution d'eau		41
..	..	..	..	..	..	10 738	11 568	Construction	F	45

Note: Data cover the fiscal year ending 30 June. ISIC 182 includes part of 191; ISIC 201 includes part of 202.
ISIC 311 includes 312 and 319;ISIC 321 includes 323. Part of ISIC 11, 403, 41, are not covered by the inquiry..

Table AS.3 EMPLOYMENT - EMPLOI

Units

ISIC revision 3				1990-91	1991-92	1992-93	1993-94	1994-95	1995-96	1996-97	1997-98
C	10_14		Mining and quarrying	..	..	84 000	78 000	82 000	85 000	86 000	..
	10		Mining of coal and lignite; extraction of peat	..	..	27 000	27 000	26 000	27 000	27 000	..
	11		Crude petroleum & natural gas; service activities	..	..	6 000	6 000	5 000	5 000	5 000	..
	13		Mining of metal ores	..	..	..	..	..	..	29 039	..
	14		Other mining and quarrying	..	..	21 000	17 000	22 000	25 000	25 000	..
D	15_37		Total manufacturing	..	..	914 835	911 132	924 263	944 847	945 386	..
	15		Food products and beverages	..	..	161 983	164 695	164 811	161 299	162 474	..
		151	Production of meat, fish, fruit, vegetables, oils & fats	..	..	62 987	62 961	63 174	62 263	63 372	..
		152	Dairy products	..	..	14 758	15 792	16 346	16 626	17 216	..
		153	Grain mill and starch products; animal feeds	..	..	7 789	8 092	7 433	6 432	6 796	..
		154	Other food products	..	..	59 157	60 729	60 810	59 167	59 804	..
		155	Beverages	..	..	17 292	17 121	17 049	16 812	15 628	..
	16		Tobacco products	..	..	2 881	2 333	1 822	1 612	1 569	..
	17		Textiles	..	..	34 211	34 266	33 755	32 251	31 170	..
		171	Spinning, weaving and finishing of textiles	..	..	14 417	14 694	14 660	13 662	13 091	..
		172	Other textiles	..	..	11 710	12 494	12 407	12 221	11 769	..
		173	Knitted and crocheted fabrics and articles	..	..	8 084	7 078	6 688	6 368	6 310	..
	18		Wearing apparel and fur	..	..	39 891	37 276	39 202	38 790	..	..
		181	Wearing apparel, except fur apparel	..	..	37 054	34 366	35 962	35 726	..	..
		182	Dressing and dyeing of fur; articles of fur	..	..	2 837	2 910	3 240	3 064	..	..
	19		Leather and articles of leather; footwear	..	..	8 509	8 378	7 098	6 976	6 747	..
		191	Leather and articles of leather	..	..	1 037	1 056	1 072	1 263	892	..
		192	Footwear	..	..	7 472	7 322	6 026	5 713	5 855	..
	20		Wood & products of wood & cork, exc. furniture	..	..	44 478	47 441	49 327	46 276	44 226	..
		201	Sawmilling and planing of wood	..	..	34 070	34 276	35 271	32 949	..	..
		202	Wood products; articles of cork, straw, etc.	..	..	10 409	13 166	14 056	13 327	..	..
	21		Paper and paper products	..	..	..	..	..	..	..	..
	22		Printing, publishing and recorded media	..	..	109 777	110 181	113 679	112 203	114 512	..
		221	Publishing	..	..	..	..	..	..	..	..
		222	Printing and related activities	..	..	..	..	..	..	..	..
		223	Reproduction of recorded media	..	..	..	..	..	..	..	..
	23		Coke and petroleum products; nuclear fuel	..	..	4 474	4 539	4 178	4 360	4 314	..
		231	Coke oven products	..	..	540	486	529	776	647	..
		232	Refined petroleum products	..	..	3 934	4 053	3 649	3 584	3 667	..
		233	Processing of nuclear fuel	..	..	..	..	..	..	..	..
	24		Chemicals and chemical products	..	..	39 861	38 757	40 618	40 462	45 922	..
		241	Basic chemicals	..	..	20 521	20 165	21 358	20 368	25 921	..
		242	Other chemical products	..	..	19 340	18 592	19 260	20 094	20 001	..
		243	Man-made fibres	..	..	..	..	..	..	..	..
	25		Rubber and plastic products	..	..	123 564	125 523	127 401	130 329	128 348	..
		251	Rubber products	..	..	8 274	8 114	8 107	7 842	8 154	..
		252	Plastic products	..	..	32 484	34 310	35 270	36 058	34 979	..
	26		Other non-metallic mineral products	..	..	39 899	38 876	39 017	36 351	37 169	..
		261	Glass and glass products	..	..	5 746	5 647	5 764	5 243	5 328	..
		269	Non-metallic mineral products, nec	..	..	34 153	33 229	33 253	31 108	31 841	..
	27		Basic metals	..	..	..	..	..	..	..	..
		271	Basic iron and steel	..	..	..	..	..	..	..	..
		272	Basic precious and non-ferrous metals	..	..	..	..	..	..	..	..
		273	Casting of metals	..	..	..	..	..	..	..	..
	28		Fabricated metal products, exc. machinery	..	..	..	..	..	..	..	..
		281	Structural metal prod., tanks, reservoirs, etc.	..	..	..	..	..	..	..	..
		289	Other fabricated metal products	..	..	..	..	..	..	..	..
	29		Machinery and equipment, nec	..	..	216 622	218 564	217 919	219 659	218 577	..
		291	General purpose machinery	..	..	..	..	..	..	..	..
		292	Special purpose machinery	..	..	..	..	..	..	..	..
		293	Domestic appliances, nec	..	..	12 782	14 366	13 550	13 088	12 909	..
	30		Office, accounting & computing machinery	..	..	2 128	2 622	3 551	3 174	2 848	..
	31		Electrical machinery & apparatus, nec	..	..	25 253	26 830	25 778	25 625	25 722	..
		311	Electric motors, generators and transformers	..	..	16 027	16 897	15 182	15 630	16 028	..
		312	Electricity distribution & control apparatus	..	..	..	..	..	..	..	..
		313	Insulated wire and cable	..	..	4 324	4 173	4 655	4 429	4 476	..
		314	Accumulators, primary cells & primary batteries	..	..	1 340	991	1 097	732	646	..
		315	Electric lamps and lighting equipment	..	..	3 562	4 769	4 844	4 834	4 572	..
		319	Other electrical equipment, nec	..	..	..	..	..	..	..	..
	32		Radio, TV, communication equipment	..	..	16 321	13 734	17 488	18 093	18 144	..
		321	Electronic valves, tubes; other electronic components	..	..	8 675	6 790	9 830	10 501	10 625	..
		322	TV/radio transmitters; line teleph./telegr. apparatus	..	..	7 646	6 944	7 658	7 592	7 519	..
		323	TV/radio receivers; sound and video app.; assoc. gds.	..	..	..	..	..	..	..	..
	33		Medical, precision & optical instrum.; watch & clocks	..	..	11 988	12 417	12 942	11 420	10 696	..
		331	Medical appl.; instruments & appl. for measuring, etc.	..	..	..	..	..	..	..	..
		332	Optical instruments and photographic equipment	..	..	..	..	..	..	..	..
		333	Watches and clocks	..	..	..	..	..	..	..	..
	34		Motor vehicles, trailers and semi-trailers	..	..	54 445	53 916	55 044	55 908	55 391	..
		341	Motor vehicles	..	..	22 896	22 559	22 477	20 875	20 915	..
		342	Bodies for motor vehicles, trailers & semi-trailers	..	..	6 234	6 424	7 345	7 721	7 692	..
		343	Parts & accessories for motor vehicles & their engines	..	..	25 315	24 932	25 222	27 313	26 783	..
	35		Other transport equipment	..	..	28 630	28 271	28 447	28 368	28 597	..
		351	Building and repairing of ships and boats	..	..	9 778	9 981	10 275	10 754	10 645	..
		352	Railway and tramway locomotives and rolling stock	..	..	4 641	5 256	5 338	5 083	5 199	..
		353	Aircraft and spacecraft	..	..	13 660	12 528	12 230	11 902	12 165	..
		359	Transport equipment, nec	..	..	552	507	604	629	589	..
	36		Furniture; manufacturing, nec	..	..	53 517	55 879	54 873	52 569	55 046	..
		361	Furniture	..	..	39 570	42 937	42 550	40 704	42 183	..
		369	Manufacturing, nec	..	..	13 947	12 942	12 323	11 865	12 863	..
	37		Recycling	..	..	..	..	..	..	..	..
E	40_41		Electricity, gas and water supply	..	..	91 000	81 000	73 000	64 000	58 000	..
	40		Electricity, gas, steam and hot water supply	..	..	63 000	55 000	51 000	45 000	41 000	..
	41		Collection, purification and distribution of water	..	..	27 000	26 000	22 000	19 000	17 000	..
F	45		Construction	..	..	267 000	274 000	306 000	340 000	373 000	..

Note: Data cover the fiscal year ending 30 June. ISIC 182 includes part of 191; ISIC 201 includes part of 202.
ISIC 311 includes 312 and 319;ISIC 321 includes 323. Part of ISIC 11, 403, 41, are not covered by the inquiry..

Table AS.4 **SALAIRES ET TRAITEMENTS, SALARIÉS - WAGES AND SALARIES, EMPLOYEES**

Millions de A$ (Prix courants)

1990-91	1991-92	1992-93	1993-94	1994-95	1995-96	1996-97	1997-98		CITI révision 3	
..	..	3 889	3 946	..	..	5 781	..	Activites extractives	C	10_14
..	..	1 779	1 950	..	..	2 121	..	Extraction de charbon, de lignite et de tourbe		10
..	..	329	322	..	..	1 502	..	Pétrole brut et gaz naturel ; activités annexes		11
..	..	..	..	..	..	1 855	..	Extraction de minerais métalliques		13
..	..	292	..	..	..	304	..	Autres activités extractives		14
..	..	27 809	28 107	29 653	30 515	32 629	..	Total activites manufacturieres	D	15_37
..	..	4 692	4 850	4 928	5 053	5 522	..	Produits alimentaires et boissons		15
..	..	1 756	1 770	1 779	1 819	2 002	..	Viande, poisson, fruits, légumes, huiles et graisses		151
..	..	491	517	548	583	646	..	Produits laitiers		152
..	..	427	445	416	434	..	..	Travail des grains; prod. amyl. ; alim. pour animaux		153
..	..	1 410	1 540	1 585	1 617	2 262	..	Autres produits alimentaires		154
..	..	608	579	601	600	612	..	Boissons		155
..	..	127	119	112	95	89	..	Produits à base de tabac		16
..	..	904	887	961	916	979	..	Textiles		17
..	..	431	440	471	442	453	..	Filature, tissage et achèvement des textiles		171
..	..	276	276	312	305	334	..	Autres articles textiles		172
..	..	197	171	178	169	192	..	Etoffes et articles de bonneterie		173
..	..	..	..	..	..	890	..	Articles d'habillement et fourrure		18
..	..	743	663	705	721	780	..	Articles d'habillement autres qu'en fourrure		181
..	..	75	82	88	117	110	..	Prép. et teinture des fourrures ; articles en fourrure		182
..	..	..	..	..	..	..	..	Cuir et articles en cuir ; chaussures		19
..	..	..	..	..	..	..	..	Cuir et articles en cuir sauf chaussures		191
..	..	163	160	146	141	146	..	Chaussures		192
..	..	1 039	1 081	1 371	1 191	1 218	..	Bois, articles en bois, sauf meubles		20
..	..	788	761	850	..	..	..	Sciage et rabotage du bois		201
..	..	251	320	368	..	..	..	Articles en bois, liège, vannerie, sparterie		202
..	..	..	..	..	..	..	..	Papier, articles en papier et carton		21
..	..	..	..	..	..	..	..	Edition, imprimerie et reproduction		22
..	..	..	..	..	..	..	..	Edition		221
..	..	..	..	..	..	..	..	Imprimerie et activités annexes		222
..	..	..	..	..	..	..	..	Reproduction de supports enregistrés		223
..	..	249	262	263	247	286	..	Coke. et prod. pétroliers ; comb. nucléaires		23
..	..	18	18	23	36	34	..	Cokéfaction		231
..	..	231	245	240	211	252	..	Produits pétroliers raffinés		232
..	..					..	..	Traitement de combustibles nucléaires		233
..	..	1 687	1 723	1 781	1 922	..	..	Produits chimiques		24
..	..	1 049	1 069	1 054	1 159	..	..	Produits chimiques de base		241
..	..	638	654	727	763	..	..	Autres produits chimiques		242
..	..	..	..	..	..	..	..	Fibres synthétiques ou artificielles		243
..	..	1 209	1 249	..	1 419	1 502	..	Art. en caoutchouc et matières plastiques		25
..	..	273	293	298	308	331	..	Articles en caoutchouc		251
..	..	935	956	..	1 111	1 171	..	Articles en matières plastiques		252
..	..	1 300	1 289	1 177	1 251	1 370	..	Autres produits minéraux non métalliques		26
..	..	207	213	210	200	217	..	Verre et articles en verre		261
..	..	1 093	1 076	967	1 050	1 154	..	Produits minéraux non métalliques, nca		269
..	..	2 807	2 751	2 868	..	..	..	Produits métallurgiques de base		27
..	..	2 747	2 701	2 820	..	..	..	Sidérurgie et première transformation de l'acier		271
..	..	..	..	..	..	..	..	Première transf. métaux précieux et non ferreux		272
..	..	60	50	48	27	..	..	Fonderie		273
..	..	2 420	..	..	..	..	..	Ouvrages en métaux, sauf machines		28
..	..	1 523	1 578	1 701	..	..	..	Constr. et menuis. métal.; réservoirs; générateurs		281
..	..	1 050	1 069	894	..	..	..	Autres ouvrages en métaux		289
..	..	1 280	1 334	1 418	..	..	..	Machines et matériels, nca		29
..	..	382	360	380	392	..	..	Machines d'usage général		291
..	..	556	588	656	..	..	..	Machines d'usage spécifique		292
..	..	342	386	382	414	404	..	Appareils domestiques, nca		293
..	..	68	82	123	119	110	..	Mach. de bureau, comptables, information		30
..	..	734	768	571	832	1 289	..	Machines et appareils électriques, nca		31
..	..	448	465	472	490	542	..	Moteurs, génératrices, transformateurs électriques		311
..	..	..	..	..	..	..	..	Matériel électrique de distribution et de commande		312
..	..	150	164	183	182	194	..	Fils et câbles électriques isolés		313
..	..	37	31	40	38	22	..	Accumulateurs et piles électriques		314
..	..	100	108	118	123	127	..	Lampes électriques et appareils d'éclairage		315
..	..	..	..	..	..	..	..	Autres matériels électriques, nca.		319
..	..	567	425	765	653	703	..	Equipements et appareils de radio, TV, etc.		32
..	..	255	172	..	333	383	..	Tubes et valves électroniques;autres composants		321
..	..	313	252	293	320	320	..	Emetteurs radio/TV; app. téléphonie et télégraphie		322
..	..	..	..	..	..	..	..	Récepteurs radio/TV, son et image; art. associés		323
..	..	359	374	211	367	360	..	Instr. médicaux, de précision ; optique ; horlog.		33
..	..	..	..	..	..	..	..	Instruments médicaux et de précision, etc.		331
..	..	..	..	..	..	..	..	Instruments d'optique ; matériel photographique		332
..	..	..	..	..	..	..	..	Horlogerie		333
..	..	1 549	1 633	1 828	1 865	2 001	..	Véhicules automobiles; remorques, etc.		34
..	..	706	782	859	804	865	..	Véhicules automobiles		341
..	..	146	148	187	207	223	..	Carross. automobiles, remorques, semi-remorques		342
..	..	698	703	781	854	913	..	Pièces et accessoires pour automobiles et moteurs		343
..	..	1 064	1 062	1 032	1 065	1 230	..	Autres matériels de transport		35
..	..	311	322	337	354	372	..	Construction et réparation de navires		351
..	..	142	169	164	174	201	..	Matériel ferroviaire roulant		352
..	..	602	558	531	522	642	..	Construction aéronautique et spatiale		353
..	..	10	13	16	15	15	..	Autres équipements de transport		359
..	..	1 110	1 139	1 166	1 246	1 335	..	Meubles ; activités de fabrication, nca		36
..	..	836	878	904	965	1 016	..	Meubles		361
..	..	274	261	262	281	319	..	Activités de fabrication, nca		369
..	..	..	..	..	..	..	..	Récupération		37
..	..	..	..	..	..	3 079	..	Electricité, gaz et eau	E	40_41
..	..	..	..	..	..	2 325	..	Prod. et dist. d'électricité gaz, vapeur d'eau chaude		40
..	..	..	..	..	..	754	..	Captage, épuration et distribution d'eau		41
..	..	6 187	7 095	7 874	8 991	..	..	Construction	F	45

Note: Data cover the fiscal year ending 30 June. ISIC 182 includes part of 191; ISIC 201 includes part of 202.
ISIC 311 includes 312 and 319;ISIC 321 includes 323. Part of ISIC 11, 403, 41, are not covered by the inquiry..

AUSTRALIA

Table AS.5 **ESTABLISHMENTS - ÉTABLISSEMENTS**

Units

ISIC revision 3				1990-91	1991-92	1992-93	1993-94	1994-95	1995-96	1996-97	1997-98
C	10_14		Mining and quarrying	..	..	1 217	641	..	..	500	..
	10		Mining of coal and lignite; extraction of peat	..	..	156	159	..	..	173	..
	11		Crude petroleum & natural gas; service activities	..	..	103	102	..	..	86	..
	13		Mining of metal ores	..	..	..	..	..	..	241	..
	14		Other mining and quarrying	..	..	613	..	..	..	..	..
D	15_37		Total manufacturing	..	..	42 707	44 585	47 332	54 340	55 373	..
	15		Food products and beverages	..	..	3 745	3 697	3 702	3 907	3 935	..
		151	Production of meat, fish, fruit, vegetables, oils & fats	..	..	1 022	987	1 045	1 078	1 095	..
		152	Dairy products	..	..	216	223	229	259	235	..
		153	Grain mill and starch products; animal feeds	..	..	272	283	303	235	280	..
		154	Other food products	..	..	1 846	1 762	1 661	1 813	1 930	..
		155	Beverages	..	..	388	442	466	521	439	..
	16		Tobacco products	..	..	6	6	5	5	5	..
	17		Textiles	..	..	1 476	1 579	1 630	1 641	1 692	..
		171	Spinning, weaving and finishing of textiles	..	..	329	304	332	287	263	..
		172	Other textiles	..	..	840	1 045	1 054	1 131	1 192	..
		173	Knitted and crocheted fabrics and articles	..	..	307	230	244	223	237	..
	18		Wearing apparel and fur	..	..	2 902	3 150	3 818	4 471	5 031	..
		181	Wearing apparel, except fur apparel	..	..	2 767	3 041	3 700	4 345	4 915	..
		182	Dressing and dyeing of fur; articles of fur	..	..	135	109	118	126	116	..
	19		Leather and articles of leather; footwear	..	..	399	405	345	433	412	..
		191	Leather and articles of leather	..	..	139	148	175	175	155	..
		192	Footwear	..	..	260	257	170	258	257	..
	20		Wood & products of wood & cork, exc. furniture	..	..	3 831	4 112	4 417	4 794	4 545	..
		201	Sawmilling and planing of wood	..	..	2 807	2 893	2 901	3 294	..	..
		202	Wood products; articles of cork, straw, etc.	..	..	1 024	1 219	1 516	1 500	..	..
	21		Paper and paper products	..	..	..	..	..	..	..	..
	22		Printing, publishing and recorded media	..	..	5 499	6 026	6 103	6 738	6 758	..
		221	Publishing	..	..	..	..	..	..	..	..
		222	Printing and related activities	..	..	..	..	..	..	..	..
		223	Reproduction of recorded media	..	..	..	..	..	..	..	..
	23		Coke and petroleum products; nuclear fuel	..	..	71	75	50	65	59	..
		231	Coke oven products	..	..	38	43	26	36	30	..
		232	Refined petroleum products	..	..	33	32	24	29	29	..
		233	Processing of nuclear fuel	..	..	..	..	..	..	..	..
	24		Chemicals and chemical products	..	..	1 045	1 085	1 126	1 210	1 369	..
		241	Basic chemicals	..	..	355	377	424	421	595	..
		242	Other chemical products	..	..	690	708	702	789	774	..
		243	Man-made fibres	..	..	..	..	..	..	..	..
	25		Rubber and plastic products	..	..	4 487	4 748	4 839	5 206	5 093	..
		251	Rubber products	..	..	287	289	274	278	296	..
		252	Plastic products	..	..	1 450	1 562	1 630	1 775	1 684	..
	26		Other non-metallic mineral products	..	..	1 879	2 020	2 030	2 138	2 172	..
		261	Glass and glass products	..	..	184	293	325	318	353	..
		269	Non-metallic mineral products, nec	..	..	1 695	1 727	1 705	1 820	1 819	..
	27		Basic metals	..	..	..	..	..	..	..	..
		271	Basic iron and steel	..	..	..	..	..	..	..	..
		272	Basic precious and non-ferrous metals	..	..	..	..	..	..	..	..
		273	Casting of metals	..	..	..	..	..	..	..	..
	28		Fabricated metal products, exc. machinery	..	..	..	..	..	..	..	..
		281	Structural metal prod., tanks, reservoirs, etc.	..	..	..	..	..	..	..	..
		289	Other fabricated metal products	..	..	..	..	..	..	..	..
	29		Machinery and equipment, nec	..	..	11 932	12 315	12 420	13 812	13 968	..
		291	General purpose machinery	..	..	..	..	..	..	..	..
		292	Special purpose machinery	..	..	..	..	..	..	..	..
		293	Domestic appliances, nec	..	..	315	377	275	363	344	..
	30		Office, accounting & computing machinery	..	..	130	170	161	150	165	..
	31		Electrical machinery & apparatus, nec	..	..	966	1 200	1 209	1 237	1 244	..
		311	Electric motors, generators and transformers	..	..	673	771	700	660	724	..
		312	Electricity distribution & control apparatus	..	..	..	..	..	..	..	..
		313	Insulated wire and cable	..	..	56	56	64	53	65	..
		314	Accumulators, primary cells & primary batteries	..	..	17	19	19	18	13	..
		315	Electric lamps and lighting equipment	..	..	220	354	426	506	442	..
		319	Other electrical equipment, nec	..	..	..	..	..	..	..	..
	32		Radio, TV, communication equipment	..	..	557	649	668	593	679	..
		321	Electronic valves, tubes; other electronic components	..	..	422	510	553	463	528	..
		322	TV/radio transmitters; line teleph./telegr. apparatus	..	..	135	139	115	130	151	..
		323	TV/radio receivers; sound and video app.; assoc. gds.	..	..	..	..	..	..	..	..
	33		Medical, precision & optical instrum.; watch & clocks	..	..	1 087	1 140	1 147	1 259	1 232	..
		331	Medical appl.; instruments & appl. for measuring, etc.	..	..	..	..	..	..	..	..
		332	Optical instruments and photographic equipment	..	..	..	..	..	..	..	..
		333	Watches and clocks	..	..	..	..	..	..	..	..
	34		Motor vehicles, trailers and semi-trailers	..	..	1 492	1 480	1 315	1 562	1 531	..
		341	Motor vehicles	..	..	83	82	82	66	44	..
		342	Bodies for motor vehicles, trailers & semi-trailers	..	..	434	478	511	596	584	..
		343	Parts & accessories for motor vehicles & their engines	..	..	975	921	722	899	904	..
	35		Other transport equipment	..	..	898	963	1 097	1 035	1 181	..
		351	Building and repairing of ships and boats	..	..	571	573	635	667	756	..
		352	Railway and tramway locomotives and rolling stock	..	..	32	40	108	24	32	..
		353	Aircraft and spacecraft	..	..	230	284	289	271	318	..
		359	Transport equipment, nec	..	..	64	65	65	73	75	..
	36		Furniture; manufacturing, nec	..	..	6 122	6 317	6 826	7 134	7 412	..
		361	Furniture	..	..	4 081	4 454	4 930	5 165	5 338	..
		369	Manufacturing, nec	..	..	2 041	1 863	1 896	1 969	2 074	..
	37		Recycling	..	..	..	..	..	..	..	..
E	40_41		Electricity, gas and water supply	..	..	..	..	..	..	82	..
	40		Electricity, gas, steam and hot water supply	..	..	..	..	..	..	82	..
	41		Collection, purification and distribution of water	..	..	..	..	..	..	..	..
F	45		Construction	..	..	60 744	68 750	..	..	..	..

Note: Data cover the fiscal year ending 30 June. ISIC 182 includes part of 191; ISIC 201 includes part of 202.
ISIC 311 includes 312 and 319;ISIC 321 includes 323. Part of ISIC 11, 403, 41, are not covered by the incuiry..

Table AU.1 **PRODUCTION - PRODUCTION**

Millions de Sch (Prix courants)

1991	1992	1993	1994	1995	1996	1997	1998			CITI révision 3	
20 665	19 999	20 338	18 906	11 473	..	13 171	13 886	Activites extractives	C	10_14	
1 630	1 628	1 431	1 134	1 027	..	965	..	Extraction de charbon, de lignite et de tourbe		10	
13 115	11 998	..	..	..	..	..	..	Pétrole brut et gaz naturel ; activités annexes		11	
705	641	..	..	..	..	..	..	Extraction de minerais métalliques		13	
5 215	5 732	5 787	6 042	9 096	..	10 543	11 468	Autres activités extractives		14	
1 072 598	1 058 828	1 020 148	1 073 536	1 100 497	..	1 241 795	1 353 266	Total activites manufacturieres	D	15_37	
149 697	155 810	158 323	156 777	154 958	..	167 479	165 196	Produits alimentaires et boissons		15	
37 407	39 081	38 354	37 789	46 741	..	53 291	51 308	Viande, poisson, fruits, légumes, huiles et graisses			151
28 712	27 363	28 935	25 604	30 880	..	27 299	25 713	Produits laitiers			152
4 768	4 675	4 512	4 477	9 051	..	9 588	10 165	Travail des grains; prod. amyl. ; alim. pour animaux			153
50 671	52 896	54 550	55 212	42 145	..	49 818	51 189	Autres produits alimentaires			154
28 140	31 795	31 972	33 696	26 141	..	27 483	26 820	Boissons			155
19 782	20 086	19 921	20 849	..	..	..	..	Produits à base de tabac		16	
40 812	41 053	38 216	37 395	39 377	..	33 777	36 372	Textiles		17	
18 209	17 819	16 147	15 818	21 479	..	15 555	16 077	Filature, tissage et achèvement des textiles			171
14 038	14 309	13 779	13 579	11 804	..	10 818	12 247	Autres articles textiles			172
8 564	8 924	8 290	7 998	6 095	..	7 404	8 049	Etoffes et articles de bonneterie			173
15 343	14 682	13 459	12 188	14 298	..	15 508	15 588	Articles d'habillement et fourrure		18	
15 113	14 450	13 298	12 014	13 992	..	15 258	15 352	Articles d'habillement autres qu'en fourrure			181
230	232	161	173	307	..	249	237	Prép. et teinture des fourrures ; articles en fourrure			182
10 970	10 166	9 407	9 024	9 409	..	11 209	10 723	Cuir et articles en cuir ; chaussures		19	
3 283	3 346	3 112	3 001	3 360	..	3 841	4 455	Cuir et articles en cuir sauf chaussures			191
7 687	6 820	6 294	6 023	6 048	..	7 367	6 268	Chaussures			192
67 512	68 963	65 885	71 603	54 689	..	61 174	67 022	Bois, articles en bois, sauf meubles		20	
23 456	22 495	19 653	21 789	20 993	..	23 380	27 180	Sciage et rabotage du bois			201
44 056	46 468	46 232	49 814	33 696	..	37 794	39 842	Articles en bois, liège, vannerie, sparterie			202
47 909	47 126	43 232	48 478	56 697	..	59 067	59 964	Papier, articles en papier et carton		21	
34 926	34 954	33 225	34 144	43 341	..	46 887	49 234	Edition, imprimerie et reproduction		22	
10 198	10 483	9 945	11 216	16 143	..	18 203	..	Edition			221
23 742	23 478	22 333	22 034	25 098	..	24 197	26 167	Imprimerie et activités annexes			222
986	993	947	894	2 101	..	4 487	..	Reproduction de supports enregistrés			223
37 748	32 910	32 796	33 047	..	..	..	..	Coke. et prod. pétroliers ; comb. nucléaires		23	
2 215	1 758	1 999	2 279	..	..	..	..	Cokéfaction			231
35 533	31 152	30 797	30 768	..	..	..	..	Produits pétroliers raffinés			232
..	..	..	..	..	..	..	..	Traitement de combustibles nucléaires			233
93 172	89 468	85 440	90 238	62 046	..	84 306	91 021	Produits chimiques		24	
51 915	48 076	42 555	47 051	26 972	..	31 170	31 511	Produits chimiques de base			241
41 257	41 392	42 885	43 187	..	..	..	..	Autres produits chimiques			242
..	..	..	..	..	..	..	..	Fibres synthétiques ou artificielles			243
29 626	29 967	29 296	30 375	42 213	..	48 478	50 729	Art. en caoutchouc et matières plastiques		25	
10 902	10 936	9 679	10 258	7 853	..	9 295	9 858	Articles en caoutchouc			251
18 723	19 032	19 617	20 117	34 360	..	39 182	40 871	Articles en matières plastiques			252
60 186	61 104	61 281	65 735	61 763	..	69 639	68 628	Autres produits minéraux non métalliques		26	
11 593	12 126	11 732	12 139	11 170	..	12 683	12 660	Verre et articles en verre			261
48 593	48 977	49 549	53 596	50 593	..	56 956	55 968	Produits minéraux non métalliques, nca			269
78 719	75 239	66 782	71 839	78 865	..	63 581	91 647	Produits métallurgiques de base		27	
42 531	40 116	36 851	40 652	50 301	..	30 689	56 716	Sidérurgie et première transformation de l'acier			271
26 433	25 575	21 372	21 880	21 628	..	25 444	26 271	Première transf. métaux précieux et non ferreux			272
9 756	9 548	8 559	9 307	6 936	..	7 448	8 660	Fonderie			273
69 192	72 052	70 224	74 868	86 967	..	87 246	97 027	Ouvrages en métaux, sauf machines		28	
27 407	28 923	29 354	31 203	43 822	..	40 206	43 900	Constr. et menuis. métal.; réservoirs; générateurs			281
41 784	43 128	40 871	43 664	43 146	..	47 039	53 127	Autres ouvrages en métaux			289
118 406	104 294	98 250	108 040	108 890	..	125 583	134 154	Machines et matériels, nca		29	
31 877	18 974	19 911	22 268	42 108	..	47 198	53 193	Machines d'usage général			291
79 774	78 262	71 724	78 872	59 253	..	69 646	71 768	Machines d'usage spécifique			292
6 755	7 057	6 615	6 900	7 528	..	8 740	9 192	Appareils domestiques, nca			293
..	..	..	..	565	..	666	3 983	Mach. de bureau, comptables, information		30	
72 165	70 992	71 862	71 569	45 437	..	50 895	56 761	Machines et appareils électriques, nca		31	
14 787	12 832	16 700	14 751	13 240	..	12 640	16 158	Moteurs, génératrices, transformateurs électriques			311
..	..	..	..	10 965	..	13 523	13 468	Matériel électrique de distribution et de commande			312
5 658	5 730	6 021	6 041	7 083	..	7 722	7 670	Fils et câbles électriques isolés			313
51 720	52 430	49 141	50 778	1 907	..	2 378	2 341	Accumulateurs et piles électriques			314
..	..	..	..	4 230	..	5 491	5 775	Lampes électriques et appareils d'éclairage			315
..	..	..	..	8 013	..	9 142	11 349	Autres matériels électriques, nca.			319
43 971	41 498	39 838	50 004	61 965	..	93 721	106 031	Equipements et appareils de radio, TV, etc.		32	
33 350	27 926	28 380	35 061	11 468	..	13 433	23 274	Tubes et valves électroniques;autres composants			321
10 621	13 572	11 457	14 943	34 943	..	62 280	64 286	Emetteurs radio/TV; app. téléphonie et télégraphie			322
..	..	..	..	15 553	..	18 008	18 471	Récepteurs radio/TV, son et image; art. associés			323
11 945	12 901	12 593	12 561	14 110	..	18 339	20 246	Instr. médicaux, de précision ; optique ; horlog.		33	
7 681	8 116	8 396	8 912	11 604	..	14 998	16 954	Instruments médicaux et de précision, etc.			331
3 683	4 174	3 550	2 995	2 452	..	3 266	3 221	Instruments d'optique ; matériel photographique			332
582	611	647	654	54	..	74	71	Horlogerie			333
55 576	60 381	54 931	60 415	64 751	..	84 084	92 699	Véhicules automobiles; remorques, etc.		34	
..	..	..	..	46 829	..	66 786	54 561	Véhicules automobiles			341
..	..	..	..	4 491	..	5 634	6 042	Carross. automobiles, remorques, semi-remorques			342
..	..	..	..	13 431	..	11 663	32 097	Pièces et accessoires pour automobiles et moteurs			343
7 594	6 808	7 173	6 049	12 253	..	11 042	15 148	Autres matériels de transport		35	
1 371	714	931	405	365	..	315	297	Construction et réparation de navires			351
4 467	4 303	4 580	3 885	10 368	..	..	12 356	Matériel ferroviaire roulant			352
..	..	..	..	214	..	281	340	Construction aéronautique et spatiale			353
..	..	..	..	1 305	..	..	2 155	Autres équipements de transport			359
7 346	8 374	8 013	8 339	52 702	..	57 039	57 073	Meubles ; activités de fabrication, nca		36	
367	445	410	419	33 762	..	37 303	37 380	Meubles			361
6 980	7 930	7 603	7 920	18 940	..	19 737	19 693	Activités de fabrication, nca			369
..	..	..	..	1 461	..	1 695	1 999	Récupération		37	
117 575	121 262	125 801	124 681	134 961	..	142 371	142 261	Electricité, gaz et eau	E	40_41	
113 673	117 031	120 953	119 671	130 750	..	138 052	137 470	Prod. et dist. d'électricité gaz, vapeur d'eau chaude		40	
3 902	4 231	4 848	5 009	4 210	..	4 319	4 791	Captage, épuration et distribution d'eau		41	
..	180 728	187 669	202 143	277 540	..	311 225	310 247	Construction	F	45	

AUSTRIA

Table AU.2 VALUE ADDED AT FACTOR COSTS - VALEUR AJOUTÉE AU COUT DES FACTEURS

Millions of Sch (Current Prices)

ISIC revision 3				1991	1992	1993	1994	1995	1996	1997	1998
C	10_14		Mining and quarrying	7 982	7 595	7 494	6 756	6 653	..	7 117	6 795
	10		Mining of coal and lignite; extraction of peat	966	903	842	600	1 004	..	898	..
	11		Crude petroleum & natural gas; service activities	4 145	3 754	..	..	..	..	..	..
	13		Mining of metal ores	222	207	..	..	..	..	..	..
	14		Other mining and quarrying	2 649	2 731	2 809	2 940	4 674	..	4 989	5 203
D	15_37		Total manufacturing	372 142	369 703	359 832	381 165	420 731	..	412 858	475 050
	15		Food products and beverages	38 162	42 325	43 810	42 886	46 233	..	44 085	45 478
		151	Production of meat, fish, fruit, vegetables, oils & fats	6 901	7 358	7 538	6 757	12 605	..	12 183	12 724
		152	Dairy products	3 802	4 323	4 457	3 646	3 140	..	4 287	3 658
		153	Grain mill and starch products; animal feeds	1 070	970	988	936	2 173	..	1 961	2 191
		154	Other food products	16 449	17 169	18 065	17 959	18 616	..	17 830	18 377
		155	Beverages	9 940	12 505	12 761	13 588	9 699	..	7 825	8 528
	16		Tobacco products	16 635	17 204	16 853	17 948	..	..	..	..
	17		Textiles	14 788	15 293	14 562	14 137	16 148	..	12 498	12 987
		171	Spinning, weaving and finishing of textiles	6 492	6 452	5 840	5 893	8 895	..	5 365	5 404
		172	Other textiles	4 849	5 141	5 130	4 875	4 790	..	4 390	5 038
		173	Knitted and crocheted fabrics and articles	3 448	3 700	3 592	3 368	2 463	..	2 744	2 545
	18		Wearing apparel and fur	6 059	5 667	5 096	4 537	5 796	..	5 021	5 141
		181	Wearing apparel, except fur apparel	5 994	5 607	5 037	4 480	5 658	..	4 917	5 057
		182	Dressing and dyeing of fur; articles of fur	65	61	59	57	139	..	105	84
	19		Leather and articles of leather; footwear	3 448	3 264	3 073	2 579	3 020	..	3 239	2 819
		191	Leather and articles of leather	979	947	907	733	799	..	795	906
		192	Footwear	2 469	2 317	2 166	1 846	2 221	..	2 444	1 913
	20		Wood & products of wood & cork, exc. furniture	23 741	24 563	23 579	24 950	19 497	..	20 506	21 342
		201	Sawmilling and planing of wood	6 111	5 786	5 221	5 702	5 677	..	6 139	6 443
		202	Wood products; articles of cork, straw, etc.	17 630	18 777	18 358	19 248	13 820	..	14 367	14 900
	21		Paper and paper products	15 255	14 319	13 100	15 441	19 690	..	18 361	18 702
	22		Printing, publishing and recorded media	14 330	13 843	13 675	14 045	19 187	..	20 680	20 835
		221	Publishing	2 763	2 260	2 765	3 267	5 110	..	5 876	..
		222	Printing and related activities	11 075	11 155	10 415	10 326	12 770	..	12 031	12 258
		223	Reproduction of recorded media	492	428	495	452	1 306	..	2 773	..
	23		Coke and petroleum products; nuclear fuel	6 130	5 453	5 756	6 900	..	..	..	..
		231	Coke oven products	805	625	704	635	..	..	..	..
		232	Refined petroleum products	5 325	4 829	5 051	6 265	..	..	..	..
		233	Processing of nuclear fuel	..	..	..	..		..		
	24		Chemicals and chemical products	27 757	27 070	27 257	28 523	21 724	..	26 268	29 600
		241	Basic chemicals	14 186	13 389	12 456	14 156	7 170	..	7 755	9 517
		242	Other chemical products	13 571	13 681	14 801	14 367	..	..	..	..
		243	Man-made fibres	..	..	..	..	..	..	..	..
	25		Rubber and plastic products	11 011	11 602	11 750	12 214	17 420	..	18 153	19 380
		251	Rubber products	3 824	3 926	3 543	3 980	3 357	..	3 014	3 482
		252	Plastic products	7 187	7 676	8 207	8 234	14 063	..	15 139	15 899
	26		Other non-metallic mineral products	26 163	26 320	25 612	28 071	28 793	..	29 204	28 458
		261	Glass and glass products	6 074	6 702	6 630	6 941	6 324	..	7 035	6 865
		269	Non-metallic mineral products, nec	20 090	19 618	18 983	21 131	22 470	..	22 168	21 593
	27		Basic metals	26 152	22 413	19 982	22 029	27 767	..	18 917	30 197
		271	Basic iron and steel	16 633	12 989	11 888	13 381	20 113	..	10 570	21 104
		272	Basic precious and non-ferrous metals	5 031	4 662	3 858	4 259	4 464	..	5 001	5 198
		273	Casting of metals	4 489	4 762	4 236	4 390	3 190	..	3 346	3 895
	28		Fabricated metal products, exc. machinery	28 957	30 477	29 392	32 310	38 850	..	36 932	40 918
		281	Structural metal prod., tanks, reservoirs, etc.	10 486	11 476	10 881	12 475	17 347	..	15 624	16 956
		289	Other fabricated metal products	18 472	19 001	18 511	19 834	21 503	..	21 308	23 961
	29		Machinery and equipment, nec	42 443	36 933	35 447	40 211	44 300	..	46 145	49 637
		291	General purpose machinery	12 745	6 585	8 146	9 167	17 809	..	18 621	20 527
		292	Special purpose machinery	27 306	27 479	24 657	28 385	23 293	..	24 347	25 488
		293	Domestic appliances, nec	2 392	2 868	2 644	2 659	3 199	..	3 177	3 622
	30		Office, accounting & computing machinery	..	..	..	..	212	..	261	607
	31		Electrical machinery & apparatus, nec	27 451	27 300	28 214	27 734	20 708	..	18 134	20 812
		311	Electric motors, generators and transformers	5 885	5 870	7 161	6 349	6 201	..	4 111	5 698
		312	Electricity distribution & control apparatus	..	..	..	..	5 048	..	5 416	5 367
		313	Insulated wire and cable	2 280	2 373	2 594	2 640	2 831	..	2 501	2 488
		314	Accumulators, primary cells & primary batteries	19 286	19 057	18 459	18 745	733	..	828	901
		315	Electric lamps and lighting equipment	..	..	..	..	2 015	..	1 913	2 347
		319	Other electrical equipment, nec	..	..	..	..	3 879	..	3 365	4 010
	32		Radio, TV, communication equipment	13 855	14 250	14 100	16 409	25 970	..	24 737	29 936
		321	Electronic valves, tubes; other electronic components	9 247	8 721	9 107	10 398	5 747	..	5 530	7 643
		322	TV/radio transmitters; line teleph./telegr. apparatus	4 608	5 529	4 993	6 012	16 629	..	15 750	18 676
		323	TV/radio receivers; sound and video app.; assoc. gds.	..	..	..	..	3 593	..	3 456	3 617
	33		Medical, precision & optical instrum.; watch & clocks	5 609	5 989	5 973	5 867	7 371	..	8 475	9 627
		331	Medical appl.; instruments & appl. for measuring, etc.	3 513	3 688	3 907	4 138	6 114	..	6 805	8 008
		332	Optical instruments and photographic equipment	1 863	2 053	1 814	1 454	1 226	..	1 633	1 581
		333	Watches and clocks	232	249	252	275	32	..	37	38
	34		Motor vehicles, trailers and semi-trailers	17 843	19 199	16 198	18 465	19 334	..	22 037	24 036
		341	Motor vehicles	..	..	..	..	12 356	..	15 459	12 630
		342	Bodies for motor vehicles, trailers & semi-trailers	..	..	..	..	1 737	..	1 842	2 046
		343	Parts & accessories for motor vehicles & their engines	..	..	..	..	5 241	..	4 736	9 360
	35		Other transport equipment	3 333	2 929	3 215	2 753	3 964	..	3 653	4 169
		351	Building and repairing of ships and boats	481	266	472	210	185	..	150	132
		352	Railway and tramway locomotives and rolling stock	2 132	2 053	2 103	1 883	3 317	..	..	3 302
		353	Aircraft and spacecraft	..	..	..	..	90	..	101	155
		359	Transport equipment, nec	..	..	..	..	372	..	..	580
	36		Furniture; manufacturing, nec	3 018	3 290	3 188	3 155	22 386	..	22 536	24 250
		361	Furniture	126	156	140	150	16 191	..	16 527	16 854
		369	Manufacturing, nec	2 892	3 134	3 047	3 005	6 195	..	6 009	7 396
	37		Recycling	..	..	..	..	631	..	650	734
E	40_41		Electricity, gas and water supply	51 528	55 342	58 609	58 548	62 268	..	66 013	64 477
	40		Electricity, gas, steam and hot water supply	48 999	52 565	55 404	55 245	59 463	..	63 147	61 415
	41		Collection, purification and distribution of water	2 529	2 776	3 205	3 303	2 805	..	2 865	3 061
F	45		Construction	..	82 904	88 204	95 979	130 184	..	141 697	145 308

Table AU.3 **INVESTISSEMENT - INVESTMENT**

Millions de Sch (Prix courants)

1991	1992	1993	1994	1995	1996	1997	1998		CITI révision 3		
1 802	1 774	1 932	1 941	2 364	..	2 070	2 010	Activites extractives	C	10_14	
63	55	69	46	111	..	63	..	Extraction de charbon, de lignite et de tourbe		10	
856	964	..	..	..	..	..	..	Pétrole brut et gaz naturel ; activités annexes		11	
48	41	..	..	..	..	..	..	Extraction de minerais métalliques		13	
835	714	838	903	1 807	..	1 716	1 658	Autres activités extractives		14	
69 911	70 599	59 098	57 422	64 346	..	67 788	71 743	Total activites manufacturieres	D	15_37	
8 418	8 622	9 484	9 515	9 313	..	10 196	8 739	Produits alimentaires et boissons		15	
1 260	1 461	1 649	1 740	2 452	..	2 490	2 898	Viande, poisson, fruits, légumes, huiles et graisses			151
679	858	1 536	1 255	1 179	..	1 222	892	Produits laitiers			152
172	306	237	153	547	..	352	346	Travail des grains; prod. amyl. ; alim. pour animaux			153
3 203	2 654	2 716	3 066	2 446	..	3 133	2 710	Autres produits alimentaires			154
3 104	3 344	3 347	3 302	2 689	..	2 999	1 893	Boissons			155
252	338	383	290	..	..	..	..	Produits à base de tabac		16	
3 298	2 196	1 649	1 743	2 028	..	1 825	2 323	Textiles		17	
2 127	1 258	682	714	1 145	..	819	1 053	Filature, tissage et achèvement des textiles			171
893	645	692	649	597	..	658	832	Autres articles textiles			172
279	294	274	380	286	..	348	438	Etoffes et articles de bonneterie			173
373	405	433	289	370	..	340	387	Articles d'habillement et fourrure		18	
368	405	431	282	362	..	321	365	Articles d'habillement autres qu'en fourrure			181
5	1	2	6	8	..	18	22	Prép. et teinture des fourrures ; articles en fourrure			182
370	337	404	250	261	..	420	376	Cuir et articles en cuir ; chaussures		19	
167	121	168	75	55	..	206	256	Cuir et articles en cuir sauf chaussures			191
203	216	237	175	206	..	214	120	Chaussures			192
4 869	4 574	4 115	4 153	4 386	..	4 608	4 226	Bois, articles en bois, sauf meubles		20	
1 818	1 477	1 238	1 344	1 691	..	1 956	1 791	Sciage et rabotage du bois			201
3 051	3 097	2 878	2 810	2 695	..	2 652	2 435	Articles en bois, liège, vannerie, sparterie			202
6 098	5 351	3 709	3 042	3 205	..	4 934	3 501	Papier, articles en papier et carton		21	
3 519	2 281	2 390	1 936	2 746	..	2 446	3 413	Edition, imprimerie et reproduction		22	
1 408	627	466	543	652	..	569	..	Edition			221
2 022	1 563	1 866	1 354	1 912	..	1 522	2 584	Imprimerie et activités annexes			222
89	92	58	38	183	..	355	..	Reproduction de supports enregistrés			223
1 265	1 504	863	828	..	..	..	..	Coke. et prod. pétroliers ; comb. nucléaires		23	
93	104	234	213	..	..	..	..	Cokéfaction			231
1 172	1 399	629	616	..	..	..	..	Produits pétroliers raffinés			232
..	..	..	..	..	..	..	..	Traitement de combustibles nucléaires			233
6 842	7 709	6 519	5 391	3 957	..	5 874	5 633	Produits chimiques		24	
4 235	4 778	3 157	2 333	1 237	..	2 433	2 166	Produits chimiques de base			241
2 607	2 931	3 362	3 058	..	..	..	..	Autres produits chimiques			242
..	..	..	..	..	..	..	..	Fibres synthétiques ou artificielles			243
2 234	2 270	1 906	2 095	2 767	..	3 254	3 672	Art. en caoutchouc et matières plastiques		25	
547	629	381	354	388	..	430	466	Articles en caoutchouc			251
1 687	1 642	1 524	1 741	2 379	..	2 824	3 206	Articles en matières plastiques			252
5 673	5 981	4 654	5 521	5 479	..	4 347	5 189	Autres produits minéraux non métalliques		26	
1 556	1 694	1 046	1 197	1 490	..	848	1 033	Verre et articles en verre			261
4 118	4 287	3 607	4 324	3 989	..	3 499	4 156	Produits minéraux non métalliques, nca			269
5 307	5 732	3 670	3 841	4 474	..	2 765	6 864	Produits métallurgiques de base		27	
2 620	2 725	1 825	2 048	3 444	..	1 737	4 846	Sidérurgie et première transformation de l'acier			271
1 730	1 898	1 182	993	493	..	534	1 235	Première transf. métaux précieux et non ferreux			272
957	1 109	662	800	537	..	494	783	Fonderie			273
4 601	4 462	4 220	4 478	5 542	..	5 431	5 918	Ouvrages en métaux, sauf machines		28	
1 224	1 342	1 277	1 595	2 219	..	2 100	2 019	Constr. et menuis. métal.; réservoirs; générateurs			281
3 377	3 120	2 943	2 884	3 322	..	3 332	3 898	Autres ouvrages en métaux			289
5 404	4 573	3 838	4 542	4 459	..	4 583	5 353	Machines et matériels, nca		29	
1 405	1 065	1 149	1 710	1 708	..	1 714	2 187	Machines d'usage général			291
3 447	3 035	2 348	2 472	2 351	..	2 528	2 769	Machines d'usage spécifique			292
552	472	341	360	401	..	341	396	Appareils domestiques, nca			293
..	..	..	..	30	..	23	50	Mach. de bureau, comptables, information		30	
3 264	3 128	2 952	2 165	2 066	..	1 962	2 628	Machines et appareils électriques, nca		31	
768	511	679	511	465	..	280	588	Moteurs, générateurs, transformateurs électriques			311
..	..	..	..	590	..	551	753	Matériel électrique de distribution et de commande			312
217	410	373	340	219	..	352	314	Fils et câbles électriques isolés			313
2 279	2 207	1 899	1 313	70	..	150	114	Accumulateurs et piles électriques			314
..	..	..	..	167	..	181	260	Lampes électriques et appareils d'éclairage			315
..	..	..	..	555	..	448	599	Autres matériels électriques, nca.			319
2 726	3 411	2 319	3 096	3 785	..	3 923	3 764	Equipements et appareils de radio, TV, etc.		32	
2 263	2 669	1 671	2 484	1 364	..	1 890	1 768	Tubes et valves électroniques;autres composants			321
463	743	648	612	1 895	..	1 541	1 531	Emetteurs radio/TV; app. téléphonie et télégraphie			322
..	..	..	..	526	..	492	465	Récepteurs radio/TV, son et image; art. associés			323
710	545	715	832	769	..	796	981	Instr. médicaux, de précision ; optique ; horlog.		33	
471	333	413	642	673	..	659	820	Instruments médicaux et de précision, etc.			331
230	204	283	170	95	..	134	157	Instruments d'optique ; matériel photographique			332
10	8	20	21	1	..	2	3	Horlogerie			333
4 200	6 245	4 227	2 847	3 586	..	4 946	4 094	Véhicules automobiles; remorques, etc.		34	
..	..	..	..	2 611	..	4 005	2 069	Véhicules automobiles			341
..	..	..	..	232	..	335	494	Carross. automobiles, remorques, semi-remorques			342
..	..	..	..	744	..	606	1 531	Pièces et accessoires pour automobiles et moteurs			343
238	311	265	227	390	..	470	544	Autres matériels de transport		35	
39	14	7	14	34	..	14	11	Construction et réparation de navires			351
182	188	208	171	294	..	..	388	Matériel ferroviaire roulant			352
..	..	..	..	39	..	6	42	Construction aéronautique et spatiale			353
..	..	..	..	24	..	..	102	Autres équipements de transport			359
250	622	382	340	3 034	..	2 680	2 756	Meubles ; activités de fabrication, nca		36	
6	15	15	10	2 276	..	2 001	2 087	Meubles			361
243	607	367	331	758	..	680	668	Activités de fabrication, nca			369
..	..	..	..	162	..	164	197	Récupération		37	
18 999	21 812	21 098	21 459	23 980	..	22 767	23 054	Electricité, gaz et eau	E	40_41	
17 684	20 561	19 747	20 149	22 606	..	21 452	21 674	Prod. et dist. d'électricité gaz, vapeur d'eau chaude		40	
1 315	1 251	1 350	1 310	1 374	..	1 315	1 380	Captage, épuration et distribution d'eau		41	
..	7 295	8 025	7 698	11 487	..	12 841	13 548	Construction	F	45	

AUSTRIA

Table AU.4 INVESTMENT IN MACHINERY & EQUIPMENT - DÉPENSES DE MACHINES ET D'ÉQUIPEMENT

Millions of Sch (Current Prices)

ISIC revision 3				1991	1992	1993	1994	1995	1996	1997	1998
C	10_14		Mining and quarrying	1 148	1 294	1 300	1 439	1 593	..	..	1 509
	10		Mining of coal and lignite; extraction of peat	17	36	26	24	94	..	48	..
	11		Crude petroleum & natural gas; service activities	749	905	..	..	..	..	..	..
	13		Mining of metal ores	15	6	..	..	..	..	..	..
	14		Other mining and quarrying	366	347	316	503	1 092	..	1 149	1 170
D	15_37		Total manufacturing	46 673	46 402	37 977	37 289	44 189	..	46 799	50 446
	15		Food products and beverages	4 613	4 722	5 265	4 926	5 588	..	5 825	5 566
		151	Production of meat, fish, fruit, vegetables, oils & fats	633	796	849	773	1 363	..	1 287	1 782
		152	Dairy products	356	576	843	595	687	..	856	684
		153	Grain mill and starch products; animal feeds	122	84	127	111	437	..	256	233
		154	Other food products	1 949	1 761	1 847	1 990	1 595	..	1 778	1 683
		155	Beverages	1 554	1 505	1 600	1 457	1 507	..	1 648	1 183
	16		Tobacco products	180	250	143	104	..	..	..	..
	17		Textiles	2 176	1 487	1 058	1 119	1 519	..	1 331	1 679
		171	Spinning, weaving and finishing of textiles	1 351	844	467	517	930	..	638	844
		172	Other textiles	646	424	415	419	384	..	499	603
		173	Knitted and crocheted fabrics and articles	180	219	176	183	204	..	193	232
	18		Wearing apparel and fur	223	256	199	166	270	..	227	256
		181	Wearing apparel, except fur apparel	221	255	199	162	265	..	220	247
		182	Dressing and dyeing of fur; articles of fur	2	1	1	4	5	..	6	9
	19		Leather and articles of leather; footwear	224	211	202	199	207	..	274	177
		191	Leather and articles of leather	91	78	58	54	46	..	118	88
		192	Footwear	133	133	143	145	161	..	156	89
	20		Wood & products of wood & cork, exc. furniture	2 846	2 772	2 408	2 203	2 852	..	2 774	2 781
		201	Sawmilling and planing of wood	1 108	872	778	761	1 174	..	1 385	1 323
		202	Wood products; articles of cork, straw, etc.	1 738	1 900	1 630	1 442	1 678	..	1 390	1 458
	21		Paper and paper products	4 513	3 927	2 677	2 451	2 550	..	4 411	2 857
	22		Printing, publishing and recorded media	2 760	1 716	1 789	1 475	2 011	..	1 770	2 744
		221	Publishing	1 109	523	326	417	428	..	356	..
		222	Printing and related activities	1 572	1 128	1 408	1 029	1 453	..	1 249	2 171
		223	Reproduction of recorded media	79	65	56	29	131	..	164	..
	23		Coke and petroleum products; nuclear fuel	876	984	588	611	..	..	..	..
		231	Coke oven products	43	83	143	155	..	..	..	..
		232	Refined petroleum products	833	901	444	456	..	..	..	..
		233	Processing of nuclear fuel	..	..	..	..	..	..	..	..
	24		Chemicals and chemical products	4 546	4 683	3 735	3 714	2 717	..	4 188	3 891
		241	Basic chemicals	3 101	3 304	1 760	1 676	930	..	2 099	1 612
		242	Other chemical products	1 445	1 379	1 974	2 037	..	..	..	..
		243	Man-made fibres	..	..	..	..	..	..	..	..
	25		Rubber and plastic products	1 654	1 498	1 304	1 608	2 073	..	2 395	2 487
		251	Rubber products	501	467	328	284	347	..	369	441
		252	Plastic products	1 153	1 031	976	1 324	1 726	..	2 026	2 046
	26		Other non-metallic mineral products	3 677	3 622	2 915	3 159	3 813	..	3 311	3 611
		261	Glass and glass products	1 198	1 268	892	714	1 015	..	729	805
		269	Non-metallic mineral products, nec	2 479	2 353	2 023	2 445	2 798	..	2 581	2 805
	27		Basic metals	4 115	4 270	2 881	3 119	3 860	..	2 163	5 345
		271	Basic iron and steel	2 065	2 284	1 543	1 741	2 998	..	1 372	3 810
		272	Basic precious and non-ferrous metals	1 461	1 194	795	791	421	..	437	929
		273	Casting of metals	589	792	543	588	441	..	353	606
	28		Fabricated metal products, exc. machinery	2 895	2 853	2 553	2 615	3 715	..	3 507	4 118
		281	Structural metal prod., tanks, reservoirs, etc.	660	747	712	782	1 246	..	1 178	1 201
		289	Other fabricated metal products	2 235	2 107	1 841	1 833	2 469	..	2 329	2 916
	29		Machinery and equipment, nec	3 309	2 788	2 393	2 668	2 996	..	2 922	3 335
		291	General purpose machinery	859	745	595	883	1 143	..	996	1 270
		292	Special purpose machinery	2 057	1 697	1 513	1 509	1 545	..	1 644	1 788
		293	Domestic appliances, nec	393	346	284	276	308	..	282	277
	30		Office, accounting & computing machinery	..	..	..	..	12	..	15	31
	31		Electrical machinery & apparatus, nec	2 125	1 854	1 987	1 562	1 529	..	1 438	1 819
		311	Electric motors, generators and transformers	553	350	525	350	323	..	241	324
		312	Electricity distribution & control apparatus	..	..	..	..	467	..	451	551
		313	Insulated wire and cable	148	272	353	255	161	..	212	205
		314	Accumulators, primary cells & primary batteries	1 423	1 231	1 110	957	55	..	85	71
		315	Electric lamps and lighting equipment	..	..	..	..	108	..	140	204
		319	Other electrical equipment, nec	..	..	..	..	416	..	309	463
	32		Radio, TV, communication equipment	2 344	2 784	1 652	2 549	3 093	..	3 277	3 371
		321	Electronic valves, tubes; other electronic components	1 983	2 298	1 249	2 093	1 153	..	1 503	1 583
		322	TV/radio transmitters; line teleph./telegr. apparatus	362	486	403	457	1 443	..	1 306	1 371
		323	TV/radio receivers; sound and video app.; assoc. gds.	..	..	..	..	496	..	469	418
	33		Medical, precision & optical instrum.; watch & clocks	470	431	556	383	436	..	548	688
		331	Medical appl.; instruments & appl. for measuring, etc.	318	252	286	241	365	..	452	555
		332	Optical instruments and photographic equipment	148	175	262	125	71	..	95	131
		333	Watches and clocks	5	4	8	17	0	..	1	2
	34		Motor vehicles, trailers and semi-trailers	2 792	4 874	3 169	2 270	1 768	..	3 036	3 065
		341	Motor vehicles	..	..	..	..	1 046	..	2 511	1 662
		342	Bodies for motor vehicles, trailers & semi-trailers	..	..	..	..	143	..	181	275
		343	Parts & accessories for motor vehicles & their engines	..	..	..	..	578	..	343	1 128
	35		Other transport equipment	130	162	207	101	211	..	316	267
		351	Building and repairing of ships and boats	15	8	4	9	16	..	8	8
		352	Railway and tramway locomotives and rolling stock	112	104	158	66	167	..	..	200
		353	Aircraft and spacecraft	..	..	..	..	13	..	4	9
		359	Transport equipment, nec	..	..	..	..	15	..	..	50
	36		Furniture; manufacturing, nec	204	258	298	287	1 753	..	1 578	1 607
		361	Furniture	2	6	5	4	1 217	..	1 123	1 113
		369	Manufacturing, nec	202	252	293	283	537	..	455	494
	37		Recycling	..	..	..	..	126	..	131	151
E	40_41		Electricity, gas and water supply	12 885	14 284	12 678	13 510	15 117	..	..	16 012
	40		Electricity, gas, steam and hot water supply	12 632	13 997	12 456	13 217	14 945	..	15 635	15 866
	41		Collection, purification and distribution of water	253	287	222	294	173	..	170	147
F	45		Construction	..	3 685	3 855	3 631	6 798	..	7 291	8 623

Table AU.5 **EMPLOI, SALARIÉS - EMPLOYMENT, EMPLOYEES**

Unités

1991	1992	1993	1994	1995	1996	1997	1998		CITI révision 3		
7 457	6 850	6 755	6 108	6 141	..	6 397	6 161	Activites extractives	C	10_14	
1 203	1 046	887	648	1 201	..	1 082	..	Extraction de charbon, de lignite et de tourbe		10	
2 320	1 999	..	..	..	..	..	..	Pétrole brut et gaz naturel ; activités annexes		11	
699	561	..	..	..	..	..	..	Extraction de minerais métalliques		13	
3 235	3 244	3 199	3 199	4 411	..	4 792	5 171	Autres activités extractives		14	
628 960	596 088	559 957	553 754	629 747	..	607 661	615 287	Total activites manufacturieres	D	15_37	
64 751	63 594	62 300	60 965	81 391	..	76 736	76 921	Produits alimentaires et boissons		15	
16 101	15 792	15 887	15 345	22 878	..	20 987	21 217	Viande, poisson, fruits, légumes, huiles et graisses			151
7 394	6 753	5 838	5 401	6 765	..	5 473	5 069	Produits laitiers			152
1 119	1 042	990	912	3 623	..	3 026	3 130	Travail des grains; prod. amyl. ; alim. pour animaux			153
27 428	27 535	27 240	27 360	36 936	..	37 031	37 825	Autres produits alimentaires			154
12 709	12 472	12 345	11 947	11 189	..	10 220	9 681	Boissons			155
1 320	1 296	1 248	1 195	..	..	..	..	Produits à base de tabac		16	
35 340	32 859	28 593	28 127	26 608	..	21 664	21 857	Textiles		17	
14 482	13 467	10 771	11 092	12 236	..	9 490	9 472	Filature, tissage et achèvement des textiles			171
10 698	9 900	9 316	9 226	8 853	..	7 438	7 697	Autres articles textiles			172
10 160	9 492	8 506	7 809	5 519	..	4 737	4 688	Etoffes et articles de bonneterie			173
21 941	18 916	15 854	13 444	15 904	..	13 342	12 373	Articles d'habillement et fourrure		18	
21 763	18 766	15 738	13 319	15 552	..	13 051	12 173	Articles d'habillement autres qu'en fourrure			181
178	150	116	125	352	..	290	200	Prép. et teinture des fourrures ; articles en fourrure			182
9 983	8 519	7 837	7 535	7 641	..	7 186	6 996	Cuir et articles en cuir ; chaussures		19	
2 593	2 566	2 242	2 383	2 382	..	2 379	2 722	Cuir et articles en cuir sauf chaussures			191
7 390	5 953	5 595	5 152	5 259	..	4 807	4 274	Chaussures			192
50 280	49 435	48 010	47 693	36 036	..	36 325	37 665	Bois, articles en bois, sauf meubles		20	
10 814	10 387	9 828	9 423	8 894	..	9 096	9 703	Sciage et rabotage du bois			201
39 466	39 048	38 182	38 270	27 142	..	27 229	27 962	Articles en bois, liège, vannerie, sparterie			202
20 346	19 718	18 860	18 306	17 861	..	17 517	17 516	Papier, articles en papier et carton		21	
24 142	23 369	21 765	20 980	26 443	..	26 142	25 969	Edition, imprimerie et reproduction		22	
4 830	5 272	4 797	4 862	7 198	..	7 919	..	Edition			221
18 360	17 233	16 253	15 356	18 451	..	17 075	17 075	Imprimerie et activités annexes			222
952	864	715	762	794	..	1 149	..	Reproduction de supports enregistrés			223
4 845	4 361	4 193	4 493	..	..	..	..	Coke. et prod. pétroliers ; comb. nucléaires		23	
1 092	840	885	941	..	..	..	..	Cokéfaction			231
3 753	3 521	3 308	3 552	..	..	..	..	Produits pétroliers raffinés			232
..	..	..	..	..	..	..	..	Traitement de combustibles nucléaires			233
38 832	37 465	35 504	34 318	23 749	..	25 751	27 140	Produits chimiques		24	
19 910	19 383	17 920	17 112	7 532	..	7 029	7 162	Produits chimiques de base			241
18 922	18 082	17 584	17 206	..	..	..	..	Autres produits chimiques			242
..	..	..	..	..	..	..	..	Fibres synthétiques ou artificielles			243
20 501	20 401	19 966	20 033	26 764	..	26 397	27 797	Art. en caoutchouc et matières plastiques		25	
6 339	5 914	5 339	5 142	5 090	..	4 119	4 512	Articles en caoutchouc			251
14 162	14 487	14 627	14 891	21 674	..	22 277	23 285	Articles en matières plastiques			252
37 521	36 248	34 234	34 471	35 601	..	34 334	33 720	Autres produits minéraux non métalliques		26	
9 224	8 933	8 375	8 641	7 966	..	7 847	8 058	Verre et articles en verre			261
28 297	27 315	25 859	25 830	27 635	..	26 487	25 662	Produits minéraux non métalliques, nca			269
41 966	38 651	34 915	34 462	33 123	..	32 413	32 885	Produits métallurgiques de base		27	
24 445	23 190	20 813	20 459	21 817	..	21 119	21 114	Sidérurgie et première transformation de l'acier			271
9 229	7 582	6 984	6 746	6 081	..	6 149	6 033	Première transf. métaux précieux et non ferreux			272
8 292	7 879	7 118	7 257	5 225	..	5 146	5 739	Fonderie			273
57 424	55 799	53 617	55 214	63 919	..	61 083	64 325	Ouvrages en métaux, sauf machines		28	
21 171	21 268	20 606	21 151	29 535	..	27 184	29 005	Constr. et menuis. métal.; réservoirs; générateurs			281
36 253	34 531	33 011	34 063	34 384	..	33 900	35 319	Autres ouvrages en métaux			289
78 697	66 699	62 514	61 987	70 859	..	70 155	70 119	Machines et matériels, nca		29	
24 390	13 791	14 163	14 522	27 759	..	27 339	28 214	Machines d'usage général			291
49 319	48 070	44 001	42 939	37 668	..	37 120	36 491	Machines d'usage spécifique			292
4 988	4 838	4 350	4 526	5 432	..	5 696	5 415	Appareils domestiques, nca			293
..	..	..	..	322	..	461	804	Mach. de bureau, comptables, information		30	
45 623	44 532	41 813	40 847	32 047	..	27 955	28 132	Machines et appareils électriques, nca		31	
12 066	11 669	11 310	10 922	8 399	..	6 573	7 149	Moteurs, génératrices, transformateurs électriques			311
..	..	..	..	8 427	..	8 247	7 285	Matériel électrique de distribution et de commande			312
3 417	3 428	3 348	3 423	4 544	..	3 827	3 746	Fils et câbles électriques isolés			313
30 140	29 435	27 155	26 502	1 203	..	999	950	Accumulateurs et piles électriques			314
..	..	..	..	3 143	..	2 913	3 207	Lampes électriques et appareils d'éclairage			315
..	..	..	..	6 331	..	5 395	5 795	Autres matériels électriques, nca.			319
25 470	24 701	22 902	23 325	32 785	..	30 509	29 502	Equipements et appareils de radio, TV, etc.		32	
17 882	16 906	15 878	15 994	6 713	..	7 658	8 877	Tubes et valves électroniques;autres composants			321
7 588	7 795	7 024	7 331	20 274	..	18 140	15 885	Emetteurs radio/TV; app. téléphonie et télégraphie			322
..	..	..	..	5 798	..	4 711	4 740	Récepteurs radio/TV, son et image; art. associés			323
11 748	11 908	10 912	10 824	13 251	..	14 577	15 666	Instr. médicaux, de précision ; optique ; horlog.		33	
7 179	7 177	6 913	7 214	10 871	..	11 782	12 984	Instruments médicaux et de précision, etc.			331
3 918	4 108	3 492	3 099	2 338	..	2 701	2 621	Instruments d'optique ; matériel photographique			332
651	623	507	511	42	..	94	61	Horlogerie			333
25 241	25 014	23 531	24 287	23 864	..	25 817	27 305	Véhicules automobiles; remorques, etc.		34	
..	..	..	..	11 370	..	16 368	13 741	Véhicules automobiles			341
..	..	..	..	3 436	..	3 556	3 891	Carross. automobiles, remorques, semi-remorques			342
..	..	..	..	9 058	..	5 893	9 673	Pièces et accessoires pour automobiles et moteurs			343
5 891	5 206	4 742	4 472	5 021	..	4 944	5 150	Autres matériels de transport		35	
986	577	376	385	322	..	315	256	Construction et réparation de navires			351
3 533	3 254	3 023	2 855	3 871	..	..	3 835	Matériel ferroviaire roulant			352
..	..	..	..	191	..	252	263	Construction aéronautique et spatiale			353
..	..	..	..	637	..	..	795	Autres équipements de transport			359
7 098	7 397	6 647	6 776	49 900	..	48 408	47 723	Meubles ; activités de fabrication, nca		36	
362	404	333	381	38 254	..	37 362	36 728	Meubles			361
6 736	6 993	6 314	6 395	11 646	..	11 046	10 994	Activités de fabrication, nca			369
..	..	..	..	967	..	805	943	Récupération		37	
37 010	36 777	36 096	35 349	37 983	..	37 557	36 950	Electricité, gaz et eau	E	40_41	
34 667	34 408	33 725	32 975	36 041	..	35 558	34 745	Prod. et dist. d'électricité gaz, vapeur d'eau chaude		40	
2 343	2 369	2 371	2 374	1 942	..	2 000	2 205	Captage, épuration et distribution d'eau		41	
..	171 370	171 704	173 452	224 674	..	235 083	235 327	Construction	F	45	

AUSTRIA

Table AU.6 EMPLOYMENT, FEMALES - EMPLOI, FEMMES

Units (Current Prices)

ISIC revision 3				1991	1992	1993	1994	1995	1996	1997	1998
C	10_14		Mining and quarrying	717	641	628	578	745	..	774	836
	10		Mining of coal and lignite; extraction of peat	57	49	54	43	119	..	112	..
	11		Crude petroleum & natural gas; service activities	244	189	..	..	..	..	..	..
	13		Mining of metal ores	80	64	..	..	..	..	..	..
	14		Other mining and quarrying	336	339	335	331	565	..	607	715
D	15_37		Total manufacturing	181 380	169 674	155 763	151 504	175 973	..	167 980	169 847
	15		Food products and beverages	23 131	22 401	22 127	21 878	34 718	..	33 791	34 320
		151	Production of meat, fish, fruit, vegetables, oils & fats	6 471	6 121	6 099	5 924	10 753	..	9 887	10 103
		152	Dairy products	1 792	1 595	1 450	1 382	2 002	..	1 589	1 537
		153	Grain mill and starch products; animal feeds	210	199	188	177	839	..	667	775
		154	Other food products	12 101	11 938	11 865	11 907	18 704	..	19 454	19 744
		155	Beverages	2 557	2 548	2 525	2 488	2 420	..	2 193	2 160
	16		Tobacco products	506	494	460	426	..	..	..	..
	17		Textiles	19 569	17 828	15 448	15 052	12 528	..	10 808	10 730
		171	Spinning, weaving and finishing of textiles	5 979	5 410	4 262	4 506	3 851	..	3 585	3 547
		172	Other textiles	5 885	5 374	5 006	4 987	4 923	..	4 090	4 153
		173	Knitted and crocheted fabrics and articles	7 705	7 044	6 180	5 559	3 754	..	3 133	3 031
	18		Wearing apparel and fur	19 277	16 545	13 779	11 518	13 270	..	11 057	10 142
		181	Wearing apparel, except fur apparel	19 164	16 443	13 708	11 441	13 030	..	10 851	9 992
		182	Dressing and dyeing of fur; articles of fur	113	102	71	77	240	..	205	149
	19		Leather and articles of leather; footwear	6 625	5 544	5 033	4 801	4 749	..	4 472	4 228
		191	Leather and articles of leather	1 559	1 536	1 332	1 405	1 410	..	1 419	1 525
		192	Footwear	5 066	4 008	3 701	3 396	3 339	..	3 052	2 703
	20		Wood & products of wood & cork, exc. furniture	9 584	9 409	9 178	9 010	5 846	..	5 969	6 166
		201	Sawmilling and planing of wood	1 431	1 399	1 340	1 122	1 100	..	1 169	1 231
		202	Wood products; articles of cork, straw, etc.	8 153	8 010	7 838	7 888	4 746	..	4 799	4 935
	21		Paper and paper products	4 086	3 830	3 624	3 314	3 296	..	3 109	3 342
	22		Printing, publishing and recorded media	8 220	8 109	7 550	7 270	9 476	..	9 868	9 857
		221	Publishing	2 394	2 578	2 436	2 421	3 606	..	3 993	..
		222	Printing and related activities	5 343	5 073	4 727	4 441	5 690	..	5 593	5 597
		223	Reproduction of recorded media	483	458	387	408	180	..	282	..
	23		Coke and petroleum products; nuclear fuel	900	809	775	866	..	..	..	..
		231	Coke oven products	121	121	130	136	..	..	..	..
		232	Refined petroleum products	779	688	645	730	..	..	..	..
		233	Processing of nuclear fuel	..	..	..	..	..	..	..	..
	24		Chemicals and chemical products	11 231	10 687	10 133	9 583	7 757	..	7 611	8 307
		241	Basic chemicals	3 661	3 455	3 176	2 866	1 279	..	1 160	1 248
		242	Other chemical products	7 570	7 232	6 957	6 717	..	..	..	..
		243	Man-made fibres	..	..	..	..	..	..	..	..
	25		Rubber and plastic products	6 426	6 224	5 753	5 847	7 166	..	7 056	7 454
		251	Rubber products	1 408	1 289	1 074	1 068	961	..	736	711
		252	Plastic products	5 018	4 935	4 679	4 779	6 205	..	6 320	6 742
	26		Other non-metallic mineral products	7 595	7 299	6 764	6 935	6 780	..	6 421	6 619
		261	Glass and glass products	3 202	2 992	2 765	2 986	2 264	..	2 213	2 390
		269	Non-metallic mineral products, nec	4 393	4 307	3 999	3 949	4 516	..	4 207	4 230
	27		Basic metals	4 941	4 601	4 135	3 977	3 916	..	3 824	3 879
		271	Basic iron and steel	2 103	2 079	1 832	1 742	1 877	..	1 751	1 787
		272	Basic precious and non-ferrous metals	1 679	1 404	1 326	1 254	1 042	..	1 098	1 040
		273	Casting of metals	1 159	1 118	977	981	997	..	975	1 053
	28		Fabricated metal products, exc. machinery	11 831	11 420	10 642	10 973	12 022	..	11 517	11 873
		281	Structural metal prod., tanks, reservoirs, etc.	2 532	2 722	2 418	2 492	3 976	..	3 546	3 770
		289	Other fabricated metal products	9 299	8 698	8 224	8 481	8 046	..	7 972	8 103
	29		Machinery and equipment, nec	11 361	9 762	9 088	9 097	10 700	..	10 737	10 951
		291	General purpose machinery	3 669	2 376	2 405	2 448	4 374	..	4 507	4 716
		292	Special purpose machinery	5 805	5 564	5 071	4 928	4 748	..	4 609	4 754
		293	Domestic appliances, nec	1 887	1 822	1 612	1 721	1 578	..	1 621	1 481
	30		Office, accounting & computing machinery	..	..	..	..	96	..	115	215
	31		Electrical machinery & apparatus, nec	13 180	12 399	11 689	11 108	9 937	..	8 888	8 718
		311	Electric motors, generators and transformers	3 364	3 202	3 139	2 987	1 742	..	1 645	1 779
		312	Electricity distribution & control apparatus	..	..	..	..	2 475	..	2 525	2 255
		313	Insulated wire and cable	953	919	951	925	1 342	..	1 092	1 070
		314	Accumulators, primary cells & primary batteries	8 863	8 278	7 599	7 196	143	..	125	113
		315	Electric lamps and lighting equipment	..	..	..	..	1 068	..	1 061	1 028
		319	Other electrical equipment, nec	..	..	..	..	3 167	..	2 440	2 472
	32		Radio, TV, communication equipment	10 607	9 868	8 407	8 506	9 852	..	8 609	8 357
		321	Electronic valves, tubes; other electronic components	7 969	7 175	6 138	6 080	2 000	..	2 178	2 418
		322	TV/radio transmitters; line teleph./telegr. apparatus	2 638	2 693	2 269	2 426	5 194	..	4 402	3 873
		323	TV/radio receivers; sound and video app.; assoc. gds.	..	..	..	..	2 658	..	2 028	2 066
	33		Medical, precision & optical instrum.; watch & clocks	5 106	5 231	4 659	4 542	5 499	..	5 839	6 198
		331	Medical appl.; instruments & appl. for measuring, etc.	2 809	2 842	2 716	2 754	4 352	..	4 514	4 908
		332	Optical instruments and photographic equipment	1 953	2 053	1 676	1 511	1 136	..	1 297	1 281
		333	Watches and clocks	344	336	267	277	11	..	28	9
	34		Motor vehicles, trailers and semi-trailers	3 970	3 994	3 658	3 940	3 326	..	3 630	3 822
		341	Motor vehicles	..	..	..	..	958	..	1 530	1 333
		342	Bodies for motor vehicles, trailers & semi-trailers	..	..	..	..	398	..	407	483
		343	Parts & accessories for motor vehicles & their engines	..	..	..	..	1 970	..	1 693	2 006
	35		Other transport equipment	643	501	463	442	678	..	691	714
		351	Building and repairing of ships and boats	78	34	31	30	81	..	61	76
		352	Railway and tramway locomotives and rolling stock	352	307	241	255	405	..	..	372
		353	Aircraft and spacecraft	..	..	..	..	34	..	85	80
		359	Transport equipment, nec	..	..	..	..	158	..	..	186
	36		Furniture; manufacturing, nec	2 591	2 719	2 398	2 419	13 133	..	12 858	12 806
		361	Furniture	96	104	83	109	7 839	..	7 647	7 615
		369	Manufacturing, nec	2 495	2 615	2 315	2 310	5 294	..	5 211	5 192
	37		Recycling	..	..	..	..	139	..	111	185
E	40_41		Electricity, gas and water supply	5 040	4 998	4 794	4 597	5 041	..	5 045	5 019
	40		Electricity, gas, steam and hot water supply	4 844	4 806	4 591	4 391	4 841	..	4 835	4 736
	41		Collection, purification and distribution of water	196	192	203	206	200	..	210	283
F	45		Construction	..	11 409	11 320	11 404	25 773	..	26 984	27 726

Table AU.7 **SALAIRES ET CHARGES SOCIALES, SALARIÉS - COMPENSATION OF LABOUR, EMPLOYEES**

Millions de Sch (Prix courants)

1991	1992	1993	1994	1995	1996	1997	1998		CITI révision 3		
4 246	4 081	4 348	4 295	3 478	..	3 592	3 688	Activites extractives	C	10_14	
672	534	511	347	692	..	637	..	Extraction de charbon, de lignite et de tourbe		10	
1 607	1 613	..	..	..	..	..	..	Pétrole brut et gaz naturel ; activités annexes		11	
484	361	..	..	..	..	..	..	Extraction de minerais métalliques		13	
1 483	1 573	1 619	1 685	2 308	..	2 500	2 687	Autres activités extractives		14	
255 316	259 287	256 899	261 543	297 250	..	291 277	307 255	Total activites manufacturieres	D	15_37	
24 566	25 643	26 424	26 761	32 858	..	31 561	31 270	Produits alimentaires et boissons		15	
5 005	5 352	5 544	5 782	8 550	..	8 021	8 091	Viande, poisson, fruits, légumes, huiles et graisses			151
3 157	3 082	2 866	2 628	3 720	..	3 020	2 604	Produits laitiers			152
510	540	523	519	1 763	..	1 419	1 430	Travail des grains; prod. amyl. ; alim. pour animaux			153
10 070	10 586	11 167	11 309	12 755	..	13 461	13 595	Autres produits alimentaires			154
5 824	6 083	6 324	6 523	6 069	..	5 640	5 549	Boissons			155
951	946	1 036	799	..	..	..	..	Produits à base de tabac		16	
11 203	11 343	10 670	10 408	11 038	..	8 794	8 976	Textiles		17	
5 162	5 131	4 545	4 420	5 875	..	4 024	4 154	Filature, tissage et achèvement des textiles			171
3 337	3 389	3 376	3 412	3 272	..	2 811	2 925	Autres articles textiles			172
2 705	2 823	2 749	2 577	1 890	..	1 958	2 103	Etoffes et articles de bonneterie			173
4 928	4 627	4 240	3 766	4 626	..	4 017	3 892	Articles d'habillement et fourrure		18	
4 880	4 582	4 203	3 728	4 542	..	3 951	3 839	Articles d'habillement autres qu'en fourrure			181
48	45	38	38	84	..	66	53	Prép. et teinture des fourrures ; articles en fourrure			182
2 559	2 428	2 315	2 173	2 342	..	2 434	2 219	Cuir et articles en cuir ; chaussures		19	
696	717	683	681	664	..	731	771	Cuir et articles en cuir sauf chaussures			191
1 863	1 711	1 632	1 493	1 678	..	1 703	1 448	Chaussures			192
15 167	16 282	16 615	17 195	12 713	..	13 838	14 374	Bois, articles en bois, sauf meubles		20	
3 192	3 158	3 119	3 142	3 138	..	3 236	3 708	Sciage et rabotage du bois			201
11 975	13 124	13 496	14 053	9 575	..	10 602	10 666	Articles en bois, liège, vannerie, sparterie			202
9 727	10 085	9 764	9 954	9 786	..	10 294	9 988	Papier, articles en papier et carton		21	
11 533	12 021	11 757	11 673	14 094	..	13 873	14 308	Edition, imprimerie et reproduction		22	
2 465	2 722	2 642	2 829	3 965	..	4 481	..	Edition			221
8 706	8 848	8 802	8 538	9 754	..	8 814	9 250	Imprimerie et activités annexes			222
362	451	313	306	375	..	578	..	Reproduction de supports enregistrés			223
3 421	3 748	3 816	4 577	..	..	..	..	Coke. et prod. pétroliers ; comb. nucléaires		23	
506	439	495	593	..	..	..	..	Cokéfaction			231
2 915	3 309	3 321	3 984	..	..	..	..	Produits pétroliers raffinés			232
..	..	..	..	..	..	..	..	Traitement de combustibles nucléaires			233
19 104	19 395	19 367	19 405	13 951	..	16 212	17 559	Produits chimiques		24	
10 351	10 400	10 085	10 052	4 957	..	5 004	5 143	Produits chimiques de base			241
8 753	8 995	9 282	9 353	..	..	..	..	Autres produits chimiques			242
..	..	..	..	..	..	..	..	Fibres synthétiques ou artificielles			243
7 569	8 177	8 264	8 561	12 156	..	12 557	13 656	Art. en caoutchouc et matières plastiques		25	
2 920	2 980	2 708	2 749	2 695	..	2 245	2 438	Articles en caoutchouc			251
4 649	5 197	5 555	5 811	9 460	..	10 313	11 217	Articles en matières plastiques			252
16 445	17 440	17 184	17 792	19 169	..	19 162	19 317	Autres produits minéraux non métalliques		26	
3 603	3 923	3 880	4 025	3 868	..	4 083	4 225	Verre et articles en verre			261
12 842	13 517	13 304	13 766	15 300	..	15 079	15 092	Produits minéraux non métalliques, nca			269
19 288	18 955	17 726	17 787	18 433	..	12 907	19 730	Produits métallurgiques de base		27	
11 961	11 872	11 062	11 134	12 858	..	7 200	13 354	Sidérurgie et première transformation de l'acier			271
4 011	3 712	3 503	3 448	3 173	..	3 317	3 551	Première transf. métaux précieux et non ferreux			272
3 315	3 371	3 161	3 205	2 402	..	2 389	2 825	Fonderie			273
21 781	23 111	22 863	23 808	28 649	..	27 738	29 835	Ouvrages en métaux, sauf machines		28	
8 512	9 142	9 148	9 473	13 846	..	12 447	13 433	Constr. et menuis. métal.; réservoirs; générateurs			281
13 270	13 968	13 715	14 335	14 803	..	15 291	16 401	Autres ouvrages en métaux			289
34 665	30 739	30 238	30 798	35 332	..	36 848	37 610	Machines et matériels, nca		29	
10 716	6 059	6 543	6 882	14 004	..	14 426	15 499	Machines d'usage général			291
22 200	22 897	21 948	22 113	19 024	..	19 945	19 524	Machines d'usage spécifique			292
1 750	1 783	1 748	1 804	2 304	..	2 477	2 587	Appareils domestiques, nca			293
..	..	..	..	155	..	185	284	Mach. de bureau, comptables, information		30	
21 233	21 989	22 623	22 620	15 992	..	14 280	14 862	Machines et appareils électriques, nca		31	
4 794	4 815	5 013	5 002	4 421	..	3 508	3 983	Moteurs, générateurs, transformateurs électriques			311
..	..	..	..	4 309	..	4 034	3 909	Matériel électrique de distribution et de commande			312
1 553	1 664	1 777	1 840	2 479	..	2 278	2 129	Fils et câbles électriques isolés			313
14 886	15 509	15 832	15 778	667	..	593	553	Accumulateurs et piles électriques			314
..	..	..	..	1 463	..	1 482	1 701	Lampes électriques et appareils d'éclairage			315
..	..	..	..	2 653	..	2 385	2 587	Autres matériels électriques, nca.			319
10 851	11 027	11 053	11 726	21 256	..	20 601	21 405	Equipements et appareils de radio, TV, etc.		32	
7 249	7 011	7 114	7 375	3 361	..	3 417	4 724	Tubes et valves électroniques;autres composants			321
3 602	4 015	3 938	4 351	14 764	..	14 459	13 885	Emetteurs radio/TV; app. téléphonie et télégraphie			322
..	..	..	..	3 131	..	2 726	2 796	Récepteurs radio/TV, son et image; art. associés			323
4 615	4 991	4 935	5 064	5 660	..	6 574	7 082	Instr. médicaux, de précision ; optique ; horlog.		33	
2 905	3 028	3 140	3 456	4 560	..	5 292	5 818	Instruments médicaux et de précision, etc.			331
1 511	1 732	1 605	1 413	1 073	..	1 260	1 236	Instruments d'optique ; matériel photographique			332
198	232	190	194	26	..	22	28	Horlogerie			333
10 441	11 253	10 909	11 857	11 633	..	13 921	15 012	Véhicules automobiles; remorques, etc.		34	
..	..	..	..	5 937	..	9 610	8 352	Véhicules automobiles			341
..	..	..	..	1 295	..	1 441	1 576	Carross. automobiles, remorques, semi-remorques			342
..	..	..	..	4 402	..	2 870	5 083	Pièces et accessoires pour automobiles et moteurs			343
2 885	2 431	2 476	2 194	3 092	..	2 997	3 290	Autres matériels de transport		35	
613	328	340	187	139	..	126	108	Construction et réparation de navires			351
1 659	1 530	1 520	1 363	2 624	..	..	2 756	Matériel ferroviaire roulant			352
..	..	..	..	86	..	85	106	Construction aéronautique et spatiale			353
..	..	..	..	243	..	..	320	Autres équipements de transport			359
2 385	2 655	2 626	2 625	17 012	..	17 182	17 398	Meubles ; activités de fabrication, nca		36	
107	123	129	132	12 356	..	12 863	12 677	Meubles			361
2 278	2 533	2 497	2 493	4 656	..	4 319	4 721	Activités de fabrication, nca			369
..	..	..	..	434	..	333	394	Récupération		37	
25 376	27 669	27 893	28 508	29 413	..	29 653	30 532	Electricité, gaz et eau	E	40_41	
24 229	26 456	26 640	27 231	28 377	..	28 557	29 314	Prod. et dist. d'électricité gaz, vapeur d'eau chaude		40	
1 146	1 214	1 253	1 277	1 036	..	1 096	1 217	Captage, épuration et distribution d'eau		41	
..	64 361	67 305	72 469	102 205	..	110 952	108 317	Construction	F	45	

Table AU.8 **WAGES AND SALARIES, EMPLOYEES - SALAIRES ET TRAITEMENTS, SALARIÉS**

Millions of Sch (Current Prices)

ISIC revision 3				1991	1992	1993	1994	1995	1996	1997	1998
C	10_14		Mining and quarrying	3 272	3 149	3 327	3 022	2 629	..	2 723	2 833
	10		Mining of coal and lignite; extraction of peat	517	415	410	276	533	..	490	..
	11		Crude petroleum & natural gas; service activities	1 226	1 231	..	..	..	..	..	..
	13		Mining of metal ores	368	262	..	..	..	..	..	..
	14		Other mining and quarrying	1 162	1 241	1 265	1 307	1 771	..	1 909	2 067
D	15_37		Total manufacturing	199 419	201 581	199 411	201 470	228 712	..	224 381	236 902
	15		Food products and beverages	19 092	19 827	20 384	20 524	25 396	..	24 261	24 123
		151	Production of meat, fish, fruit, vegetables, oils & fats	3 941	4 175	4 309	4 443	6 606	..	6 197	6 233
		152	Dairy products	2 441	2 386	2 204	2 039	2 861	..	2 311	1 993
		153	Grain mill and starch products; animal feeds	409	422	410	403	1 364	..	1 086	1 103
		154	Other food products	7 821	8 201	8 595	8 710	9 896	..	10 302	10 455
		155	Beverages	4 480	4 643	4 866	4 928	4 670	..	4 364	4 338
	16		Tobacco products	585	584	626	627	..	..	..	..
	17		Textiles	8 826	8 897	8 303	8 090	8 510	..	6 779	6 905
		171	Spinning, weaving and finishing of textiles	4 075	4 020	3 556	3 449	4 538	..	3 095	3 196
		172	Other textiles	2 619	2 642	2 598	2 626	2 510	..	2 182	2 259
		173	Knitted and crocheted fabrics and articles	2 132	2 236	2 149	2 014	1 462	..	1 502	1 449
	18		Wearing apparel and fur	3 871	3 590	3 302	2 921	3 549	..	3 088	3 017
		181	Wearing apparel, except fur apparel	3 833	3 553	3 271	2 892	3 482	..	3 036	2 974
		182	Dressing and dyeing of fur; articles of fur	38	37	31	29	67	..	52	43
	19		Leather and articles of leather; footwear	2 026	1 911	1 798	1 681	1 819	..	1 871	1 724
		191	Leather and articles of leather	538	558	529	531	514	..	562	611
		192	Footwear	1 488	1 354	1 269	1 151	1 305	..	1 309	1 114
	20		Wood & products of wood & cork, exc. furniture	11 995	12 796	13 025	13 385	9 823	..	10 619	11 082
		201	Sawmilling and planing of wood	2 550	2 494	2 448	2 451	2 422	..	2 489	2 851
		202	Wood products; articles of cork, straw, etc.	9 445	10 302	10 577	10 934	7 401	..	8 130	8 231
	21		Paper and paper products	7 562	7 887	7 562	7 621	7 490	..	7 865	7 635
	22		Printing, publishing and recorded media	9 348	9 796	9 489	9 308	11 120	..	10 905	11 239
		221	Publishing	2 001	2 233	2 178	2 276	3 128	..	3 558	..
		222	Printing and related activities	7 056	7 189	7 062	6 791	7 698	..	6 893	7 263
		223	Reproduction of recorded media	290	374	250	241	294	..	453	..
	23		Coke and petroleum products; nuclear fuel	2 611	2 928	2 929	2 993	..	..	..	..
		231	Coke oven products	409	352	394	467	..	..	..	..
		232	Refined petroleum products	2 203	2 576	2 536	2 525	..	..	..	..
		233	Processing of nuclear fuel	..	..	..	..	..	..	..	..
	24		Chemicals and chemical products	14 484	14 659	14 698	14 695	10 748	..	12 535	13 620
		241	Basic chemicals	7 843	7 817	7 606	7 494	3 832	..	3 895	4 015
		242	Other chemical products	6 641	6 842	7 092	7 201	..	..	..	..
		243	Man-made fibres	..	..	..	..	..	..	..	..
	25		Rubber and plastic products	6 009	6 437	6 489	6 699	9 401	..	9 626	10 486
		251	Rubber products	2 325	2 346	2 136	2 168	2 094	..	1 691	1 840
		252	Plastic products	3 683	4 091	4 353	4 531	7 307	..	7 935	8 647
	26		Other non-metallic mineral products	12 718	13 393	13 128	13 431	14 624	..	14 483	14 522
		261	Glass and glass products	2 815	3 030	2 955	3 056	2 922	..	3 043	3 136
		269	Non-metallic mineral products, nec	9 903	10 363	10 173	10 375	11 702	..	11 440	11 386
	27		Basic metals	14 586	14 196	13 362	13 540	14 057	..	9 835	15 072
		271	Basic iron and steel	8 902	8 681	8 217	8 389	9 834	..	5 485	10 226
		272	Basic precious and non-ferrous metals	3 099	2 907	2 719	2 690	2 394	..	2 533	2 690
		273	Casting of metals	2 586	2 608	2 426	2 461	1 828	..	1 818	2 156
	28		Fabricated metal products, exc. machinery	17 128	18 108	17 858	18 491	22 316	..	21 531	23 121
		281	Structural metal prod., tanks, reservoirs, etc.	6 603	7 115	7 116	7 363	10 807	..	9 671	10 432
		289	Other fabricated metal products	10 524	10 993	10 742	11 128	11 509	..	11 861	12 688
	29		Machinery and equipment, nec	27 332	24 066	23 722	23 935	27 373	..	28 524	29 197
		291	General purpose machinery	8 389	4 756	5 117	5 348	10 814	..	11 121	12 005
		292	Special purpose machinery	17 564	17 914	17 248	17 198	14 781	..	15 498	15 198
		293	Domestic appliances, nec	1 379	1 396	1 358	1 389	1 778	..	1 905	1 994
	30		Office, accounting & computing machinery	..	..	..	..	123	..	143	219
	31		Electrical machinery & apparatus, nec	16 612	17 109	17 749	17 579	12 338	..	10 965	11 442
		311	Electric motors, generators and transformers	3 781	3 774	3 907	3 864	3 420	..	2 720	3 104
		312	Electricity distribution & control apparatus	..	..	..	..	3 334	..	3 119	3 019
		313	Insulated wire and cable	1 232	1 304	1 380	1 400	1 895	..	1 710	1 595
		314	Accumulators, primary cells & primary batteries	11 599	12 031	12 462	12 315	513	..	459	429
		315	Electric lamps and lighting equipment	..	..	..	..	1 135	..	1 130	1 310
		319	Other electrical equipment, nec	..	..	..	..	2 043	..	1 827	1 985
	32		Radio, TV, communication equipment	8 638	8 686	8 674	9 167	16 464	..	15 886	16 414
		321	Electronic valves, tubes; other electronic components	5 765	5 529	5 566	5 718	2 600	..	2 571	3 655
		322	TV/radio transmitters; line teleph./telegr. apparatus	2 873	3 157	3 108	3 448	11 443	..	11 221	10 600
		323	TV/radio receivers; sound and video app.; assoc. gds.	..	..	..	..	2 422	..	2 094	2 158
	33		Medical, precision & optical instrum.; watch & clocks	3 679	3 947	3 864	3 917	4 391	..	5 097	5 513
		331	Medical appl.; instruments & appl. for measuring, etc.	2 331	2 413	2 490	2 700	3 550	..	4 107	4 535
		332	Optical instruments and photographic equipment	1 188	1 355	1 223	1 066	821	..	972	956
		333	Watches and clocks	160	178	151	150	21	..	18	22
	34		Motor vehicles, trailers and semi-trailers	8 136	8 758	8 417	9 102	8 930	..	10 604	11 436
		341	Motor vehicles	..	..	..	..	4 524	..	7 306	6 371
		342	Bodies for motor vehicles, trailers & semi-trailers	..	..	..	..	1 007	..	1 104	1 206
		343	Parts & accessories for motor vehicles & their engines	..	..	..	..	3 399	..	2 193	3 859
	35		Other transport equipment	2 288	1 923	1 970	1 723	2 414	..	2 340	2 591
		351	Building and repairing of ships and boats	491	269	289	144	110	..	99	86
		352	Railway and tramway locomotives and rolling stock	1 336	1 203	1 189	1 066	2 048	..	..	2 175
		353	Aircraft and spacecraft	..	..	..	..	67	..	66	84
		359	Transport equipment, nec	..	..	..	..	189	..	..	247
	36		Furniture; manufacturing, nec	1 891	2 082	2 062	2 040	13 170	..	13 264	13 429
		361	Furniture	82	95	100	101	9 563	..	9 923	9 785
		369	Manufacturing, nec	1 809	1 987	1 962	1 939	3 607	..	3 342	3 644
	37		Recycling	..	..	..	..	334	..	255	301
E	40_41		Electricity, gas and water supply	16 014	16 966	17 448	17 556	19 921	..	20 312	20 993
	40		Electricity, gas, steam and hot water supply	15 206	16 112	16 563	16 646	19 165	..	19 539	20 146
	41		Collection, purification and distribution of water	808	854	885	910	756	..	772	847
F	45		Construction	..	51 010	53 115	57 127	79 787	..	86 401	84 783

Table AU.9 CHARGES SOCIALES DES EMPLOYEURS - SUPPLEMENTS TO WAGES & SALARIES

Millions de Sch (Prix courants)

1991	1992	1993	1994	1995	1996	1997	1998		CITI révision 3		
974	932	1 021	1 273	849	..	869	855	Activites extractives	C	10_14	
156	118	101	71	160	..	147	..	Extraction de charbon, de lignite et de tourbe		10	
381	382	..	..	..	..	..	..	Pétrole brut et gaz naturel ; activités annexes		11	
117	100	..	..	..	..	..	..	Extraction de minerais métalliques		13	
321	332	355	378	537	..	592	620	Autres activités extractives		14	
55 897	57 706	57 488	60 073	68 538	..	66 896	70 353	Total activites manufacturieres	D	15_37	
5 474	5 816	6 040	6 236	7 462	..	7 299	7 147	Produits alimentaires et boissons		15	
1 063	1 177	1 235	1 339	1 944	..	1 823	1 858	Viande, poisson, fruits, légumes, huiles et graisses			151
716	696	662	589	859	..	709	611	Produits laitiers			152
101	118	113	116	400	..	333	327	Travail des grains; prod. amyl. ; alim. pour animaux			153
2 249	2 385	2 572	2 599	2 860	..	3 159	3 140	Autres produits alimentaires			154
1 344	1 440	1 458	1 594	1 399	..	1 275	1 211	Boissons			155
366	362	411	172	..	..	..	..	Produits à base de tabac		16	
2 378	2 446	2 367	2 319	2 528	..	2 014	2 071	Textiles		17	
1 088	1 111	989	970	1 337	..	929	958	Filature, tissage et achèvement des textiles			171
718	747	778	786	762	..	629	666	Autres articles textiles			172
572	588	600	562	429	..	456	447	Etoffes et articles de bonneterie			173
1 056	1 037	939	845	1 077	..	928	875	Articles d'habillement et fourrure		18	
1 047	1 029	932	836	1 060	..	914	865	Articles d'habillement autres qu'en fourrure			181
9	8	7	9	18	..	14	11	Prép. et teinture des fourrures ; articles en fourrure			182
533	516	517	492	523	..	563	494	Cuir et articles en cuir ; chaussures		19	
158	159	154	150	150	..	169	160	Cuir et articles en cuir sauf chaussures			191
375	357	363	342	373	..	394	334	Chaussures			192
3 172	3 486	3 590	3 810	2 890	..	3 218	3 292	Bois, articles en bois, sauf meubles		20	
642	663	671	691	716	..	747	857	Sciage et rabotage du bois			201
2 530	2 822	2 919	3 119	2 174	..	2 471	2 435	Articles en bois, liège, vannerie, sparterie			202
2 164	2 198	2 202	2 333	2 296	..	2 429	2 352	Papier, articles en papier et carton		21	
2 185	2 226	2 268	2 365	2 974	..	2 968	3 069	Edition, imprimerie et reproduction		22	
464	489	464	553	837	..	923	..	Edition			221
1 650	1 659	1 740	1 747	2 056	..	1 921	1 988	Imprimerie et activités annexes			222
72	77	64	65	81	..	124	..	Reproduction de supports enregistrés			223
809	820	887	1 584	..	..	..	..	Coke. et prod. pétroliers ; comb. nucléaires		23	
97	87	101	125	..	..	..	..	Cokéfaction			231
712	733	786	1 459	..	..	..	..	Produits pétroliers raffinés			232
..	..	..	..	..	..	..	..	Traitement de combustibles nucléaires			233
4 621	4 737	4 669	4 710	3 203	..	3 677	3 939	Produits chimiques		24	
2 508	2 583	2 479	2 559	1 125	..	1 109	1 128	Produits chimiques de base			241
2 113	2 153	2 190	2 151	..	..	..	..	Autres produits chimiques			242
..	..	..	..	..	..	..	..	Fibres synthétiques ou artificielles			243
1 560	1 741	1 774	1 861	2 754	..	2 931	3 169	Art. en caoutchouc et matières plastiques		25	
595	634	572	581	601	..	554	599	Articles en caoutchouc			251
965	1 106	1 202	1 281	2 153	..	2 378	2 571	Articles en matières plastiques			252
3 727	4 047	4 056	4 360	4 545	..	4 679	4 795	Autres produits minéraux non métalliques		26	
788	893	925	969	947	..	1 040	1 090	Verre et articles en verre			261
2 939	3 154	3 131	3 392	3 598	..	3 639	3 706	Produits minéraux non métalliques, nca			269
4 701	4 758	4 363	4 247	4 376	..	3 072	4 658	Produits métallurgiques de base		27	
3 059	3 191	2 845	2 744	3 023	..	1 716	3 128	Sidérurgie et première transformation de l'acier			271
913	805	783	758	778	..	784	861	Première transf. métaux précieux et non ferreux			272
730	762	735	745	574	..	572	669	Fonderie			273
4 654	5 003	5 005	5 317	6 333	..	6 207	6 714	Ouvrages en métaux, sauf machines		28	
1 908	2 028	2 032	2 110	3 039	..	2 776	3 001	Constr. et menuis. métal.; réservoirs; générateurs			281
2 745	2 975	2 973	3 207	3 294	..	3 430	3 713	Autres ouvrages en métaux			289
7 333	6 672	6 516	6 863	7 959	..	8 324	8 413	Machines et matériels, nca		29	
2 327	1 302	1 426	1 534	3 190	..	3 305	3 493	Machines d'usage général			291
4 636	4 983	4 700	4 915	4 243	..	4 447	4 326	Machines d'usage spécifique			292
371	387	390	415	526	..	572	593	Appareils domestiques, nca			293
..	..	..	..	32	..	41	64	Mach. de bureau, comptables, information		30	
4 620	4 879	4 874	5 041	3 653	..	3 315	3 421	Machines et appareils électriques, nca		31	
1 014	1 041	1 106	1 138	1 001	..	788	879	Moteurs, génératrices, transformateurs électriques			311
..	..	..	..	975	..	915	891	Matériel électrique de distribution et de commande			312
320	360	397	440	584	..	568	534	Fils et câbles électriques isolés			313
3 287	3 478	3 371	3 463	154	..	134	124	Accumulateurs et piles électriques			314
..	..	..	..	328	..	352	391	Lampes électriques et appareils d'éclairage			315
..	..	..	..	611	..	558	602	Autres matériels électriques, nca.			319
2 213	2 340	2 378	2 560	4 791	..	4 715	4 991	Equipements et appareils de radio, TV, etc.		32	
1 484	1 483	1 548	1 657	761	..	845	1 068	Tubes et valves électroniques;autres composants			321
729	858	830	903	3 321	..	3 238	3 284	Emetteurs radio/TV; app. téléphonie et télégraphie			322
..	..	..	..	710	..	632	638	Récepteurs radio/TV, son et image; art. associés			323
935	1 044	1 070	1 147	1 269	..	1 477	1 569	Instr. médicaux, de précision ; optique ; horlog.		33	
574	615	650	756	1 011	..	1 185	1 284	Instruments médicaux et de précision, etc.			331
323	376	382	347	252	..	288	279	Instruments d'optique ; matériel photographique			332
38	54	39	44	6	..	4	6	Horlogerie			333
2 305	2 496	2 492	2 755	2 703	..	3 317	3 576	Véhicules automobiles; remorques, etc.		34	
..	..	..	..	1 413	..	2 304	1 981	Véhicules automobiles			341
..	..	..	..	288	..	336	370	Carross. automobiles, remorques, semi-remorques			342
..	..	..	..	1 002	..	677	1 225	Pièces et accessoires pour automobiles et moteurs			343
597	507	507	471	678	..	657	700	Autres matériels de transport		35	
122	60	52	43	30	..	27	22	Construction et réparation de navires			351
323	327	331	297	575	..	..	582	Matériel ferroviaire roulant			352
..	..	..	..	19	..	19	22	Construction aéronautique et spatiale			353
..	..	..	..	54	..	..	73	Autres équipements de transport			359
493	573	565	585	3 842	..	3 918	3 969	Meubles ; activités de fabrication, nca		36	
24	28	30	31	2 793	..	2 940	2 892	Meubles			361
469	545	535	554	1 049	..	977	1 078	Activités de fabrication, nca			369
..	..	..	..	100	..	78	93	Récupération		37	
9 362	10 703	10 446	10 952	9 492	..	9 342	9 539	Electricité, gaz et eau	E	40_41	
9 024	10 344	10 077	10 585	9 211	..	9 018	9 168	Prod. et dist. d'électricité gaz, vapeur d'eau chaude		40	
338	360	368	368	280	..	323	370	Captage, épuration et distribution d'eau		41	
..	13 350	14 190	15 342	22 418	..	24 551	23 534	Construction	F	45	

Table AU.10 **HOURS WORKED BY EMPLOYEES - HEURES OUVRÉES, SALARIÉS**

Thousands

ISIC revision 3				1991	1992	1993	1994	1995	1996	1997	1998
C	10_14		Mining and quarrying	..	..	..	..	10 791	..	11 442	11 054
	10		Mining of coal and lignite; extraction of peat	..	..	..	..	2 072	..	1 804	..
	11		Crude petroleum & natural gas; service activities	..	..	..	..	..	..	..	..
	13		Mining of metal ores	..	..	..	..	..	..	..	..
	14		Other mining and quarrying	..	..	..	..	7 909	..	8 781	9 412
D	15_37		Total manufacturing	..	..	..	..	1057 207	..	1024 102	1035 612
	15		Food products and beverages	..	..	..	..	139 536	..	130 780	130 091
		151	Production of meat, fish, fruit, vegetables, oils & fats	..	..	..	..	39 041	..	35 931	36 128
		152	Dairy products	..	..	..	..	11 792	..	9 275	8 610
		153	Grain mill and starch products; animal feeds	..	..	..	..	6 469	..	5 137	5 361
		154	Other food products	..	..	..	..	62 765	..	63 170	63 751
		155	Beverages	..	..	..	..	19 469	..	17 267	16 242
	16		Tobacco products	..	..	..	..	..	..	..	..
	17		Textiles	..	..	..	..	43 392	..	35 692	36 168
		171	Spinning, weaving and finishing of textiles	..	..	..	..	20 986	..	15 734	15 441
		172	Other textiles	..	..	..	..	13 640	..	12 218	12 795
		173	Knitted and crocheted fabrics and articles	..	..	..	..	8 765	..	7 740	7 932
	18		Wearing apparel and fur	..	..	..	..	25 874	..	21 438	19 871
		181	Wearing apparel, except fur apparel	..	..	..	..	25 354	..	20 947	19 556
		182	Dressing and dyeing of fur; articles of fur	..	..	..	..	520	..	490	315
	19		Leather and articles of leather; footwear	..	..	..	..	11 668	..	12 014	11 607
		191	Leather and articles of leather	..	..	..	..	3 831	..	4 131	4 803
		192	Footwear	..	..	..	..	7 837	..	7 883	6 803
	20		Wood & products of wood & cork, exc. furniture	..	..	..	..	61 479	..	62 087	64 270
		201	Sawmilling and planing of wood	..	..	..	..	15 319	..	15 410	16 265
		202	Wood products; articles of cork, straw, etc.	..	..	..	..	46 160	..	46 677	48 004
	21		Paper and paper products	..	..	..	..	30 003	..	29 226	29 372
	22		Printing, publishing and recorded media	..	..	..	..	42 250	..	42 350	42 123
		221	Publishing	..	..	..	..	10 496	..	11 991	..
		222	Printing and related activities	..	..	..	..	30 445	..	28 290	28 341
		223	Reproduction of recorded media	..	..	..	..	1 309	..	2 068	..
	23		Coke and petroleum products; nuclear fuel	..	..	..	..	..	..	..	..
		231	Coke oven products	..	..	..	..	..	..		..
		232	Refined petroleum products	..	..	..	..	..	..	..	..
		233	Processing of nuclear fuel	..	..	..	..	..	..		..
	24		Chemicals and chemical products	..	..	..	..	39 594	..	42 786	45 121
		241	Basic chemicals	..	..	..	..	12 618	..	11 500	11 871
		242	Other chemical products	..	..	..	..	..	..	..	..
		243	Man-made fibres	..	..	..	..	..	..	..	..
	25		Rubber and plastic products	..	..	..	..	45 041	..	44 352	46 816
		251	Rubber products	..	..	..	..	8 141	..	6 793	7 388
		252	Plastic products	..	..	..	..	36 901	..	37 558	39 428
	26		Other non-metallic mineral products	..	..	..	..	60 892	..	58 075	56 818
		261	Glass and glass products	..	..	..	..	12 998	..	12 885	13 076
		269	Non-metallic mineral products, nec	..	..	..	..	47 893	..	45 190	43 743
	27		Basic metals	..	..	..	..	55 883	..	54 872	54 817
		271	Basic iron and steel	..	..	..	..	36 612	..	35 558	35 124
		272	Basic precious and non-ferrous metals	..	..	..	..	10 316	..	10 345	9 910
		273	Casting of metals	..	..	..	..	8 956	..	8 969	9 783
	28		Fabricated metal products, exc. machinery	..	..	..	..	108 118	..	102 689	107 563
		281	Structural metal prod., tanks, reservoirs, etc.	..	..	..	..	50 410	..	45 950	49 213
		289	Other fabricated metal products	..	..	..	..	57 709	..	56 740	58 350
	29		Machinery and equipment, nec	..	..	..	..	119 782	..	119 082	119 719
		291	General purpose machinery	..	..	..	..	47 129	..	47 006	48 186
		292	Special purpose machinery	..	..	..	..	63 789	..	62 560	62 529
		293	Domestic appliances, nec	..	..	..	..	8 864	..	9 516	9 004
	30		Office, accounting & computing machinery	..	..	..	..	549	..	756	1 245
	31		Electrical machinery & apparatus, nec	..	..	..	..	52 896	..	47 009	47 682
		311	Electric motors, generators and transformers	..	..	..	..	13 835	..	10 917	11 967
		312	Electricity distribution & control apparatus	..	..	..	..	13 870	..	14 168	12 561
		313	Insulated wire and cable	..	..	..	..	7 520	..	6 201	6 166
		314	Accumulators, primary cells & primary batteries	..	..	..	..	2 037	..	1 665	1 569
		315	Electric lamps and lighting equipment	..	..	..	..	4 934	..	4 769	5 391
		319	Other electrical equipment, nec	..	..	..	..	10 699	..	9 288	10 029
	32		Radio, TV, communication equipment	..	..	..	..	55 921	..	54 958	52 872
		321	Electronic valves, tubes; other electronic components	..	..	..	..	11 267	..	13 518	15 370
		322	TV/radio transmitters; line teleph./telegr. apparatus	..	..	..	..	34 793	..	33 557	28 905
		323	TV/radio receivers; sound and video app.; assoc. gds.	..	..	..	..	9 862	..	7 883	8 597
	33		Medical, precision & optical instrum.; watch & clocks	..	..	..	..	21 858	..	24 305	26 220
		331	Medical appl.; instruments & appl. for measuring, etc.	..	..	..	..	18 012	..	19 796	21 838
		332	Optical instruments and photographic equipment	..	..	..	..	3 779	..	4 350	4 276
		333	Watches and clocks	..	..	..	..	67	..	159	106
	34		Motor vehicles, trailers and semi-trailers	..	..	..	..	39 837	..	42 714	45 352
		341	Motor vehicles	..	..	..	..	19 077	..	26 858	22 816
		342	Bodies for motor vehicles, trailers & semi-trailers	..	..	..	..	5 618	..	5 812	6 247
		343	Parts & accessories for motor vehicles & their engines	..	..	..	..	15 141	..	10 045	16 289
	35		Other transport equipment	..	..	..	..	8 704	..	8 552	9 032
		351	Building and repairing of ships and boats	..	..	..	..	545	..	498	424
		352	Railway and tramway locomotives and rolling stock	..	..	..	..	6 815	..	..	6 793
		353	Aircraft and spacecraft	..	..	..	..	291	..	378	427
		359	Transport equipment, nec	..	..	..	..	1 054	..	..	1 387
	36		Furniture; manufacturing, nec	..	..	..	..	82 539	..	80 732	79 546
		361	Furniture	..	..	..	..	63 831	..	62 501	61 500
		369	Manufacturing, nec	..	..	..	..	18 708	..	18 231	18 046
	37		Recycling	..	..	..	..	1 709	..	1 362	1 636
E	40_41		Electricity, gas and water supply	..	..	..	..	64 599	..	65 515	63 617
	40		Electricity, gas, steam and hot water supply	..	..	..	..	61 470	..	61 861	59 551
	41		Collection, purification and distribution of water	..	..	..	..	..	..	3 653	4 067
F	45		Construction	..	..	..	..	..	..	392 975	388 712

Table AU.11 ÉTABLISSEMENTS - ESTABLISHMENTS

Unités

1991	1992	1993	1994	1995	1996	1997	1998	CITI révision 3			
159	156	156	154	..	..	..	..	Activites extractives	C	10_14	
10	12	10	10	..	..	..	..	Extraction de charbon, de lignite et de tourbe		10	
8	6	7	7	..	..	..	..	Pétrole brut et gaz naturel ; activités annexes		11	
6	5	4	3	..	..	..	..	Extraction de minerais métalliques		13	
135	133	135	134	..	..	..	..	Autres activités extractives		14	
8 920	8 598	8 385	8 185	..	..	..	..	Total activites manufacturieres	D	15_37	
1 047	1 034	1 018	986	..	..	..	..	Produits alimentaires et boissons		15	
224	225	218	202	..	..	..	..	Viande, poisson, fruits, légumes, huiles et graisses			151
125	116	119	103	..	..	..	..	Produits laitiers			152
46	46	41	40	..	..	..	..	Travail des grains; prod. amyl. ; alim. pour animaux			153
418	421	418	416	..	..	..	..	Autres produits alimentaires			154
234	226	222	225	..	..	..	..	Boissons			155
14	14	14	13	..	..	..	..	Produits à base de tabac		16	
432	410	400	387	..	..	..	..	Textiles		17	
112	104	101	99	..	..	..	..	Filature, tissage et achèvement des textiles			171
222	210	200	200	..	..	..	..	Autres articles textiles			172
98	96	99	88	..	..	..	..	Etoffes et articles de bonneterie			173
369	342	308	280	..	..	..	..	Articles d'habillement et fourrure		18	
361	336	303	275	..	..	..	..	Articles d'habillement autres qu'en fourrure			181
8	6	5	5	..	..	..	..	Prép. et teinture des fourrures ; articles en fourrure			182
97	92	90	82	..	..	..	..	Cuir et articles en cuir ; chaussures		19	
40	41	41	35	..	..	..	..	Cuir et articles en cuir sauf chaussures			191
57	51	49	47	..	..	..	..	Chaussures			192
2 528	2 465	2 367	2 306	..	..	..	..	Bois, articles en bois, sauf meubles		20	
1 807	1 737	1 672	1 638	..	..	..	..	Sciage et rabotage du bois			201
721	728	695	668	..	..	..	..	Articles en bois, liège, vannerie, sparterie			202
147	141	141	136	..	..	..	..	Papier, articles en papier et carton		21	
324	329	324	313	..	..	..	..	Edition, imprimerie et reproduction		22	
50	55	55	58	..	..	..	..	Edition			221
257	257	253	239	..	..	..	..	Imprimerie et activités annexes			222
17	17	16	16	..	..	..	..	Reproduction de supports enregistrés			223
58	54	56	59	..	..	..	..	Coke. et prod. pétroliers ; comb. nucléaires		23	
29	27	29	34	..	..	..	..	Cokéfaction			231
29	27	27	25	..	..	..	..	Produits pétroliers raffinés			232
..	..	..	..	..	..	..	..	Traitement de combustibles nucléaires			233
426	400	396	378	..	..	..	..	Produits chimiques		24	
145	143	143	136	..	..	..	..	Produits chimiques de base			241
281	257	253	242	..	..	..	..	Autres produits chimiques			242
..	..	..	..	..	..	..	..	Fibres synthétiques ou artificielles			243
286	291	290	285	..	..	..	..	Art. en caoutchouc et matières plastiques		25	
37	35	33	33	..	..	..	..	Articles en caoutchouc			251
249	256	257	252	..	..	..	..	Articles en matières plastiques			252
535	537	526	523	..	..	..	..	Autres produits minéraux non métalliques		26	
73	73	72	75	..	..	..	..	Verre et articles en verre			261
462	464	454	448	..	..	..	..	Produits minéraux non métalliques, nca			269
172	163	154	154	..	..	..	..	Produits métallurgiques de base		27	
41	39	37	37	..	..	..	..	Sidérurgie et première transformation de l'acier			271
57	54	51	49	..	..	..	..	Première transf. métaux précieux et non ferreux			272
74	70	66	68	..	..	..	..	Fonderie			273
869	888	878	883	..	..	..	..	Ouvrages en métaux, sauf machines		28	
318	323	322	328	..	..	..	..	Constr. et menuis. métal.; réservoirs; générateurs			281
551	565	556	555	..	..	..	..	Autres ouvrages en métaux			289
889	694	697	677	..	..	..	..	Machines et matériels, nca		29	
338	144	143	146	..	..	..	..	Machines d'usage général			291
518	521	525	503	..	..	..	..	Machines d'usage spécifique			292
33	29	29	28	..	..	..	..	Appareils domestiques, nca			293
..	..	..	..	..	..	..	..	Mach. de bureau, comptables, information		30	
271	291	293	290	..	..	..	..	Machines et appareils électriques, nca		31	
74	85	87	85	..	..	..	..	Moteurs, génératrices, transformateurs électriques			311
..	..	..	..	..	..	..	..	Matériel électrique de distribution et de commande			312
22	23	22	21	..	..	..	..	Fils et câbles électriques isolés			313
175	183	184	184	..	..	..	..	Accumulateurs et piles électriques			314
..	..	..	..	..	..	..	..	Lampes électriques et appareils d'éclairage			315
..	..	..	..	..	..	..	..	Autres matériels électriques, nca.			319
127	118	114	114	..	..	..	..	Equipements et appareils de radio, TV, etc.		32	
63	58	59	53	..	..	..	..	Tubes et valves électroniques;autres composants			321
64	60	55	61	..	..	..	..	Emetteurs radio/TV; app. téléphonie et télégraphie			322
..	..	..	..	..	..	..	..	Récepteurs radio/TV, son et image; art. associés			323
136	134	134	132	..	..	..	..	Instr. médicaux, de précision ; optique ; horlog.		33	
98	96	98	100	..	..	..	..	Instruments médicaux et de précision, etc.			331
22	21	21	20	..	..	..	..	Instruments d'optique ; matériel photographique			332
16	17	15	12	..	..	..	..	Horlogerie			333
117	119	109	113	..	..	..	..	Véhicules automobiles; remorques, etc.		34	
..	..	..	..	..	..	..	..	Véhicules automobiles			341
..	..	..	..	..	..	..	..	Carross. automobiles, remorques, semi-remorques			342
..	..	..	..	..	..	..	..	Pièces et accessoires pour automobiles et moteurs			343
18	21	21	21	..	..	..	..	Autres matériels de transport		35	
5	5	5	5	..	..	..	..	Construction et réparation de navires			351
7	8	8	8	..	..	..	..	Matériel ferroviaire roulant			352
4	4	4	5	..	..	..	..	Construction aéronautique et spatiale			353
2	4	4	3	..	..	..	..	Autres équipements de transport			359
58	61	55	53	..	..	..	..	Meubles ; activités de fabrication, nca		36	
9	9	9	9	..	..	..	..	Meubles			361
49	52	46	44	..	..	..	..	Activités de fabrication, nca			369
..	..	..	..	..	..	..	..	Récupération		37	
589	606	603	609	..	..	..	..	Electricité, gaz et eau	E	40_41	
402	418	415	419	..	..	..	..	Prod. et dist. d'électricité gaz, vapeur d'eau chaude		40	
187	188	188	190	..	..	..	..	Captage, épuration et distribution d'eau		41	
..	4 033	3 955	4 006	..	..	..	..	Construction	F	45	

Table AU.12 **ENTERPRISES - ENTREPRISES**

Units

ISIC revision 3				1991	1992	1993	1994	1995	1996	1997	1998
C	10_14		Mining and quarrying	..	..	..	..	303	..	302	318
	10		Mining of coal and lignite; extraction of peat	..	..	..	..	11	..	12	14
	11		Crude petroleum & natural gas; service activities	..	..	..	..	3	..	4	2
	13		Mining of metal ores	..	..	..	..	1	..	1	1
	14		Other mining and quarrying	..	..	..	..	288	..	285	301
D	15_37		Total manufacturing	..	..	..	..	25 509	..	25 465	26 321
	15		Food products and beverages	..	..	..	..	4 736	..	4 619	4 541
		151	Production of meat, fish, fruit, vegetables, oils & fats	..	..	..	..	1 561	..	1 499	1 487
		152	Dairy products	..	..	..	..	106	..	115	121
		153	Grain mill and starch products; animal feeds	..	..	..	..	227	..	203	199
		154	Other food products	..	..	..	..	2 580	..	2 556	2 478
		155	Beverages	..	..	..	..	262	..	246	258
	16		Tobacco products	..	..	..	..	1	..	1	1
	17		Textiles	..	..	..	..	962	..	897	923
		171	Spinning, weaving and finishing of textiles	..	..	..	..	176	..	202	133
		172	Other textiles	..	..	..	..	662	..	607	674
		173	Knitted and crocheted fabrics and articles	..	..	..	..	124	..	87	116
	18		Wearing apparel and fur	..	..	..	..	1 354	..	1 175	1 105
		181	Wearing apparel, except fur apparel	..	..	..	..	1 232	..	1 057	1 016
		182	Dressing and dyeing of fur; articles of fur	..	..	..	..	122	..	119	89
	19		Leather and articles of leather; footwear	..	..	..	..	243	..	232	255
		191	Leather and articles of leather	..	..	..	..	140	..	132	156
		192	Footwear	..	..	..	..	103	..	99	100
	20		Wood & products of wood & cork, exc. furniture	..	..	..	..	3 487	..	3 368	3 481
		201	Sawmilling and planing of wood	..	..	..	..	1 311	..	1 285	1 247
		202	Wood products; articles of cork, straw, etc.	..	..	..	..	2 176	..	2 084	2 234
	21		Paper and paper products	..	..	..	..	143	..	144	172
	22		Printing, publishing and recorded media	..	..	..	..	1 436	..	1 493	1 575
		221	Publishing	..	..	..	..	364	..	376	443
		222	Printing and related activities	..	..	..	..	1 065	..	1 110	1 129
		223	Reproduction of recorded media	..	..	..	..	7	..	7	2
	23		Coke and petroleum products; nuclear fuel	..	..	..	..	8	..	5	5
		231	Coke oven products	..	..	..	..	..	..	..	..
		232	Refined petroleum products	..	..	..	..	8	..	5	5
		233	Processing of nuclear fuel	..	..	..	..	..	..	..	..
	24		Chemicals and chemical products	..	..	..	..	378	..	346	353
		241	Basic chemicals	..	..	..	..	66	..	69	70
		242	Other chemical products	..	..	..	..	308	..	273	278
		243	Man-made fibres	..	..	..	..	4	..	4	5
	25		Rubber and plastic products	..	..	..	..	605	..	538	574
		251	Rubber products	..	..	..	..	45	..	44	44
		252	Plastic products	..	..	..	..	560	..	493	529
	26		Other non-metallic mineral products	..	..	..	..	1 198	..	1 202	1 314
		261	Glass and glass products	..	..	..	..	181	..	176	225
		269	Non-metallic mineral products, nec	..	..	..	..	1 017	..	1 026	1 089
	27		Basic metals	..	..	..	..	152	..	158	180
		271	Basic iron and steel	..	..	..	..	41	..	38	68
		272	Basic precious and non-ferrous metals	..	..	..	..	41	..	46	49
		273	Casting of metals	..	..	..	..	70	..	75	64
	28		Fabricated metal products, exc. machinery	..	..	..	..	2 968	..	3 098	3 373
		281	Structural metal prod., tanks, reservoirs, etc.	..	..	..	..	832	..	927	1 039
		289	Other fabricated metal products	..	..	..	..	2 136	..	2 171	2 334
	29		Machinery and equipment, nec	..	..	..	..	1 727	..	1 797	1 614
		291	General purpose machinery	..	..	..	..	589	..	663	600
		292	Special purpose machinery	..	..	..	..	1 058	..	1 044	915
		293	Domestic appliances, nec	..	..	..	..	80	..	89	98
	30		Office, accounting & computing machinery	..	..	..	..	49	..	52	47
	31		Electrical machinery & apparatus, nec	..	..	..	..	385	..	434	452
		311	Electric motors, generators and transformers	..	..	..	..	87	..	70	107
		312	Electricity distribution & control apparatus	..	..	..	..	114	..	214	132
		313	Insulated wire and cable	..	..	..	..	23	..	17	20
		314	Accumulators, primary cells & primary batteries	..	..	..	..	9	..	6	5
		315	Electric lamps and lighting equipment	..	..	..	..	74	..	71	88
		319	Other electrical equipment, nec	..	..	..	..	78	..	56	100
	32		Radio, TV, communication equipment	..	..	..	..	129	..	141	161
		321	Electronic valves, tubes; other electronic components	..	..	..	..	58	..	27	100
		322	TV/radio transmitters; line teleph./telegr. apparatus	..	..	..	..	24	..	88	13
		323	TV/radio receivers; sound and video app.; assoc. gds.	..	..	..	..	47	..	26	48
	33		Medical, precision & optical instrum.; watch & clocks	..	..	..	..	954	..	1 040	1 111
		331	Medical appl.; instruments & appl. for measuring, etc.	..	..	..	..	855	..	902	998
		332	Optical instruments and photographic equipment	..	..	..	..	89	..	120	109
		333	Watches and clocks	..	..	..	..	10	..	18	5
	34		Motor vehicles, trailers and semi-trailers	..	..	..	..	201	..	235	226
		341	Motor vehicles	..	..	..	..	22	..	20	14
		342	Bodies for motor vehicles, trailers & semi-trailers	..	..	..	..	134	..	134	162
		343	Parts & accessories for motor vehicles & their engines	..	..	..	..	45	..	81	49
	35		Other transport equipment	..	..	..	..	79	..	88	85
		351	Building and repairing of ships and boats	..	..	..	..	34	..	33	25
		352	Railway and tramway locomotives and rolling stock	..	..	..	..	11	..	10	8
		353	Aircraft and spacecraft	..	..	..	..	23	..	43	32
		359	Transport equipment, nec	..	..	..	..	11	..	2	20
	36		Furniture; manufacturing, nec	..	..	..	..	4 223	..	4 322	4 681
		361	Furniture	..	..	..	..	3 191	..	3 255	3 568
		369	Manufacturing, nec	..	..	..	..	1 032	..	1 066	1 114
	37		Recycling	..	..	..	..	91	..	81	94
E	40_41		Electricity, gas and water supply	..	..	..	..	613	..	699	656
	40		Electricity, gas, steam and hot water supply	..	..	..	..	422	..	517	464
	41		Collection, purification and distribution of water	..	..	..	..	191	..	181	192
F	45		Construction	..	..	..	..	15 817	..	17 403	17 327

Table CZ.1 **PRODUCTION - PRODUCTION**

Millions de CK (Prix courants)

1991	1992	1993	1994	1995	1996	1997	1998			CITI révision 3	
..	..	..	..	86 391	88 273	63 755	61 444	Activites extractives	C	10_14	
..	..	..	..	73 154	72 317	50 588	48 427	Extraction de charbon, de lignite et de tourbe		10	
..	..	..	..	1 860	1 905	1 547	1 372	Pétrole brut et gaz naturel ; activités annexes		11	
..	..	..	..	..	..	..	..	Extraction de minerais métalliques		13	
..	..	..	..	7 723	9 993	9 646	10 044	Autres activités extractives		14	
..	..	..	..	1 296 686	1 419 906	1 519 765	1 617 062	Total activites manufacturieres	D	15_37	
..	..	..	..	194 309	232 056	250 905	255 963	Produits alimentaires et boissons		15	
..	..	..	..	..	..	80 320	78 809	Viande, poisson, fruits, légumes, huiles et graisses			151
..	..	..	..	..	..	34 814	38 942	Produits laitiers			152
..	..	..	..	..	..	32 699	31 451	Travail des grains; prod. amyl. ; alim. pour animaux			153
..	..	..	..	..	..	59 194	62 748	Autres produits alimentaires			154
..	..	..	..	..	..	43 878	44 013	Boissons			155
..	..	..	..	..	..	..	..	Produits à base de tabac		16	
..	..	..	..	55 374	55 590	48 863	52 236	Textiles		17	
..	..	..	..	..	..	29 914	33 349	Filature, tissage et achèvement des textiles			171
..	..	..	..	..	..	13 501	13 900	Autres articles textiles			172
..	..	..	..	..	..	5 448	4 987	Etoffes et articles de bonneterie			173
..	..	..	..	16 878	16 528	20 574	18 602	Articles d'habillement et fourrure		18	
..	..	..	..	..	..	19 693	18 021	Articles d'habillement autres qu'en fourrure			181
..	..	..	..	..	..	881	581	Prép. et teinture des fourrures ; articles en fourrure			182
..	..	..	..	17 203	16 380	12 110	11 664	Cuir et articles en cuir ; chaussures		19	
..	..	..	..	..	..	4 519	4 319	Cuir et articles en cuir sauf chaussures			191
..	..	..	..	..	..	7 591	7 345	Chaussures			192
..	..	..	..	36 331	40 349	44 047	47 599	Bois, articles en bois, sauf meubles		20	
..	..	..	..	..	..	12 284	13 187	Sciage et rabotage du bois			201
..	..	..	..	..	..	31 763	34 412	Articles en bois, liège, vannerie, sparterie			202
..	..	..	..	36 797	33 693	34 505	38 893	Papier, articles en papier et carton		21	
..	..	..	..	32 198	37 599	43 633	48 020	Edition, imprimerie et reproduction		22	
..	..	..	..	..	..	25 749	24 934	Edition			221
..	..	..	..	..	..	16 689	20 925	Imprimerie et activités annexes			222
..	..	..	..	..	..	1 195	2 161	Reproduction de supports enregistrés			223
..	..	..	..	..	..	..	..	Coke. et prod. pétroliers ; comb. nucléaires		23	
..	..	..	..	..	..	..	..	Cokéfaction			231
..	..	..	..	..	..	..	..	Produits pétroliers raffinés			232
..	..	..	..	..	..	..	..	Traitement de combustibles nucléaires			233
..	..	..	..	72 696	100 559	97 516	100 684	Produits chimiques		24	
..	..	..	..	..	..	61 210	61 385	Produits chimiques de base			241
..	..	..	..	..	..	34 679	37 496	Autres produits chimiques			242
..	..	..	..	..	..	1 627	1 803	Fibres synthétiques ou artificielles			243
..	..	..	..	37 966	48 027	58 770	64 818	Art. en caoutchouc et matières plastiques		25	
..	..	..	..	..	..	18 474	20 129	Articles en caoutchouc			251
..	..	..	..	..	..	40 296	44 689	Articles en matières plastiques			252
..	..	..	..	64 207	75 712	86 217	90 133	Autres produits minéraux non métalliques		26	
..	..	..	..	..	..	31 797	32 565	Verre et articles en verre			261
..	..	..	..	..	..	54 420	57 568	Produits minéraux non métalliques, nca			269
..	..	..	..	219 590	179 985	146 968	153 987	Produits métallurgiques de base		27	
..	..	..	..	..	..	120 203	124 784	Sidérurgie et première transformation de l'acier			271
..	..	..	..	..	..	12 121	13 011	Première transf. métaux précieux et non ferreux			272
..	..	..	..	..	..	14 644	16 192	Fonderie			273
..	..	..	..	93 512	108 352	126 247	139 915	Ouvrages en métaux, sauf machines		28	
..	..	..	..	..	..	40 699	57 001	Constr. et menuis. métal.; réservoirs; générateurs			281
..	..	..	..	..	..	85 548	82 914	Autres ouvrages en métaux			289
..	..	..	..	112 000	127 827	140 141	157 288	Machines et matériels, nca		29	
..	..	..	..	..	..	57 399	74 759	Machines d'usage général			291
..	..	..	..	..	..	73 977	74 444	Machines d'usage spécifique			292
..	..	..	..	..	..	8 765	8 085	Appareils domestiques, nca			293
..	..	..	..	2 270	2 280	1 255	3 649	Mach. de bureau, comptables, information		30	
..	..	..	..	53 688	61 849	72 575	81 909	Machines et appareils électriques, nca		31	
..	..	..	..	..	..	16 826	21 045	Moteurs, génératrices, transformateurs électriques			311
..	..	..	..	..	..	22 550	24 280	Matériel électrique de distribution et de commande			312
..	..	..	..	..	..	7 993	8 548	Fils et câbles électriques isolés			313
..	..	..	..	..	..	2 917	3 412	Accumulateurs et piles électriques			314
..	..	..	..	..	..	4 448	4 138	Lampes électriques et appareils d'éclairage			315
..	..	..	..	..	..	17 841	20 486	Autres matériels électriques, nca.			319
..	..	..	..	14 037	18 849	19 673	23 688	Equipements et appareils de radio, TV, etc.		32	
..	..	..	..	..	..	6 752	8 411	Tubes et valves électroniques;autres composants			321
..	..	..	..	..	..	8 461	10 352	Emetteurs radio/TV; app. téléphonie et télégraphie			322
..	..	..	..	..	..	4 460	4 925	Récepteurs radio/TV, son et image; art. associés			323
..	..	..	..	15 263	18 655	21 632	26 387	Instr. médicaux, de précision ; optique ; horlog.		33	
..	..	..	..	..	..	19 143	23 309	Instruments médicaux et de précision, etc.			331
..	..	..	..	..	..	2 211	2 645	Instruments d'optique ; matériel photographique			332
..	..	..	..	..	..	278	433	Horlogerie			333
..	..	..	..	76 444	98 407	141 353	168 445	Véhicules automobiles; remorques, etc.		34	
..	..	..	..	..	..	102 272	118 352	Véhicules automobiles			341
..	..	..	..	..	..	4 116	5 920	Carross. automobiles, remorques, semi-remorques			342
..	..	..	..	..	..	34 965	44 173	Pièces et accessoires pour automobiles et moteurs			343
..	..	..	..	22 487	27 000	24 899	18 609	Autres matériels de transport		35	
..	..	..	..	..	..	400	393	Construction et réparation de navires			351
..	..	..	..	..	..	15 631	10 941	Matériel ferroviaire roulant			352
..	..	..	..	..	..	4 122	4 014	Construction aéronautique et spatiale			353
..	..	..	..	..	..	4 746	3 261	Autres équipements de transport			359
..	..	..	..	39 948	48 177	50 627	51 833	Meubles ; activités de fabrication, nca		36	
..	..	..	..	..	..	27 797	33 297	Meubles			361
..	..	..	..	..	..	22 830	18 536	Activités de fabrication, nca			369
..	..	..	..	11 116	9 450	10 393	9 803	Récupération		37	
..	..	..	..	231 723	225 518	225 524	237 440	Electricité, gaz et eau	E	40_41	
..	..	..	..	218 506	210 389	208 666	218 820	Prod. et dist. d'électricité gaz, vapeur d'eau chaude		40	
..	..	..	..	13 217	15 129	16 858	18 620	Captage, épuration et distribution d'eau		41	
..	..	..	..	322 994	360 886	404 857	397 172	Construction	F	45	

CZECH REPUBLIC

Table CZ.2 VALUE ADDED - VALEUR AJOUTÉE

Millions of CK (Current Prices)

ISIC revision 3				1991	1992	1993	1994	1995	1996	1997	1998
C	10_14		Mining and quarrying	..	..	..	..	28 648	29 493	30 326	29 050
	10		Mining of coal and lignite; extraction of peat	..	..	..	..	23 675	24 625	25 354	24 131
	11		Crude petroleum & natural gas; service activities	..	..	..	..	880	821	677	662
	13		Mining of metal ores	..	..	..	..	..	..	..	..
	14		Other mining and quarrying	..	..	..	..	3 093	4 074	3 894	4 081
D	15_37		Total manufacturing	..	..	..	..	337 157	362 696	400 511	417 068
	15		Food products and beverages	..	..	..	..	41 918	44 588	49 517	49 355
		151	Production of meat, fish, fruit, vegetables, oils & fats	..	..	..	..	..	..	12 553	11 745
		152	Dairy products	..	..	..	..	..	..	4 376	4 754
		153	Grain mill and starch products; animal feeds	..	..	..	..	..	..	5 440	5 929
		154	Other food products	..	..	..	..	..	..	14 931	14 978
		155	Beverages	..	..	..	..	..	..	12 217	11 949
	16		Tobacco products	..	..	..	..	..	..	..	..
	17		Textiles	..	..	..	..	14 149	15 136	14 914	15 735
		171	Spinning, weaving and finishing of textiles	..	..	..	..	..	..	8 901	9 672
		172	Other textiles	..	..	..	..	..	..	4 105	4 297
		173	Knitted and crocheted fabrics and articles	..	..	..	..	..	..	1 908	1 766
	18		Wearing apparel and fur	..	..	..	..	7 485	6 035	7 496	7 427
		181	Wearing apparel, except fur apparel	..	..	..	..	..	..	7 302	7 271
		182	Dressing and dyeing of fur; articles of fur	..	..	..	..	..	..	194	156
	19		Leather and articles of leather; footwear	..	..	..	..	5 225	5 118	3 730	3 551
		191	Leather and articles of leather	..	..	..	..	..	..	1 213	1 081
		192	Footwear	..	..	..	..	..	..	2 517	2 470
	20		Wood & products of wood & cork, exc. furniture	..	..	..	..	11 133	12 183	11 470	11 669
		201	Sawmilling and planing of wood	..	..	..	..	..	..	3 037	3 560
		202	Wood products; articles of cork, straw, etc.	..	..	..	..	..	..	8 433	8 109
	21		Paper and paper products	..	..	..	..	9 745	7 676	7 653	7 916
	22		Printing, publishing and recorded media	..	..	..	..	9 078	11 804	12 884	14 349
		221	Publishing	..	..	..	..	..	..	6 792	7 027
		222	Printing and related activities	..	..	..	..	..	..	5 658	6 781
		223	Reproduction of recorded media	..	..	..	..	..	..	434	541
	23		Coke and petroleum products; nuclear fuel	..	..	..	..	..	..	..	..
		231	Coke oven products	..	..	..	..	..	..	..	..
		232	Refined petroleum products	..	..	..	..	..	..	..	..
		233	Processing of nuclear fuel	..	..	..	..	..	..	..	..
	24		Chemicals and chemical products	..	..	..	..	20 303	25 156	26 042	26 716
		241	Basic chemicals	..	..	..	..	..	..	15 547	16 787
		242	Other chemical products	..	..	..	..	..	..	10 111	9 483
		243	Man-made fibres	..	..	..	..	..	..	384	446
	25		Rubber and plastic products	..	..	..	..	10 689	14 188	17 173	18 386
		251	Rubber products	..	..	..	..	..	..	5 931	6 352
		252	Plastic products	..	..	..	..	..	..	11 242	12 034
	26		Other non-metallic mineral products	..	..	..	..	23 374	27 605	33 020	34 623
		261	Glass and glass products	..	..	..	..	..	..	12 422	12 922
		269	Non-metallic mineral products, nec	..	..	..	..	..	..	20 598	21 701
	27		Basic metals	..	..	..	..	30 445	29 369	30 943	32 742
		271	Basic iron and steel	..	..	..	..	..	..	24 659	25 536
		272	Basic precious and non-ferrous metals	..	..	..	..	..	..	1 715	2 152
		273	Casting of metals	..	..	..	..	..	..	4 569	5 054
	28		Fabricated metal products, exc. machinery	..	..	..	..	32 832	35 918	35 892	40 191
		281	Structural metal prod., tanks, reservoirs, etc.	..	..	..	..	..	..	9 127	13 128
		289	Other fabricated metal products	..	..	..	..	..	..	26 765	27 063
	29		Machinery and equipment, nec	..	..	..	..	37 868	40 862	45 742	46 970
		291	General purpose machinery	..	..	..	..	..	..	18 227	18 799
		292	Special purpose machinery	..	..	..	..	..	..	24 752	25 797
		293	Domestic appliances, nec	..	..	..	..	..	..	2 763	2 374
	30		Office, accounting & computing machinery	..	..	..	..	618	743	481	551
	31		Electrical machinery & apparatus, nec	..	..	..	..	16 254	19 284	22 462	24 255
		311	Electric motors, generators and transformers	..	..	..	..	..	..	5 294	5 425
		312	Electricity distribution & control apparatus	..	..	..	..	..	..	7 298	7 752
		313	Insulated wire and cable	..	..	..	..	..	..	2 086	2 160
		314	Accumulators, primary cells & primary batteries	..	..	..	..	..	..	896	940
		315	Electric lamps and lighting equipment	..	..	..	..	..	..	1 553	1 336
		319	Other electrical equipment, nec	..	..	..	..	..	..	5 335	6 642
	32		Radio, TV, communication equipment	..	..	..	..	5 315	6 100	5 967	6 828
		321	Electronic valves, tubes; other electronic components	..	..	..	..	..	..	2 427	3 104
		322	TV/radio transmitters; line teleph./telegr. apparatus	..	..	..	..	..	..	2 758	3 090
		323	TV/radio receivers; sound and video app.; assoc. gds.	..	..	..	..	..	..	782	634
	33		Medical, precision & optical instrum.; watch & clocks	..	..	..	..	6 256	7 354	7 855	9 254
		331	Medical appl.; instruments & appl. for measuring, etc.	..	..	..	..	..	..	6 793	7 997
		332	Optical instruments and photographic equipment	..	..	..	..	..	..	993	1 128
		333	Watches and clocks	..	..	..	..	..	..	69	129
	34		Motor vehicles, trailers and semi-trailers	..	..	..	..	15 928	21 018	30 571	34 173
		341	Motor vehicles	..	..	..	..	..	..	19 613	20 521
		342	Bodies for motor vehicles, trailers & semi-trailers	..	..	..	..	..	..	889	1 087
		343	Parts & accessories for motor vehicles & their engines	..	..	..	..	..	..	10 069	12 565
	35		Other transport equipment	..	..	..	..	6 766	7 308	8 976	5 544
		351	Building and repairing of ships and boats	..	..	..	..	..	..	181	136
		352	Railway and tramway locomotives and rolling stock	..	..	..	..	..	..	6 596	3 626
		353	Aircraft and spacecraft	..	..	..	..	..	..	881	1 060
		359	Transport equipment, nec	..	..	..	..	..	..	1 318	722
	36		Furniture; manufacturing, nec	..	..	..	..	14 166	16 326	14 510	14 908
		361	Furniture	..	..	..	..	..	..	6 976	8 256
		369	Manufacturing, nec	..	..	..	..	..	..	7 534	6 652
	37		Recycling	..	..	..	..	2 119	1 788	2 135	1 412
E	40_41		Electricity, gas and water supply	..	..	..	..	68 521	65 366	63 578	68 806
	40		Electricity, gas, steam and hot water supply	..	..	..	..	62 883	59 018	56 337	60 952
	41		Collection, purification and distribution of water	..	..	..	..	5 638	6 348	7 241	7 854
F	45		Construction	..	..	..	..	91 069	91 362	97 226	90 804

Table CZ.3 **INVESTISSEMENT - INVESTMENT**

Millions de CK (Prix courants)

1991	1992	1993	1994	1995	1996	1997	1998		CITI révision 3		
..	..	..	..	8 983	12 013	8 206	8 542	Activites extractives	C	10_14	
..	..	..	..	6 135	9 577	6 326	6 044	Extraction de charbon, de lignite et de tourbe		10	
..	..	..	..	232	457	246	239	Pétrole brut et gaz naturel ; activités annexes		11	
..	..	..	..	..	..	..	..	Extraction de minerais métalliques		13	
..	..	..	..	1 459	1 541	1 408	1 999	Autres activités extractives		14	
..	..	..	..	94 691	115 260	134 995	124 721	Total activites manufacturieres	D	15_37	
..	..	..	..	12 246	16 390	17 885	15 426	Produits alimentaires et boissons		15	
..	..	..	..	..	..	3 761	3 600	Viande, poisson, fruits, légumes, huiles et graisses			151
..	..	..	..	..	..	839	1 391	Produits laitiers			152
..	..	..	..	..	..	1 308	837	Travail des grains; prod. amyl. ; alim. pour animaux			153
..	..	..	..	..	..	4 812	5 268	Autres produits alimentaires			154
..	..	..	..	..	..	7 165	4 330	Boissons			155
..	..	..	..	..	..	..	..	Produits à base de tabac		16	
..	..	..	..	3 696	4 366	4 697	5 455	Textiles		17	
..	..	..	..	..	..	3 003	3 432	Filature, tissage et achèvement des textiles			171
..	..	..	..	..	..	1 406	1 804	Autres articles textiles			172
..	..	..	..	..	..	288	219	Etoffes et articles de bonneterie			173
..	..	..	..	758	963	1 147	758	Articles d'habillement et fourrure		18	
..	..	..	..	..	..	1 052	723	Articles d'habillement autres qu'en fourrure			181
..	..	..	..	..	..	95	35	Prép. et teinture des fourrures ; articles en fourrure			182
..	..	..	..	626	533	429	368	Cuir et articles en cuir ; chaussures		19	
..	..	..	..	..	..	120	195	Cuir et articles en cuir sauf chaussures			191
..	..	..	..	..	..	309	173	Chaussures			192
..	..	..	..	2 268	4 466	4 198	2 804	Bois, articles en bois, sauf meubles		20	
..	..	..	..	..	..	650	498	Sciage et rabotage du bois			201
..	..	..	..	..	..	3 548	2 306	Articles en bois, liège, vannerie, sparterie			202
..	..	..	..	3 515	5 196	3 511	5 633	Papier, articles en papier et carton		21	
..	..	..	..	3 446	2 880	4 049	3 227	Edition, imprimerie et reproduction		22	
..	..	..	..	..	..	1 732	1 465	Edition			221
..	..	..	..	..	..	2 155	1 706	Imprimerie et activités annexes			222
..	..	..	..	..	..	162	56	Reproduction de supports enregistrés			223
..	..	..	..	..	..	..	..	Coke. et prod. pétroliers ; comb. nucléaires		23	
..	..	..	..	..	..	..	..	Cokéfaction			231
..	..	..	..	..	..	..	..	Produits pétroliers raffinés			232
..	..	..	..	..	..	..	..	Traitement de combustibles nucléaires			233
..	..	..	..	7 516	10 905	10 311	12 059	Produits chimiques		24	
..	..	..	..	..	..	7 932	8 319	Produits chimiques de base			241
..	..	..	..	..	..	2 197	3 557	Autres produits chimiques			242
..	..	..	..	..	..	182	183	Fibres synthétiques ou artificielles			243
..	..	..	..	3 542	3 603	5 729	6 527	Art. en caoutchouc et matières plastiques		25	
..	..	..	..	..	..	2 049	2 912	Articles en caoutchouc			251
..	..	..	..	..	..	3 680	3 615	Articles en matières plastiques			252
..	..	..	..	11 885	12 219	12 544	13 270	Autres produits minéraux non métalliques		26	
..	..	..	..	..	..	4 202	2 810	Verre et articles en verre			261
..	..	..	..	..	..	8 342	10 460	Produits minéraux non métalliques, nca			269
..	..	..	..	6 727	10 387	15 537	7 828	Produits métallurgiques de base		27	
..	..	..	..	..	..	13 952	5 974	Sidérurgie et première transformation de l'acier			271
..	..	..	..	..	..	542	649	Première transf. métaux précieux et non ferreux			272
..	..	..	..	..	..	1 043	1 205	Fonderie			273
..	..	..	..	6 193	7 162	11 555	10 515	Ouvrages en métaux, sauf machines		28	
..	..	..	..	..	..	3 854	3 726	Constr. et menuis. métal.; réservoirs; générateurs			281
..	..	..	..	..	..	7 701	6 789	Autres ouvrages en métaux			289
..	..	..	..	6 636	7 386	12 014	7 760	Machines et matériels, nca		29	
..	..	..	..	..	..	6 231	3 466	Machines d'usage général			291
..	..	..	..	..	..	5 304	3 735	Machines d'usage spécifique			292
..	..	..	..	..	..	479	559	Appareils domestiques, nca			293
..	..	..	..	69	83	67	207	Mach. de bureau, comptables, information		30	
..	..	..	..	3 660	4 370	6 391	8 709	Machines et appareils électriques, nca		31	
..	..	..	..	..	..	1 210	1 175	Moteurs, génératrices, transformateurs électriques			311
..	..	..	..	..	..	2 157	4 057	Matériel électrique de distribution et de commande			312
..	..	..	..	..	..	564	807	Fils et câbles électriques isolés			313
..	..	..	..	..	..	386	279	Accumulateurs et piles électriques			314
..	..	..	..	..	..	446	225	Lampes électriques et appareils d'éclairage			315
..	..	..	..	..	..	1 628	2 166	Autres matériels électriques, nca.			319
..	..	..	..	892	1 185	1 935	2 402	Equipements et appareils de radio, TV, etc.		32	
..	..	..	..	..	..	1 057	1 571	Tubes et valves électroniques;autres composants			321
..	..	..	..	..	..	451	580	Emetteurs radio/TV; app. téléphonie et télégraphie			322
..	..	..	..	..	..	427	251	Récepteurs radio/TV, son et image; art. associés			323
..	..	..	..	1 419	1 402	1 414	1 442	Instr. médicaux, de précision ; optique ; horlog.		33	
..	..	..	..	..	..	1 165	1 069	Instruments médicaux et de précision, etc.			331
..	..	..	..	..	..	243	321	Instruments d'optique ; matériel photographique			332
..	..	..	..	..	..	6	52	Horlogerie			333
..	..	..	..	8 330	14 598	13 342	14 281	Véhicules automobiles; remorques, etc.		34	
..	..	..	..	..	..	8 144	9 627	Véhicules automobiles			341
..	..	..	..	..	..	353	248	Carross. automobiles, remorques, semi-remorques			342
..	..	..	..	..	..	4 845	4 406	Pièces et accessoires pour automobiles et moteurs			343
..	..	..	..	1 795	1 937	1 711	1 118	Autres matériels de transport		35	
..	..	..	..	..	..	6	9	Construction et réparation de navires			351
..	..	..	..	..	..	1 094	649	Matériel ferroviaire roulant			352
..	..	..	..	..	..	445	375	Construction aéronautique et spatiale			353
..	..	..	..	..	..	166	85	Autres équipements de transport			359
..	..	..	..	2 516	2 437	3 441	2 584	Meubles ; activités de fabrication, nca		36	
..	..	..	..	..	..	1 754	1 458	Meubles			361
..	..	..	..	..	..	1 687	1 126	Activités de fabrication, nca			369
..	..	..	..	959	371	696	357	Récupération		37	
..	..	..	..	68 959	52 339	49 491	54 021	Electricité, gaz et eau	E	40_41	
..	..	..	..	65 442	48 618	44 855	49 034	Prod. et dist. d'électricité gaz, vapeur d'eau chaude		40	
..	..	..	..	3 517	3 721	4 636	4 987	Captage, épuration et distribution d'eau		41	
..	..	..	..	14 239	12 410	16 579	14 982	Construction	F	45	

Table CZ.4 **EMPLOYMENT, EMPLOYEES - EMPLOI, SALARIÉS**

Units

ISIC revision 3				1991	1992	1993	1994	1995	1996	1997	1998
C	10_14		Mining and quarrying	..	..	..	..	96 000	84 000	78 000	70 000
	10		Mining of coal and lignite; extraction of peat	..	..	..	..	72 000	67 000	62 000	57 000
	11		Crude petroleum & natural gas; service activities	..	..	..	..	2 000	2 000	2 000	1 000
	13		Mining of metal ores	..	..	..	..	..	..	..	..
	14		Other mining and quarrying	..	..	..	..	18 000	12 000	9 000	8 000
D	15_37		Total manufacturing	..	..	..	..	1347 000	1341 000	1345 000	1304 000
	15		Food products and beverages	..	..	..	..	139 000	152 000	162 000	155 000
		151	Production of meat, fish, fruit, vegetables, oils & fats	..	..	..	..	..	..	46 000	42 000
		152	Dairy products	..	..	..	..	..	..	15 000	15 000
		153	Grain mill and starch products; animal feeds	..	..	..	..	..	..	15 000	13 000
		154	Other food products	..	..	..	..	..	..	60 000	61 000
		155	Beverages	..	..	..	..	..	..	26 000	24 000
	16		Tobacco products	..	..	..	..	..	..	..	..
	17		Textiles	..	..	..	..	92 000	88 000	79 000	76 000
		171	Spinning, weaving and finishing of textiles	..	..	..	..	..	..	46 000	47 000
		172	Other textiles	..	..	..	..	..	..	20 000	19 000
		173	Knitted and crocheted fabrics and articles	..	..	..	..	..	..	13 000	10 000
	18		Wearing apparel and fur	..	..	..	..	56 000	53 000	54 000	51 000
		181	Wearing apparel, except fur apparel	..	..	..	..	..	..	53 000	50 000
		182	Dressing and dyeing of fur; articles of fur	..	..	..	..	..	..	1 000	1 000
	19		Leather and articles of leather; footwear	..	..	..	..	43 000	31 000	27 000	25 000
		191	Leather and articles of leather	..	..	..	..	..	..	9 000	7 000
		192	Footwear	..	..	..	..	..	..	18 000	18 000
	20		Wood & products of wood & cork, exc. furniture	..	..	..	..	53 000	49 000	53 000	52 000
		201	Sawmilling and planing of wood	..	..	..	..	..	..	15 000	15 000
		202	Wood products; articles of cork, straw, etc.	..	..	..	..	..	..	38 000	37 000
	21		Paper and paper products	..	..	..	..	31 000	25 000	23 000	22 000
	22		Printing, publishing and recorded media	..	..	..	..	26 000	34 000	34 000	34 000
		221	Publishing	..	..	..	..	..	..	17 000	15 000
		222	Printing and related activities	..	..	..	..	..	..	16 000	18 000
		223	Reproduction of recorded media	..	..	..	..	..	..	1 000	1 000
	23		Coke and petroleum products; nuclear fuel	..	..	..	..	..	..	..	..
		231	Coke oven products	..	..	..	..	..	..	..	..
		232	Refined petroleum products	..	..	..	..	..	..	..	..
		233	Processing of nuclear fuel	..	..	..	..	..	..	..	..
	24		Chemicals and chemical products	..	..	..	..	49 000	54 000	49 000	46 000
		241	Basic chemicals	..	..	..	..	..	..	28 000	26 000
		242	Other chemical products	..	..	..	..	..	..	20 000	19 000
		243	Man-made fibres	..	..	..	..	..	..	1 000	1 000
	25		Rubber and plastic products	..	..	..	..	40 000	44 000	52 000	51 000
		251	Rubber products	..	..	..	..	..	..	16 000	14 000
		252	Plastic products	..	..	..	..	..	..	36 000	37 000
	26		Other non-metallic mineral products	..	..	..	..	76 000	80 000	84 000	82 000
		261	Glass and glass products	..	..	..	..	..	..	34 000	33 000
		269	Non-metallic mineral products, nec	..	..	..	..	..	..	50 000	49 000
	27		Basic metals	..	..	..	..	122 000	107 000	98 000	91 000
		271	Basic iron and steel	..	..	..	..	..	..	71 000	66 000
		272	Basic precious and non-ferrous metals	..	..	..	..	..	..	5 000	5 000
		273	Casting of metals	..	..	..	..	..	..	22 000	20 000
	28		Fabricated metal products, exc. machinery	..	..	..	..	121 000	132 000	134 000	145 000
		281	Structural metal prod., tanks, reservoirs, etc.	..	..	..	..	..	..	40 000	46 000
		289	Other fabricated metal products	..	..	..	..	..	..	94 000	99 000
	29		Machinery and equipment, nec	..	..	..	..	184 000	182 000	179 000	162 000
		291	General purpose machinery	..	..	..	..	..	..	66 000	58 000
		292	Special purpose machinery	..	..	..	..	..	..	103 000	95 000
		293	Domestic appliances, nec	..	..	..	..	..	..	10 000	9 000
	30		Office, accounting & computing machinery	..	..	..	..	8 000	4 000	2 000	2 000
	31		Electrical machinery & apparatus, nec	..	..	..	..	67 000	73 000	80 000	78 000
		311	Electric motors, generators and transformers	..	..	..	..	..	..	20 000	18 000
		312	Electricity distribution & control apparatus	..	..	..	..	..	..	23 000	22 000
		313	Insulated wire and cable	..	..	..	..	..	..	5 000	6 000
		314	Accumulators, primary cells & primary batteries	..	..	..	..	..	..	2 000	2 000
		315	Electric lamps and lighting equipment	..	..	..	..	..	..	7 000	6 000
		319	Other electrical equipment, nec	..	..	..	..	..	..	23 000	24 000
	32		Radio, TV, communication equipment	..	..	..	..	24 000	23 000	23 000	26 000
		321	Electronic valves, tubes; other electronic components	..	..	..	..	..	..	12 000	14 000
		322	TV/radio transmitters; line teleph./telegr. apparatus	..	..	..	..	..	..	9 000	10 000
		323	TV/radio receivers; sound and video app.; assoc. gds.	..	..	..	..	..	..	2 000	2 000
	33		Medical, precision & optical instrum.; watch & clocks	..	..	..	..	25 000	29 000	29 000	29 000
		331	Medical appl.; instruments & appl. for measuring, etc.	..	..	..	..	..	..	25 000	23 000
		332	Optical instruments and photographic equipment	..	..	..	..	..	..	4 000	5 000
		333	Watches and clocks	..	..	..	..	..	..		1 000
	34		Motor vehicles, trailers and semi-trailers	..	..	..	..	56 000	57 000	62 000	66 000
		341	Motor vehicles	..	..	..	..	..	..	32 000	31 000
		342	Bodies for motor vehicles, trailers & semi-trailers	..	..	..	..	..	..	3 000	4 000
		343	Parts & accessories for motor vehicles & their engines	..	..	..	..	..	..	27 000	31 000
	35		Other transport equipment	..	..	..	..	42 000	36 000	35 000	27 000
		351	Building and repairing of ships and boats	..	..	..	..	..	..		
		352	Railway and tramway locomotives and rolling stock	..	..	..	..	..	..	18 000	15 000
		353	Aircraft and spacecraft	..	..	..	..	..	..	9 000	8 000
		359	Transport equipment, nec	..	..	..	..	..	..	8 000	4 000
	36		Furniture; manufacturing, nec	..	..	..	..	67 000	74 000	73 000	70 000
		361	Furniture	..	..	..	..	..	..	36 000	38 000
		369	Manufacturing, nec	..	..	..	..	..	..	37 000	32 000
	37		Recycling	..	..	..	..	7 000	7 000	6 000	6 000
E	40_41		Electricity, gas and water supply	..	..	..	..	87 000	84 000	84 000	81 000
	40		Electricity, gas, steam and hot water supply	..	..	..	..	65 000	63 000	60 000	57 000
	41		Collection, purification and distribution of water	..	..	..	..	22 000	21 000	24 000	24 000
F	45		Construction	..	..	..	..	..	..	366 000	332 000

Table CZ.5 **SALAIRES ET TRAITEMENTS, SALARIÉS - WAGES AND SALARIES, EMPLOYEES**

Millions de CK (Prix courants)

1991	1992	1993	1994	1995	1996	1997	1998		CITI révision 3		
..	..	..	..	10 642	11 657	12 412	12 777	Activites extractives	C	10_14	
..	..	..	..	8 913	9 571	10 075	10 407	Extraction de charbon, de lignite et de tourbe		10	
..	..	..	..	285	320	254	232	Pétrole brut et gaz naturel ; activités annexes		11	
..	..	..	..	..	..	..	..	Extraction de minerais métalliques		13	
..	..	..	..	1 088	1 378	1 367	1 425	Autres activités extractives		14	
..	..	..	..	125 039	144 650	164 442	176 263	Total activites manufacturieres	D	15_37	
..	..	..	..	12 661	15 381	18 811	19 321	Produits alimentaires et boissons		15	
..	..	..	..	..	..	5 110	4 980	Viande, poisson, fruits, légumes, huiles et graisses			151
..	..	..	..	..	..	1 738	1 936	Produits laitiers			152
..	..	..	..	..	..	1 757	1 833	Travail des grains; prod. amyl. ; alim. pour animaux			153
..	..	..	..	..	..	6 554	6 817	Autres produits alimentaires			154
..	..	..	..	..	..	3 652	3 755	Boissons			155
..	..	..	..	..	..	..	..	Produits à base de tabac		16	
..	..	..	..	6 452	7 102	7 243	7 824	Textiles		17	
..	..	..	..	..	..	4 260	4 807	Filature, tissage et achèvement des textiles			171
..	..	..	..	..	..	1 835	2 043	Autres articles textiles			172
..	..	..	..	..	..	1 148	974	Etoffes et articles de bonneterie			173
..	..	..	..	3 685	3 523	4 166	4 365	Articles d'habillement et fourrure		18	
..	..	..	..	..	..	4 031	4 267	Articles d'habillement autres qu'en fourrure			181
..	..	..	..	..	..	135	98	Prép. et teinture des fourrures ; articles en fourrure			182
..	..	..	..	2 562	2 582	2 413	2 358	Cuir et articles en cuir ; chaussures		19	
..	..	..	..	..	..	854	668	Cuir et articles en cuir sauf chaussures			191
..	..	..	..	..	..	1 559	1 690	Chaussures			192
..	..	..	..	4 919	4 668	5 227	5 370	Bois, articles en bois, sauf meubles		20	
..	..	..	..	..	..	1 458	1 433	Sciage et rabotage du bois			201
..	..	..	..	..	..	3 769	3 937	Articles en bois, liège, vannerie, sparterie			202
..	..	..	..	2 244	2 674	2 860	3 157	Papier, articles en papier et carton		21	
..	..	..	..	3 701	4 376	5 081	5 630	Edition, imprimerie et reproduction		22	
..	..	..	..	..	..	2 839	2 975	Edition			221
..	..	..	..	..	..	2 122	2 463	Imprimerie et activités annexes			222
..	..	..	..	..	..	120	192	Reproduction de supports enregistrés			223
..	..	..	..	..	..	..	..	Coke. et prod. pétroliers ; comb. nucléaires		23	
..	..	..	..	..	..	..	..	Cokéfaction			231
..	..	..	..	..	..	..	..	Produits pétroliers raffinés			232
..	..	..	..	..	..	..	..	Traitement de combustibles nucléaires			233
..	..	..	..	5 064	6 945	7 279	7 712	Produits chimiques		24	
..	..	..	..	..	..	4 263	4 526	Produits chimiques de base			241
..	..	..	..	..	..	2 862	2 998	Autres produits chimiques			242
..	..	..	..	..	..	154	188	Fibres synthétiques ou artificielles			243
..	..	..	..	3 878	5 061	6 446	7 253	Art. en caoutchouc et matières plastiques		25	
..	..	..	..	..	..	2 147	2 292	Articles en caoutchouc			251
..	..	..	..	..	..	4 299	4 961	Articles en matières plastiques			252
..	..	..	..	7 769	9 029	10 835	11 831	Autres produits minéraux non métalliques		26	
..	..	..	..	..	..	4 229	4 670	Verre et articles en verre			261
..	..	..	..	..	..	6 606	7 161	Produits minéraux non métalliques, nca			269
..	..	..	..	12 089	13 773	14 248	14 604	Produits métallurgiques de base		27	
..	..	..	..	..	..	10 897	11 179	Sidérurgie et première transformation de l'acier			271
..	..	..	..	..	..	713	767	Première transf. métaux précieux et non ferreux			272
..	..	..	..	..	..	2 638	2 658	Fonderie			273
..	..	..	..	12 259	14 675	16 709	19 804	Ouvrages en métaux, sauf machines		28	
..	..	..	..	..	..	5 455	6 897	Constr. et menuis. métal.; réservoirs; générateurs			281
..	..	..	..	..	..	11 254	12 907	Autres ouvrages en métaux			289
..	..	..	..	18 259	20 447	23 289	23 584	Machines et matériels, nca		29	
..	..	..	..	..	..	9 019	8 888	Machines d'usage général			291
..	..	..	..	..	..	13 111	13 541	Machines d'usage spécifique			292
..	..	..	..	..	..	1 159	1 155	Appareils domestiques, nca			293
..	..	..	..	306	361	212	286	Mach. de bureau, comptables, information		30	
..	..	..	..	6 710	8 339	10 235	10 894	Machines et appareils électriques, nca		31	
..	..	..	..	..	..	2 446	2 391	Moteurs, génératrices, transformateurs électriques			311
..	..	..	..	..	..	3 264	3 409	Matériel électrique de distribution et de commande			312
..	..	..	..	..	..	683	892	Fils et câbles électriques isolés			313
..	..	..	..	..	..	309	338	Accumulateurs et piles électriques			314
..	..	..	..	..	..	737	736	Lampes électriques et appareils d'éclairage			315
..	..	..	..	..	..	2 796	3 128	Autres matériels électriques, nca.			319
..	..	..	..	2 179	2 451	2 888	3 487	Equipements et appareils de radio, TV, etc.		32	
..	..	..	..	..	..	1 342	1 626	Tubes et valves électroniques;autres composants			321
..	..	..	..	..	..	1 246	1 532	Emetteurs radio/TV; app. téléphonie et télégraphie			322
..	..	..	..	..	..	300	329	Récepteurs radio/TV, son et image; art. associés			323
..	..	..	..	2 686	3 261	3 576	3 921	Instr. médicaux, de précision ; optique ; horlog.		33	
..	..	..	..	..	..	3 093	3 243	Instruments médicaux et de précision, etc.			331
..	..	..	..	..	..	433	592	Instruments d'optique ; matériel photographique			332
..	..	..	..	..	..	50	86	Horlogerie			333
..	..	..	..	5 839	7 412	9 188	11 079	Véhicules automobiles; remorques, etc.		34	
..	..	..	..	..	..	5 167	5 904	Véhicules automobiles			341
..	..	..	..	..	..	401	621	Carross. automobiles, remorques, semi-remorques			342
..	..	..	..	..	..	3 620	4 554	Pièces et accessoires pour automobiles et moteurs			343
..	..	..	..	3 478	4 114	4 534	3 916	Autres matériels de transport		35	
..	..	..	..	..	..	43	51	Construction et réparation de navires			351
..	..	..	..	..	..	2 372	2 144	Matériel ferroviaire roulant			352
..	..	..	..	..	..	1 300	1 296	Construction aéronautique et spatiale			353
..	..	..	..	..	..	819	425	Autres équipements de transport			359
..	..	..	..	5 650	6 617	7 009	7 476	Meubles ; activités de fabrication, nca		36	
..	..	..	..	..	..	3 564	4 223	Meubles			361
..	..	..	..	..	..	3 445	3 253	Activités de fabrication, nca			369
..	..	..	..	697	704	808	884	Récupération		37	
..	..	..	..	10 391	11 703	12 992	13 888	Electricité, gaz et eau	E	40_41	
..	..	..	..	8 284	9 373	9 978	10 549	Prod. et dist. d'électricité gaz, vapeur d'eau chaude		40	
..	..	..	..	2 107	2 330	3 014	3 339	Captage, épuration et distribution d'eau		41	
..	..	..	..	..	..	47 167	46 033	Construction	F	45	

CZECH REPUBLIC

Table CZ.6 HOURS WORKED BY OPERATIVES - HEURES OUVRÉES, OUVRIERS

Thousands

ISIC revision 3				1991	1992	1993	1994	1995	1996	1997	1998
C	10_14		Mining and quarrying	..	..	..	..	117 000	110 000	108 000	101 000
	10		Mining of coal and lignite; extraction of peat	..	..	..	..	101 000	94 000	88 000	82 000
	11		Crude petroleum & natural gas; service activities	..	..	..	..	3 000	3 000	2 000	2 000
	13		Mining of metal ores	..	..	..	..	..	..	..	..
	14		Other mining and quarrying	..	..	..	..	10 000	10 000	12 000	12 000
D	15_37		Total manufacturing	..	..	..	..	1252 000	1192 000	1462 000	1471 000
	15		Food products and beverages	..	..	..	..	135 834	134 000	192 000	184 000
		151	Production of meat, fish, fruit, vegetables, oils & fats	..	..	..	..	..	..	53 000	52 000
		152	Dairy products	..	..	..	..	..	..	20 000	20 000
		153	Grain mill and starch products; animal feeds	..	..	..	..	..	..	18 000	15 000
		154	Other food products	..	..	..	..	..	..	72 000	71 000
		155	Beverages	..	..	..	..	..	..	29 000	26 000
	16		Tobacco products	..	..	..	..	..	..	..	..
	17		Textiles	..	..	..	..	100 003	90 000	98 000	92 000
		171	Spinning, weaving and finishing of textiles	..	..	..	..	..	..	60 000	60 000
		172	Other textiles	..	..	..	..	..	..	24 000	19 000
		173	Knitted and crocheted fabrics and articles	..	..	..	..	..	..	14 000	13 000
	18		Wearing apparel and fur	..	..	..	..	39 189	34 000	54 000	57 000
		181	Wearing apparel, except fur apparel	..	..	..	..	..	..	53 000	56 000
		182	Dressing and dyeing of fur; articles of fur	..	..	..	..	..	..	1 000	1 000
	19		Leather and articles of leather; footwear	..	..	..	..	37 941	32 000	36 000	30 000
		191	Leather and articles of leather	..	..	..	..	..	..	9 000	8 000
		192	Footwear	..	..	..	..	..	..	27 000	22 000
	20		Wood & products of wood & cork, exc. furniture	..	..	..	..	29 266	26 000	47 000	51 000
		201	Sawmilling and planing of wood	..	..	..	..	..	..	16 000	19 000
		202	Wood products; articles of cork, straw, etc.	..	..	..	..	..	..	31 000	32 000
	21		Paper and paper products	..	..	..	..	27 417	27 000	27 000	28 000
	22		Printing, publishing and recorded media	..	..	..	..	15 772	15 000	23 000	23 000
		221	Publishing	..	..	..	..	..	..	10 000	9 000
		222	Printing and related activities	..	..	..	..	..	..	12 000	13 000
		223	Reproduction of recorded media	..	..	..	..	..	..	1 000	1 000
	23		Coke and petroleum products; nuclear fuel	..	..	..	..	..	..	..	..
		231	Coke oven products	..	..	..	..	..	..	..	..
		232	Refined petroleum products	..	..	..	..	..	..	..	..
		233	Processing of nuclear fuel	..	..	..	..	..	..	..	..
	24		Chemicals and chemical products	..	..	..	..	46 332	52 000	51 000	48 000
		241	Basic chemicals	..	..	..	..	..	..	32 000	30 000
		242	Other chemical products	..	..	..	..	..	..	18 000	16 000
		243	Man-made fibres	..	..	..	..	..	..	1 000	2 000
	25		Rubber and plastic products	..	..	..	..	36 087	38 000	56 000	61 000
		251	Rubber products	..	..	..	..	..	..	18 000	19 000
		252	Plastic products	..	..	..	..	..	..	38 000	42 000
	26		Other non-metallic mineral products	..	..	..	..	85 658	80 000	103 000	101 000
		261	Glass and glass products	..	..	..	..	..	..	41 000	40 000
		269	Non-metallic mineral products, nec	..	..	..	..	..	..	62 000	61 000
	27		Basic metals	..	..	..	..	132 945	129 000	119 000	116 000
		271	Basic iron and steel	..	..	..	..	..	..	87 000	85 000
		272	Basic precious and non-ferrous metals	..	..	..	..	..	..	6 000	6 000
		273	Casting of metals	..	..	..	..	..	..	26 000	25 000
	28		Fabricated metal products, exc. machinery	..	..	..	..	91 678	88 000	135 000	149 000
		281	Structural metal prod., tanks, reservoirs, etc.	..	..	..	..	..	..	46 000	46 000
		289	Other fabricated metal products	..	..	..	..	..	..	89 000	103 000
	29		Machinery and equipment, nec	..	..	..	..	186 126	170 000	187 000	184 000
		291	General purpose machinery	..	..	..	..	..	..	65 000	59 000
		292	Special purpose machinery	..	..	..	..	..	..	112 000	115 000
		293	Domestic appliances, nec	..	..	..	..	..	..	10 000	10 000
	30		Office, accounting & computing machinery	..	..	..	..	3 335	3 000	2 000	2 000
	31		Electrical machinery & apparatus, nec	..	..	..	..	62 756	62 000	83 000	89 000
		311	Electric motors, generators and transformers	..	..	..	..	..	..	21 000	18 000
		312	Electricity distribution & control apparatus	..	..	..	..	..	..	24 000	24 000
		313	Insulated wire and cable	..	..	..	..	..	..	6 000	8 000
		314	Accumulators, primary cells & primary batteries	..	..	..	..	..	..	2 000	2 000
		315	Electric lamps and lighting equipment	..	..	..	..	..	..	7 000	8 000
		319	Other electrical equipment, nec	..	..	..	..	..	..	23 000	29 000
	32		Radio, TV, communication equipment	..	..	..	..	18 004	19 000	24 000	28 000
		321	Electronic valves, tubes; other electronic components	..	..	..	..	..	..	15 000	18 000
		322	TV/radio transmitters; line teleph./telegr. apparatus	..	..	..	..	..	..	7 000	8 000
		323	TV/radio receivers; sound and video app.; assoc. gds.	..	..	..	..	..	..	2 000	2 000
	33		Medical, precision & optical instrum.; watch & clocks	..	..	..	..	21 314	22 000	25 000	26 000
		331	Medical appl.; instruments & appl. for measuring, etc.	..	..	..	..	..	..	20 000	19 000
		332	Optical instruments and photographic equipment	..	..	..	..	..	..	4 000	6 000
		333	Watches and clocks	..	..	..	..	..	..	1 000	1 000
	34		Motor vehicles, trailers and semi-trailers	..	..	..	..	63 631	67 000	75 000	84 000
		341	Motor vehicles	..	..	..	..	..	..	35 000	37 000
		342	Bodies for motor vehicles, trailers & semi-trailers	..	..	..	..	..	..	4 000	6 000
		343	Parts & accessories for motor vehicles & their engines	..	..	..	..	..	..	36 000	41 000
	35		Other transport equipment	..	..	..	..	38 165	37 000	37 000	29 000
		351	Building and repairing of ships and boats	..	..	..	..	..	..	..	..
		352	Railway and tramway locomotives and rolling stock	..	..	..	..	..	..	20 000	17 000
		353	Aircraft and spacecraft	..	..	..	..	..	..	9 000	8 000
		359	Transport equipment, nec	..	..	..	..	..	..	8 000	4 000
	36		Furniture; manufacturing, nec	..	..	..	..	58 218	54 000	73 000	72 000
		361	Furniture	..	..	..	..	..	..	42 000	41 000
		369	Manufacturing, nec	..	..	..	..	..	..	31 000	31 000
	37		Recycling	..	..	..	..	4 778	4 000	7 000	7 000
E	40_41		Electricity, gas and water supply	..	..	..	..	91 000	85 000	87 000	81 000
	40		Electricity, gas, steam and hot water supply	..	..	..	..	68 000	63 000	61 000	56 000
	41		Collection, purification and distribution of water	..	..	..	..	23 000	22 000	26 000	25 000
F	45		Construction	..	..	..	..	379 000	..	467 000	427 000

Table CZ.7 **ENTREPRISES - ENTERPRISES**

Unités

1991	1992	1993	1994	1995	1996	1997	1998			CITI révision 3	
..	..	..	..	287	290	307	269	Activites extractives	C	10_14	
..	..	..	..	30	29	33	49	Extraction de charbon, de lignite et de tourbe		10	
..	..	..	..	19	10	5	9	Pétrole brut et gaz naturel ; activités annexes		11	
..	..	..	..	..	..	..	..	Extraction de minerais métalliques		13	
..	..	..	..	228	244	266	208	Autres activités extractives		14	
..	..	..	..	100 889	110 163	145 814	128 333	Total activites manufacturieres	D	15_37	
..	..	..	..	5 595	5 623	6 423	6 465	Produits alimentaires et boissons		15	
..	..	..	..	..	..	2 638	1 560	Viande, poisson, fruits, légumes, huiles et graisses			151
..	..	..	..	..	..	175	169	Produits laitiers			152
..	..	..	..	..	..	467	359	Travail des grains; prod. amyl. ; alim. pour animaux			153
..	..	..	..	..	..	2 514	3 234	Autres produits alimentaires			154
..	..	..	..	..	..	629	1 143	Boissons			155
..	..	..	..	..	..	..	..	Produits à base de tabac		16	
..	..	..	..	3 525	3 195	3 855	3 925	Textiles		17	
..	..	..	..	..	..	410	459	Filature, tissage et achèvement des textiles			171
..	..	..	..	..	..	2 142	2 392	Autres articles textiles			172
..	..	..	..	..	..	1 303	1 074	Etoffes et articles de bonneterie			173
..	..	..	..	11 383	8 768	12 726	9 170	Articles d'habillement et fourrure		18	
..	..	..	..	..	..	12 379	8 915	Articles d'habillement autres qu'en fourrure			181
..	..	..	..	..	..	347	255	Prép. et teinture des fourrures ; articles en fourrure			182
..	..	..	..	1 064	1 107	1 423	1 534	Cuir et articles en cuir ; chaussures		19	
..	..	..	..	..	..	928	1 234	Cuir et articles en cuir sauf chaussures			191
..	..	..	..	..	..	495	300	Chaussures			192
..	..	..	..	15 093	17 370	23 509	21 995	Bois, articles en bois, sauf meubles		20	
..	..	..	..	..	..	1 731	1 835	Sciage et rabotage du bois			201
..	..	..	..	..	..	21 778	20 160	Articles en bois, liège, vannerie, sparterie			202
..	..	..	..	366	391	562	401	Papier, articles en papier et carton		21	
..	..	..	..	4 166	4 202	5 702	5 950	Edition, imprimerie et reproduction		22	
..	..	..	..	..	..	2 652	1 652	Edition			221
..	..	..	..	..	..	2 875	3 945	Imprimerie et activités annexes			222
..	..	..	..	..	..	175	353	Reproduction de supports enregistrés			223
..	..	..	..	..	..	..	..	Coke. et prod. pétroliers ; comb. nucléaires		23	
..	..	..	..	..	..	..	..	Cokéfaction			231
..	..	..	..	..	..	..	..	Produits pétroliers raffinés			232
..	..	..	..	..	..	..	..	Traitement de combustibles nucléaires			233
..	..	..	..	1 024	893	1 063	1 594	Produits chimiques		24	
..	..	..	..	..	..	224	616	Produits chimiques de base			241
..	..	..	..	..	..	828	968	Autres produits chimiques			242
..	..	..	..	..	..	11	10	Fibres synthétiques ou artificielles			243
..	..	..	..	1 637	1 988	3 678	2 339	Art. en caoutchouc et matières plastiques		25	
..	..	..	..	..	..	867	508	Articles en caoutchouc			251
..	..	..	..	..	..	2 811	1 831	Articles en matières plastiques			252
..	..	..	..	4 573	5 587	7 011	6 377	Autres produits minéraux non métalliques		26	
..	..	..	..	..	..	2 298	1 909	Verre et articles en verre			261
..	..	..	..	..	..	4 713	4 468	Produits minéraux non métalliques, nca			269
..	..	..	..	364	310	497	423	Produits métallurgiques de base		27	
..	..	..	..	..	..	151	112	Sidérurgie et première transformation de l'acier			271
..	..	..	..	..	..	80	54	Première transf. métaux précieux et non ferreux			272
..	..	..	..	..	..	266	257	Fonderie			273
..	..	..	..	21 562	28 181	36 241	28 111	Ouvrages en métaux, sauf machines		28	
..	..	..	..	..	..	1 953	2 853	Constr. et menuis. métal.; réservoirs; générateurs			281
..	..	..	..	..	..	34 288	25 258	Autres ouvrages en métaux			289
..	..	..	..	5 858	6 351	7 752	6 127	Machines et matériels, nca		29	
..	..	..	..	..	..	4 054	3 754	Machines d'usage général			291
..	..	..	..	..	..	2 998	2 019	Machines d'usage spécifique			292
..	..	..	..	..	..	700	354	Appareils domestiques, nca			293
..	..	..	..	160	152	192	152	Mach. de bureau, comptables, information		30	
..	..	..	..	9 936	9 641	9 996	13 025	Machines et appareils électriques, nca		31	
..	..	..	..	..	..	4 378	9 094	Moteurs, génératrices, transformateurs électriques			311
..	..	..	..	..	..	2 564	1 478	Matériel électrique de distribution et de commande			312
..	..	..	..	..	..	71	68	Fils et câbles électriques isolés			313
..	..	..	..	..	..	69	35	Accumulateurs et piles électriques			314
..	..	..	..	..	..	91	156	Lampes électriques et appareils d'éclairage			315
..	..	..	..	..	..	2 823	2 194	Autres matériels électriques, nca.			319
..	..	..	..	2 548	1 796	4 139	2 211	Equipements et appareils de radio, TV, etc.		32	
..	..	..	..	..	..	908	603	Tubes et valves électroniques;autres composants			321
..	..	..	..	..	..	1 095	930	Emetteurs radio/TV; app. téléphonie et télégraphie			322
..	..	..	..	..	..	2 136	678	Récepteurs radio/TV, son et image; art. associés			323
..	..	..	..	2 740	3 422	4 544	3 639	Instr. médicaux, de précision ; optique ; horlog.		33	
..	..	..	..	..	..	3 709	3 324	Instruments médicaux et de précision, etc.			331
..	..	..	..	..	..	660	230	Instruments d'optique ; matériel photographique			332
..	..	..	..	..	..	175	85	Horlogerie			333
..	..	..	..	548	311	341	688	Véhicules automobiles; remorques, etc.		34	
..	..	..	..	..	..	130	57	Véhicules automobiles			341
..	..	..	..	..	..	89	77	Carross. automobiles, remorques, semi-remorques			342
..	..	..	..	..	..	122	554	Pièces et accessoires pour automobiles et moteurs			343
..	..	..	..	305	422	390	383	Autres matériels de transport		35	
..	..	..	..	..	..	55	98	Construction et réparation de navires			351
..	..	..	..	..	..	41	59	Matériel ferroviaire roulant			352
..	..	..	..	..	..	53	68	Construction aéronautique et spatiale			353
..	..	..	..	..	..	241	158	Autres équipements de transport			359
..	..	..	..	7 992	10 058	14 450	13 177	Meubles ; activités de fabrication, nca		36	
..	..	..	..	..	..	4 923	5 438	Meubles			361
..	..	..	..	..	..	9 527	7 739	Activités de fabrication, nca			369
..	..	..	..	434	386	1 301	627	Récupération		37	
..	..	..	..	890	810	779	687	Electricité, gaz et eau	E	40_41	
..	..	..	..	706	671	619	556	Prod. et dist. d'électricité gaz, vapeur d'eau chaude		40	
..	..	..	..	184	138	160	131	Captage, épuration et distribution d'eau		41	
..	..	..	..	..	..	106 661	111 948	Construction	F	45	

DENMARK

Table DN.1 **EMPLOYMENT, EMPLOYEES - EMPLOI, SALARIÉS**

Units

ISIC revision 3				1991	1992	1993	1994	1995	1996	1997	1998
C	10_14		Mining and quarrying	..	..	3 626	3 736	3 771	..	..	..
	10		Mining of coal and lignite; extraction of peat	..	..	165	144	144	149	150	146
	11		Crude petroleum & natural gas; service activities	..	..	1 284	1 239	1 351	1 241	1 201	1 288
	13		Mining of metal ores	..	..	..	..	..	..	..	..
	14		Other mining and quarrying	..	..	..	..	..	2 256	2 288	1 729
D	15_37		Total manufacturing	..	..	464 695	484 882	490 634	482 280	486 124	488 701
	15		Food products and beverages	..	..	90 271	90 836	89 625	87 766	86 056	84 590
		151	Production of meat, fish, fruit, vegetables, oils & fats	..	..	38 406	37 900	36 379	35 830	34 879	34 775
		152	Dairy products	..	..	11 988	12 119	12 356	12 216	12 236	11 592
		153	Grain mill and starch products; animal feeds	..	..	2 770	2 963	3 175	3 122	3 021	2 843
		154	Other food products	..	..	30 311	31 036	30 790	30 135	29 301	28 914
		155	Beverages	..	..	6 796	6 818	6 925	6 463	6 619	6 466
	16		Tobacco products	..	..	1 461	1 460	1 472	1 404	1 354	1 415
	17		Textiles	..	..	12 222	12 215	11 163	10 314	10 021	9 601
		171	Spinning, weaving and finishing of textiles	..	..	3 084	3 013	2 233	2 145	2 199	1 973
		172	Other textiles	..	..	5 615	6 010	5 891	5 667	5 636	5 542
		173	Knitted and crocheted fabrics and articles	..	..	3 523	3 192	3 039	2 502	2 186	2 086
	18		Wearing apparel and fur	..	..	9 865	9 107	7 721	6 504	5 614	5 389
		181	Wearing apparel, except fur apparel	..	..	9 640	8 867	7 520	6 294	5 417	5 194
		182	Dressing and dyeing of fur; articles of fur	..	..	225	240	201	210	197	195
	19		Leather and articles of leather; footwear	..	..	1 978	2 042	1 875	1 663	1 628	1 570
		191	Leather and articles of leather	..	..	611	601	527	535	488	422
		192	Footwear	..	..	1 367	1 441	1 348	1 128	1 140	1 148
	20		Wood & products of wood & cork, exc. furniture	..	..	13 768	15 853	15 711	16 367	16 486	16 821
		201	Sawmilling and planing of wood	..	..	1 931	2 171	1 998	2 044	2 129	2 093
		202	Wood products; articles of cork, straw, etc.	..	..	11 837	13 682	13 713	14 323	14 357	14 728
	21		Paper and paper products	..	..	9 578	10 039	10 304	9 992	10 354	10 125
	22		Printing, publishing and recorded media	..	..	49 280	48 874	50 709	52 028	52 311	52 469
		221	Publishing	..	..	31 202	30 940	33 522	35 015	35 711	35 261
		222	Printing and related activities	..	..	18 017	17 852	16 963	16 740	16 053	16 525
		223	Reproduction of recorded media	..	..	61	82	224	273	547	683
	23		Coke and petroleum products; nuclear fuel	..	..	901	977	989	928	815	762
		231	Coke oven products	..	..	..	..	..	..	..	..
		232	Refined petroleum products	..	..	901	977	989	928	815	762
		233	Processing of nuclear fuel	..	..	..	..	..	..	..	..
	24		Chemicals and chemical products	..	..	25 233	27 039	27 975	28 021	28 240	28 303
		241	Basic chemicals	..	..	6 418	5 783	5 955	6 674	5 105	4 973
		242	Other chemical products	..	..	18 815	21 256	22 020	20 235	21 978	22 128
		243	Man-made fibres	..	..	..	..	..	1 112	1 157	1 202
	25		Rubber and plastic products	..	..	20 595	21 024	21 100	21 512	22 433	23 093
		251	Rubber products	..	..	2 809	2 708	2 540	2 469	2 555	2 276
		252	Plastic products	..	..	17 786	18 316	18 560	19 043	19 878	20 817
	26		Other non-metallic mineral products	..	..	18 260	19 719	20 128	19 966	20 187	21 737
		261	Glass and glass products	..	..	3 386	4 148	4 615	4 353	4 555	5 265
		269	Non-metallic mineral products, nec	..	..	14 874	15 571	15 513	..	15 632	14 330
	27		Basic metals	..	..	7 658	7 203	7 753	7 543	8 231	10 516
		271	Basic iron and steel	..	..	4 439	4 895	5 466	5 306	5 912	6 174
		272	Basic precious and non-ferrous metals	..	..	2 129	1 879	2 183	2 116	2 187	2 412
		273	Casting of metals	..	..	1 090	429	104	121	132	1 930
	28		Fabricated metal products, exc. machinery	..	..	42 751	46 429	47 338	45 961	46 541	45 503
		281	Structural metal prod., tanks, reservoirs, etc.	..	..	14 001	15 050	15 525	15 766	17 052	16 820
		289	Other fabricated metal products	..	..	28 750	31 379	31 813	30 195	29 489	28 683
	29		Machinery and equipment, nec	..	..	67 066	72 277	75 301	72 247	74 054	73 925
		291	General purpose machinery	..	..	37 493	41 705	44 602	43 157	44 302	45 025
		292	Special purpose machinery	..	..	22 195	23 248	23 807	23 117	23 737	23 414
		293	Domestic appliances, nec	..	..	7 378	7 324	6 892	5 973	6 015	5 486
	30		Office, accounting & computing machinery	..	..	2 017	2 145	2 009	1 985	1 900	2 194
	31		Electrical machinery & apparatus, nec	..	..	14 613	15 260	16 129	16 603	16 371	17 500
		311	Electric motors, generators and transformers	..	..	3 087	3 434	4 410	4 351	5 220	6 028
		312	Electricity distribution & control apparatus	..	..	4 174	4 183	3 861	3 977	4 062	4 180
		313	Insulated wire and cable	..	..	1 487	1 445	1 411	1 537	1 796	1 999
		314	Accumulators, primary cells & primary batteries	..	..	322	276	273	260	271	208
		315	Electric lamps and lighting equipment	..	..	2 254	2 689	2 719	2 768	2 528	2 516
		319	Other electrical equipment, nec	..	..	3 289	3 233	3 455	3 710	2 494	2 569
	32		Radio, TV, communication equipment	..	..	10 894	11 139	11 451	11 912	12 550	13 188
		321	Electronic valves, tubes; other electronic components	..	..	3 057	2 661	2 801	2 985	3 638	3 530
		322	TV/radio transmitters; line teleph./telegr. apparatus	..	..	2 749	3 126	2 979	3 062	2 605	3 274
		323	TV/radio receivers; sound and video app.; assoc. gds.	..	..	5 088	5 352	5 671	5 865	6 307	6 384
	33		Medical, precision & optical instrum.; watch & clocks	..	..	14 273	14 457	15 577	14 936	15 096	14 873
		331	Medical appl.; instruments & appl. for measuring, etc.	..	..	13 212	13 068	14 152	13 360	13 473	13 266
		332	Optical instruments and photographic equipment	..	..	1 047	1 376	1 415	1 570	1 615	1 591
		333	Watches and clocks	..	..	14	13	10	6	8	16
	34		Motor vehicles, trailers and semi-trailers	..	..	5 488	7 051	7 844	7 811	8 053	8 010
		341	Motor vehicles	..	..	83	55	..	..	..	..
		342	Bodies for motor vehicles, trailers & semi-trailers	..	..	3 339	3 890	4 320	4 463	4 676	4 467
		343	Parts & accessories for motor vehicles & their engines	..	..	2 066	3 106	3 524	3 348	3 377	3 543
	35		Other transport equipment	..	..	14 124	15 698	14 814	13 432	13 142	12 333
		351	Building and repairing of ships and boats	..	..	11 605	13 092	12 108	10 842	10 246	9 447
		352	Railway and tramway locomotives and rolling stock	..	..	895	774	717	672	688	434
		353	Aircraft and spacecraft	..	..	560	687	734	787	996	1 066
		359	Transport equipment, nec	..	..	1 064	1 145	1 255	1 131	1 212	1 386
	36		Furniture; manufacturing, nec	..	..	31 905	33 675	33 257	33 031	34 321	34 427
		361	Furniture	..	..	23 991	25 720	25 038	24 828	25 347	25 628
		369	Manufacturing, nec	..	..	7 914	7 955	8 219	8 203	8 974	8 799
	37		Recycling	..	..	494	363	389	354	366	357
E	40_41		Electricity, gas and water supply	..	..	19 048	20 412	20 352	19 837	..	18 917
	40		Electricity, gas, steam and hot water supply	..	..	16 101	16 736	16 811	16 503	16 199	15 401
	41		Collection, purification and distribution of water	..	..	2 947	3 676	3 541	3 334	3 283	3 516
F	45		Construction	..	..	130 468	143 926	146 758	148 722	153 732	157 431

Note:ISIC 242 includes 243.

Table DN.2 **EMPLOI, FEMMES - EMPLOYMENT, FEMALES**

Unités

1991	1992	1993	1994	1995	1996	1997	1998			CITI révision 3
..	..	584	576	539	..	..	..	Activites extractives	C	10_14
..	..	23	22	22	22	21	..	Extraction de charbon, de lignite et de tourbe		10
..	..	182	181	184	175	164	167	Pétrole brut et gaz naturel ; activités annexes		11
..	..	..	..	..	..	..	..	Extraction de minerais métalliques		13
..	..	..	..	..	311	316	..	Autres activités extractives		14
..	..	155 623	156 833	158 585	..	..	..	Total activites manufacturieres	D	15_37
..	..	37 808	37 398	37 294	36 821	36 030	35 421	Produits alimentaires et boissons		15
..	..	13 279	12 612	12 360	12 232	11 845	11 807	Viande, poisson, fruits, légumes, huiles et graisses		151
..	..	3 713	3 712	3 821	3 852	3 988	3 823	Produits laitiers		152
..	..	729	743	873	876	910	808	Travail des grains; prod. amyl. ; alim. pour animaux		153
..	..	18 431	18 703	18 636	18 341	17 654	17 370	Autres produits alimentaires		154
..	..	1 656	1 628	1 604	1 520	1 633	1 613	Boissons		155
..	..	752	741	736	661	654	705	Produits à base de tabac		16
..	..	6 963	6 624	5 958	5 358	4 991	4 789	Textiles		17
..	..	1 385	1 273	921	865	880	864	Filature, tissage et achèvement des textiles		171
..	..	3 063	3 156	3 100	2 949	2 859	2 803	Autres articles textiles		172
..	..	2 515	2 195	1 937	1 544	1 252	1 122	Etoffes et articles de bonneterie		173
..	..	8 877	7 969	6 691	5 562	4 711	4 425	Articles d'habillement et fourrure		18
..	..	8 740	7 836	6 569	5 425	4 586	4 301	Articles d'habillement autres qu'en fourrure		181
..	..	137	133	122	137	125	124	Prép. et teinture des fourrures ; articles en fourrure		182
..	..	1 022	1 036	969	837	864	842	Cuir et articles en cuir ; chaussures		19
..	..	311	299	279	238	235	209	Cuir et articles en cuir sauf chaussures		191
..	..	711	737	690	599	629	633	Chaussures		192
..	..	2 535	2 837	2 849	3 049	3 157	3 335	Bois, articles en bois, sauf meubles		20
..	..	300	307	297	315	354	342	Sciage et rabotage du bois		201
..	..	2 235	2 530	2 552	2 734	2 803	2 993	Articles en bois, liège, vannerie, sparterie		202
..	..	2 716	2 749	2 774	2 683	2 774	2 743	Papier, articles en papier et carton		21
..	..	18 818	18 526	19 210	20 049	20 309	20 622	Edition, imprimerie et reproduction		22
..	..	12 102	12 036	13 042	13 911	14 300	14 453	Edition		221
..	..	6 690	6 459	6 082	6 013	5 760	5 861	Imprimerie et activités annexes		222
..	..	26	31	86	125	249	308	Reproduction de supports enregistrés		223
..	..	124	136	149	147	128	119	Coke. et prod. pétroliers ; comb. nucléaires		23
..	..	..	..	..	..	..	..	Cokéfaction		231
..	..	124	136	149	147	128	119	Produits pétroliers raffinés		232
..	..	..	..	..	..	..	..	Traitement de combustibles nucléaires		233
..	..	10 723	11 456	12 034	11 976	12 064	12 381	Produits chimiques		24
..	..	2 122	1 401	1 511	2 366	1 362	1 368	Produits chimiques de base		241
..	..	8 601	10 055	10 523	9 263	10 326	10 612	Autres produits chimiques		242
..	..	..	..	..	347	376	401	Fibres synthétiques ou artificielles		243
..	..	7 215	7 270	7 384	7 626	7 974	8 332	Art. en caoutchouc et matières plastiques		25
..	..	714	676	606	590	602	530	Articles en caoutchouc		251
..	..	6 501	6 594	6 778	7 036	7 372	7 802	Articles en matières plastiques		252
..	..	3 885	3 875	4 035	3 894	4 060	4 394	Autres produits minéraux non métalliques		26
..	..	901	1 017	1 169	1 085	1 190	1 378	Verre et articles en verre		261
..	..	2 984	2 858	2 866	..	2 870	3 016	Produits minéraux non métalliques, nca		269
..	..	1 491	1 200	1 276	1 303	1 423	2 011	Produits métallurgiques de base		27
..	..	713	760	812	826	907	1 009	Sidérurgie et première transformation de l'acier		271
..	..	555	350	444	451	487	662	Première transf. métaux précieux et non ferreux		272
..	..	223	90	20	26	29	340	Fonderie		273
..	..	8 062	8 543	8 971	8 621	8 828	8 499	Ouvrages en métaux, sauf machines		28
..	..	1 608	1 662	1 729	1 820	2 047	1 974	Constr. et menuis. métal.; réservoirs; générateurs		281
..	..	6 454	6 881	7 242	6 801	6 781	6 525	Autres ouvrages en métaux		289
..	..	14 890	15 651	16 224	15 407	15 917	15 905	Machines et matériels, nca		29
..	..	9 311	10 167	10 819	10 233	10 573	10 733	Machines d'usage général		291
..	..	3 531	3 485	3 481	3 440	3 479	3 436	Machines d'usage spécifique		292
..	..	2 048	1 999	1 924	1 734	1 865	1 736	Appareils domestiques, nca		293
..	..	715	706	588	556	578	704	Mach. de bureau, comptables, information		30
..	..	5 093	5 368	5 565	5 537	5 469	5 974	Machines et appareils électriques, nca		31
..	..	1 038	1 169	1 500	1 436	1 691	2 031	Moteurs, génératrices, transformateurs électriques		311
..	..	1 456	1 550	1 247	1 262	1 246	1 272	Matériel électrique de distribution et de commande		312
..	..	488	446	474	555	616	679	Fils et câbles électriques isolés		313
..	..	120	111	105	106	112	92	Accumulateurs et piles électriques		314
..	..	942	1 061	1 129	1 192	1 123	1 096	Lampes électriques et appareils d'éclairage		315
..	..	1 049	1 031	1 110	986	681	804	Autres matériels électriques, nca.		319
..	..	4 606	4 848	5 024	5 306	5 683	6 280	Equipements et appareils de radio, TV, etc.		32
..	..	1 388	1 270	1 359	1 530	1 889	1 864	Tubes et valves électroniques;autres composants		321
..	..	1 041	1 301	1 248	1 250	1 022	1 508	Emetteurs radio/TV; app. téléphonie et télégraphie		322
..	..	2 177	2 277	2 417	2 526	2 772	2 908	Récepteurs radio/TV, son et image; art. associés		323
..	..	6 443	6 455	6 962	7 012	7 003	6 908	Instr. médicaux, de précision ; optique ; horlog.		33
..	..	6 122	6 031	6 501	6 494	6 417	6 323	Instruments médicaux et de précision, etc.		331
..	..	317	420	460	517	583	582	Instruments d'optique ; matériel photographique		332
..	..	4	4	1	1	3	3	Horlogerie		333
..	..	893	1 047	1 258	1 288	1 371	1 432	Véhicules automobiles; remorques, etc.		34
..	..	13	7	..	..	..	..	Véhicules automobiles		341
..	..	361	371	455	494	540	529	Carross. automobiles, remorques, semi-remorques		342
..	..	519	669	803	794	831	903	Pièces et accessoires pour automobiles et moteurs		343
..	..	1 358	1 420	1 518	1 366	1 385	1 376	Autres matériels de transport		35
..	..	907	971	989	878	824	835	Construction et réparation de navires		351
..	..	118	97	100	94	99	64	Matériel ferroviaire roulant		352
..	..	95	105	117	123	194	180	Construction aéronautique et spatiale		353
..	..	238	247	312	271	268	297	Autres équipements de transport		359
..	..	10 519	10 903	11 062	11 129	11 913	12 039	Meubles ; activités de fabrication, nca		36
..	..	6 250	6 658	6 752	6 823	7 212	7 473	Meubles		361
..	..	4 269	4 245	4 310	4 306	4 701	4 566	Activités de fabrication, nca		369
..	..	115	75	54	54	54	55	Récupération		37
..	..	3 936	4 072	4 131	4 080	..	3 914	Electricité, gaz et eau	E	40_41
..	..	3 478	3 517	3 617	3 591	3 549	3 405	Prod. et dist. d'électricité gaz, vapeur d'eau chaude		40
..	..	458	555	514	489	498	509	Captage, épuration et distribution d'eau		41
..	..	18 044	17 427	17 924	17 449	17 720	18 269	Construction	F	45

DENMARK

Table DN.3 WAGES AND SALARIES, EMPLOYEES - SALAIRES ET TRAITEMENTS, SALARIÉS

Millions of DKr (Current Prices)

ISIC revision 3				1991	1992	1993	1994	1995	1996	1997	1998
C	10_14		Mining and quarrying	..	..	977	993	1 019	..	..	..
	10		Mining of coal and lignite; extraction of peat	..	..	36	34	34	34	38	39
	11		Crude petroleum & natural gas; service activities	..	..	482	459	495	496	487	512
	13		Mining of metal ores	..	..	..	..	..	..	..	..
	14		Other mining and quarrying	..	..	..	..	..	501	519	410
D	15_37		Total manufacturing	..	..	91 258	96 073	101 688	103 228	106 980	111 027
	15		Food products and beverages	..	..	16 662	17 097	17 496	17 452	17 610	17 991
		151	Production of meat, fish, fruit, vegetables, oils & fats	..	..	7 667	7 735	7 829	7 728	7 783	8 069
		152	Dairy products	..	..	2 642	2 643	2 791	2 809	2 877	2 778
		153	Grain mill and starch products; animal feeds	..	..	582	626	706	705	697	674
		154	Other food products	..	..	4 095	4 409	4 419	4 496	4 506	4 677
		155	Beverages	..	..	1 676	1 685	1 751	1 715	1 749	1 793
	16		Tobacco products	..	..	316	312	324	325	325	344
	17		Textiles	..	..	2 061	2 122	2 047	1 979	1 992	1 978
		171	Spinning, weaving and finishing of textiles	..	..	561	554	447	436	442	426
		172	Other textiles	..	..	974	1 042	1 065	1 083	1 113	1 114
		173	Knitted and crocheted fabrics and articles	..	..	526	527	535	460	437	438
	18		Wearing apparel and fur	..	..	1 392	1 324	1 206	1 084	1 007	973
		181	Wearing apparel, except fur apparel	..	..	1 370	1 301	1 188	1 066	986	951
		182	Dressing and dyeing of fur; articles of fur	..	..	22	23	18	19	21	22
	19		Leather and articles of leather; footwear	..	..	341	372	350	320	313	327
		191	Leather and articles of leather	..	..	104	101	91	94	83	79
		192	Footwear	..	..	237	272	259	226	230	247
	20		Wood & products of wood & cork, exc. furniture	..	..	2 363	2 803	2 998	3 079	3 278	3 445
		201	Sawmilling and planing of wood	..	..	295	354	343	353	391	402
		202	Wood products; articles of cork, straw, etc.	..	..	2 068	2 449	2 654	2 726	2 888	3 043
	21		Paper and paper products	..	..	2 162	2 286	2 457	2 458	2 632	2 652
	22		Printing, publishing and recorded media	..	..	8 731	8 885	9 110	9 629	9 790	10 198
		221	Publishing	..	..	4 805	4 979	5 245	5 565	5 726	5 986
		222	Printing and related activities	..	..	3 912	3 888	3 822	4 009	3 955	4 073
		223	Reproduction of recorded media	..	..	14	18	43	55	109	139
	23		Coke and petroleum products; nuclear fuel	..	..	257	283	302	304	291	273
		231	Coke oven products	..	..	..	..	..	..	..	..
		232	Refined petroleum products	..	..	257	283	302	304	291	273
		233	Processing of nuclear fuel	..	..	..	..	..	..	..	..
	24		Chemicals and chemical products	..	..	6 288	6 831	7 391	7 794	8 280	8 605
		241	Basic chemicals	..	..	1 653	1 512	1 622	1 948	1 527	1 536
		242	Other chemical products	..	..	4 634	5 319	5 769	5 565	6 456	6 757
		243	Man-made fibres	..	..	..	..	..	281	297	312
	25		Rubber and plastic products	..	..	4 109	4 237	4 350	4 628	4 945	5 252
		251	Rubber products	..	..	569	552	543	542	565	559
		252	Plastic products	..	..	3 540	3 686	3 807	4 086	4 380	4 693
	26		Other non-metallic mineral products	..	..	3 791	4 104	4 485	4 458	4 667	5 068
		261	Glass and glass products	..	..	648	772	918	881	957	1 073
		269	Non-metallic mineral products, nec	..	..	3 144	3 333	3 567	..	3 709	3 995
	27		Basic metals	..	..	1 643	1 557	1 744	1 791	1 973	2 504
		271	Basic iron and steel	..	..	958	1 058	1 210	1 246	1 409	1 459
		272	Basic precious and non-ferrous metals	..	..	501	430	516	523	542	619
		273	Casting of metals	..	..	184	69	18	22	22	427
	28		Fabricated metal products, exc. machinery	..	..	8 151	8 867	9 557	9 682	10 064	10 027
		281	Structural metal prod., tanks, reservoirs, etc.	..	..	2 939	3 147	3 414	3 577	3 998	4 017
		289	Other fabricated metal products	..	..	5 212	5 719	6 143	6 107	6 066	6 010
	29		Machinery and equipment, nec	..	..	14 328	15 212	16 724	16 951	17 489	18 201
		291	General purpose machinery	..	..	8 115	8 849	9 969	10 248	10 628	11 060
		292	Special purpose machinery	..	..	4 724	4 884	5 260	5 347	5 538	5 836
		293	Domestic appliances, nec	..	..	1 488	1 479	1 495	1 356	1 322	1 305
	30		Office, accounting & computing machinery	..	..	536	586	551	547	486	573
	31		Electrical machinery & apparatus, nec	..	..	2 921	3 015	3 332	3 523	3 594	3 982
		311	Electric motors, generators and transformers	..	..	617	706	936	981	1 175	1 403
		312	Electricity distribution & control apparatus	..	..	893	882	835	900	913	988
		313	Insulated wire and cable	..	..	319	296	320	347	409	476
		314	Accumulators, primary cells & primary batteries	..	..	61	56	58	64	61	49
		315	Electric lamps and lighting equipment	..	..	385	473	524	539	517	528
		319	Other electrical equipment, nec	..	..	645	603	659	692	519	540
	32		Radio, TV, communication equipment	..	..	2 304	2 263	2 408	2 575	2 854	3 036
		321	Electronic valves, tubes; other electronic components	..	..	647	560	592	637	820	842
		322	TV/radio transmitters; line teleph./telegr. apparatus	..	..	744	736	717	771	706	826
		323	TV/radio receivers; sound and video app.; assoc. gds.	..	..	914	967	1 098	1 167	1 329	1 368
	33		Medical, precision & optical instrum.; watch & clocks	..	..	3 154	3 223	3 569	3 583	3 720	3 734
		331	Medical appl.; instruments & appl. for measuring, etc.	..	..	2 898	2 924	3 237	3 235	3 302	3 303
		332	Optical instruments and photographic equipment	..	..	255	297	330	346	417	430
		333	Watches and clocks	..	..	2	2	2	1	1	2
	34		Motor vehicles, trailers and semi-trailers	..	..	1 045	1 303	1 543	1 650	1 717	1 816
		341	Motor vehicles	..	..	17	10	..	..	..	..
		342	Bodies for motor vehicles, trailers & semi-trailers	..	..	649	737	854	958	1 010	1 033
		343	Parts & accessories for motor vehicles & their engines	..	..	379	555	689	693	707	784
	35		Other transport equipment	..	..	3 300	3 466	3 558	3 249	3 255	3 089
		351	Building and repairing of ships and boats	..	..	2 781	2 940	2 983	2 649	2 578	2 360
		352	Railway and tramway locomotives and rolling stock	..	..	217	180	165	169	170	135
		353	Aircraft and spacecraft	..	..	116	138	166	191	261	290
		359	Transport equipment, nec	..	..	186	208	243	240	247	303
	36		Furniture; manufacturing, nec	..	..	5 313	5 846	6 099	6 080	6 610	6 873
		361	Furniture	..	..	3 919	4 420	4 584	4 525	4 830	5 021
		369	Manufacturing, nec	..	..	1 394	1 427	1 514	1 553	1 780	1 852
	37		Recycling	..	..	90	77	86	87	88	86
E	40_41		Electricity, gas and water supply	..	..	4 236	4 430	4 536	4 579	..	4 684
	40		Electricity, gas, steam and hot water supply	..	..	3 805	3 926	4 047	4 120	4 182	4 196
	41		Collection, purification and distribution of water	..	..	431	504	489	459	472	488
F	45		Construction	..	..	24 120	26 586	28 619	29 055	31 182	32 949

Note:ISIC 242 includes 243.

Table DN.4 **ÉTABLISSEMENTS - ESTABLISHMENTS**

Unités

1991	1992	1993	1994	1995	1996	1997	1998		CITI révision 3		
..	..	327	341	312	..	..	..	Activites extractives	C	10_14	
..	..	16	13	12	11	12	11	Extraction de charbon, de lignite et de tourbe		10	
..	..	13	17	17	13	13	24	Pétrole brut et gaz naturel ; activités annexes		11	
..	..	..	..	..	..	..	..	Extraction de minerais métalliques		13	
..	..	..	..	..	275	258	264	Autres activités extractives		14	
..	..	25 477	24 603	24 024	23 552	..	..	Total activites manufacturieres	D	15_37	
..	..	2 747	2 663	2 595	2 484	2 396	2 285	Produits alimentaires et boissons		15	
..	..	655	634	631	594	550	522	Viande, poisson, fruits, légumes, huiles et graisses			151
..	..	184	181	180	162	153	135	Produits laitiers			152
..	..	102	105	103	106	100	99	Travail des grains; prod. amyl. ; alim. pour animaux			153
..	..	1 737	1 678	1 616	1 559	1 534	1 472	Autres produits alimentaires			154
..	..	69	65	65	63	59	57	Boissons			155
..	..	17	15	13	13	12	12	Produits à base de tabac		16	
..	..	961	924	876	845	804	731	Textiles		17	
..	..	178	173	156	153	136	131	Filature, tissage et achèvement des textiles			171
..	..	579	558	535	543	530	481	Autres articles textiles			172
..	..	204	193	185	149	138	119	Etoffes et articles de bonneterie			173
..	..	1 400	1 217	1 149	1 076	974	884	Articles d'habillement et fourrure		18	
..	..	1 311	1 131	1 069	998	897	812	Articles d'habillement autres qu'en fourrure			181
..	..	89	86	80	78	77	72	Prép. et teinture des fourrures ; articles en fourrure			182
..	..	205	196	189	173	150	140	Cuir et articles en cuir ; chaussures		19	
..	..	97	92	83	72	62	62	Cuir et articles en cuir sauf chaussures			191
..	..	108	104	106	101	88	78	Chaussures			192
..	..	997	964	938	911	888	853	Bois, articles en bois, sauf meubles		20	
..	..	218	217	196	196	188	179	Sciage et rabotage du bois			201
..	..	779	747	742	715	700	674	Articles en bois, liège, vannerie, sparterie			202
..	..	269	260	252	244	249	244	Papier, articles en papier et carton		21	
..	..	3 605	3 484	3 409	3 440	3 364	3 227	Edition, imprimerie et reproduction		22	
..	..	1 418	1 393	1 391	1 464	1 437	1 401	Edition			221
..	..	2 168	2 071	1 991	1 937	1 876	1 776	Imprimerie et activités annexes			222
..	..	19	20	27	39	51	50	Reproduction de supports enregistrés			223
..	..	17	17	18	16	15	14	Coke. et prod. pétroliers ; comb. nucléaires		23	
..	..	..	..	..	..	..	..	Cokéfaction			231
..	..	17	17	18	16	15	14	Produits pétroliers raffinés			232
..	..	..	..	..	..	..	..	Traitement de combustibles nucléaires			233
..	..	478	475	460	456	444	418	Produits chimiques		24	
..	..	95	92	94	96	98	90	Produits chimiques de base			241
..	..	383	383	366	357	343	324	Autres produits chimiques			242
..	..	..	..	..	3	3	4	Fibres synthétiques ou artificielles			243
..	..	778	755	730	736	726	709	Art. en caoutchouc et matières plastiques		25	
..	..	134	124	118	112	107	93	Articles en caoutchouc			251
..	..	644	631	612	624	619	616	Articles en matières plastiques			252
..	..	1 429	1 382	1 318	1 279	1 257	1 222	Autres produits minéraux non métalliques		26	
..	..	176	192	191	189	187	184	Verre et articles en verre			261
..	..	1 253	1 190	1 127	1 090	1 070	1 038	Produits minéraux non métalliques, nca			269
..	..	191	190	189	190	195	223	Produits métallurgiques de base		27	
..	..	85	93	101	104	113	121	Sidérurgie et première transformation de l'acier			271
..	..	39	45	46	44	44	43	Première transf. métaux précieux et non ferreux			272
..	..	67	52	42	42	38	59	Fonderie			273
..	..	4 250	4 115	4 017	3 953	3 849	3 766	Ouvrages en métaux, sauf machines		28	
..	..	524	528	502	521	557	556	Constr. et menuis. métal.; réservoirs; générateurs			281
..	..	3 726	3 587	3 515	3 432	3 292	3 210	Autres ouvrages en métaux			289
..	..	2 299	2 296	2 328	2 266	2 304	2 290	Machines et matériels, nca		29	
..	..	1 014	1 003	1 029	1 005	1 009	995	Machines d'usage général			291
..	..	1 182	1 202	1 210	1 185	1 223	1 229	Machines d'usage spécifique			292
..	..	103	91	89	76	72	66	Appareils domestiques, nca			293
..	..	138	145	140	128	111	115	Mach. de bureau, comptables, information		30	
..	..	1 097	1 083	1 063	1 013	959	933	Machines et appareils électriques, nca		31	
..	..	75	77	79	80	83	78	Moteurs, génératrices, transformateurs électriques			311
..	..	172	167	150	158	158	153	Matériel électrique de distribution et de commande			312
..	..	30	30	31	31	30	32	Fils et câbles électriques isolés			313
..	..	19	18	20	17	14	12	Accumulateurs et piles électriques			314
..	..	221	205	201	173	168	161	Lampes électriques et appareils d'éclairage			315
..	..	580	586	582	554	506	497	Autres matériels électriques, nca.			319
..	..	297	272	262	255	246	240	Equipements et appareils de radio, TV, etc.		32	
..	..	123	104	104	99	98	94	Tubes et valves électroniques;autres composants			321
..	..	57	55	51	52	46	43	Emetteurs radio/TV; app. téléphonie et télégraphie			322
..	..	117	113	107	104	102	103	Récepteurs radio/TV, son et image; art. associés			323
..	..	678	682	673	683	691	640	Instr. médicaux, de précision ; optique ; horlog.		33	
..	..	624	628	621	633	641	596	Instruments médicaux et de précision, etc.			331
..	..	51	49	49	47	45	39	Instruments d'optique ; matériel photographique			332
..	..	3	5	3	3	5	5	Horlogerie			333
..	..	216	221	214	210	197	184	Véhicules automobiles; remorques, etc.		34	
..	..	11	9	..	..	..	..	Véhicules automobiles			341
..	..	137	131	134	136	129	118	Carross. automobiles, remorques, semi-remorques			342
..	..	68	81	80	74	68	66	Pièces et accessoires pour automobiles et moteurs			343
..	..	549	524	511	495	491	483	Autres matériels de transport		35	
..	..	437	415	396	379	379	367	Construction et réparation de navires			351
..	..	4	5	7	6	5	6	Matériel ferroviaire roulant			352
..	..	45	42	44	46	45	46	Construction aéronautique et spatiale			353
..	..	63	62	64	64	62	64	Autres équipements de transport			359
..	..	2 832	2 696	2 650	2 652	2 591	2 465	Meubles ; activités de fabrication, nca		36	
..	..	1 670	1 603	1 583	1 572	1 534	1 460	Meubles			361
..	..	1 162	1 093	1 067	1 080	1 057	1 005	Activités de fabrication, nca			369
..	..	27	27	30	34	36	30	Récupération		37	
..	..	1 664	1 869	1 896	1 908	..	2 018	Electricité, gaz et eau	E	40_41	
..	..	835	866	877	900	950	973	Prod. et dist. d'électricité gaz, vapeur d'eau chaude		40	
..	..	829	1 003	1 019	1 008	1 014	1 045	Captage, épuration et distribution d'eau		41	
..	..	23 352	23 135	23 273	23 656	24 108	23 790	Construction	F	45	

Note:ISIC 242 includes 243.

FINLAND

Table FN.1 **PRODUCTION - PRODUCTION**

Millions of Mk (Current Prices)

ISIC revision 3				1991	1992	1993	1994	1995	1996	1997	1998
C	10_14		Mining and quarrying	..	..	..	..	4 142	4 485	4 933	4 677
	10		Mining of coal and lignite; extraction of peat	..	..	..	..	1 406	1 523	1 642	1 319
	11		Crude petroleum & natural gas; service activities	..	..	..	..	..	..	..	..
	13		Mining of metal ores	..	..	..	..	560	580	644	495
	14		Other mining and quarrying	..	..	..	..	2 176	2 382	2 647	2 863
D	15_37		Total manufacturing	..	..	..	..	446 664	464 971	524 246	475 981
	15		Food products and beverages	..	..	..	..	45 880	49 024	48 888	48 964
		151	Production of meat, fish, fruit, vegetables, oils & fats	..	..	..	..	15 668	17 197	16 952	16 890
		152	Dairy products	..	..	..	..	11 220	11 628	10 454	10 306
		153	Grain mill and starch products; animal feeds	..	..	..	..	4 430	4 661	4 972	4 891
		154	Other food products	..	..	..	..	10 429	10 987	11 773	11 846
		155	Beverages	..	..	..	..	4 133	4 550	4 736	5 030
	16		Tobacco products	..	..	..	..	914	687	642	684
	17		Textiles	..	..	..	..	3 501	3 705	3 938	4 077
		171	Spinning, weaving and finishing of textiles	..	..	..	..	805	746	794	862
		172	Other textiles	..	..	..	..	2 096	2 308	2 509	2 593
		173	Knitted and crocheted fabrics and articles	..	..	..	..	600	652	635	622
	18		Wearing apparel and fur	..	..	..	..	3 106	3 176	3 343	3 534
		181	Wearing apparel, except fur apparel	..	..	..	..	2 969	3 005	3 176	3 364
		182	Dressing and dyeing of fur; articles of fur	..	..	..	..	138	171	167	170
	19		Leather and articles of leather; footwear	..	..	..	..	1 382	1 368	1 376	1 444
		191	Leather and articles of leather	..	..	..	..	462	444	443	402
		192	Footwear	..	..	..	..	920	924	933	1 042
	20		Wood & products of wood & cork, exc. furniture	..	..	..	..	21 506	20 752	26 502	28 231
		201	Sawmilling and planing of wood	..	..	..	..	12 630	11 502	15 428	16 154
		202	Wood products; articles of cork, straw, etc.	..	..	..	..	8 876	9 251	11 075	12 077
	21		Paper and paper products	..	..	..	..	76 834	67 003	74 207	78 538
	22		Printing, publishing and recorded media	..	..	..	..	19 519	21 028	21 587	23 063
		221	Publishing	..	..	..	..	10 516	11 276	11 738	12 819
		222	Printing and related activities	..	..	..	..	8 910	9 651	9 748	10 092
		223	Reproduction of recorded media	..	..	..	..	93	101	101	152
	23		Coke and petroleum products; nuclear fuel	..	..	..	..	11 388	13 954	13 643	12 865
		231	Coke oven products	..	..	..	..	..	..	..	..
		232	Refined petroleum products	..	..	..	..	10 761	13 269	12 946	12 145
		233	Processing of nuclear fuel	..	..	..	..	..	..	..	..
	24		Chemicals and chemical products	..	..	..	..	22 937	23 482	26 538	27 281
		241	Basic chemicals	..	..	..	..	15 535	15 779	17 485	17 742
		242	Other chemical products	..	..	..	..	6 692	7 179	8 388	8 865
		243	Man-made fibres	..	..	..	..	..	..	..	..
	25		Rubber and plastic products	..	..	..	..	9 109	9 485	10 857	12 504
		251	Rubber products	..	..	..	..	1 651	1 584	1 698	1 901
		252	Plastic products	..	..	..	..	7 458	7 902	9 159	10 604
	26		Other non-metallic mineral products	..	..	..	..	7 986	8 827	10 087	11 190
		261	Glass and glass products	..	..	..	..	1 934	2 107	2 306	2 598
		269	Non-metallic mineral products, nec	..	..	..	..	5 981	6 664	7 718	8 592
	27		Basic metals	..	..	..	..	30 936	29 937	33 101	31 426
		271	Basic iron and steel	..	..	..	..	19 203	18 244	19 513	20 056
		272	Basic precious and non-ferrous metals	..	..	..	..	9 320	10 219	11 968	9 745
		273	Casting of metals	..	..	..	..	1 506	1 473	1 621	1 625
	28		Fabricated metal products, exc. machinery	..	..	..	..	18 791	20 219	21 789	23 440
		281	Structural metal prod., tanks, reservoirs, etc.	..	..	..	..	9 874	10 080	10 341	11 361
		289	Other fabricated metal products	..	..	..	..	8 917	10 139	11 448	12 079
	29		Machinery and equipment, nec	..	..	..	..	40 008	48 167	51 799	53 732
		291	General purpose machinery	..	..	..	..	17 123	20 134	22 673	24 334
		292	Special purpose machinery	..	..	..	..	22 086	27 140	28 114	28 310
		293	Domestic appliances, nec	..	..	..	..	800	893	1 012	1 088
	30		Office, accounting & computing machinery	..	..	..	..	66 432	72 078	87 974	5 829
	31		Electrical machinery & apparatus, nec	..	..	..	..	12 583	13 416	14 303	14 579
		311	Electric motors, generators and transformers	..	..	..	..	5 177	6 010	6 283	6 492
		312	Electricity distribution & control apparatus	..	..	..	..	2 526	2 532	3 066	3 403
		313	Insulated wire and cable	..	..	..	..	1 689	1 891	2 168	2 095
		314	Accumulators, primary cells & primary batteries	..	..	..	..	..	..	..	10
		315	Electric lamps and lighting equipment	..	..	..	..	823	874	908	1 069
		319	Other electrical equipment, nec	..	..	..	..	1 176	1 287	1 462	1 511
	32		Radio, TV, communication equipment	..	..	..	..	26 785	31 021	42 261	59 704
		321	Electronic valves, tubes; other electronic components	..	..	..	..	1 492	1 785	2 082	2 258
		322	TV/radio transmitters; line teleph./telegr. apparatus	..	..	..	..	24 065	28 151	38 818	56 274
		323	TV/radio receivers; sound and video app.; assoc. gds.	..	..	..	..	1 228	1 085	1 361	1 172
	33		Medical, precision & optical instrum.; watch & clocks	..	..	..	..	5 559	6 291	7 433	8 112
		331	Medical appl.; instruments & appl. for measuring, etc.	..	..	..	..	5 403	6 120	7 232	7 955
		332	Optical instruments and photographic equipment	..	..	..	..	150	165	194	156
		333	Watches and clocks	..	..	..	..	7	5	8	1
	34		Motor vehicles, trailers and semi-trailers	..	..	..	..	3 952	4 063	4 637	5 171
		341	Motor vehicles	..	..	..	..	1 643	1 552	1 799	1 965
		342	Bodies for motor vehicles, trailers & semi-trailers	..	..	..	..	1 726	1 961	2 241	2 678
		343	Parts & accessories for motor vehicles & their engines	..	..	..	..	584	551	597	528
	35		Other transport equipment	..	..	..	..	10 501	10 088	11 575	12 650
		351	Building and repairing of ships and boats	..	..	..	..	8 389	7 670	8 652	9 773
		352	Railway and tramway locomotives and rolling stock	..	..	..	..	795	944	1 420	1 353
		353	Aircraft and spacecraft	..	..	..	..	1 052	..	..	..
		359	Transport equipment, nec	..	..	..	..	264	358	287	234
	36		Furniture; manufacturing, nec	..	..	..	..	6 484	6 647	7 171	8 503
		361	Furniture	..	..	..	..	4 460	4 561	5 035	6 112
		369	Manufacturing, nec	..	..	..	..	2 024	2 087	2 136	2 391
	37		Recycling	..	..	..	..	569	552	594	460
E	40_41		Electricity, gas and water supply	..	..	..	..	22 947	28 122	24 916	1 647
	40		Electricity, gas, steam and hot water supply	..	..	..	..	21 518	26 676	23 462	29
	41		Collection, purification and distribution of water	..	..	..	..	1 384	1 333	..	1 617
F	45		Construction	..	..	..	..	50 530	53 206	58 937	73 699

Note:Data are provisional.
ISIC 313 includes 314.

Table FN.2 VALEUR AJOUTÉE AU COUT DES FACTEURS - VALUE ADDED AT FACTOR COSTS

Millions de Mk (Prix courants)

1991	1992	1993	1994	1995	1996	1997	1998			CITI révision 3	
..	..	..	..	1 608	1 621	1 675	1 624	Activites extractives	C	10_14	
..	..	..	..	592	611	621	597	Extraction de charbon, de lignite et de tourbe		10	
..	..	..	..	..	..	..	..	Pétrole brut et gaz naturel ; activités annexes		11	
..	..	..	..	175	170	185	103	Extraction de minerais métalliques		13	
..	..	..	..	841	840	869	924	Autres activités extractives		14	
..	..	..	..	119 979	120 328	135 192	149 951	Total activites manufacturieres	D	15_37	
..	..	..	..	10 499	11 467	11 323	11 501	Produits alimentaires et boissons		15	
..	..	..	..	3 317	3 474	3 643	3 805	Viande, poisson, fruits, légumes, huiles et graisses			151
..	..	..	..	1 455	1 644	1 460	1 444	Produits laitiers			152
..	..	..	..	889	831	874	896	Travail des grains; prod. amyl. ; alim. pour animaux			153
..	..	..	..	3 727	4 007	3 818	3 794	Autres produits alimentaires			154
..	..	..	..	1 111	1 511	1 529	1 562	Boissons			155
..	..	..	..	224	215	199	196	Produits à base de tabac		16	
..	..	..	..	1 349	1 504	1 657	1 729	Textiles		17	
..	..	..	..	227	246	277	294	Filature, tissage et achèvement des textiles			171
..	..	..	..	855	953	1 080	1 141	Autres articles textiles			172
..	..	..	..	267	306	299	295	Etoffes et articles de bonneterie			173
..	..	..	..	1 265	1 266	1 272	1 328	Articles d'habillement et fourrure		18	
..	..	..	..	1 225	1 204	1 209	1 256	Articles d'habillement autres qu'en fourrure			181
..	..	..	..	40	62	63	73	Prép. et teinture des fourrures ; articles en fourrure			182
..	..	..	..	532	508	524	549	Cuir et articles en cuir ; chaussures		19	
..	..	..	..	156	148	141	137	Cuir et articles en cuir sauf chaussures			191
..	..	..	..	376	361	384	412	Chaussures			192
..	..	..	..	..	..	..	7 497	Bois, articles en bois, sauf meubles		20	
..	..	..	..	3 017	2 144	3 816	3 072	Sciage et rabotage du bois			201
..	..	..	..	3 080	3 319	3 964	4 424	Articles en bois, liège, vannerie, sparterie			202
..	..	..	..	25 275	18 755	20 891	24 304	Papier, articles en papier et carton		21	
..	..	..	..	7 725	8 340	8 777	9 369	Edition, imprimerie et reproduction		22	
..	..	..	..	3 799	4 114	4 460	4 961	Edition			221
..	..	..	..	3 896	4 186	4 277	4 340	Imprimerie et activités annexes			222
..	..	..	..	30	40	40	68	Reproduction de supports enregistrés			223
..	..	..	..	1 434	1 723	1 507	1 770	Coke. et prod. pétroliers ; comb. nucléaires		23	
..	..	..	..	..	..	..	..	Cokéfaction			231
..	..	..	..	1 257	1 506	1 285	1 555	Produits pétroliers raffinés			232
..	..	..	..	..	..	..	..	Traitement de combustibles nucléaires			233
..	..	..	..	7 940	7 813	8 683	9 104	Produits chimiques		24	
..	..	..	..	4 905	4 681	5 084	5 468	Produits chimiques de base			241
..	..	..	..	2 896	3 036	3 438	3 483	Autres produits chimiques			242
..	..	..	..	..	..	..	..	Fibres synthétiques ou artificielles			243
..	..	..	..	3 406	3 949	4 418	5 029	Art. en caoutchouc et matières plastiques		25	
..	..	..	..	698	721	806	901	Articles en caoutchouc			251
..	..	..	..	2 708	3 229	3 611	4 128	Articles en matières plastiques			252
..	..	..	..	3 265	3 491	4 012	4 647	Autres produits minéraux non métalliques		26	
..	..	..	..	833	882	901	1 064	Verre et articles en verre			261
..	..	..	..	2 406	2 574	3 075	3 583	Produits minéraux non métalliques, nca			269
..	..	..	..	7 446	6 075	6 862	6 394	Produits métallurgiques de base		27	
..	..	..	..	5 365	4 125	4 326	4 303	Sidérurgie et première transformation de l'acier			271
..	..	..	..	1 005	1 236	1 750	1 303	Première transf. métaux précieux et non ferreux			272
..	..	..	..	734	714	786	788	Fonderie			273
..	..	..	..	6 955	7 952	8 625	9 483	Ouvrages en métaux, sauf machines		28	
..	..	..	..	3 023	3 401	3 518	3 904	Constr. et menuis. métal.; réservoirs; générateurs			281
..	..	..	..	3 932	4 551	5 107	5 579	Autres ouvrages en métaux			289
..	..	..	..	13 604	16 128	16 590	17 398	Machines et matériels, nca		29	
..	..	..	..	6 080	7 045	7 228	7 607	Machines d'usage général			291
..	..	..	..	7 286	8 777	9 021	9 410	Machines d'usage spécifique			292
..	..	..	..	238	306	342	382	Appareils domestiques, nca			293
..	..	..	..	..	..	..	466	Mach. de bureau, comptables, information		30	
..	..	..	..	4 111	4 584	5 089	5 221	Machines et appareils électriques, nca		31	
..	..	..	..	1 674	2 050	2 302	2 290	Moteurs, génératrices, transformateurs électriques			311
..	..	..	..	978	990	1 197	1 342	Matériel électrique de distribution et de commande			312
..	..	..	..	462	544	584	622	Fils et câbles électriques isolés			313
..	..	..	..	..	..	..	2	Accumulateurs et piles électriques			314
..	..	..	..	316	342	373	400	Lampes électriques et appareils d'éclairage			315
..	..	..	..	476	535	563	565	Autres matériels électriques, nca.			319
..	..	..	..	8 085	9 431	14 103	21 275	Equipements et appareils de radio, TV, etc.		32	
..	..	..	..	661	781	870	922	Tubes et valves électroniques;autres composants			321
..	..	..	..	7 179	8 419	12 875	20 036	Emetteurs radio/TV; app. téléphonie et télégraphie			322
..	..	..	..	245	231	358	317	Récepteurs radio/TV, son et image; art. associés			323
..	..	..	..	2 299	2 695	3 042	3 343	Instr. médicaux, de précision ; optique ; horlog.		33	
..	..	..	..	2 225	2 614	2 954	3 273	Instruments médicaux et de précision, etc.			331
..	..	..	..	71	79	86	70	Instruments d'optique ; matériel photographique			332
..	..	..	..	3	2	2		Horlogerie			333
..	..	..	..	1 398	1 544	1 743	1 976	Véhicules automobiles; remorques, etc.		34	
..	..	..	..	556	604	694	799	Véhicules automobiles			341
..	..	..	..	609	707	792	952	Carross. automobiles, remorques, semi-remorques			342
..	..	..	..	233	234	257	225	Pièces et accessoires pour automobiles et moteurs			343
..	..	..	..	3 487	3 614	3 980	3 852	Autres matériels de transport		35	
..	..	..	..	2 519	2 486	2 780	2 583	Construction et réparation de navires			351
..	..	..	..	258	342	329	401	Matériel ferroviaire roulant			352
..	..	..	..	605	..	..	..	Construction aéronautique et spatiale			353
..	..	..	..	105	118	112	78	Autres équipements de transport			359
..	..	..	..	2 819	2 903	3 136	3 428	Meubles ; activités de fabrication, nca		36	
..	..	..	..	1 929	1 987	2 225	2 452	Meubles			361
..	..	..	..	890	916	911	976	Activités de fabrication, nca			369
..	..	..	..	105	92	113	93	Récupération		37	
..	..	..	..	11 799	12 897	12 823	1 195	Electricité, gaz et eau	E	40_41	
..	..	..	..	10 780	11 874	11 791	19	Prod. et dist. d'électricité gaz, vapeur d'eau chaude		40	
..	..	..	..	1 018	999	..	1 176	Captage, épuration et distribution d'eau		41	
..	..	..	..	15 171	15 640	19 343	25 438	Construction	F	45	

Note:Data are provisional.
ISIC313 includes 314.

FINLAND

Table FN.3 **INVESTMENT - INVESTISSEMENT**

Millions of Mk (Current Prices)

ISIC revision 3				1991	1992	1993	1994	1995	1996	1997	1998
C	10_14		Mining and quarrying	..	..	..	..	402	369	324	418
	10		Mining of coal and lignite; extraction of peat	..	..	..	..	193	153	136	152
	11		Crude petroleum & natural gas; service activities	..	..	..	..	..	..	..	..
	13		Mining of metal ores	..	..	..	..	57	61	13	27
	14		Other mining and quarrying	..	..	..	..	152	155	176	239
D	15_37		Total manufacturing	..	..	..	..	23 751	26 017	28 320	25 073
	15		Food products and beverages	..	..	..	..	2 030	1 987	1 942	1 979
		151	Production of meat, fish, fruit, vegetables, oils & fats	..	..	..	..	513	739	610	682
		152	Dairy products	..	..	..	..	374	344	228	355
		153	Grain mill and starch products; animal feeds	..	..	..	..	258	178	193	136
		154	Other food products	..	..	..	..	555	430	507	502
		155	Beverages	..	..	..	..	328	295	404	304
	16		Tobacco products	..	..	..	..	95	64	28	6
	17		Textiles	..	..	..	..	255	124	229	242
		171	Spinning, weaving and finishing of textiles	..	..	..	..	14	19	16	36
		172	Other textiles	..	..	..	..	216	84	190	180
		173	Knitted and crocheted fabrics and articles	..	..	..	..	24	21	23	27
	18		Wearing apparel and fur	..	..	..	..	64	80	104	73
		181	Wearing apparel, except fur apparel	..	..	..	..	62	76	98	69
		182	Dressing and dyeing of fur; articles of fur	..	..	..	..	2	4	7	3
	19		Leather and articles of leather; footwear	..	..	..	..	47	50	62	50
		191	Leather and articles of leather	..	..	..	..	20	16	20	15
		192	Footwear	..	..	..	..	27	34	42	35
	20		Wood & products of wood & cork, exc. furniture	..	..	..	..	1 366	972	1 341	1 736
		201	Sawmilling and planing of wood	..	..	..	..	798	602	932	949
		202	Wood products; articles of cork, straw, etc.	..	..	..	..	568	371	409	787
	21		Paper and paper products	..	..	..	..	4 156	9 021	5 662	5 579
	22		Printing, publishing and recorded media	..	..	..	..	953	877	1 336	1 471
		221	Publishing	..	..	..	..	295	241	484	604
		222	Printing and related activities	..	..	..	..	651	630	842	850
		223	Reproduction of recorded media	..	..	..	..	7	6	11	18
	23		Coke and petroleum products; nuclear fuel	..	..	..	..	252	271	863	235
		231	Coke oven products	..	..	..	..	..	..	..	..
		232	Refined petroleum products	..	..	..	..	230	264	839	216
		233	Processing of nuclear fuel	..	..	..	..	..	..	..	..
	24		Chemicals and chemical products	..	..	..	..	2 227	1 240	2 342	1 951
		241	Basic chemicals	..	..	..	..	1 829	857	1 548	1 525
		242	Other chemical products	..	..	..	..	356	369	790	419
		243	Man-made fibres	..	..	..	..	..	..	..	..
	25		Rubber and plastic products	..	..	..	..	642	655	723	1 174
		251	Rubber products	..	..	..	..	84	140	142	325
		252	Plastic products	..	..	..	..	558	515	581	849
	26		Other non-metallic mineral products	..	..	..	..	450	451	550	741
		261	Glass and glass products	..	..	..	..	188	186	177	152
		269	Non-metallic mineral products, nec	..	..	..	..	261	265	371	589
	27		Basic metals	..	..	..	..	2 163	2 395	2 761	1 956
		271	Basic iron and steel	..	..	..	..	1 016	1 688	2 216	1 139
		272	Basic precious and non-ferrous metals	..	..	..	..	1 027	609	396	698
		273	Casting of metals	..	..	..	..	96	98	149	118
	28		Fabricated metal products, exc. machinery	..	..	..	..	915	1 016	1 184	1 551
		281	Structural metal prod., tanks, reservoirs, etc.	..	..	..	..	291	352	410	521
		289	Other fabricated metal products	..	..	..	..	624	664	775	1 030
	29		Machinery and equipment, nec	..	..	..	..	1 617	1 775	1 689	1 723
		291	General purpose machinery	..	..	..	..	825	815	891	891
		292	Special purpose machinery	..	..	..	..	753	938	767	788
		293	Domestic appliances, nec	..	..	..	..	39	22	31	44
	30		Office, accounting & computing machinery	..	..	..	..	3 260	2 519	3 750	42
	31		Electrical machinery & apparatus, nec	..	..	..	..	507	433	581	662
		311	Electric motors, generators and transformers	..	..	..	..	226	180	254	263
		312	Electricity distribution & control apparatus	..	..	..	..	75	83	117	112
		313	Insulated wire and cable	..	..	..	..	58	64	87	166
		314	Accumulators, primary cells & primary batteries	..	..	..	..	..	..	..	0
		315	Electric lamps and lighting equipment	..	..	..	..	57	30	41	39
		319	Other electrical equipment, nec	..	..	..	..	73	72	66	83
	32		Radio, TV, communication equipment	..	..	..	..	1 761	1 090	1 835	2 374
		321	Electronic valves, tubes; other electronic components	..	..	..	..	247	154	321	273
		322	TV/radio transmitters; line teleph./telegr. apparatus	..	..	..	..	1 438	909	1 475	2 075
		323	TV/radio receivers; sound and video app.; assoc. gds.	..	..	..	..	76	27	39	27
	33		Medical, precision & optical instrum.; watch & clocks	..	..	..	..	177	194	334	323
		331	Medical appl.; instruments & appl. for measuring, etc.	..	..	..	..	171	185	316	297
		332	Optical instruments and photographic equipment	..	..	..	..	6	9	17	26
		333	Watches and clocks	..	..	..	..	0	0	0	0
	34		Motor vehicles, trailers and semi-trailers	..	..	..	..	152	205	219	208
		341	Motor vehicles	..	..	..	..	64	118	135	87
		342	Bodies for motor vehicles, trailers & semi-trailers	..	..	..	..	40	65	34	92
		343	Parts & accessories for motor vehicles & their engines	..	..	..	..	49	22	50	29
	35		Other transport equipment	..	..	..	..	331	314	370	521
		351	Building and repairing of ships and boats	..	..	..	..	168	139	131	226
		352	Railway and tramway locomotives and rolling stock	..	..	..	..	9	20	15	10
		353	Aircraft and spacecraft	..	..	..	..	144	..	..	..
		359	Transport equipment, nec	..	..	..	..	10	5	20	10
	36		Furniture; manufacturing, nec	..	..	..	..	315	245	358	435
		361	Furniture	..	..	..	..	184	177	262	329
		369	Manufacturing, nec	..	..	..	..	131	67	96	106
	37		Recycling	..	..	..	..	16	38	54	39
E	40_41		Electricity, gas and water supply	..	..	..	..	4 631	4 310	11 261	480
	40		Electricity, gas, steam and hot water supply	..	..	..	..	4 237	3 884	10 790	10
	41		Collection, purification and distribution of water	..	..	..	..	418	..	..	470
F	45		Construction	..	..	..	..	2 581	3 073	4 467	2 934

Note:Data are provisional.
ISIC 313 includes 314.

Table FN.4 DÉPENSES DE MACHINES ET D'ÉQUIPEMENT - INVESTMENT IN MACHINERY & EQUIPMENT

Millions de Mk (Prix courants)

1991	1992	1993	1994	1995	1996	1997	1998	CITI révision 3			
..	..	..	..	315	296	277	354	Activites extractives	C	10_14	
..	..	..	..	152	112	111	126	Extraction de charbon, de lignite et de tourbe		10	
..	..	..	..	..	..	..	..	Pétrole brut et gaz naturel ; activités annexes		11	
..	..	..	..	23	52	7	23	Extraction de minerais métalliques		13	
..	..	..	..	140	132	159	206	Autres activités extractives		14	
..	..	..	..	18 552	20 459	21 993	20 555	Total activites manufacturieres	D	15_37	
..	..	..	..	1 521	1 452	1 352	1 517	Produits alimentaires et boissons		15	
..	..	..	..	354	465	330	459	Viande, poisson, fruits, légumes, huiles et graisses			151
..	..	..	..	308	316	188	277	Produits laitiers			152
..	..	..	..	172	124	138	102	Travail des grains; prod. amyl. ; alim. pour animaux			153
..	..	..	..	447	325	430	445	Autres produits alimentaires			154
..	..	..	..	240	222	267	233	Boissons			155
..	..	..	..	73	22	25	6	Produits à base de tabac		16	
..	..	..	..	199	112	203	203	Textiles		17	
..	..	..	..	12	18	14	22	Filature, tissage et achèvement des textiles			171
..	..	..	..	164	75	169	161	Autres articles textiles			172
..	..	..	..	23	19	19	21	Etoffes et articles de bonneterie			173
..	..	..	..	48	54	75	56	Articles d'habillement et fourrure		18	
..	..	..	..	46	50	72	53	Articles d'habillement autres qu'en fourrure			181
..	..	..	..	1	3	3	3	Prép. et teinture des fourrures ; articles en fourrure			182
..	..	..	..	36	44	46	39	Cuir et articles en cuir ; chaussures		19	
..	..	..	..	15	13	16	12	Cuir et articles en cuir sauf chaussures			191
..	..	..	..	21	31	30	26	Chaussures			192
..	..	..	..	1 021	697	1 010	1 269	Bois, articles en bois, sauf meubles		20	
..	..	..	..	581	394	707	658	Sciage et rabotage du bois			201
..	..	..	..	440	303	303	611	Articles en bois, liège, vannerie, sparterie			202
..	..	..	..	3 206	7 495	4 578	4 962	Papier, articles en papier et carton		21	
..	..	..	..	806	761	1 019	1 176	Edition, imprimerie et reproduction		22	
..	..	..	..	201	200	311	363	Edition			221
..	..	..	..	598	556	698	795	Imprimerie et activités annexes			222
..	..	..	..	7	6	10	18	Reproduction de supports enregistrés			223
..	..	..	..	203	193	745	153	Coke. et prod. pétroliers ; comb. nucléaires		23	
..	..	..	..	..	..	..	..	Cokéfaction			231
..	..	..	..	184	186	720	136	Produits pétroliers raffinés			232
..	..	..	..	..	..	..	..	Traitement de combustibles nucléaires			233
..	..	..	..	1 760	965	1 808	1 606	Produits chimiques		24	
..	..	..	..	1 477	697	1 353	1 292	Produits chimiques de base			241
..	..	..	..	248	256	450	307	Autres produits chimiques			242
..	..	..	..	..	..	..	..	Fibres synthétiques ou artificielles			243
..	..	..	..	536	542	590	953	Art. en caoutchouc et matières plastiques		25	
..	..	..	..	80	104	107	232	Articles en caoutchouc			251
..	..	..	..	456	438	483	722	Articles en matières plastiques			252
..	..	..	..	331	329	439	561	Autres produits minéraux non métalliques		26	
..	..	..	..	149	153	144	128	Verre et articles en verre			261
..	..	..	..	181	176	294	433	Produits minéraux non métalliques, nca			269
..	..	..	..	1 762	1 679	2 131	1 666	Produits métallurgiques de base		27	
..	..	..	..	880	1 236	1 702	998	Sidérurgie et première transformation de l'acier			271
..	..	..	..	787	381	307	574	Première transf. métaux précieux et non ferreux			272
..	..	..	..	81	62	122	94	Fonderie			273
..	..	..	..	712	748	921	1 177	Ouvrages en métaux, sauf machines		28	
..	..	..	..	225	251	288	388	Constr. et menuis. métal.; réservoirs; générateurs			281
..	..	..	..	487	497	633	790	Autres ouvrages en métaux			289
..	..	..	..	1 023	1 304	1 229	1 326	Machines et matériels, nca		29	
..	..	..	..	540	592	638	663	Machines d'usage général			291
..	..	..	..	454	694	562	628	Machines d'usage spécifique			292
..	..	..	..	28	18	29	34	Appareils domestiques, nca			293
..	..	..	..	2 657	2 031	2 911	38	Mach. de bureau, comptables, information		30	
..	..	..	..	409	389	472	556	Machines et appareils électriques, nca		31	
..	..	..	..	204	172	196	222	Moteurs, génératrices, transformateurs électriques			311
..	..	..	..	69	76	100	90	Matériel électrique de distribution et de commande			312
..	..	..	..	37	58	80	149	Fils et câbles électriques isolés			313
..	..	..	..	..	..	..	0	Accumulateurs et piles électriques			314
..	..	..	..	28	21	23	30	Lampes électriques et appareils d'éclairage			315
..	..	..	..	59	58	57	65	Autres matériels électriques, nca.			319
..	..	..	..	1 486	847	1 417	2 140	Equipements et appareils de radio, TV, etc.		32	
..	..	..	..	200	144	296	255	Tubes et valves électroniques;autres composants			321
..	..	..	..	1 218	677	1 088	1 860	Emetteurs radio/TV; app. téléphonie et télégraphie			322
..	..	..	..	68	26	33	24	Récepteurs radio/TV, son et image; art. associés			323
..	..	..	..	158	165	196	263	Instr. médicaux, de précision ; optique ; horlog.		33	
..	..	..	..	152	157	186	248	Instruments médicaux et de précision, etc.			331
..	..	..	..	6	9	10	15	Instruments d'optique ; matériel photographique			332
..	..	..	..	0	0	0	0	Horlogerie			333
..	..	..	..	100	166	178	156	Véhicules automobiles; remorques, etc.		34	
..	..	..	..	33	112	123	83	Véhicules automobiles			341
..	..	..	..	31	34	28	51	Carross. automobiles, remorques, semi-remorques			342
..	..	..	..	35	20	27	22	Pièces et accessoires pour automobiles et moteurs			343
..	..	..	..	256	235	318	387	Autres matériels de transport		35	
..	..	..	..	104	82	96	133	Construction et réparation de navires			351
..	..	..	..	8	15	14	10	Matériel ferroviaire roulant			352
..	..	..	..	135	..	..	..	Construction aéronautique et spatiale			353
..	..	..	..	9	4	15	8	Autres équipements de transport			359
..	..	..	..	237	197	288	310	Meubles ; activités de fabrication, nca		36	
..	..	..	..	123	141	205	219	Meubles			361
..	..	..	..	114	55	83	91	Activités de fabrication, nca			369
..	..	..	..	12	32	41	36	Récupération		37	
..	..	..	..	1 853	1 851	3 980	58	Electricité, gaz et eau	E	40_41	
..	..	..	..	1 812	1 804	3 940	9	Prod. et dist. d'électricité gaz, vapeur d'eau chaude		40	
..	..	..	..	42	50	..	48	Captage, épuration et distribution d'eau		41	
..	..	..	..	2 054	2 846	3 094	2 632	Construction	F	45	

Note:Data are provisional.

ISIC 313 includes 314.

FINLAND

Table FN.5 EMPLOYMENT, EMPLOYEES - EMPLOI, SALARIÉS

Units

ISIC revision 3				1991	1992	1993	1994	1995	1996	1997	1998
C	10_14		Mining and quarrying	..	..	..	..	4 108	4 120	4 346	4 219
	10		Mining of coal and lignite; extraction of peat	..	..	..	..	1 037	1 167	1 229	1 096
	11		Crude petroleum & natural gas; service activities	..	..	..	..	..	..	..	..
	13		Mining of metal ores	..	..	..	..	616	594	597	584
	14		Other mining and quarrying	..	..	..	..	2 456	2 359	2 520	2 539
D	15_37		Total manufacturing	..	..	..	..	463 174	468 452	482 781	414 261
	15		Food products and beverages	..	..	..	..	42 912	42 359	41 506	42 132
		151	Production of meat, fish, fruit, vegetables, oils & fats	..	..	..	..	14 442	14 810	14 612	14 962
		152	Dairy products	..	..	..	..	6 520	5 649	5 292	5 805
		153	Grain mill and starch products; animal feeds	..	..	..	..	2 149	2 224	2 253	2 271
		154	Other food products	..	..	..	..	15 954	16 057	15 839	15 553
		155	Beverages	..	..	..	..	3 848	3 619	3 513	3 541
	16		Tobacco products	..	..	..	..	915	761	723	682
	17		Textiles	..	..	..	..	6 395	6 584	6 537	6 551
		171	Spinning, weaving and finishing of textiles	..	..	..	..	1 233	1 189	1 164	1 101
		172	Other textiles	..	..	..	..	3 368	3 478	3 603	3 731
		173	Knitted and crocheted fabrics and articles	..	..	..	..	1 794	1 917	1 771	1 719
	18		Wearing apparel and fur	..	..	..	..	8 337	7 333	7 372	7 373
		181	Wearing apparel, except fur apparel	..	..	..	..	8 008	7 007	7 071	7 034
		182	Dressing and dyeing of fur; articles of fur	..	..	..	..	330	327	301	339
	19		Leather and articles of leather; footwear	..	..	..	..	3 360	3 081	2 800	2 916
		191	Leather and articles of leather	..	..	..	..	932	783	707	665
		192	Footwear	..	..	..	..	2 428	2 298	2 093	2 251
	20		Wood & products of wood & cork, exc. furniture	..	..	..	..	26 628	25 586	26 292	27 641
		201	Sawmilling and planing of wood	..	..	..	..	10 440	9 599	9 548	10 135
		202	Wood products; articles of cork, straw, etc.	..	..	..	..	16 188	15 987	16 744	17 506
	21		Paper and paper products	..	..	..	..	38 943	38 582	38 700	38 254
	22		Printing, publishing and recorded media	..	..	..	..	30 531	29 182	29 144	30 192
		221	Publishing	..	..	..	..	15 880	15 159	14 550	15 418
		222	Printing and related activities	..	..	..	..	14 548	13 924	14 475	14 634
		223	Reproduction of recorded media	..	..	..	..	103	99	119	140
	23		Coke and petroleum products; nuclear fuel	..	..	..	..	3 344	3 422	3 400	3 325
		231	Coke oven products	..	..	..	..	..	..	..	..
		232	Refined petroleum products	..	..	..	..	3 146	3 219	3 201	3 138
		233	Processing of nuclear fuel	..	..	..	..	..	..	..	..
	24		Chemicals and chemical products	..	..	..	..	18 461	18 111	18 042	18 617
		241	Basic chemicals	..	..	..	..	8 547	8 376	8 228	8 502
		242	Other chemical products	..	..	..	..	9 180	9 183	9 291	9 583
		243	Man-made fibres	..	..	..	..	..	..	..	..
	25		Rubber and plastic products	..	..	..	..	13 781	14 053	15 018	16 365
		251	Rubber products	..	..	..	..	2 577	2 331	2 386	2 535
		252	Plastic products	..	..	..	..	11 204	11 722	12 633	13 831
	26		Other non-metallic mineral products	..	..	..	..	12 706	12 690	12 964	14 003
		261	Glass and glass products	..	..	..	..	3 301	3 355	3 494	3 710
		269	Non-metallic mineral products, nec	..	..	..	..	9 252	9 204	9 366	10 294
	27		Basic metals	..	..	..	..	16 207	16 754	16 748	17 266
		271	Basic iron and steel	..	..	..	..	9 710	10 107	10 149	10 493
		272	Basic precious and non-ferrous metals	..	..	..	..	3 371	3 794	3 728	3 830
		273	Casting of metals	..	..	..	..	2 773	2 853	2 871	2 943
	28		Fabricated metal products, exc. machinery	..	..	..	..	28 548	30 312	32 089	34 355
		281	Structural metal prod., tanks, reservoirs, etc.	..	..	..	..	12 686	13 462	13 521	14 624
		289	Other fabricated metal products	..	..	..	..	15 861	16 851	18 569	19 731
	29		Machinery and equipment, nec	..	..	..	..	49 630	53 508	55 004	57 740
		291	General purpose machinery	..	..	..	..	21 801	23 408	24 301	24 708
		292	Special purpose machinery	..	..	..	..	26 342	28 683	29 210	31 547
		293	Domestic appliances, nec	..	..	..	..	1 489	1 417	1 493	1 485
	30		Office, accounting & computing machinery	..	..	..	..	81 238	83 067	88 221	2 362
	31		Electrical machinery & apparatus, nec	..	..	..	..	15 750	15 920	16 342	16 128
		311	Electric motors, generators and transformers	..	..	..	..	5 675	6 525	6 115	6 121
		312	Electricity distribution & control apparatus	..	..	..	..	4 029	3 642	4 259	4 474
		313	Insulated wire and cable	..	..	..	..	1 691	1 695	1 914	1 885
		314	Accumulators, primary cells & primary batteries	..	..	..	..	..	..	..	16
		315	Electric lamps and lighting equipment	..	..	..	..	1 421	1 419	1 416	1 502
		319	Other electrical equipment, nec	..	..	..	..	2 198	2 203	2 245	2 131
	32		Radio, TV, communication equipment	..	..	..	..	21 350	23 421	26 570	30 702
		321	Electronic valves, tubes; other electronic components	..	..	..	..	3 290	3 500	3 563	3 795
		322	TV/radio transmitters; line teleph./telegr. apparatus	..	..	..	..	16 682	18 583	21 601	25 664
		323	TV/radio receivers; sound and video app.; assoc. gds.	..	..	..	..	1 377	1 338	1 406	1 243
	33		Medical, precision & optical instrum.; watch & clocks	..	..	..	..	8 113	8 453	8 944	9 998
		331	Medical appl.; instruments & appl. for measuring, etc.	..	..	..	..	7 811	8 143	8 612	9 717
		332	Optical instruments and photographic equipment	..	..	..	..	290	299	321	280
		333	Watches and clocks	..	..	..	..	11	11	11	1
	34		Motor vehicles, trailers and semi-trailers	..	..	..	..	6 308	6 331	7 007	7 264
		341	Motor vehicles	..	..	..	..	2 432	2 227	2 592	2 686
		342	Bodies for motor vehicles, trailers & semi-trailers	..	..	..	..	2 916	3 137	3 417	3 771
		343	Parts & accessories for motor vehicles & their engines	..	..	..	..	960	967	998	807
	35		Other transport equipment	..	..	..	..	15 599	15 151	14 839	15 008
		351	Building and repairing of ships and boats	..	..	..	..	10 147	9 834	9 301	9 617
		352	Railway and tramway locomotives and rolling stock	..	..	..	..	2 381	2 166	2 174	2 099
		353	Aircraft and spacecraft	..	..	..	..	2 613	..	..	..
		359	Transport equipment, nec	..	..	..	..	458	444	395	348
	36		Furniture; manufacturing, nec	..	..	..	..	13 860	13 493	14 198	15 063
		361	Furniture	..	..	..	..	9 763	9 567	10 336	11 153
		369	Manufacturing, nec	..	..	..	..	4 097	3 926	3 862	3 909
	37		Recycling	..	..	..	..	258	298	321	323
E	40_41		Electricity, gas and water supply	..	..	..	..	19 152	19 227	18 965	2 071
	40		Electricity, gas, steam and hot water supply	..	..	..	..	16 902	17 011	16 847	46
	41		Collection, purification and distribution of water	..	..	..	..	2 250	2 139	..	2 025
F	45		Construction	..	..	..	..	71 769	74 330	79 878	98 724

Note:Data are provisional.
ISIC 313 includes 314.

Table FN.6 SALAIRES ET CHARGES SOCIALES, SALARIÉS - COMPENSATION OF LABOUR, EMPLOYEES

Millions de Mk (Prix courants)

1991	1992	1993	1994	1995	1996	1997	1998		CITI révision 3		
..	..	..	..	736	723	790	809	Activites extractives	C	10_14	
..	..	..	..	178	176	196	186	Extraction de charbon, de lignite et de tourbe		10	
..	..	..	..	..	..	..	..	Pétrole brut et gaz naturel ; activités annexes		11	
..	..	..	..	140	134	137	136	Extraction de minerais métalliques		13	
..	..	..	..	418	414	457	488	Autres activités extractives		14	
..	..	..	..	82 897	86 317	90 516	82 180	Total activites manufacturieres	D	15_37	
..	..	..	..	7 078	7 201	7 134	7 439	Produits alimentaires et boissons		15	
..	..	..	..	2 315	2 465	2 436	2 597	Viande, poisson, fruits, légumes, huiles et graisses			151
..	..	..	..	1 132	996	935	1 055	Produits laitiers			152
..	..	..	..	427	458	473	489	Travail des grains; prod. amyl. ; alim. pour animaux			153
..	..	..	..	2 421	2 517	2 537	2 515	Autres produits alimentaires			154
..	..	..	..	783	765	753	783	Boissons			155
..	..	..	..	178	151	153	157	Produits à base de tabac		16	
..	..	..	..	902	957	975	1 042	Textiles		17	
..	..	..	..	162	163	160	164	Filature, tissage et achèvement des textiles			171
..	..	..	..	511	561	583	640	Autres articles textiles			172
..	..	..	..	229	233	231	238	Etoffes et articles de bonneterie			173
..	..	..	..	958	907	923	955	Articles d'habillement et fourrure		18	
..	..	..	..	921	860	877	907	Articles d'habillement autres qu'en fourrure			181
..	..	..	..	37	46	46	48	Prép. et teinture des fourrures ; articles en fourrure			182
..	..	..	..	394	383	354	386	Cuir et articles en cuir ; chaussures		19	
..	..	..	..	107	101	82	84	Cuir et articles en cuir sauf chaussures			191
..	..	..	..	288	283	272	302	Chaussures			192
..	..	..	..	4 091	3 974	4 247	4 647	Bois, articles en bois, sauf meubles		20	
..	..	..	..	1 684	1 555	1 609	1 794	Sciage et rabotage du bois			201
..	..	..	..	2 407	2 419	2 639	2 853	Articles en bois, liège, vannerie, sparterie			202
..	..	..	..	8 805	8 976	9 328	9 838	Papier, articles en papier et carton		21	
..	..	..	..	5 467	5 553	5 620	6 121	Edition, imprimerie et reproduction		22	
..	..	..	..	2 891	2 873	2 921	3 249	Edition			221
..	..	..	..	2 557	2 661	2 679	2 838	Imprimerie et activités annexes			222
..	..	..	..	19	19	20	33	Reproduction de supports enregistrés			223
..	..	..	..	782	804	838	825	Coke. et prod. pétroliers ; comb. nucléaires		23	
..	..	..	..	..	..	..	..	Cokéfaction			231
..	..	..	..	739	760	795	782	Produits pétroliers raffinés			232
..	..	..	..	..	..	..	..	Traitement de combustibles nucléaires			233
..	..	..	..	3 676	3 669	3 793	4 040	Produits chimiques		24	
..	..	..	..	1 899	1 903	1 957	2 107	Produits chimiques de base			241
..	..	..	..	1 624	1 653	1 720	1 818	Autres produits chimiques			242
..	..	..	..	..	..	..	..	Fibres synthétiques ou artificielles			243
..	..	..	..	2 280	2 380	2 645	3 014	Art. en caoutchouc et matières plastiques		25	
..	..	..	..	444	435	469	552	Articles en caoutchouc			251
..	..	..	..	1 836	1 945	2 176	2 462	Articles en matières plastiques			252
..	..	..	..	2 150	2 241	2 335	2 583	Autres produits minéraux non métalliques		26	
..	..	..	..	586	635	659	728	Verre et articles en verre			261
..	..	..	..	1 537	1 585	1 654	1 855	Produits minéraux non métalliques, nca			269
..	..	..	..	3 274	3 415	3 533	3 787	Produits métallurgiques de base		27	
..	..	..	..	1 983	2 121	2 215	2 373	Sidérurgie et première transformation de l'acier			271
..	..	..	..	714	793	807	855	Première transf. métaux précieux et non ferreux			272
..	..	..	..	506	502	512	559	Fonderie			273
..	..	..	..	4 738	5 184	5 524	6 150	Ouvrages en métaux, sauf machines		28	
..	..	..	..	2 201	2 381	2 458	2 762	Constr. et menuis. métal.; réservoirs; générateurs			281
..	..	..	..	2 538	2 803	3 066	3 389	Autres ouvrages en métaux			289
..	..	..	..	9 595	10 702	11 315	12 194	Machines et matériels, nca		29	
..	..	..	..	4 210	4 619	4 978	5 203	Machines d'usage général			291
..	..	..	..	5 180	5 869	6 106	6 737	Machines d'usage spécifique			292
..	..	..	..	205	215	232	255	Appareils domestiques, nca			293
..	..	..	..	14 264	14 909	15 899	480	Mach. de bureau, comptables, information		30	
..	..	..	..	2 736	2 901	3 047	3 146	Machines et appareils électriques, nca		31	
..	..	..	..	1 060	1 285	1 231	1 261	Moteurs, génératrices, transformateurs électriques			311
..	..	..	..	681	642	777	855	Matériel électrique de distribution et de commande			312
..	..	..	..	326	325	379	389	Fils et câbles électriques isolés			313
..	..	..	..	..	..	..	3	Accumulateurs et piles électriques			314
..	..	..	..	216	218	231	253	Lampes électriques et appareils d'éclairage			315
..	..	..	..	353	365	375	385	Autres matériels électriques, nca.			319
..	..	..	..	3 936	4 407	4 855	6 472	Equipements et appareils de radio, TV, etc.		32	
..	..	..	..	487	524	584	630	Tubes et valves électroniques;autres composants			321
..	..	..	..	3 214	3 657	4 033	5 616	Emetteurs radio/TV; app. téléphonie et télégraphie			322
..	..	..	..	234	226	239	226	Récepteurs radio/TV, son et image; art. associés			323
..	..	..	..	1 527	1 616	1 813	2 059	Instr. médicaux, de précision ; optique ; horlog.		33	
..	..	..	..	1 476	1 564	1 754	2 012	Instruments médicaux et de précision, etc.			331
..	..	..	..	49	50	57	47	Instruments d'optique ; matériel photographique			332
..	..	..	..	2	1	2	0	Horlogerie			333
..	..	..	..	1 059	1 095	1 224	1 344	Véhicules automobiles; remorques, etc.		34	
..	..	..	..	436	413	502	525	Véhicules automobiles			341
..	..	..	..	462	514	552	674	Carross. automobiles, remorques, semi-remorques			342
..	..	..	..	161	167	171	144	Pièces et accessoires pour automobiles et moteurs			343
..	..	..	..	3 008	2 839	2 815	3 115	Autres matériels de transport		35	
..	..	..	..	1 907	1 777	1 664	1 940	Construction et réparation de navires			351
..	..	..	..	455	367	414	410	Matériel ferroviaire roulant			352
..	..	..	..	573	..	..	..	Construction aéronautique et spatiale			353
..	..	..	..	73	84	69	62	Autres équipements de transport			359
..	..	..	..	1 953	2 004	2 086	2 328	Meubles ; activités de fabrication, nca		36	
..	..	..	..	1 358	1 412	1 499	1 716	Meubles			361
..	..	..	..	595	592	587	612	Activités de fabrication, nca			369
..	..	..	..	45	48	59	58	Récupération		37	
..	..	..	..	3 691	3 969	3 925	361	Electricité, gaz et eau	E	40_41	
..	..	..	..	3 354	3 614	3 573	11	Prod. et dist. d'électricité gaz, vapeur d'eau chaude		40	
..	..	..	..	322	..	..	350	Captage, épuration et distribution d'eau		41	
..	..	..	..	11 655	12 469	13 525	16 886	Construction	F	45	

Note:Data are provisional.
ISIC 313 includes 314.

FINLAND

Table FN.7 WAGES AND SALARIES, EMPLOYEES - SALAIRES ET TRAITEMENTS, SALARIÉS

Millions of Mk (Current Prices)

ISIC revision 3				1991	1992	1993	1994	1995	1996	1997	1998
C	10_14		Mining and quarrying	..	..	..	..	560	552	604	621
	10		Mining of coal and lignite; extraction of peat	..	..	..	..	136	136	152	145
	11		Crude petroleum & natural gas; service activities	..	..	..	..	..	..	..	..
	13		Mining of metal ores	..	..	..	..	105	104	106	107
	14		Other mining and quarrying	..	..	..	..	320	311	346	369
D	15_37		Total manufacturing	..	..	..	..	63 907	67 394	70 839	64 291
	15		Food products and beverages	..	..	..	..	5 422	5 567	5 525	5 787
		151	Production of meat, fish, fruit, vegetables, oils & fats	..	..	..	..	1 769	1 922	1 880	2 011
		152	Dairy products	..	..	..	..	857	767	719	822
		153	Grain mill and starch products; animal feeds	..	..	..	..	328	352	365	376
		154	Other food products	..	..	..	..	1 864	1 951	1 986	1 979
		155	Beverages	..	..	..	..	604	574	575	599
	16		Tobacco products	..	..	..	..	135	120	118	123
	17		Textiles	..	..	..	..	695	753	772	814
		171	Spinning, weaving and finishing of textiles	..	..	..	..	126	129	126	129
		172	Other textiles	..	..	..	..	392	439	463	499
		173	Knitted and crocheted fabrics and articles	..	..	..	..	177	184	183	186
	18		Wearing apparel and fur	..	..	..	..	758	719	733	761
		181	Wearing apparel, except fur apparel	..	..	..	..	729	683	697	723
		182	Dressing and dyeing of fur; articles of fur	..	..	..	..	29	36	36	38
	19		Leather and articles of leather; footwear	..	..	..	..	305	298	279	302
		191	Leather and articles of leather	..	..	..	..	83	79	65	65
		192	Footwear	..	..	..	..	222	219	215	237
	20		Wood & products of wood & cork, exc. furniture	..	..	..	..	3 131	3 093	3 315	3 614
		201	Sawmilling and planing of wood	..	..	..	..	1 281	1 195	1 249	1 388
		202	Wood products; articles of cork, straw, etc.	..	..	..	..	1 850	1 898	2 065	2 226
	21		Paper and paper products	..	..	..	..	6 730	6 958	7 213	7 654
	22		Printing, publishing and recorded media	..	..	..	..	4 217	4 358	4 405	4 813
		221	Publishing	..	..	..	..	2 224	2 259	2 288	2 556
		222	Printing and related activities	..	..	..	..	1 978	2 085	2 102	2 231
		223	Reproduction of recorded media	..	..	..	..	15	15	15	26
	23		Coke and petroleum products; nuclear fuel	..	..	..	..	602	633	664	663
		231	Coke oven products	..	..	..	..	..	..	..	..
		232	Refined petroleum products	..	..	..	..	568	598	629	629
		233	Processing of nuclear fuel	..	..	..	..	..	..	..	..
	24		Chemicals and chemical products	..	..	..	..	2 825	2 866	2 977	3 145
		241	Basic chemicals	..	..	..	..	1 460	1 482	1 533	1 608
		242	Other chemical products	..	..	..	..	1 255	1 299	1 356	1 449
		243	Man-made fibres	..	..	..	..	..	..	..	..
	25		Rubber and plastic products	..	..	..	..	1 764	1 862	2 074	2 369
		251	Rubber products	..	..	..	..	345	333	367	429
		252	Plastic products	..	..	..	..	1 419	1 529	1 707	1 939
	26		Other non-metallic mineral products	..	..	..	..	1 652	1 725	1 809	1 995
		261	Glass and glass products	..	..	..	..	457	496	509	566
		269	Non-metallic mineral products, nec	..	..	..	..	1 174	1 213	1 283	1 429
	27		Basic metals	..	..	..	..	2 530	2 689	2 769	2 961
		271	Basic iron and steel	..	..	..	..	1 546	1 674	1 750	1 850
		272	Basic precious and non-ferrous metals	..	..	..	..	544	622	626	669
		273	Casting of metals	..	..	..	..	388	392	393	441
	28		Fabricated metal products, exc. machinery	..	..	..	..	3 650	4 031	4 333	4 789
		281	Structural metal prod., tanks, reservoirs, etc.	..	..	..	..	1 696	1 851	1 930	2 143
		289	Other fabricated metal products	..	..	..	..	1 953	2 180	2 403	2 646
	29		Machinery and equipment, nec	..	..	..	..	7 325	8 293	8 831	9 580
		291	General purpose machinery	..	..	..	..	3 231	3 609	3 897	4 114
		292	Special purpose machinery	..	..	..	..	3 938	4 514	4 750	5 264
		293	Domestic appliances, nec	..	..	..	..	156	170	183	202
	30		Office, accounting & computing machinery	..	..	..	..	11 083	11 714	12 511	391
	31		Electrical machinery & apparatus, nec	..	..	..	..	2 128	2 255	2 413	2 509
		311	Electric motors, generators and transformers	..	..	..	..	804	978	965	1 007
		312	Electricity distribution & control apparatus	..	..	..	..	540	513	626	694
		313	Insulated wire and cable	..	..	..	..	249	251	296	304
		314	Accumulators, primary cells & primary batteries	..	..	..	..	..	..	..	3
		315	Electric lamps and lighting equipment	..	..	..	..	168	170	182	200
		319	Other electrical equipment, nec	..	..	..	..	284	287	297	303
	32		Radio, TV, communication equipment	..	..	..	..	3 069	3 466	3 817	5 019
		321	Electronic valves, tubes; other electronic components	..	..	..	..	380	415	457	500
		322	TV/radio transmitters; line teleph./telegr. apparatus	..	..	..	..	2 511	2 872	3 170	4 340
		323	TV/radio receivers; sound and video app.; assoc. gds.	..	..	..	..	178	179	190	179
	33		Medical, precision & optical instrum.; watch & clocks	..	..	..	..	1 178	1 274	1 430	1 625
		331	Medical appl.; instruments & appl. for measuring, etc.	..	..	..	..	1 138	1 234	1 383	1 589
		332	Optical instruments and photographic equipment	..	..	..	..	38	39	45	37
		333	Watches and clocks	..	..	..	..	1	1	1	0
	34		Motor vehicles, trailers and semi-trailers	..	..	..	..	820	855	956	1 052
		341	Motor vehicles	..	..	..	..	335	319	390	416
		342	Bodies for motor vehicles, trailers & semi-trailers	..	..	..	..	360	405	431	522
		343	Parts & accessories for motor vehicles & their engines	..	..	..	..	125	131	135	114
	35		Other transport equipment	..	..	..	..	2 329	2 251	2 210	2 448
		351	Building and repairing of ships and boats	..	..	..	..	1 475	1 414	1 312	1 533
		352	Railway and tramway locomotives and rolling stock	..	..	..	..	352	295	322	319
		353	Aircraft and spacecraft	..	..	..	..	447	..	..	..
		359	Transport equipment, nec	..	..	..	..	55	67	55	49
	36		Furniture; manufacturing, nec	..	..	..	..	1 525	1 574	1 642	1 828
		361	Furniture	..	..	..	..	1 061	1 111	1 182	1 346
		369	Manufacturing, nec	..	..	..	..	464	463	460	482
	37		Recycling	..	..	..	..	35	38	46	45
E	40_41		Electricity, gas and water supply	..	..	..	..	2 783	3 022	3 029	277
	40		Electricity, gas, steam and hot water supply	..	..	..	..	2 530	2 754	2 764	8
	41		Collection, purification and distribution of water	..	..	..	..	245	..	..	268
F	45		Construction	..	..	..	..	8 922	9 560	10 421	13 075

Note:Data are provisional.

ISIC 313 includes 314.

FINLANDE

Table FN.8 **CHARGES SOCIALES DES EMPLOYEURS - SUPPLEMENTS TO WAGES & SALARIES**

Millions de Mk (Prix courants)

1991	1992	1993	1994	1995	1996	1997	1998			CITI révision 3	
..	..	..	..	176	171	185	188	Activites extractives	C	10_14	
..	..	..	..	42	39	44	41	Extraction de charbon, de lignite et de tourbe		10	
..	..	..	..	..	..	..	..	Pétrole brut et gaz naturel ; activités annexes		11	
..	..	..	..	36	29	30	29	Extraction de minerais métalliques		13	
..	..	..	..	98	103	111	119	Autres activités extractives		14	
..	..	..	..	15 922	15 861	16 386	17 888	Total activites manufacturieres	D	15_37	
..	..	..	..	1 656	1 634	1 609	1 652	Produits alimentaires et boissons		15	
..	..	..	..	546	543	556	585	Viande, poisson, fruits, légumes, huiles et graisses			151
..	..	..	..	275	229	215	233	Produits laitiers			152
..	..	..	..	99	105	108	113	Travail des grains; prod. amyl. ; alim. pour animaux			153
..	..	..	..	557	565	551	536	Autres produits alimentaires			154
..	..	..	..	180	191	178	185	Boissons			155
..	..	..	..	43	31	35	34	Produits à base de tabac		16	
..	..	..	..	207	205	203	228	Textiles		17	
..	..	..	..	36	34	34	35	Filature, tissage et achèvement des textiles			171
..	..	..	..	119	122	120	141	Autres articles textiles			172
..	..	..	..	51	49	48	52	Etoffes et articles de bonneterie			173
..	..	..	..	200	188	190	194	Articles d'habillement et fourrure		18	
..	..	..	..	192	178	181	184	Articles d'habillement autres qu'en fourrure			181
..	..	..	..	8	10	9	10	Prép. et teinture des fourrures ; articles en fourrure			182
..	..	..	..	89	85	75	84	Cuir et articles en cuir ; chaussures		19	
..	..	..	..	24	22	18	18	Cuir et articles en cuir sauf chaussures			191
..	..	..	..	65	63	57	65	Chaussures			192
..	..	..	..	959	881	933	1 033	Bois, articles en bois, sauf meubles		20	
..	..	..	..	403	359	359	406	Sciage et rabotage du bois			201
..	..	..	..	556	521	574	627	Articles en bois, liège, vannerie, sparterie			202
..	..	..	..	2 074	2 018	2 115	2 184	Papier, articles en papier et carton		21	
..	..	..	..	1 251	1 195	1 215	1 307	Edition, imprimerie et reproduction		22	
..	..	..	..	667	614	633	693	Edition			221
..	..	..	..	579	576	577	607	Imprimerie et activités annexes			222
..	..	..	..	5	4	5	7	Reproduction de supports enregistrés			223
..	..	..	..	179	171	175	162	Coke. et prod. pétroliers ; comb. nucléaires		23	
..	..	..	..	..	..	..	..	Cokéfaction			231
..	..	..	..	170	162	166	153	Produits pétroliers raffinés			232
..	..	..	..	..	..	..	..	Traitement de combustibles nucléaires			233
..	..	..	..	851	803	816	895	Produits chimiques		24	
..	..	..	..	439	422	424	499	Produits chimiques de base			241
..	..	..	..	369	354	363	369	Autres produits chimiques			242
..	..	..	..	..	..	..	..	Fibres synthétiques ou artificielles			243
..	..	..	..	516	518	571	645	Art. en caoutchouc et matières plastiques		25	
..	..	..	..	99	102	102	123	Articles en caoutchouc			251
..	..	..	..	417	416	469	523	Articles en matières plastiques			252
..	..	..	..	498	516	526	588	Autres produits minéraux non métalliques		26	
..	..	..	..	128	139	150	162	Verre et articles en verre			261
..	..	..	..	364	372	371	426	Produits minéraux non métalliques, nca			269
..	..	..	..	743	726	764	826	Produits métallurgiques de base		27	
..	..	..	..	437	447	465	523	Sidérurgie et première transformation de l'acier			271
..	..	..	..	170	170	181	186	Première transf. métaux précieux et non ferreux			272
..	..	..	..	119	109	119	118	Fonderie			273
..	..	..	..	1 089	1 153	1 191	1 361	Ouvrages en métaux, sauf machines		28	
..	..	..	..	504	530	528	618	Constr. et menuis. métal.; réservoirs; générateurs			281
..	..	..	..	584	623	663	743	Autres ouvrages en métaux			289
..	..	..	..	2 270	2 409	2 485	2 614	Machines et matériels, nca		29	
..	..	..	..	980	1 011	1 081	1 089	Machines d'usage général			291
..	..	..	..	1 241	1 354	1 355	1 473	Machines d'usage spécifique			292
..	..	..	..	49	44	49	52	Appareils domestiques, nca			293
..	..	..	..	3 182	3 196	3 388	89	Mach. de bureau, comptables, information		30	
..	..	..	..	609	646	634	636	Machines et appareils électriques, nca		31	
..	..	..	..	256	308	265	254	Moteurs, génératrices, transformateurs électriques			311
..	..	..	..	141	129	151	161	Matériel électrique de distribution et de commande			312
..	..	..	..	76	73	83	85	Fils et câbles électriques isolés			313
..	..	..	..	..	..	..	1	Accumulateurs et piles électriques			314
..	..	..	..	48	48	49	53	Lampes électriques et appareils d'éclairage			315
..	..	..	..	69	78	78	83	Autres matériels électriques, nca.			319
..	..	..	..	867	940	1 039	1 453	Equipements et appareils de radio, TV, etc.		32	
..	..	..	..	107	109	127	130	Tubes et valves électroniques;autres composants			321
..	..	..	..	703	784	863	1 276	Emetteurs radio/TV; app. téléphonie et télégraphie			322
..	..	..	..	57	47	49	47	Récepteurs radio/TV, son et image; art. associés			323
..	..	..	..	349	341	383	433	Instr. médicaux, de précision ; optique ; horlog.		33	
..	..	..	..	338	330	371	423	Instruments médicaux et de précision, etc.			331
..	..	..	..	11	11	12	10	Instruments d'optique ; matériel photographique			332
..	..	..	..	0	0	0	0	Horlogerie			333
..	..	..	..	240	239	268	291	Véhicules automobiles; remorques, etc.		34	
..	..	..	..	101	94	112	109	Véhicules automobiles			341
..	..	..	..	103	109	121	152	Carross. automobiles, remorques, semi-remorques			342
..	..	..	..	36	36	36	30	Pièces et accessoires pour automobiles et moteurs			343
..	..	..	..	679	588	606	667	Autres matériels de transport		35	
..	..	..	..	432	363	353	406	Construction et réparation de navires			351
..	..	..	..	103	73	92	91	Matériel ferroviaire roulant			352
..	..	..	..	127	..	..	..	Construction aéronautique et spatiale			353
..	..	..	..	18	17	14	13	Autres équipements de transport			359
..	..	..	..	428	430	444	499	Meubles ; activités de fabrication, nca		36	
..	..	..	..	297	301	317	370	Meubles			361
..	..	..	..	130	130	128	129	Activités de fabrication, nca			369
..	..	..	..	11	10	13	13	Récupération		37	
..	..	..	..	908	948	897	84	Electricité, gaz et eau	E	40_41	
..	..	..	..	824	860	809	2	Prod. et dist. d'électricité gaz, vapeur d'eau chaude		40	
..	..	..	..	78	..	..	82	Captage, épuration et distribution d'eau		41	
..	..	..	..	2 732	2 909	3 104	3 812	Construction	F	45	

Note:Data are provisional.

ISIC 313 includes 314.

FINLAND

Table FN.9 HOURS WORKED BY OPERATIVES - HEURES OUVRÉES, OUVRIERS

Thousands

ISIC revision 3				1991	1992	1993	1994	1995	1996	1997	1998
C	10_14		Mining and quarrying	..	..	..	..	4 371	4 363	4 619	4 411
	10		Mining of coal and lignite; extraction of peat	..	..	..	..	787	914	955	871
	11		Crude petroleum & natural gas; service activities	..	..	..	..	..	..	..	..
	13		Mining of metal ores	..	..	..	..	729	692	714	678
	14		Other mining and quarrying	..	..	..	..	2 855	2 757	2 950	2 862
D	15_37		Total manufacturing	..	..	..	..	490 483	486 132	502 811	442 412
	15		Food products and beverages	..	..	..	..	47 854	47 681	47 149	47 455
		151	Production of meat, fish, fruit, vegetables, oils & fats	..	..	..	..	16 930	17 675	17 873	18 306
		152	Dairy products	..	..	..	..	6 834	6 256	5 896	6 081
		153	Grain mill and starch products; animal feeds	..	..	..	..	1 993	1 890	1 916	1 879
		154	Other food products	..	..	..	..	18 398	18 316	18 114	17 697
		155	Beverages	..	..	..	..	3 699	3 544	3 350	3 493
	16		Tobacco products	..	..	..	..	763	737	699	648
	17		Textiles	..	..	..	..	7 926	8 206	8 024	8 148
		171	Spinning, weaving and finishing of textiles	..	..	..	..	1 562	1 487	1 435	1 408
		172	Other textiles	..	..	..	..	4 098	4 387	4 359	4 493
		173	Knitted and crocheted fabrics and articles	..	..	..	..	2 266	2 332	2 230	2 246
	18		Wearing apparel and fur	..	..	..	..	9 710	8 507	8 575	8 900
		181	Wearing apparel, except fur apparel	..	..	..	..	9 288	8 092	8 192	8 457
		182	Dressing and dyeing of fur; articles of fur	..	..	..	..	422	415	383	443
	19		Leather and articles of leather; footwear	..	..	..	..	4 466	4 007	3 847	4 367
		191	Leather and articles of leather	..	..	..	..	1 250	1 058	985	949
		192	Footwear	..	..	..	..	3 216	2 949	2 862	3 418
	20		Wood & products of wood & cork, exc. furniture	..	..	..	..	34 888	33 062	35 250	36 886
		201	Sawmilling and planing of wood	..	..	..	..	14 173	12 880	13 153	14 201
		202	Wood products; articles of cork, straw, etc.	..	..	..	..	20 715	20 182	22 097	22 685
	21		Paper and paper products	..	..	..	..	45 715	45 106	45 569	46 463
	22		Printing, publishing and recorded media	..	..	..	..	25 537	23 699	23 359	24 309
		221	Publishing	..	..	..	..	7 410	6 371	5 297	6 173
		222	Printing and related activities	..	..	..	..	18 047	17 265	17 960	18 033
		223	Reproduction of recorded media	..	..	..	..	80	63	102	103
	23		Coke and petroleum products; nuclear fuel	..	..	..	..	2 266	2 215	2 231	2 023
		231	Coke oven products	..	..	..	..	..	..	..	..
		232	Refined petroleum products	..	..	..	..	1 991	1 948	1 970	1 781
		233	Processing of nuclear fuel	..	..	..	..	..	..	..	..
	24		Chemicals and chemical products	..	..	..	..	14 842	14 462	13 973	14 212
		241	Basic chemicals	..	..	..	..	7 862	7 615	7 105	7 292
		242	Other chemical products	..	..	..	..	6 181	6 227	6 266	6 267
		243	Man-made fibres	..	..	..	..	..	..	..	..
	25		Rubber and plastic products	..	..	..	..	16 495	16 470	17 906	20 107
		251	Rubber products	..	..	..	..	3 152	2 792	2 880	3 105
		252	Plastic products	..	..	..	..	13 343	13 678	15 026	17 002
	26		Other non-metallic mineral products	..	..	..	..	15 132	14 794	15 749	17 431
		261	Glass and glass products	..	..	..	..	4 027	4 240	4 425	4 802
		269	Non-metallic mineral products, nec	..	..	..	..	10 904	10 415	11 227	12 629
	27		Basic metals	..	..	..	..	18 885	19 165	19 058	19 735
		271	Basic iron and steel	..	..	..	..	10 819	11 144	11 195	11 509
		272	Basic precious and non-ferrous metals	..	..	..	..	3 713	4 159	4 039	4 195
		273	Casting of metals	..	..	..	..	3 916	3 862	3 824	4 031
	28		Fabricated metal products, exc. machinery	..	..	..	..	35 286	37 115	39 774	43 309
		281	Structural metal prod., tanks, reservoirs, etc.	..	..	..	..	14 433	15 230	15 395	16 761
		289	Other fabricated metal products	..	..	..	..	20 853	21 885	24 379	26 548
	29		Machinery and equipment, nec	..	..	..	..	50 022	52 654	54 562	57 823
		291	General purpose machinery	..	..	..	..	22 040	23 029	24 085	24 941
		292	Special purpose machinery	..	..	..	..	26 252	27 943	28 594	30 948
		293	Domestic appliances, nec	..	..	..	..	1 730	1 682	1 883	1 934
	30		Office, accounting & computing machinery	..	..	..	..	80 348	79 126	83 543	2 155
	31		Electrical machinery & apparatus, nec	..	..	..	..	16 775	16 378	17 337	16 873
		311	Electric motors, generators and transformers	..	..	..	..	5 692	6 106	6 268	6 104
		312	Electricity distribution & control apparatus	..	..	..	..	4 171	3 795	4 396	4 713
		313	Insulated wire and cable	..	..	..	..	1 909	1 895	2 116	2 037
		314	Accumulators, primary cells & primary batteries	..	..	..	..	..	..	..	17
		315	Electric lamps and lighting equipment	..	..	..	..	1 770	1 698	1 723	1 764
		319	Other electrical equipment, nec	..	..	..	..	2 239	2 308	2 269	2 239
	32		Radio, TV, communication equipment	..	..	..	..	16 063	16 558	17 815	19 541
		321	Electronic valves, tubes; other electronic components	..	..	..	..	4 298	4 414	4 615	4 988
		322	TV/radio transmitters; line teleph./telegr. apparatus	..	..	..	..	10 337	10 858	11 697	13 364
		323	TV/radio receivers; sound and video app.; assoc. gds.	..	..	..	..	1 428	1 286	1 503	1 189
	33		Medical, precision & optical instrum.; watch & clocks	..	..	..	..	5 234	5 234	5 529	6 504
		331	Medical appl.; instruments & appl. for measuring, etc.	..	..	..	..	4 918	4 930	5 268	6 229
		332	Optical instruments and photographic equipment	..	..	..	..	312	298	256	275
		333	Watches and clocks	..	..	..	..	4	6	5	1
	34		Motor vehicles, trailers and semi-trailers	..	..	..	..	7 617	8 288	9 380	9 669
		341	Motor vehicles	..	..	..	..	2 465	3 016	3 417	3 556
		342	Bodies for motor vehicles, trailers & semi-trailers	..	..	..	..	3 908	4 073	4 617	5 026
		343	Parts & accessories for motor vehicles & their engines	..	..	..	..	1 244	1 199	1 346	1 088
	35		Other transport equipment	..	..	..	..	18 413	16 902	16 067	17 504
		351	Building and repairing of ships and boats	..	..	..	..	12 728	11 321	10 429	11 852
		352	Railway and tramway locomotives and rolling stock	..	..	..	..	2 822	2 374	2 382	2 386
		353	Aircraft and spacecraft	..	..	..	..	2 335	..	..	..
		359	Transport equipment, nec	..	..	..	..	528	528	466	424
	36		Furniture; manufacturing, nec	..	..	..	..	15 969	15 628	17 132	18 064
		361	Furniture	..	..	..	..	11 202	11 243	12 587	13 569
		369	Manufacturing, nec	..	..	..	..	4 767	4 385	4 545	4 495
	37		Recycling	..	..	..	..	277	138	283	284
E	40_41		Electricity, gas and water supply	..	..	..	..	15 505	15 335	14 441	2 205
	40		Electricity, gas, steam and hot water supply	..	..	..	..	13 141	13 024	12 235	38
	41		Collection, purification and distribution of water	..	..	..	..	..	..	..	2 166
F	45		Construction	..	..	..	..	82 435	84 383	95 612	122 834

Note:Data are provisional.

ISIC 313 includes 314.

FINLANDE

Table FN.10 **ÉTABLISSEMENTS - ESTABLISHMENTS**

Unités

1991	1992	1993	1994	1995	1996	1997	1998			CITI révision 3	
..	..	..	..	1 262	1 372	1 319	1 289	Activites extractives	C	**10_14**	
..	..	..	..	775	851	822	750	Extraction de charbon, de lignite et de tourbe		**10**	
..	..	..	..	..	..	..	..	Pétrole brut et gaz naturel ; activités annexes		**11**	
..	..	..	..	10	8	6	9	Extraction de minerais métalliques		**13**	
..	..	..	..	477	513	491	530	Autres activités extractives		**14**	
..	..	..	..	30 017	31 210	30 451	27 990	Total activites manufacturieres	D	**15_37**	
..	..	..	..	2 027	2 146	2 034	2 226	Produits alimentaires et boissons		**15**	
..	..	..	..	537	587	556	652	Viande, poisson, fruits, légumes, huiles et graisses			151
..	..	..	..	111	106	97	94	Produits laitiers			152
..	..	..	..	188	196	185	194	Travail des grains; prod. amyl. ; alim. pour animaux			153
..	..	..	..	1 119	1 181	1 117	1 197	Autres produits alimentaires			154
..	..	..	..	72	76	79	89	Boissons			155
..	..	..	..	5	4	4	5	Produits à base de tabac		**16**	
..	..	..	..	1 050	1 061	1 021	1 131	Textiles		**17**	
..	..	..	..	186	195	183	201	Filature, tissage et achèvement des textiles			171
..	..	..	..	673	683	650	720	Autres articles textiles			172
..	..	..	..	191	183	188	210	Etoffes et articles de bonneterie			173
..	..	..	..	1 339	1 352	1 325	1 480	Articles d'habillement et fourrure		**18**	
..	..	..	..	1 194	1 223	1 203	1 352	Articles d'habillement autres qu'en fourrure			181
..	..	..	..	145	129	122	128	Prép. et teinture des fourrures ; articles en fourrure			182
..	..	..	..	390	387	375	397	Cuir et articles en cuir ; chaussures		**19**	
..	..	..	..	289	285	279	288	Cuir et articles en cuir sauf chaussures			191
..	..	..	..	101	102	96	109	Chaussures			192
..	..	..	..	2 966	3 058	2 979	3 174	Bois, articles en bois, sauf meubles		**20**	
..	..	..	..	1 237	1 284	1 267	1 340	Sciage et rabotage du bois			201
..	..	..	..	1 729	1 774	1 712	1 834	Articles en bois, liège, vannerie, sparterie			202
..	..	..	..	330	306	299	303	Papier, articles en papier et carton		**21**	
..	..	..	..	2 653	2 790	2 724	2 980	Edition, imprimerie et reproduction		**22**	
..	..	..	..	1 155	1 219	1 196	1 390	Edition			221
..	..	..	..	1 438	1 520	1 473	1 523	Imprimerie et activités annexes			222
..	..	..	..	60	51	55	67	Reproduction de supports enregistrés			223
..	..	..	..	22	22	20	22	Coke. et prod. pétroliers ; comb. nucléaires		**23**	
..	..	..	..	2	1	1	2	Cokéfaction			231
..	..	..	..	20	21	19	20	Produits pétroliers raffinés			232
..	..	..	..	..	..	..	..	Traitement de combustibles nucléaires			233
..	..	..	..	361	380	378	398	Produits chimiques		**24**	
..	..	..	..	158	159	147	155	Produits chimiques de base			241
..	..	..	..	201	219	229	240	Autres produits chimiques			242
..	..	..	..	2	2	2	3	Fibres synthétiques ou artificielles			243
..	..	..	..	679	701	692	739	Art. en caoutchouc et matières plastiques		**25**	
..	..	..	..	77	69	66	78	Articles en caoutchouc			251
..	..	..	..	602	632	626	661	Articles en matières plastiques			252
..	..	..	..	999	1 031	985	1 102	Autres produits minéraux non métalliques		**26**	
..	..	..	..	131	135	126	149	Verre et articles en verre			261
..	..	..	..	868	896	859	953	Produits minéraux non métalliques, nca			269
..	..	..	..	162	177	179	200	Produits métallurgiques de base		**27**	
..	..	..	..	78	82	78	95	Sidérurgie et première transformation de l'acier			271
..	..	..	..	25	29	31	32	Première transf. métaux précieux et non ferreux			272
..	..	..	..	59	66	70	73	Fonderie			273
..	..	..	..	4 030	4 227	4 166	4 549	Ouvrages en métaux, sauf machines		**28**	
..	..	..	..	1 384	1 422	1 405	1 525	Constr. et menuis. métal.; réservoirs; générateurs			281
..	..	..	..	2 646	2 805	2 761	3 024	Autres ouvrages en métaux			289
..	..	..	..	3 406	3 566	3 538	3 734	Machines et matériels, nca		**29**	
..	..	..	..	1 642	1 706	1 704	1 771	Machines d'usage général			291
..	..	..	..	1 715	1 817	1 791	1 914	Machines d'usage spécifique			292
..	..	..	..	49	43	43	49	Appareils domestiques, nca			293
..	..	..	..	4 799	5 001	4 866	65	Mach. de bureau, comptables, information		**30**	
..	..	..	..	494	498	490	505	Machines et appareils électriques, nca		**31**	
..	..	..	..	102	104	98	91	Moteurs, génératrices, transformateurs électriques			311
..	..	..	..	107	103	102	110	Matériel électrique de distribution et de commande			312
..	..	..	..	35	37	36	38	Fils et câbles électriques isolés			313
..	..	..	..	5	5	5	6	Accumulateurs et piles électriques			314
..	..	..	..	91	94	95	98	Lampes électriques et appareils d'éclairage			315
..	..	..	..	154	155	154	162	Autres matériels électriques, nca.			319
..	..	..	..	306	323	317	373	Equipements et appareils de radio, TV, etc.		**32**	
..	..	..	..	173	185	183	223	Tubes et valves électroniques;autres composants			321
..	..	..	..	88	90	88	102	Emetteurs radio/TV; app. téléphonie et télégraphie			322
..	..	..	..	45	48	46	48	Récepteurs radio/TV, son et image; art. associés			323
..	..	..	..	747	751	669	852	Instr. médicaux, de précision ; optique ; horlog.		**33**	
..	..	..	..	704	710	624	808	Instruments médicaux et de précision, etc.			331
..	..	..	..	36	35	40	40	Instruments d'optique ; matériel photographique			332
..	..	..	..	7	6	5	4	Horlogerie			333
..	..	..	..	279	289	278	304	Véhicules automobiles; remorques, etc.		**34**	
..	..	..	..	32	34	35	32	Véhicules automobiles			341
..	..	..	..	176	182	177	201	Carross. automobiles, remorques, semi-remorques			342
..	..	..	..	71	73	66	71	Pièces et accessoires pour automobiles et moteurs			343
..	..	..	..	521	546	554	639	Autres matériels de transport		**35**	
..	..	..	..	464	491	499	575	Construction et réparation de navires			351
..	..	..	..	16	13	14	17	Matériel ferroviaire roulant			352
..	..	..	..	14	13	15	15	Construction aéronautique et spatiale			353
..	..	..	..	27	29	26	32	Autres équipements de transport			359
..	..	..	..	2 372	2 489	2 449	2 702	Meubles ; activités de fabrication, nca		**36**	
..	..	..	..	1 504	1 550	1 521	1 670	Meubles			361
..	..	..	..	868	939	928	1 032	Activités de fabrication, nca			369
..	..	..	..	80	105	109	110	Récupération		**37**	
..	..	..	..	952	1 017	1 032	376	Electricité, gaz et eau	E	**40_41**	
..	..	..	..	631	664	692	5	Prod. et dist. d'électricité gaz, vapeur d'eau chaude		**40**	
..	..	..	..	244	188	..	371	Captage, épuration et distribution d'eau		**41**	
..	..	..	..	23 914	25 056	24 916	28 544	Construction	F	**45**	

Note:Data are provisional.

FINLAND

Table FN.11 ENTERPRISES - ENTREPRISES

Units

ISIC revision 3				1991	1992	1993	1994	1995	1996	1997	1998
C	10_14		Mining and quarrying	..	..	..	..	1 178	1 293	1 241	1 206
	10		Mining of coal and lignite; extraction of peat	..	..	..	..	755	829	800	730
	11		Crude petroleum & natural gas; service activities	..	..	..	..	..	..	..	..
	12		Mining of uranium and thorium ores	..	..	..	..	..	..	..	..
	13		Mining of metal ores	..	..	..	..	5	5	4	7
	14		Other mining and quarrying	..	..	..	..	418	459	437	469
D	15_37		Total manufacturing	..	..	..	..	29 077	30 312	29 609	27 159
	15		Food products and beverages	..	..	..	..	1 859	1 990	1 892	2 089
		151	Production of meat, fish, fruit, vegetables, oils & fats	..	..	..	..	497	553	531	623
		152	Dairy products	..	..	..	..	54	56	55	58
		153	Grain mill and starch products; animal feeds	..	..	..	..	176	185	173	182
		154	Other food products	..	..	..	..	1 063	1 122	1 056	1 140
		155	Beverages	..	..	..	..	69	74	77	86
	16		Tobacco products	..	..	..	..	4	3	3	4
	17		Textiles	..	..	..	..	1 042	1 052	1 014	1 121
		171	Spinning, weaving and finishing of textiles	..	..	..	..	185	194	182	199
		172	Other textiles	..	..	..	..	670	679	647	717
		173	Knitted and crocheted fabrics and articles	..	..	..	..	187	179	185	205
	18		Wearing apparel and fur	..	..	..	..	1 329	1 342	1 316	1 467
		181	Wearing apparel, except fur apparel	..	..	..	..	1 184	1 213	1 194	1 340
		182	Dressing and dyeing of fur; articles of fur	..	..	..	..	145	129	122	127
	19		Leather and articles of leather; footwear	..	..	..	..	388	386	373	395
		191	Leather and articles of leather	..	..	..	..	288	285	278	287
		192	Footwear	..	..	..	..	100	101	95	108
	20		Wood & products of wood & cork, exc. furniture	..	..	..	..	2 879	2 974	2 906	3 094
		201	Sawmilling and planing of wood	..	..	..	..	1 200	1 242	1 230	1 304
		202	Wood products; articles of cork, straw, etc.	..	..	..	..	1 679	1 732	1 676	1 790
	21		Paper and paper products	..	..	..	..	229	220	226	229
	22		Printing, publishing and recorded media	..	..	..	..	2 608	2 737	2 665	2 916
		221	Publishing	..	..	..	..	1 127	1 189	1 166	1 356
		222	Printing and related activities	..	..	..	..	1 421	1 497	1 444	1 493
		223	Reproduction of recorded media	..	..	..	..	60	51	55	67
	23		Coke and petroleum products; nuclear fuel	..	..	..	..	15	16	14	14
		231	Coke oven products	..	..	..	..	2	1	1	2
		232	Refined petroleum products	..	..	..	..	13	15	13	12
		233	Processing of nuclear fuel	..	..	..	..	..	..	..	..
	24		Chemicals and chemical products	..	..	..	..	293	313	317	335
		241	Basic chemicals	..	..	..	..	107	109	101	110
		242	Other chemical products	..	..	..	..	184	202	214	222
		243	Man-made fibres	..	..	..	..	2	2	2	3
	25		Rubber and plastic products	..	..	..	..	651	673	667	708
		251	Rubber products	..	..	..	..	76	68	65	76
		252	Plastic products	..	..	..	..	575	605	602	632
	26		Other non-metallic mineral products	..	..	..	..	902	936	895	995
		261	Glass and glass products	..	..	..	..	128	130	119	142
		269	Non-metallic mineral products, nec	..	..	..	..	774	806	776	853
	27		Basic metals	..	..	..	..	137	149	148	166
		271	Basic iron and steel	..	..	..	..	60	65	61	76
		272	Basic precious and non-ferrous metals	..	..	..	..	21	21	19	19
		273	Casting of metals	..	..	..	..	56	63	68	71
	28		Fabricated metal products, exc. machinery	..	..	..	..	4 001	4 197	4 135	4 513
		281	Structural metal prod., tanks, reservoirs, etc.	..	..	..	..	1 369	1 409	1 393	1 511
	29		Machinery and equipment, nec	..	..	..	..	3 340	3 500	3 468	3 652
		291	General purpose machinery	..	..	..	..	1 620	1 685	1 684	1 746
		292	Special purpose machinery	..	..	..	..	1 672	1 773	1 742	1 858
		293	Domestic appliances, nec	..	..	..	..	48	42	42	48
	30		Office, accounting & computing machinery	..	..	..	..	4 700	4 912	4 785	64
	31		Electrical machinery & apparatus, nec	..	..	..	..	483	486	480	496
		311	Electric motors, generators and transformers	..	..	..	..	97	99	95	89
		312	Electricity distribution & control apparatus	..	..	..	..	106	101	100	107
		313	Insulated wire and cable	..	..	..	..	31	33	32	35
		314	Accumulators, primary cells & primary batteries	..	..	..	..	5	5	5	6
		315	Electric lamps and lighting equipment	..	..	..	..	91	94	95	98
		319	Other electrical equipment, nec	..	..	..	..	153	154	153	161
	32		Radio, TV, communication equipment	..	..	..	..	279	294	289	341
		321	Electronic valves, tubes; other electronic components	..	..	..	..	168	179	177	217
		322	TV/radio transmitters; line teleph./telegr. apparatus	..	..	..	..	67	68	67	78
		323	TV/radio receivers; sound and video app.; assoc. gds.	..	..	..	..	44	47	45	46
	33		Medical, precision & optical instrum.; watch & clocks	..	..	..	..	732	740	658	841
		331	Medical appl.; instruments & appl. for measuring, etc.	..	..	..	..	690	699	613	797
		332	Optical instruments and photographic equipment	..	..	..	..	35	35	40	40
		333	Watches and clocks	..	..	..	..	7	6	5	4
	34		Motor vehicles, trailers and semi-trailers	..	..	..	..	277	287	276	301
		341	Motor vehicles	..	..	..	..	32	34	35	32
		342	Bodies for motor vehicles, trailers & semi-trailers	..	..	..	..	174	180	175	198
		343	Parts & accessories for motor vehicles & their engines	..	..	..	..	71	73	66	71
	35		Other transport equipment	..	..	..	..	501	533	543	628
		351	Building and repairing of ships and boats	..	..	..	..	457	485	495	571
		352	Railway and tramway locomotives and rolling stock	..	..	..	..	4	7	8	11
		353	Aircraft and spacecraft	..	..	..	..	13	12	14	14
		359	Transport equipment, nec	..	..	..	..	27	29	26	32
	36		Furniture; manufacturing, nec	..	..	..	..	2 356	2 475	2 437	2 687
		361	Furniture	..	..	..	..	1 495	1 541	1 511	1 657
		369	Manufacturing, nec	..	..	..	..	861	934	926	1 030
	37		Recycling	..	..	..	..	72	97	102	103
E	40_41		Electricity, gas and water supply	..	..	..	..	862	908	896	373
	40		Electricity, gas, steam and hot water supply	..	..	..	..	542	556	557	3
	41		Collection, purification and distribution of water	..	..	..	..	..	..	..	370
F	45		Construction	..	..	..	..	23 839	24 956	24 808	28 412

Note:Data are provisional.

FINLANDE

Table FN.12 EXPORTATIONS - EXPORTS

Millions de Mk (Prix courants)

1991	1992	1993	1994	1995	1996	1997	1998		CITI révision 3		
..	..	..	..	559	543	600	626	Activites extractives	C	10_14	
..	..	..	..	117	63	66	61	Extraction de charbon, de lignite et de tourbe		10	
..	..	..	..	..	..	..	..	Pétrole brut et gaz naturel ; activités annexes		11	
..	..	..	..	..	..	..	..	Extraction de Minerais d'Uranium et de Thorium		12	
..	..	..	..	98	122	131	151	Extraction de minerais métalliques		13	
..	..	..	..	344	358	403	415	Autres activités extractives		14	
..	..	..	..	207 093	226 021	263 548	236 555	Total activites manufacturieres	D	15_37	
..	..	..	..	4 543	5 306	5 936	5 665	Produits alimentaires et boissons		15	
..	..	..	..	837	1 050	1 169	1 241	Viande, poisson, fruits, légumes, huiles et graisses			151
..	..	..	..	1 349	1 556	1 859	1 742	Produits laitiers			152
..	..	..	..	145	303	470	453	Travail des grains; prod. amyl. ; alim. pour animaux			153
..	..	..	..	1 424	1 592	1 566	1 452	Autres produits alimentaires			154
..	..	..	..	789	804	872	778	Boissons			155
..	..	..	..	145	33	38	49	Produits à base de tabac		16	
..	..	..	..	1 295	1 401	1 548	1 647	Textiles		17	
..	..	..	..	189	192	196	286	Filature, tissage et achèvement des textiles			171
..	..	..	..	997	1 084	1 210	1 217	Autres articles textiles			172
..	..	..	..	109	125	142	144	Etoffes et articles de bonneterie			173
..	..	..	..	1 042	1 095	1 005	1 081	Articles d'habillement et fourrure		18	
..	..	..	..	988	1 022	934	1 001	Articles d'habillement autres qu'en fourrure			181
..	..	..	..	54	73	70	81	Prép. et teinture des fourrures ; articles en fourrure			182
..	..	..	..	467	459	471	543	Cuir et articles en cuir ; chaussures		19	
..	..	..	..	107	95	85	95	Cuir et articles en cuir sauf chaussures			191
..	..	..	..	361	364	386	448	Chaussures			192
..	..	..	..	11 599	11 658	14 246	14 551	Bois, articles en bois, sauf meubles		20	
..	..	..	..	7 252	7 136	8 954	9 277	Sciage et rabotage du bois			201
..	..	..	..	4 347	4 522	5 292	5 274	Articles en bois, liège, vannerie, sparterie			202
..	..	..	..	51 010	45 292	51 079	56 063	Papier, articles en papier et carton		21	
..	..	..	..	1 590	1 699	1 925	2 012	Edition, imprimerie et reproduction		22	
..	..	..	..	101	137	124	152	Edition			221
..	..	..	..	1 488	1 559	1 799	1 857	Imprimerie et activités annexes			222
..	..	..	..	1	2	2	3	Reproduction de supports enregistrés			223
..	..	..	..	3 336	4 901	4 595	4 348	Coke. et prod. pétroliers ; comb. nucléaires		23	
..	..	..	..	..	..	..	..	Cokéfaction			231
..	..	..	..	3 309	4 869	4 563	4 304	Produits pétroliers raffinés			232
..	..	..	..	..	..	..	..	Traitement de combustibles nucléaires			233
..	..	..	..	9 834	10 158	12 594	12 207	Produits chimiques		24	
..	..	..	..	7 151	7 153	9 029	8 361	Produits chimiques de base			241
..	..	..	..	2 106	2 565	2 989	3 323	Autres produits chimiques			242
..	..	..	..	..	..	..	..	Fibres synthétiques ou artificielles			243
..	..	..	..	3 415	3 675	4 302	4 941	Art. en caoutchouc et matières plastiques		25	
..	..	..	..	783	772	839	978	Articles en caoutchouc			251
..	..	..	..	2 632	2 902	3 463	3 964	Articles en matières plastiques			252
..	..	..	..	2 055	2 324	2 697	2 729	Autres produits minéraux non métalliques		26	
..	..	..	..	1 032	1 114	1 359	1 325	Verre et articles en verre			261
..	..	..	..	987	1 183	1 318	1 404	Produits minéraux non métalliques, nca			269
..	..	..	..	12 889	12 664	14 165	13 511	Produits métallurgiques de base		27	
..	..	..	..	8 764	8 912	9 234	9 451	Sidérurgie et première transformation de l'acier			271
..	..	..	..	3 300	3 342	4 417	3 604	Première transf. métaux précieux et non ferreux			272
..	..	..	..	413	409	514	457	Fonderie			273
..	..	..	..	5 205	5 536	6 099	6 080	Ouvrages en métaux, sauf machines		28	
..	..	..	..	3 244	3 315	3 610	3 635	Constr. et menuis. métal.; réservoirs; générateurs			281
..	..	..	..	22 816	25 057	28 642	30 094	Machines et matériels, nca		29	
..	..	..	..	10 292	11 367	12 956	14 214	Machines d'usage général			291
..	..	..	..	12 186	13 281	15 279	15 429	Machines d'usage spécifique			292
..	..	..	..	338	409	407	450	Appareils domestiques, nca			293
..	..	..	..	37 925	47 382	57 103	4 701	Mach. de bureau, comptables, information		30	
..	..	..	..	6 374	7 139	7 809	8 412	Machines et appareils électriques, nca		31	
..	..	..	..	3 028	3 552	3 942	4 300	Moteurs, génératrices, transformateurs électriques			311
..	..	..	..	1 131	1 147	1 296	1 497	Matériel électrique de distribution et de commande			312
..	..	..	..	646	927	1 113	1 224	Fils et câbles électriques isolés			313
..	..	..	..	..	..	..	..	Accumulateurs et piles électriques			314
..	..	..	..	417	441	456	528	Lampes électriques et appareils d'éclairage			315
..	..	..	..	712	712	835	864	Autres matériels électriques, nca.			319
..	..	..	..	19 388	23 759	32 929	51 209	Equipements et appareils de radio, TV, etc.		32	
..	..	..	..	409	521	575	700	Tubes et valves électroniques;autres composants			321
..	..	..	..	17 934	22 407	31 378	49 696	Emetteurs radio/TV; app. téléphonie et télégraphie			322
..	..	..	..	1 045	831	977	814	Récepteurs radio/TV, son et image; art. associés			323
..	..	..	..	3 006	3 663	4 420	4 943	Instr. médicaux, de précision ; optique ; horlog.		33	
..	..	..	..	2 996	3 662	4 387	4 918	Instruments médicaux et de précision, etc.			331
..	..	..	..	9	1	33	25	Instruments d'optique ; matériel photographique			332
..	..	..	..	2		0	..	Horlogerie			333
..	..	..	..	1 679	1 894	2 030	2 277	Véhicules automobiles; remorques, etc.		34	
..	..	..	..	936	1 058	1 098	1 225	Véhicules automobiles			341
..	..	..	..	559	672	737	903	Carross. automobiles, remorques, semi-remorques			342
..	..	..	..	184	164	194	149	Pièces et accessoires pour automobiles et moteurs			343
..	..	..	..	5 356	8 924	7 803	7 300	Autres matériels de transport		35	
..	..	..	..	4 905	8 460	7 149	6 576	Construction et réparation de navires			351
..	..	..	..	127	150	72	156	Matériel ferroviaire roulant			352
..	..	..	..	284	..	..	..	Construction aéronautique et spatiale			353
..	..	..	..	40	30	68	22	Autres équipements de transport			359
..	..	..	..	1 921	1 907	1 965	2 133	Meubles ; activités de fabrication, nca		36	
..	..	..	..	1 095	1 076	1 117	1 231	Meubles			361
..	..	..	..	826	831	848	903	Activités de fabrication, nca			369
..	..	..	..	201	95	146	60	Récupération		37	
..	..	..	..	148	949	251	2	Electricité, gaz et eau	E	40_41	
..	..	..	..	148	949	251	0	Prod. et dist. d'électricité gaz, vapeur d'eau chaude		40	
..	..	..	..	0	..	..	2	Captage, épuration et distribution d'eau		41	
..	..	..	..	4 092	3 190	3 705	4 075	Construction	F	45	

Note:Data are provisional.
ISIC 313 includes 314.

FINLAND

Table FN.13 IMPORTS - IMPORTATIONS

Millions of Mk (Current Prices)

ISIC revision 3				1991	1992	1993	1994	1995	1996	1997	1998
C	10_14		Mining and quarrying	..	..	..	..	50	75	64	133
	10		Mining of coal and lignite; extraction of peat	..	..	..	..	1	21	8	1
	11		Crude petroleum & natural gas; service activities	..	..	..	..	..	..	..	..
	12		Mining of uranium and thorium ores	..	..	..	..	..	..	..	..
	13		Mining of metal ores	..	..	..	..	8	11	12	6
	14		Other mining and quarrying	..	..	..	..	41	43	43	126
D	15_37		Total manufacturing	..	..	..	..	72 894	78 265	88 292	59 924
	15		Food products and beverages	..	..	..	..	4 099	5 048	5 322	4 431
		151	Production of meat, fish, fruit, vegetables, oils & fats	..	..	..	..	1 095	1 735	1 449	1 299
		152	Dairy products	..	..	..	..	358	79	94	380
		153	Grain mill and starch products; animal feeds	..	..	..	..	609	487	615	536
		154	Other food products	..	..	..	..	1 818	2 174	2 556	1 417
		155	Beverages	..	..	..	..	219	573	607	798
	16		Tobacco products	..	..	..	..	90	163	179	197
	17		Textiles	..	..	..	..	935	758	975	831
		171	Spinning, weaving and finishing of textiles	..	..	..	..	140	110	161	174
		172	Other textiles	..	..	..	..	665	501	657	525
		173	Knitted and crocheted fabrics and articles	..	..	..	..	130	146	157	132
	18		Wearing apparel and fur	..	..	..	..	923	613	851	883
		181	Wearing apparel, except fur apparel	..	..	..	..	897	570	809	857
		182	Dressing and dyeing of fur; articles of fur	..	..	..	..	26	42	42	26
	19		Leather and articles of leather; footwear	..	..	..	..	232	261	271	162
		191	Leather and articles of leather	..	..	..	..	75	81	85	58
		192	Footwear	..	..	..	..	156	181	186	104
	20		Wood & products of wood & cork, exc. furniture	..	..	..	..	306	298	478	464
		201	Sawmilling and planing of wood	..	..	..	..	14	18	71	44
		202	Wood products; articles of cork, straw, etc.	..	..	..	..	292	280	407	419
	21		Paper and paper products	..	..	..	..	2 756	2 722	3 187	3 177
	22		Printing, publishing and recorded media	..	..	..	..	185	217	392	425
		221	Publishing	..	..	..	..	66	46	115	130
		222	Printing and related activities	..	..	..	..	117	168	265	295
		223	Reproduction of recorded media	..	..	..	..	2	3	13	0
	23		Coke and petroleum products; nuclear fuel	..	..	..	..	6 954	9 743	10 050	8 379
		231	Coke oven products	..	..	..	..	..	..	..	..
		232	Refined petroleum products	..	..	..	..	6 537	9 370	9 666	7 957
		233	Processing of nuclear fuel	..	..	..	..	..	..	..	..
	24		Chemicals and chemical products	..	..	..	..	4 344	4 049	5 355	5 934
		241	Basic chemicals	..	..	..	..	2 506	2 326	3 349	3 782
		242	Other chemical products	..	..	..	..	1 419	1 474	1 690	1 825
		243	Man-made fibres	..	..	..	..	..	..	..	..
	25		Rubber and plastic products	..	..	..	..	2 238	2 243	2 359	2 206
		251	Rubber products	..	..	..	..	609	590	356	310
		252	Plastic products	..	..	..	..	1 629	1 653	2 003	1 895
	26		Other non-metallic mineral products	..	..	..	..	689	734	1 093	1 149
		261	Glass and glass products	..	..	..	..	332	322	539	570
		269	Non-metallic mineral products, nec	..	..	..	..	330	378	514	579
	27		Basic metals	..	..	..	..	4 870	6 500	6 848	5 926
		271	Basic iron and steel	..	..	..	..	2 868	2 846	3 475	3 291
		272	Basic precious and non-ferrous metals	..	..	..	..	1 845	3 580	3 261	2 575
		273	Casting of metals	..	..	..	..	69	74	112	61
	28		Fabricated metal products, exc. machinery	..	..	..	..	1 034	1 151	1 245	1 196
		281	Structural metal prod., tanks, reservoirs, etc.	..	..	..	..	426	547	587	532
	29		Machinery and equipment, nec	..	..	..	..	4 926	5 238	6 458	7 350
		291	General purpose machinery	..	..	..	..	2 263	2 793	4 146	3 970
		292	Special purpose machinery	..	..	..	..	2 452	2 202	2 112	3 158
		293	Domestic appliances, nec	..	..	..	..	211	242	201	222
	30		Office, accounting & computing machinery	..	..	..	..	19 157	19 265	21 615	2 455
	31		Electrical machinery & apparatus, nec	..	..	..	..	3 592	3 561	3 744	3 108
		311	Electric motors, generators and transformers	..	..	..	..	1 462	1 322	1 398	1 607
		312	Electricity distribution & control apparatus	..	..	..	..	356	514	671	627
		313	Insulated wire and cable	..	..	..	..	565	697	902	233
		314	Accumulators, primary cells & primary batteries	..	..	..	..	..	..	..	0
		315	Electric lamps and lighting equipment	..	..	..	..	224	234	226	272
		319	Other electrical equipment, nec	..	..	..	..	225	281	392	369
	32		Radio, TV, communication equipment	..	..	..	..	11 838	11 852	13 523	7 478
		321	Electronic valves, tubes; other electronic components	..	..	..	..	241	224	331	234
		322	TV/radio transmitters; line teleph./telegr. apparatus	..	..	..	..	10 840	11 062	12 537	6 738
		323	TV/radio receivers; sound and video app.; assoc. gds.	..	..	..	..	757	566	654	506
	33		Medical, precision & optical instrum.; watch & clocks	..	..	..	..	502	598	980	856
		331	Medical appl.; instruments & appl. for measuring, etc.	..	..	..	..	449	558	908	813
		332	Optical instruments and photographic equipment	..	..	..	..	52	40	71	44
		333	Watches and clocks	..	..	..	..	1	0	1	..
	34		Motor vehicles, trailers and semi-trailers	..	..	..	..	584	754	832	722
		341	Motor vehicles	..	..	..	..	363	519	544	451
		342	Bodies for motor vehicles, trailers & semi-trailers	..	..	..	..	171	187	230	226
		343	Parts & accessories for motor vehicles & their engines	..	..	..	..	50	48	58	45
	35		Other transport equipment	..	..	..	..	1 846	1 740	1 797	1 936
		351	Building and repairing of ships and boats	..	..	..	..	1 331	1 066	794	1 355
		352	Railway and tramway locomotives and rolling stock	..	..	..	..	78	85	524	311
		353	Aircraft and spacecraft	..	..	..	..	395	..	..	..
		359	Transport equipment, nec	..	..	..	..	42	94	73	48
	36		Furniture; manufacturing, nec	..	..	..	..	724	723	723	633
		361	Furniture	..	..	..	..	341	355	398	285
		369	Manufacturing, nec	..	..	..	..	383	368	326	348
	37		Recycling	..	..	..	..	70	37	16	28
E	40_41		Electricity, gas and water supply	..	..	..	..	642	2 116	1 393	7
	40		Electricity, gas, steam and hot water supply	..	..	..	..	641	2 114	1 392	3
	41		Collection, purification and distribution of water	..	..	..	..	0	1	..	3
F	45		Construction	..	..	..	..	278	248	271	368

Note:Data are provisional.
ISIC 313 includes 314.

Table FR.1 **PRODUCTION - PRODUCTION**

Millions de FF (Prix courants)

1991	1992	1993	1994	1995	1996	1997	1998			CITI révision 3	
47 206	43 557	42 541	41 387	39 576	37 940	29 279	26 935	Activites extractives	C	10_14	
..	..	..	..	..	..	..	..	Extraction de charbon, de lignite et de tourbe		10	
..	..	..	..	..	..	..	..	Pétrole brut et gaz naturel ; activités annexes		11	
..	..	..	..	..	..	..	..	Extraction de Minerais d'Uranium et de Thorium		12	
709	612	620	416	431	411	342	274	Extraction de minerais métalliques		13	
23 164	21 007	19 907	20 449	21 124	19 569	20 097	19 981	Autres activités extractives		14	
2 784 593	2 754 794	3 285 968	2 815 444	2 994 536	3 048 550	3 619 644	3 439 898	Total activites manufacturieres	D	15_37	
..	..	61 652	..	61 786	54 874	58 425	54 266	Produits alimentaires et boissons		15	
..	..	..	..	..	..	..	..	Viande, poisson, fruits, légumes, huiles et graisses			151
..	..	..	..	..	..	..	..	Produits laitiers			152
..	..	..	..	..	..	..	..	Travail des grains; prod. amyl. ; alim. pour animaux			153
..	..	..	..	..	..	..	..	Autres produits alimentaires			154
..	..	..	..	..	..	..	..	Boissons			155
..	..	..	..	..	..	..	..	Produits à base de tabac		16	
90 633	91 823	82 264	87 901	90 238	87 260	90 050	94 471	Textiles		17	
50 272	48 667	42 625	46 472	47 082	45 330	47 937	50 593	Filature, tissage et achèvement des textiles			171
23 838	25 936	24 280	26 650	28 720	28 242	27 905	29 696	Autres articles textiles			172
16 523	17 221	15 359	14 779	14 436	13 689	14 208	14 182	Etoffes et articles de bonneterie			173
62 331	62 015	59 420	59 799	62 335	62 890	61 478	61 993	Articles d'habillement et fourrure		18	
61 624	61 304	58 968	59 349	61 805	62 323	60 941	61 539	Articles d'habillement autres qu'en fourrure			181
708	710	452	450	530	567	537	454	Prép. et teinture des fourrures ; articles en fourrure			182
26 191	24 633	24 312	23 054	23 001	21 172	21 230	22 127	Cuir et articles en cuir ; chaussures		19	
8 467	8 399	7 226	8 693	8 409	7 851	8 081	8 431	Cuir et articles en cuir sauf chaussures			191
17 724	16 234	17 086	14 361	14 592	13 321	13 150	13 696	Chaussures			192
31 909	31 534	29 514	31 707	33 124	31 919	33 682	36 319	Bois, articles en bois, sauf meubles		20	
299	334	411	439	526	349	353	304	Sciage et rabotage du bois			201
31 610	31 200	29 103	31 268	32 598	31 570	33 328	36 015	Articles en bois, liège, vannerie, sparterie			202
96 200	95 171	89 983	96 634	111 661	101 023	104 543	106 257	Papier, articles en papier et carton		21	
132 236	131 965	131 019	136 253	141 368	144 550	145 248	151 791	Edition, imprimerie et reproduction		22	
77 196	76 390	78 039	81 550	83 963	88 149	87 788	91 827	Edition			221
53 296	53 591	51 229	52 873	55 320	53 974	55 006	57 358	Imprimerie et activités annexes			222
1 744	1 984	1 751	1 830	2 085	2 428	2 454	2 606	Reproduction de supports enregistrés			223
223 052	201 773	220 117	222 724	234 245	263 675	274 000	248 126	Coke. et prod. pétroliers ; comb. nucléaires		23	
24 155	24 892	22 876	26 819	30 228	32 712	32 707	31 194	Cokéfaction			231
198 897	176 880	197 242	195 905	204 016	230 963	241 293	216 932	Produits pétroliers raffinés			232
..	..	..	..	..	..	..	..	Traitement de combustibles nucléaires			233
389 892	399 416	398 844	422 907	455 275	463 193	504 456	509 875	Produits chimiques		24	
134 181	130 233	122 724	135 161	156 316	160 900	170 124	162 159	Produits chimiques de base			241
252 599	266 165	273 381	284 864	295 666	299 179	330 650	343 802	Autres produits chimiques			242
3 111	3 017	2 739	2 882	3 293	3 115	3 682	3 914	Fibres synthétiques ou artificielles			243
122 666	124 550	119 144	128 359	138 386	138 676	152 160	169 031	Art. en caoutchouc et matières plastiques		25	
42 014	42 550	39 653	40 671	44 730	44 937	49 616	55 965	Articles en caoutchouc			251
80 651	82 000	79 491	87 688	93 655	93 739	102 544	113 066	Articles en matières plastiques			252
107 002	104 188	99 145	104 389	107 781	105 130	107 511	112 556	Autres produits minéraux non métalliques		26	
34 762	35 264	33 486	35 903	38 336	37 473	39 006	40 589	Verre et articles en verre			261
72 240	68 924	65 659	68 486	69 446	67 658	68 505	71 966	Produits minéraux non métalliques, nca			269
153 321	143 854	129 544	146 727	167 447	164 757	181 360	187 621	Produits métallurgiques de base		27	
93 896	87 495	79 891	90 604	103 845	103 993	112 860	117 079	Sidérurgie et première transformation de l'acier			271
39 976	37 826	33 893	38 483	43 882	41 399	48 016	47 655	Première transf. métaux précieux et non ferreux			272
19 448	18 533	15 760	17 639	19 720	19 365	20 484	22 887	Fonderie			273
193 574	188 255	173 844	183 802	196 485	198 391	210 235	220 214	Ouvrages en métaux, sauf machines		28	
79 521	76 598	72 058	71 940	71 395	69 193	72 012	71 488	Constr. et menuis. métal.; réservoirs; générateurs			281
201 173	204 050	189 630	197 633	211 888	218 710	224 629	241 246	Machines et matériels, nca		29	
107 822	112 159	105 608	108 874	112 673	117 005	119 962	131 630	Machines d'usage général			291
72 828	70 238	63 389	67 753	76 914	78 896	82 099	86 857	Machines d'usage spécifique			292
20 523	21 653	20 633	21 006	22 302	22 808	22 569	22 759	Appareils domestiques, nca			293
63 898	56 911	60 433	63 555	71 768	69 563	72 222	77 390	Mach. de bureau, comptables, information		30	
121 978	108 787	114 542	122 493	131 043	131 006	136 612	144 844	Machines et appareils électriques, nca		31	
18 050	17 025	16 120	16 004	18 136	17 818	19 294	21 409	Moteurs, génératrices, transformateurs électriques			311
53 339	38 789	47 009	49 431	52 162	51 534	54 335	58 849	Matériel électrique de distribution et de commande			312
21 768	22 421	21 853	22 697	23 306	20 012	19 868	17 683	Fils et câbles électriques isolés			313
6 535	6 447	6 648	6 976	7 295	7 329	7 105	7 323	Accumulateurs et piles électriques			314
7 743	7 859	7 226	7 870	8 314	7 868	7 946	8 914	Lampes électriques et appareils d'éclairage			315
14 543	16 246	15 685	19 514	21 829	26 445	28 063	30 665	Autres matériels électriques, nca.			319
131 520	126 171	123 174	131 030	147 476	166 588	181 362	205 173	Equipements et appareils de radio, TV, etc.		32	
65 838	62 263	56 928	59 565	64 838	73 108	75 332	75 674	Tubes et valves électroniques;autres composants			321
50 600	48 016	51 742	56 666	66 297	73 159	82 252	104 003	Emetteurs radio/TV; app. téléphonie et télégraphie			322
15 082	15 891	14 503	14 799	16 341	20 320	23 778	25 496	Récepteurs radio/TV, son et image; art. associés			323
53 930	53 759	51 628	53 256	56 403	60 541	61 781	65 046	Instr. médicaux, de précision ; optique ; horlog.		33	
42 976	43 218	41 666	43 008	45 387	49 967	51 199	53 719	Instruments médicaux et de précision, etc.			331
7 498	7 336	7 061	7 439	7 956	7 595	7 885	8 634	Instruments d'optique ; matériel photographique			332
3 456	3 204	2 902	2 809	3 059	2 980	2 698	2 693	Horlogerie			333
386 431	405 533	363 608	416 172	426 833	424 945	479 485	538 333	Véhicules automobiles; remorques, etc.		34	
303 856	320 773	287 348	329 401	331 726	333 828	381 906	429 345	Véhicules automobiles			341
16 574	14 852	12 293	12 971	15 474	15 526	15 669	18 715	Carross. automobiles, remorques, semi-remorques			342
66 002	69 908	63 967	73 800	79 632	75 591	81 910	90 272	Pièces et accessoires pour automobiles et moteurs			343
123 287	126 513	110 035	112 697	112 858	119 204	132 792	149 472	Autres matériels de transport		35	
11 904	10 897	8 925	12 390	11 613	10 313	10 333	13 459	Construction et réparation de navires			351
5 755	14 495	13 155	11 870	11 880	13 152	13 772	14 710	Matériel ferroviaire roulant			352
99 542	95 161	82 501	82 549	82 509	89 312	101 977	114 161	Construction aéronautique et spatiale			353
6 086	5 959	5 453	5 889	6 855	6 427	6 710	7 142	Autres équipements de transport			359
73 370	73 893	72 489	74 352	74 923	75 358	76 825	82 751	Meubles ; activités de fabrication, nca		36	
48 394	48 711	46 075	48 134	48 737	49 685	49 825	53 852	Meubles			361
24 977	25 182	26 414	26 218	26 186	25 674	26 999	28 899	Activités de fabrication, nca			369
..	..	..	..	..	..	..	15 263	Récupération		37	
288 260	298 428	311 990	312 719	328 440	340 212	338 703	350 452	Electricité, gaz et eau	E	40_41	
252 467	258 923	269 280	266 757	280 359	290 797	287 686	294 568	Prod. et dist. d'électricité gaz, vapeur d'eau chaude		40	
35 793	39 505	42 710	45 962	48 082	49 415	51 017	55 883	Captage, épuration et distribution d'eau		41	
..	..	..	..	..	..	..	..	Construction	F	45	

Note:Parts of ISIC 10 and 11, 16, part of 23 are not covered by the inquiry and do not figure in the subtotals and totals.

FRANCE

Table FR.2 VALUE ADDED AT FACTOR COSTS - VALEUR AJOUTÉE AU COUT DES FACTEURS

Millions of FF (Current Prices)

ISIC revision 3				1991	1992	1993	1994	1995	1996	1997	1998
C	10_14		Mining and quarrying	20 001	17 811	17 706	17 661	15 337	14 625	10 844	7 627
	10		Mining of coal and lignite; extraction of peat	..	..	..	..	..	..	..	..
	11		Crude petroleum & natural gas; service activities	..	..	..	..	..	..	..	..
	12		Mining of uranium and thorium ores	..	..	..	..	..	..	..	..
	13		Mining of metal ores	340	275	137	176	180	175	136	115
	14		Other mining and quarrying	9 485	8 693	7 862	8 160	8 354	7 642	7 665	7 512
D	15_37		Total manufacturing	903 034	882 429	831 946	889 206	920 018	908 693	1103 267	983 858
	15		Food products and beverages	..	..	33 884	..	33 177	34 476	36 698	35 182
		151	Production of meat, fish, fruit, vegetables, oils & fats	..	..	..	..	..	..	..	..
		152	Dairy products	..	..	..	..	..	..	..	..
		153	Grain mill and starch products; animal feeds	..	..	..	..	..	..	..	..
		154	Other food products	..	..	..	..	..	..	..	..
		155	Beverages	..	..	..	..	..	..	..	..
	16		Tobacco products	..	..	..	..	..	..	..	..
	17		Textiles	30 263	30 207	27 293	29 345	28 450	26 400	27 415	27 366
		171	Spinning, weaving and finishing of textiles	15 115	14 629	12 991	14 486	13 630	12 459	13 066	13 095
		172	Other textiles	9 029	9 520	8 909	9 442	9 623	9 084	9 267	9 259
		173	Knitted and crocheted fabrics and articles	6 119	6 058	5 393	5 417	5 197	4 857	5 082	5 012
	18		Wearing apparel and fur	23 390	22 725	21 690	20 501	19 723	19 099	18 332	18 030
		181	Wearing apparel, except fur apparel	23 137	22 509	21 562	20 369	19 572	18 946	18 195	17 950
		182	Dressing and dyeing of fur; articles of fur	253	216	128	133	151	154	137	80
	19		Leather and articles of leather; footwear	11 014	10 234	9 832	9 798	9 425	8 841	8 792	8 796
		191	Leather and articles of leather	3 628	3 599	3 222	3 840	3 574	3 487	3 591	3 705
		192	Footwear	7 386	6 635	6 610	5 958	5 851	5 355	5 201	5 090
	20		Wood & products of wood & cork, exc. furniture	10 344	10 295	9 521	9 768	9 852	9 537	10 063	10 472
		201	Sawmilling and planing of wood	152	170	225	254	290	211	223	186
		202	Wood products; articles of cork, straw, etc.	10 193	10 125	9 297	9 514	9 562	9 326	9 840	10 287
	21		Paper and paper products	29 871	28 320	26 831	28 467	31 811	28 519	30 015	30 717
	22		Printing, publishing and recorded media	51 570	51 371	50 917	53 232	54 016	55 193	55 482	55 977
		221	Publishing	28 753	28 247	29 081	30 853	31 282	32 975	33 161	33 862
		222	Printing and related activities	22 235	22 439	21 202	21 764	22 026	21 476	21 472	21 182
		223	Reproduction of recorded media	582	685	634	614	708	742	848	933
	23		Coke and petroleum products; nuclear fuel	24 761	21 637	24 490	23 861	25 060	31 133	28 261	25 912
		231	Coke oven products	9 872	9 771	9 854	10 497	12 902	14 386	14 870	14 265
		232	Refined petroleum products	14 888	11 866	14 636	13 364	12 159	16 747	13 392	11 646
		233	Processing of nuclear fuel	..	..	..	..	..	..	..	..
	24		Chemicals and chemical products	118 316	119 544	120 197	131 982	145 357	141 831	152 376	150 669
		241	Basic chemicals	34 467	33 256	30 704	39 478	47 882	42 791	47 087	45 483
		242	Other chemical products	82 995	85 506	88 904	91 866	96 851	98 376	104 515	104 312
		243	Man-made fibres	855	782	589	638	625	664	773	875
	25		Rubber and plastic products	47 745	48 980	47 437	50 287	51 076	51 940	53 718	59 365
		251	Rubber products	18 558	18 746	17 817	18 268	18 888	19 343	19 920	21 730
		252	Plastic products	29 187	30 233	29 620	32 019	32 188	32 596	33 798	37 635
	26		Other non-metallic mineral products	45 329	43 129	41 537	43 759	44 707	42 767	43 139	45 102
		261	Glass and glass products	15 790	15 176	14 406	15 302	16 061	15 651	16 038	16 425
		269	Non-metallic mineral products, nec	29 539	27 953	27 131	28 457	28 646	27 115	27 101	28 677
	27		Basic metals	41 648	38 148	32 926	39 487	45 094	40 438	43 554	43 713
		271	Basic iron and steel	24 984	22 959	19 518	24 498	28 984	24 878	26 421	26 403
		272	Basic precious and non-ferrous metals	8 481	7 523	6 725	7 792	8 700	8 109	9 265	8 865
		273	Casting of metals	8 183	7 665	6 683	7 198	7 411	7 451	7 868	8 444
	28		Fabricated metal products, exc. machinery	79 630	77 573	70 645	74 634	77 236	76 351	82 877	83 023
		281	Structural metal prod., tanks, reservoirs, etc.	29 918	29 646	27 498	27 750	26 114	24 451	26 843	25 208
	29		Machinery and equipment, nec	75 508	75 490	71 508	73 461	75 364	75 949	78 006	80 545
		291	General purpose machinery	41 161	41 722	39 853	40 738	41 470	42 462	42 847	44 910
		292	Special purpose machinery	26 693	25 644	23 699	24 641	25 821	25 698	27 236	28 281
		293	Domestic appliances, nec	7 654	8 125	7 956	8 081	8 074	7 789	7 923	7 354
	30		Office, accounting & computing machinery	31 688	27 891	20 778	22 726	23 778	22 408	21 364	24 284
	31		Electrical machinery & apparatus, nec	46 588	42 042	43 213	46 051	46 890	46 198	47 877	48 616
		311	Electric motors, generators and transformers	7 025	6 608	6 207	6 091	6 621	6 357	6 916	7 441
		312	Electricity distribution & control apparatus	21 926	16 871	18 747	19 898	20 095	19 423	19 976	20 448
		313	Insulated wire and cable	6 105	6 513	6 456	6 680	6 309	5 803	6 077	5 415
		314	Accumulators, primary cells & primary batteries	2 669	2 543	2 640	2 518	2 555	2 376	2 067	1 974
		315	Electric lamps and lighting equipment	3 403	3 273	2 916	3 171	3 313	3 055	3 306	3 565
		319	Other electrical equipment, nec	5 461	6 233	6 246	7 693	7 997	9 184	9 536	9 773
	32		Radio, TV, communication equipment	51 202	47 719	45 619	47 518	50 651	53 276	55 718	58 872
		321	Electronic valves, tubes; other electronic components	26 665	23 980	22 225	24 137	24 171	25 759	26 726	26 620
		322	TV/radio transmitters; line teleph./telegr. apparatus	20 329	19 866	19 420	19 551	22 614	23 254	23 829	27 808
		323	TV/radio receivers; sound and video app.; assoc. gds.	4 208	3 873	3 973	3 830	3 866	4 264	5 163	4 444
	33		Medical, precision & optical instrum.; watch & clocks	23 080	22 833	23 319	23 827	24 822	24 349	25 047	24 915
		331	Medical appl.; instruments & appl. for measuring, etc.	18 192	18 029	18 660	19 131	19 982	19 823	20 471	20 317
		332	Optical instruments and photographic equipment	3 461	3 385	3 308	3 436	3 501	3 342	3 429	3 479
		333	Watches and clocks	1 427	1 419	1 351	1 259	1 339	1 184	1 147	1 120
	34		Motor vehicles, trailers and semi-trailers	92 010	94 333	78 965	92 653	91 873	89 344	95 658	108 149
		341	Motor vehicles	62 806	64 975	51 978	61 928	60 029	58 934	64 629	75 717
		342	Bodies for motor vehicles, trailers & semi-trailers	5 053	4 616	4 011	4 000	4 483	4 515	4 545	5 366
		343	Parts & accessories for motor vehicles & their engines	24 151	24 742	22 976	26 725	27 360	25 895	26 483	27 066
	35		Other transport equipment	40 287	41 465	37 924	40 272	37 823	38 227	44 588	49 058
		351	Building and repairing of ships and boats	3 547	3 829	2 998	3 710	3 424	2 910	3 057	4 174
		352	Railway and tramway locomotives and rolling stock	2 130	5 483	4 784	4 773	4 762	4 748	4 722	3 974
		353	Aircraft and spacecraft	32 572	30 271	28 442	29 857	27 486	28 548	34 724	38 760
		359	Transport equipment, nec	2 037	1 882	1 699	1 932	2 150	2 021	2 085	2 149
	36		Furniture; manufacturing, nec	28 789	28 493	27 305	27 579	27 009	26 894	26 753	27 142
		361	Furniture	18 303	18 074	17 029	17 201	16 849	16 740	16 625	17 248
		369	Manufacturing, nec	10 487	10 419	10 276	10 378	10 160	10 153	10 128	9 894
	37		Recycling	..	..	..	..	..	..	..	3 134
E	40_41		Electricity, gas and water supply	146 742	156 421	157 891	150 819	153 771	153 823	143 387	153 574
	40		Electricity, gas, steam and hot water supply	135 305	144 660	145 261	137 765	140 271	140 405	130 318	141 486
	41		Collection, purification and distribution of water	11 437	11 761	12 630	13 054	13 500	13 417	13 069	12 088
F	45		Construction	..	..	..	..	..	..	..	..

Note:Parts of ISIC 10 and 11, 16, part of 23 are not covered by the inquiry and do not figure in the subtotals and totals.

Table FR.3 INVESTISSEMENT - INVESTMENT

Millions de FF (Prix courants)

1991	1992	1993	1994	1995	1996	1997	1998			CITI révision 3
2 380	1 991	1 665	1 783	2 052	1 887	2 283	2 239	Activites extractives	C	10_14
..	..	..	..	..	..	..	..	Extraction de charbon, de lignite et de tourbe		10
..	..	..	..	..	..	..	..	Pétrole brut et gaz naturel ; activités annexes		11
..	..	..	..	..	..	..	..	Extraction de Minerais d'Uranium et de Thorium		12
77	46	80	147	60	41	19	20	Extraction de minerais métalliques		13
2 303	1 945	1 585	1 636	1 991	1 846	1 458	1 612	Autres activités extractives		14
147 107	135 133	111 990	110 476	121 403	128 785	125 951	134 624	Total activites manufacturieres	D	15_37
..	..	2 477	..	2 805	2 391	1 772	2 759	Produits alimentaires et boissons		15
..	..	..	..	..	..	..	..	Viande, poisson, fruits, légumes, huiles et graisses		151
..	..	..	..	..	..	..	..	Produits laitiers		152
..	..	..	..	..	..	..	..	Travail des grains; prod. amyl. ; alim. pour animaux		153
..	..	..	..	..	..	..	..	Autres produits alimentaires		154
..	..	..	..	..	..	..	..	Boissons		155
..	..	..	..	..	..	..	..	Produits à base de tabac		16
4 040	3 879	3 015	3 081	3 519	3 383	3 506	3 751	Textiles		17
2 234	2 037	1 413	1 560	1 855	1 714	2 012	2 085	Filature, tissage et achèvement des textiles		171
1 345	1 337	1 180	1 123	1 233	1 192	1 081	1 243	Autres articles textiles		172
461	506	422	398	431	476	412	423	Etoffes et articles de bonneterie		173
1 294	1 448	1 256	1 031	1 216	1 105	1 122	1 055	Articles d'habillement et fourrure		18
1 278	1 431	1 252	1 030	1 211	1 095	1 115	1 051	Articles d'habillement autres qu'en fourrure		181
16	17	3	2	4	9	7	4	Prép. et teinture des fourrures ; articles en fourrure		182
739	674	625	527	623	543	501	556	Cuir et articles en cuir ; chaussures		19
302	269	184	182	210	172	162	191	Cuir et articles en cuir sauf chaussures		191
437	406	441	345	413	371	339	364	Chaussures		192
1 655	1 493	1 001	1 534	1 327	1 375	1 227	1 611	Bois, articles en bois, sauf meubles		20
10	8	8	11	10	6	7	8	Sciage et rabotage du bois		201
1 645	1 485	993	1 522	1 317	1 369	1 220	1 603	Articles en bois, liège, vannerie, sparterie		202
7 625	7 641	4 265	3 770	4 706	5 245	4 920	6 811	Papier, articles en papier et carton		21
4 189	3 721	3 261	3 774	4 009	4 667	4 273	4 327	Edition, imprimerie et reproduction		22
1 684	1 346	1 150	1 488	1 322	1 461	1 527	1 357	Edition		221
2 339	2 269	1 988	2 175	2 483	3 031	2 588	2 667	Imprimerie et activités annexes		222
166	106	122	111	204	175	157	304	Reproduction de supports enregistrés		223
11 480	11 907	12 342	11 817	9 140	7 720	7 215	7 089	Coke. et prod. pétroliers ; comb. nucléaires		23
5 197	5 861	7 270	6 224	4 684	3 789	3 800	3 522	Cokéfaction		231
6 283	6 047	5 072	5 593	4 456	3 931	3 415	3 567	Produits pétroliers raffinés		232
..	..	..	..	..	..	..	..	Traitement de combustibles nucléaires		233
22 592	21 485	18 600	16 229	17 879	20 525	20 022	22 259	Produits chimiques		24
11 257	9 176	6 988	6 139	7 321	9 156	8 138	9 432	Produits chimiques de base		241
11 046	12 006	11 328	9 866	10 173	11 239	11 624	12 649	Autres produits chimiques		242
289	303	284	224	386	130	260	178	Fibres synthétiques ou artificielles		243
6 852	6 827	5 874	5 629	7 096	7 760	8 613	10 164	Art. en caoutchouc et matières plastiques		25
1 947	1 968	1 571	1 332	2 001	2 442	3 216	2 991	Articles en caoutchouc		251
4 904	4 859	4 303	4 297	5 094	5 318	5 397	7 174	Articles en matières plastiques		252
6 163	5 931	4 770	4 982	6 135	6 429	5 907	6 530	Autres produits minéraux non métalliques		26
1 853	1 858	1 659	2 032	2 227	2 314	1 782	2 347	Verre et articles en verre		261
4 310	4 073	3 112	2 950	3 908	4 115	4 124	4 183	Produits minéraux non métalliques, nca		269
10 496	6 563	5 234	4 505	5 213	6 423	7 101	7 280	Produits métallurgiques de base		27
5 168	4 293	3 534	2 740	3 176	4 152	4 874	4 516	Sidérurgie et première transformation de l'acier		271
4 292	1 437	1 149	1 094	1 166	1 282	1 415	1 493	Première transf. métaux précieux et non ferreux		272
1 036	833	551	671	871	988	812	1 271	Fonderie		273
8 793	8 556	6 862	6 322	7 207	7 531	8 091	9 476	Ouvrages en métaux, sauf machines		28
2 397	2 453	2 055	1 628	1 422	1 561	1 719	1 709	Constr. et menuis. métal.; réservoirs; générateurs		281
7 977	6 549	5 370	5 062	6 579	6 992	6 575	7 362	Machines et matériels, nca		29
3 864	3 586	2 707	2 543	3 078	3 494	3 475	3 910	Machines d'usage général		291
2 843	1 884	1 489	1 544	2 187	2 365	2 058	2 240	Machines d'usage spécifique		292
1 270	1 078	1 174	975	1 315	1 133	1 042	1 212	Appareils domestiques, nca		293
4 358	3 745	3 077	2 754	2 598	2 758	1 928	2 203	Mach. de bureau, comptables, information		30
6 685	5 673	4 749	4 486	4 919	5 507	5 245	6 186	Machines et appareils électriques, nca		31
642	635	454	455	673	581	520	585	Moteurs, génératrices, transformateurs électriques		311
2 730	2 228	1 729	1 852	1 826	2 100	1 950	2 628	Matériel électrique de distribution et de commande		312
1 429	814	723	641	647	639	597	818	Fils et câbles électriques isolés		313
439	328	491	328	332	349	420	353	Accumulateurs et piles électriques		314
383	369	274	292	329	341	376	323	Lampes électriques et appareils d'éclairage		315
1 062	1 299	1 078	919	1 112	1 498	1 382	1 479	Autres matériels électriques, nca.		319
6 588	5 557	5 516	6 882	7 770	8 020	8 199	7 954	Equipements et appareils de radio, TV, etc.		32
3 363	2 826	3 205	4 445	4 550	4 962	5 210	4 531	Tubes et valves électroniques;autres composants		321
2 242	1 684	1 786	1 804	2 267	1 990	2 278	2 808	Emetteurs radio/TV; app. téléphonie et télégraphie		322
983	1 046	525	633	954	1 068	712	615	Récepteurs radio/TV, son et image; art. associés		323
2 186	1 938	1 807	1 780	1 976	2 109	1 950	2 218	Instr. médicaux, de précision ; optique ; horlog.		33
1 725	1 543	1 434	1 415	1 516	1 766	1 642	1 823	Instruments médicaux et de précision, etc.		331
278	240	240	265	289	235	203	292	Instruments d'optique ; matériel photographique		332
183	155	133	100	171	108	105	103	Horlogerie		333
24 022	23 716	18 849	21 211	23 818	24 489	23 007	19 480	Véhicules automobiles; remorques, etc.		34
18 802	18 574	14 748	17 202	18 904	19 580	16 558	13 823	Véhicules automobiles		341
546	401	251	230	338	326	348	434	Carross. automobiles, remorques, semi-remorques		342
4 675	4 741	3 850	3 779	4 576	4 583	6 101	5 223	Pièces et accessoires pour automobiles et moteurs		343
6 280	5 032	3 228	2 870	3 192	3 591	3 873	4 667	Autres matériels de transport		35
300	384	293	302	290	445	359	358	Construction et réparation de navires		351
263	497	397	376	334	383	334	293	Matériel ferroviaire roulant		352
5 462	3 905	2 317	2 004	2 334	2 515	2 901	3 762	Construction aéronautique et spatiale		353
255	245	221	188	234	247	279	255	Autres équipements de transport		359
3 092	2 798	2 288	2 230	2 482	2 613	2 677	2 889	Meubles ; activités de fabrication, nca		36
1 892	1 742	1 427	1 328	1 529	1 504	1 568	1 647	Meubles		361
1 200	1 055	861	902	953	1 109	1 109	1 241	Activités de fabrication, nca		369
..	..	..	..	..	..	..	754	Récupération		37
43 262	44 396	45 989	44 806	45 766	45 131	42 782	39 212	Electricité, gaz et eau	E	40_41
41 551	42 277	43 671	41 387	41 988	41 732	39 222	34 859	Prod. et dist. d'électricité gaz, vapeur d'eau chaude		40
1 710	2 120	2 318	3 419	3 778	3 399	3 560	4 354	Captage, épuration et distribution d'eau		41
..	..	..	..	..	..	..	..	Construction	F	45

Note: Parts of ISIC 10 and 11, 16, part of 23 are not covered by the inquiry and do not figure in the subtotals and totals.

Table FR.4 **EMPLOYMENT, EMPLOYEES - EMPLOI, SALARIÉS**

Units

ISIC revision 3				1991	1992	1993	1994	1995	1996	1997	1998
C	10_14		Mining and quarrying	59 112	55 122	51 884	49 267	47 951	45 266	38 009	35 063
	10		Mining of coal and lignite; extraction of peat	..	..	..	..	..	..	..	..
	11		Crude petroleum & natural gas; service activities	..	..	..	..	..	..	..	..
	12		Mining of uranium and thorium ores	..	..	..	..	..	..	..	..
	13		Mining of metal ores	1 231	1 110	819	584	513	497	438	330
	14		Other mining and quarrying	29 554	26 689	25 105	24 157	24 063	22 729	22 486	21 145
D	15_37		Total manufacturing	3135 067	3024 167	3255 148	2811 061	2963 341	2943 533	2960 296	2943 769
	15		Food products and beverages	..	..	160 806	..	160 736	181 261	192 478	167 651
		151	Production of meat, fish, fruit, vegetables, oils & fats	..	..	..	..	..	..	..	..
		152	Dairy products	..	..	..	..	..	..	..	..
		153	Grain mill and starch products; animal feeds	..	..	..	..	..	..	..	..
		154	Other food products	..	..	..	..	..	..	..	..
		155	Beverages	..	..	..	..	..	..	..	..
	16		Tobacco products	..	..	..	..	..	..	..	..
	17		Textiles	147 728	139 486	127 467	127 383	126 124	117 234	113 561	110 630
		171	Spinning, weaving and finishing of textiles	72 732	65 677	59 621	59 024	58 026	54 364	52 157	51 743
		172	Other textiles	42 244	41 696	39 518	40 167	41 313	38 522	37 042	35 839
		173	Knitted and crocheted fabrics and articles	32 752	32 113	28 328	28 192	26 785	24 348	24 362	23 048
	18		Wearing apparel and fur	138 907	131 776	120 814	112 709	107 113	102 020	96 061	90 879
		181	Wearing apparel, except fur apparel	137 787	130 715	120 014	111 992	106 412	101 253	95 414	90 272
		182	Dressing and dyeing of fur; articles of fur	1 120	1 061	800	717	701	767	647	607
	19		Leather and articles of leather; footwear	58 820	53 768	51 541	48 979	46 719	44 473	42 341	40 721
		191	Leather and articles of leather	18 779	17 181	15 512	15 926	14 787	14 521	14 099	13 758
		192	Footwear	40 041	36 587	36 029	33 053	31 932	29 952	28 242	26 963
	20		Wood & products of wood & cork, exc. furniture	47 917	46 516	42 997	42 767	42 468	40 704	40 511	40 394
		201	Sawmilling and planing of wood	782	795	957	980	1 271	894	891	715
		202	Wood products; articles of cork, straw, etc.	47 135	45 721	42 040	41 787	41 197	39 810	39 620	39 679
	21		Paper and paper products	98 686	97 386	94 220	92 205	90 780	88 569	87 371	86 202
	22		Printing, publishing and recorded media	159 427	152 284	146 839	147 076	147 711	142 963	141 639	140 584
		221	Publishing	73 001	68 535	67 609	68 061	68 593	65 338	64 347	64 274
		222	Printing and related activities	84 560	81 746	77 422	77 196	76 930	75 497	74 942	73 698
		223	Reproduction of recorded media	1 866	2 003	1 808	1 819	2 188	2 128	2 350	2 612
	23		Coke and petroleum products; nuclear fuel	35 411	35 362	34 959	33 681	33 259	31 258	29 988	29 436
		231	Coke oven products	13 024	12 600	12 154	12 032	11 897	11 397	11 110	11 417
		232	Refined petroleum products	22 387	22 762	22 805	21 649	21 362	19 861	18 878	18 019
		233	Processing of nuclear fuel	..	..	..	..	..	..	..	..
	24		Chemicals and chemical products	287 818	284 142	275 695	272 944	270 795	267 310	269 491	271 694
		241	Basic chemicals	80 133	79 353	76 013	74 747	75 176	74 013	73 510	73 142
		242	Other chemical products	204 084	201 369	196 373	195 189	192 849	190 885	193 543	196 158
		243	Man-made fibres	3 601	3 420	3 309	3 008	2 770	2 412	2 438	2 394
	25		Rubber and plastic products	190 610	187 891	184 019	182 540	185 754	185 971	188 105	195 951
		251	Rubber products	77 592	73 131	70 654	67 317	68 088	67 484	67 343	67 469
		252	Plastic products	113 018	114 760	113 365	115 223	117 666	118 487	120 762	128 482
	26		Other non-metallic mineral products	144 704	140 410	134 339	130 777	129 348	128 381	125 057	123 631
		261	Glass and glass products	51 853	52 532	50 454	49 825	50 398	49 086	48 765	48 798
		269	Non-metallic mineral products, nec	92 851	87 878	83 885	80 952	78 950	79 295	76 292	74 833
	27		Basic metals	144 473	140 624	132 615	127 774	127 558	126 049	122 709	121 162
		271	Basic iron and steel	76 562	75 683	71 811	69 112	68 145	67 860	65 711	63 886
		272	Basic precious and non-ferrous metals	28 661	27 364	26 430	25 590	25 212	24 708	23 670	22 930
		273	Casting of metals	39 250	37 577	34 374	33 072	34 201	33 481	33 328	34 346
	28		Fabricated metal products, exc. machinery	329 084	319 071	297 581	294 009	297 190	299 222	304 872	303 944
		281	Structural metal prod., tanks, reservoirs, etc.	118 522	117 989	112 148	106 305	99 315	97 981	98 974	96 532
	29		Machinery and equipment, nec	290 859	284 576	268 625	251 140	252 262	254 815	252 377	253 616
		291	General purpose machinery	153 238	152 570	146 868	136 709	136 286	140 014	139 937	142 483
		292	Special purpose machinery	104 132	98 616	90 071	84 356	85 185	84 975	84 791	84 852
		293	Domestic appliances, nec	33 489	33 390	31 686	30 075	30 791	29 826	27 649	26 281
	30		Office, accounting & computing machinery	59 515	52 201	45 729	42 420	40 610	36 195	37 357	43 078
	31		Electrical machinery & apparatus, nec	167 983	149 701	151 433	151 710	152 346	151 161	150 426	152 455
		311	Electric motors, generators and transformers	27 501	26 908	24 867	23 076	23 770	22 663	22 865	22 787
		312	Electricity distribution & control apparatus	76 914	58 509	64 573	62 716	62 490	62 308	61 877	63 690
		313	Insulated wire and cable	20 035	20 082	19 026	20 864	18 849	16 375	15 722	15 895
		314	Accumulators, primary cells & primary batteries	9 376	9 114	9 014	8 510	8 432	7 929	7 477	7 221
		315	Electric lamps and lighting equipment	12 982	12 556	11 570	11 637	11 527	11 141	11 099	11 402
		319	Other electrical equipment, nec	21 175	22 532	22 383	24 907	27 278	30 745	31 386	31 460
	32		Radio, TV, communication equipment	162 681	152 455	143 284	141 746	145 624	152 651	151 675	153 636
		321	Electronic valves, tubes; other electronic components	87 035	82 492	76 889	73 163	69 954	70 383	69 001	71 294
		322	TV/radio transmitters; line teleph./telegr. apparatus	60 519	56 135	52 803	53 843	60 799	66 057	64 636	66 424
		323	TV/radio receivers; sound and video app.; assoc. gds.	15 127	13 828	13 592	14 740	14 871	16 211	18 038	15 918
	33		Medical, precision & optical instrum.; watch & clocks	88 943	86 989	83 148	83 818	84 291	81 623	80 565	81 332
		331	Medical appl.; instruments & appl. for measuring, etc.	66 280	65 312	63 376	64 954	65 327	63 636	63 459	64 382
		332	Optical instruments and photographic equipment	15 443	14 767	13 586	13 244	13 537	13 094	12 286	12 556
		333	Watches and clocks	7 220	6 910	6 186	5 620	5 427	4 893	4 820	4 394
	34		Motor vehicles, trailers and semi-trailers	319 373	309 959	297 562	288 317	290 390	282 992	270 183	261 072
		341	Motor vehicles	206 886	199 178	193 418	185 198	184 407	183 497	172 064	162 019
		342	Bodies for motor vehicles, trailers & semi-trailers	23 971	22 609	20 731	19 885	20 598	20 174	20 482	21 888
		343	Parts & accessories for motor vehicles & their engines	88 516	88 172	83 413	83 234	85 385	79 321	77 637	77 165
	35		Other transport equipment	133 367	135 856	129 173	124 619	120 702	117 345	116 424	114 646
		351	Building and repairing of ships and boats	17 413	15 911	14 436	14 864	14 196	14 131	13 561	13 378
		352	Railway and tramway locomotives and rolling stock	8 706	18 133	17 726	16 685	15 593	15 616	15 250	14 237
		353	Aircraft and spacecraft	97 492	92 743	88 432	84 690	82 340	79 341	78 855	78 614
		359	Transport equipment, nec	9 756	9 069	8 579	8 380	8 573	8 257	8 758	8 417
	36		Furniture; manufacturing, nec	128 761	123 714	115 351	114 447	111 561	111 336	107 166	106 835
		361	Furniture	83 508	81 004	75 568	75 104	73 950	73 622	70 290	70 553
		369	Manufacturing, nec	45 253	42 710	39 783	39 343	37 611	37 714	36 876	36 282
	37		Recycling	..	..	..	..	..	..	..	11 372
E	40_41		Electricity, gas and water supply	191 672	190 875	195 695	196 537	198 188	196 720	194 408	199 648
	40		Electricity, gas, steam and hot water supply	160 964	159 174	163 807	164 759	166 226	165 503	163 447	163 127
	41		Collection, purification and distribution of water	30 708	31 701	31 888	31 778	31 962	31 217	30 961	31 522
F	45		Construction	..	..	..	..	..	..	..	..

Note: Parts of ISIC 10 and 11, 16, part of 23 are not covered by the inquiry and do not figure in the subtotals and totals.

Table FR.5 **SALAIRES ET TRAITEMENTS, SALARIÉS - WAGES AND SALARIES, EMPLOYEES**

Millions de FF (Prix courants)

1991	1992	1993	1994	1995	1996	1997	1998			CITI révision 3	
10 027	9 896	9 547	9 248	9 091	8 895	6 722	6 404	Activites extractives	C	10_14	
..	..	..	..	..	..	..	..	Extraction de charbon, de lignite et de tourbe		10	
..	..	..	..	..	..	..	..	Pétrole brut et gaz naturel ; activités annexes		11	
..	..	..	..	..	..	..	..	Extraction de Minerais d'Uranium et de Thorium		12	
182	165	131	89	83	83	78	59	Extraction de minerais métalliques		13	
4 128	3 886	3 718	3 634	3 702	3 622	3 685	3 612	Autres activités extractives		14	
459 492	461 367	449 106	450 186	458 287	463 188	518 344	482 736	Total activites manufacturieres	D	15_37	
..	..	3 934	..	3 620	13 482	14 288	12 924	Produits alimentaires et boissons		15	
..	..	..	..	..	..	..	..	Viande, poisson, fruits, légumes, huiles et graisses			151
..	..	..	..	..	..	..	..	Produits laitiers			152
..	..	..	..	..	..	..	..	Travail des grains; prod. amyl. ; alim. pour animaux			153
..	..	..	..	..	..	..	..	Autres produits alimentaires			154
..	..	..	..	..	..	..	..	Boissons			155
..	..	..	..	..	..	..	..	Produits à base de tabac		16	
16 686	16 289	15 056	15 306	15 583	14 830	14 770	14 732	Textiles		17	
8 518	7 951	7 309	7 374	7 326	6 975	6 985	7 025	Filature, tissage et achèvement des textiles			171
4 768	4 883	4 627	4 850	5 204	4 973	4 843	4 848	Autres articles textiles			172
3 400	3 454	3 119	3 081	3 053	2 882	2 942	2 859	Etoffes et articles de bonneterie			173
13 936	13 700	12 889	12 218	11 968	11 661	11 209	10 900	Articles d'habillement et fourrure		18	
13 796	13 566	12 792	12 138	11 879	11 562	11 120	10 807	Articles d'habillement autres qu'en fourrure			181
140	134	97	80	89	100	89	94	Prép. et teinture des fourrures ; articles en fourrure			182
6 078	5 711	5 548	5 328	5 182	4 948	4 951	4 867	Cuir et articles en cuir ; chaussures		19	
2 039	1 946	1 796	1 915	1 797	1 757	1 803	1 808	Cuir et articles en cuir sauf chaussures			191
4 039	3 765	3 752	3 413	3 386	3 191	3 148	3 059	Chaussures			192
5 259	5 326	4 996	5 031	5 099	5 013	5 157	5 253	Bois, articles en bois, sauf meubles		20	
102	105	129	152	178	139	151	128	Sciage et rabotage du bois			201
5 157	5 221	4 867	4 879	4 922	4 874	5 006	5 125	Articles en bois, liège, vannerie, sparterie			202
14 177	14 405	14 590	14 453	14 569	14 579	14 645	14 648	Papier, articles en papier et carton		21	
27 800	27 064	26 681	27 187	27 588	27 722	28 131	28 414	Edition, imprimerie et reproduction		22	
14 743	14 204	14 268	14 824	15 023	15 237	15 468	15 816	Edition			221
12 830	12 600	12 179	12 121	12 270	12 200	12 320	12 219	Imprimerie et activités annexes			222
227	259	234	241	294	285	343	379	Reproduction de supports enregistrés			223
8 363	8 431	8 473	8 513	8 444	8 181	8 077	8 174	Coke. et prod. pétroliers ; comb. nucléaires		23	
2 501	2 451	2 439	2 511	2 527	2 487	2 500	2 631	Cokéfaction			231
5 862	5 980	6 034	6 002	5 917	5 694	5 576	5 543	Produits pétroliers raffinés			232
..	..	..	..	..	..	..	..	Traitement de combustibles nucléaires			233
52 082	53 475	54 405	54 740	56 406	56 694	59 234	60 804	Produits chimiques		24	
14 963	15 208	15 152	15 110	15 706	15 998	16 220	16 210	Produits chimiques de base			241
36 552	37 699	38 684	39 083	40 242	40 287	42 585	44 190	Autres produits chimiques			242
567	569	570	548	457	409	429	404	Fibres synthétiques ou artificielles			243
24 338	24 812	25 122	25 383	26 311	26 751	28 172	29 912	Art. en caoutchouc et matières plastiques		25	
10 401	10 205	10 244	9 867	10 197	10 337	10 614	10 919	Articles en caoutchouc			251
13 937	14 606	14 877	15 516	16 114	16 414	17 558	18 994	Articles en matières plastiques			252
19 991	20 045	19 834	19 836	20 160	20 326	20 563	20 624	Autres produits minéraux non métalliques		26	
7 290	7 549	7 519	7 702	8 012	8 030	8 199	8 264	Verre et articles en verre			261
12 701	12 496	12 315	12 133	12 148	12 296	12 364	12 360	Produits minéraux non métalliques, nca			269
23 091	22 713	21 609	21 330	22 068	21 751	21 968	21 784	Produits métallurgiques de base		27	
13 983	13 694	13 071	12 781	13 155	12 757	12 813	12 531	Sidérurgie et première transformation de l'acier			271
4 474	4 480	4 350	4 367	4 456	4 531	4 577	4 473	Première transf. métaux précieux et non ferreux			272
4 634	4 538	4 187	4 181	4 457	4 463	4 578	4 780	Fonderie			273
43 454	43 606	41 204	41 134	42 669	43 706	45 528	46 148	Ouvrages en métaux, sauf machines		28	
16 270	16 762	16 118	15 413	14 860	14 832	15 258	15 024	Constr. et menuis. métal.; réservoirs; générateurs			281
41 857	42 671	41 408	39 239	40 450	41 715	42 268	43 544	Machines et matériels, nca		29	
22 304	23 266	22 841	21 738	22 382	23 383	23 934	24 954	Machines d'usage général			291
15 564	15 241	14 372	13 556	14 002	14 276	14 415	14 802	Machines d'usage spécifique			292
3 989	4 164	4 196	3 945	4 066	4 056	3 919	3 788	Appareils domestiques, nca			293
13 816	12 896	11 545	12 029	10 852	10 493	10 243	11 979	Mach. de bureau, comptables, information		30	
23 667	21 347	22 911	23 789	24 488	24 985	25 566	26 434	Machines et appareils électriques, nca		31	
3 786	3 782	3 606	3 414	3 607	3 505	3 631	3 795	Moteurs, génératrices, transformateurs électriques			311
11 292	8 556	10 268	10 438	10 630	10 940	11 103	11 442	Matériel électrique de distribution et de commande			312
2 782	2 872	2 901	3 214	3 123	2 776	2 781	2 933	Fils et câbles électriques isolés			313
1 346	1 358	1 359	1 350	1 356	1 352	1 382	1 345	Accumulateurs et piles électriques			314
1 682	1 676	1 577	1 623	1 674	1 644	1 660	1 776	Lampes électriques et appareils d'éclairage			315
2 779	3 103	3 199	3 751	4 099	4 766	5 008	5 143	Autres matériels électriques, nca.			319
28 286	30 018	26 927	27 085	28 366	31 042	31 127	32 906	Equipements et appareils de radio, TV, etc.		32	
15 016	17 316	14 275	13 860	13 284	13 936	13 798	14 446	Tubes et valves électroniques;autres composants			321
11 268	10 887	10 847	11 138	12 948	14 640	14 763	15 954	Emetteurs radio/TV; app. téléphonie et télégraphie			322
2 002	1 815	1 806	2 088	2 134	2 466	2 566	2 507	Récepteurs radio/TV, son et image; art. associés			323
13 358	13 730	13 474	13 837	14 337	14 041	14 401	14 696	Instr. médicaux, de précision ; optique ; horlog.		33	
10 519	10 838	10 746	11 222	11 613	11 422	11 854	12 105	Instruments médicaux et de précision, etc.			331
1 999	2 051	1 953	1 896	2 001	1 952	1 881	1 960	Instruments d'optique ; matériel photographique			332
840	841	774	719	724	666	665	632	Horlogerie			333
43 712	44 363	43 908	44 667	45 514	45 932	45 855	46 068	Véhicules automobiles; remorques, etc.		34	
29 289	29 499	29 308	29 780	30 002	30 978	30 689	30 130	Véhicules automobiles			341
2 870	2 835	2 692	2 524	2 690	2 710	2 770	3 038	Carross. automobiles, remorques, semi-remorques			342
11 554	12 029	11 908	12 363	12 822	12 245	12 395	12 900	Pièces et accessoires pour automobiles et moteurs			343
24 226	25 572	24 038	24 445	23 628	23 966	24 414	24 397	Autres matériels de transport		35	
2 295	2 182	1 990	2 213	2 013	2 001	1 997	2 084	Construction et réparation de navires			351
1 191	2 760	2 763	2 719	2 628	2 709	2 723	2 692	Matériel ferroviaire roulant			352
19 644	19 557	18 245	18 423	17 822	18 105	18 466	18 401	Construction aéronautique et spatiale			353
1 095	1 073	1 041	1 090	1 166	1 151	1 227	1 220	Autres équipements de transport			359
15 313	15 192	14 488	14 635	14 604	14 853	14 801	14 883	Meubles ; activités de fabrication, nca		36	
9 601	9 651	9 235	9 358	9 439	9 531	9 487	9 591	Meubles			361
5 712	5 540	5 253	5 278	5 164	5 322	5 313	5 291	Activités de fabrication, nca			369
..	..	..	..	..	..	..	1 569	Récupération		37	
33 298	34 531	36 055	37 950	38 423	40 074	40 544	41 419	Electricité, gaz et eau	E	40_41	
28 382	29 290	30 497	32 296	32 583	34 127	34 450	35 189	Prod. et dist. d'électricité gaz, vapeur d'eau chaude		40	
4 916	5 241	5 558	5 653	5 840	5 947	6 093	6 230	Captage, épuration et distribution d'eau		41	
..	..	..	..	..	..	..	..	Construction	F	45	

Note: Parts of ISIC 10 and 11, 16, part of 23 are not covered by the inquiry and do not figure in the subtotals and totals.

FRANCE

Table FR.6 SUPPLEMENTS TO WAGES & SALARIES - CHARGES SOCIALES DES EMPLOYEURS

Millions of FF (Current Prices)

ISIC revision 3				1991	1992	1993	1994	1995	1996	1997	1998
C	10_14		Mining and quarrying	4 600	4 554	5 723	4 238	4 247	4 195	3 170	2 871
	10		Mining of coal and lignite; extraction of peat	..	..	1 045	..	..	..	..	..
	11		Crude petroleum & natural gas; service activities	..	..	1 301	..	..	..	..	..
	12		Mining of uranium and thorium ores	..	..	59	..	..	..	..	..
	13		Mining of metal ores	126	117	121	48	46	45	41	29
	14		Other mining and quarrying	1 998	1 904	1 837	1 818	1 837	1 792	1 835	1 656
D	15_37		Total manufacturing	194 824	195 379	195 726	197 854	199 972	200 503	231 160	208 077
	15		Food products and beverages	..	..	..	..	..	..	..	..
		151	Production of meat, fish, fruit, vegetables, oils & fats	..	..	..	..	..	..	..	..
		152	Dairy products	..	..	..	..	..	..	..	..
		153	Grain mill and starch products; animal feeds	..	..	..	..	..	..	..	..
		154	Other food products	..	..	..	..	..	..	..	..
		155	Beverages	..	..	..	..	..	..	..	..
	16		Tobacco products	..	..	..	..	..	..	..	..
	17		Textiles	6 668	6 566	6 166	6 279	6 245	5 566	5 229	5 517
		171	Spinning, weaving and finishing of textiles	3 442	3 201	3 017	3 050	2 959	2 662	2 500	2 678
		172	Other textiles	1 910	1 994	1 901	2 000	2 086	1 894	1 749	1 841
		173	Knitted and crocheted fabrics and articles	1 316	1 371	1 248	1 229	1 200	1 010	979	999
	18		Wearing apparel and fur	5 372	5 373	5 076	4 730	4 499	4 003	3 583	3 728
		181	Wearing apparel, except fur apparel	5 317	5 321	5 037	4 698	4 463	3 966	3 552	3 691
		182	Dressing and dyeing of fur; articles of fur	55	52	38	32	35	37	31	37
	19		Leather and articles of leather; footwear	2 375	2 285	2 223	2 129	2 018	1 773	1 640	1 732
		191	Leather and articles of leather	808	786	732	784	725	668	648	673
		192	Footwear	1 568	1 499	1 490	1 345	1 294	1 105	991	1 058
	20		Wood & products of wood & cork, exc. furniture	2 250	2 307	2 169	2 184	2 170	2 079	2 105	2 153
		201	Sawmilling and planing of wood	40	42	49	56	68	51	54	47
		202	Wood products; articles of cork, straw, etc.	2 211	2 264	2 119	2 128	2 102	2 028	2 051	2 106
	21		Paper and paper products	5 956	6 142	6 044	6 201	6 225	6 212	6 202	6 241
	22		Printing, publishing and recorded media	11 276	11 232	11 296	11 569	11 703	11 682	11 836	12 006
		221	Publishing	5 870	5 813	5 953	6 257	6 307	6 348	6 457	6 628
		222	Printing and related activities	5 313	5 312	5 244	5 210	5 276	5 218	5 238	5 224
		223	Reproduction of recorded media	93	107	99	103	120	116	140	154
	23		Coke and petroleum products; nuclear fuel	4 665	4 850	5 020	5 180	5 162	5 070	5 105	5 070
		231	Coke oven products	1 374	1 506	1 536	1 610	1 603	1 606	1 550	1 601
		232	Refined petroleum products	3 292	3 344	3 484	3 569	3 559	3 464	3 556	3 469
		233	Processing of nuclear fuel	..	..	..	..	..	..	..	..
	24		Chemicals and chemical products	23 136	24 230	24 818	25 280	25 859	26 112	27 435	27 938
		241	Basic chemicals	7 262	7 623	7 592	7 605	7 919	8 094	8 640	8 343
		242	Other chemical products	15 550	16 264	16 853	17 297	17 689	17 795	18 563	19 394
		243	Man-made fibres	324	343	372	378	251	223	232	202
	25		Rubber and plastic products	9 998	10 356	10 621	10 775	11 114	11 172	11 704	12 431
		251	Rubber products	4 306	4 300	4 357	4 227	4 345	4 382	4 507	4 616
		252	Plastic products	5 691	6 056	6 264	6 549	6 769	6 790	7 196	7 815
	26		Other non-metallic mineral products	8 583	8 690	8 640	8 721	8 736	8 779	8 809	8 907
		261	Glass and glass products	2 966	3 127	3 135	3 209	3 291	3 314	3 358	3 392
		269	Non-metallic mineral products, nec	5 616	5 563	5 505	5 512	5 446	5 465	5 450	5 515
	27		Basic metals	10 049	10 137	9 854	9 723	9 868	9 831	9 767	9 647
		271	Basic iron and steel	6 063	6 135	5 966	5 813	5 825	5 818	5 765	5 613
		272	Basic precious and non-ferrous metals	2 016	2 066	2 089	2 089	2 116	2 111	2 105	2 045
		273	Casting of metals	1 970	1 935	1 799	1 821	1 927	1 901	1 897	1 990
	28		Fabricated metal products, exc. machinery	18 224	18 422	17 659	17 787	18 297	18 573	19 169	19 531
		281	Structural metal prod., tanks, reservoirs, etc.	7 048	7 270	7 120	6 916	6 617	6 539	6 658	6 585
	29		Machinery and equipment, nec	17 430	18 038	17 653	17 177	17 580	18 075	18 123	18 670
		291	General purpose machinery	9 294	9 869	9 802	9 589	9 817	10 240	10 387	10 795
		292	Special purpose machinery	6 514	6 443	6 137	5 914	6 048	6 125	6 183	6 379
		293	Domestic appliances, nec	1 623	1 726	1 713	1 674	1 715	1 710	1 553	1 496
	30		Office, accounting & computing machinery	6 078	5 731	5 249	5 154	4 801	4 481	4 433	5 240
	31		Electrical machinery & apparatus, nec	9 881	9 072	9 864	10 352	10 555	10 764	11 012	11 452
		311	Electric motors, generators and transformers	1 545	1 550	1 499	1 454	1 509	1 456	1 507	1 572
		312	Electricity distribution & control apparatus	4 718	3 663	4 465	4 541	4 599	4 750	4 866	5 036
		313	Insulated wire and cable	1 184	1 260	1 249	1 433	1 378	1 242	1 250	1 321
		314	Accumulators, primary cells & primary batteries	553	569	585	589	590	572	560	552
		315	Electric lamps and lighting equipment	697	709	674	698	716	708	705	746
		319	Other electrical equipment, nec	1 184	1 321	1 392	1 639	1 763	2 035	2 125	2 224
	32		Radio, TV, communication equipment	12 201	9 875	11 825	11 873	12 503	13 484	13 369	14 215
		321	Electronic valves, tubes; other electronic components	6 685	4 420	6 302	6 192	5 813	5 993	5 821	6 087
		322	TV/radio transmitters; line teleph./telegr. apparatus	4 727	4 680	4 739	4 784	5 782	6 458	6 481	7 083
		323	TV/radio receivers; sound and video app.; assoc. gds.	789	775	784	897	908	1 034	1 066	1 046
	33		Medical, precision & optical instrum.; watch & clocks	5 564	5 769	5 753	5 997	6 170	6 051	6 132	6 342
		331	Medical appl.; instruments & appl. for measuring, etc.	4 411	4 589	4 614	4 865	5 017	4 938	5 051	5 211
		332	Optical instruments and photographic equipment	823	847	832	837	865	855	823	879
		333	Watches and clocks	330	332	307	295	288	257	258	251
	34		Motor vehicles, trailers and semi-trailers	18 489	18 981	18 977	19 817	19 677	19 921	19 655	19 467
		341	Motor vehicles	12 503	12 771	12 879	13 415	13 070	13 562	13 253	12 804
		342	Bodies for motor vehicles, trailers & semi-trailers	1 204	1 195	1 129	1 084	1 141	1 144	1 154	1 260
		343	Parts & accessories for motor vehicles & their engines	4 781	5 015	4 969	5 318	5 466	5 216	5 247	5 403
	35		Other transport equipment	10 459	11 071	10 801	10 800	10 721	10 827	11 029	11 157
		351	Building and repairing of ships and boats	973	930	877	985	899	880	872	914
		352	Railway and tramway locomotives and rolling stock	506	1 206	1 212	1 195	1 152	1 190	1 209	1 229
		353	Aircraft and spacecraft	8 533	8 488	8 275	8 161	8 192	8 286	8 458	8 515
		359	Transport equipment, nec	448	447	438	458	477	471	490	499
	36		Furniture; manufacturing, nec	6 170	6 253	6 018	6 126	6 069	6 046	5 932	5 957
		361	Furniture	3 868	3 983	3 852	3 921	3 929	3 892	3 796	3 844
		369	Manufacturing, nec	2 302	2 270	2 166	2 206	2 141	2 155	2 136	2 114
	37		Recycling	..	..	..	..	..	..	..	674
E	40_41		Electricity, gas and water supply	21 157	22 278	23 317	24 093	24 660	25 439	..	26 686
	40		Electricity, gas, steam and hot water supply	18 727	19 661	20 518	21 219	21 657	22 340	22 566	23 447
	41		Collection, purification and distribution of water	2 430	2 617	2 798	2 873	3 003	3 100	3 156	3 239
F	45		Construction	..	..	..	..	..	..	..	..

Note: Parts of ISIC 10 and 11, 16, part of 23 are not covered by the inquiry and do not figure in the subtotals and totals.

FRANCE

Table FR.7 ENTREPRISES - ENTERPRISES

Unités (Prix courants)

1991	1992	1993	1994	1995	1996	1997	1998			CITI révision 3	
447	388	372	371	378	346	332	328	Activites extractives	C	10_14	
..	..	..	..	..	..	..	..	Extraction de charbon, de lignite et de tourbe		10	
..	..	..	..	..	..	..	..	Pétrole brut et gaz naturel ; activités annexes		11	
..	..	..	..	..	..	..	..	Extraction de Minerais d'Uranium et de Thorium		12	
5	6	6	5	5	5	4	3	Extraction de minerais métalliques		13	
423	363	346	347	354	321	307	306	Autres activités extractives		14	
24 484	23 880	22 438	22 332	22 066	21 803	22 231	21 792	Total activites manufacturieres	D	15_37	
..	..	..	..	..	..	..	..	Produits alimentaires et boissons		15	
..	..	..	..	..	..	..	..	Viande, poisson, fruits, légumes, huiles et graisses			151
..	..	..	..	..	..	..	..	Produits laitiers			152
..	..	..	..	..	..	..	..	Travail des grains; prod. amyl. ; alim. pour animaux			153
..	..	..	..	..	..	..	..	Autres produits alimentaires			154
..	..	..	..	..	..	..	..	Boissons			155
..	..	..	..	..	..	..	..	Produits à base de tabac		16	
1 691	1 646	1 546	1 567	1 528	1 427	1 398	1 390	Textiles		17	
826	773	732	729	730	702	671	671	Filature, tissage et achèvement des textiles			171
533	547	519	537	525	489	495	494	Autres articles textiles			172
332	326	295	301	273	236	232	225	Etoffes et articles de bonneterie			173
2 160	2 091	1 913	1 829	1 715	1 669	1 560	1 383	Articles d'habillement et fourrure		18	
2 131	2 064	1 891	1 809	1 694	1 649	1 543	1 369	Articles d'habillement autres qu'en fourrure			181
29	27	22	20	21	20	17	14	Prép. et teinture des fourrures ; articles en fourrure			182
602	559	517	492	470	441	423	403	Cuir et articles en cuir ; chaussures		19	
284	253	223	217	203	191	185	176	Cuir et articles en cuir sauf chaussures			191
318	306	294	275	267	250	238	227	Chaussures			192
755	741	686	675	668	656	649	657	Bois, articles en bois, sauf meubles		20	
12	13	17	18	23	18	16	16	Sciage et rabotage du bois			201
743	728	669	657	645	638	633	641	Articles en bois, liège, vannerie, sparterie			202
711	688	655	667	658	648	660	656	Papier, articles en papier et carton		21	
2 112	2 044	1 931	2 002	2 012	1 980	1 884	1 862	Edition, imprimerie et reproduction		22	
596	569	554	601	622	650	599	635	Edition			221
1 502	1 458	1 362	1 384	1 371	1 312	1 266	1 206	Imprimerie et activités annexes			222
14	17	15	17	19	18	19	21	Reproduction de supports enregistrés			223
57	57	54	53	64	60	60	61	Coke. et prod. pétroliers ; comb. nucléaires		23	
9	9	8	9	11	9	9	10	Cokéfaction			231
48	48	46	44	53	51	51	51	Produits pétroliers raffinés			232
..	..	..	..	..	..	..	..	Traitement de combustibles nucléaires			233
1 201	1 205	1 188	1 225	1 196	1 184	1 188	1 226	Produits chimiques		24	
241	256	243	256	272	281	283	287	Produits chimiques de base			241
954	942	935	956	911	891	893	927	Autres produits chimiques			242
6	7	10	13	13	12	12	12	Fibres synthétiques ou artificielles			243
1 380	1 401	1 364	1 382	1 412	1 419	1 448	1 465	Art. en caoutchouc et matières plastiques		25	
168	169	165	154	168	163	170	174	Articles en caoutchouc			251
1 212	1 232	1 199	1 228	1 244	1 256	1 278	1 291	Articles en matières plastiques			252
990	984	990	986	939	918	919	926	Autres produits minéraux non métalliques		26	
195	210	207	193	197	198	195	197	Verre et articles en verre			261
795	774	783	793	742	720	724	729	Produits minéraux non métalliques, nca			269
526	503	477	484	489	481	474	463	Produits métallurgiques de base		27	
132	132	125	143	147	150	149	145	Sidérurgie et première transformation de l'acier			271
107	100	98	94	94	96	91	86	Première transf. métaux précieux et non ferreux			272
287	271	254	247	248	235	234	232	Fonderie			273
4 884	4 774	4 396	4 398	4 410	4 458	4 577	4 580	Ouvrages en métaux, sauf machines		28	
1 692	1 701	1 616	1 567	1 477	1 443	1 494	1 477	Constr. et menuis. métal.; réservoirs; générateurs			281
2 578	2 562	2 419	2 198	2 146	2 196	2 235	2 237	Machines et matériels, nca		29	
1 256	1 257	1 198	1 040	1 012	1 079	1 123	1 143	Machines d'usage général			291
1 240	1 221	1 144	1 083	1 056	1 043	1 043	1 027	Machines d'usage spécifique			292
82	84	77	75	78	74	69	67	Appareils domestiques, nca			293
117	110	74	81	88	73	76	75	Mach. de bureau, comptables, information		30	
795	809	763	761	756	735	723	741	Machines et appareils électriques, nca		31	
200	205	181	170	182	157	162	167	Moteurs, génératrices, transformateurs électriques			311
300	283	264	269	251	247	247	246	Matériel électrique de distribution et de commande			312
64	69	65	63	63	61	51	55	Fils et câbles électriques isolés			313
12	13	12	14	15	13	15	14	Accumulateurs et piles électriques			314
109	108	99	112	108	102	100	103	Lampes électriques et appareils d'éclairage			315
110	131	142	133	137	155	148	156	Autres matériels électriques, nca.			319
641	583	540	558	606	601	599	603	Equipements et appareils de radio, TV, etc.		32	
391	327	319	315	332	340	337	346	Tubes et valves électroniques;autres composants			321
191	199	174	189	221	208	208	203	Emetteurs radio/TV; app. téléphonie et télégraphie			322
59	57	47	54	53	53	54	54	Récepteurs radio/TV, son et image; art. associés			323
942	897	856	891	870	863	866	859	Instr. médicaux, de précision ; optique ; horlog.		33	
729	694	666	708	700	700	709	708	Instruments médicaux et de précision, etc.			331
131	125	115	112	106	100	98	98	Instruments d'optique ; matériel photographique			332
82	78	75	71	64	63	59	53	Horlogerie			333
604	569	540	545	541	522	531	536	Véhicules automobiles; remorques, etc.		34	
52	52	46	49	45	49	55	53	Véhicules automobiles			341
287	263	248	251	255	245	251	260	Carross. automobiles, remorques, semi-remorques			342
265	254	246	245	241	228	225	223	Pièces et accessoires pour automobiles et moteurs			343
328	320	297	291	289	273	288	287	Autres matériels de transport		35	
135	130	108	101	97	87	88	93	Construction et réparation de navires			351
38	43	41	39	38	40	41	37	Matériel ferroviaire roulant			352
93	92	92	92	93	91	99	100	Construction aéronautique et spatiale			353
62	55	56	59	61	55	60	57	Autres équipements de transport			359
1 410	1 337	1 232	1 247	1 209	1 199	1 170	1 138	Meubles ; activités de fabrication, nca		36	
862	813	759	765	750	743	717	693	Meubles			361
548	524	473	482	459	456	453	445	Activités de fabrication, nca			369
..	..	..	..	..	..	..	244	Récupération		37	
154	144	164	180	178	170	171	185	Electricité, gaz et eau	E	40_41	
81	74	86	98	104	105	104	109	Prod. et dist. d'électricité gaz, vapeur d'eau chaude		40	
73	70	78	82	74	65	67	76	Captage, épuration et distribution d'eau		41	
..	..	..	..	..	..	..	..	Construction	F	45	

Note:Parts of ISIC 10 and 11, 16, part of 23 are not covered by the inquiry and do not figure in the subtotals and totals.

Table FR.8 EXPORTS - EXPORTATIONS

Millions of FF (Current Prices)

ISIC revision 3				1991	1992	1993	1994	1995	1996	1997	1998
C	10_14		Mining and quarrying	6 868	7 125	7 365	7 451	7 358	6 688	3 671	2 956
	10		Mining of coal and lignite; extraction of peat	..	..	..	..	..	..	..	..
	11		Crude petroleum & natural gas; service activities	..	..	..	..	..	..	..	..
	12		Mining of uranium and thorium ores	..	..	..	..	..	..	..	..
	13		Mining of metal ores	167	156	162	160	248	248	210	164
	14		Other mining and quarrying	2 290	2 170	2 197	2 383	2 528	2 324	2 373	2 389
D	15_37		Total manufacturing	852 783	871 485	933 883	926 814	1007 520	1064 025	1224 875	1314 563
	15		Food products and beverages	..	..	..	..	..	..	..	..
		151	Production of meat, fish, fruit, vegetables, oils & fats	..	..	..	..	..	..	..	..
		152	Dairy products	..	..	..	..	..	..	..	..
		153	Grain mill and starch products; animal feeds	..	..	..	..	..	..	..	..
		154	Other food products	..	..	..	..	..	..	..	..
		155	Beverages	..	..	..	..	..	..	..	..
	16		Tobacco products	..	..	..	..	..	..	..	..
	17		Textiles	28 399	28 494	25 124	28 054	30 101	29 560	31 325	35 310
		171	Spinning, weaving and finishing of textiles	18 523	17 409	15 431	17 351	18 031	17 879	19 892	21 485
		172	Other textiles	6 568	7 434	6 805	7 855	9 257	8 874	8 187	10 510
		173	Knitted and crocheted fabrics and articles	3 307	3 651	2 889	2 848	2 813	2 807	3 245	3 315
	18		Wearing apparel and fur	14 699	15 235	14 396	15 013	16 545	16 872	17 826	18 018
		181	Wearing apparel, except fur apparel	14 446	14 989	14 225	14 847	16 263	16 547	17 534	17 763
		182	Dressing and dyeing of fur; articles of fur	253	246	170	166	281	324	292	255
	19		Leather and articles of leather; footwear	7 871	7 442	7 176	7 845	7 399	7 453	7 290	7 589
		191	Leather and articles of leather	2 929	3 038	2 870	4 328	3 774	3 855	3 580	3 703
		192	Footwear	4 942	4 405	4 305	3 517	3 625	3 598	3 710	3 885
	20		Wood & products of wood & cork, exc. furniture	3 960	4 443	4 265	5 267	5 751	5 420	5 999	7 179
		201	Sawmilling and planing of wood	1	13	2	1	12	2	1	1
		202	Wood products; articles of cork, straw, etc.	3 959	4 431	4 263	5 266	5 739	5 418	5 999	7 178
	21		Paper and paper products	22 964	24 821	24 512	27 032	32 993	29 926	33 715	35 243
	22		Printing, publishing and recorded media	8 237	8 542	9 002	9 757	9 960	10 602	10 906	12 020
		221	Publishing	5 210	4 915	5 318	5 579	5 683	6 275	5 940	6 370
		222	Printing and related activities	2 816	3 260	3 325	3 780	3 798	3 722	4 326	5 093
		223	Reproduction of recorded media	210	366	359	398	480	605	639	557
	23		Coke and petroleum products; nuclear fuel	32 873	28 339	26 137	25 695	26 619	36 130	35 127	32 115
		231	Coke oven products	4 936	5 613	6 747	8 400	9 965	13 139	11 332	11 647
		232	Refined petroleum products	27 937	22 726	19 390	17 295	16 653	22 990	23 795	20 469
		233	Processing of nuclear fuel	..	..	..	..	..	..	..	..
	24		Chemicals and chemical products	135 843	138 830	140 774	156 373	172 886	179 078	205 334	206 535
		241	Basic chemicals	63 174	61 318	61 374	70 896	77 838	76 914	87 626	85 356
		242	Other chemical products	70 834	75 741	77 622	83 836	93 136	100 385	114 999	118 418
		243	Man-made fibres	1 835	1 772	1 778	1 640	1 912	1 779	2 709	2 760
	25		Rubber and plastic products	33 279	34 189	33 466	35 951	38 851	41 830	46 502	54 047
		251	Rubber products	19 130	19 292	18 429	18 995	20 339	22 375	24 899	28 127
		252	Plastic products	14 149	14 897	15 037	16 956	18 512	19 455	21 603	25 920
	26		Other non-metallic mineral products	21 871	21 603	22 095	23 472	25 281	25 880	28 422	29 183
		261	Glass and glass products	11 581	11 374	11 710	12 822	13 646	13 601	15 320	15 469
		269	Non-metallic mineral products, nec	10 290	10 229	10 385	10 650	11 636	12 279	13 102	13 715
	27		Basic metals	69 439	66 002	61 358	67 778	80 554	77 655	88 039	89 311
		271	Basic iron and steel	43 951	40 623	38 663	42 709	49 997	48 265	54 198	55 902
		272	Basic precious and non-ferrous metals	21 423	21 404	18 990	20 993	25 515	24 031	28 159	26 834
		273	Casting of metals	4 065	3 976	3 705	4 077	5 042	5 358	5 683	6 575
	28		Fabricated metal products, exc. machinery	29 084	30 424	29 103	39 731	35 870	36 708	39 854	43 791
		281	Structural metal prod., tanks, reservoirs, etc.	11 131	12 316	11 751	19 490	12 295	11 091	11 292	11 394
	29		Machinery and equipment, nec	76 283	82 576	78 252	90 222	98 576	101 770	105 042	117 232
		291	General purpose machinery	38 663	43 714	40 218	47 603	49 236	50 793	50 601	59 599
		292	Special purpose machinery	29 634	29 973	29 344	33 859	39 665	40 437	43 802	47 379
		293	Domestic appliances, nec	7 985	8 889	8 690	8 759	9 676	10 539	10 639	10 254
	30		Office, accounting & computing machinery	30 341	29 666	28 538	28 394	35 115	37 528	37 905	40 962
	31		Electrical machinery & apparatus, nec	37 629	34 276	38 723	41 163	46 754	50 429	56 344	61 937
		311	Electric motors, generators and transformers	5 921	5 434	5 473	5 282	5 928	6 297	7 049	8 547
		312	Electricity distribution & control apparatus	16 812	12 715	16 171	17 526	18 985	19 947	22 259	27 347
		313	Insulated wire and cable	5 357	6 002	6 673	6 182	7 291	7 202	8 043	6 829
		314	Accumulators, primary cells & primary batteries	2 206	2 083	2 322	2 455	2 764	2 821	2 797	2 705
		315	Electric lamps and lighting equipment	2 194	2 407	2 628	3 089	3 412	3 762	4 498	5 002
		319	Other electrical equipment, nec	5 139	5 634	5 455	6 629	8 374	10 401	11 697	11 508
	32		Radio, TV, communication equipment	46 018	49 143	43 610	52 008	59 629	76 042	96 337	110 027
		321	Electronic valves, tubes; other electronic components	26 626	27 479	21 605	25 011	28 242	36 688	45 539	42 551
		322	TV/radio transmitters; line teleph./telegr. apparatus	9 978	11 550	13 293	18 020	20 363	24 469	32 951	49 136
		323	TV/radio receivers; sound and video app.; assoc. gds.	9 415	10 113	8 712	8 977	11 023	14 885	17 847	18 340
	33		Medical, precision & optical instrum.; watch & clocks	19 086	19 351	20 199	22 214	23 941	25 232	27 109	28 757
		331	Medical appl.; instruments & appl. for measuring, etc.	14 549	15 010	16 416	18 182	19 438	20 813	22 530	23 471
		332	Optical instruments and photographic equipment	2 835	2 743	2 361	2 614	2 969	2 986	3 205	3 883
		333	Watches and clocks	1 703	1 599	1 423	1 418	1 534	1 433	1 374	1 402
	34		Motor vehicles, trailers and semi-trailers	158 990	169 384	156 424	177 062	189 323	200 132	235 134	272 548
		341	Motor vehicles	129 340	138 258	127 116	145 049	151 299	161 609	195 208	226 882
		342	Bodies for motor vehicles, trailers & semi-trailers	3 881	3 328	2 431	2 661	3 041	3 197	3 668	5 037
		343	Parts & accessories for motor vehicles & their engines	25 769	27 799	26 878	29 352	34 983	35 327	36 258	40 630
	35		Other transport equipment	59 461	61 822	51 490	56 081	52 611	56 032	72 389	86 296
		351	Building and repairing of ships and boats	7 329	5 713	4 167	7 812	6 923	4 673	6 754	5 799
		352	Railway and tramway locomotives and rolling stock	1 453	3 798	3 406	3 738	3 462	3 870	5 364	6 290
		353	Aircraft and spacecraft	48 093	49 759	41 562	41 748	39 111	44 319	56 675	70 146
		359	Transport equipment, nec	2 586	2 552	2 354	2 783	3 116	3 170	3 596	4 062
	36		Furniture; manufacturing, nec	16 456	16 900	16 687	17 701	18 761	19 747	20 956	22 466
		361	Furniture	6 558	6 924	6 664	7 119	7 591	8 058	8 512	9 580
		369	Manufacturing, nec	9 898	9 976	10 023	10 582	11 170	11 689	12 444	12 886
	37		Recycling	..	..	..	..	..	..	..	3 996
E	40_41		Electricity, gas and water supply	15 861	16 452	17 944	19 363	21 551	20 802	19 649	17 796
	40		Electricity, gas, steam and hot water supply	15 604	16 112	17 572	18 936	21 238	20 481	19 122	17 304
	41		Collection, purification and distribution of water	258	340	373	427	313	321	527	492
F	45		Construction	..	..	..	..	..	..	..	..

Note:Parts of ISIC 10 and 11, 16, part of 23 are not covered by the inquiry and do not figure in the subtotals and totals.

ALLEMAGNE

Table GE.1 **PRODUCTION - PRODUCTION**

Millions de DM (Prix courants)

1991	1992	1993	1994	1995	1996	1997	1998			CITI révision 3
44 318	41 759	40 670	41 116	40 518	29 347	25 614	23 286	Activites extractives	C	10_14
29 978	26 477	26 311	24 872	23 861	13 630	13 029	11 187	Extraction de charbon, de lignite et de tourbe		10
5 063	5 384	4 962	5 232	5 685	5 798	4 272	4 100	Pétrole brut et gaz naturel ; activités annexes		11
..	..	..	..	..	..	..	..	Extraction de Minerais d'Uranium et de Thorium		12
494	503	552	608	653	..	..	..	Extraction de minerais métalliques		13
8 784	9 395	8 845	10 404	10 319	9 918	8 313	7 999	Autres activités extractives		14
2 023 736	2 025 012	1 899 201	1 966 634	2 046 645	2 063 339	2 174 872	2 247 538	Total activites manufacturieres	D	15_37
219 437	223 643	221 681	223 553	225 617	226 970	231 050	228 559	Produits alimentaires et boissons		15
..	..	..	..	60 725	63 831	65 905	64 769	Viande, poisson, fruits, légumes, huiles et graisses		151
..	..	..	..	40 031	39 794	39 396	40 111	Produits laitiers		152
..	..	..	..	16 011	16 653	16 672	16 085	Travail des grains; prod. amyl. ; alim. pour animaux		153
..	..	..	..	65 456	65 104	67 625	67 353	Autres produits alimentaires		154
..	..	..	..	43 394	39 306	38 571	40 241	Boissons		155
29 327	28 007	29 340	29 343	29 918	29 633	29 593	30 053	Produits à base de tabac		16
40 017	38 284	34 313	33 107	32 137	31 550	31 697	32 055	Textiles		17
..	..	..	..	15 917	3 381	3 377	..	Filature, tissage et achèvement des textiles		171
..	..	..	..	11 661	12 117	12 456	13 112	Autres articles textiles		172
..	..	..	..	4 559	4 485	4 057	4 008	Etoffes et articles de bonneterie		173
30 181	28 277	26 928	24 752	23 518	22 462	22 543	22 608	Articles d'habillement et fourrure		18
..	..	..	..	23 456	22 401	22 483	22 555	Articles d'habillement autres qu'en fourrure		181
..	..	..	..	61	62	60	53	Prép. et teinture des fourrures ; articles en fourrure		182
9 753	9 136	8 775	8 589	8 363	8 545	8 615	8 697	Cuir et articles en cuir ; chaussures		19
..	..	..	..	2 412	2 309	2 111	2 107	Cuir et articles en cuir sauf chaussures		191
..	..	..	..	5 952	6 236	6 505	6 590	Chaussures		192
28 238	29 477	28 670	31 755	32 718	30 943	31 981	32 378	Bois, articles en bois, sauf meubles		20
..	..	..	..	7 988	7 325	7 919	8 069	Sciage et rabotage du bois		201
..	..	..	..	24 729	23 617	24 061	24 310	Articles en bois, liège, vannerie, sparterie		202
50 094	48 746	44 522	46 911	54 033	50 319	50 445	51 544	Papier, articles en papier et carton		21
..	..	..	..	68 768	71 744	74 298	77 508	Edition, imprimerie et reproduction		22
..	..	..	..	35 117	38 573	42 475	44 729	Edition		221
..	..	..	..	32 386	31 889	29 884	31 278	Imprimerie et activités annexes		222
..	..	..	..	..	..	..	704	Reproduction de supports enregistrés		223
..	..	..	..	111 145	120 445	132 396	103 346	Coke. et prod. pétroliers ; comb. nucléaires		23
..	..	..	..	..	..	..	1 376	Cokéfaction		231
..	..	..	..	109 300	118 433	130 189	101 452	Produits pétroliers raffinés		232
..	..	..	..	..	..	..	517	Traitement de combustibles nucléaires		233
206 922	204 704	194 521	223 110	219 652	215 653	231 091	228 455	Produits chimiques		24
..	..	..	..	111 159	107 900	117 242	101 837	Produits chimiques de base		241
..	..	..	..	101 248	100 704	106 357	119 981	Autres produits chimiques		242
..	..	..	..	7 244	7 050	7 492	6 638	Fibres synthétiques ou artificielles		243
80 248	82 543	78 203	82 643	89 007	86 682	90 267	93 320	Art. en caoutchouc et matières plastiques		25
..	..	..	..	21 512	20 490	21 308	21 839	Articles en caoutchouc		251
..	..	..	..	67 495	66 192	68 959	71 481	Articles en matières plastiques		252
60 351	65 069	66 603	73 379	73 706	70 277	70 204	69 289	Autres produits minéraux non métalliques		26
..	..	..	..	15 421	15 117	15 394	15 861	Verre et articles en verre		261
..	..	..	..	58 283	55 159	54 810	53 430	Produits minéraux non métalliques, nca		269
104 250	97 862	84 104	93 485	101 610	92 433	99 911	102 784	Produits métallurgiques de base		27
..	..	..	..	58 275	51 581	55 749	57 599	Sidérurgie et première transformation de l'acier		271
..	..	..	..	31 441	29 098	31 540	30 956	Première transf. métaux précieux et non ferreux		272
..	..	..	..	11 895	11 753	12 622	14 229	Fonderie		273
120 255	125 835	117 825	118 910	128 124	126 571	130 936	139 637	Ouvrages en métaux, sauf machines		28
..	..	..	..	48 486	48 493	48 456	50 004	Constr. et menuis. métal.; réservoirs; générateurs		281
..	..	..	..	79 636	78 078	82 480	89 631	Autres ouvrages en métaux		289
261 592	266 007	243 583	247 237	252 712	260 183	270 433	285 250	Machines et matériels, nca		29
..	..	..	..	120 614	123 317	125 258	133 052	Machines d'usage général		291
..	..	..	..	109 165	113 659	120 566	127 816	Machines d'usage spécifique		292
..	..	..	..	22 931	23 206	24 608	24 381	Appareils domestiques, nca		293
25 798	25 329	23 203	27 058	28 158	26 442	27 405	27 489	Mach. de bureau, comptables, information		30
106 646	102 495	97 600	119 372	123 719	127 021	118 501	127 182	Machines et appareils électriques, nca		31
..	..	..	..	15 681	14 599	17 094	19 029	Moteurs, génératrices, transformateurs électriques		311
..	..	..	..	68 464	72 486	59 924	62 618	Matériel électrique de distribution et de commande		312
..	..	..	..	6 871	6 242	7 664	7 578	Fils et câbles électriques isolés		313
..	..	..	..	3 145	2 462	2 572	2 516	Accumulateurs et piles électriques		314
..	..	..	..	9 184	9 055	9 466	9 812	Lampes électriques et appareils d'éclairage		315
..	..	..	..	20 374	22 177	21 780	25 628	Autres matériels électriques, nca.		319
55 085	57 555	40 525	40 756	39 553	39 269	57 840	60 937	Equipements et appareils de radio, TV, etc.		32
..	..	..	..	10 500	11 243	19 972	21 965	Tubes et valves électroniques;autres composants		321
..	..	..	..	11 957	15 256	25 714	26 131	Emetteurs radio/TV; app. téléphonie et télégraphie		322
..	..	..	..	17 095	12 770	12 154	12 841	Récepteurs radio/TV, son et image; art. associés		323
53 319	55 606	68 250	50 087	49 956	51 509	53 568	56 136	Instr. médicaux, de précision ; optique ; horlog.		33
..	..	..	..	41 908	42 491	44 741	47 512	Instruments médicaux et de précision, etc.		331
..	..	..	..	6 580	7 667	7 526	7 295	Instruments d'optique ; matériel photographique		332
..	..	..	..	1 468	1 350	1 303	1 329	Horlogerie		333
261 215	259 158	218 809	234 524	262 731	283 553	315 948	371 592	Véhicules automobiles; remorques, etc.		34
..	..	..	..	197 704	217 160	239 458	286 787	Véhicules automobiles		341
..	..	..	..	10 703	9 935	12 890	12 064	Carross. automobiles, remorques, semi-remorques		342
..	..	..	..	54 325	56 457	63 600	72 741	Pièces et accessoires pour automobiles et moteurs		343
38 080	35 543	35 251	34 546	33 400	33 238	38 316	39 262	Autres matériels de transport		35
..	..	..	..	10 108	8 490	8 225	7 790	Construction et réparation de navires		351
..	..	..	..	6 071	5 575	6 618	6 260	Matériel ferroviaire roulant		352
..	..	..	..	13 792	16 037	20 365	21 919	Construction aéronautique et spatiale		353
..	..	..	..	3 429	3 136	3 108	3 293	Autres équipements de transport		359
55 928	56 660	55 839	55 219	55 987	55 711	54 963	56 637	Meubles ; activités de fabrication, nca		36
..	..	..	..	43 273	42 890	42 055	43 736	Meubles		361
..	..	..	..	12 715	12 820	11 347	11 293	Activités de fabrication, nca		369
..	..	..	..	2 114	2 186	2 872	2 818	Récupération		37
..	..	..	..	..	..	..	..	Electricité, gaz et eau	E	40_41
..	..	..	..	..	..	..	..	Prod. et dist. d'électricité gaz, vapeur d'eau chaude		40
..	..	..	..	..	..	..	..	Construction	F	45

Table GE.2 EMPLOYMENT - EMPLOI

Units

ISIC revision 3				1991	1992	1993	1994	1995	1996	1997	1998
C	10_14		Mining and quarrying	308 784	252 723	220 178	197 649	185 736	166 722	147 972	137 767
	10		Mining of coal and lignite; extraction of peat	241 332	199 760	172 960	150 982	139 644	122 988	111 823	102 533
	11		Crude petroleum & natural gas; service activities	7 709	7 032	6 750	6 473	6 328	6 036	5 741	5 787
	12		Mining of uranium and thorium ores	..	..	..	..				
	13		Mining of metal ores	2 977	1 168	427	399	352			
	14		Other mining and quarrying	56 765	44 764	40 041	39 795	39 412	37 698	30 408	29 447
D	15_37		Total manufacturing	9124 419	8159 693	7434 831	6942 979	6700 505	6460 097	6267 669	6266 788
	15		Food products and beverages	685 728	631 609	601 333	585 818	573 581	568 626	551 650	544 082
		151	Production of meat, fish, fruit, vegetables, oils & fats	..	..	..	..	162 754	162 413	153 586	151 459
		152	Dairy products	..	..	..	..	49 241	46 994	44 745	43 866
		153	Grain mill and starch products; animal feeds	..	..	..	..	21 754	21 788	21 870	21 433
		154	Other food products	..	..	..	..	254 008	254 729	253 291	251 979
		155	Beverages	..	..	..	..	85 415	82 130	77 793	75 346
	16		Tobacco products	19 293	17 076	15 850	15 455	14 633	13 794	12 969	12 595
	17		Textiles	275 560	209 291	180 319	164 229	151 215	139 555	132 057	128 973
		171	Spinning, weaving and finishing of textiles	..	..	..	..	71 194	63 776	59 099	..
		172	Other textiles	..	..	..	..	55 006	53 562	52 280	52 530
		173	Knitted and crocheted fabrics and articles	..	..	..	..	24 981	22 194	20 651	18 892
	18		Wearing apparel and fur	216 791	167 805	139 123	121 482	105 948	93 667	84 398	79 023
		181	Wearing apparel, except fur apparel	..	..	..	..	105 412	93 224	83 942	78 589
		182	Dressing and dyeing of fur; articles of fur	..	..	..	..	536	442	456	434
	19		Leather and articles of leather; footwear	76 716	52 405	42 839	38 447	34 543	31 479	28 945	28 060
		191	Leather and articles of leather	..	..	..	..	12 573	11 246	9 982	9 470
		192	Footwear	..	..	..	..	21 972	20 233	18 964	18 590
	20		Wood & products of wood & cork, exc. furniture	151 592	143 434	132 923	132 211	130 431	124 093	119 363	116 319
		201	Sawmilling and planing of wood	..	..	..	..	26 714	25 107	24 538	24 076
		202	Wood products; articles of cork, straw, etc.	..	..	..	..	103 718	98 987	94 826	92 244
	21		Paper and paper products	181 918	171 306	159 427	153 127	154 747	152 145	148 304	145 972
	22		Printing, publishing and recorded media	..	..	..	..	268 124	263 362	259 734	258 674
		221	Publishing	..	..	..	..	107 565	112 432	115 821	114 863
		222	Printing and related activities	..	..	..	..	155 503	146 264	138 866	135 871
		223	Reproduction of recorded media	..	..	..	..	..	774	..	5 396
	23		Coke and petroleum products; nuclear fuel	..	..	..	..	26 831	24 461	22 854	21 608
		231	Coke oven products	..	..	..	..	2 278	2 158	..	1 947
		232	Refined petroleum products	..	..	..	..	23 155	21 246	19 896	18 782
		233	Processing of nuclear fuel	..	..	..	..	1 398	1 058	..	878
	24		Chemicals and chemical products	709 347	649 837	595 673	561 075	535 896	517 532	500 546	484 639
		241	Basic chemicals	..	..	..	..	246 873	239 666	226 806	184 223
		242	Other chemical products	..	..	..	..	264 736	255 212	252 640	280 793
		243	Man-made fibres	..	..	..	..	24 288	22 655	21 102	19 623
	25		Rubber and plastic products	416 279	401 096	373 394	360 901	365 278	350 856	346 070	353 009
		251	Rubber products	..	..	..	..	84 140	78 108	76 738	78 341
		252	Plastic products	..	..	..	..	280 960	272 580	269 168	274 667
	26		Other non-metallic mineral products	345 971	310 198	295 660	289 299	284 868	270 198	258 377	252 750
		261	Glass and glass products	..	..	..	..	68 891	66 951	65 057	64 341
		269	Non-metallic mineral products, nec	..	..	..	..	215 976	203 248	193 320	188 408
	27		Basic metals	476 857	408 954	353 938	319 068	294 968	279 543	271 352	270 719
		271	Basic iron and steel	..	..	..	..	161 904	149 956	144 422	140 908
		272	Basic precious and non-ferrous metals	..	..	..	..	65 135	62 290	61 366	61 982
		273	Casting of metals	..	..	..	..	..	67 270	65 538	67 829
	28		Fabricated metal products, exc. machinery	726 526	690 010	643 951	603 026	613 906	590 740	574 954	592 322
		281	Structural metal prod., tanks, reservoirs, etc.	..	..	..	..	219 239	208 023	198 237	197 457
	29		Machinery and equipment, nec	1571 708	1382 307	1224 429	1118 485	1055 269	1031 355	991 985	986 921
		291	General purpose machinery	..	..	..	..	504 962	493 741	476 035	475 212
		292	Special purpose machinery	..	..	..	..	471 372	462 217	443 603	442 465
		293	Domestic appliances, nec	..	..	..	..	78 935	75 397	72 348	69 245
	30		Office, accounting & computing machinery	108 561	86 896	62 501	55 564	49 685	45 129	45 636	42 187
	31		Electrical machinery & apparatus, nec	656 462	568 624	514 302	499 818	498 063	472 785	448 678	452 945
		311	Electric motors, generators and transformers	..	..	..	..	83 410	75 990	70 990	72 872
		312	Electricity distribution & control apparatus	..	..	..	..	249 399	240 561	227 481	225 030
		313	Insulated wire and cable	..	..	..	..	26 248	23 237	22 225	23 025
		314	Accumulators, primary cells & primary batteries	..	..	..	..	10 563	9 367	8 600	8 362
		315	Electric lamps and lighting equipment	..	..	..	..	37 192	34 981	34 688	34 804
		319	Other electrical equipment, nec	..	..	..	..	91 252	88 650	84 693	88 852
	32		Radio, TV, communication equipment	280 004	235 024	192 130	175 783	158 861	152 483	146 135	148 918
		321	Electronic valves, tubes; other electronic components	..	..	..	..	59 421	61 677	61 448	61 378
		322	TV/radio transmitters; line teleph./telegr. apparatus	..	..	..	..	55 314	53 363	53 434	56 471
		323	TV/radio receivers; sound and video app.; assoc. gds.	..	..	..	..	44 125	37 443	31 253	31 068
	33		Medical, precision & optical instrum.; watch & clocks	361 992	318 677	319 020	268 724	244 072	235 294	228 791	223 914
		331	Medical appl.; instruments & appl. for measuring, etc.	..	..	..	..	205 620	196 361	192 105	189 553
		332	Optical instruments and photographic equipment	..	..	..	..	31 356	32 563	31 064	29 058
		333	Watches and clocks	..	..	..	..	7 096	6 370	5 623	5 303
	34		Motor vehicles, trailers and semi-trailers	823 955	769 062	706 180	661 530	691 902	684 420	698 278	739 446
		341	Motor vehicles	..	..	..	..	399 022	396 272	400 614	425 414
		342	Bodies for motor vehicles, trailers & semi-trailers	..	..	..	..	45 917	41 831	41 359	39 061
		343	Parts & accessories for motor vehicles & their engines	..	..	..	..	246 963	246 318	256 305	274 971
	35		Other transport equipment	256 975	229 584	219 102	197 928	182 561	164 568	153 849	145 781
		351	Building and repairing of ships and boats	..	..	..	..	36 887	31 998	28 422	25 510
		352	Railway and tramway locomotives and rolling stock	..	..	..	..	69 681	61 189	55 353	48 667
		353	Aircraft and spacecraft	..	..	..	..	62 905	59 115	58 558	59 927
		359	Transport equipment, nec	..	..	..	..	13 088	12 267	11 516	11 676
	36		Furniture; manufacturing, nec	334 574	302 439	282 290	268 145	258 585	246 525	235 208	230 487
		361	Furniture	..	..	..	..	188 647	179 722	171 666	169 460
		369	Manufacturing, nec	..	..	..	..	69 567	66 455	63 216	53 699
	37		Recycling	..	..	..	..	6 540	7 487	7 536	7 444
E	40_41		Electricity, gas and water supply	..	..	..	..	..	..	..	..
	40		Electricity, gas, steam and hot water supply	..	..	..	..	..	..	..	..
	41		Collection, purification and distribution of water	..	..	..	..	..	..	..	..
F	45		Construction	..	..	..	..	..	..	..	..

Table GE.3 **SALAIRES ET TRAITEMENTS - TOTAL WAGES AND SALARIES**

Millions de DM (Prix courants)

1991	1992	1993	1994	1995	1996	1997	1998		CITI révision 3		
13 578	13 701	12 919	11 541	11 437	10 751	9 788	9 182	Activites extractives	C	10_14	
10 750	10 872	10 219	8 747	8 567	7 973	7 465	6 884	Extraction de charbon, de lignite et de tourbe		10	
522	543	554	564	568	570	553	562	Pétrole brut et gaz naturel ; activités annexes		11	
..	..	..	..	..	..	..	..	Extraction de Minerais d'Uranium et de Thorium		12	
100	62	26	27	25	..	..	..	Extraction de minerais métalliques		13	
2 206	2 223	2 119	2 203	2 276	2 207	1 771	1 736	Autres activités extractives		14	
423 802	433 461	413 855	403 607	406 823	404 612	398 828	407 147	Total activites manufacturieres	D	15_37	
25 056	26 022	26 335	26 324	26 198	26 433	25 915	25 761	Produits alimentaires et boissons		15	
..	..	..	..	6 820	6 935	6 563	6 538	Viande, poisson, fruits, légumes, huiles et graisses			151
..	..	..	..	2 689	2 631	2 538	2 521	Produits laitiers			152
..	..	..	..	1 382	1 401	1 439	1 413	Travail des grains; prod. amyl. ; alim. pour animaux			153
..	..	..	..	10 179	10 351	10 495	10 469	Autres produits alimentaires			154
..	..	..	..	5 131	4 894	4 698	4 819	Boissons			155
1 049	1 012	1 072	1 121	1 115	1 076	1 007	1 005	Produits à base de tabac		16	
8 877	8 366	7 657	7 244	6 908	6 569	6 298	6 228	Textiles		17	
..	..	..	..	3 438	916	866	814	Filature, tissage et achèvement des textiles			171
..	..	..	..	2 473	2 503	2 475	2 547	Autres articles textiles			172
..	..	..	..	997	910	853	790	Etoffes et articles de bonneterie			173
5 963	5 483	4 938	4 452	4 078	3 780	3 517	3 377	Articles d'habillement et fourrure		18	
..	..	..	..	4 058	3 760	3 497	3 359	Articles d'habillement autres qu'en fourrure			181
..	..	..	..	20	19	19	19	Prép. et teinture des fourrures ; articles en fourrure			182
2 075	1 870	1 663	1 570	1 480	1 407	1 332	1 335	Cuir et articles en cuir ; chaussures		19	
..	..	..	..	499	460	414	398	Cuir et articles en cuir sauf chaussures			191
..	..	..	..	981	947	919	937	Chaussures			192
6 040	6 431	6 342	6 477	6 622	6 384	6 215	6 142	Bois, articles en bois, sauf meubles		20	
..	..	..	..	1 254	1 194	1 191	1 174	Sciage et rabotage du bois			201
..	..	..	..	5 368	5 190	5 023	4 969	Articles en bois, liège, vannerie, sparterie			202
8 544	8 868	8 627	8 535	8 916	8 962	8 886	8 942	Papier, articles en papier et carton		21	
..	..	..	..	17 295	17 622	17 392	17 660	Edition, imprimerie et reproduction		22	
..	..	..	..	7 433	8 117	8 227	8 534	Edition			221
..	..	..	..	9 582	9 241	8 735	8 638	Imprimerie et activités annexes			222
..	..	..	..	..	..	..	352	Reproduction de supports enregistrés			223
..	..	..	..	2 064	1 980	1 945	1 920	Coke. et prod. pétroliers ; comb. nucléaires		23	
..	..	..	..	144	146	..	133	Cokéfaction			231
..	..	..	..	1 801	1 739	1 725	1 704	Produits pétroliers raffinés			232
..	..	..	..	120	95	..	82	Traitement de combustibles nucléaires			233
40 800	42 283	40 835	39 874	39 908	39 645	38 902	38 410	Produits chimiques		24	
..	..	..	..	19 625	19 646	19 877	15 397	Produits chimiques de base			241
..	..	..	..	18 762	18 545	18 549	21 696	Autres produits chimiques			242
..	..	..	..	1 521	1 453	1 376	1 318	Fibres synthétiques ou artificielles			243
18 884	19 811	19 132	19 191	19 777	19 325	19 385	20 079	Art. en caoutchouc et matières plastiques		25	
..	..	..	..	5 000	4 713	4 746	4 920	Articles en caoutchouc			251
..	..	..	..	14 778	14 613	14 639	15 159	Articles en matières plastiques			252
14 432	15 049	15 216	15 708	15 879	15 414	14 811	14 781	Autres produits minéraux non métalliques		26	
..	..	..	..	3 804	3 825	3 706	3 692	Verre et articles en verre			261
..	..	..	..	12 075	11 588	11 103	11 088	Produits minéraux non métalliques, nca			269
22 014	21 659	19 260	18 571	18 167	17 414	17 557	17 846	Produits métallurgiques de base		27	
..	..	..	..	9 913	9 265	9 353	9 249	Sidérurgie et première transformation de l'acier			271
..	..	..	..	4 276	4 187	4 274	4 413	Première transf. métaux précieux et non ferreux			272
..	..	..	..	3 978	3 962	3 930	4 183	Fonderie			273
32 904	34 325	32 937	32 011	34 200	33 634	33 087	34 692	Ouvrages en métaux, sauf machines		28	
..	..	..	..	12 445	12 054	11 517	11 674	Constr. et menuis. métal.; réservoirs; générateurs			281
..	..	..	..	21 756	21 579	21 571	23 018	Autres ouvrages en métaux			289
74 292	75 488	70 782	68 501	68 337	68 711	67 360	68 647	Machines et matériels, nca		29	
..	..	..	..	33 006	33 244	32 503	33 272	Machines d'usage général			291
..	..	..	..	30 628	30 817	30 305	30 919	Machines d'usage spécifique			292
..	..	..	..	4 703	4 648	4 551	4 457	Appareils domestiques, nca			293
6 167	6 070	4 691	4 331	3 911	3 800	3 608	3 261	Mach. de bureau, comptables, information		30	
30 563	30 495	29 436	30 350	31 589	31 520	30 205	31 271	Machines et appareils électriques, nca		31	
..	..	..	..	4 875	4 722	4 407	4 544	Moteurs, génératrices, transformateurs électriques			311
..	..	..	..	17 087	17 284	16 571	16 984	Matériel électrique de distribution et de commande			312
..	..	..	..	1 624	1 489	1 457	1 532	Fils et câbles électriques isolés			313
..	..	..	..	645	607	547	543	Accumulateurs et piles électriques			314
..	..	..	..	2 031	2 003	2 014	2 064	Lampes électriques et appareils d'éclairage			315
..	..	..	..	5 327	5 415	5 210	5 604	Autres matériels électriques, nca.			319
13 419	13 973	11 981	11 600	10 910	10 765	10 767	11 195	Equipements et appareils de radio, TV, etc.		32	
..	..	..	..	3 972	4 235	4 313	4 349	Tubes et valves électroniques;autres composants			321
..	..	..	..	4 294	4 208	4 449	4 814	Emetteurs radio/TV; app. téléphonie et télégraphie			322
..	..	..	..	2 645	2 322	2 005	2 032	Récepteurs radio/TV, son et image; art. associés			323
16 545	17 410	18 281	15 577	14 912	15 009	14 716	14 706	Instr. médicaux, de précision ; optique ; horlog.		33	
..	..	..	..	12 720	12 604	12 399	12 497	Instruments médicaux et de précision, etc.			331
..	..	..	..	1 805	2 046	1 994	1 902	Instruments d'optique ; matériel photographique			332
..	..	..	..	386	359	322	308	Horlogerie			333
48 563	49 849	46 144	45 301	49 637	51 240	52 707	56 959	Véhicules automobiles; remorques, etc.		34	
..	..	..	..	31 043	32 157	32 919	35 841	Véhicules automobiles			341
..	..	..	..	2 447	2 267	2 333	2 220	Carross. automobiles, remorques, semi-remorques			342
..	..	..	..	16 146	16 815	17 455	18 897	Pièces et accessoires pour automobiles et moteurs			343
11 663	11 948	12 323	11 835	11 532	10 801	10 583	10 374	Autres matériels de transport		35	
..	..	..	..	2 240	1 902	1 756	1 553	Construction et réparation de navires			351
..	..	..	..	3 592	3 419	3 227	3 015	Matériel ferroviaire roulant			352
..	..	..	..	5 005	4 814	4 931	5 104	Construction aéronautique et spatiale			353
..	..	..	..	695	667	670	703	Autres équipements de transport			359
13 289	13 570	13 358	13 100	13 065	12 751	12 261	12 177	Meubles ; activités de fabrication, nca		36	
..	..	..	..	9 844	9 577	9 185	9 183	Meubles			361
..	..	..	..	3 223	3 173	2 720	2 625	Activités de fabrication, nca			369
..	..	..	..	322	370	372	380	Récupération		37	
..	..	..	..	..	..	..	..	Electricité, gaz et eau	E	40_41	
..	..	..	..	..	..	..	..	Prod. et dist. d'électricité gaz, vapeur d'eau chaude		40	
..	..	..	..	..	..	..	..	Construction	F	45	

GERMANY

Table GE.4 **HOURS WORKED BY OPERATIVES - HEURES OUVRÉES, SALARIÉS**

Thousands

ISIC revision 3				1991	1992	1993	1994	1995	1996	1997	1998
C	10_14		Mining and quarrying	337 051	291 229	244 362	212 466	200 899	175 869	154 485	144 036
	10		Mining of coal and lignite; extraction of peat	261 969	225 745	186 269	153 923	143 906	123 106	110 638	101 755
	11		Crude petroleum & natural gas; service activities	5 632	4 453	4 390	4 214	4 034	3 667	3 502	3 404
	12		Mining of uranium and thorium ores	..	..	..	..	..	..	..	..
	13		Mining of metal ores	3 016	1 408	526	521	439			
	14		Other mining and quarrying	66 434	59 623	53 177	53 808	52 520	49 096	40 345	38 877
D	15_37		Total manufacturing	9123 556	8403 517	7382 293	7013 665	6807 442	6428 411	6245 787	6296 293
	15		Food products and beverages	780 757	728 508	679 633	656 266	635 066	620 679	594 038	579 423
		151	Production of meat, fish, fruit, vegetables, oils & fats	..	..	..	..	195 852	194 210	183 414	180 921
		152	Dairy products	..	..	..	..	61 963	58 264	54 275	52 008
		153	Grain mill and starch products; animal feeds	..	..	..	..	22 933	22 758	22 375	21 857
		154	Other food products	..	..	..	..	257 887	255 933	251 173	246 294
		155	Beverages	..	..	..	..	96 431	..	..	78 343
	16		Tobacco products	19 114	17 221	15 260	14 673	13 588	12 685	11 360	10 972
	17		Textiles	281 030	232 804	197 526	180 124	162 549	150 132	144 259	141 742
		171	Spinning, weaving and finishing of textiles	..	..	..	..	77 003	20 197	19 245	18 083
		172	Other textiles	..	..	..	..	59 280	57 460	56 300	57 537
		173	Knitted and crocheted fabrics and articles	..	..	..	..	26 266	23 737	22 127	20 218
	18		Wearing apparel and fur	222 102	176 567	144 118	119 258	99 092	84 967	74 691	69 735
		181	Wearing apparel, except fur apparel	..	..	..	..	98 601	84 512	74 240	69 308
		182	Dressing and dyeing of fur; articles of fur	..	..	..	..	491	455	451	427
	19		Leather and articles of leather; footwear	74 705	56 870	45 868	40 857	36 969	32 599	30 352	29 550
		191	Leather and articles of leather	..	..	..	..	13 326	11 638	10 564	10 019
		192	Footwear	..	..	..	..	23 643	20 961	19 788	19 531
	20		Wood & products of wood & cork, exc. furniture	194 280	187 134	172 168	173 364	168 819	156 344	151 581	148 204
		201	Sawmilling and planing of wood	..	..	..	..	36 315	33 503	33 645	32 854
		202	Wood products; articles of cork, straw, etc.	..	..	..	..	132 504	122 841	117 936	115 350
	21		Paper and paper products	210 817	198 053	183 769	177 676	177 530	171 686	169 003	165 807
	22		Printing, publishing and recorded media	..	..	..	..	200 699	188 562	182 470	179 047
		221	Publishing	..	..	..	..	..	31 038	..	..
		222	Printing and related activities	..	..	..	..	163 558	153 059	146 587	145 181
		223	Reproduction of recorded media	..	..	..	..	..	..	..	5 103
	23		Coke and petroleum products; nuclear fuel	..	..	..	..	19 814	18 208	16 608	15 546
		231	Coke oven products	..	..	..	..	2 773	2 663	..	2 374
		232	Refined petroleum products	..	..	..	..	16 275	14 946	13 457	12 615
		233	Processing of nuclear fuel	..	..	..	..	766	599	..	557
	24		Chemicals and chemical products	562 548	520 358	453 156	420 377	403 998	380 542	369 888	362 087
		241	Basic chemicals	..	..	..	..	..	..	..	155 018
		242	Other chemical products	..	..	..	..	..	..	165 610	182 933
		243	Man-made fibres	..	..	..	..	28 897	26 678	25 176	24 136
	25		Rubber and plastic products	483 494	472 104	425 343	420 910	425 653	404 784	402 006	417 158
		251	Rubber products	..	..	..	..	96 041	87 641	85 972	89 744
		252	Plastic products	..	..	..	..	329 612	317 143	316 034	327 414
	26		Other non-metallic mineral products	408 301	383 598	362 592	359 224	348 827	320 594	304 415	297 001
		261	Glass and glass products	..	..	..	..	83 309	80 237	77 368	76 334
		269	Non-metallic mineral products, nec	..	..	..	..	265 518	240 357	227 047	220 667
	27		Basic metals	519 530	460 496	380 488	360 456	340 786	310 572	309 953	313 276
		271	Basic iron and steel	..	..	..	..	184 127	163 908	161 975	159 706
		272	Basic precious and non-ferrous metals	..	..	..	..	71 273	65 017	66 565	67 101
		273	Casting of metals	..	..	..	..	85 386	81 647	81 413	86 469
	28		Fabricated metal products, exc. machinery	838 878	813 441	728 688	701 368	712 505	671 879	659 274	689 205
		281	Structural metal prod., tanks, reservoirs, etc.	..	..	..	..	250 401	233 730	221 594	221 284
	29		Machinery and equipment, nec	1460 645	1333 289	1127 905	1054 510	1006 350	956 308	922 536	926 900
		291	General purpose machinery	..	..	..	..	478 426	453 661	440 273	442 026
		292	Special purpose machinery	..	..	..	..	449 820	429 581	414 003	417 061
		293	Domestic appliances, nec	..	..	..	..	78 104	73 066	68 260	67 813
	30		Office, accounting & computing machinery	54 282	42 434	29 475	26 580	23 848	21 050	24 018	22 784
	31		Electrical machinery & apparatus, nec	590 612	535 969	462 861	448 865	452 137	416 152	390 193	397 246
		311	Electric motors, generators and transformers	..	..	..	..	85 250	74 675	71 439	74 802
		312	Electricity distribution & control apparatus	..	..	..	..	199 254	185 246	168 548	165 130
		313	Insulated wire and cable	..	..	..	..	28 077	26 965	24 641	25 362
		314	Accumulators, primary cells & primary batteries	..	..	..	..	10 296	9 092	8 486	8 141
		315	Electric lamps and lighting equipment	..	..	..	..	38 322	35 433	35 812	36 251
		319	Other electrical equipment, nec	..	..	..	..	90 938	84 741	81 267	87 560
	32		Radio, TV, communication equipment	215 485	185 303	149 542	136 854	126 999	119 070	113 103	116 218
		321	Electronic valves, tubes; other electronic components	..	..	..	..	53 835	54 057	54 232	55 972
		322	TV/radio transmitters; line teleph./telegr. apparatus	..	..	..	..	34 926	34 117	33 940	34 644
		323	TV/radio receivers; sound and video app.; assoc. gds.	..	..	..	..	38 238	30 896	24 931	25 602
	33		Medical, precision & optical instrum.; watch & clocks	293 717	269 538	249 041	223 787	199 665	190 103	183 960	179 728
		331	Medical appl.; instruments & appl. for measuring, etc.	..	..	..	..	167 544	157 803	153 599	151 050
		332	Optical instruments and photographic equipment	..	..	..	..	24 992	25 900	24 781	23 410
		333	Watches and clocks	..	..	..	..	7 129	6 400	5 580	5 268
	34		Motor vehicles, trailers and semi-trailers	940 146	885 524	744 731	727 887	775 691	757 981	777 209	830 180
		341	Motor vehicles	..	..	..	..	440 984	437 186	441 273	468 679
		342	Bodies for motor vehicles, trailers & semi-trailers	..	..	..	..	52 456	45 988	47 100	45 110
		343	Parts & accessories for motor vehicles & their engines	..	..	..	..	282 251	274 807	288 836	316 391
	35		Other transport equipment	233 929	218 562	203 396	186 074	174 003	157 264	145 139	139 720
		351	Building and repairing of ships and boats	..	..	..	..	39 948	34 494	29 707	29 713
		352	Railway and tramway locomotives and rolling stock	..	..	..	..	80 309	69 383	62 040	53 928
		353	Aircraft and spacecraft	..	..	..	..	39 120	40 276	40 640	42 926
		359	Transport equipment, nec	..	..	..	..	14 626	13 111	12 752	13 153
	36		Furniture; manufacturing, nec	384 733	355 217	327 979	308 258	293 478	275 684	258 773	254 204
		361	Furniture	..	..	..	..	219 983	206 678	193 192	191 009
		369	Manufacturing, nec	..	..	..	..	73 495	69 006	..	..
	37		Recycling	..	..	..	..	9 376	10 566	10 958	10 560
E	40_41		Electricity, gas and water supply	..	..	..	..	..	..	..	..
	40		Electricity, gas, steam and hot water supply	..	..	..	..	..	..	..	..
	41		Collection, purification and distribution of water	..	..	..	..	..	..	..	..
F	45		Construction	..	..	..	..	..	..	..	..

Table GE.5 **ÉTABLISSEMENTS - ESTABLISHMENTS**

Unités

1991	1992	1993	1994	1995	1996	1997	1998			CITI révision 3	
1 485	1 461	1 441	1 475	1 521	1 523	1 520	1 473	Activites extractives	C	10_14	
169	160	151	146	150	143	138	124	Extraction de charbon, de lignite et de tourbe		10	
50	51	49	48	50	49	47	45	Pétrole brut et gaz naturel ; activités annexes		11	
..	..	..	..	..	..	..	..	Extraction de Minerais d'Uranium et de Thorium		12	
14	9	6	6	5	..	..	..	Extraction de minerais métalliques		13	
1 252	1 241	1 235	1 275	1 316	1 331	1 335	1 304	Autres activités extractives		14	
54 617	54 072	52 223	50 999	49 468	48 850	47 459	46 885	Total activites manufacturieres	D	15_37	
7 153	6 918	6 720	6 652	6 480	6 431	6 144	5 911	Produits alimentaires et boissons		15	
..	..	..	..	1 918	1 907	1 800	1 718	Viande, poisson, fruits, légumes, huiles et graisses			151
..	..	..	..	340	325	314	305	Produits laitiers			152
..	..	..	..	339	331	325	314	Travail des grains; prod. amyl. ; alim. pour animaux			153
..	..	..	..	2 962	3 018	2 858	2 763	Autres produits alimentaires			154
..	..	..	..	921	871	836	811	Boissons			155
57	53	46	44	41	36	33	34	Produits à base de tabac		16	
1 844	1 701	1 593	1 516	1 460	1 404	1 305	1 253	Textiles		17	
..	..	..	..	533	508	463	..	Filature, tissage et achèvement des textiles			171
..	..	..	..	645	637	604	586	Autres articles textiles			172
..	..	..	..	281	257	237	218	Etoffes et articles de bonneterie			173
2 252	1 957	1 674	1 434	1 256	1 105	938	846	Articles d'habillement et fourrure		18	
..	..	..	..	1 244	1 095	928	837	Articles d'habillement autres qu'en fourrure			181
..	..	..	..	12	10	10	9	Prép. et teinture des fourrures ; articles en fourrure			182
633	568	503	454	385	346	315	299	Cuir et articles en cuir ; chaussures		19	
..	..	..	..	186	166	150	138	Cuir et articles en cuir sauf chaussures			191
..	..	..	..	199	180	165	161	Chaussures			192
3 476	3 462	2 497	2 404	2 305	2 243	2 122	2 043	Bois, articles en bois, sauf meubles		20	
..	..	..	..	923	889	840	797	Sciage et rabotage du bois			201
..	..	..	..	..	1 354	1 282	1 247	Articles en bois, liège, vannerie, sparterie			202
1 147	1 143	1 105	1 076	1 073	1 059	1 041	1 036	Papier, articles en papier et carton		21	
..	..	..	..	2 979	2 909	2 863	2 817	Edition, imprimerie et reproduction		22	
..	..	..	..	822	831	842	827	Edition			221
..	..	..	..	2 137	2 058	2 001	1 928	Imprimerie et activités annexes			222
..	..	..	..	..	..	..	25	Reproduction de supports enregistrés			223
..	..	..	..	85	83	76	72	Coke. et prod. pétroliers ; comb. nucléaires		23	
..	..	..	..	8	8	..	6	Cokéfaction			231
..	..	..	..	74	72	66	63	Produits pétroliers raffinés			232
..	..	..	..	3	3	..	3	Traitement de combustibles nucléaires			233
1 774	1 768	1 744	1 740	1 717	1 718	1 709	1 722	Produits chimiques		24	
..	..	..	..	434	436	451	469	Produits chimiques de base			241
..	..	..	..	1 237	1 236	1 215	1 211	Autres produits chimiques			242
..	..	..	..	45	44	43	43	Fibres synthétiques ou artificielles			243
2 822	2 918	2 954	2 998	3 038	3 031	3 005	3 003	Art. en caoutchouc et matières plastiques		25	
..	..	..	..	292	290	293	291	Articles en caoutchouc			251
..	..	..	..	2 741	2 737	2 707	2 712	Articles en matières plastiques			252
3 467	3 524	3 606	3 733	3 783	3 829	3 782	3 776	Autres produits minéraux non métalliques		26	
..	..	..	..	420	425	415	414	Verre et articles en verre			261
..	..	..	..	3 363	3 404	3 337	3 362	Produits minéraux non métalliques, nca			269
1 314	1 318	1 271	1 217	1 140	1 120	1 079	1 069	Produits métallurgiques de base		27	
..	..	..	..	455	433	415	405	Sidérurgie et première transformation de l'acier			271
..	..	..	..	225	224	215	213	Première transf. métaux précieux et non ferreux			272
..	..	..	..	460	463	450	452	Fonderie			273
6 860	6 984	7 018	6 872	6 975	7 008	6 913	6 971	Ouvrages en métaux, sauf machines		28	
..	..	..	..	2 686	2 639	2 555	2 525	Constr. et menuis. métal.; réservoirs; générateurs			281
7 868	7 767	7 563	7 260	6 966	6 896	6 753	6 719	Machines et matériels, nca		29	
..	..	..	..	3 133	3 204	3 140	3 141	Machines d'usage général			291
..	..	..	..	3 494	3 487	3 414	3 384	Machines d'usage spécifique			292
..	..	..	..	217	205	201	195	Appareils domestiques, nca			293
220	222	211	217	207	204	213	200	Mach. de bureau, comptables, information		30	
2 388	2 385	2 366	2 351	2 412	2 375	2 316	2 336	Machines et appareils électriques, nca		31	
..	..	..	..	521	520	498	492	Moteurs, génératrices, transformateurs électriques			311
..	..	..	..	1 037	1 026	987	998	Matériel électrique de distribution et de commande			312
..	..	..	..	114	114	116	122	Fils et câbles électriques isolés			313
..	..	..	..	34	31	29	30	Accumulateurs et piles électriques			314
..	..	..	..	301	282	277	273	Lampes électriques et appareils d'éclairage			315
..	..	..	..	405	402	409	421	Autres matériels électriques, nca.			319
756	749	737	711	602	600	579	615	Equipements et appareils de radio, TV, etc.		32	
..	..	..	..	264	281	276	272	Tubes et valves électroniques;autres composants			321
..	..	..	..	157	139	141	178	Emetteurs radio/TV; app. téléphonie et télégraphie			322
..	..	..	..	181	180	162	165	Récepteurs radio/TV, son et image; art. associés			323
2 542	2 580	2 659	2 611	2 353	2 348	2 311	2 301	Instr. médicaux, de précision ; optique ; horlog.		33	
..	..	..	..	2 093	2 093	2 074	2 080	Instruments médicaux et de précision, etc.			331
..	..	..	..	187	181	173	165	Instruments d'optique ; matériel photographique			332
..	..	..	..	73	74	64	57	Horlogerie			333
1 012	1 055	1 077	1 092	1 109	1 072	1 055	1 072	Véhicules automobiles; remorques, etc.		34	
..	..	..	..	90	92	87	93	Véhicules automobiles			341
..	..	..	..	434	389	358	354	Carross. automobiles, remorques, semi-remorques			342
..	..	..	..	585	591	610	625	Pièces et accessoires pour automobiles et moteurs			343
446	451	461	452	441	419	412	407	Autres matériels de transport		35	
..	..	..	..	112	106	105	109	Construction et réparation de navires			351
..	..	..	..	158	144	138	131	Matériel ferroviaire roulant			352
..	..	..	..	85	85	90	95	Construction aéronautique et spatiale			353
..	..	..	..	78	78	73	73	Autres équipements de transport			359
2 925	2 863	2 804	2 690	2 549	2 479	2 331	2 220	Meubles ; activités de fabrication, nca		36	
..	..	..	..	1 734	1 693	1 600	1 512	Meubles			361
..	..	..	..	815	776	722	..	Activités de fabrication, nca			369
..	..	..	..	112	135	164	163	Récupération		37	
..	..	..	..	..	..	..	..	Electricité, gaz et eau	E	40_41	
..	..	..	..	..	..	..	..	Prod. et dist. d'électricité gaz, vapeur d'eau chaude		40	
..	..	..	..	..	..	..	..	Captage, épuration et distribution d'eau		41	
..	..	..	..	..	..	..	..	Construction	F	45	

GERMANY

Table GE.6 **ENTERPRISES - ENTREPRISES**

Units

ISIC revision 3				1991	1992	1993	1994	1995	1996	1997	1998
C	10_14		Mining and quarrying	..	..	..	..	783	788	787	774
	10		Mining of coal and lignite; extraction of peat	..	..	..	..	64	55	53	47
	11		Crude petroleum & natural gas; service activities	..	..	..	..	12	12	12	12
	12		Mining of uranium and thorium ores	..	..	..	..	..	..	..	..
	13		Mining of metal ores	..	..	..	..	..	..	..	..
	14		Other mining and quarrying	..	..	..	..	703	721	722	715
D	15_37		Total manufacturing	..	..	..	..	42 296	41 732	40 502	39 806
	15		Food products and beverages	..	..	..	..	5 780	5 749	5 488	5 256
		151	Production of meat, fish, fruit, vegetables, oils & fats	..	..	..	..	1 731	1 724	1 633	1 547
		152	Dairy products	..	..	..	..	224	217	210	202
		153	Grain mill and starch products; animal feeds	..	..	..	..	257	248	238	231
		154	Other food products	..	..	..	..	2 731	2 614	2 642	2 549
		155	Beverages	..	..	..	..	837	796	727	728
	16		Tobacco products	..	..	..	..	28	26	22	23
	17		Textiles	..	..	..	..	1 304	1 279	1 192	1 146
		171	Spinning, weaving and finishing of textiles	..	..	..	..	..	..	..	..
		172	Other textiles	..	..	..	..	591	594	559	546
		173	Knitted and crocheted fabrics and articles	..	..	..	..	243	231	215	199
	18		Wearing apparel and fur	..	..	..	..	1 081	956	816	736
		181	Wearing apparel, except fur apparel	..	..	..	..	1 068	946	806	727
		182	Dressing and dyeing of fur; articles of fur	..	..	..	..	13	10	10	9
	19		Leather and articles of leather; footwear	..	..	..	..	357	319	291	276
		191	Leather and articles of leather	..	..	..	..	182	161	145	133
		192	Footwear	..	..	..	..	175	158	146	143
	20		Wood & products of wood & cork, exc. furniture	..	..	..	..	2 305	2 108	2 001	1 923
		201	Sawmilling and planing of wood	..	..	..	..	812	845	802	762
		202	Wood products; articles of cork, straw, etc.	..	..	..	..	..	1 263	1 200	1 162
	21		Paper and paper products	..	..	..	..	906	889	874	873
	22		Printing, publishing and recorded media	..	..	..	..	2 812	2 751	2 704	2 655
		221	Publishing	..	..	..	..	744	789	..	813
		222	Printing and related activities	..	..	..	..	2 050	1 943	1 885	1 819
		223	Reproduction of recorded media	..	..	..	..	..	..	..	24
	23		Coke and petroleum products; nuclear fuel	..	..	..	..	54	53	49	45
		231	Coke oven products	..	..	..	..	..	..	..	..
		232	Refined petroleum products	..	..	..	..	..	..	..	..
		233	Processing of nuclear fuel	..	..	..	..	..	..	..	..
	24		Chemicals and chemical products	..	..	..	..	1 293	1 290	1 273	1 279
		241	Basic chemicals	..	..	..	..	244	248	..	..
		242	Other chemical products	..	..	..	..	1 013	1 009	982	972
		243	Man-made fibres	..	..	..	..	36	33	32	33
	25		Rubber and plastic products	..	..	..	..	2 644	2 652	2 630	2 627
		251	Rubber products	..	..	..	..	242	236	238	226
		252	Plastic products	..	..	..	..	2 402	2 416	2 393	2 402
	26		Other non-metallic mineral products	..	..	..	..	2 437	2 431	2 366	2 322
		261	Glass and glass products	..	..	..	..	350	350	343	347
		269	Non-metallic mineral products, nec	..	..	..	..	2 087	2 081	2 022	1 975
	27		Basic metals	..	..	..	..	968	948	915	904
		271	Basic iron and steel	..	..	..	..	386	355	338	328
		272	Basic precious and non-ferrous metals	..	..	..	..	172	175	171	170
		273	Casting of metals	..	..	..	..	410	418	406	405
	28		Fabricated metal products, exc. machinery	..	..	..	..	6 158	6 126	6 040	6 043
		281	Structural metal prod., tanks, reservoirs, etc.	..	..	..	..	2 462	2 421	2 355	2 335
		289	Other fabricated metal products	..	..	..	..	3 696	3 705	3 684	3 710
	29		Machinery and equipment, nec	..	..	..	..	6 087	6 026	5 919	5 866
		291	General purpose machinery	..	..	..	..	2 749	2 679	2 642	2 629
		292	Special purpose machinery	..	..	..	..	3 157	3 174	3 110	3 079
		293	Domestic appliances, nec	..	..	..	..	181	173	167	159
	30		Office, accounting & computing machinery	..	..	..	..	207	162	172	164
	31		Electrical machinery & apparatus, nec	..	..	..	..	1 892	1 891	1 860	1 872
		311	Electric motors, generators and transformers	..	..	..	..	427	437	422	421
		312	Electricity distribution & control apparatus	..	..	..	..	759	766	744	747
		313	Insulated wire and cable	..	..	..	..	89	91	97	101
		314	Accumulators, primary cells & primary batteries	..	..	..	..	23	21	23	24
		315	Electric lamps and lighting equipment	..	..	..	..	284	260	257	253
		319	Other electrical equipment, nec	..	..	..	..	310	316	318	327
	32		Radio, TV, communication equipment	..	..	..	..	451	470	457	471
		321	Electronic valves, tubes; other electronic components	..	..	..	..	199	213	216	222
		322	TV/radio transmitters; line teleph./telegr. apparatus	..	..	..	..	103	100	101	111
		323	TV/radio receivers; sound and video app.; assoc. gds.	..	..	..	..	149	157	141	138
	33		Medical, precision & optical instrum.; watch & clocks	..	..	..	..	2 081	2 069	2 040	2 020
		331	Medical appl.; instruments & appl. for measuring, etc.	..	..	..	..	1 864	1 853	1 841	1 832
		332	Optical instruments and photographic equipment	..	..	..	..	150	148	141	135
		333	Watches and clocks	..	..	..	..	67	68	59	53
	34		Motor vehicles, trailers and semi-trailers	..	..	..	..	889	858	832	855
		341	Motor vehicles	..	..	..	..	57	60	54	59
		342	Bodies for motor vehicles, trailers & semi-trailers	..	..	..	..	410	368	331	328
		343	Parts & accessories for motor vehicles & their engines	..	..	..	..	422	430	447	467
	35		Other transport equipment	..	..	..	..	309	283	280	279
		351	Building and repairing of ships and boats	..	..	..	..	104	100	101	98
		352	Railway and tramway locomotives and rolling stock	..	..	..	..	76	57	55	52
		353	Aircraft and spacecraft	..	..	..	..	57	56	59	64
		359	Transport equipment, nec	..	..	..	..	72	70	66	66
	36		Furniture; manufacturing, nec	..	..	..	..	2 355	2 298	2 161	2 053
		361	Furniture	..	..	..	..	1 605	1 567	1 478	1 392
		369	Manufacturing, nec	..	..	..	..	750	731	..	..
	37		Recycling	..	..	..	..	81	98	118	118
E	40_41		Electricity, gas and water supply	..	..	..	..	..	..	..	..
	40		Electricity, gas, steam and hot water supply	..	..	..	..	..	..	..	..
F	45		Construction	..	..	..	..	..	..	..	..

Table GR.1 PRODUCTION - PRODUCTION

Millions de Dr (Prix courants)

1991	1992	1993	1994	1995	1996	1997	1998			CITI révision 3	
..	..	..	..	..	..	..	..	Activites extractives	C	10_14	
..	..	..	..	..	..	..	..	Extraction de charbon, de lignite et de tourbe		10	
..	..	..	..	..	..	..	..	Pétrole brut et gaz naturel ; activités annexes		11	
..	..	..	..	..	..	..	..	Extraction de Minerais d'Uranium et de Thorium		12	
..	..	..	..	..	..	..	..	Extraction de minerais métalliques		13	
..	..	..	..	..	..	..	..	Autres activités extractives		14	
..	..	5 734 980	6 292 839	7 026 424	7 438 210	..	..	Total activites manufacturieres	D	15_37	
..	..	1 499 763	1 653 302	1 785 575	1 880 111	..	..	Produits alimentaires et boissons		15	
..	..	423 353	469 704	531 154	560 393	..	..	Viande, poisson, fruits, légumes, huiles et graisses			151
..	..	249 894	300 757	317 593	324 290	..	..	Produits laitiers			152
..	..	208 915	199 249	206 371	206 637	..	..	Travail des grains; prod. amyl. ; alim. pour animaux			153
..	..	303 015	331 583	354 795	384 548	..	..	Autres produits alimentaires			154
..	..	314 586	352 009	375 664	404 243	..	..	Boissons			155
..	..	93 769	113 629	129 878	136 920	..	..	Produits à base de tabac		16	
..	..	384 903	413 311	439 973	434 676	..	..	Textiles		17	
..	..	221 813	243 266	270 362	259 103	..	..	Filature, tissage et achèvement des textiles			171
..	..	59 394	64 677	68 762	74 801	..	..	Autres articles textiles			172
..	..	103 697	105 367	100 850	100 773	..	..	Etoffes et articles de bonneterie			173
..	..	321 704	319 539	345 668	333 767	..	..	Articles d'habillement et fourrure		18	
..	..	310 040	306 320	333 047	316 219	..	..	Articles d'habillement autres qu'en fourrure			181
..	..	11 664	13 219	12 621	17 548	..	..	Prép. et teinture des fourrures ; articles en fourrure			182
..	..	82 547	84 747	88 260	88 165	..	..	Cuir et articles en cuir ; chaussures		19	
..	..	26 530	23 860	23 507	24 022	..	..	Cuir et articles en cuir sauf chaussures			191
..	..	56 017	60 887	64 753	64 143	..	..	Chaussures			192
..	..	107 459	97 197	111 832	114 399	..	..	Bois, articles en bois, sauf meubles		20	
..	..	14 692	14 970	14 328	14 951	..	..	Sciage et rabotage du bois			201
..	..	92 767	82 228	97 504	99 448	..	..	Articles en bois, liège, vannerie, sparterie			202
..	..	155 032	177 871	228 518	220 923	..	..	Papier, articles en papier et carton		21	
..	..	132 061	150 669	175 868	216 752	..	..	Edition, imprimerie et reproduction		22	
..	..	88 071	98 939	120 290	157 888	..	..	Edition			221
..	..	43 990	51 730	55 578	58 864	..	..	Imprimerie et activités annexes			222
..	..	..	..	..	..	..	..	Reproduction de supports enregistrés			223
..	..	567 731	658 902	735 162	917 306	..	..	Coke. et prod. pétroliers ; comb. nucléaires		23	
..	..	..	..	..	..	..	..	Cokéfaction			231
..	..	567 731	658 902	735 162	917 306	..	..	Produits pétroliers raffinés			232
..	..	..	..	..	..	..	..	Traitement de combustibles nucléaires			233
..	..	548 544	635 786	689 507	743 825	..	..	Produits chimiques		24	
..	..	124 741	143 781	174 527	190 491	..	..	Produits chimiques de base			241
..	..	411 013	476 010	496 377	535 787	..	..	Autres produits chimiques			242
..	..	12 790	15 995	18 604	17 547	..	..	Fibres synthétiques ou artificielles			243
..	..	178 935	208 853	260 864	261 443	..	..	Art. en caoutchouc et matières plastiques		25	
..	..	32 744	35 784	40 473	37 338	..	..	Articles en caoutchouc			251
..	..	146 191	173 069	220 391	224 105	..	..	Articles en matières plastiques			252
..	..	339 298	353 009	380 034	434 300	..	..	Autres produits minéraux non métalliques		26	
..	..	17 111	17 665	18 908	19 020	..	..	Verre et articles en verre			261
..	..	322 187	335 344	361 125	415 280	..	..	Produits minéraux non métalliques, nca			269
..	..	397 759	478 112	634 066	583 656	..	..	Produits métallurgiques de base		27	
..	..	207 461	235 380	309 038	274 606	..	..	Sidérurgie et première transformation de l'acier			271
..	..	188 707	240 795	321 765	305 047	..	..	Première transf. métaux précieux et non ferreux			272
..	..	1 590	1 937	3 263	4 003	..	..	Fonderie			273
..	..	189 700	207 014	232 780	258 061	..	..	Ouvrages en métaux, sauf machines		28	
..	..	67 710	64 245	75 163	91 850	..	..	Constr. et menuis. métal.; réservoirs; générateurs			281
..	..	121 990	142 769	157 618	166 211	..	..	Autres ouvrages en métaux			289
..	..	157 786	175 337	196 952	230 792	..	..	Machines et matériels, nca		29	
..	..	51 518	59 115	65 547	69 828	..	..	Machines d'usage général			291
..	..	58 071	59 813	71 535	84 138	..	..	Machines d'usage spécifique			292
..	..	48 197	56 408	59 870	76 826	..	..	Appareils domestiques, nca			293
..	..	1 106	909	1 473	2 428	..	..	Mach. de bureau, comptables, information		30	
..	..	146 340	146 447	177 639	161 914	..	..	Machines et appareils électriques, nca		31	
..	..	19 293	20 118	23 389	27 480	..	..	Moteurs, génératrices, transformateurs électriques			311
..	..	24 285	26 504	25 258	25 380	..	..	Matériel électrique de distribution et de commande			312
..	..	82 114	77 236	108 287	88 426	..	..	Fils et câbles électriques isolés			313
..	..	10 714	11 298	7 980	7 498	..	..	Accumulateurs et piles électriques			314
..	..	7 804	8 826	10 294	10 367	..	..	Lampes électriques et appareils d'éclairage			315
..	..	2 130	2 464	2 430	2 763	..	..	Autres matériels électriques, nca.			319
..	..	131 100	108 395	91 186	82 403	..	..	Equipements et appareils de radio, TV, etc.		32	
..	..	3 242	5 000	4 365	4 325	..	..	Tubes et valves électroniques;autres composants			321
..	..	126 218	101 823	85 267	76 423	..	..	Emetteurs radio/TV; app. téléphonie et télégraphie			322
..	..	1 640	1 572	1 553	1 656	..	..	Récepteurs radio/TV, son et image; art. associés			323
..	..	12 445	14 225	18 756	23 086	..	..	Instr. médicaux, de précision ; optique ; horlog.		33	
..	..	10 395	11 788	15 124	18 672	..	..	Instruments médicaux et de précision, etc.			331
..	..	2 007	2 400	3 596	4 385	..	..	Instruments d'optique ; matériel photographique			332
..	..	43	38	36	29	..	..	Horlogerie			333
..	..	62 203	52 282	48 203	47 026	..	..	Véhicules automobiles; remorques, etc.		34	
..	..	52 319	42 792	38 357	34 700	..	..	Véhicules automobiles			341
..	..	6 750	4 970	5 429	7 407	..	..	Carross. automobiles, remorques, semi-remorques			342
..	..	3 133	4 520	4 417	4 920	..	..	Pièces et accessoires pour automobiles et moteurs			343
..	..	127 083	139 686	138 927	151 130	..	..	Autres matériels de transport		35	
..	..	66 907	71 978	69 301	75 745	..	..	Construction et réparation de navires			351
..	..	14 828	16 982	17 198	22 122	..	..	Matériel ferroviaire roulant			352
..	..	42 735	47 906	49 246	49 291	..	..	Construction aéronautique et spatiale			353
..	..	2 613	2 821	3 183	3 972	..	..	Autres équipements de transport			359
..	..	92 894	97 992	107 661	110 436	..	..	Meubles ; activités de fabrication, nca		36	
..	..	70 762	75 030	79 599	83 476	..	..	Meubles			361
..	..	22 133	22 962	28 062	26 960	..	..	Activités de fabrication, nca			369
..	..	4 819	5 624	7 641	4 690	..	..	Récupération		37	
..	..	679 884	734 446	..	..	..	..	Electricité, gaz et eau	E	40_41	
..	..	623 673	671 402	..	..	..	..	Prod. et dist. d'électricité gaz, vapeur d'eau chaude		40	
..	..	..	..	..	..	..	..	Construction	F	45	

GREECE

Table GR.2 VALUE ADDED - VALEUR AJOUTÉE

Millions of Dr (Current Prices)

ISIC revision 3				1991	1992	1993	1994	1995	1996	1997	1998
C	10_14		Mining and quarrying	..	..	..	..	..	..	..	..
	10		Mining of coal and lignite; extraction of peat	..	..	..	..	..	..	..	..
	11		Crude petroleum & natural gas; service activities	..	..	..	..	..	..	..	..
	12		Mining of uranium and thorium ores	..	..	..	..	..	..	..	..
	13		Mining of metal ores	..	..	..	..	..	..	..	..
	14		Other mining and quarrying	..	..	..	..	..	..	..	..
D	15_37		Total manufacturing	..	..	2100 613	2278 580	2488 186	2636 239	..	..
	15		Food products and beverages	..	..	507 817	569 956	611 712	656 158	..	..
		151	Production of meat, fish, fruit, vegetables, oils & fats	..	..	118 542	128 941	150 053	156 749	..	..
		152	Dairy products	..	..	77 494	94 330	105 360	115 602	..	..
		153	Grain mill and starch products; animal feeds	..	..	41 215	43 982	45 790	44 817	..	..
		154	Other food products	..	..	132 573	146 289	153 957	165 455	..	..
		155	Beverages	..	..	137 992	156 414	156 552	173 535	..	..
	16		Tobacco products	..	..	38 381	44 908	48 358	52 383	..	..
	17		Textiles	..	..	162 707	171 950	174 958	172 856	..	..
		171	Spinning, weaving and finishing of textiles	..	..	97 602	105 206	109 641	104 165	..	..
		172	Other textiles	..	..	24 779	26 509	26 283	29 558	..	..
		173	Knitted and crocheted fabrics and articles	..	..	40 325	40 234	39 035	39 133	..	..
	18		Wearing apparel and fur	..	..	141 847	136 583	143 799	135 746	..	..
		181	Wearing apparel, except fur apparel	..	..	138 128	132 799	140 595	131 127	..	..
		182	Dressing and dyeing of fur; articles of fur	..	..	3 719	3 784	3 204	4 618	..	..
	19		Leather and articles of leather; footwear	..	..	35 446	33 418	34 560	33 206	..	..
		191	Leather and articles of leather	..	..	12 127	8 056	7 595	7 875	..	..
		192	Footwear	..	..	23 319	25 362	26 964	25 331	..	..
	20		Wood & products of wood & cork, exc. furniture	..	..	39 305	33 261	38 854	40 439	..	..
		201	Sawmilling and planing of wood	..	..	3 878	4 611	4 576	4 811	..	..
		202	Wood products; articles of cork, straw, etc.	..	..	35 427	28 650	34 278	35 628	..	..
	21		Paper and paper products	..	..	60 873	66 986	79 246	80 432	..	..
	22		Printing, publishing and recorded media	..	..	74 634	82 293	92 300	111 511	..	..
		221	Publishing	..	..	51 604	55 966	64 733	80 599	..	..
		222	Printing and related activities	..	..	23 030	26 327	27 566	30 912	..	..
		223	Reproduction of recorded media	..	..	..	..	..	..	..	..
	23		Coke and petroleum products; nuclear fuel	..	..	105 191	105 745	129 570	114 638	..	..
		231	Coke oven products	..	..	..	..	..	..	..	..
		232	Refined petroleum products	..	..	105 191	105 745	129 570	114 638	..	..
		233	Processing of nuclear fuel	..	..	..	..	..	..	..	..
	24		Chemicals and chemical products	..	..	221 893	258 414	280 154	314 269	..	..
		241	Basic chemicals	..	..	47 252	50 897	61 018	73 574	..	..
		242	Other chemical products	..	..	168 937	200 739	211 370	233 899	..	..
		243	Man-made fibres	..	..	5 703	6 778	7 766	6 796	..	..
	25		Rubber and plastic products	..	..	71 631	80 875	93 079	97 471	..	..
		251	Rubber products	..	..	12 314	13 527	14 042	12 604	..	..
		252	Plastic products	..	..	59 317	67 347	79 037	84 868	..	..
	26		Other non-metallic mineral products	..	..	142 544	149 696	155 809	191 272	..	..
		261	Glass and glass products	..	..	7 937	8 706	9 203	9 693	..	..
		269	Non-metallic mineral products, nec	..	..	134 607	140 990	146 606	181 579	..	..
	27		Basic metals	..	..	98 675	118 137	165 363	147 708	..	..
		271	Basic iron and steel	..	..	45 256	46 251	69 525	59 202	..	..
		272	Basic precious and non-ferrous metals	..	..	52 697	70 945	94 171	86 332	..	..
		273	Casting of metals	..	..	722	941	1 667	2 175	..	..
	28		Fabricated metal products, exc. machinery	..	..	74 028	81 758	89 617	100 047	..	..
		281	Structural metal prod., tanks, reservoirs, etc.	..	..	26 488	25 222	28 631	34 750	..	..
		289	Other fabricated metal products	..	..	47 540	56 535	60 986	65 297	..	..
	29		Machinery and equipment, nec	..	..	63 608	73 376	80 216	95 277	..	..
		291	General purpose machinery	..	..	23 532	27 287	29 232	33 157	..	..
		292	Special purpose machinery	..	..	21 069	23 843	27 825	33 727	..	..
		293	Domestic appliances, nec	..	..	19 007	22 246	23 158	28 392	..	..
	30		Office, accounting & computing machinery	..	..	484	473	932	1 568	..	..
	31		Electrical machinery & apparatus, nec	..	..	50 053	50 187	54 452	54 684	..	..
		311	Electric motors, generators and transformers	..	..	8 785	9 126	9 763	12 009	..	..
		312	Electricity distribution & control apparatus	..	..	10 495	12 332	11 262	11 436	..	..
		313	Insulated wire and cable	..	..	21 405	19 205	23 892	21 803	..	..
		314	Accumulators, primary cells & primary batteries	..	..	5 061	4 409	4 034	3 742	..	..
		315	Electric lamps and lighting equipment	..	..	3 429	3 989	4 468	4 453	..	..
		319	Other electrical equipment, nec	..	..	878	1 125	1 033	1 242	..	..
	32		Radio, TV, communication equipment	..	..	52 520	51 281	42 035	44 164	..	..
		321	Electronic valves, tubes; other electronic components	..	..	1 514	2 010	1 895	2 044	..	..
		322	TV/radio transmitters; line teleph./telegr. apparatus	..	..	50 125	48 392	39 205	41 170	..	..
		323	TV/radio receivers; sound and video app.; assoc. gds.	..	..	881	879	935	951	..	..
	33		Medical, precision & optical instrum.; watch & clocks	..	..	5 932	6 560	7 994	9 620	..	..
		331	Medical appl.; instruments & appl. for measuring, etc.	..	..	4 972	5 287	6 299	7 638	..	..
		332	Optical instruments and photographic equipment	..	..	935	1 256	1 673	1 966	..	..
		333	Watches and clocks	..	..	24	16	22	17	..	..
	34		Motor vehicles, trailers and semi-trailers	..	..	16 723	13 871	15 267	16 743	..	..
		341	Motor vehicles	..	..	12 839	10 011	10 985	11 390	..	..
		342	Bodies for motor vehicles, trailers & semi-trailers	..	..	2 258	1 547	2 088	2 948	..	..
		343	Parts & accessories for motor vehicles & their engines	..	..	1 627	2 313	2 194	2 404	..	..
	35		Other transport equipment	..	..	91 323	102 850	100 458	112 910	..	..
		351	Building and repairing of ships and boats	..	..	47 425	51 805	47 786	55 947	..	..
		352	Railway and tramway locomotives and rolling stock	..	..	11 247	12 888	13 097	16 865	..	..
		353	Aircraft and spacecraft	..	..	31 810	37 191	38 530	38 837	..	..
		359	Transport equipment, nec	..	..	841	966	1 044	1 261	..	..
	36		Furniture; manufacturing, nec	..	..	43 583	44 954	48 296	52 262	..	..
		361	Furniture	..	..	32 692	34 638	36 044	39 812	..	..
		369	Manufacturing, nec	..	..	10 891	10 316	12 252	12 450	..	..
	37		Recycling	..	..	1 414	1 051	1 157	875	..	..
E	40_41		Electricity, gas and water supply	..	..	233 468	284 713	..	..	..	..
	40		Electricity, gas, steam and hot water supply	..	..	193 157	245 646	..	..	..	..
F	45		Construction	..	..	..	..	..	..	..	..

Table GR.3 **INVESTISSEMENT - INVESTMENT**

Millions de Dr (Prix courants)

1991	1992	1993	1994	1995	1996	1997	1998		CITI révision 3		
..	..	..	..	..	..	..	..	Activites extractives	C	10_14	
..	..	..	..	..	..	..	..	Extraction de charbon, de lignite et de tourbe		10	
..	..	..	..	..	..	..	..	Pétrole brut et gaz naturel ; activités annexes		11	
..	..	..	..	..	..	..	..	Extraction de Minerais d'Uranium et de Thorium		12	
..	..	..	..	..	..	..	..	Extraction de minerais métalliques		13	
..	..	..	..	..	..	..	..	Autres activités extractives		14	
..	..	302 020	286 936	344 414	360 456	..	..	Total activites manufacturieres	D	15_37	
..	..	99 330	90 021	111 305	106 762	..	..	Produits alimentaires et boissons		15	
..	..	21 073	22 704	23 298	29 610	..	..	Viande, poisson, fruits, légumes, huiles et graisses			151
..	..	31 356	18 856	36 849	15 056	..	..	Produits laitiers			152
..	..	11 553	6 101	5 076	9 703	..	..	Travail des grains; prod. amyl. ; alim. pour animaux			153
..	..	18 895	22 956	22 538	20 870	..	..	Autres produits alimentaires			154
..	..	16 454	19 403	23 544	31 523	..	..	Boissons			155
..	..	2 705	5 567	4 859	8 270	..	..	Produits à base de tabac		16	
..	..	23 906	28 919	27 060	37 950	..	..	Textiles		17	
..	..	16 562	17 066	16 623	23 054	..	..	Filature, tissage et achèvement des textiles			171
..	..	2 349	7 206	5 050	10 365	..	..	Autres articles textiles			172
..	..	4 995	4 647	5 387	4 531	..	..	Etoffes et articles de bonneterie			173
..	..	9 169	11 218	12 406	13 182	..	..	Articles d'habillement et fourrure		18	
..	..	9 022	10 999	12 231	12 873	..	..	Articles d'habillement autres qu'en fourrure			181
..	..	147	219	175	309	..	..	Prép. et teinture des fourrures ; articles en fourrure			182
..	..	2 682	1 516	2 721	5 064	..	..	Cuir et articles en cuir ; chaussures		19	
..	..	1 260	328	581	703	..	..	Cuir et articles en cuir sauf chaussures			191
..	..	1 422	1 187	2 140	4 360	..	..	Chaussures			192
..	..	3 870	4 790	3 637	5 626	..	..	Bois, articles en bois, sauf meubles		20	
..	..	303	306	470	625	..	..	Sciage et rabotage du bois			201
..	..	3 567	4 484	3 167	5 001	..	..	Articles en bois, liège, vannerie, sparterie			202
..	..	6 843	9 734	16 448	14 702	..	..	Papier, articles en papier et carton		21	
..	..	12 836	11 135	8 541	10 618	..	..	Edition, imprimerie et reproduction		22	
..	..	8 351	8 749	5 557	6 110	..	..	Edition			221
..	..	4 485	2 387	2 984	4 507	..	..	Imprimerie et activités annexes			222
..	..				..	..	..	Reproduction de supports enregistrés			223
..	..	10 073	8 968	16 973	14 345	..	..	Coke. et prod. pétroliers ; comb. nucléaires		23	
..	..			..	..	..	..	Cokéfaction			231
..	..	10 073	8 968	16 973	14 345	..	..	Produits pétroliers raffinés			232
..	..			..	..	..	..	Traitement de combustibles nucléaires			233
..	..	19 838	20 357	27 942	30 868	..	..	Produits chimiques		24	
..	..	5 544	5 446	8 843	5 461	..	..	Produits chimiques de base			241
..	..	12 427	13 186	17 445	24 397	..	..	Autres produits chimiques			242
..	..	1 867	1 725	1 654	1 010	..	..	Fibres synthétiques ou artificielles			243
..	..	10 498	12 389	18 656	17 249	..	..	Art. en caoutchouc et matières plastiques		25	
..	..	619	819	1 480	369	..	..	Articles en caoutchouc			251
..	..	9 879	11 570	17 176	16 880	..	..	Articles en matières plastiques			252
..	..	18 227	17 848	16 406	26 014	..	..	Autres produits minéraux non métalliques		26	
..	..	1 667	721	896	2 764	..	..	Verre et articles en verre			261
..	..	16 560	17 127	15 510	23 250	..	..	Produits minéraux non métalliques, nca			269
..	..	22 947	15 747	20 962	23 796	..	..	Produits métallurgiques de base		27	
..	..	10 894	5 099	10 813	9 708	..	..	Sidérurgie et première transformation de l'acier			271
..	..	11 983	10 548	10 059	13 820	..	..	Première transf. métaux précieux et non ferreux			272
..	..	70	100	90	268	..	..	Fonderie			273
..	..	8 345	8 687	10 832	13 393	..	..	Ouvrages en métaux, sauf machines		28	
..	..	3 504	4 306	3 707	3 442	..	..	Constr. et menuis. métal.; réservoirs; générateurs			281
..	..	4 841	4 381	7 125	9 951	..	..	Autres ouvrages en métaux			289
..	..	7 794	9 727	9 608	10 746	..	..	Machines et matériels, nca		29	
..	..	3 924	3 605	3 841	4 038	..	..	Machines d'usage général			291
..	..	2 126	2 712	2 863	2 913	..	..	Machines d'usage spécifique			292
..	..	1 743	3 410	2 904	3 795	..	..	Appareils domestiques, nca			293
..	..	27	27	59	216	..	..	Mach. de bureau, comptables, information		30	
..	..	7 192	10 143	8 468	8 443	..	..	Machines et appareils électriques, nca		31	
..	..	337	697	714	707	..	..	Moteurs, génératrices, transformateurs électriques			311
..	..	1 247	2 592	2 746	3 158	..	..	Matériel électrique de distribution et de commande			312
..	..	4 956	3 718	2 680	3 297	..	..	Fils et câbles électriques isolés			313
..	..	175	2 768	1 447	521	..	..	Accumulateurs et piles électriques			314
..	..	314	227	520	707	..	..	Lampes électriques et appareils d'éclairage			315
..	..	163	141	361	53	..	..	Autres matériels électriques, nca.			319
..	..	8 547	5 031	4 512	3 427	..	..	Equipements et appareils de radio, TV, etc.		32	
..	..	407	742	395	400	..	..	Tubes et valves électroniques;autres composants			321
..	..	8 110	4 276	4 044	2 912	..	..	Emetteurs radio/TV; app. téléphonie et télégraphie			322
..	..	30	13	73	115	..	..	Récepteurs radio/TV, son et image; art. associés			323
..	..	590	2 009	1 194	947	..	..	Instr. médicaux, de précision ; optique ; horlog.		33	
..	..	555	460	661	910	..	..	Instruments médicaux et de précision, etc.			331
..	..	34	1 549	533	37	..	..	Instruments d'optique ; matériel photographique			332
..	..	1	0	..	..	..	..	Horlogerie			333
..	..	1 994	1 231	502	664	..	..	Véhicules automobiles; remorques, etc.		34	
..	..	1 542	935	289	432	..	..	Véhicules automobiles			341
..	..	214	125	103	68	..	..	Carross. automobiles, remorques, semi-remorques			342
..	..	239	170	110	164	..	..	Pièces et accessoires pour automobiles et moteurs			343
..	..	19 874	7 438	16 170	3 285	..	..	Autres matériels de transport		35	
..	..	3 126	353	1 165	1 643	..	..	Construction et réparation de navires			351
..	..	12	2	12	26	..	..	Matériel ferroviaire roulant			352
..	..	16 649	7 049	13 803	1 445	..	..	Construction aéronautique et spatiale			353
..	..	88	35	1 189	171	..	..	Autres équipements de transport			359
..	..	4 441	4 364	4 672	4 786	..	..	Meubles ; activités de fabrication, nca		36	
..	..	2 603	2 830	3 548	3 964	..	..	Meubles			361
..	..	1 837	1 534	1 124	822	..	..	Activités de fabrication, nca			369
..	..	293	69	484	105	..	..	Récupération		37	
..	..	160 850	163 934	..	..	..	..	Electricité, gaz et eau	E	40_41	
..	..	142 715	151 489	..	..	..	..	Prod. et dist. d'électricité gaz, vapeur d'eau chaude		40	
..	..	..	..	..	..	..	..	Construction	F	45	

GREECE

Table GR.4 INVESTMENT IN MACHINERY & EQUIPMENT - DÉPENSES DE MACHINES ET D'ÉQUIPEMENT

Millions of Dr (Current Prices)

ISIC revision 3				1991	1992	1993	1994	1995	1996	1997	1998
C	10_14		Mining and quarrying	..	..	..	..	..	..	..	..
	10		Mining of coal and lignite; extraction of peat	..	..	..	..	..	..	..	..
	11		Crude petroleum & natural gas; service activities	..	..	..	..	..	..	..	..
	12		Mining of uranium and thorium ores	..	..	..	..	..	..	..	..
	13		Mining of metal ores	..	..	..	..	..	..	..	..
	14		Other mining and quarrying	..	..	..	..	..	..	..	..
D	15_37		Total manufacturing	..	..	183 428	181 973	222 105	249 218	..	..
	15		Food products and beverages	..	..	62 667	55 757	78 151	74 681	..	..
		151	Production of meat, fish, fruit, vegetables, oils & fats	..	..	12 604	13 492	15 880	20 374	..	..
		152	Dairy products	..	..	20 547	10 582	25 651	9 203	..	..
		153	Grain mill and starch products; animal feeds	..	..	7 296	3 201	3 252	6 614	..	..
		154	Other food products	..	..	10 004	12 432	13 895	13 725	..	..
		155	Beverages	..	..	12 216	16 049	19 473	24 765	..	..
	16		Tobacco products	..	..	2 397	3 604	3 623	5 971	..	..
	17		Textiles	..	..	13 901	19 427	19 136	27 286	..	..
		171	Spinning, weaving and finishing of textiles	..	..	8 144	10 009	11 120	16 760	..	..
		172	Other textiles	..	..	1 883	5 534	3 595	7 406	..	..
		173	Knitted and crocheted fabrics and articles	..	..	3 874	3 885	4 421	3 120	..	..
	18		Wearing apparel and fur	..	..	5 726	6 228	6 820	8 724	..	..
		181	Wearing apparel, except fur apparel	..	..	5 655	6 137	6 721	8 535	..	..
		182	Dressing and dyeing of fur; articles of fur	..	..	72	90	99	189	..	..
	19		Leather and articles of leather; footwear	..	..	1 098	1 268	1 814	2 724	..	..
		191	Leather and articles of leather	..	..	354	195	407	458	..	..
		192	Footwear	..	..	743	1 073	1 407	2 266	..	..
	20		Wood & products of wood & cork, exc. furniture	..	..	2 046	1 337	1 826	3 646	..	..
		201	Sawmilling and planing of wood	..	..	153	235	214	529	..	..
		202	Wood products; articles of cork, straw, etc.	..	..	1 893	1 101	1 611	3 117	..	..
	21		Paper and paper products	..	..	5 494	5 916	8 282	10 328	..	..
	22		Printing, publishing and recorded media	..	..	7 712	9 286	4 751	8 158	..	..
		221	Publishing	..	..	4 355	5 993	2 508	4 646	..	..
		222	Printing and related activities	..	..	3 357	3 293	2 243	3 512	..	..
		223	Reproduction of recorded media	..	..	..	..	..	..	..	..
	23		Coke and petroleum products; nuclear fuel	..	..	5 482	7 271	10 705	13 515	..	..
		231	Coke oven products	..	..	..	..	..	..	..	..
		232	Refined petroleum products	..	..	5 482	7 271	10 705	13 515	..	..
		233	Processing of nuclear fuel	..	..	..	..	..	..	..	..
	24		Chemicals and chemical products	..	..	13 164	15 552	18 743	17 042	..	..
		241	Basic chemicals	..	..	4 027	5 030	7 671	4 623	..	..
		242	Other chemical products	..	..	7 425	8 945	9 460	11 366	..	..
		243	Man-made fibres	..	..	1 712	1 577	1 612	1 053	..	..
	25		Rubber and plastic products	..	..	8 184	9 476	14 804	11 436	..	..
		251	Rubber products	..	..	497	700	1 422	179	..	..
		252	Plastic products	..	..	7 687	8 776	13 383	11 257	..	..
	26		Other non-metallic mineral products	..	..	13 583	11 847	11 718	17 940	..	..
		261	Glass and glass products	..	..	975	394	737	2 169	..	..
		269	Non-metallic mineral products, nec	..	..	12 608	11 453	10 981	15 771	..	..
	27		Basic metals	..	..	15 410	10 279	15 701	16 900	..	..
		271	Basic iron and steel	..	..	7 138	3 225	7 981	7 275	..	..
		272	Basic precious and non-ferrous metals	..	..	8 218	6 963	7 661	9 416	..	..
		273	Casting of metals	..	..	53	90	59	209	..	..
	28		Fabricated metal products, exc. machinery	..	..	5 762	5 915	7 499	11 484	..	..
		281	Structural metal prod., tanks, reservoirs, etc.	..	..	1 989	2 352	2 009	3 395	..	..
		289	Other fabricated metal products	..	..	3 773	3 563	5 490	8 089	..	..
	29		Machinery and equipment, nec	..	..	4 826	5 087	5 398	6 817	..	..
		291	General purpose machinery	..	..	2 003	1 602	2 241	2 539	..	..
		292	Special purpose machinery	..	..	1 078	1 406	1 386	1 790	..	..
		293	Domestic appliances, nec	..	..	1 744	2 079	1 771	2 488	..	..
	30		Office, accounting & computing machinery	..	..	5	27	49	95	..	..
	31		Electrical machinery & apparatus, nec	..	..	5 803	6 634	5 804	4 403	..	..
		311	Electric motors, generators and transformers	..	..	241	591	666	592	..	..
		312	Electricity distribution & control apparatus	..	..	1 100	1 042	1 587	886	..	..
		313	Insulated wire and cable	..	..	3 976	2 165	2 065	1 965	..	..
		314	Accumulators, primary cells & primary batteries	..	..	149	2 550	1 077	470	..	..
		315	Electric lamps and lighting equipment	..	..	273	163	299	443	..	..
		319	Other electrical equipment, nec	..	..	64	123	110	47	..	..
	32		Radio, TV, communication equipment	..	..	2 866	1 721	1 872	2 050	..	..
		321	Electronic valves, tubes; other electronic components	..	..	336	316	315	322	..	..
		322	TV/radio transmitters; line teleph./telegr. apparatus	..	..	2 501	1 400	1 485	1 612	..	..
		323	TV/radio receivers; sound and video app.; assoc. gds.	..	..	29	4	73	115	..	..
	33		Medical, precision & optical instrum.; watch & clocks	..	..	519	554	766	357	..	..
		331	Medical appl.; instruments & appl. for measuring, etc.	..	..	484	270	497	321	..	..
		332	Optical instruments and photographic equipment	..	..	34	283	269	37	..	..
		333	Watches and clocks	..	..	1	0		..	..	..
	34		Motor vehicles, trailers and semi-trailers	..	..	810	525	213	387	..	..
		341	Motor vehicles	..	..	654	309	106	235	..	..
		342	Bodies for motor vehicles, trailers & semi-trailers	..	..	71	91	24	65	..	..
		343	Parts & accessories for motor vehicles & their engines	..	..	85	124	82	87	..	..
	35		Other transport equipment	..	..	2 957	1 700	1 462	1 772	..	..
		351	Building and repairing of ships and boats	..	..	2 023	1 216	834	965	..	..
		352	Railway and tramway locomotives and rolling stock	..	..	12	2	12	26	..	..
		353	Aircraft and spacecraft	..	..	845	458	319	682	..	..
		359	Transport equipment, nec	..	..	78	24	297	98	..	..
	36		Furniture; manufacturing, nec	..	..	2 738	2 511	2 614	3 403	..	..
		361	Furniture	..	..	1 855	1 526	2 016	2 824	..	..
		369	Manufacturing, nec	..	..	883	985	598	579	..	..
	37		Recycling	..	..	280	51	356	100	..	..
E	40_41		Electricity, gas and water supply	..	..	..	..	..	..	..	..
	40		Electricity, gas, steam and hot water supply	..	..	..	..	..	..	..	..
F	45		Construction	..	..	..	..	..	..	..	..

Table GR.5 **EMPLOI, SALARIÉS - EMPLOYMENT, EMPLOYEES**

Unités

1991	1992	1993	1994	1995	1996	1997	1998	CITI révision 3			
..	..	..	..	..	..	..	..	Activites extractives	C	10_14	
..	..	..	..	..	..	..	..	Extraction de charbon, de lignite et de tourbe		10	
..	..	..	..	..	..	..	..	Pétrole brut et gaz naturel ; activités annexes		11	
..	..	..	..	..	..	..	..	Extraction de Minerais d'Uranium et de Thorium		12	
..	..	..	..	..	..	..	..	Extraction de minerais métalliques		13	
..	..	..	..	..	..	..	..	Autres activités extractives		14	
..	..	269 357	257 057	250 460	238 860	..	..	Total activites manufacturieres	D	15_37	
..	..	51 328	51 253	50 586	49 396	..	..	Produits alimentaires et boissons		15	
..	..	16 428	16 474	16 137	15 237	..	..	Viande, poisson, fruits, légumes, huiles et graisses			151
..	..	6 794	6 938	7 059	7 130	..	..	Produits laitiers			152
..	..	4 037	3 509	3 499	3 366	..	..	Travail des grains; prod. amyl. ; alim. pour animaux			153
..	..	14 964	15 415	15 283	15 114	..	..	Autres produits alimentaires			154
..	..	9 105	8 917	8 608	8 549	..	..	Boissons			155
..	..	3 315	2 996	2 925	2 784	..	..	Produits à base de tabac		16	
..	..	27 687	24 586	23 670	21 340	..	..	Textiles		17	
..	..	15 929	14 176	13 981	12 183	..	..	Filature, tissage et achèvement des textiles			171
..	..	4 479	3 988	4 076	4 077	..	..	Autres articles textiles			172
..	..	7 279	6 422	5 613	5 080	..	..	Etoffes et articles de bonneterie			173
..	..	35 292	30 834	28 773	25 227	..	..	Articles d'habillement et fourrure		18	
..	..	34 487	30 063	28 084	24 498	..	..	Articles d'habillement autres qu'en fourrure			181
..	..	805	771	689	729	..	..	Prép. et teinture des fourrures ; articles en fourrure			182
..	..	6 209	5 680	5 351	4 815	..	..	Cuir et articles en cuir ; chaussures		19	
..	..	1 144	964	865	758	..	..	Cuir et articles en cuir sauf chaussures			191
..	..	5 065	4 716	4 486	4 057	..	..	Chaussures			192
..	..	5 625	5 028	4 742	4 666	..	..	Bois, articles en bois, sauf meubles		20	
..	..	838	761	754	751	..	..	Sciage et rabotage du bois			201
..	..	4 787	4 267	3 988	3 915	..	..	Articles en bois, liège, vannerie, sparterie			202
..	..	7 851	8 022	8 293	8 256	..	..	Papier, articles en papier et carton		21	
..	..	10 042	10 166	10 088	10 022	..	..	Edition, imprimerie et reproduction		22	
..	..	6 685	6 787	6 872	6 848	..	..	Edition			221
..	..	3 357	3 379	3 216	3 174	..	..	Imprimerie et activités annexes			222
..	..	..	..	..	..	..	..	Reproduction de supports enregistrés			223
..	..	4 182	4 261	4 117	3 710	..	..	Coke. et prod. pétroliers ; comb. nucléaires		23	
..	..	..	..	..	..	..	..	Cokéfaction			231
..	..	4 182	4 261	4 117	3 710	..	..	Produits pétroliers raffinés			232
..	..	..	..	..	..	..	..	Traitement de combustibles nucléaires			233
..	..	20 598	20 146	19 208	18 837	..	..	Produits chimiques		24	
..	..	4 593	4 430	4 342	4 396	..	..	Produits chimiques de base			241
..	..	14 949	14 761	13 929	13 531	..	..	Autres produits chimiques			242
..	..	1 056	955	937	910	..	..	Fibres synthétiques ou artificielles			243
..	..	8 968	9 060	9 392	8 861	..	..	Art. en caoutchouc et matières plastiques		25	
..	..	1 005	998	1 012	803	..	..	Articles en caoutchouc			251
..	..	7 963	8 062	8 380	8 058	..	..	Articles en matières plastiques			252
..	..	18 713	17 580	16 993	16 455	..	..	Autres produits minéraux non métalliques		26	
..	..	1 089	1 018	906	890	..	..	Verre et articles en verre			261
..	..	17 624	16 562	16 087	15 565	..	..	Produits minéraux non métalliques, nca			269
..	..	10 964	10 639	10 571	10 616	..	..	Produits métallurgiques de base		27	
..	..	6 077	5 639	5 421	5 251	..	..	Sidérurgie et première transformation de l'acier			271
..	..	4 787	4 836	4 895	5 111	..	..	Première transf. métaux précieux et non ferreux			272
..	..	100	164	255	254	..	..	Fonderie			273
..	..	11 079	10 636	10 410	10 252	..	..	Ouvrages en métaux, sauf machines		28	
..	..	4 315	3 744	3 713	3 892	..	..	Constr. et menuis. métal.; réservoirs; générateurs			281
..	..	6 764	6 892	6 697	6 360	..	..	Autres ouvrages en métaux			289
..	..	11 284	11 108	11 167	11 273	..	..	Machines et matériels, nca		29	
..	..	4 316	4 260	4 132	4 201	..	..	Machines d'usage général			291
..	..	4 702	4 638	4 873	4 895	..	..	Machines d'usage spécifique			292
..	..	2 266	2 210	2 162	2 177	..	..	Appareils domestiques, nca			293
..	..	72	58	102	235	..	..	Mach. de bureau, comptables, information		30	
..	..	5 484	5 292	5 244	5 714	..	..	Machines et appareils électriques, nca		31	
..	..	1 161	1 116	1 108	1 193	..	..	Moteurs, génératrices, transformateurs électriques			311
..	..	1 525	1 535	1 518	1 444	..	..	Matériel électrique de distribution et de commande			312
..	..	1 563	1 524	1 537	2 014	..	..	Fils et câbles électriques isolés			313
..	..	571	430	400	377	..	..	Accumulateurs et piles électriques			314
..	..	556	533	541	557	..	..	Lampes électriques et appareils d'éclairage			315
..	..	108	154	140	129	..	..	Autres matériels électriques, nca.			319
..	..	2 574	2 566	2 519	2 461	..	..	Equipements et appareils de radio, TV, etc.		32	
..	..	308	365	338	256	..	..	Tubes et valves électroniques;autres composants			321
..	..	2 051	2 032	2 032	2 068	..	..	Emetteurs radio/TV; app. téléphonie et télégraphie			322
..	..	215	169	149	137	..	..	Récepteurs radio/TV, son et image; art. associés			323
..	..	1 070	1 122	1 250	1 304	..	..	Instr. médicaux, de précision ; optique ; horlog.		33	
..	..	898	893	923	945	..	..	Instruments médicaux et de précision, etc.			331
..	..	166	223	322	355	..	..	Instruments d'optique ; matériel photographique			332
..	..	6	6	5	4	..	..	Horlogerie			333
..	..	2 342	2 224	2 127	1 820	..	..	Véhicules automobiles; remorques, etc.		34	
..	..	1 711	1 599	1 481	1 144	..	..	Véhicules automobiles			341
..	..	345	249	282	323	..	..	Carross. automobiles, remorques, semi-remorques			342
..	..	286	376	364	353	..	..	Pièces et accessoires pour automobiles et moteurs			343
..	..	16 175	15 855	15 469	13 773	..	..	Autres matériels de transport		35	
..	..	8 803	8 560	8 509	6 978	..	..	Construction et réparation de navires			351
..	..	2 373	2 398	2 214	2 238	..	..	Matériel ferroviaire roulant			352
..	..	4 892	4 778	4 595	4 395	..	..	Construction aéronautique et spatiale			353
..	..	107	119	151	162	..	..	Autres équipements de transport			359
..	..	8 318	7 846	7 340	6 928	..	..	Meubles ; activités de fabrication, nca		36	
..	..	6 414	6 011	5 569	5 233	..	..	Meubles			361
..	..	1 904	1 835	1 771	1 695	..	..	Activités de fabrication, nca			369
..	..	185	99	123	115	..	..	Récupération		37	
..	..	29 053	29 454	..	..	..	..	Electricité, gaz et eau	E	40_41	
..	..	22 488	22 755	..	..	..	..	Prod. et dist. d'électricité gaz, vapeur d'eau chaude		40	
..	..	..	..	..	..	..	..	Construction	F	45	

GREECE

Table GR.6 COMPENSATION OF LABOUR, EMPLOYEES - SALAIRES ET CHARGES SOCIALES, SALARIÉS

Millions of Dr (Current Prices)

ISIC revision 3				1991	1992	1993	1994	1995	1996	1997	1998
C	10_14		Mining and quarrying	..	..	..	..	..	..	..	..
	10		Mining of coal and lignite; extraction of peat	..	..	..	..	..	..	..	..
	11		Crude petroleum & natural gas; service activities	..	..	..	..	..	..	..	..
	12		Mining of uranium and thorium ores	..	..	..	..	..	..	..	..
	13		Mining of metal ores	..	..	..	..	..	..	..	..
	14		Other mining and quarrying	..	..	..	..	..	..	..	..
D	15_37		Total manufacturing	..	..	1034 794	1130 253	1238 658	..	..	..
	15		Food products and beverages	..	..	193 941	225 832	245 536	..	..	..
		151	Production of meat, fish, fruit, vegetables, oils & fats	..	..	54 164	60 755	64 933	..	..	..
		152	Dairy products	..	..	26 775	33 384	37 090	..	..	..
		153	Grain mill and starch products; animal feeds	..	..	15 529	15 742	17 476	..	..	..
		154	Other food products	..	..	52 248	64 200	69 957	..	..	..
		155	Beverages	..	..	45 225	51 751	56 079	..	..	..
	16		Tobacco products	..	..	13 972	18 138	19 435	..	..	..
	17		Textiles	..	..	87 628	89 416	95 601	..	..	..
		171	Spinning, weaving and finishing of textiles	..	..	56 341	57 763	63 270	..	..	..
		172	Other textiles	..	..	12 323	12 861	14 084	..	..	..
		173	Knitted and crocheted fabrics and articles	..	..	18 963	18 792	18 247	..	..	..
	18		Wearing apparel and fur	..	..	84 979	83 760	88 451	..	..	..
		181	Wearing apparel, except fur apparel	..	..	83 242	81 972	86 665	..	..	..
		182	Dressing and dyeing of fur; articles of fur	..	..	1 737	1 788	1 786	..	..	..
	19		Leather and articles of leather; footwear	..	..	16 549	17 977	19 343	..	..	..
		191	Leather and articles of leather	..	..	3 650	3 608	3 599	..	..	..
		192	Footwear	..	..	12 899	14 369	15 744	..	..	..
	20		Wood & products of wood & cork, exc. furniture	..	..	19 899	19 085	20 694	..	..	..
		201	Sawmilling and planing of wood	..	..	2 569	2 605	2 691	..	..	..
		202	Wood products; articles of cork, straw, etc.	..	..	17 330	16 480	18 003	..	..	..
	21		Paper and paper products	..	..	32 645	35 590	41 210	..	..	..
	22		Printing, publishing and recorded media	..	..	41 295	47 571	50 940	..	..	..
		221	Publishing	..	..	28 947	32 815	35 821	..	..	..
		222	Printing and related activities	..	..	12 347	14 756	15 119	..	..	..
		223	Reproduction of recorded media	..	..	..	..	..	..	..	..
	23		Coke and petroleum products; nuclear fuel	..	..	26 333	32 809	38 214	..	..	..
		231	Coke oven products	..	..	..	..	..	..	..	..
		232	Refined petroleum products	..	..	26 333	32 809	38 214	..	..	..
		233	Processing of nuclear fuel	..	..	..	..	..	..	..	..
	24		Chemicals and chemical products	..	..	102 652	114 283	124 141	..	..	..
		241	Basic chemicals	..	..	28 650	31 491	35 553	..	..	..
		242	Other chemical products	..	..	70 439	78 469	84 074	..	..	..
		243	Man-made fibres	..	..	3 563	4 323	4 514	..	..	..
	25		Rubber and plastic products	..	..	32 878	37 098	42 345	..	..	..
		251	Rubber products	..	..	5 578	6 158	6 970	..	..	..
		252	Plastic products	..	..	27 300	30 940	35 375	..	..	..
	26		Other non-metallic mineral products	..	..	81 782	87 322	95 370	..	..	..
		261	Glass and glass products	..	..	3 715	3 858	4 139	..	..	..
		269	Non-metallic mineral products, nec	..	..	78 068	83 464	91 231	..	..	..
	27		Basic metals	..	..	63 806	63 874	74 164	..	..	..
		271	Basic iron and steel	..	..	32 193	32 671	36 509	..	..	..
		272	Basic precious and non-ferrous metals	..	..	31 245	30 545	36 577	..	..	..
		273	Casting of metals	..	..	368	659	1 078	..	..	..
	28		Fabricated metal products, exc. machinery	..	..	38 833	41 786	46 262	..	..	..
		281	Structural metal prod., tanks, reservoirs, etc.	..	..	14 459	13 742	15 788	..	..	..
		289	Other fabricated metal products	..	..	24 373	28 044	30 474	..	..	..
	29		Machinery and equipment, nec	..	..	42 394	47 962	54 624	..	..	..
		291	General purpose machinery	..	..	14 584	16 145	17 940	..	..	..
		292	Special purpose machinery	..	..	18 815	21 723	25 875	..	..	..
		293	Domestic appliances, nec	..	..	8 994	10 094	10 810	..	..	..
	30		Office, accounting & computing machinery	..	..	304	242	638	..	..	..
	31		Electrical machinery & apparatus, nec	..	..	24 426	23 654	26 580	..	..	..
		311	Electric motors, generators and transformers	..	..	5 497	5 703	6 502	..	..	..
		312	Electricity distribution & control apparatus	..	..	5 468	6 086	6 364	..	..	..
		313	Insulated wire and cable	..	..	7 863	7 695	8 993	..	..	..
		314	Accumulators, primary cells & primary batteries	..	..	3 800	1 994	2 278	..	..	..
		315	Electric lamps and lighting equipment	..	..	1 434	1 631	1 896	..	..	..
		319	Other electrical equipment, nec	..	..	365	545	547	..	..	..
	32		Radio, TV, communication equipment	..	..	12 380	13 202	14 695	..	..	..
		321	Electronic valves, tubes; other electronic components	..	..	1 006	1 240	1 253	..	..	..
		322	TV/radio transmitters; line teleph./telegr. apparatus	..	..	10 799	11 383	12 894	..	..	..
		323	TV/radio receivers; sound and video app.; assoc. gds.	..	..	575	578	548	..	..	..
	33		Medical, precision & optical instrum.; watch & clocks	..	..	3 283	3 747	4 798	..	..	..
		331	Medical appl.; instruments & appl. for measuring, etc.	..	..	2 822	3 050	3 647	..	..	..
		332	Optical instruments and photographic equipment	..	..	446	682	1 137	..	..	..
		333	Watches and clocks	..	..	15	16	14	..	..	..
	34		Motor vehicles, trailers and semi-trailers	..	..	9 897	10 237	10 863	..	..	..
		341	Motor vehicles	..	..	7 836	7 998	8 315	..	..	..
		342	Bodies for motor vehicles, trailers & semi-trailers	..	..	1 212	929	1 147	..	..	..
		343	Parts & accessories for motor vehicles & their engines	..	..	849	1 310	1 400	..	..	..
	35		Other transport equipment	..	..	81 881	92 664	99 373	..	..	..
		351	Building and repairing of ships and boats	..	..	44 454	47 455	52 398	..	..	..
		352	Railway and tramway locomotives and rolling stock	..	..	9 696	11 085	11 301	..	..	..
		353	Aircraft and spacecraft	..	..	27 421	33 761	35 104	..	..	..
		359	Transport equipment, nec	..	..	310	364	569	..	..	..
	36		Furniture; manufacturing, nec	..	..	22 410	23 719	24 932	..	..	..
		361	Furniture	..	..	17 289	18 166	18 838	..	..	..
		369	Manufacturing, nec	..	..	5 121	5 553	6 094	..	..	..
	37		Recycling	..	..	629	287	449	..	..	..
E	40_41		Electricity, gas and water supply	..	..	143 441	172 839	..	..	..	..
	40		Electricity, gas, steam and hot water supply	..	..	110 146	133 895	..	..	..	..
F	45		Construction	..	..	..	..	..	..	..	..

Table GR.7 **SALAIRES ET TRAITEMENTS, SALARIÉS - WAGES AND SALARIES, EMPLOYEES**

Millions de Dr (Prix courants)

1991	1992	1993	1994	1995	1996	1997	1998			CITI révision 3
..	..	..	..	..	..	..	..	Activites extractives	C	**10_14**
..	..	..	..	..	..	..	..	Extraction de charbon, de lignite et de tourbe		**10**
..	..	..	..	..	..	..	..	Pétrole brut et gaz naturel ; activités annexes		**11**
..	..	..	..	..	..	..	..	Extraction de Minerais d'Uranium et de Thorium		**12**
..	..	..	..	..	..	..	..	Extraction de minerais métalliques		**13**
..	..	..	..	..	..	..	..	Autres activités extractives		**14**
..	..	791 644	865 853	947 536	..	..	..	Total activites manufacturieres	D	**15_37**
..	..	148 713	173 365	188 161	..	..	..	Produits alimentaires et boissons		**15**
..	..	41 636	46 999	49 924	..	..	..	Viande, poisson, fruits, légumes, huiles et graisses		151
..	..	20 396	25 715	28 565	..	..	..	Produits laitiers		152
..	..	11 969	12 079	13 370	..	..	..	Travail des grains; prod. amyl. ; alim. pour animaux		153
..	..	40 283	49 264	53 858	..	..	..	Autres produits alimentaires		154
..	..	34 431	39 308	42 443	..	..	..	Boissons		155
..	..	10 289	13 475	14 663	..	..	..	Produits à base de tabac		**16**
..	..	67 005	68 374	72 806	..	..	..	Textiles		**17**
..	..	42 749	43 851	47 871	..	..	..	Filature, tissage et achèvement des textiles		171
..	..	9 499	9 943	10 813	..	..	..	Autres articles textiles		172
..	..	14 757	14 580	14 121	..	..	..	Etoffes et articles de bonneterie		173
..	..	66 200	65 271	68 478	..	..	..	Articles d'habillement et fourrure		**18**
..	..	64 840	63 876	67 077	..	..	..	Articles d'habillement autres qu'en fourrure		181
..	..	1 361	1 395	1 400	..	..	..	Prép. et teinture des fourrures ; articles en fourrure		182
..	..	12 748	13 814	14 742	..	..	..	Cuir et articles en cuir ; chaussures		**19**
..	..	2 816	2 791	2 776	..	..	..	Cuir et articles en cuir sauf chaussures		191
..	..	9 932	11 023	11 966	..	..	..	Chaussures		192
..	..	15 445	14 705	15 882	..	..	..	Bois, articles en bois, sauf meubles		**20**
..	..	1 972	1 975	2 029	..	..	..	Sciage et rabotage du bois		201
..	..	13 472	12 730	13 852	..	..	..	Articles en bois, liège, vannerie, sparterie		202
..	..	24 874	27 001	31 116	..	..	..	Papier, articles en papier et carton		**21**
..	..	31 940	37 600	40 117	..	..	..	Edition, imprimerie et reproduction		**22**
..	..	22 518	26 301	28 683	..	..	..	Edition		221
..	..	9 422	11 299	11 434	..	..	..	Imprimerie et activités annexes		222
..	..	..	..	..	..	..	..	Reproduction de supports enregistrés		223
..	..	18 783	23 716	28 334	..	..	..	Coke. et prod. pétroliers ; comb. nucléaires		**23**
..	..	..	..	..	..	..	..	Cokéfaction		231
..	..	18 783	23 716	28 334	..	..	..	Produits pétroliers raffinés		232
..	..	..	..	..	..	..	..	Traitement de combustibles nucléaires		233
..	..	78 686	87 368	94 741	..	..	..	Produits chimiques		**24**
..	..	21 990	23 997	26 932	..	..	..	Produits chimiques de base		241
..	..	54 009	60 058	64 344	..	..	..	Autres produits chimiques		242
..	..	2 687	3 313	3 465	..	..	..	Fibres synthétiques ou artificielles		243
..	..	25 303	28 820	32 806	..	..	..	Art. en caoutchouc et matières plastiques		**25**
..	..	4 185	4 805	5 492	..	..	..	Articles en caoutchouc		251
..	..	21 118	24 015	27 314	..	..	..	Articles en matières plastiques		252
..	..	62 410	66 685	72 548	..	..	..	Autres produits minéraux non métalliques		**26**
..	..	2 859	2 973	3 089	..	..	..	Verre et articles en verre		261
..	..	59 551	63 712	69 460	..	..	..	Produits minéraux non métalliques, nca		269
..	..	47 385	48 413	56 256	..	..	..	Produits métallurgiques de base		**27**
..	..	24 035	24 618	27 433	..	..	..	Sidérurgie et première transformation de l'acier		271
..	..	23 068	23 287	28 005	..	..	..	Première transf. métaux précieux et non ferreux		272
..	..	281	508	817	..	..	..	Fonderie		273
..	..	29 950	32 075	35 438	..	..	..	Ouvrages en métaux, sauf machines		**28**
..	..	11 190	10 516	12 078	..	..	..	Constr. et menuis. métal.; réservoirs; générateurs		281
..	..	18 760	21 559	23 361	..	..	..	Autres ouvrages en métaux		289
..	..	32 342	36 515	41 712	..	..	..	Machines et matériels, nca		**29**
..	..	11 244	12 406	13 733	..	..	..	Machines d'usage général		291
..	..	14 084	16 243	19 545	..	..	..	Machines d'usage spécifique		292
..	..	7 014	7 865	8 434	..	..	..	Appareils domestiques, nca		293
..	..	239	186	498	..	..	..	Mach. de bureau, comptables, information		**30**
..	..	19 152	18 306	20 502	..	..	..	Machines et appareils électriques, nca		**31**
..	..	4 208	4 355	4 949	..	..	..	Moteurs, génératrices, transformateurs électriques		311
..	..	4 276	4 763	4 926	..	..	..	Matériel électrique de distribution et de commande		312
..	..	6 135	5 977	6 996	..	..	..	Fils et câbles électriques isolés		313
..	..	3 133	1 530	1 728	..	..	..	Accumulateurs et piles électriques		314
..	..	1 113	1 265	1 474	..	..	..	Lampes électriques et appareils d'éclairage		315
..	..	286	417	429	..	..	..	Autres matériels électriques, nca.		319
..	..	9 311	10 357	11 610	..	..	..	Equipements et appareils de radio, TV, etc.		**32**
..	..	787	957	969	..	..	..	Tubes et valves électroniques;autres composants		321
..	..	8 073	8 943	10 207	..	..	..	Emetteurs radio/TV; app. téléphonie et télégraphie		322
..	..	450	457	434	..	..	..	Récepteurs radio/TV, son et image; art. associés		323
..	..	2 528	2 856	3 681	..	..	..	Instr. médicaux, de précision ; optique ; horlog.		**33**
..	..	2 171	2 324	2 767	..	..	..	Instruments médicaux et de précision, etc.		331
..	..	346	519	902	..	..	..	Instruments d'optique ; matériel photographique		332
..	..	11	12	11	..	..	..	Horlogerie		333
..	..	7 323	7 704	8 252	..	..	..	Véhicules automobiles; remorques, etc.		**34**
..	..	5 749	5 976	6 306	..	..	..	Véhicules automobiles		341
..	..	919	718	871	..	..	..	Carross. automobiles, remorques, semi-remorques		342
..	..	656	1 009	1 075	..	..	..	Pièces et accessoires pour automobiles et moteurs		343
..	..	63 156	70 711	75 633	..	..	..	Autres matériels de transport		**35**
..	..	34 344	36 414	40 175	..	..	..	Construction et réparation de navires		351
..	..	7 554	8 634	8 815	..	..	..	Matériel ferroviaire roulant		352
..	..	21 016	25 385	26 200	..	..	..	Construction aéronautique et spatiale		353
..	..	242	278	443	..	..	..	Autres équipements de transport		359
..	..	17 365	18 314	19 212	..	..	..	Meubles ; activités de fabrication, nca		**36**
..	..	13 402	14 027	14 513	..	..	..	Meubles		361
..	..	3 963	4 287	4 699	..	..	..	Activités de fabrication, nca		369
..	..	496	222	348	..	..	..	Récupération		**37**
..	..	111 078	128 937	..	..	..	..	Electricité, gaz et eau	E	**40_41**
..	..	86 438	99 875	..	..	..	..	Prod. et dist. d'électricité gaz, vapeur d'eau chaude		**40**
..	..	..	..	..	..	..	..	Construction	F	**45**

GREECE

Table GR.8 SUPPLEMENTS TO WAGES & SALARIES - CHARGES SOCIALES DES EMPLOYEURS

Millions of Dr (Current Prices)

ISIC revision 3				1991	1992	1993	1994	1995	1996	1997	1998
C	10_14		Mining and quarrying	..	..	..	..	..	..	..	..
	10		Mining of coal and lignite; extraction of peat	..	..	..	..	..	..	..	..
	11		Crude petroleum & natural gas; service activities	..	..	..	..	..	..	..	..
	12		Mining of uranium and thorium ores	..	..	..	..	..	..	..	..
	13		Mining of metal ores	..	..	..	..	..	..	..	..
	14		Other mining and quarrying	..	..	..	..	..	..	..	..
D	15_37		Total manufacturing	..	..	243 150	264 400	291 122	297 310	..	..
	15		Food products and beverages	..	..	45 227	52 467	57 375	60 887	..	..
		151	Production of meat, fish, fruit, vegetables, oils & fats	..	..	12 528	13 756	15 009	15 321	..	..
		152	Dairy products	..	..	6 379	7 669	8 525	8 994	..	..
		153	Grain mill and starch products; animal feeds	..	..	3 560	3 663	4 106	4 301	..	..
		154	Other food products	..	..	11 965	14 936	16 099	17 210	..	..
		155	Beverages	..	..	10 795	12 443	13 636	15 061	..	..
	16		Tobacco products	..	..	3 683	4 663	4 772	4 800	..	..
	17		Textiles	..	..	20 622	21 041	22 795	18 152	..	..
		171	Spinning, weaving and finishing of textiles	..	..	13 592	13 912	15 399	14 589	..	..
		172	Other textiles	..	..	2 824	2 917	3 270	4 288	..	..
		173	Knitted and crocheted fabrics and articles	..	..	4 206	4 212	4 126	..	..	..
	18		Wearing apparel and fur	..	..	18 779	18 490	19 973	19 150	..	..
		181	Wearing apparel, except fur apparel	..	..	18 402	18 096	19 588	18 701	..	..
		182	Dressing and dyeing of fur; articles of fur	..	..	376	394	385	449	..	..
	19		Leather and articles of leather; footwear	..	..	3 800	4 163	4 601	4 428	..	..
		191	Leather and articles of leather	..	..	834	817	823	857	..	..
		192	Footwear	..	..	2 966	3 346	3 778	3 571	..	..
	20		Wood & products of wood & cork, exc. furniture	..	..	4 455	4 380	4 812	5 199	..	..
		201	Sawmilling and planing of wood	..	..	597	630	661	734	..	..
		202	Wood products; articles of cork, straw, etc.	..	..	3 858	3 750	4 150	4 464	..	..
	21		Paper and paper products	..	..	7 771	8 588	10 094	10 912	..	..
	22		Printing, publishing and recorded media	..	..	9 355	9 971	10 822	11 123	..	..
		221	Publishing	..	..	6 429	6 514	7 138	7 163	..	..
		222	Printing and related activities	..	..	2 925	3 457	3 684	3 959	..	..
		223	Reproduction of recorded media	..	..	..	..	..	..	..	..
	23		Coke and petroleum products; nuclear fuel	..	..	7 550	9 093	9 880	8 636	..	..
		231	Coke oven products	..	..	..	..	..	..	..	..
		232	Refined petroleum products	..	..	7 550	9 093	9 880	8 636	..	..
		233	Processing of nuclear fuel	..	..	..	..	..	..	..	..
	24		Chemicals and chemical products	..	..	23 966	26 914	29 400	30 754	..	..
		241	Basic chemicals	..	..	6 659	7 493	8 621	9 407	..	..
		242	Other chemical products	..	..	16 431	18 411	19 731	20 260	..	..
		243	Man-made fibres	..	..	876	1 010	1 049	1 087	..	..
	25		Rubber and plastic products	..	..	7 575	8 278	9 539	9 930	..	..
		251	Rubber products	..	..	1 393	1 353	1 478	1 281	..	..
		252	Plastic products	..	..	6 183	6 925	8 061	8 649	..	..
	26		Other non-metallic mineral products	..	..	19 372	20 637	22 821	24 451	..	..
		261	Glass and glass products	..	..	856	885	1 050	1 077	..	..
		269	Non-metallic mineral products, nec	..	..	18 517	19 752	21 771	23 373	..	..
	27		Basic metals	..	..	16 421	15 461	17 908	18 996	..	..
		271	Basic iron and steel	..	..	8 158	8 053	9 076	9 363	..	..
		272	Basic precious and non-ferrous metals	..	..	8 177	7 257	8 572	9 356	..	..
		273	Casting of metals	..	..	86	151	260	277	..	..
	28		Fabricated metal products, exc. machinery	..	..	8 883	9 712	10 823	11 605	..	..
		281	Structural metal prod., tanks, reservoirs, etc.	..	..	3 270	3 226	3 710	4 169	..	..
		289	Other fabricated metal products	..	..	5 613	6 485	7 113	7 436	..	..
	29		Machinery and equipment, nec	..	..	10 052	11 447	12 912	14 939	..	..
		291	General purpose machinery	..	..	3 341	3 739	4 207	4 946	..	..
		292	Special purpose machinery	..	..	4 731	5 480	6 330	7 281	..	..
		293	Domestic appliances, nec	..	..	1 980	2 229	2 376	2 713	..	..
	30		Office, accounting & computing machinery	..	..	64	55	140	222	..	..
	31		Electrical machinery & apparatus, nec	..	..	5 274	5 348	6 079	6 237	..	..
		311	Electric motors, generators and transformers	..	..	1 289	1 348	1 552	1 698	..	..
		312	Electricity distribution & control apparatus	..	..	1 191	1 323	1 438	1 550	..	..
		313	Insulated wire and cable	..	..	1 727	1 718	1 998	1 911	..	..
		314	Accumulators, primary cells & primary batteries	..	..	667	464	550	482	..	..
		315	Electric lamps and lighting equipment	..	..	321	366	423	479	..	..
		319	Other electrical equipment, nec	..	..	79	128	118	117	..	..
	32		Radio, TV, communication equipment	..	..	3 069	2 845	3 085	3 254	..	..
		321	Electronic valves, tubes; other electronic components	..	..	219	284	283	250	..	..
		322	TV/radio transmitters; line teleph./telegr. apparatus	..	..	2 725	2 440	2 687	2 887	..	..
		323	TV/radio receivers; sound and video app.; assoc. gds.	..	..	125	121	114	118	..	..
	33		Medical, precision & optical instrum.; watch & clocks	..	..	754	892	1 117	1 316	..	..
		331	Medical appl.; instruments & appl. for measuring, etc.	..	..	651	725	880	1 045	..	..
		332	Optical instruments and photographic equipment	..	..	100	163	234	268	..	..
		333	Watches and clocks	..	..	3	3	3	3	..	..
	34		Motor vehicles, trailers and semi-trailers	..	..	2 574	2 533	2 610	2 517	..	..
		341	Motor vehicles	..	..	2 087	2 021	2 009	1 849	..	..
		342	Bodies for motor vehicles, trailers & semi-trailers	..	..	293	211	276	350	..	..
		343	Parts & accessories for motor vehicles & their engines	..	..	194	300	325	317	..	..
	35		Other transport equipment	..	..	18 725	21 953	23 740	23 852	..	..
		351	Building and repairing of ships and boats	..	..	10 110	11 040	12 223	10 760	..	..
		352	Railway and tramway locomotives and rolling stock	..	..	2 142	2 451	2 486	3 176	..	..
		353	Aircraft and spacecraft	..	..	6 406	8 376	8 905	9 761	..	..
		359	Transport equipment, nec	..	..	68	86	126	154	..	..
	36		Furniture; manufacturing, nec	..	..	5 045	5 405	5 721	5 843	..	..
		361	Furniture	..	..	3 887	4 139	4 325	4 402	..	..
		369	Manufacturing, nec	..	..	1 158	1 266	1 396	1 441	..	..
	37		Recycling	..	..	133	65	101	107	..	..
E	40_41		Electricity, gas and water supply	..	..	30 160	41 611	..	..	..	..
	40		Electricity, gas, steam and hot water supply	..	..	23 708	34 821	..	..	..	..
F	45		Construction	..	..	..	..	..	..	..	..

Table GR.9 **JOURS OUVRÉS, OUVRIERS - DAYS WORKED BY OPERATIVES**

Milliers

1991	1992	1993	1994	1995	1996	1997	1998			CITI révision 3
..	..	..	..	..	..	..	..	Activites extractives	C	10_14
..	..	..	..	..	..	..	..	Extraction de charbon, de lignite et de tourbe		10
..	..	..	..	..	..	..	..	Pétrole brut et gaz naturel ; activités annexes		11
..	..	..	..	..	..	..	..	Extraction de Minerais d'Uranium et de Thorium		12
..	..	..	..	..	..	..	..	Extraction de minerais métalliques		13
..	..	..	..	..	..	..	..	Autres activités extractives		14
..	..	47 149	44 986	42 471	39 502	..	..	Total activites manufacturieres	D	15_37
..	..	8 228	8 179	7 776	7 442	..	..	Produits alimentaires et boissons		15
..	..	3 179	3 134	3 035	2 869	..	..	Viande, poisson, fruits, légumes, huiles et graisses		151
..	..	1 005	1 076	981	911	..	..	Produits laitiers		152
..	..	645	536	501	472	..	..	Travail des grains; prod. amyl. ; alim. pour animaux		153
..	..	2 609	2 680	2 621	2 552	..	..	Autres produits alimentaires		154
..	..	791	753	638	638	..	..	Boissons		155
..	..	579	464	390	342	..	..	Produits à base de tabac		16
..	..	6 669	6 178	5 810	5 005	..	..	Textiles		17
..	..	3 717	3 624	3 484	2 796	..	..	Filature, tissage et achèvement des textiles		171
..	..	1 076	933	980	1 000	..	..	Autres articles textiles		172
..	..	1 876	1 621	1 346	1 209	..	..	Etoffes et articles de bonneterie		173
..	..	8 776	8 012	7 153	6 337	..	..	Articles d'habillement et fourrure		18
..	..	8 555	7 820	6 980	6 138	..	..	Articles d'habillement autres qu'en fourrure		181
..	..	221	192	173	199	..	..	Prép. et teinture des fourrures ; articles en fourrure		182
..	..	1 509	1 372	1 297	1 176	..	..	Cuir et articles en cuir ; chaussures		19
..	..	267	217	190	161	..	..	Cuir et articles en cuir sauf chaussures		191
..	..	1 242	1 155	1 106	1 014	..	..	Chaussures		192
..	..	1 078	921	884	869	..	..	Bois, articles en bois, sauf meubles		20
..	..	171	150	153	157	..	..	Sciage et rabotage du bois		201
..	..	906	771	731	712	..	..	Articles en bois, liège, vannerie, sparterie		202
..	..	1 036	1 061	1 086	1 064	..	..	Papier, articles en papier et carton		21
..	..	982	991	959	984	..	..	Edition, imprimerie et reproduction		22
..	..	366	369	338	360	..	..	Edition		221
..	..	616	622	621	625	..	..	Imprimerie et activités annexes		222
..	..	..	..	..	..	..	..	Reproduction de supports enregistrés		223
..	..	133	91	54	15	..	..	Coke. et prod. pétroliers ; comb. nucléaires		23
..	..	..	..	..	..	..	..	Cokéfaction		231
..	..	133	91	54	15	..	..	Produits pétroliers raffinés		232
..	..	..	..	..	..	..	..	Traitement de combustibles nucléaires		233
..	..	1 917	1 944	1 843	1 839	..	..	Produits chimiques		24
..	..	284	314	294	317	..	..	Produits chimiques de base		241
..	..	1 393	1 397	1 328	1 309	..	..	Autres produits chimiques		242
..	..	240	232	222	213	..	..	Fibres synthétiques ou artificielles		243
..	..	1 734	1 773	1 829	1 763	..	..	Art. en caoutchouc et matières plastiques		25
..	..	203	203	204	170	..	..	Articles en caoutchouc		251
..	..	1 531	1 570	1 625	1 593	..	..	Articles en matières plastiques		252
..	..	2 620	2 472	2 189	2 079	..	..	Autres produits minéraux non métalliques		26
..	..	268	257	205	199	..	..	Verre et articles en verre		261
..	..	2 352	2 215	1 984	1 880	..	..	Produits minéraux non métalliques, nca		269
..	..	2 079	1 807	1 857	1 809	..	..	Produits métallurgiques de base		27
..	..	1 169	914	912	859	..	..	Sidérurgie et première transformation de l'acier		271
..	..	883	854	879	889	..	..	Première transf. métaux précieux et non ferreux		272
..	..	27	39	66	61	..	..	Fonderie		273
..	..	2 180	2 117	2 147	2 087	..	..	Ouvrages en métaux, sauf machines		28
..	..	879	773	811	843	..	..	Constr. et menuis. métal.; réservoirs; générateurs		281
..	..	1 301	1 344	1 336	1 245	..	..	Autres ouvrages en métaux		289
..	..	1 932	1 851	1 910	1 916	..	..	Machines et matériels, nca		29
..	..	755	669	676	654	..	..	Machines d'usage général		291
..	..	742	721	799	816	..	..	Machines d'usage spécifique		292
..	..	436	461	435	445	..	..	Appareils domestiques, nca		293
..	..	8	4	2	11	..	..	Mach. de bureau, comptables, information		30
..	..	868	827	833	798	..	..	Machines et appareils électriques, nca		31
..	..	102	86	81	92	..	..	Moteurs, génératrices, transformateurs électriques		311
..	..	277	284	291	276	..	..	Matériel électrique de distribution et de commande		312
..	..	289	274	275	235	..	..	Fils et câbles électriques isolés		313
..	..	67	51	52	61	..	..	Accumulateurs et piles électriques		314
..	..	115	112	112	112	..	..	Lampes électriques et appareils d'éclairage		315
..	..	17	20	22	22	..	..	Autres matériels électriques, nca.		319
..	..	293	274	245	220	..	..	Equipements et appareils de radio, TV, etc.		32
..	..	64	67	67	56	..	..	Tubes et valves électroniques;autres composants		321
..	..	195	183	155	145	..	..	Emetteurs radio/TV; app. téléphonie et télégraphie		322
..	..	33	24	24	19	..	..	Récepteurs radio/TV, son et image; art. associés		323
..	..	170	199	190	224	..	..	Instr. médicaux, de précision ; optique ; horlog.		33
..	..	136	152	142	169	..	..	Instruments médicaux et de précision, etc.		331
..	..	32	45	48	55	..	..	Instruments d'optique ; matériel photographique		332
..	..	1	1	1	1	..	..	Horlogerie		333
..	..	495	456	439	364	..	..	Véhicules automobiles; remorques, etc.		34
..	..	352	329	308	235	..	..	Véhicules automobiles		341
..	..	80	43	48	51	..	..	Carross. automobiles, remorques, semi-remorques		342
..	..	63	84	83	79	..	..	Pièces et accessoires pour automobiles et moteurs		343
..	..	1 978	2 048	1 850	1 545	..	..	Autres matériels de transport		35
..	..	1 961	2 009	1 809	1 502	..	..	Construction et réparation de navires		351
..	..		3	2	3	..	..	Matériel ferroviaire roulant		352
..	..	9	11	13	12	..	..	Construction aéronautique et spatiale		353
..	..	8	24	25	27	..	..	Autres équipements de transport		359
..	..	1 850	1 931	1 710	1 593	..	..	Meubles ; activités de fabrication, nca		36
..	..	1 436	1 409	1 302	1 204	..	..	Meubles		361
..	..	414	523	408	390	..	..	Activités de fabrication, nca		369
..	..	35	14	19	19	..	..	Récupération		37
..	..	..	..	..	..	..	..	Electricité, gaz et eau	E	40_41
..	..	..	..	..	..	..	..	Prod. et dist. d'électricité gaz, vapeur d'eau chaude		40
..	..	..	..	..	..	..	..	Construction	F	45

Table GR.10 ESTABLISHMENTS - ÉTABLISSEMENTS

Units

ISIC revision 3				1991	1992	1993	1994	1995	1996	1997	1998
C	10_14		Mining and quarrying	..	..	..	..	..	..	..	..
	10		Mining of coal and lignite; extraction of peat	..	..	..	..	..	..	..	..
	11		Crude petroleum & natural gas; service activities	..	..	..	..	..	..	..	..
	12		Mining of uranium and thorium ores	..	..	..	..	..	..	..	..
	13		Mining of metal ores	..	..	..	..	..	..	..	..
	14		Other mining and quarrying	..	..	..	..	..	..	..	..
D	15_37		Total manufacturing	..	..	6 119	5 945	5 813	..	..	..
	15		Food products and beverages	..	..	1 126	1 106	1 091	..	..	..
		151	Production of meat, fish, fruit, vegetables, oils & fats	..	..	400	387	380	..	..	..
		152	Dairy products	..	..	88	89	89	..	..	..
		153	Grain mill and starch products; animal feeds	..	..	134	130	129	..	..	..
		154	Other food products	..	..	346	342	339	..	..	..
		155	Beverages	..	..	158	158	154	..	..	..
	16		Tobacco products	..	..	14	14	14	..	..	..
	17		Textiles	..	..	525	499	480	..	..	..
		171	Spinning, weaving and finishing of textiles	..	..	220	201	194	..	..	..
		172	Other textiles	..	..	133	126	124	..	..	..
		173	Knitted and crocheted fabrics and articles	..	..	172	172	162	..	..	..
	18		Wearing apparel and fur	..	..	903	843	797	..	..	..
		181	Wearing apparel, except fur apparel	..	..	863	805	759	..	..	..
		182	Dressing and dyeing of fur; articles of fur	..	..	40	38	38	..	..	..
	19		Leather and articles of leather; footwear	..	..	245	235	228	..	..	..
		191	Leather and articles of leather	..	..	57	50	47	..	..	..
		192	Footwear	..	..	188	185	181	..	..	..
	20		Wood & products of wood & cork, exc. furniture	..	..	177	166	160	..	..	..
		201	Sawmilling and planing of wood	..	..	36	32	32	..	..	..
		202	Wood products; articles of cork, straw, etc.	..	..	141	134	128	..	..	..
	21		Paper and paper products	..	..	138	138	139	..	..	..
	22		Printing, publishing and recorded media	..	..	238	234	236	..	..	..
		221	Publishing	..	..	112	112	111	..	..	..
		222	Printing and related activities	..	..	126	122	125	..	..	..
		223	Reproduction of recorded media	..	..	..	..	..	..	..	..
	23		Coke and petroleum products; nuclear fuel	..	..	18	17	17	..	..	..
		231	Coke oven products	..	..	..	..	..	..	..	..
		232	Refined petroleum products	..	..	18	17	17	..	..	..
		233	Processing of nuclear fuel	..	..	..	..	..	..	..	..
	24		Chemicals and chemical products	..	..	337	327	323	..	..	..
		241	Basic chemicals	..	..	66	65	65	..	..	..
		242	Other chemical products	..	..	266	257	253	..	..	..
		243	Man-made fibres	..	..	5	5	5	..	..	..
	25		Rubber and plastic products	..	..	257	263	268	..	..	..
		251	Rubber products	..	..	25	25	25	..	..	..
		252	Plastic products	..	..	232	238	243	..	..	..
	26		Other non-metallic mineral products	..	..	514	505	496	..	..	..
		261	Glass and glass products	..	..	25	22	20	..	..	..
		269	Non-metallic mineral products, nec	..	..	489	483	476	..	..	..
	27		Basic metals	..	..	129	126	125	..	..	..
		271	Basic iron and steel	..	..	75	70	69	..	..	..
		272	Basic precious and non-ferrous metals	..	..	48	51	50	..	..	..
		273	Casting of metals	..	..	6	5	6	..	..	..
	28		Fabricated metal products, exc. machinery	..	..	379	362	361	..	..	..
		281	Structural metal prod., tanks, reservoirs, etc.	..	..	156	140	139	..	..	..
		289	Other fabricated metal products	..	..	223	222	222	..	..	..
	29		Machinery and equipment, nec	..	..	342	339	332	..	..	..
		291	General purpose machinery	..	..	150	149	140	..	..	..
		292	Special purpose machinery	..	..	127	127	129	..	..	..
		293	Domestic appliances, nec	..	..	65	63	63	..	..	..
	30		Office, accounting & computing machinery	..	..	5	4	4	..	..	..
	31		Electrical machinery & apparatus, nec	..	..	118	117	114	..	..	..
		311	Electric motors, generators and transformers	..	..	17	16	15	..	..	..
		312	Electricity distribution & control apparatus	..	..	35	37	35	..	..	..
		313	Insulated wire and cable	..	..	21	21	22	..	..	..
		314	Accumulators, primary cells & primary batteries	..	..	10	8	8	..	..	..
		315	Electric lamps and lighting equipment	..	..	29	28	28	..	..	..
		319	Other electrical equipment, nec	..	..	6	7	6	..	..	..
	32		Radio, TV, communication equipment	..	..	31	31	29	..	..	..
		321	Electronic valves, tubes; other electronic components	..	..	11	13	11	..	..	..
		322	TV/radio transmitters; line teleph./telegr. apparatus	..	..	6	6	6	..	..	..
		323	TV/radio receivers; sound and video app.; assoc. gds.	..	..	14	12	12	..	..	..
	33		Medical, precision & optical instrum.; watch & clocks	..	..	36	31	31	..	..	..
		331	Medical appl.; instruments & appl. for measuring, etc.	..	..	27	23	23	..	..	..
		332	Optical instruments and photographic equipment	..	..	8	7	7	..	..	..
		333	Watches and clocks	..	..	1	1	1	..	..	..
	34		Motor vehicles, trailers and semi-trailers	..	..	47	47	47	..	..	..
		341	Motor vehicles	..	..	12	11	11	..	..	..
		342	Bodies for motor vehicles, trailers & semi-trailers	..	..	19	16	16	..	..	..
		343	Parts & accessories for motor vehicles & their engines	..	..	16	20	20	..	..	..
	35		Other transport equipment	..	..	114	126	127	..	..	..
		351	Building and repairing of ships and boats	..	..	92	101	101	..	..	..
		352	Railway and tramway locomotives and rolling stock	..	..	12	13	13	..	..	..
		353	Aircraft and spacecraft	..	..	5	6	6	..	..	..
		359	Transport equipment, nec	..	..	5	6	7	..	..	..
	36		Furniture; manufacturing, nec	..	..	413	405	384	..	..	..
		361	Furniture	..	..	319	313	295	..	..	..
		369	Manufacturing, nec	..	..	94	92	89	..	..	..
	37		Recycling	..	..	13	10	10	..	..	..
E	40_41		Electricity, gas and water supply	..	..	49	49	..	..	..	..
	40		Electricity, gas, steam and hot water supply	..	..	2	1	..	..	..	..
F	45		Construction	..	..	..	..	..	..	..	..

Table HU.1 **PRODUCTION - PRODUCTION**

Millions de Ft (Prix courants)

1991	1992	1993	1994	1995	1996	1997	1998			CITI révision 3	
..	57 943	29 750	25 699	27 308	36 299	43 692	35 365	Activites extractives	C	**10_14**	
..	40 595	22 516	15 312	15 096	19 931	22 309	11 618	Extraction de charbon, de lignite et de tourbe		**10**	
..	6 260	12 322	11 088	15 330	15 780	19 640	22 193	Pétrole brut et gaz naturel ; activités annexes		**11**	
..	2 109	..	..	..	..	..	..	Extraction de Minerais d'Uranium et de Thorium		**12**	
..	3 513	3 590	2 459	3 539	4 542	4 933	5 544	Extraction de minerais métalliques		**13**	
..	5 466	7 234	10 387	12 212	16 368	21 383	23 747	Autres activités extractives		**14**	
..	1 665 207	1 899 059	2 401 682	3 305 463	4 113 852	5 646 532	6 947 401	Total activites manufacturieres	D	**15_37**	
..	437 178	505 677	668 379	862 559	1 066 692	1 224 687	1 388 730	Produits alimentaires et boissons		**15**	
..	166 324	204 786	278 288	358 879	432 755	543 720	613 819	Viande, poisson, fruits, légumes, huiles et graisses			151
..	55 200	48 150	73 520	92 990	110 342	130 561	158 723	Produits laitiers			152
..	76 085	70 289	94 199	126 261	180 401	193 725	193 078	Travail des grains; prod. amyl. ; alim. pour animaux			153
..	86 821	109 994	134 730	168 971	220 854	224 514	255 375	Autres produits alimentaires			154
..	52 748	72 458	87 642	115 458	122 340	132 167	167 735	Boissons			155
..	14 462	18 403	23 755	26 634	35 853	39 624	46 850	Produits à base de tabac		**16**	
..	50 041	58 060	66 412	86 148	101 868	126 548	122 941	Textiles		**17**	
..	25 340	29 196	30 717	36 365	41 475	52 473	56 829	Filature, tissage et achèvement des textiles			171
..	16 302	21 254	26 353	38 221	45 735	55 735	54 521	Autres articles textiles			172
..	8 399	7 610	9 342	11 562	14 658	18 340	11 591	Etoffes et articles de bonneterie			173
..	35 888	39 827	47 958	60 590	71 911	85 650	128 812	Articles d'habillement et fourrure		**18**	
..	..	1 972	2 277	2 491	3 269	4 217	127 752	Articles d'habillement autres qu'en fourrure			181
..	..	36 863	44 667	56 743	67 174	79 863	1 060	Prép. et teinture des fourrures ; articles en fourrure			182
..	23 876	26 964	26 349	32 183	35 058	46 479	54 118	Cuir et articles en cuir ; chaussures		**19**	
..	9 357	10 777	8 802	11 243	10 243	13 785	13 276	Cuir et articles en cuir sauf chaussures			191
..	14 519	16 187	17 547	20 940	24 815	32 694	40 842	Chaussures			192
..	30 702	34 390	42 363	60 244	72 055	89 534	100 505	Bois, articles en bois, sauf meubles		**20**	
..	15 452	16 253	20 044	29 321	35 946	47 405	20 899	Sciage et rabotage du bois			201
..	15 250	18 137	22 319	30 923	36 109	42 129	79 606	Articles en bois, liège, vannerie, sparterie			202
..	39 125	35 970	50 160	77 802	82 693	109 823	149 994	Papier, articles en papier et carton		**21**	
..	70 449	80 769	100 213	130 770	149 809	185 782	201 872	Edition, imprimerie et reproduction		**22**	
..	37 932	41 281	48 455	60 199	69 607	89 899	95 917	Edition			221
..	30 602	36 786	48 076	65 329	74 148	86 763	101 172	Imprimerie et activités annexes			222
..	1 915	2 702	3 682	5 242	6 054	9 120	4 783	Reproduction de supports enregistrés			223
..	174 905	193 586	198 788	249 414	316 643	400 240	416 126	Coke. et prod. pétroliers ; comb. nucléaires		**23**	
..	..	191 136	196 116	245 953	312 434	394 497	7 153	Cokéfaction			231
..	..	..	..	..	..	..	408 973	Produits pétroliers raffinés			232
..	..	..	..	..	..	..	..	Traitement de combustibles nucléaires			233
..	176 081	202 599	255 445	357 423	404 092	521 808	561 524	Produits chimiques		**24**	
..	68 060	80 072	100 072	161 895	176 284	230 470	250 575	Produits chimiques de base			241
..	103 528	119 653	149 417	187 579	217 510	277 364	296 906	Autres produits chimiques			242
..	4 493	2 874	5 956	7 949	10 298	13 974	14 043	Fibres synthétiques ou artificielles			243
..	48 345	56 671	86 493	128 336	159 921	204 119	252 038	Art. en caoutchouc et matières plastiques		**25**	
..	12 864	7 956	15 886	23 436	31 182	42 117	55 831	Articles en caoutchouc			251
..	35 481	48 715	70 607	104 900	128 739	162 002	196 207	Articles en matières plastiques			252
..	60 249	71 510	91 761	125 195	148 060	187 729	232 085	Autres produits minéraux non métalliques		**26**	
..	..	18 504	22 470	31 514	36 621	42 769	46 427	Verre et articles en verre			261
..	..	53 006	69 291	93 681	141 223	144 960	185 658	Produits minéraux non métalliques, nca			269
..	83 885	79 645	110 995	195 396	231 800	312 891	345 206	Produits métallurgiques de base		**27**	
..	50 667	51 332	76 140	120 013	133 527	175 792	198 245	Sidérurgie et première transformation de l'acier			271
..	27 742	22 099	27 450	45 879	85 915	120 320	125 132	Première transf. métaux précieux et non ferreux			272
..	5 476	6 214	7 405	29 504	12 358	16 779	21 829	Fonderie			273
..	85 363	102 260	140 015	198 319	186 941	253 891	292 353	Ouvrages en métaux, sauf machines		**28**	
..	16 776	21 799	31 546	47 958	56 593	76 237	137 432	Constr. et menuis. métal.; réservoirs; générateurs			281
..	68 587	..	92 966	131 562	108 998	150 225	154 921	Autres ouvrages en métaux			289
..	114 234	121 854	148 670	200 871	244 360	312 954	359 679	Machines et matériels, nca		**29**	
..	..	..	..	..	..	..	142 780	Machines d'usage général			291
..	..	16 935	19 997	26 862	33 166	46 841	144 945	Machines d'usage spécifique			292
..	..	9 634	16 003	19 322	21 587	30 593	71 954	Appareils domestiques, nca			293
..	6 688	12 728	10 030	12 098	93 343	336 492	496 559	Mach. de bureau, comptables, information		**30**	
..	52 160	66 252	92 986	139 504	203 735	244 952	338 786	Machines et appareils électriques, nca		**31**	
..	7 044	8 841	10 162	20 094	25 193	28 359	35 376	Moteurs, génératrices, transformateurs électriques			311
..	3 487	4 331	4 908	9 109	13 325	16 548	45 754	Matériel électrique de distribution et de commande			312
..	11 859	9 735	16 960	27 929	42 670	44 748	36 579	Fils et câbles électriques isolés			313
..	1 593	3 308	3 306	3 669	3 959	3 580	6 453	Accumulateurs et piles électriques			314
..	19 032	27 786	36 649	49 381	68 957	83 100	114 011	Lampes électriques et appareils d'éclairage			315
..	9 145	..	17 002	24 242	40 741	58 009	100 613	Autres matériels électriques, nca.			319
..	32 503	40 440	53 582	85 073	138 880	282 757	330 978	Equipements et appareils de radio, TV, etc.		**32**	
..	4 674	6 177	9 055	15 506	32 914	73 243	89 013	Tubes et valves électroniques;autres composants			321
..	17 726	21 369	25 147	44 060	49 484	64 125	50 407	Emetteurs radio/TV; app. téléphonie et télégraphie			322
..	10 103	12 894	19 380	25 507	56 482	145 389	191 558	Récepteurs radio/TV, son et image; art. associés			323
..	27 840	29 953	38 372	53 100	62 816	77 700	84 041	Instr. médicaux, de précision ; optique ; horlog.		**33**	
..	..	24 147	30 442	43 147	51 272	63 458	74 387	Instruments médicaux et de précision, etc.			331
..	..	2 480	3 154	4 678	5 572	6 894	9 442	Instruments d'optique ; matériel photographique			332
..	..	8	13	40	55	52	212	Horlogerie			333
..	60 175	75 015	94 617	158 298	229 462	506 725	897 364	Véhicules automobiles; remorques, etc.		**34**	
..	33 889	40 106	53 151	99 284	134 869	332 101	630 064	Véhicules automobiles			341
..	1 053	1 213	881	1 395	4 465	8 726	14 007	Carross. automobiles, remorques, semi-remorques			342
..	25 233	33 696	40 585	57 619	90 128	165 898	253 293	Pièces et accessoires pour automobiles et moteurs			343
..	7 907	8 967	11 441	13 051	16 799	22 596	45 641	Autres matériels de transport		**35**	
..	1 537	404	274	385	587	649	1 081	Construction et réparation de navires			351
..	2 741	1 828	4 228	4 153	5 562	7 212	28 021	Matériel ferroviaire roulant			352
..	2 009	1 801	1 396	1 639	1 722	2 246	7 406	Construction aéronautique et spatiale			353
..	1 620	4 934	5 543	6 874	8 928	12 489	9 133	Autres équipements de transport			359
..	32 468	36 814	41 806	50 181	56 758	67 225	89 351	Meubles ; activités de fabrication, nca		**36**	
..	25 240	28 306	31 210	36 842	41 721	49 475	65 387	Meubles			361
..	7 228	8 508	10 596	13 339	15 037	17 750	23 964	Activités de fabrication, nca			369
..	683	705	1 092	2 274	4 303	6 326	11 848	Récupération		**37**	
..	297 965	287 060	327 905	429 666	432 276	565 834	678 450	Electricité, gaz et eau	E	**40_41**	
..	260 102	248 060	283 604	381 053	374 469	496 552	585 857	Prod. et dist. d'électricité gaz, vapeur d'eau chaude		**40**	
..	223 789	258 171	340 493	386 142	458 152	565 250	698 837	Construction	F	**45**	

HUNGARY

Table HU.2 VALUE ADDED - VALEUR AJOUTÉE

Millions of Ft (Current Prices)

ISIC revision 3				1991	1992	1993	1994	1995	1996	1997	1998
C	10_14		Mining and quarrying	..	29 883	10 584	9 288	11 471	15 500	19 512	14 695
	10		Mining of coal and lignite; extraction of peat	..	22 746	7 871	5 128	6 140	9 208	10 610	4 661
	11		Crude petroleum & natural gas; service activities	..	2 402	5 961	6 044	7 136	6 307	8 867	9 828
	12		Mining of uranium and thorium ores	..	840	..	..	..	..	..	..
	13		Mining of metal ores	..	1 988	2 161	1 397	2 345	2 632	2 942	3 009
	14		Other mining and quarrying	..	1 907	2 713	4 160	5 331	6 292	8 902	10 034
D	15_37		Total manufacturing	..	490 418	597 669	759 966	993 143	1199 562	1666 785	2005 009
	15		Food products and beverages	..	105 001	122 141	154 515	180 715	208 287	248 399	287 617
		151	Production of meat, fish, fruit, vegetables, oils & fats	..	32 031	42 877	57 418	69 969	77 451	96 392	114 088
		152	Dairy products	..	12 297	9 718	14 112	15 178	19 138	22 762	21 131
		153	Grain mill and starch products; animal feeds	..	17 430	14 657	22 957	29 369	35 146	40 761	44 026
		154	Other food products	..	26 939	31 807	35 902	41 359	55 075	59 105	63 314
		155	Beverages	..	16 304	23 082	24 126	24 840	21 477	29 379	45 058
	16		Tobacco products	..	3 325	3 816	6 286	4 853	9 538	11 891	13 298
	17		Textiles	..	15 205	18 917	22 396	29 799	35 966	45 488	44 703
		171	Spinning, weaving and finishing of textiles	..	7 102	8 973	9 691	10 481	13 108	16 034	18 905
		172	Other textiles	..	5 805	7 267	8 938	14 664	16 818	21 734	21 726
		173	Knitted and crocheted fabrics and articles	..	2 298	2 677	3 767	4 654	6 040	7 720	4 072
	18		Wearing apparel and fur	..	18 386	21 019	26 125	32 059	38 598	47 999	67 043
		181	Wearing apparel, except fur apparel	..	..	985	1 085	1 121	1 434	1 959	66 835
		182	Dressing and dyeing of fur; articles of fur	..	..	19 682	24 675	30 459	36 728	45 469	208
	19		Leather and articles of leather; footwear	..	8 077	9 691	12 250	15 213	16 881	21 603	23 018
		191	Leather and articles of leather	..	1 934	2 281	3 931	4 777	4 250	5 587	4 786
		192	Footwear	..	6 143	7 410	8 319	10 436	12 631	16 016	18 232
	20		Wood & products of wood & cork, exc. furniture	..	8 064	9 854	12 448	16 409	20 451	26 973	29 996
		201	Sawmilling and planing of wood	..	3 602	4 029	5 254	7 420	8 876	13 053	5 907
		202	Wood products; articles of cork, straw, etc.	..	4 462	5 825	7 194	8 989	11 575	13 920	24 089
	21		Paper and paper products	..	9 243	10 200	14 516	21 472	23 748	33 270	42 135
	22		Printing, publishing and recorded media	..	20 647	27 424	35 535	37 668	44 953	66 399	67 808
		221	Publishing	..	9 311	12 344	16 412	16 662	19 301	31 272	29 049
		222	Printing and related activities	..	10 596	14 047	17 331	19 920	23 234	31 068	36 881
		223	Reproduction of recorded media	..	740	1 033	1 792	1 086	2 418	4 059	1 878
	23		Coke and petroleum products; nuclear fuel	..	79 694	91 635	83 140	98 109	111 064	155 241	169 317
		231	Coke oven products	..	79 694	90 963	82 208	96 910	109 586	153 021	2 196
		232	Refined petroleum products	..	..	..	..	..	..	..	167 121
		233	Processing of nuclear fuel	..	..	..	..	..	..	..	..
	24		Chemicals and chemical products	..	46 300	61 156	87 840	127 302	133 331	184 097	205 839
		241	Basic chemicals	..	10 848	15 905	28 667	56 902	48 614	66 752	77 507
		242	Other chemical products	..	34 935	44 495	58 310	70 198	82 307	113 510	125 052
		243	Man-made fibres	..	517	756	863	202	2 410	3 835	3 280
	25		Rubber and plastic products	..	13 298	17 801	25 710	36 421	46 463	62 390	76 856
		251	Rubber products	..	2 972	2 741	5 141	7 343	9 772	12 277	17 010
		252	Plastic products	..	10 326	15 060	20 569	29 078	36 691	50 113	59 846
	26		Other non-metallic mineral products	..	19 922	26 687	36 176	49 766	56 205	73 080	89 262
		261	Glass and glass products	..	..	6 536	8 087	12 524	12 907	16 007	17 451
		269	Non-metallic mineral products, nec	..	..	20 151	28 089	37 242	43 298	57 073	71 811
	27		Basic metals	..	10 219	11 567	20 709	37 455	39 214	46 823	58 247
		271	Basic iron and steel	..	5 973	6 447	12 612	25 500	22 281	22 278	26 993
		272	Basic precious and non-ferrous metals	..	2 909	3 373	5 500	8 246	12 667	19 066	23 251
		273	Casting of metals	..	1 337	1 747	2 597	3 709	4 266	5 479	8 003
	28		Fabricated metal products, exc. machinery	..	23 092	30 081	41 994	58 574	64 455	90 036	103 640
		281	Structural metal prod., tanks, reservoirs, etc.	..	5 331	7 346	10 760	17 029	20 747	28 535	45 513
		289	Other fabricated metal products	..	17 761	19 254	25 418	35 073	35 881	51 767	58 127
	29		Machinery and equipment, nec	..	46 288	46 534	53 747	75 025	90 400	123 086	134 870
		291	General purpose machinery	..	..	..	3 325	5 356	6 610	8 654	58 674
		292	Special purpose machinery	..	..	..	..	..	..	..	57 706
		293	Domestic appliances, nec	..	..	..	..	..	..	..	18 490
	30		Office, accounting & computing machinery	..	1 675	..	..	..	..	..	89 216
	31		Electrical machinery & apparatus, nec	..	13 367	19 247	36 579	42 241	75 729	97 561	126 460
		311	Electric motors, generators and transformers	..	1 610	2 302	3 145	7 200	9 250	11 276	10 657
		312	Electricity distribution & control apparatus	..	756	1 497	1 344	2 763	4 123	5 919	15 277
		313	Insulated wire and cable	..	3 326	3 891	5 070	6 589	11 037	12 419	8 101
		314	Accumulators, primary cells & primary batteries	..	587	1 402	1 418	1 422	1 426	817	2 152
		315	Electric lamps and lighting equipment	..	3 550	5 103	18 224	14 035	31 248	40 155	52 045
		319	Other electrical equipment, nec	..	3 538	..	..	..	..	..	38 228
	32		Radio, TV, communication equipment	..	8 420	13 168	18 757	28 096	35 501	58 476	73 809
		321	Electronic valves, tubes; other electronic components	..	1 708	2 541	3 913	7 469	11 052	19 663	24 136
		322	TV/radio transmitters; line teleph./telegr. apparatus	..	4 858	7 714	8 913	15 116	18 051	29 265	22 064
		323	TV/radio receivers; sound and video app.; assoc. gds.	..	1 854	2 913	5 931	5 511	6 398	9 548	27 609
	33		Medical, precision & optical instrum.; watch & clocks	..	11 509	13 141	16 671	22 030	26 419	34 096	35 943
		331	Medical appl.; instruments & appl. for measuring, etc.	..	..	10 351	13 054	..	..	..	31 745
		332	Optical instruments and photographic equipment	..	..	1 293	1 519	2 085	2 285	2 916	4 135
		333	Watches and clocks	..	..	..	..	11	9	7	63
	34		Motor vehicles, trailers and semi-trailers	..	15 057	24 814	32 254	54 695	77 610	142 208	216 100
		341	Motor vehicles	..	4 605	11 379	16 669	34 851	48 324	88 459	135 088
		342	Bodies for motor vehicles, trailers & semi-trailers	..	453	475	339	457	1 227	2 054	5 066
		343	Parts & accessories for motor vehicles & their engines	..	9 999	12 960	15 246	19 387	28 059	51 695	75 946
	35		Other transport equipment	..	2 368	2 925	5 001	5 270	6 621	8 351	15 517
		351	Building and repairing of ships and boats	..	218	..	19	82	202	227	412
		352	Railway and tramway locomotives and rolling stock	..	892	..	1 004	1 320	2 140	2 460	8 949
		353	Aircraft and spacecraft	..	432	432	393	303	325	375	4 313
		359	Transport equipment, nec	..	826	2 788	3 585	3 565	3 954	5 289	1 843
	36		Furniture; manufacturing, nec	..	11 093	12 937	15 012	17 184	19 346	24 135	31 642
		361	Furniture	..	8 231	9 462	10 451	11 486	13 060	16 253	22 503
		369	Manufacturing, nec	..	2 862	3 475	4 561	5 698	6 286	7 882	9 139
	37		Recycling	..	168	98	227	701	1 049	1 833	2 673
E	40_41		Electricity, gas and water supply	..	102 874	108 430	123 602	161 632	200 429	284 064	341 742
	40		Electricity, gas, steam and hot water supply	..	81 944	87 200	98 607	134 493	168 890	248 016	298 248
F	45		Construction	..	90 649	102 881	126 217	131 194	144 659	201 159	244 640

HONGRIE

Table HU.3 **EMPLOI, SALARIÉS - EMPLOYMENT, EMPLOYEES**

Unités

1991	1992	1993	1994	1995	1996	1997	1998			CITI révision 3
..	46 987	30 872	20 036	..	..	..	..	Activites extractives	C	10_14
..	35 098	19 134	9 695	8 337	7 554	5 987	..	Extraction de charbon, de lignite et de tourbe		10
..	2 670	3 855	3 472	3 368	3 046	2 671	..	Pétrole brut et gaz naturel ; activités annexes		11
..	2 006	1 901	1 910	1 405	1 251	111	..	Extraction de Minerais d'Uranium et de Thorium		12
..	2 820	2 238	1 847	1 337	1 441	1 275	..	Extraction de minerais métalliques		13
..	4 393	3 744	3 112	3 433	3 327	3 384	..	Autres activités extractives		14
..	835 136	726 719	679 320	774 559	769 064	775 429	..	Total activites manufacturieres	D	15_37
..	180 238	151 625	138 956	147 220	140 406	133 226	..	Produits alimentaires et boissons		15
..	..	..	59 467	130 341	126 143	119 814	..	Viande, poisson, fruits, légumes, huiles et graisses		151
..	..	..	16 315	16 879	14 263	13 412	..	Produits laitiers		152
..	..	..	15 073	..	..	..	..	Travail des grains; prod. amyl. ; alim. pour animaux		153
..	..	..	32 045	..	..	..	..	Autres produits alimentaires		154
..	..	..	16 056	..	..	..	..	Boissons		155
..	3 392	3 016	2 740	2 415	2 327	2 129	..	Produits à base de tabac		16
..	55 764	49 226	44 417	45 640	44 637	46 087	..	Textiles		17
..	..	..	20 035	45 640	44 637	46 087	..	Filature, tissage et achèvement des textiles		171
..	..	..	14 663	..	..	..	..	Autres articles textiles		172
..	..	..	9 719	..	..	..	..	Etoffes et articles de bonneterie		173
..	62 087	58 509	57 387	61 504	67 090	62 011	..	Articles d'habillement et fourrure		18
..	..	..	56 680	60 609	66 372	61 349	..	Articles d'habillement autres qu'en fourrure		181
..	..	..	707	895	718	662	..	Prép. et teinture des fourrures ; articles en fourrure		182
..	33 644	30 409	23 723	25 362	25 711	27 586	..	Cuir et articles en cuir ; chaussures		19
..	..	..	5 103	25 362	25 711	27 586	..	Cuir et articles en cuir sauf chaussures		191
..	..	..	18 620	..	..	..	..	Chaussures		192
..	21 932	18 117	17 457	24 588	23 899	23 607	..	Bois, articles en bois, sauf meubles		20
..	..	..	3 665	8 791	8 743	9 174	..	Sciage et rabotage du bois		201
..	..	..	13 792	..	..	..	..	Articles en bois, liège, vannerie, sparterie		202
..	13 491	10 696	10 318	10 996	10 392	11 044	..	Papier, articles en papier et carton		21
..	24 385	24 424	21 747	32 360	31 143	31 436	..	Edition, imprimerie et reproduction		22
..	..	..	7 035	10 780	10 348	10 393	..	Edition		221
..	..	..	14 077	20 323	19 456	19 640	..	Imprimerie et activités annexes		222
..	..	..	635	1 257	1 339	1 403	..	Reproduction de supports enregistrés		223
..	22 170	20 996	18 484	17 797	16 757	15 509	..	Coke. et prod. pétroliers ; comb. nucléaires		23
..	..	..	1 073	16 736	15 702	14 469	..	Cokéfaction		231
..	..	..	17 411	1 061	1 055	1 040	..	Produits pétroliers raffinés		232
..	..	..	..	..	..	..	..	Traitement de combustibles nucléaires		233
..	52 308	49 328	46 224	46 507	43 912	42 461	..	Produits chimiques		24
..	..	..	14 865	15 125	14 454	14 083	..	Produits chimiques de base		241
..	..	..	29 415	..	..	..	..	Autres produits chimiques		242
..	..	..	1 944	..	..	..	..	Fibres synthétiques ou artificielles		243
..	22 370	20 148	19 635	27 345	28 054	30 717	..	Art. en caoutchouc et matières plastiques		25
..	..	..	4 876	6 129	6 500	6 642	..	Articles en caoutchouc		251
..	..	..	14 759	21 216	21 554	24 075	..	Articles en matières plastiques		252
..	39 744	35 153	31 639	35 966	34 884	34 036	..	Autres produits minéraux non métalliques		26
..	..	..	9 152	10 912	10 144	9 481	..	Verre et articles en verre		261
..	..	..	22 487	..	..	..	..	Produits minéraux non métalliques, nca		269
..	38 495	30 441	29 008	29 754	29 948	26 982	..	Produits métallurgiques de base		27
..	..	..	13 532	29 754	29 948	26 982	..	Sidérurgie et première transformation de l'acier		271
..	..	..	6 508	..	..	..	..	Première transf. métaux précieux et non ferreux		272
..	..	..	8 968	..	..	..	..	Fonderie		273
..	49 807	49 030	42 150	59 411	54 777	57 536	..	Ouvrages en métaux, sauf machines		28
..	..	..	18 470	25 137	24 069	26 002	..	Constr. et menuis. métal.; réservoirs; générateurs		281
..	..	..	23 680	34 274	30 708	31 534	..	Autres ouvrages en métaux		289
..	86 515	65 807	56 644	69 033	69 614	71 242	..	Machines et matériels, nca		29
..	..	..	13 649	14 363	14 289	15 750	..	Machines d'usage général		291
..	..	..	34 268	46 216	45 847	47 208	..	Machines d'usage spécifique		292
..	..	..	8 727	8 454	9 478	8 284	..	Appareils domestiques, nca		293
..	2 540	2 904	2 273	2 583	4 701	6 974	..	Mach. de bureau, comptables, information		30
..	33 417	29 610	30 294	36 422	39 728	42 692	..	Machines et appareils électriques, nca		31
..	..	..	5 852	5 465	5 869	6 100	..	Moteurs, génératrices, transformateurs électriques		311
..	..	..	1 651	2 065	2 395	2 644	..	Matériel électrique de distribution et de commande		312
..	..	..	3 519	4 455	5 800	6 211	..	Fils et câbles électriques isolés		313
..	..	..	878	827	694	505	..	Accumulateurs et piles électriques		314
..	..	..	12 356	13 250	13 778	14 294	..	Lampes électriques et appareils d'éclairage		315
..	..	..	6 038	7 588	7 693	9 019	..	Autres matériels électriques, nca.		319
..	26 073	21 453	18 741	23 926	25 102	30 373	..	Equipements et appareils de radio, TV, etc.		32
..	..	..	5 617	7 947	9 949	13 852	..	Tubes et valves électroniques;autres composants		321
..	..	..	9 173	11 268	10 833	12 228	..	Emetteurs radio/TV; app. téléphonie et télégraphie		322
..	..	..	3 951	4 711	4 320	4 293	..	Récepteurs radio/TV, son et image; art. associés		323
..	21 861	16 299	15 098	19 306	18 889	18 740	..	Instr. médicaux, de précision ; optique ; horlog.		33
..	..	..	13 714	17 427	16 960	17 015	..	Instruments médicaux et de précision, etc.		331
..	..	..	1 384	1 879	1 929	1 725	..	Instruments d'optique ; matériel photographique		332
..	..	..	..	..	..	..	..	Horlogerie		333
..	31 846	27 476	24 496	24 959	25 905	30 753	..	Véhicules automobiles; remorques, etc.		34
..	..	..	8 181	8 419	7 598	9 158	..	Véhicules automobiles		341
..	..	..	468	643	920	1 160	..	Carross. automobiles, remorques, semi-remorques		342
..	..	..	15 847	15 897	17 387	20 435	..	Pièces et accessoires pour automobiles et moteurs		343
..	4 889	5 805	4 722	4 228	3 987	4 080	..	Autres matériels de transport		35
..	..	..	356	283	249	263	..	Construction et réparation de navires		351
..	..	..	818	821	775	1 010	..	Matériel ferroviaire roulant		352
..	..	..	..	408	422	394	..	Construction aéronautique et spatiale		353
..	..	..	3 548	..	..	..	..	Autres équipements de transport		359
..	8 135	6 195	23 088	26 442	26 301	25 287	..	Meubles ; activités de fabrication, nca		36
..	..	..	17 664	..	..	..	..	Meubles		361
..	..	..	5 424	..	..	..	..	Activités de fabrication, nca		369
..	33	52	79	795	900	921	..	Récupération		37
..	92 781	93 456	98 839	96 863	89 780	84 488	..	Electricité, gaz et eau	E	40_41
..	59 109	62 967	70 385	69 260	64 877	60 405	..	Prod. et dist. d'électricité gaz, vapeur d'eau chaude		40
..	140 622	125 895	104 591	151 503	142 662	137 643	..	Construction	F	45

Table HU.4 **EMPLOYMENT, FEMALES - EMPLOI, FEMMES**

Units

ISIC revision 3				1991	1992	1993	1994	1995	1996	1997	1998
C	10_14		Mining and quarrying	..	..	..	..	2 317	2 049	..	..
	10		Mining of coal and lignite; extraction of peat	..	..	..	..	1 140	1 056	..	..
	11		Crude petroleum & natural gas; service activities	..	..	..	..	425	363	..	..
	12		Mining of uranium and thorium ores	..	..	..	..	105	86	..	..
	13		Mining of metal ores	..	..	..	..	183	173	..	..
	14		Other mining and quarrying	..	..	..	..	464	371	..	..
D	15_37		Total manufacturing	..	..	..	284 749	271 606	264 942	..	..
	15		Food products and beverages	..	..	..	56 329	52 560	48 561	..	..
		151	Production of meat, fish, fruit, vegetables, oils & fats	..	..	..	25 428	23 791	22 323	..	..
		152	Dairy products	..	..	..	6 046	5 894	5 307	..	..
		153	Grain mill and starch products; animal feeds	..	..	..	4 600	4 488	4 176	..	..
		154	Other food products	..	..	..	14 494	13 372	12 734	..	..
		155	Beverages	..	..	..	5 761	5 015	4 021	..	..
	16		Tobacco products	..	..	..	1 350	1 147	1 005	..	..
	17		Textiles	..	..	..	28 956	26 238	24 959	..	..
		171	Spinning, weaving and finishing of textiles	..	..	..	10 982	8 942	8 177	..	..
		172	Other textiles	..	..	..	10 347	10 220	10 243	..	..
		173	Knitted and crocheted fabrics and articles	..	..	..	7 627	7 076	6 539	..	..
	18		Wearing apparel and fur	..	..	..	44 528	41 187	42 994	..	..
		181	Wearing apparel, except fur apparel	..	..	..	44 151	40 853	42 716	..	..
		182	Dressing and dyeing of fur; articles of fur	..	..	..	377	334	278	..	..
	19		Leather and articles of leather; footwear	..	..	..	16 085	15 453	15 406	..	..
		191	Leather and articles of leather	..	..	..	2 787	2 492	2 510	..	..
		192	Footwear	..	..	..	13 298	12 961	12 896	..	..
	20		Wood & products of wood & cork, exc. furniture	..	..	..	4 826	4 674	4 198	..	..
		201	Sawmilling and planing of wood	..	..	..	915	952	851	..	..
		202	Wood products; articles of cork, straw, etc.	..	..	..	3 911	3 722	3 347	..	..
	21		Paper and paper products	..	..	..	4 078	3 747	3 431	..	..
	22		Printing, publishing and recorded media	..	..	..	10 537	9 886	8 709	..	..
		221	Publishing	..	..	..	3 827	3 560	3 018	..	..
		222	Printing and related activities	..	..	..	6 376	6 033	5 415	..	..
		223	Reproduction of recorded media	..	..	..	334	293	276	..	..
	23		Coke and petroleum products; nuclear fuel	..	..	..	4 627	5 010	4 986	..	..
		231	Coke oven products	..	..	..	275	266	255	..	..
		232	Refined petroleum products	..	..	..	4 352	4 744	4 731	..	..
		233	Processing of nuclear fuel	..	..	..	..	..	..	..	..
	24		Chemicals and chemical products	..	..	..	19 748	19 336	18 838	..	..
		241	Basic chemicals	..	..	..	4 992	4 969	4 705	..	..
		242	Other chemical products	..	..	..	13 954	13 637	13 478	..	..
		243	Man-made fibres	..	..	..	802	730	655	..	..
	25		Rubber and plastic products	..	..	..	7 755	7 822	8 005	..	..
		251	Rubber products	..	..	..	1 575	1 614	1 606	..	..
		252	Plastic products	..	..	..	6 180	6 208	6 399	..	..
	26		Other non-metallic mineral products	..	..	..	10 678	10 059	10 177	..	..
		261	Glass and glass products	..	..	..	3 364	3 055	3 207	..	..
		269	Non-metallic mineral products, nec	..	..	..	7 314	7 004	6 970	..	..
	27		Basic metals	..	..	..	6 895	6 677	6 347	..	..
		271	Basic iron and steel	..	..	..	3 322	3 163	3 940	..	..
		272	Basic precious and non-ferrous metals	..	..	..	1 541	1 511	1 861	..	..
		273	Casting of metals	..	..	..	2 032	2 003	546	..	..
	28		Fabricated metal products, exc. machinery	..	..	..	10 649	10 008	8 307	..	..
		281	Structural metal prod., tanks, reservoirs, etc.	..	..	..	3 426	3 373	3 071	..	..
		289	Other fabricated metal products	..	..	..	7 223	6 635	5 236	..	..
	29		Machinery and equipment, nec	..	..	..	12 550	11 168	10 958	..	..
		291	General purpose machinery	..	..	..	3 190	2 633	2 481	..	..
		292	Special purpose machinery	..	..	..	5 980	5 367	5 094	..	..
		293	Domestic appliances, nec	..	..	..	3 380	3 168	3 383	..	..
	30		Office, accounting & computing machinery	..	..	..	1 130	1 011	2 071	..	..
	31		Electrical machinery & apparatus, nec	..	..	..	13 084	14 192	14 857	..	..
		311	Electric motors, generators and transformers	..	..	..	2 011	2 143	2 103	..	..
		312	Electricity distribution & control apparatus	..	..	..	522	494	835	..	..
		313	Insulated wire and cable	..	..	..	1 547	2 344	2 708	..	..
		314	Accumulators, primary cells & primary batteries	..	..	..	185	129	112	..	..
		315	Electric lamps and lighting equipment	..	..	..	5 505	5 430	5 216	..	..
		319	Other electrical equipment, nec	..	..	..	3 314	3 652	3 883	..	..
	32		Radio, TV, communication equipment	..	..	..	9 532	11 210	11 250	..	..
		321	Electronic valves, tubes; other electronic components	..	..	..	3 442	4 477	5 078	..	..
		322	TV/radio transmitters; line teleph./telegr. apparatus	..	..	..	3 990	4 396	3 954	..	..
		323	TV/radio receivers; sound and video app.; assoc. gds.	..	..	..	2 100	2 337	2 218	..	..
	33		Medical, precision & optical instrum.; watch & clocks	..	..	..	6 498	6 833	6 213	..	..
		331	Medical appl.; instruments & appl. for measuring, etc.	..	..	..	5 667	5 833	5 356	..	..
		332	Optical instruments and photographic equipment	..	..	..	831	1 000	857	..	..
		333	Watches and clocks	..	..	..				..	..
	34		Motor vehicles, trailers and semi-trailers	..	..	..	6 592	5 989	6 377	..	..
		341	Motor vehicles	..	..	..	2 126	1 899	1 534	..	..
		342	Bodies for motor vehicles, trailers & semi-trailers	..	..	..	68	65	170	..	..
		343	Parts & accessories for motor vehicles & their engines	..	..	..	4 398	4 025	4 673	..	..
	35		Other transport equipment	..	..	..	926	682	713	..	..
		351	Building and repairing of ships and boats	..	..	..	74	45	18	..	..
		352	Railway and tramway locomotives and rolling stock	..	..	..	116	95	97	..	..
		353	Aircraft and spacecraft	..	..	..	..	..	..	..	..
		359	Transport equipment, nec	..	..	..	736	542	598	..	..
	36		Furniture; manufacturing, nec	..	..	..	7 374	6 638	6 492	..	..
		361	Furniture	..	..	..	5 017	4 244	3 678	..	..
		369	Manufacturing, nec	..	..	..	2 357	2 394	2 814	..	..
	37		Recycling	..	..	..	22	79	88	..	..
E	40_41		Electricity, gas and water supply	..	..	..	..	24 224	22 792	..	..
	40		Electricity, gas, steam and hot water supply	..	..	..	..	18 505	17 457	..	..
F	45		Construction	..	..	..	..	10 937	9 663	..	..

Table HU.5 **SALAIRES ET TRAITEMENTS, SALARIÉS - WAGES AND SALARIES, EMPLOYEES**

Millions de Ft (Prix courants)

1991	1992	1993	1994	1995	1996	1997	1998			CITI révision 3	
..	..	..	..	..	..	..	..	Activites extractives	C	10_14	
..	..	..	..	4 528	5 020	5 037	..	Extraction de charbon, de lignite et de tourbe		10	
..	..	..	..	2 862	2 870	3 181	..	Pétrole brut et gaz naturel ; activités annexes		11	
..	..	..	..	831	904	109	..	Extraction de Minerais d'Uranium et de Thorium		12	
..	..	..	..	871	1 128	1 205	..	Extraction de minerais métalliques		13	
..	..	..	..	1 772	2 132	2 614	..	Autres activités extractives		14	
..	..	..	264 429	345 335	408 676	501 561	..	Total activites manufacturieres	D	15_37	
..	..	..	54 974	66 063	75 687	84 563	..	Produits alimentaires et boissons		15	
..	..	..	21 737	56 974	65 819	73 334	..	Viande, poisson, fruits, légumes, huiles et graisses			151
..	..	..	6 393	9 089	9 868	11 229	..	Produits laitiers			152
..	..	..	6 496	..	..	..	..	Travail des grains; prod. amyl. ; alim. pour animaux			153
..	..	..	12 909	..	..	..	..	Autres produits alimentaires			154
..	..	..	7 439	..	..	..	..	Boissons			155
..	..	..	1 805	2 155	2 164	2 697	..	Produits à base de tabac		16	
..	..	..	12 133	14 066	16 412	20 245	..	Textiles		17	
..	..	..	5 912	14 066	16 412	20 245	..	Filature, tissage et achèvement des textiles			171
..	..	..	3 949	..	..	..	..	Autres articles textiles			172
..	..	..	2 272	..	..	..	..	Etoffes et articles de bonneterie			173
..	..	..	13 054	16 560	20 126	24 419	..	Articles d'habillement et fourrure		18	
..	..	..	12 872	16 320	19 898	24 174	..	Articles d'habillement autres qu'en fourrure			181
..	..	..	182	240	228	245	..	Prép. et teinture des fourrures ; articles en fourrure			182
..	..	..	5 787	7 255	8 865	11 042	..	Cuir et articles en cuir ; chaussures		19	
..	..	..	1 342	7 255	8 865	11 042	..	Cuir et articles en cuir sauf chaussures			191
..	..	..	4 445	..	..	..	..	Chaussures			192
..	..	..	5 226	7 815	8 654	10 020	..	Bois, articles en bois, sauf meubles		20	
..	..	..	1 010	2 957	3 537	4 264	..	Sciage et rabotage du bois			201
..	..	..	4 216	..	..	..	..	Articles en bois, liège, vannerie, sparterie			202
..	..	..	4 611	6 249	6 996	9 391	..	Papier, articles en papier et carton		21	
..	..	..	11 057	15 987	17 359	19 960	..	Edition, imprimerie et reproduction		22	
..	..	..	4 725	6 860	7 316	8 290	..	Edition			221
..	..	..	6 010	8 477	9 204	10 474	..	Imprimerie et activités annexes			222
..	..	..	322	650	839	1 196	..	Reproduction de supports enregistrés			223
..	..	..	13 248	15 614	18 221	20 937	..	Coke. et prod. pétroliers ; comb. nucléaires		23	
..	..	..	522	14 918	17 381	19 947	..	Cokéfaction			231
..	..	..	12 726	696	840	990	..	Produits pétroliers raffinés			232
..	..	..	..	..	..	..	..	Traitement de combustibles nucléaires			233
..	..	..	24 964	31 221	36 771	44 644	..	Produits chimiques		24	
..	..	..	7 512	9 737	11 474	13 722	..	Produits chimiques de base			241
..	..	..	16 772	..	..	..	..	Autres produits chimiques			242
..	..	..	680	..	..	..	..	Fibres synthétiques ou artificielles			243
..	..	..	7 670	12 092	14 856	19 339	..	Art. en caoutchouc et matières plastiques		25	
..	..	..	2 082	2 924	3 852	4 933	..	Articles en caoutchouc			251
..	..	..	5 588	9 168	11 004	14 406	..	Articles en matières plastiques			252
..	..	..	12 678	16 386	19 378	23 509	..	Autres produits minéraux non métalliques		26	
..	..	..	3 290	4 055	4 761	5 739	..	Verre et articles en verre			261
..	..	..	9 388	..	..	..	..	Produits minéraux non métalliques, nca			269
..	..	..	13 315	16 656	20 810	23 477	..	Produits métallurgiques de base		27	
..	..	..	6 434	16 656	20 810	23 477	..	Sidérurgie et première transformation de l'acier			271
..	..	..	3 024	..	..	..	..	Première transf. métaux précieux et non ferreux			272
..	..	..	3 857	..	..	..	..	Fonderie			273
..	..	..	16 041	24 531	25 582	32 803	..	Ouvrages en métaux, sauf machines		28	
..	..	..	7 195	10 635	11 661	15 018	..	Constr. et menuis. métal.; réservoirs; générateurs			281
..	..	..	8 846	13 896	13 921	17 785	..	Autres ouvrages en métaux			289
..	..	..	21 779	30 586	36 619	45 718	..	Machines et matériels, nca		29	
..	..	..	5 328	6 792	8 142	10 835	..	Machines d'usage général			291
..	..	..	13 490	20 265	23 769	29 665	..	Machines d'usage spécifique			292
..	..	..	2 961	3 529	4 708	5 218	..	Appareils domestiques, nca			293
..	..	..	821	1 114	2 597	4 704	..	Mach. de bureau, comptables, information		30	
..	..	..	12 865	17 634	23 330	30 463	..	Machines et appareils électriques, nca		31	
..	..	..	2 456	2 800	3 724	4 582	..	Moteurs, génératrices, transformateurs électriques			311
..	..	..	851	1 247	1 720	2 198	..	Matériel électrique de distribution et de commande			312
..	..	..	1 499	2 227	3 483	4 778	..	Fils et câbles électriques isolés			313
..	..	..	504	515	480	439	..	Accumulateurs et piles électriques			314
..	..	..	5 349	6 736	8 334	10 705	..	Lampes électriques et appareils d'éclairage			315
..	..	..	2 206	3 324	4 258	5 990	..	Autres matériels électriques, nca.			319
..	..	..	6 899	10 330	13 855	19 942	..	Equipements et appareils de radio, TV, etc.		32	
..	..	..	1 746	3 088	4 997	8 535	..	Tubes et valves électroniques;autres composants			321
..	..	..	3 450	5 154	6 360	8 305	..	Emetteurs radio/TV; app. téléphonie et télégraphie			322
..	..	..	1 703	2 088	2 498	3 102	..	Récepteurs radio/TV, son et image; art. associés			323
..	..	..	6 124	8 820	10 064	12 268	..	Instr. médicaux, de précision ; optique ; horlog.		33	
..	..	..	5 578	8 012	9 202	11 100	..	Instruments médicaux et de précision, etc.			331
..	..	..	546	808	862	1 168	..	Instruments d'optique ; matériel photographique			332
..	..	..	..	..	..	..	..	Horlogerie			333
..	..	..	10 375	13 135	17 581	26 240	..	Véhicules automobiles; remorques, etc.		34	
..	..	..	3 837	4 935	6 063	9 365	..	Véhicules automobiles			341
..	..	..	172	241	441	702	..	Carross. automobiles, remorques, semi-remorques			342
..	..	..	6 366	7 959	11 077	16 173	..	Pièces et accessoires pour automobiles et moteurs			343
..	..	..	2 380	2 618	2 960	3 745	..	Autres matériels de transport		35	
..	..	..	126	112	83	112	..	Construction et réparation de navires			351
..	..	..	464	486	583	940	..	Matériel ferroviaire roulant			352
..	..	..	..	199	248	236	..	Construction aéronautique et spatiale			353
..	..	..	1 790	..	..	..	..	Autres équipements de transport			359
..	..	..	6 595	8 137	9 265	10 775	..	Meubles ; activités de fabrication, nca		36	
..	..	..	5 040	..	..	..	..	Meubles			361
..	..	..	1 555	..	..	..	..	Activités de fabrication, nca			369
..	..	..	28	311	524	660	..	Récupération		37	
..	..	..	..	58 490	67 672	77 265	..	Electricité, gaz et eau	E	40_41	
..	..	..	..	45 417	53 141	60 243	..	Prod. et dist. d'électricité gaz, vapeur d'eau chaude		40	
..	..	..	..	53 767	56 841	65 056	..	Construction	F	45	

HUNGARY

Table HU.6 **ESTABLISHMENTS - ÉTABLISSEMENTS**

Units

ISIC revision 3				1991	1992	1993	1994	1995	1996	1997	1998
C	10_14		Mining and quarrying	..	177	187	192	220	258	284	297
	10		Mining of coal and lignite; extraction of peat	..	30	31	30	29	33	39	36
	11		Crude petroleum & natural gas; service activities	..	8	10	12	17	19	16	20
	12		Mining of uranium and thorium ores	..	2	..	..	..	..	..	1
	13		Mining of metal ores	..	4	6	7	8	8	7	9
	14		Other mining and quarrying	..	133	156	162	191	225	245	261
D	15_37		Total manufacturing	..	22 226	25 637	28 064	30 197	33 581	34 644	34 197
	15		Food products and beverages	..	2 416	3 110	3 459	3 667	3 864	3 862	3 678
		151	Production of meat, fish, fruit, vegetables, oils & fats	..	640	770	842	905	944	960	925
		152	Dairy products	..	92	123	128	133	129	119	118
		153	Grain mill and starch products; animal feeds	..	205	334	337	365	369	367	341
		154	Other food products	..	1 017	1 207	1 385	1 467	1 597	1 531	1 490
		155	Beverages	..	462	676	767	797	825	885	804
	16		Tobacco products	..	13	13	14	9	10	8	8
	17		Textiles	..	910	1 123	1 269	1 394	1 576	1 654	1 431
		171	Spinning, weaving and finishing of textiles	..	146	158	169	170	181	191	270
		172	Other textiles	..	542	721	850	948	1 099	1 171	881
		173	Knitted and crocheted fabrics and articles	..	222	244	250	276	296	292	280
	18		Wearing apparel and fur	..	1 411	1 483	1 543	1 653	1 800	1 810	2 202
		181	Wearing apparel, except fur apparel	..	..	1 432	1 495	1 607	1 755	1 765	2 165
		182	Dressing and dyeing of fur; articles of fur	..	..	..	48	46	45	45	37
	19		Leather and articles of leather; footwear	..	421	465	507	527	561	574	569
		191	Leather and articles of leather	..	139	165	185	192	209	211	205
		192	Footwear	..	282	300	322	335	352	363	364
	20		Wood & products of wood & cork, exc. furniture	..	1 258	1 487	1 665	1 841	2 111	2 201	2 209
		201	Sawmilling and planing of wood	..	351	448	519	598	689	779	683
		202	Wood products; articles of cork, straw, etc.	..	907	1 039	1 146	1 243	1 422	1 422	1 526
	21		Paper and paper products	..	207	248	291	311	355	364	395
	22		Printing, publishing and recorded media	..	2 579	3 218	3 685	4 183	4 919	5 380	5 549
		221	Publishing	..	990	1 338	1 591	1 858	2 234	2 427	2 489
		222	Printing and related activities	..	1 334	1 604	1 831	2 037	2 388	2 621	2 768
		223	Reproduction of recorded media	..	255	276	263	288	297	332	292
	23		Coke and petroleum products; nuclear fuel	..	10	15	13	13	13	14	12
		231	Coke oven products	..	9	14	12	12	12	13	1
		232	Refined petroleum products	..	1	..	..	..	..	..	11
		233	Processing of nuclear fuel	..		..	..	..	..	..	..
	24		Chemicals and chemical products	..	576	613	619	655	704	716	680
		241	Basic chemicals	..	180	190	209	228	255	260	223
		242	Other chemical products	..	386	414	402	418	440	447	450
		243	Man-made fibres	..	10	9	8	9	9	9	7
	25		Rubber and plastic products	..	1 107	1 256	1 350	1 419	1 512	1 565	1 563
		251	Rubber products	..	135	157	167	167	183	193	198
		252	Plastic products	..	972	1 099	1 183	1 252	1 329	1 372	1 365
	26		Other non-metallic mineral products	..	660	781	867	923	1 046	1 082	1 137
		261	Glass and glass products	..	..	144	161	180	209	209	207
		269	Non-metallic mineral products, nec	..	..	637	706	743	847	873	930
	27		Basic metals	..	237	250	253	272	289	280	315
		271	Basic iron and steel	..	78	80	76	77	83	85	107
		272	Basic precious and non-ferrous metals	..	41	41	44	49	54	53	63
		273	Casting of metals	..	118	129	133	146	152	142	145
	28		Fabricated metal products, exc. machinery	..	2 900	3 220	3 497	3 749	4 185	4 206	4 407
		281	Structural metal prod., tanks, reservoirs, etc.	..	601	827	987	1 128	1 338	1 491	1 862
		289	Other fabricated metal products	..	2 299	..	2 392	2 503	2 725	2 586	2 545
	29		Machinery and equipment, nec	..	3 191	3 475	3 609	3 781	4 106	4 259	3 372
		291	General purpose machinery	..	..	..	..	..	..	..	1 994
		292	Special purpose machinery	..	..	..	..	..	..	..	1 228
		293	Domestic appliances, nec	..	..	82	81	87	103	102	150
	30		Office, accounting & computing machinery	..	159	187	208	217	259	274	261
	31		Electrical machinery & apparatus, nec	..	871	1 037	1 100	1 171	1 318	1 423	1 297
		311	Electric motors, generators and transformers	..	72	89	86	98	103	105	421
		312	Electricity distribution & control apparatus	..	80	93	87	98	106	122	156
		313	Insulated wire and cable	..	20	24	24	30	39	39	47
		314	Accumulators, primary cells & primary batteries	..	20	24	19	21	23	23	23
		315	Electric lamps and lighting equipment	..	86	102	106	119	131	129	126
		319	Other electrical equipment, nec	..	593	..	198	190	198	200	524
	32		Radio, TV, communication equipment	..	728	769	816	882	949	1 030	925
		321	Electronic valves, tubes; other electronic components	..	214	219	223	239	248	271	277
		322	TV/radio transmitters; line teleph./telegr. apparatus	..	404	442	482	529	591	636	478
		323	TV/radio receivers; sound and video app.; assoc. gds.	..	110	108	111	114	110	123	170
	33		Medical, precision & optical instrum.; watch & clocks	..	1 067	1 156	1 357	1 427	1 610	1 487	1 433
		331	Medical appl.; instruments & appl. for measuring, etc.	..	..	696	808	842	954	988	1 251
		332	Optical instruments and photographic equipment	..	..	71	87	119	141	125	169
		333	Watches and clocks	..	..	8	5	4	8	7	13
	34		Motor vehicles, trailers and semi-trailers	..	203	229	234	246	264	281	308
		341	Motor vehicles	..	32	35	33	42	38	38	35
		342	Bodies for motor vehicles, trailers & semi-trailers	..	15	21	29	33	39	38	45
		343	Parts & accessories for motor vehicles & their engines	..	156	173	172	171	187	205	228
	35		Other transport equipment	..	139	162	183	202	230	204	217
		351	Building and repairing of ships and boats	..	45	52	56	62	73	84	86
		352	Railway and tramway locomotives and rolling stock	..	8	6	9	10	10	10	28
		353	Aircraft and spacecraft	..	5	10	14	15	20	17	25
		359	Transport equipment, nec	..	81	..	..	..	..	..	78
	36		Furniture; manufacturing, nec	..	1 099	1 266	1 450	1 569	1 798	1 866	2 073
		361	Furniture	..	534	666	771	872	1 071	1 192	1 350
		369	Manufacturing, nec	..	565	..	..	..	..	..	723
	37		Recycling	..	64	74	75	86	102	104	156
E	40_41		Electricity, gas and water supply	..	255	337	366	393	420	450	479
	40		Electricity, gas, steam and hot water supply	..	163	199	202	207	210	226	244
F	45		Construction	..	10 854	12 670	14 562	15 836	17 912	18 855	20 510

Table IC.1 **EXPORTATIONS - EXPORTS**

Millions de IKr (Prix courants)

1991	1992	1993	1994	1995	1996	1997	1998			CITI révision 3	
..	..	..	..	..	..	..	..	Activites extractives	C	10_14	
..	..	..	..	..	..	..	..	Extraction de charbon, de lignite et de tourbe		10	
..	..	..	..	..	..	..	..	Pétrole brut et gaz naturel ; activités annexes		11	
..	..	..	..	..	..	..	..	Extraction de Minerais d'Uranium et de Thorium		12	
..	..	..	..	..	..	..	..	Extraction de minerais métalliques		13	
..	..	..	..	1 298	978	916	844	Autres activités extractives		14	
..	..	..	..	..	..	..	..	Total activites manufacturieres	D	15_37	
..	..	..	..	82 763	91 645	92 814	98 313	Produits alimentaires et boissons		15	
..	..	..	..	81 927	91 227	92 467	97 966	Viande, poisson, fruits, légumes, huiles et graisses			151
..	..	..	..	17	23	16	12	Produits laitiers			152
..	..	..	..	534	162	130	149	Travail des grains; prod. amyl. ; alim. pour animaux			153
..	..	..	..	62	61	48	58	Autres produits alimentaires			154
..	..	..	..	224	172	153	128	Boissons			155
..	..	..	..	..	..	..	..	Produits à base de tabac		16	
..	..	..	..	996	1 005	1 106	1 111	Textiles		17	
..	..	..	..	..	..	..	..	Filature, tissage et achèvement des textiles			171
..	..	..	..	549	538	534	568	Autres articles textiles			172
..	..	..	..	162	150	169	185	Etoffes et articles de bonneterie			173
..	..	..	..	1 068	1 286	1 292	729	Articles d'habillement et fourrure		18	
..	..	..	..	..	..	..	103	Articles d'habillement autres qu'en fourrure			181
..	..	..	..	977	1 196	1 205	626	Prép. et teinture des fourrures ; articles en fourrure			182
..	..	..	..	23	35	39	66	Cuir et articles en cuir ; chaussures		19	
..	..	..	..	21	34	36	57	Cuir et articles en cuir sauf chaussures			191
..	..	..	..	3	2	4	9	Chaussures			192
..	..	..	..	32	13	28	48	Bois, articles en bois, sauf meubles		20	
..	..	..	..	26	..	16	32	Sciage et rabotage du bois			201
..	..	..	..	6	13	..	15	Articles en bois, liège, vannerie, sparterie			202
..	..	..	..	169	249	208	188	Papier, articles en papier et carton		21	
..	..	..	..	72	54	74	119	Edition, imprimerie et reproduction		22	
..	..	..	..	51	46	64	110	Edition			221
..	..	..	..	3	2	4	3	Imprimerie et activités annexes			222
..	..	..	..	18	5	6	7	Reproduction de supports enregistrés			223
..	..	..	..	56	76	78	220	Coke. et prod. pétroliers ; comb. nucléaires		23	
..	..	..	..	..	..	..	..	Cokéfaction			231
..	..	..	..	56	76	78	220	Produits pétroliers raffinés			232
..	..	..	..	..	..	..	..	Traitement de combustibles nucléaires			233
..	..	..	..	764	629	547	587	Produits chimiques		24	
..	..	..	..	140	147	70	78	Produits chimiques de base			241
..	..	..	..	..	..	477	510	Autres produits chimiques			242
..	..	..	..	2	..	..	..	Fibres synthétiques ou artificielles			243
..	..	..	..	292	383	524	498	Art. en caoutchouc et matières plastiques		25	
..	..	..	..	16	24	31	33	Articles en caoutchouc			251
..	..	..	..	275	359	493	466	Articles en matières plastiques			252
..	..	..	..	144	153	189	148	Autres produits minéraux non métalliques		26	
..	..	..	..	2	9	9	8	Verre et articles en verre			261
..	..	..	..	..	..	..	..	Produits minéraux non métalliques, nca			269
..	..	..	..	15 833	16 460	19 554	22 261	Produits métallurgiques de base		27	
..	..	..	..	3 227	3 835	3 733	3 242	Sidérurgie et première transformation de l'acier			271
..	..	..	..	12 605	12 343	15 487	18 658	Première transf. métaux précieux et non ferreux			272
..	..	..	..	..	282	335	362	Fonderie			273
..	..	..	..	572	211	258	283	Ouvrages en métaux, sauf machines		28	
..	..	..	..	..	..	..	10	Constr. et menuis. métal.; réservoirs; générateurs			281
..	..	..	..	..	..	..	..	Autres ouvrages en métaux			289
..	..	..	..	1 754	1 928	1 888	2 007	Machines et matériels, nca		29	
..	..	..	..	992	1 487	1 457	1 402	Machines d'usage général			291
..	..	..	..	..	..	..	605	Machines d'usage spécifique			292
..	..	..	..	92	..	..	1	Appareils domestiques, nca			293
..	..	..	..	18	37	16	77	Mach. de bureau, comptables, information		30	
..	..	..	..	66	93	39	136	Machines et appareils électriques, nca		31	
..	..	..	..	2	7	11	82	Moteurs, génératrices, transformateurs électriques			311
..	..	..	..	4	8	6	40	Matériel électrique de distribution et de commande			312
..	..	..	..	1	..	1	2	Fils et câbles électriques isolés			313
..	..	..	..	4	8	5	1	Accumulateurs et piles électriques			314
..	..	..	..	8	10	5	5	Lampes électriques et appareils d'éclairage			315
..	..	..	..	47	59	10	6	Autres matériels électriques, nca.			319
..	..	..	..	..	3	3	10	Equipements et appareils de radio, TV, etc.		32	
..	..	..	..	..	..	..	1	Tubes et valves électroniques;autres composants			321
..	..	..	..	..	..	1	4	Emetteurs radio/TV; app. téléphonie et télégraphie			322
..	..	..	..	..	2	2	5	Récepteurs radio/TV, son et image; art. associés			323
..	..	..	..	304	293	485	707	Instr. médicaux, de précision ; optique ; horlog.		33	
..	..	..	..	..	..	..	..	Instruments médicaux et de précision, etc.			331
..	..	..	..	2	1		3	Instruments d'optique ; matériel photographique			332
..	..	..	..	2	4	3	4	Horlogerie			333
..	..	..	..	11	13	32	35	Véhicules automobiles; remorques, etc.		34	
..	..	..	..	10	8	26	22	Véhicules automobiles			341
..	..	..	..	..	1		1	Carross. automobiles, remorques, semi-remorques			342
..	..	..	..	1	5	6	12	Pièces et accessoires pour automobiles et moteurs			343
..	..	..	..	4 133	4 171	5 305	2 351	Autres matériels de transport		35	
..	..	..	..	2 225	1 625	2 641	563	Construction et réparation de navires			351
..	..	..	..	..	..	..	..	Matériel ferroviaire roulant			352
..	..	..	..	1 905	2 546	2 664	1 788	Construction aéronautique et spatiale			353
..	..	..	..	..	1	..	..	Autres équipements de transport			359
..	..	..	..	17	12	12	10	Meubles ; activités de fabrication, nca		36	
..	..	..	..	15	6	9	3	Meubles			361
..	..	..	..	2	5	..	6	Activités de fabrication, nca			369
..	..	..	..	..	..	..	..	Récupération		37	
..	..	..	..	..	..	..	..	Electricité, gaz et eau	E	40_41	
..	..	..	..	..	..	..	..	Prod. et dist. d'électricité gaz, vapeur d'eau chaude		40	
..	..	..	..	..	..	..	..	Construction	F	45	

IRELAND

Table IR.1 PRODUCTION - PRODUCTION

Millions of Ir£ (Current Prices)

ISIC revision 3				1991	1992	1993	1994	1995	1996	1997	1998
C	10_14		Mining and quarrying	389	365	362	434	512	517	512	..
	10		Mining of coal and lignite; extraction of peat	..	..	..	..	..	..	..	..
	11		Crude petroleum & natural gas; service activities	..	..	..	..	..	..	..	..
	13		Mining of metal ores	..	..	..	..	..	..	..	..
	14		Other mining and quarrying	..	..	..	..	..	..	..	..
D	15_37		Total manufacturing	20 898	22 361	24 904	27 787	33 583	36 296	41 394	..
	15		Food products and beverages	7 664	8 017	8 559	8 768	9 631	..	..	..
		151	Production of meat, fish, fruit, vegetables, oils & fats	2 486	2 570	2 696	2 651	2 918	..	4 938	..
		152	Dairy products	2 107	2 108	2 200	2 116	2 258	2 292	2 190	..
		153	Grain mill and starch products; animal feeds	584	610	581	680	758	..	621	..
		154	Other food products	1 593	1 779	2 144	2 365	2 569	..	..	..
		155	Beverages	895	950	937	955	1 128	1 204	1 264	..
	16		Tobacco products	143	155	148	165	173	187	189	..
	17		Textiles	452	421	423	445	442	430	396	..
		171	Spinning, weaving and finishing of textiles	240	211	223	233	229	..	..	..
		172	Other textiles	137	133	120	131	138	..	..	..
		173	Knitted and crocheted fabrics and articles	75	77	80	81	76	..	..	..
	18		Wearing apparel and fur	291	294	301	320	336	347	342	..
		181	Wearing apparel, except fur apparel	289	293	299	317	333	..	..	..
		182	Dressing and dyeing of fur; articles of fur	2	1	2	3	3	3	2	..
	19		Leather and articles of leather; footwear	65	66	65	88	83	51	58	..
		191	Leather and articles of leather	44	50	50	70	67	..	..	..
		192	Footwear	21	16	15	18	16	16	14	..
	20		Wood & products of wood & cork, exc. furniture	230	237	241	282	322	343	424	..
		201	Sawmilling and planing of wood	120	125	131	158	180	196	126	..
		202	Wood products; articles of cork, straw, etc.	110	112	109	124	142	..	..	..
	21		Paper and paper products	335	343	347	369	467	469	487	..
	22		Printing, publishing and recorded media	1 311	1 497	1 715	2 235	2 573	3 133	3 653	..
		221	Publishing	530	598	644	678	702	789	..	..
		222	Printing and related activities	..	..	..	..	..	..	..	..
		223	Reproduction of recorded media	781	899	1 071	1 557	1 871	2 344	2 804	..
	23		Coke and petroleum products; nuclear fuel	..	..	..	..	..	..	..	..
		231	Coke oven products	..	..	..	..	..	..	..	..
		232	Refined petroleum products	..	..	..	..	..	..	..	..
		233	Processing of nuclear fuel	..	..	..	..	..	..	..	..
	24		Chemicals and chemical products	2 685	3 229	3 809	4 614	5 446	6 326	8 075	..
		241	Basic chemicals	1 407	1 627	1 969	2 282	2 792	..	..	..
		242	Other chemical products	1 159	1 471	1 704	2 181	2 467	..	2 361	..
		243	Man-made fibres	119	131	136	150	187	173	306	..
	25		Rubber and plastic products	515	546	559	628	734	774	776	..
		251	Rubber products	133	141	134	165	186	149	95	..
		252	Plastic products	382	405	425	463	548	..	681	..
	26		Other non-metallic mineral products	582	556	584	653	716	757	861	..
		261	Glass and glass products	152	141	150	171	194	..	..	..
		269	Non-metallic mineral products, nec	430	415	434	482	523	..	..	..
	27		Basic metals	278	222	253	272	303	311	343	..
		271	Basic iron and steel	89	74	81	92	97	..	..	..
		272	Basic precious and non-ferrous metals	189	148	172	180	206	218	234	..
		273	Casting of metals	..	..	..	..	..	..	4	..
	28		Fabricated metal products, exc. machinery	591	584	585	633	690	781	834	..
		281	Structural metal prod., tanks, reservoirs, etc.	252	243	242	274	304	..	..	..
		289	Other fabricated metal products	339	341	343	359	387	..	..	..
	29		Machinery and equipment, nec	732	756	769	923	1 072	1 106	1 146	..
		291	General purpose machinery	388	393	396	481	588	..	650	..
		292	Special purpose machinery	150	173	173	207	246	..	..	..
		293	Domestic appliances, nec	193	190	200	235	239	254	227	..
	30		Office, accounting & computing machinery	2 177	2 471	3 136	3 559	5 994	6 451	7 072	..
	31		Electrical machinery & apparatus, nec	543	587	611	680	1 010	1 103	1 250	..
		311	Electric motors, generators and transformers	110	124	151	186	230	336	390	..
		312	Electricity distribution & control apparatus	124	132	139	153	209	221	265	..
		313	Insulated wire and cable	109	122	117	134	230	279	248	..
		314	Accumulators, primary cells & primary batteries	74	76	82	94	234	178	10	..
		315	Electric lamps and lighting equipment	12	14	13	12	13	16	19	..
		319	Other electrical equipment, nec	115	118	109	101	95	..	318	..
	32		Radio, TV, communication equipment	470	489	737	978	1 151	1 209	2 428	..
		321	Electronic valves, tubes; other electronic components	160	204	337	475	681	601	1 784	..
		322	TV/radio transmitters; line teleph./telegr. apparatus	273	240	337	436	405	540	578	..
		323	TV/radio receivers; sound and video app.; assoc. gds.	37	45	63	68	66	69	65	..
	33		Medical, precision & optical instrum.; watch & clocks	658	699	837	904	1 005	1 186	1 306	..
		331	Medical appl.; instruments & appl. for measuring, etc.	535	561	639	682	765	..	1 038	..
		332	Optical instruments and photographic equipment	123	138	198	222	240	201	268	..
		333	Watches and clocks	..	..	..	..	..	..	..	..
	34		Motor vehicles, trailers and semi-trailers	167	178	187	232	280	288	319	..
		341	Motor vehicles	59	54	44	44	61	61	60	..
		342	Bodies for motor vehicles, trailers & semi-trailers	35	35	36	45	50	47	57	..
		343	Parts & accessories for motor vehicles & their engines	72	89	106	143	169	180	202	..
	35		Other transport equipment	239	233	229	221	245	290	343	..
		351	Building and repairing of ships and boats	27	25	23	23	29	..	22	..
		352	Railway and tramway locomotives and rolling stock	210	207	205	197	213	258	314	..
		353	Aircraft and spacecraft	..	..	..	..	..	..	..	..
		359	Transport equipment, nec	1	1	1	1	3	..	..	..
	36		Furniture; manufacturing, nec	771	781	809	818	907	..	..	..
		361	Furniture	152	159	166	173	191	..	300	..
		369	Manufacturing, nec	619	622	643	645	717	..	..	..
	37		Recycling	..	..	..	..	..	..	22	..
E	40_41		Electricity, gas and water supply	1 256	1 330	1 396	1 402	1 472	1 644	1 833	..
	40		Electricity, gas, steam and hot water supply	1 188	1 258	1 315	1 319	1 387	1 549	1 749	..
	41		Collection, purification and distribution of water	68	72	81	83	84	94	85	..
F	45		Construction	..	..	..	..	..	..	..	..

Note:ISIC 171 includes part of 172. Parts of ISIC 171, 172 and 173 are not covered. ISIC 201 includes part of 202. Parts of ISIC 202 and 21 are not covered. ISIC 221 includes 222
Parts of ISIC 23 and 241 are not covered. ISIC 272 includes 273. Part of ISIC292 is not covered. ISIC 332 includes 333. ISIC 369 includes part of 23 and 37. ISIC 402 includes 403.

Table IR.2 VALEUR AJOUTÉE - VALUE ADDED

Millions de Ir£ (Prix courants)

1991	1992	1993	1994	1995	1996	1997	1998			CITI révision 3
262	246	239	314	371	352	346	..	Activites extractives	C	10_14
..	..	..	..	..	..	..	..	Extraction de charbon, de lignite et de tourbe		10
..	..	..	..	..	..	..	..	Pétrole brut et gaz naturel ; activités annexes		11
..	..	..	..	..	..	..	..	Extraction de minerais métalliques		13
..	..	..	..	..	..	..	..	Autres activités extractives		14
9 664	10 485	11 722	13 325	16 237	18 135	21 210	..	Total activites manufacturieres	D	15_37
2 624	2 852	3 155	3 378	3 663	..	..	..	Produits alimentaires et boissons		15
477	459	500	502	548	..	..	..	Viande, poisson, fruits, légumes, huiles et graisses		151
389	415	449	399	405	462	449	..	Produits laitiers		152
129	127	127	155	161	..	..	..	Travail des grains; prod. amyl. ; alim. pour animaux		153
1 064	1 225	1 501	1 676	1 814	..	..	..	Autres produits alimentaires		154
565	626	578	646	735	778	837	..	Boissons		155
99	104	102	113	126	135	142	..	Produits à base de tabac		16
195	185	199	203	189	190	187	..	Textiles		17
88	76	91	95	84	..	..	..	Filature, tissage et achèvement des textiles		171
68	68	64	65	63	..	..	..	Autres articles textiles		172
39	40	43	43	42	..	..	..	Etoffes et articles de bonneterie		173
136	134	148	151	159	158	147	..	Articles d'habillement et fourrure		18
136	134	147	150	158	..	..	..	Articles d'habillement autres qu'en fourrure		181
1	1	1	1	2	2	1	..	Prép. et teinture des fourrures ; articles en fourrure		182
20	18	21	26	26	18	20	..	Cuir et articles en cuir ; chaussures		19
10	12	14	17	16	..	..	..	Cuir et articles en cuir sauf chaussures		191
10	7	7	9	9	8	7	..	Chaussures		192
87	96	95	108	115	126	162	..	Bois, articles en bois, sauf meubles		20
41	46	49	63	62	69	36	..	Sciage et rabotage du bois		201
46	50	45	45	53	..	..	..	Articles en bois, liège, vannerie, sparterie		202
144	158	170	174	222	213	247	..	Papier, articles en papier et carton		21
928	1 043	1 198	1 617	1 960	2 486	2 933	..	Edition, imprimerie et reproduction		22
341	371	399	419	430	478	..	..	Edition		221
..	..	..	..	..	..	..	..	Imprimerie et activités annexes		222
586	672	799	1 198	1 531	2 009	2 404	..	Reproduction de supports enregistrés		223
..	..	..	..	..	..	..	..	Coke. et prod. pétroliers ; comb. nucléaires		23
..	..	..	..	..	..	..	..	Cokéfaction		231
..	..	..	..	..	..	..	..	Produits pétroliers raffinés		232
..	..	..	..	..	..	..	..	Traitement de combustibles nucléaires		233
1 741	2 190	2 652	3 253	3 857	4 508	6 067	..	Produits chimiques		24
952	1 152	1 453	1 733	2 112	..	..	..	Produits chimiques de base		241
737	987	1 142	1 456	1 664	..	..	..	Autres produits chimiques		242
52	51	57	64	81	58	149	..	Fibres synthétiques ou artificielles		243
252	277	281	323	349	365	376	..	Art. en caoutchouc et matières plastiques		25
75	80	75	93	107	72	47	..	Articles en caoutchouc		251
177	197	206	231	242	..	..	..	Articles en matières plastiques		252
325	317	351	380	402	422	480	..	Autres produits minéraux non métalliques		26
88	88	97	115	120	..	..	..	Verre et articles en verre		261
237	229	253	266	282	..	..	..	Produits minéraux non métalliques, nca		269
101	52	87	63	82	85	106	..	Produits métallurgiques de base		27
32	26	36	18	21	..	..	..	Sidérurgie et première transformation de l'acier		271
68	26	52	46	61	60	72	..	Première transf. métaux précieux et non ferreux		272
..	..	..	..	..	..	2	..	Fonderie		273
258	264	269	284	307	340	377	..	Ouvrages en métaux, sauf machines		28
106	102	103	116	119	..	..	..	Constr. et menuis. métal.; réservoirs; générateurs		281
153	162	166	168	188	..	..	..	Autres ouvrages en métaux		289
372	387	401	457	519	537	547	..	Machines et matériels, nca		29
199	204	201	225	283	..	..	..	Machines d'usage général		291
87	94	100	114	124	..	..	..	Machines d'usage spécifique		292
86	89	100	118	112	115	99	..	Appareils domestiques, nca		293
998	933	923	1 001	2 163	2 146	2 003	..	Mach. de bureau, comptables, information		30
275	304	303	355	524	615	667	..	Machines et appareils électriques, nca		31
67	77	97	121	147	225	261	..	Moteurs, génératrices, transformateurs électriques		311
67	67	74	83	116	131	160	..	Matériel électrique de distribution et de commande		312
34	42	37	44	103	126	95	..	Fils et câbles électriques isolés		313
40	44	34	49	104	88	3	..	Accumulateurs et piles électriques		314
6	7	6	5	6	9	10	..	Lampes électriques et appareils d'éclairage		315
61	66	55	54	48	..	..	..	Autres matériels électriques, nca.		319
214	191	267	312	341	488	1 158	..	Equipements et appareils de radio, TV, etc.		32
77	86	106	113	152	233	916	..	Tubes et valves électroniques;autres composants		321
120	91	136	168	167	227	215	..	Emetteurs radio/TV; app. téléphonie et télégraphie		322
17	14	25	30	22	28	27	..	Récepteurs radio/TV, son et image; art. associés		323
419	461	559	572	618	721	775	..	Instr. médicaux, de précision ; optique ; horlog.		33
342	366	412	430	480	..	609	..	Instruments médicaux et de précision, etc.		331
77	94	147	142	138	109	166	..	Instruments d'optique ; matériel photographique		332
..	..	..	..	..	..	..	..	Horlogerie		333
72	83	84	96	116	105	110	..	Véhicules automobiles; remorques, etc.		34
24	27	22	12	25	26	21	..	Véhicules automobiles		341
14	14	13	18	18	16	19	..	Carross. automobiles, remorques, semi-remorques		342
34	42	48	66	73	63	70	..	Pièces et accessoires pour automobiles et moteurs		343
112	130	125	115	133	155	176	..	Autres matériels de transport		35
12	11	10	10	16	..	..	..	Construction et réparation de navires		351
99	118	114	104	116	137	161	..	Matériel ferroviaire roulant		352
..	..	..	..	..	..	..	..	Construction aéronautique et spatiale		353
1	0		1	2	..	..	..	Autres équipements de transport		359
293	306	333	344	364	..	..	..	Meubles ; activités de fabrication, nca		36
71	74	78	81	86	..	..	..	Meubles		361
222	233	255	262	278	..	..	..	Activités de fabrication, nca		369
..	..	..	..	..	..	5	..	Récupération		37
775	798	853	825	836	995	1 081	..	Electricité, gaz et eau	E	40_41
728	749	795	765	776	928	..	..	Prod. et dist. d'électricité gaz, vapeur d'eau chaude		40
47	49	59	60	60	67	..	..	Captage, épuration et distribution d'eau		41
..	..	..	..	..	..	..	..	Construction	F	45

Note:ISIC 171 includes part of 172. Parts of ISIC 171, 172 and 173 are not covered. ISIC 201 includes part of 202. Parts of ISIC 202 and 21 are not covered. ISIC 221 includes 222
Parts of ISIC 23 and 241 are not covered. ISIC 272 includes 273. Part of ISIC292 is not covered. ISIC 332 includes 333. ISIC 369 includes part of 23 and 37. ISIC 402 includes 403.

Table IR.3 **INVESTMENT - INVESTISSEMENT**

Millions of Ir£ (Current Prices)

ISIC revision 3				1991	1992	1993	1994	1995	1996	1997	1998
C	10_14		Mining and quarrying	41	44	29	15	39	33	39	..
	10		Mining of coal and lignite; extraction of peat	..	..	..	..	..	..	..	..
	11		Crude petroleum & natural gas; service activities	..	..	..	..	..	..	..	..
	13		Mining of metal ores	..	..	..	..	..	..	..	..
	14		Other mining and quarrying	..	..	..	..	..	..	..	..
D	15_37		Total manufacturing	1 024	1 109	1 163	1 084	1 283	1 728	2 047	..
	15		Food products and beverages	244	223	247	283	229	..	..	..
		151	Production of meat, fish, fruit, vegetables, oils & fats	64	46	65	63	64	..	..	..
		152	Dairy products	60	49	47	94	49	66	42	..
		153	Grain mill and starch products; animal feeds	8	10	12	32	14	..	..	..
		154	Other food products	42	66	64	36	65	..	..	..
		155	Beverages	70	52	59	57	37	44	68	..
	16		Tobacco products	2	5	1		5	4	6	..
	17		Textiles	9	10	19	12	25	10	18	..
		171	Spinning, weaving and finishing of textiles	3	3	11	5	12	..	..	..
		172	Other textiles	4	3	4	3	8	..	..	..
		173	Knitted and crocheted fabrics and articles	3	4	4	4	5	..	..	..
	18		Wearing apparel and fur	13	11	20	11	14	7	11	..
		181	Wearing apparel, except fur apparel	13	11	20	11	14	..	..	..
		182	Dressing and dyeing of fur; articles of fur	0	0	0	0	0			..
	19		Leather and articles of leather; footwear	1	0	0	4	2	1	2	..
		191	Leather and articles of leather	1	0	0	4	2	..	..	..
		192	Footwear	0		0		0			..
	20		Wood & products of wood & cork, exc. furniture	10	12	17	26	12	22	16	..
		201	Sawmilling and planing of wood	7	7	12	23	6	16	5	..
		202	Wood products; articles of cork, straw, etc.	4	5	5	3	6	..	..	..
	21		Paper and paper products	16	14	12	17	18	19	27	..
	22		Printing, publishing and recorded media	45	47	58	42	75	160	160	..
		221	Publishing	30	32	32	20	26	53	..	..
		222	Printing and related activities	..	..	..	..	..	..	..	..
		223	Reproduction of recorded media	15	15	25	22	48	107	84	..
	23		Coke and petroleum products; nuclear fuel	..	..	..	..	..	..	..	..
		231	Coke oven products	..	..	..	..	..	..	..	..
		232	Refined petroleum products	..	..	..	..	..	..	..	..
		233	Processing of nuclear fuel	..	..	..	..	..	..	..	..
	24		Chemicals and chemical products	233	376	295	259	272	415	467	..
		241	Basic chemicals	53	101	121	118	100	..	..	..
		242	Other chemical products	178	251	164	137	150	..	..	..
		243	Man-made fibres	3	23	9	4	22	8	7	..
	25		Rubber and plastic products	56	25	27	37	49	61	50	..
		251	Rubber products	33	7	8	7	16	5	3	..
		252	Plastic products	23	18	19	30	33	..	..	..
	26		Other non-metallic mineral products	31	16	24	26	42	39	71	..
		261	Glass and glass products	10	6	9	10	17	..	..	..
		269	Non-metallic mineral products, nec	21	10	15	16	25	..	..	..
	27		Basic metals	13	12	9	9	11	13	36	..
		271	Basic iron and steel	2	5	5	1	2	..	..	..
		272	Basic precious and non-ferrous metals	10	8	5	8	10	9	30	..
		273	Casting of metals	..	..	..	..	..	..	..	..
	28		Fabricated metal products, exc. machinery	25	13	13	25	28	40	41	..
		281	Structural metal prod., tanks, reservoirs, etc.	8	6	5	9	9	..	..	..
		289	Other fabricated metal products	17	7	8	15	18	..	..	..
	29		Machinery and equipment, nec	32	28	26	30	41	40	60	..
		291	General purpose machinery	11	9	11	12	16	..	..	..
		292	Special purpose machinery	9	5	7	10	11	..	..	..
		293	Domestic appliances, nec	12	14	8	9	14	14	17	..
	30		Office, accounting & computing machinery	79	147	254	117	144	283	226	..
	31		Electrical machinery & apparatus, nec	31	28	24	43	125	78	82	..
		311	Electric motors, generators and transformers	11	15	13	27	62	50	39	..
		312	Electricity distribution & control apparatus	8	4	6	9	14	13	22	..
		313	Insulated wire and cable	2	3	4	2	30	6	11	..
		314	Accumulators, primary cells & primary batteries	7	3	..	4	15	6		..
		315	Electric lamps and lighting equipment	0	0	0	0	0	1	1	..
		319	Other electrical equipment, nec	3	2	3	1	4	..	..	..
	32		Radio, TV, communication equipment	17	32	24	57	67	71	324	..
		321	Electronic valves, tubes; other electronic components	7	7	9	35	46	44	297	..
		322	TV/radio transmitters; line teleph./telegr. apparatus	8	19	12	21	18	25	25	..
		323	TV/radio receivers; sound and video app.; assoc. gds.	2	6	2	1	3	2	2	..
	33		Medical, precision & optical instrum.; watch & clocks	47	41	45	43	61	109	86	..
		331	Medical appl.; instruments & appl. for measuring, etc.	26	27	38	30	55	..	71	..
		332	Optical instruments and photographic equipment	20	14	7	12	6	42	16	..
		333	Watches and clocks	..	..	..	..	..	..	..	..
	34		Motor vehicles, trailers and semi-trailers	6	6	6	10	14	18	18	..
		341	Motor vehicles	1	0	0	2	1	5	9	..
		342	Bodies for motor vehicles, trailers & semi-trailers	0	1	1	1	1	2	2	..
		343	Parts & accessories for motor vehicles & their engines	5	6	5	6	13	11	7	..
	35		Other transport equipment	79	33	11		10	15	44	..
		351	Building and repairing of ships and boats	4	0	0	1	1	..	..	..
		352	Railway and tramway locomotives and rolling stock	76	32	10		6	13	43	..
		353	Aircraft and spacecraft	..	..	..	..	..	..	..	..
		359	Transport equipment, nec	0	0	0		4	..	..	..
	36		Furniture; manufacturing, nec	34	31	31	32	39	..	..	..
		361	Furniture	8	9	7	9	8	..	..	..
		369	Manufacturing, nec	26	22	24	24	31	..	..	..
	37		Recycling	..	..	..	..	..	..	3	..
E	40_41		Electricity, gas and water supply	153	252	346	199	161	212	217	..
	40		Electricity, gas, steam and hot water supply	138	236	329	186	154	190	..	..
	41		Collection, purification and distribution of water	15	16	18	13	8	22	..	..
F	45		Construction	..	..	..	..	..	..	..	..

Note:ISIC 171 includes part of 172. Parts of ISIC 171, 172 and 173 are not covered. ISIC 201 includes part of 202. Parts of ISIC 202 and 21 are not covered. ISIC 221 includes 222
Parts of ISIC 23 and 241 are not covered. ISIC 272 includes 273. Part of ISIC292 is not covered. ISIC 332 includes 333. ISIC 369 includes part of 23 and 37. ISIC 402 includes 403.

Table IR.4 **EMPLOI, SALARIÉS - EMPLOYMENT, EMPLOYEES**

Unités

1991	1992	1993	1994	1995	1996	1997	1998			CITI révision 3	
5 359	5 023	5 264	5 258	5 030	5 202	5 271	..	Activites extractives	C	10_14	
..	..	..	..	..	..	..	..	Extraction de charbon, de lignite et de tourbe		10	
..	..	..	..	..	..	..	..	Pétrole brut et gaz naturel ; activités annexes		11	
..	..	..	..	..	..	..	..	Extraction de Minerais d'Uranium et de Thorium		12	
..	..	..	..	..	..	..	..	Extraction de minerais métalliques		13	
..	..	..	..	..	..	..	..	Autres activités extractives		14	
195 607	197 658	198 757	203 921	218 972	225 053	238 893	..	Total activites manufacturieres	D	15_37	
43 172	43 160	43 160	43 001	44 285	..	..	..	Produits alimentaires et boissons		15	
16 013	16 759	16 979	16 613	17 601	..	..	..	Viande, poisson, fruits, légumes, huiles et graisses			151
8 292	7 728	7 704	7 567	7 702	8 087	7 594	..	Produits laitiers			152
2 524	2 362	2 293	2 411	2 874	..	..	..	Travail des grains; prod. amyl. ; alim. pour animaux			153
10 970	11 118	11 116	11 222	11 238	..	..	..	Autres produits alimentaires			154
5 373	5 193	5 068	5 188	4 870	4 838	4 742	..	Boissons			155
1 301	1 276	1 178	1 027	1 018	1 001	961	..	Produits à base de tabac		16	
9 319	8 755	8 429	8 227	7 952	7 689	7 295	..	Textiles		17	
4 175	3 732	3 559	3 646	3 471	..	..	..	Filature, tissage et achèvement des textiles			171
2 816	2 634	2 442	2 276	2 345	..	..	..	Autres articles textiles			172
2 328	2 389	2 428	2 305	2 136	..	..	..	Etoffes et articles de bonneterie			173
11 561	11 734	11 014	11 339	11 197	10 151	9 414	..	Articles d'habillement et fourrure		18	
11 501	11 689	10 958	11 275	11 125	..	..	..	Articles d'habillement autres qu'en fourrure			181
60	45	56	64	72	64	50	..	Prép. et teinture des fourrures ; articles en fourrure			182
1 262	1 171	1 151	1 193	1 199	975	1 025	..	Cuir et articles en cuir ; chaussures		19	
519	532	547	651	688	..	..	..	Cuir et articles en cuir sauf chaussures			191
743	639	604	542	511	500	467	..	Chaussures			192
4 198	4 165	4 034	3 993	4 401	4 474	5 056	..	Bois, articles en bois, sauf meubles		20	
1 502	1 485	1 495	1 611	1 700	1 803	1 168	..	Sciage et rabotage du bois			201
2 696	2 680	2 539	2 382	2 701	..	..	..	Articles en bois, liège, vannerie, sparterie			202
4 237	4 122	4 081	4 126	4 351	4 505	4 727	..	Papier, articles en papier et carton		21	
12 396	13 090	14 001	14 728	15 587	16 067	17 771	..	Edition, imprimerie et reproduction		22	
10 671	11 055	11 411	11 679	11 567	11 732	..	..	Edition			221
..	..	..	..	..	..	..	..	Imprimerie et activités annexes			222
1 725	2 035	2 590	3 049	4 020	4 335	5 310	..	Reproduction de supports enregistrés			223
..	..	..	..	..	..	..	..	Coke. et prod. pétroliers ; comb. nucléaires		23	
..	..	..	..	..	..	..	..	Cokéfaction			231
..	..	..	..	..	..	..	..	Produits pétroliers raffinés			232
..	..	..	..	..	..	..	..	Traitement de combustibles nucléaires			233
14 648	15 288	16 242	17 422	18 116	19 423	20 974	..	Produits chimiques		24	
4 336	4 351	4 465	4 702	4 846	..	..	..	Produits chimiques de base			241
8 988	9 544	10 435	11 358	11 843	..	..	..	Autres produits chimiques			242
1 324	1 393	1 342	1 362	1 427	1 524	2 196	..	Fibres synthétiques ou artificielles			243
8 079	8 399	8 484	8 802	9 496	10 307	10 014	..	Art. en caoutchouc et matières plastiques		25	
2 149	2 235	2 221	2 286	2 434	2 409	1 780	..	Articles en caoutchouc			251
5 930	6 164	6 263	6 516	7 062	..	..	..	Articles en matières plastiques			252
10 042	9 507	9 135	9 193	9 347	9 470	10 027	..	Autres produits minéraux non métalliques		26	
4 114	3 804	3 284	3 348	3 607	..	..	..	Verre et articles en verre			261
5 928	5 703	5 851	5 845	5 740	..	..	..	Produits minéraux non métalliques, nca			269
2 455	2 338	2 366	2 294	2 187	2 095	2 301	..	Produits métallurgiques de base		27	
1 015	996	1 079	1 086	939	..	..	..	Sidérurgie et première transformation de l'acier			271
1 440	1 342	1 287	1 208	1 248	1 128	1 087	..	Première transf. métaux précieux et non ferreux			272
..	..	..	..	..	..	126	..	Fonderie			273
10 218	10 010	9 696	9 945	10 372	11 476	11 664	..	Ouvrages en métaux, sauf machines		28	
4 493	4 419	4 259	4 395	4 576	..	..	..	Constr. et menuis. métal.; réservoirs; générateurs			281
5 725	5 591	5 437	5 550	5 796	..	..	..	Autres ouvrages en métaux			289
12 300	12 526	12 428	13 473	14 104	14 220	14 948	..	Machines et matériels, nca		29	
5 634	5 803	5 672	6 117	6 722	..	..	..	Machines d'usage général			291
3 272	3 481	3 436	3 862	3 716	..	..	..	Machines d'usage spécifique			292
3 394	3 242	3 320	3 494	3 666	3 712	3 594	..	Appareils domestiques, nca			293
8 016	7 939	8 886	10 235	14 413	15 231	15 168	..	Mach. de bureau, comptables, information		30	
10 244	10 597	10 204	10 064	12 358	11 982	13 686	..	Machines et appareils électriques, nca		31	
1 974	2 090	2 116	2 251	2 700	3 332	3 595	..	Moteurs, génératrices, transformateurs électriques			311
2 270	2 356	2 350	2 428	3 283	3 058	3 576	..	Matériel électrique de distribution et de commande			312
1 343	1 519	1 476	1 438	2 059	2 184	3 110	..	Fils et câbles électriques isolés			313
1 492	1 397	1 231	1 374	2 098	1 840	149	..	Accumulateurs et piles électriques			314
320	358	328	336	354	384	398	..	Lampes électriques et appareils d'éclairage			315
2 845	2 877	2 703	2 237	1 864	..	..	..	Autres matériels électriques, nca.			319
4 883	5 197	6 093	7 184	7 228	8 254	13 512	..	Equipements et appareils de radio, TV, etc.		32	
2 506	2 717	3 297	3 706	4 072	4 461	8 920	..	Tubes et valves électroniques;autres composants			321
1 811	1 836	2 083	2 547	2 139	2 896	3 603	..	Emetteurs radio/TV; app. téléphonie et télégraphie			322
566	644	713	931	1 017	897	989	..	Récepteurs radio/TV, son et image; art. associés			323
9 269	9 643	10 169	10 606	11 796	12 873	14 513	..	Instr. médicaux, de précision ; optique ; horlog.		33	
7 373	7 508	7 872	8 403	9 281	..	11 705	..	Instruments médicaux et de précision, etc.			331
1 896	2 135	2 297	2 203	2 515	2 386	2 808	..	Instruments d'optique ; matériel photographique			332
..	..	..	..	..	..	..	..	Horlogerie			333
3 780	3 855	3 272	3 648	4 403	4 344	4 224	..	Véhicules automobiles; remorques, etc.		34	
1 573	1 481	826	849	1 444	1 439	1 367	..	Véhicules automobiles			341
812	758	660	690	687	653	697	..	Carross. automobiles, remorques, semi-remorques			342
1 395	1 616	1 786	2 109	2 272	2 252	2 160	..	Pièces et accessoires pour automobiles et moteurs			343
5 109	5 736	5 457	3 988	5 140	5 118	5 161	..	Autres matériels de transport		35	
591	561	460	468	496	..	..	..	Construction et réparation de navires			351
4 481	5 142	4 967	3 492	4 504	4 491	4 532	..	Matériel ferroviaire roulant			352
..	..	..	..	..	..	..	..	Construction aéronautique et spatiale			353
37	33	30	28	140	..	..	..	Autres équipements de transport			359
9 118	9 150	9 277	9 433	10 022	..	..	..	Meubles ; activités de fabrication, nca		36	
3 701	3 703	3 835	3 906	4 064	..	..	..	Meubles			361
5 417	5 447	5 442	5 527	5 958	..	..	..	Activités de fabrication, nca			369
..	..	..	..	..	..	214	..	Récupération		37	
14 342	14 442	14 287	13 977	13 268	12 737	12 193	..	Electricité, gaz et eau	E	40_41	
12 104	12 161	12 038	11 827	11 153	10 708	..	..	Prod. et dist. d'électricité gaz, vapeur d'eau chaude		40	
..	..	..	..	..	..	..	..	Construction	F	45	

Note:ISIC 171 includes part of 172. Parts of ISIC 171, 172 and 173 are not covered. ISIC 201 includes part of 202. Parts of ISIC 202 and 21 are not covered. ISIC 221 includes 222
Parts of ISIC 23 and 241 are not covered. ISIC 272 includes 273. Part of ISIC292 is not covered. ISIC 332 includes 333. ISIC 369 includes part of 23 and 37. ISIC 402 includes 403.

IRELAND

Table IR.5 EMPLOYMENT, FEMALES - EMPLOI, FEMMES

Units

ISIC revision 3				1991	1992	1993	1994	1995	1996	1997	1998
C	10_14		Mining and quarrying	302	312	321	337	361	418	368	..
	10		Mining of coal and lignite; extraction of peat	..	..	..	..	..	..	..	..
	11		Crude petroleum & natural gas; service activities	..	..	..	..	..	..	..	..
	12		Mining of uranium and thorium ores	..	..	..	..	..	..	..	..
	13		Mining of metal ores	..	..	..	..	..	..	..	..
	14		Other mining and quarrying	..	..	..	..	..	..	..	..
D	15_37		Total manufacturing	63 198	64 217	66 340	68 957	73 804	75 749	81 933	..
	15		Food products and beverages	10 210	10 099	10 788	10 443	10 940	..	..	..
		151	Production of meat, fish, fruit, vegetables, oils & fats	4 029	3 884	4 444	4 097	4 369	..	..	..
		152	Dairy products	1 489	1 414	1 435	1 324	1 453	1 442	1 457	..
		153	Grain mill and starch products; animal feeds	350	353	380	424	498	..	..	..
		154	Other food products	3 469	3 577	3 644	3 711	3 658	..	..	..
		155	Beverages	873	871	885	887	962	916	885	..
	16		Tobacco products	516	500	449	418	425	409	401	..
	17		Textiles	4 088	4 054	3 903	3 766	3 649	3 426	3 350	..
		171	Spinning, weaving and finishing of textiles	1 490	1 464	1 371	1 372	1 354	..	..	..
		172	Other textiles	983	918	851	816	814	..	..	..
		173	Knitted and crocheted fabrics and articles	1 615	1 672	1 681	1 578	1 481	..	..	..
	18		Wearing apparel and fur	9 346	9 257	8 714	8 942	8 571	7 712	7 017	..
		181	Wearing apparel, except fur apparel	9 325	9 244	8 691	8 928	8 559	..	..	..
		182	Dressing and dyeing of fur; articles of fur	21	13	23	14	12	12	11	..
	19		Leather and articles of leather; footwear	494	490	497	461	478	451	499	..
		191	Leather and articles of leather	203	203	235	255	278	..	..	..
		192	Footwear	291	287	262	206	200	197	195	..
	20		Wood & products of wood & cork, exc. furniture	449	446	447	418	502	404	582	..
		201	Sawmilling and planing of wood	115	99	93	103	113	121	91	..
		202	Wood products; articles of cork, straw, etc.	334	347	354	315	389	..	..	..
	21		Paper and paper products	1 132	1 082	1 059	1 040	1 063	1 106	1 206	..
	22		Printing, publishing and recorded media	3 772	4 209	4 565	4 847	5 361	5 512	6 288	..
		221	Publishing	3 193	3 438	3 608	3 698	3 822	3 897	..	..
		222	Printing and related activities	..	..	..	..	..	..	..	..
		223	Reproduction of recorded media	579	771	957	1 149	1 539	1 615	2 072	..
	23		Coke and petroleum products; nuclear fuel	..	..	..	..	..	..	..	..
		231	Coke oven products	..	..	..	..	..	..	..	..
		232	Refined petroleum products	..	..	..	..	..	..	..	..
		233	Processing of nuclear fuel	..	..	..	..	..	..	..	..
	24		Chemicals and chemical products	4 807	5 157	5 611	5 804	6 210	6 893	7 592	..
		241	Basic chemicals	589	654	812	767	831	..	..	..
		242	Other chemical products	3 991	4 239	4 547	4 777	5 102	..	..	..
		243	Man-made fibres	227	264	252	260	277	309	599	..
	25		Rubber and plastic products	2 069	2 133	2 290	2 289	2 581	2 865	2 866	..
		251	Rubber products	462	514	595	598	662	701	661	..
		252	Plastic products	1 607	1 619	1 695	1 691	1 919	..	..	..
	26		Other non-metallic mineral products	1 505	1 378	1 326	1 471	1 463	1 484	1 637	..
		261	Glass and glass products	792	660	554	674	691	..	..	..
		269	Non-metallic mineral products, nec	713	718	772	797	772	..	..	..
	27		Basic metals	261	225	254	337	253	261	264	..
		271	Basic iron and steel	65	54	79	153	67	..	..	..
		272	Basic precious and non-ferrous metals	196	171	175	184	186	186	172	..
		273	Casting of metals	..	..	..	..	..	..	8	..
	28		Fabricated metal products, exc. machinery	1 267	1 193	1 174	1 399	1 336	1 510	1 568	..
		281	Structural metal prod., tanks, reservoirs, etc.	377	365	356	388	412	..	..	..
		289	Other fabricated metal products	890	828	818	1 011	924	..	..	..
	29		Machinery and equipment, nec	3 320	3 236	3 215	3 567	3 469	3 421	3 617	..
		291	General purpose machinery	824	840	768	871	905	..	..	..
		292	Special purpose machinery	572	673	731	981	650	..	..	..
		293	Domestic appliances, nec	1 924	1 723	1 716	1 715	1 914	1 885	1 989	..
	30		Office, accounting & computing machinery	3 317	3 234	3 831	4 484	6 214	6 196	6 578	..
	31		Electrical machinery & apparatus, nec	5 069	5 158	4 742	4 875	5 921	5 788	6 495	..
		311	Electric motors, generators and transformers	830	892	880	985	1 174	1 556	1 637	..
		312	Electricity distribution & control apparatus	923	925	889	1 210	1 597	1 569	1 674	..
		313	Insulated wire and cable	599	675	579	611	929	929	1 533	..
		314	Accumulators, primary cells & primary batteries	827	754	558	664	1 081	991	58	..
		315	Electric lamps and lighting equipment	140	172	152	123	139	135	152	..
		319	Other electrical equipment, nec	1 750	1 740	1 684	1 282	1 001	..	..	..
	32		Radio, TV, communication equipment	2 656	2 914	3 506	3 743	3 650	4 437	6 702	..
		321	Electronic valves, tubes; other electronic components	1 480	1 656	1 994	1 884	2 010	2 574	4 338	..
		322	TV/radio transmitters; line teleph./telegr. apparatus	813	857	1 033	1 237	974	1 325	1 766	..
		323	TV/radio receivers; sound and video app.; assoc. gds.	363	401	479	622	666	538	598	..
	33		Medical, precision & optical instrum.; watch & clocks	5 059	5 371	5 766	6 068	6 688	7 261	8 230	..
		331	Medical appl.; instruments & appl. for measuring, etc.	3 858	4 031	4 306	4 715	5 215	..	6 540	..
		332	Optical instruments and photographic equipment	1 201	1 340	1 460	1 353	1 473	1 478	1 690	..
		333	Watches and clocks	..	..	..	..	..	..	..	..
	34		Motor vehicles, trailers and semi-trailers	752	907	996	1 208	1 286	1 265	1 166	..
		341	Motor vehicles	114	102	53	54	103	93	88	..
		342	Bodies for motor vehicles, trailers & semi-trailers	67	67	60	55	54	45	61	..
		343	Parts & accessories for motor vehicles & their engines	571	738	883	1 099	1 129	1 127	1 017	..
	35		Other transport equipment	353	408	398	325	462	923	507	..
		351	Building and repairing of ships and boats	43	37	37	42	47	..	..	..
		352	Railway and tramway locomotives and rolling stock	300	364	357	276	307	769	369	..
		353	Aircraft and spacecraft	..	..	..	..	..	..	..	..
		359	Transport equipment, nec	10	7	4	7	108	..	..	..
	36		Furniture; manufacturing, nec	2 756	2 766	2 809	3 052	3 282	..	..	..
		361	Furniture	430	459	474	489	508	..	..	..
		369	Manufacturing, nec	2 326	2 307	2 335	2 563	2 774	..	..	..
	37		Recycling	..	..	..	..	..	..	43	..
E	40_41		Electricity, gas and water supply	2 021	2 083	2 106	2 064	1 938	1 853	1 918	..
	40		Electricity, gas, steam and hot water supply	1 866	1 899	1 919	1 898	1 768	1 709	..	..
F	45		Construction	..	..	..	..	..	..	..	..

Note:ISIC 171 includes part of 172. Parts of ISIC 171, 172 and 173 are not covered. ISIC 201 includes part of 202. Parts of ISIC 202 and 21 are not covered. ISIC 221 includes 222
Parts of ISIC 23 and 241 are not covered. ISIC 272 includes 273. Part of ISIC292 is not covered. ISIC 332 includes 333. ISIC 369 includes part of 23 and 37. ISIC 402 includes 403.

Table IR.6 SALAIRES ET TRAITEMENTS, SALARIÉS - WAGES AND SALARIES, EMPLOYEES

Millions de Ir£ (Prix courants)

1991	1992	1993	1994	1995	1996	1997	1998			CITI révision 3	
95	95	99	106	114	115	..	..	Activites extractives	C	10_14	
..	..	..	..	..	..	..	..	Extraction de charbon, de lignite et de tourbe		10	
..	..	..	..	..	..	..	..	Pétrole brut et gaz naturel ; activités annexes		11	
..	..	..	..	..	..	..	..	Extraction de Minerais d'Uranium et de Thorium		12	
..	..	..	..	..	..	..	..	Extraction de minerais métalliques		13	
..	..	..	..	..	..	..	..	Autres activités extractives		14	
2 598	2 760	2 883	3 051	3 335	3 588	..	..	Total activites manufacturieres	D	15_37	
585	616	639	655	694	..	..	..	Produits alimentaires et boissons		15	
168	184	193	191	205	..	..	..	Viande, poisson, fruits, légumes, huiles et graisses			151
132	135	135	137	145	151	..	..	Produits laitiers			152
37	36	36	39	46	..	..	..	Travail des grains; prod. amyl. ; alim. pour animaux			153
139	150	159	170	176	..	..	..	Autres produits alimentaires			154
109	111	116	119	123	122	..	..	Boissons			155
25	26	24	23	24	24	..	..	Produits à base de tabac		16	
98	96	95	98	99	99	..	..	Textiles		17	
45	41	42	43	43	..	..	..	Filature, tissage et achèvement des textiles			171
34	34	31	33	34	..	..	..	Autres articles textiles			172
20	21	22	22	22	..	..	..	Etoffes et articles de bonneterie			173
89	95	94	98	103	93	..	..	Articles d'habillement et fourrure		18	
89	94	93	98	103	..	..	..	Articles d'habillement autres qu'en fourrure			181
0	0	0	1	1	1	..	..	Prép. et teinture des fourrures ; articles en fourrure			182
12	10	11	13	11	10	..	..	Cuir et articles en cuir ; chaussures		19	
5	5	6	8	6	..	..	..	Cuir et articles en cuir sauf chaussures			191
7	5	5	6	5	5	..	..	Chaussures			192
42	43	43	44	50	54	..	..	Bois, articles en bois, sauf meubles		20	
18	19	20	21	23	26	..	..	Sciage et rabotage du bois			201
25	24	24	23	27	..	..	..	Articles en bois, liège, vannerie, sparterie			202
63	65	68	70	78	80	..	..	Papier, articles en papier et carton		21	
199	224	250	266	284	308	..	..	Edition, imprimerie et reproduction		22	
173	189	201	208	210	224	..	..	Edition			221
..	..	..	..	..	..	..	..	Imprimerie et activités annexes			222
26	35	49	58	74	84	..	..	Reproduction de supports enregistrés			223
..	..	..	..	..	..	..	..	Coke. et prod. pétroliers ; comb. nucléaires		23	
..	..	..	..	..	..	..	..	Cokéfaction			231
..	..	..	..	..	..	..	..	Produits pétroliers raffinés			232
..	..	..	..	..	..	..	..	Traitement de combustibles nucléaires			233
253	279	308	339	369	400	..	..	Produits chimiques		24	
99	105	108	115	128	..	..	..	Produits chimiques de base			241
135	151	177	198	215	..	..	..	Autres produits chimiques			242
19	23	23	25	26	26	..	..	Fibres synthétiques ou artificielles			243
103	112	116	126	139	153	..	..	Art. en caoutchouc et matières plastiques		25	
33	36	34	37	42	38	..	..	Articles en caoutchouc			251
70	77	82	88	97	..	..	..	Articles en matières plastiques			252
140	139	146	151	153	163	..	..	Autres produits minéraux non métalliques		26	
58	57	61	65	63	..	..	..	Verre et articles en verre			261
82	82	86	86	90	..	..	..	Produits minéraux non métalliques, nca			269
42	38	43	41	38	38	..	..	Produits métallurgiques de base		27	
17	16	20	20	16	..	..	..	Sidérurgie et première transformation de l'acier			271
25	22	23	21	22	22	..	..	Première transf. métaux précieux et non ferreux			272
..	..	..	..	..	..	..	..	Fonderie			273
117	119	117	128	139	159	..	..	Ouvrages en métaux, sauf machines		28	
50	51	49	54	57	..	..	..	Constr. et menuis. métal.; réservoirs; générateurs			281
67	69	68	74	82	..	..	..	Autres ouvrages en métaux			289
150	164	165	190	206	217	..	..	Machines et matériels, nca		29	
74	81	81	93	108	..	..	..	Machines d'usage général			291
38	44	44	53	53	..	..	..	Machines d'usage spécifique			292
38	39	40	44	46	51	..	..	Appareils domestiques, nca			293
131	136	153	175	211	248	..	..	Mach. de bureau, comptables, information		30	
125	137	134	139	177	178	..	..	Machines et appareils électriques, nca		31	
28	30	32	36	43	59	..	..	Moteurs, génératrices, transformateurs électriques			311
26	27	30	32	44	41	..	..	Matériel électrique de distribution et de commande			312
18	20	19	22	31	35	..	..	Fils et câbles électriques isolés			313
19	21	17	18	29	23	..	..	Accumulateurs et piles électriques			314
3	3	4	3	4	4	..	..	Lampes électriques et appareils d'éclairage			315
31	35	32	28	26	..	..	..	Autres matériels électriques, nca.			319
66	72	84	104	112	137	..	..	Equipements et appareils de radio, TV, etc.		32	
33	35	40	50	57	68	..	..	Tubes et valves électroniques;autres composants			321
27	31	35	43	44	57	..	..	Emetteurs radio/TV; app. téléphonie et télégraphie			322
6	7	9	11	11	12	..	..	Récepteurs radio/TV, son et image; art. associés			323
121	132	143	155	172	192	..	..	Instr. médicaux, de précision ; optique ; horlog.		33	
98	104	113	121	135	..	..	..	Instruments médicaux et de précision, etc.			331
23	28	31	34	37	38	..	..	Instruments d'optique ; matériel photographique			332
..	..	..	..	..	..	..	..	Horlogerie			333
45	47	38	47	60	64	..	..	Véhicules automobiles; remorques, etc.		34	
22	21	12	14	22	23	..	..	Véhicules automobiles			341
8	8	7	7	8	7	..	..	Carross. automobiles, remorques, semi-remorques			342
15	17	20	26	30	34	..	..	Pièces et accessoires pour automobiles et moteurs			343
86	99	91	71	92	108	..	..	Autres matériels de transport		35	
7	6	7	6	7	..	..	..	Construction et réparation de navires			351
79	92	84	64	84	99	..	..	Matériel ferroviaire roulant			352
..	..	..	..	..	..	..	..	Construction aéronautique et spatiale			353
0	0	0	0	1	..	..	..	Autres équipements de transport			359
104	112	118	118	124	..	..	..	Meubles ; activités de fabrication, nca		36	
31	33	37	38	39	..	..	..	Meubles			361
74	78	81	80	86	..	..	..	Activités de fabrication, nca			369
..	..	..	..	..	..	..	..	Récupération		37	
252	270	274	287	288	281	..	..	Electricité, gaz et eau	E	40_41	
228	244	246	259	259	252	..	..	Prod. et dist. d'électricité gaz, vapeur d'eau chaude		40	
..	..	..	..	..	..	..	..	Construction	F	45	

Note:ISIC 171 includes part of 172. Parts of ISIC 171, 172 and 173 are not covered. ISIC 201 includes part of 202. Parts of ISIC 202 and 21 are not covered. ISIC 221 includes 222
Parts of ISIC 23 and 241 are not covered. ISIC 272 includes 273. Part of ISIC292 is not covered. ISIC 332 includes 333. ISIC 369 includes part of 23 and 37. ISIC 402 includes 403.

Table IR.7 **ESTABLISHMENTS - ÉTABLISSEMENTS**

Units

ISIC revision 3				1991	1992	1993	1994	1995	1996	1997	1998
C	10_14		Mining and quarrying	102	108	99	106	117	120	127	..
	10		Mining of coal and lignite; extraction of peat	..	..	..	..	..	..	..	..
	11		Crude petroleum & natural gas; service activities	..	..	..	..	..	..	..	..
	12		Mining of uranium and thorium ores	..	..	..	..	..	..	..	..
	13		Mining of metal ores	..	..	..	..	..	..	..	..
	14		Other mining and quarrying	..	..	..	..	..	..	..	..
D	15_37		Total manufacturing	4 546	4 542	4 544	4 603	4 604	4 599	4 739	..
	15		Food products and beverages	790	796	775	799	830	..	..	..
		151	Production of meat, fish, fruit, vegetables, oils & fats	255	267	273	276	288	..	..	..
		152	Dairy products	95	99	98	104	116	109	105	..
		153	Grain mill and starch products; animal feeds	107	100	94	94	103	..	..	..
		154	Other food products	270	265	249	262	258	..	..	..
		155	Beverages	63	65	61	63	65	63	62	..
	16		Tobacco products	6	6	6	6	6	6	6	..
	17		Textiles	200	195	195	198	190	188	190	..
		171	Spinning, weaving and finishing of textiles	96	90	88	94	92	..	..	..
		172	Other textiles	49	49	45	44	41	..	..	..
		173	Knitted and crocheted fabrics and articles	55	56	62	60	57	..	..	..
	18		Wearing apparel and fur	246	232	222	209	199	182	182	..
		181	Wearing apparel, except fur apparel	238	226	215	202	192	..	..	..
		182	Dressing and dyeing of fur; articles of fur	8	6	7	7	7	6	6	..
	19		Leather and articles of leather; footwear	42	40	41	39	33	31	30	..
		191	Leather and articles of leather	24	25	26	24	19	..	..	..
		192	Footwear	18	15	15	15	14	12	10	..
	20		Wood & products of wood & cork, exc. furniture	225	229	220	214	208	215	229	..
		201	Sawmilling and planing of wood	49	48	48	45	43	46	41	..
		202	Wood products; articles of cork, straw, etc.	176	181	172	169	165	..	..	..
	21		Paper and paper products	100	106	110	104	102	106	109	..
	22		Printing, publishing and recorded media	384	396	427	440	443	433	463	..
		221	Publishing	362	371	392	401	400	398	..	..
		222	Printing and related activities	..	..	..	..	..	..	..	..
		223	Reproduction of recorded media	22	25	35	39	43	35	38	..
	23		Coke and petroleum products; nuclear fuel	..	..	..	..	..	..	..	..
		231	Coke oven products	..	..	..	..	..	..	..	..
		232	Refined petroleum products	..	..	..	..	..	..	..	..
		233	Processing of nuclear fuel	..	..	..	..	..	..	..	..
	24		Chemicals and chemical products	239	232	239	245	243	237	250	..
		241	Basic chemicals	74	70	72	73	73	..	..	..
		242	Other chemical products	160	157	162	167	165	..	..	..
		243	Man-made fibres	5	5	5	5	5	6	24	..
	25		Rubber and plastic products	242	238	238	238	231	241	262	..
		251	Rubber products	44	41	41	43	42	39	35	..
		252	Plastic products	198	197	197	195	189	..	..	..
	26		Other non-metallic mineral products	280	275	280	292	283	282	290	..
		261	Glass and glass products	49	49	50	50	52	..	..	..
		269	Non-metallic mineral products, nec	231	226	230	242	231	..	..	..
	27		Basic metals	42	40	46	52	51	49	50	..
		271	Basic iron and steel	19	18	20	22	20	..	..	..
		272	Basic precious and non-ferrous metals	23	22	26	30	31	25	15	..
		273	Casting of metals	..	..	..	..	..	..	12	..
	28		Fabricated metal products, exc. machinery	518	502	496	503	485	485	483	..
		281	Structural metal prod., tanks, reservoirs, etc.	260	251	240	245	237	..	..	..
		289	Other fabricated metal products	258	251	256	258	248	..	..	..
	29		Machinery and equipment, nec	316	325	324	335	337	352	373	..
		291	General purpose machinery	157	161	161	164	163	..	..	..
		292	Special purpose machinery	144	148	148	155	157	..	..	..
		293	Domestic appliances, nec	15	16	15	16	17	18	17	..
	30		Office, accounting & computing machinery	67	66	69	71	71	70	70	..
	31		Electrical machinery & apparatus, nec	153	150	150	148	155	166	179	..
		311	Electric motors, generators and transformers	28	27	25	26	28	30	30	..
		312	Electricity distribution & control apparatus	47	48	50	48	49	47	54	..
		313	Insulated wire and cable	22	23	22	21	22	24	27	..
		314	Accumulators, primary cells & primary batteries	25	22	23	24	25	30	8	..
		315	Electric lamps and lighting equipment	20	19	20	19	20	24	25	..
		319	Other electrical equipment, nec	11	11	10	10	11	..	..	..
	32		Radio, TV, communication equipment	52	54	53	53	53	53	55	..
		321	Electronic valves, tubes; other electronic components	21	24	24	23	24	25	30	..
		322	TV/radio transmitters; line teleph./telegr. apparatus	20	17	17	18	17	19	18	..
		323	TV/radio receivers; sound and video app.; assoc. gds.	11	13	12	12	12	9	7	..
	33		Medical, precision & optical instrum.; watch & clocks	133	137	129	131	135	139	139	..
		331	Medical appl.; instruments & appl. for measuring, etc.	111	115	108	109	112	..	119	..
		332	Optical instruments and photographic equipment	22	22	21	22	23	21	20	..
		333	Watches and clocks	..	..	..	..	..	..	..	..
	34		Motor vehicles, trailers and semi-trailers	104	103	94	92	99	95	85	..
		341	Motor vehicles	23	22	14	15	23	22	20	..
		342	Bodies for motor vehicles, trailers & semi-trailers	44	44	47	46	44	41	37	..
		343	Parts & accessories for motor vehicles & their engines	37	37	33	31	32	32	28	..
	35		Other transport equipment	47	46	48	47	48	48	45	..
		351	Building and repairing of ships and boats	29	26	27	27	26	..	..	..
		352	Railway and tramway locomotives and rolling stock	14	16	17	16	16	17	16	..
		353	Aircraft and spacecraft	..	..	..	..	..	..	..	..
		359	Transport equipment, nec	4	4	4	4	6	..	..	..
	36		Furniture; manufacturing, nec	360	374	382	387	402	..	..	..
		361	Furniture	245	255	259	269	282	..	..	..
		369	Manufacturing, nec	115	119	123	118	120	..	..	..
	37		Recycling	..	..	..	..	..	..	14	..
E	40_41		Electricity, gas and water supply	96	94	95	96	96	84	90	..
	40		Electricity, gas, steam and hot water supply	41	41	41	41	41	33	..	..
F	45		Construction	..	..	..	..	..	..	..	..

Note:ISIC 171 includes part of 172. Parts of ISIC 171, 172 and 173 are not covered. ISIC 201 includes part of 202. Parts of ISIC 202 and 21 are not covered. ISIC 221 includes 222. Parts of ISIC 23 and 241 are not covered. ISIC 272 includes 273. Part of ISIC292 is not covered. ISIC 332 includes 333. ISIC 369 includes part of 23 and 37. ISIC 402 includes 403.

Table IT.1 PRODUCTION - PRODUCTION

Milliards de L (Prix courants)

1991	1992	1993	1994	1995	1996	1997	1998		CITI révision 3		
..	13 676	7 487	8 061	..	..	19 187	..	Activites extractives	C	10_14	
..	..	..	..	..	..	7	..	Extraction de charbon, de lignite et de tourbe		10	
..	11 352	5 279	5 636	..	..	13 161	..	Pétrole brut et gaz naturel ; activités annexes		11	
..	39	53	32	..	..	..	..	Extraction de minerais métalliques		13	
..	2 286	2 155	2 393	..	5 827	..	..	Autres activités extractives		14	
..	677 387	691 196	812 281	..	..	1 344 723	..	Total activites manufacturieres	D	15_37	
..	92 684	95 335	99 797	..	162 420	153 732	..	Produits alimentaires et boissons		15	
..	32 739	33 175	33 113	..	48 818	48 341	..	Viande, poisson, fruits, légumes, huiles et graisses			151
..	15 466	15 037	17 914	..	28 366	25 744	..	Produits laitiers			152
..	9 954	10 544	10 500	..	19 813	15 313	..	Travail des grains; prod. amyl. ; alim. pour animaux			153
..	22 354	23 100	23 383	..	41 517	41 237	..	Autres produits alimentaires			154
..	12 170	13 479	14 887	..	23 903	23 097	..	Boissons			155
..	6 224	6 424	6 750	..	15 015	15 684	..	Produits à base de tabac		16	
..	40 981	42 818	50 184	..	73 826	73 037	..	Textiles		17	
..	24 987	26 254	32 170	..	43 977	43 831	..	Filature, tissage et achèvement des textiles			171
..	5 479	5 540	6 155	..	13 241	13 540	..	Autres articles textiles			172
..	10 515	11 024	11 859	..	16 608	15 667	..	Etoffes et articles de bonneterie			173
..	26 459	26 949	29 636	..	51 043	50 196	..	Articles d'habillement et fourrure		18	
..	26 259	26 758	29 495	..	50 287	49 265	..	Articles d'habillement autres qu'en fourrure			181
..	200	192	141	..	756	930	..	Prép. et teinture des fourrures ; articles en fourrure			182
..	20 507	22 974	27 942	..	46 062	44 424	..	Cuir et articles en cuir ; chaussures		19	
..	8 510	9 392	11 696	..	18 801	19 947	..	Cuir et articles en cuir sauf chaussures			191
..	11 997	13 582	16 245	..	27 260	24 477	..	Chaussures			192
..	8 107	8 546	9 756	..	24 613	26 618	..	Bois, articles en bois, sauf meubles		20	
..	1 011	966	1 078	..	4 100	4 648	..	Sciage et rabotage du bois			201
..	7 096	7 579	8 678	..	20 513	21 970	..	Articles en bois, liège, vannerie, sparterie			202
..	16 128	16 989	20 898	..	30 074	29 974	..	Papier, articles en papier et carton		21	
..	19 381	19 806	20 981	..	38 000	40 146	..	Edition, imprimerie et reproduction		22	
..	10 800	11 138	11 686	..	18 380	20 223	..	Edition			221
..	8 142	8 292	9 024	..	19 023	19 272	..	Imprimerie et activités annexes			222
..	439	377	271	..	597	650	..	Reproduction de supports enregistrés			223
..	26 802	34 177	66 329	..	71 303	75 428	..	Coke. et prod. pétroliers ; comb. nucléaires		23	
..	..	..	..	..	..	..	..	Cokéfaction			231
..	26 644	34 031	66 195	..	..	75 295	..	Produits pétroliers raffinés			232
..	..	..	..	..	..	..	..	Traitement de combustibles nucléaires			233
..	64 272	67 265	72 832	..	112 550	114 680	..	Produits chimiques		24	
..	20 205	22 063	25 949	..	35 229	38 967	..	Produits chimiques de base			241
..	40 724	41 527	43 274	..	72 832	71 395	..	Autres produits chimiques			242
..	3 343	3 675	3 609	..	4 488	4 318	..	Fibres synthétiques ou artificielles			243
..	22 242	24 423	29 119	..	51 566	52 427	..	Art. en caoutchouc et matières plastiques		25	
..	5 752	6 028	7 458	..	11 350	11 771	..	Articles en caoutchouc			251
..	16 490	18 395	21 662	..	40 216	40 656	..	Articles en matières plastiques			252
..	32 953	31 742	33 340	..	51 653	55 456	..	Autres produits minéraux non métalliques		26	
..	5 350	5 321	5 919	..	8 497	8 377	..	Verre et articles en verre			261
..	27 603	26 421	27 421	..	43 156	47 080	..	Produits minéraux non métalliques, nca			269
..	38 985	35 068	48 605	..	57 747	67 615	..	Produits métallurgiques de base		27	
..	26 232	22 217	33 105	..	36 476	42 790	..	Sidérurgie et première transformation de l'acier			271
..	8 207	8 278	9 354	..	11 701	14 779	..	Première transf. métaux précieux et non ferreux			272
..	4 547	4 573	6 146	..	9 570	10 046	..	Fonderie			273
..	41 296	41 734	46 291	..	104 830	110 648	..	Ouvrages en métaux, sauf machines		28	
..	15 622	14 982	14 574	..	31 881	32 286	..	Constr. et menuis. métal.; réservoirs; générateurs			281
..	25 674	26 752	31 718	..	72 946	78 362	..	Autres ouvrages en métaux			289
..	75 287	78 414	90 820	..	151 320	158 881	..	Machines et matériels, nca		29	
..	28 581	29 040	35 174	..	71 335	75 506	..	Machines d'usage général			291
..	35 312	36 971	39 862	..	60 907	63 949	..	Machines d'usage spécifique			292
..	11 395	12 403	15 784	..	19 076	19 426	..	Appareils domestiques, nca			293
..	6 797	7 351	8 046	..	11 828	12 861	..	Mach. de bureau, comptables, information		30	
..	32 060	28 811	31 405	..	46 524	51 056	..	Machines et appareils électriques, nca		31	
..	6 226	6 704	6 470	..	9 967	9 035	..	Moteurs, génératrices, transformateurs électriques			311
..	4 576	4 362	5 417	..	9 377	10 876	..	Matériel électrique de distribution et de commande			312
..	4 758	4 040	4 370	..	4 748	5 779	..	Fils et câbles électriques isolés			313
..	918	824	1 015	..	1 797	1 413	..	Accumulateurs et piles électriques			314
..	2 166	2 441	2 680	..	4 426	4 996	..	Lampes électriques et appareils d'éclairage			315
..	13 415	10 440	11 454	..	16 209	18 958	..	Autres matériels électriques, nca.			319
..	15 274	15 858	17 873	..	26 224	29 645	..	Equipements et appareils de radio, TV, etc.		32	
..	3 658	4 420	5 380	..	8 242	7 964	..	Tubes et valves électroniques;autres composants			321
..	10 076	9 964	10 475	..	15 554	18 952	..	Emetteurs radio/TV; app. téléphonie et télégraphie			322
..	1 540	1 474	2 018	..	2 429	2 729	..	Récepteurs radio/TV, son et image; art. associés			323
..	9 308	10 654	11 127	..	21 757	19 210	..	Instr. médicaux, de précision ; optique ; horlog.		33	
..	7 185	8 466	8 620	..	16 208	15 105	..	Instruments médicaux et de précision, etc.			331
..	1 838	1 950	2 243	..	5 008	3 906	..	Instruments d'optique ; matériel photographique			332
..	284	238	264	..	540	199	..	Horlogerie			333
..	39 197	31 729	41 531	..	69 710	80 300	..	Véhicules automobiles; remorques, etc.		34	
..	26 824	19 754	27 185	..	46 113	53 804	..	Véhicules automobiles			341
..	2 151	2 115	2 452	..	4 146	4 045	..	Carross. automobiles, remorques, semi-remorques			342
..	10 222	9 861	11 894	..	19 451	22 451	..	Pièces et accessoires pour automobiles et moteurs			343
..	17 051	17 303	18 190	..	24 979	25 490	..	Autres matériels de transport		35	
..	4 333	4 109	4 424	..	6 236	7 637	..	Construction et réparation de navires			351
..	2 818	3 498	3 263	..	2 075	2 480	..	Matériel ferroviaire roulant			352
..	6 266	5 860	5 981	..	9 715	9 284	..	Construction aéronautique et spatiale			353
..	3 634	3 836	4 521	..	6 953	6 090	..	Autres équipements de transport			359
..	24 943	26 326	30 210	..	56 682	55 106	..	Meubles ; activités de fabrication, nca		36	
..	17 161	17 136	19 959	..	38 440	36 686	..	Meubles			361
..	7 782	9 190	10 251	..	18 242	18 420	..	Activités de fabrication, nca			369
..	446	497	616	..	2 064	2 109	..	Récupération		37	
..	64 081	67 804	73 562	..	80 532	85 356	..	Electricité, gaz et eau	E	40_41	
..	60 923	64 533	70 124	..	76 726	81 292	..	Prod. et dist. d'électricité gaz, vapeur d'eau chaude		40	
..	3 158	3 272	3 438	..	3 806	4 064	..	Captage, épuration et distribution d'eau		41	
..	82 951	76 870	64 738	..	193 160	212 061	..	Construction	F	45	

Table IT.2 VALUE ADDED - VALEUR AJOUTÉE

Billions of L (Current Prices)

ISIC revision 3				1991	1992	1993	1994	1995	1996	1997	1998
C	10_14		Mining and quarrying	..	4 664	4 977	5 529	..	..	7 075	..
	10		Mining of coal and lignite; extraction of peat	..	..	..	..	..	..	1	..
	11		Crude petroleum & natural gas; service activities	..	3 556	3 937	4 514	..	..	4 611	..
	13		Mining of metal ores	..	14	16	5	..	..	..	..
	14		Other mining and quarrying	..	1 093	1 024	1 010	..	2 355	..	..
D	15_37		Total manufacturing	..	209 926	214 355	233 824	..	..	..	..
	15		Food products and beverages	..	19 546	20 704	20 377	..	31 503	30 347	..
		151	Production of meat, fish, fruit, vegetables, oils & fats	..	5 737	5 948	5 388	..	7 741	7 620	..
		152	Dairy products	..	2 920	2 895	3 442	..	4 808	4 570	..
		153	Grain mill and starch products; animal feeds	..	1 488	1 492	1 520	..	2 419	2 135	..
		154	Other food products	..	6 438	6 801	6 489	..	11 975	11 822	..
		155	Beverages	..	2 963	3 567	3 538	..	4 561	4 201	..
	16		Tobacco products	..	829	643	1 147	..	1 188	1 080	..
	17		Textiles	..	13 125	13 901	15 506	..	21 534	21 039	..
		171	Spinning, weaving and finishing of textiles	..	8 408	8 865	10 412	..	13 094	13 060	..
		172	Other textiles	..	1 798	1 845	1 901	..	3 877	3 774	..
		173	Knitted and crocheted fabrics and articles	..	2 919	3 191	3 194	..	4 562	4 205	..
	18		Wearing apparel and fur	..	8 535	8 496	9 134	..	15 907	14 852	..
		181	Wearing apparel, except fur apparel	..	8 446	8 410	9 065	..	15 687	14 642	..
		182	Dressing and dyeing of fur; articles of fur	..	89	86	70	..	220	210	..
	19		Leather and articles of leather; footwear	..	5 708	6 216	6 869	..	11 873	10 584	..
		191	Leather and articles of leather	..	2 205	2 257	2 536	..	4 463	3 973	..
		192	Footwear	..	3 502	3 959	4 332	..	7 410	6 611	..
	20		Wood & products of wood & cork, exc. furniture	..	2 544	2 673	2 866	..	7 896	7 745	..
		201	Sawmilling and planing of wood	..	279	291	323	..	946	1 032	..
		202	Wood products; articles of cork, straw, etc.	..	2 266	2 383	2 543	..	6 950	6 713	..
	21		Paper and paper products	..	4 980	5 241	6 163	..	8 629	8 037	..
	22		Printing, publishing and recorded media	..	8 570	8 689	8 307	..	13 455	13 438	..
		221	Publishing	..	4 863	4 884	4 466	..	5 293	5 681	..
		222	Printing and related activities	..	3 476	3 638	3 762	..	8 004	7 605	..
		223	Reproduction of recorded media	..	231	167	79	..	158	152	..
	23		Coke and petroleum products; nuclear fuel	..	3 273	5 002	5 794	..	4 466	4 678	..
		231	Coke oven products	..	..	..	..	..	..	..	..
		232	Refined petroleum products	..	3 210	4 945	5 750	..	..	4 644	..
		233	Processing of nuclear fuel	..	..	..	..	..	..	..	..
	24		Chemicals and chemical products	..	20 687	21 840	23 653	..	29 045	28 688	..
		241	Basic chemicals	..	5 245	5 799	6 871	..	7 807	8 346	..
		242	Other chemical products	..	14 368	14 822	15 574	..	20 092	19 237	..
		243	Man-made fibres	..	1 075	1 219	1 209	..	1 146	1 106	..
	25		Rubber and plastic products	..	8 359	9 125	10 735	..	16 749	16 083	..
		251	Rubber products	..	2 794	2 954	3 826	..	3 968	3 968	..
		252	Plastic products	..	5 565	6 171	6 909	..	12 781	12 115	..
	26		Other non-metallic mineral products	..	13 319	12 895	13 441	..	18 390	18 518	..
		261	Glass and glass products	..	2 184	2 150	2 586	..	3 543	3 113	..
		269	Non-metallic mineral products, nec	..	11 136	10 745	10 854	..	14 848	15 405	..
	27		Basic metals	..	9 290	8 209	11 823	..	14 277	15 287	..
		271	Basic iron and steel	..	5 976	4 884	7 799	..	8 562	9 195	..
		272	Basic precious and non-ferrous metals	..	1 510	1 597	1 961	..	2 413	2 561	..
		273	Casting of metals	..	1 803	1 728	2 062	..	3 303	3 532	..
	28		Fabricated metal products, exc. machinery	..	15 425	15 689	16 820	..	41 106	42 449	..
		281	Structural metal prod., tanks, reservoirs, etc.	..	5 491	5 344	5 014	..	11 649	11 272	..
		289	Other fabricated metal products	..	9 934	10 346	11 806	..	29 456	31 177	..
	29		Machinery and equipment, nec	..	26 496	27 722	30 951	..	48 479	48 246	..
		291	General purpose machinery	..	10 360	10 594	12 997	..	24 303	24 268	..
		292	Special purpose machinery	..	12 458	13 040	13 199	..	18 795	18 823	..
		293	Domestic appliances, nec	..	3 678	4 087	4 755	..	5 380	5 155	..
	30		Office, accounting & computing machinery	..	1 588	1 234	765	..	2 087	2 644	..
	31		Electrical machinery & apparatus, nec	..	12 860	11 078	11 046	..	14 770	16 113	..
		311	Electric motors, generators and transformers	..	2 554	2 324	2 135	..	2 784	2 942	..
		312	Electricity distribution & control apparatus	..	2 092	2 094	2 329	..	3 258	3 593	..
		313	Insulated wire and cable	..	1 331	1 179	1 141	..	1 156	1 225	..
		314	Accumulators, primary cells & primary batteries	..	291	215	245	..	547	397	..
		315	Electric lamps and lighting equipment	..	788	878	867	..	1 270	1 413	..
		319	Other electrical equipment, nec	..	5 803	4 387	4 329	..	5 755	6 544	..
	32		Radio, TV, communication equipment	..	6 261	6 965	7 361	..	8 406	8 737	..
		321	Electronic valves, tubes; other electronic components	..	1 495	1 925	2 130	..	2 914	2 657	..
		322	TV/radio transmitters; line teleph./telegr. apparatus	..	4 388	4 662	4 678	..	4 908	5 429	..
		323	TV/radio receivers; sound and video app.; assoc. gds.	..	378	377	554	..	584	651	..
	33		Medical, precision & optical instrum.; watch & clocks	..	4 385	4 943	5 124	..	8 525	7 630	..
		331	Medical appl.; instruments & appl. for measuring, etc.	..	3 459	3 990	4 075	..	6 554	6 142	..
		332	Optical instruments and photographic equipment	..	848	886	974	..	1 828	1 419	..
		333	Watches and clocks	..	78	68	75	..	143	69	..
	34		Motor vehicles, trailers and semi-trailers	..	10 873	8 562	11 210	..	14 455	17 792	..
		341	Motor vehicles	..	6 610	4 452	6 325	..	7 047	9 754	..
		342	Bodies for motor vehicles, trailers & semi-trailers	..	643	657	743	..	1 239	1 002	..
		343	Parts & accessories for motor vehicles & their engines	..	3 620	3 453	4 143	..	6 169	7 035	..
	35		Other transport equipment	..	5 914	6 698	6 466	..	7 700	7 559	..
		351	Building and repairing of ships and boats	..	962	985	751	..	2 227	1 767	..
		352	Railway and tramway locomotives and rolling stock	..	1 228	1 885	1 713	..	689	807	..
		353	Aircraft and spacecraft	..	2 577	2 687	2 618	..	3 070	3 158	..
		359	Transport equipment, nec	..	1 147	1 140	1 384	..	1 713	1 827	..
	36		Furniture; manufacturing, nec	..	7 280	7 713	8 066	..	15 641	15 092	..
		361	Furniture	..	5 111	5 354	5 586	..	10 715	10 425	..
		369	Manufacturing, nec	..	2 169	2 359	2 480	..	4 927	4 668	..
	37		Recycling	..	78	116	201	..	397	523	..
E	40_41		Electricity, gas and water supply	..	28 291	28 945	33 433	..	37 776	36 789	..
	40		Electricity, gas, steam and hot water supply	..	27 021	27 567	31 993	..	36 008	35 080	..
	41		Collection, purification and distribution of water	..	1 270	1 378	1 440	..	1 768	1 710	..
F	45		Construction	..	25 816	24 285	20 951	..	68 082	67 732	..

Table IT.3 **INVESTISSEMENT - INVESTMENT**

Milliards de L (Prix courants)

1991	1992	1993	1994	1995	1996	1997	1998	CITI révision 3			
..	1 667	2 177	1 719	..	911	1 863	..	Activites extractives	C	10_14	
..	..	..	..	..	..	0	..	Extraction de charbon, de lignite et de tourbe		10	
..	1 384	1 947	1 558	..	..	1 020	..	Pétrole brut et gaz naturel ; activités annexes		11	
..	8	11	2	..	..	..	..	Extraction de minerais métalliques		13	
..	274	219	159	..	711	640	..	Autres activités extractives		14	
..	35 287	32 828	34 369	..	65 414	59 945	..	Total activites manufacturieres	D	15_37	
..	4 065	3 827	3 935	..	6 427	6 503	..	Produits alimentaires et boissons		15	
..	1 248	1 220	1 204	..	1 775	1 772	..	Viande, poisson, fruits, légumes, huiles et graisses			151
..	559	589	638	..	896	1 008	..	Produits laitiers			152
..	349	313	303	..	621	522	..	Travail des grains; prod. amyl. ; alim. pour animaux			153
..	1 146	1 020	1 112	..	2 141	2 022	..	Autres produits alimentaires			154
..	763	685	678	..	994	1 179	..	Boissons			155
..	222	84	68	..	59	59	..	Produits à base de tabac		16	
..	1 735	1 723	2 066	..	3 843	3 337	..	Textiles		17	
..	1 148	1 162	1 473	..	2 334	2 065	..	Filature, tissage et achèvement des textiles			171
..	273	294	263	..	556	590	..	Autres articles textiles			172
..	314	267	330	..	953	682	..	Etoffes et articles de bonneterie			173
..	538	502	606	..	1 573	1 666	..	Articles d'habillement et fourrure		18	
..	536	499	603	..	1 552	1 645	..	Articles d'habillement autres qu'en fourrure			181
..	2	3	2	..	21	21	..	Prép. et teinture des fourrures ; articles en fourrure			182
..	486	470	615	..	1 393	1 548	..	Cuir et articles en cuir ; chaussures		19	
..	204	156	222	..	596	502	..	Cuir et articles en cuir sauf chaussures			191
..	282	314	393	..	797	1 046	..	Chaussures			192
..	473	376	372	..	1 700	1 784	..	Bois, articles en bois, sauf meubles		20	
..	45	28	32	..	170	388	..	Sciage et rabotage du bois			201
..	428	348	340	..	1 530	1 396	..	Articles en bois, liège, vannerie, sparterie			202
..	1 097	1 185	1 063	..	2 084	2 154	..	Papier, articles en papier et carton		21	
..	702	822	842	..	1 885	1 727	..	Edition, imprimerie et reproduction		22	
..	305	400	281	..	478	394	..	Edition			221
..	390	414	550	..	1 385	1 315	..	Imprimerie et activités annexes			222
..	6	8	10	..	22	18	..	Reproduction de supports enregistrés			223
..	1 584	1 761	1 227	..	1 244	1 471	..	Coke. et prod. pétroliers ; comb. nucléaires		23	
..	..	..	..	..	..	..	..	Cokéfaction			231
..	1 576	1 752	1 221	..	..	1 470	..	Produits pétroliers raffinés			232
..	..	..	..	..	..	..	..	Traitement de combustibles nucléaires			233
..	3 900	3 236	3 191	..	5 511	5 385	..	Produits chimiques		24	
..	1 696	1 142	933	..	1 298	2 073	..	Produits chimiques de base			241
..	1 944	1 859	1 990	..	4 063	3 128	..	Autres produits chimiques			242
..	260	235	268	..	150	184	..	Fibres synthétiques ou artificielles			243
..	1 360	1 434	1 535	..	3 773	3 050	..	Art. en caoutchouc et matières plastiques		25	
..	288	339	385	..	809	549	..	Articles en caoutchouc			251
..	1 072	1 096	1 151	..	2 965	2 502	..	Articles en matières plastiques			252
..	2 588	2 299	2 387	..	3 754	3 581	..	Autres produits minéraux non métalliques		26	
..	462	361	410	..	782	906	..	Verre et articles en verre			261
..	2 126	1 939	1 976	..	2 972	2 675	..	Produits minéraux non métalliques, nca			269
..	2 870	1 458	2 027	..	2 459	2 784	..	Produits métallurgiques de base		27	
..	2 208	867	1 357	..	1 313	1 470	..	Sidérurgie et première transformation de l'acier			271
..	348	307	314	..	489	549	..	Première transf. métaux précieux et non ferreux			272
..	314	284	356	..	658	766	..	Fonderie			273
..	2 139	1 972	2 318	..	7 712	7 077	..	Ouvrages en métaux, sauf machines		28	
..	674	497	466	..	1 674	1 863	..	Constr. et menuis. métal.; réservoirs; générateurs			281
..	1 465	1 475	1 853	..	6 038	5 214	..	Autres ouvrages en métaux			289
..	2 528	2 544	3 086	..	9 739	6 034	..	Machines et matériels, nca		29	
..	1 047	959	1 314	..	3 348	3 291	..	Machines d'usage général			291
..	998	930	1 081	..	2 787	2 172	..	Machines d'usage spécifique			292
..	484	654	690	..	3 604	571	..	Appareils domestiques, nca			293
..	247	214	195	..	292	401	..	Mach. de bureau, comptables, information		30	
..	1 361	1 205	1 195	..	2 592	2 145	..	Machines et appareils électriques, nca		31	
..	287	217	210	..	376	428	..	Moteurs, génératrices, transformateurs électriques			311
..	200	198	224	..	424	451	..	Matériel électrique de distribution et de commande			312
..	179	153	117	..	213	259	..	Fils et câbles électriques isolés			313
..	52	51	38	..	50	47	..	Accumulateurs et piles électriques			314
..	104	96	120	..	187	204	..	Lampes électriques et appareils d'éclairage			315
..	541	490	486	..	1 342	756	..	Autres matériels électriques, nca.			319
..	936	1 012	1 109	..	1 724	1 876	..	Equipements et appareils de radio, TV, etc.		32	
..	371	563	710	..	962	1 154	..	Tubes et valves électroniques;autres composants			321
..	512	390	353	..	682	630	..	Emetteurs radio/TV; app. téléphonie et télégraphie			322
..	53	60	46	..	80	93	..	Récepteurs radio/TV, son et image; art. associés			323
..	405	385	418	..	947	755	..	Instr. médicaux, de précision ; optique ; horlog.		33	
..	315	299	314	..	708	575	..	Instruments médicaux et de précision, etc.			331
..	84	83	99	..	218	173	..	Instruments d'optique ; matériel photographique			332
..	5	3	5	..	22	8	..	Horlogerie			333
..	3 599	3 726	4 019	..	2 699	3 211	..	Véhicules automobiles; remorques, etc.		34	
..	2 905	2 977	3 124	..	1 166	1 791	..	Véhicules automobiles			341
..	72	98	101	..	122	102	..	Carross. automobiles, remorques, semi-remorques			342
..	622	651	794	..	1 410	1 318	..	Pièces et accessoires pour automobiles et moteurs			343
..	1 606	1 755	1 157	..	1 446	952	..	Autres matériels de transport		35	
..	181	95	100	..	273	372	..	Construction et réparation de navires			351
..	677	1 030	563	..	87	80	..	Matériel ferroviaire roulant			352
..	538	394	307	..	531	181	..	Construction aéronautique et spatiale			353
..	210	236	188	..	556	319	..	Autres équipements de transport			359
..	825	805	905	..	2 436	2 297	..	Meubles ; activités de fabrication, nca		36	
..	583	570	637	..	1 772	1 576	..	Meubles			361
..	242	236	268	..	664	721	..	Activités de fabrication, nca			369
..	21	32	31	..	121	149	..	Récupération		37	
..	15 340	13 668	11 680	..	8 465	11 337	..	Electricité, gaz et eau	E	40_41	
..	14 864	13 159	11 231	..	7 868	10 771	..	Prod. et dist. d'électricité gaz, vapeur d'eau chaude		40	
..	476	509	448	..	596	566	..	Captage, épuration et distribution d'eau		41	
..	2 088	1 941	2 056	..	16 245	9 532	..	Construction	F	45	

ITALY

Table IT.4 INVESTMENT IN MACHINERY & EQUIPMENT - DÉPENSES DE MACHINES ET D'ÉQUIPEMENT

Billions of L (Current Prices)

ISIC revision 3				1991	1992	1993	1994	1995	1996	1997	1998
C	10_14		Mining and quarrying	..	1 604	2 100	1 691	..	..	1 619	..
	10		Mining of coal and lignite; extraction of peat	..	..	..	..	..	..	0	..
	11		Crude petroleum & natural gas; service activities	..	1 376	1 930	1 535	..	..	1 017	..
	13		Mining of metal ores	..	6	10	2	..	..	..	..
	14		Other mining and quarrying	..	222	159	154	..	..	443	..
D	15_37		Total manufacturing	..	28 378	26 188	29 230	..	..	49 081	..
	15		Food products and beverages	..	3 146	2 960	3 205	..	..	5 122	..
		151	Production of meat, fish, fruit, vegetables, oils & fats	..	930	873	909	..	..	1 292	..
		152	Dairy products	..	450	465	552	..	..	782	..
		153	Grain mill and starch products; animal feeds	..	257	237	234	..	..	383	..
		154	Other food products	..	904	849	931	..	..	1 750	..
		155	Beverages	..	605	536	579	..	..	915	..
	16		Tobacco products	..	153	39	48	..	..	43	..
	17		Textiles	..	1 400	1 380	1 768	..	..	2 751	..
		171	Spinning, weaving and finishing of textiles	..	966	943	1 272	..	..	1 730	..
		172	Other textiles	..	201	237	222	..	..	463	..
		173	Knitted and crocheted fabrics and articles	..	233	200	275	..	..	558	..
	18		Wearing apparel and fur	..	362	349	439	..	..	1 176	..
		181	Wearing apparel, except fur apparel	..	360	347	436	..	..	1 160	..
		182	Dressing and dyeing of fur; articles of fur	..	2	2	2	..	..	15	..
	19		Leather and articles of leather; footwear	..	375	370	523	..	..	1 213	..
		191	Leather and articles of leather	..	140	114	169	..	..	382	..
		192	Footwear	..	234	256	354	..	..	830	..
	20		Wood & products of wood & cork, exc. furniture	..	354	292	330	..	..	1 336	..
		201	Sawmilling and planing of wood	..	34	23	34	..	..	230	..
		202	Wood products; articles of cork, straw, etc.	..	320	270	296	..	..	1 105	..
	21		Paper and paper products	..	922	1 059	943	..	..	1 799	..
	22		Printing, publishing and recorded media	..	522	572	670	..	..	1 394	..
		221	Publishing	..	217	220	213	..	..	352	..
		222	Printing and related activities	..	299	345	450	..	..	1 025	..
		223	Reproduction of recorded media	..	6	7	7	..	..	16	..
	23		Coke and petroleum products; nuclear fuel	..	1 439	1 703	1 125	..	..	1 386	..
		231	Coke oven products	..	..	..	..	..	..	..	..
		232	Refined petroleum products	..	1 432	1 695	1 119	..	..	1 385	..
		233	Processing of nuclear fuel	..	..	..	..	..	..	..	..
	24		Chemicals and chemical products	..	3 216	2 668	2 725	..	..	4 800	..
		241	Basic chemicals	..	1 513	1 018	841	..	..	1 884	..
		242	Other chemical products	..	1 472	1 446	1 639	..	..	2 745	..
		243	Man-made fibres	..	231	204	245	..	..	171	..
	25		Rubber and plastic products	..	1 168	1 209	1 355	..	..	2 597	..
		251	Rubber products	..	262	292	364	..	..	500	..
		252	Plastic products	..	906	917	991	..	..	2 097	..
	26		Other non-metallic mineral products	..	2 093	1 872	2 128	..	..	2 841	..
		261	Glass and glass products	..	397	328	360	..	..	762	..
		269	Non-metallic mineral products, nec	..	1 696	1 545	1 767	..	..	2 079	..
	27		Basic metals	..	2 453	1 289	1 833	..	..	2 390	..
		271	Basic iron and steel	..	1 876	750	1 159	..	..	1 254	..
		272	Basic precious and non-ferrous metals	..	311	282	266	..	..	485	..
		273	Casting of metals	..	267	257	408	..	..	651	..
	28		Fabricated metal products, exc. machinery	..	1 639	1 588	2 055	..	..	5 527	..
		281	Structural metal prod., tanks, reservoirs, etc.	..	454	372	408	..	..	1 336	..
		289	Other fabricated metal products	..	1 185	1 216	1 647	..	..	4 191	..
	29		Machinery and equipment, nec	..	2 015	2 001	2 575	..	..	4 765	..
		291	General purpose machinery	..	849	778	1 108	..	..	2 619	..
		292	Special purpose machinery	..	741	721	866	..	..	1 620	..
		293	Domestic appliances, nec	..	425	502	601	..	..	527	..
	30		Office, accounting & computing machinery	..	222	194	182	..	..	360	..
	31		Electrical machinery & apparatus, nec	..	1 085	956	1 031	..	..	1 727	..
		311	Electric motors, generators and transformers	..	217	196	195	..	..	373	..
		312	Electricity distribution & control apparatus	..	167	152	203	..	..	365	..
		313	Insulated wire and cable	..	159	130	106	..	..	222	..
		314	Accumulators, primary cells & primary batteries	..	45	29	33	..	..	43	..
		315	Electric lamps and lighting equipment	..	87	74	95	..	..	163	..
		319	Other electrical equipment, nec	..	409	376	399	..	..	562	..
	32		Radio, TV, communication equipment	..	801	903	1 008	..	..	1 690	..
		321	Electronic valves, tubes; other electronic components	..	356	530	648	..	..	1 053	..
		322	TV/radio transmitters; line teleph./telegr. apparatus	..	402	342	325	..	..	573	..
		323	TV/radio receivers; sound and video app.; assoc. gds.	..	43	31	35	..	..	64	..
	33		Medical, precision & optical instrum.; watch & clocks	..	340	317	359	..	..	612	..
		331	Medical appl.; instruments & appl. for measuring, etc.	..	262	253	269	..	..	443	..
		332	Optical instruments and photographic equipment	..	73	61	85	..	..	161	..
		333	Watches and clocks	..	5	3	5	..	..	7	..
	34		Motor vehicles, trailers and semi-trailers	..	3 349	3 245	3 661	..	..	2 952	..
		341	Motor vehicles	..	2 752	2 585	2 875	..	..	1 704	..
		342	Bodies for motor vehicles, trailers & semi-trailers	..	44	70	81	..	..	74	..
		343	Parts & accessories for motor vehicles & their engines	..	553	590	706	..	..	1 174	..
	35		Other transport equipment	..	680	595	531	..	..	807	..
		351	Building and repairing of ships and boats	..	121	77	81	..	..	316	..
		352	Railway and tramway locomotives and rolling stock	..	90	91	53	..	..	72	..
		353	Aircraft and spacecraft	..	316	217	231	..	..	153	..
		359	Transport equipment, nec	..	153	209	165	..	..	265	..
	36		Furniture; manufacturing, nec	..	626	605	705	..	..	1 689	..
		361	Furniture	..	436	417	481	..	..	1 107	..
		369	Manufacturing, nec	..	190	188	223	..	..	582	..
	37		Recycling	..	17	19	32	..	..	106	..
E	40_41		Electricity, gas and water supply	..	9 656	8 743	7 390	..	..	10 753	..
	40		Electricity, gas, steam and hot water supply	..	9 240	8 392	7 037	..	..	10 211	..
	41		Collection, purification and distribution of water	..	416	351	353	..	..	542	..
F	45		Construction	..	1 556	1 119	1 681	..	..	6 770	..

Table IT.5 **EMPLOI - EMPLOYMENT**

Unités

1991	1992	1993	1994	1995	1996	1997	1998	CITI révision 3			
..	24 077	23 828	22 134	..	35 508	35 933	..	Activites extractives	C	10_14	
..	..	..	..	..	..	14	..	Extraction de charbon, de lignite et de tourbe		10	
..	9 070	8 354	7 994	..	..	6 986	..	Pétrole brut et gaz naturel ; activités annexes		11	
..	511	778	697	..	..	..	..	Extraction de minerais métalliques		13	
..	14 496	14 696	13 443	..	28 901	24 481	..	Autres activités extractives		14	
..	3 043 817	2 920 620	2 887 601	..	4 847 292	4 817 085	..	Total activites manufacturieres	D	15_37	
..	222 732	219 970	217 085	..	425 794	418 411	..	Produits alimentaires et boissons		15	
..	81 178	79 278	74 638	..	100 348	98 099	..	Viande, poisson, fruits, légumes, huiles et graisses			151
..	30 983	30 881	32 961	..	59 576	53 970	..	Produits laitiers			152
..	15 193	14 382	13 533	..	21 904	21 988	..	Travail des grains; prod. amyl. ; alim. pour animaux			153
..	67 191	65 648	66 728	..	203 950	204 720	..	Autres produits alimentaires			154
..	28 187	29 781	29 225	..	40 016	39 634	..	Boissons			155
..	16 964	16 238	14 144	..	13 814	12 535	..	Produits à base de tabac		16	
..	221 461	215 387	216 250	..	333 253	324 488	..	Textiles		17	
..	135 638	131 505	132 780	..	177 300	174 276	..	Filature, tissage et achèvement des textiles			171
..	30 274	29 955	28 907	..	63 764	63 164	..	Autres articles textiles			172
..	55 549	53 927	54 563	..	92 189	87 048	..	Etoffes et articles de bonneterie			173
..	191 797	185 072	174 728	..	363 471	346 963	..	Articles d'habillement et fourrure		18	
..	190 021	183 405	173 276	..	355 723	340 396	..	Articles d'habillement autres qu'en fourrure			181
..	1 776	1 667	1 452	..	7 748	6 567	..	Prép. et teinture des fourrures ; articles en fourrure			182
..	126 298	121 536	121 165	..	229 837	226 307	..	Cuir et articles en cuir ; chaussures		19	
..	36 850	32 602	32 212	..	70 489	69 507	..	Cuir et articles en cuir sauf chaussures			191
..	89 448	88 934	88 953	..	159 348	156 800	..	Chaussures			192
..	42 813	42 162	43 192	..	172 476	170 173	..	Bois, articles en bois, sauf meubles		20	
..	4 064	4 155	4 520	..	17 588	19 125	..	Sciage et rabotage du bois			201
..	38 749	38 007	38 672	..	154 888	151 048	..	Articles en bois, liège, vannerie, sparterie			202
..	64 326	63 347	63 828	..	83 503	83 029	..	Papier, articles en papier et carton		21	
..	91 688	90 998	87 367	..	176 066	172 726	..	Edition, imprimerie et reproduction		22	
..	38 846	37 564	35 755	..	50 002	48 922	..	Edition			221
..	51 873	52 408	50 739	..	124 833	122 257	..	Imprimerie et activités annexes			222
..	969	1 026	873	..	1 231	1 547	..	Reproduction de supports enregistrés			223
..	23 273	21 324	25 245	..	24 764	24 161	..	Coke. et prod. pétroliers ; comb. nucléaires		23	
..	..	..	..	..	..	..	..	Cokéfaction			231
..	22 462	20 632	24 612	..	..	23 790	..	Produits pétroliers raffinés			232
..	..	..	..	..	..	..	..	Traitement de combustibles nucléaires			233
..	206 879	199 279	189 598	..	212 511	213 219	..	Produits chimiques		24	
..	62 303	60 878	56 889	..	54 831	56 885	..	Produits chimiques de base			241
..	130 139	124 191	120 449	..	146 549	145 543	..	Autres produits chimiques			242
..	14 437	14 210	12 260	..	11 131	10 791	..	Fibres synthétiques ou artificielles			243
..	112 449	112 839	117 139	..	196 280	201 869	..	Art. en caoutchouc et matières plastiques		25	
..	34 914	32 252	33 845	..	44 929	46 911	..	Articles en caoutchouc			251
..	77 535	80 587	83 294	..	151 351	154 958	..	Articles en matières plastiques			252
..	166 344	159 138	152 588	..	247 932	242 533	..	Autres produits minéraux non métalliques		26	
..	31 327	29 921	28 053	..	44 547	41 151	..	Verre et articles en verre			261
..	135 017	129 217	124 535	..	203 385	201 382	..	Produits minéraux non métalliques, nca			269
..	147 988	114 301	126 881	..	137 619	142 709	..	Produits métallurgiques de base		27	
..	95 337	64 722	77 287	..	75 794	77 512	..	Sidérurgie et première transformation de l'acier			271
..	22 853	22 429	22 045	..	21 739	24 281	..	Première transf. métaux précieux et non ferreux			272
..	29 798	27 150	27 549	..	40 086	40 916	..	Fonderie			273
..	246 278	241 466	242 043	..	611 326	620 786	..	Ouvrages en métaux, sauf machines		28	
..	90 776	86 831	81 929	..	192 785	190 466	..	Constr. et menuis. métal.; réservoirs; générateurs			281
..	155 502	154 635	160 114	..	418 541	430 320	..	Autres ouvrages en métaux			289
..	375 215	367 990	371 241	..	550 053	568 589	..	Machines et matériels, nca		29	
..	142 204	141 770	151 380	..	272 513	287 118	..	Machines d'usage général			291
..	182 363	174 548	164 988	..	214 911	220 742	..	Machines d'usage spécifique			292
..	50 648	51 672	54 873	..	62 629	60 729	..	Appareils domestiques, nca			293
..	23 249	19 651	16 494	..	21 632	24 757	..	Mach. de bureau, comptables, information		30	
..	173 734	150 813	139 806	..	200 014	207 418	..	Machines et appareils électriques, nca		31	
..	38 388	33 878	28 833	..	38 542	37 636	..	Moteurs, génératrices, transformateurs électriques			311
..	28 068	27 498	28 730	..	41 198	42 019	..	Matériel électrique de distribution et de commande			312
..	15 838	14 418	12 752	..	11 538	13 789	..	Fils et câbles électriques isolés			313
..	3 449	3 123	3 113	..	4 958	4 280	..	Accumulateurs et piles électriques			314
..	11 509	10 889	9 856	..	18 167	18 517	..	Lampes électriques et appareils d'éclairage			315
..	76 482	61 007	56 522	..	85 611	91 177	..	Autres matériels électriques, nca.			319
..	83 672	85 232	89 238	..	107 073	105 784	..	Equipements et appareils de radio, TV, etc.		32	
..	23 645	23 735	24 755	..	32 241	29 978	..	Tubes et valves électroniques;autres composants			321
..	52 366	55 446	57 301	..	66 163	66 787	..	Emetteurs radio/TV; app. téléphonie et télégraphie			322
..	7 661	6 051	7 182	..	8 669	9 019	..	Récepteurs radio/TV, son et image; art. associés			323
..	58 067	61 031	60 762	..	125 744	115 837	..	Instr. médicaux, de précision ; optique ; horlog.		33	
..	43 913	46 952	46 499	..	97 177	92 734	..	Instruments médicaux et de précision, etc.			331
..	12 942	13 190	13 560	..	27 358	22 029	..	Instruments d'optique ; matériel photographique			332
..	1 212	889	703	..	1 209	1 074	..	Horlogerie			333
..	201 242	185 576	181 178	..	192 841	190 946	..	Véhicules automobiles; remorques, etc.		34	
..	129 570	120 838	115 953	..	103 874	101 086	..	Véhicules automobiles			341
..	14 182	12 880	12 922	..	15 967	14 654	..	Carross. automobiles, remorques, semi-remorques			342
..	57 490	51 858	52 303	..	73 000	75 206	..	Pièces et accessoires pour automobiles et moteurs			343
..	113 427	116 283	103 606	..	106 166	102 895	..	Autres matériels de transport		35	
..	25 050	23 275	22 083	..	30 489	28 989	..	Construction et réparation de navires			351
..	27 900	28 090	23 065	..	10 768	10 545	..	Matériel ferroviaire roulant			352
..	42 048	46 813	41 226	..	40 297	39 058	..	Construction aéronautique et spatiale			353
..	18 429	18 105	17 232	..	24 612	24 303	..	Autres équipements de transport			359
..	132 779	129 789	131 323	..	308 118	292 541	..	Meubles ; activités de fabrication, nca		36	
..	94 972	91 303	92 794	..	216 935	204 040	..	Meubles			361
..	37 807	38 486	38 529	..	91 183	88 501	..	Activités de fabrication, nca			369
..	1 142	1 198	2 700	..	7 005	8 409	..	Récupération		37	
..	168 150	167 157	161 694	..	167 795	160 490	..	Electricité, gaz et eau	E	40_41	
..	151 259	150 192	144 713	..	150 095	143 110	..	Prod. et dist. d'électricité gaz, vapeur d'eau chaude		40	
..	16 891	16 965	16 981	..	17 700	17 380	..	Captage, épuration et distribution d'eau		41	
..	394 643	368 302	320 154	..	1 366 504	1 297 943	..	Construction	F	45	

ITALY

Table IT.6 EMPLOYMENT, FEMALES - EMPLOI, FEMMES

Units

ISIC revision 3				1991	1992	1993	1994	1995	1996	1997	1998
C	10_14		Mining and quarrying	..	1 006	1 766	1 770	..	..	2 872	..
	10		Mining of coal and lignite; extraction of peat	..	..	..	..	..	..	3	..
	11		Crude petroleum & natural gas; service activities	..	89	1 024	954	..	..	1 302	..
	13		Mining of metal ores	..	42	58	30	..	..	..	..
	14		Other mining and quarrying	..	875	684	786	..	..	1 327	..
D	15_37		Total manufacturing	..	759 736	731 346	721 135	..	..	1186 060	..
	15		Food products and beverages	..	56 063	52 889	52 880	..	..	100 251	..
		151	Production of meat, fish, fruit, vegetables, oils & fats	..	22 901	21 347	22 008	..	..	24 675	..
		152	Dairy products	..	5 327	4 696	4 308	..	..	10 452	..
		153	Grain mill and starch products; animal feeds	..	2 101	2 100	1 899	..	..	2 568	..
		154	Other food products	..	20 090	18 834	18 823	..	..	55 377	..
		155	Beverages	..	5 644	5 912	5 842	..	..	7 179	..
	16		Tobacco products	..	1 947	2 041	918	..	..	3 272	..
	17		Textiles	..	108 546	103 831	104 364	..	..	144 789	..
		171	Spinning, weaving and finishing of textiles	..	53 669	51 358	52 193	..	..	67 275	..
		172	Other textiles	..	15 528	15 084	14 267	..	..	27 450	..
		173	Knitted and crocheted fabrics and articles	..	39 349	37 389	37 904	..	..	50 064	..
	18		Wearing apparel and fur	..	146 720	141 003	132 450	..	..	223 264	..
		181	Wearing apparel, except fur apparel	..	145 747	139 994	131 698	..	..	221 022	..
		182	Dressing and dyeing of fur; articles of fur	..	973	1 009	752	..	..	2 242	..
	19		Leather and articles of leather; footwear	..	60 240	58 101	56 548	..	..	94 776	..
		191	Leather and articles of leather	..	15 655	15 074	13 801	..	..	25 064	..
		192	Footwear	..	44 585	43 027	42 747	..	..	69 712	..
	20		Wood & products of wood & cork, exc. furniture	..	8 737	8 400	8 727	..	..	20 797	..
		201	Sawmilling and planing of wood	..	939	993	934	..	..	2 976	..
		202	Wood products; articles of cork, straw, etc.	..	7 798	7 407	7 793	..	..	17 821	..
	21		Paper and paper products	..	11 701	12 235	12 172	..	..	17 211	..
	22		Printing, publishing and recorded media	..	24 062	23 588	21 221	..	..	41 149	..
		221	Publishing	..	12 621	12 465	10 108	..	..	15 901	..
		222	Printing and related activities	..	11 128	10 929	10 953	..	..	24 883	..
		223	Reproduction of recorded media	..	313	194	160	..	..	365	..
	23		Coke and petroleum products; nuclear fuel	..	1 986	1 617	2 411	..	..	2 571	..
		231	Coke oven products	..	..	..	..	..	..	..	..
		232	Refined petroleum products	..	1 944	1 579	2 376	..	..	2 544	..
		233	Processing of nuclear fuel	..	..	..	..	..	..	..	..
	24		Chemicals and chemical products	..	42 240	39 739	37 958	..	..	47 331	..
		241	Basic chemicals	..	5 555	4 627	4 757	..	..	6 988	..
		242	Other chemical products	..	34 783	33 314	31 463	..	..	38 523	..
		243	Man-made fibres	..	1 902	1 798	1 738	..	..	1 820	..
	25		Rubber and plastic products	..	25 537	25 712	27 567	..	..	49 331	..
		251	Rubber products	..	4 901	5 077	5 223	..	..	9 897	..
		252	Plastic products	..	20 636	20 635	22 344	..	..	39 434	..
	26		Other non-metallic mineral products	..	25 586	25 094	24 543	..	..	36 781	..
		261	Glass and glass products	..	4 493	4 577	4 139	..	..	6 487	..
		269	Non-metallic mineral products, nec	..	21 093	20 517	20 404	..	..	30 294	..
	27		Basic metals	..	8 419	8 085	8 695	..	..	11 110	..
		271	Basic iron and steel	..	3 480	3 383	3 980	..	..	4 277	..
		272	Basic precious and non-ferrous metals	..	2 302	2 200	2 293	..	..	2 613	..
		273	Casting of metals	..	2 637	2 502	2 422	..	..	4 220	..
	28		Fabricated metal products, exc. machinery	..	39 273	37 926	39 791	..	..	93 016	..
		281	Structural metal prod., tanks, reservoirs, etc.	..	7 235	7 074	6 903	..	..	13 917	..
		289	Other fabricated metal products	..	32 038	30 852	32 888	..	..	79 099	..
	29		Machinery and equipment, nec	..	50 453	51 945	54 472	..	..	78 651	..
		291	General purpose machinery	..	20 271	20 955	23 082	..	..	40 479	..
		292	Special purpose machinery	..	18 204	18 263	17 312	..	..	24 426	..
		293	Domestic appliances, nec	..	11 978	12 727	14 078	..	..	13 746	..
	30		Office, accounting & computing machinery	..	2 303	2 436	1 885	..	..	5 704	..
	31		Electrical machinery & apparatus, nec	..	37 054	34 349	33 022	..	..	47 053	..
		311	Electric motors, generators and transformers	..	8 427	8 528	8 115	..	..	9 491	..
		312	Electricity distribution & control apparatus	..	6 384	5 780	7 141	..	..	9 434	..
		313	Insulated wire and cable	..	2 745	2 112	2 363	..	..	2 743	..
		314	Accumulators, primary cells & primary batteries	..	352	286	252	..	..	442	..
		315	Electric lamps and lighting equipment	..	4 363	3 903	3 552	..	..	5 545	..
		319	Other electrical equipment, nec	..	14 783	13 740	11 599	..	..	19 398	..
	32		Radio, TV, communication equipment	..	28 148	23 355	22 499	..	..	30 004	..
		321	Electronic valves, tubes; other electronic components	..	8 751	8 223	8 553	..	..	10 805	..
		322	TV/radio transmitters; line teleph./telegr. apparatus	..	15 953	12 847	10 307	..	..	14 989	..
		323	TV/radio receivers; sound and video app.; assoc. gds.	..	3 444	2 285	3 639	..	..	4 210	..
	33		Medical, precision & optical instrum.; watch & clocks	..	19 359	19 344	18 845	..	..	27 123	..
		331	Medical appl.; instruments & appl. for measuring, etc.	..	12 584	12 558	11 826	..	..	18 146	..
		332	Optical instruments and photographic equipment	..	6 318	6 496	6 756	..	..	8 641	..
		333	Watches and clocks	..	457	290	263	..	..	336	..
	34		Motor vehicles, trailers and semi-trailers	..	15 966	15 120	15 870	..	..	31 126	..
		341	Motor vehicles	..	3 003	3 011	3 658	..	..	14 441	..
		342	Bodies for motor vehicles, trailers & semi-trailers	..	1 210	1 123	969	..	..	1 032	..
		343	Parts & accessories for motor vehicles & their engines	..	11 753	10 986	11 243	..	..	15 653	..
	35		Other transport equipment	..	6 346	6 244	5 951	..	..	11 566	..
		351	Building and repairing of ships and boats	..	785	756	647	..	..	1 734	..
		352	Railway and tramway locomotives and rolling stock	..	502	614	482	..	..	682	..
		353	Aircraft and spacecraft	..	2 251	2 062	2 015	..	..	4 161	..
		359	Transport equipment, nec	..	2 808	2 812	2 807	..	..	4 989	..
	36		Furniture; manufacturing, nec	..	38 827	38 016	37 968	..	..	68 108	..
		361	Furniture	..	22 765	21 539	22 282	..	..	38 068	..
		369	Manufacturing, nec	..	16 062	16 477	15 686	..	..	30 040	..
	37		Recycling	..	223	276	378	..	..	1 076	..
E	40_41		Electricity, gas and water supply	..	15 116	15 258	15 178	..	..	20 194	..
	40		Electricity, gas, steam and hot water supply	..	13 768	13 869	13 823	..	..	17 887	..
	41		Collection, purification and distribution of water	..	1 348	1 389	1 355	..	..	2 307	..
F	45		Construction	..	22 696	22 910	21 250	..	..	50 368	..

Table IT.7 **SALAIRES ET CHARGES SOCIALES - COMPENSATION OF LABOUR**

Milliards de L (Prix courants)

1991	1992	1993	1994	1995	1996	1997	1998			CITI révision 3	
..	1 282	1 285	1 359	..	..	1 923	..	Activites extractives	C	10_14	
..	..	..	..	..	..	0	..	Extraction de charbon, de lignite et de tourbe		10	
..	592	586	652	..	..	703	..	Pétrole brut et gaz naturel ; activités annexes		11	
..	22	31	32	..	..	..	..	Extraction de minerais métalliques		13	
..	668	668	675	..	1 161	..	..	Autres activités extractives		14	
..	143 821	142 119	148 035	..	213 655	214 389	..	Total activites manufacturieres	D	15_37	
..	11 438	11 868	12 200	..	16 636	16 871	..	Produits alimentaires et boissons		15	
..	3 807	3 946	3 795	..	4 364	4 369	..	Viande, poisson, fruits, légumes, huiles et graisses			151
..	1 643	1 684	1 824	..	2 921	2 776	..	Produits laitiers			152
..	860	841	843	..	1 056	1 077	..	Travail des grains; prod. amyl. ; alim. pour animaux			153
..	3 515	3 563	3 809	..	5 953	6 320	..	Autres produits alimentaires			154
..	1 613	1 834	1 929	..	2 341	2 329	..	Boissons			155
..	660	623	561	..	740	720	..	Produits à base de tabac		16	
..	8 569	8 590	9 113	..	12 717	12 539	..	Textiles		17	
..	5 682	5 646	6 067	..	7 774	7 901	..	Filature, tissage et achèvement des textiles			171
..	1 109	1 132	1 155	..	2 167	2 071	..	Autres articles textiles			172
..	1 779	1 812	1 891	..	2 776	2 566	..	Etoffes et articles de bonneterie			173
..	5 663	5 512	5 619	..	11 199	9 569	..	Articles d'habillement et fourrure		18	
..	5 606	5 456	5 568	..	11 081	9 460	..	Articles d'habillement autres qu'en fourrure			181
..	58	55	51	..	118	109	..	Prép. et teinture des fourrures ; articles en fourrure			182
..	3 835	3 868	4 128	..	7 268	6 838	..	Cuir et articles en cuir ; chaussures		19	
..	1 343	1 216	1 286	..	2 347	2 257	..	Cuir et articles en cuir sauf chaussures			191
..	2 492	2 652	2 841	..	4 921	4 581	..	Chaussures			192
..	1 588	1 626	1 724	..	3 986	3 908	..	Bois, articles en bois, sauf meubles		20	
..	153	152	181	..	477	508	..	Sciage et rabotage du bois			201
..	1 436	1 474	1 543	..	3 509	3 400	..	Articles en bois, liège, vannerie, sparterie			202
..	3 166	3 212	3 446	..	4 363	4 444	..	Papier, articles en papier et carton		21	
..	5 728	6 020	6 021	..	8 623	8 580	..	Edition, imprimerie et reproduction		22	
..	3 103	3 246	3 162	..	3 685	3 675	..	Edition			221
..	2 571	2 712	2 810	..	4 882	4 842	..	Imprimerie et activités annexes			222
..	54	62	48	..	56	63	..	Reproduction de supports enregistrés			223
..	1 647	1 533	1 966	..	2 047	2 104	..	Coke. et prod. pétroliers ; comb. nucléaires		23	
..	..	..	..	..	..	..	..	Cokéfaction			231
..	1 599	1 489	1 925	..	..	2 075	..	Produits pétroliers raffinés			232
..	..	..	..	..	..	..	..	Traitement de combustibles nucléaires			233
..	13 518	13 555	13 464	..	16 510	16 397	..	Produits chimiques		24	
..	3 913	4 007	3 875	..	4 091	4 351	..	Produits chimiques de base			241
..	8 820	8 792	8 882	..	11 656	11 335	..	Autres produits chimiques			242
..	786	756	708	..	762	711	..	Fibres synthétiques ou artificielles			243
..	5 201	5 421	5 879	..	9 258	9 423	..	Art. en caoutchouc et matières plastiques		25	
..	1 725	1 666	1 843	..	2 435	2 537	..	Articles en caoutchouc			251
..	3 476	3 755	4 035	..	6 823	6 886	..	Articles en matières plastiques			252
..	7 958	7 954	7 955	..	11 351	11 099	..	Autres produits minéraux non métalliques		26	
..	1 523	1 540	1 526	..	2 009	1 854	..	Verre et articles en verre			261
..	6 434	6 414	6 429	..	9 342	9 245	..	Produits minéraux non métalliques, nca			269
..	7 640	5 995	7 253	..	8 464	9 075	..	Produits métallurgiques de base		27	
..	4 978	3 399	4 510	..	4 907	5 207	..	Sidérurgie et première transformation de l'acier			271
..	1 280	1 252	1 286	..	1 425	1 607	..	Première transf. métaux précieux et non ferreux			272
..	1 383	1 343	1 457	..	2 132	2 261	..	Fonderie			273
..	10 644	10 815	11 269	..	22 568	23 325	..	Ouvrages en métaux, sauf machines		28	
..	3 920	3 881	3 727	..	6 647	6 807	..	Constr. et menuis. métal.; réservoirs; générateurs			281
..	6 724	6 934	7 542	..	15 921	16 518	..	Autres ouvrages en métaux			289
..	18 907	19 133	20 405	..	29 496	30 358	..	Machines et matériels, nca		29	
..	7 192	7 314	8 287	..	14 418	15 030	..	Machines d'usage général			291
..	9 341	9 326	9 332	..	11 758	11 988	..	Machines d'usage spécifique			292
..	2 374	2 493	2 787	..	3 320	3 340	..	Appareils domestiques, nca			293
..	1 455	1 218	1 010	..	1 688	1 859	..	Mach. de bureau, comptables, information		30	
..	8 749	7 493	7 287	..	9 420	10 014	..	Machines et appareils électriques, nca		31	
..	1 890	1 716	1 506	..	1 785	1 850	..	Moteurs, génératrices, transformateurs électriques			311
..	1 348	1 358	1 478	..	2 112	2 221	..	Matériel électrique de distribution et de commande			312
..	827	734	696	..	615	819	..	Fils et câbles électriques isolés			313
..	170	153	169	..	305	263	..	Accumulateurs et piles électriques			314
..	501	499	474	..	742	785	..	Lampes électriques et appareils d'éclairage			315
..	4 013	3 034	2 964	..	3 860	4 075	..	Autres matériels électriques, nca.			319
..	4 446	4 764	5 114	..	5 823	5 916	..	Equipements et appareils de radio, TV, etc.		32	
..	1 125	1 228	1 315	..	1 794	1 668	..	Tubes et valves électroniques;autres composants			321
..	2 985	3 224	3 483	..	3 657	3 849	..	Emetteurs radio/TV; app. téléphonie et télégraphie			322
..	337	311	315	..	373	400	..	Récepteurs radio/TV, son et image; art. associés			323
..	2 870	3 301	3 358	..	4 842	4 443	..	Instr. médicaux, de précision ; optique ; horlog.		33	
..	2 296	2 725	2 765	..	3 787	3 607	..	Instruments médicaux et de précision, etc.			331
..	518	538	560	..	1 006	792	..	Instruments d'optique ; matériel photographique			332
..	55	38	33	..	49	44	..	Horlogerie			333
..	9 272	8 377	9 009	..	10 773	11 331	..	Véhicules automobiles; remorques, etc.		34	
..	6 149	5 486	5 783	..	5 976	6 323	..	Véhicules automobiles			341
..	581	536	616	..	847	729	..	Carross. automobiles, remorques, semi-remorques			342
..	2 542	2 354	2 611	..	3 950	4 280	..	Pièces et accessoires pour automobiles et moteurs			343
..	5 791	6 068	5 812	..	6 244	6 322	..	Autres matériels de transport		35	
..	1 166	1 128	1 069	..	1 553	1 491	..	Construction et réparation de navires			351
..	1 411	1 554	1 403	..	652	684	..	Matériel ferroviaire roulant			352
..	2 414	2 574	2 521	..	2 908	2 989	..	Construction aéronautique et spatiale			353
..	800	812	820	..	1 132	1 158	..	Autres équipements de transport			359
..	5 029	5 113	5 311	..	9 432	8 975	..	Meubles ; activités de fabrication, nca		36	
..	3 600	3 613	3 742	..	6 599	6 266	..	Meubles			361
..	1 429	1 500	1 569	..	2 833	2 709	..	Activités de fabrication, nca			369
..	45	61	132	..	207	278	..	Récupération		37	
..	12 896	12 826	13 306	..	14 890	14 119	..	Electricité, gaz et eau	E	40_41	
..	11 725	11 662	12 146	..	13 592	12 865	..	Prod. et dist. d'électricité gaz, vapeur d'eau chaude		40	
..	1 171	1 164	1 161	..	1 298	1 255	..	Captage, épuration et distribution d'eau		41	
..	18 666	17 556	15 775	..	35 684	33 097	..	Construction	F	45	

ITALY

Table IT.8 SUPPLEMENTS TO WAGES & SALARIES - CHARGES SOCIALES DES EMPLOYEURS

Billions of L (Current Prices)

ISIC revision 3				1991	1992	1993	1994	1995	1996	1997	1998
C	10_14		Mining and quarrying	..	405	388	396	..	..	..	..
	10		Mining of coal and lignite; extraction of peat	..	..	..	..	..	..	..	..
	11		Crude petroleum & natural gas; service activities	..	180	168	170	..	..	..	..
	13		Mining of metal ores	..	7	9	9	..	..	..	..
	14		Other mining and quarrying	..	218	210	217	..	..	..	..
D	15_37		Total manufacturing	..	45 692	45 026	46 766	..	..	..	..
	15		Food products and beverages	..	3 580	3 675	3 734	..	..	..	..
		151	Production of meat, fish, fruit, vegetables, oils & fats	..	1 178	1 215	1 146	..	..	..	..
		152	Dairy products	..	504	510	544	..	..	..	..
		153	Grain mill and starch products; animal feeds	..	277	270	263	..	..	..	..
		154	Other food products	..	1 123	1 127	1 195	..	..	..	..
		155	Beverages	..	498	553	586	..	..	..	..
	16		Tobacco products	..	92	96	78	..	..	..	..
	17		Textiles	..	2 785	2 800	2 965	..	..	..	..
		171	Spinning, weaving and finishing of textiles	..	1 859	1 848	1 986	..	..	..	..
		172	Other textiles	..	361	367	375	..	..	..	..
		173	Knitted and crocheted fabrics and articles	..	565	585	604	..	..	..	..
	18		Wearing apparel and fur	..	1 771	1 742	1 759	..	..	..	..
		181	Wearing apparel, except fur apparel	..	1 752	1 724	1 743	..	..	..	..
		182	Dressing and dyeing of fur; articles of fur	..	19	18	16	..	..	..	..
	19		Leather and articles of leather; footwear	..	1 217	1 212	1 302	..	..	..	..
		191	Leather and articles of leather	..	429	385	407	..	..	..	..
		192	Footwear	..	789	827	895	..	..	..	..
	20		Wood & products of wood & cork, exc. furniture	..	526	544	566	..	..	..	..
		201	Sawmilling and planing of wood	..	51	52	60	..	..	..	..
		202	Wood products; articles of cork, straw, etc.	..	475	493	506	..	..	..	..
	21		Paper and paper products	..	1 035	1 036	1 116	..	..	..	..
	22		Printing, publishing and recorded media	..	1 818	1 890	1 898	..	..	..	..
		221	Publishing	..	970	1 009	989	..	..	..	..
		222	Printing and related activities	..	831	863	895	..	..	..	..
		223	Reproduction of recorded media	..	16	18	14	..	..	..	..
	23		Coke and petroleum products; nuclear fuel	..	476	436	575	..	..	..	..
		231	Coke oven products	..	..	..	..	..	..	..	..
		232	Refined petroleum products	..	460	422	561	..	..	..	..
		233	Processing of nuclear fuel	..	..	..	..	..	..	..	..
	24		Chemicals and chemical products	..	4 104	4 172	4 167	..	..	..	..
		241	Basic chemicals	..	1 183	1 214	1 185	..	..	..	..
		242	Other chemical products	..	2 677	2 721	2 758	..	..	..	..
		243	Man-made fibres	..	243	237	224	..	..	..	..
	25		Rubber and plastic products	..	1 706	1 779	1 909	..	..	..	..
		251	Rubber products	..	570	553	593	..	..	..	..
		252	Plastic products	..	1 136	1 227	1 316	..	..	..	..
	26		Other non-metallic mineral products	..	2 650	2 638	2 626	..	..	..	..
		261	Glass and glass products	..	503	510	498	..	..	..	..
		269	Non-metallic mineral products, nec	..	2 147	2 128	2 129	..	..	..	..
	27		Basic metals	..	2 481	1 936	2 330	..	..	..	..
		271	Basic iron and steel	..	1 599	1 103	1 453	..	..	..	..
		272	Basic precious and non-ferrous metals	..	404	383	390	..	..	..	..
		273	Casting of metals	..	478	451	487	..	..	..	..
	28		Fabricated metal products, exc. machinery	..	3 391	3 485	3 588	..	..	..	..
		281	Structural metal prod., tanks, reservoirs, etc.	..	1 187	1 239	1 143	..	..	..	..
		289	Other fabricated metal products	..	2 203	2 246	2 445	..	..	..	..
	29		Machinery and equipment, nec	..	6 171	6 175	6 527	..	..	..	..
		291	General purpose machinery	..	2 304	2 353	2 647	..	..	..	..
		292	Special purpose machinery	..	3 072	3 010	2 990	..	..	..	..
		293	Domestic appliances, nec	..	795	812	890	..	..	..	..
	30		Office, accounting & computing machinery	..	505	442	368	..	..	..	..
	31		Electrical machinery & apparatus, nec	..	2 698	2 263	2 244	..	..	..	..
		311	Electric motors, generators and transformers	..	580	521	478	..	..	..	..
		312	Electricity distribution & control apparatus	..	416	413	464	..	..	..	..
		313	Insulated wire and cable	..	251	217	213	..	..	..	..
		314	Accumulators, primary cells & primary batteries	..	52	48	51	..	..	..	..
		315	Electric lamps and lighting equipment	..	162	161	154	..	..	..	..
		319	Other electrical equipment, nec	..	1 237	903	883	..	..	..	..
	32		Radio, TV, communication equipment	..	1 349	1 405	1 499	..	..	..	..
		321	Electronic valves, tubes; other electronic components	..	341	365	387	..	..	..	..
		322	TV/radio transmitters; line teleph./telegr. apparatus	..	900	933	1 013	..	..	..	..
		323	TV/radio receivers; sound and video app.; assoc. gds.	..	108	107	99	..	..	..	..
	33		Medical, precision & optical instrum.; watch & clocks	..	906	1 022	1 043	..	..	..	..
		331	Medical appl.; instruments & appl. for measuring, etc.	..	722	844	859	..	..	..	..
		332	Optical instruments and photographic equipment	..	168	167	174	..	..	..	..
		333	Watches and clocks	..	17	12	11	..	..	..	..
	34		Motor vehicles, trailers and semi-trailers	..	2 999	2 729	2 929	..	..	..	..
		341	Motor vehicles	..	1 970	1 799	1 910	..	..	..	..
		342	Bodies for motor vehicles, trailers & semi-trailers	..	191	174	197	..	..	..	..
		343	Parts & accessories for motor vehicles & their engines	..	838	756	823	..	..	..	..
	35		Other transport equipment	..	1 773	1 872	1 791	..	..	..	..
		351	Building and repairing of ships and boats	..	397	381	361	..	..	..	..
		352	Railway and tramway locomotives and rolling stock	..	419	468	415	..	..	..	..
		353	Aircraft and spacecraft	..	698	759	748	..	..	..	..
		359	Transport equipment, nec	..	260	264	266	..	..	..	..
	36		Furniture; manufacturing, nec	..	1 643	1 655	1 712	..	..	..	..
		361	Furniture	..	1 183	1 172	1 207	..	..	..	..
		369	Manufacturing, nec	..	460	483	505	..	..	..	..
	37		Recycling	..	15	20	39	..	..	..	..
E	40_41		Electricity, gas and water supply	..	4 078	4 157	4 406	..	..	..	..
	40		Electricity, gas, steam and hot water supply	..	3 724	3 796	4 053	..	..	..	..
	41		Collection, purification and distribution of water	..	354	360	353	..	..	..	..
F	45		Construction	..	6 396	5 920	5 355	..	..	..	..

Table IT.9 HEURES OUVRÉES, OUVRIERS - HOURS WORKED BY OPERATIVES

Milliers

1991	1992	1993	1994	1995	1996	1997	1998		CITI révision 3	
..	25 267	24 495	23 724	..	..	37 348	..	Activites extractives	C	10_14
..	..	..	..	..	..	13	..	Extraction de charbon, de lignite et de tourbe		10
..	5 375	4 873	5 309	..	..	2 330	..	Pétrole brut et gaz naturel ; activités annexes		11
..	568	886	872	..	..	..	..	Extraction de minerais métalliques		13
..	19 324	18 736	17 543	..	..	..	..	Autres activités extractives		14
..	3 402 887	3 218 757	3 380 003	..	..	5 075 294	..	Total activites manufacturieres	D	15_37
..	271 216	266 595	261 424	..	..	405 774	..	Produits alimentaires et boissons		15
..	105 320	101 794	97 757	..	..	111 459	..	Viande, poisson, fruits, légumes, huiles et graisses		151
..	38 921	37 891	37 885	..	..	58 662	..	Produits laitiers		152
..	15 455	15 013	14 391	..	..	19 968	..	Travail des grains; prod. amyl. ; alim. pour animaux		153
..	81 046	80 588	80 589	..	..	178 232	..	Autres produits alimentaires		154
..	30 474	31 309	30 802	..	..	37 453	..	Boissons		155
..	20 135	19 379	15 631	..	..	13 477	..	Produits à base de tabac		16
..	275 250	266 591	274 374	..	..	374 453	..	Textiles		17
..	170 212	165 287	171 361	..	..	211 121	..	Filature, tissage et achèvement des textiles		171
..	36 349	35 412	35 215	..	..	67 184	..	Autres articles textiles		172
..	68 689	65 892	67 798	..	..	96 147	..	Etoffes et articles de bonneterie		173
..	233 039	221 157	215 095	..	..	378 733	..	Articles d'habillement et fourrure		18
..	230 902	219 232	213 285	..	..	373 977	..	Articles d'habillement autres qu'en fourrure		181
..	2 137	1 925	1 810	..	..	4 756	..	Prép. et teinture des fourrures ; articles en fourrure		182
..	159 988	152 864	157 811	..	..	267 369	..	Cuir et articles en cuir ; chaussures		19
..	46 330	40 094	42 811	..	..	78 761	..	Cuir et articles en cuir sauf chaussures		191
..	113 658	112 770	115 000	..	..	188 608	..	Chaussures		192
..	57 777	55 843	58 269	..	..	144 447	..	Bois, articles en bois, sauf meubles		20
..	5 346	5 391	6 002	..	..	17 996	..	Sciage et rabotage du bois		201
..	52 431	50 452	52 267	..	..	126 451	..	Articles en bois, liège, vannerie, sparterie		202
..	82 341	83 093	82 713	..	..	101 987	..	Papier, articles en papier et carton		21
..	82 318	82 683	80 763	..	..	146 361	..	Edition, imprimerie et reproduction		22
..	16 910	16 416	14 494	..	..	15 739	..	Edition		221
..	64 635	65 322	65 353	..	..	129 732	..	Imprimerie et activités annexes		222
..	773	945	916	..	..	891	..	Reproduction de supports enregistrés		223
..	19 463	17 447	19 100	..	..	18 148	..	Coke. et prod. pétroliers ; comb. nucléaires		23
..	..	..	..	..	..	..	..	Cokéfaction		231
..	18 575	16 694	18 484	..	..	17 837	..	Produits pétroliers raffinés		232
..	..	..	..	..	..	..	..	Traitement de combustibles nucléaires		233
..	153 124	142 387	139 137	..	..	159 918	..	Produits chimiques		24
..	54 904	50 708	49 445	..	..	48 553	..	Produits chimiques de base		241
..	82 994	75 990	74 568	..	..	97 224	..	Autres produits chimiques		242
..	15 226	15 689	15 124	..	..	14 141	..	Fibres synthétiques ou artificielles		243
..	138 618	139 587	146 237	..	..	244 656	..	Art. en caoutchouc et matières plastiques		25
..	42 944	38 624	40 212	..	..	56 607	..	Articles en caoutchouc		251
..	95 674	100 963	106 025	..	..	188 049	..	Articles en matières plastiques		252
..	215 806	202 964	194 455	..	..	274 493	..	Autres produits minéraux non métalliques		26
..	38 599	37 874	35 551	..	..	44 521	..	Verre et articles en verre		261
..	177 207	165 090	158 904	..	..	229 972	..	Produits minéraux non métalliques, nca		269
..	176 527	142 394	159 668	..	..	185 578	..	Produits métallurgiques de base		27
..	113 492	80 495	96 197	..	..	103 507	..	Sidérurgie et première transformation de l'acier		271
..	25 519	27 440	26 126	..	..	27 492	..	Première transf. métaux précieux et non ferreux		272
..	37 516	34 459	37 345	..	..	54 579	..	Fonderie		273
..	310 236	301 498	417 819	..	..	726 453	..	Ouvrages en métaux, sauf machines		28
..	114 227	104 366	101 820	..	..	215 364	..	Constr. et menuis. métal.; réservoirs; générateurs		281
..	196 009	197 132	315 999	..	..	511 090	..	Autres ouvrages en métaux		289
..	396 792	385 835	405 378	..	..	602 845	..	Machines et matériels, nca		29
..	147 672	145 488	161 440	..	..	293 959	..	Machines d'usage général		291
..	181 409	173 068	170 303	..	..	226 232	..	Machines d'usage spécifique		292
..	67 711	67 279	73 635	..	..	82 654	..	Appareils domestiques, nca		293
..	9 778	9 610	9 679	..	..	9 891	..	Mach. de bureau, comptables, information		30
..	177 050	155 832	149 972	..	..	216 993	..	Machines et appareils électriques, nca		31
..	38 871	33 585	30 756	..	..	40 874	..	Moteurs, génératrices, transformateurs électriques		311
..	28 431	27 506	27 253	..	..	42 261	..	Matériel électrique de distribution et de commande		312
..	19 027	16 302	14 997	..	..	16 225	..	Fils et câbles électriques isolés		313
..	4 272	3 352	3 816	..	..	5 641	..	Accumulateurs et piles électriques		314
..	12 501	11 746	11 014	..	..	18 796	..	Lampes électriques et appareils d'éclairage		315
..	73 948	63 341	62 136	..	..	93 197	..	Autres matériels électriques, nca.		319
..	66 936	64 644	66 997	..	..	78 019	..	Equipements et appareils de radio, TV, etc.		32
..	22 022	21 092	24 393	..	..	29 623	..	Tubes et valves électroniques;autres composants		321
..	38 094	37 820	34 842	..	..	39 890	..	Emetteurs radio/TV; app. téléphonie et télégraphie		322
..	6 820	5 732	7 762	..	..	8 506	..	Récepteurs radio/TV, son et image; art. associés		323
..	51 173	51 909	49 362	..	..	83 278	..	Instr. médicaux, de précision ; optique ; horlog.		33
..	34 057	33 696	30 479	..	..	57 739	..	Instruments médicaux et de précision, etc.		331
..	15 852	17 039	17 939	..	..	24 318	..	Instruments d'optique ; matériel photographique		332
..	1 264	1 174	944	..	..	1 220	..	Horlogerie		333
..	220 921	188 215	208 426	..	..	237 738	..	Véhicules automobiles; remorques, etc.		34
..	139 525	113 800	129 000	..	..	125 386	..	Véhicules automobiles		341
..	15 370	14 431	15 244	..	..	18 534	..	Carross. automobiles, remorques, semi-remorques		342
..	66 026	59 984	64 182	..	..	93 818	..	Pièces et accessoires pour automobiles et moteurs		343
..	118 906	107 388	100 718	..	..	102 493	..	Autres matériels de transport		35
..	27 497	24 735	21 571	..	..	33 266	..	Construction et réparation de navires		351
..	35 466	32 573	30 201	..	..	11 142	..	Matériel ferroviaire roulant		352
..	33 524	28 870	27 565	..	..	28 246	..	Construction aéronautique et spatiale		353
..	22 419	21 210	21 381	..	..	29 839	..	Autres équipements de transport		359
..	164 019	159 335	163 826	..	..	294 393	..	Meubles ; activités de fabrication, nca		36
..	118 282	114 505	118 645	..	..	209 321	..	Meubles		361
..	45 737	44 830	45 181	..	..	85 073	..	Activités de fabrication, nca		369
..	1 474	1 507	3 149	..	..	7 797	..	Récupération		37
..	121 966	118 232	112 401	..	..	106 982	..	Electricité, gaz et eau	E	40_41
..	109 510	105 489	99 732	..	..	93 804	..	Prod. et dist. d'électricité gaz, vapeur d'eau chaude		40
..	12 456	12 743	12 669	..	..	13 177	..	Captage, épuration et distribution d'eau		41
..	482 194	528 841	382 000	..	..	1 026 362	..	Construction	F	45

ITALY

Table IT.10 ESTABLISHMENTS - ÉTABLISSEMENTS

Units

ISIC revision 3				1991	1992	1993	1994	1995	1996	1997	1998
C	10_14		Mining and quarrying	..	386	384	367	..	..	..	..
	10		Mining of coal and lignite; extraction of peat	..	2	2	..	..	..	..	..
	11		Crude petroleum & natural gas; service activities	..	17	21	18	..	..	..	..
	13		Mining of metal ores	..	4	6	7	..	..	..	..
	14		Other mining and quarrying	..	362	355	342	..	..	..	..
D	15_37		Total manufacturing	..	38 708	39 118	38 983	..	..	..	..
	15		Food products and beverages	..	2 696	2 734	2 698	..	..	..	..
		151	Production of meat, fish, fruit, vegetables, oils & fats	..	977	1 003	965	..	..	..	..
		152	Dairy products	..	357	375	368	..	..	..	..
		153	Grain mill and starch products; animal feeds	..	256	243	243	..	..	..	..
		154	Other food products	..	718	727	746	..	..	..	..
		155	Beverages	..	388	386	376	..	..	..	..
	16		Tobacco products	..	42	40	32	..	..	..	..
	17		Textiles	..	3 465	3 514	3 511	..	..	..	..
		171	Spinning, weaving and finishing of textiles	..	1 778	1 804	1 812	..	..	..	..
		172	Other textiles	..	600	611	584	..	..	..	..
		173	Knitted and crocheted fabrics and articles	..	1 087	1 099	1 115	..	..	..	..
	18		Wearing apparel and fur	..	3 946	4 063	3 888	..	..	..	..
		181	Wearing apparel, except fur apparel	..	3 899	4 020	3 852	..	..	..	..
		182	Dressing and dyeing of fur; articles of fur	..	47	43	36	..	..	..	..
	19		Leather and articles of leather; footwear	..	2 721	2 747	2 740	..	..	..	..
		191	Leather and articles of leather	..	835	784	778	..	..	..	..
		192	Footwear	..	1 886	1 963	1 962	..	..	..	..
	20		Wood & products of wood & cork, exc. furniture	..	979	983	1 018	..	..	..	..
		201	Sawmilling and planing of wood	..	105	113	124	..	..	..	..
		202	Wood products; articles of cork, straw, etc.	..	874	870	894	..	..	..	..
	21		Paper and paper products	..	782	793	806	..	..	..	..
	22		Printing, publishing and recorded media	..	1 370	1 383	1 344	..	..	..	..
		221	Publishing	..	298	306	289	..	..	..	..
		222	Printing and related activities	..	1 056	1 058	1 037	..	..	..	..
		223	Reproduction of recorded media	..	16	19	18	..	..	..	..
	23		Coke and petroleum products; nuclear fuel	..	117	119	113	..	..	..	..
		231	Coke oven products	..	..	..	..	..	..	..	..
		232	Refined petroleum products	..	..	..	..	..	..	..	..
		233	Processing of nuclear fuel	..	..	..	..	..	..	..	..
	24		Chemicals and chemical products	..	1 220	1 215	1 214	..	..	..	..
		241	Basic chemicals	..	284	297	297	..	..	..	..
		242	Other chemical products	..	895	873	876	..	..	..	..
		243	Man-made fibres	..	41	45	41	..	..	..	..
	25		Rubber and plastic products	..	1 592	1 677	1 742	..	..	..	..
		251	Rubber products	..	258	266	272	..	..	..	..
		252	Plastic products	..	1 334	1 411	1 470	..	..	..	..
	26		Other non-metallic mineral products	..	2 330	2 312	2 222	..	..	..	..
		261	Glass and glass products	..	320	318	307	..	..	..	..
		269	Non-metallic mineral products, nec	..	2 010	1 994	1 915	..	..	..	..
	27		Basic metals	..	1 008	993	995	..	..	..	..
		271	Basic iron and steel	..	521	510	495	..	..	..	..
		272	Basic precious and non-ferrous metals	..	128	135	139	..	..	..	..
		273	Casting of metals	..	359	348	361	..	..	..	..
	28		Fabricated metal products, exc. machinery	..	4 879	4 948	5 084	..	..	..	..
		281	Structural metal prod., tanks, reservoirs, etc.	..	1 641	1 645	1 624	..	..	..	..
		289	Other fabricated metal products	..	3 238	3 303	3 460	..	..	..	..
	29		Machinery and equipment, nec	..	4 511	4 560	4 605	..	..	..	..
		291	General purpose machinery	..	1 784	1 852	1 972	..	..	..	..
		292	Special purpose machinery	..	2 493	2 467	2 398	..	..	..	..
		293	Domestic appliances, nec	..	234	241	235	..	..	..	..
	30		Office, accounting & computing machinery	..	116	126	112	..	..	..	..
	31		Electrical machinery & apparatus, nec	..	1 672	1 647	1 632	..	..	..	..
		311	Electric motors, generators and transformers	..	426	407	396	..	..	..	..
		312	Electricity distribution & control apparatus	..	345	341	331	..	..	..	..
		313	Insulated wire and cable	..	126	121	116	..	..	..	..
		314	Accumulators, primary cells & primary batteries	..	25	25	24	..	..	..	..
		315	Electric lamps and lighting equipment	..	171	168	175	..	..	..	..
		319	Other electrical equipment, nec	..	579	585	590	..	..	..	..
	32		Radio, TV, communication equipment	..	497	514	504	..	..	..	..
		321	Electronic valves, tubes; other electronic components	..	208	216	223	..	..	..	..
		322	TV/radio transmitters; line teleph./telegr. apparatus	..	238	251	232	..	..	..	..
		323	TV/radio receivers; sound and video app.; assoc. gds.	..	51	47	49	..	..	..	..
	33		Medical, precision & optical instrum.; watch & clocks	..	802	802	789	..	..	..	..
		331	Medical appl.; instruments & appl. for measuring, etc.	..	609	609	609	..	..	..	..
		332	Optical instruments and photographic equipment	..	174	173	166	..	..	..	..
		333	Watches and clocks	..	19	20	14	..	..	..	..
	34		Motor vehicles, trailers and semi-trailers	..	736	695	686	..	..	..	..
		341	Motor vehicles	..	47	42	40	..	..	..	..
		342	Bodies for motor vehicles, trailers & semi-trailers	..	199	188	185	..	..	..	..
		343	Parts & accessories for motor vehicles & their engines	..	490	465	461	..	..	..	..
	35		Other transport equipment	..	489	485	466	..	..	..	..
		351	Building and repairing of ships and boats	..	218	217	208	..	..	..	..
		352	Railway and tramway locomotives and rolling stock	..	51	45	32	..	..	..	..
		353	Aircraft and spacecraft	..	45	43	41	..	..	..	..
		359	Transport equipment, nec	..	175	180	185	..	..	..	..
	36		Furniture; manufacturing, nec	..	2 705	2 729	2 740	..	..	..	..
		361	Furniture	..	1 924	1 927	1 952	..	..	..	..
		369	Manufacturing, nec	..	781	802	788	..	..	..	..
	37		Recycling	..	33	39	42	..	..	..	..
E	40_41		Electricity, gas and water supply	..	311	325	315	..	..	..	..
	40		Electricity, gas, steam and hot water supply	..	200	209	208	..	..	..	..
	41		Collection, purification and distribution of water	..	111	116	107	..	..	..	..
F	45		Construction	..	7 436	7 659	6 994	..	..	..	..

Table IT.11 **ENTREPRISES - ENTERPRISES**

Unités

1991	1992	1993	1994	1995	1996	1997	1998			CITI révision 3	
..	..	..	..	..	4 574	4 336	..	Activites extractives	C	**10_14**	
..	..	..	..	..	1	3	..	Extraction de charbon, de lignite et de tourbe		**10**	
..	..	..	..	..	21	20	..	Pétrole brut et gaz naturel ; activités annexes		**11**	
..	..	..	..	..	4	2	..	Extraction de minerais métalliques		**13**	
..	..	..	..	..	4 548	4 311	..	Autres activités extractives		**14**	
..	..	..	..	..	568 768	547 496	..	Total activites manufacturieres	D	**15_37**	
..	..	..	..	..	69 950	65 205	..	Produits alimentaires et boissons		**15**	
..	..	..	..	..	10 669	10 942	..	Viande, poisson, fruits, légumes, huiles et graisses			151
..	..	..	..	..	6 602	5 552	..	Produits laitiers			152
..	..	..	..	..	3 062	2 867	..	Travail des grains; prod. amyl. ; alim. pour animaux			153
..	..	..	..	..	46 327	42 662	..	Autres produits alimentaires			154
..	..	..	..	..	3 290	3 182	..	Boissons			155
..	..	..	..	..	70	97	..	Produits à base de tabac		**16**	
..	..	..	..	..	33 432	31 860	..	Textiles		**17**	
..	..	..	..	..	11 621	11 302	..	Filature, tissage et achèvement des textiles			171
..	..	..	..	..	9 797	9 514	..	Autres articles textiles			172
..	..	..	..	..	12 014	11 044	..	Etoffes et articles de bonneterie			173
..	..	..	..	..	51 906	49 925	..	Articles d'habillement et fourrure		**18**	
..	..	..	..	..	48 644	47 210	..	Articles d'habillement autres qu'en fourrure			181
..	..	..	..	..	3 262	2 715	..	Prép. et teinture des fourrures ; articles en fourrure			182
..	..	..	..	..	25 661	24 788	..	Cuir et articles en cuir ; chaussures		**19**	
..	..	..	..	..	9 737	9 704	..	Cuir et articles en cuir sauf chaussures			191
..	..	..	..	..	15 924	15 084	..	Chaussures			192
..	..	..	..	..	51 106	50 068	..	Bois, articles en bois, sauf meubles		**20**	
..	..	..	..	..	4 254	4 801	..	Sciage et rabotage du bois			201
..	..	..	..	..	46 852	45 267	..	Articles en bois, liège, vannerie, sparterie			202
..	..	..	..	..	4 767	4 764	..	Papier, articles en papier et carton		**21**	
..	..	..	..	..	26 202	26 380	..	Edition, imprimerie et reproduction		**22**	
..	..	..	..	..	6 481	6 727	..	Edition			221
..	..	..	..	..	19 366	19 292	..	Imprimerie et activités annexes			222
..	..	..	..	..	355	361	..	Reproduction de supports enregistrés			223
..	..	..	..	..	439	440	..	Coke. et prod. pétroliers ; comb. nucléaires		**23**	
..	..	..	..	..	1	3	..	Cokéfaction			231
..	..	..	..	..	437	436	..	Produits pétroliers raffinés			232
..	..	..	..	..	1	1	..	Traitement de combustibles nucléaires			233
..	..	..	..	..	5 612	5 911	..	Produits chimiques		**24**	
..	..	..	..	..	891	928	..	Produits chimiques de base			241
..	..	..	..	..	4 669	4 894	..	Autres produits chimiques			242
..	..	..	..	..	52	89	..	Fibres synthétiques ou artificielles			243
..	..	..	..	..	13 042	13 719	..	Art. en caoutchouc et matières plastiques		**25**	
..	..	..	..	..	1 582	1 709	..	Articles en caoutchouc			251
..	..	..	..	..	11 460	12 010	..	Articles en matières plastiques			252
..	..	..	..	..	28 478	26 373	..	Autres produits minéraux non métalliques		**26**	
..	..	..	..	..	5 369	5 206	..	Verre et articles en verre			261
..	..	..	..	..	23 109	21 167	..	Produits minéraux non métalliques, nca			269
..	..	..	..	..	3 370	3 825	..	Produits métallurgiques de base		**27**	
..	..	..	..	..	1 342	1 616	..	Sidérurgie et première transformation de l'acier			271
..	..	..	..	..	324	538	..	Première transf. métaux précieux et non ferreux			272
..	..	..	..	..	1 704	1 671	..	Fonderie			273
..	..	..	..	..	92 851	89 921	..	Ouvrages en métaux, sauf machines		**28**	
..	..	..	..	..	33 995	31 467	..	Constr. et menuis. métal.; réservoirs; générateurs			281
..	..	..	..	..	58 856	58 454	..	Autres ouvrages en métaux			289
..	..	..	..	..	41 818	40 270	..	Machines et matériels, nca		**29**	
..	..	..	..	..	24 436	23 435	..	Machines d'usage général			291
..	..	..	..	..	16 343	15 969	..	Machines d'usage spécifique			292
..	..	..	..	..	1 039	866	..	Appareils domestiques, nca			293
..	..	..	..	..	505	893	..	Mach. de bureau, comptables, information		**30**	
..	..	..	..	..	16 385	17 397	..	Machines et appareils électriques, nca		**31**	
..	..	..	..	..	3 028	2 599	..	Moteurs, génératrices, transformateurs électriques			311
..	..	..	..	..	2 062	2 260	..	Matériel électrique de distribution et de commande			312
..	..	..	..	..	361	406	..	Fils et câbles électriques isolés			313
..	..	..	..	..	88	141	..	Accumulateurs et piles électriques			314
..	..	..	..	..	2 265	2 123	..	Lampes électriques et appareils d'éclairage			315
..	..	..	..	..	8 581	9 868	..	Autres matériels électriques, nca.			319
..	..	..	..	..	9 525	9 262	..	Equipements et appareils de radio, TV, etc.		**32**	
..	..	..	..	..	1 311	1 012	..	Tubes et valves électroniques;autres composants			321
..	..	..	..	..	7 682	7 751	..	Emetteurs radio/TV; app. téléphonie et télégraphie			322
..	..	..	..	..	532	499	..	Récepteurs radio/TV, son et image; art. associés			323
..	..	..	..	..	27 544	25 586	..	Instr. médicaux, de précision ; optique ; horlog.		**33**	
..	..	..	..	..	24 409	22 548	..	Instruments médicaux et de précision, etc.			331
..	..	..	..	..	3 005	2 907	..	Instruments d'optique ; matériel photographique			332
..	..	..	..	..	130	131	..	Horlogerie			333
..	..	..	..	..	1 939	2 068	..	Véhicules automobiles; remorques, etc.		**34**	
..	..	..	..	..	51	61	..	Véhicules automobiles			341
..	..	..	..	..	573	607	..	Carross. automobiles, remorques, semi-remorques			342
..	..	..	..	..	1 315	1 400	..	Pièces et accessoires pour automobiles et moteurs			343
..	..	..	..	..	4 056	4 072	..	Autres matériels de transport		**35**	
..	..	..	..	..	2 788	2 866	..	Construction et réparation de navires			351
..	..	..	..	..	56	54	..	Matériel ferroviaire roulant			352
..	..	..	..	..	68	69	..	Construction aéronautique et spatiale			353
..	..	..	..	..	1 144	1 083	..	Autres équipements de transport			359
..	..	..	..	..	58 156	52 808	..	Meubles ; activités de fabrication, nca		**36**	
..	..	..	..	..	40 750	35 893	..	Meubles			361
..	..	..	..	..	17 406	16 915	..	Activités de fabrication, nca			369
..	..	..	..	..	1 954	1 864	..	Récupération		**37**	
..	..	..	..	..	1 980	2 301	..	Electricité, gaz et eau	E	**40_41**	
..	..	..	..	..	1 042	1 178	..	Prod. et dist. d'électricité gaz, vapeur d'eau chaude		**40**	
..	..	..	..	..	938	1 123	..	Captage, épuration et distribution d'eau		**41**	
..	..	..	..	..	461 641	452 881	..	Construction	F	**45**	

Table JN.1 **PRODUCTION - PRODUCTION**

Billions of Y (Current Prices)

ISIC revision 3				1991	1992	1993	1994	1995	1996	1997	1998
C	10_14		Mining and quarrying	..	..	..	..	..	..	..	..
	10		Mining of coal and lignite; extraction of peat	..	..	..	..	..	..	..	..
	11		Crude petroleum & natural gas; service activities	..	..	..	..	..	..	..	..
	13		Mining of metal ores	..	..	..	..	..	..	..	..
	14		Other mining and quarrying	..	..	..	..	..	..	..	..
D	15_37		Total manufacturing	..	..	..	298 112	306 754	313 645	324 038	305 801
	15		Food products and beverages	..	..	..	32 108	32 203	32 394	32 646	33 039
		151	Production of meat, fish, fruit, vegetables, oils & fats	..	..	..	7 830	7 888	7 869	7 972	8 063
		152	Dairy products	..	..	..	2 261	2 299	2 302	2 286	2 296
		153	Grain mill and starch products; animal feeds	..	..	..	2 823	2 609	2 659	2 652	2 694
		154	Other food products	..	..	..	12 472	12 706	12 903	12 931	13 185
		155	Beverages	..	..	..	6 722	6 702	6 661	6 805	6 801
	16		Tobacco products	..	..	..	4 579	4 505	4 390	4 576	4 617
	17		Textiles	..	..	..	7 311	6 944	6 621	6 510	5 951
		171	Spinning, weaving and finishing of textiles	..	..	..	3 298	3 044	2 851	2 821	2 457
		172	Other textiles	..	..	..	2 223	2 196	2 147	2 136	2 022
		173	Knitted and crocheted fabrics and articles	..	..	..	1 789	1 704	1 623	1 553	1 472
	18		Wearing apparel and fur	..	..	..	2 718	2 581	2 437	2 301	2 149
		181	Wearing apparel, except fur apparel	..	..	..	2 704	2 568	2 420	2 290	2 138
		182	Dressing and dyeing of fur; articles of fur	..	..	..	14	13	18	11	11
	19		Leather and articles of leather; footwear	..	..	..	1 253	1 311	1 294	1 173	1 070
		191	Leather and articles of leather	..	..	..	615	602	596	516	510
		192	Footwear	..	..	..	638	709	698	657	559
	20		Wood & products of wood & cork, exc. furniture	..	..	..	5 099	4 978	5 048	4 924	4 124
		201	Sawmilling and planing of wood	..	..	..	2 094	1 975	1 928	1 835	1 540
		202	Wood products; articles of cork, straw, etc.	..	..	..	3 005	3 003	3 120	3 089	2 584
	21		Paper and paper products	..	..	..	8 068	8 467	8 616	8 621	8 164
	22		Printing, publishing and recorded media	..	..	..	12 715	13 324	13 711	14 223	13 848
		221	Publishing	..	..	..	4 620	4 826	5 043	5 092	5 151
		222	Printing and related activities	..	..	..	7 869	8 232	8 431	8 595	8 455
		223	Reproduction of recorded media	..	..	..	225	266	237	536	241
	23		Coke and petroleum products; nuclear fuel	..	..	..	7 868	7 688	8 515	9 132	8 259
		231	Coke oven products	..	..	..	633	632	688	708	692
		232	Refined petroleum products	..	..	..	7 153	6 961	7 718	8 345	7 465
		233	Processing of nuclear fuel	..	..	..	82	95	108	79	103
	24		Chemicals and chemical products	..	..	..	22 413	23 390	23 465	24 640	23 228
		241	Basic chemicals	..	..	..	8 390	8 994	9 035	9 932	9 151
		242	Other chemical products	..	..	..	13 143	13 512	13 569	13 834	13 320
		243	Man-made fibres	..	..	..	880	884	862	875	758
	25		Rubber and plastic products	..	..	..	13 103	13 404	13 458	13 889	13 283
		251	Rubber products	..	..	..	2 841	3 002	3 056	3 132	3 001
		252	Plastic products	..	..	..	10 263	10 402	10 402	10 757	10 282
	26		Other non-metallic mineral products	..	..	..	10 230	10 130	10 200	10 207	9 355
		261	Glass and glass products	..	..	..	1 687	1 593	1 615	1 633	1 481
		269	Non-metallic mineral products, nec	..	..	..	8 544	8 537	8 585	8 574	7 875
	27		Basic metals	..	..	..	16 680	17 600	17 454	18 521	16 522
		271	Basic iron and steel	..	..	..	11 693	12 057	11 854	12 506	11 098
		272	Basic precious and non-ferrous metals	..	..	..	3 176	3 625	3 706	4 005	3 667
		273	Casting of metals	..	..	..	1 811	1 918	1 895	2 010	1 756
	28		Fabricated metal products, exc. machinery	..	..	..	19 643	19 936	20 281	20 220	18 881
		281	Structural metal prod., tanks, reservoirs, etc.	..	..	..	10 052	10 152	10 461	10 206	9 636
		289	Other fabricated metal products	..	..	..	9 591	9 784	9 820	10 014	9 245
	29		Machinery and equipment, nec	..	..	..	29 232	31 704	33 518	34 414	31 765
		291	General purpose machinery	..	..	..	12 056	12 519	12 860	13 672	12 861
		292	Special purpose machinery	..	..	..	12 193	13 831	15 279	15 635	14 311
		293	Domestic appliances, nec	..	..	..	4 984	5 354	5 379	5 106	4 592
	30		Office, accounting & computing machinery	..	..	..	21 129	22 020	24 386	25 517	23 406
	31		Electrical machinery & apparatus, nec	..	..	..	15 104	15 488	15 590	16 471	14 635
		311	Electric motors, generators and transformers	..	..	..	2 207	2 371	2 329	2 429	2 116
		312	Electricity distribution & control apparatus	..	..	..	4 316	4 326	4 399	4 421	4 161
		313	Insulated wire and cable	..	..	..	1 797	1 927	2 052	2 216	1 834
		314	Accumulators, primary cells & primary batteries	..	..	..	750	730	831	927	1 005
		315	Electric lamps and lighting equipment	..	..	..	1 327	1 276	1 295	1 331	1 238
		319	Other electrical equipment, nec	..	..	..	4 708	4 858	4 683	5 147	4 281
	32		Radio, TV, communication equipment	..	..	..	24 779	27 211	28 526	30 250	28 806
		321	Electronic valves, tubes; other electronic components	..	..	..	10 533	11 840	11 897	12 395	11 707
		322	TV/radio transmitters; line teleph./telegr. apparatus	..	..	..	3 576	4 160	5 441	5 504	5 037
		323	TV/radio receivers; sound and video app.; assoc. gds.	..	..	..	10 669	11 210	11 188	12 351	12 062
	33		Medical, precision & optical instrum.; watch & clocks	..	..	..	5 333	5 469	5 633	6 113	6 138
		331	Medical appl.; instruments & appl. for measuring, etc.	..	..	..	3 417	3 710	3 895	4 180	4 266
		332	Optical instruments and photographic equipment	..	..	..	1 188	1 091	1 124	1 290	1 262
		333	Watches and clocks	..	..	..	727	669	614	643	609
	34		Motor vehicles, trailers and semi-trailers	..	..	..	39 507	39 556	40 642	42 518	40 291
		341	Motor vehicles	..	..	..	19 283	19 548	20 563	21 310	21 098
		342	Bodies for motor vehicles, trailers & semi-trailers	..	..	..	2 868	2 824	2 821	2 958	2 534
		343	Parts & accessories for motor vehicles & their engines	..	..	..	17 355	17 184	17 257	18 250	16 659
	35		Other transport equipment	..	..	..	3 828	3 817	3 543	4 128	4 357
		351	Building and repairing of ships and boats	..	..	..	2 162	2 109	1 905	2 288	2 478
		352	Railway and tramway locomotives and rolling stock	..	..	..	366	428	388	488	340
		353	Aircraft and spacecraft	..	..	..	837	839	854	972	1 140
		359	Transport equipment, nec	..	..	..	463	441	396	380	400
	36		Furniture; manufacturing, nec	..	..	..	7 879	7 839	7 878	7 630	7 526
		361	Furniture	..	..	..	2 919	2 909	2 972	2 891	2 515
		369	Manufacturing, nec	..	..	..	4 960	4 930	4 907	4 739	5 011
	37		Recycling	..	..	..	385	451	432	461	399
E	40_41		Electricity, gas and water supply	..	..	..	..	..	..	..	..
	40		Electricity, gas, steam and hot water supply	..	..	..	..	..	..	..	..
	41		Collection, purification and distribution of water	..	..	..	..	..	..	..	..
F	45		Construction	..	..	..	..	..	..	..	..

Table JN.2 **VALEUR AJOUTÉE - VALUE ADDED**

Milliards de Y (Prix courants)

1991	1992	1993	1994	1995	1996	1997	1998			CITI révision 3	
..	..	..	..	..	..	..	..	Activites extractives	C	10_14	
..	..	..	..	..	..	..	..	Extraction de charbon, de lignite et de tourbe		10	
..	..	..	..	..	..	..	..	Pétrole brut et gaz naturel ; activités annexes		11	
..	..	..	..	..	..	..	..	Extraction de minerais métalliques		13	
..	..	..	..	..	..	..	..	Autres activités extractives		14	
..	..	..	112 349	117 204	119 304	119 873	113 193	Total activites manufacturieres	D	15_37	
..	..	..	11 621	11 962	11 933	11 818	12 049	Produits alimentaires et boissons		15	
..	..	..	2 436	2 509	2 443	2 471	2 533	Viande, poisson, fruits, légumes, huiles et graisses			151
..	..	..	717	742	713	675	694	Produits laitiers			152
..	..	..	568	579	523	503	525	Travail des grains; prod. amyl. ; alim. pour animaux			153
..	..	..	5 608	5 802	5 854	5 756	5 850	Autres produits alimentaires			154
..	..	..	2 292	2 330	2 401	2 413	2 447	Boissons			155
..	..	..	702	762	794	904	869	Produits à base de tabac		16	
..	..	..	3 059	2 934	2 811	2 775	2 538	Textiles		17	
..	..	..	1 417	1 316	1 248	1 244	1 086	Filature, tissage et achèvement des textiles			171
..	..	..	916	923	897	893	841	Autres articles textiles			172
..	..	..	726	695	666	638	612	Etoffes et articles de bonneterie			173
..	..	..	1 433	1 365	1 289	1 198	1 127	Articles d'habillement et fourrure		18	
..	..	..	1 427	1 360	1 281	1 193	1 122	Articles d'habillement autres qu'en fourrure			181
..	..	..	6	5	8	5	4	Prép. et teinture des fourrures ; articles en fourrure			182
..	..	..	508	539	522	477	410	Cuir et articles en cuir ; chaussures		19	
..	..	..	230	231	224	195	194	Cuir et articles en cuir sauf chaussures			191
..	..	..	278	308	298	282	216	Chaussures			192
..	..	..	1 904	1 882	1 884	1 804	1 551	Bois, articles en bois, sauf meubles		20	
..	..	..	737	696	676	628	544	Sciage et rabotage du bois			201
..	..	..	1 166	1 186	1 208	1 176	1 007	Articles en bois, liège, vannerie, sparterie			202
..	..	..	3 040	3 205	3 260	3 225	3 032	Papier, articles en papier et carton		21	
..	..	..	6 618	6 893	7 054	7 211	7 115	Edition, imprimerie et reproduction		22	
..	..	..	2 808	2 922	3 031	3 082	3 093	Edition			221
..	..	..	3 658	3 813	3 872	3 911	3 886	Imprimerie et activités annexes			222
..	..	..	151	158	151	217	136	Reproduction de supports enregistrés			223
..	..	..	1 715	1 376	1 162	855	940	Coke. et prod. pétroliers ; comb. nucléaires		23	
..	..	..	185	185	197	184	166	Cokéfaction			231
..	..	..	1 507	1 149	910	652	730	Produits pétroliers raffinés			232
..	..	..	22	42	55	18	44	Traitement de combustibles nucléaires			233
..	..	..	11 365	11 971	11 885	12 110	11 315	Produits chimiques		24	
..	..	..	3 682	4 031	3 873	4 035	3 729	Produits chimiques de base			241
..	..	..	7 324	7 583	7 667	7 713	7 292	Autres produits chimiques			242
..	..	..	359	357	344	361	293	Fibres synthétiques ou artificielles			243
..	..	..	5 440	5 562	5 630	5 772	5 541	Art. en caoutchouc et matières plastiques		25	
..	..	..	1 349	1 401	1 433	1 499	1 417	Articles en caoutchouc			251
..	..	..	4 091	4 161	4 197	4 273	4 125	Articles en matières plastiques			252
..	..	..	5 011	4 946	4 986	4 944	4 470	Autres produits minéraux non métalliques		26	
..	..	..	927	841	848	876	769	Verre et articles en verre			261
..	..	..	4 084	4 105	4 138	4 068	3 701	Produits minéraux non métalliques, nca			269
..	..	..	5 600	5 979	5 976	6 245	5 367	Produits métallurgiques de base		27	
..	..	..	3 885	4 132	4 079	4 296	3 656	Sidérurgie et première transformation de l'acier			271
..	..	..	871	960	1 007	1 037	920	Première transf. métaux précieux et non ferreux			272
..	..	..	843	887	889	913	792	Fonderie			273
..	..	..	8 826	9 059	9 228	9 013	8 468	Ouvrages en métaux, sauf machines		28	
..	..	..	4 431	4 503	4 611	4 377	4 155	Constr. et menuis. métal.; réservoirs; générateurs			281
..	..	..	4 395	4 556	4 617	4 635	4 313	Autres ouvrages en métaux			289
..	..	..	12 041	13 076	13 670	13 961	12 845	Machines et matériels, nca		29	
..	..	..	5 000	5 186	5 301	5 593	5 154	Machines d'usage général			291
..	..	..	4 914	5 637	6 200	6 341	5 876	Machines d'usage spécifique			292
..	..	..	2 127	2 252	2 169	2 027	1 815	Appareils domestiques, nca			293
..	..	..	5 981	5 510	6 036	5 725	5 579	Mach. de bureau, comptables, information		30	
..	..	..	5 635	5 895	5 872	6 023	5 388	Machines et appareils électriques, nca		31	
..	..	..	816	899	869	900	764	Moteurs, génératrices, transformateurs électriques			311
..	..	..	1 783	1 783	1 787	1 752	1 711	Matériel électrique de distribution et de commande			312
..	..	..	463	559	630	670	537	Fils et câbles électriques isolés			313
..	..	..	332	312	321	337	353	Accumulateurs et piles électriques			314
..	..	..	513	495	508	487	448	Lampes électriques et appareils d'éclairage			315
..	..	..	1 729	1 846	1 757	1 877	1 575	Autres matériels électriques, nca.			319
..	..	..	8 437	9 586	9 917	10 161	9 335	Equipements et appareils de radio, TV, etc.		32	
..	..	..	3 867	4 445	4 273	4 263	3 834	Tubes et valves électroniques;autres composants			321
..	..	..	1 222	1 464	1 841	1 705	1 602	Emetteurs radio/TV; app. téléphonie et télégraphie			322
..	..	..	3 348	3 677	3 802	4 193	3 900	Récepteurs radio/TV, son et image; art. associés			323
..	..	..	2 151	2 296	2 428	2 548	2 493	Instr. médicaux, de précision ; optique ; horlog.		33	
..	..	..	1 546	1 716	1 822	1 890	1 853	Instruments médicaux et de précision, etc.			331
..	..	..	453	414	425	474	473	Instruments d'optique ; matériel photographique			332
..	..	..	153	166	181	184	168	Horlogerie			333
..	..	..	9 750	10 707	11 788	11 615	11 176	Véhicules automobiles; remorques, etc.		34	
..	..	..	3 549	4 432	5 418	5 196	5 587	Véhicules automobiles			341
..	..	..	597	608	649	680	564	Carross. automobiles, remorques, semi-remorques			342
..	..	..	5 603	5 668	5 721	5 739	5 024	Pièces et accessoires pour automobiles et moteurs			343
..	..	..	1 475	1 479	1 260	1 535	1 642	Autres matériels de transport		35	
..	..	..	795	762	584	825	915	Construction et réparation de navires			351
..	..	..	157	183	161	187	121	Matériel ferroviaire roulant			352
..	..	..	309	333	354	381	453	Construction aéronautique et spatiale			353
..	..	..	215	201	160	142	153	Autres équipements de transport			359
..	..	..	3 241	3 185	3 170	3 087	3 016	Meubles ; activités de fabrication, nca		36	
..	..	..	1 242	1 243	1 277	1 228	1 059	Meubles			361
..	..	..	1 999	1 942	1 893	1 858	1 956	Activités de fabrication, nca			369
..	..	..	137	172	166	182	151	Récupération		37	
..	..	..	..	..	..	..	..	Electricité, gaz et eau	E	40_41	
..	..	..	..	..	..	..	..	Prod. et dist. d'électricité gaz, vapeur d'eau chaude		40	
..	..	..	..	..	..	..	..	Captage, épuration et distribution d'eau		41	
..	..	..	..	..	..	..	..	Construction	F	45	

Table JN.3 **INVESTMENT - INVESTISSEMENT**

Billions of Y (Current Prices)

ISIC revision 3				1991	1992	1993	1994	1995	1996	1997	1998
C	10_14		Mining and quarrying	..	..	..	..	..	..	..	..
	10		Mining of coal and lignite; extraction of peat	..	..	..	..	..	..	..	..
	11		Crude petroleum & natural gas; service activities	..	..	..	..	..	..	..	..
	13		Mining of metal ores	..	..	..	..	..	..	..	..
	14		Other mining and quarrying	..	..	..	..	..	..	..	..
D	15_37		Total manufacturing	..	..	..	12 382	12 098	13 202	14 070	13 913
	15		Food products and beverages	..	..	..	1 184	1 059	1 100	1 178	1 106
		151	Production of meat, fish, fruit, vegetables, oils & fats	..	..	..	206	196	201	216	177
		152	Dairy products	..	..	..	104	96	83	77	96
		153	Grain mill and starch products; animal feeds	..	..	..	99	63	71	68	67
		154	Other food products	..	..	..	528	478	513	475	439
		155	Beverages	..	..	..	246	227	233	342	326
	16		Tobacco products	..	..	..	59	34	80	22	32
	17		Textiles	..	..	..	214	188	202	175	162
		171	Spinning, weaving and finishing of textiles	..	..	..	119	104	109	96	92
		172	Other textiles	..	..	..	56	56	64	54	48
		173	Knitted and crocheted fabrics and articles	..	..	..	39	27	28	26	22
	18		Wearing apparel and fur	..	..	..	43	29	33	30	22
		181	Wearing apparel, except fur apparel	..	..	..	43	29	33	30	22
		182	Dressing and dyeing of fur; articles of fur	..	..	..	..	..	..	..	..
	19		Leather and articles of leather; footwear	..	..	..	25	21	22	19	17
		191	Leather and articles of leather	..	..	..	16	13	16	12	12
		192	Footwear	..	..	..	9	9	6	7	5
	20		Wood & products of wood & cork, exc. furniture	..	..	..	131	125	131	144	103
		201	Sawmilling and planing of wood	..	..	..	49	44	45	40	32
		202	Wood products; articles of cork, straw, etc.	..	..	..	82	81	86	104	72
	21		Paper and paper products	..	..	..	415	381	489	582	527
	22		Printing, publishing and recorded media	..	..	..	474	456	491	570	455
		221	Publishing	..	..	..	160	119	137	159	111
		222	Printing and related activities	..	..	..	306	327	332	396	324
		223	Reproduction of recorded media	..	..	..	8	11	22	16	20
	23		Coke and petroleum products; nuclear fuel	..	..	..	383	456	466	332	216
		231	Coke oven products	..	..	..	22	17	19	16	15
		232	Refined petroleum products	..	..	..	297	344	346	244	143
		233	Processing of nuclear fuel	..	..	..	64	95	100	73	58
	24		Chemicals and chemical products	..	..	..	1 280	1 189	1 170	1 347	1 496
		241	Basic chemicals	..	..	..	646	522	528	627	669
		242	Other chemical products	..	..	..	550	593	551	656	720
		243	Man-made fibres	..	..	..	83	74	90	64	107
	25		Rubber and plastic products	..	..	..	618	632	664	726	770
		251	Rubber products	..	..	..	142	140	147	161	193
		252	Plastic products	..	..	..	476	492	516	565	577
	26		Other non-metallic mineral products	..	..	..	596	553	567	619	595
		261	Glass and glass products	..	..	..	104	104	110	123	121
		269	Non-metallic mineral products, nec	..	..	..	491	450	457	496	474
	27		Basic metals	..	..	..	1 147	1 211	1 203	1 018	865
		271	Basic iron and steel	..	..	..	873	906	890	674	584
		272	Basic precious and non-ferrous metals	..	..	..	167	215	208	211	182
		273	Casting of metals	..	..	..	107	90	105	133	99
	28		Fabricated metal products, exc. machinery	..	..	..	722	600	719	799	679
		281	Structural metal prod., tanks, reservoirs, etc.	..	..	..	326	231	269	313	243
		289	Other fabricated metal products	..	..	..	396	369	450	486	436
	29		Machinery and equipment, nec	..	..	..	946	909	1 041	1 117	1 098
		291	General purpose machinery	..	..	..	394	389	414	458	472
		292	Special purpose machinery	..	..	..	359	366	476	489	479
		293	Domestic appliances, nec	..	..	..	192	153	151	170	147
	30		Office, accounting & computing machinery	..	..	..	821	691	724	776	835
	31		Electrical machinery & apparatus, nec	..	..	..	553	629	770	827	688
		311	Electric motors, generators and transformers	..	..	..	52	74	76	69	63
		312	Electricity distribution & control apparatus	..	..	..	116	115	103	122	106
		313	Insulated wire and cable	..	..	..	79	100	99	130	106
		314	Accumulators, primary cells & primary batteries	..	..	..	55	54	72	70	78
		315	Electric lamps and lighting equipment	..	..	..	39	40	49	56	46
		319	Other electrical equipment, nec	..	..	..	213	246	370	381	289
	32		Radio, TV, communication equipment	..	..	..	1 255	1 688	1 940	2 031	2 109
		321	Electronic valves, tubes; other electronic components	..	..	..	782	1 108	1 156	1 180	1 207
		322	TV/radio transmitters; line teleph./telegr. apparatus	..	..	..	84	100	122	126	165
		323	TV/radio receivers; sound and video app.; assoc. gds.	..	..	..	390	480	662	724	737
	33		Medical, precision & optical instrum.; watch & clocks	..	..	..	172	155	173	192	261
		331	Medical appl.; instruments & appl. for measuring, etc.	..	..	..	102	91	108	122	184
		332	Optical instruments and photographic equipment	..	..	..	40	36	44	45	46
		333	Watches and clocks	..	..	..	30	28	21	24	31
	34		Motor vehicles, trailers and semi-trailers	..	..	..	1 386	1 157	1 314	1 626	1 964
		341	Motor vehicles	..	..	..	458	398	436	466	596
		342	Bodies for motor vehicles, trailers & semi-trailers	..	..	..	90	72	90	115	99
		343	Parts & accessories for motor vehicles & their engines	..	..	..	838	687	788	1 046	1 269
	35		Other transport equipment	..	..	..	154	116	103	103	125
		351	Building and repairing of ships and boats	..	..	..	98	52	50	47	50
		352	Railway and tramway locomotives and rolling stock	..	..	..	10	11	18	13	26
		353	Aircraft and spacecraft	..	..	..	31	39	21	29	38
		359	Transport equipment, nec	..	..	..	15	14	14	14	12
	36		Furniture; manufacturing, nec	..	..	..	231	167	183	217	203
		361	Furniture	..	..	..	78	58	66	70	59
		369	Manufacturing, nec	..	..	..	153	109	117	147	144
	37		Recycling	..	..	..	12	16	21	18	17
E	40_41		Electricity, gas and water supply	..	..	..	..	..	..	..	..
	40		Electricity, gas, steam and hot water supply	..	..	..	..	..	..	..	..
	41		Collection, purification and distribution of water	..	..	..	..	..	..	..	..
F	45		Construction	..	..	..	..	..	..	..	..

Table JN.4 DÉPENSES DE MACHINES ET D'ÉQUIPEMENT - INVESTMENT IN MACHINERY & EQUIPMENT

Milliards de Y (Prix courants)

1991	1992	1993	1994	1995	1996	1997	1998		CITI révision 3		
..	..	..	..	..	..	..	..	Activites extractives	C	10_14	
..	..	..	..	..	..	..	..	Extraction de charbon, de lignite et de tourbe		10	
..	..	..	..	..	..	..	..	Pétrole brut et gaz naturel ; activités annexes		11	
..	..	..	..	..	..	..	..	Extraction de minerais métalliques		13	
..	..	..	..	..	..	..	..	Autres activités extractives		14	
..	..	..	6 457	6 903	7 265	7 683	7 602	Total activites manufacturieres	D	15_37	
..	..	..	547	539	524	573	556	Produits alimentaires et boissons		15	
..	..	..	73	78	75	85	65	Viande, poisson, fruits, légumes, huiles et graisses			151
..	..	..	64	62	54	51	61	Produits laitiers			152
..	..	..	45	33	33	29	31	Travail des grains; prod. amyl. ; alim. pour animaux			153
..	..	..	232	238	237	217	206	Autres produits alimentaires			154
..	..	..	133	128	125	191	193	Boissons			155
..	..	..	50	27	64	16	22	Produits à base de tabac		16	
..	..	..	95	106	98	86	78	Textiles		17	
..	..	..	58	64	57	52	50	Filature, tissage et achèvement des textiles			171
..	..	..	18	27	26	22	19	Autres articles textiles			172
..	..	..	19	15	15	12	9	Etoffes et articles de bonneterie			173
..	..	..	12	9	9	8	6	Articles d'habillement et fourrure		18	
..	..	..	12	9	9	8	6	Articles d'habillement autres qu'en fourrure			181
..	..	..	..	..	..	..	..	Prép. et teinture des fourrures ; articles en fourrure			182
..	..	..	7	8	7	8	5	Cuir et articles en cuir ; chaussures		19	
..	..	..	4	6	5	6	4	Cuir et articles en cuir sauf chaussures			191
..	..	..	3	2	2	2	1	Chaussures			192
..	..	..	41	53	41	51	39	Bois, articles en bois, sauf meubles		20	
..	..	..	11	12	10	11	8	Sciage et rabotage du bois			201
..	..	..	30	42	31	40	30	Articles en bois, liège, vannerie, sparterie			202
..	..	..	282	266	332	397	375	Papier, articles en papier et carton		21	
..	..	..	185	223	243	296	229	Edition, imprimerie et reproduction		22	
..	..	..	34	41	54	76	50	Edition			221
..	..	..	146	175	175	209	167	Imprimerie et activités annexes			222
..	..	..	5	8	14	10	11	Reproduction de supports enregistrés			223
..	..	..	233	363	294	251	150	Coke. et prod. pétroliers ; comb. nucléaires		23	
..	..	..	11	9	9	8	7	Cokéfaction			231
..	..	..	189	274	262	187	92	Produits pétroliers raffinés			232
..	..	..	33	79	24	56	51	Traitement de combustibles nucléaires			233
..	..	..	791	730	715	801	917	Produits chimiques		24	
..	..	..	434	364	375	423	472	Produits chimiques de base			241
..	..	..	291	310	284	331	369	Autres produits chimiques			242
..	..	..	67	55	56	48	75	Fibres synthétiques ou artificielles			243
..	..	..	290	325	338	338	375	Art. en caoutchouc et matières plastiques		25	
..	..	..	68	80	81	80	104	Articles en caoutchouc			251
..	..	..	222	245	258	259	271	Articles en matières plastiques			252
..	..	..	261	267	257	308	294	Autres produits minéraux non métalliques		26	
..	..	..	69	72	76	70	77	Verre et articles en verre			261
..	..	..	192	196	181	239	217	Produits minéraux non métalliques, nca			269
..	..	..	779	827	826	710	603	Produits métallurgiques de base		27	
..	..	..	612	646	655	511	440	Sidérurgie et première transformation de l'acier			271
..	..	..	115	134	123	134	118	Première transf. métaux précieux et non ferreux			272
..	..	..	52	46	48	65	45	Fonderie			273
..	..	..	286	288	325	341	299	Ouvrages en métaux, sauf machines		28	
..	..	..	102	98	102	113	86	Constr. et menuis. métal.; réservoirs; générateurs			281
..	..	..	184	189	223	229	213	Autres ouvrages en métaux			289
..	..	..	366	392	432	477	458	Machines et matériels, nca		29	
..	..	..	172	192	189	227	222	Machines d'usage général			291
..	..	..	119	145	186	190	183	Machines d'usage spécifique			292
..	..	..	75	56	57	60	53	Appareils domestiques, nca			293
..	..	..	316	270	221	314	317	Mach. de bureau, comptables, information		30	
..	..	..	299	365	468	465	378	Machines et appareils électriques, nca		31	
..	..	..	24	38	37	34	31	Moteurs, génératrices, transformateurs électriques			311
..	..	..	41	41	36	41	37	Matériel électrique de distribution et de commande			312
..	..	..	47	54	65	81	65	Fils et câbles électriques isolés			313
..	..	..	41	42	54	52	56	Accumulateurs et piles électriques			314
..	..	..	15	18	20	19	16	Lampes électriques et appareils d'éclairage			315
..	..	..	130	171	257	238	173	Autres matériels électriques, nca.			319
..	..	..	810	1 158	1 327	1 320	1 390	Equipements et appareils de radio, TV, etc.		32	
..	..	..	567	834	860	828	888	Tubes et valves électroniques;autres composants			321
..	..	..	28	38	39	41	58	Emetteurs radio/TV; app. téléphonie et télégraphie			322
..	..	..	215	287	428	450	444	Récepteurs radio/TV, son et image; art. associés			323
..	..	..	60	61	63	64	98	Instr. médicaux, de précision ; optique ; horlog.		33	
..	..	..	33	33	36	39	64	Instruments médicaux et de précision, etc.			331
..	..	..	12	12	16	13	17	Instruments d'optique ; matériel photographique			332
..	..	..	14	15	11	13	17	Horlogerie			333
..	..	..	796	660	714	901	1 060	Véhicules automobiles; remorques, etc.		34	
..	..	..	253	226	239	237	313	Véhicules automobiles			341
..	..	..	47	34	40	56	51	Carross. automobiles, remorques, semi-remorques			342
..	..	..	496	400	434	608	696	Pièces et accessoires pour automobiles et moteurs			343
..	..	..	47	45	39	42	46	Autres matériels de transport		35	
..	..	..	28	23	25	21	23	Construction et réparation de navires			351
..	..	..	3	5	3	6	6	Matériel ferroviaire roulant			352
..	..	..	12	11	9	12	14	Construction aéronautique et spatiale			353
..	..	..	5	6	2	3	3	Autres équipements de transport			359
..	..	..	84	66	64	77	74	Meubles ; activités de fabrication, nca		36	
..	..	..	23	23	22	24	17	Meubles			361
..	..	..	61	43	42	53	56	Activités de fabrication, nca			369
..	..	..	2	4	5	3	4	Récupération		37	
..	..	..	..	..	..	..	..	Electricité, gaz et eau	E	40_41	
..	..	..	..	..	..	..	..	Prod. et dist. d'électricité gaz, vapeur d'eau chaude		40	
..	..	..	..	..	..	..	..	Captage, épuration et distribution d'eau		41	
..	..	..	..	..	..	..	..	Construction	F	45	

Table JN.5 EMPLOYMENT, EMPLOYEES - EMPLOI, SALARIÉS

Thousands

ISIC revision 3				1991	1992	1993	1994	1995	1996	1997	1998
C	10_14		Mining and quarrying	..	..	..	..	..	..	..	..
	10		Mining of coal and lignite; extraction of peat	..	..	..	..	..	..	..	..
	11		Crude petroleum & natural gas; service activities	..	..	..	..	..	..	..	..
	13		Mining of metal ores	..	..	..	..	..	..	..	..
	14		Other mining and quarrying	..	..	..	..	..	..	..	..
D	15_37		Total manufacturing	..	..	..	10 289	10 187	9 987	9 831	9 725
	15		Food products and beverages	..	..	..	1 204	1 226	1 219	1 205	1 247
		151	Production of meat, fish, fruit, vegetables, oils & fats	..	..	..	333	335	330	329	341
		152	Dairy products	..	..	..	45	45	45	43	44
		153	Grain mill and starch products; animal feeds	..	..	..	30	30	30	29	30
		154	Other food products	..	..	..	707	728	728	719	747
		155	Beverages	..	..	..	88	89	86	85	84
	16		Tobacco products	..	..	..	16	15	15	14	14
	17		Textiles	..	..	..	499	473	448	427	408
		171	Spinning, weaving and finishing of textiles	..	..	..	210	193	179	169	155
		172	Other textiles	..	..	..	134	132	127	122	123
		173	Knitted and crocheted fabrics and articles	..	..	..	156	149	142	135	130
	18		Wearing apparel and fur	..	..	..	397	364	337	312	287
		181	Wearing apparel, except fur apparel	..	..	..	396	363	336	311	286
		182	Dressing and dyeing of fur; articles of fur	..	..	..	1	1	1	1	1
	19		Leather and articles of leather; footwear	..	..	..	79	81	78	71	66
		191	Leather and articles of leather	..	..	..	34	32	31	27	27
		192	Footwear	..	..	..	45	49	47	44	39
	20		Wood & products of wood & cork, exc. furniture	..	..	..	263	258	253	240	224
		201	Sawmilling and planing of wood	..	..	..	104	100	94	88	81
		202	Wood products; articles of cork, straw, etc.	..	..	..	159	158	159	152	143
	21		Paper and paper products	..	..	..	266	265	260	255	256
	22		Printing, publishing and recorded media	..	..	..	536	541	535	533	545
		221	Publishing	..	..	..	115	114	115	117	120
		222	Printing and related activities	..	..	..	416	421	414	409	417
		223	Reproduction of recorded media	..	..	..	6	6	7	7	8
	23		Coke and petroleum products; nuclear fuel	..	..	..	37	37	36	35	34
		231	Coke oven products	..	..	..	10	10	10	10	10
		232	Refined petroleum products	..	..	..	24	23	23	22	21
		233	Processing of nuclear fuel	..	..	..	4	4	4	4	2
	24		Chemicals and chemical products	..	..	..	397	391	388	383	382
		241	Basic chemicals	..	..	..	124	120	119	118	118
		242	Other chemical products	..	..	..	249	248	247	243	245
		243	Man-made fibres	..	..	..	25	23	22	21	19
	25		Rubber and plastic products	..	..	..	561	565	555	552	549
		251	Rubber products	..	..	..	127	127	125	123	120
		252	Plastic products	..	..	..	433	439	430	429	429
	26		Other non-metallic mineral products	..	..	..	424	421	410	401	389
		261	Glass and glass products	..	..	..	56	55	53	52	51
		269	Non-metallic mineral products, nec	..	..	..	368	366	358	349	338
	27		Basic metals	..	..	..	384	372	358	349	331
		271	Basic iron and steel	..	..	..	231	218	207	197	189
		272	Basic precious and non-ferrous metals	..	..	..	67	68	67	68	65
		273	Casting of metals	..	..	..	86	86	83	83	78
	28		Fabricated metal products, exc. machinery	..	..	..	870	869	856	844	826
		281	Structural metal prod., tanks, reservoirs, etc.	..	..	..	377	378	372	361	359
		289	Other fabricated metal products	..	..	..	493	491	485	482	467
	29		Machinery and equipment, nec	..	..	..	1 120	1 128	1 131	1 123	1 115
		291	General purpose machinery	..	..	..	454	452	455	458	458
		292	Special purpose machinery	..	..	..	512	521	527	519	518
		293	Domestic appliances, nec	..	..	..	155	155	149	146	139
	30		Office, accounting & computing machinery	..	..	..	448	438	427	428	431
	31		Electrical machinery & apparatus, nec	..	..	..	615	600	580	578	556
		311	Electric motors, generators and transformers	..	..	..	108	108	102	103	95
		312	Electricity distribution & control apparatus	..	..	..	181	176	172	169	165
		313	Insulated wire and cable	..	..	..	51	51	51	50	47
		314	Accumulators, primary cells & primary batteries	..	..	..	23	22	23	22	24
		315	Electric lamps and lighting equipment	..	..	..	51	50	49	48	47
		319	Other electrical equipment, nec	..	..	..	201	193	182	186	178
	32		Radio, TV, communication equipment	..	..	..	869	868	845	837	837
		321	Electronic valves, tubes; other electronic components	..	..	..	349	351	345	344	347
		322	TV/radio transmitters; line teleph./telegr. apparatus	..	..	..	87	90	93	87	90
		323	TV/radio receivers; sound and video app.; assoc. gds.	..	..	..	432	427	407	405	400
	33		Medical, precision & optical instrum.; watch & clocks	..	..	..	254	246	236	238	241
		331	Medical appl.; instruments & appl. for measuring, etc.	..	..	..	156	156	152	156	159
		332	Optical instruments and photographic equipment	..	..	..	68	62	59	60	60
		333	Watches and clocks	..	..	..	30	27	24	22	22
	34		Motor vehicles, trailers and semi-trailers	..	..	..	787	768	769	768	753
		341	Motor vehicles	..	..	..	184	183	181	183	184
		342	Bodies for motor vehicles, trailers & semi-trailers	..	..	..	58	56	55	55	50
		343	Parts & accessories for motor vehicles & their engines	..	..	..	544	528	532	531	518
	35		Other transport equipment	..	..	..	115	115	107	107	110
		351	Building and repairing of ships and boats	..	..	..	57	56	51	50	52
		352	Railway and tramway locomotives and rolling stock	..	..	..	15	17	16	17	15
		353	Aircraft and spacecraft	..	..	..	27	26	26	26	28
		359	Transport equipment, nec	..	..	..	17	16	15	14	15
	36		Furniture; manufacturing, nec	..	..	..	371	361	354	342	337
		361	Furniture	..	..	..	157	152	155	148	139
		369	Manufacturing, nec	..	..	..	214	209	199	194	198
	37		Recycling	..	..	..	10	11	11	11	11
E	40_41		Electricity, gas and water supply	..	..	..	..	..	..	..	..
	40		Electricity, gas, steam and hot water supply	..	..	..	..	..	..	..	..
	41		Collection, purification and distribution of water	..	..	..	..	..	..	..	..
F	45		Construction	..	..	..	..	..	..	..	..

Table JN.6 **EMPLOI, FEMMES - EMPLOYMENT, FEMALES**

Milliers

1991	1992	1993	1994	1995	1996	1997	1998	CITI révision 3			
..	..	..	..	..	..	..	..	Activites extractives	C	10_14	
..	..	..	..	..	..	..	..	Extraction de charbon, de lignite et de tourbe		10	
..	..	..	..	..	..	..	..	Pétrole brut et gaz naturel ; activités annexes		11	
..	..	..	..	..	..	..	..	Extraction de minerais métalliques		13	
..	..	..	..	..	..	..	..	Autres activités extractives		14	
..	..	..	3 707	3 646	3 540	3 462	3 400	Total activites manufacturieres	D	15_37	
..	..	..	682	695	694	686	712	Produits alimentaires et boissons		15	
..	..	..	217	217	214	212	219	Viande, poisson, fruits, légumes, huiles et graisses			151
..	..	..	15	15	15	14	14	Produits laitiers			152
..	..	..	8	8	8	8	8	Travail des grains; prod. amyl. ; alim. pour animaux			153
..	..	..	414	427	430	426	445	Autres produits alimentaires			154
..	..	..	28	28	27	26	26	Boissons			155
..	..	..	6	5	5	5	5	Produits à base de tabac		16	
..	..	..	303	288	272	258	246	Textiles		17	
..	..	..	103	95	87	82	74	Filature, tissage et achèvement des textiles			171
..	..	..	79	78	75	72	72	Autres articles textiles			172
..	..	..	121	115	110	105	101	Etoffes et articles de bonneterie			173
..	..	..	334	306	284	263	240	Articles d'habillement et fourrure		18	
..	..	..	333	305	283	262	240	Articles d'habillement autres qu'en fourrure			181
..	..	..	1	1	1	..	..	Prép. et teinture des fourrures ; articles en fourrure			182
..	..	..	42	44	41	38	35	Cuir et articles en cuir ; chaussures		19	
..	..	..	17	16	16	14	13	Cuir et articles en cuir sauf chaussures			191
..	..	..	25	27	26	24	21	Chaussures			192
..	..	..	80	78	76	71	65	Bois, articles en bois, sauf meubles		20	
..	..	..	28	27	25	23	21	Sciage et rabotage du bois			201
..	..	..	52	51	51	48	45	Articles en bois, liège, vannerie, sparterie			202
..	..	..	87	86	83	81	82	Papier, articles en papier et carton		21	
..	..	..	165	166	162	160	164	Edition, imprimerie et reproduction		22	
..	..	..	25	25	25	26	28	Edition			221
..	..	..	138	138	134	131	133	Imprimerie et activités annexes			222
..	..	..	2	2	2	3	3	Reproduction de supports enregistrés			223
..	..	..	4	4	4	4	4	Coke. et prod. pétroliers ; comb. nucléaires		23	
..	..	..	2	2	2	2	2	Cokéfaction			231
..	..	..	2	2	2	2	2	Produits pétroliers raffinés			232
..	..	..	..	..	..	..	..	Traitement de combustibles nucléaires			233
..	..	..	100	98	98	96	96	Produits chimiques		24	
..	..	..	13	12	12	12	12	Produits chimiques de base			241
..	..	..	82	82	82	81	81	Autres produits chimiques			242
..	..	..	4	4	4	3	3	Fibres synthétiques ou artificielles			243
..	..	..	208	211	208	207	206	Art. en caoutchouc et matières plastiques		25	
..	..	..	37	37	36	36	36	Articles en caoutchouc			251
..	..	..	172	173	172	171	171	Articles en matières plastiques			252
..	..	..	104	103	99	97	92	Autres produits minéraux non métalliques		26	
..	..	..	13	13	13	13	12	Verre et articles en verre			261
..	..	..	91	90	87	84	80	Produits minéraux non métalliques, nca			269
..	..	..	48	47	45	44	42	Produits métallurgiques de base		27	
..	..	..	21	20	19	18	17	Sidérurgie et première transformation de l'acier			271
..	..	..	10	10	10	10	10	Première transf. métaux précieux et non ferreux			272
..	..	..	17	17	16	16	15	Fonderie			273
..	..	..	230	232	225	222	212	Ouvrages en métaux, sauf machines		28	
..	..	..	77	80	75	73	72	Constr. et menuis. métal.; réservoirs; générateurs			281
..	..	..	153	152	149	148	140	Autres ouvrages en métaux			289
..	..	..	242	246	244	243	238	Machines et matériels, nca		29	
..	..	..	85	86	86	87	86	Machines d'usage général			291
..	..	..	93	97	97	97	96	Machines d'usage spécifique			292
..	..	..	64	64	61	59	56	Appareils domestiques, nca			293
..	..	..	170	167	162	161	160	Mach. de bureau, comptables, information		30	
..	..	..	227	222	213	211	202	Machines et appareils électriques, nca		31	
..	..	..	36	36	34	34	31	Moteurs, génératrices, transformateurs électriques			311
..	..	..	60	59	57	56	54	Matériel électrique de distribution et de commande			312
..	..	..	11	11	11	10	10	Fils et câbles électriques isolés			313
..	..	..	7	7	7	7	7	Accumulateurs et piles électriques			314
..	..	..	21	21	21	21	20	Lampes électriques et appareils d'éclairage			315
..	..	..	91	88	84	84	81	Autres matériels électriques, nca.			319
..	..	..	364	352	335	329	321	Equipements et appareils de radio, TV, etc.		32	
..	..	..	117	115	112	111	109	Tubes et valves électroniques;autres composants			321
..	..	..	28	29	29	28	28	Emetteurs radio/TV; app. téléphonie et télégraphie			322
..	..	..	219	209	194	191	184	Récepteurs radio/TV, son et image; art. associés			323
..	..	..	94	91	90	90	90	Instr. médicaux, de précision ; optique ; horlog.		33	
..	..	..	51	51	51	52	51	Instruments médicaux et de précision, etc.			331
..	..	..	30	28	28	28	28	Instruments d'optique ; matériel photographique			332
..	..	..	13	12	11	10	10	Horlogerie			333
..	..	..	144	137	136	134	131	Véhicules automobiles; remorques, etc.		34	
..	..	..	11	11	10	10	10	Véhicules automobiles			341
..	..	..	5	4	4	4	4	Carross. automobiles, remorques, semi-remorques			342
..	..	..	128	122	121	120	118	Pièces et accessoires pour automobiles et moteurs			343
..	..	..	15	15	14	14	14	Autres matériels de transport		35	
..	..	..	5	5	4	4	4	Construction et réparation de navires			351
..	..	..	2	2	2	2	2	Matériel ferroviaire roulant			352
..	..	..	3	3	3	3	3	Construction aéronautique et spatiale			353
..	..	..	6	5	5	5	5	Autres équipements de transport			359
..	..	..	143	138	132	127	124	Meubles ; activités de fabrication, nca		36	
..	..	..	48	46	45	43	40	Meubles			361
..	..	..	95	92	87	84	84	Activités de fabrication, nca			369
..	..	..	2	2	2	2	2	Récupération		37	
..	..	..	..	..	..	..	..	Electricité, gaz et eau	E	40_41	
..	..	..	..	..	..	..	..	Prod. et dist. d'électricité gaz, vapeur d'eau chaude		40	
..	..	..	..	..	..	..	..	Captage, épuration et distribution d'eau		41	
..	..	..	..	..	..	..	..	Construction	F	45	

Table JN.7 **WAGES AND SALARIES, EMPLOYEES - SALAIRES ET TRAITEMENTS, SALARIÉS**

Billions of Y (Current Prices)

ISIC revision 3				1991	1992	1993	1994	1995	1996	1997	1998
C	10_14		Mining and quarrying	..	..	..	..	..	..	..	..
	10		Mining of coal and lignite; extraction of peat	..	..	..	..	..	..	..	..
	11		Crude petroleum & natural gas; service activities	..	..	..	..	..	..	..	..
	13		Mining of metal ores	..	..	..	..	..	..	..	..
	14		Other mining and quarrying	..	..	..	..	..	..	..	..
D	15_37		Total manufacturing	..	..	..	32 174	32 417	32 668	32 967	32 186
	15		Food products and beverages	..	..	..	2 583	2 675	2 684	2 677	2 702
		151	Production of meat, fish, fruit, vegetables, oils & fats	..	..	..	566	581	585	586	596
		152	Dairy products	..	..	..	167	174	174	172	170
		153	Grain mill and starch products; animal feeds	..	..	..	91	91	92	89	90
		154	Other food products	..	..	..	1 474	1 539	1 550	1 540	1 565
		155	Beverages	..	..	..	284	291	283	289	282
	16		Tobacco products	..	..	..	105	107	105	103	100
	17		Textiles	..	..	..	878	838	796	768	708
		171	Spinning, weaving and finishing of textiles	..	..	..	475	436	402	386	346
		172	Other textiles	..	..	..	194	199	200	196	184
		173	Knitted and crocheted fabrics and articles	..	..	..	209	203	193	187	178
	18		Wearing apparel and fur	..	..	..	495	454	421	395	353
		181	Wearing apparel, except fur apparel	..	..	..	494	453	419	395	352
		182	Dressing and dyeing of fur; articles of fur	..	..	..	1	1	2	1	1
	19		Leather and articles of leather; footwear	..	..	..	139	140	136	117	104
		191	Leather and articles of leather	..	..	..	52	51	52	40	39
		192	Footwear	..	..	..	87	90	83	77	65
	20		Wood & products of wood & cork, exc. furniture	..	..	..	300	298	310	302	264
		201	Sawmilling and planing of wood	..	..	..	70	69	66	63	56
		202	Wood products; articles of cork, straw, etc.	..	..	..	230	230	244	239	208
	21		Paper and paper products	..	..	..	816	819	824	830	817
	22		Printing, publishing and recorded media	..	..	..	1 822	1 869	1 907	1 946	1 940
		221	Publishing	..	..	..	809	827	847	865	867
		222	Printing and related activities	..	..	..	987	1 015	1 030	1 048	1 041
		223	Reproduction of recorded media	..	..	..	26	26	30	33	32
	23		Coke and petroleum products; nuclear fuel	..	..	..	203	204	202	200	182
		231	Coke oven products	..	..	..	16	14	15	15	15
		232	Refined petroleum products	..	..	..	165	166	163	161	153
		233	Processing of nuclear fuel	..	..	..	23	24	24	24	14
	24		Chemicals and chemical products	..	..	..	2 038	2 031	2 048	2 058	2 028
		241	Basic chemicals	..	..	..	704	688	694	705	705
		242	Other chemical products	..	..	..	1 206	1 220	1 234	1 240	1 219
		243	Man-made fibres	..	..	..	128	123	120	112	104
	25		Rubber and plastic products	..	..	..	1 622	1 652	1 605	1 622	1 585
		251	Rubber products	..	..	..	482	481	483	478	458
		252	Plastic products	..	..	..	1 140	1 171	1 122	1 144	1 126
	26		Other non-metallic mineral products	..	..	..	1 071	1 074	1 061	1 075	1 015
		261	Glass and glass products	..	..	..	234	228	226	224	218
		269	Non-metallic mineral products, nec	..	..	..	837	846	835	851	797
	27		Basic metals	..	..	..	1 839	1 776	1 747	1 712	1 582
		271	Basic iron and steel	..	..	..	1 243	1 158	1 124	1 076	1 004
		272	Basic precious and non-ferrous metals	..	..	..	323	330	336	342	315
		273	Casting of metals	..	..	..	273	289	287	294	264
	28		Fabricated metal products, exc. machinery	..	..	..	2 157	2 166	2 189	2 192	2 082
		281	Structural metal prod., tanks, reservoirs, etc.	..	..	..	963	977	978	958	922
		289	Other fabricated metal products	..	..	..	1 194	1 189	1 212	1 234	1 160
	29		Machinery and equipment, nec	..	..	..	3 905	3 969	4 152	4 174	4 060
		291	General purpose machinery	..	..	..	1 682	1 680	1 769	1 808	1 762
		292	Special purpose machinery	..	..	..	1 661	1 706	1 800	1 786	1 751
		293	Domestic appliances, nec	..	..	..	563	583	584	580	547
	30		Office, accounting & computing machinery	..	..	..	1 818	1 835	1 863	1 909	1 921
	31		Electrical machinery & apparatus, nec	..	..	..	2 196	2 186	2 169	2 224	2 102
		311	Electric motors, generators and transformers	..	..	..	384	401	390	405	359
		312	Electricity distribution & control apparatus	..	..	..	645	632	648	646	630
		313	Insulated wire and cable	..	..	..	228	229	235	238	218
		314	Accumulators, primary cells & primary batteries	..	..	..	105	104	110	108	113
		315	Electric lamps and lighting equipment	..	..	..	171	170	170	170	168
		319	Other electrical equipment, nec	..	..	..	663	650	616	658	614
	32		Radio, TV, communication equipment	..	..	..	3 286	3 445	3 491	3 561	3 618
		321	Electronic valves, tubes; other electronic components	..	..	..	1 489	1 549	1 573	1 611	1 652
		322	TV/radio transmitters; line teleph./telegr. apparatus	..	..	..	409	444	475	448	469
		323	TV/radio receivers; sound and video app.; assoc. gds.	..	..	..	1 388	1 452	1 443	1 503	1 498
	33		Medical, precision & optical instrum.; watch & clocks	..	..	..	897	882	846	882	896
		331	Medical appl.; instruments & appl. for measuring, etc.	..	..	..	574	585	577	611	628
		332	Optical instruments and photographic equipment	..	..	..	217	201	183	191	188
		333	Watches and clocks	..	..	..	107	97	87	81	80
	34		Motor vehicles, trailers and semi-trailers	..	..	..	3 710	3 726	3 848	3 974	3 913
		341	Motor vehicles	..	..	..	1 157	1 189	1 203	1 238	1 288
		342	Bodies for motor vehicles, trailers & semi-trailers	..	..	..	310	307	315	326	291
		343	Parts & accessories for motor vehicles & their engines	..	..	..	2 243	2 230	2 330	2 410	2 334
	35		Other transport equipment	..	..	..	510	508	474	488	499
		351	Building and repairing of ships and boats	..	..	..	262	250	230	231	244
		352	Railway and tramway locomotives and rolling stock	..	..	..	60	73	65	74	63
		353	Aircraft and spacecraft	..	..	..	144	143	145	147	155
		359	Transport equipment, nec	..	..	..	43	42	33	36	37
	36		Furniture; manufacturing, nec	..	..	..	735	716	761	748	709
		361	Furniture	..	..	..	304	300	351	341	310
		369	Manufacturing, nec	..	..	..	431	416	410	407	399
	37		Recycling	..	..	..	13	17	16	16	15
E	40_41		Electricity, gas and water supply	..	..	..	..	..	..	..	..
	40		Electricity, gas, steam and hot water supply	..	..	..	..	..	..	..	..
	41		Collection, purification and distribution of water	..	..	..	..	..	..	..	..
F	45		Construction	..	..	..	..	..	..	..	..

Table JN.8 **ÉTABLISSEMENTS - ESTABLISHMENTS**

Unités

1991	1992	1993	1994	1995	1996	1997	1998		CITI révision 3		
..	..	..	..	..	..	..	..	Activites extractives	C	10_14	
..	..	..	..	..	..	..	..	Extraction de charbon, de lignite et de tourbe		10	
..	..	..	..	..	..	..	..	Pétrole brut et gaz naturel ; activités annexes		11	
..	..	..	..	..	..	..	..	Extraction de minerais métalliques		13	
..	..	..	..	..	..	..	..	Autres activités extractives		14	
..	..	..	382 825	387 726	369 612	358 246	373 713	Total activites manufacturieres	D	15_37	
..	..	..	46 225	47 245	45 405	44 111	47 034	Produits alimentaires et boissons		15	
..	..	..	13 908	13 949	13 543	13 216	14 052	Viande, poisson, fruits, légumes, huiles et graisses			151
..	..	..	797	800	793	770	809	Produits laitiers			152
..	..	..	1 407	1 399	1 348	1 315	1 386	Travail des grains; prod. amyl. ; alim. pour animaux			153
..	..	..	26 726	27 687	26 424	25 579	27 474	Autres produits alimentaires			154
..	..	..	3 387	3 410	3 297	3 231	3 313	Boissons			155
..	..	..	52	52	50	50	50	Produits à base de tabac		16	
..	..	..	31 279	30 450	28 575	27 089	27 007	Textiles		17	
..	..	..	12 326	11 735	10 958	10 312	9 680	Filature, tissage et achèvement des textiles			171
..	..	..	10 055	10 031	9 279	8 834	9 418	Autres articles textiles			172
..	..	..	8 898	8 684	8 338	7 943	7 909	Etoffes et articles de bonneterie			173
..	..	..	20 242	19 283	17 644	16 248	16 120	Articles d'habillement et fourrure		18	
..	..	..	20 166	19 205	17 573	16 182	16 058	Articles d'habillement autres qu'en fourrure			181
..	..	..	76	78	71	66	62	Prép. et teinture des fourrures ; articles en fourrure			182
..	..	..	5 412	5 800	5 643	5 293	5 144	Cuir et articles en cuir ; chaussures		19	
..	..	..	2 985	2 851	2 742	2 571	2 526	Cuir et articles en cuir sauf chaussures			191
..	..	..	2 427	2 949	2 901	2 722	2 618	Chaussures			192
..	..	..	23 970	23 844	22 573	21 470	21 272	Bois, articles en bois, sauf meubles		20	
..	..	..	9 887	9 628	8 960	8 428	8 121	Sciage et rabotage du bois			201
..	..	..	14 083	14 216	13 613	13 042	13 151	Articles en bois, liège, vannerie, sparterie			202
..	..	..	10 392	10 538	10 186	9 826	10 349	Papier, articles en papier et carton		21	
..	..	..	26 536	27 235	25 850	25 277	27 562	Edition, imprimerie et reproduction		22	
..	..	..	2 307	2 308	2 286	2 365	2 767	Edition			221
..	..	..	24 154	24 837	23 477	22 809	24 644	Imprimerie et activités annexes			222
..	..	..	75	90	87	103	151	Reproduction de supports enregistrés			223
..	..	..	1 076	1 094	1 094	1 097	1 188	Coke. et prod. pétroliers ; comb. nucléaires		23	
..	..	..	801	813	812	823	910	Cokéfaction			231
..	..	..	268	274	275	267	272	Produits pétroliers raffinés			232
..	..	..	7	7	7	7	6	Traitement de combustibles nucléaires			233
..	..	..	5 232	5 302	5 291	5 255	5 501	Produits chimiques		24	
..	..	..	1 217	1 231	1 265	1 288	1 346	Produits chimiques de base			241
..	..	..	3 928	3 979	3 941	3 886	4 069	Autres produits chimiques			242
..	..	..	87	92	85	81	86	Fibres synthétiques ou artificielles			243
..	..	..	22 171	22 909	22 222	21 705	23 046	Art. en caoutchouc et matières plastiques		25	
..	..	..	3 339	3 505	3 341	3 255	3 422	Articles en caoutchouc			251
..	..	..	18 832	19 404	18 881	18 450	19 624	Articles en matières plastiques			252
..	..	..	19 199	19 297	18 740	18 211	18 690	Autres produits minéraux non métalliques		26	
..	..	..	1 094	1 104	1 060	1 031	1 147	Verre et articles en verre			261
..	..	..	18 105	18 193	17 680	17 180	17 543	Produits minéraux non métalliques, nca			269
..	..	..	7 333	7 358	6 954	6 983	7 280	Produits métallurgiques de base		27	
..	..	..	3 143	3 156	2 999	3 001	3 224	Sidérurgie et première transformation de l'acier			271
..	..	..	1 131	1 155	1 121	1 183	1 320	Première transf. métaux précieux et non ferreux			272
..	..	..	3 059	3 047	2 834	2 799	2 736	Fonderie			273
..	..	..	48 450	49 830	47 482	46 253	48 425	Ouvrages en métaux, sauf machines		28	
..	..	..	21 681	22 505	21 326	20 793	22 036	Constr. et menuis. métal.; réservoirs; générateurs			281
..	..	..	26 769	27 325	26 156	25 460	26 389	Autres ouvrages en métaux			289
..	..	..	41 101	42 990	41 291	40 751	43 074	Machines et matériels, nca		29	
..	..	..	15 060	15 801	15 168	15 090	16 053	Machines d'usage général			291
..	..	..	22 705	23 841	22 923	22 637	23 945	Machines d'usage spécifique			292
..	..	..	3 336	3 348	3 200	3 024	3 076	Appareils domestiques, nca			293
..	..	..	6 248	6 150	5 910	5 818	6 112	Mach. de bureau, comptables, information		30	
..	..	..	14 798	14 943	14 246	13 775	14 326	Machines et appareils électriques, nca		31	
..	..	..	2 665	2 708	2 611	2 486	2 517	Moteurs, génératrices, transformateurs électriques			311
..	..	..	5 162	5 189	4 941	4 713	5 048	Matériel électrique de distribution et de commande			312
..	..	..	923	960	918	890	922	Fils et câbles électriques isolés			313
..	..	..	197	212	210	210	209	Accumulateurs et piles électriques			314
..	..	..	1 374	1 386	1 291	1 247	1 278	Lampes électriques et appareils d'éclairage			315
..	..	..	4 477	4 488	4 275	4 229	4 352	Autres matériels électriques, nca.			319
..	..	..	11 961	11 832	11 158	10 832	10 917	Equipements et appareils de radio, TV, etc.		32	
..	..	..	2 592	2 697	2 700	2 744	2 817	Tubes et valves électroniques;autres composants			321
..	..	..	482	507	514	503	523	Emetteurs radio/TV; app. téléphonie et télégraphie			322
..	..	..	8 887	8 628	7 944	7 585	7 577	Récepteurs radio/TV, son et image; art. associés			323
..	..	..	7 051	7 064	6 805	6 720	7 177	Instr. médicaux, de précision ; optique ; horlog.		33	
..	..	..	4 525	4 608	4 424	4 416	4 710	Instruments médicaux et de précision, etc.			331
..	..	..	2 018	1 984	1 956	1 891	2 064	Instruments d'optique ; matériel photographique			332
..	..	..	508	472	425	413	403	Horlogerie			333
..	..	..	10 467	10 648	10 126	9 964	10 436	Véhicules automobiles; remorques, etc.		34	
..	..	..	49	51	49	50	59	Véhicules automobiles			341
..	..	..	262	272	285	276	267	Carross. automobiles, remorques, semi-remorques			342
..	..	..	10 156	10 325	9 792	9 638	10 110	Pièces et accessoires pour automobiles et moteurs			343
..	..	..	2 756	2 805	2 679	2 593	2 791	Autres matériels de transport		35	
..	..	..	1 334	1 365	1 298	1 234	1 308	Construction et réparation de navires			351
..	..	..	406	436	418	403	420	Matériel ferroviaire roulant			352
..	..	..	217	221	217	228	269	Construction aéronautique et spatiale			353
..	..	..	799	783	746	728	794	Autres équipements de transport			359
..	..	..	23 332	23 444	21 987	21 190	22 604	Meubles ; activités de fabrication, nca		36	
..	..	..	9 537	9 433	9 085	8 732	8 848	Meubles			361
..	..	..	13 795	14 011	12 902	12 458	13 756	Activités de fabrication, nca			369
..	..	..	692	714	681	669	689	Récupération		37	
..	..	..	..	..	..	..	..	Electricité, gaz et eau	E	40_41	
..	..	..	..	..	..	..	..	Prod. et dist. d'électricité gaz, vapeur d'eau chaude		40	
..	..	..	..	..	..	..	..	Captage, épuration et distribution d'eau		41	
..	..	..	..	..	..	..	..	Construction	F	45	

KOREA

Table KR.1 **PRODUCTION - PRODUCTION**

Millions of Won (Current Prices)

ISIC revision 3				1991	1992	1993	1994	1995	1996	1997	1998
C	10_14		Mining and quarrying	2206 188	1979 918	1849 908	1767 686	1740 513	1825 026	1818 787	..
	10		Mining of coal and lignite; extraction of peat	1284 128	1012 542	756 991	595 910	424 301	369 455	359 823	267 018
	11		Crude petroleum & natural gas; service activities	..	..	..	..	..	..	..	..
	13		Mining of metal ores	62 219	54 781	37 498	35 431	34 398	37 893	36 444	26 469
	14		Other mining and quarrying	859 841	912 595	1055 419	1136 345	1281 814	1417 678	1422 520	1261 242
D	15_37		Total manufacturing	205699 006	226816 626	255926 066	299246 191	364821 270	401952 973	434894 981	425007 060
	15		Food products and beverages	17167 280	19000 744	20904 331	23545 088	26429 987	29858 839	31425 829	33852 463
		151	Production of meat, fish, fruit, vegetables, oils & fats	3883 970	4063 952	4332 140	4876 745	5759 627	5911 411	6150 893	6891 355
		152	Dairy products	2225 280	2325 005	2488 185	3017 891	3318 680	3518 163	3520 369	3728 233
		153	Grain mill and starch products; animal feeds	3421 952	3890 683	4627 555	4936 900	5953 735	7515 962	8351 444	8718 170
		154	Other food products	4573 656	5467 654	6092 947	6434 926	7103 244	7866 402	8398 517	9623 950
		155	Beverages	3062 422	3253 450	3363 504	4278 626	4294 701	5046 901	5004 606	4890 755
	16		Tobacco products	2974 982	3037 239	3223 976	3254 118	3379 356	4049 236	4195 409	3727 537
	17		Textiles	14311 117	15885 288	17284 429	18174 008	19765 258	20339 085	21314 136	23183 034
		171	Spinning, weaving and finishing of textiles	11024 834	11987 611	11900 792	13038 662	14757 763	14635 482	15565 441	16451 716
		172	Other textiles	1785 600	2064 389	2747 012	3043 129	3272 369	3387 127	3306 591	3575 838
		173	Knitted and crocheted fabrics and articles	1500 683	1833 288	2636 625	2092 217	1735 126	2316 476	2442 104	3155 480
	18		Wearing apparel and fur	5921 649	6047 414	7767 903	8813 922	10812 604	10375 808	10313 991	7209 566
		181	Wearing apparel, except fur apparel	5603 214	5759 637	7478 540	8440 956	10391 772	9840 981	9962 914	6927 851
		182	Dressing and dyeing of fur; articles of fur	318 435	287 777	289 363	372 966	420 832	534 827	351 077	281 715
	19		Leather and articles of leather; footwear	6642 851	6368 120	5585 091	5553 985	5441 433	5546 083	5108 235	5128 795
		191	Leather and articles of leather	2510 717	2621 000	2658 225	2910 750	2883 132	2863 318	2853 276	3148 608
		192	Footwear	4132 134	3747 120	2926 866	2643 235	2558 301	2682 765	2254 959	1980 187
	20		Wood & products of wood & cork, exc. furniture	2150 883	2171 009	2731 767	2913 500	3121 322	3271 172	3320 333	2575 620
		201	Sawmilling and planing of wood	832 971	814 404	1065 618	1165 350	1185 360	1214 620	1000 312	744 413
		202	Wood products; articles of cork, straw, etc.	1317 912	1356 605	1666 149	1748 150	1935 962	2056 552	2320 021	1831 207
	21		Paper and paper products	5193 134	5811 911	6115 412	7508 571	9614 828	9934 263	10246 857	10401 802
	22		Printing, publishing and recorded media	3190 461	3874 895	4942 764	5683 990	6874 598	8104 625	7865 769	6936 424
		221	Publishing	2056 885	2506 493	3376 007	3783 006	4673 304	5529 282	5131 568	4673 297
		222	Printing and related activities	1064 755	1236 087	1423 996	1727 937	2012 311	2353 111	2508 993	2085 246
		223	Reproduction of recorded media	68 821	132 315	142 761	173 047	188 983	222 232	225 208	177 881
	23		Coke and petroleum products; nuclear fuel	9567 178	11938 107	12766 450	13230 778	16415 970	21045 051	28038 455	30049 161
		231	Coke oven products	37 167	33 208	29 297	25 490	22 697	30 555	30 022	13 282
		232	Refined petroleum products	9484 642	11864 424	12689 660	13141 173	16331 436	20955 674	27901 939	29919 654
		233	Processing of nuclear fuel	45 369	40 475	47 493	64 115	61 837	58 822	106 494	116 225
	24		Chemicals and chemical products	18122 323	21105 790	23592 873	27738 282	33518 623	35913 492	41681 366	46989 107
		241	Basic chemicals	8105 575	10402 930	10794 566	13094 946	17387 167	17889 848	21565 625	25919 897
		242	Other chemical products	8135 476	8912 446	10258 410	11336 508	11649 041	13812 799	15116 942	16080 193
		243	Man-made fibres	1881 272	1790 414	2539 897	3306 828	4482 415	4210 845	4998 799	4989 017
	25		Rubber and plastic products	7707 767	8243 781	9708 630	11174 981	14069 213	14896 606	15221 677	15455 937
		251	Rubber products	2143 157	2378 890	2742 489	3060 868	3508 206	4024 416	4046 092	4696 131
		252	Plastic products	5564 610	5864 891	6966 141	8114 113	10561 007	10872 190	11175 585	10759 806
	26		Other non-metallic mineral products	10081 216	11165 830	12015 022	13062 244	15290 894	16409 426	17034 045	14519 643
		261	Glass and glass products	1496 252	1754 011	1948 761	2242 890	2744 959	2993 803	3028 786	3268 468
		269	Non-metallic mineral products, nec	8584 964	9411 819	10066 261	10819 354	12545 935	13415 623	14005 259	11251 175
	27		Basic metals	18168 098	19136 516	21435 077	24911 038	31007 526	32528 347	35204 200	38844 689
		271	Basic iron and steel	13791 537	14735 772	16917 007	19234 063	22875 681	24602 326	26221 380	29622 892
		272	Basic precious and non-ferrous metals	3575 952	3636 521	3749 571	4713 810	7015 842	6618 638	7700 308	8124 146
		273	Casting of metals	800 609	764 223	768 499	963 165	1116 003	1307 383	1282 512	1097 651
	28		Fabricated metal products, exc. machinery	8574 259	9206 270	10801 665	14008 030	16957 734	20187 467	19971 920	16140 330
		281	Structural metal prod., tanks, reservoirs, etc.	3537 337	4080 900	5158 260	6983 935	8681 388	11110 107	10969 690	8068 737
		289	Other fabricated metal products	5036 922	5125 370	5643 405	7024 095	8276 346	9077 360	9002 230	8071 593
	29		Machinery and equipment, nec	17689 163	17866 339	20078 513	24582 100	30924 278	34551 298	36286 989	27214 507
		291	General purpose machinery	5533 592	6612 789	7376 159	9425 332	13455 047	15522 177	15584 596	11067 649
		292	Special purpose machinery	8475 000	7623 271	8547 687	10338 662	13282 147	14734 710	15131 231	11292 948
		293	Domestic appliances, nec	3680 571	3630 279	4154 667	4818 106	4187 084	4294 411	5571 162	4853 910
	30		Office, accounting & computing machinery	1485 854	1576 276	3087 347	3720 556	5263 691	6117 763	13312 502	13029 317
	31		Electrical machinery & apparatus, nec	6584 394	7866 935	8546 351	9880 754	12554 570	16235 215	15627 318	14224 459
		311	Electric motors, generators and transformers	1565 385	1959 179	2362 984	2547 672	3113 386	3857 188	3736 490	3639 884
		312	Electricity distribution & control apparatus	1914 127	2210 169	2334 837	2789 599	3618 851	4287 052	4169 754	3205 873
		313	Insulated wire and cable	1795 563	2040 861	2049 946	2582 198	3001 767	4775 327	4100 125	4125 553
		314	Accumulators, primary cells & primary batteries	328 705	474 215	367 386	401 982	539 514	656 885	855 896	953 261
		315	Electric lamps and lighting equipment	389 983	474 425	612 094	759 908	955 621	1074 626	1053 563	975 961
		319	Other electrical equipment, nec	590 631	708 086	819 104	799 395	1325 431	1584 137	1711 490	1323 927
	32		Radio, TV, communication equipment	19732 279	21523 566	25811 970	32923 365	45890 917	47181 274	44972 880	52896 566
		321	Electronic valves, tubes; other electronic components	6775 519	7459 524	10957 791	14790 701	24897 450	24551 388	24422 456	32541 835
		322	TV/radio transmitters; line teleph./telegr. apparatus	2421 019	2861 252	2727 629	4176 478	4627 686	6511 424	11866 155	13821 581
		323	TV/radio receivers; sound and video app.; assoc. gds.	10535 741	11202 790	12126 550	13956 186	16365 781	16118 462	8684 269	6533 150
	33		Medical, precision & optical instrum.; watch & clocks	2054 848	2158 663	2610 832	3220 095	3807 622	4140 677	4803 301	4008 521
		331	Medical appl.; instruments & appl. for measuring, etc.	1010 104	1088 416	1321 317	1756 280	2375 880	2658 721	2797 880	2511 078
		332	Optical instruments and photographic equipment	493 400	514 020	740 857	923 324	939 496	1044 458	1507 646	999 795
		333	Watches and clocks	551 344	556 227	548 658	540 491	492 246	437 498	497 775	497 648
	34		Motor vehicles, trailers and semi-trailers	17745 997	20248 773	23457 427	28788 727	35089 516	40855 066	44181 398	32786 716
		341	Motor vehicles	11041 008	13260 815	15007 290	18356 736	22161 271	25557 644	29084 006	21132 727
		342	Bodies for motor vehicles, trailers & semi-trailers	1268 482	941 920	844 013	963 373	1424 982	1267 105	713 616	887 456
		343	Parts & accessories for motor vehicles & their engines	5436 507	6046 038	7606 124	9468 618	11503 263	14030 317	14383 776	10766 533
	35		Other transport equipment	5530 912	7104 406	7480 792	9900 494	11032 792	12618 843	16996 884	18850 128
		351	Building and repairing of ships and boats	4372 836	5887 570	5918 061	7534 843	9199 091	10538 723	13740 927	15732 468
		352	Railway and tramway locomotives and rolling stock	193 017	384 912	759 885	1141 533	616 982	737 941	1593 747	1257 976
		353	Aircraft and spacecraft	440 669	341 668	330 341	594 804	564 503	660 124	984 418	1376 448
		359	Transport equipment, nec	524 390	490 256	472 505	629 314	652 216	682 055	677 792	483 236
	36		Furniture; manufacturing, nec	5069 761	5436 716	5858 246	6453 851	7104 511	7279 143	7099 649	6245 305
		361	Furniture	2268 107	2693 877	2955 946	3345 800	3636 358	3844 170	3697 537	2691 338
		369	Manufacturing, nec	2801 654	2742 839	2902 300	3108 051	3468 153	3434 973	3402 112	3553 967
	37		Recycling	32 600	42 038	119 198	203 714	454 027	514 194	671 838	737 433
E	40_41		Electricity, gas and water supply	..	..	..	..	..	..	..	..
	40		Electricity, gas, steam and hot water supply	..	..	..	..	..	..	..	..
	41		Collection, purification and distribution of water	..	..	..	..	..	..	..	..
F	45		Construction	..	..	..	..	..	..	..	..

Table KR.2 **VALEUR AJOUTÉE - VALUE ADDED**

Millions de Won (Prix courants)

1991	1992	1993	1994	1995	1996	1997	1998			CITI révision 3
1 265 130	1 237 648	1 193 393	1 208 790	1 244 548	1 326 437	1 284 815	1 079 161	Activites extractives	C	10_14
575 528	510 105	348 935	306 464	236 495	213 514	215 219	127 568	Extraction de charbon, de lignite et de tourbe		10
..	..	..	..	..	..	..	..	Pétrole brut et gaz naturel ; activités annexes		11
45 156	44 187	30 270	30 057	29 583	32 125	30 854	21 039	Extraction de minerais métalliques		13
644 446	683 356	814 188	872 269	978 470	1 080 798	1 038 742	930 554	Autres activités extractives		14
86 366 481	96 018 026	108 521 742	129 610 868	159 448 345	174 215 220	181 085 369	176 729 615	Total activites manufacturieres	D	15_37
6 925 216	8 005 391	8 746 335	9 948 876	10 884 639	12 409 929	12 585 717	13 054 904	Produits alimentaires et boissons		15
1 313 293	1 506 846	1 716 000	1 803 977	2 280 219	2 293 553	2 203 514	2 353 774	Viande, poisson, fruits, légumes, huiles et graisses		151
880 638	881 441	1 058 390	1 196 196	1 336 674	1 324 222	1 407 371	1 498 542	Produits laitiers		152
1 061 908	1 178 702	1 331 627	1 471 068	1 672 832	1 977 180	2 180 624	2 232 125	Travail des grains; prod. amyl. ; alim. pour animaux		153
2 076 277	2 658 882	2 877 380	3 156 161	3 382 817	3 967 445	4 152 378	4 335 294	Autres produits alimentaires		154
1 593 100	1 779 520	1 762 938	2 321 474	2 212 097	2 847 529	2 641 830	2 635 169	Boissons		155
2 210 145	2 134 273	2 176 722	2 225 032	2 527 308	3 051 063	3 011 703	2 661 016	Produits à base de tabac		16
6 143 505	7 219 959	7 726 945	8 216 740	8 820 887	9 066 799	8 933 730	9 762 149	Textiles		17
4 781 156	5 437 592	5 364 497	5 913 416	6 683 152	6 504 870	6 625 609	6 893 955	Filature, tissage et achèvement des textiles		171
743 941	935 557	1 260 276	1 385 307	1 415 985	1 486 730	1 397 797	1 651 414	Autres articles textiles		172
618 408	846 810	1 102 172	918 017	721 750	1 075 199	910 324	1 216 780	Etoffes et articles de bonneterie		173
2 833 333	2 945 354	3 855 188	4 296 242	5 250 871	5 195 016	5 087 106	3 584 504	Articles d'habillement et fourrure		18
2 688 735	2 811 758	3 758 431	4 157 425	5 104 408	4 989 379	4 946 118	3 449 139	Articles d'habillement autres qu'en fourrure		181
144 598	133 596	96 757	138 817	146 463	205 637	140 988	135 365	Prép. et teinture des fourrures ; articles en fourrure		182
2 616 980	2 613 670	2 293 992	2 337 979	2 160 517	2 319 970	1 915 106	1 935 477	Cuir et articles en cuir ; chaussures		19
764 981	877 831	904 197	1 062 075	964 915	927 464	835 804	1 034 135	Cuir et articles en cuir sauf chaussures		191
1 851 999	1 735 839	1 389 795	1 275 904	1 195 602	1 392 506	1 079 302	901 342	Chaussures		192
901 561	860 537	1 065 791	1 176 435	1 298 164	1 366 162	1 344 720	1 108 780	Bois, articles en bois, sauf meubles		20
316 933	317 811	383 084	461 796	491 680	529 003	384 319	294 123	Sciage et rabotage du bois		201
584 628	542 726	682 707	714 639	806 484	837 159	960 401	814 657	Articles en bois, liège, vannerie, sparterie		202
2 015 431	2 339 049	2 529 781	3 107 959	3 669 364	4 039 778	4 103 757	4 348 404	Papier, articles en papier et carton		21
1 919 130	2 325 740	3 119 794	3 509 289	4 161 717	4 793 035	4 578 759	4 115 490	Edition, imprimerie et reproduction		22
1 328 326	1 588 284	2 229 319	2 480 297	2 964 912	3 397 876	3 142 194	2 864 878	Edition		221
556 273	656 273	806 936	951 041	1 089 757	1 283 433	1 321 864	1 133 072	Imprimerie et activités annexes		222
34 531	81 183	83 539	77 951	107 048	111 726	114 701	117 540	Reproduction de supports enregistrés		223
2 589 889	3 231 117	3 285 770	4 028 940	5 383 395	5 109 266	8 041 360	7 447 774	Coke. et prod. pétroliers ; comb. nucléaires		23
13 482	16 751	12 312	12 157	7 929	17 960	23 693	4 138	Cokéfaction		231
2 544 361	3 188 712	3 240 315	3 970 040	5 329 463	5 059 354	7 958 401	7 376 912	Produits pétroliers raffinés		232
32 046	25 654	33 143	46 743	46 003	31 952	59 266	66 724	Traitement de combustibles nucléaires		233
7 855 822	9 242 148	10 719 576	12 657 456	14 853 890	16 089 697	17 789 163	20 275 484	Produits chimiques		24
2 753 225	3 721 922	4 033 078	5 175 956	6 939 026	7 174 348	7 974 964	9 527 597	Produits chimiques de base		241
4 195 964	4 652 703	5 491 707	6 049 922	6 099 412	7 144 907	7 847 496	8 686 181	Autres produits chimiques		242
906 633	867 523	1 194 791	1 431 578	1 815 452	1 770 442	1 966 703	2 061 706	Fibres synthétiques ou artificielles		243
3 203 730	3 685 331	4 241 116	4 897 704	6 282 508	6 679 528	6 610 865	6 851 928	Art. en caoutchouc et matières plastiques		25
1 000 438	1 148 346	1 250 449	1 434 948	1 672 485	2 039 073	1 898 523	2 195 585	Articles en caoutchouc		251
2 203 292	2 536 985	2 990 667	3 462 756	4 610 023	4 640 455	4 712 342	4 656 343	Articles en matières plastiques		252
5 067 890	5 444 796	6 026 923	6 460 618	7 706 751	8 012 126	8 002 748	6 986 801	Autres produits minéraux non métalliques		26
894 011	1 027 066	1 044 496	1 330 481	1 653 141	1 727 097	1 689 099	1 866 806	Verre et articles en verre		261
4 173 879	4 417 730	4 982 427	5 130 137	6 053 610	6 285 029	6 313 649	5 119 995	Produits minéraux non métalliques, nca		269
6 667 396	7 059 148	7 436 022	8 798 750	10 920 647	10 942 609	11 260 064	13 957 755	Produits métallurgiques de base		27
5 231 812	5 637 117	5 936 175	6 934 928	8 340 739	8 553 009	8 767 681	10 357 807	Sidérurgie et première transformation de l'acier		271
1 041 251	1 032 544	1 105 432	1 373 565	2 038 064	1 739 655	1 911 141	3 086 534	Première transf. métaux précieux et non ferreux		272
394 333	389 487	394 415	490 257	541 844	649 945	581 242	513 414	Fonderie		273
3 650 636	4 205 286	5 135 666	6 689 555	7 453 567	9 497 884	9 004 550	7 805 215	Ouvrages en métaux, sauf machines		28
1 270 690	1 717 457	2 273 578	3 147 924	3 388 840	4 963 891	4 638 780	3 702 406	Constr. et menuis. métal.; réservoirs; générateurs		281
2 379 946	2 487 829	2 862 088	3 541 631	4 064 727	4 533 993	4 365 770	4 102 809	Autres ouvrages en métaux		289
7 330 608	7 568 090	8 610 322	10 354 247	12 890 390	14 766 203	14 766 201	12 049 270	Machines et matériels, nca		29
2 360 620	2 807 020	3 349 053	4 110 250	5 793 785	6 795 273	6 391 605	4 938 383	Machines d'usage général		291
3 545 954	3 372 785	3 716 057	4 483 249	5 576 386	6 386 319	6 379 233	5 127 218	Machines d'usage spécifique		292
1 424 034	1 388 285	1 545 212	1 760 748	1 520 219	1 584 611	1 995 363	1 983 669	Appareils domestiques, nca		293
578 088	593 465	994 599	1 273 915	1 634 955	2 117 207	4 265 706	3 429 662	Mach. de bureau, comptables, information		30
2 501 941	3 152 200	3 427 822	4 006 084	5 082 025	6 397 981	6 181 781	5 754 478	Machines et appareils électriques, nca		31
595 227	799 261	961 848	1 050 143	1 255 826	1 512 011	1 496 978	1 471 629	Moteurs, génératrices, transformateurs électriques		311
762 547	942 702	943 066	1 149 861	1 536 892	1 862 622	1 766 460	1 350 238	Matériel électrique de distribution et de commande		312
542 130	600 448	676 454	917 897	1 034 115	1 526 638	1 345 459	1 505 978	Fils et câbles électriques isolés		313
147 591	261 342	190 131	169 879	232 370	265 274	352 104	365 225	Accumulateurs et piles électriques		314
186 110	224 715	293 707	365 794	462 628	506 051	487 934	461 731	Lampes électriques et appareils d'éclairage		315
268 336	323 732	362 616	352 510	560 194	725 385	732 846	599 677	Autres matériels électriques, nca.		319
8 549 805	8 848 544	11 431 469	16 029 479	24 993 880	26 242 371	23 013 300	26 014 555	Equipements et appareils de radio, TV, etc.		32
3 515 331	3 712 301	6 666 354	9 348 707	17 090 457	16 454 439	13 987 909	18 738 010	Tubes et valves électroniques;autres composants		321
951 779	1 228 591	1 041 415	1 900 779	1 848 407	2 811 320	5 562 260	5 109 693	Emetteurs radio/TV; app. téléphonie et télégraphie		322
4 082 695	3 907 652	3 723 700	4 779 993	6 055 016	6 976 612	3 463 131	2 166 852	Récepteurs radio/TV, son et image; art. associés		323
929 893	983 913	1 177 389	1 434 516	1 749 224	1 906 316	2 054 730	1 774 904	Instr. médicaux, de précision ; optique ; horlog.		33
462 268	494 793	603 062	715 706	1 052 964	1 189 120	1 219 216	1 144 376	Instruments médicaux et de précision, etc.		331
233 095	242 103	332 976	489 240	475 263	508 721	672 256	467 192	Instruments d'optique ; matériel photographique		332
234 530	247 017	241 351	229 570	220 997	208 475	163 258	163 336	Horlogerie		333
6 637 398	7 358 174	8 414 378	10 054 689	13 081 433	15 057 427	17 789 762	12 714 504	Véhicules automobiles; remorques, etc.		34
3 904 088	4 385 033	4 853 214	5 708 542	7 762 515	8 799 844	11 719 076	7 858 167	Véhicules automobiles		341
413 474	283 126	314 729	331 803	448 319	416 498	282 526	360 775	Carross. automobiles, remorques, semi-remorques		342
2 319 836	2 690 015	3 246 435	4 014 344	4 870 599	5 841 085	5 788 160	4 495 562	Pièces et accessoires pour automobiles et moteurs		343
2 719 908	3 494 616	3 176 821	4 740 094	4 940 410	5 400 333	7 224 680	7 932 277	Autres matériels de transport		35
2 124 909	3 029 205	2 657 396	3 677 076	4 093 883	4 627 993	5 911 922	6 929 912	Construction et réparation de navires		351
99 754	148 104	224 112	441 400	244 395	203 076	713 317	444 899	Matériel ferroviaire roulant		352
291 720	139 454	137 674	386 256	339 334	313 754	357 200	383 790	Construction aéronautique et spatiale		353
203 525	177 853	157 639	235 362	262 798	255 510	242 241	173 676	Autres équipements de transport		359
2 501 231	2 685 064	2 883 946	3 292 537	3 581 417	3 551 136	3 293 656	2 940 317	Meubles ; activités de fabrication, nca		36
1 179 547	1 355 529	1 462 449	1 741 095	1 798 022	1 871 736	1 706 295	1 248 966	Meubles		361
1 321 684	1 329 535	1 421 497	1 551 442	1 783 395	1 679 400	1 587 361	1 691 351	Activités de fabrication, nca		369
16 945	22 161	45 375	73 732	120 386	203 384	226 205	223 967	Récupération		37
..	..	..	..	..	..	..	..	Electricité, gaz et eau	E	40_41
..	..	..	..	..	..	..	..	Prod. et dist. d'électricité gaz, vapeur d'eau chaude		40
..	..	..	..	..	..	..	..	Captage, épuration et distribution d'eau		41
..	..	..	..	..	..	..	..	Construction	F	45

Table KR.3 **INVESTMENT - INVESTISSEMENT**

Millions of Won (Current Prices)

ISIC revision 3				1991	1992	1993	1994	1995	1996	1997	1998
C	10_14		Mining and quarrying	300 719	357 142	336 720	236 127	280 726	274 367	172 547	361 687
	10		Mining of coal and lignite; extraction of peat	102 565	138 917	117 266	79 818	61 486	76 663	43 195	150 597
	11		Crude petroleum & natural gas; service activities	..	..	..	..	..	..	..	..
	13		Mining of metal ores	10 582	8 434	6 449	8 071	4 253	3 889	3 709	5 991
	14		Other mining and quarrying	187 572	209 791	213 005	148 238	214 987	193 815	125 643	205 099
D	15_37		Total manufacturing	32365 767	26192 079	28738 026	31855 089	43312 249	94 493	52399 542	53459 326
	15		Food products and beverages	1571 621	1636 749	1801 477	2241 527	2645 663	2831 348	2692 802	2745 538
		151	Production of meat, fish, fruit, vegetables, oils & fats	325 756	313 006	396 355	336 925	490 643	579 640	346 111	408 566
		152	Dairy products	162 767	150 535	132 985	118 429	247 831	191 340	169 574	133 821
		153	Grain mill and starch products; animal feeds	216 742	274 468	295 543	261 703	333 233	344 876	337 642	448 279
		154	Other food products	537 508	519 783	614 437	705 293	818 196	820 075	667 894	756 491
		155	Beverages	328 848	378 957	362 157	819 177	755 760	895 417	1171 581	998 381
	16		Tobacco products	49 877	45 292	32 670	32 472	37 865	52 860	62 608	142 583
	17		Textiles	2276 665	2327 977	1926 502	2072 077	2209 131	1800 615	1238 900	1958 185
		171	Spinning, weaving and finishing of textiles	1963 302	2028 625	1518 907	1612 467	1824 858	1259 302	906 097	1535 862
		172	Other textiles	196 285	159 315	300 724	359 178	311 407	452 530	255 304	301 647
		173	Knitted and crocheted fabrics and articles	117 078	140 037	106 871	100 432	72 866	88 783	77 499	120 676
	18		Wearing apparel and fur	231 535	181 588	241 229	255 404	317 345	287 488	314 431	175 544
		181	Wearing apparel, except fur apparel	218 190	171 300	211 508	248 562	288 992	270 368	307 355	172 704
		182	Dressing and dyeing of fur; articles of fur	13 345	10 288	29 721	6 842	28 353	17 120	7 076	2 840
	19		Leather and articles of leather; footwear	402 998	303 713	216 092	273 159	211 410	195 952	146 126	118 392
		191	Leather and articles of leather	103 638	93 742	80 761	145 435	85 575	95 203	60 498	61 119
		192	Footwear	299 360	209 971	135 331	127 724	125 835	100 749	85 628	57 273
	20		Wood & products of wood & cork, exc. furniture	172 906	176 953	178 055	156 954	326 251	220 074	228 248	300 484
		201	Sawmilling and planing of wood	56 512	59 736	47 845	58 689	77 674	50 227	33 047	25 719
		202	Wood products; articles of cork, straw, etc.	116 394	117 217	130 210	98 265	248 577	169 847	195 201	274 765
	21		Paper and paper products	901 709	866 482	775 321	846 086	1250 761	1262 321	1941 700	1671 593
	22		Printing, publishing and recorded media	285 794	362 178	410 137	424 745	542 645	528 013	490 635	750 170
		221	Publishing	174 442	268 147	307 089	315 495	378 999	393 500	297 756	594 305
		222	Printing and related activities	103 807	84 740	93 671	98 997	141 188	122 573	176 225	141 119
		223	Reproduction of recorded media	7 545	9 291	9 377	10 253	22 458	11 940	16 654	14 746
	23		Coke and petroleum products; nuclear fuel	1441 529	823 161	604 541	886 916	1929 341	3088 087	2822 832	1702 047
		231	Coke oven products	4 473	14 322	1 471	1 162	209	5 435	9 893	425
		232	Refined petroleum products	1436 094	803 249	598 984	873 116	1906 184	3033 597	2762 885	1677 432
		233	Processing of nuclear fuel	962	5 590	4 086	12 638	22 948	49 055	50 054	24 190
	24		Chemicals and chemical products	5416 119	3783 799	3542 437	3822 553	4878 961	5730 115	7694 520	8378 323
		241	Basic chemicals	4104 548	2396 645	1782 741	1949 098	2750 441	4123 655	5991 592	6434 140
		242	Other chemical products	859 544	737 748	1186 278	1116 450	1502 294	1007 430	1173 849	1173 464
		243	Man-made fibres	452 027	649 406	573 418	757 005	626 226	599 030	529 079	770 719
	25		Rubber and plastic products	1163 460	1124 792	1305 158	1323 576	1737 650	2034 142	1520 535	1226 935
		251	Rubber products	323 911	411 469	355 734	364 416	477 521	694 788	434 867	391 795
		252	Plastic products	839 549	713 323	949 424	959 160	1260 129	1339 354	1085 668	835 140
	26		Other non-metallic mineral products	1936 831	2112 971	1826 272	1905 489	2061 801	2514 828	1958 459	2575 652
		261	Glass and glass products	367 065	308 472	299 646	403 738	691 167	677 288	582 452	523 741
		269	Non-metallic mineral products, nec	1569 766	1804 499	1526 626	1501 751	1370 634	1837 540	1376 007	2051 911
	27		Basic metals	3215 471	2958 595	2302 204	2996 560	6020 974	4825 660	7408 316	4898 648
		271	Basic iron and steel	2745 949	2145 523	1168 850	2436 245	5259 362	4159 889	6786 519	3726 501
		272	Basic precious and non-ferrous metals	377 906	694 480	1014 722	442 278	650 249	510 805	533 412	1089 829
		273	Casting of metals	91 616	118 592	118 632	118 037	111 363	154 966	88 385	82 318
	28		Fabricated metal products, exc. machinery	817 901	916 986	1001 693	1353 528	1425 596	1645 068	1511 791	1140 526
		281	Structural metal prod., tanks, reservoirs, etc.	293 310	333 919	426 284	622 182	632 766	847 961	784 487	593 278
		289	Other fabricated metal products	524 591	583 067	575 409	731 346	792 830	797 107	727 304	547 248
	29		Machinery and equipment, nec	1708 511	1856 638	2437 306	2123 387	2597 647	2509 808	2893 978	2511 044
		291	General purpose machinery	552 224	808 970	690 124	731 904	1241 684	1217 546	1216 678	899 998
		292	Special purpose machinery	896 336	794 030	1273 592	1035 867	1013 425	999 384	1306 196	1360 518
		293	Domestic appliances, nec	259 951	253 638	473 590	355 616	342 538	292 878	371 104	250 528
	30		Office, accounting & computing machinery	136 546	87 390	260 522	210 727	343 779	427 121	449 650	514 167
	31		Electrical machinery & apparatus, nec	511 503	601 961	835 452	744 728	947 872	1324 982	1083 871	1029 295
		311	Electric motors, generators and transformers	91 243	129 672	237 778	159 160	183 667	221 288	199 858	272 939
		312	Electricity distribution & control apparatus	137 400	140 503	261 288	185 066	199 666	280 853	252 000	292 531
		313	Insulated wire and cable	111 409	111 264	174 431	201 950	223 375	539 069	312 633	253 412
		314	Accumulators, primary cells & primary batteries	44 763	60 786	63 924	51 668	59 600	89 154	94 802	88 109
		315	Electric lamps and lighting equipment	53 619	79 613	40 429	45 703	100 992	71 081	99 841	54 942
		319	Other electrical equipment, nec	73 069	80 123	57 602	101 181	180 572	123 537	124 737	67 362
	32		Radio, TV, communication equipment	2727 261	2456 979	3240 714	4817 443	7317 454	8276 312	7568 832	9551 880
		321	Electronic valves, tubes; other electronic components	1770 182	1505 398	2376 288	3989 767	6130 582	6357 544	6434 723	8833 530
		322	TV/radio transmitters; line teleph./telegr. apparatus	177 202	147 543	175 956	263 246	366 028	300 826	561 066	416 541
		323	TV/radio receivers; sound and video app.; assoc. gds.	779 877	804 038	688 470	564 430	820 844	1617 942	573 043	301 809
	33		Medical, precision & optical instrum.; watch & clocks	178 761	185 644	186 185	209 698	289 205	343 776	1058 936	896 765
		331	Medical appl.; instruments & appl. for measuring, etc.	72 589	129 022	90 971	109 192	148 443	157 107	144 390	192 382
		332	Optical instruments and photographic equipment	85 095	46 283	61 860	90 371	126 374	171 505	902 174	696 065
		333	Watches and clocks	21 077	10 339	33 354	10 135	14 388	15 164	12 372	8 318
	34		Motor vehicles, trailers and semi-trailers	2984 493	2214 268	3058 428	3118 907	3975 915	6358 061	7023 559	7150 428
		341	Motor vehicles	1759 799	1186 647	2043 497	1927 103	2296 195	4509 348	5050 932	5387 530
		342	Bodies for motor vehicles, trailers & semi-trailers	267 991	70 434	86 947	67 306	134 053	116 430	77 750	40 865
		343	Parts & accessories for motor vehicles & their engines	956 703	957 187	927 984	1124 498	1545 667	1732 283	1894 877	1722 033
	35		Other transport equipment	836 256	807 672	2119 939	1626 070	1759 610	2261 316	1858 325	3591 780
		351	Building and repairing of ships and boats	635 992	696 843	1819 463	1375 048	1494 103	1782 002	1503 156	2425 733
		352	Railway and tramway locomotives and rolling stock	54 135	25 774	158 525	69 242	131 097	86 728	154 448	314 487
		353	Aircraft and spacecraft	99 593	54 312	119 105	153 593	73 156	277 295	170 292	805 347
		359	Transport equipment, nec	46 536	30 743	22 846	28 187	61 254	115 291	30 429	46 213
	36		Furniture; manufacturing, nec	3386 584	347 312	416 521	383 735	433 869	389 663	362 808	375 886
		361	Furniture	3192 316	209 926	192 307	205 747	252 817	223 233	181 642	147 586
		369	Manufacturing, nec	194 268	137 386	224 214	177 988	181 052	166 430	181 166	228 300
	37		Recycling	11 436	12 979	19 171	29 348	51 504	46 452	67 680	53 461
E	40_41		Electricity, gas and water supply	..	..	..	..	..	..	..	..
	40		Electricity, gas, steam and hot water supply	..	..	..	..	..	..	..	..
	41		Collection, purification and distribution of water	..	..	..	..	..	..	..	..
F	45		Construction	..	..	..	..	..	..	..	..

Table KR.4 DÉPENSES DE MACHINES ET D'ÉQUIPEMENT - INVESTMENT IN MACHINERY & EQUIPMENT

Millions de Won (Prix courants)

1991	1992	1993	1994	1995	1996	1997	1998			CITI révision 3	
115 888	124 435	72 399	87 779	83 125	94 493	66 213	92 114	Activites extractives	C	10_14	
26 789	49 483	14 591	17 367	12 847	15 272	9 226	22 662	Extraction de charbon, de lignite et de tourbe		10	
..	..	..	..	..	..	..	..	Pétrole brut et gaz naturel ; activités annexes		11	
4 074	2 883	1 733	2 924	418	572	868	436	Extraction de minerais métalliques		13	
85 025	72 069	56 075	67 488	69 860	78 649	56 119	69 016	Autres activités extractives		14	
17 820 795	17 181 351	13 889 159	14 793 274	21 672 766	22 096 663	28 046 486	34 686 144	Total activites manufacturieres	D	15_37	
802 532	698 250	808 023	928 825	1 105 202	1 245 147	1 062 191	1 368 599	Produits alimentaires et boissons		15	
140 178	130 294	156 746	131 929	162 723	202 182	124 690	151 870	Viande, poisson, fruits, légumes, huiles et graisses			151
119 164	81 069	74 113	58 730	99 917	83 701	100 895	78 656	Produits laitiers			152
114 019	107 852	147 374	148 358	160 391	199 977	182 296	233 904	Travail des grains; prod. amyl. ; alim. pour animaux			153
267 084	264 257	264 010	331 509	412 268	433 927	315 553	415 288	Autres produits alimentaires			154
162 087	114 778	165 780	258 299	269 903	325 360	338 757	488 881	Boissons			155
35 840	18 597	43 382	22 845	20 032	41 571	31 216	91 666	Produits à base de tabac		16	
1 179 221	1 358 357	963 410	1 172 248	1 254 310	921 832	724 337	861 462	Textiles		17	
1 019 706	1 243 331	770 560	929 528	1 073 574	699 871	560 757	601 649	Filature, tissage et achèvement des textiles			171
108 147	71 363	151 224	199 128	141 814	174 570	118 775	181 610	Autres articles textiles			172
51 368	43 663	41 626	43 592	38 922	47 391	44 805	78 203	Etoffes et articles de bonneterie			173
97 079	76 384	62 377	72 553	82 684	85 963	62 427	38 339	Articles d'habillement et fourrure		18	
91 729	74 799	60 174	70 247	75 789	81 959	59 317	37 127	Articles d'habillement autres qu'en fourrure			181
5 350	1 585	2 203	2 306	6 895	4 004	3 110	1 212	Prép. et teinture des fourrures ; articles en fourrure			182
164 366	83 737	82 130	97 598	69 941	68 607	51 189	44 803	Cuir et articles en cuir ; chaussures		19	
41 634	34 137	30 475	57 928	34 602	37 976	25 250	19 418	Cuir et articles en cuir sauf chaussures			191
122 732	49 600	51 655	39 670	35 339	30 631	25 939	25 385	Chaussures			192
61 516	54 725	71 647	53 862	136 918	95 990	121 128	99 622	Bois, articles en bois, sauf meubles		20	
14 850	15 533	15 338	15 955	24 939	14 496	11 506	10 365	Sciage et rabotage du bois			201
46 666	39 192	56 309	37 907	111 979	81 494	109 622	89 257	Articles en bois, liège, vannerie, sparterie			202
509 066	445 139	655 415	549 498	653 908	700 195	987 312	1 161 769	Papier, articles en papier et carton		21	
181 569	142 562	172 622	131 120	286 408	200 783	263 538	350 985	Edition, imprimerie et reproduction		22	
103 767	88 623	113 022	73 999	189 911	112 955	154 353	262 984	Edition			221
72 270	49 909	55 009	53 814	88 394	81 500	99 271	85 183	Imprimerie et activités annexes			222
5 532	4 030	4 591	3 307	8 103	6 328	9 914	2 818	Reproduction de supports enregistrés			223
681 003	731 291	396 515	136 745	1 278 701	2 386 681	1 742 227	1 720 996	Coke. et prod. pétroliers ; comb. nucléaires		23	
1 264	9 242	80	287	12	615	5 573	146	Cokéfaction			231
679 155	720 586	393 563	136 179	1 275 170	2 384 467	1 697 079	1 691 024	Produits pétroliers raffinés			232
584	1 463	2 872	279	3 519	1 599	39 575	29 826	Traitement de combustibles nucléaires			233
3 552 742	4 171 703	1 913 988	1 914 264	2 329 088	2 371 762	4 104 535	7 950 182	Produits chimiques		24	
2 763 469	3 420 006	985 580	1 000 570	1 333 845	1 630 185	3 091 796	6 530 901	Produits chimiques de base			241
359 211	277 953	493 562	330 812	534 940	365 954	497 467	617 048	Autres produits chimiques			242
430 062	473 744	434 846	582 882	460 303	375 623	515 272	802 233	Fibres synthétiques ou artificielles			243
796 192	578 677	631 578	661 157	873 971	1 004 002	711 259	806 485	Art. en caoutchouc et matières plastiques		25	
269 882	206 668	156 917	144 367	222 399	299 377	89 697	348 457	Articles en caoutchouc			251
526 310	372 009	474 661	516 790	651 572	704 625	621 562	458 028	Articles en matières plastiques			252
926 157	1 207 800	745 633	879 876	945 788	996 979	1 077 974	995 579	Autres produits minéraux non métalliques		26	
300 450	204 266	169 293	152 373	416 768	280 103	483 435	272 005	Verre et articles en verre			261
625 707	1 003 534	576 340	727 503	529 020	716 876	594 539	723 574	Produits minéraux non métalliques, nca			269
994 000	3 323 037	1 573 106	1 505 701	2 664 698	2 670 899	4 001 624	3 226 031	Produits métallurgiques de base		27	
767 059	2 970 876	829 791	1 018 869	2 319 628	2 380 267	3 721 801	2 519 151	Sidérurgie et première transformation de l'acier			271
177 069	309 493	704 592	436 358	303 371	211 538	238 007	657 806	Première transf. métaux précieux et non ferreux			272
49 872	42 668	38 723	50 474	41 699	79 094	41 816	49 074	Fonderie			273
462 185	385 245	397 809	500 998	583 673	668 050	652 157	483 709	Ouvrages en métaux, sauf machines		28	
122 628	140 443	124 822	172 056	210 739	292 663	286 905	215 673	Constr. et menuis. métal.; réservoirs; générateurs			281
339 557	244 802	272 987	328 942	372 934	375 387	365 252	268 036	Autres ouvrages en métaux			289
818 843	697 000	696 748	739 545	891 223	1 010 274	1 181 088	887 183	Machines et matériels, nca		29	
281 987	262 143	245 183	272 609	380 869	445 735	508 111	363 901	Machines d'usage général			291
378 350	353 487	332 485	342 730	380 246	482 802	507 341	444 416	Machines d'usage spécifique			292
158 506	81 370	119 080	124 206	130 108	81 737	165 636	78 866	Appareils domestiques, nca			293
57 923	35 772	85 477	97 023	126 386	214 166	184 878	216 567	Mach. de bureau, comptables, information		30	
313 355	289 285	279 900	368 817	398 919	671 047	510 323	457 842	Machines et appareils électriques, nca		31	
52 298	67 296	59 782	76 238	61 661	91 264	96 511	95 703	Moteurs, génératrices, transformateurs électriques			311
92 994	52 666	70 362	55 639	84 383	119 673	95 685	88 524	Matériel électrique de distribution et de commande			312
67 962	62 286	72 076	146 056	113 202	347 073	155 332	158 666	Fils et câbles électriques isolés			313
28 598	32 459	25 686	24 373	26 569	33 645	43 949	46 957	Accumulateurs et piles électriques			314
29 022	29 963	24 910	22 932	34 098	36 770	58 513	35 904	Lampes électriques et appareils d'éclairage			315
42 481	44 615	27 084	43 579	79 006	42 622	60 333	32 088	Autres matériels électriques, nca.			319
2 074 979	1 207 270	2 416 611	2 831 711	5 449 743	3 861 601	6 670 808	6 721 987	Equipements et appareils de radio, TV, etc.		32	
1 573 231	758 100	2 058 710	2 507 600	4 838 614	3 118 666	6 073 467	6 440 050	Tubes et valves électroniques;autres composants			321
105 151	69 355	60 147	103 447	160 727	151 909	304 711	204 972	Emetteurs radio/TV; app. téléphonie et télégraphie			322
396 597	379 815	297 754	220 664	450 402	591 026	292 630	76 965	Récepteurs radio/TV, son et image; art. associés			323
98 262	56 758	74 874	67 897	87 743	120 659	200 552	614 388	Instr. médicaux, de précision ; optique ; horlog.		33	
37 043	37 463	36 196	31 464	37 926	51 479	44 281	53 012	Instruments médicaux et de précision, etc.			331
45 787	15 054	32 162	32 567	45 650	61 987	152 744	558 477	Instruments d'optique ; matériel photographique			332
15 432	4 241	6 516	3 866	4 167	7 193	3 527	2 899	Horlogerie			333
1 477 968	1 155 238	1 334 422	1 561 529	1 969 343	1 938 142	3 210 045	5 315 897	Véhicules automobiles; remorques, etc.		34	
934 089	683 705	763 643	1 054 732	1 212 741	1 073 878	2 170 335	3 894 111	Véhicules automobiles			341
79 058	25 631	40 581	16 191	41 247	49 313	36 268	30 558	Carross. automobiles, remorques, semi-remorques			342
464 821	445 902	530 198	490 606	715 355	814 951	1 003 442	1 391 228	Pièces et accessoires pour automobiles et moteurs			343
352 212	343 896	356 687	360 058	311 301	669 672	355 494	1 135 054	Autres matériels de transport		35	
262 622	311 274	282 147	268 189	212 065	586 681	257 183	755 601	Construction et réparation de navires			351
13 632	6 610	23 859	30 613	31 850	13 709	22 278	59 186	Matériel ferroviaire roulant			352
45 153	13 583	40 317	49 934	43 787	32 538	62 449	298 817	Construction aéronautique et spatiale			353
30 805	12 429	10 364	11 322	23 599	36 744	13 584	21 450	Autres équipements de transport			359
2 176 956	115 411	119 468	126 137	138 000	133 099	113 165	109 180	Meubles ; activités de fabrication, nca		36	
2 076 202	63 513	66 558	73 920	83 174	71 051	62 523	47 316	Meubles			361
100 754	51 898	52 910	52 217	54 826	62 048	50 642	61 864	Activités de fabrication, nca			369
6 829	5 217	7 337	13 267	14 786	19 542	27 019	27 819	Récupération		37	
..	..	..	..	..	..	..	..	Electricité, gaz et eau	E	40_41	
..	..	..	..	..	..	..	..	Prod. et dist. d'électricité gaz, vapeur d'eau chaude		40	
..	..	..	..	..	..	..	..	Captage, épuration et distribution d'eau		41	
..	..	..	..	..	..	..	..	Construction	F	45	

Table KR.5 EMPLOYMENT, EMPLOYEES - EMPLOI, SALARIÉS

Units

ISIC revision 3				1991	1992	1993	1994	1995	1996	1997	1998
C	10_14		Mining and quarrying	59 268	49 405	44 080	34 576	29 339	26 860	23 920	19 525
	10		Mining of coal and lignite; extraction of peat	37 623	29 309	24 528	16 651	12 634	11 313	9 990	8 604
	11		Crude petroleum & natural gas; service activities	..	..	..	..	..	..	..	..
	13		Mining of metal ores	2 053	1 612	991	715	579	582	613	291
	14		Other mining and quarrying	19 592	18 484	18 561	17 210	16 126	14 965	13 317	10 630
D	15_37		Total manufacturing	2853 563	2734 179	2804 591	2848 789	2865 221	2812 089	2618 792	2258 359
	15		Food products and beverages	194 800	190 895	191 941	194 235	199 617	191 258	180 420	164 239
		151	Production of meat, fish, fruit, vegetables, oils & fats	63 562	61 549	60 370	61 627	62 870	61 527	60 688	56 206
		152	Dairy products	14 887	14 176	13 589	15 493	15 153	14 048	13 410	10 747
		153	Grain mill and starch products; animal feeds	18 491	17 745	19 376	19 713	20 507	20 486	19 190	18 943
		154	Other food products	77 611	78 393	79 752	78 491	79 579	74 978	68 485	63 391
		155	Beverages	20 249	19 032	18 854	18 911	21 508	20 219	18 647	14 952
	16		Tobacco products	6 855	6 473	5 823	5 463	5 043	4 909	4 305	3 752
	17		Textiles	331 828	315 210	307 127	294 928	270 876	247 456	223 792	201 252
		171	Spinning, weaving and finishing of textiles	251 132	235 416	218 756	215 688	199 743	177 855	160 082	137 674
		172	Other textiles	43 212	43 870	48 256	46 388	44 720	42 611	38 347	36 502
		173	Knitted and crocheted fabrics and articles	37 484	35 924	40 115	32 852	26 413	26 990	25 363	27 076
	18		Wearing apparel and fur	204 105	191 042	200 082	193 309	189 420	171 082	153 951	123 250
		181	Wearing apparel, except fur apparel	198 090	186 952	196 705	190 198	185 638	167 165	151 506	121 365
		182	Dressing and dyeing of fur; articles of fur	6 015	4 090	3 377	3 111	3 782	3 917	2 445	1 885
	19		Leather and articles of leather; footwear	177 054	139 719	108 925	93 096	78 406	66 614	53 722	48 068
		191	Leather and articles of leather	33 539	29 993	28 454	27 557	24 906	22 911	19 458	17 800
		192	Footwear	143 515	109 726	80 471	65 539	53 500	43 703	34 264	30 268
	20		Wood & products of wood & cork, exc. furniture	40 515	37 328	40 478	39 597	38 069	35 140	29 306	22 503
		201	Sawmilling and planing of wood	14 958	13 797	14 006	14 817	13 372	12 111	9 005	6 570
		202	Wood products; articles of cork, straw, etc.	25 557	23 531	26 472	24 780	24 697	23 029	20 301	15 933
	21		Paper and paper products	63 278	62 413	63 225	66 540	66 873	64 866	60 393	51 127
	22		Printing, publishing and recorded media	71 250	75 953	86 309	88 243	89 475	88 692	84 119	68 110
		221	Publishing	39 194	42 972	48 685	49 977	50 088	49 275	48 095	38 545
		222	Printing and related activities	30 760	31 279	35 646	36 435	37 403	37 529	34 064	28 136
		223	Reproduction of recorded media	1 296	1 702	1 978	1 831	1 984	1 888	1 960	1 429
	23		Coke and petroleum products; nuclear fuel	9 821	9 909	10 004	10 572	12 419	12 207	12 485	12 139
		231	Coke oven products	374	453	326	284	242	243	108	49
		232	Refined petroleum products	9 066	9 067	9 258	9 839	11 661	11 413	11 730	11 446
		233	Processing of nuclear fuel	381	389	420	449	516	551	647	644
	24		Chemicals and chemical products	141 936	144 599	148 515	153 102	151 960	149 215	143 690	134 190
		241	Basic chemicals	48 664	50 454	49 468	50 675	52 535	50 791	48 624	48 910
		242	Other chemical products	80 970	81 832	79 495	80 201	77 961	80 741	78 702	71 676
		243	Man-made fibres	12 302	12 313	19 552	22 226	21 464	17 683	16 364	13 604
	25		Rubber and plastic products	121 934	119 059	131 236	133 268	137 002	138 738	129 821	119 389
		251	Rubber products	31 254	31 445	32 850	33 443	34 901	34 696	33 122	31 759
		252	Plastic products	90 680	87 614	98 386	99 825	102 101	104 042	96 699	87 630
	26		Other non-metallic mineral products	134 197	134 253	135 237	132 281	128 632	122 257	109 013	82 272
		261	Glass and glass products	24 573	25 313	24 840	25 352	25 587	25 458	22 663	18 339
		269	Non-metallic mineral products, nec	109 624	108 940	110 397	106 929	103 045	96 799	86 350	63 933
	27		Basic metals	119 525	114 900	115 812	116 032	121 174	115 775	108 878	98 003
		271	Basic iron and steel	72 884	71 343	73 894	73 652	76 207	74 461	70 348	64 505
		272	Basic precious and non-ferrous metals	29 961	28 339	27 348	27 217	31 171	27 168	25 405	21 787
		273	Casting of metals	16 680	15 218	14 570	15 163	13 796	14 146	13 125	11 711
	28		Fabricated metal products, exc. machinery	149 086	146 442	165 262	179 188	188 079	195 141	176 346	149 316
		281	Structural metal prod., tanks, reservoirs, etc.	47 334	49 644	59 984	66 836	72 708	78 951	71 113	57 423
		289	Other fabricated metal products	101 752	96 798	105 278	112 352	115 371	116 190	105 233	91 893
	29		Machinery and equipment, nec	261 293	252 036	269 418	284 458	291 280	295 546	279 961	223 329
		291	General purpose machinery	84 930	88 558	99 516	111 476	122 416	127 265	120 186	92 467
		292	Special purpose machinery	128 189	118 959	126 455	128 076	130 315	133 866	125 582	102 132
		293	Domestic appliances, nec	48 174	44 519	43 447	44 906	38 549	34 415	34 193	28 730
	30		Office, accounting & computing machinery	25 121	21 066	27 718	29 931	31 769	31 356	53 930	41 368
	31		Electrical machinery & apparatus, nec	105 739	109 836	121 194	122 354	130 279	136 505	124 261	106 689
		311	Electric motors, generators and transformers	31 265	32 085	36 324	35 415	36 162	37 510	34 274	32 707
		312	Electricity distribution & control apparatus	30 640	31 946	36 997	36 880	40 026	40 613	36 953	28 450
		313	Insulated wire and cable	15 949	15 864	16 488	18 012	16 970	20 881	17 859	15 843
		314	Accumulators, primary cells & primary batteries	4 907	5 596	4 376	4 698	5 114	5 067	4 939	4 895
		315	Electric lamps and lighting equipment	11 719	11 831	13 228	14 090	14 420	14 324	12 647	11 385
		319	Other electrical equipment, nec	11 259	12 514	13 781	13 259	17 587	18 110	17 589	13 409
	32		Radio, TV, communication equipment	271 427	258 305	248 451	261 558	269 419	269 088	238 747	219 637
		321	Electronic valves, tubes; other electronic components	104 458	102 338	108 154	117 645	134 987	137 954	138 921	138 395
		322	TV/radio transmitters; line teleph./telegr. apparatus	37 364	33 732	30 098	34 870	31 816	36 059	39 203	38 632
		323	TV/radio receivers; sound and video app.; assoc. gds.	129 605	122 235	110 199	109 043	102 616	95 075	60 623	42 610
	33		Medical, precision & optical instrum.; watch & clocks	44 483	41 121	45 282	47 403	48 603	48 324	45 330	39 785
		331	Medical appl.; instruments & appl. for measuring, etc.	21 607	20 285	23 225	26 719	28 236	28 973	26 972	25 033
		332	Optical instruments and photographic equipment	12 946	11 690	12 728	13 087	13 975	14 190	14 205	10 699
		333	Watches and clocks	9 930	9 146	9 329	7 597	6 392	5 161	4 153	4 053
	34		Motor vehicles, trailers and semi-trailers	181 264	182 003	190 684	207 592	218 235	233 515	221 854	186 923
		341	Motor vehicles	69 704	74 776	77 403	89 859	93 143	102 882	98 140	88 047
		342	Bodies for motor vehicles, trailers & semi-trailers	12 982	8 075	8 385	7 304	8 763	6 706	4 910	4 642
		343	Parts & accessories for motor vehicles & their engines	98 578	99 152	104 896	110 429	116 329	123 927	118 804	94 234
	35		Other transport equipment	67 957	65 071	73 500	82 972	89 563	92 279	95 145	88 498
		351	Building and repairing of ships and boats	52 659	50 015	55 007	63 959	73 077	75 142	75 473	68 697
		352	Railway and tramway locomotives and rolling stock	3 256	4 387	7 992	8 148	6 461	5 738	8 796	8 685
		353	Aircraft and spacecraft	5 200	4 607	4 879	5 187	4 952	5 141	6 148	6 928
		359	Transport equipment, nec	6 842	6 062	5 622	5 678	5 073	6 258	4 728	4 188
	36		Furniture; manufacturing, nec	129 018	115 450	116 674	110 693	106 876	99 350	85 961	70 683
		361	Furniture	52 002	50 727	55 174	53 785	52 869	49 241	40 542	29 837
		369	Manufacturing, nec	77 016	64 723	61 500	56 908	54 007	50 109	45 419	40 846
	37		Recycling	1 077	1 096	1 694	1 974	2 152	2 776	3 362	3 837
E	40_41		Electricity, gas and water supply	..	..	..	..	..	..	..	..
	40		Electricity, gas, steam and hot water supply	..	..	..	..	..	..	..	..
	41		Collection, purification and distribution of water	..	..	..	..	..	..	..	..
F	45		Construction	..	..	..	..	..	..	..	..

Table KR.6 EMPLOI, FEMMES - EMPLOYMENT, FEMALES

Unités (Prix courants)

1991	1992	1993	1994	1995	1996	1997	1998			CITI révision 3	
3 564	3 002	2 543	2 335	2 118	2 125	1 844	1 695	Activites extractives	C	10_14	
1 945	1 455	972	796	618	611	531	617	Extraction de charbon, de lignite et de tourbe		10	
..	..	..	..	..	..	..	..	Pétrole brut et gaz naturel ; activités annexes		11	
119	106	64	44	31	30	39	18	Extraction de minerais métalliques		13	
1 500	1 441	1 507	1 495	1 469	1 484	1 274	1 060	Autres activités extractives		14	
1 062 976	982 587	970 456	963 599	948 915	923 279	803 178	666 736	Total activites manufacturieres	D	15_37	
88 768	86 878	88 410	91 506	94 771	90 104	83 722	76 511	Produits alimentaires et boissons		15	
38 601	37 627	36 264	38 208	37 629	37 643	36 453	33 174	Viande, poisson, fruits, légumes, huiles et graisses			151
4 621	4 045	3 955	5 295	4 715	4 683	3 831	3 118	Produits laitiers			152
4 281	4 306	4 935	5 192	5 700	5 513	4 956	5 528	Travail des grains; prod. amyl. ; alim. pour animaux			153
36 796	36 572	38 512	38 262	40 902	36 876	34 079	31 373	Autres produits alimentaires			154
4 469	4 328	4 744	4 549	5 825	5 389	4 403	3 318	Boissons			155
1 637	1 553	1 281	1 208	1 068	1 027	847	758	Produits à base de tabac		16	
191 452	171 394	164 494	154 313	139 527	123 619	107 427	96 198	Textiles		17	
143 588	126 076	115 375	110 412	100 989	86 816	73 751	63 467	Filature, tissage et achèvement des textiles			171
23 655	22 826	24 455	23 742	22 415	20 782	18 578	16 995	Autres articles textiles			172
24 209	22 492	24 664	20 159	16 123	16 021	15 098	15 736	Etoffes et articles de bonneterie			173
146 691	140 555	140 963	135 788	131 765	120 304	108 875	88 835	Articles d'habillement et fourrure		18	
143 961	138 670	139 103	134 334	129 951	118 458	107 692	88 033	Articles d'habillement autres qu'en fourrure			181
2 730	1 885	1 860	1 454	1 814	1 846	1 183	802	Prép. et teinture des fourrures ; articles en fourrure			182
100 899	78 628	57 701	47 513	38 444	33 343	25 246	22 769	Cuir et articles en cuir ; chaussures		19	
12 898	11 864	11 154	10 747	10 089	9 587	7 868	7 311	Cuir et articles en cuir sauf chaussures			191
88 001	66 764	46 547	36 766	28 355	23 756	17 378	15 458	Chaussures			192
10 720	9 910	10 153	9 915	9 192	8 829	6 985	5 085	Bois, articles en bois, sauf meubles		20	
2 774	2 469	2 532	2 742	2 446	2 405	1 566	1 150	Sciage et rabotage du bois			201
7 946	7 441	7 621	7 173	6 746	6 424	5 419	3 935	Articles en bois, liège, vannerie, sparterie			202
15 917	15 764	15 587	15 782	15 836	15 088	12 966	10 291	Papier, articles en papier et carton		21	
20 957	22 459	24 732	25 264	26 884	26 688	25 788	20 147	Edition, imprimerie et reproduction		22	
10 488	12 287	13 349	13 704	14 668	14 849	15 333	11 936	Edition			221
9 984	9 513	10 670	10 942	11 473	11 204	9 761	7 771	Imprimerie et activités annexes			222
485	659	713	618	743	635	694	440	Reproduction de supports enregistrés			223
692	686	579	584	623	4 450	615	475	Coke. et prod. pétroliers ; comb. nucléaires		23	
22	21	70	13	11	13	12	7	Cokéfaction			231
631	625	477	549	577	4 393	560	454	Produits pétroliers raffinés			232
39	40	32	22	35	44	43	14	Traitement de combustibles nucléaires			233
38 724	37 969	38 371	38 827	38 010	49 654	30 762	27 002	Produits chimiques		24	
5 338	5 430	5 419	5 121	5 552	4 763	4 160	4 446	Produits chimiques de base			241
29 670	28 647	26 971	26 638	26 861	41 163	23 465	20 251	Autres produits chimiques			242
3 716	3 892	5 981	7 068	5 597	3 728	3 137	2 305	Fibres synthétiques ou artificielles			243
32 487	31 122	32 852	33 969	35 064	34 760	31 251	28 256	Art. en caoutchouc et matières plastiques		25	
7 330	7 091	6 522	6 735	7 275	7 002	6 556	6 005	Articles en caoutchouc			251
25 157	24 031	26 330	27 234	27 789	27 758	24 695	22 251	Articles en matières plastiques			252
31 292	30 026	29 819	28 574	27 315	24 689	20 729	14 497	Autres produits minéraux non métalliques		26	
5 382	5 358	5 620	5 288	5 405	4 848	4 032	2 880	Verre et articles en verre			261
25 910	24 668	24 199	23 286	21 910	19 841	16 697	11 617	Produits minéraux non métalliques, nca			269
11 784	10 481	10 978	11 176	11 785	10 723	9 217	8 123	Produits métallurgiques de base		27	
5 338	4 683	5 292	5 345	5 809	5 092	4 456	4 220	Sidérurgie et première transformation de l'acier			271
4 124	3 795	3 576	3 686	4 092	3 687	3 058	2 414	Première transf. métaux précieux et non ferreux			272
2 322	2 003	2 110	2 145	1 884	1 944	1 703	1 489	Fonderie			273
32 752	31 969	34 780	36 903	37 997	39 317	34 390	28 168	Ouvrages en métaux, sauf machines		28	
6 363	6 646	7 636	8 595	8 995	9 812	8 770	6 273	Constr. et menuis. métal.; réservoirs; générateurs			281
26 389	25 323	27 144	28 308	29 002	29 505	25 620	21 895	Autres ouvrages en métaux			289
46 607	42 960	46 662	49 452	49 513	49 522	44 392	34 344	Machines et matériels, nca		29	
14 443	14 089	17 024	18 831	20 498	21 810	19 245	14 982	Machines d'usage général			291
15 756	14 424	15 944	16 433	16 306	16 488	14 728	11 607	Machines d'usage spécifique			292
16 408	14 447	13 694	14 188	12 709	11 224	10 419	7 755	Appareils domestiques, nca			293
12 956	9 499	12 299	12 251	13 615	13 199	18 000	14 138	Mach. de bureau, comptables, information		30	
40 125	42 215	45 273	46 651	47 984	48 734	42 167	34 721	Machines et appareils électriques, nca		31	
14 748	15 114	14 403	15 045	14 235	14 375	12 370	11 403	Moteurs, génératrices, transformateurs électriques			311
10 575	11 800	14 497	13 695	14 093	14 001	11 787	8 744	Matériel électrique de distribution et de commande			312
3 726	3 297	3 972	4 848	5 076	4 979	4 567	3 776	Fils et câbles électriques isolés			313
1 638	1 617	1 035	1 129	1 181	1 047	1 094	1 126	Accumulateurs et piles électriques			314
5 210	5 240	5 484	5 857	5 608	5 364	4 689	3 996	Lampes électriques et appareils d'éclairage			315
4 228	5 147	5 882	6 077	7 791	8 968	7 660	5 676	Autres matériels électriques, nca.			319
135 762	125 505	120 870	129 029	135 307	136 983	107 808	90 674	Equipements et appareils de radio, TV, etc.		32	
51 082	48 671	52 563	57 468	71 825	77 784	61 824	55 612	Tubes et valves électroniques;autres composants			321
18 255	14 596	13 204	16 752	13 887	15 469	15 136	14 136	Emetteurs radio/TV; app. téléphonie et télégraphie			322
66 425	62 238	55 103	54 809	49 595	43 730	30 848	20 926	Récepteurs radio/TV, son et image; art. associés			323
19 352	17 258	18 657	18 567	18 541	17 746	16 299	14 711	Instr. médicaux, de précision ; optique ; horlog.		33	
8 519	7 863	8 600	9 285	9 503	9 266	8 500	7 771	Instruments médicaux et de précision, etc.			331
5 849	5 386	5 470	5 688	5 984	6 091	5 897	5 033	Instruments d'optique ; matériel photographique			332
4 984	4 009	4 587	3 594	3 054	2 389	1 902	1 907	Horlogerie			333
25 657	25 488	26 757	28 777	30 272	32 739	38 998	20 594	Véhicules automobiles; remorques, etc.		34	
2 616	2 499	2 999	3 768	3 269	4 467	3 065	1 928	Véhicules automobiles			341
1 033	699	782	741	975	609	400	383	Carross. automobiles, remorques, semi-remorques			342
22 008	22 290	22 976	24 268	26 028	27 663	35 533	18 283	Pièces et accessoires pour automobiles et moteurs			343
5 483	5 567	5 668	7 046	7 109	7 429	6 862	5 713	Autres matériels de transport		35	
3 300	3 433	3 554	4 765	5 086	5 296	4 900	4 037	Construction et réparation de navires			351
347	384	575	592	556	549	581	541	Matériel ferroviaire roulant			352
291	409	257	370	312	328	367	337	Construction aéronautique et spatiale			353
1 545	1 341	1 282	1 319	1 155	1 256	1 014	798	Autres équipements de transport			359
51 978	44 436	43 228	40 107	37 842	33 819	29 205	23 926	Meubles ; activités de fabrication, nca		36	
14 637	14 341	15 414	14 875	14 049	12 476	9 746	7 014	Meubles			361
37 341	30 095	27 814	25 232	23 793	21 343	19 459	16 912	Activités de fabrication, nca			369
284	265	342	397	451	513	627	800	Récupération		37	
..	..	..	..	..	..	..	..	Electricité, gaz et eau	E	40_41	
..	..	..	..	..	..	..	..	Prod. et dist. d'électricité gaz, vapeur d'eau chaude		40	
..	..	..	..	..	..	..	..	Captage, épuration et distribution d'eau		41	
..	..	..	..	..	..	..	..	Construction	F	45	

KOREA

Table KR.7 COMPENSATION OF LABOUR, EMPLOYEES - SALAIRES ET CHARGES SOCIALES, SALARIÉS

Millions of Won (Current Prices)

ISIC revision 3				1991	1992	1993	1994	1995	1996	1997	1998
C	10_14		Mining and quarrying	529 380	546 213	524 051	466 500	469 133	486 382	475 445	..
	10		Mining of coal and lignite; extraction of peat	342 241	339 822	303 110	231 695	212 902	215 447	214 834	..
	11		Crude petroleum & natural gas; service activities	..	..	..	..	..	..	..	..
	13		Mining of metal ores	21 218	17 114	10 167	7 380	7 977	8 972	10 266	..
	14		Other mining and quarrying	165 921	189 277	210 774	227 425	248 254	261 963	250 345	..
D	15_37		Total manufacturing	25656 260	28304 044	32316 858	36879 031	42738 567	47804 802	46958 936	..
	15		Food products and beverages	1513 620	1667 139	1920 527	2124 398	2447 969	2639 421	2647 974	..
		151	Production of meat, fish, fruit, vegetables, oils & fats	418 224	445 359	503 439	555 656	643 567	694 176	707 501	..
		152	Dairy products	142 988	158 291	180 320	220 167	238 129	262 373	262 540	..
		153	Grain mill and starch products; animal feeds	193 047	206 166	258 501	275 094	318 353	364 253	379 943	..
		154	Other food products	539 418	629 218	735 605	793 411	907 324	958 480	944 879	..
		155	Beverages	219 943	228 105	242 662	280 070	340 596	360 139	353 111	..
	16		Tobacco products	111 581	124 967	124 839	122 729	133 759	162 219	147 711	..
	17		Textiles	2390 750	2692 543	2870 951	3048 318	3249 934	3265 719	3136 963	..
		171	Spinning, weaving and finishing of textiles	1846 555	2051 964	2084 037	2255 697	2447 393	2386 996	2299 173	..
		172	Other textiles	295 049	353 069	434 982	477 609	520 343	548 593	511 547	..
		173	Knitted and crocheted fabrics and articles	249 146	287 510	351 932	315 012	282 198	330 130	326 243	..
	18		Wearing apparel and fur	1235 917	1324 391	1597 994	1758 518	1993 347	1976 917	1842 642	..
		181	Wearing apparel, except fur apparel	1187 080	1287 246	1564 366	1720 192	1940 040	1917 015	1806 173	..
		182	Dressing and dyeing of fur; articles of fur	48 837	37 145	33 628	38 326	53 307	59 902	36 469	..
	19		Leather and articles of leather; footwear	1222 251	1065 577	953 491	896 174	853 992	843 269	710 209	..
		191	Leather and articles of leather	267 530	285 150	293 410	311 794	315 110	333 527	295 533	..
		192	Footwear	954 721	780 427	660 081	584 380	538 882	509 742	414 676	..
	20		Wood & products of wood & cork, exc. furniture	338 884	352 887	411 776	456 303	488 836	516 370	459 015	..
		201	Sawmilling and planing of wood	120 857	124 519	138 325	164 414	164 880	166 800	128 633	..
		202	Wood products; articles of cork, straw, etc.	218 027	228 368	273 451	291 889	323 956	349 570	330 382	..
	21		Paper and paper products	585 168	667 863	728 885	859 544	1002 505	1082 586	1092 605	..
	22		Printing, publishing and recorded media	700 322	872 607	1131 819	1275 819	1402 242	1569 629	1562 974	..
		221	Publishing	452 164	570 859	762 125	848 790	909 695	1015 255	1020 179	..
		222	Printing and related activities	236 006	283 075	348 611	402 242	465 916	524 761	511 730	..
		223	Reproduction of recorded media	12 152	18 673	21 083	24 787	26 631	29 613	31 065	..
	23		Coke and petroleum products; nuclear fuel	176 798	162 138	189 501	236 522	300 872	368 147	438 364	..
		231	Coke oven products	4 499	5 314	4 506	4 689	4 345	5 579	1 601	..
		232	Refined petroleum products	166 589	150 655	177 484	224 723	288 285	347 151	418 665	..
		233	Processing of nuclear fuel	5 710	6 169	7 511	7 110	8 242	15 417	18 098	..
	24		Chemicals and chemical products	1585 603	1827 463	2121 026	2455 156	2805 103	3155 059	3194 143	..
		241	Basic chemicals	630 563	737 453	804 151	947 595	1113 941	1219 285	1252 131	..
		242	Other chemical products	811 149	926 688	1020 812	1147 132	1241 184	1487 675	1507 794	..
		243	Man-made fibres	143 891	163 322	296 063	360 429	449 978	448 099	434 218	..
	25		Rubber and plastic products	1047 908	1156 759	1430 062	1594 937	1904 517	2169 542	2127 815	..
		251	Rubber products	287 190	320 461	380 304	426 268	523 008	587 739	623 423	..
		252	Plastic products	760 718	836 298	1049 758	1168 669	1381 509	1581 803	1504 392	..
	26		Other non-metallic mineral products	1298 748	1519 386	1659 889	1800 370	2043 677	2146 489	2027 174	..
		261	Glass and glass products	273 082	327 545	353 436	387 643	458 960	512 418	486 205	..
		269	Non-metallic mineral products, nec	1025 666	1191 841	1306 453	1412 727	1584 717	1634 071	1540 969	..
	27		Basic metals	1519 390	1635 445	1785 950	2079 491	2409 495	2583 313	2490 411	..
		271	Basic iron and steel	1052 630	1133 660	1272 066	1481 365	1673 202	1846 450	1764 397	..
		272	Basic precious and non-ferrous metals	300 760	328 717	337 736	387 403	520 394	493 652	490 292	..
		273	Casting of metals	166 000	173 068	176 148	210 723	215 899	243 211	235 722	..
	28		Fabricated metal products, exc. machinery	1364 184	1495 351	1892 305	2313 183	2742 407	3211 300	3036 066	..
		281	Structural metal prod., tanks, reservoirs, etc.	482 135	565 753	770 178	952 688	1198 517	1448 217	1368 215	..
		289	Other fabricated metal products	882 049	929 598	1122 127	1360 495	1543 890	1763 083	1667 851	..
	29		Machinery and equipment, nec	2550 512	2832 012	3277 162	3829 759	4513 870	5197 888	5229 504	..
		291	General purpose machinery	799 176	1039 964	1204 668	1476 367	1903 718	2212 867	2265 748	..
		292	Special purpose machinery	1335 473	1354 681	1588 251	1781 011	2083 380	2443 315	2427 227	..
		293	Domestic appliances, nec	415 863	437 367	484 243	572 381	526 772	541 706	536 529	..
	30		Office, accounting & computing machinery	196 877	198 944	306 000	367 335	452 438	473 972	1090 815	..
	31		Electrical machinery & apparatus, nec	915 261	1068 514	1300 955	1411 246	1776 431	2130 915	2054 836	..
		311	Electric motors, generators and transformers	248 993	275 788	388 895	381 963	466 793	565 524	554 232	..
		312	Electricity distribution & control apparatus	275 733	328 621	389 580	434 485	563 858	638 160	619 415	..
		313	Insulated wire and cable	168 576	199 024	208 673	250 365	276 560	410 771	348 703	..
		314	Accumulators, primary cells & primary batteries	50 783	61 878	56 901	65 124	80 876	91 329	102 649	..
		315	Electric lamps and lighting equipment	80 987	94 912	117 932	142 423	175 989	182 314	174 437	..
		319	Other electrical equipment, nec	90 189	108 291	138 974	136 886	212 355	242 817	255 400	..
	32		Radio, TV, communication equipment	2487 534	2698 227	2756 228	3253 881	4113 803	4763 459	4267 885	..
		321	Electronic valves, tubes; other electronic components	1010 124	1123 596	1309 837	1589 502	2246 595	2536 297	2617 845	..
		322	TV/radio transmitters; line teleph./telegr. apparatus	335 141	358 981	332 374	452 924	507 799	598 273	666 536	..
		323	TV/radio receivers; sound and video app.; assoc. gds.	1142 269	1215 650	1114 017	1211 455	1359 409	1628 889	983 504	..
	33		Medical, precision & optical instrum.; watch & clocks	353 834	375 260	452 260	524 546	637 036	701 845	726 468	..
		331	Medical appl.; instruments & appl. for measuring, etc.	171 095	183 090	230 993	300 900	385 263	442 731	437 678	..
		332	Optical instruments and photographic equipment	101 864	101 404	124 038	138 619	172 610	192 126	226 847	..
		333	Watches and clocks	80 875	90 766	97 229	85 027	79 163	66 988	61 943	..
	34		Motor vehicles, trailers and semi-trailers	2150 157	2400 950	2832 872	3548 317	4119 561	5017 070	4936 820	..
		341	Motor vehicles	1046 159	1233 382	1441 358	1923 719	2181 992	2728 209	2698 860	..
		342	Bodies for motor vehicles, trailers & semi-trailers	156 912	111 448	132 660	114 507	159 439	152 582	88 048	..
		343	Parts & accessories for motor vehicles & their engines	947 086	1056 120	1258 854	1510 091	1778 130	2136 279	2149 912	..
	35		Other transport equipment	922 467	1137 156	1406 983	1711 734	2013 358	2426 880	2430 089	..
		351	Building and repairing of ships and boats	755 922	944 370	1114 759	1384 158	1691 086	2067 663	1992 216	..
		352	Railway and tramway locomotives and rolling stock	34 497	63 124	149 898	153 486	121 272	107 498	202 086	..
		353	Aircraft and spacecraft	73 014	67 003	80 946	106 130	130 375	155 265	158 951	..
		359	Transport equipment, nec	59 034	62 659	61 380	67 960	70 625	96 454	76 836	..
	36		Furniture; manufacturing, nec	981 428	1019 368	1148 551	1189 206	1307 080	1362 515	1255 272	..
		361	Furniture	423 818	467 786	559 833	588 840	668 714	699 735	622 718	..
		369	Manufacturing, nec	557 610	551 582	588 718	600 366	638 366	662 780	632 554	..
	37		Recycling	7 066	9 097	16 832	21 545	26 335	40 278	53 181	..
E	40_41		Electricity, gas and water supply	..	..	..	..	..	..	..	..
	40		Electricity, gas, steam and hot water supply	..	..	..	..	..	..	..	..
	41		Collection, purification and distribution of water	..	..	..	..	..	..	..	..
F	45		Construction	..	..	..	..	..	..	..	..

Table KR.8 SALAIRES ET TRAITEMENTS, SALARIÉS - WAGES AND SALARIES, EMPLOYEES

Millions de Won (Prix courants)

1991	1992	1993	1994	1995	1996	1997	1998			CITI révision 3	
474 466	486 208	474 717	409 712	412 822	425 298	415 698	344 350	Activites extractives	C	10_14	
307 185	300 994	274 751	200 795	185 464	185 672	184 786	179 038	Extraction de charbon, de lignite et de tourbe		10	
..	..	..	..	..	..	..	..	Pétrole brut et gaz naturel ; activités annexes		11	
19 416	15 695	9 528	6 892	7 427	8 255	9 631	3 610	Extraction de minerais métalliques		13	
147 865	169 519	190 438	202 025	219 931	231 371	221 281	161 702	Autres activités extractives		14	
22 830 419	25 234 409	28 834 306	32 791 213	37 844 431	42 329 152	41 489 165	34 693 726	Total activites manufacturieres	D	15_37	
1 338 969	1 471 187	1 700 148	1 876 719	2 150 178	2 334 932	2 348 228	2 084 397	Produits alimentaires et boissons		15	
376 625	400 061	451 248	498 478	577 742	624 560	638 207	585 239	Viande, poisson, fruits, légumes, huiles et graisses			151
124 529	137 929	159 727	191 025	200 684	229 586	234 400	192 316	Produits laitiers			152
168 160	178 724	222 362	240 853	280 541	317 010	327 676	285 694	Travail des grains; prod. amyl. ; alim. pour animaux			153
481 396	559 725	654 619	706 687	802 605	856 931	842 752	777 213	Autres produits alimentaires			154
188 259	194 748	212 192	239 676	288 606	306 845	305 193	243 935	Boissons			155
90 407	96 805	98 072	94 832	104 475	129 951	114 340	96 186	Produits à base de tabac		16	
2 180 703	2 445 466	2 620 686	2 780 040	2 963 269	2 976 148	2 857 443	2 478 888	Textiles		17	
1 683 964	1 864 057	1 904 537	2 057 129	2 229 133	2 171 105	2 090 768	1 721 440	Filature, tissage et achèvement des textiles			171
267 825	317 215	393 667	433 121	472 614	500 861	465 155	434 796	Autres articles textiles			172
228 914	264 194	322 482	289 790	261 522	304 182	301 520	322 652	Etoffes et articles de bonneterie			173
1 135 558	1 223 068	1 474 682	1 619 693	1 841 478	1 816 280	1 704 683	1 267 738	Articles d'habillement et fourrure		18	
1 090 314	1 188 623	1 444 705	1 584 729	1 792 961	1 761 427	1 671 125	1 241 888	Articles d'habillement autres qu'en fourrure			181
45 244	34 445	29 977	34 964	48 517	54 853	33 558	25 850	Prép. et teinture des fourrures ; articles en fourrure			182
1 120 506	976 683	873 497	819 654	781 635	767 470	641 583	553 815	Cuir et articles en cuir ; chaussures		19	
242 282	257 780	265 275	281 040	285 511	300 639	266 654	235 439	Cuir et articles en cuir sauf chaussures			191
878 224	718 903	608 222	538 614	496 124	466 831	374 929	318 376	Chaussures			192
310 888	323 807	378 054	418 383	447 587	470 210	418 310	290 548	Bois, articles en bois, sauf meubles		20	
111 867	115 526	129 074	152 448	153 559	154 053	119 555	78 614	Sciage et rabotage du bois			201
199 021	208 281	248 980	265 935	294 028	316 157	298 755	211 934	Articles en bois, liège, vannerie, sparterie			202
515 855	590 942	650 678	760 377	885 549	957 960	969 185	792 660	Papier, articles en papier et carton		21	
633 442	786 902	1 021 646	1 141 758	1 254 869	1 400 673	1 400 432	1 045 151	Edition, imprimerie et reproduction		22	
404 675	508 980	678 107	747 287	797 423	888 865	902 428	656 690	Edition			221
217 911	261 255	324 891	372 471	433 659	485 733	470 396	366 752	Imprimerie et activités annexes			222
10 856	16 667	18 648	22 000	23 787	26 075	27 608	21 709	Reproduction de supports enregistrés			223
130 856	145 191	164 920	205 785	258 217	318 049	349 763	338 710	Coke. et prod. pétroliers ; comb. nucléaires		23	
4 048	4 866	4 076	4 285	3 931	4 237	1 374	958	Cokéfaction			231
122 182	135 428	154 587	195 840	247 908	301 045	334 875	323 101	Produits pétroliers raffinés			232
4 626	4 897	6 257	5 660	6 378	12 767	13 514	14 651	Traitement de combustibles nucléaires			233
1 398 365	1 606 516	1 870 388	2 161 919	2 454 942	2 740 787	2 769 649	2 574 106	Produits chimiques		24	
561 480	652 026	706 332	830 054	969 223	1 037 496	1 064 592	1 096 795	Produits chimiques de base			241
710 472	810 653	902 112	1 013 489	1 086 735	1 304 064	1 326 841	1 148 675	Autres produits chimiques			242
126 413	143 837	261 944	318 376	398 984	399 227	378 216	328 636	Fibres synthétiques ou artificielles			243
935 579	1 037 187	1 288 092	1 432 077	1 697 547	1 935 741	1 905 910	1 677 056	Art. en caoutchouc et matières plastiques		25	
256 601	288 031	344 389	381 113	467 494	525 730	559 383	493 240	Articles en caoutchouc			251
678 978	749 156	943 703	1 050 964	1 230 053	1 410 011	1 346 527	1 183 816	Articles en matières plastiques			252
1 154 925	1 352 672	1 475 461	1 596 684	1 795 101	1 894 126	1 784 492	1 310 101	Autres produits minéraux non métalliques		26	
245 865	294 979	313 209	346 566	408 489	449 066	423 200	335 913	Verre et articles en verre			261
909 060	1 057 693	1 162 252	1 250 118	1 386 612	1 445 060	1 361 292	974 188	Produits minéraux non métalliques, nca			269
1 294 699	1 385 232	1 487 621	1 694 347	1 954 375	2 117 799	2 056 608	1 759 068	Produits métallurgiques de base		27	
880 507	935 220	1 027 759	1 162 650	1 303 006	1 462 272	1 413 213	1 250 048	Sidérurgie et première transformation de l'acier			271
266 118	294 348	301 226	343 034	457 866	437 760	432 058	337 969	Première transf. métaux précieux et non ferreux			272
148 074	155 664	158 636	188 663	193 503	217 767	211 337	171 051	Fonderie			273
1 217 818	1 349 374	1 704 624	2 078 683	2 471 457	2 892 692	2 739 948	2 179 340	Ouvrages en métaux, sauf machines		28	
430 552	515 194	690 851	857 374	1 079 857	1 306 108	1 232 912	957 125	Constr. et menuis. métal.; réservoirs; générateurs			281
787 266	834 180	1 013 773	1 221 309	1 391 600	1 586 584	1 507 036	1 222 215	Autres ouvrages en métaux			289
2 264 371	2 523 626	2 928 188	3 392 294	3 992 799	4 600 348	4 630 394	3 484 231	Machines et matériels, nca		29	
714 396	923 039	1 078 204	1 305 256	1 690 466	1 964 246	2 010 141	1 406 739	Machines d'usage général			291
1 183 834	1 213 386	1 423 946	1 590 357	1 837 912	2 157 767	2 154 212	1 660 074	Machines d'usage spécifique			292
366 141	387 201	426 038	496 681	464 421	478 335	466 041	417 418	Appareils domestiques, nca			293
176 941	178 267	273 709	329 176	394 563	420 721	910 344	667 520	Mach. de bureau, comptables, information		30	
814 283	944 461	1 167 762	1 261 723	1 583 479	1 893 386	1 834 587	1 452 838	Machines et appareils électriques, nca		31	
225 003	244 639	351 386	344 198	420 760	509 705	501 156	447 586	Moteurs, génératrices, transformateurs électriques			311
243 915	285 690	350 041	386 871	501 601	569 284	554 870	390 656	Matériel électrique de distribution et de commande			312
144 039	174 185	182 777	221 156	241 518	348 115	297 489	227 957	Fils et câbles électriques isolés			313
45 604	55 550	50 538	57 671	71 380	82 503	92 095	76 614	Accumulateurs et piles électriques			314
73 620	86 505	108 032	129 203	159 819	164 181	158 941	141 066	Lampes électriques et appareils d'éclairage			315
82 102	97 892	124 988	122 624	188 401	219 598	230 036	168 959	Autres matériels électriques, nca.			319
2 202 877	2 343 643	2 401 830	2 816 779	3 518 523	4 081 073	3 648 665	3 554 480	Equipements et appareils de radio, TV, etc.		32	
887 776	983 381	1 135 995	1 366 718	1 930 224	2 193 709	2 245 673	2 396 375	Tubes et valves électroniques;autres composants			321
291 103	306 791	286 029	395 207	430 707	513 650	557 826	605 342	Emetteurs radio/TV; app. téléphonie et télégraphie			322
1 023 998	1 053 471	979 806	1 054 854	1 157 592	1 373 714	845 166	552 763	Récepteurs radio/TV, son et image; art. associés			323
318 879	338 453	406 723	464 745	567 681	625 359	640 905	528 131	Instr. médicaux, de précision ; optique ; horlog.		33	
153 460	163 480	207 432	264 500	339 164	392 516	385 989	342 008	Instruments médicaux et de précision, etc.			331
92 239	92 629	112 483	125 454	156 393	171 766	198 514	130 208	Instruments d'optique ; matériel photographique			332
73 180	82 344	86 808	74 791	72 124	61 077	56 402	55 915	Horlogerie			333
1 884 215	2 179 582	2 538 401	3 227 348	3 729 235	4 537 481	4 450 738	3 513 911	Véhicules automobiles; remorques, etc.		34	
904 450	1 136 012	1 287 216	1 772 725	2 002 559	2 497 243	2 454 664	2 047 859	Véhicules automobiles			341
132 647	99 294	120 386	101 559	139 941	133 807	75 746	70 999	Carross. automobiles, remorques, semi-remorques			342
847 118	944 276	1 130 799	1 353 064	1 586 735	1 906 431	1 920 328	1 395 053	Pièces et accessoires pour automobiles et moteurs			343
808 617	1 004 875	1 243 807	1 511 754	1 786 238	2 142 408	2 121 439	2 107 802	Autres matériels de transport		35	
661 357	835 356	989 810	1 231 569	1 512 008	1 834 441	1 749 082	1 697 729	Construction et réparation de navires			351
30 488	55 708	129 278	128 101	101 776	90 863	173 051	181 439	Matériel ferroviaire roulant			352
64 450	58 757	70 125	91 140	111 056	131 668	131 899	174 538	Construction aéronautique et spatiale			353
52 322	55 054	54 594	60 944	61 398	85 436	67 407	54 096	Autres équipements de transport			359
895 201	922 137	1 050 085	1 086 963	1 187 269	1 239 273	1 144 064	885 947	Meubles ; activités de fabrication, nca		36	
388 538	422 938	512 908	536 865	605 862	634 961	565 756	377 074	Meubles			361
506 663	499 199	537 177	550 098	581 407	604 312	578 308	508 873	Activités de fabrication, nca			369
6 465	8 333	15 232	19 480	23 965	36 285	47 455	51 102	Récupération		37	
..	..	..	..	..	..	..	..	Electricité, gaz et eau	E	40_41	
..	..	..	..	..	..	..	..	Prod. et dist. d'électricité gaz, vapeur d'eau chaude		40	
..	..	..	..	..	..	..	..	Captage, épuration et distribution d'eau		41	
..	..	..	..	..	..	..	..	Construction	F	45	

KOREA

Table KR.9 SUPPLEMENTS TO WAGES & SALARIES - CHARGES SOCIALES DES EMPLOYEURS

Millions of Won (Current Prices)

ISIC revision 3				1991	1992	1993	1994	1995	1996	1997	1998
C	10_14		Mining and quarrying	54 914	60 005	49 334	56 788	56 311	61 084	59 747	..
	10		Mining of coal and lignite; extraction of peat	35 056	38 828	28 359	30 900	27 438	29 775	30 048	..
	11		Crude petroleum & natural gas; service activities	..	..	..	..	..	..	..	..
	13		Mining of metal ores	1 802	1 419	639	488	550	717	635	..
	14		Other mining and quarrying	18 056	19 758	20 336	25 400	28 323	30 592	29 064	..
D	15_37		Total manufacturing	2825 841	3069 635	3482 552	4087 818	4894 136	5475 650	5469 771	..
	15		Food products and beverages	174 651	195 952	220 379	247 679	297 791	304 489	299 746	..
		151	Production of meat, fish, fruit, vegetables, oils & fats	41 599	45 298	52 191	57 178	65 825	69 616	69 294	..
		152	Dairy products	18 459	20 362	20 593	29 142	37 445	32 787	28 140	..
		153	Grain mill and starch products; animal feeds	24 887	27 442	36 139	34 241	37 812	47 243	52 267	..
		154	Other food products	58 022	69 493	80 986	86 724	104 719	101 549	102 127	..
		155	Beverages	31 684	33 357	30 470	40 394	51 990	53 294	47 918	..
	16		Tobacco products	21 174	28 162	26 767	27 897	29 284	32 268	33 371	..
	17		Textiles	210 047	247 077	250 265	268 278	286 665	289 571	279 520	..
		171	Spinning, weaving and finishing of textiles	162 591	187 907	179 500	198 568	218 260	215 891	208 405	..
		172	Other textiles	27 224	35 854	41 315	44 488	47 729	47 732	46 392	..
		173	Knitted and crocheted fabrics and articles	20 232	23 316	29 450	25 222	20 676	25 948	24 723	..
	18		Wearing apparel and fur	100 359	101 323	123 312	138 825	151 869	160 637	137 959	..
		181	Wearing apparel, except fur apparel	96 766	98 623	119 661	135 463	147 079	155 588	135 048	..
		182	Dressing and dyeing of fur; articles of fur	3 593	2 700	3 651	3 362	4 790	5 049	2 911	..
	19		Leather and articles of leather; footwear	101 745	88 894	79 994	76 520	72 357	75 799	68 626	..
		191	Leather and articles of leather	25 248	27 370	28 135	30 754	29 599	32 888	28 879	..
		192	Footwear	76 497	61 524	51 859	45 766	42 758	42 911	39 747	..
	20		Wood & products of wood & cork, exc. furniture	27 996	29 080	33 722	37 920	41 249	46 160	40 705	..
		201	Sawmilling and planing of wood	8 990	8 993	9 251	11 966	11 321	12 747	9 078	..
		202	Wood products; articles of cork, straw, etc.	19 006	20 087	24 471	25 954	29 928	33 413	31 627	..
	21		Paper and paper products	69 313	76 921	78 207	99 167	116 956	124 626	123 420	..
	22		Printing, publishing and recorded media	66 880	85 705	110 173	134 061	147 373	168 956	162 542	..
		221	Publishing	47 489	61 879	84 018	101 503	112 272	126 390	117 751	..
		222	Printing and related activities	18 095	21 820	23 720	29 771	32 257	39 028	41 334	..
		223	Reproduction of recorded media	1 296	2 006	2 435	2 787	2 844	3 538	3 457	..
	23		Coke and petroleum products; nuclear fuel	45 942	16 947	24 581	30 737	42 655	50 098	88 601	..
		231	Coke oven products	451	448	430	404	414	1 342	227	..
		232	Refined petroleum products	44 407	15 227	22 897	28 883	40 377	46 106	83 790	..
		233	Processing of nuclear fuel	1 084	1 272	1 254	1 450	1 864	2 650	4 584	..
	24		Chemicals and chemical products	187 238	220 947	250 638	293 237	350 161	414 272	424 494	..
		241	Basic chemicals	69 083	85 427	97 819	117 541	144 718	181 789	187 539	..
		242	Other chemical products	100 677	116 035	118 700	133 643	154 449	183 611	180 953	..
		243	Man-made fibres	17 478	19 485	34 119	42 053	50 994	48 872	56 002	..
	25		Rubber and plastic products	112 329	119 572	141 970	162 860	206 970	233 801	221 905	..
		251	Rubber products	30 589	32 430	35 915	45 155	55 514	62 009	64 040	..
		252	Plastic products	81 740	87 142	106 055	117 705	151 456	171 792	157 865	..
	26		Other non-metallic mineral products	143 823	166 714	184 428	203 686	248 576	252 363	242 682	..
		261	Glass and glass products	27 217	32 566	40 227	41 077	50 471	63 352	63 005	..
		269	Non-metallic mineral products, nec	116 606	134 148	144 201	162 609	198 105	189 011	179 677	..
	27		Basic metals	224 691	250 213	298 329	385 144	455 120	465 514	433 803	..
		271	Basic iron and steel	172 123	198 440	244 307	318 715	370 196	384 178	351 184	..
		272	Basic precious and non-ferrous metals	34 642	34 369	36 510	44 369	62 528	55 892	58 234	..
		273	Casting of metals	17 926	17 404	17 512	22 060	22 396	25 444	24 385	..
	28		Fabricated metal products, exc. machinery	146 366	145 977	187 681	234 500	270 950	318 608	296 118	..
		281	Structural metal prod., tanks, reservoirs, etc.	51 583	50 559	79 327	95 314	118 660	142 109	135 303	..
		289	Other fabricated metal products	94 783	95 418	108 354	139 186	152 290	176 499	160 815	..
	29		Machinery and equipment, nec	286 141	308 386	348 974	437 465	521 071	597 540	599 110	..
		291	General purpose machinery	84 780	116 925	126 464	171 111	213 252	248 621	255 607	..
		292	Special purpose machinery	151 639	141 295	164 305	190 654	245 468	285 548	273 015	..
		293	Domestic appliances, nec	49 722	50 166	58 205	75 700	62 351	63 371	70 488	..
	30		Office, accounting & computing machinery	19 936	20 677	32 291	38 159	57 875	53 251	180 471	..
	31		Electrical machinery & apparatus, nec	100 978	124 053	133 193	149 523	192 952	237 529	220 249	..
		311	Electric motors, generators and transformers	23 990	31 149	37 509	37 765	46 033	55 819	53 076	..
		312	Electricity distribution & control apparatus	31 818	42 931	39 539	47 614	62 257	68 876	64 545	..
		313	Insulated wire and cable	24 537	24 839	25 896	29 209	35 042	62 656	51 214	..
		314	Accumulators, primary cells & primary batteries	5 179	6 328	6 363	7 453	9 496	8 826	10 554	..
		315	Electric lamps and lighting equipment	7 367	8 407	9 900	13 220	16 170	18 133	15 496	..
		319	Other electrical equipment, nec	8 087	10 399	13 986	14 262	23 954	23 219	25 364	..
	32		Radio, TV, communication equipment	284 657	354 584	354 398	437 102	595 280	682 386	619 220	..
		321	Electronic valves, tubes; other electronic components	122 348	140 215	173 842	222 784	316 371	342 588	372 172	..
		322	TV/radio transmitters; line teleph./telegr. apparatus	44 038	52 190	46 345	57 717	77 092	84 623	108 710	..
		323	TV/radio receivers; sound and video app.; assoc. gds.	118 271	162 179	134 211	156 601	201 817	255 175	138 338	..
	33		Medical, precision & optical instrum.; watch & clocks	34 955	36 807	45 537	59 801	69 355	76 486	85 563	..
		331	Medical appl.; instruments & appl. for measuring, etc.	17 635	19 610	23 561	36 400	46 099	50 215	51 689	..
		332	Optical instruments and photographic equipment	9 625	8 775	11 555	13 165	16 217	20 360	28 333	..
		333	Watches and clocks	7 695	8 422	10 421	10 236	7 039	5 911	5 541	..
	34		Motor vehicles, trailers and semi-trailers	265 942	221 368	294 471	320 969	390 326	479 589	486 082	..
		341	Motor vehicles	141 709	97 370	154 142	150 994	179 433	230 966	244 196	..
		342	Bodies for motor vehicles, trailers & semi-trailers	24 265	12 154	12 274	12 948	19 498	18 775	12 302	..
		343	Parts & accessories for motor vehicles & their engines	99 968	111 844	128 055	157 027	191 395	229 848	229 584	..
	35		Other transport equipment	113 850	132 281	163 176	199 980	227 120	284 472	308 650	..
		351	Building and repairing of ships and boats	94 565	109 014	124 949	152 589	179 078	233 222	243 134	..
		352	Railway and tramway locomotives and rolling stock	4 009	7 416	20 620	25 385	19 496	16 635	29 035	..
		353	Aircraft and spacecraft	8 564	8 246	10 821	14 990	19 319	23 597	27 052	..
		359	Transport equipment, nec	6 712	7 605	6 786	7 016	9 227	11 018	9 429	..
	36		Furniture; manufacturing, nec	86 227	97 231	98 466	102 243	119 811	123 242	111 208	..
		361	Furniture	35 280	44 848	46 925	51 975	62 852	64 774	56 962	..
		369	Manufacturing, nec	50 947	52 383	51 541	50 268	56 959	58 468	54 246	..
	37		Recycling	601	764	1 600	2 065	2 370	3 993	5 726	..
E	40_41		Electricity, gas and water supply	..	..	..	..	..	..	..	..
	40		Electricity, gas, steam and hot water supply	..	..	..	..	..	..	..	..
	41		Collection, purification and distribution of water	..	..	..	..	..	..	..	..
F	45		Construction	..	..	..	..	..	..	..	..

Table KR.10 ÉTABLISSEMENTS - ESTABLISHMENTS

Unités

1991	1992	1993	1994	1995	1996	1997	1998			CITI révision 3	
1 788	1 678	1 642	1 206	1 082	975	852	771	Activites extractives	C	10_14	
307	261	204	168	130	116	101	102	Extraction de charbon, de lignite et de tourbe		10	
..	..	..	..	..	..	..	..	Pétrole brut et gaz naturel ; activités annexes		11	
20	12	10	13	9	10	9	7	Extraction de minerais métalliques		13	
1 461	1 405	1 428	1 025	943	849	742	662	Autres activités extractives		14	
72 213	74 679	88 864	91 372	96 202	97 144	92 138	79 544	Total activites manufacturieres	D	15_37	
4 753	5 044	5 792	5 858	6 248	6 239	6 166	5 824	Produits alimentaires et boissons		15	
1 311	1 421	1 711	1 753	1 901	1 931	1 907	1 784	Viande, poisson, fruits, légumes, huiles et graisses			151
107	103	110	116	118	113	103	100	Produits laitiers			152
445	533	690	770	869	910	972	953	Travail des grains; prod. amyl. ; alim. pour animaux			153
2 240	2 417	2 750	2 711	2 847	2 787	2 714	2 561	Autres produits alimentaires			154
650	570	531	508	513	498	470	426	Boissons			155
20	20	16	16	16	16	14	14	Produits à base de tabac		16	
7 982	8 106	9 617	9 838	9 740	9 471	8 954	8 103	Textiles		17	
4 502	4 556	5 228	5 436	5 314	5 108	4 764	4 352	Filature, tissage et achèvement des textiles			171
1 815	1 932	2 337	2 417	2 572	2 509	2 374	2 026	Autres articles textiles			172
1 665	1 618	2 052	1 985	1 854	1 854	1 816	1 725	Etoffes et articles de bonneterie			173
6 638	6 573	8 141	8 460	8 732	8 691	8 243	6 781	Articles d'habillement et fourrure		18	
6 507	6 468	7 987	8 293	8 545	8 514	8 109	6 676	Articles d'habillement autres qu'en fourrure			181
131	105	154	167	187	177	134	105	Prép. et teinture des fourrures ; articles en fourrure			182
2 924	2 721	3 130	3 085	3 027	2 922	2 619	2 284	Cuir et articles en cuir ; chaussures		19	
1 176	1 101	1 245	1 231	1 218	1 153	1 060	893	Cuir et articles en cuir sauf chaussures			191
1 748	1 620	1 885	1 854	1 809	1 769	1 559	1 391	Chaussures			192
2 194	2 150	2 556	2 505	2 490	2 392	2 056	1 677	Bois, articles en bois, sauf meubles		20	
1 149	1 121	1 254	1 229	1 202	1 123	934	726	Sciage et rabotage du bois			201
1 045	1 029	1 302	1 276	1 288	1 269	1 122	951	Articles en bois, liège, vannerie, sparterie			202
2 236	2 309	2 496	2 600	2 671	2 713	2 639	2 300	Papier, articles en papier et carton		21	
3 189	3 406	4 192	4 366	4 819	4 796	4 502	3 962	Edition, imprimerie et reproduction		22	
1 105	1 014	1 274	1 276	1 320	1 352	1 293	1 186	Edition			221
2 036	2 341	2 841	3 011	3 423	3 370	3 139	2 709	Imprimerie et activités annexes			222
48	51	77	79	76	74	70	67	Reproduction de supports enregistrés			223
65	77	77	76	70	71	61	55	Coke. et prod. pétroliers ; comb. nucléaires		23	
7	8	8	6	6	8	6	4	Cokéfaction			231
57	68	68	69	63	62	54	50	Produits pétroliers raffinés			232
1	1	1	1	1	1	1	1	Traitement de combustibles nucléaires			233
2 168	2 353	2 553	2 644	2 796	2 672	2 663	2 694	Produits chimiques		24	
925	1 023	1 091	1 112	1 197	1 050	942	1 047	Produits chimiques de base			241
1 158	1 228	1 388	1 417	1 488	1 578	1 613	1 562	Autres produits chimiques			242
85	102	74	115	111	44	108	85	Fibres synthétiques ou artificielles			243
4 057	4 313	5 251	5 416	5 671	6 019	5 819	5 139	Art. en caoutchouc et matières plastiques		25	
650	686	772	801	858	897	858	814	Articles en caoutchouc			251
3 407	3 627	4 479	4 615	4 813	5 122	4 961	4 325	Articles en matières plastiques			252
4 204	4 405	4 708	4 657	4 709	4 559	4 184	3 378	Autres produits minéraux non métalliques		26	
389	431	534	573	599	646	620	510	Verre et articles en verre			261
3 815	3 974	4 174	4 084	4 110	3 913	3 564	2 868	Produits minéraux non métalliques, nca			269
1 798	1 835	1 988	1 921	2 075	2 076	1 992	1 908	Produits métallurgiques de base		27	
713	757	847	832	911	903	848	898	Sidérurgie et première transformation de l'acier			271
645	668	685	640	708	704	685	581	Première transf. métaux précieux et non ferreux			272
440	410	456	449	456	469	459	429	Fonderie			273
5 636	6 129	8 229	8 790	9 616	10 126	9 915	8 038	Ouvrages en métaux, sauf machines		28	
1 704	1 884	2 865	3 134	3 365	3 734	3 599	2 760	Constr. et menuis. métal.; réservoirs; générateurs			281
3 932	4 245	5 364	5 656	6 251	6 392	6 316	5 278	Autres ouvrages en métaux			289
8 759	9 302	11 246	11 582	12 408	12 916	12 252	10 251	Machines et matériels, nca		29	
2 879	3 169	4 251	4 437	4 857	5 156	4 852	4 051	Machines d'usage général			291
4 754	4 959	5 854	5 988	6 389	6 714	6 422	5 411	Machines d'usage spécifique			292
1 126	1 174	1 141	1 157	1 162	1 046	978	789	Appareils domestiques, nca			293
400	428	530	599	657	620	604	571	Mach. de bureau, comptables, information		30	
2 706	2 993	3 780	4 043	4 542	4 674	4 385	3 811	Machines et appareils électriques, nca		31	
768	798	987	1 074	1 157	1 199	1 122	1 032	Moteurs, génératrices, transformateurs électriques			311
874	1 018	1 311	1 404	1 556	1 604	1 487	1 243	Matériel électrique de distribution et de commande			312
247	273	375	410	451	475	431	386	Fils et câbles électriques isolés			313
45	45	41	34	41	45	50	53	Accumulateurs et piles électriques			314
424	466	605	639	749	775	684	594	Lampes électriques et appareils d'éclairage			315
348	393	461	482	588	576	611	503	Autres matériels électriques, nca.			319
3 181	3 132	3 393	3 434	3 546	3 573	3 238	2 829	Equipements et appareils de radio, TV, etc.		32	
936	963	1 186	1 251	1 383	1 441	1 329	1 109	Tubes et valves électroniques;autres composants			321
578	581	618	620	667	726	732	732	Emetteurs radio/TV; app. téléphonie et télégraphie			322
1 667	1 588	1 589	1 563	1 496	1 406	1 177	988	Récepteurs radio/TV, son et image; art. associés			323
1 248	1 345	1 668	1 801	1 982	2 000	1 971	1 779	Instr. médicaux, de précision ; optique ; horlog.		33	
639	717	904	1 002	1 121	1 140	1 140	1 100	Instruments médicaux et de précision, etc.			331
367	383	487	517	581	600	602	505	Instruments d'optique ; matériel photographique			332
242	245	277	282	280	260	229	174	Horlogerie			333
2 291	2 392	2 655	2 815	3 070	3 362	3 083	2 604	Véhicules automobiles; remorques, etc.		34	
25	23	40	78	60	62	51	73	Véhicules automobiles			341
89	94	177	178	204	113	103	105	Carross. automobiles, remorques, semi-remorques			342
2 177	2 275	2 438	2 559	2 806	3 187	2 929	2 426	Pièces et accessoires pour automobiles et moteurs			343
656	638	767	808	954	1 045	990	936	Autres matériels de transport		35	
365	346	459	503	609	670	648	634	Construction et réparation de navires			351
61	85	97	98	143	143	127	124	Matériel ferroviaire roulant			352
15	29	31	39	39	40	40	41	Construction aéronautique et spatiale			353
215	178	180	168	163	192	175	137	Autres équipements de transport			359
5 010	4 917	5 947	5 896	6 183	5 996	5 523	4 311	Meubles ; activités de fabrication, nca		36	
2 360	2 445	3 061	3 080	3 314	3 173	2 749	2 017	Meubles			361
2 650	2 472	2 886	2 816	2 869	2 823	2 774	2 294	Activités de fabrication, nca			369
98	91	132	162	180	195	265	295	Récupération		37	
..	..	..	..	..	..	..	..	Electricité, gaz et eau	E	40_41	
..	..	..	..	..	..	..	..	Prod. et dist. d'électricité gaz, vapeur d'eau chaude		40	
..	..	..	..	..	..	..	..	Captage, épuration et distribution d'eau		41	
..	..	..	..	..	..	..	..	Construction	F	45	

LUXEMBOURG

Table LX.1 **PRODUCTION - PRODUCTION**

Millions of LF (Current Prices)

ISIC revision 3				1991	1992	1993	1994	1995	1996	1997	1998
C	10_14		Mining and quarrying	..	..	..	..	..	..	..	..
	10		Mining of coal and lignite; extraction of peat	..	..	..	..	..	..	..	..
	11		Crude petroleum & natural gas; service activities	..	..	..	..	..	..	..	..
	13		Mining of metal ores	..	..	..	..	..	..	..	..
	14		Other mining and quarrying	..	..	..	..	..	..	..	..
D	15_37		Total manufacturing	202 384	201 472	214 883	229 499	239 044	226 251	257 060	..
	15		Food products and beverages	19 639	21 115	20 374	20 629	20 844	21 045	22 136	..
		151	Production of meat, fish, fruit, vegetables, oils & fats	..	..	..	..	..	..	..	..
		152	Dairy products	..	..	..	..	..	..	..	..
		153	Grain mill and starch products; animal feeds	..	..	..	..	..	..	..	..
		154	Other food products	3 384	4 053	4 115	4 329	4 432	4 565	4 670	..
		155	Beverages	..	..	..	..	..	..	..	..
	16		Tobacco products	..	..	..	..	..	..	..	..
	17		Textiles	11 417	12 817	13 324	14 972	15 088	13 380	15 456	..
		171	Spinning, weaving and finishing of textiles	..	..	..	..	..	..	..	..
		172	Other textiles	..	..	..	..	..	..	..	..
		173	Knitted and crocheted fabrics and articles	..	..	..	..	..	..	..	..
	18		Wearing apparel and fur	..	..	..	..	..	..	..	..
		181	Wearing apparel, except fur apparel	..	..	..	..	..	..	..	..
		182	Dressing and dyeing of fur; articles of fur	..	..	..	..	..	..	..	..
	19		Leather and articles of leather; footwear	..	..	..	..	..	..	..	..
		191	Leather and articles of leather	..	..	..	..	..	..	..	..
		192	Footwear	..	..	..	..	..	..	..	..
	20		Wood & products of wood & cork, exc. furniture	..	..	..	..	..	..	..	..
		201	Sawmilling and planing of wood	..	..	..	..	..	..	..	..
		202	Wood products; articles of cork, straw, etc.	..	..	..	..	..	..	..	..
	21		Paper and paper products	7 337	8 515	8 786	9 273	11 002	10 655	11 974	..
	22		Printing, publishing and recorded media	..	..	..	..	..	..	..	..
		221	Publishing	..	..	..	..	..	..	..	..
		222	Printing and related activities	..	..	..	..	..	..	..	..
		223	Reproduction of recorded media	..	..	..	..	..	..	..	..
	23		Coke and petroleum products; nuclear fuel	..	..	..	..	..	..	..	..
		231	Coke oven products	..	..	..	..	..	..	..	..
		232	Refined petroleum products	..	..	..	..	..	..	..	..
		233	Processing of nuclear fuel	..	..	..	..	..	..	..	..
	24		Chemicals and chemical products	10 544	14 561	14 821	15 785	16 134	17 318	18 970	..
		241	Basic chemicals	..	..	..	..	..	..	..	..
		242	Other chemical products	..	..	..	..	..	..	..	..
		243	Man-made fibres	..	..	..	..	..	..	..	..
	25		Rubber and plastic products	24 139	24 255	23 627	26 391	31 028	31 054	30 659	..
		251	Rubber products	..	..	..	..	..	..	..	..
		252	Plastic products	..	..	..	..	..	..	..	..
	26		Other non-metallic mineral products	16 168	15 280	15 211	17 370	18 467	17 242	19 998	..
		261	Glass and glass products	..	..	..	..	..	..	..	..
		269	Non-metallic mineral products, nec	..	..	..	..	..	..	..	..
	27		Basic metals	67 166	60 363	75 085	75 419	70 499	63 371	73 602	..
		271	Basic iron and steel	..	..	..	..	..	..	..	..
		272	Basic precious and non-ferrous metals	..	..	..	..	..	..	..	..
		273	Casting of metals	..	..	..	..	..	..	..	..
	28		Fabricated metal products, exc. machinery	22 831	23 784	23 098	26 670	29 342	27 123	31 397	..
		281	Structural metal prod., tanks, reservoirs, etc.	..	..	..	..	..	..	..	..
		289	Other fabricated metal products	..	..	..	..	..	..	..	..
	29		Machinery and equipment, nec	13 885	12 976	12 742	13 931	15 304	15 826	17 743	..
		291	General purpose machinery	..	..	..	..	..	..	..	..
		292	Special purpose machinery	..	..	..	..	..	..	..	..
		293	Domestic appliances, nec	..	..	..	..	..	..	..	..
	30		Office, accounting & computing machinery	5 269	4 068	4 127	4 777	5 848	6 105	6 450	..
	31		Electrical machinery & apparatus, nec	..	..	..	..	..	..	..	..
		311	Electric motors, generators and transformers	..	..	..	..	..	..	..	..
		312	Electricity distribution & control apparatus	..	..	..	..	..	..	..	..
		313	Insulated wire and cable	..	..	..	..	..	..	..	..
		314	Accumulators, primary cells & primary batteries	..	..	..	..	..	..	..	..
		315	Electric lamps and lighting equipment	..	..	..	..	..	..	..	..
		319	Other electrical equipment, nec	..	..	..	..	..	..	..	..
	32		Radio, TV, communication equipment	..	..	..	..	..	..	..	..
		321	Electronic valves, tubes; other electronic components	..	..	..	..	..	..	..	..
		322	TV/radio transmitters; line teleph./telegr. apparatus	..	..	..	..	..	..	..	..
		323	TV/radio receivers; sound and video app.; assoc. gds.	..	..	..	..	..	..	..	..
	33		Medical, precision & optical instrum.; watch & clocks	..	..	..	..	..	..	..	..
		331	Medical appl.; instruments & appl. for measuring, etc.	..	..	..	..	..	..	..	..
		332	Optical instruments and photographic equipment	..	..	..	..	..	..	..	..
		333	Watches and clocks	..	..	..	..	..	..	..	..
	34		Motor vehicles, trailers and semi-trailers	1 397	1 120	844	1 031	945	891	1 134	..
		341	Motor vehicles	..	..	..	..	..	..	..	..
		342	Bodies for motor vehicles, trailers & semi-trailers	..	..	..	..	..	..	..	..
		343	Parts & accessories for motor vehicles & their engines	..	..	..	..	..	..	..	..
	35		Other transport equipment	..	..	..	..	..	..	..	..
		351	Building and repairing of ships and boats	..	..	..	..	..	..	..	..
		352	Railway and tramway locomotives and rolling stock	..	..	..	..	..	..	..	..
		353	Aircraft and spacecraft	..	..	..	..	..	..	..	..
		359	Transport equipment, nec	..	..	..	..	..	..	..	..
	36		Furniture; manufacturing, nec	1 567	1 575	1 668	1 938	2 901	3 962	5 685	..
		361	Furniture	..	..	..	..	..	..	..	..
		369	Manufacturing, nec	..	..	..	..	..	..	..	..
	37		Recycling	1 024	1 044	1 175	1 312	1 642	1 505	1 856	..
E	40_41		Electricity, gas and water supply	11 402	11 552	12 007	12 601	12 874	13 688	14 156	..
	40		Electricity, gas, steam and hot water supply	..	..	..	..	..	..	..	..
	41		Collection, purification and distribution of water	..	..	..	..	..	..	..	..
F	45		Construction	..	..	..	..	..	..	..	..

Note: ISIC 15 includes ISIC 16; ISIC 21 includes ISIC 22; ISIC 30 includes ISIC 31, 32 and 33; ISIC 34 ncludes ISIC 35; ISIC 36 includes ISIC 20. Data for 1997 are provisional.

Table LX.2 **VALEUR AJOUTÉE AU COUT DES FACTEURS - VALUE ADDED AT FACTOR COSTS**

Millions de LF (Prix courants)

1991	1992	1993	1994	1995	1996	1997	1998			CITI révision 3	
..	..	..	..	..	..	..	..	Activites extractives	C	10_14	
..	..	..	..	..	..	..	..	Extraction de charbon, de lignite et de tourbe		10	
..	..	..	..	..	..	..	..	Pétrole brut et gaz naturel ; activités annexes		11	
..	..	..	..	..	..	..	..	Extraction de minerais métalliques		13	
..	..	..	..	..	..	..	..	Autres activités extractives		14	
68 856	68 312	68 875	72 744	76 401	71 557	82 873	..	Total activites manufacturieres	D	15_37	
5 701	6 346	6 406	6 614	6 591	6 653	7 057	..	Produits alimentaires et boissons		15	
..	..	..	..	..	..	..	..	Viande, poisson, fruits, légumes, huiles et graisses			151
..	..	..	..	..	..	..	..	Produits laitiers			152
..	..	..	..	..	..	..	..	Travail des grains; prod. amyl. ; alim. pour animaux			153
1 905	2 074	2 134	2 237	2 281	2 345	2 391	..	Autres produits alimentaires			154
..	..	..	..	..	..	..	..	Boissons			155
..	..	..	..	..	..	..	..	Produits à base de tabac		16	
4 674	5 545	6 098	7 173	6 363	5 522	6 650	..	Textiles		17	
..	..	..	..	..	..	..	..	Filature, tissage et achèvement des textiles			171
..	..	..	..	..	..	..	..	Autres articles textiles			172
..	..	..	..	..	..	..	..	Etoffes et articles de bonneterie			173
..	..	..	..	..	..	..	..	Articles d'habillement et fourrure		18	
..	..	..	..	..	..	..	..	Articles d'habillement autres qu'en fourrure			181
..	..	..	..	..	..	..	..	Prép. et teinture des fourrures ; articles en fourrure			182
..	..	..	..	..	..	..	..	Cuir et articles en cuir ; chaussures		19	
..	..	..	..	..	..	..	..	Cuir et articles en cuir sauf chaussures			191
..	..	..	..	..	..	..	..	Chaussures			192
..	..	..	..	..	..	..	..	Bois, articles en bois, sauf meubles		20	
..	..	..	..	..	..	..	..	Sciage et rabotage du bois			201
..	..	..	..	..	..	..	..	Articles en bois, liège, vannerie, sparterie			202
3 135	3 600	3 865	4 169	4 695	4 487	4 986	..	Papier, articles en papier et carton		21	
..	..	..	..	..	..	..	..	Edition, imprimerie et reproduction		22	
..	..	..	..	..	..	..	..	Edition			221
..	..	..	..	..	..	..	..	Imprimerie et activités annexes			222
..	..	..	..	..	..	..	..	Reproduction de supports enregistrés			223
..	..	..	..	..	..	..	..	Coke. et prod. pétroliers ; comb. nucléaires		23	
..	..	..	..	..	..	..	..	Cokéfaction			231
..	..	..	..	..	..	..	..	Produits pétroliers raffinés			232
..	..	..	..	..	..	..	..	Traitement de combustibles nucléaires			233
3 267	4 678	4 844	5 263	5 266	5 614	6 086	..	Produits chimiques		24	
..	..	..	..	..	..	..	..	Produits chimiques de base			241
..	..	..	..	..	..	..	..	Autres produits chimiques			242
..	..	..	..	..	..	..	..	Fibres synthétiques ou artificielles			243
9 453	9 390	9 456	10 626	11 714	11 593	11 559	..	Art. en caoutchouc et matières plastiques		25	
..	..	..	..	..	..	..	..	Articles en caoutchouc			251
..	..	..	..	..	..	..	..	Articles en matières plastiques			252
7 711	6 730	6 945	7 246	8 632	7 821	9 221	..	Autres produits minéraux non métalliques		26	
..	..	..	..	..	..	..	..	Verre et articles en verre			261
..	..	..	..	..	..	..	..	Produits minéraux non métalliques, nca			269
18 706	15 626	15 184	14 793	14 691	12 660	16 832	..	Produits métallurgiques de base		27	
..	..	..	..	..	..	..	..	Sidérurgie et première transformation de l'acier			271
..	..	..	..	..	..	..	..	Première transf. métaux précieux et non ferreux			272
..	..	..	..	..	..	..	..	Fonderie			273
6 784	6 959	7 205	7 479	8 025	7 694	8 701	..	Ouvrages en métaux, sauf machines		28	
..	..	..	..	..	..	..	..	Constr. et menuis. métal.; réservoirs; générateurs			281
..	..	..	..	..	..	..	..	Autres ouvrages en métaux			289
6 076	6 062	5 485	5 690	6 106	6 143	6 680	..	Machines et matériels, nca		29	
..	..	..	..	..	..	..	..	Machines d'usage général			291
..	..	..	..	..	..	..	..	Machines d'usage spécifique			292
..	..	..	..	..	..	..	..	Appareils domestiques, nca			293
1 973	2 111	2 176	2 294	2 582	2 676	2 823	..	Mach. de bureau, comptables, information		30	
..	..	..	..	..	..	..	..	Machines et appareils électriques, nca		31	
..	..	..	..	..	..	..	..	Moteurs, génératrices, transformateurs électriques			311
..	..	..	..	..	..	..	..	Matériel électrique de distribution et de commande			312
..	..	..	..	..	..	..	..	Fils et câbles électriques isolés			313
..	..	..	..	..	..	..	..	Accumulateurs et piles électriques			314
..	..	..	..	..	..	..	..	Lampes électriques et appareils d'éclairage			315
..	..	..	..	..	..	..	..	Autres matériels électriques, nca.			319
..	..	..	..	..	..	..	..	Equipements et appareils de radio, TV, etc.		32	
..	..	..	..	..	..	..	..	Tubes et valves électroniques;autres composants			321
..	..	..	..	..	..	..	..	Emetteurs radio/TV; app. téléphonie et télégraphie			322
..	..	..	..	..	..	..	..	Récepteurs radio/TV, son et image; art. associés			323
..	..	..	..	..	..	..	..	Instr. médicaux, de précision ; optique ; horlog.		33	
..	..	..	..	..	..	..	..	Instruments médicaux et de précision, etc.			331
..	..	..	..	..	..	..	..	Instruments d'optique ; matériel photographique			332
..	..	..	..	..	..	..	..	Horlogerie			333
449	372	285	387	370	347	442	..	Véhicules automobiles; remorques, etc.		34	
..	..	..	..	..	..	..	..	Véhicules automobiles			341
..	..	..	..	..	..	..	..	Carross. automobiles, remorques, semi-remorques			342
..	..	..	..	..	..	..	..	Pièces et accessoires pour automobiles et moteurs			343
..	..	..	..	..	..	..	..	Autres matériels de transport		35	
..	..	..	..	..	..	..	..	Construction et réparation de navires			351
..	..	..	..	..	..	..	..	Matériel ferroviaire roulant			352
..	..	..	..	..	..	..	..	Construction aéronautique et spatiale			353
..	..	..	..	..	..	..	..	Autres équipements de transport			359
661	624	646	722	1 005	1 044	1 424	..	Meubles ; activités de fabrication, nca		36	
..	..	..	..	..	..	..	..	Meubles			361
..	..	..	..	..	..	..	..	Activités de fabrication, nca			369
267	269	279	289	362	335	412	..	Récupération		37	
6 121	6 460	7 074	7 518	7 752	8 005	8 053	..	Electricité, gaz et eau	E	40_41	
..	..	..	..	..	..	..	..	Prod. et dist. d'électricité gaz, vapeur d'eau chaude		40	
..	..	..	..	..	..	..	..	Captage, épuration et distribution d'eau		41	
..	..	..	..	..	..	..	..	Construction	F	45	

Note: ISIC 15 includes ISIC 16; ISIC 21 includes ISIC 22; ISIC 30 includes ISIC 31, 32 and 33; ISIC 34 includes ISIC 35; ISIC 36 includes ISIC 20. Data for 1997 are provisional.

LUXEMBOURG

Table LX.3 EMPLOYMENT, EMPLOYEES - EMPLOI, SALARIÉS

Units

ISIC revision 3				1991	1992	1993	1994	1995	1996	1997	1998
C	10_14		Mining and quarrying	..	..	..	..	..	..	..	..
	10		Mining of coal and lignite; extraction of peat	..	..	..	..	..	..	..	..
	11		Crude petroleum & natural gas; service activities	..	..	..	..	..	..	..	..
	13		Mining of metal ores	..	..	..	..	..	..	..	..
	14		Other mining and quarrying	..	..	..	..	..	..	..	..
D	15_37		Total manufacturing	34 775	34 071	32 782	32 003	32 379	31 974	32 158	..
	15		Food products and beverages	3 583	3 600	3 614	3 641	3 723	3 733	3 810	..
		151	Production of meat, fish, fruit, vegetables, oils & fats	..	..	..	..	..	..	..	..
		152	Dairy products	..	..	..	..	..	..	..	..
		153	Grain mill and starch products; animal feeds	..	..	..	..	..	..	..	..
		154	Other food products	1 672	1 707	1 756	1 822	1 857	1 911	1 928	..
		155	Beverages	..	..	..	..	..	..	..	..
	16		Tobacco products	..	..	..	..	..	..	..	..
	17		Textiles	1 057	1 080	1 017	1 025	1 001	1 052	1 157	..
		171	Spinning, weaving and finishing of textiles	..	..	..	..	..	..	..	..
		172	Other textiles	..	..	..	..	..	..	..	..
		173	Knitted and crocheted fabrics and articles	..	..	..	..	..	..	..	..
	18		Wearing apparel and fur	..	..	..	..	..	..	..	..
		181	Wearing apparel, except fur apparel	..	..	..	..	..	..	..	..
		182	Dressing and dyeing of fur; articles of fur	..	..	..	..	..	..	..	..
	19		Leather and articles of leather; footwear	..	..	..	..	..	..	..	..
		191	Leather and articles of leather	..	..	..	..	..	..	..	..
		192	Footwear	..	..	..	..	..	..	..	..
	20		Wood & products of wood & cork, exc. furniture	..	..	..	..	..	..	..	..
		201	Sawmilling and planing of wood	..	..	..	..	..	..	..	..
		202	Wood products; articles of cork, straw, etc.	..	..	..	..	..	..	..	..
	21		Paper and paper products	1 971	1 901	1 888	1 926	1 984	1 989	2 060	..
	22		Printing, publishing and recorded media	..	..	..	..	..	..	..	..
		221	Publishing	..	..	..	..	..	..	..	..
		222	Printing and related activities	..	..	..	..	..	..	..	..
		223	Reproduction of recorded media	..	..	..	..	..	..	..	..
	23		Coke and petroleum products; nuclear fuel	..	..	..	..	..	..	..	..
		231	Coke oven products	..	..	..	..	..	..	..	..
		232	Refined petroleum products	..	..	..	..	..	..	..	..
		233	Processing of nuclear fuel	..	..	..	..	..	..	..	..
	24		Chemicals and chemical products	1 005	1 346	1 323	1 279	1 302	1 300	1 417	..
		241	Basic chemicals	..	..	..	..	..	..	..	..
		242	Other chemical products	..	..	..	..	..	..	..	..
		243	Man-made fibres	..	..	..	..	..	..	..	..
	25		Rubber and plastic products	3 866	3 736	3 651	3 592	3 714	3 878	3 917	..
		251	Rubber products	..	..	..	..	..	..	..	..
		252	Plastic products	..	..	..	..	..	..	..	..
	26		Other non-metallic mineral products	3 125	3 034	3 051	3 054	2 973	2 890	2 888	..
		261	Glass and glass products	..	..	..	..	..	..	..	..
		269	Non-metallic mineral products, nec	..	..	..	..	..	..	..	..
	27		Basic metals	10 427	9 695	8 731	8 108	7 753	7 113	6 543	..
		271	Basic iron and steel	..	..	..	..	..	..	..	..
		272	Basic precious and non-ferrous metals	..	..	..	..	..	..	..	..
		273	Casting of metals	..	..	..	..	..	..	..	..
	28		Fabricated metal products, exc. machinery	3 944	3 991	3 886	3 794	3 961	3 873	3 865	..
		281	Structural metal prod., tanks, reservoirs, etc.	..	..	..	..	..	..	..	..
		289	Other fabricated metal products	..	..	..	..	..	..	..	..
	29		Machinery and equipment, nec	3 237	3 168	3 145	2 992	2 976	2 981	3 148	..
		291	General purpose machinery	..	..	..	..	..	..	..	..
		292	Special purpose machinery	..	..	..	..	..	..	..	..
		293	Domestic appliances, nec	..	..	..	..	..	..	..	..
	30		Office, accounting & computing machinery	1 475	1 534	1 532	1 614	1 760	1 870	1 935	..
	31		Electrical machinery & apparatus, nec	..	..	..	..	..	..	..	..
		311	Electric motors, generators and transformers	..	..	..	..	..	..	..	..
		312	Electricity distribution & control apparatus	..	..	..	..	..	..	..	..
		313	Insulated wire and cable	..	..	..	..	..	..	..	..
		314	Accumulators, primary cells & primary batteries	..	..	..	..	..	..	..	..
		315	Electric lamps and lighting equipment	..	..	..	..	..	..	..	..
		319	Other electrical equipment, nec	..	..	..	..	..	..	..	..
	32		Radio, TV, communication equipment	..	..	..	..	..	..	..	..
		321	Electronic valves, tubes; other electronic components	..	..	..	..	..	..	..	..
		322	TV/radio transmitters; line teleph./telegr. apparatus	..	..	..	..	..	..	..	..
		323	TV/radio receivers; sound and video app.; assoc. gds.	..	..	..	..	..	..	..	..
	33		Medical, precision & optical instrum.; watch & clocks	..	..	..	..	..	..	..	..
		331	Medical appl.; instruments & appl. for measuring, etc.	..	..	..	..	..	..	..	..
		332	Optical instruments and photographic equipment	..	..	..	..	..	..	..	..
		333	Watches and clocks	..	..	..	..	..	..	..	..
	34		Motor vehicles, trailers and semi-trailers	428	380	312	276	262	256	270	..
		341	Motor vehicles	..	..	..	..	..	..	..	..
		342	Bodies for motor vehicles, trailers & semi-trailers	..	..	..	..	..	..	..	..
		343	Parts & accessories for motor vehicles & their engines	..	..	..	..	..	..	..	..
	35		Other transport equipment	..	..	..	..	..	..	..	..
		351	Building and repairing of ships and boats	..	..	..	..	..	..	..	..
		352	Railway and tramway locomotives and rolling stock	..	..	..	..	..	..	..	..
		353	Aircraft and spacecraft	..	..	..	..	..	..	..	..
		359	Transport equipment, nec	..	..	..	..	..	..	..	..
	36		Furniture; manufacturing, nec	462	427	451	494	725	794	882	..
		361	Furniture	..	..	..	..	..	..	..	..
		369	Manufacturing, nec	..	..	..	..	..	..	..	..
	37		Recycling	195	179	181	207	246	245	265	..
E	40_41		Electricity, gas and water supply	1 348	1 361	1 361	1 472	1 455	1 456	1 452	..
	40		Electricity, gas, steam and hot water supply	..	..	..	..	..	..	..	..
	41		Collection, purification and distribution of water	..	..	..	..	..	..	..	..
F	45		Construction	..	..	..	..	..	..	..	..

Note: ISIC 15 includes ISIC 16; ISIC 21 includes ISIC 22; ISIC 30 includes ISIC 31, 32 and 33; ISIC 34 includes ISIC 35; ISIC 36 includes ISIC 20. Data for 1997 are provisional.

Table LX.4 **SALAIRES ET CHARGES SOCIALES, SALARIÉS - COMPENSATION OF LABOUR, EMPLOYEES**

Millions de LF (Prix courants)

1991	1992	1993	1994	1995	1996	1997	1998			CITI révision 3	
..	..	..	..	..	..	..	..	Activites extractives	C	10_14	
..	..	..	..	..	..	..	..	Extraction de charbon, de lignite et de tourbe		10	
..	..	..	..	..	..	..	..	Pétrole brut et gaz naturel ; activités annexes		11	
..	..	..	..	..	..	..	..	Extraction de minerais métalliques		13	
..	..	..	..	..	..	..	..	Autres activités extractives		14	
43 282	44 920	45 810	46 512	48 507	48 321	50 054	..	Total activites manufacturieres	D	15_37	
3 138	3 400	3 535	3 650	3 789	3 854	3 984	..	Produits alimentaires et boissons		15	
..	..	..	..	..	..	..	..	Viande, poisson, fruits, légumes, huiles et graisses			151
..	..	..	..	..	..	..	..	Produits laitiers			152
..	..	..	..	..	..	..	..	Travail des grains; prod. amyl. ; alim. pour animaux			153
1 146	1 257	1 348	1 425	1 505	1 626	1 682	..	Autres produits alimentaires			154
..	..	..	..	..	..	..	..	Boissons			155
..	..	..	..	..	..	..	..	Produits à base de tabac		16	
1 778	1 873	2 083	2 173	2 169	2 370	2 569	..	Textiles		17	
..	..	..	..	..	..	..	..	Filature, tissage et achèvement des textiles			171
..	..	..	..	..	..	..	..	Autres articles textiles			172
..	..	..	..	..	..	..	..	Etoffes et articles de bonneterie			173
..	..	..	..	..	..	..	..	Articles d'habillement et fourrure		18	
..	..	..	..	..	..	..	..	Articles d'habillement autres qu'en fourrure			181
..	..	..	..	..	..	..	..	Prép. et teinture des fourrures ; articles en fourrure			182
..	..	..	..	..	..	..	..	Cuir et articles en cuir ; chaussures		19	
..	..	..	..	..	..	..	..	Cuir et articles en cuir sauf chaussures			191
..	..	..	..	..	..	..	..	Chaussures			192
..	..	..	..	..	..	..	..	Bois, articles en bois, sauf meubles		20	
..	..	..	..	..	..	..	..	Sciage et rabotage du bois			201
..	..	..	..	..	..	..	..	Articles en bois, liège, vannerie, sparterie			202
2 315	2 368	2 599	2 742	2 941	3 013	3 224	..	Papier, articles en papier et carton		21	
..	..	..	..	..	..	..	..	Edition, imprimerie et reproduction		22	
..	..	..	..	..	..	..	..	Edition			221
..	..	..	..	..	..	..	..	Imprimerie et activités annexes			222
..	..	..	..	..	..	..	..	Reproduction de supports enregistrés			223
..	..	..	..	..	..	..	..	Coke. et prod. pétroliers ; comb. nucléaires		23	
..	..	..	..	..	..	..	..	Cokéfaction			231
..	..	..	..	..	..	..	..	Produits pétroliers raffinés			232
..	..	..	..	..	..	..	..	Traitement de combustibles nucléaires			233
1 357	1 742	1 855	1 892	1 909	1 998	2 162	..	Produits chimiques		24	
..	..	..	..	..	..	..	..	Produits chimiques de base			241
..	..	..	..	..	..	..	..	Autres produits chimiques			242
..	..	..	..	..	..	..	..	Fibres synthétiques ou artificielles			243
5 527	5 956	6 227	6 290	6 592	6 772	6 921	..	Art. en caoutchouc et matières plastiques		25	
..	..	..	..	..	..	..	..	Articles en caoutchouc			251
..	..	..	..	..	..	..	..	Articles en matières plastiques			252
3 380	3 430	3 632	3 856	3 863	3 765	3 894	..	Autres produits minéraux non métalliques		26	
..	..	..	..	..	..	..	..	Verre et articles en verre			261
..	..	..	..	..	..	..	..	Produits minéraux non métalliques, nca			269
14 621	14 382	13 867	13 527	13 761	12 715	12 447	..	Produits métallurgiques de base		27	
..	..	..	..	..	..	..	..	Sidérurgie et première transformation de l'acier			271
..	..	..	..	..	..	..	..	Première transf. métaux précieux et non ferreux			272
..	..	..	..	..	..	..	..	Fonderie			273
4 363	4 701	4 788	4 915	5 366	5 429	5 654	..	Ouvrages en métaux, sauf machines		28	
..	..	..	..	..	..	..	..	Constr. et menuis. métal.; réservoirs; générateurs			281
..	..	..	..	..	..	..	..	Autres ouvrages en métaux			289
4 601	4 786	4 863	4 852	5 050	5 084	5 533	..	Machines et matériels, nca		29	
..	..	..	..	..	..	..	..	Machines d'usage général			291
..	..	..	..	..	..	..	..	Machines d'usage spécifique			292
..	..	..	..	..	..	..	..	Appareils domestiques, nca			293
1 274	1 378	1 493	1 633	1 820	1 986	2 181	..	Mach. de bureau, comptables, information		30	
..	..	..	..	..	..	..	..	Machines et appareils électriques, nca		31	
..	..	..	..	..	..	..	..	Moteurs, génératrices, transformateurs électriques			311
..	..	..	..	..	..	..	..	Matériel électrique de distribution et de commande			312
..	..	..	..	..	..	..	..	Fils et câbles électriques isolés			313
..	..	..	..	..	..	..	..	Accumulateurs et piles électriques			314
..	..	..	..	..	..	..	..	Lampes électriques et appareils d'éclairage			315
..	..	..	..	..	..	..	..	Autres matériels électriques, nca.			319
..	..	..	..	..	..	..	..	Equipements et appareils de radio, TV, etc.		32	
..	..	..	..	..	..	..	..	Tubes et valves électroniques;autres composants			321
..	..	..	..	..	..	..	..	Emetteurs radio/TV; app. téléphonie et télégraphie			322
..	..	..	..	..	..	..	..	Récepteurs radio/TV, son et image; art. associés			323
..	..	..	..	..	..	..	..	Instr. médicaux, de précision ; optique ; horlog.		33	
..	..	..	..	..	..	..	..	Instruments médicaux et de précision, etc.			331
..	..	..	..	..	..	..	..	Instruments d'optique ; matériel photographique			332
..	..	..	..	..	..	..	..	Horlogerie			333
365	356	267	296	269	261	271	..	Véhicules automobiles; remorques, etc.		34	
..	..	..	..	..	..	..	..	Véhicules automobiles			341
..	..	..	..	..	..	..	..	Carross. automobiles, remorques, semi-remorques			342
..	..	..	..	..	..	..	..	Pièces et accessoires pour automobiles et moteurs			343
..	..	..	..	..	..	..	..	Autres matériels de transport		35	
..	..	..	..	..	..	..	..	Construction et réparation de navires			351
..	..	..	..	..	..	..	..	Matériel ferroviaire roulant			352
..	..	..	..	..	..	..	..	Construction aéronautique et spatiale			353
..	..	..	..	..	..	..	..	Autres équipements de transport			359
383	373	410	461	711	802	912	..	Meubles ; activités de fabrication, nca		36	
..	..	..	..	..	..	..	..	Meubles			361
..	..	..	..	..	..	..	..	Activités de fabrication, nca			369
180	174	191	225	268	273	302	..	Récupération		37	
2 296	2 477	2 608	3 007	3 021	3 100	3 210	..	Electricité, gaz et eau	E	40_41	
..	..	..	..	..	..	..	..	Prod. et dist. d'électricité gaz, vapeur d'eau chaude		40	
..	..	..	..	..	..	..	..	Captage, épuration et distribution d'eau		41	
..	..	..	..	..	..	..	..	Construction	F	45	

Note: ISIC 15 includes ISIC 16; ISIC 21 includes ISIC 22; ISIC 30 includes ISIC 31, 32 and 33; ISIC 34 includes ISIC 35; ISIC 36 includes ISIC 20. Data for 1997 are provisional.

LUXEMBOURG

Table LX.5 WAGES AND SALARIES, EMPLOYEES - SALAIRES ET TRAITEMENTS, SALARIÉS

Millions of LF (Current Prices)

ISIC revision 3				1991	1992	1993	1994	1995	1996	1997	1998
C	10_14		Mining and quarrying	..	..	..	..	..	..	..	..
	10		Mining of coal and lignite; extraction of peat	..	..	..	..	..	..	..	..
	11		Crude petroleum & natural gas; service activities	..	..	..	..	..	..	..	..
	13		Mining of metal ores	..	..	..	..	..	..	..	..
	14		Other mining and quarrying	..	..	..	..	..	..	..	..
D	15_37		Total manufacturing	37 556	38 193	39 041	39 873	41 423	41 429	43 073	..
	15		Food products and beverages	2 741	2 961	3 075	3 192	3 314	3 371	3 489	..
		151	Production of meat, fish, fruit, vegetables, oils & fats	..	..	..	..	..	..	..	..
		152	Dairy products	..	..	..	..	..	..	..	..
		153	Grain mill and starch products; animal feeds	..	..	..	..	..	..	..	..
		154	Other food products	1 002	1 094	1 170	1 242	1 311	1 418	1 469	..
		155	Beverages	..	..	..	..	..	..	..	..
	16		Tobacco products	..	..	..	..	..	..	..	..
	17		Textiles	1 557	1 616	1 784	1 887	1 898	2 103	2 248	..
		171	Spinning, weaving and finishing of textiles	..	..	..	..	..	..	..	..
		172	Other textiles	..	..	..	..	..	..	..	..
		173	Knitted and crocheted fabrics and articles	..	..	..	..	..	..	..	..
	18		Wearing apparel and fur	..	..	..	..	..	..	..	..
		181	Wearing apparel, except fur apparel	..	..	..	..	..	..	..	..
		182	Dressing and dyeing of fur; articles of fur	..	..	..	..	..	..	..	..
	19		Leather and articles of leather; footwear	..	..	..	..	..	..	..	..
		191	Leather and articles of leather	..	..	..	..	..	..	..	..
		192	Footwear	..	..	..	..	..	..	..	..
	20		Wood & products of wood & cork, exc. furniture	..	..	..	..	..	..	..	..
		201	Sawmilling and planing of wood	..	..	..	..	..	..	..	..
		202	Wood products; articles of cork, straw, etc.	..	..	..	..	..	..	..	..
	21		Paper and paper products	2 046	2 083	2 281	2 412	2 579	2 646	2 835	..
	22		Printing, publishing and recorded media	..	..	..	..	..	..	..	..
		221	Publishing	..	..	..	..	..	..	..	..
		222	Printing and related activities	..	..	..	..	..	..	..	..
		223	Reproduction of recorded media	..	..	..	..	..	..	..	..
	23		Coke and petroleum products; nuclear fuel	..	..	..	..	..	..	..	..
		231	Coke oven products	..	..	..	..	..	..	..	..
		232	Refined petroleum products	..	..	..	..	..	..	..	..
		233	Processing of nuclear fuel	..	..	..	..	..	..	..	..
	24		Chemicals and chemical products	1 190	1 534	1 617	1 660	1 692	1 773	1 923	..
		241	Basic chemicals	..	..	..	..	..	..	..	..
		242	Other chemical products	..	..	..	..	..	..	..	..
		243	Man-made fibres	..	..	..	..	..	..	..	..
	25		Rubber and plastic products	4 804	5 154	5 378	5 474	5 789	5 946	6 092	..
		251	Rubber products	..	..	..	..	..	..	..	..
		252	Plastic products	..	..	..	..	..	..	..	..
	26		Other non-metallic mineral products	2 903	2 916	3 106	3 327	3 324	3 248	3 372	..
		261	Glass and glass products	..	..	..	..	..	..	..	..
		269	Non-metallic mineral products, nec	..	..	..	..	..	..	..	..
	27		Basic metals	12 618	11 732	11 409	11 149	11 017	10 225	10 080	..
		271	Basic iron and steel	..	..	..	..	..	..	..	..
		272	Basic precious and non-ferrous metals	..	..	..	..	..	..	..	..
		273	Casting of metals	..	..	..	..	..	..	..	..
	28		Fabricated metal products, exc. machinery	3 791	4 069	4 155	4 285	4 680	4 740	4 952	..
		281	Structural metal prod., tanks, reservoirs, etc.	..	..	..	..	..	..	..	..
		289	Other fabricated metal products	..	..	..	..	..	..	..	..
	29		Machinery and equipment, nec	3 992	4 148	4 187	4 207	4 466	4 494	4 891	..
		291	General purpose machinery	..	..	..	..	..	..	..	..
		292	Special purpose machinery	..	..	..	..	..	..	..	..
		293	Domestic appliances, nec	..	..	..	..	..	..	..	..
	30		Office, accounting & computing machinery	1 120	1 203	1 303	1 429	1 588	1 734	1 907	..
	31		Electrical machinery & apparatus, nec	..	..	..	..	..	..	..	..
		311	Electric motors, generators and transformers	..	..	..	..	..	..	..	..
		312	Electricity distribution & control apparatus	..	..	..	..	..	..	..	..
		313	Insulated wire and cable	..	..	..	..	..	..	..	..
		314	Accumulators, primary cells & primary batteries	..	..	..	..	..	..	..	..
		315	Electric lamps and lighting equipment	..	..	..	..	..	..	..	..
		319	Other electrical equipment, nec	..	..	..	..	..	..	..	..
	32		Radio, TV, communication equipment	..	..	..	..	..	..	..	..
		321	Electronic valves, tubes; other electronic components	..	..	..	..	..	..	..	..
		322	TV/radio transmitters; line teleph./telegr. apparatus	..	..	..	..	..	..	..	..
		323	TV/radio receivers; sound and video app.; assoc. gds.	..	..	..	..	..	..	..	..
	33		Medical, precision & optical instrum.; watch & clocks	..	..	..	..	..	..	..	..
		331	Medical appl.; instruments & appl. for measuring, etc.	..	..	..	..	..	..	..	..
		332	Optical instruments and photographic equipment	..	..	..	..	..	..	..	..
		333	Watches and clocks	..	..	..	..	..	..	..	..
	34		Motor vehicles, trailers and semi-trailers	316	306	230	257	232	224	233	..
		341	Motor vehicles	..	..	..	..	..	..	..	..
		342	Bodies for motor vehicles, trailers & semi-trailers	..	..	..	..	..	..	..	..
		343	Parts & accessories for motor vehicles & their engines	..	..	..	..	..	..	..	..
	35		Other transport equipment	..	..	..	..	..	..	..	..
		351	Building and repairing of ships and boats	..	..	..	..	..	..	..	..
		352	Railway and tramway locomotives and rolling stock	..	..	..	..	..	..	..	..
		353	Aircraft and spacecraft	..	..	..	..	..	..	..	..
		359	Transport equipment, nec	..	..	..	..	..	..	..	..
	36		Furniture; manufacturing, nec	328	323	354	399	614	692	792	..
		361	Furniture	..	..	..	..	..	..	..	..
		369	Manufacturing, nec	..	..	..	..	..	..	..	..
	37		Recycling	152	148	163	196	231	234	260	..
E	40_41		Electricity, gas and water supply	1 994	2 127	2 253	2 634	2 685	2 763	2 865	..
	40		Electricity, gas, steam and hot water supply	..	..	..	..	..	..	..	..
	41		Collection, purification and distribution of water	..	..	..	..	..	..	..	..
F	45		Construction	..	..	..	..	..	..	..	..

Note: ISIC 15 includes ISIC 16; ISIC 21 includes ISIC 22; ISIC 30 includes ISIC 31, 32 and 33; ISIC 34 includes ISIC 35; ISIC 36 includes ISIC 20. Data for 1997 are provisional.

Table LX.6 CHARGES SOCIALES DES EMPLOYEURS - SUPPLEMENTS TO WAGES & SALARIES

Millions de LF (Prix courants)

1991	1992	1993	1994	1995	1996	1997	1998			CITI révision 3	
..	..	..	..	..	..	..	..	Activites extractives	C	10_14	
..	..	..	..	..	..	..	..	Extraction de charbon, de lignite et de tourbe		10	
..	..	..	..	..	..	..	..	Pétrole brut et gaz naturel ; activités annexes		11	
..	..	..	..	..	..	..	..	Extraction de minerais métalliques		13	
..	..	..	..	..	..	..	..	Autres activités extractives		14	
5 726	6 726	6 770	6 639	7 084	6 892	6 981	..	Total activites manufacturieres	D	15_37	
397	439	460	458	475	483	494	..	Produits alimentaires et boissons		15	
..	..	..	..	..	..	..	..	Viande, poisson, fruits, légumes, huiles et graisses			151
..	..	..	..	..	..	..	..	Produits laitiers			152
..	..	..	..	..	..	..	..	Travail des grains; prod. amyl. ; alim. pour animaux			153
144	163	178	183	194	207	213	..	Autres produits alimentaires			154
..	..	..	..	..	..	..	..	Boissons			155
..	..	..	..	..	..	..	..	Produits à base de tabac		16	
221	257	300	286	271	268	321	..	Textiles		17	
..	..	..	..	..	..	..	..	Filature, tissage et achèvement des textiles			171
..	..	..	..	..	..	..	..	Autres articles textiles			172
..	..	..	..	..	..	..	..	Etoffes et articles de bonneterie			173
..	..	..	..	..	..	..	..	Articles d'habillement et fourrure		18	
..	..	..	..	..	..	..	..	Articles d'habillement autres qu'en fourrure			181
..	..	..	..	..	..	..	..	Prép. et teinture des fourrures ; articles en fourrure			182
..	..	..	..	..	..	..	..	Cuir et articles en cuir ; chaussures		19	
..	..	..	..	..	..	..	..	Cuir et articles en cuir sauf chaussures			191
..	..	..	..	..	..	..	..	Chaussures			192
..	..	..	..	..	..	..	..	Bois, articles en bois, sauf meubles		20	
..	..	..	..	..	..	..	..	Sciage et rabotage du bois			201
..	..	..	..	..	..	..	..	Articles en bois, liège, vannerie, sparterie			202
269	285	318	330	363	367	389	..	Papier, articles en papier et carton		21	
..	..	..	..	..	..	..	..	Edition, imprimerie et reproduction		22	
..	..	..	..	..	..	..	..	Edition			221
..	..	..	..	..	..	..	..	Imprimerie et activités annexes			222
..	..	..	..	..	..	..	..	Reproduction de supports enregistrés			223
..	..	..	..	..	..	..	..	Coke. et prod. pétroliers ; comb. nucléaires		23	
..	..	..	..	..	..	..	..	Cokéfaction			231
..	..	..	..	..	..	..	..	Produits pétroliers raffinés			232
..	..	..	..	..	..	..	..	Traitement de combustibles nucléaires			233
167	208	238	232	217	225	239	..	Produits chimiques		24	
..	..	..	..	..	..	..	..	Produits chimiques de base			241
..	..	..	..	..	..	..	..	Autres produits chimiques			242
..	..	..	..	..	..	..	..	Fibres synthétiques ou artificielles			243
723	802	849	817	803	826	828	..	Art. en caoutchouc et matières plastiques		25	
..	..	..	..	..	..	..	..	Articles en caoutchouc			251
..	..	..	..	..	..	..	..	Articles en matières plastiques			252
477	514	526	529	539	517	522	..	Autres produits minéraux non métalliques		26	
..	..	..	..	..	..	..	..	Verre et articles en verre			261
..	..	..	..	..	..	..	..	Produits minéraux non métalliques, nca			269
2 004	2 650	2 458	2 378	2 744	2 490	2 367	..	Produits métallurgiques de base		27	
..	..	..	..	..	..	..	..	Sidérurgie et première transformation de l'acier			271
..	..	..	..	..	..	..	..	Première transf. métaux précieux et non ferreux			272
..	..	..	..	..	..	..	..	Fonderie			273
573	632	632	630	686	690	703	..	Ouvrages en métaux, sauf machines		28	
..	..	..	..	..	..	..	..	Constr. et menuis. métal.; réservoirs; générateurs			281
..	..	..	..	..	..	..	..	Autres ouvrages en métaux			289
610	638	676	645	584	590	642	..	Machines et matériels, nca		29	
..	..	..	..	..	..	..	..	Machines d'usage général			291
..	..	..	..	..	..	..	..	Machines d'usage spécifique			292
..	..	..	..	..	..	..	..	Appareils domestiques, nca			293
154	175	190	204	231	252	274	..	Mach. de bureau, comptables, information		30	
..	..	..	..	..	..	..	..	Machines et appareils électriques, nca		31	
..	..	..	..	..	..	..	..	Moteurs, génératrices, transformateurs électriques			311
..	..	..	..	..	..	..	..	Matériel électrique de distribution et de commande			312
..	..	..	..	..	..	..	..	Fils et câbles électriques isolés			313
..	..	..	..	..	..	..	..	Accumulateurs et piles électriques			314
..	..	..	..	..	..	..	..	Lampes électriques et appareils d'éclairage			315
..	..	..	..	..	..	..	..	Autres matériels électriques, nca.			319
..	..	..	..	..	..	..	..	Equipements et appareils de radio, TV, etc.		32	
..	..	..	..	..	..	..	..	Tubes et valves électroniques;autres composants			321
..	..	..	..	..	..	..	..	Emetteurs radio/TV; app. téléphonie et télégraphie			322
..	..	..	..	..	..	..	..	Récepteurs radio/TV, son et image; art. associés			323
..	..	..	..	..	..	..	..	Instr. médicaux, de précision ; optique ; horlog.		33	
..	..	..	..	..	..	..	..	Instruments médicaux et de précision, etc.			331
..	..	..	..	..	..	..	..	Instruments d'optique ; matériel photographique			332
..	..	..	..	..	..	..	..	Horlogerie			333
50	50	38	39	37	37	38	..	Véhicules automobiles; remorques, etc.		34	
..	..	..	..	..	..	..	..	Véhicules automobiles			341
..	..	..	..	..	..	..	..	Carross. automobiles, remorques, semi-remorques			342
..	..	..	..	..	..	..	..	Pièces et accessoires pour automobiles et moteurs			343
..	..	..	..	..	..	..	..	Autres matériels de transport		35	
..	..	..	..	..	..	..	..	Construction et réparation de navires			351
..	..	..	..	..	..	..	..	Matériel ferroviaire roulant			352
..	..	..	..	..	..	..	..	Construction aéronautique et spatiale			353
..	..	..	..	..	..	..	..	Autres équipements de transport			359
55	50	56	63	98	109	120	..	Meubles ; activités de fabrication, nca		36	
..	..	..	..	..	..	..	..	Meubles			361
..	..	..	..	..	..	..	..	Activités de fabrication, nca			369
28	26	28	29	37	39	43	..	Récupération		37	
302	350	355	374	335	337	344	..	Electricité, gaz et eau	E	40_41	
..	..	..	..	..	..	..	..	Prod. et dist. d'électricité gaz, vapeur d'eau chaude		40	
..	..	..	..	..	..	..	..	Captage, épuration et distribution d'eau		41	
..	..	..	..	..	..	..	..	Construction	F	45	

Note: ISIC 15 includes ISIC 16; ISIC 21 includes ISIC 22; ISIC 30 includes ISIC 31, 32 and 33; ISIC 34 includes ISIC 35; ISIC 36 includes ISIC 20. Data for 1997 are provisional.

Table LX.7 **ESTABLISHMENTS - ÉTABLISSEMENTS**

Units

ISIC revision 3				1991	1992	1993	1994	1995	1996	1997	1998
C	**10_14**		Mining and quarrying	..	..	..	..	..	..	..	..
	10		Mining of coal and lignite; extraction of peat	..	..	..	..	..	..	..	..
	11		Crude petroleum & natural gas; service activities	..	..	..	..	..	..	..	..
	13		Mining of metal ores	..	..	..	..	..	..	..	..
	14		Other mining and quarrying	..	..	..	..	..	..	..	..
D	**15_37**		Total manufacturing	909	897	910	919	909	908	909	..
	15		Food products and beverages	248	236	230	227	218	218	218	..
		151	Production of meat, fish, fruit, vegetables, oils & fats	..	..	..	..	..	..	..	..
		152	Dairy products	..	..	..	..	..	..	..	..
		153	Grain mill and starch products; animal feeds	..	..	..	..	..	..	..	..
		154	Other food products	206	195	188	184	175	175	175	..
		155	Beverages	..	..	..	..	..	..	..	..
	16		Tobacco products	..	..	..	..	..	..	..	..
	17		Textiles	58	56	58	58	56	56	56	..
		171	Spinning, weaving and finishing of textiles	..	..	..	..	..	..	..	..
		172	Other textiles	..	..	..	..	..	..	..	..
		173	Knitted and crocheted fabrics and articles	..	..	..	..	..	..	..	..
	18		Wearing apparel and fur	..	..	..	..	..	..	..	..
		181	Wearing apparel, except fur apparel	..	..	..	..	..	..	..	..
		182	Dressing and dyeing of fur; articles of fur	..	..	..	..	..	..	..	..
	19		Leather and articles of leather; footwear	..	..	..	..	..	..	..	..
		191	Leather and articles of leather	..	..	..	..	..	..	..	..
		192	Footwear	..	..	..	..	..	..	..	..
	20		Wood & products of wood & cork, exc. furniture	..	..	..	..	..	..	..	..
		201	Sawmilling and planing of wood	..	..	..	..	..	..	..	..
		202	Wood products; articles of cork, straw, etc.	..	..	..	..	..	..	..	..
	21		Paper and paper products	150	155	162	164	164	164	164	..
	22		Printing, publishing and recorded media	..	..	..	..	..	..	..	..
		221	Publishing	..	..	..	..	..	..	..	..
		222	Printing and related activities	..	..	..	..	..	..	..	..
		223	Reproduction of recorded media	..	..	..	..	..	..	..	..
	23		Coke and petroleum products; nuclear fuel	..	..	..	..	..	..	..	..
		231	Coke oven products	..	..	..	..	..	..	..	..
		232	Refined petroleum products	..	..	..	..	..	..	..	..
		233	Processing of nuclear fuel	..	..	..	..	..	..	..	..
	24		Chemicals and chemical products	22	21	21	20	20	20	20	..
		241	Basic chemicals	..	..	..	..	..	..	..	..
		242	Other chemical products	..	..	..	..	..	..	..	..
		243	Man-made fibres	..	..	..	..	..	..	..	..
	25		Rubber and plastic products	22	21	23	23	21	21	21	..
		251	Rubber products	..	..	..	..	..	..	..	..
		252	Plastic products	..	..	..	..	..	..	..	..
	26		Other non-metallic mineral products	42	46	46	42	37	37	37	..
		261	Glass and glass products	..	..	..	..	..	..	..	..
		269	Non-metallic mineral products, nec	..	..	..	..	..	..	..	..
	27		Basic metals	13	12	14	15	15	16	16	..
		271	Basic iron and steel	..	..	..	..	..	..	..	..
		272	Basic precious and non-ferrous metals	..	..	..	..	..	..	..	..
		273	Casting of metals	..	..	..	..	..	..	..	..
	28		Fabricated metal products, exc. machinery	138	135	140	147	147	147	147	..
		281	Structural metal prod., tanks, reservoirs, etc.	..	..	..	..	..	..	..	..
		289	Other fabricated metal products	..	..	..	..	..	..	..	..
	29		Machinery and equipment, nec	42	41	42	40	43	43	43	..
		291	General purpose machinery	..	..	..	..	..	..	..	..
		292	Special purpose machinery	..	..	..	..	..	..	..	..
		293	Domestic appliances, nec	..	..	..	..	..	..	..	..
	30		Office, accounting & computing machinery	54	52	58	60	64	64	64	..
	31		Electrical machinery & apparatus, nec	..	..	..	..	..	..	..	..
		311	Electric motors, generators and transformers	..	..	..	..	..	..	..	..
		312	Electricity distribution & control apparatus	..	..	..	..	..	..	..	..
		313	Insulated wire and cable	..	..	..	..	..	..	..	..
		314	Accumulators, primary cells & primary batteries	..	..	..	..	..	..	..	..
		315	Electric lamps and lighting equipment	..	..	..	..	..	..	..	..
		319	Other electrical equipment, nec	..	..	..	..	..	..	..	..
	32		Radio, TV, communication equipment	..	..	..	..	..	..	..	..
		321	Electronic valves, tubes; other electronic components	..	..	..	..	..	..	..	..
		322	TV/radio transmitters; line teleph./telegr. apparatus	..	..	..	..	..	..	..	..
		323	TV/radio receivers; sound and video app.; assoc. gds.	..	..	..	..	..	..	..	..
	33		Medical, precision & optical instrum.; watch & clocks	..	..	..	..	..	..	..	..
		331	Medical appl.; instruments & appl. for measuring, etc.	..	..	..	..	..	..	..	..
		332	Optical instruments and photographic equipment	..	..	..	..	..	..	..	..
		333	Watches and clocks	..	..	..	..	..	..	..	..
	34		Motor vehicles, trailers and semi-trailers	17	17	16	14	16	14	14	..
		341	Motor vehicles	..	..	..	..	..	..	..	..
		342	Bodies for motor vehicles, trailers & semi-trailers	..	..	..	..	..	..	..	..
		343	Parts & accessories for motor vehicles & their engines	..	..	..	..	..	..	..	..
	35		Other transport equipment	..	..	..	..	..	..	..	..
		351	Building and repairing of ships and boats	..	..	..	..	..	..	..	..
		352	Railway and tramway locomotives and rolling stock	..	..	..	..	..	..	..	..
		353	Aircraft and spacecraft	..	..	..	..	..	..	..	..
		359	Transport equipment, nec	..	..	..	..	..	..	..	..
	36		Furniture; manufacturing, nec	94	95	89	95	92	92	92	..
		361	Furniture	..	..	..	..	..	..	..	..
		369	Manufacturing, nec	..	..	..	..	..	..	..	..
	37		Recycling	9	10	11	14	16	16	17	..
E	**40_41**		Electricity, gas and water supply	35	35	36	38	39	38	38	..
	40		Electricity, gas, steam and hot water supply	..	..	..	..	..	..	..	..
	41		Collection, purification and distribution of water	..	..	..	..	..	..	..	..
F	**45**		Construction	..	..	..	..	..	..	..	..

Note: ISIC 15 includes ISIC 16; ISIC 21 includes ISIC 22; ISIC 30 includes ISIC 31, 32 and 33; ISIC 34 includes ISIC 35; ISIC 36 includes ISIC 20. Data for 1997 are provisional.

Table LX.8 **EXPORTATIONS - EXPORTS**

Millions de LF (Prix courants)

1991	1992	1993	1994	1995	1996	1997	1998			CITI révision 3	
..	..	..	..	..	..	..	..	Activites extractives	C	10_14	
..	..	..	..	..	..	..	..	Extraction de charbon, de lignite et de tourbe		10	
..	..	..	..	..	..	..	..	Pétrole brut et gaz naturel ; activités annexes		11	
..	..	..	..	..	..	..	..	Extraction de minerais métalliques		13	
..	..	..	..	..	..	..	..	Autres activités extractives		14	
158 468	161 057	166 311	185 687	187 397	180 571	200 461	..	Total activites manufacturieres	D	15_37	
6 041	6 111	5 762	6 619	6 257	6 543	6 961	..	Produits alimentaires et boissons		15	
..	..	..	..	..	..	..	..	Viande, poisson, fruits, légumes, huiles et graisses			151
..	..	..	..	..	..	..	..	Produits laitiers			152
..	..	..	..	..	..	..	..	Travail des grains; prod. amyl. ; alim. pour animaux			153
29	141	180	194	224	202	226	..	Autres produits alimentaires			154
..	..	..	..	..	..	..	..	Boissons			155
..	..	..	..	..	..	..	..	Produits à base de tabac		16	
10 817	12 273	12 713	14 408	14 559	13 039	14 870	..	Textiles		17	
..	..	..	..	..	..	..	..	Filature, tissage et achèvement des textiles			171
..	..	..	..	..	..	..	..	Autres articles textiles			172
..	..	..	..	..	..	..	..	Etoffes et articles de bonneterie			173
..	..	..	..	..	..	..	..	Articles d'habillement et fourrure		18	
..	..	..	..	..	..	..	..	Articles d'habillement autres qu'en fourrure			181
..	..	..	..	..	..	..	..	Prép. et teinture des fourrures ; articles en fourrure			182
..	..	..	..	..	..	..	..	Cuir et articles en cuir ; chaussures		19	
..	..	..	..	..	..	..	..	Cuir et articles en cuir sauf chaussures			191
..	..	..	..	..	..	..	..	Chaussures			192
..	..	..	..	..	..	..	..	Bois, articles en bois, sauf meubles		20	
..	..	..	..	..	..	..	..	Sciage et rabotage du bois			201
..	..	..	..	..	..	..	..	Articles en bois, liège, vannerie, sparterie			202
2 460	3 890	3 309	3 465	4 453	4 411	5 002	..	Papier, articles en papier et carton		21	
..	..	..	..	..	..	..	..	Edition, imprimerie et reproduction		22	
..	..	..	..	..	..	..	..	Edition			221
..	..	..	..	..	..	..	..	Imprimerie et activités annexes			222
..	..	..	..	..	..	..	..	Reproduction de supports enregistrés			223
..	..	..	..	..	..	..	..	Coke. et prod. pétroliers ; comb. nucléaires		23	
..	..	..	..	..	..	..	..	Cokéfaction			231
..	..	..	..	..	..	..	..	Produits pétroliers raffinés			232
..	..	..	..	..	..	..	..	Traitement de combustibles nucléaires			233
15 433	21 074	20 476	20 963	21 254	18 178	20 014	..	Produits chimiques		24	
..	..	..	..	..	..	..	..	Produits chimiques de base			241
..	..	..	..	..	..	..	..	Autres produits chimiques			242
..	..	..	..	..	..	..	..	Fibres synthétiques ou artificielles			243
24 467	26 932	25 649	29 910	32 710	32 797	33 995	..	Art. en caoutchouc et matières plastiques		25	
..	..	..	..	..	..	..	..	Articles en caoutchouc			251
..	..	..	..	..	..	..	..	Articles en matières plastiques			252
9 929	9 353	9 290	12 102	11 699	10 468	11 644	..	Autres produits minéraux non métalliques		26	
..	..	..	..	..	..	..	..	Verre et articles en verre			261
..	..	..	..	..	..	..	..	Produits minéraux non métalliques, nca			269
53 892	47 161	56 183	59 499	54 272	51 272	56 365	..	Produits métallurgiques de base		27	
..	..	..	..	..	..	..	..	Sidérurgie et première transformation de l'acier			271
..	..	..	..	..	..	..	..	Première transf. métaux précieux et non ferreux			272
..	..	..	..	..	..	..	..	Fonderie			273
15 768	16 713	16 311	19 209	21 135	19 581	22 609	..	Ouvrages en métaux, sauf machines		28	
..	..	..	..	..	..	..	..	Constr. et menuis. métal.; réservoirs; générateurs			281
..	..	..	..	..	..	..	..	Autres ouvrages en métaux			289
13 002	12 771	12 088	13 937	13 789	14 863	18 125	..	Machines et matériels, nca		29	
..	..	..	..	..	..	..	..	Machines d'usage général			291
..	..	..	..	..	..	..	..	Machines d'usage spécifique			292
..	..	..	..	..	..	..	..	Appareils domestiques, nca			293
4 937	3 341	3 291	3 854	4 916	5 110	5 746	..	Mach. de bureau, comptables, information		30	
..	..	..	..	..	..	..	..	Machines et appareils électriques, nca		31	
..	..	..	..	..	..	..	..	Moteurs, génératrices, transformateurs électriques			311
..	..	..	..	..	..	..	..	Matériel électrique de distribution et de commande			312
..	..	..	..	..	..	..	..	Fils et câbles électriques isolés			313
..	..	..	..	..	..	..	..	Accumulateurs et piles électriques			314
..	..	..	..	..	..	..	..	Lampes électriques et appareils d'éclairage			315
..	..	..	..	..	..	..	..	Autres matériels électriques, nca.			319
..	..	..	..	..	..	..	..	Equipements et appareils de radio, TV, etc.		32	
..	..	..	..	..	..	..	..	Tubes et valves électroniques;autres composants			321
..	..	..	..	..	..	..	..	Emetteurs radio/TV; app. téléphonie et télégraphie			322
..	..	..	..	..	..	..	..	Récepteurs radio/TV, son et image; art. associés			323
..	..	..	..	..	..	..	..	Instr. médicaux, de précision ; optique ; horlog.		33	
..	..	..	..	..	..	..	..	Instruments médicaux et de précision, etc.			331
..	..	..	..	..	..	..	..	Instruments d'optique ; matériel photographique			332
..	..	..	..	..	..	..	..	Horlogerie			333
1 016	857	603	796	696	646	812	..	Véhicules automobiles; remorques, etc.		34	
..	..	..	..	..	..	..	..	Véhicules automobiles			341
..	..	..	..	..	..	..	..	Carross. automobiles, remorques, semi-remorques			342
..	..	..	..	..	..	..	..	Pièces et accessoires pour automobiles et moteurs			343
..	..	..	..	..	..	..	..	Autres matériels de transport		35	
..	..	..	..	..	..	..	..	Construction et réparation de navires			351
..	..	..	..	..	..	..	..	Matériel ferroviaire roulant			352
..	..	..	..	..	..	..	..	Construction aéronautique et spatiale			353
..	..	..	..	..	..	..	..	Autres équipements de transport			359
533	392	506	805	1 462	2 541	4 125	..	Meubles ; activités de fabrication, nca		36	
..	..	..	..	..	..	..	..	Meubles			361
..	..	..	..	..	..	..	..	Activités de fabrication, nca			369
172	191	131	120	197	157	193	..	Récupération		37	
916	926	952	1 168	963	1 010	1 015	..	Electricité, gaz et eau	E	40_41	
..	..	..	..	..	..	..	..	Prod. et dist. d'électricité gaz, vapeur d'eau chaude		40	
..	..	..	..	..	..	..	..	Captage, épuration et distribution d'eau		41	
..	..	..	..	..	..	..	..	Construction	F	45	

Note: ISIC 15 includes ISIC 16; ISIC 21 includes ISIC 22; ISIC 30 includes ISIC 31, 32 and 33; ISIC 34 includes ISIC 35; ISIC 36 includes ISIC 20. Data for 1997 are provisional.

NETHERLANDS

Table NL.1 PRODUCTION - PRODUCTION

Millions of Gld (Current Prices)

ISIC revision 3				1991	1992	1993	1994	1995	1996	1997	1998
C	10_14		Mining and quarrying	..	..	..	..	..	..	..	..
	10		Mining of coal and lignite; extraction of peat	..	..	..	..	..	..	..	..
	11		Crude petroleum & natural gas; service activities	..	..	..	..	..	..	..	..
	13		Mining of metal ores	..	..	..	..	..	..	..	..
	14		Other mining and quarrying	..	..	316	556	539	671	880	895
D	15_37		Total manufacturing	..	277 484	266 641	278 045	296 860	306 095	333 445	..
	15		Food products and beverages	..	71 694	70 236	71 229	71 810	74 099	77 333	79 863
		151	Production of meat, fish, fruit, vegetables, oils & fats	..	24 065	23 082	23 710	24 301	25 668	26 979	28 097
		152	Dairy products	..	15 513	14 210	13 952	13 549	13 049	13 539	14 254
		153	Grain mill and starch products; animal feeds	..	..	..	..	..	..	..	..
		154	Other food products	..	..	..	..	..	..	..	..
		155	Beverages	..	6 366	6 391	6 909	7 047	7 146	7 449	7 556
	16		Tobacco products	..	6 229	6 621	6 932	7 289	7 932	8 687	9 050
	17		Textiles	..	4 801	4 726	4 645	4 684	4 566	4 925	5 045
		171	Spinning, weaving and finishing of textiles	..	1 928	1 721	1 732	1 604	1 527	1 529	1 584
		172	Other textiles	..	..	..	..	..	..	..	..
		173	Knitted and crocheted fabrics and articles	..	..	..	..	..	..	..	..
	18		Wearing apparel and fur	..	1 288	1 260	1 289	1 158	988	924	1 000
		181	Wearing apparel, except fur apparel	..	1 288	1 260	1 289	1 158	..	..	..
		182	Dressing and dyeing of fur; articles of fur	..					..	..	..
	19		Leather and articles of leather; footwear	..	710	652	585	560	504	564	643
		191	Leather and articles of leather	..	371	352	326	288	267	319	381
		192	Footwear	..	338	300	259	272	237	245	263
	20		Wood & products of wood & cork, exc. furniture	..	2 369	2 371	2 452	2 736	2 687	2 930	3 161
		201	Sawmilling and planing of wood	..	..	..	..	..	..	..	..
		202	Wood products; articles of cork, straw, etc.	..	..	..	..	..	..	..	..
	21		Paper and paper products	..	8 472	7 521	8 067	9 368	8 940	9 683	10 470
	22		Printing, publishing and recorded media	..	17 316	16 375	16 935	17 373	18 220	18 378	19 584
		221	Publishing	..	7 648	7 812	8 101	8 675	9 202	9 623	10 745
		222	Printing and related activities	..	8 503	7 283	7 407	7 794	8 073	7 829	7 868
		223	Reproduction of recorded media	..	1 165	1 280	1 427	904	945	926	971
	23		Coke and petroleum products; nuclear fuel	..	17 291	19 029	18 164	17 518	22 466	20 700	16 452
		231	Coke oven products	..	..	..	..	..	..	..	..
		232	Refined petroleum products	..	..	..	..	..	..	..	..
		233	Processing of nuclear fuel	..	..	..	..	..	..	..	..
	24		Chemicals and chemical products	..	41 217	40 584	45 826	51 493	51 130	59 687	57 333
		241	Basic chemicals	..	25 119	23 979	28 321	32 193	31 436	36 181	33 708
		242	Other chemical products	..	..	..	..	..	..	..	..
		243	Man-made fibres	..	..	..	..	..	..	..	..
	25		Rubber and plastic products	..	8 206	7 687	8 196	8 693	8 586	9 393	9 855
		251	Rubber products	..	1 097	1 043	1 039	919	952	966	991
		252	Plastic products	..	7 109	6 644	7 156	7 774	7 634	8 427	8 863
	26		Other non-metallic mineral products	..	7 801	7 559	8 268	8 455	8 568	8 857	9 468
		261	Glass and glass products	..	1 451	1 466	1 559	1 631	1 752	1 753	1 889
		269	Non-metallic mineral products, nec	..	6 350	6 093	6 709	6 824	6 816	7 104	7 579
	27		Basic metals	..	8 390	7 874	9 045	10 477	9 696	10 943	11 542
		271	Basic iron and steel	..	..	..	..	..	..	..	..
		272	Basic precious and non-ferrous metals	..	..	..	..	..	..	..	..
		273	Casting of metals	..	806	737	905	979	997	1 087	1 178
	28		Fabricated metal products, exc. machinery	..	16 766	15 197	16 001	17 307	17 837	19 413	21 314
		281	Structural metal prod., tanks, reservoirs, etc.	..	9 585	8 618	8 839	9 398	9 624	10 571	11 592
		289	Other fabricated metal products	..	7 181	6 579	7 162	7 909	8 213	8 842	9 722
	29		Machinery and equipment, nec	..	16 585	15 708	16 671	18 532	20 171	21 592	23 176
		291	General purpose machinery	..	9 738	9 125	9 498	10 087	10 954	11 414	12 660
		292	Special purpose machinery	..	6 336	6 093	6 728	8 007	9 218	10 179	10 515
		293	Domestic appliances, nec	..	511	489	446	438	529	682	682
	30		Office, accounting & computing machinery	..	2 809	2 382	2 353	3 333	3 155	5 174	5 784
	31		Electrical machinery & apparatus, nec	..	..	..	..	..	..	..	..
		311	Electric motors, generators and transformers	..	1 050	1 109	740	641	604	676	883
		312	Electricity distribution & control apparatus	..	1 209	1 138	1 620	1 863	1 894	2 069	1 878
		313	Insulated wire and cable	..	922	799	848	1 113	1 327	1 384	1 318
		314	Accumulators, primary cells & primary batteries	..	..	..	..	..	..	..	..
		315	Electric lamps and lighting equipment	..	..	..	..	..	..	..	..
		319	Other electrical equipment, nec	..	..	..	..	..	..	..	..
	32		Radio, TV, communication equipment	..	..	..	..	..	..	..	..
		321	Electronic valves, tubes; other electronic components	..	..	..	..	..	..	..	..
		322	TV/radio transmitters; line teleph./telegr. apparatus	..	..	..	..	..	..	..	..
		323	TV/radio receivers; sound and video app.; assoc. gds.	..	..	..	..	..	..	..	..
	33		Medical, precision & optical instrum.; watch & clocks	..	..	..	..	..	..	..	..
		331	Medical appl.; instruments & appl. for measuring, etc.	..	..	..	..	..	..	..	..
		332	Optical instruments and photographic equipment	..	..	..	..	..	..	..	..
		333	Watches and clocks	..	..	..	..	..	..	..	..
	34		Motor vehicles, trailers and semi-trailers	..	..	..	..	..	..	..	..
		341	Motor vehicles	..	..	..	..	..	..	..	..
		342	Bodies for motor vehicles, trailers & semi-trailers	..	1 770	1 552	1 645	1 833	1 870	1 795	2 147
		343	Parts & accessories for motor vehicles & their engines	..	817	897	1 051	1 273	1 498	1 694	1 732
	35		Other transport equipment	..	..	..	..	..	..	..	..
		351	Building and repairing of ships and boats	..	4 139	3 589	3 808	3 973	4 235	4 461	5 474
		352	Railway and tramway locomotives and rolling stock	..	..	..	..	..	..	..	..
		353	Aircraft and spacecraft	..	..	..	..	..	..	..	..
		359	Transport equipment, nec	..	701	659	699	709	756	899	1 109
	36		Furniture; manufacturing, nec	..	4 437	4 556	4 404	4 848	5 047	5 579	5 967
		361	Furniture	..	3 599	3 663	3 618	3 873	3 982	4 397	4 788
		369	Manufacturing, nec	..	838	893	787	975	1 065	1 181	1 179
	37		Recycling	..		..	63	130	407	472	480
E	40_41		Electricity, gas and water supply	..	..	33 497	34 631	36 639	..	..	..
	40		Electricity, gas, steam and hot water supply	..	..	31 212	32 242	33 859	..	..	..
	41		Collection, purification and distribution of water	..	..	2 285	2 389	2 780	..	..	..
F	45		Construction	..	..	84 057	87 232	91 126	..	..	..

Table NL.2 **VALEUR AJOUTÉE - VALUE ADDED**

Millions de Gld (Prix courants)

1991	1992	1993	1994	1995	1996	1997	1998			CITI révision 3	
..	..	..	..	..	..	..	..	Activites extractives	C	10_14	
..	..	..	..	..	..	..	..	Extraction de charbon, de lignite et de tourbe		10	
..	..	..	..	..	..	..	..	Pétrole brut et gaz naturel ; activités annexes		11	
..	..	..	..	..	..	..	..	Extraction de minerais métalliques		13	
..	..	133	226	219	270	343	321	Autres activités extractives		14	
..	83 251	82 350	88 444	91 276	91 569	97 768	99 106	Total activites manufacturieres	D	15_37	
..	15 224	15 353	15 769	15 436	16 049	16 560	16 462	Produits alimentaires et boissons		15	
..	3 773	3 818	3 642	3 807	4 037	4 147	4 453	Viande, poisson, fruits, légumes, huiles et graisses			151
..	2 133	1 851	1 966	1 834	1 675	1 809	1 623	Produits laitiers			152
..	..	..	..	..	..	..	..	Travail des grains; prod. amyl. ; alim. pour animaux			153
..	..	..	..	..	..	..	..	Autres produits alimentaires			154
..	3 161	3 137	3 440	3 299	3 399	3 476	3 405	Boissons			155
..	3 784	3 860	4 131	4 422	4 818	5 388	5 758	Produits à base de tabac		16	
..	1 663	1 641	1 598	1 535	1 533	1 542	1 601	Textiles		17	
..	801	731	718	630	611	570	600	Filature, tissage et achèvement des textiles			171
..	..	..	..	..	..	..	..	Autres articles textiles			172
..	..	..	..	..	..	..	..	Etoffes et articles de bonneterie			173
..	394	388	372	316	268	237	262	Articles d'habillement et fourrure		18	
..	394	388	372	316	..	..	..	Articles d'habillement autres qu'en fourrure			181
..	..	..	..	..	..	..	..	Prép. et teinture des fourrures ; articles en fourrure			182
..	252	221	186	179	169	176	191	Cuir et articles en cuir ; chaussures		19	
..	107	90	78	69	71	75	87	Cuir et articles en cuir sauf chaussures			191
..	146	131	109	110	98	101	104	Chaussures			192
..	819	845	835	912	903	939	1 027	Bois, articles en bois, sauf meubles		20	
..	..	..	..	..	..	..	..	Sciage et rabotage du bois			201
..	..	..	..	..	..	..	..	Articles en bois, liège, vannerie, sparterie			202
..	2 974	2 825	2 866	3 050	3 082	3 312	3 640	Papier, articles en papier et carton		21	
..	6 983	7 233	7 513	7 489	7 827	8 102	8 387	Edition, imprimerie et reproduction		22	
..	2 868	3 492	3 662	3 874	4 153	4 493	4 867	Edition			221
..	3 797	3 257	3 306	3 285	3 387	3 311	3 220	Imprimerie et activités annexes			222
..	318	483	546	330	287	298	300	Reproduction de supports enregistrés			223
..	1 622	2 137	2 130	1 945	2 468	2 699	2 635	Coke. et prod. pétroliers ; comb. nucléaires		23	
..	..	..	..	..	..	..	..	Cokéfaction			231
..	..	..	..	..	..	..	..	Produits pétroliers raffinés			232
..	..	..	..	..	..	..	..	Traitement de combustibles nucléaires			233
..	11 470	11 429	14 984	16 945	14 817	16 138	14 524	Produits chimiques		24	
..	6 455	5 716	8 794	10 217	9 078	9 246	8 129	Produits chimiques de base			241
..	..	..	..	..	..	..	..	Autres produits chimiques			242
..	..	..	..	..	..	..	..	Fibres synthétiques ou artificielles			243
..	3 005	2 862	3 022	2 926	3 155	3 364	3 605	Art. en caoutchouc et matières plastiques		25	
..	498	450	442	390	402	396	414	Articles en caoutchouc			251
..	2 507	2 412	2 579	2 536	2 753	2 968	3 191	Articles en matières plastiques			252
..	3 290	3 316	3 661	3 763	3 753	3 879	4 014	Autres produits minéraux non métalliques		26	
..	668	685	729	779	788	736	771	Verre et articles en verre			261
..	2 622	2 631	2 932	2 984	2 965	3 143	3 243	Produits minéraux non métalliques, nca			269
..	3 053	3 001	3 565	4 047	3 522	3 851	3 880	Produits métallurgiques de base		27	
..	..	..	..	..	..	..	..	Sidérurgie et première transformation de l'acier			271
..	..	..	..	..	..	..	..	Première transf. métaux précieux et non ferreux			272
..	384	340	402	412	411	410	446	Fonderie			273
..	6 098	5 645	5 846	6 086	6 326	6 814	7 323	Ouvrages en métaux, sauf machines		28	
..	3 135	2 847	2 878	2 945	3 035	3 299	3 579	Constr. et menuis. métal.; réservoirs; générateurs			281
..	2 964	2 797	2 969	3 140	3 291	3 515	3 744	Autres ouvrages en métaux			289
..	6 432	6 081	6 377	6 742	7 074	7 457	7 694	Machines et matériels, nca		29	
..	3 825	3 493	3 532	3 591	3 643	3 841	4 276	Machines d'usage général			291
..	2 435	2 422	2 696	3 008	3 275	3 407	3 207	Machines d'usage spécifique			292
..	172	166	150	143	156	209	211	Appareils domestiques, nca			293
..	937	714	749	856	826	1 002	1 203	Mach. de bureau, comptables, information		30	
..	..	..	..	..	..	..	..	Machines et appareils électriques, nca		31	
..	399	401	277	188	203	220	266	Moteurs, génératrices, transformateurs électriques			311
..	563	479	646	721	787	796	701	Matériel électrique de distribution et de commande			312
..	410	381	385	425	481	479	509	Fils et câbles électriques isolés			313
..	..	..	..	..	..	..	..	Accumulateurs et piles électriques			314
..	..	..	..	..	..	..	..	Lampes électriques et appareils d'éclairage			315
..	..	..	..	..	..	..	..	Autres matériels électriques, nca.			319
..	..	..	..	..	..	..	..	Equipements et appareils de radio, TV, etc.		32	
..	..	..	..	..	..	..	..	Tubes et valves électroniques;autres composants			321
..	..	..	..	..	..	..	..	Emetteurs radio/TV; app. téléphonie et télégraphie			322
..	..	..	..	..	..	..	..	Récepteurs radio/TV, son et image; art. associés			323
..	..	..	..	..	..	..	..	Instr. médicaux, de précision ; optique ; horlog.		33	
..	..	..	..	..	..	..	..	Instruments médicaux et de précision, etc.			331
..	..	..	..	..	..	..	..	Instruments d'optique ; matériel photographique			332
..	..	..	..	..	..	..	..	Horlogerie			333
..	..	..	..	..	..	..	..	Véhicules automobiles; remorques, etc.		34	
..	..	..	..	..	..	..	..	Véhicules automobiles			341
..	487	476	467	521	527	509	571	Carross. automobiles, remorques, semi-remorques			342
..	309	313	357	427	399	468	481	Pièces et accessoires pour automobiles et moteurs			343
..	..	..	..	..	..	..	..	Autres matériels de transport		35	
..	1 162	1 108	1 099	1 056	993	904	1 159	Construction et réparation de navires			351
..	..	..	..	..	..	..	..	Matériel ferroviaire roulant			352
..	..	..	..	..	..	..	..	Construction aéronautique et spatiale			353
..	209	195	206	204	229	270	311	Autres équipements de transport			359
..	1 698	1 779	1 739	1 816	1 836	2 012	2 116	Meubles ; activités de fabrication, nca		36	
..	1 349	1 387	1 381	1 414	1 396	1 523	1 635	Meubles			361
..	349	392	358	403	440	489	481	Activités de fabrication, nca			369
..	..	..	21	44	112	163	151	Récupération		37	
..	..	10 514	10 859	11 933	12 704	..	..	Electricité, gaz et eau	E	40_41	
..	..	8 970	9 225	9 944	10 559	..	..	Prod. et dist. d'électricité gaz, vapeur d'eau chaude		40	
..	..	1 544	1 634	1 989	2 145	..	..	Captage, épuration et distribution d'eau		41	
..	..	29 054	29 475	30 270	31 208	32 460	..	Construction	F	45	

NETHERLANDS

Table NL.3 INVESTMENT - INVESTISSEMENT

Millions of Gld (Current Prices)

ISIC revision 3				1991	1992	1993	1994	1995	1996	1997	1998
C	10_14		Mining and quarrying	..	..	3 885	2 028	1 719	1 634	4 147	2 508
	10		Mining of coal and lignite; extraction of peat	..	..	..	..	..	..	..	..
	11		Crude petroleum & natural gas; service activities	..	..	..	..	..	..	..	..
	13		Mining of metal ores	..	..	..	..	..	..	..	..
	14		Other mining and quarrying	..	..	..	..	..	..	..	..
D	15_37		Total manufacturing	..	..	13 029	12 371	15 425	17 057	19 710	18 828
	15		Food products and beverages	..	..	2 913	2 809	2 717	2 919	3 279	3 002
		151	Production of meat, fish, fruit, vegetables, oils & fats	..	..	..	778	567	..	..	..
		152	Dairy products	..	..	..	447	347	354	377	310
		153	Grain mill and starch products; animal feeds	..	..	..	398	357	..	..	..
		154	Other food products	..	..	..	826	918	928	1 040	934
		155	Beverages	..	..	..	360	439	533	571	549
	16		Tobacco products	..	..	249	157	146	214	191	140
	17		Textiles	..	..	212	186	208	211	192	219
		171	Spinning, weaving and finishing of textiles	..	..	..	81	54	35	23	..
		172	Other textiles	..	..	..	86	118	..	..	..
		173	Knitted and crocheted fabrics and articles	..	..	..	19	15	27	19	..
	18		Wearing apparel and fur	..	..	26	27	67	39	52	45
		181	Wearing apparel, except fur apparel	..	..	..	..	67	39	52	45
		182	Dressing and dyeing of fur; articles of fur	..	..	..	..				
	19		Leather and articles of leather; footwear	..	..	17	23	25	18	24	30
		191	Leather and articles of leather	..	..	..	14	9	..	..	..
		192	Footwear	..	..	..	8	14	9	10	14
	20		Wood & products of wood & cork, exc. furniture	..	..	109	99	226	233	259	255
		201	Sawmilling and planing of wood	..	..	..	6	19	19	16	16
		202	Wood products; articles of cork, straw, etc.	..	..	..	93	124	..	..	..
	21		Paper and paper products	..	..	504	474	621	739	635	817
	22		Printing, publishing and recorded media	..	..	995	1 223	1 310	1 429	1 482	1 883
		221	Publishing	..	..	..	363	360	263	386	568
		222	Printing and related activities	..	..	..	825	903	1 099	1 048	1 207
		223	Reproduction of recorded media	..	..	..	35	48	66	47	107
	23		Coke and petroleum products; nuclear fuel	..	..	765	707	493	494	3 576	807
		231	Coke oven products	..	..	..	..	..	..	..	..
		232	Refined petroleum products	..	..	..	..	..	..	..	..
		233	Processing of nuclear fuel	..	..	..	..	..	..	..	..
	24		Chemicals and chemical products	..	..	2 469	2 460	2 556	3 228	3 232	3 763
		241	Basic chemicals	..	..	..	1 631	1 511	2 395	2 279	2 803
		242	Other chemical products	..	..	..	830	883	..	..	..
		243	Man-made fibres	..	..	..	..	142	145	146	33
	25		Rubber and plastic products	..	..	737	562	664	807	649	774
		251	Rubber products	..	..	..	66	84	109	75	73
		252	Plastic products	..	..	..	496	580	699	575	701
	26		Other non-metallic mineral products	..	..	550	661	849	814	831	796
		261	Glass and glass products	..	..	..	142	150	167	196	126
		269	Non-metallic mineral products, nec	..	..	..	519	629	124	100	..
	27		Basic metals	..	..	284	349	375	500	508	662
		271	Basic iron and steel	..	..	..	218	..	..	..	..
		272	Basic precious and non-ferrous metals	..	..	..	69	111	179	258	306
		273	Casting of metals	..	..	..	62	79	57	61	79
	28		Fabricated metal products, exc. machinery	..	..	714	628	1 054	1 132	1 250	1 383
		281	Structural metal prod., tanks, reservoirs, etc.	..	..	..	264	279	..	..	..
		289	Other fabricated metal products	..	..	..	364	548	..	..	..
	29		Machinery and equipment, nec	..	..	577	579	820	1 103	997	1 219
		291	General purpose machinery	..	..	..	334	366	..	..	..
		292	Special purpose machinery	..	..	..	245	328	..	..	..
		293	Domestic appliances, nec	..	..	..	..	..	25	29	..
	30		Office, accounting & computing machinery	..	..	160	91	111	122	122	174
	31		Electrical machinery & apparatus, nec	..	..	218	167	254	255	253	401
		311	Electric motors, generators and transformers	..	..	..	19	18	25	34	35
		312	Electricity distribution & control apparatus	..	..	..	53	86	79	84	155
		313	Insulated wire and cable	..	..	..	25	62	52	47	58
		314	Accumulators, primary cells & primary batteries	..	..	..	17	2	3	1	2
		315	Electric lamps and lighting equipment	..	..	..	..	18	35	20	29
		319	Other electrical equipment, nec	..	..	..	53	59	48	56	..
	32		Radio, TV, communication equipment	..	..	652	508	698	1 444	886	868
		321	Electronic valves, tubes; other electronic components	..	..	..	..	..	..	..	..
		322	TV/radio transmitters; line teleph./telegr. apparatus	..	..	..	..	..	..	..	..
		323	TV/radio receivers; sound and video app.; assoc. gds.	..	..	..	..	..	..	..	..
	33		Medical, precision & optical instrum.; watch & clocks	..	..	262	146	211	301	241	243
		331	Medical appl.; instruments & appl. for measuring, etc.	..	..	..	127	130	..	..	..
		332	Optical instruments and photographic equipment	..	..	..	19	37	52	37	23
		333	Watches and clocks	..	..	..	..	1	1	1	2
	34		Motor vehicles, trailers and semi-trailers	..	..	337	323	..	449	305	505
		341	Motor vehicles	..	..	..	115	..	313	..	..
		342	Bodies for motor vehicles, trailers & semi-trailers	..	..	..	53	52	58	..	..
		343	Parts & accessories for motor vehicles & their engines	..	..	..	42	47	78	56	..
	35		Other transport equipment	..	..	..	..	..	206	269	259
		351	Building and repairing of ships and boats	..	..	..	101	165	158	184	184
		352	Railway and tramway locomotives and rolling stock	..	..	..	2	..	..	..	1
		353	Aircraft and spacecraft	..	..	..	..	..	..	..	..
		359	Transport equipment, nec	..	..	..	..	..	..	..	..
	36		Furniture; manufacturing, nec	..	..	280	192	296	336	393	429
		361	Furniture	..	..	..	146	225	228	280	321
		369	Manufacturing, nec	..	..	..	47	49	..	..	..
	37		Recycling	..	..	..	..	39	65	88	156
E	40_41		Electricity, gas and water supply	..	..	4 815	6 289	5 052	7 742	4 314	3 916
	40		Electricity, gas, steam and hot water supply	..	..	4 113	5 291	4 172	6 935	3 527	3 065
	41		Collection, purification and distribution of water	..	..	702	998	880	807	787	852
F	45		Construction	..	..	1 813	1 722	2 649	3 404	3 497	3 605

Note:When available: ISIC 241 includes 243; 292 includes 293; 332includes part of 331 and 333; 34 includes 353; 352 includes part of 359.

Table NL.4 **DÉPENSES DE MACHINES ET D'ÉQUIPEMENT - INVESTMENT IN MACHINERY & EQUIPMENT**

Millions de Gld (Prix courants)

1991	1992	1993	1994	1995	1996	1997	1998			CITI révision 3
..	..	2 434	1 203	985	544	2 699	1 238	Activites extractives	C	10_14
..	..	..	..	..	..	..	..	Extraction de charbon, de lignite et de tourbe		10
..	..	..	..	..	..	..	..	Pétrole brut et gaz naturel ; activités annexes		11
..	..	..	..	..	..	..	..	Extraction de minerais métalliques		13
..	..	..	..	..	..	..	..	Autres activités extractives		14
..	..	9 957	9 759	12 739	14 048	16 395	15 590	Total activites manufacturieres	D	15_37
..	..	2 074	2 096	2 202	2 358	2 490	2 461	Produits alimentaires et boissons		15
..	..	..	571	437	..	..	..	Viande, poisson, fruits, légumes, huiles et graisses		151
..	..	..	353	294	307	323	242	Produits laitiers		152
..	..	..	288	270	..	..	..	Travail des grains; prod. amyl. ; alim. pour animaux		153
..	..	..	607	715	723	772	779	Autres produits alimentaires		154
..	..	..	277	371	475	393	500	Boissons		155
..	..	125	147	138	198	159	108	Produits à base de tabac		16
..	..	176	156	172	191	162	190	Textiles		17
..	..	..	71	39	33	20	..	Filature, tissage et achèvement des textiles		171
..	..	..	68	92	..	..	..	Autres articles textiles		172
..	..	..	17	14	25	18	..	Etoffes et articles de bonneterie		173
..	..	17	16	44	29	37	37	Articles d'habillement et fourrure		18
..	..	..	..	44	29	37	37	Articles d'habillement autres qu'en fourrure		181
..	..	..	..					Prép. et teinture des fourrures ; articles en fourrure		182
..	..	13	14	24	16	21	21	Cuir et articles en cuir ; chaussures		19
..	..	..	9	7	..	..	..	Cuir et articles en cuir sauf chaussures		191
..	..	..	5	13	9	9	12	Chaussures		192
..	..	68	68	173	151	169	176	Bois, articles en bois, sauf meubles		20
..	..	..	5	13	13	11	12	Sciage et rabotage du bois		201
..	..	..	63	83	..	..	..	Articles en bois, liège, vannerie, sparterie		202
..	..	446	392	524	640	562	715	Papier, articles en papier et carton		21
..	..	763	928	1 137	1 225	1 302	1 463	Edition, imprimerie et reproduction		22
..	..	..	247	272	222	326	369	Edition		221
..	..	..	653	822	941	933	1 029	Imprimerie et activités annexes		222
..	..	..	28	43	62	43	66	Reproduction de supports enregistrés		223
..	..	666	651	441	459	3 386	768	Coke. et prod. pétroliers ; comb. nucléaires		23
..	..	..	..	..	..	..	..	Cokéfaction		231
..	..	..	..	..	..	..	..	Produits pétroliers raffinés		232
..	..	..	..	..	..	..	..	Traitement de combustibles nucléaires		233
..	..	2 053	2 071	2 143	2 879	2 816	3 377	Produits chimiques		24
..	..	..	1 488	1 343	2 189	2 071	2 623	Produits chimiques de base		241
..	..	..	583	607	..	..	..	Autres produits chimiques		242
..	..	..	..	134	137	129	28	Fibres synthétiques ou artificielles		243
..	..	598	483	540	658	555	675	Art. en caoutchouc et matières plastiques		25
..	..	..	55	62	98	70	69	Articles en caoutchouc		251
..	..	..	428	478	559	485	607	Articles en matières plastiques		252
..	..	435	498	661	640	651	652	Autres produits minéraux non métalliques		26
..	..	..	132	128	140	176	110	Verre et articles en verre		261
..	..	..	366	443	98	76	..	Produits minéraux non métalliques, nca		269
..	..	235	300	340	393	441	556	Produits métallurgiques de base		27
..	..	..	194	..	..	..	..	Sidérurgie et première transformation de l'acier		271
..	..	..	61	103	146	230	256	Première transf. métaux précieux et non ferreux		272
..	..	..	45	63	47	51	70	Fonderie		273
..	..	481	455	828	870	974	1 084	Ouvrages en métaux, sauf machines		28
..	..	..	163	171	..	..	..	Constr. et menuis. métal.; réservoirs; générateurs		281
..	..	..	292	402	..	..	..	Autres ouvrages en métaux		289
..	..	361	390	625	758	701	935	Machines et matériels, nca		29
..	..	..	232	234	..	..	..	Machines d'usage général		291
..	..	..	158	222	..	..	..	Machines d'usage spécifique		292
..	..	..	..	..	17	20	..	Appareils domestiques, nca		293
..	..	124	73	92	97	103	133	Mach. de bureau, comptables, information		30
..	..	152	122	208	213	218	273	Machines et appareils électriques, nca		31
..	..	..	12	14	18	31	27	Moteurs, génératrices, transformateurs électriques		311
..	..	..	46	74	74	71	97	Matériel électrique de distribution et de commande		312
..	..	..	21	56	42	42	53	Fils et câbles électriques isolés		313
..	..	..	8	2	3	1	1	Accumulateurs et piles électriques		314
..	..	..	..	15	27	17	23	Lampes électriques et appareils d'éclairage		315
..	..	..	35	37	33	43	..	Autres matériels électriques, nca.		319
..	..	597	453	636	1 281	742	783	Equipements et appareils de radio, TV, etc.		32
..	..	..	..	..	..	..	..	Tubes et valves électroniques;autres composants		321
..	..	..	..	..	..	..	..	Emetteurs radio/TV; app. téléphonie et télégraphie		322
..	..	..	..	..	..	..	..	Récepteurs radio/TV, son et image; art. associés		323
..	..	145	108	165	201	182	185	Instr. médicaux, de précision ; optique ; horlog.		33
..	..	..	93	92	..	..	..	Instruments médicaux et de précision, etc.		331
..	..	..	15	28	35	29	19	Instruments d'optique ; matériel photographique		332
..	..	..	..	1	1	1	1	Horlogerie		333
..	..	217	209	..	373	233	411	Véhicules automobiles; remorques, etc.		34
..	..	..	82	..	266	..	..	Véhicules automobiles		341
..	..	..	19	39	42	..	..	Carross. automobiles, remorques, semi-remorques		342
..	..	..	37	44	65	52	..	Pièces et accessoires pour automobiles et moteurs		343
..	..	..	..	..	129	155	149	Autres matériels de transport		35
..	..	..	61	91	93	109	92	Construction et réparation de navires		351
..	..	..	2	..	..	..	1	Matériel ferroviaire roulant		352
..	..	..	..	..	..	..	..	Construction aéronautique et spatiale		353
..	..	..	..	..	..	..	..	Autres équipements de transport		359
..	..	211	129	220	251	271	322	Meubles ; activités de fabrication, nca		36
..	..	..	96	168	169	191	243	Meubles		361
..	..	..	34	36	..	..	..	Activités de fabrication, nca		369
..	..	..	..	23	42	65	117	Récupération		37
..	..	2 406	3 465	2 040	4 390	1 908	1 567	Electricité, gaz et eau	E	40_41
..	..	2 282	3 285	1 866	4 214	1 715	1 368	Prod. et dist. d'électricité gaz, vapeur d'eau chaude		40
..	..	124	179	173	176	193	199	Captage, épuration et distribution d'eau		41
..	..	1 106	1 027	2 084	2 567	2 821	2 945	Construction	F	45

Note:When available: ISIC 241 includes 243; 292 includes 293; 332includes part of 331 and 333; 34 includes 353; 352 includes part of 359.

NETHERLANDS

Table NL.5 **EMPLOYMENT, EMPLOYEES - EMPLOI, SALARIÉS**

ISIC revision 3				1991	1992	1993	1994	1995	1996	1997	1998
C	10_14		Mining and quarrying	..	..	..	..	..	..	..	..
	10		Mining of coal and lignite; extraction of peat	..	..	..	..	..	..	..	..
	11		Crude petroleum & natural gas; service activities	..	..	..	..	..	..	..	..
	13		Mining of metal ores	..	..	..	..	..	..	..	..
	14		Other mining and quarrying	..	..	807	1 005	1 011	1 308	1 697	1 724
D	15_37		Total manufacturing	..	765 848	724 000	694 000	681 000	666 793	660 650	667 033
	15		Food products and beverages	..	121 135	119 000	114 000	109 000	106 931	105 914	106 491
		151	Production of meat, fish, fruit, vegetables, oils & fats	..	38 618	37 361	35 433	33 685	33 604	33 882	35 297
		152	Dairy products	..	16 875	16 646	15 474	14 272	11 900	11 787	11 409
		153	Grain mill and starch products; animal feeds	..	..	..	..	..	..	..	..
		154	Other food products	..	..	..	..	..	..	..	..
		155	Beverages	..	11 403	11 217	11 172	10 976	11 231	11 290	11 016
	16		Tobacco products	..	5 923	5 740	5 524	5 452	5 424	5 391	5 354
	17		Textiles	..	19 566	18 730	17 413	16 499	15 141	14 508	14 345
		171	Spinning, weaving and finishing of textiles	..	8 814	8 118	7 203	6 254	5 515	5 173	5 133
		172	Other textiles	..	..	..	..	..	..	..	..
		173	Knitted and crocheted fabrics and articles	..	..	..	..	..	..	..	..
	18		Wearing apparel and fur	..	6 891	6 216	5 910	5 062	4 076	3 649	3 831
		181	Wearing apparel, except fur apparel	..	6 891	6 216	5 910	5 062	..	..	..
		182	Dressing and dyeing of fur; articles of fur	..	..	..	..	..	..	..	..
	19		Leather and articles of leather; footwear	..	3 909	3 473	2 767	2 453	2 091	1 978	2 092
		191	Leather and articles of leather	..	1 530	1 358	1 080	918	874	777	886
		192	Footwear	..	2 379	2 115	1 687	1 535	1 217	1 201	1 206
	20		Wood & products of wood & cork, exc. furniture	..	11 297	11 061	10 716	11 379	10 666	10 617	11 125
		201	Sawmilling and planing of wood	..	..	..	..	..	..	..	..
		202	Wood products; articles of cork, straw, etc.	..	..	..	..	..	..	..	..
	21		Paper and paper products	..	25 329	23 976	22 542	22 838	22 467	23 140	25 098
	22		Printing, publishing and recorded media	..	64 428	63 806	61 058	59 051	58 362	57 141	56 130
		221	Publishing	..	23 197	27 937	26 556	26 947	26 837	26 835	26 773
		222	Printing and related activities	..	39 159	33 559	32 061	30 395	29 920	28 612	27 375
		223	Reproduction of recorded media	..	2 072	2 310	2 441	1 709	1 605	1 694	1 982
	23		Coke and petroleum products; nuclear fuel	..	7 349	7 653	7 210	6 984	6 812	6 619	6 630
		231	Coke oven products	..	..	..	..	..	..	..	..
		232	Refined petroleum products	..	..	..	..	..	..	..	..
		233	Processing of nuclear fuel	..	..	..	..	..	..	..	..
	24		Chemicals and chemical products	..	86 880	79 333	76 407	73 009	72 131	72 165	68 436
		241	Basic chemicals	..	44 544	37 097	35 192	33 459	35 362	30 995	29 455
		242	Other chemical products	..	..	..	..	..	..	..	..
		243	Man-made fibres	..	..	..	..	..	..	..	..
	25		Rubber and plastic products	..	30 554	29 349	28 428	27 156	27 006	27 606	27 936
		251	Rubber products	..	4 854	4 703	4 311	3 649	3 764	3 740	3 562
		252	Plastic products	..	25 700	24 646	24 117	23 507	23 242	23 866	24 374
	26		Other non-metallic mineral products	..	29 085	27 910	27 310	27 351	26 961	26 351	27 070
		261	Glass and glass products	..	5 697	5 977	5 861	5 664	5 682	5 495	5 766
		269	Non-metallic mineral products, nec	..	23 388	21 933	21 449	21 687	21 279	20 856	21 304
	27		Basic metals	..	29 580	26 605	25 966	25 994	24 626	24 337	24 348
		271	Basic iron and steel	..	..	..	..	..	..	..	..
		272	Basic precious and non-ferrous metals	..	..	..	..	..	..	..	..
		273	Casting of metals	..	4 839	4 505	4 441	4 517	4 494	4 209	4 253
	28		Fabricated metal products, exc. machinery	..	71 457	67 488	65 721	65 691	64 823	66 299	69 646
		281	Structural metal prod., tanks, reservoirs, etc.	..	37 307	35 701	34 004	33 937	32 959	34 292	35 925
		289	Other fabricated metal products	..	34 150	31 787	31 717	31 754	31 864	32 007	33 721
	29		Machinery and equipment, nec	..	72 797	69 130	67 793	68 140	68 665	68 124	70 949
		291	General purpose machinery	..	41 490	39 346	38 191	37 946	37 762	37 512	39 365
		292	Special purpose machinery	..	29 304	27 809	27 994	28 508	29 087	28 583	29 563
		293	Domestic appliances, nec	..	2 003	1 975	1 608	1 686	1 816	2 029	2 021
	30		Office, accounting & computing machinery	..	8 949	7 128	6 110	7 221	6 963	6 410	6 451
	31		Electrical machinery & apparatus, nec	..	..	..	..	..	..	..	..
		311	Electric motors, generators and transformers	..	4 332	4 213	2 679	2 239	2 203	2 264	2 875
		312	Electricity distribution & control apparatus	..	5 583	5 114	6 840	6 723	6 588	6 834	6 261
		313	Insulated wire and cable	..	2 768	2 747	2 554	3 130	3 100	2 943	2 945
		314	Accumulators, primary cells & primary batteries	..	..	..	..	..	..	..	..
		315	Electric lamps and lighting equipment	..	..	..	..	..	..	..	..
		319	Other electrical equipment, nec	..	..	..	..	..	..	..	..
	32		Radio, TV, communication equipment	..	..	..	..	..	..	..	..
		321	Electronic valves, tubes; other electronic components	..	..	..	..	..	..	..	..
		322	TV/radio transmitters; line teleph./telegr. apparatus	..	..	..	..	..	..	..	..
		323	TV/radio receivers; sound and video app.; assoc. gds.	..	..	..	..	..	..	..	..
	33		Medical, precision & optical instrum.; watch & clocks	..	..	..	..	..	..	..	..
		331	Medical appl.; instruments & appl. for measuring, etc.	..	..	..	..	..	..	..	..
		332	Optical instruments and photographic equipment	..	..	..	..	..	..	..	..
		333	Watches and clocks	..	..	..	..	..	..	..	..
	34		Motor vehicles, trailers and semi-trailers	..	..	..	..	..	..	..	..
		341	Motor vehicles	..	..	..	..	..	..	..	..
		342	Bodies for motor vehicles, trailers & semi-trailers	..	7 188	6 549	6 037	6 540	6 321	6 205	6 427
		343	Parts & accessories for motor vehicles & their engines	..	3 417	3 374	3 400	3 996	4 304	4 685	4 837
	35		Other transport equipment	..	..	..	..	..	..	..	..
		351	Building and repairing of ships and boats	..	15 156	13 980	13 425	12 696	12 104	11 556	11 903
		352	Railway and tramway locomotives and rolling stock	..	..	..	..	..	..	..	..
		353	Aircraft and spacecraft	..	..	..	..	..	..	..	..
		359	Transport equipment, nec	..	2 536	2 344	2 381	2 264	2 235	2 709	2 962
	36		Furniture; manufacturing, nec	..	21 774	21 627	20 309	20 856	20 433	21 067	21 359
		361	Furniture	..	18 081	17 854	17 014	17 065	16 631	17 035	17 406
		369	Manufacturing, nec	..	3 693	3 773	3 295	3 791	3 802	4 032	3 953
	37		Recycling	..	..	..	110	342	888	1 173	1 253
E	40_41		Electricity, gas and water supply	..	..	41 803	40 334	39 079	..	..	..
	40		Electricity, gas, steam and hot water supply	..	..	34 387	32 988	31 846	..	..	..
	41		Collection, purification and distribution of water	..	..	7 416	7 346	7 233	..	..	..
F	45		Construction	..	..	372	371	373	..	..	..

Table NL.6 **SALAIRES ET CHARGES SOCIALES, SALARIÉS - COMPENSATION OF LABOUR, EMPLOYEES**

Millions de Gld (Prix courants)

1991	1992	1993	1994	1995	1996	1997	1998		CITI révision 3		
..	..	..	..	..	..	..	..	Activites extractives	C	10_14	
..	..	..	..	..	..	..	..	Extraction de charbon, de lignite et de tourbe		10	
..	..	..	..	..	..	..	..	Pétrole brut et gaz naturel ; activités annexes		11	
..	..	..	..	..	..	..	..	Extraction de minerais métalliques		13	
..	..	67	88	82	106	149	151	Autres activités extractives		14	
..	53 548	53 212	51 847	52 092	52 578	53 616	55 757	Total activites manufacturieres	D	15_37	
..	8 152	8 566	8 501	8 269	8 355	8 373	8 601	Produits alimentaires et boissons		15	
..	2 286	2 394	2 332	2 266	2 300	2 374	2 527	Viande, poisson, fruits, légumes, huiles et graisses			151
..	1 409	1 470	1 412	1 324	1 169	1 144	1 082	Produits laitiers			152
..	..	..	..	..	..	..	..	Travail des grains; prod. amyl. ; alim. pour animaux			153
..	..	..	..	..	..	..	..	Autres produits alimentaires			154
..	956	1 010	1 072	1 039	1 108	1 074	1 142	Boissons			155
..	487	526	513	519	525	544	553	Produits à base de tabac		16	
..	1 173	1 166	1 089	1 041	999	999	1 020	Textiles		17	
..	566	543	492	431	405	375	391	Filature, tissage et achèvement des textiles			171
..	..	..	..	..	..	..	..	Autres articles textiles			172
..	..	..	..	..	..	..	..	Etoffes et articles de bonneterie			173
..	312	301	297	266	218	202	212	Articles d'habillement et fourrure		18	
..	312	301	297	266	..	..	..	Articles d'habillement autres qu'en fourrure			181
..	..	..	..	..	..	..	..	Prép. et teinture des fourrures ; articles en fourrure			182
..	200	181	151	142	123	123	133	Cuir et articles en cuir ; chaussures		19	
..	80	72	62	55	52	52	57	Cuir et articles en cuir sauf chaussures			191
..	120	109	89	87	71	71	76	Chaussures			192
..	646	657	644	699	667	694	752	Bois, articles en bois, sauf meubles		20	
..	..	..	..	..	..	..	..	Sciage et rabotage du bois			201
..	..	..	..	..	..	..	..	Articles en bois, liège, vannerie, sparterie			202
..	1 799	1 775	1 746	1 771	1 806	1 887	2 110	Papier, articles en papier et carton		21	
..	4 712	4 649	4 640	4 628	4 789	4 802	4 873	Edition, imprimerie et reproduction		22	
..	1 777	2 107	2 142	2 245	2 353	2 415	2 520	Edition			221
..	2 765	2 376	2 321	2 270	2 304	2 260	2 209	Imprimerie et activités annexes			222
..	169	167	177	114	132	127	144	Reproduction de supports enregistrés			223
..	834	1 032	881	868	884	897	934	Coke. et prod. pétroliers ; comb. nucléaires		23	
..	..	..	..	..	..	..	..	Cokéfaction			231
..	..	..	..	..	..	..	..	Produits pétroliers raffinés			232
..	..	..	..	..	..	..	..	Traitement de combustibles nucléaires			233
..	7 432	7 529	7 068	7 001	7 064	7 248	7 068	Produits chimiques		24	
..	4 141	4 054	3 547	3 591	3 930	3 584	3 466	Produits chimiques de base			241
..	..	..	..	..	..	..	..	Autres produits chimiques			242
..	..	..	..	..	..	..	..	Fibres synthétiques ou artificielles			243
..	1 941	1 941	1 941	1 902	1 939	2 020	2 146	Art. en caoutchouc et matières plastiques		25	
..	322	336	317	273	286	278	287	Articles en caoutchouc			251
..	1 619	1 605	1 624	1 630	1 653	1 742	1 859	Articles en matières plastiques			252
..	1 966	1 942	1 984	2 039	2 040	2 036	2 174	Autres produits minéraux non métalliques		26	
..	409	443	456	472	459	444	472	Verre et articles en verre			261
..	1 557	1 499	1 528	1 567	1 581	1 592	1 702	Produits minéraux non métalliques, nca			269
..	2 350	2 254	2 128	2 219	2 175	2 268	2 326	Produits métallurgiques de base		27	
..	..	..	..	..	..	..	..	Sidérurgie et première transformation de l'acier			271
..	..	..	..	..	..	..	..	Première transf. métaux précieux et non ferreux			272
..	292	276	290	296	305	297	316	Fonderie			273
..	4 416	4 314	4 293	4 398	4 466	4 693	5 090	Ouvrages en métaux, sauf machines		28	
..	2 305	2 276	2 247	2 283	2 292	2 440	2 649	Constr. et menuis. métal.; réservoirs; générateurs			281
..	2 111	2 039	2 045	2 114	2 174	2 253	2 441	Autres ouvrages en métaux			289
..	4 862	4 733	4 770	4 915	5 060	5 148	5 652	Machines et matériels, nca		29	
..	2 765	2 717	2 708	2 737	2 768	2 837	3 139	Machines d'usage général			291
..	1 983	1 905	1 964	2 079	2 183	2 184	2 377	Machines d'usage spécifique			292
..	115	111	99	98	109	127	136	Appareils domestiques, nca			293
..	607	535	491	563	550	556	590	Mach. de bureau, comptables, information		30	
..	..	..	..	..	..	..	..	Machines et appareils électriques, nca		31	
..	275	293	181	151	154	161	223	Moteurs, génératrices, transformateurs électriques			311
..	393	359	490	491	515	544	500	Matériel électrique de distribution et de commande			312
..	211	208	207	238	262	256	257	Fils et câbles électriques isolés			313
..	..	..	..	..	..	..	..	Accumulateurs et piles électriques			314
..	..	..	..	..	..	..	..	Lampes électriques et appareils d'éclairage			315
..	..	..	..	..	..	..	..	Autres matériels électriques, nca.			319
..	..	..	..	..	..	..	..	Equipements et appareils de radio, TV, etc.		32	
..	..	..	..	..	..	..	..	Tubes et valves électroniques;autres composants			321
..	..	..	..	..	..	..	..	Emetteurs radio/TV; app. téléphonie et télégraphie			322
..	..	..	..	..	..	..	..	Récepteurs radio/TV, son et image; art. associés			323
..	..	..	..	..	..	..	..	Instr. médicaux, de précision ; optique ; horlog.		33	
..	..	..	..	..	..	..	..	Instruments médicaux et de précision, etc.			331
..	..	..	..	..	..	..	..	Instruments d'optique ; matériel photographique			332
..	..	..	..	..	..	..	..	Horlogerie			333
..	..	..	..	..	..	..	..	Véhicules automobiles; remorques, etc.		34	
..	..	..	..	..	..	..	..	Véhicules automobiles			341
..	417	402	364	407	407	395	437	Carross. automobiles, remorques, semi-remorques			342
..	204	207	216	262	296	335	349	Pièces et accessoires pour automobiles et moteurs			343
..	..	..	..	..	..	..	..	Autres matériels de transport		35	
..	1 026	974	959	929	909	888	940	Construction et réparation de navires			351
..	..	..	..	..	..	..	..	Matériel ferroviaire roulant			352
..	..	..	..	..	..	..	..	Construction aéronautique et spatiale			353
..	136	135	140	142	146	173	190	Autres équipements de transport			359
..	1 253	1 283	1 223	1 282	1 261	1 347	1 423	Meubles ; activités de fabrication, nca		36	
..	1 029	1 038	1 008	1 041	1 015	1 081	1 145	Meubles			361
..	225	245	214	241	246	266	278	Activités de fabrication, nca			369
..		..	5	23	63	82	94	Récupération		37	
..	..	..	..	..	..	..	..	Electricité, gaz et eau	E	40_41	
..	..	..	..	..	..	..	..	Prod. et dist. d'électricité gaz, vapeur d'eau chaude		40	
..	..	..	..	..	..	..	..	Captage, épuration et distribution d'eau		41	
..	..	23 401	23 467	23 851	..	..	..	Construction	F	45	

NETHERLANDS

Table NL.7 **WAGES AND SALARIES, EMPLOYEES - SALAIRES ET TRAITEMENTS, SALARIÉS**

Millions of Gld (Current Prices)

ISIC revision 3				1991	1992	1993	1994	1995	1996	1997	1998
C	10_14		Mining and quarrying	..	..	..	..	..	..	..	..
	10		Mining of coal and lignite; extraction of peat	..	..	..	..	..	..	..	..
	11		Crude petroleum & natural gas; service activities	..	..	..	..	..	..	..	..
	13		Mining of metal ores	..	..	..	..	..	..	..	..
	14		Other mining and quarrying	..	..	54	73	67	88	123	120
D	15_37		Total manufacturing	..	44 844	43 604	43 803	44 292	44 405	45 434	45 106
	15		Food products and beverages	..	6 703	7 004	6 969	6 829	6 859	6 904	6 797
		151	Production of meat, fish, fruit, vegetables, oils & fats	..	1 927	2 012	1 994	1 943	1 959	2 036	2 055
		152	Dairy products	..	1 097	1 149	1 085	1 013	901	878	787
		153	Grain mill and starch products; animal feeds	..	..	..	..	..	..	..	..
		154	Other food products	..	..	..	..	..	..	..	..
		155	Beverages	..	807	819	843	848	889	839	875
	16		Tobacco products	..	390	432	407	431	437	440	440
	17		Textiles	..	945	961	909	878	838	848	818
		171	Spinning, weaving and finishing of textiles	..	449	447	409	362	335	317	310
		172	Other textiles	..	..	..	..	..	..	..	..
		173	Knitted and crocheted fabrics and articles	..	..	..	..	..	..	..	..
	18		Wearing apparel and fur	..	264	253	253	227	187	175	175
		181	Wearing apparel, except fur apparel	..	264	253	253	227	..	..	..
		182	Dressing and dyeing of fur; articles of fur	..					..	..	..
	19		Leather and articles of leather; footwear	..	164	148	127	122	106	106	110
		191	Leather and articles of leather	..	64	59	52	47	44	44	48
		192	Footwear	..	99	89	75	75	62	62	62
	20		Wood & products of wood & cork, exc. furniture	..	535	541	544	598	572	600	613
		201	Sawmilling and planing of wood	..	..	..	..	..	..	..	..
		202	Wood products; articles of cork, straw, etc.	..	..	..	..	..	..	..	..
	21		Paper and paper products	..	1 516	1 482	1 482	1 522	1 542	1 627	1 715
	22		Printing, publishing and recorded media	..	3 910	3 887	3 930	3 943	4 069	4 066	3 940
		221	Publishing	..	1 502	1 795	1 830	1 924	2 014	2 046	2 046
		222	Printing and related activities	..	2 269	1 950	1 950	1 920	1 938	1 908	1 774
		223	Reproduction of recorded media	..	140	142	151	99	117	112	120
	23		Coke and petroleum products; nuclear fuel	..	..	737	722	705	717	749	742
		231	Coke oven products	..	..	..	..	..	..	..	..
		232	Refined petroleum products	..	..	..	..	..	..	..	..
		233	Processing of nuclear fuel	..	..	..	..	..	..	..	..
	24		Chemicals and chemical products	..	6 180	5 980	5 875	5 826	5 828	6 012	5 674
		241	Basic chemicals	..	3 406	3 112	2 955	2 990	3 225	2 954	2 807
		242	Other chemical products	..	..	..	..	..	..	..	..
		243	Man-made fibres	..	..	..	..	..	..	..	..
	25		Rubber and plastic products	..	1 647	1 646	1 677	1 641	1 666	1 742	1 744
		251	Rubber products	..	274	283	273	235	239	236	230
		252	Plastic products	..	1 373	1 363	1 404	1 407	1 427	1 506	1 514
	26		Other non-metallic mineral products	..	1 620	1 609	1 672	1 743	1 721	1 724	1 738
		261	Glass and glass products	..	341	373	384	411	387	379	395
		269	Non-metallic mineral products, nec	..	1 279	1 236	1 288	1 332	1 334	1 345	1 343
	27		Basic metals	..	1 904	1 767	1 799	1 826	1 783	1 808	1 769
		271	Basic iron and steel	..	..	..	..	..	..	..	..
		272	Basic precious and non-ferrous metals	..	..	..	..	..	..	..	..
		273	Casting of metals	..	245	231	251	259	265	259	262
	28		Fabricated metal products, exc. machinery	..	3 715	3 606	3 701	3 819	3 857	4 080	4 188
		281	Structural metal prod., tanks, reservoirs, etc.	..	1 944	1 912	1 945	1 988	1 979	2 128	2 186
		289	Other fabricated metal products	..	1 770	1 694	1 757	1 832	1 877	1 952	2 002
	29		Machinery and equipment, nec	..	4 116	3 984	4 097	4 234	4 321	4 407	4 611
		291	General purpose machinery	..	2 334	2 290	2 324	2 354	2 361	2 432	2 570
		292	Special purpose machinery	..	1 685	1 602	1 689	1 796	1 867	1 865	1 929
		293	Domestic appliances, nec	..	97	93	85	85	93	110	112
	30		Office, accounting & computing machinery	..	517	439	431	507	477	487	490
	31		Electrical machinery & apparatus, nec	..	..	..	..	..	..	..	..
		311	Electric motors, generators and transformers	..	236	244	158	132	134	139	183
		312	Electricity distribution & control apparatus	..	340	307	416	416	438	460	396
		313	Insulated wire and cable	..	185	180	181	202	220	215	213
		314	Accumulators, primary cells & primary batteries	..	..	..	..	..	..	..	..
		315	Electric lamps and lighting equipment	..	..	..	..	..	..	..	..
		319	Other electrical equipment, nec	..	..	..	..	..	..	..	..
	32		Radio, TV, communication equipment	..	..	..	..	..	..	..	..
		321	Electronic valves, tubes; other electronic components	..	..	..	..	..	..	..	..
		322	TV/radio transmitters; line teleph./telegr. apparatus	..	..	..	..	..	..	..	..
		323	TV/radio receivers; sound and video app.; assoc. gds.	..	..	..	..	..	..	..	..
	33		Medical, precision & optical instrum.; watch & clocks	..	..	..	..	..	..	..	..
		331	Medical appl.; instruments & appl. for measuring, etc.	..	..	..	..	..	..	..	..
		332	Optical instruments and photographic equipment	..	..	..	..	..	..	..	..
		333	Watches and clocks	..	..	..	..	..	..	..	..
	34		Motor vehicles, trailers and semi-trailers	..	..	..	..	..	..	..	..
		341	Motor vehicles	..	..	..	..	..	..	..	..
		342	Bodies for motor vehicles, trailers & semi-trailers	..	356	340	317	357	355	346	360
		343	Parts & accessories for motor vehicles & their engines	..	170	171	187	227	255	287	284
	35		Other transport equipment	..	..	..	..	..	..	..	..
		351	Building and repairing of ships and boats	..	866	813	819	796	773	762	765
		352	Railway and tramway locomotives and rolling stock	..	..	..	..	..	..	..	..
		353	Aircraft and spacecraft	..	..	..	..	..	..	..	..
		359	Transport equipment, nec	..	116	114	121	123	127	151	159
	36		Furniture; manufacturing, nec	..	1 040	1 059	1 042	1 106	1 091	1 170	1 173
		361	Furniture	..	855	861	862	901	882	942	945
		369	Manufacturing, nec	..	185	198	180	205	209	228	228
	37		Recycling	..		..	4	21	56	75	78
E	40_41		Electricity, gas and water supply	..	..	2 949	2 967	2 992	..	..	..
	40		Electricity, gas, steam and hot water supply	..	..	2 454	2 469	2 486	..	..	..
	41		Collection, purification and distribution of water	..	..	495	498	506	..	..	..
F	45		Construction	..	..	17 278	17 327	17 610	..	..	..

Table NL.8 CHARGES SOCIALES DES EMPLOYEURS - SUPPLEMENTS TO WAGES & SALARIES

Millions de Gld (Prix courants)

1991	1992	1993	1994	1995	1996	1997	1998			CITI révision 3	
..	..	..	..	..	..	..	..	Activites extractives	C	10_14	
..	..	..	..	..	..	..	..	Extraction de charbon, de lignite et de tourbe		10	
..	..	..	..	..	..	..	..	Pétrole brut et gaz naturel ; activités annexes		11	
..	..	..	..	..	..	..	..	Extraction de minerais métalliques		13	
..	..	13	15	15	18	26	31	Autres activités extractives		14	
..	8 704	9 379	8 041	7 802	8 173	8 182	10 651	Total activites manufacturieres	D	15_37	
..	1 449	1 562	1 533	1 441	1 497	1 469	1 804	Produits alimentaires et boissons		15	
..	359	382	337	322	341	339	472	Viande, poisson, fruits, légumes, huiles et graisses			151
..	312	321	327	311	268	266	295	Produits laitiers			152
..	..	..	..	..	..	..	..	Travail des grains; prod. amyl. ; alim. pour animaux			153
..	..	..	..	..	..	..	..	Autres produits alimentaires			154
..	149	190	229	191	219	234	267	Boissons			155
..	98	94	105	88	88	104	113	Produits à base de tabac		16	
..	228	205	180	163	161	151	202	Textiles		17	
..	117	96	83	69	70	59	82	Filature, tissage et achèvement des textiles			171
..	..	..	..	..	..	..	..	Autres articles textiles			172
..	..	..	..	..	..	..	..	Etoffes et articles de bonneterie			173
..	48	48	44	39	32	27	37	Articles d'habillement et fourrure		18	
..	48	48	44	39	..	..	..	Articles d'habillement autres qu'en fourrure			181
..					..	..	..	Prép. et teinture des fourrures ; articles en fourrure			182
..	36	33	24	20	17	16	23	Cuir et articles en cuir ; chaussures		19	
..	15	13	10	8	7	7	10	Cuir et articles en cuir sauf chaussures			191
..	21	20	14	12	10	9	13	Chaussures			192
..	111	116	100	101	96	94	139	Bois, articles en bois, sauf meubles		20	
..	..	..	..	..	..	..	..	Sciage et rabotage du bois			201
..	..	..	..	..	..	..	..	Articles en bois, liège, vannerie, sparterie			202
..	283	294	263	249	264	260	396	Papier, articles en papier et carton		21	
..	802	763	709	686	720	736	933	Edition, imprimerie et reproduction		22	
..	275	312	312	321	339	369	474	Edition			221
..	496	426	371	350	365	352	436	Imprimerie et activités annexes			222
..	29	25	26	15	15	15	23	Reproduction de supports enregistrés			223
..	..	295	159	163	167	149	192	Coke. et prod. pétroliers ; comb. nucléaires		23	
..	..	..	..	..	..	..	..	Cokéfaction			231
..	..	..	..	..	..	..	..	Produits pétroliers raffinés			232
..	..	..	..	..	..	..	..	Traitement de combustibles nucléaires			233
..	1 252	1 549	1 192	1 176	1 235	1 236	1 395	Produits chimiques		24	
..	735	941	592	601	705	630	659	Produits chimiques de base			241
..	..	..	..	..	..	..	..	Autres produits chimiques			242
..	..	..	..	..	..	..	..	Fibres synthétiques ou artificielles			243
..	294	295	264	261	273	277	402	Art. en caoutchouc et matières plastiques		25	
..	48	54	43	38	47	41	58	Articles en caoutchouc			251
..	246	242	221	223	226	236	345	Articles en matières plastiques			252
..	346	333	312	296	319	312	436	Autres produits minéraux non métalliques		26	
..	68	69	72	60	73	65	77	Verre et articles en verre			261
..	278	264	240	236	246	247	359	Produits minéraux non métalliques, nca			269
..	446	487	330	393	392	460	557	Produits métallurgiques de base		27	
..	..	..	..	..	..	..	..	Sidérurgie et première transformation de l'acier			271
..	..	..	..	..	..	..	..	Première transf. métaux précieux et non ferreux			272
..	47	45	39	37	40	37	54	Fonderie			273
..	701	708	592	579	609	613	903	Ouvrages en métaux, sauf machines		28	
..	361	364	303	296	312	312	464	Constr. et menuis. métal.; réservoirs; générateurs			281
..	341	344	289	283	297	301	438	Autres ouvrages en métaux			289
..	746	749	673	680	739	740	1 041	Machines et matériels, nca		29	
..	431	428	382	385	408	405	569	Machines d'usage général			291
..	298	303	277	281	315	318	448	Machines d'usage spécifique			292
..	18	18	14	14	16	17	24	Appareils domestiques, nca			293
..	90	96	59	56	73	69	100	Mach. de bureau, comptables, information		30	
..	..	..	..	..	..	..	..	Machines et appareils électriques, nca		31	
..	39	49	23	19	20	22	40	Moteurs, génératrices, transformateurs électriques			311
..	53	52	74	75	77	84	103	Matériel électrique de distribution et de commande			312
..	26	29	26	36	42	41	44	Fils et câbles électriques isolés			313
..	..	..	..	..	..	..	..	Accumulateurs et piles électriques			314
..	..	..	..	..	..	..	..	Lampes électriques et appareils d'éclairage			315
..	..	..	..	..	..	..	..	Autres matériels électriques, nca.			319
..	..	..	..	..	..	..	..	Equipements et appareils de radio, TV, etc.		32	
..	..	..	..	..	..	..	..	Tubes et valves électroniques;autres composants			321
..	..	..	..	..	..	..	..	Emetteurs radio/TV; app. téléphonie et télégraphie			322
..	..	..	..	..	..	..	..	Récepteurs radio/TV, son et image; art. associés			323
..	..	..	..	..	..	..	..	Instr. médicaux, de précision ; optique ; horlog.		33	
..	..	..	..	..	..	..	..	Instruments médicaux et de précision, etc.			331
..	..	..	..	..	..	..	..	Instruments d'optique ; matériel photographique			332
..	..	..	..	..	..	..	..	Horlogerie			333
..	..	..	..	..	..	..	..	Véhicules automobiles; remorques, etc.		34	
..	..	..	..	..	..	..	..	Véhicules automobiles			341
..	61	62	47	50	52	49	76	Carross. automobiles, remorques, semi-remorques			342
..	34	36	29	34	41	48	64	Pièces et accessoires pour automobiles et moteurs			343
..	..	..	..	..	..	..	..	Autres matériels de transport		35	
..	160	161	140	133	136	126	175	Construction et réparation de navires			351
..	..	..	..	..	..	..	..	Matériel ferroviaire roulant			352
..	..	..	..	..	..	..	..	Construction aéronautique et spatiale			353
..	20	21	19	19	19	22	32	Autres équipements de transport			359
..	213	224	181	177	170	177	250	Meubles ; activités de fabrication, nca		36	
..	174	177	146	140	133	140	200	Meubles			361
..	40	47	35	37	37	37	50	Activités de fabrication, nca			369
..		..	1	2	7	8	15	Récupération		37	
..	..	453	457	506	..	..	..	Electricité, gaz et eau	E	40_41	
..	..	380	378	418	..	..	..	Prod. et dist. d'électricité gaz, vapeur d'eau chaude		40	
..	..	73	79	88	..	..	..	Captage, épuration et distribution d'eau		41	
..	..	6 126	6 143	6 244	..	..	..	Construction	F	45	

Table NL.9 **ENTERPRISES - ENTREPRISES**

Units

ISIC revision 3				1991	1992	1993	1994	1995	1996	1997	1998
C	10_14		Mining and quarrying	..	..	..	..	..	..	..	..
	10		Mining of coal and lignite; extraction of peat	..	..	..	..	..	..	..	..
	11		Crude petroleum & natural gas; service activities	..	..	..	..	..	..	..	..
	13		Mining of metal ores	..	..	..	..	..	..	..	..
	14		Other mining and quarrying	..	..	23	29	29	32	39	36
D	15_37		Total manufacturing	..	6 601	6 681	6 716	6 404	6 212	6 350	6 433
	15		Food products and beverages	..	884	901	922	887	870	871	880
		151	Production of meat, fish, fruit, vegetables, oils & fats	..	312	313	318	319	306	321	322
		152	Dairy products	..	38	40	37	39	38	38	43
		153	Grain mill and starch products; animal feeds	..	..	..	..	..	..	..	..
		154	Other food products	..	..	..	..	..	..	..	..
		155	Beverages	..	34	34	34	35	33	33	34
	16		Tobacco products	..	15	13	12	12	11	11	11
	17		Textiles	..	221	224	222	224	201	187	178
		171	Spinning, weaving and finishing of textiles	..	78	77	70	64	59	54	50
		172	Other textiles	..	..	..	..	..	..	..	..
		173	Knitted and crocheted fabrics and articles	..	..	..	..	..	..	..	..
	18		Wearing apparel and fur	..	124	108	126	92	83	71	77
		181	Wearing apparel, except fur apparel	..	124	108	126	92	..	..	..
		182	Dressing and dyeing of fur; articles of fur	..					..	..	..
	19		Leather and articles of leather; footwear	..	74	72	63	53	46	44	41
		191	Leather and articles of leather	..	34	31	27	24	23	20	20
		192	Footwear	..	40	41	36	29	23	24	21
	20		Wood & products of wood & cork, exc. furniture	..	193	199	183	192	183	185	194
		201	Sawmilling and planing of wood	..	..	..	..	..	..	..	..
		202	Wood products; articles of cork, straw, etc.	..	..	..	..	..	..	..	..
	21		Paper and paper products	..	165	156	156	166	159	176	203
	22		Printing, publishing and recorded media	..	761	753	754	693	674	658	654
		221	Publishing	..	170	173	162	157	150	161	162
		222	Printing and related activities	..	578	565	578	523	510	484	477
		223	Reproduction of recorded media	..	13	15	14	13	14	13	15
	23		Coke and petroleum products; nuclear fuel	..	18	19	18	18	17	19	21
		231	Coke oven products	..	..	..	..	..	..	..	..
		232	Refined petroleum products	..	..	..	..	..	..	..	..
		233	Processing of nuclear fuel	..	..	..	..	..	..	..	..
	24		Chemicals and chemical products	..	282	306	309	304	308	323	327
		241	Basic chemicals	..	101	121	124	118	122	113	116
		242	Other chemical products	..	..	..	..	..	..	..	..
		243	Man-made fibres	..	..	..	..	..	..	..	..
	25		Rubber and plastic products	..	327	336	344	344	344	348	351
		251	Rubber products	..	30	35	32	25	27	24	23
		252	Plastic products	..	297	301	312	319	317	324	328
	26		Other non-metallic mineral products	..	307	303	281	271	263	269	276
		261	Glass and glass products	..	25	29	34	32	30	30	35
		269	Non-metallic mineral products, nec	..	282	274	257	239	233	239	241
	27		Basic metals	..	84	88	97	87	85	88	84
		271	Basic iron and steel	..	..	..	..	..	..	..	..
		272	Basic precious and non-ferrous metals	..	..	..	..	..	..	..	..
		273	Casting of metals	..	45	47	47	45	45	42	40
	28		Fabricated metal products, exc. machinery	..	1 066	1 089	1 110	1 023	988	1 054	1 093
		281	Structural metal prod., tanks, reservoirs, etc.	..	581	593	585	525	509	544	561
		289	Other fabricated metal products	..	485	496	525	498	479	510	532
	29		Machinery and equipment, nec	..	944	953	986	918	890	891	915
		291	General purpose machinery	..	519	519	528	488	478	497	519
		292	Special purpose machinery	..	406	413	438	410	390	370	371
		293	Domestic appliances, nec	..	19	21	20	20	22	24	25
	30		Office, accounting & computing machinery	..	34	37	29	26	23	26	24
	31		Electrical machinery & apparatus, nec	..	..	..	..	..	..	..	..
		311	Electric motors, generators and transformers	..	50	52	44	41	40	41	43
		312	Electricity distribution & control apparatus	..	27	28	34	34	32	36	35
		313	Insulated wire and cable	..	10	11	13	17	15	17	14
		314	Accumulators, primary cells & primary batteries	..	..	..	..	..	..	..	..
		315	Electric lamps and lighting equipment	..	..	..	..	..	..	..	..
		319	Other electrical equipment, nec	..	..	..	..	..	..	..	..
	32		Radio, TV, communication equipment	..	..	..	..	..	..	..	..
		321	Electronic valves, tubes; other electronic components	..	..	..	..	..	..	..	..
		322	TV/radio transmitters; line teleph./telegr. apparatus	..	..	..	..	..	..	..	..
		323	TV/radio receivers; sound and video app.; assoc. gds.	..	..	..	..	..	..	..	..
	33		Medical, precision & optical instrum.; watch & clocks	..	..	..	..	..	..	..	..
		331	Medical appl.; instruments & appl. for measuring, etc.	..	..	..	..	..	..	..	..
		332	Optical instruments and photographic equipment	..	..	..	..	..	..	..	..
		333	Watches and clocks	..	..	..	..	..	..	..	..
	34		Motor vehicles, trailers and semi-trailers	..	..	..	..	..	..	..	..
		341	Motor vehicles	..	..	..	..	..	..	..	..
		342	Bodies for motor vehicles, trailers & semi-trailers	..	104	109	103	112	102	109	111
		343	Parts & accessories for motor vehicles & their engines	..	21	20	21	23	26	30	30
	35		Other transport equipment	..	..	..	..	..	..	..	..
		351	Building and repairing of ships and boats	..	143	142	137	137	132	140	130
		352	Railway and tramway locomotives and rolling stock	..	..	..	..	..	..	..	..
		353	Aircraft and spacecraft	..	..	..	..	..	..	..	..
		359	Transport equipment, nec	..	29	28	30	29	28	31	30
	36		Furniture; manufacturing, nec	..	384	409	393	383	360	370	355
		361	Furniture	..	321	343	333	319	303	309	295
		369	Manufacturing, nec	..	63	66	60	64	57	61	60
	37		Recycling	..	..	..	6	11	22	25	27
E	40_41		Electricity, gas and water supply	..	..	105	98	102	..	..	..
	40		Electricity, gas, steam and hot water supply	..	..	78	73	78	..	..	..
	41		Collection, purification and distribution of water	..	..	27	25	24	..	..	..
F	45		Construction	..	..	..	..	..	..	..	..

Table NL.10 **EXPORTATIONS - EXPORTS**

Millions de Gld (Prix courants)

1991	1992	1993	1994	1995	1996	1997	1998		CITI révision 3		
..	..	..	..	..	..	..	..	Activites extractives	C	10_14	
..	..	..	..	..	..	..	..	Extraction de charbon, de lignite et de tourbe		10	
..	..	..	..	..	..	..	..	Pétrole brut et gaz naturel ; activités annexes		11	
..	..	..	..	..	..	..	..	Extraction de minerais métalliques		13	
..	..	90	206	110	181	333	..	Autres activités extractives		14	
..	124 062	125 179	132 681	149 213	166 305	143 840	..	Total activites manufacturieres	D	15_37	
..	26 706	28 216	28 233	29 327	30 049	32 350	..	Produits alimentaires et boissons		15	
..	11 890	11 731	11 634	12 290	12 266	13 276	..	Viande, poisson, fruits, légumes, huiles et graisses			151
..	5 623	6 152	5 629	5 408	5 403	5 400	..	Produits laitiers			152
..	..	..	..	..	..	..	..	Travail des grains; prod. amyl. ; alim. pour animaux			153
..	..	..	..	..	..	..	..	Autres produits alimentaires			154
..	1 836	2 037	2 363	2 461	2 596	2 892	..	Boissons			155
..	3 256	3 267	3 711	4 053	4 621	5 274	..	Produits à base de tabac		16	
..	2 407	2 341	2 821	3 039	2 745	2 935	..	Textiles		17	
..	1 194	960	1 167	1 184	1 116	1 048	..	Filature, tissage et achèvement des textiles			171
..	..	..	..	..	..	..	..	Autres articles textiles			172
..	..	..	..	..	..	..	..	Etoffes et articles de bonneterie			173
..	456	508	484	445	352	331	..	Articles d'habillement et fourrure		18	
..	456	508	484	445	352	331	..	Articles d'habillement autres qu'en fourrure			181
..							..	Prép. et teinture des fourrures ; articles en fourrure			182
..	239	238	255	248	232	276	..	Cuir et articles en cuir ; chaussures		19	
..	161	162	182	162	165	205	..	Cuir et articles en cuir sauf chaussures			191
..	78	76	73	86	68	71	..	Chaussures			192
..	358	413	374	468	455	470	..	Bois, articles en bois, sauf meubles		20	
..	..	..	..	..	..	..	..	Sciage et rabotage du bois			201
..	..	..	..	..	..	..	..	Articles en bois, liège, vannerie, sparterie			202
..	3 766	3 613	4 431	5 016	4 755	5 259	..	Papier, articles en papier et carton		21	
..	2 329	2 580	2 410	2 357	2 527	2 786	..	Edition, imprimerie et reproduction		22	
..	980	1 172	1 024	..	988	1 165	..	Edition			221
..	802	772	791	841	868	914	..	Imprimerie et activités annexes			222
..	546	637	596	..	671	706	..	Reproduction de supports enregistrés			223
..	7 054	9 106	8 299	8 389	10 855	11 923	..	Coke. et prod. pétroliers ; comb. nucléaires		23	
..	..	..	..	..	..	..	..	Cokéfaction			231
..	..	..	..	..	..	..	..	Produits pétroliers raffinés			232
..	..	..	..	..	..	..	..	Traitement de combustibles nucléaires			233
..	29 055	29 147	33 700	37 981	37 290	43 077	..	Produits chimiques		24	
..	19 098	19 528	22 061	24 825	24 528	27 545	..	Produits chimiques de base			241
..	..	..	..	..	..	..	..	Autres produits chimiques			242
..	..	..	..	..	..	..	..	Fibres synthétiques ou artificielles			243
..	3 270	3 337	3 688	4 137	4 037	4 143	..	Art. en caoutchouc et matières plastiques		25	
..	618	535	616	602	602	582	..	Articles en caoutchouc			251
..	2 651	2 803	3 072	3 536	3 435	3 562	..	Articles en matières plastiques			252
..	1 798	1 714	1 935	2 028	2 027	2 041	..	Autres produits minéraux non métalliques		26	
..	732	675	746	819	822	833	..	Verre et articles en verre			261
..	1 066	..	..	1 209	1 205	1 208	..	Produits minéraux non métalliques, nca			269
..	6 107	5 824	6 924	7 872	7 395	8 270	..	Produits métallurgiques de base		27	
..	..	..	..	..	..	..	..	Sidérurgie et première transformation de l'acier			271
..	..	..	..	..	..	..	..	Première transf. métaux précieux et non ferreux			272
..	437	408	552	587	601	688	..	Fonderie			273
..	4 947	4 431	4 869	5 334	5 641	6 690	..	Ouvrages en métaux, sauf machines		28	
..	2 221	1 992	2 113	2 264	2 496	3 143	..	Constr. et menuis. métal.; réservoirs; générateurs			281
..	2 727	2 439	2 756	3 069	3 145	3 546	..	Autres ouvrages en métaux			289
..	7 959	7 835	8 190	9 816	10 551	12 063	..	Machines et matériels, nca		29	
..	4 245	4 467	4 180	4 606	5 087	5 838	..	Machines d'usage général			291
..	3 501	3 186	3 873	5 073	5 281	6 007	..	Machines d'usage spécifique			292
..	213	183	136	138	182	217	..	Appareils domestiques, nca			293
..	1 911	1 571	1 781	2 310	2 331	4 493	..	Mach. de bureau, comptables, information		30	
..	..	..	..	..	..	..	..	Machines et appareils électriques, nca		31	
..	385	420	264	210	222	198	..	Moteurs, génératrices, transformateurs électriques			311
..	539	444	582	754	761	890	..	Matériel électrique de distribution et de commande			312
..	194	183	226	287	396	468	..	Fils et câbles électriques isolés			313
..	..	..	..	..	..	..	..	Accumulateurs et piles électriques			314
..	..	..	..	..	..	..	..	Lampes électriques et appareils d'éclairage			315
..	..	..	..	..	..	..	..	Autres matériels électriques, nca.			319
..	..	..	..	..	..	..	..	Equipements et appareils de radio, TV, etc.		32	
..	..	..	..	..	..	..	..	Tubes et valves électroniques;autres composants			321
..	..	..	..	..	..	..	..	Emetteurs radio/TV; app. téléphonie et télégraphie			322
..	..	..	..	..	..	..	..	Récepteurs radio/TV, son et image; art. associés			323
..	..	..	..	..	..	..	..	Instr. médicaux, de précision ; optique ; horlog.		33	
..	..	..	..	..	..	..	..	Instruments médicaux et de précision, etc.			331
..	..	..	..	..	..	..	..	Instruments d'optique ; matériel photographique			332
..	..	..	..	..	..	..	..	Horlogerie			333
..	..	..	..	..	..	..	..	Véhicules automobiles; remorques, etc.		34	
..	..	..	..	..	..	..	..	Véhicules automobiles			341
..	446	427	494	502	622	573	..	Carross. automobiles, remorques, semi-remorques			342
..	478	600	683	888	1 114	1 244	..	Pièces et accessoires pour automobiles et moteurs			343
..	..	..	..	..	..	..	..	Autres matériels de transport		35	
..	2 028	1 692	1 927	2 004	2 034	2 163	..	Construction et réparation de navires			351
..	..	..	..	..	..	..	..	Matériel ferroviaire roulant			352
..	..	..	..	..	..	..	..	Construction aéronautique et spatiale			353
..	190	210	241	245	248	316	..	Autres équipements de transport			359
..	1 173	1 182	1 119	1 303	1 353	1 459	..	Meubles ; activités de fabrication, nca		36	
..	730	720	717	824	792	840	..	Meubles			361
..	444	462	402	479	561	619	..	Activités de fabrication, nca			369
..		..			25	..	..	Récupération		37	
..	..	..	..	..	..	..	..	Electricité, gaz et eau	E	40_41	
..	..	..	..	..	..	..	..	Prod. et dist. d'électricité gaz, vapeur d'eau chaude		40	
..	..	..	..	..	..	..	..	Captage, épuration et distribution d'eau		41	
..	..	..	..	..	..	..	..	Construction	F	45	

Table NZ.1 **EMPLOYMENT, EMPLOYEES - EMPLOI, SALARIÉS**

Units

ISIC revision 3				1990-91	1991-92	1992-93	1993-94	1994-95	1995-96	1996-97	1997-98
C	10_14		Mining and quarrying	3 830	3 730	3 450	3 750	3 990	4 020	3 500	3 164
	10		Mining of coal and lignite; extraction of peat	990	860	780	810	860	830	878	848
	11		Crude petroleum & natural gas; service activities	740	760	640	630	680	640	504	490
	12		Mining of uranium and thorium ores	..	..	..	..	..	..	..	..
	13		Mining of metal ores	470	620	730	600	640	780	801	509
	14		Other mining and quarrying	1 640	1 490	1 300	1 720	1 820	1 760	1 260	1 291
D	15_37		Total manufacturing	213 580	202 940	208 040	220 010	228 250	228 590	228 197	224 180
	15		Food products and beverages	58 270	56 600	57 890	57 570	56 760	58 650	62 553	62 641
		151	Production of meat, fish, fruit, vegetables, oils & fats	35 180	35 080	36 000	34 280	33 610	34 970	43 202	45 822
		152	Dairy products	8 270	7 550	7 950	7 700	7 740	8 070	6 859	7 072
		153	Grain mill and starch products; animal feeds	1 600	1 670	1 610	1 410	1 460	1 310	3 388	964
		154	Other food products	10 270	9 740	9 820	11 450	11 100	11 370	11 878	11 766
		155	Beverages	2 965	2 560	2 495	2 730	2 860	2 930	4 086	4 090
	16		Tobacco products	650	560	480	520	500	490	446	481
	17		Textiles	10 350	9 350	9 450	9 250	9 810	9 160	8 089	8 364
		171	Spinning, weaving and finishing of textiles	2 340	2 250	4 300	3 320	3 630	3 310	2 902	2 486
		172	Other textiles	5 230	4 660	3 140	3 860	4 090	4 130	3 655	4 353
		173	Knitted and crocheted fabrics and articles	2 770	2 420	2 000	2 070	2 080	1 730	1 532	1 526
	18		Wearing apparel and fur	12 220	11 170	11 340	11 510	11 220	10 070	9 896	8 698
		181	Wearing apparel, except fur apparel	..	..	..	..	..	..	9 376	8 218
		182	Dressing and dyeing of fur; articles of fur	..	..	..	..	..	..	520	480
	19		Leather and articles of leather; footwear	3 920	3 580	3 700	4 560	4 550	4 260	3 324	2 692
		191	Leather and articles of leather	2 060	2 220	2 270	2 800	2 910	2 790	2 018	1 733
		192	Footwear	1 860	1 370	1 420	1 770	1 640	1 470	1 306	960
	20		Wood & products of wood & cork, exc. furniture	12 100	11 970	13 000	14 650	16 010	15 870	17 001	17 201
		201	Sawmilling and planing of wood	6 250	6 530	7 150	7 250	7 560	7 510	7 955	8 340
		202	Wood products; articles of cork, straw, etc.	5 860	5 440	5 850	7 400	8 450	8 360	9 046	8 861
	21		Paper and paper products	9 490	8 850	8 630	8 480	8 540	8 660	8 659	8 056
	22		Printing, publishing and recorded media	18 040	17 390	17 730	18 180	19 230	19 280	19 837	19 893
		221	Publishing	8 890	8 620	9 100	9 080	9 600	9 410	9 448	9 853
		222	Printing and related activities	9 160	8 760	8 630	9 100	9 620	9 860	10 380	10 026
		223	Reproduction of recorded media	..	3	6	..	..	..	..	14
	23		Coke and petroleum products; nuclear fuel	950	930	890	1 000	970	920	760	660
		231	Coke oven products	200	180	220	420	400	380	166	176
		232	Refined petroleum products	750	750	670	580	570	550	594	484
		233	Processing of nuclear fuel	..	..	..	..	..	..	..	..
	24		Chemicals and chemical products	10 000	10 730	10 690	9 700	10 330	9 730	10 288	9 846
		241	Basic chemicals	3 620	4 210	4 110	4 020	4 330	4 010	4 028	3 877
		242	Other chemical products	6 390	6 520	6 570	5 670	5 990	5 710	6 261	5 969
		243	Man-made fibres	..	..	..	..	..	..	206	172
	25		Rubber and plastic products	8 440	8 250	9 220	9 070	9 770	10 000	8 269	8 758
		251	Rubber products	1 940	1 780	2 870	1 840	1 850	1 880	968	1 593
		252	Plastic products	6 510	6 470	6 360	7 240	7 910	8 130	7 301	7 165
	26		Other non-metallic mineral products	5 550	5 340	5 640	5 400	5 790	6 000	7 430	7 040
		261	Glass and glass products	1 070	950	1 110	1 070	1 130	1 140	1 596	1 063
		269	Non-metallic mineral products, nec	4 490	4 390	4 530	4 330	4 650	4 860	5 834	5 978
	27		Basic metals	6 630	5 880	6 330	5 600	5 810	5 860	4 062	3 765
		271	Basic iron and steel	2 680	2 300	2 300	3 130	3 220	3 110	2 255	2 055
		272	Basic precious and non-ferrous metals	1 500	1 300	1 300	1 730	1 740	1 780	1 066	1 064
		273	Casting of metals	2 440	2 290	2 730	740	840	970	1 233	1 140
	28		Fabricated metal products, exc. machinery	14 800	13 900	14 000	16 650	17 950	18 510	20 760	21 518
		281	Structural metal prod., tanks, reservoirs, etc.	..	..	..	6 480	7 230	7 830	7 989	9 174
		289	Other fabricated metal products	..	..	..	10 170	10 730	10 670	12 771	12 344
	29		Machinery and equipment, nec	14 790	13 940	14 700	16 990	17 970	18 410	13 649	13 047
		291	General purpose machinery	..	..	..	1 130	1 330	1 330	1 242	1 295
		292	Special purpose machinery	..	..	..	11 730	12 240	12 740	12 408	11 752
		293	Domestic appliances, nec	4 050	3 870	4 140	4 120	4 400	4 330	4 942	4 417
	30		Office, accounting & computing machinery	350	280	240	310	340	380	574	654
	31		Electrical machinery & apparatus, nec	4 560	4 020	4 080	4 170	4 850	4 580	4 149	4 024
		311	Electric motors, generators and transformers	..	..	..	..	..	..	2 963	2 825
		312	Electricity distribution & control apparatus	..	..	..	2 970	3 560	3 320	2 890	..
		313	Insulated wire and cable	720	580	500	540	650	640	685	585
		314	Accumulators, primary cells & primary batteries	480	420	390	220	190	190	98	131
		315	Electric lamps and lighting equipment	610	490	450	440	440	430	405	484
		319	Other electrical equipment, nec	2 750	2 530	2 730	..	..	..	..	..
	32		Radio, TV, communication equipment	2 300	1 730	1 530	1 880	2 190	2 270	2 835	2 892
		321	Electronic valves, tubes; other electronic components	1 180	890	640	..	..	..	..	1 528
		322	TV/radio transmitters; line teleph./telegr. apparatus	1 120	850	890	940	1 160	1 270	1 404	1 364
		323	TV/radio receivers; sound and video app.; assoc. gds.	..	..	..	940	1 040	990	1 490	..
	33		Medical, precision & optical instrum.; watch & clocks	1 200	1 260	1 210	800	840	920	1 069	1 014
		331	Medical appl.; instruments & appl. for measuring, etc.	..	..	..	650	690	760	917	854
		332	Optical instruments and photographic equipment	320	270	210	160	160	150	152	160
		333	Watches and clocks	..	..	..	..	..	..	..	..
	34		Motor vehicles, trailers and semi-trailers	6 210	4 860	4 980	5 200	5 600	5 420	4 831	4 585
		341	Motor vehicles	3 210	2 310	2 340	2 300	2 290	2 190	1 601	1 460
		342	Bodies for motor vehicles, trailers & semi-trailers	610	710	770	750	960	1 010	1 063	964
		343	Parts & accessories for motor vehicles & their engines	2 410	1 850	1 860	2 160	2 350	2 230	2 168	2 161
	35		Other transport equipment	1 740	1 780	1 730	7 900	7 960	8 270	3 819	3 568
		351	Building and repairing of ships and boats	1 620	1 690	1 670	2 710	2 840	2 870	2 964	2 645
		352	Railway and tramway locomotives and rolling stock	..	..	..	2 210	2 100	2 240	..	..
		353	Aircraft and spacecraft	540	600	610	2 640	2 700	2 870	736	781
		359	Transport equipment, nec	200	210	120	340	320	290	120	142
	36		Furniture; manufacturing, nec	9 790	9 170	9 150	10 450	11 100	10 870	11 262	10 657
		361	Furniture	6 860	6 430	6 490	7 200	7 900	7 860	7 403	7 226
		369	Manufacturing, nec	2 930	2 740	2 660	3 250	3 210	3 000	3 063	2 652
	37		Recycling	..	..	..	160	170	30	35	..
E	40_41		Electricity, gas and water supply	13 050	12 640	11 340	10 930	10 140	9 830	9 809	8 457
	40		Electricity, gas, steam and hot water supply	12 940	12 050	10 720	9 810	9 060	8 790	9 363	7 893
F	45		Construction	44 780	39 010	38 070	46 120	51 630	56 200	58 918	60 437

Table NZ.2 **EMPLOI, FEMMES - EMPLOYMENT, FEMALES**

Unités

1990-91	1991-92	1992-93	1993-94	1994-95	1995-96	1996-97	1997-98			CITI révision 3	
450	460	370	380	400	410	370	336	Activites extractives	C	10_14	
90	85	55	70	80	85	84	85	Extraction de charbon, de lignite et de tourbe		10	
150	170	120	110	120	140	107	98	Pétrole brut et gaz naturel ; activités annexes		11	
..	..	..	..	..	..	..	..	Extraction de Minerais d'Uranium et de Thorium		12	
55	70	80	60	70	65	78	53	Extraction de minerais métalliques		13	
130	130	110	140	130	130	99	99	Autres activités extractives		14	
62 100	58 430	59 290	62 530	64 910	63 620	63 679	62 047	Total activites manufacturieres	D	15_37	
15 430	14 760	15 000	16 760	17 110	17 420	18 642	18 743	Produits alimentaires et boissons		15	
7 710	7 690	7 720	8 460	8 950	9 080	10 651	11 942	Viande, poisson, fruits, légumes, huiles et graisses			151
1 970	1 660	1 820	1 920	1 970	1 980	1 705	1 662	Produits laitiers			152
470	490	480	340	410	320	1 594	290	Travail des grains; prod. amyl. ; alim. pour animaux			153
4 450	4 140	4 210	5 240	4 900	5 100	5 154	5 218	Autres produits alimentaires			154
815	790	765	800	880	940	1 244	1 293	Boissons			155
280	240	200	240	230	210	197	206	Produits à base de tabac		16	
4 580	4 100	4 200	4 380	4 580	4 280	3 788	3 796	Textiles		17	
750	690	1 280	1 110	1 240	1 080	869	797	Filature, tissage et achèvement des textiles			171
2 230	2 050	1 700	1 990	2 100	2 110	1 922	2 017	Autres articles textiles			172
1 590	1 370	1 210	1 280	1 240	1 100	997	982	Etoffes et articles de bonneterie			173
10 000	9 180	9 120	9 420	9 160	8 260	7 502	6 454	Articles d'habillement et fourrure		18	
..	..	..	..	..	..	7 468	6 418	Articles d'habillement autres qu'en fourrure			181
..	..	..	..	..	..	34	36	Prép. et teinture des fourrures ; articles en fourrure			182
1 790	1 480	1 560	1 830	1 780	1 630	1 329	999	Cuir et articles en cuir ; chaussures		19	
690	700	730	780	820	770	598	486	Cuir et articles en cuir sauf chaussures			191
1 090	780	830	1 040	950	850	731	514	Chaussures			192
1 500	1 350	1 420	1 610	1 770	1 700	2 137	2 142	Bois, articles en bois, sauf meubles		20	
620	600	700	720	760	730	1 052	1 047	Sciage et rabotage du bois			201
880	750	720	900	1 020	960	1 085	1 096	Articles en bois, liège, vannerie, sparterie			202
1 610	1 490	1 560	1 530	1 600	1 680	1 540	1 430	Papier, articles en papier et carton		21	
7 660	7 450	7 640	7 700	8 240	8 220	8 486	8 646	Edition, imprimerie et reproduction		22	
4 300	4 320	4 640	4 560	4 910	4 850	4 860	5 155	Edition			221
3 360	3 130	2 980	3 130	3 340	3 380	3 622	3 487	Imprimerie et activités annexes			222
..	..	6	..	..	..	5	5	Reproduction de supports enregistrés			223
120	120	120	150	140	140	110	97	Coke. et prod. pétroliers ; comb. nucléaires		23	
45	40	50	85	80	80	43	43	Cokéfaction			231
75	85	70	65	60	60	68	54	Produits pétroliers raffinés			232
..	..	..	..	..	..	..	..	Traitement de combustibles nucléaires			233
3 220	3 350	3 410	3 040	3 330	3 060	3 147	3 020	Produits chimiques		24	
730	920	860	780	840	830	794	752	Produits chimiques de base			241
2 480	2 450	2 570	2 260	2 470	2 240	2 354	2 269	Autres produits chimiques			242
..	..	..	..	..	..	80	64	Fibres synthétiques ou artificielles			243
2 090	2 050	2 100	2 150	2 340	2 460	2 104	2 023	Art. en caoutchouc et matières plastiques		25	
360	340	430	290	300	320	251	288	Articles en caoutchouc			251
1 730	1 710	1 670	1 870	2 040	2 150	1 853	1 735	Articles en matières plastiques			252
750	750	810	840	850	960	1 149	1 041	Autres produits minéraux non métalliques		26	
170	150	170	150	160	190	326	196	Verre et articles en verre			261
580	600	640	680	690	770	823	845	Produits minéraux non métalliques, nca			269
820	670	750	570	580	560	337	327	Produits métallurgiques de base		27	
310	260	270	330	320	270	199	189	Sidérurgie et première transformation de l'acier			271
100	70	85	160	160	160	77	74	Première transf. métaux précieux et non ferreux			272
410	340	400	90	110	130	129	131	Fonderie			273
2 450	2 420	2 330	2 450	2 670	2 760	3 151	3 300	Ouvrages en métaux, sauf machines		28	
..	..	..	850	950	1 000	1 075	1 359	Constr. et menuis. métal.; réservoirs; générateurs			281
..	..	..	1 610	1 720	1 760	2 076	1 941	Autres ouvrages en métaux			289
2 640	2 520	2 690	3 070	3 190	3 160	1 616	1 548	Machines et matériels, nca		29	
..	..	..	130	160	150	125	140	Machines d'usage général			291
..	..	..	1 420	1 470	1 560	1 491	1 408	Machines d'usage spécifique			292
1 340	1 260	1 370	1 510	1 570	1 460	1 691	1 477	Appareils domestiques, nca			293
75	70	45	70	85	110	138	140	Mach. de bureau, comptables, information		30	
1 490	1 380	1 320	1 370	1 690	1 450	1 107	1 115	Machines et appareils électriques, nca		31	
..	..	..	..	..	..	801	826	Moteurs, génératrices, transformateurs électriques			311
..	..	..	1 110	1 390	1 150	810	..	Matériel électrique de distribution et de commande			312
110	90	70	70	110	120	128	105	Fils et câbles électriques isolés			313
85	75	75	30	25	20	11	19	Accumulateurs et piles électriques			314
230	190	160	150	170	150	167	166	Lampes électriques et appareils d'éclairage			315
1 080	1 030	1 010	..	..	..	..	..	Autres matériels électriques, nca.			319
870	630	580	660	790	860	1 121	1 109	Equipements et appareils de radio, TV, etc.		32	
380	280	200	..			534	578	Tubes et valves électroniques;autres composants			321
490	360	380	370	470	490	588	531	Emetteurs radio/TV; app. téléphonie et télégraphie			322
..	..	..	300	310	370	580	..	Récepteurs radio/TV, son et image; art. associés			323
530	610	590	320	340	390	395	478	Instr. médicaux, de précision ; optique ; horlog.		33	
..	..	..	260	270	310	334	408	Instruments médicaux et de précision, etc.			331
130	130	95	70	75	75	62	70	Instruments d'optique ; matériel photographique			332
..	..	..	..	..	..	..	..	Horlogerie			333
1 220	950	1 020	1 030	1 100	1 030	1 078	1 178	Véhicules automobiles; remorques, etc.		34	
610	450	450	410	390	370	282	301	Véhicules automobiles			341
70	85	90	80	90	90	112	99	Carross. automobiles, remorques, semi-remorques			342
540	420	480	550	620	570	685	779	Pièces et accessoires pour automobiles et moteurs			343
200	210	240	660	640	780	357	376	Autres matériels de transport		35	
120	140	155	230	210	210	225	229	Construction et réparation de navires			351
..	..	..	210	230	220	..	..	Matériel ferroviaire roulant			352
90	90	100	160	150	280	98	89	Construction aéronautique et spatiale			353
40	45	55	65	65	60	35	59	Autres équipements de transport			359
2 660	2 470	2 390	2 630	2 700	2 530	2 574	2 417	Meubles ; activités de fabrication, nca		36	
1 510	1 420	1 390	1 470	1 550	1 510	1 424	1 411	Meubles			361
1 150	1 050	1 000	1 160	1 150	1 020	1 072	929	Activités de fabrication, nca			369
..	..	..	25	20	..	3	..	Récupération		37	
2 260	2 230	2 110	1 940	1 930	1 960	2 164	1 931	Electricité, gaz et eau	E	40_41	
2 240	2 170	2 040	1 840	1 820	1 840	2 077	1 805	Prod. et dist. d'électricité gaz, vapeur d'eau chaude		40	
4 340	3 910	3 790	4 040	4 400	4 880	5 127	5 339	Construction	F	45	

Table NZ.3 WAGES AND SALARIES, EMPLOYEES - SALAIRES ET TRAITEMENTS, SALARIÉS

Millions of NZ$ (Current Prices)

ISIC revision 3				1990-91	1991-92	1992-93	1993-94	1994-95	1995-96	1996-97	1997-98
C	10_14		Mining and quarrying	..	..	..	139	160	157	150	139
	10		Mining of coal and lignite; extraction of peat	..	..	..	33	42	37	39	37
	11		Crude petroleum & natural gas; service activities	..	..	..	35	39	40	39	42
	12		Mining of uranium and thorium ores	..	..	..	..	..	..	..	..
	13		Mining of metal ores	..	..	..	24	25	28	25	15
	14		Other mining and quarrying	..	..	..	46	55	52	46	46
D	15_37		Total manufacturing	..	..	..	6 525	6 885	7 099	7 044	6 875
	15		Food products and beverages	..	..	..	1 588	1 676	1 766	1 785	1 727
		151	Production of meat, fish, fruit, vegetables, oils & fats	..	..	..	902	945	988	1 003	962
		152	Dairy products	..	..	..	267	273	318	312	306
		153	Grain mill and starch products; animal feeds	..	..	..	46	55	51	52	46
		154	Other food products	..	..	..	292	311	311	307	308
		155	Beverages	..	..	..	81	92	98	111	106
	16		Tobacco products	..	..	..	18	18	19	18	20
	17		Textiles	..	..	..	226	226	208	193	197
		171	Spinning, weaving and finishing of textiles	..	..	..	97	97	95	84	82
		172	Other textiles	..	..	..	90	90	83	83	90
		173	Knitted and crocheted fabrics and articles	..	..	..	39	38	31	27	25
	18		Wearing apparel and fur	..	..	..	200	198	172	154	149
		181	Wearing apparel, except fur apparel	..	..	..	200	198	172	154	149
		182	Dressing and dyeing of fur; articles of fur	..	..	..	..	..	..	..	..
	19		Leather and articles of leather; footwear	..	..	..	110	107	100	89	83
		191	Leather and articles of leather	..	..	..	67	69	66	60	60
		192	Footwear	..	..	..	43	38	34	28	23
	20		Wood & products of wood & cork, exc. furniture	..	..	..	420	445	454	466	451
		201	Sawmilling and planing of wood	..	..	..	190	204	209	219	218
		202	Wood products; articles of cork, straw, etc.	..	..	..	230	241	245	247	233
	21		Paper and paper products	..	..	..	336	351	353	332	316
	22		Printing, publishing and recorded media	..	..	..	615	654	692	705	734
		221	Publishing	..	..	..	283	296	314	316	317
		222	Printing and related activities	..	..	..	332	358	378	389	418
		223	Reproduction of recorded media	..	..	..	..	..	..	..	..
	23		Coke and petroleum products; nuclear fuel	..	..	..	48	51	49	50	41
		231	Coke oven products	..	..	..	..	..	..	..	..
		232	Refined petroleum products	..	..	..	..	..	..	..	..
		233	Processing of nuclear fuel	..	..	..	..	..	..	..	..
	24		Chemicals and chemical products	..	..	..	351	363	355	341	325
		241	Basic chemicals	..	..	..	146	155	152	154	149
		242	Other chemical products	..	..	..	205	208	203	187	176
		243	Man-made fibres	..	..	..	..	..	..	..	..
	25		Rubber and plastic products	..	..	..	292	305	305	296	290
		251	Rubber products	..	..	..	66	69	68	64	55
		252	Plastic products	..	..	..	226	236	237	233	235
	26		Other non-metallic mineral products	..	..	..	158	177	184	192	203
		261	Glass and glass products	..	..	..	36	38	40	40	38
		269	Non-metallic mineral products, nec	..	..	..	122	139	144	152	165
	27		Basic metals	..	..	..	240	254	253	230	217
		271	Basic iron and steel	..	..	..	139	141	130	116	116
		272	Basic precious and non-ferrous metals	..	..	..	75	82	90	94	83
		273	Casting of metals	..	..	..	26	31	33	19	19
	28		Fabricated metal products, exc. machinery	..	..	..	457	479	544	557	545
		281	Structural metal prod., tanks, reservoirs, etc.	..	..	..	161	185	237	236	228
		289	Other fabricated metal products	..	..	..	296	293	307	321	317
	29		Machinery and equipment, nec	..	..	..	532	574	601	592	547
		291	General purpose machinery	..	..	..	42	51	53	50	57
		292	Special purpose machinery	..	..	..	374	400	425	426	380
		293	Domestic appliances, nec	..	..	..	116	123	124	116	111
	30		Office, accounting & computing machinery	..	..	..	14	12	12	13	16
	31		Electrical machinery & apparatus, nec	..	..	..	177	205	223	223	231
		311	Electric motors, generators and transformers	..	..	..	..	..	..	..	..
		312	Electricity distribution & control apparatus	..	..	..	139	166	185	183	191
		313	Insulated wire and cable	..	..	..	21	23	24	25	24
		314	Accumulators, primary cells & primary batteries	..	..	..	7	..	..	..	6
		315	Electric lamps and lighting equipment	..	..	..	10	..	..	..	10
		319	Other electrical equipment, nec	..	..	..	..	..	..	..	..
	32		Radio, TV, communication equipment	..	..	..	52	59	67	83	88
		321	Electronic valves, tubes; other electronic components	..	..	..	..	..	..	..	..
		322	TV/radio transmitters; line teleph./telegr. apparatus	..	..	..	23	27	31	34	39
		323	TV/radio receivers; sound and video app.; assoc. gds.	..	..	..	29	31	36	49	49
	33		Medical, precision & optical instrum.; watch & clocks	..	..	..	17	19	22	34	33
		331	Medical appl.; instruments & appl. for measuring, etc.	..	..	..	14	15	19	31	30
		332	Optical instruments and photographic equipment	..	..	..	3	3	3	4	3
		333	Watches and clocks	..	..	..	..	..	..	..	..
	34		Motor vehicles, trailers and semi-trailers	..	..	..	153	164	165	143	121
		341	Motor vehicles	..	..	..	71	72	75	58	39
		342	Bodies for motor vehicles, trailers & semi-trailers	..	..	..	30	40	37	31	31
		343	Parts & accessories for motor vehicles & their engines	..	..	..	51	52	54	55	51
	35		Other transport equipment	..	..	..	267	284	299	318	321
		351	Building and repairing of ships and boats	..	..	..	66	81	79	79	76
		352	Railway and tramway locomotives and rolling stock	..	..	..	74	76	79	84	87
		353	Aircraft and spacecraft	..	..	..	116	116	131	147	152
		359	Transport equipment, nec	..	..	..	11	11	11	8	7
	36		Furniture; manufacturing, nec	..	..	..	248	260	254	230	219
		361	Furniture	..	..	..	174	189	186	170	163
		369	Manufacturing, nec	..	..	..	74	71	68	61	56
	37		Recycling	..	..	..	..	..	..	..	..
E	40_41		Electricity, gas and water supply	..	..	..	381	379	377	383	339
	40		Electricity, gas, steam and hot water supply	..	..	..	342	342	341	347	297
F	45		Construction	..	..	..	1 007	1 139	1 243	1 255	1 384

Table NZ.4 **HEURES OUVRÉES, SALARIÉS - HOURS WORKED BY EMPLOYEES**

Thousands

1990-91	1991-92	1992-93	1993-94	1994-95	1995-96	1996-97	1997-98			CITI révision 3
..	..	..	7 850	8 774	8 190	7 450	6 498	Activites extractives	C	10_14
..	..	..	1 643	2 085	1 749	1 779	1 621	Extraction de charbon, de lignite et de tourbe		10
..	..	..	1 443	1 464	1 360	1 314	1 181	Pétrole brut et gaz naturel ; activités annexes		11
..	..	..	..	..	..	..	..	Extraction de Minerais d'Uranium et de Thorium		12
..	..	..	1 375	1 375	1 506	1 281	725	Extraction de minerais métalliques		13
..	..	..	3 388	3 850	3 575	3 075	2 971	Autres activités extractives		14
..	..	..	449 182	461 590	458 983	439 367	416 007	Total activites manufacturieres	D	15_37
..	..	..	107 660	111 151	113 846	112 043	105 783	Produits alimentaires et boissons		15
..	..	..	60 648	61 970	63 427	63 083	58 906	Viande, poisson, fruits, légumes, huiles et graisses		151
..	..	..	16 843	16 895	19 133	18 436	17 406	Produits laitiers		152
..	..	..	3 279	3 727	3 376	3 090	2 886	Travail des grains; prod. amyl. ; alim. pour animaux		153
..	..	..	22 267	23 322	22 733	21 948	21 117	Autres produits alimentaires		154
..	..	..	4 623	5 236	5 176	5 487	5 467	Boissons		155
..	..	..	1 117	1 046	997	912	958	Produits à base de tabac		16
..	..	..	17 692	17 396	15 830	14 321	14 130	Textiles		17
..	..	..	7 106	7 181	6 872	6 072	5 741	Filature, tissage et achèvement des textiles		171
..	..	..	7 370	7 109	6 393	6 125	6 401	Autres articles textiles		172
..	..	..	3 216	3 107	2 565	2 124	1 988	Etoffes et articles de bonneterie		173
..	..	..	19 184	18 764	15 839	13 567	12 665	Articles d'habillement et fourrure		18
..	..	..	19 184	18 765	15 839	13 567	12 665	Articles d'habillement autres qu'en fourrure		181
..	..	..	..	..	..	..	..	Prép. et teinture des fourrures ; articles en fourrure		182
..	..	..	8 865	8 205	7 452	6 361	5 880	Cuir et articles en cuir ; chaussures		19
..	..	..	5 066	4 986	4 660	4 124	4 122	Cuir et articles en cuir sauf chaussures		191
..	..	..	3 799	3 219	2 792	2 237	1 758	Chaussures		192
..	..	..	32 824	33 913	33 410	33 054	30 927	Bois, articles en bois, sauf meubles		20
..	..	..	15 078	15 820	15 817	15 887	15 172	Sciage et rabotage du bois		201
..	..	..	17 746	18 093	17 593	17 167	15 754	Articles en bois, liège, vannerie, sparterie		202
..	..	..	18 746	19 037	18 640	16 803	15 478	Papier, articles en papier et carton		21
..	..	..	37 134	38 578	39 626	39 617	39 601	Edition, imprimerie et reproduction		22
..	..	..	16 454	16 734	17 424	16 907	16 298	Edition		221
..	..	..	20 680	21 843	22 202	22 710	23 303	Imprimerie et activités annexes		222
..	..	..	..	..	..	..	..	Reproduction de supports enregistrés		223
..	..	..	2 022	2 040	1 847	1 811	1 434	Coke. et prod. pétroliers ; comb. nucléaires		23
..	..	..	..	..	..	..	..	Cokéfaction		231
..	..	..	..	..	..	..	..	Produits pétroliers raffinés		232
..	..	..						Traitement de combustibles nucléaires		233
..	..	..	21 543	21 575	20 526	19 092	17 361	Produits chimiques		24
..	..	..	8 855	9 234	8 817	8 606	7 928	Produits chimiques de base		241
..	..	..	12 688	12 341	11 710	10 486	9 433	Autres produits chimiques		242
..	..	..	..	..	..	..	..	Fibres synthétiques ou artificielles		243
..	..	..	21 824	22 380	21 699	20 199	19 123	Art. en caoutchouc et matières plastiques		25
..	..	..	4 496	4 754	4 451	4 041	3 441	Articles en caoutchouc		251
..	..	..	17 328	17 627	17 248	16 158	15 682	Articles en matières plastiques		252
..	..	..	10 859	11 812	11 796	12 147	12 462	Autres produits minéraux non métalliques		26
..	..	..	2 099	2 163	2 205	2 265	2 017	Verre et articles en verre		261
..	..	..	8 760	9 649	9 591	9 883	10 445	Produits minéraux non métalliques, nca		269
..	..	..	13 115	13 408	13 424	11 585	10 464	Produits métallurgiques de base		27
..	..	..	7 586	7 407	7 001	5 953	5 559	Sidérurgie et première transformation de l'acier		271
..	..	..	3 549	3 760	4 036	4 239	3 551	Première transf. métaux précieux et non ferreux		272
..	..	..	1 980	2 241	2 388	1 393	1 354	Fonderie		273
..	..	..	33 415	33 930	37 017	36 382	34 866	Ouvrages en métaux, sauf machines		28
..	..	..	11 739	12 942	15 637	14 981	14 433	Constr. et menuis. métal.; réservoirs; générateurs		281
..	..	..	21 676	20 987	21 380	21 401	20 434	Autres ouvrages en métaux		289
..	..	..	36 422	38 196	38 038	36 074	32 488	Machines et matériels, nca		29
..	..	..	2 575	2 978	2 974	2 656	2 922	Machines d'usage général		291
..	..	..	25 494	26 648	26 956	25 934	22 463	Machines d'usage spécifique		292
..	..	..	8 353	8 571	8 107	7 484	7 103	Appareils domestiques, nca		293
..	..	..	798	655	658	689	788	Mach. de bureau, comptables, information		30
..	..	..	12 807	14 486	14 845	13 909	13 906	Machines et appareils électriques, nca		31
..	..	..	..	..	..	..	..	Moteurs, génératrices, transformateurs électriques		311
..	..	..	10 420	12 092	12 501	11 535	11 626	Matériel électrique de distribution et de commande		312
..	..	..	1 226	..	..	..	1 295	Fils et câbles électriques isolés		313
..	..	..	452	..	..	..	351	Accumulateurs et piles électriques		314
..	..	..	709	670	578	587	635	Lampes électriques et appareils d'éclairage		315
..	..	..	..	..	..	..	..	Autres matériels électriques, nca.		319
..	..	..	3 669	4 015	4 383	5 075	5 125	Equipements et appareils de radio, TV, etc.		32
..	..	..	..	..	..	..	..	Tubes et valves électroniques;autres composants		321
..	..	..	1 645	1 856	1 963	2 012	2 213	Emetteurs radio/TV; app. téléphonie et télégraphie		322
..	..	..	2 024	2 160	2 420	3 063	2 912	Récepteurs radio/TV, son et image; art. associés		323
..	..	..	1 141	1 221	1 401	2 124	1 937	Instr. médicaux, de précision ; optique ; horlog.		33
..	..	..	889	987	1 177	1 887	1 723	Instruments médicaux et de précision, etc.		331
..	..	..	252	234	224	238	214	Instruments d'optique ; matériel photographique		332
..	..	..	..	..	..	..	..	Horlogerie		333
..	..	..	11 240	11 881	11 496	9 773	8 098	Véhicules automobiles; remorques, etc.		34
..	..	..	4 799	4 847	4 851	3 645	2 403	Véhicules automobiles		341
..	..	..	2 333	2 926	2 605	2 147	2 102	Carross. automobiles, remorques, semi-remorques		342
..	..	..	4 107	4 108	4 040	3 981	3 594	Pièces et accessoires pour automobiles et moteurs		343
..	..	..	16 122	16 615	16 611	16 533	16 249	Autres matériels de transport		35
..	..	..	4 711	5 339	4 821	4 661	4 330	Construction et réparation de navires		351
..	..	..	4 550	4 583	4 631	4 751	4 741	Matériel ferroviaire roulant		352
..	..	..	5 907	5 796	6 293	6 569	6 661	Construction aéronautique et spatiale		353
..	..	..	954	896	866	553	518	Autres équipements de transport		359
..	..	..	20 566	20 955	19 529	17 221	16 207	Meubles ; activités de fabrication, nca		36
..	..	..	14 658	15 336	14 391	12 730	12 104	Meubles		361
..	..	..	5 908	5 619	5 138	4 490	4 103	Activités de fabrication, nca		369
..	..	..	..	..	..	..	..	Récupération		37
..	..	..	21 557	20 490	19 277	18 222	15 288	Electricité, gaz et eau	E	40_41
..	..	..	19 319	18 315	17 244	16 188	13 093	Prod. et dist. d'électricité gaz, vapeur d'eau chaude		40
..	..	..	74 319	81 497	85 420	83 664	88 981	Construction	F	45

Table NZ.5 ESTABLISHMENTS - ÉTABLISSEMENTS

Units

ISIC revision 3				1990-91	1991-92	1992-93	1993-94	1994-95	1995-96	1996-97	1997-98
C	10_14		Mining and quarrying	..	387	357	505	512	524	502	489
	10		Mining of coal and lignite; extraction of peat	..	39	37	60	59	54	53	48
	11		Crude petroleum & natural gas; service activities	..	20	17	25	32	33	38	39
	12		Mining of uranium and thorium ores	..	..	..	..	..	..	..	..
	13		Mining of metal ores	..	119	102	108	93	108	88	84
	14		Other mining and quarrying	..	209	201	312	328	329	323	318
D	15_37		Total manufacturing	..	17 779	17 958	20 057	21 553	21 677	21 711	22 313
	15		Food products and beverages	..	1 268	1 274	1 718	1 786	1 730	1 748	1 784
		151	Production of meat, fish, fruit, vegetables, oils & fats	..	483	478	633	654	631	632	629
		152	Dairy products	..	75	76	138	143	148	149	145
		153	Grain mill and starch products; animal feeds	..	70	65	113	111	104	114	113
		154	Other food products	..	511	511	624	649	604	605	631
		155	Beverages	..	129	144	210	229	243	248	266
	16		Tobacco products	..	4	3	8	7	6	7	7
	17		Textiles	..	702	717	785	836	830	830	845
		171	Spinning, weaving and finishing of textiles	..	116	130	152	164	156	152	153
		172	Other textiles	..	449	459	509	548	567	567	587
		173	Knitted and crocheted fabrics and articles	..	137	128	124	124	107	111	105
	18		Wearing apparel and fur	..	1 137	1 153	1 169	1 215	1 188	1 114	1 104
		181	Wearing apparel, except fur apparel	..	1 111	1 127	..	..	..	..	..
		182	Dressing and dyeing of fur; articles of fur	..	26	26	..	..	..	..	..
	19		Leather and articles of leather; footwear	..	235	234	288	282	272	253	245
		191	Leather and articles of leather	..	160	162	208	207	203	191	188
		192	Footwear	..	75	72	80	75	69	62	57
	20		Wood & products of wood & cork, exc. furniture	..	1 563	1 622	1 836	1 982	2 014	2 010	2 103
		201	Sawmilling and planing of wood	..	389	402	557	586	600	565	578
		202	Wood products; articles of cork, straw, etc.	..	1 174	1 220	1 279	1 396	1 414	1 445	1 525
	21		Paper and paper products	..	118	114	166	162	145	140	136
	22		Printing, publishing and recorded media	..	1 810	1 875	2 055	2 265	2 233	2 216	2 221
		221	Publishing	..	579	605	700	798	784	798	822
		222	Printing and related activities	..	1 231	1 270	1 355	1 467	1 449	1 418	1 399
		223	Reproduction of recorded media	..	..	..	..	..	..	..	..
	23		Coke and petroleum products; nuclear fuel	..	26	26	43	45	41	39	40
		231	Coke oven products	..	19	23	41	43	39	35	38
		232	Refined petroleum products	..	7	3	2	2	2	4	2
		233	Processing of nuclear fuel	..	..	..	..	..	..	..	..
	24		Chemicals and chemical products	..	389	398	597	652	656	628	647
		241	Basic chemicals	..	140	143	221	240	243	226	229
		242	Other chemical products	..	249	255	376	412	413	402	418
		243	Man-made fibres	..	..	..	..	..	..	..	..
	25		Rubber and plastic products	..	554	511	598	621	627	633	640
		251	Rubber products	..	95	92	121	120	105	110	112
		252	Plastic products	..	459	419	477	501	522	523	528
	26		Other non-metallic mineral products	..	619	654	799	858	843	859	854
		261	Glass and glass products	..	129	147	140	154	130	130	135
		269	Non-metallic mineral products, nec	..	490	507	659	704	713	729	719
	27		Basic metals	..	154	160	219	223	228	221	272
		271	Basic iron and steel	..	68	70	111	125	127	126	175
		272	Basic precious and non-ferrous metals	..	25	31	39	33	35	35	42
		273	Casting of metals	..	61	59	69	65	66	60	55
	28		Fabricated metal products, exc. machinery	..	2 256	2 278	2 439	2 602	2 651	2 798	3 091
		281	Structural metal prod., tanks, reservoirs, etc.	..	784	790	853	902	898	948	1 053
		289	Other fabricated metal products	..	1 472	1 488	1 586	1 700	1 753	1 850	2 038
	29		Machinery and equipment, nec	..	2 493	2 444	2 614	2 875	3 062	3 043	3 022
		291	General purpose machinery	..	91	98	140	171	179	187	199
		292	Special purpose machinery	..	2 402	2 346	2 413	2 634	2 812	2 786	2 764
		293	Domestic appliances, nec	..	..	..	61	70	71	70	59
	30		Office, accounting & computing machinery	..	67	56	82	78	73	92	109
	31		Electrical machinery & apparatus, nec	..	366	349	368	366	355	351	355
		311	Electric motors, generators and transformers	..	..	..	..	..	..	..	..
		312	Electricity distribution & control apparatus	..	299	280	300	291	279	269	270
		313	Insulated wire and cable	..	10	7	11	14	13	14	15
		314	Accumulators, primary cells & primary batteries	..	5	6	6	6	6	7	8
		315	Electric lamps and lighting equipment	..	52	56	51	55	57	61	62
		319	Other electrical equipment, nec	..	..	..	..	..	..	..	..
	32		Radio, TV, communication equipment	..	179	172	196	225	232	227	240
		321	Electronic valves, tubes; other electronic components	..	..	..	..	..	..	..	..
		322	TV/radio transmitters; line teleph./telegr. apparatus	..	62	52	58	56	58	57	63
		323	TV/radio receivers; sound and video app.; assoc. gds.	..	117	120	138	169	174	170	177
	33		Medical, precision & optical instrum.; watch & clocks	..	237	228	234	255	270	281	306
		331	Medical appl.; instruments & appl. for measuring, etc.	..	213	203	214	232	244	255	279
		332	Optical instruments and photographic equipment	..	24	25	20	23	26	26	27
		333	Watches and clocks	..	..	..	..	..	..	..	..
	34		Motor vehicles, trailers and semi-trailers	..	542	569	594	620	600	604	609
		341	Motor vehicles	..	93	104	100	96	94	75	75
		342	Bodies for motor vehicles, trailers & semi-trailers	..	145	150	151	183	172	192	184
		343	Parts & accessories for motor vehicles & their engines	..	304	315	343	341	334	337	350
	35		Other transport equipment	..	632	620	676	748	782	778	830
		351	Building and repairing of ships and boats	..	532	515	487	536	577	577	627
		352	Railway and tramway locomotives and rolling stock	..	..	..	53	53	52	54	54
		353	Aircraft and spacecraft	..	75	82	103	121	120	117	120
		359	Transport equipment, nec	..	25	23	33	38	33	30	29
	36		Furniture; manufacturing, nec	..	2 428	2 501	2 570	2 846	2 837	2 836	2 852
		361	Furniture	..	1 471	1 496	1 589	1 769	1 767	1 799	1 848
		369	Manufacturing, nec	..	957	1 005	981	1 077	1 070	1 037	1 004
	37		Recycling	..	..	..	3	4	2	3	1
E	40_41		Electricity, gas and water supply	..	85	92	536	548	533	515	524
	40		Electricity, gas, steam and hot water supply	..	73	78	392	407	384	375	384
F	45		Construction	..	26 127	25 533	24 996	29 028	31 305	32 869	34 960

Table NZ.6 **ENTREPRISES - ENTERPRISES**

Unités

1990-91	1991-92	1992-93	1993-94	1994-95	1995-96	1996-97	1997-98		CITI révision 3		
455	449	422	394	401	402	331	325	Activites extractives	C	10_14	
41	39	37	38	37	32	28	26	Extraction de charbon, de lignite et de tourbe		10	
67	78	77	18	25	26	14	12	Pétrole brut et gaz naturel ; activités annexes		11	
..	..	..	..	..	..	..	..	Extraction de Minerais d'Uranium et de Thorium		12	
124	119	102	101	89	100	78	77	Extraction de minerais métalliques		13	
223	213	206	239	252	248	195	191	Autres activités extractives		14	
17 344	17 779	17 958	18 252	19 730	19 916	19 754	20 409	Total activites manufacturieres	D	15_37	
1 229	1 268	1 274	1 388	1 468	1 419	1 375	1 417	Produits alimentaires et boissons		15	
468	483	478	513	547	528	581	574	Viande, poisson, fruits, légumes, huiles et graisses			151
74	75	76	85	86	85	32	34	Produits laitiers			152
86	81	75	89	87	76	80	92	Travail des grains; prod. amyl. ; alim. pour animaux			153
473	500	501	551	574	541	506	523	Autres produits alimentaires			154
128	129	144	175	200	214	208	228	Boissons			155
4	4	3	3	2	2	3	3	Produits à base de tabac		16	
683	702	717	753	804	796	768	800	Textiles		17	
116	116	130	140	151	148	132	139	Filature, tissage et achèvement des textiles			171
431	449	459	493	532	548	529	559	Autres articles textiles			172
136	137	128	124	124	107	107	102	Etoffes et articles de bonneterie			173
1 071	1 111	1 127	1 117	1 167	1 144	1 076	1 079	Articles d'habillement et fourrure		18	
..	..	..	..	..	..	1 062	1 066	Articles d'habillement autres qu'en fourrure			181
..	..	..	..	..	..	14	13	Prép. et teinture des fourrures ; articles en fourrure			182
278	261	260	264	261	251	213	208	Cuir et articles en cuir ; chaussures		19	
191	186	188	192	189	185	154	153	Cuir et articles en cuir sauf chaussures			191
87	75	72	73	72	67	59	55	Chaussures			192
1 504	1 520	1 579	1 741	1 881	1 919	1 883	1 979	Bois, articles en bois, sauf meubles		20	
372	389	402	507	534	549	422	433	Sciage et rabotage du bois			201
1 132	1 131	1 177	1 238	1 352	1 375	1 461	1 546	Articles en bois, liège, vannerie, sparterie			202
118	118	114	127	115	98	92	95	Papier, articles en papier et carton		21	
1 724	1 810	1 875	1 897	2 097	2 058	2 033	2 028	Edition, imprimerie et reproduction		22	
535	576	603	583	674	661	663	677	Edition			221
1 187	1 231	1 270	1 321	1 431	1 403	1 358	1 335	Imprimerie et activités annexes			222
2	3	2	..	..	..	12	16	Reproduction de supports enregistrés			223
28	26	26	35	41	37	26	27	Coke. et prod. pétroliers ; comb. nucléaires		23	
21	19	23	33	39	35	23	25	Cokéfaction			231
7	7	3	2	2	2	3	2	Produits pétroliers raffinés			232
..	..	..	..	..	..	..	..	Traitement de combustibles nucléaires			233
373	389	398	444	495	502	454	481	Produits chimiques		24	
..	..	..	178	192	197	174	182	Produits chimiques de base			241
240	249	255	270	307	309	280	299	Autres produits chimiques			242
..	..	..	..	..	..	33	34	Fibres synthétiques ou artificielles			243
517	554	511	530	555	565	554	566	Art. en caoutchouc et matières plastiques		25	
84	95	92	93	93	80	80	84	Articles en caoutchouc			251
433	459	419	440	465	487	474	482	Articles en matières plastiques			252
634	619	654	633	694	674	660	658	Autres produits minéraux non métalliques		26	
128	129	147	130	144	122	119	125	Verre et articles en verre			261
506	490	507	504	551	553	541	533	Produits minéraux non métalliques, nca			269
157	162	166	202	211	215	136	183	Produits métallurgiques de base		27	
14	10	12	108	122	125	52	85	Sidérurgie et première transformation de l'acier			271
13	13	17	32	29	31	14	15	Première transf. métaux précieux et non ferreux			272
130	139	137	64	62	61	126	136	Fonderie			273
2 184	2 222	2 235	2 265	2 424	2 492	2 610	2 908	Ouvrages en métaux, sauf machines		28	
..	..	..	813	861	861	890	992	Constr. et menuis. métal.; réservoirs; générateurs			281
..	..	..	1 460	1 569	1 638	1 720	1 916	Autres ouvrages en métaux			289
2 429	2 472	2 421	2 506	2 762	2 955	2 840	2 829	Machines et matériels, nca		29	
..	..	..	123	155	167	171	184	Machines d'usage général			291
..	..	..	2 337	2 552	2 731	2 669	2 645	Machines d'usage spécifique			292
45	46	43	48	56	58	55	49	Appareils domestiques, nca			293
57	67	56	64	61	59	75	77	Mach. de bureau, comptables, information		30	
361	366	349	341	339	332	320	331	Machines et appareils électriques, nca		31	
..	..	..	..	..	..	245	251	Moteurs, génératrices, transformateurs électriques			311
..	..	..	278	270	263	253	..	Matériel électrique de distribution et de commande			312
12	10	7	8	11	10	10	12	Fils et câbles électriques isolés			313
4	5	6	6	6	6	6	7	Accumulateurs et piles électriques			314
51	52	56	50	53	55	59	61	Lampes électriques et appareils d'éclairage			315
294	299	280	..	..	..	..	..	Autres matériels électriques, nca.			319
178	179	172	188	218	223	219	231	Equipements et appareils de radio, TV, etc.		32	
121	117	120	..	..	..	167	174	Tubes et valves électroniques;autres composants			321
57	62	52	56	55	56	52	57	Emetteurs radio/TV; app. téléphonie et télégraphie			322
..	..	..	132	163	167	169	..	Récepteurs radio/TV, son et image; art. associés			323
241	237	228	226	246	262	273	296	Instr. médicaux, de précision ; optique ; horlog.		33	
..	..	..	206	223	236	247	270	Instruments médicaux et de précision, etc.			331
26	24	25	20	23	26	26	26	Instruments d'optique ; matériel photographique			332
..	..	..	..	..	..	..	..	Horlogerie			333
529	542	569	580	605	584	571	581	Véhicules automobiles; remorques, etc.		34	
89	93	104	97	92	91	68	68	Véhicules automobiles			341
136	145	150	147	177	166	183	179	Carross. automobiles, remorques, semi-remorques			342
304	304	315	336	336	328	320	334	Pièces et accessoires pour automobiles et moteurs			343
409	403	411	606	677	714	697	750	Autres matériels de transport		35	
543	532	515	482	530	572	566	615	Construction et réparation de navires			351
..	..	..	1	1	1	..	..	Matériel ferroviaire roulant			352
71	75	82	94	112	110	104	108	Construction aéronautique et spatiale			353
29	25	23	30	34	31	27	27	Autres équipements de transport			359
2 312	2 428	2 501	2 516	2 790	2 786	2 850	2 873	Meubles ; activités de fabrication, nca		36	
1 428	1 471	1 496	1 556	1 738	1 733	1 754	1 802	Meubles			361
884	957	1 005	962	1 055	1 053	1 011	978	Activités de fabrication, nca			369
..	..	..	3	4	2	3	..	Récupération		37	
135	152	160	169	173	180	166	190	Electricité, gaz et eau	E	40_41	
68	73	78	84	88	85	87	100	Prod. et dist. d'électricité gaz, vapeur d'eau chaude		40	
25 702	26 127	25 533	24 416	28 466	30 759	32 267	34 351	Construction	F	45	

NEW ZEALAND

Table NZ.7 **EXPORTS - EXPORTATIONS**

Millions of NZ$ (Current Prices)

ISIC revision 3				1990-91	1991-92	1992-93	1993-94	1994-95	1995-96	1996-97	1997-98
C	10_14		Mining and quarrying	..	..	..	..	..	..	..	..
	10		Mining of coal and lignite; extraction of peat	..	..	..	..	..	..	..	..
	11		Crude petroleum & natural gas; service activities	..	..	..	..	..	..	..	..
	12		Mining of uranium and thorium ores	..	..	..	..	..	..	..	..
	13		Mining of metal ores	..	..	27	25	19	22	11	19
	14		Other mining and quarrying	..	..	27	28	24	25	27	25
D	15_37		Total manufacturing	..	9 769	11 707	11 936	11 954	12 461	12 674	13 674
	15		Food products and beverages	..	8 399	8 760	8 804	8 716	9 376	9 727	10 419
		151	Production of meat, fish, fruit, vegetables, oils & fats	..	5 071	5 034	4 910	4 907	4 861	4 888	5 024
		152	Dairy products	..	3 022	3 410	3 500	3 376	4 034	4 349	4 866
		153	Grain mill and starch products; animal feeds	..	48	36	48	48	54	46	50
		154	Other food products	..	162	185	240	282	298	285	298
		155	Beverages	..	96	95	105	102	129	159	181
	16		Tobacco products	..	4	4	5	6	5	7	8
	17		Textiles	..	..	..	..	..	..	..	..
		171	Spinning, weaving and finishing of textiles	..	859	826	1 020	979	905	780	..
		172	Other textiles	..	..	..	..	..	..	122	133
		173	Knitted and crocheted fabrics and articles	..	71	78	79	93	82	85	87
	18		Wearing apparel and fur	..	161	178	177	187	176	165	162
		181	Wearing apparel, except fur apparel	..	99	105	94	101	88	104	106
		182	Dressing and dyeing of fur; articles of fur	..	63	72	83	86	89	61	56
	19		Leather and articles of leather; footwear	..	243	261	285	288	315	305	341
		191	Leather and articles of leather	..	212	229	243	247	275	264	298
		192	Footwear	..	31	33	42	42	40	41	42
	20		Wood & products of wood & cork, exc. furniture	..	..	890	980	982	880	923	970
		201	Sawmilling and planing of wood	..	401	563	585	554	466	532	640
		202	Wood products; articles of cork, straw, etc.	..	279	327	395	428	414	391	330
	21		Paper and paper products	..	764	763	824	979	878	724	910
	22		Printing, publishing and recorded media	..	..	..	..	..	..	..	..
		221	Publishing	..	..	..	..	..	..	..	..
		222	Printing and related activities	..	..	..	..	..	..	..	..
		223	Reproduction of recorded media	..	..	..	..	..	..	..	..
	23		Coke and petroleum products; nuclear fuel	..	..	..	..	..	..	..	..
		231	Coke oven products	..	0	0	0	0	0	0	0
		232	Refined petroleum products	..	104	117	73	16	34	57	36
		233	Processing of nuclear fuel	..	..	..	..	..	..	..	..
	24		Chemicals and chemical products	..	..	..	..	..	..	..	..
		241	Basic chemicals	..	261	313	590	620	521	648	499
		242	Other chemical products	..	263	271	344	378	378	341	406
		243	Man-made fibres	..	..	..	..	..	..	..	..
	25		Rubber and plastic products	..	198	240	267	266	250	256	268
		251	Rubber products	..	81	98	112	111	101	94	102
		252	Plastic products	..	117	141	155	155	149	162	166
	26		Other non-metallic mineral products	..	..	..	..	..	..	..	..
		261	Glass and glass products	..	26	29	31	27	29	30	18
		269	Non-metallic mineral products, nec	..	..	..	..	..	..	..	..
	27		Basic metals	..	..	..	..	..	..	..	..
		271	Basic iron and steel	..	332	314	283	307	272	307	341
		272	Basic precious and non-ferrous metals	..	922	970	1 034	1 240	1 078	1 169	1 177
		273	Casting of metals	..	..	..	..	..	..	..	..
	28		Fabricated metal products, exc. machinery	..	..	..	..	..	..	..	..
		281	Structural metal prod., tanks, reservoirs, etc.	..	..	..	..	..	..	..	..
		289	Other fabricated metal products	..	..	..	..	..	..	..	..
	29		Machinery and equipment, nec	..	..	..	..	..	..	..	..
		291	General purpose machinery	..	..	..	..	..	..	..	..
		292	Special purpose machinery	..	..	..	..	..	..	..	..
		293	Domestic appliances, nec	..	145	178	215	192	207	155	181
	30		Office, accounting & computing machinery	..	..	..	..	..	..	..	..
	31		Electrical machinery & apparatus, nec	..	..	..	..	..	..	..	..
		311	Electric motors, generators and transformers	..	..	..	..	..	..	..	..
		312	Electricity distribution & control apparatus	..	116	128	182	205	227	242	271
		313	Insulated wire and cable	..	48	55	59	54	30	41	52
		314	Accumulators, primary cells & primary batteries	..	2	1	2	1	1	1	1
		315	Electric lamps and lighting equipment	..	17	19	22	26	22	25	23
		319	Other electrical equipment, nec	..	..	..	..	..	..	..	..
	32		Radio, TV, communication equipment	..	..	..	..	..	..	..	..
		321	Electronic valves, tubes; other electronic components	..	..	..	..	..	..	..	..
		322	TV/radio transmitters; line teleph./telegr. apparatus	..	..	..	..	..	..	..	..
		323	TV/radio receivers; sound and video app.; assoc. gds.	..	123	147	183	209	206	273	303
	33		Medical, precision & optical instrum.; watch & clocks	..	..	..	..	..	..	..	..
		331	Medical appl.; instruments & appl. for measuring, etc.	..	..	..	..	..	..	..	..
		332	Optical instruments and photographic equipment	..	19	40	35	27	16	19	22
		333	Watches and clocks	..	..	..	..	..	..	..	..
	34		Motor vehicles, trailers and semi-trailers	..	..	128	151	111	128	131	165
		341	Motor vehicles	..	22	23	14	13	14	15	44
		342	Bodies for motor vehicles, trailers & semi-trailers	..	7	7	12	12	13	15	13
		343	Parts & accessories for motor vehicles & their engines	..	85	98	124	87	100	100	109
	35		Other transport equipment	..	..	..	..	..	..	..	..
		351	Building and repairing of ships and boats	..	..	..	..	..	..	..	..
		352	Railway and tramway locomotives and rolling stock	..	1	0	1	1	0	6	9
		353	Aircraft and spacecraft	..	247	184	87	226	125	212	73
		359	Transport equipment, nec	..	..	8	10	8	8	10	10
	36		Furniture; manufacturing, nec	..	..	485	443	420	453	436	431
		361	Furniture	..	60	69	78	61	57	54	68
		369	Manufacturing, nec	..	..	416	365	359	395	382	363
	37		Recycling	..	..	..	..	..	..	..	..
E	40_41		Electricity, gas and water supply	..	..	..	..	..	..	..	..
	40		Electricity, gas, steam and hot water supply	..	..	..	..	..	..	..	..
F	45		Construction	..	..	..	..	..	..	..	..

Table NZ.8 **IMPORTATIONS - IMPORTS**

Millions de NZ$ (Prix courants)

1990-91	1991-92	1992-93	1993-94	1994-95	1995-96	1996-97	1997-98			CITI révision 3	
..	..	..	..	..	..	..	..	Activites extractives	C	10_14	
..	..	..	..	..	..	..	..	Extraction de charbon, de lignite et de tourbe		10	
..	..	..	..	..	..	..	..	Pétrole brut et gaz naturel ; activités annexes		11	
..	..	..	..	..	..	..	..	Extraction de Minerais d'Uranium et de Thorium		12	
..	..	2	1	2	1	2	2	Extraction de minerais métalliques		13	
..	173	166	150	154	138	137	168	Autres activités extractives		14	
..	5 112	5 778	6 886	7 510	8 047	7 073	7 126	Total activites manufacturieres	D	15_37	
..	1 032	1 145	1 286	1 353	1 382	1 477	1 718	Produits alimentaires et boissons		15	
..	392	437	457	497	483	493	589	Viande, poisson, fruits, légumes, huiles et graisses			151
..	19	22	33	38	39	44	43	Produits laitiers			152
..	89	96	129	131	131	148	164	Travail des grains; prod. amyl. ; alim. pour animaux			153
..	380	424	478	502	544	586	685	Autres produits alimentaires			154
..	152	166	189	184	185	207	238	Boissons			155
..	6	6	7	8	9	8	11	Produits à base de tabac		16	
..	..	..	..	..	..	..	..	Textiles		17	
..	..	..	..	..	..	..	..	Filature, tissage et achèvement des textiles			171
..	..	..	..	..	..	..	..	Autres articles textiles			172
..	215	214	272	266	286	342	369	Etoffes et articles de bonneterie			173
..	177	193	231	249	270	312	351	Articles d'habillement et fourrure		18	
..	175	189	225	243	262	308	344	Articles d'habillement autres qu'en fourrure			181
..	3	4	6	5	8	4	6	Prép. et teinture des fourrures ; articles en fourrure			182
..	148	170	192	194	198	227	230	Cuir et articles en cuir ; chaussures		19	
..	60	64	74	74	71	77	82	Cuir et articles en cuir sauf chaussures			191
..	88	106	118	119	127	150	148	Chaussures			192
..	..	..	..	..	..	..	..	Bois, articles en bois, sauf meubles		20	
..	34	44	42	45	48	48	50	Sciage et rabotage du bois			201
..	45	46	52	61	67	74	86	Articles en bois, liège, vannerie, sparterie			202
..	552	476	508	637	571	635	725	Papier, articles en papier et carton		21	
..	..	..	..	..	..	..	..	Edition, imprimerie et reproduction		22	
..	..	..	..	..	..	..	..	Edition			221
..	..	..	..	..	..	..	..	Imprimerie et activités annexes			222
..	..	..	..	..	..	..	..	Reproduction de supports enregistrés			223
..	..	..	..	..	..	..	..	Coke. et prod. pétroliers ; comb. nucléaires		23	
..	1	2	2	3	3	3	4	Cokéfaction			231
..	298	294	333	362	550	612	512	Produits pétroliers raffinés			232
..	..	..	..	..	..	..	..	Traitement de combustibles nucléaires			233
..	..	..	..	..	..	..	..	Produits chimiques		24	
..	1 444	1 580	1 679	1 786	1 684	1 628	1 775	Produits chimiques de base			241
..	1 106	1 214	1 255	1 303	1 371	1 457	1 621	Autres produits chimiques			242
..	..	..	..	..	..	..	..	Fibres synthétiques ou artificielles			243
..	397	450	469	523	521	513	614	Art. en caoutchouc et matières plastiques		25	
..	204	241	235	270	269	239	312	Articles en caoutchouc			251
..	192	209	233	253	252	274	302	Articles en matières plastiques			252
..	..	..	..	..	..	..	..	Autres produits minéraux non métalliques		26	
..	125	123	147	154	159	146	158	Verre et articles en verre			261
..	..	..	..	..	..	..	..	Produits minéraux non métalliques, nca			269
..	..	..	..	..	..	..	..	Produits métallurgiques de base		27	
..	370	409	479	503	533	480	454	Sidérurgie et première transformation de l'acier			271
..	304	321	345	397	333	345	362	Première transf. métaux précieux et non ferreux			272
..	..	..	..	..	..	..	..	Fonderie			273
..	..	..	..	..	..	..	..	Ouvrages en métaux, sauf machines		28	
..	..	..	..	..	..	..	..	Constr. et menuis. métal.; réservoirs; générateurs			281
..	..	..	..	..	..	..	..	Autres ouvrages en métaux			289
..	..	..	..	..	..	..	..	Machines et matériels, nca		29	
..	..	..	..	..	..	..	..	Machines d'usage général			291
..	..	..	..	..	..	..	..	Machines d'usage spécifique			292
..	194	191	211	213	219	220	218	Appareils domestiques, nca			293
..	..	..	..	..	..	..	..	Mach. de bureau, comptables, information		30	
..	..	..	..	..	..	..	..	Machines et appareils électriques, nca		31	
..	..	..	..	..	..	..	..	Moteurs, génératrices, transformateurs électriques			311
..	354	375	423	471	516	539	522	Matériel électrique de distribution et de commande			312
..	56	64	68	67	70	65	75	Fils et câbles électriques isolés			313
..	13	16	19	18	16	23	27	Accumulateurs et piles électriques			314
..	94	95	100	110	106	114	118	Lampes électriques et appareils d'éclairage			315
..	..	..	..	..	..	..	..	Autres matériels électriques, nca.			319
..	..	..	..	..	..	..	..	Equipements et appareils de radio, TV, etc.		32	
..	..	..	..	..	..	..	..	Tubes et valves électroniques;autres composants			321
..	..	..	..	..	..	..	..	Emetteurs radio/TV; app. téléphonie et télégraphie			322
..	1 015	1 071	1 229	1 459	1 383	1 405	1 499	Récepteurs radio/TV, son et image; art. associés			323
..	..	..	..	..	..	..	..	Instr. médicaux, de précision ; optique ; horlog.		33	
..	..	..	..	..	..	..	..	Instruments médicaux et de précision, etc.			331
..	153	193	198	209	218	234	243	Instruments d'optique ; matériel photographique			332
..	..	..	..	..	..	..	..	Horlogerie			333
..	2 799	3 338	4 194	4 547	5 095	3 901	2 713	Véhicules automobiles; remorques, etc.		34	
..	1 399	1 669	2 097	2 273	2 548	1 950	2 212	Véhicules automobiles			341
..	14	19	25	32	30	33	25	Carross. automobiles, remorques, semi-remorques			342
..	328	406	390	379	357	390	475	Pièces et accessoires pour automobiles et moteurs			343
..	..	..	..	..	..	..	..	Autres matériels de transport		35	
..	..	..	..	..	..	..	..	Construction et réparation de navires			351
..	6	13	4	8	10	17	20	Matériel ferroviaire roulant			352
..	881	414	774	670	213	468	926	Construction aéronautique et spatiale			353
..	72	77	74	78	82	88	96	Autres équipements de transport			359
..	..	528	551	596	637	704	765	Meubles ; activités de fabrication, nca		36	
..	65	71	92	107	132	147	163	Meubles			361
..	..	457	459	489	506	557	602	Activités de fabrication, nca			369
..	..	..	..	..	..	..	..	Récupération		37	
..	..	..	..	..	..	..	..	Electricité, gaz et eau	E	40_41	
..	..	..	..	..	..	..	..	Prod. et dist. d'électricité gaz, vapeur d'eau chaude		40	
..	..	..	..	..	..	..	..	Construction	F	45	

Table NY.1 **PRODUCTION - PRODUCTION**

Millions of NKr (Current Prices)

ISIC revision 3				1991	1992	1993	1994	1995	1996	1997	1998
C	10_14		Mining and quarrying	..	..	..	..	..	..	..	..
	10		Mining of coal and lignite; extraction of peat	258	226	163	191	165	125	152	153
	11		Crude petroleum & natural gas; service activities	..	..	..	..	..	..	..	..
	12		Mining of uranium and thorium ores	..	..	..	..	..	..	..	..
	13		Mining of metal ores	1 243	1 152	1 029	1 161	1 118	901	742	537
	14		Other mining and quarrying	2 357	1 973	1 974	3 123	3 336	4 138	4 592	5 167
D	15_37		Total manufacturing	296 888	286 087	291 964	341 234	367 958	390 959	427 715	459 599
	15		Food products and beverages	79 083	79 743	..	..	..	..	..	..
		151	Production of meat, fish, fruit, vegetables, oils & fats	36 823	36 895	37 302	41 445	43 281	46 301	49 199	54 577
		152	Dairy products	12 194	13 478	13 624	13 814	13 455	12 715	13 888	13 716
		153	Grain mill and starch products; animal feeds	10 000	9 161	8 895	10 566	10 994	11 849	11 777	12 207
		154	Other food products	9 134	8 985	9 506	11 010	11 770	11 589	11 859	12 950
		155	Beverages	10 932	11 224	..	..	..	..	..	..
	16		Tobacco products	..	..	..	..	..	..	..	..
	17		Textiles	3 302	3 247	3 205	3 887	3 520	3 867	4 102	4 247
		171	Spinning, weaving and finishing of textiles	1 037	1 039	989	1 106	1 022	1 068	1 207	1 108
		172	Other textiles	1 732	1 598	1 586	2 037	2 038	2 281	2 328	2 619
		173	Knitted and crocheted fabrics and articles	534	611	630	744	459	517	566	520
	18		Wearing apparel and fur	942	964	1 033	1 287	1 431	1 526	1 432	1 371
		181	Wearing apparel, except fur apparel	884	906	983	1 218	1 377	1 447	1 352	1 294
		182	Dressing and dyeing of fur; articles of fur	58	58	51	69	54	79	80	77
	19		Leather and articles of leather; footwear	506	505	516	626	527	479	521	466
		191	Leather and articles of leather	318	336	340	417	283	251	266	248
		192	Footwear	188	169	176	209	244	228	255	217
	20		Wood & products of wood & cork, exc. furniture	11 452	9 405	9 406	13 993	14 737	14 678	16 708	17 309
		201	Sawmilling and planing of wood	5 747	4 800	4 541	6 766	6 823	6 539	7 787	7 910
		202	Wood products; articles of cork, straw, etc.	5 704	4 605	4 865	7 228	7 914	8 139	8 921	9 399
	21		Paper and paper products	17 332	15 364	15 028	17 052	22 230	19 375	18 268	19 713
	22		Printing, publishing and recorded media	20 469	18 857	19 295	24 716	25 927	28 982	30 573	33 038
		221	Publishing	13 534	13 329	13 725	16 082	16 664	19 017	20 261	21 408
		222	Printing and related activities	6 850	5 443	5 486	8 260	8 902	9 544	9 945	11 245
		223	Reproduction of recorded media	85	85	85	374	361	421	366	384
	23		Coke and petroleum products; nuclear fuel	15 430	14 766	16 518	15 868	13 425	18 845	18 959	14 930
		231	Coke oven products	..	..	..	..	..	..	..	..
		232	Refined petroleum products	15 430	14 766	16 518	15 868	13 425	18 845	18 959	14 930
		233	Processing of nuclear fuel	..	..	..	..	..	..	..	..
	24		Chemicals and chemical products	22 990	22 775	24 320	27 305	29 532	30 126	33 490	35 124
		241	Basic chemicals	16 326	15 331	16 340	19 053	20 785	19 814	22 846	23 036
		242	Other chemical products	6 663	7 443	7 979	8 252	8 748	10 313	10 644	12 088
		243	Man-made fibres	..	..	..	..	..	..	..	..
	25		Rubber and plastic products	5 231	4 333	4 433	5 793	6 507	7 052	7 504	7 809
		251	Rubber products	799	428	409	691	676	791	770	815
		252	Plastic products	4 432	3 905	4 023	5 102	5 831	6 262	6 734	6 994
	26		Other non-metallic mineral products	7 462	6 616	6 672	9 188	9 906	12 497	12 920	13 477
		261	Glass and glass products	..	..	1 189	1 508	1 505	1 920	2 167	2 224
		269	Non-metallic mineral products, nec	..	..	5 483	7 680	8 401	10 577	10 752	..
	27		Basic metals	27 774	25 217	25 408	31 885	37 113	35 546	37 871	40 141
		271	Basic iron and steel	6 864	6 722	..	7 722	8 892	9 031	8 840	11 183
		272	Basic precious and non-ferrous metals	20 711	18 350	18 429	22 931	26 816	25 100	27 569	27 339
		273	Casting of metals	199	145	..	1 232	1 406	1 415	1 462	1 618
	28		Fabricated metal products, exc. machinery	10 919	9 560	9 474	11 986	13 793	14 698	16 615	20 490
		281	Structural metal prod., tanks, reservoirs, etc.	..	..	..	5 561	6 894	6 845	7 739	8 764
		289	Other fabricated metal products	..	..	..	6 425	6 899	7 853	8 876	11 726
	29		Machinery and equipment, nec	17 527	16 599	16 463	20 985	23 812	26 236	27 730	32 155
		291	General purpose machinery	10 745	10 444	9 954	12 696	14 543	15 968	17 982	21 637
		292	Special purpose machinery	5 900	5 298	5 623	7 241	8 305	9 161	8 545	9 327
		293	Domestic appliances, nec	882	856	886	1 048	963	1 107	1 202	1 191
	30		Office, accounting & computing machinery	1 088	1 005	928	1 450	1 292	1 129	1 516	2 053
	31		Electrical machinery & apparatus, nec	9 190	8 811	8 709	10 083	10 673	11 264	12 316	13 859
		311	Electric motors, generators and transformers	2 710	2 545	2 986	3 138	3 058	3 614	4 063	4 236
		312	Electricity distribution & control apparatus	1 845	1 770	1 924	2 361	2 685	2 261	2 775	3 351
		313	Insulated wire and cable	2 396	2 451	1 939	2 205	2 595	2 605	2 429	2 571
		314	Accumulators, primary cells & primary batteries	102	..	..	138	171	160	149	42
		315	Electric lamps and lighting equipment	1 033	968	914	1 009	1 140	1 140	1 147	1 193
		319	Other electrical equipment, nec	1 104	..	..	1 231	1 024	1 484	1 753	2 466
	32		Radio, TV, communication equipment	3 489	3 701	3 766	4 633	5 672	6 411	7 243	5 985
		321	Electronic valves, tubes; other electronic components	401	526	570	652	804	945	1 456	2 398
		322	TV/radio transmitters; line teleph./telegr. apparatus	2 562	2 764	2 583	3 410	4 487	4 873	5 264	2 870
		323	TV/radio receivers; sound and video app.; assoc. gds.	526	411	612	571	380	593	523	717
	33		Medical, precision & optical instrum.; watch & clocks	3 367	3 530	3 842	4 809	5 859	6 720	7 814	8 538
		331	Medical appl.; instruments & appl. for measuring, etc.	3 179	..	..	4 616	5 630	6 384	7 460	8 390
		332	Optical instruments and photographic equipment	187	..	..	193	229	336	..	..
		333	Watches and clocks	..	..	..	..	..	..	..	..
	34		Motor vehicles, trailers and semi-trailers	2 317	2 481	2 566	3 225	4 404	4 888	5 088	5 263
		341	Motor vehicles	31	..	..	316	..	461	393	407
		342	Bodies for motor vehicles, trailers & semi-trailers	675	..	..	884	..	1 241	1 122	1 252
		343	Parts & accessories for motor vehicles & their engines	1 610	1 690	1 771	2 024	2 846	3 186	3 573	3 604
	35		Other transport equipment	30 090	32 016	33 327	34 506	36 097	39 484	53 297	62 264
		351	Building and repairing of ships and boats	26 170	28 093	29 445	29 595	31 314	34 550	48 483	60 548
		352	Railway and tramway locomotives and rolling stock	1 272	1 352	1 449	1 335	1 156	1 248	1 379	439
		353	Aircraft and spacecraft	2 292	2 180	2 041	3 090	3 147	3 215	2 981	805
		359	Transport equipment, nec	357	391	393	486	480	471	454	471
	36		Furniture; manufacturing, nec	6 795	6 460	6 407	8 241	8 835	9 967	11 529	12 637
		361	Furniture	5 059	4 663	4 667	6 154	6 611	7 680	9 123	10 196
		369	Manufacturing, nec	1 735	1 798	1 740	2 087	2 224	2 286	2 405	2 442
	37		Recycling	136	132	132	484	461	580	893	1 504
E	40_41		Electricity, gas and water supply	..	..	..	..	..	..	..	..
	40		Electricity, gas, steam and hot water supply	..	..	..	..	..	..	..	..
F	45		Construction	..	..	..	..	..	..	..	..

Note:Unless given separately, ISIC 155 includes 16.

Table NY.2 VALEUR AJOUTÉE - VALUE ADDED

Millions de NKr (Prix courants)

1991	1992	1993	1994	1995	1996	1997	1998			CITI révision 3	
..	..	..	..	..	..	..	..	Activites extractives	C	10_14	
55	76	32	30	32	9	45	49	Extraction de charbon, de lignite et de tourbe		10	
..	..	..	..	..	..	..	..	Pétrole brut et gaz naturel ; activités annexes		11	
..	..	..	..	..	..	..	..	Extraction de Minerais d'Uranium et de Thorium		12	
377	411	271	440	348	269	307	176	Extraction de minerais métalliques		13	
1 055	891	910	1 357	1 464	1 598	1 742	1 827	Autres activités extractives		14	
85 187	84 969	88 899	103 518	114 575	115 565	125 071	133 531	Total activites manufacturieres	D	15_37	
17 022	19 183	..	..	..	..	..	..	Produits alimentaires et boissons		15	
5 718	5 983	5 975	6 347	6 453	7 101	7 971	8 885	Viande, poisson, fruits, légumes, huiles et graisses			151
		239	244	511	1 158	1 437	1 695	Produits laitiers			152
1 437	1 834	1 814	2 057	2 206	1 402	1 672	1 391	Travail des grains; prod. amyl. ; alim. pour animaux			153
3 323	3 370	3 530	4 149	4 311	3 776	4 183	4 587	Autres produits alimentaires			154
8 022	8 450	..	..	..	..	..	..	Boissons			155
..	1 801	..	..	..	..	..	..	Produits à base de tabac		16	
1 285	1 294	1 252	1 452	1 298	1 481	1 521	1 645	Textiles		17	
410	409	391	426	389	406	469	424	Filature, tissage et achèvement des textiles			171
650	609	582	715	714	873	837	990	Autres articles textiles			172
224	276	279	311	194	203	215	231	Etoffes et articles de bonneterie			173
369	378	408	484	514	542	523	520	Articles d'habillement et fourrure		18	
347	355	389	455	490	503	492	485	Articles d'habillement autres qu'en fourrure			181
22	23	20	29	24	39	32	35	Prép. et teinture des fourrures ; articles en fourrure			182
185	167	173	208	179	161	159	148	Cuir et articles en cuir ; chaussures		19	
104	92	96	113	83	74	66	67	Cuir et articles en cuir sauf chaussures			191
81	75	76	96	96	87	93	81	Chaussures			192
3 371	2 810	2 916	4 256	4 173	3 895	4 616	5 001	Bois, articles en bois, sauf meubles		20	
1 549	1 286	1 366	1 848	1 573	1 350	1 864	2 081	Sciage et rabotage du bois			201
1 821	1 524	1 549	2 408	2 601	2 545	2 752	2 920	Articles en bois, liège, vannerie, sparterie			202
4 260	3 634	3 906	4 599	7 236	5 884	4 990	5 705	Papier, articles en papier et carton		21	
8 961	8 466	8 911	11 252	11 616	12 208	13 048	13 245	Edition, imprimerie et reproduction		22	
5 977	6 038	6 435	7 373	7 700	8 072	8 701	8 855	Edition			221
2 935	2 380	2 428	3 666	3 712	3 938	4 172	4 225	Imprimerie et activités annexes			222
48	48	48	213	205	199	175	165	Reproduction de supports enregistrés			223
1 569	940	1 806	1 591	870	1 627	1 925	1 418	Coke. et prod. pétroliers ; comb. nucléaires		23	
..	..	..	..	..	..	..	..	Cokéfaction			231
1 569	940	1 806	1 591	870	1 627	1 925	1 418	Produits pétroliers raffinés			232
..	..	..	..	..	..	..	..	Traitement de combustibles nucléaires			233
6 998	7 079	7 834	8 850	10 743	10 361	10 421	10 298	Produits chimiques		24	
4 623	4 316	4 772	5 889	7 251	6 282	7 050	6 568	Produits chimiques de base			241
2 375	2 828	3 062	2 960	3 492	4 079	3 371	3 730	Autres produits chimiques			242
..	..	..	..	..	..	..	..	Fibres synthétiques ou artificielles			243
1 860	1 680	1 701	2 105	2 294	2 472	2 552	2 721	Art. en caoutchouc et matières plastiques		25	
334	176	180	270	287	314	264	328	Articles en caoutchouc			251
1 526	1 505	1 521	1 835	2 007	2 157	2 288	2 393	Articles en matières plastiques			252
2 983	2 407	2 755	3 525	3 840	4 176	4 312	4 682	Autres produits minéraux non métalliques		26	
..	..	501	622	646	748	829	832	Verre et articles en verre			261
..	..	2 254	2 903	3 194	3 428	3 483	..	Produits minéraux non métalliques, nca			269
5 942	5 560	6 017	7 611	9 867	8 376	8 851	9 938	Produits métallurgiques de base		27	
1 534	1 528	..	1 907	2 408	2 114	2 072	2 541	Sidérurgie et première transformation de l'acier			271
4 322	3 978	4 182	5 133	6 827	5 631	6 117	6 666	Première transf. métaux précieux et non ferreux			272
86	54	..	571	632	632	662	730	Fonderie			273
4 322	3 805	3 896	5 040	5 564	6 209	6 805	8 095	Ouvrages en métaux, sauf machines		28	
..	..	..	2 077	2 540	2 670	3 068	3 408	Constr. et menuis. métal.; réservoirs; générateurs			281
..	..	..	2 963	3 023	3 540	3 737	4 687	Autres ouvrages en métaux			289
6 421	5 989	5 912	7 291	8 084	8 418	9 374	10 902	Machines et matériels, nca		29	
3 781	3 534	3 341	4 118	4 702	4 791	5 719	6 846	Machines d'usage général			291
2 286	2 132	2 223	2 768	3 031	3 192	3 165	3 580	Machines d'usage spécifique			292
354	322	348	405	351	434	490	476	Appareils domestiques, nca			293
337	334	283	404	352	402	580	425	Mach. de bureau, comptables, information		30	
3 294	3 141	3 272	3 636	3 828	3 806	4 658	5 102	Machines et appareils électriques, nca		31	
920	859	1 018	955	1 195	1 183	1 692	1 678	Moteurs, génératrices, transformateurs électriques			311
758	710	736	867	972	869	1 122	1 268	Matériel électrique de distribution et de commande			312
858	802	848	858	899	848	793	820	Fils et câbles électriques isolés			313
23	..	..	38	63	35	29	23	Accumulateurs et piles électriques			314
392	365	359	402	425	421	434	447	Lampes électriques et appareils d'éclairage			315
343	..	..	517	274	450	587	867	Autres matériels électriques, nca.			319
1 520	1 338	1 477	1 695	1 976	2 165	2 341	2 301	Equipements et appareils de radio, TV, etc.		32	
164	214	203	250	312	336	505	800	Tubes et valves électroniques;autres composants			321
1 181	956	1 030	1 217	1 537	1 575	1 594	1 204	Emetteurs radio/TV; app. téléphonie et télégraphie			322
175	169	244	229	127	254	242	297	Récepteurs radio/TV, son et image; art. associés			323
1 328	1 351	1 410	1 877	2 187	2 178	2 476	2 881	Instr. médicaux, de précision ; optique ; horlog.		33	
1 260	..	..	1 801	2 132	2 119	2 457	2 816	Instruments médicaux et de précision, etc.			331
68	..	..	77	55	59	..	..	Instruments d'optique ; matériel photographique			332
..	..	..	..	..	..	..	..	Horlogerie			333
800	851	955	1 136	1 464	1 565	1 614	1 833	Véhicules automobiles; remorques, etc.		34	
12	..	..	66	..	34	20	96	Véhicules automobiles			341
246	..	..	321	..	391	350	413	Carross. automobiles, remorques, semi-remorques			342
543	598	686	749	997	1 140	1 244	1 324	Pièces et accessoires pour automobiles et moteurs			343
9 834	10 297	11 569	11 359	12 109	12 255	14 141	15 479	Autres matériels de transport		35	
8 237	8 735	9 951	9 623	10 601	10 283	12 197	14 986	Construction et réparation de navires			351
486	520	566	507	429	487	558	93	Matériel ferroviaire roulant			352
984	896	928	1 062	911	1 343	1 226	277	Construction aéronautique et spatiale			353
127	144	124	167	168	142	159	123	Autres équipements de transport			359
2 496	2 436	2 453	3 236	3 277	3 557	4 245	4 386	Meubles ; activités de fabrication, nca		36	
1 753	1 634	1 649	2 282	2 333	2 570	3 199	3 443	Meubles			361
742	802	804	954	944	988	1 046	943	Activités de fabrication, nca			369
31	29	29	188	163	147	274	386	Récupération		37	
..	..	..	..	..	..	..	..	Electricité, gaz et eau	E	40_41	
..	..	..	..	..	..	..	..	Prod. et dist. d'électricité gaz, vapeur d'eau chaude		40	
..	..	..	..	..	..	..	..	Construction	F	45	

Note:Unless given separately, ISIC 155 includes 16.

NORWAY

Table NY.3 **INVESTMENT - INVESTISSEMENT**

Millions of NKr (Current Prices)

ISIC revision 3				1991	1992	1993	1994	1995	1996	1997	1998
C	10_14		Mining and quarrying	..	..	..	..	..	..	..	..
	10		Mining of coal and lignite; extraction of peat	30	20	36	11	19	47	11	13
	11		Crude petroleum & natural gas; service activities	..	..	..	..	..	..	..	..
	12		Mining of uranium and thorium ores	..	..	..	..	..	..	..	..
	13		Mining of metal ores	77	72	56	83	111	10		32
	14		Other mining and quarrying	190	168	136	169	253	307	267	304
D	15_37		Total manufacturing	11 899	11 501	9 358	10 881	15 756	17 593	17 315	20 919
	15		Food products and beverages	2 520	2 314	..	..	..	..	..	..
		151	Production of meat, fish, fruit, vegetables, oils & fats	870	777	943	1 201	1 313	1 740	1 454	1 628
		152	Dairy products	472	443	333	421	367	405	295	391
		153	Grain mill and starch products; animal feeds	283	201	163	196	311	488	366	435
		154	Other food products	404	410	381	426	346	438	658	615
		155	Beverages	491	483	..	..	..	..	..	..
	16		Tobacco products	..	37	..	..	..	..	..	..
	17		Textiles	117	78	160	152	171	180	159	173
		171	Spinning, weaving and finishing of textiles	16	14	36	61	54	69	82	63
		172	Other textiles	86	40	103	62	95	71	55	95
		173	Knitted and crocheted fabrics and articles	15	25	22	29	22	40	21	15
	18		Wearing apparel and fur	11	9	13	28	34	25	36	7
		181	Wearing apparel, except fur apparel	10	9	11	27	32	24	20	6
		182	Dressing and dyeing of fur; articles of fur	1	0	1	1	2	1	16	1
	19		Leather and articles of leather; footwear	10	15	6	18	18	10	10	4
		191	Leather and articles of leather	5	13	3	8	16	1	3	2
		192	Footwear	5	2	2	10	2	10	8	3
	20		Wood & products of wood & cork, exc. furniture	472	418	294	773	824	960	609	649
		201	Sawmilling and planing of wood	327	287	157	509	471	359	313	384
		202	Wood products; articles of cork, straw, etc.	145	131	137	264	353	601	296	265
	21		Paper and paper products	1 240	2 973	938	621	1 669	1 337	1 500	2 085
	22		Printing, publishing and recorded media	623	644	619	955	1 243	1 404	1 504	2 424
		221	Publishing	336	384	298	585	495	803	858	1 096
		222	Printing and related activities	286	259	319	360	727	585	587	1 336
		223	Reproduction of recorded media	1	1	1	10	21	17	59	
	23		Coke and petroleum products; nuclear fuel	263	206	345	263	176	483	215	723
		231	Coke oven products	..	..	..	..	..	..	..	..
		232	Refined petroleum products	263	206	345	263	176	483	215	723
		233	Processing of nuclear fuel	..	..	..	..	..	..	..	..
	24		Chemicals and chemical products	1 907	1 164	1 025	1 181	3 134	3 167	2 811	3 158
		241	Basic chemicals	1 430	830	656	759	2 884	2 704	2 097	2 616
		242	Other chemical products	476	334	369	423	250	463	714	542
		243	Man-made fibres	..	..	..	..	..	..	..	..
	25		Rubber and plastic products	270	155	204	253	360	366	463	482
		251	Rubber products		10	12	..	24	28	55	56
		252	Plastic products	323	145	192	264	336	338	408	426
	26		Other non-metallic mineral products	706	35	287	514	813	972	596	800
		261	Glass and glass products	..	..	27	37	78	107	138	187
		269	Non-metallic mineral products, nec	..	..	260	477	735	865	458	..
	27		Basic metals	1 208	898	519	905	1 134	2 085	2 310	1 544
		271	Basic iron and steel	308	152	..	306	420	425	381	535
		272	Basic precious and non-ferrous metals	887	732	398	556	647	1 523	1 842	938
		273	Casting of metals	13	14	..	43	67	137	87	71
	28		Fabricated metal products, exc. machinery	338	436	246	240	443	611	597	919
		281	Structural metal prod., tanks, reservoirs, etc.	..	..	..	58	177	198	242	292
		289	Other fabricated metal products	..	..	..	182	266	413	355	627
	29		Machinery and equipment, nec	556	483	422	453	662	462	553	960
		291	General purpose machinery	323	291	263	246	317	96	541	656
		292	Special purpose machinery	191	163	132	174	308	337		283
		293	Domestic appliances, nec	42	29	27	32	37	28	49	21
	30		Office, accounting & computing machinery	46	7	11	28	56		71	21
	31		Electrical machinery & apparatus, nec	405	295	228	222	272	121	336	590
		311	Electric motors, generators and transformers	110	44	43	33	40	69	54	134
		312	Electricity distribution & control apparatus	57	63	55	65	100	82	78	78
		313	Insulated wire and cable	166	121	112	93	106	66	81	
		314	Accumulators, primary cells & primary batteries		..	..	1	2	5	3	..
		315	Electric lamps and lighting equipment	67	45	23	12	16	31	20	3
		319	Other electrical equipment, nec	5	..	..	18	9		99	524
	32		Radio, TV, communication equipment	164	110	134	176	178	143	191	134
		321	Electronic valves, tubes; other electronic components	15	10	16	32	35	47	87	68
		322	TV/radio transmitters; line teleph./telegr. apparatus	140	91	106	135	137	91	94	39
		323	TV/radio receivers; sound and video app.; assoc. gds.	9	9	12	9	6	6	10	26
	33		Medical, precision & optical instrum.; watch & clocks	79	118	117	179	273	289	168	291
		331	Medical appl.; instruments & appl. for measuring, etc.	77	..	..	177	263	281	156	284
		332	Optical instruments and photographic equipment	3	..	..	2	10	8	..	..
		333	Watches and clocks	..	..	..	..	..	..	..	..
	34		Motor vehicles, trailers and semi-trailers	121	85	155	134	201	271	445	427
		341	Motor vehicles	2	..	..	3	..	13	13	9
		342	Bodies for motor vehicles, trailers & semi-trailers	9	..	..	13	..	30	26	39
		343	Parts & accessories for motor vehicles & their engines	111	77	146	118	178	228	406	379
	35		Other transport equipment	584	839	912	821	846	770	1 069	1 155
		351	Building and repairing of ships and boats	519	744	729	698	698	718	916	1 064
		352	Railway and tramway locomotives and rolling stock	13	42	128	57	21		57	2
		353	Aircraft and spacecraft	40	46	37	55	118	74	76	76
		359	Transport equipment, nec	11	7	18	11	9	23	21	12
	36		Furniture; manufacturing, nec	259	180	211	250	342	392	431	440
		361	Furniture	148	94	129	191	266	284	331	327
		369	Manufacturing, nec	111	86	83	59	77	108	100	113
	37		Recycling	1	1	1	27	18	113	111	25
E	40_41		Electricity, gas and water supply	..	..	..	..	..	..	..	..
	40		Electricity, gas, steam and hot water supply	..	..	..	..	..	..	..	..
F	45		Construction	..	..	..	..	..	..	..	..

Note:Unless given separately, ISIC 155 includes 16.

Table NY.4 DÉPENSES DE MACHINES ET D'ÉQUIPEMENT - INVESTMENT IN MACHINERY & EQUIPMENT

Millions de NKr (Prix courants)

1991	1992	1993	1994	1995	1996	1997	1998			CITI révision 3	
..	..	..	..	..	..	..	..	Activites extractives	C	10_14	
20	14	7		6	62	7		Extraction de charbon, de lignite et de tourbe		10	
..	..	..	..	..	..	..	..	Pétrole brut et gaz naturel ; activités annexes		11	
..	..	..	..	..	..	..	..	Extraction de Minerais d'Uranium et de Thorium		12	
43	21	11	34	96	12		14	Extraction de minerais métalliques		13	
139	108	105	118	112	142	173	116	Autres activités extractives		14	
8 309	6 658	6 093	7 198	8 396	10 872	11 558	14 317	Total activites manufacturieres	D	15_37	
1 760	1 605	..	..	..	..	..	..	Produits alimentaires et boissons		15	
561	506	521	658	657	942	837	949	Viande, poisson, fruits, légumes, huiles et graisses			151
359	348	248	270	238	268	194	256	Produits laitiers			152
197	113	97	126	173	202	164	200	Travail des grains; prod. amyl. ; alim. pour animaux			153
271	279	245	287	244	308	450	430	Autres produits alimentaires			154
373	358	..	..	..	..	..	..	Boissons			155
..	19	..	..	..	..	..	..	Produits à base de tabac		16	
109	66	153	117	108	144	112	110	Textiles		17	
16	17	36	46	40	62	55	47	Filature, tissage et achèvement des textiles			171
82	26	98	47	52	56	36	48	Autres articles textiles			172
11	23	20	25	17	25	21	15	Etoffes et articles de bonneterie			173
9	9	10	19	22	21	15	7	Articles d'habillement et fourrure		18	
8	9	9	20	21	21	16	5	Articles d'habillement autres qu'en fourrure			181
1	0	1	..	2	..	..	1	Prép. et teinture des fourrures ; articles en fourrure			182
8	11	4	12	3	9	5	4	Cuir et articles en cuir ; chaussures		19	
4	9	1	4	..	1	1	1	Cuir et articles en cuir sauf chaussures			191
4	2	2	8	2	8	4	2	Chaussures			192
321	307	180	482	499	395	382	391	Bois, articles en bois, sauf meubles		20	
212	179	92	291	253	168	222	188	Sciage et rabotage du bois			201
109	128	88	191	246	227	160	203	Articles en bois, liège, vannerie, sparterie			202
576	582	513	484	770	1 093	1 098	1 482	Papier, articles en papier et carton		21	
559	499	634	800	801	1 017	1 093	2 022	Edition, imprimerie et reproduction		22	
294	258	369	417	373	511	573	809	Edition			221
263	241	264	370	408	492	514	1 219	Imprimerie et activités annexes			222
1	1	1	13	20	15	6		Reproduction de supports enregistrés			223
198	32	61	85	107	49	39	53	Coke. et prod. pétroliers ; comb. nucléaires		23	
..	..	..	..	..	..	..	..	Cokéfaction			231
198	32	61	85	107	49	39	53	Produits pétroliers raffinés			232
..	..	..	..	..	..	..	..	Traitement de combustibles nucléaires			233
1 170	787	650	771	1 222	1 908	1 495	2 018	Produits chimiques		24	
821	522	355	495	1 031	1 611	1 054	1 796	Produits chimiques de base			241
348	266	295	277	191	297	442	222	Autres produits chimiques			242
..	..	..	..	..	..	..	..	Fibres synthétiques ou artificielles			243
268	123	170	207	249	284	364	378	Art. en caoutchouc et matières plastiques		25	
4	8	13	10	21	24	37	39	Articles en caoutchouc			251
264	115	157	196	228	260	327	339	Articles en matières plastiques			252
503	257	162	192	259	345	366	449	Autres produits minéraux non métalliques		26	
..	..	22	28	68	96	85	139	Verre et articles en verre			261
..	..	140	163	191	249	280	..	Produits minéraux non métalliques, nca			269
920	590	324	534	536	1 029	1 499	1 025	Produits métallurgiques de base		27	
212	51	..	89	107	147	266	332	Sidérurgie et première transformation de l'acier			271
699	527	290	411	382	779	1 153	638	Première transf. métaux précieux et non ferreux			272
9	12	..	33	47	103	81	55	Fonderie			273
246	290	140	223	300	421	372	652	Ouvrages en métaux, sauf machines		28	
..	..	..	80	111	138	164	178	Constr. et menuis. métal.; réservoirs; générateurs			281
..	..	..	143	189	281	208	473	Autres ouvrages en métaux			289
448	424	310	332	471	565	703	807	Machines et matériels, nca		29	
249	261	180	161	219	245	403	459	Machines d'usage général			291
164	141	106	153	232	296	260	311	Machines d'usage spécifique			292
35	22	25	18	20	24	41	36	Appareils domestiques, nca			293
46	6	11	24	56	35	71	16	Mach. de bureau, comptables, information		30	
295	179	181	236	213	58	293	679	Machines et appareils électriques, nca		31	
92	38	34	68	31	40	36	76	Moteurs, génératrices, transformateurs électriques			311
51	55	37	58	65	72	59	70	Matériel électrique de distribution et de commande			312
103	45	83	78	95	51	77	68	Fils et câbles électriques isolés			313
..	..	..	1	2	3	3	..	Accumulateurs et piles électriques			314
49	27	21	12	15	30	33	12	Lampes électriques et appareils d'éclairage			315
0	..	..	19	6		86	452	Autres matériels électriques, nca.			319
158	108	128	166	187	135	167	113	Equipements et appareils de radio, TV, etc.		32	
13	9	14	25	26	44	65	65	Tubes et valves électroniques;autres composants			321
137	91	103	133	155	88	92	32	Emetteurs radio/TV; app. téléphonie et télégraphie			322
9	8	11	7	6	4	10	16	Récepteurs radio/TV, son et image; art. associés			323
69	111	94	157	229	257	262	241	Instr. médicaux, de précision ; optique ; horlog.		33	
67	..	..	155	220	251	253	234	Instruments médicaux et de précision, etc.			331
3	..	..	2	9	6	..	..	Instruments d'optique ; matériel photographique			332
..	..	..	..	..	..	..	..	Horlogerie			333
129	82	141	108	149	235	294	372	Véhicules automobiles; remorques, etc.		34	
0	..	..	3	..	11	13	6	Véhicules automobiles			341
9	..	..	7	..	28	18	34	Carross. automobiles, remorques, semi-remorques			342
120	74	134	98	129	196	264	332	Pièces et accessoires pour automobiles et moteurs			343
309	411	506	385	406	478	654	852	Autres matériels de transport		35	
257	359	376	293	280	410	568	771	Construction et réparation de navires			351
11	10	84	47	10	11	12	2	Matériel ferroviaire roulant			352
31	35	29	38	107	48	58	73	Construction aéronautique et spatiale			353
10	6	18	6	8	8	17	6	Autres équipements de transport			359
208	158	185	206	243	306	281	335	Meubles ; activités de fabrication, nca		36	
112	82	109	160	188	212	201	213	Meubles			361
96	76	76	46	55	95	80	122	Activités de fabrication, nca			369
1	1	1	20	4	80	65	16	Récupération		37	
..	..	..	..	..	..	..	..	Electricité, gaz et eau	E	40_41	
..	..	..	..	..	..	..	..	Prod. et dist. d'électricité gaz, vapeur d'eau chaude		40	
..	..	..	..	..	..	..	..	Construction	F	45	

Note:Unless given separately, ISIC 155 includes 16.

NORWAY

Table NY.5 **EMPLOYMENT, EMPLOYEES - EMPLOI, SALARIÉS**

Units

ISIC revision 3				1991	1992	1993	1994	1995	1996	1997	1998
C	10_14		Mining and quarrying	..	..	..	..	..	..	..	..
	10		Mining of coal and lignite; extraction of peat	491	458	331	352	318	290	273	246
	11		Crude petroleum & natural gas; service activities	..	..	..	..	..	..	..	..
	12		Mining of uranium and thorium ores	..	..	..	..	..	..	..	..
	13		Mining of metal ores	1 885	1 659	1 424	1 356	1 161	1 011	708	547
	14		Other mining and quarrying	2 345	1 672	1 785	2 725	2 821	2 980	3 078	3 163
D	15_37		Total manufacturing	255 787	234 742	234 670	267 346	275 336	280 743	290 177	294 168
	15		Food products and beverages	47 272	43 602	..	..	..	..	..	..
		151	Production of meat, fish, fruit, vegetables, oils & fats	22 367	21 372	22 505	24 921	25 361	25 945	26 532	26 612
		152	Dairy products	5 926	5 709	5 625	5 910	5 654	5 507	5 635	5 750
		153	Grain mill and starch products; animal feeds	2 476	1 940	1 831	2 505	2 536	2 665	2 618	2 595
		154	Other food products	11 201	9 536	9 669	11 299	11 651	11 918	11 572	12 149
		155	Beverages	5 302	5 045	..	..	..	..	..	..
	16		Tobacco products	..	..	..	..	..	..	..	..
	17		Textiles	5 024	4 769	4 604	5 403	4 708	5 031	5 219	5 232
		171	Spinning, weaving and finishing of textiles	1 557	1 454	1 452	1 462	1 409	1 428	1 547	1 360
		172	Other textiles	2 397	2 155	2 005	2 551	2 482	2 798	2 747	3 056
		173	Knitted and crocheted fabrics and articles	1 070	1 160	1 147	1 390	817	805	925	816
	18		Wearing apparel and fur	1 902	1 724	1 644	1 832	2 419	2 378	2 194	1 970
		181	Wearing apparel, except fur apparel	1 794	1 621	1 549	1 687	2 300	2 250	2 080	1 861
		182	Dressing and dyeing of fur; articles of fur	108	103	95	145	119	128	114	109
	19		Leather and articles of leather; footwear	794	753	726	877	766	725	765	621
		191	Leather and articles of leather	398	382	366	431	331	288	335	234
		192	Footwear	396	371	360	446	435	437	430	387
	20		Wood & products of wood & cork, exc. furniture	13 681	10 615	10 289	13 842	14 224	14 326	14 970	15 266
		201	Sawmilling and planing of wood	5 637	4 660	4 432	5 316	5 238	5 228	5 435	5 447
		202	Wood products; articles of cork, straw, etc.	8 044	5 955	5 857	8 526	8 986	9 098	9 535	9 819
	21		Paper and paper products	11 029	10 421	10 150	10 366	10 684	10 550	10 250	9 826
	22		Printing, publishing and recorded media	31 528	29 058	29 229	35 288	35 838	36 296	37 569	37 904
		221	Publishing	22 999	22 376	22 553	25 029	25 556	25 738	26 619	26 028
		222	Printing and related activities	8 392	6 545	6 539	9 906	9 865	10 157	10 598	11 524
		223	Reproduction of recorded media	137	137	137	353	417	401	352	352
	23		Coke and petroleum products; nuclear fuel	1 411	1 315	1 574	1 736	1 636	1 206	1 158	1 189
		231	Coke oven products	5	..	..	..	..	..	..	..
		232	Refined petroleum products	1 406	1 315	1 574	1 736	1 636	1 206	1 158	1 189
		233	Processing of nuclear fuel	..	..	..	..	..	..	..	..
	24		Chemicals and chemical products	13 638	13 287	13 214	13 518	13 856	13 856	13 687	13 748
		241	Basic chemicals	8 404	8 198	8 231	8 397	8 439	8 245	7 980	7 934
		242	Other chemical products	5 234	5 089	4 983	5 121	5 417	5 611	5 707	5 814
		243	Man-made fibres	..	..	..	..	..	..	..	..
	25		Rubber and plastic products	6 070	4 873	5 014	6 225	6 280	6 583	6 936	7 005
		251	Rubber products	1 229	590	630	871	847	840	836	843
		252	Plastic products	4 841	4 283	4 384	5 354	5 433	5 743	6 100	6 162
	26		Other non-metallic mineral products	8 003	6 756	6 515	8 076	8 287	9 394	9 596	9 935
		261	Glass and glass products	1 713	1 576	1 522	1 739	1 752	1 952	2 078	2 219
		269	Non-metallic mineral products, nec	6 290	5 180	4 993	6 337	6 535	7 442	7 518	..
	27		Basic metals	14 962	13 863	13 257	14 790	15 103	15 136	14 956	14 656
		271	Basic iron and steel	4 754	4 497	4 444	4 285	4 267	4 259	4 108	4 592
		272	Basic precious and non-ferrous metals	9 918	9 216	8 790	8 643	8 805	8 923	8 959	8 062
		273	Casting of metals	290	150	23	1 862	2 031	1 954	1 889	2 002
	28		Fabricated metal products, exc. machinery	15 457	13 051	12 705	15 523	16 693	17 660	18 919	20 746
		281	Structural metal prod., tanks, reservoirs, etc.	7 409	6 525	6 240	7 156	7 879	8 317	8 715	9 172
		289	Other fabricated metal products	8 048	6 526	6 465	8 367	8 814	9 343	10 204	11 574
	29		Machinery and equipment, nec	19 885	18 285	18 111	21 428	21 758	22 703	24 328	24 337
		291	General purpose machinery	11 204	10 715	10 268	12 142	12 122	12 670	13 923	14 394
		292	Special purpose machinery	7 439	6 470	6 744	8 124	8 393	8 796	9 155	8 648
		293	Domestic appliances, nec	1 242	1 100	1 099	1 162	1 243	1 237	1 250	1 295
	30		Office, accounting & computing machinery	989	918	774	816	889	746	767	814
	31		Electrical machinery & apparatus, nec	9 922	8 602	8 271	8 810	9 295	9 576	9 930	10 402
		311	Electric motors, generators and transformers	2 814	2 493	2 662	2 366	2 886	3 082	2 917	2 962
		312	Electricity distribution & control apparatus	2 510	2 128	2 176	2 461	2 582	2 477	2 690	2 767
		313	Insulated wire and cable	1 589	1 329	1 376	1 405	1 438	1 401	1 456	1 538
		314	Accumulators, primary cells & primary batteries	144	141	125	160	155	166	153	35
		315	Electric lamps and lighting equipment	1 431	1 318	1 157	1 278	1 313	1 376	1 344	1 361
		319	Other electrical equipment, nec	1 434	1 193	775	1 140	921	1 074	1 370	1 739
	32		Radio, TV, communication equipment	4 347	3 610	3 856	4 184	4 170	4 468	4 862	4 526
		321	Electronic valves, tubes; other electronic components	618	668	687	812	942	1 074	1 461	1 922
		322	TV/radio transmitters; line teleph./telegr. apparatus	3 098	2 418	2 508	2 730	2 836	2 979	2 806	1 883
		323	TV/radio receivers; sound and video app.; assoc. gds.	631	524	661	642	392	415	595	721
	33		Medical, precision & optical instrum.; watch & clocks	3 204	3 256	3 409	4 413	5 057	5 326	5 497	5 829
		331	Medical appl.; instruments & appl. for measuring, etc.	3 103	3 162	3 310	4 273	4 927	5 207	5 383	5 733
		332	Optical instruments and photographic equipment	101	94	99	140	130	119	..	..
		333	Watches and clocks	..	..	..	..	..	..	..	..
	34		Motor vehicles, trailers and semi-trailers	3 224	3 150	3 199	3 463	4 544	4 822	4 842	4 729
		341	Motor vehicles	33	167	165	197	..	317	301	298
		342	Bodies for motor vehicles, trailers & semi-trailers	1 011	846	826	1 051	..	1 211	1 112	1 191
		343	Parts & accessories for motor vehicles & their engines	2 180	2 137	2 208	2 215	3 131	3 294	3 429	3 240
	35		Other transport equipment	32 949	33 515	34 564	35 202	35 761	34 942	37 078	37 318
		351	Building and repairing of ships and boats	27 274	28 343	29 408	30 117	30 814	30 340	32 764	35 743
		352	Railway and tramway locomotives and rolling stock	2 180	1 987	1 959	1 971	1 921	1 844	1 765	492
		353	Aircraft and spacecraft	2 988	2 668	2 623	2 484	2 457	2 149	2 104	591
		359	Transport equipment, nec	507	517	574	630	569	609	445	492
	36		Furniture; manufacturing, nec	10 421	9 249	8 952	11 038	11 903	12 317	13 772	14 225
		361	Furniture	7 461	6 372	6 405	7 921	8 504	9 063	10 302	10 747
		369	Manufacturing, nec	2 960	2 877	2 547	3 117	3 399	3 254	3 470	3 478
	37		Recycling	75	70	70	407	323	467	610	923
E	40_41		Electricity, gas and water supply	..	..	..	..	..	..	..	..
	40		Electricity, gas, steam and hot water supply	..	..	..	..	..	..	..	..
F	45		Construction	..	..	..	..	..	..	..	..

Note:Unless given separately, ISIC 155 includes 16.

Table NY.6 **SALAIRES ET CHARGES SOCIALES, SALARIÉS - COMPENSATION OF LABOUR, EMPLOYEES**

Millions de NKr (Prix courants)

1991	1992	1993	1994	1995	1996	1997	1998			CITI révision 3	
..	..	..	..	..	..	..	..	Activites extractives	C	**10_14**	
123	141	113	128	114	108	106	96	Extraction de charbon, de lignite et de tourbe		**10**	
..	..	..	..	..	..	..	..	Pétrole brut et gaz naturel ; activités annexes		**11**	
..	..	..	..	..	..	..	..	Extraction de Minerais d'Uranium et de Thorium		**12**	
431	388	373	417	319	296	212	176	Extraction de minerais métalliques		**13**	
554	440	471	718	789	801	929	1 029	Autres activités extractives		**14**	
60 933	58 206	59 502	69 708	74 172	78 035	84 852	91 180	Total activites manufacturieres	D	**15_37**	
..	..	..	..	..	..	..	..	Produits alimentaires et boissons		**15**	
4 344	4 353	4 640	5 334	5 595	5 944	6 412	6 711	Viande, poisson, fruits, légumes, huiles et graisses			151
1 390	1 386	1 394	1 466	1 477	1 520	1 607	1 704	Produits laitiers			152
625	515	525	723	748	809	853	900	Travail des grains; prod. amyl. ; alim. pour animaux			153
2 134	1 978	2 033	2 435	2 646	2 672	2 752	3 102	Autres produits alimentaires			154
1 394	1 385	..	..	..	..	..	..	Boissons			155
..	..	..	..	..	..	..	..	Produits à base de tabac		**16**	
923	906	884	1 083	963	1 066	1 185	1 247	Textiles		**17**	
296	297	291	305	297	314	360	329	Filature, tissage et achèvement des textiles			171
445	410	393	532	516	593	645	742	Autres articles textiles			172
182	200	199	247	149	159	180	175	Etoffes et articles de bonneterie			173
300	273	261	310	431	451	438	421	Articles d'habillement et fourrure		**18**	
282	258	243	281	405	423	409	392	Articles d'habillement autres qu'en fourrure			181
18	16	18	29	25	27	30	29	Prép. et teinture des fourrures ; articles en fourrure			182
137	130	133	160	144	136	158	138	Cuir et articles en cuir ; chaussures		**19**	
72	68	70	83	64	61	76	61	Cuir et articles en cuir sauf chaussures			191
65	62	63	77	80	75	82	77	Chaussures			192
2 730	2 206	2 158	3 027	3 227	3 343	3 702	3 969	Bois, articles en bois, sauf meubles		**20**	
1 151	992	951	1 164	1 206	1 257	1 402	1 457	Sciage et rabotage du bois			201
1 579	1 214	1 206	1 863	2 021	2 086	2 300	2 512	Articles en bois, liège, vannerie, sparterie			202
2 659	2 585	2 569	2 761	3 033	3 084	3 135	3 286	Papier, articles en papier et carton		**21**	
7 103	6 465	6 637	8 463	8 778	9 301	9 966	10 798	Edition, imprimerie et reproduction		**22**	
4 768	4 549	4 724	5 486	5 729	6 069	6 591	6 841	Edition			221
2 292	1 873	1 871	2 873	2 925	3 093	3 274	3 851	Imprimerie et activités annexes			222
42	42	42	104	125	139	100	105	Reproduction de supports enregistrés			223
541	496	584	654	645	496	526	528	Coke. et prod. pétroliers ; comb. nucléaires		**23**	
..	..	..	..	..	..	..	..	Cokéfaction			231
541	496	584	654	645	496	526	528	Produits pétroliers raffinés			232
..	..	..	..	..	..	..	..	Traitement de combustibles nucléaires			233
4 063	4 123	4 182	4 426	4 850	4 934	5 149	5 399	Produits chimiques		**24**	
2 620	2 661	2 658	2 821	2 994	3 009	3 038	3 123	Produits chimiques de base			241
1 442	1 462	1 524	1 605	1 856	1 925	2 111	2 276	Autres produits chimiques			242
..	..	..	..	..	..	..	..	Fibres synthétiques ou artificielles			243
1 327	1 113	1 173	1 481	1 611	1 750	1 947	2 056	Art. en caoutchouc et matières plastiques		**25**	
271	134	146	219	214	213	230	249	Articles en caoutchouc			251
1 056	979	1 027	1 262	1 396	1 537	1 717	1 807	Articles en matières plastiques			252
1 928	1 747	1 667	2 125	2 273	2 616	2 829	3 093	Autres produits minéraux non métalliques		**26**	
..	..	364	425	451	529	613	656	Verre et articles en verre			261
..	..	1 304	1 700	1 822	2 087	2 217	..	Produits minéraux non métalliques, nca			269
4 289	4 069	4 011	4 590	4 806	4 980	5 131	5 377	Produits métallurgiques de base		**27**	
1 292	1 242	..	1 257	1 316	1 370	1 385	1 668	Sidérurgie et première transformation de l'acier			271
2 934	2 790	2 750	2 863	2 966	3 086	3 198	3 113	Première transf. métaux précieux et non ferreux			272
62	37	..	470	524	524	548	596	Fonderie			273
3 477	3 063	3 017	3 915	4 331	4 701	5 300	6 225	Ouvrages en métaux, sauf machines		**28**	
..	..	..	1 810	2 080	2 263	2 440	2 698	Constr. et menuis. métal.; réservoirs; générateurs			281
..	..	..	2 106	2 252	2 438	2 860	3 528	Autres ouvrages en métaux			289
5 040	4 925	4 954	6 066	6 393	6 844	7 698	8 339	Machines et matériels, nca		**29**	
2 965	2 979	2 891	3 553	3 684	3 988	4 564	5 053	Machines d'usage général			291
1 813	1 697	1 816	2 220	2 407	2 553	2 769	2 912	Machines d'usage spécifique			292
262	250	248	293	302	303	365	374	Appareils domestiques, nca			293
298	264	260	308	272	249	287	323	Mach. de bureau, comptables, information		**30**	
2 541	2 414	2 401	2 605	2 853	3 045	3 287	3 692	Machines et appareils électriques, nca		**31**	
733	712	792	732	906	1 031	1 062	1 161	Moteurs, génératrices, transformateurs électriques			311
628	579	608	703	784	754	848	956	Matériel électrique de distribution et de commande			312
521	499	473	517	561	547	564	627	Fils et câbles électriques isolés			313
38	..	..	44	48	49	44	12	Accumulateurs et piles électriques			314
305	296	283	303	319	345	359	375	Lampes électriques et appareils d'éclairage			315
316	..	..	305	234	319	411	561	Autres matériels électriques, nca.			319
1 283	1 045	1 119	1 292	1 395	1 620	1 755	1 658	Equipements et appareils de radio, TV, etc.		**32**	
129	159	156	194	245	284	390	586	Tubes et valves électroniques;autres composants			321
1 010	760	799	923	1 042	1 214	1 197	862	Emetteurs radio/TV; app. téléphonie et télégraphie			322
144	126	165	175	108	122	168	210	Récepteurs radio/TV, son et image; art. associés			323
1 016	1 075	1 139	1 472	1 720	1 842	2 100	2 182	Instr. médicaux, de précision ; optique ; horlog.		**33**	
990	..	..	1 434	1 685	1 813	2 064	2 157	Instruments médicaux et de précision, etc.			331
26	..	..	38	35	30	..	..	Instruments d'optique ; matériel photographique			332
..	..	..	..	..	..	..	..	Horlogerie			333
709	718	761	844	1 194	1 312	1 407	1 428	Véhicules automobiles; remorques, etc.		**34**	
7	..	..	49	..	85	89	91	Véhicules automobiles			341
206	..	..	246	..	293	293	332	Carross. automobiles, remorques, semi-remorques			342
496	498	538	549	838	933	1 025	1 005	Pièces et accessoires pour automobiles et moteurs			343
8 618	9 050	9 674	10 005	10 526	10 569	11 923	12 926	Autres matériels de transport		**35**	
7 070	7 617	8 211	8 418	9 104	9 088	10 503	12 403	Construction et réparation de navires			351
492	493	506	524	506	524	528	191	Matériel ferroviaire roulant			352
957	829	832	921	792	798	777	212	Construction aéronautique et spatiale			353
99	110	124	143	124	160	115	120	Autres équipements de transport			359
2 050	1 907	1 892	2 401	2 531	2 805	3 345	3 647	Meubles ; activités de fabrication, nca		**36**	
1 409	1 276	1 290	1 676	1 740	2 029	2 450	2 736	Meubles			361
641	631	603	725	791	777	895	911	Activités de fabrication, nca			369
18	17	17	109	94	129	181	295	Récupération		**37**	
..	..	..	..	..	..	..	..	Electricité, gaz et eau	E	**40_41**	
..	..	..	..	..	..	..	..	Prod. et dist. d'électricité gaz, vapeur d'eau chaude		**40**	
..	..	..	..	..	..	..	..	Construction	F	**45**	

Note:Unless given separately, ISIC 155 includes 16.

Table NY.7 **WAGES AND SALARIES, EMPLOYEES - SALAIRES ET TRAITEMENTS, SALARIÉS**

Millions of NKr (Current Prices)

ISIC revision 3				1991	1992	1993	1994	1995	1996	1997	1998
C	10_14		Mining and quarrying	..	..	..	..	..	..	..	..
	10		Mining of coal and lignite; extraction of peat	121	138	113	89	82	86	85	80
	11		Crude petroleum & natural gas; service activities	..	..	..	..	..	..	..	..
	12		Mining of uranium and thorium ores	..	..	..	..	..	..	..	..
	13		Mining of metal ores	402	362	352	305	275	249	177	144
	14		Other mining and quarrying	485	388	422	615	672	688	791	877
D	15_37		Total manufacturing	52 917	50 574	52 688	58 459	62 221	65 753	71 453	76 786
	15		Food products and beverages	..	..	..	..	..	..	..	..
		151	Production of meat, fish, fruit, vegetables, oils & fats	3 842	3 865	4 193	4 612	4 842	5 135	5 557	5 828
		152	Dairy products	1 213	1 213	1 245	1 213	1 220	1 267	1 335	1 413
		153	Grain mill and starch products; animal feeds	544	447	466	605	627	664	710	746
		154	Other food products	1 844	1 708	1 791	2 052	2 228	2 274	2 342	2 635
		155	Beverages	1 203	1 196	..	..	..	..	..	..
	16		Tobacco products	..	..	..	..	..	..	..	..
	17		Textiles	802	787	782	936	838	926	1 026	1 078
		171	Spinning, weaving and finishing of textiles	258	259	259	267	259	271	311	285
		172	Other textiles	386	355	347	453	449	516	558	644
		173	Knitted and crocheted fabrics and articles	158	173	175	216	130	139	156	149
	18		Wearing apparel and fur	260	238	232	273	379	393	382	366
		181	Wearing apparel, except fur apparel	245	225	217	247	357	370	357	341
		182	Dressing and dyeing of fur; articles of fur	15	13	15	26	22	24	26	25
	19		Leather and articles of leather; footwear	118	112	117	139	125	117	137	120
		191	Leather and articles of leather	62	59	62	72	55	52	64	53
		192	Footwear	56	53	55	67	70	65	72	67
	20		Wood & products of wood & cork, exc. furniture	2 374	1 920	1 916	2 606	2 774	2 872	3 176	3 425
		201	Sawmilling and planing of wood	1 001	864	845	993	1 031	1 077	1 192	1 254
		202	Wood products; articles of cork, straw, etc.	1 373	1 055	1 070	1 614	1 743	1 795	1 984	2 171
	21		Paper and paper products	2 291	2 226	2 253	2 299	2 525	2 585	2 607	2 724
	22		Printing, publishing and recorded media	6 133	5 593	5 859	6 976	7 257	7 766	8 316	8 934
		221	Publishing	4 126	3 943	4 179	4 484	4 689	5 047	5 474	5 619
		222	Printing and related activities	1 971	1 613	1 643	2 403	2 462	2 604	2 756	3 227
		223	Reproduction of recorded media	37	37	37	89	106	114	86	88
	23		Coke and petroleum products; nuclear fuel	474	434	520	533	534	408	422	461
		231	Coke oven products	..	..	..	..	..	..	..	..
		232	Refined petroleum products	474	434	520	533	534	408	422	461
		233	Processing of nuclear fuel	..	..	..	..	..	..	..	..
	24		Chemicals and chemical products	3 540	3 584	3 696	3 577	3 893	3 947	4 082	4 339
		241	Basic chemicals	2 292	2 322	2 359	2 281	2 415	2 395	2 372	2 514
		242	Other chemical products	1 248	1 261	1 338	1 296	1 478	1 552	1 710	1 826
		243	Man-made fibres	..	..	..	..	..	..	..	..
	25		Rubber and plastic products	1 150	966	1 038	1 245	1 357	1 468	1 642	1 732
		251	Rubber products	235	116	129	183	178	177	191	210
		252	Plastic products	916	849	909	1 062	1 179	1 290	1 451	1 522
	26		Other non-metallic mineral products	1 658	1 494	1 467	1 774	1 893	2 211	2 396	2 615
		261	Glass and glass products	..	..	320	350	361	445	518	553
		269	Non-metallic mineral products, nec	..	..	1 148	1 424	1 532	1 766	1 878	..
	27		Basic metals	3 788	3 593	3 604	3 770	3 970	4 216	4 332	4 521
		271	Basic iron and steel	1 151	1 105	..	1 059	1 103	1 149	1 167	1 407
		272	Basic precious and non-ferrous metals	2 583	2 457	2 461	2 322	2 426	2 629	2 708	2 617
		273	Casting of metals	53	32	..	390	441	438	457	497
	28		Fabricated metal products, exc. machinery	3 012	2 659	2 666	3 334	3 698	4 026	4 543	5 342
		281	Structural metal prod., tanks, reservoirs, etc.	..	..	..	1 550	1 781	1 950	2 099	2 318
		289	Other fabricated metal products	..	..	..	1 784	1 917	2 076	2 444	3 024
	29		Machinery and equipment, nec	4 359	4 263	4 376	5 089	5 362	5 780	6 511	7 046
		291	General purpose machinery	2 564	2 585	2 556	2 997	3 091	3 382	3 885	4 273
		292	Special purpose machinery	1 569	1 463	1 602	1 843	2 013	2 135	2 312	2 457
		293	Domestic appliances, nec	226	215	218	249	259	263	314	316
	30		Office, accounting & computing machinery	256	226	226	251	232	204	232	262
	31		Electrical machinery & apparatus, nec	2 204	2 094	2 117	2 136	2 317	2 524	2 737	3 096
		311	Electric motors, generators and transformers	635	613	695	591	725	849	872	959
		312	Electricity distribution & control apparatus	542	500	533	588	653	627	705	802
		313	Insulated wire and cable	456	439	420	416	455	456	469	527
		314	Accumulators, primary cells & primary batteries	33	..	..	37	39	41	37	11
		315	Electric lamps and lighting equipment	263	255	249	249	259	286	302	318
		319	Other electrical equipment, nec	275	..	..	256	186	265	353	480
	32		Radio, TV, communication equipment	1 113	902	990	1 070	1 159	1 334	1 442	1 352
		321	Electronic valves, tubes; other electronic components	111	137	137	164	206	239	328	498
		322	TV/radio transmitters; line teleph./telegr. apparatus	878	656	709	765	862	994	974	681
		323	TV/radio receivers; sound and video app.; assoc. gds.	124	109	144	141	90	101	140	174
	33		Medical, precision & optical instrum.; watch & clocks	876	928	1 000	1 217	1 436	1 505	1 732	1 804
		331	Medical appl.; instruments & appl. for measuring, etc.	854	..	..	1 185	1 406	1 481	1 702	1 783
		332	Optical instruments and photographic equipment	22	..	..	32	30	24	..	..
		333	Watches and clocks	..	..	..	..	..	..	..	..
	34		Motor vehicles, trailers and semi-trailers	613	621	674	713	1 021	1 118	1 195	1 208
		341	Motor vehicles	6	..	..	41	..	72	75	77
		342	Bodies for motor vehicles, trailers & semi-trailers	179	..	..	208	..	249	247	280
		343	Parts & accessories for motor vehicles & their engines	429	432	476	463	717	798	872	851
	35		Other transport equipment	7 459	7 835	8 525	8 521	8 924	8 993	10 116	10 900
		351	Building and repairing of ships and boats	6 129	6 612	7 257	7 201	7 738	7 761	8 931	10 465
		352	Railway and tramway locomotives and rolling stock	422	409	426	426	417	419	424	155
		353	Aircraft and spacecraft	823	719	733	775	667	681	658	175
		359	Transport equipment, nec	85	94	109	119	103	131	102	106
	36		Furniture; manufacturing, nec	1 776	1 655	1 673	2 038	2 163	2 400	2 865	3 142
		361	Furniture	1 223	1 108	1 142	1 438	1 497	1 751	2 111	2 372
		369	Manufacturing, nec	553	548	531	600	665	649	754	769
	37		Recycling	15	15	15	94	81	108	151	247
E	40_41		Electricity, gas and water supply	..	..	..	..	..	..	..	..
	40		Electricity, gas, steam and hot water supply	..	..	..	..	..	..	..	..
F	45		Construction	..	..	..	..	..	..	..	..

Note:Unless given separately, ISIC 155 includes 16.

Table NY.8 **CHARGES SOCIALES DES EMPLOYEURS - SUPPLEMENTS TO WAGES & SALARIES**

Millions de NKr (Prix courants)

1991	1992	1993	1994	1995	1996	1997	1998			CITI révision 3
..	..	..	..	..	..	..	..	Activites extractives	C	10_14
3	3	1	39	32	21	21	16	Extraction de charbon, de lignite et de tourbe		10
..	..	..	..	..	..	..	..	Pétrole brut et gaz naturel ; activités annexes		11
..	..	..	..	..	..	..	..	Extraction de Minerais d'Uranium et de Thorium		12
29	26	21	112	44	46	35	31	Extraction de minerais métalliques		13
69	53	49	103	116	113	138	152	Autres activités extractives		14
8 016	7 632	6 814	11 249	11 951	12 282	13 398	14 394	Total activites manufacturieres	D	15_37
1 241	1 188	..	..	..	..	..	..	Produits alimentaires et boissons		15
501	488	447	721	753	808	856	883	Viande, poisson, fruits, légumes, huiles et graisses		151
178	172	149	253	258	252	273	291	Produits laitiers		152
82	69	59	118	122	145	143	153	Travail des grains; prod. amyl. ; alim. pour animaux		153
290	270	242	383	418	398	410	467	Autres produits alimentaires		154
191	189	..	..	..	..	..	..	Boissons		155
..	..	..	..	..	..	..	..	Produits à base de tabac		16
121	119	102	147	125	139	160	168	Textiles		17
38	38	32	38	39	43	50	44	Filature, tissage et achèvement des textiles		171
59	55	46	78	67	77	87	98	Autres articles textiles		172
24	27	24	31	19	21	24	26	Etoffes et articles de bonneterie		173
39	35	29	37	52	57	56	55	Articles d'habillement et fourrure		18
37	33	27	34	48	54	52	51	Articles d'habillement autres qu'en fourrure		181
2	2	2	4	3	3	3	3	Prép. et teinture des fourrures ; articles en fourrure		182
18	18	16	21	19	19	21	19	Cuir et articles en cuir ; chaussures		19
10	9	8	11	9	9	11	8	Cuir et articles en cuir sauf chaussures		191
9	8	8	10	10	10	10	10	Chaussures		192
356	286	242	421	453	471	526	544	Bois, articles en bois, sauf meubles		20
150	128	106	171	175	181	210	203	Sciage et rabotage du bois		201
206	158	136	250	279	290	316	341	Articles en bois, liège, vannerie, sparterie		202
368	359	316	462	507	499	529	563	Papier, articles en papier et carton		21
969	872	778	1 487	1 521	1 536	1 650	1 864	Edition, imprimerie et reproduction		22
643	606	545	1 003	1 040	1 021	1 116	1 222	Edition		221
321	260	227	470	462	489	519	625	Imprimerie et activités annexes		222
5	5	5	15	19	26	15	17	Reproduction de supports enregistrés		223
68	62	64	121	111	88	105	67	Coke. et prod. pétroliers ; comb. nucléaires		23
..	..	..	..	..	..	..	..	Cokéfaction		231
68	62	64	121	111	88	105	67	Produits pétroliers raffinés		232
..	..	..	..	..	..	..	..	Traitement de combustibles nucléaires		233
523	539	486	849	957	987	1 067	1 061	Produits chimiques		24
328	338	300	540	579	614	665	610	Produits chimiques de base		241
195	201	186	309	378	373	402	451	Autres produits chimiques		242
..	..	..	..	..	..	..	..	Fibres synthétiques ou artificielles		243
177	147	135	236	254	283	304	324	Art. en caoutchouc et matières plastiques		25
36	17	17	35	36	35	39	38	Articles en caoutchouc		251
140	130	118	200	217	247	266	284	Articles en matières plastiques		252
270	253	200	352	380	405	433	479	Autres produits minéraux non métalliques		26
..	..	44	76	90	84	95	102	Verre et articles en verre		261
..	..	156	276	290	322	338	..	Produits minéraux non métalliques, nca		269
501	476	408	820	836	764	800	856	Produits métallurgiques de base		27
141	137	..	198	213	221	219	262	Sidérurgie et première transformation de l'acier		271
351	334	289	542	541	457	490	495	Première transf. métaux précieux et non ferreux		272
9	5	..	80	82	86	92	100	Fonderie		273
465	405	351	581	633	676	757	884	Ouvrages en métaux, sauf machines		28
..	..	..	260	299	313	341	379	Constr. et menuis. métal.; réservoirs; générateurs		281
..	..	..	322	334	363	415	504	Autres ouvrages en métaux		289
681	662	579	978	1 030	1 065	1 187	1 293	Machines et matériels, nca		29
400	394	335	557	593	607	679	780	Machines d'usage général		291
244	233	214	377	394	419	458	455	Machines d'usage spécifique		292
37	35	30	45	43	41	51	59	Appareils domestiques, nca		293
42	38	34	56	39	45	55	60	Mach. de bureau, comptables, information		30
337	320	284	468	536	520	550	595	Machines et appareils électriques, nca		31
97	98	97	141	181	182	190	202	Moteurs, génératrices, transformateurs électriques		311
86	79	75	114	132	127	142	154	Matériel électrique de distribution et de commande		312
65	59	52	102	105	90	95	101	Fils et câbles électriques isolés		313
5	..	..	7	9	8	7	1	Accumulateurs et piles électriques		314
42	41	34	54	61	58	58	57	Lampes électriques et appareils d'éclairage		315
41	..	..	50	48	54	59	81	Autres matériels électriques, nca.		319
170	143	129	221	236	286	313	305	Equipements et appareils de radio, TV, etc.		32
18	22	19	30	38	45	61	88	Tubes et valves électroniques;autres composants		321
132	104	90	158	179	219	224	181	Emetteurs radio/TV; app. téléphonie et télégraphie		322
19	17	21	34	18	21	28	36	Récepteurs radio/TV, son et image; art. associés		323
140	147	139	255	285	337	368	378	Instr. médicaux, de précision ; optique ; horlog.		33
136	..	..	249	279	331	362	373	Instruments médicaux et de précision, etc.		331
4	..	..	6	5	6	..	..	Instruments d'optique ; matériel photographique		332
..	..	..	..	..	..	..	..	Horlogerie		333
96	96	87	131	172	194	212	219	Véhicules automobiles; remorques, etc.		34
1	..	..	8	..	13	13	14	Véhicules automobiles		341
28	..	..	37	..	45	46	51	Carross. automobiles, remorques, semi-remorques		342
67	66	62	86	120	135	152	154	Pièces et accessoires pour automobiles et moteurs		343
1 159	1 215	1 149	1 484	1 602	1 577	1 808	2 026	Autres matériels de transport		35
941	1 005	954	1 216	1 367	1 327	1 572	1 938	Construction et réparation de navires		351
69	84	81	98	89	105	103	36	Matériel ferroviaire roulant		352
135	110	99	146	125	117	119	38	Construction aéronautique et spatiale		353
14	16	15	24	21	28	13	15	Autres équipements de transport		359
274	251	219	363	368	406	480	505	Meubles ; activités de fabrication, nca		36
186	168	147	237	243	278	339	364	Meubles		361
88	83	72	125	125	128	141	142	Activités de fabrication, nca		369
2	2	2	15	13	21	30	48	Récupération		37
..	..	..	..	..	..	..	..	Electricité, gaz et eau	E	40_41
..	..	..	..	..	..	..	..	Prod. et dist. d'électricité gaz, vapeur d'eau chaude		40
..	..	..	..	..	..	..	..	Construction	F	45

Note:Unless given separately, ISIC 155 includes 16.

NORWAY

Table NY.9 HOURS WORKED BY EMPLOYEES - HEURES OUVRÉES, SALARIÉS

Thousands

ISIC revision 3				1991	1992	1993	1994	1995	1996	1997	1998
C	10_14		Mining and quarrying	..	..	..	..	..	..	..	..
	10		Mining of coal and lignite; extraction of peat	835	739	551	595	532	489	472	384
	11		Crude petroleum & natural gas; service activities	..	..	..	..	..	..	..	..
	12		Mining of uranium and thorium ores	..	..	..	..	..	..	..	..
	13		Mining of metal ores	6 913	5 474	3 716	2 149	1 849	1 621	1 127	888
	14		Other mining and quarrying	1 301	1 126	5 174	4 609	4 823	5 083	5 236	5 460
D	15_37		Total manufacturing	401 723	371 114	371 973	427 080	441 515	450 574	465 929	474 148
	15		Food products and beverages	..	..	..	..	..	..	..	..
		151	Production of meat, fish, fruit, vegetables, oils & fats	34 771	33 492	35 202	38 544	39 383	40 337	41 425	41 411
		152	Dairy products	10 151	9 771	9 664	9 938	9 598	9 635	9 568	9 641
		153	Grain mill and starch products; animal feeds	4 163	3 326	3 194	4 358	4 430	4 827	4 780	4 601
		154	Other food products	16 660	14 732	14 572	16 561	17 500	17 985	17 618	18 851
		155	Beverages	..	..	..	..	..	..	..	..
	16		Tobacco products	..	..	..	..	..	..	..	..
	17		Textiles	7 627	7 086	6 797	8 108	6 989	7 624	7 908	7 918
		171	Spinning, weaving and finishing of textiles	2 344	2 156	2 118	2 169	2 033	2 114	2 317	1 983
		172	Other textiles	3 707	3 256	3 036	3 989	3 814	4 359	4 243	4 766
		173	Knitted and crocheted fabrics and articles	1 576	1 674	1 643	1 950	1 142	1 151	1 348	1 169
	18		Wearing apparel and fur	2 816	2 420	2 305	2 613	3 615	3 544	3 283	2 936
		181	Wearing apparel, except fur apparel	2 660	2 285	2 162	2 387	3 429	3 345	3 099	2 766
		182	Dressing and dyeing of fur; articles of fur	156	136	143	226	186	199	184	170
	19		Leather and articles of leather; footwear	1 172	1 079	1 099	1 299	1 110	1 005	1 032	897
		191	Leather and articles of leather	600	566	575	667	489	449	458	359
		192	Footwear	572	513	524	632	621	556	574	538
	20		Wood & products of wood & cork, exc. furniture	22 243	17 340	16 903	22 981	23 726	23 800	24 642	25 260
		201	Sawmilling and planing of wood	9 252	7 742	7 401	8 786	8 833	8 864	9 007	9 270
		202	Wood products; articles of cork, straw, etc.	12 990	9 598	9 502	14 195	14 893	14 936	15 635	15 990
	21		Paper and paper products	18 046	16 829	16 721	17 324	17 939	17 602	16 995	16 490
	22		Printing, publishing and recorded media	40 914	35 798	36 525	45 531	46 376	46 899	48 477	49 591
		221	Publishing	26 876	24 702	25 674	28 719	29 364	29 526	30 288	29 637
		222	Printing and related activities	13 793	10 851	10 606	16 185	16 300	16 658	17 582	19 368
		223	Reproduction of recorded media	245	245	245	627	712	715	607	586
	23		Coke and petroleum products; nuclear fuel	2 337	2 106	2 555	2 858	2 709	2 003	1 936	1 982
		231	Coke oven products	9	..	..	..	..	..	..	..
		232	Refined petroleum products	2 329	2 106	2 555	2 858	2 709	2 003	1 936	1 982
		233	Processing of nuclear fuel	..	..	..	..	..	..	..	..
	24		Chemicals and chemical products	22 596	22 565	22 601	23 381	23 925	23 979	23 392	24 370
		241	Basic chemicals	14 129	14 265	14 296	14 664	14 752	14 544	13 961	14 074
		242	Other chemical products	8 468	8 300	8 306	8 717	9 173	9 435	9 431	10 296
		243	Man-made fibres	..	..	..	..	..	..	..	..
	25		Rubber and plastic products	9 934	8 013	8 286	10 342	10 594	10 893	11 556	11 852
		251	Rubber products	1 955	909	977	1 414	1 382	1 379	1 402	1 508
		252	Plastic products	7 978	7 104	7 309	8 928	9 212	9 515	10 154	10 344
	26		Other non-metallic mineral products	13 036	10 979	10 615	13 359	13 866	15 738	16 260	16 985
		261	Glass and glass products	2 691	2 408	2 348	2 698	2 765	3 134	3 375	3 604
		269	Non-metallic mineral products, nec	10 345	8 571	8 267	10 661	11 101	12 604	12 885	..
	27		Basic metals	25 068	23 438	22 741	25 256	25 507	26 012	26 087	24 864
		271	Basic iron and steel	8 029	7 695	7 650	7 407	7 271	7 370	6 992	7 429
		272	Basic precious and non-ferrous metals	16 592	15 508	15 054	14 987	14 990	15 532	15 979	14 183
		273	Casting of metals	447	235	37	2 861	3 246	3 110	3 115	3 252
	28		Fabricated metal products, exc. machinery	24 823	21 243	20 594	25 332	27 826	29 410	31 449	35 057
		281	Structural metal prod., tanks, reservoirs, etc.	12 080	10 756	10 179	11 845	13 249	13 896	14 642	15 468
		289	Other fabricated metal products	12 743	10 487	10 415	13 487	14 577	15 514	16 807	19 588
	29		Machinery and equipment, nec	32 487	30 437	30 387	36 353	37 521	38 774	41 243	41 468
		291	General purpose machinery	18 665	18 081	17 380	20 908	21 195	22 100	24 233	24 661
		292	Special purpose machinery	11 883	10 542	11 239	13 514	14 311	14 759	14 906	14 785
		293	Domestic appliances, nec	1 940	1 814	1 768	1 931	2 015	1 915	2 104	2 022
	30		Office, accounting & computing machinery	1 641	1 501	1 317	1 418	1 473	1 286	1 287	1 465
	31		Electrical machinery & apparatus, nec	16 171	14 079	13 795	14 357	15 194	15 786	16 677	17 729
		311	Electric motors, generators and transformers	4 720	4 110	4 418	3 846	4 807	5 124	4 970	4 939
		312	Electricity distribution & control apparatus	4 139	3 423	3 589	3 984	4 253	3 986	4 552	4 704
		313	Insulated wire and cable	2 612	2 245	2 401	2 295	2 301	2 358	2 447	2 810
		314	Accumulators, primary cells & primary batteries	231	229	212	264	268	267	247	61
		315	Electric lamps and lighting equipment	2 206	2 095	1 854	2 007	1 948	2 192	2 148	2 204
		319	Other electrical equipment, nec	2 263	1 978	1 322	1 961	1 618	1 859	2 313	3 011
	32		Radio, TV, communication equipment	7 055	6 107	6 488	7 085	7 057	7 551	8 176	7 639
		321	Electronic valves, tubes; other electronic components	1 047	1 125	1 145	1 365	1 608	1 794	2 464	3 271
		322	TV/radio transmitters; line teleph./telegr. apparatus	5 021	4 190	4 259	4 665	4 816	5 099	4 839	3 314
		323	TV/radio receivers; sound and video app.; assoc. gds.	987	792	1 084	1 055	633	658	873	1 054
	33		Medical, precision & optical instrum.; watch & clocks	5 577	5 656	5 755	7 663	8 942	9 520	9 684	9 693
		331	Medical appl.; instruments & appl. for measuring, etc.	5 406	5 502	5 596	7 441	8 733	9 318	9 491	9 537
		332	Optical instruments and photographic equipment	171	154	159	222	209	202	..	..
		333	Watches and clocks	..	..	..	..	..	..	..	..
	34		Motor vehicles, trailers and semi-trailers	5 135	5 036	5 133	5 720	7 808	8 045	7 992	7 771
		341	Motor vehicles	57	316	305	352	..	547	532	521
		342	Bodies for motor vehicles, trailers & semi-trailers	1 602	1 356	1 311	1 687	..	1 981	1 875	1 994
		343	Parts & accessories for motor vehicles & their engines	3 476	3 364	3 517	3 681	5 483	5 517	5 586	5 255
	35		Other transport equipment	51 866	54 552	55 568	58 351	58 440	57 206	61 410	61 485
		351	Building and repairing of ships and boats	42 735	46 198	47 077	49 910	50 270	49 666	54 261	58 905
		352	Railway and tramway locomotives and rolling stock	3 427	3 140	3 111	3 229	3 098	2 971	2 849	821
		353	Aircraft and spacecraft	5 005	4 451	4 506	4 255	4 195	3 594	3 614	996
		359	Transport equipment, nec	698	764	874	958	879	976	687	763
	36		Furniture; manufacturing, nec	16 055	14 486	14 193	17 812	19 161	19 635	22 008	22 574
		361	Furniture	11 600	9 945	10 122	12 757	13 737	14 434	16 465	17 244
		369	Manufacturing, nec	4 454	4 541	4 071	5 055	5 424	5 201	5 543	5 330
	37		Recycling	131	124	124	719	543	771	1 136	1 712
E	40_41		Electricity, gas and water supply	700	666	809	..	..	..	..	..
	40		Electricity, gas, steam and hot water supply	..	..	..	..	..	..	..	..
F	45		Construction	..	..	..	..	..	..	..	..

Note:Unless given separately, ISIC 155 includes 16.

Table NY.10 ÉTABLISSEMENTS - ESTABLISHMENTS

Unités

1991	1992	1993	1994	1995	1996	1997	1998		CITI révision 3		
..	..	..	..	..	..	..	..	Activites extractives	C	10_14	
7	4	3	12	12	8	10	11	Extraction de charbon, de lignite et de tourbe		10	
..	..	..	..	..	..	..	..	Pétrole brut et gaz naturel ; activités annexes		11	
..	..	..	..	..	..	..	..	Extraction de Minerais d'Uranium et de Thorium		12	
10	8	8	8	8	7	6	8	Extraction de minerais métalliques		13	
143	62	73	332	333	346	336	348	Autres activités extractives		14	
6 075	3 900	3 916	10 957	10 668	11 019	11 644	11 737	Total activites manufacturieres	D	15_37	
1 345	904	895	1 810	1 766	1 727	1 678	1 653	Produits alimentaires et boissons		15	
633	454	466	835	816	814	836	831	Viande, poisson, fruits, légumes, huiles et graisses			151
140	122	116	124	114	99	99	94	Produits laitiers			152
116	59	57	179	170	174	183	176	Travail des grains; prod. amyl. ; alim. pour animaux			153
407	227	220	623	612	588	510	506	Autres produits alimentaires			154
49	42	36	49	54	52	50	46	Boissons			155
..	6	..	2	2	2	2	2	Produits à base de tabac		16	
182	120	115	323	307	315	332	331	Textiles		17	
31	26	24	49	53	46	47	44	Filature, tissage et achèvement des textiles			171
121	70	68	221	214	228	242	247	Autres articles textiles			172
30	24	23	53	40	41	43	40	Etoffes et articles de bonneterie			173
82	61	58	141	154	160	169	165	Articles d'habillement et fourrure		18	
74	56	53	119	136	140	152	149	Articles d'habillement autres qu'en fourrure			181
8	5	5	22	18	20	17	16	Prép. et teinture des fourrures ; articles en fourrure			182
30	19	18	44	45	43	49	46	Cuir et articles en cuir ; chaussures		19	
16	10	9	23	22	22	25	23	Cuir et articles en cuir sauf chaussures			191
14	9	9	21	23	21	24	23	Chaussures			192
571	303	286	987	959	987	1 008	1 051	Bois, articles en bois, sauf meubles		20	
212	125	122	306	297	312	327	330	Sciage et rabotage du bois			201
359	178	164	681	662	675	681	721	Articles en bois, liège, vannerie, sparterie			202
102	84	82	119	115	108	108	108	Papier, articles en papier et carton		21	
805	453	452	1 831	1 785	1 870	2 099	2 062	Edition, imprimerie et reproduction		22	
289	204	207	594	577	615	754	758	Edition			221
511	244	240	1 171	1 135	1 177	1 272	1 236	Imprimerie et activités annexes			222
5	5	5	66	73	78	73	68	Reproduction de supports enregistrés			223
22	11	59	80	78	8	9	6	Coke. et prod. pétroliers ; comb. nucléaires		23	
1	..	..	..	..	..	..	..	Cokéfaction			231
21	11	59	80	78	8	9	6	Produits pétroliers raffinés			232
..	..	..	..	..	..	..	..	Traitement de combustibles nucléaires			233
127	100	99	161	162	177	181	197	Produits chimiques		24	
54	47	50	65	65	65	72	82	Produits chimiques de base			241
73	53	49	96	97	112	109	115	Autres produits chimiques			242
..	..	..	..	..	..	..	..	Fibres synthétiques ou artificielles			243
197	116	126	335	337	356	363	354	Art. en caoutchouc et matières plastiques		25	
31	15	16	65	63	67	64	58	Articles en caoutchouc			251
166	101	110	270	274	289	299	296	Articles en matières plastiques			252
274	143	177	521	500	600	607	615	Autres produits minéraux non métalliques		26	
35	29	30	70	57	65	68	74	Verre et articles en verre			261
239	114	147	451	443	535	539	541	Produits minéraux non métalliques, nca			269
81	65	67	127	116	121	120	121	Produits métallurgiques de base		27	
39	33	37	56	54	57	57	62	Sidérurgie et première transformation de l'acier			271
29	28	28	28	26	28	31	26	Première transf. métaux précieux et non ferreux			272
13	4	2	43	36	36	32	33	Fonderie			273
654	410	389	1 174	1 150	1 178	1 299	1 344	Ouvrages en métaux, sauf machines		28	
284	206	193	406	387	397	403	426	Constr. et menuis. métal.; réservoirs; générateurs			281
370	204	196	768	763	781	896	918	Autres ouvrages en métaux			289
471	300	294	1 142	1 123	1 164	1 268	1 295	Machines et matériels, nca		29	
280	196	187	607	606	638	707	727	Machines d'usage général			291
171	91	94	507	489	497	531	537	Machines d'usage spécifique			292
20	13	13	28	28	29	30	31	Appareils domestiques, nca			293
12	6	6	20	21	25	25	18	Mach. de bureau, comptables, information		30	
182	127	119	314	304	316	354	351	Machines et appareils électriques, nca		31	
36	31	31	62	65	63	63	64	Moteurs, génératrices, transformateurs électriques			311
51	31	33	75	74	73	83	77	Matériel électrique de distribution et de commande			312
16	12	14	16	15	18	18	18	Fils et câbles électriques isolés			313
3	2	2	5	5	4	3	3	Accumulateurs et piles électriques			314
20	16	15	35	38	37	38	46	Lampes électriques et appareils d'éclairage			315
56	35	24	121	107	121	149	143	Autres matériels électriques, nca.			319
52	42	44	72	66	75	73	85	Equipements et appareils de radio, TV, etc.		32	
14	13	15	30	29	34	34	44	Tubes et valves électroniques;autres composants			321
19	15	14	17	17	20	19	14	Emetteurs radio/TV; app. téléphonie et télégraphie			322
19	14	15	25	20	21	20	27	Récepteurs radio/TV, son et image; art. associés			323
76	55	55	269	285	333	323	330	Instr. médicaux, de précision ; optique ; horlog.		33	
72	52	52	259	278	327	315	322	Instruments médicaux et de précision, etc.			331
4	3	3	10	7	6	7	7	Instruments d'optique ; matériel photographique			332
..	..	..	..	..	..	1	1	Horlogerie			333
85	57	52	103	91	97	104	108	Véhicules automobiles; remorques, etc.		34	
3	2	2	8	1	4	4	4	Véhicules automobiles			341
36	20	20	52	49	53	52	54	Carross. automobiles, remorques, semi-remorques			342
46	35	30	43	41	40	48	50	Pièces et accessoires pour automobiles et moteurs			343
354	283	300	613	582	615	629	617	Autres matériels de transport		35	
319	255	268	566	539	564	582	591	Construction et réparation de navires			351
9	9	9	8	9	12	14	4	Matériel ferroviaire roulant			352
19	13	14	23	20	23	21	12	Construction aéronautique et spatiale			353
7	6	9	16	14	16	12	10	Autres équipements de transport			359
367	232	220	700	678	680	780	802	Meubles ; activités de fabrication, nca		36	
270	165	169	483	461	469	540	565	Meubles			361
97	67	51	217	217	211	240	237	Activités de fabrication, nca			369
4	3	3	69	42	62	64	76	Récupération		37	
..	..	..	..	..	..	..	..	Electricité, gaz et eau	E	40_41	
..	..	..	..	..	..	..	..	Prod. et dist. d'électricité gaz, vapeur d'eau chaude		40	
..	..	..	..	..	..	..	..	Construction	F	45	

Note:Unless given separately, ISIC 155 includes 16.

POLAND

Table PL .1 PRODUCTION - PRODUCTION

Millions of Zl (Current Prices)

ISIC revision 3				1991	1992	1993	1994	1995	1996	1997	1998
C	10_14		Mining and quarrying	..	9 624	10 872	15 775	19 617	22 421	26 276	24 565
	10		Mining of coal and lignite; extraction of peat	..	5 804	..	..	..	..	..	..
	11		Crude petroleum & natural gas; service activities	..	..	..	..	..	..	..	..
	12		Mining of uranium and thorium ores	..	..	..	..	..	..	..	..
	13		Mining of metal ores	..	..	..	..	..	..	..	..
	14		Other mining and quarrying	..	..	..	..	..	..	..	..
D	15_37		Total manufacturing	..	78 975	108 727	144 442	206 128	251 826	309 719	349 248
	15		Food products and beverages	..	20 592	..	..	..	..	..	..
		151	Production of meat, fish, fruit, vegetables, oils & fats	..	..	..	..	..	..	..	..
		152	Dairy products	..	..	..	..	..	..	..	..
		153	Grain mill and starch products; animal feeds	..	..	..	..	..	..	..	..
		154	Other food products	..	..	..	..	..	..	..	..
		155	Beverages	..	..	..	..	..	..	..	..
	16		Tobacco products	..	1 601	..	..	..	..	..	..
	17		Textiles	..	2 857	..	..	..	..	..	..
		171	Spinning, weaving and finishing of textiles	..	..	..	..	..	..	..	..
		172	Other textiles	..	..	..	..	..	..	..	..
		173	Knitted and crocheted fabrics and articles	..	..	..	..	..	..	..	..
	18		Wearing apparel and fur	..	3 022	..	..	..	..	..	..
		181	Wearing apparel, except fur apparel	..	..	..	..	..	..	..	..
		182	Dressing and dyeing of fur; articles of fur	..	..	..	..	..	..	..	..
	19		Leather and articles of leather; footwear	..	1 295	..	..	..	..	..	..
		191	Leather and articles of leather	..	..	..	..	..	..	..	..
		192	Footwear	..	..	..	..	..	..	..	..
	20		Wood & products of wood & cork, exc. furniture	..	2 208	..	..	..	..	..	..
		201	Sawmilling and planing of wood	..	..	..	..	..	..	..	..
		202	Wood products; articles of cork, straw, etc.	..	..	..	..	..	..	..	..
	21		Paper and paper products	..	1 296	..	..	..	..	..	..
	22		Printing, publishing and recorded media	..	2 001	..	..	..	..	..	..
		221	Publishing	..	..	..	..	..	..	..	..
		222	Printing and related activities	..	..	..	..	..	..	..	..
		223	Reproduction of recorded media	..	..	..	..	..	..	..	..
	23		Coke and petroleum products; nuclear fuel	..	6 423	..	..	..	..	..	..
		231	Coke oven products	..	..	..	..	..	..	..	..
		232	Refined petroleum products	..	..	..	..	..	..	..	..
		233	Processing of nuclear fuel	..	..	..	..	..	..	..	..
	24		Chemicals and chemical products	..	5 653	..	..	..	..	..	..
		241	Basic chemicals	..	..	..	..	..	..	..	..
		242	Other chemical products	..	..	..	..	..	..	..	..
		243	Man-made fibres	..	..	..	..	..	..	..	..
	25		Rubber and plastic products	..	2 396	..	..	..	..	..	..
		251	Rubber products	..	..	..	..	..	..	..	..
		252	Plastic products	..	..	..	..	..	..	..	..
	26		Other non-metallic mineral products	..	3 518	..	..	..	..	..	..
		261	Glass and glass products	..	..	..	..	..	..	..	..
		269	Non-metallic mineral products, nec	..	..	..	..	..	..	..	..
	27		Basic metals	..	5 864	..	..	..	..	..	..
		271	Basic iron and steel	..	..	..	..	..	..	..	..
		272	Basic precious and non-ferrous metals	..	..	..	..	..	..	..	..
		273	Casting of metals	..	..	..	..	..	..	..	..
	28		Fabricated metal products, exc. machinery	..	3 637	..	..	..	..	..	..
		281	Structural metal prod., tanks, reservoirs, etc.	..	..	..	..	..	..	..	..
		289	Other fabricated metal products	..	..	..	..	..	..	..	..
	29		Machinery and equipment, nec	..	5 028	..	..	..	..	..	..
		291	General purpose machinery	..	..	..	..	..	..	..	..
		292	Special purpose machinery	..	..	..	..	..	..	..	..
		293	Domestic appliances, nec	..	..	..	..	..	..	..	..
	30		Office, accounting & computing machinery	..	117	..	..	..	..	..	..
	31		Electrical machinery & apparatus, nec	..	2 200	..	..	..	..	..	..
		311	Electric motors, generators and transformers	..	..	..	..	..	..	..	..
		312	Electricity distribution & control apparatus	..	..	..	..	..	..	..	..
		313	Insulated wire and cable	..	..	..	..	..	..	..	..
		314	Accumulators, primary cells & primary batteries	..	..	..	..	..	..	..	..
		315	Electric lamps and lighting equipment	..	..	..	..	..	..	..	..
		319	Other electrical equipment, nec	..	..	..	..	..	..	..	..
	32		Radio, TV, communication equipment	..	1 053	..	..	..	..	..	..
		321	Electronic valves, tubes; other electronic components	..	..	..	..	..	..	..	..
		322	TV/radio transmitters; line teleph./telegr. apparatus	..	..	..	..	..	..	..	..
		323	TV/radio receivers; sound and video app.; assoc. gds.	..	..	..	..	..	..	..	..
	33		Medical, precision & optical instrum.; watch & clocks	..	738	..	..	..	..	..	..
		331	Medical appl.; instruments & appl. for measuring, etc.	..	..	..	..	..	..	..	..
		332	Optical instruments and photographic equipment	..	..	..	..	..	..	..	..
		333	Watches and clocks	..	..	..	..	..	..	..	..
	34		Motor vehicles, trailers and semi-trailers	..	2 639	..	..	..	..	..	..
		341	Motor vehicles	..	..	..	..	..	..	..	..
		342	Bodies for motor vehicles, trailers & semi-trailers	..	..	..	..	..	..	..	..
		343	Parts & accessories for motor vehicles & their engines	..	..	..	..	..	..	..	..
	35		Other transport equipment	..	2 115	..	..	..	..	..	..
		351	Building and repairing of ships and boats	..	..	..	..	..	..	..	..
		352	Railway and tramway locomotives and rolling stock	..	..	..	..	..	..	..	..
		353	Aircraft and spacecraft	..	..	..	..	..	..	..	..
		359	Transport equipment, nec	..	..	..	..	..	..	..	..
	36		Furniture; manufacturing, nec	..	2 456	..	..	..	..	..	..
		361	Furniture	..	..	..	..	..	..	..	..
		369	Manufacturing, nec	..	..	..	..	..	..	..	..
	37		Recycling	..	269	..	..	..	..	..	..
E	40_41		Electricity, gas and water supply	..	11 162	15 383	19 742	24 788	28 616	33 064	36 250
	40		Electricity, gas, steam and hot water supply	..	10 037	..	..	..	..	..	..
F	45		Construction	..	..	..	34 536	47 525	61 093	76 966	99 425

Table PL.2 VALEUR AJOUTÉE - VALUE ADDED

Millions de Zl (Prix courants)

1991	1992	1993	1994	1995	1996	1997	1998			CITI révision 3
..	3 871	5 721	8 882	11 101	12 759	14 631	13 897	Activites extractives	C	10_14
..	2 861	..	..	..	..	..	..	Extraction de charbon, de lignite et de tourbe		10
..	..	..	..	..	..	..	..	Pétrole brut et gaz naturel ; activités annexes		11
..	..	..	..	..	..	..	..	Extraction de Minerais d'Uranium et de Thorium		12
..	..	..	..	..	..	..	..	Extraction de minerais métalliques		13
..	..	..	..	..	..	..	..	Autres activités extractives		14
..	30 925	39 294	43 514	63 322	75 788	92 321	104 340	Total activites manufacturieres	D	15_37
..	7 924	..	..	..	..	..	..	Produits alimentaires et boissons		15
..	..	..	..	..	..	..	..	Viande, poisson, fruits, légumes, huiles et graisses		151
..	..	..	..	..	..	..	..	Produits laitiers		152
..	..	..	..	..	..	..	..	Travail des grains; prod. amyl. ; alim. pour animaux		153
..	..	..	..	..	..	..	..	Autres produits alimentaires		154
..	..	..	..	..	..	..	..	Boissons		155
..	1 149	..	..	..	..	..	..	Produits à base de tabac		16
..	1 194	..	..	..	..	..	..	Textiles		17
..	..	..	..	..	..	..	..	Filature, tissage et achèvement des textiles		171
..	..	..	..	..	..	..	..	Autres articles textiles		172
..	..	..	..	..	..	..	..	Etoffes et articles de bonneterie		173
..	1 465	..	..	..	..	..	..	Articles d'habillement et fourrure		18
..	..	..	..	..	..	..	..	Articles d'habillement autres qu'en fourrure		181
..	..	..	..	..	..	..	..	Prép. et teinture des fourrures ; articles en fourrure		182
..	533	..	..	..	..	..	..	Cuir et articles en cuir ; chaussures		19
..	..	..	..	..	..	..	..	Cuir et articles en cuir sauf chaussures		191
..	..	..	..	..	..	..	..	Chaussures		192
..	812	..	..	..	..	..	..	Bois, articles en bois, sauf meubles		20
..	..	..	..	..	..	..	..	Sciage et rabotage du bois		201
..	..	..	..	..	..	..	..	Articles en bois, liège, vannerie, sparterie		202
..	387	..	..	..	..	..	..	Papier, articles en papier et carton		21
..	765	..	..	..	..	..	..	Edition, imprimerie et reproduction		22
..	..	..	..	..	..	..	..	Edition		221
..	..	..	..	..	..	..	..	Imprimerie et activités annexes		222
..	..	..	..	..	..	..	..	Reproduction de supports enregistrés		223
..	2 665	..	..	..	..	..	..	Coke. et prod. pétroliers ; comb. nucléaires		23
..	..	..	..	..	..	..	..	Cokéfaction		231
..	..	..	..	..	..	..	..	Produits pétroliers raffinés		232
..	..	..	..	..	..	..	..	Traitement de combustibles nucléaires		233
..	1 838	..	..	..	..	..	..	Produits chimiques		24
..	..	..	..	..	..	..	..	Produits chimiques de base		241
..	..	..	..	..	..	..	..	Autres produits chimiques		242
..	..	..	..	..	..	..	..	Fibres synthétiques ou artificielles		243
..	926	..	..	..	..	..	..	Art. en caoutchouc et matières plastiques		25
..	..	..	..	..	..	..	..	Articles en caoutchouc		251
..	..	..	..	..	..	..	..	Articles en matières plastiques		252
..	1 412	..	..	..	..	..	..	Autres produits minéraux non métalliques		26
..	..	..	..	..	..	..	..	Verre et articles en verre		261
..	..	..	..	..	..	..	..	Produits minéraux non métalliques, nca		269
..	1 517	..	..	..	..	..	..	Produits métallurgiques de base		27
..	..	..	..	..	..	..	..	Sidérurgie et première transformation de l'acier		271
..	..	..	..	..	..	..	..	Première transf. métaux précieux et non ferreux		272
..	..	..	..	..	..	..	..	Fonderie		273
..	1 668	..	..	..	..	..	..	Ouvrages en métaux, sauf machines		28
..	..	..	..	..	..	..	..	Constr. et menuis. métal.; réservoirs; générateurs		281
..	..	..	..	..	..	..	..	Autres ouvrages en métaux		289
..	2 309	..	..	..	..	..	..	Machines et matériels, nca		29
..	..	..	..	..	..	..	..	Machines d'usage général		291
..	..	..	..	..	..	..	..	Machines d'usage spécifique		292
..	..	..	..	..	..	..	..	Appareils domestiques, nca		293
..	63	..	..	..	..	..	..	Mach. de bureau, comptables, information		30
..	846	..	..	..	..	..	..	Machines et appareils électriques, nca		31
..	..	..	..	..	..	..	..	Moteurs, génératrices, transformateurs électriques		311
..	..	..	..	..	..	..	..	Matériel électrique de distribution et de commande		312
..	..	..	..	..	..	..	..	Fils et câbles électriques isolés		313
..	..	..	..	..	..	..	..	Accumulateurs et piles électriques		314
..	..	..	..	..	..	..	..	Lampes électriques et appareils d'éclairage		315
..	..	..	..	..	..	..	..	Autres matériels électriques, nca.		319
..	329	..	..	..	..	..	..	Equipements et appareils de radio, TV, etc.		32
..	..	..	..	..	..	..	..	Tubes et valves électroniques;autres composants		321
..	..	..	..	..	..	..	..	Emetteurs radio/TV; app. téléphonie et télégraphie		322
..	..	..	..	..	..	..	..	Récepteurs radio/TV, son et image; art. associés		323
..	396	..	..	..	..	..	..	Instr. médicaux, de précision ; optique ; horlog.		33
..	..	..	..	..	..	..	..	Instruments médicaux et de précision, etc.		331
..	..	..	..	..	..	..	..	Instruments d'optique ; matériel photographique		332
..	..	..	..	..	..	..	..	Horlogerie		333
..	832	..	..	..	..	..	..	Véhicules automobiles; remorques, etc.		34
..	..	..	..	..	..	..	..	Véhicules automobiles		341
..	..	..	..	..	..	..	..	Carross. automobiles, remorques, semi-remorques		342
..	..	..	..	..	..	..	..	Pièces et accessoires pour automobiles et moteurs		343
..	751	..	..	..	..	..	..	Autres matériels de transport		35
..	..	..	..	..	..	..	..	Construction et réparation de navires		351
..	..	..	..	..	..	..	..	Matériel ferroviaire roulant		352
..	..	..	..	..	..	..	..	Construction aéronautique et spatiale		353
..	..	..	..	..	..	..	..	Autres équipements de transport		359
..	1 052	..	..	..	..	..	..	Meubles ; activités de fabrication, nca		36
..	..	..	..	..	..	..	..	Meubles		361
..	..	..	..	..	..	..	..	Activités de fabrication, nca		369
..	92	..	..	..	..	..	..	Récupération		37
..	4 331	6 193	7 635	10 539	12 736	14 073	15 726	Electricité, gaz et eau	E	40_41
..	3 715	..	..	..	..	..	..	Prod. et dist. d'électricité gaz, vapeur d'eau chaude		40
..	8 932	10 150	14 371	19 506	25 049	32 769	41 989	Construction	F	45

POLAND

Table PL.3 INVESTMENT - INVESTISSEMENT

Millions of Zl (Current Prices)

ISIC revision 3				1991	1992	1993	1994	1995	1996	1997	1998
C	10_14		Mining and quarrying	..	888	1 032	1 444	1 674	1 840	2 057	2 299
	10		Mining of coal and lignite; extraction of peat	..	725	..	..	..	..	..	..
	11		Crude petroleum & natural gas; service activities	..	..	..	..	..	..	..	..
	12		Mining of uranium and thorium ores	..	..	..	..	..	..	..	..
	13		Mining of metal ores	..	..	..	..	..	..	..	..
	14		Other mining and quarrying	..	..	..	..	..	..	..	..
D	15_37		Total manufacturing	..	4 580	4 965	8 396	11 733	17 215	23 084	29 054
	15		Food products and beverages	..	1 236	1 176	1 708	2 695	3 731	4 348	4 617
		151	Production of meat, fish, fruit, vegetables, oils & fats	..	..	..	..	..	..	..	..
		152	Dairy products	..	..	..	..	..	..	..	..
		153	Grain mill and starch products; animal feeds	..	..	..	..	..	..	..	..
		154	Other food products	..	..	..	..	..	..	..	..
		155	Beverages	..	..	..	..	..	..	..	..
	16		Tobacco products	..	43	110	84	122	322	434	334
	17		Textiles	..	109	151	280	327	385	473	738
		171	Spinning, weaving and finishing of textiles	..	..	..	..	..	..	..	..
		172	Other textiles	..	..	..	..	..	..	..	..
		173	Knitted and crocheted fabrics and articles	..	..	..	..	..	..	..	..
	18		Wearing apparel and fur	..	115	124	191	233	258	323	309
		181	Wearing apparel, except fur apparel	..	..	..	..	..	..	..	..
		182	Dressing and dyeing of fur; articles of fur	..	..	..	..	..	..	..	..
	19		Leather and articles of leather; footwear	..	42	36	51	80	98	133	101
		191	Leather and articles of leather	..	..	..	..	..	..	..	..
		192	Footwear	..	..	..	..	..	..	..	..
	20		Wood & products of wood & cork, exc. furniture	..	100	157	321	510	562	1 136	1 850
		201	Sawmilling and planing of wood	..	..	..	..	..	..	..	..
		202	Wood products; articles of cork, straw, etc.	..	..	..	..	..	..	..	..
	21		Paper and paper products	..	94	239	453	489	687	848	1 010
	22		Printing, publishing and recorded media	..	98	162	321	255	421	718	827
		221	Publishing	..	..	..	..	..	..	..	..
		222	Printing and related activities	..	..	..	..	..	..	..	..
		223	Reproduction of recorded media	..	..	..	..	..	..	..	..
	23		Coke and petroleum products; nuclear fuel	..	183	306	561	709	1 254	1 482	2 304
		231	Coke oven products	..	..	..	..	..	..	..	..
		232	Refined petroleum products	..	..	..	..	..	..	..	..
		233	Processing of nuclear fuel	..	..	..	..	..	..	..	..
	24		Chemicals and chemical products	..	388	451	911	1 181	1 554	2 366	2 832
		241	Basic chemicals	..	..	..	..	..	..	..	..
		242	Other chemical products	..	..	..	..	..	..	..	..
		243	Man-made fibres	..	..	..	..	..	..	..	..
	25		Rubber and plastic products	..	128	221	293	543	855	1 172	1 460
		251	Rubber products	..	..	..	..	..	..	..	..
		252	Plastic products	..	..	..	..	..	..	..	..
	26		Other non-metallic mineral products	..	212	308	609	845	1 279	1 767	2 673
		261	Glass and glass products	..	..	..	..	..	..	..	..
		269	Non-metallic mineral products, nec	..	..	..	..	..	..	..	..
	27		Basic metals	..	480	428	705	1 101	1 913	1 612	1 671
		271	Basic iron and steel	..	..	..	..	..	..	..	..
		272	Basic precious and non-ferrous metals	..	..	..	..	..	..	..	..
		273	Casting of metals	..	..	..	..	..	..	..	..
	28		Fabricated metal products, exc. machinery	..	126	171	357	326	580	855	1 031
		281	Structural metal prod., tanks, reservoirs, etc.	..	..	..	..	..	..	..	..
		289	Other fabricated metal products	..	..	..	..	..	..	..	..
	29		Machinery and equipment, nec	..	200	217	378	584	773	995	1 008
		291	General purpose machinery	..	..	..	..	..	..	..	..
		292	Special purpose machinery	..	..	..	..	..	..	..	..
		293	Domestic appliances, nec	..	..	..	..	..	..	..	..
	30		Office, accounting & computing machinery	..	4	16	14	16	17	24	31
	31		Electrical machinery & apparatus, nec	..	179	190	261	381	557	764	794
		311	Electric motors, generators and transformers	..	..	..	..	..	..	..	..
		312	Electricity distribution & control apparatus	..	..	..	..	..	..	..	..
		313	Insulated wire and cable	..	..	..	..	..	..	..	..
		314	Accumulators, primary cells & primary batteries	..	..	..	..	..	..	..	..
		315	Electric lamps and lighting equipment	..	..	..	..	..	..	..	..
		319	Other electrical equipment, nec	..	..	..	..	..	..	..	..
	32		Radio, TV, communication equipment	..	118	118	155	151	327	332	499
		321	Electronic valves, tubes; other electronic components	..	..	..	..	..	..	..	..
		322	TV/radio transmitters; line teleph./telegr. apparatus	..	..	..	..	..	..	..	..
		323	TV/radio receivers; sound and video app.; assoc. gds.	..	..	..	..	..	..	..	..
	33		Medical, precision & optical instrum.; watch & clocks	..	33	34	60	91	133	183	198
		331	Medical appl.; instruments & appl. for measuring, etc.	..	..	..	..	..	..	..	..
		332	Optical instruments and photographic equipment	..	..	..	..	..	..	..	..
		333	Watches and clocks	..	..	..	..	..	..	..	..
	34		Motor vehicles, trailers and semi-trailers	..	543	147	188	494	808	2 191	3 508
		341	Motor vehicles	..	..	..	..	..	..	..	..
		342	Bodies for motor vehicles, trailers & semi-trailers	..	..	..	..	..	..	..	..
		343	Parts & accessories for motor vehicles & their engines	..	..	..	..	..	..	..	..
	35		Other transport equipment	..	55	88	207	197	247	274	304
		351	Building and repairing of ships and boats	..	..	..	..	..	..	..	..
		352	Railway and tramway locomotives and rolling stock	..	..	..	..	..	..	..	..
		353	Aircraft and spacecraft	..	..	..	..	..	..	..	..
		359	Transport equipment, nec	..	..	..	..	..	..	..	..
	36		Furniture; manufacturing, nec	..	85	103	250	378	423	575	854
		361	Furniture	..	..	..	..	..	..	..	..
		369	Manufacturing, nec	..	..	..	..	..	..	..	..
	37		Recycling	..	11	12	39	27	31	81	101
E	40_41		Electricity, gas and water supply	..	2 507	3 571	4 869	6 768	8 770	10 027	10 255
	40		Electricity, gas, steam and hot water supply	..	1 603	..	..	..	..	..	..
F	45		Construction	..	956	1 025	1 556	2 157	3 845	5 547	7 739

Table PL.4 DÉPENSES DE MACHINES ET D'ÉQUIPEMENT - INVESTMENT IN MACHINERY & EQUIPMENT

Millions de Zl (Prix courants)

1991	1992	1993	1994	1995	1996	1997	1998			CITI révision 3	
..	..	488	751	952	1 199	1 225	1 289	Activites extractives	C	10_14	
..	..	..	..	..	..	..	..	Extraction de charbon, de lignite et de tourbe		10	
..	..	..	..	..	..	..	..	Pétrole brut et gaz naturel ; activités annexes		11	
..	..	..	..	..	..	..	..	Extraction de Minerais d'Uranium et de Thorium		12	
..	..	..	..	..	..	..	..	Extraction de minerais métalliques		13	
..	..	..	..	..	..	..	..	Autres activités extractives		14	
..	..	2 939	5 113	7 215	10 918	14 367	18 154	Total activites manufacturieres	D	15_37	
..	..	661	1 006	1 684	2 304	2 729	2 749	Produits alimentaires et boissons		15	
..	..	..	..	..	..	..	..	Viande, poisson, fruits, légumes, huiles et graisses			151
..	..	..	..	..	..	..	..	Produits laitiers			152
..	..	..	..	..	..	..	..	Travail des grains; prod. amyl. ; alim. pour animaux			153
..	..	..	..	..	..	..	..	Autres produits alimentaires			154
..	..	..	..	..	..	..	..	Boissons			155
..	..	74	60	93	258	293	216	Produits à base de tabac		16	
..	..	109	202	259	280	373	529	Textiles		17	
..	..	..	..	..	..	..	..	Filature, tissage et achèvement des textiles			171
..	..	..	..	..	..	..	..	Autres articles textiles			172
..	..	..	..	..	..	..	..	Etoffes et articles de bonneterie			173
..	..	71	102	134	139	123	145	Articles d'habillement et fourrure		18	
..	..	..	..	..	..	..	..	Articles d'habillement autres qu'en fourrure			181
..	..	..	..	..	..	..	..	Prép. et teinture des fourrures ; articles en fourrure			182
..	..	17	24	41	60	79	53	Cuir et articles en cuir ; chaussures		19	
..	..	..	..	..	..	..	..	Cuir et articles en cuir sauf chaussures			191
..	..	..	..	..	..	..	..	Chaussures			192
..	..	92	222	369	384	832	1 391	Bois, articles en bois, sauf meubles		20	
..	..	..	..	..	..	..	..	Sciage et rabotage du bois			201
..	..	..	..	..	..	..	..	Articles en bois, liège, vannerie, sparterie			202
..	..	177	368	393	516	664	777	Papier, articles en papier et carton		21	
..	..	112	260	197	315	502	551	Edition, imprimerie et reproduction		22	
..	..	..	..	..	..	..	..	Edition			221
..	..	..	..	..	..	..	..	Imprimerie et activités annexes			222
..	..	..	..	..	..	..	..	Reproduction de supports enregistrés			223
..	..	148	314	338	697	728	1 265	Coke. et prod. pétroliers ; comb. nucléaires		23	
..	..	..	..	..	..	..	..	Cokéfaction			231
..	..	..	..	..	..	..	..	Produits pétroliers raffinés			232
..	..	..	..	..	..	..	..	Traitement de combustibles nucléaires			233
..	..	222	380	663	896	1 363	1 509	Produits chimiques		24	
..	..	..	..	..	..	..	..	Produits chimiques de base			241
..	..	..	..	..	..	..	..	Autres produits chimiques			242
..	..	..	..	..	..	..	..	Fibres synthétiques ou artificielles			243
..	..	163	208	383	603	809	1 002	Art. en caoutchouc et matières plastiques		25	
..	..	..	..	..	..	..	..	Articles en caoutchouc			251
..	..	..	..	..	..	..	..	Articles en matières plastiques			252
..	..	183	309	552	847	1 031	1 803	Autres produits minéraux non métalliques		26	
..	..	..	..	..	..	..	..	Verre et articles en verre			261
..	..	..	..	..	..	..	..	Produits minéraux non métalliques, nca			269
..	..	248	452	668	1 194	1 052	1 146	Produits métallurgiques de base		27	
..	..	..	..	..	..	..	..	Sidérurgie et première transformation de l'acier			271
..	..	..	..	..	..	..	..	Première transf. métaux précieux et non ferreux			272
..	..	..	..	..	..	..	..	Fonderie			273
..	..	114	240	195	340	492	586	Ouvrages en métaux, sauf machines		28	
..	..	..	..	..	..	..	..	Constr. et menuis. métal.; réservoirs; générateurs			281
..	..	..	..	..	..	..	..	Autres ouvrages en métaux			289
..	..	127	249	362	507	597	654	Machines et matériels, nca		29	
..	..	..	..	..	..	..	..	Machines d'usage général			291
..	..	..	..	..	..	..	..	Machines d'usage spécifique			292
..	..	..	..	..	..	..	..	Appareils domestiques, nca			293
..	..	6	9	8	9	9	11	Mach. de bureau, comptables, information		30	
..	..	113	178	243	402	528	548	Machines et appareils électriques, nca		31	
..	..	..	..	..	..	..	..	Moteurs, génératrices, transformateurs électriques			311
..	..	..	..	..	..	..	..	Matériel électrique de distribution et de commande			312
..	..	..	..	..	..	..	..	Fils et câbles électriques isolés			313
..	..	..	..	..	..	..	..	Accumulateurs et piles électriques			314
..	..	..	..	..	..	..	..	Lampes électriques et appareils d'éclairage			315
..	..	..	..	..	..	..	..	Autres matériels électriques, nca.			319
..	..	91	109	84	237	202	291	Equipements et appareils de radio, TV, etc.		32	
..	..	..	..	..	..	..	..	Tubes et valves électroniques;autres composants			321
..	..	..	..	..	..	..	..	Emetteurs radio/TV; app. téléphonie et télégraphie			322
..	..	..	..	..	..	..	..	Récepteurs radio/TV, son et image; art. associés			323
..	..	22	39	61	77	106	118	Instr. médicaux, de précision ; optique ; horlog.		33	
..	..	..	..	..	..	..	..	Instruments médicaux et de précision, etc.			331
..	..	..	..	..	..	..	..	Instruments d'optique ; matériel photographique			332
..	..	..	..	..	..	..	..	Horlogerie			333
..	..	77	114	187	501	1 397	2 180	Véhicules automobiles; remorques, etc.		34	
..	..	..	..	..	..	..	..	Véhicules automobiles			341
..	..	..	..	..	..	..	..	Carross. automobiles, remorques, semi-remorques			342
..	..	..	..	..	..	..	..	Pièces et accessoires pour automobiles et moteurs			343
..	..	52	118	110	126	121	155	Autres matériels de transport		35	
..	..	..	..	..	..	..	..	Construction et réparation de navires			351
..	..	..	..	..	..	..	..	Matériel ferroviaire roulant			352
..	..	..	..	..	..	..	..	Construction aéronautique et spatiale			353
..	..	..	..	..	..	..	..	Autres équipements de transport			359
..	..	55	128	177	210	289	416	Meubles ; activités de fabrication, nca		36	
..	..	..	..	..	..	..	..	Meubles			361
..	..	..	..	..	..	..	..	Activités de fabrication, nca			369
..	..	7	25	14	17	51	58	Récupération		37	
..	..	714	939	1 913	2 725	3 325	3 691	Electricité, gaz et eau	E	40_41	
..	..	..	..	..	..	..	..	Prod. et dist. d'électricité gaz, vapeur d'eau chaude		40	
..	..	179	241	405	649	977	1 256	Construction	F	45	

POLAND

Table PL.5 **EMPLOYMENT, EMPLOYEES - EMPLOI, SALARIÉS**

Thousands

ISIC revision 3				1991	1992	1993	1994	1995	1996	1997	1998
C	10_14		Mining and quarrying	..	46	42	39	37	36	34	31
	10		Mining of coal and lignite; extraction of peat	..	37	35	33	31	29	28	26
	11		Crude petroleum & natural gas; service activities	..	..	0	0	0	0	0	0
	12		Mining of uranium and thorium ores	..	..	..	..	..	..	..	..
	13		Mining of metal ores	..	..	3	3	3	3	2	2
	14		Other mining and quarrying	..	..	4	4	4	3	3	3
D	15_37		Total manufacturing	..	277	270	269	281	280	282	280
	15		Food products and beverages	..	46	47	45	48	49	50	51
		151	Production of meat, fish, fruit, vegetables, oils & fats	..	..	..	..	..	..	..	..
		152	Dairy products	..	..	..	..	..	..	..	..
		153	Grain mill and starch products; animal feeds	..	..	..	..	..	..	..	..
		154	Other food products	..	..	..	..	..	..	..	..
		155	Beverages	..	..	..	..	..	..	..	..
	16		Tobacco products	..	1	1	1	1	1	1	1
	17		Textiles	..	18	17	17	17	15	15	13
		171	Spinning, weaving and finishing of textiles	..	..	..	..	..	..	..	..
		172	Other textiles	..	..	..	..	..	..	..	..
		173	Knitted and crocheted fabrics and articles	..	..	..	..	..	..	..	..
	18		Wearing apparel and fur	..	21	21	23	27	26	25	26
		181	Wearing apparel, except fur apparel	..	..	..	..	..	..	..	..
		182	Dressing and dyeing of fur; articles of fur	..	..	..	..	..	..	..	..
	19		Leather and articles of leather; footwear	..	9	8	8	8	8	8	6
		191	Leather and articles of leather	..	..	..	..	..	..	..	..
		192	Footwear	..	..	..	..	..	..	..	..
	20		Wood & products of wood & cork, exc. furniture	..	10	10	10	11	12	12	12
		201	Sawmilling and planing of wood	..	..	..	..	..	..	..	..
		202	Wood products; articles of cork, straw, etc.	..	..	..	..	..	..	..	..
	21		Paper and paper products	..	4	4	4	4	4	4	4
	22		Printing, publishing and recorded media	..	6	6	6	7	7	8	8
		221	Publishing	..	..	..	..	..	..	..	..
		222	Printing and related activities	..	..	..	..	..	..	..	..
		223	Reproduction of recorded media	..	..	..	..	..	..	..	..
	23		Coke and petroleum products; nuclear fuel	..	2	2	2	2	2	2	2
		231	Coke oven products	..	..	..	..	..	..	..	..
		232	Refined petroleum products	..	..	..	..	..	..	..	..
		233	Processing of nuclear fuel	..	..	..	..	..	..	..	..
	24		Chemicals and chemical products	..	14	14	14	14	14	14	13
		241	Basic chemicals	..	..	..	..	..	..	..	..
		242	Other chemical products	..	..	..	..	..	..	..	..
		243	Man-made fibres	..	..	..	..	..	..	..	..
	25		Rubber and plastic products	..	7	8	9	9	10	10	11
		251	Rubber products	..	..	..	..	..	..	..	..
		252	Plastic products	..	..	..	..	..	..	..	..
	26		Other non-metallic mineral products	..	17	17	16	17	17	17	17
		261	Glass and glass products	..	..	..	..	..	..	..	..
		269	Non-metallic mineral products, nec	..	..	..	..	..	..	..	..
	27		Basic metals	..	17	17	15	15	15	15	14
		271	Basic iron and steel	..	..	..	..	..	..	..	..
		272	Basic precious and non-ferrous metals	..	..	..	..	..	..	..	..
		273	Casting of metals	..	..	..	..	..	..	..	..
	28		Fabricated metal products, exc. machinery	..	15	14	15	16	17	18	19
		281	Structural metal prod., tanks, reservoirs, etc.	..	..	..	..	..	..	..	..
		289	Other fabricated metal products	..	..	..	..	..	..	..	..
	29		Machinery and equipment, nec	..	33	31	30	30	29	28	27
		291	General purpose machinery	..	..	..	..	..	..	..	..
		292	Special purpose machinery	..	..	..	..	..	..	..	..
		293	Domestic appliances, nec	..	..	..	..	..	..	..	..
	30		Office, accounting & computing machinery	..	1	1	0	0	0	1	1
	31		Electrical machinery & apparatus, nec	..	9	9	9	9	9	10	10
		311	Electric motors, generators and transformers	..	..	..	..	..	..	..	..
		312	Electricity distribution & control apparatus	..	..	..	..	..	..	..	..
		313	Insulated wire and cable	..	..	..	..	..	..	..	..
		314	Accumulators, primary cells & primary batteries	..	..	..	..	..	..	..	..
		315	Electric lamps and lighting equipment	..	..	..	..	..	..	..	..
		319	Other electrical equipment, nec	..	..	..	..	..	..	..	..
	32		Radio, TV, communication equipment	..	6	5	5	5	4	4	4
		321	Electronic valves, tubes; other electronic components	..	..	..	..	..	..	..	..
		322	TV/radio transmitters; line teleph./telegr. apparatus	..	..	..	..	..	..	..	..
		323	TV/radio receivers; sound and video app.; assoc. gds.	..	..	..	..	..	..	..	..
	33		Medical, precision & optical instrum.; watch & clocks	..	4	4	4	4	4	4	4
		331	Medical appl.; instruments & appl. for measuring, etc.	..	..	..	..	..	..	..	..
		332	Optical instruments and photographic equipment	..	..	..	..	..	..	..	..
		333	Watches and clocks	..	..	..	..	..	..	..	..
	34		Motor vehicles, trailers and semi-trailers	..	11	10	10	10	10	11	11
		341	Motor vehicles	..	..	..	..	..	..	..	..
		342	Bodies for motor vehicles, trailers & semi-trailers	..	..	..	..	..	..	..	..
		343	Parts & accessories for motor vehicles & their engines	..	..	..	..	..	..	..	..
	35		Other transport equipment	..	12	12	12	11	11	10	9
		351	Building and repairing of ships and boats	..	..	..	..	..	..	..	..
		352	Railway and tramway locomotives and rolling stock	..	..	..	..	..	..	..	..
		353	Aircraft and spacecraft	..	..	..	..	..	..	..	..
		359	Transport equipment, nec	..	..	..	..	..	..	..	..
	36		Furniture; manufacturing, nec	..	12	13	13	15	16	17	18
		361	Furniture	..	..	..	..	..	..	..	..
		369	Manufacturing, nec	..	..	..	..	..	..	..	..
	37		Recycling	..	1	1	1	1	1	1	1
E	40_41		Electricity, gas and water supply	..	24	27	28	28	28	28	27
	40		Electricity, gas, steam and hot water supply	..	..	22	22	22	22	22	21
F	45		Construction	..	82	72	68	69	68	72	74

Table PL.6 **EMPLOI, FEMMES - EMPLOYMENT, FEMALES**

Milliers

1991	1992	1993	1994	1995	1996	1997	1998			CITI révision 3
..	..	5	4	4	4	4	4	Activites extractives	C	**10_14**
..	..	4	3	3	3	3	3	Extraction de charbon, de lignite et de tourbe		**10**
..	..	0	0	0	0	0	0	Pétrole brut et gaz naturel ; activités annexes		**11**
..	..	..	..	..	..	..	..	Extraction de Minerais d'Uranium et de Thorium		**12**
..	..	0	0	0	0	0	0	Extraction de minerais métalliques		**13**
..	..	1	1	1	1	1	1	Autres activités extractives		**14**
..	..	115	121	122	122	123	120	Total activites manufacturieres	D	**15_37**
..	..	23	24	24	25	25	25	Produits alimentaires et boissons		**15**
..	..	..	..	..	..	..	..	Viande, poisson, fruits, légumes, huiles et graisses		151
..	..	..	..	..	..	..	..	Produits laitiers		152
..	..	..	..	..	..	..	..	Travail des grains; prod. amyl. ; alim. pour animaux		153
..	..	..	..	..	..	..	..	Autres produits alimentaires		154
..	..	..	..	..	..	..	..	Boissons		155
..	..	1	1	1	1	1	1	Produits à base de tabac		**16**
..	..	11	12	10	10	10	8	Textiles		**17**
..	..	..	..	..	..	..	..	Filature, tissage et achèvement des textiles		171
..	..	..	..	..	..	..	..	Autres articles textiles		172
..	..	..	..	..	..	..	..	Etoffes et articles de bonneterie		173
..	..	22	24	27	26	25	26	Articles d'habillement et fourrure		**18**
..	..	..	..	..	..	..	..	Articles d'habillement autres qu'en fourrure		181
..	..	..	..	..	..	..	..	Prép. et teinture des fourrures ; articles en fourrure		182
..	..	5	5	5	5	6	5	Cuir et articles en cuir ; chaussures		**19**
..	..	..	..	..	..	..	..	Cuir et articles en cuir sauf chaussures		191
..	..	..	..	..	..	..	..	Chaussures		192
..	..	2	2	2	3	3	3	Bois, articles en bois, sauf meubles		**20**
..	..	..	..	..	..	..	..	Sciage et rabotage du bois		201
..	..	..	..	..	..	..	..	Articles en bois, liège, vannerie, sparterie		202
..	..	2	2	2	2	2	2	Papier, articles en papier et carton		**21**
..	..	3	4	4	4	5	5	Edition, imprimerie et reproduction		**22**
..	..	..	..	..	..	..	..	Edition		221
..	..	..	..	..	..	..	..	Imprimerie et activités annexes		222
..	..	..	..	..	..	..	..	Reproduction de supports enregistrés		223
..	..	1	1	1	1	1	1	Coke. et prod. pétroliers ; comb. nucléaires		**23**
..	..	..	..	..	..	..	..	Cokéfaction		231
..	..	..	..	..	..	..	..	Produits pétroliers raffinés		232
..	..	..	..	..	..	..	..	Traitement de combustibles nucléaires		233
..	..	5	6	6	6	6	5	Produits chimiques		**24**
..	..	..	..	..	..	..	..	Produits chimiques de base		241
..	..	..	..	..	..	..	..	Autres produits chimiques		242
..	..	..	..	..	..	..	..	Fibres synthétiques ou artificielles		243
..	..	3	3	4	4	4	4	Art. en caoutchouc et matières plastiques		**25**
..	..	..	..	..	..	..	..	Articles en caoutchouc		251
..	..	..	..	..	..	..	..	Articles en matières plastiques		252
..	..	5	5	5	5	5	5	Autres produits minéraux non métalliques		**26**
..	..	..	..	..	..	..	..	Verre et articles en verre		261
..	..	..	..	..	..	..	..	Produits minéraux non métalliques, nca		269
..	..	4	4	4	3	3	3	Produits métallurgiques de base		**27**
..	..	..	..	..	..	..	..	Sidérurgie et première transformation de l'acier		271
..	..	..	..	..	..	..	..	Première transf. métaux précieux et non ferreux		272
..	..	..	..	..	..	..	..	Fonderie		273
..	..	4	4	4	4	4	4	Ouvrages en métaux, sauf machines		**28**
..	..	..	..	..	..	..	..	Constr. et menuis. métal.; réservoirs; générateurs		281
..	..	..	..	..	..	..	..	Autres ouvrages en métaux		289
..	..	7	7	7	6	6	6	Machines et matériels, nca		**29**
..	..	..	..	..	..	..	..	Machines d'usage général		291
..	..	..	..	..	..	..	..	Machines d'usage spécifique		292
..	..	..	..	..	..	..	..	Appareils domestiques, nca		293
..	..	0	0	0	0	0	0	Mach. de bureau, comptables, information		**30**
..	..	3	4	4	4	4	4	Machines et appareils électriques, nca		**31**
..	..	..	..	..	..	..	..	Moteurs, génératrices, transformateurs électriques		311
..	..	..	..	..	..	..	..	Matériel électrique de distribution et de commande		312
..	..	..	..	..	..	..	..	Fils et câbles électriques isolés		313
..	..	..	..	..	..	..	..	Accumulateurs et piles électriques		314
..	..	..	..	..	..	..	..	Lampes électriques et appareils d'éclairage		315
..	..	..	..	..	..	..	..	Autres matériels électriques, nca.		319
..	..	2	2	2	2	2	2	Equipements et appareils de radio, TV, etc.		**32**
..	..	..	..	..	..	..	..	Tubes et valves électroniques;autres composants		321
..	..	..	..	..	..	..	..	Emetteurs radio/TV; app. téléphonie et télégraphie		322
..	..	..	..	..	..	..	..	Récepteurs radio/TV, son et image; art. associés		323
..	..	2	2	2	2	2	2	Instr. médicaux, de précision ; optique ; horlog.		**33**
..	..	..	..	..	..	..	..	Instruments médicaux et de précision, etc.		331
..	..	..	..	..	..	..	..	Instruments d'optique ; matériel photographique		332
..	..	..	..	..	..	..	..	Horlogerie		333
..	..	3	3	3	3	3	3	Véhicules automobiles; remorques, etc.		**34**
..	..	..	..	..	..	..	..	Véhicules automobiles		341
..	..	..	..	..	..	..	..	Carross. automobiles, remorques, semi-remorques		342
..	..	..	..	..	..	..	..	Pièces et accessoires pour automobiles et moteurs		343
..	..	2	2	2	2	2	2	Autres matériels de transport		**35**
..	..	..	..	..	..	..	..	Construction et réparation de navires		351
..	..	..	..	..	..	..	..	Matériel ferroviaire roulant		352
..	..	..	..	..	..	..	..	Construction aéronautique et spatiale		353
..	..	..	..	..	..	..	..	Autres équipements de transport		359
..	..	5	5	5	6	6	6	Meubles ; activités de fabrication, nca		**36**
..	..	..	..	..	..	..	..	Meubles		361
..	..	..	..	..	..	..	..	Activités de fabrication, nca		369
..	..	0	0	0	0	0	0	Récupération		**37**
..	..	6	6	6	6	6	6	Electricité, gaz et eau	E	**40_41**
..	..	4	5	5	5	5	4	Prod. et dist. d'électricité gaz, vapeur d'eau chaude		**40**
..	..	10	10	9	10	10	10	Construction	F	**45**

POLAND

Table PL.7 COMPENSATION OF LABOUR, EMPLOYEES - SALAIRES ET CHARGES SOCIALES, SALARIÉS

Millions of Zl (Current Prices)

ISIC revision 3				1991	1992	1993	1994	1995	1996	1997	1998
C	10_14		Mining and quarrying	..	..	4 929	6 924	8 766	10 529	12 104	12 808
	10		Mining of coal and lignite; extraction of peat	..	..	..	..	..	..	..	..
	11		Crude petroleum & natural gas; service activities	..	..	..	..	..	..	..	..
	12		Mining of uranium and thorium ores	..	..	..	..	..	..	..	..
	13		Mining of metal ores	..	..	..	..	..	..	..	..
	14		Other mining and quarrying	..	..	..	..	..	..	..	..
D	15_37		Total manufacturing	..	..	17 236	24 657	34 276	44 943	55 222	64 166
	15		Food products and beverages	..	..	..	..	..	..	..	..
		151	Production of meat, fish, fruit, vegetables, oils & fats	..	..	..	..	..	..	..	..
		152	Dairy products	..	..	..	..	..	..	..	..
		153	Grain mill and starch products; animal feeds	..	..	..	..	..	..	..	..
		154	Other food products	..	..	..	..	..	..	..	..
		155	Beverages	..	..	..	..	..	..	..	..
	16		Tobacco products	..	..	..	..	..	..	..	..
	17		Textiles	..	..	..	..	..	..	..	..
		171	Spinning, weaving and finishing of textiles	..	..	..	..	..	..	..	..
		172	Other textiles	..	..	..	..	..	..	..	..
		173	Knitted and crocheted fabrics and articles	..	..	..	..	..	..	..	..
	18		Wearing apparel and fur	..	..	..	..	..	..	..	..
		181	Wearing apparel, except fur apparel	..	..	..	..	..	..	..	..
		182	Dressing and dyeing of fur; articles of fur	..	..	..	..	..	..	..	..
	19		Leather and articles of leather; footwear	..	..	..	..	..	..	..	..
		191	Leather and articles of leather	..	..	..	..	..	..	..	..
		192	Footwear	..	..	..	..	..	..	..	..
	20		Wood & products of wood & cork, exc. furniture	..	..	..	..	..	..	..	..
		201	Sawmilling and planing of wood	..	..	..	..	..	..	..	..
		202	Wood products; articles of cork, straw, etc.	..	..	..	..	..	..	..	..
	21		Paper and paper products	..	..	..	..	..	..	..	..
	22		Printing, publishing and recorded media	..	..	..	..	..	..	..	..
		221	Publishing	..	..	..	..	..	..	..	..
		222	Printing and related activities	..	..	..	..	..	..	..	..
		223	Reproduction of recorded media	..	..	..	..	..	..	..	..
	23		Coke and petroleum products; nuclear fuel	..	..	..	..	..	..	..	..
		231	Coke oven products	..	..	..	..	..	..	..	..
		232	Refined petroleum products	..	..	..	..	..	..	..	..
		233	Processing of nuclear fuel	..	..	..	..	..	..	..	..
	24		Chemicals and chemical products	..	..	..	..	..	..	..	..
		241	Basic chemicals	..	..	..	..	..	..	..	..
		242	Other chemical products	..	..	..	..	..	..	..	..
		243	Man-made fibres	..	..	..	..	..	..	..	..
	25		Rubber and plastic products	..	..	..	..	..	..	..	..
		251	Rubber products	..	..	..	..	..	..	..	..
		252	Plastic products	..	..	..	..	..	..	..	..
	26		Other non-metallic mineral products	..	..	..	..	..	..	..	..
		261	Glass and glass products	..	..	..	..	..	..	..	..
		269	Non-metallic mineral products, nec	..	..	..	..	..	..	..	..
	27		Basic metals	..	..	..	..	..	..	..	..
		271	Basic iron and steel	..	..	..	..	..	..	..	..
		272	Basic precious and non-ferrous metals	..	..	..	..	..	..	..	..
		273	Casting of metals	..	..	..	..	..	..	..	..
	28		Fabricated metal products, exc. machinery	..	..	..	..	..	..	..	..
		281	Structural metal prod., tanks, reservoirs, etc.	..	..	..	..	..	..	..	..
		289	Other fabricated metal products	..	..	..	..	..	..	..	..
	29		Machinery and equipment, nec	..	..	..	..	..	..	..	..
		291	General purpose machinery	..	..	..	..	..	..	..	..
		292	Special purpose machinery	..	..	..	..	..	..	..	..
		293	Domestic appliances, nec	..	..	..	..	..	..	..	..
	30		Office, accounting & computing machinery	..	..	..	..	..	..	..	..
	31		Electrical machinery & apparatus, nec	..	..	..	..	..	..	..	..
		311	Electric motors, generators and transformers	..	..	..	..	..	..	..	..
		312	Electricity distribution & control apparatus	..	..	..	..	..	..	..	..
		313	Insulated wire and cable	..	..	..	..	..	..	..	..
		314	Accumulators, primary cells & primary batteries	..	..	..	..	..	..	..	..
		315	Electric lamps and lighting equipment	..	..	..	..	..	..	..	..
		319	Other electrical equipment, nec	..	..	..	..	..	..	..	..
	32		Radio, TV, communication equipment	..	..	..	..	..	..	..	..
		321	Electronic valves, tubes; other electronic components	..	..	..	..	..	..	..	..
		322	TV/radio transmitters; line teleph./telegr. apparatus	..	..	..	..	..	..	..	..
		323	TV/radio receivers; sound and video app.; assoc. gds.	..	..	..	..	..	..	..	..
	33		Medical, precision & optical instrum.; watch & clocks	..	..	..	..	..	..	..	..
		331	Medical appl.; instruments & appl. for measuring, etc.	..	..	..	..	..	..	..	..
		332	Optical instruments and photographic equipment	..	..	..	..	..	..	..	..
		333	Watches and clocks	..	..	..	..	..	..	..	..
	34		Motor vehicles, trailers and semi-trailers	..	..	..	..	..	..	..	..
		341	Motor vehicles	..	..	..	..	..	..	..	..
		342	Bodies for motor vehicles, trailers & semi-trailers	..	..	..	..	..	..	..	..
		343	Parts & accessories for motor vehicles & their engines	..	..	..	..	..	..	..	..
	35		Other transport equipment	..	..	..	..	..	..	..	..
		351	Building and repairing of ships and boats	..	..	..	..	..	..	..	..
		352	Railway and tramway locomotives and rolling stock	..	..	..	..	..	..	..	..
		353	Aircraft and spacecraft	..	..	..	..	..	..	..	..
		359	Transport equipment, nec	..	..	..	..	..	..	..	..
	36		Furniture; manufacturing, nec	..	..	..	..	..	..	..	..
		361	Furniture	..	..	..	..	..	..	..	..
		369	Manufacturing, nec	..	..	..	..	..	..	..	..
	37		Recycling	..	..	..	..	..	..	..	..
E	40_41		Electricity, gas and water supply	..	..	2 214	3 468	4 179	5 894	6 703	7 385
	40		Electricity, gas, steam and hot water supply	..	..	..	..	..	..	..	..
F	45		Construction	..	..	5 396	6 081	7 887	11 509	14 397	18 189

Table PL.8 SALAIRES ET TRAITEMENTS, SALARIÉS - WAGES AND SALARIES, EMPLOYEES

Millions de Zl (Prix courants)

1991	1992	1993	1994	1995	1996	1997	1998			CITI révision 3
..	2 613	3 444	4 918	6 006	7 171	7 876	8 426	Activites extractives	C	10_14
..	2 214	2 999	4 265	5 094	6 085	6 735	7 177	Extraction de charbon, de lignite et de tourbe		10
..	40	4	5	8	10	12	12	Pétrole brut et gaz naturel ; activités annexes		11
..	..	..	..	..	..	..	..	Extraction de Minerais d'Uranium et de Thorium		12
..	207	257	399	559	653	634	685	Extraction de minerais métalliques		13
..	153	185	249	345	423	495	552	Autres activités extractives		14
..	8 625	11 709	15 970	22 097	27 970	34 291	39 075	Total activites manufacturieres	D	15_37
..	1 505	1 967	2 516	3 535	4 535	5 679	6 568	Produits alimentaires et boissons		15
..	..	..	..	..	..	..	..	Viande, poisson, fruits, légumes, huiles et graisses		151
..	..	..	..	..	..	..	..	Produits laitiers		152
..	..	..	..	..	..	..	..	Travail des grains; prod. amyl. ; alim. pour animaux		153
..	..	..	..	..	..	..	..	Autres produits alimentaires		154
..	..	..	..	..	..	..	..	Boissons		155
..	44	72	107	147	230	281	318	Produits à base de tabac		16
..	492	635	861	1 073	1 235	1 425	1 473	Textiles		17
..	..	..	..	..	..	..	..	Filature, tissage et achèvement des textiles		171
..	..	..	..	..	..	..	..	Autres articles textiles		172
..	..	..	..	..	..	..	..	Etoffes et articles de bonneterie		173
..	507	768	1 024	1 461	1 676	1 953	2 254	Articles d'habillement et fourrure		18
..	..	..	..	..	..	..	..	Articles d'habillement autres qu'en fourrure		181
..	..	..	..	..	..	..	..	Prép. et teinture des fourrures ; articles en fourrure		182
..	212	265	332	436	551	704	607	Cuir et articles en cuir ; chaussures		19
..	..	..	..	..	..	..	..	Cuir et articles en cuir sauf chaussures		191
..	..	..	..	..	..	..	..	Chaussures		192
..	243	349	492	714	914	1 106	1 300	Bois, articles en bois, sauf meubles		20
..	..	..	..	..	..	..	..	Sciage et rabotage du bois		201
..	..	..	..	..	..	..	..	Articles en bois, liège, vannerie, sparterie		202
..	127	172	258	372	485	602	673	Papier, articles en papier et carton		21
..	230	312	433	648	856	1 187	1 461	Edition, imprimerie et reproduction		22
..	..	..	..	..	..	..	..	Edition		221
..	..	..	..	..	..	..	..	Imprimerie et activités annexes		222
..	..	..	..	..	..	..	..	Reproduction de supports enregistrés		223
..	130	203	289	390	462	521	573	Coke. et prod. pétroliers ; comb. nucléaires		23
..	..	..	..	..	..	..	..	Cokéfaction		231
..	..	..	..	..	..	..	..	Produits pétroliers raffinés		232
..	..	..	..	..	..	..	..	Traitement de combustibles nucléaires		233
..	503	706	1 064	1 512	1 943	2 320	2 683	Produits chimiques		24
..	..	..	..	..	..	..	..	Produits chimiques de base		241
..	..	..	..	..	..	..	..	Autres produits chimiques		242
..	..	..	..	..	..	..	..	Fibres synthétiques ou artificielles		243
..	228	357	539	766	996	1 299	1 557	Art. en caoutchouc et matières plastiques		25
..	..	..	..	..	..	..	..	Articles en caoutchouc		251
..	..	..	..	..	..	..	..	Articles en matières plastiques		252
..	510	689	932	1 302	1 642	2 066	2 392	Autres produits minéraux non métalliques		26
..	..	..	..	..	..	..	..	Verre et articles en verre		261
..	..	..	..	..	..	..	..	Produits minéraux non métalliques, nca		269
..	640	860	1 173	1 594	2 092	2 382	2 522	Produits métallurgiques de base		27
..	..	..	..	..	..	..	..	Sidérurgie et première transformation de l'acier		271
..	..	..	..	..	..	..	..	Première transf. métaux précieux et non ferreux		272
..	..	..	..	..	..	..	..	Fonderie		273
..	452	594	829	1 208	1 655	2 091	2 567	Ouvrages en métaux, sauf machines		28
..	..	..	..	..	..	..	..	Constr. et menuis. métal.; réservoirs; générateurs		281
..	..	..	..	..	..	..	..	Autres ouvrages en métaux		289
..	1 039	1 368	1 795	2 425	2 994	3 551	3 932	Machines et matériels, nca		29
..	..	..	..	..	..	..	..	Machines d'usage général		291
..	..	..	..	..	..	..	..	Machines d'usage spécifique		292
..	..	..	..	..	..	..	..	Appareils domestiques, nca		293
..	22	27	28	38	49	77	94	Mach. de bureau, comptables, information		30
..	283	406	580	789	1 023	1 311	1 542	Machines et appareils électriques, nca		31
..	..	..	..	..	..	..	..	Moteurs, génératrices, transformateurs électriques		311
..	..	..	..	..	..	..	..	Matériel électrique de distribution et de commande		312
..	..	..	..	..	..	..	..	Fils et câbles électriques isolés		313
..	..	..	..	..	..	..	..	Accumulateurs et piles électriques		314
..	..	..	..	..	..	..	..	Lampes électriques et appareils d'éclairage		315
..	..	..	..	..	..	..	..	Autres matériels électriques, nca.		319
..	185	244	310	399	480	586	663	Equipements et appareils de radio, TV, etc.		32
..	..	..	..	..	..	..	..	Tubes et valves électroniques;autres composants		321
..	..	..	..	..	..	..	..	Emetteurs radio/TV; app. téléphonie et télégraphie		322
..	..	..	..	..	..	..	..	Récepteurs radio/TV, son et image; art. associés		323
..	140	186	268	373	473	577	661	Instr. médicaux, de précision ; optique ; horlog.		33
..	..	..	..	..	..	..	..	Instruments médicaux et de précision, etc.		331
..	..	..	..	..	..	..	..	Instruments d'optique ; matériel photographique		332
..	..	..	..	..	..	..	..	Horlogerie		333
..	359	459	656	855	1 159	1 506	1 768	Véhicules automobiles; remorques, etc.		34
..	..	..	..	..	..	..	..	Véhicules automobiles		341
..	..	..	..	..	..	..	..	Carross. automobiles, remorques, semi-remorques		342
..	..	..	..	..	..	..	..	Pièces et accessoires pour automobiles et moteurs		343
..	422	567	781	1 009	1 207	1 372	1 500	Autres matériels de transport		35
..	..	..	..	..	..	..	..	Construction et réparation de navires		351
..	..	..	..	..	..	..	..	Matériel ferroviaire roulant		352
..	..	..	..	..	..	..	..	Construction aéronautique et spatiale		353
..	..	..	..	..	..	..	..	Autres équipements de transport		359
..	328	473	647	987	1 230	1 592	1 874	Meubles ; activités de fabrication, nca		36
..	..	..	..	..	..	..	..	Meubles		361
..	..	..	..	..	..	..	..	Activités de fabrication, nca		369
..	25	31	59	66	81	107	129	Récupération		37
..	1 101	1 800	2 609	3 363	4 177	4 929	5 416	Electricité, gaz et eau	E	40_41
..	896	1 514	2 213	2 851	3 522	4 083	4 451	Prod. et dist. d'électricité gaz, vapeur d'eau chaude		40
..	2 625	3 075	3 670	4 763	6 000	8 061	9 727	Construction	F	45

POLAND

Table PL.9 HOURS WORKED BY EMPLOYEES - HEURES OUVRÉES, SALARIÉS

Units

ISIC revision 3				1991	1992	1993	1994	1995	1996	1997	1998
C	10_14		Mining and quarrying	..	..	..	..	..	598744	559304	515791
	10		Mining of coal and lignite; extraction of peat	..	..	..	..	..	486051	460252	422622
	11		Crude petroleum & natural gas; service activities	..	..	..	..	..	1162	1246	1208
	12		Mining of uranium and thorium ores	..	..	..	..	..	..	..	..
	13		Mining of metal ores	..	..	..	..	..	49262	40385	36368
	14		Other mining and quarrying	..	..	..	..	..	62270	57421	55593
D	15_37		Total manufacturing	..	..	..	..	..	4725119	4713084	4640043
	15		Food products and beverages	..	..	..	..	..	865403	868609	879806
		151	Production of meat, fish, fruit, vegetables, oils & fats	..	..	..	..	..	..	210910	215895
		152	Dairy products	..	..	..	..	..	..	24119	24030
		153	Grain mill and starch products; animal feeds	..	..	..	..	..	..	85500	86675
		154	Other food products	..	..	..	..	..	..	9325	
		155	Beverages	..	..	..	..	..	..	116922	114307
	16		Tobacco products	..	..	..	..	..	21896	21049	19400
	17		Textiles	..	..	..	..	..	266844	238923	211329
		171	Spinning, weaving and finishing of textiles	..	..	..	..	..	..	42819	37691
		172	Other textiles	..	..	..	..	..	..	66071	57609
		173	Knitted and crocheted fabrics and articles	..	..	..	..	..	..	5499	5480
	18		Wearing apparel and fur	..	..	..	..	..	403256	402449	389316
		181	Wearing apparel, except fur apparel	..	..	..	..	..	..	6528	5077
		182	Dressing and dyeing of fur; articles of fur	..	..	..	..	..	..	392716	381438
	19		Leather and articles of leather; footwear	..	..	..	..	..	115998	115277	99223
		191	Leather and articles of leather	..	..	..	..	..	..	13048	10292
		192	Footwear	..	..	..	..	..	..	14202	14247
	20		Wood & products of wood & cork, exc. furniture	..	..	..	..	..	177678	180767	188604
		201	Sawmilling and planing of wood	..	..	..	..	..	..	65222	65142
		202	Wood products; articles of cork, straw, etc.	..	..	..	..	..	..	25440	26115
	21		Paper and paper products	..	..	..	..	..	69108	70485	70652
	22		Printing, publishing and recorded media	..	..	..	..	..	107177	118465	128253
		221	Publishing	..	..	..	..	..	..	52838	58178
		222	Printing and related activities	..	..	..	..	..	..	64284	68611
		223	Reproduction of recorded media	..	..	..	..	..	..	1342	1464
	23		Coke and petroleum products; nuclear fuel	..	..	..	..	..	45613	40310	36270
		231	Coke oven products	..	..	..	..	..	..	11602	..
		232	Refined petroleum products	..	..	..	..	..	..	28708	..
		233	Processing of nuclear fuel	..	..	..	..	..	..	..	..
	24		Chemicals and chemical products	..	..	..	..	..	238670	233475	224144
		241	Basic chemicals	..	..	..	..	..	..	90359	81229
		242	Other chemical products	..	..	..	..	..	..	4937	4073
		243	Man-made fibres	..	..	..	..	..	..	14018	14426
	25		Rubber and plastic products	..	..	..	..	..	163676	174667	183314
		251	Rubber products	..	..	..	..	..	..	44617	43813
		252	Plastic products	..	..	..	..	..	..	130050	139501
	26		Other non-metallic mineral products	..	..	..	..	..	280275	277682	277361
		261	Glass and glass products	..	..	..	..	..		72421	73742
		269	Non-metallic mineral products, nec	..	..	..	..	..	..	..	..
	27		Basic metals	..	..	..	..	..	259549	249265	230457
		271	Basic iron and steel	..	..	..	..	..	..	131780	119181
		272	Basic precious and non-ferrous metals	..	..	..	..	..	..	16681	14318
		273	Casting of metals	..	..	..	..	..	..	17533	14461
	28		Fabricated metal products, exc. machinery	..	..	..	..	..	270497	285309	300755
		281	Structural metal prod., tanks, reservoirs, etc.	..	..	..	..	..	..	73714	82479
		289	Other fabricated metal products	..	..	..	..	..	..	..	..
	29		Machinery and equipment, nec	..	..	..	..	..	499833	482675	451915
		291	General purpose machinery	..	..	..	..	..	..	79065	76635
		292	Special purpose machinery	..	..	..	..	..	..	79824	76444
		293	Domestic appliances, nec	..	..	..	..	..	..	68336	57204
	30		Office, accounting & computing machinery	..	..	..	..	..	7303	8143	8368
	31		Electrical machinery & apparatus, nec	..	..	..	..	..	157311	163929	166836
		311	Electric motors, generators and transformers	..	..	..	..	..	..	31112	28301
		312	Electricity distribution & control apparatus	..	..	..	..	..	..	37087	39417
		313	Insulated wire and cable	..	..	..	..	..	..	23492	23804
		314	Accumulators, primary cells & primary batteries	..	..	..	..	..	..	5873	5650
		315	Electric lamps and lighting equipment	..	..	..	..	..	..	29380	29754
		319	Other electrical equipment, nec	..	..	..	..	..	..	..	..
	32		Radio, TV, communication equipment	..	..	..	..	..	72798	72025	66424
		321	Electronic valves, tubes; other electronic components	..	..	..	..	..	..	18432	16289
		322	TV/radio transmitters; line teleph./telegr. apparatus	..	..	..	..	..	..	22740	22978
		323	TV/radio receivers; sound and video app.; assoc. gds.	..	..	..	..	..	..	30852	27157
	33		Medical, precision & optical instrum.; watch & clocks	..	..	..	..	..	70828	71553	69720
		331	Medical appl.; instruments & appl. for measuring, etc.	..	..	..	..	..	..	20079	20292
		332	Optical instruments and photographic equipment	..	..	..	..	..	..	25431	23807
		333	Watches and clocks	..	..	..	..	..	..	20384	20348
	34		Motor vehicles, trailers and semi-trailers	..	..	..	..	..	180390	187951	191871
		341	Motor vehicles	..	..	..	..	..	..	100099	94330
		342	Bodies for motor vehicles, trailers & semi-trailers	..	..	..	..	..	..	10639	11259
		343	Parts & accessories for motor vehicles & their engines	..	..	..	..	..	..	77213	86282
	35		Other transport equipment	..	..	..	..	..	190699	175548	162019
		351	Building and repairing of ships and boats	..	..	..	..	..	..	79867	78167
		352	Railway and tramway locomotives and rolling stock	..	..	..	..	..	..	46570	39703
		353	Aircraft and spacecraft	..	..	..	..	..	..	37018	33527
		359	Transport equipment, nec	..	..	..	..	..	..	..	..
	36		Furniture; manufacturing, nec	..	..	..	..	..	248512	262429	271044
		361	Furniture	..	..	..	..	..	..	212006	218976
		369	Manufacturing, nec	..	..	..	..	..	..	..	..
	37		Recycling	..	..	..	..	..	11806	12098	12963
E	40_41		Electricity, gas and water supply	..	..	..	..	..	428730	489884	468929
	40		Electricity, gas, steam and hot water supply	..	..	..	..	..	326901	380897	362904
F	45		Construction	..	..	..	..	..	1048576	1063596	1091038

Table PL.10 **ENTREPRISES - ENTERPRISES**

Unités

1991	1992	1993	1994	1995	1996	1997	1998		CITI révision 3		
..	..	..	..	914	1 050	1 135	1 274	Activites extractives	C	10_14	
..	..	..	..	73	88	91	102	Extraction de charbon, de lignite et de tourbe		10	
..	..	..	..	19	20	24	27	Pétrole brut et gaz naturel ; activités annexes		11	
..	..	..	..	..	..	..	..	Extraction de Minerais d'Uranium et de Thorium		12	
..	..	..	..	4	4	3	3	Extraction de minerais métalliques		13	
..	..	..	..	794	912	992	1 142	Autres activités extractives		14	
..	..	..	..	291 368	309 973	316 973	332 628	Total activites manufacturieres	D	15_37	
..	..	..	..	29 989	30 249	30 061	30 253	Produits alimentaires et boissons		15	
..	..	..	..	9 433	9 340	9 361	9 295	Viande, poisson, fruits, légumes, huiles et graisses			151
..	..	..	..	1 308	1 286	1 253	1 232	Produits laitiers			152
..	..	..	..	2 880	3 072	3 055	3 132	Travail des grains; prod. amyl. ; alim. pour animaux			153
..	..	..	..	14 368	14 494	14 503	14 799	Autres produits alimentaires			154
..	..	..	..	2 000	2 057	1 889	1 795	Boissons			155
..	..	..	..	18	20	21	19	Produits à base de tabac		16	
..	..	..	..	12 538	12 001	11 892	11 594	Textiles		17	
..	..	..	..	1 596	1 786	1 877	1 860	Filature, tissage et achèvement des textiles			171
..	..	..	..	3 548	3 652	3 568	3 968	Autres articles textiles			172
..	..	..	..	7 393	6 562	6 447	5 227	Etoffes et articles de bonneterie			173
..	..	..	..	45 750	46 756	45 548	45 506	Articles d'habillement et fourrure		18	
..	..	..	..	43 559	44 426	43 288	43 298	Articles d'habillement autres qu'en fourrure			181
..	..	..	..	2 191	2 329	2 260	2 208	Prép. et teinture des fourrures ; articles en fourrure			182
..	..	..	..	14 023	15 690	14 810	14 131	Cuir et articles en cuir ; chaussures		19	
..	..	..	..	4 136	4 259	4 060	3 987	Cuir et articles en cuir sauf chaussures			191
..	..	..	..	9 887	11 434	10 750	10 144	Chaussures			192
..	..	..	..	33 504	35 620	36 034	37 542	Bois, articles en bois, sauf meubles		20	
..	..	..	..	5 017	5 601	6 097	6 660	Sciage et rabotage du bois			201
..	..	..	..	28 487	30 089	29 937	30 882	Articles en bois, liège, vannerie, sparterie			202
..	..	..	..	2 051	2 218	2 428	2 637	Papier, articles en papier et carton		21	
..	..	..	..	17 214	19 319	21 068	23 338	Edition, imprimerie et reproduction		22	
..	..	..	..	6 090	6 473	6 832	7 373	Edition			221
..	..	..	..	10 739	12 529	13 961	15 695	Imprimerie et activités annexes			222
..	..	..	..	385	313	275	270	Reproduction de supports enregistrés			223
..	..	..	..	50	54	66	76	Coke. et prod. pétroliers ; comb. nucléaires		23	
..	..	..	..	6	7	8	7	Cokéfaction			231
..	..	..	..	44	47	58	69	Produits pétroliers raffinés			232
..	..	..	..	..	..	..	..	Traitement de combustibles nucléaires			233
..	..	..	..	3 147	3 174	3 134	3 205	Produits chimiques		24	
..	..	..	..	630	659	677	723	Produits chimiques de base			241
..	..	..	..	2 502	2 496	2 439	2 462	Autres produits chimiques			242
..	..	..	..	15	19	18	20	Fibres synthétiques ou artificielles			243
..	..	..	..	11 562	11 936	12 139	12 765	Art. en caoutchouc et matières plastiques		25	
..	..	..	..	1 464	1 354	1 293	1 288	Articles en caoutchouc			251
..	..	..	..	10 098	10 582	10 846	11 477	Articles en matières plastiques			252
..	..	..	..	18 124	18 790	18 998	19 946	Autres produits minéraux non métalliques		26	
..	..	..	..	2 593	2 630	2 602	2 664	Verre et articles en verre			261
..	..	..	..	15 531	16 128	16 396	17 410	Produits minéraux non métalliques, nca			269
..	..	..	..	747	780	826	855	Produits métallurgiques de base		27	
..	..	..	..	179	195	224	234	Sidérurgie et première transformation de l'acier			271
..	..	..	..	78	84	91	99	Première transf. métaux précieux et non ferreux			272
..	..	..	..	490	501	511	522	Fonderie			273
..	..	..	..	31 856	35 006	37 315	41 043	Ouvrages en métaux, sauf machines		28	
..	..	..	..	6 241	6 907	7 444	7 936	Constr. et menuis. métal.; réservoirs; générateurs			281
..	..	..	..	25 615	28 099	29 871	33 107	Autres ouvrages en métaux			289
..	..	..	..	13 173	15 277	16 658	18 299	Machines et matériels, nca		29	
..	..	..	..	5 891	6 842	7 293	7 991	Machines d'usage général			291
..	..	..	..	6 286	7 592	8 388	9 456	Machines d'usage spécifique			292
..	..	..	..	956	901	877	851	Appareils domestiques, nca			293
..	..	..	..	545	677	896	1 170	Mach. de bureau, comptables, information		30	
..	..	..	..	6 082	6 265	6 306	6 648	Machines et appareils électriques, nca		31	
..	..	..	..	1 299	1 714	1 878	2 018	Moteurs, génératrices, transformateurs électriques			311
..	..	..	..	719	764	789	892	Matériel électrique de distribution et de commande			312
..	..	..	..	140	147	154	147	Fils et câbles électriques isolés			313
..	..	..	..	165	171	157	157	Accumulateurs et piles électriques			314
..	..	..	..	1 217	1 141	1 107	1 143	Lampes électriques et appareils d'éclairage			315
..	..	..	..	2 542	2 328	2 221	2 291	Autres matériels électriques, nca.			319
..	..	..	..	5 270	5 465	5 253	5 405	Equipements et appareils de radio, TV, etc.		32	
..	..	..	..	935	832	753	755	Tubes et valves électroniques;autres composants			321
..	..	..	..	1 807	2 363	2 502	2 570	Emetteurs radio/TV; app. téléphonie et télégraphie			322
..	..	..	..	2 528	2 270	1 998	1 980	Récepteurs radio/TV, son et image; art. associés			323
..	..	..	..	8 110	10 104	12 080	13 201	Instr. médicaux, de précision ; optique ; horlog.		33	
..	..	..	..	6 270	8 285	10 249	11 464	Instruments médicaux et de précision, etc.			331
..	..	..	..	1 568	1 597	1 531	1 541	Instruments d'optique ; matériel photographique			332
..	..	..	..	272	222	200	196	Horlogerie			333
..	..	..	..	1 964	2 039	2 062	2 155	Véhicules automobiles; remorques, etc.		34	
..	..	..	..	83	89	97	106	Véhicules automobiles			341
..	..	..	..	148	181	203	226	Carross. automobiles, remorques, semi-remorques			342
..	..	..	..	1 733	1 769	1 762	1 823	Pièces et accessoires pour automobiles et moteurs			343
..	..	..	..	1 615	2 832	3 504	4 144	Autres matériels de transport		35	
..	..	..	..	1 233	2 402	3 022	3 639	Construction et réparation de navires			351
..	..	..	..	96	109	125	144	Matériel ferroviaire roulant			352
..	..	..	..	40	40	46	54	Construction aéronautique et spatiale			353
..	..	..	..	246	281	311	307	Autres équipements de transport			359
..	..	..	..	33 431	35 045	35 626	37 901	Meubles ; activités de fabrication, nca		36	
..	..	..	..	17 632	18 940	19 585	20 911	Meubles			361
..	..	..	..	15 799	16 105	16 041	16 990	Activités de fabrication, nca			369
..	..	..	..	605	656	705	795	Récupération		37	
..	..	..	..	1 402	1 741	2 090	2 381	Electricité, gaz et eau	E	40_41	
..	..	..	..	772	983	1 225	1 463	Prod. et dist. d'électricité gaz, vapeur d'eau chaude		40	
..	..	..	..	193 490	222 135	252 375	290 877	Construction	F	45	

PORTUGAL

Table PO.1 **PRODUCTION - PRODUCTION**

Millions of Esc (Current Prices)

ISIC revision 3				1991	1992	1993	1994	1995	1996	1997	1998
C	10_14		Mining and quarrying	..	..	..	..	..	169 655	200 715	..
	10		Mining of coal and lignite; extraction of peat	..	..	..	..	..	..	..	..
	11		Crude petroleum & natural gas; service activities	..	..	..	..	..	..	..	..
	12		Mining of uranium and thorium ores	..	..	..	..	..	..	..	..
	13		Mining of metal ores	..	..	..	..	..	30 417	..	..
	14		Other mining and quarrying	..	..	..	..	..	139 210	169 679	164 677
D	15_37		Total manufacturing	..	..	..	..	..	11312 627	12046 197	..
	15		Food products and beverages	..	..	..	..	..	1860 094	1957 865	..
		151	Production of meat, fish, fruit, vegetables, oils & fats	..	..	..	..	..	569 349	603 747	576 725
		152	Dairy products	..	..	..	..	..	247 998	286 282	251 103
		153	Grain mill and starch products; animal feeds	..	..	..	..	..	318 592	311 230	..
		154	Other food products	..	..	..	..	..	375 283	395 471	377 625
		155	Beverages	..	..	..	..	..	348 872	361 135	..
	16		Tobacco products	..	..	..	..	..	146 510	203 571	43 834
	17		Textiles	..	..	..	..	..	899 869	920 529	..
		171	Spinning, weaving and finishing of textiles	..	..	..	..	..	440 180	432 906	..
		172	Other textiles	..	..	..	..	..	200 068	215 437	242 671
		173	Knitted and crocheted fabrics and articles	..	..	..	..	..	259 621	272 186	..
	18		Wearing apparel and fur	..	..	..	..	..	699 904	828 751	772 809
		181	Wearing apparel, except fur apparel	..	..	..	..	..	693 836	822 525	768 340
		182	Dressing and dyeing of fur; articles of fur	..	..	..	..	..	6 068	6 226	4 469
	19		Leather and articles of leather; footwear	..	..	..	..	..	531 932	537 658	504 403
		191	Leather and articles of leather	..	..	..	..	..	77 717	78 951	79 190
		192	Footwear	..	..	..	..	..	454 215	458 707	425 213
	20		Wood & products of wood & cork, exc. furniture	..	..	..	..	..	515 633	602 104	624 563
		201	Sawmilling and planing of wood	..	..	..	..	..	102 349	111 441	119 531
		202	Wood products; articles of cork, straw, etc.	..	..	..	..	..	413 284	490 663	505 032
	21		Paper and paper products	..	..	..	..	..	371 000	372 412	..
	22		Printing, publishing and recorded media	..	..	..	..	..	451 019	461 321	..
		221	Publishing	..	..	..	..	..	200 407	218 270	209 592
		222	Printing and related activities	..	..	..	..	..	245 761	241 125	253 245
		223	Reproduction of recorded media	..	..	..	..	..	4 851	1 926	..
	23		Coke and petroleum products; nuclear fuel	..	..	..	..	..	827 910	812 814	..
		231	Coke oven products	..	..	..	..	..	..	..	..
		232	Refined petroleum products	..	..	..	..	..	827 910	812 814	773 284
		233	Processing of nuclear fuel	..	..	..	..	..	..	..	..
	24		Chemicals and chemical products	..	..	..	..	..	643 400	701 292	..
		241	Basic chemicals	..	..	..	..	..	233 487	267 386	..
		242	Other chemical products	..	..	..	..	..	389 471	408 567	..
		243	Man-made fibres	..	..	..	..	..	20 442	25 339	21 540
	25		Rubber and plastic products	..	..	..	..	..	251 915	280 356	..
		251	Rubber products	..	..	..	..	..	59 736	62 260	..
		252	Plastic products	..	..	..	..	..	192 179	218 096	238 686
	26		Other non-metallic mineral products	..	..	..	..	..	671 986	784 046	..
		261	Glass and glass products	..	..	..	..	..	94 643	106 363	116 678
		269	Non-metallic mineral products, nec	..	..	..	..	..	577 343	677 683	..
	27		Basic metals	..	..	..	..	..	228 292	259 099	..
		271	Basic iron and steel	..	..	..	..	..	133 117	157 165	..
		272	Basic precious and non-ferrous metals	..	..	..	..	..	56 946	46 915	..
		273	Casting of metals	..	..	..	..	..	38 229	55 019	59 166
	28		Fabricated metal products, exc. machinery	..	..	..	..	..	576 111	630 980	674 277
		281	Structural metal prod., tanks, reservoirs, etc.	..	..	..	..	..	222 623	237 786	254 659
		289	Other fabricated metal products	..	..	..	..	..	353 488	393 194	419 618
	29		Machinery and equipment, nec	..	..	..	..	..	444 823	467 302	501 685
		291	General purpose machinery	..	..	..	..	..	185 880	176 121	178 976
		292	Special purpose machinery	..	..	..	..	..	200 488	219 697	242 616
		293	Domestic appliances, nec	..	..	..	..	..	58 455	71 484	80 093
	30		Office, accounting & computing machinery	..	..	..	..	..	8 995	7 261	..
	31		Electrical machinery & apparatus, nec	..	..	..	..	..	315 457	350 471	392 838
		311	Electric motors, generators and transformers	..	..	..	..	..	35 149	39 213	30 361
		312	Electricity distribution & control apparatus	..	..	..	..	..	23 798	29 060	33 081
		313	Insulated wire and cable	..	..	..	..	..	95 101	114 104	137 211
		314	Accumulators, primary cells & primary batteries	..	..	..	..	..	14 725	13 778	14 264
		315	Electric lamps and lighting equipment	..	..	..	..	..	23 903	22 945	27 397
		319	Other electrical equipment, nec	..	..	..	..	..	122 781	131 371	150 523
	32		Radio, TV, communication equipment	..	..	..	..	..	334 697	361 512	415 159
		321	Electronic valves, tubes; other electronic components	..	..	..	..	..	161 062	164 025	75 364
		322	TV/radio transmitters; line teleph./telegr. apparatus	..	..	..	..	..	30 335	39 580	156 849
		323	TV/radio receivers; sound and video app.; assoc. gds.	..	..	..	..	..	143 300	157 907	182 947
	33		Medical, precision & optical instrum.; watch & clocks	..	..	..	..	..	55 677	58 378	73 319
		331	Medical appl.; instruments & appl. for measuring, etc.	..	..	..	..	..	39 262	41 597	55 544
		332	Optical instruments and photographic equipment	..	..	..	..	..	14 287	15 734	16 681
		333	Watches and clocks	..	..	..	..	..	2 128	1 047	1 094
	34		Motor vehicles, trailers and semi-trailers	..	..	..	..	..	820 376	834 569	859 435
		341	Motor vehicles	..	..	..	..	..	561 827	618 849	630 861
		342	Bodies for motor vehicles, trailers & semi-trailers	..	..	..	..	..	24 520	28 798	34 433
		343	Parts & accessories for motor vehicles & their engines	..	..	..	..	..	234 029	186 922	194 141
	35		Other transport equipment	..	..	..	..	..	163 067	132 430	..
		351	Building and repairing of ships and boats	..	..	..	..	..	92 334	63 943	134 164
		352	Railway and tramway locomotives and rolling stock	..	..	..	..	..	41 742	36 458	41 585
		353	Aircraft and spacecraft	..	..	..	..	..	13 047	16 183	18 123
		359	Transport equipment, nec	..	..	..	..	..	15 944	15 846	..
	36		Furniture; manufacturing, nec	..	..	..	..	..	481 938	465 514	497 131
		361	Furniture	..	..	..	..	..	267 904	283 029	318 186
		369	Manufacturing, nec	..	..	..	..	..	214 034	182 485	178 945
	37		Recycling	..	..	..	..	..	12 022	15 962	16 073
E	40_41		Electricity, gas and water supply	..	..	..	..	..	1173 540	1287 493	..
	40		Electricity, gas, steam and hot water supply	..	..	..	..	..	1148 878	1261 606	..
F	45		Construction	..	..	..	..	..	3643 562	4580 534	5771 572

Table PO.2 VALEUR AJOUTÉE - VALUE ADDED

Millions de Esc (Prix courants)

1991	1992	1993	1994	1995	1996	1997	1998	CITI révision 3			
..	..	..	..	..	72 246	80 276	..	Activites extractives	C	10_14	
..	..	..	..	..	..	..	..	Extraction de charbon, de lignite et de tourbe		10	
..	..	..	..	..	..	..	..	Pétrole brut et gaz naturel ; activités annexes		11	
..	..	..	..	..	..	..	..	Extraction de Minerais d'Uranium et de Thorium		12	
..	..	..	..	..	17 467	..	..	Extraction de minerais métalliques		13	
..	..	..	..	..	55 077	61 521	66 584	Autres activités extractives		14	
..	..	..	..	..	3 543 799	3 756 966	..	Total activites manufacturieres	D	15_37	
..	..	..	..	..	377 704	384 533	..	Produits alimentaires et boissons		15	
..	..	..	..	..	86 596	83 288	82 101	Viande, poisson, fruits, légumes, huiles et graisses			151
..	..	..	..	..	39 067	42 290	42 826	Produits laitiers			152
..	..	..	..	..	38 798	39 693	..	Travail des grains; prod. amyl. ; alim. pour animaux			153
..	..	..	..	..	113 324	122 204	117 528	Autres produits alimentaires			154
..	..	..	..	..	99 919	97 058	..	Boissons			155
..	..	..	..	..	132 637	185 144	20 525	Produits à base de tabac		16	
..	..	..	..	..	274 680	275 861	..	Textiles		17	
..	..	..	..	..	144 592	137 721	..	Filature, tissage et achèvement des textiles			171
..	..	..	..	..	63 617	68 480	77 399	Autres articles textiles			172
..	..	..	..	..	66 471	69 660	..	Etoffes et articles de bonneterie			173
..	..	..	..	..	239 988	256 637	251 902	Articles d'habillement et fourrure		18	
..	..	..	..	..	238 517	255 135	250 724	Articles d'habillement autres qu'en fourrure			181
..	..	..	..	..	1 471	1 502	1 178	Prép. et teinture des fourrures ; articles en fourrure			182
..	..	..	..	..	144 726	155 066	141 927	Cuir et articles en cuir ; chaussures		19	
..	..	..	..	..	18 223	18 129	15 816	Cuir et articles en cuir sauf chaussures			191
..	..	..	..	..	126 503	136 937	126 111	Chaussures			192
..	..	..	..	..	132 372	144 846	146 562	Bois, articles en bois, sauf meubles		20	
..	..	..	..	..	26 640	29 233	29 941	Sciage et rabotage du bois			201
..	..	..	..	..	105 732	115 613	116 621	Articles en bois, liège, vannerie, sparterie			202
..	..	..	..	..	103 300	119 032	..	Papier, articles en papier et carton		21	
..	..	..	..	..	166 782	182 626	..	Edition, imprimerie et reproduction		22	
..	..	..	..	..	63 067	69 381	67 449	Edition			221
..	..	..	..	..	102 593	112 648	115 668	Imprimerie et activités annexes			222
..	..	..	..	..	1 122	597	..	Reproduction de supports enregistrés			223
..	..	..	..	..	456 933	395 609	..	Coke. et prod. pétroliers ; comb. nucléaires		23	
..	..	..	..	..	..	..	..	Cokéfaction			231
..	..	..	..	..	456 933	395 609	436 542	Produits pétroliers raffinés			232
..	..	..	..	..	..	..	..	Traitement de combustibles nucléaires			233
..	..	..	..	..	191 357	216 235	..	Produits chimiques		24	
..	..	..	..	..	62 612	74 348	..	Produits chimiques de base			241
..	..	..	..	..	125 055	136 130	..	Autres produits chimiques			242
..	..	..	..	..	3 690	5 757	5 971	Fibres synthétiques ou artificielles			243
..	..	..	..	..	78 858	88 780	..	Art. en caoutchouc et matières plastiques		25	
..	..	..	..	..	22 004	23 890	..	Articles en caoutchouc			251
..	..	..	..	..	56 854	64 890	76 346	Articles en matières plastiques			252
..	..	..	..	..	266 422	312 613	..	Autres produits minéraux non métalliques		26	
..	..	..	..	..	39 055	44 364	50 252	Verre et articles en verre			261
..	..	..	..	..	227 367	268 249	..	Produits minéraux non métalliques, nca			269
..	..	..	..	..	52 080	60 853	..	Produits métallurgiques de base		27	
..	..	..	..	..	22 447	29 071	..	Sidérurgie et première transformation de l'acier			271
..	..	..	..	..	13 716	10 947	..	Première transf. métaux précieux et non ferreux			272
..	..	..	..	..	15 917	20 835	22 317	Fonderie			273
..	..	..	..	..	208 810	215 880	231 578	Ouvrages en métaux, sauf machines		28	
..	..	..	..	..	77 000	72 485	78 119	Constr. et menuis. métal.; réservoirs; générateurs			281
..	..	..	..	..	131 810	143 395	153 459	Autres ouvrages en métaux			289
..	..	..	..	..	155 079	164 949	175 680	Machines et matériels, nca		29	
..	..	..	..	..	58 332	57 801	56 730	Machines d'usage général			291
..	..	..	..	..	81 472	88 711	97 809	Machines d'usage spécifique			292
..	..	..	..	..	15 275	18 437	21 141	Appareils domestiques, nca			293
..	..	..	..	..	1 935	1 714	..	Mach. de bureau, comptables, information		30	
..	..	..	..	..	104 384	112 215	128 070	Machines et appareils électriques, nca		31	
..	..	..	..	..	11 888	12 523	9 806	Moteurs, génératrices, transformateurs électriques			311
..	..	..	..	..	9 611	11 782	13 366	Matériel électrique de distribution et de commande			312
..	..	..	..	..	27 810	30 232	40 326	Fils et câbles électriques isolés			313
..	..	..	..	..	5 261	4 953	4 347	Accumulateurs et piles électriques			314
..	..	..	..	..	8 466	8 311	9 795	Lampes électriques et appareils d'éclairage			315
..	..	..	..	..	41 348	44 414	50 430	Autres matériels électriques, nca.			319
..	..	..	..	..	75 850	93 416	111 152	Equipements et appareils de radio, TV, etc.		32	
..	..	..	..	..	35 600	39 294	19 454	Tubes et valves électroniques;autres composants			321
..	..	..	..	..	10 004	14 782	44 869	Emetteurs radio/TV; app. téléphonie et télégraphie			322
..	..	..	..	..	30 246	39 340	46 829	Récepteurs radio/TV, son et image; art. associés			323
..	..	..	..	..	22 414	21 354	26 205	Instr. médicaux, de précision ; optique ; horlog.		33	
..	..	..	..	..	15 618	14 449	19 013	Instruments médicaux et de précision, etc.			331
..	..	..	..	..	6 124	6 480	6 727	Instruments d'optique ; matériel photographique			332
..	..	..	..	..	672	425	466	Horlogerie			333
..	..	..	..	..	167 164	178 678	167 913	Véhicules automobiles; remorques, etc.		34	
..	..	..	..	..	108 880	119 075	109 125	Véhicules automobiles			341
..	..	..	..	..	6 834	8 039	9 774	Carross. automobiles, remorques, semi-remorques			342
..	..	..	..	..	51 450	51 564	49 014	Pièces et accessoires pour automobiles et moteurs			343
..	..	..	..	..	56 989	53 559	..	Autres matériels de transport		35	
..	..	..	..	..	32 415	25 311	36 363	Construction et réparation de navires			351
..	..	..	..	..	12 880	15 983	16 342	Matériel ferroviaire roulant			352
..	..	..	..	..	6 255	7 233	8 186	Construction aéronautique et spatiale			353
..	..	..	..	..	5 439	5 032	..	Autres équipements de transport			359
..	..	..	..	..	131 035	133 741	144 709	Meubles ; activités de fabrication, nca		36	
..	..	..	..	..	87 587	93 909	104 235	Meubles			361
..	..	..	..	..	43 448	39 832	40 474	Activités de fabrication, nca			369
..	..	..	..	..	2 300	3 625	3 374	Récupération		37	
..	..	..	..	..	452 442	475 273	..	Electricité, gaz et eau	E	40_41	
..	..	..	..	..	437 070	458 923	507 298	Prod. et dist. d'électricité gaz, vapeur d'eau chaude		40	
..	..	..	..	..	865 277	1 027 169	1 204 722	Construction	F	45	

PORTUGAL

Table PO.3 INVESTMENT - INVESTISSEMENT

Millions of Esc (Current Prices)

ISIC revision 3				1991	1992	1993	1994	1995	1996	1997	1998
C	10_14		Mining and quarrying	..	..	..	..	..	27 445	23 724	..
	10		Mining of coal and lignite; extraction of peat	..	..	..	..	..	..	..	..
	11		Crude petroleum & natural gas; service activities	..	..	..	..	..	..	..	..
	12		Mining of uranium and thorium ores	..	..	..	..	..	..	..	..
	13		Mining of metal ores	..	..	..	..	..	2 697	..	..
	14		Other mining and quarrying	..	..	..	..	..	24 789	21 596	28 990
D	15_37		Total manufacturing	..	..	..	..	..	..	..	..
	15		Food products and beverages	..	..	..	..	..	..	..	..
		151	Production of meat, fish, fruit, vegetables, oils & fats	..	..	..	..	..	21 590	8 534	35 122
		152	Dairy products	..	..	..	..	..	10 239	..	12 600
		153	Grain mill and starch products; animal feeds	..	..	..	..	..	5 406	6 330	..
		154	Other food products	..	..	..	..	..	28 264	27 796	25 407
		155	Beverages	..	..	..	..	..	24 207	18 654	..
	16		Tobacco products	..	..	..	..	..	..	..	2 652
	17		Textiles	..	..	..	..	..	57 170	39 083	..
		171	Spinning, weaving and finishing of textiles	..	..	..	..	..	28 482	20 145	..
		172	Other textiles	..	..	..	..	..	12 702	4 626	21 791
		173	Knitted and crocheted fabrics and articles	..	..	..	..	..	15 986	14 312	..
	18		Wearing apparel and fur	..	..	..	..	..	25 296	30 486	38 023
		181	Wearing apparel, except fur apparel	..	..	..	..	..	25 114	30 324	37 830
		182	Dressing and dyeing of fur; articles of fur	..	..	..	..	..	182	162	193
	19		Leather and articles of leather; footwear	..	..	..	..	..	19 783	21 770	25 549
		191	Leather and articles of leather	..	..	..	..	..	2 387	3 572	2 919
		192	Footwear	..	..	..	..	..	17 396	18 198	22 630
	20		Wood & products of wood & cork, exc. furniture	..	..	..	..	..	21 128	30 620	50 107
		201	Sawmilling and planing of wood	..	..	..	..	..	5 409	6 885	8 909
		202	Wood products; articles of cork, straw, etc.	..	..	..	..	..	15 719	23 735	41 198
	21		Paper and paper products	..	..	..	..	..	25 179	31 340	..
	22		Printing, publishing and recorded media	..	..	..	..	..	26 052	36 986	48 466
		221	Publishing	..	..	..	..	..	3 971	7 149	8 499
		222	Printing and related activities	..	..	..	..	..	21 831	29 773	39 967
		223	Reproduction of recorded media	..	..	..	..	..	250	64	..
	23		Coke and petroleum products; nuclear fuel	..	..	..	..	..	20 997	23 672	..
		231	Coke oven products	..	..	..	..	..	..	..	..
		232	Refined petroleum products	..	..	..	..	..	20 997	23 672	66 859
		233	Processing of nuclear fuel	..	..	..	..	..	..	..	..
	24		Chemicals and chemical products	..	..	..	..	..	25 527	71 371	..
		241	Basic chemicals	..	..	..	..	..	11 492	51 568	..
		242	Other chemical products	..	..	..	..	..	13 892	18 165	..
		243	Man-made fibres	..	..	..	..	..	143	1 638	4 384
	25		Rubber and plastic products	..	..	..	..	..	17 946	22 538	..
		251	Rubber products	..	..	..	..	..	3 340	3 867	..
		252	Plastic products	..	..	..	..	..	14 606	18 671	28 838
	26		Other non-metallic mineral products	..	..	..	..	..	..	67 002	..
		261	Glass and glass products	..	..	..	..	..	9 773	13 520	24 583
		269	Non-metallic mineral products, nec	..	..	..	..	..	..	53 482	..
	27		Basic metals	..	..	..	..	..	8 883	8 620	..
		271	Basic iron and steel	..	..	..	..	..	5 174	5 271	..
		272	Basic precious and non-ferrous metals	..	..	..	..	..	1 450	1 370	..
		273	Casting of metals	..	..	..	..	..	2 259	1 979	4 618
	28		Fabricated metal products, exc. machinery	..	..	..	..	..	33 414	37 149	51 452
		281	Structural metal prod., tanks, reservoirs, etc.	..	..	..	..	..	9 871	14 216	15 609
		289	Other fabricated metal products	..	..	..	..	..	23 543	22 933	35 843
	29		Machinery and equipment, nec	..	..	..	..	..	22 183	30 167	38 798
		291	General purpose machinery	..	..	..	..	..	5 543	9 509	11 332
		292	Special purpose machinery	..	..	..	..	..	14 453	16 515	23 826
		293	Domestic appliances, nec	..	..	..	..	..	2 187	4 143	3 640
	30		Office, accounting & computing machinery	..	..	..	..	..	331	102	..
	31		Electrical machinery & apparatus, nec	..	..	..	..	..	14 334	14 042	54 296
		311	Electric motors, generators and transformers	..	..	..	..	..	1 937	1 274	11 280
		312	Electricity distribution & control apparatus	..	..	..	..	..	408	1 245	3 057
		313	Insulated wire and cable	..	..	..	..	..	3 867	4 191	31 194
		314	Accumulators, primary cells & primary batteries	..	..	..	..	..	1 086	876	986
		315	Electric lamps and lighting equipment	..	..	..	..	..	2 177	1 036	1 518
		319	Other electrical equipment, nec	..	..	..	..	..	4 859	5 420	6 262
	32		Radio, TV, communication equipment	..	..	..	..	..	14 396	28 032	33 808
		321	Electronic valves, tubes; other electronic components	..	..	..	..	..	3 814	3 713	2 815
		322	TV/radio transmitters; line teleph./telegr. apparatus	..	..	..	..	..	2 353	1 558	5 092
		323	TV/radio receivers; sound and video app.; assoc. gds.	..	..	..	..	..	8 229	22 761	25 901
	33		Medical, precision & optical instrum.; watch & clocks	..	..	..	..	..	2 420	2 342	3 980
		331	Medical appl.; instruments & appl. for measuring, etc.	..	..	..	..	..	1 702	1 908	2 837
		332	Optical instruments and photographic equipment	..	..	..	..	..	667	426	1 123
		333	Watches and clocks	..	..	..	..	..	51	8	20
	34		Motor vehicles, trailers and semi-trailers	..	..	..	..	..	12 378	22 073	35 261
		341	Motor vehicles	..	..	..	..	..	..	10 464	19 192
		342	Bodies for motor vehicles, trailers & semi-trailers	..	..	..	..	..	1 090	790	1 715
		343	Parts & accessories for motor vehicles & their engines	..	..	..	..	..	18 586	10 819	14 354
	35		Other transport equipment	..	..	..	..	..	6 928	9 964	..
		351	Building and repairing of ships and boats	..	..	..	..	..	5 071	6 411	6 947
		352	Railway and tramway locomotives and rolling stock	..	..	..	..	..	682	2 426	4 205
		353	Aircraft and spacecraft	..	..	..	..	..	203	792	2 898
		359	Transport equipment, nec	..	..	..	..	..	972	335	..
	36		Furniture; manufacturing, nec	..	..	..	..	..	23 080	23 899	43 685
		361	Furniture	..	..	..	..	..	18 338	19 178	34 011
		369	Manufacturing, nec	..	..	..	..	..	4 742	4 721	9 674
	37		Recycling	..	..	..	..	..	1 155	1 056	3 401
E	40_41		Electricity, gas and water supply	..	..	..	..	..	153 307	161 322	..
	40		Electricity, gas, steam and hot water supply	..	..	..	..	..	147 222	154 146	..
F	45		Construction	..	..	..	..	..	137 203	140 633	285 262

Table PO.4 **EMPLOI, SALARIÉS - EMPLOYMENT, EMPLOYEES**

Unités

1991	1992	1993	1994	1995	1996	1997	1998	CITI révision 3			
..	..	..	..	..	15 173	15 000	..	Activites extractives	C	10_14	
..	..	..	..	..	..	..	..	Extraction de charbon, de lignite et de tourbe		10	
..	..	..	..	..	..	..	..	Pétrole brut et gaz naturel ; activités annexes		11	
..	..	..	..	..	..	..	..	Extraction de Minerais d'Uranium et de Thorium		12	
..	..	..	..	..	1 303	..	..	Extraction de minerais métalliques		13	
..	..	..	..	..	13 803	13 583	13 755	Autres activités extractives		14	
..	..	..	..	..	984 861	964 064	..	Total activites manufacturieres	D	15_37	
..	..	..	..	..	113 400	113 734	..	Produits alimentaires et boissons		15	
..	..	..	..	..	30 916	28 726	28 294	Viande, poisson, fruits, légumes, huiles et graisses			151
..	..	..	..	..	9 840	9 720	8 217	Produits laitiers			152
..	..	..	..	..	7 912	7 505	..	Travail des grains; prod. amyl. ; alim. pour animaux			153
..	..	..	..	..	49 366	52 415	46 388	Autres produits alimentaires			154
..	..	..	..	..	15 366	15 368	..	Boissons			155
..	..	..	..	..	1 227	1 199	1 277	Produits à base de tabac		16	
..	..	..	..	..	119 958	111 999	..	Textiles		17	
..	..	..	..	..	58 565	50 069	..	Filature, tissage et achèvement des textiles			171
..	..	..	..	..	30 246	30 660	29 264	Autres articles textiles			172
..	..	..	..	..	31 147	31 270	..	Etoffes et articles de bonneterie			173
..	..	..	..	..	154 108	149 014	139 918	Articles d'habillement et fourrure		18	
..	..	..	..	..	153 633	148 483	139 494	Articles d'habillement autres qu'en fourrure			181
..	..	..	..	..	475	531	424	Prép. et teinture des fourrures ; articles en fourrure			182
..	..	..	..	..	80 807	78 228	72 590	Cuir et articles en cuir ; chaussures		19	
..	..	..	..	..	7 611	7 189	6 773	Cuir et articles en cuir sauf chaussures			191
..	..	..	..	..	73 196	71 039	65 817	Chaussures			192
..	..	..	..	..	54 673	53 732	51 881	Bois, articles en bois, sauf meubles		20	
..	..	..	..	..	14 080	12 849	13 261	Sciage et rabotage du bois			201
..	..	..	..	..	40 593	40 883	38 620	Articles en bois, liège, vannerie, sparterie			202
..	..	..	..	..	15 217	15 341	..	Papier, articles en papier et carton		21	
..	..	..	..	..	38 306	37 601	..	Edition, imprimerie et reproduction		22	
..	..	..	..	..	10 038	9 310	10 612	Edition			221
..	..	..	..	..	28 002	28 219	26 774	Imprimerie et activités annexes			222
..	..	..	..	..	266	72	..	Reproduction de supports enregistrés			223
..	..	..	..	..	3 382	3 145	..	Coke. et prod. pétroliers ; comb. nucléaires		23	
..	..	..	..	..	..	..	..	Cokéfaction			231
..	..	..	..	..	3 382	3 145	2 793	Produits pétroliers raffinés			232
..	..	..	..	..	..	..	..	Traitement de combustibles nucléaires			233
..	..	..	..	..	27 471	25 680	..	Produits chimiques		24	
..	..	..	..	..	6 574	6 299	..	Produits chimiques de base			241
..	..	..	..	..	20 148	18 716	..	Autres produits chimiques			242
..	..	..	..	..	749	665	670	Fibres synthétiques ou artificielles			243
..	..	..	..	..	19 613	20 823	..	Art. en caoutchouc et matières plastiques		25	
..	..	..	..	..	5 130	5 042	..	Articles en caoutchouc			251
..	..	..	..	..	14 483	15 781	16 349	Articles en matières plastiques			252
..	..	..	..	..	66 896	70 114	..	Autres produits minéraux non métalliques		26	
..	..	..	..	..	8 889	8 881	9 521	Verre et articles en verre			261
..	..	..	..	..	58 007	61 233	..	Produits minéraux non métalliques, nca			269
..	..	..	..	..	14 812	14 266	..	Produits métallurgiques de base		27	
..	..	..	..	..	5 660	4 981	..	Sidérurgie et première transformation de l'acier			271
..	..	..	..	..	3 418	2 524	..	Première transf. métaux précieux et non ferreux			272
..	..	..	..	..	5 734	6 761	6 738	Fonderie			273
..	..	..	..	..	73 156	71 482	74 676	Ouvrages en métaux, sauf machines		28	
..	..	..	..	..	27 091	26 852	26 834	Constr. et menuis. métal.; réservoirs; générateurs			281
..	..	..	..	..	46 065	44 630	47 842	Autres ouvrages en métaux			289
..	..	..	..	..	45 935	45 419	45 034	Machines et matériels, nca		29	
..	..	..	..	..	17 195	16 879	15 478	Machines d'usage général			291
..	..	..	..	..	24 335	24 233	24 956	Machines d'usage spécifique			292
..	..	..	..	..	4 405	4 307	4 600	Appareils domestiques, nca			293
..	..	..	..	..	317	449	..	Mach. de bureau, comptables, information		30	
..	..	..	..	..	28 411	29 835	33 673	Machines et appareils électriques, nca		31	
..	..	..	..	..	2 840	2 750	2 259	Moteurs, génératrices, transformateurs électriques			311
..	..	..	..	..	2 638	2 872	3 371	Matériel électrique de distribution et de commande			312
..	..	..	..	..	5 810	6 414	8 107	Fils et câbles électriques isolés			313
..	..	..	..	..	885	899	873	Accumulateurs et piles électriques			314
..	..	..	..	..	3 304	3 159	3 045	Lampes électriques et appareils d'éclairage			315
..	..	..	..	..	12 934	13 741	16 018	Autres matériels électriques, nca.			319
..	..	..	..	..	17 386	17 228	17 509	Equipements et appareils de radio, TV, etc.		32	
..	..	..	..	..	7 672	7 270	4 115	Tubes et valves électroniques;autres composants			321
..	..	..	..	..	1 786	2 205	5 436	Emetteurs radio/TV; app. téléphonie et télégraphie			322
..	..	..	..	..	7 928	7 753	7 958	Récepteurs radio/TV, son et image; art. associés			323
..	..	..	..	..	6 191	6 090	6 750	Instr. médicaux, de précision ; optique ; horlog.		33	
..	..	..	..	..	4 052	4 100	4 825	Instruments médicaux et de précision, etc.			331
..	..	..	..	..	1 862	1 762	1 678	Instruments d'optique ; matériel photographique			332
..	..	..	..	..	277	228	247	Horlogerie			333
..	..	..	..	..	24 971	24 067	23 451	Véhicules automobiles; remorques, etc.		34	
..	..	..	..	..	11 106	10 063	10 135	Véhicules automobiles			341
..	..	..	..	..	3 225	3 241	3 481	Carross. automobiles, remorques, semi-remorques			342
..	..	..	..	..	10 640	10 763	9 835	Pièces et accessoires pour automobiles et moteurs			343
..	..	..	..	..	17 139	12 765	..	Autres matériels de transport		35	
..	..	..	..	..	10 009	5 866	6 447	Construction et réparation de navires			351
..	..	..	..	..	3 068	3 121	3 109	Matériel ferroviaire roulant			352
..	..	..	..	..	1 714	1 774	1 864	Construction aéronautique et spatiale			353
..	..	..	..	..	2 348	2 004	..	Autres équipements de transport			359
..	..	..	..	..	60 757	61 037	62 560	Meubles ; activités de fabrication, nca		36	
..	..	..	..	..	45 776	47 764	48 644	Meubles			361
..	..	..	..	..	14 981	13 273	13 916	Activités de fabrication, nca			369
..	..	..	..	..	728	816	759	Récupération		37	
..	..	..	..	..	17 080	19 195	..	Electricité, gaz et eau	E	40_41	
..	..	..	..	..	15 536	17 617	17 118	Prod. et dist. d'électricité gaz, vapeur d'eau chaude		40	
..	..	..	..	..	256 326	276 527	300 484	Construction	F	45	

PORTUGAL

Table PO.5 COMPENSATION OF LABOUR, EMPLOYEES - SALAIRES ET CHARGES SOCIALES, SALARIÉS

Millions of Esc (Current Prices)

ISIC revision 3				1991	1992	1993	1994	1995	1996	1997	1998
C	10_14		Mining and quarrying	..	..	..	..	..	32 368	35 154	..
	10		Mining of coal and lignite; extraction of peat	..	..	..	..	..	..	..	..
	11		Crude petroleum & natural gas; service activities	..	..	..	..	..	..	..	..
	12		Mining of uranium and thorium ores	..	..	..	..	..	..	..	..
	13		Mining of metal ores	..	..	..	..	..	5 711	..	..
	14		Other mining and quarrying	..	..	..	..	..	26 402	28 870	30 958
D	15_37		Total manufacturing	..	..	..	..	..	1880 502	1923 131	..
	15		Food products and beverages	..	..	..	..	..	215 651	222 227	..
		151	Production of meat, fish, fruit, vegetables, oils & fats	..	..	..	..	..	56 941	55 386	57 736
		152	Dairy products	..	..	..	..	..	22 387	23 590	20 649
		153	Grain mill and starch products; animal feeds	..	..	..	..	..	19 772	19 318	..
		154	Other food products	..	..	..	..	..	74 499	77 953	75 689
		155	Beverages	..	..	..	..	..	42 052	45 980	..
	16		Tobacco products	..	..	..	..	..	5 751	5 742	8 346
	17		Textiles	..	..	..	..	..	180 720	175 281	..
		171	Spinning, weaving and finishing of textiles	..	..	..	..	..	95 222	85 606	..
		172	Other textiles	..	..	..	..	..	42 693	44 819	48 587
		173	Knitted and crocheted fabrics and articles	..	..	..	..	..	42 805	44 856	..
	18		Wearing apparel and fur	..	..	..	..	..	191 416	194 271	192 569
		181	Wearing apparel, except fur apparel	..	..	..	..	..	190 608	193 300	191 845
		182	Dressing and dyeing of fur; articles of fur	..	..	..	..	..	808	971	724
	19		Leather and articles of leather; footwear	..	..	..	..	..	105 272	109 580	108 059
		191	Leather and articles of leather	..	..	..	..	..	12 565	12 351	12 457
		192	Footwear	..	..	..	..	..	92 707	97 229	95 602
	20		Wood & products of wood & cork, exc. furniture	..	..	..	..	..	82 021	86 797	88 215
		201	Sawmilling and planing of wood	..	..	..	..	..	18 209	18 118	19 657
		202	Wood products; articles of cork, straw, etc.	..	..	..	..	..	63 812	68 679	68 558
	21		Paper and paper products	..	..	..	..	..	45 392	49 240	..
	22		Printing, publishing and recorded media	..	..	..	..	..	100 409	101 827	..
		221	Publishing	..	..	..	..	..	37 967	37 295	42 958
		222	Printing and related activities	..	..	..	..	..	61 903	64 315	65 974
		223	Reproduction of recorded media	..	..	..	..	..	539	217	..
	23		Coke and petroleum products; nuclear fuel	..	..	..	..	..	24 601	22 534	..
		231	Coke oven products	..	..	..	..	..	..	..	..
		232	Refined petroleum products	..	..	..	..	..	24 601	22 534	21 316
		233	Processing of nuclear fuel	..	..	..	..	..	..	..	..
	24		Chemicals and chemical products	..	..	..	..	..	106 356	105 168	..
		241	Basic chemicals	..	..	..	..	..	29 306	31 126	..
		242	Other chemical products	..	..	..	..	..	74 086	71 187	..
		243	Man-made fibres	..	..	..	..	..	2 964	2 855	2 961
	25		Rubber and plastic products	..	..	..	..	..	41 632	47 149	..
		251	Rubber products	..	..	..	..	..	11 519	12 136	..
		252	Plastic products	..	..	..	..	..	30 113	35 013	39 858
	26		Other non-metallic mineral products	..	..	..	..	..	141 087	154 381	..
		261	Glass and glass products	..	..	..	..	..	23 139	24 857	28 412
		269	Non-metallic mineral products, nec	..	..	..	..	..	117 948	129 524	..
	27		Basic metals	..	..	..	..	..	36 256	37 122	..
		271	Basic iron and steel	..	..	..	..	..	16 148	15 559	..
		272	Basic precious and non-ferrous metals	..	..	..	..	..	7 930	6 354	..
		273	Casting of metals	..	..	..	..	..	12 178	15 209	15 459
	28		Fabricated metal products, exc. machinery	..	..	..	..	..	131 911	142 227	151 169
		281	Structural metal prod., tanks, reservoirs, etc.	..	..	..	..	..	49 458	51 779	53 210
		289	Other fabricated metal products	..	..	..	..	..	82 453	90 448	97 959
	29		Machinery and equipment, nec	..	..	..	..	..	104 848	108 239	114 992
		291	General purpose machinery	..	..	..	..	..	40 035	39 262	38 846
		292	Special purpose machinery	..	..	..	..	..	55 952	58 444	64 471
		293	Domestic appliances, nec	..	..	..	..	..	8 861	10 533	11 675
	30		Office, accounting & computing machinery	..	..	..	..	..	768	1 018	..
	31		Electrical machinery & apparatus, nec	..	..	..	..	..	67 253	75 454	85 146
		311	Electric motors, generators and transformers	..	..	..	..	..	8 878	9 391	7 309
		312	Electricity distribution & control apparatus	..	..	..	..	..	6 168	7 605	8 570
		313	Insulated wire and cable	..	..	..	..	..	15 308	18 701	23 797
		314	Accumulators, primary cells & primary batteries	..	..	..	..	..	3 507	3 402	3 452
		315	Electric lamps and lighting equipment	..	..	..	..	..	5 981	5 833	6 467
		319	Other electrical equipment, nec	..	..	..	..	..	27 411	30 522	35 550
	32		Radio, TV, communication equipment	..	..	..	..	..	58 348	63 184	67 078
		321	Electronic valves, tubes; other electronic components	..	..	..	..	..	27 815	28 246	13 665
		322	TV/radio transmitters; line teleph./telegr. apparatus	..	..	..	..	..	7 658	9 562	28 131
		323	TV/radio receivers; sound and video app.; assoc. gds.	..	..	..	..	..	22 875	25 376	25 282
	33		Medical, precision & optical instrum.; watch & clocks	..	..	..	..	..	14 173	14 157	16 365
		331	Medical appl.; instruments & appl. for measuring, etc.	..	..	..	..	..	9 816	9 720	11 826
		332	Optical instruments and photographic equipment	..	..	..	..	..	3 936	3 925	4 105
		333	Watches and clocks	..	..	..	..	..	421	512	435
	34		Motor vehicles, trailers and semi-trailers	..	..	..	..	..	83 586	75 889	82 347
		341	Motor vehicles	..	..	..	..	..	51 998	43 256	50 004
		342	Bodies for motor vehicles, trailers & semi-trailers	..	..	..	..	..	5 729	6 247	6 963
		343	Parts & accessories for motor vehicles & their engines	..	..	..	..	..	25 859	26 386	25 380
	35		Other transport equipment	..	..	..	..	..	54 819	39 377	..
		351	Building and repairing of ships and boats	..	..	..	..	..	34 598	17 937	27 686
		352	Railway and tramway locomotives and rolling stock	..	..	..	..	..	10 771	11 900	12 173
		353	Aircraft and spacecraft	..	..	..	..	..	5 254	5 868	6 659
		359	Transport equipment, nec	..	..	..	..	..	4 196	3 672	..
	36		Furniture; manufacturing, nec	..	..	..	..	..	86 791	90 437	95 902
		361	Furniture	..	..	..	..	..	61 856	66 668	70 836
		369	Manufacturing, nec	..	..	..	..	..	24 935	23 769	25 066
	37		Recycling	..	..	..	..	..	1 441	1 830	1 813
E	40_41		Electricity, gas and water supply	..	..	..	..	..	86 969	104 016	..
	40		Electricity, gas, steam and hot water supply	..	..	..	..	..	81 340	97 952	101 348
F	45		Construction	..	..	..	..	..	511 064	568 752	644 134

Table PO.6 **SALAIRES ET TRAITEMENTS, SALARIÉS - WAGES AND SALARIES, EMPLOYEES**

Milions de Esc (Prix courants)

1991	1992	1993	1994	1995	1996	1997	1998		CITI révision 3		
..	..	..	..	..	21 803	23 798	..	Activites extractives	C	10_14	
..	..	..	..	..	..	..	..	Extraction de charbon, de lignite et de tourbe		10	
..	..	..	..	..	..	..	..	Pétrole brut et gaz naturel ; activités annexes		11	
..	..	..	..	..	..	..	..	Extraction de Minerais d'Uranium et de Thorium		12	
..	..	..	..	..	3 963	..	..	Extraction de minerais métalliques		13	
..	..	..	..	..	17 714	19 355	..	Autres activités extractives		14	
..	..	..	..	..	1 286 787	1 320 851	..	Total activites manufacturieres	D	15_37	
..	..	..	..	..	148 612	153 983	..	Produits alimentaires et boissons		15	
..	..	..	..	..	40 272	39 328	..	Viande, poisson, fruits, légumes, huiles et graisses			151
..	..	..	..	..	16 496	17 562	..	Produits laitiers			152
..	..	..	..	..	13 434	13 378	..	Travail des grains; prod. amyl. ; alim. pour animaux			153
..	..	..	..	..	49 601	52 277	..	Autres produits alimentaires			154
..	..	..	..	..	28 809	31 438	..	Boissons			155
..	..	..	..	..	2 847	2 980	..	Produits à base de tabac		16	
..	..	..	..	..	129 908	125 410	..	Textiles		17	
..	..	..	..	..	69 234	61 829	..	Filature, tissage et achèvement des textiles			171
..	..	..	..	..	30 027	31 597	..	Autres articles textiles			172
..	..	..	..	..	30 647	31 984	..	Etoffes et articles de bonneterie			173
..	..	..	..	..	135 539	134 875	..	Articles d'habillement et fourrure		18	
..	..	..	..	..	134 947	134 188	..	Articles d'habillement autres qu'en fourrure			181
..	..	..	..	..	592	687	..	Prép. et teinture des fourrures ; articles en fourrure			182
..	..	..	..	..	75 892	79 019	..	Cuir et articles en cuir ; chaussures		19	
..	..	..	..	..	8 744	8 677	..	Cuir et articles en cuir sauf chaussures			191
..	..	..	..	..	67 148	70 342	..	Chaussures			192
..	..	..	..	..	53 930	57 094	..	Bois, articles en bois, sauf meubles		20	
..	..	..	..	..	11 821	11 674	..	Sciage et rabotage du bois			201
..	..	..	..	..	42 109	45 420	..	Articles en bois, liège, vannerie, sparterie			202
..	..	..	..	..	30 518	32 873	..	Papier, articles en papier et carton		21	
..	..	..	..	..	66 705	66 628	..	Edition, imprimerie et reproduction		22	
..	..	..	..	..	26 497	25 836	..	Edition			221
..	..	..	..	..	39 821	40 643	..	Imprimerie et activités annexes			222
..	..	..	..	..	387	149	..	Reproduction de supports enregistrés			223
..	..	..	..	..	15 192	15 548	..	Coke. et prod. pétroliers ; comb. nucléaires		23	
..	..	..	..	..	..	..	..	Cokéfaction			231
..	..	..	..	..	15 192	15 548	..	Produits pétroliers raffinés			232
..	..	..	..	..	..	..	..	Traitement de combustibles nucléaires			233
..	..	..	..	..	71 263	71 691	..	Produits chimiques		24	
..	..	..	..	..	18 690	20 477	..	Produits chimiques de base			241
..	..	..	..	..	50 652	49 297	..	Autres produits chimiques			242
..	..	..	..	..	1 921	1 917	..	Fibres synthétiques ou artificielles			243
..	..	..	..	..	28 649	32 592	..	Art. en caoutchouc et matières plastiques		25	
..	..	..	..	..	7 982	8 495	..	Articles en caoutchouc			251
..	..	..	..	..	20 667	24 097	..	Articles en matières plastiques			252
..	..	..	..	..	97 905	106 276	..	Autres produits minéraux non métalliques		26	
..	..	..	..	..	16 630	17 571	..	Verre et articles en verre			261
..	..	..	..	..	81 275	88 705	..	Produits minéraux non métalliques, nca			269
..	..	..	..	..	24 653	25 530	..	Produits métallurgiques de base		27	
..	..	..	..	..	10 884	10 596	..	Sidérurgie et première transformation de l'acier			271
..	..	..	..	..	5 438	4 416	..	Première transf. métaux précieux et non ferreux			272
..	..	..	..	..	8 331	10 518	..	Fonderie			273
..	..	..	..	..	84 904	92 063	..	Ouvrages en métaux, sauf machines		28	
..	..	..	..	..	31 198	35 038	..	Constr. et menuis. métal.; réservoirs; générateurs			281
..	..	..	..	..	53 706	57 025	..	Autres ouvrages en métaux			289
..	..	..	..	..	70 651	72 015	..	Machines et matériels, nca		29	
..	..	..	..	..	26 640	26 175	..	Machines d'usage général			291
..	..	..	..	..	37 639	38 175	..	Machines d'usage spécifique			292
..	..	..	..	..	6 372	7 665	..	Appareils domestiques, nca			293
..	..	..	..	..	514	622	..	Mach. de bureau, comptables, information		30	
..	..	..	..	..	46 927	54 464	..	Machines et appareils électriques, nca		31	
..	..	..	..	..	6 526	7 008	..	Moteurs, génératrices, transformateurs électriques			311
..	..	..	..	..	4 327	5 156	..	Matériel électrique de distribution et de commande			312
..	..	..	..	..	11 153	13 053	..	Fils et câbles électriques isolés			313
..	..	..	..	..	2 480	2 412	..	Accumulateurs et piles électriques			314
..	..	..	..	..	4 046	3 843	..	Lampes électriques et appareils d'éclairage			315
..	..	..	..	..	18 395	22 992	..	Autres matériels électriques, nca.			319
..	..	..	..	..	41 731	44 949	..	Equipements et appareils de radio, TV, etc.		32	
..	..	..	..	..	19 851	20 462	..	Tubes et valves électroniques;autres composants			321
..	..	..	..	..	5 688	7 195	..	Emetteurs radio/TV; app. téléphonie et télégraphie			322
..	..	..	..	..	16 192	17 292	..	Récepteurs radio/TV, son et image; art. associés			323
..	..	..	..	..	9 771	9 802	..	Instr. médicaux, de précision ; optique ; horlog.		33	
..	..	..	..	..	6 559	6 505	..	Instruments médicaux et de précision, etc.			331
..	..	..	..	..	2 932	3 023	..	Instruments d'optique ; matériel photographique			332
..	..	..	..	..	280	274	..	Horlogerie			333
..	..	..	..	..	54 409	53 604	..	Véhicules automobiles; remorques, etc.		34	
..	..	..	..	..	32 360	30 619	..	Véhicules automobiles			341
..	..	..	..	..	3 894	4 332	..	Carross. automobiles, remorques, semi-remorques			342
..	..	..	..	..	18 155	18 653	..	Pièces et accessoires pour automobiles et moteurs			343
..	..	..	..	..	39 158	28 089	..	Autres matériels de transport		35	
..	..	..	..	..	24 361	12 454	..	Construction et réparation de navires			351
..	..	..	..	..	7 800	8 538	..	Matériel ferroviaire roulant			352
..	..	..	..	..	4 148	4 582	..	Construction aéronautique et spatiale			353
..	..	..	..	..	2 849	2 515	..	Autres équipements de transport			359
..	..	..	..	..	56 221	59 594	..	Meubles ; activités de fabrication, nca		36	
..	..	..	..	..	40 474	44 588	..	Meubles			361
..	..	..	..	..	15 747	15 006	..	Activités de fabrication, nca			369
..	..	..	..	..	888	1 150	..	Récupération		37	
..	..	..	..	..	57 472	68 368	..	Electricité, gaz et eau	E	40_41	
..	..	..	..	..	53 594	64 125	..	Prod. et dist. d'électricité gaz, vapeur d'eau chaude		40	
..	..	..	..	..	338 063	385 863	..	Construction	F	45	

PORTUGAL

Table PO.7 SUPPLEMENTS TO WAGES & SALARIES - CHARGES SOCIALES DES EMPLOYEURS

Millions of Esc (Current Prices)

ISIC revision 3				1991	1992	1993	1994	1995	1996	1997	1998
C	10_14		Mining and quarrying	..	..	..	..	..	8 347	8 712	..
	10		Mining of coal and lignite; extraction of peat	..	..	..	..	..	..	..	..
	11		Crude petroleum & natural gas; service activities	..	..	..	..	..	..	..	..
	12		Mining of uranium and thorium ores	..	..	..	..	..	..	..	..
	13		Mining of metal ores	..	..	..	..	..	1 661	..	..
	14		Other mining and quarrying	..	..	..	..	..	6 581	6 979	..
D	15_37		Total manufacturing	..	..	..	..	..	469 859	469 604	..
	15		Food products and beverages	..	..	..	..	..	53 579	54 465	..
		151	Production of meat, fish, fruit, vegetables, oils & fats	..	..	..	..	..	13 805	13 274	..
		152	Dairy products	..	..	..	..	..	5 378	5 569	..
		153	Grain mill and starch products; animal feeds	..	..	..	..	..	4 997	4 684	..
		154	Other food products	..	..	..	..	..	18 139	18 725	..
		155	Beverages	..	..	..	..	..	11 260	12 213	..
	16		Tobacco products	..	..	..	..	..	2 816	2 702	..
	17		Textiles	..	..	..	..	..	40 154	39 217	..
		171	Spinning, weaving and finishing of textiles	..	..	..	..	..	21 939	19 843	..
		172	Other textiles	..	..	..	..	..	9 281	9 917	..
		173	Knitted and crocheted fabrics and articles	..	..	..	..	..	8 934	9 457	..
	18		Wearing apparel and fur	..	..	..	..	..	41 787	42 557	..
		181	Wearing apparel, except fur apparel	..	..	..	..	..	41 616	42 343	..
		182	Dressing and dyeing of fur; articles of fur	..	..	..	..	..	171	214	..
	19		Leather and articles of leather; footwear	..	..	..	..	..	22 642	23 501	..
		191	Leather and articles of leather	..	..	..	..	..	2 600	2 567	..
		192	Footwear	..	..	..	..	..	20 042	20 934	..
	20		Wood & products of wood & cork, exc. furniture	..	..	..	..	..	19 594	21 092	..
		201	Sawmilling and planing of wood	..	..	..	..	..	4 500	4 448	..
		202	Wood products; articles of cork, straw, etc.	..	..	..	..	..	15 094	16 644	..
	21		Paper and paper products	..	..	..	..	..	12 622	13 941	..
	22		Printing, publishing and recorded media	..	..	..	..	..	23 510	24 276	..
		221	Publishing	..	..	..	..	..	8 987	9 122	..
		222	Printing and related activities	..	..	..	..	..	14 410	15 099	..
		223	Reproduction of recorded media	..	..	..	..	..	113	55	..
	23		Coke and petroleum products; nuclear fuel	..	..	..	..	..	9 211	6 784	..
		231	Coke oven products	..	..	..	..	..	..	..	..
		232	Refined petroleum products	..	..	..	..	..	9 211	6 784	..
		233	Processing of nuclear fuel	..	..	..	..	..	..	..	..
	24		Chemicals and chemical products	..	..	..	..	..	30 915	29 348	..
		241	Basic chemicals	..	..	..	..	..	9 851	9 764	..
		242	Other chemical products	..	..	..	..	..	20 127	18 753	..
		243	Man-made fibres	..	..	..	..	..	937	831	..
	25		Rubber and plastic products	..	..	..	..	..	9 629	11 061	..
		251	Rubber products	..	..	..	..	..	2 784	2 997	..
		252	Plastic products	..	..	..	..	..	6 845	8 064	..
	26		Other non-metallic mineral products	..	..	..	..	..	34 752	38 752	..
		261	Glass and glass products	..	..	..	..	..	5 462	6 078	..
		269	Non-metallic mineral products, nec	..	..	..	..	..	29 290	32 674	..
	27		Basic metals	..	..	..	..	..	9 908	9 953	..
		271	Basic iron and steel	..	..	..	..	..	4 703	4 455	..
		272	Basic precious and non-ferrous metals	..	..	..	..	..	1 990	1 525	..
		273	Casting of metals	..	..	..	..	..	3 215	3 973	..
	28		Fabricated metal products, exc. machinery	..	..	..	..	..	33 514	34 811	..
		281	Structural metal prod., tanks, reservoirs, etc.	..	..	..	..	..	13 517	11 512	..
		289	Other fabricated metal products	..	..	..	..	..	19 997	23 299	..
	29		Machinery and equipment, nec	..	..	..	..	..	25 281	25 995	..
		291	General purpose machinery	..	..	..	..	..	10 157	9 654	..
		292	Special purpose machinery	..	..	..	..	..	13 002	13 822	..
		293	Domestic appliances, nec	..	..	..	..	..	2 122	2 519	..
	30		Office, accounting & computing machinery	..	..	..	..	..	196	264	..
	31		Electrical machinery & apparatus, nec	..	..	..	..	..	18 268	18 623	..
		311	Electric motors, generators and transformers	..	..	..	..	..	2 085	2 055	..
		312	Electricity distribution & control apparatus	..	..	..	..	..	1 513	2 089	..
		313	Insulated wire and cable	..	..	..	..	..	3 764	5 223	..
		314	Accumulators, primary cells & primary batteries	..	..	..	..	..	924	875	..
		315	Electric lamps and lighting equipment	..	..	..	..	..	1 340	1 361	..
		319	Other electrical equipment, nec	..	..	..	..	..	8 642	7 020	..
	32		Radio, TV, communication equipment	..	..	..	..	..	15 667	17 410	..
		321	Electronic valves, tubes; other electronic components	..	..	..	..	..	7 446	7 294	..
		322	TV/radio transmitters; line teleph./telegr. apparatus	..	..	..	..	..	1 708	2 164	..
		323	TV/radio receivers; sound and video app.; assoc. gds.	..	..	..	..	..	6 513	7 952	..
	33		Medical, precision & optical instrum.; watch & clocks	..	..	..	..	..	3 403	3 240	..
		331	Medical appl.; instruments & appl. for measuring, etc.	..	..	..	..	..	2 388	2 198	..
		332	Optical instruments and photographic equipment	..	..	..	..	..	894	822	..
		333	Watches and clocks	..	..	..	..	..	121	220	..
	34		Motor vehicles, trailers and semi-trailers	..	..	..	..	..	26 942	20 201	..
		341	Motor vehicles	..	..	..	..	..	19 029	12 115	..
		342	Bodies for motor vehicles, trailers & semi-trailers	..	..	..	..	..	1 350	1 419	..
		343	Parts & accessories for motor vehicles & their engines	..	..	..	..	..	6 563	6 667	..
	35		Other transport equipment	..	..	..	..	..	14 330	10 158	..
		351	Building and repairing of ships and boats	..	..	..	..	..	9 424	4 842	..
		352	Railway and tramway locomotives and rolling stock	..	..	..	..	..	2 900	3 265	..
		353	Aircraft and spacecraft	..	..	..	..	..	1 035	1 204	..
		359	Transport equipment, nec	..	..	..	..	..	971	847	..
	36		Furniture; manufacturing, nec	..	..	..	..	..	20 791	20 785	..
		361	Furniture	..	..	..	..	..	15 094	15 302	..
		369	Manufacturing, nec	..	..	..	..	..	5 697	5 483	..
	37		Recycling	..	..	..	..	..	348	468	..
E	40_41		Electricity, gas and water supply	..	..	..	..	..	28 627	34 708	..
	40		Electricity, gas, steam and hot water supply	..	..	..	..	..	27 086	33 158	..
F	45		Construction	..	..	..	..	..	122 852	129 226	..

Table PO.8 **ENTREPRISES - ENTERPRISES**

Unités

1991	1992	1993	1994	1995	1996	1997	1998			CITI révision 3
..	..	..	..	..	1 541	1 431	..	Activites extractives	C	10_14
..	..	..	..	..	..	..	..	Extraction de charbon, de lignite et de tourbe		10
..	..	..	..	..	..	..	..	Pétrole brut et gaz naturel ; activités annexes		11
..	..	..	..	..	..	..	..	Extraction de Minerais d'Uranium et de Thorium		12
..	..	..	..	..	9	..	..	Extraction de minerais métalliques		13
..	..	..	..	..	1 529	1 424	..	Autres activités extractives		14
..	..	..	..	..	84 551	82 995	..	Total activites manufacturieres	D	15_37
..	..	..	..	..	9 702	10 200	..	Produits alimentaires et boissons		15
..	..	..	..	..	1 484	1 441	1 272	Viande, poisson, fruits, légumes, huiles et graisses		151
..	..	..	..	..	318	310	304	Produits laitiers		152
..	..	..	..	..	729	760	..	Travail des grains; prod. amyl. ; alim. pour animaux		153
..	..	..	..	..	6 712	7 240	5 079	Autres produits alimentaires		154
..	..	..	..	..	459	449	..	Boissons		155
..	..	..	..	..	4	4	4	Produits à base de tabac		16
..	..	..	..	..	4 949	4 849	..	Textiles		17
..	..	..	..	..	945	863	..	Filature, tissage et achèvement des textiles		171
..	..	..	..	..	2 429	2 359	2 060	Autres articles textiles		172
..	..	..	..	..	1 575	1 627	..	Etoffes et articles de bonneterie		173
..	..	..	..	..	11 952	12 358	9 904	Articles d'habillement et fourrure		18
..	..	..	..	..	11 891	12 306	9 861	Articles d'habillement autres qu'en fourrure		181
..	..	..	..	..	61	52	43	Prép. et teinture des fourrures ; articles en fourrure		182
..	..	..	..	..	4 011	3 609	3 364	Cuir et articles en cuir ; chaussures		19
..	..	..	..	..	846	809	806	Cuir et articles en cuir sauf chaussures		191
..	..	..	..	..	3 165	2 800	2 558	Chaussures		192
..	..	..	..	..	9 640	9 535	8 277	Bois, articles en bois, sauf meubles		20
..	..	..	..	..	1 359	1 175	1 176	Sciage et rabotage du bois		201
..	..	..	..	..	8 281	8 360	7 101	Articles en bois, liège, vannerie, sparterie		202
..	..	..	..	..	486	476	..	Papier, articles en papier et carton		21
..	..	..	..	..	4 377	4 174	..	Edition, imprimerie et reproduction		22
..	..	..	..	..	954	800	812	Edition		221
..	..	..	..	..	3 396	3 358	2 865	Imprimerie et activités annexes		222
..	..	..	..	..	27	16	..	Reproduction de supports enregistrés		223
..	..	..	..	..	1	1	..	Coke. et prod. pétroliers ; comb. nucléaires		23
..	..	..	..	..	..	..	..	Cokéfaction		231
..	..	..	..	..	1	1	..	Produits pétroliers raffinés		232
..	..	..	..	..	..	..	..	Traitement de combustibles nucléaires		233
..	..	..	..	..	1 095	996	..	Produits chimiques		24
..	..	..	..	..	228	195	..	Produits chimiques de base		241
..	..	..	..	..	852	788	..	Autres produits chimiques		242
..	..	..	..	..	15	13	12	Fibres synthétiques ou artificielles		243
..	..	..	..	..	1 035	985	..	Art. en caoutchouc et matières plastiques		25
..	..	..	..	..	223	182	..	Articles en caoutchouc		251
..	..	..	..	..	812	803	833	Articles en matières plastiques		252
..	..	..	..	..	4 692	4 578	..	Autres produits minéraux non métalliques		26
..	..	..	..	..	447	424	463	Verre et articles en verre		261
..	..	..	..	..	4 245	4 154	..	Produits minéraux non métalliques, nca		269
..	..	..	..	..	747	621	..	Produits métallurgiques de base		27
..	..	..	..	..	247	177	..	Sidérurgie et première transformation de l'acier		271
..	..	..	..	..	149	115	..	Première transf. métaux précieux et non ferreux		272
..	..	..	..	..	351	329	279	Fonderie		273
..	..	..	..	..	14 731	14 064	13 045	Ouvrages en métaux, sauf machines		28
..	..	..	..	..	4 330	4 221	3 962	Constr. et menuis. métal.; réservoirs; générateurs		281
..	..	..	..	..	10 401	9 843	9 083	Autres ouvrages en métaux		289
..	..	..	..	..	3 909	3 597	3 351	Machines et matériels, nca		29
..	..	..	..	..	1 610	1 431	1 320	Machines d'usage général		291
..	..	..	..	..	2 164	2 068	1 926	Machines d'usage spécifique		292
..	..	..	..	..	135	98	105	Appareils domestiques, nca		293
..	..	..	..	..	54	55	..	Mach. de bureau, comptables, information		30
..	..	..	..	..	1 012	936	929	Machines et appareils électriques, nca		31
..	..	..	..	..	275	259	239	Moteurs, génératrices, transformateurs électriques		311
..	..	..	..	..	174	168	147	Matériel électrique de distribution et de commande		312
..	..	..	..	..	32	29	26	Fils et câbles électriques isolés		313
..	..	..	..	..	4	3	3	Accumulateurs et piles électriques		314
..	..	..	..	..	354	308	304	Lampes électriques et appareils d'éclairage		315
..	..	..	..	..	173	169	210	Autres matériels électriques, nca.		319
..	..	..	..	..	303	276	274	Equipements et appareils de radio, TV, etc.		32
..	..	..	..	..	164	132	109	Tubes et valves électroniques;autres composants		321
..	..	..	..	..	55	69	87	Emetteurs radio/TV; app. téléphonie et télégraphie		322
..	..	..	..	..	84	75	78	Récepteurs radio/TV, son et image; art. associés		323
..	..	..	..	..	712	669	666	Instr. médicaux, de précision ; optique ; horlog.		33
..	..	..	..	..	645	611	613	Instruments médicaux et de précision, etc.		331
..	..	..	..	..	38	34	31	Instruments d'optique ; matériel photographique		332
..	..	..	..	..	29	24	22	Horlogerie		333
..	..	..	..	..	470	442	392	Véhicules automobiles; remorques, etc.		34
..	..	..	..	..	55	44	29	Véhicules automobiles		341
..	..	..	..	..	177	167	161	Carross. automobiles, remorques, semi-remorques		342
..	..	..	..	..	238	231	202	Pièces et accessoires pour automobiles et moteurs		343
..	..	..	..	..	462	459	..	Autres matériels de transport		35
..	..	..	..	..	331	353	272	Construction et réparation de navires		351
..	..	..	..	..	5	5	4	Matériel ferroviaire roulant		352
..	..	..	..	..	56	43	27	Construction aéronautique et spatiale		353
..	..	..	..	..	70	58	..	Autres équipements de transport		359
..	..	..	..	..	10 119	10 038	9 709	Meubles ; activités de fabrication, nca		36
..	..	..	..	..	7 311	7 414	7 303	Meubles		361
..	..	..	..	..	2 808	2 624	2 406	Activités de fabrication, nca		369
..	..	..	..	..	88	73	69	Récupération		37
..	..	..	..	..	239	247	..	Electricité, gaz et eau	E	40_41
..	..	..	..	..	131	155	147	Prod. et dist. d'électricité gaz, vapeur d'eau chaude		40
..	..	..	..	..	68 718	64 308	68 522	Construction	F	45

PORTUGAL

Table PO.9 EXPORTS - EXPORTATIONS

Millions of Esc (Current Prices)

ISIC revision 3				1991	1992	1993	1994	1995	1996	1997	1998
C	10_14		Mining and quarrying	..	..	..	..	..	46 066	51 171	..
	10		Mining of coal and lignite; extraction of peat	..	..	..	..	..	..	..	..
	11		Crude petroleum & natural gas; service activities	..	..	..	..	..	..	..	..
	12		Mining of uranium and thorium ores	..	..	..	..	..	..	..	..
	13		Mining of metal ores	..	..	..	..	..	29 793	..	..
	14		Other mining and quarrying	..	..	..	..	..	15 863	20 931	..
D	15_37		Total manufacturing	..	..	..	..	..	3353 866	3759 798	..
	15		Food products and beverages	..	..	..	..	..	208 147	222 341	..
		151	Production of meat, fish, fruit, vegetables, oils & fats	..	..	..	..	..	76 326	82 048	..
		152	Dairy products	..	..	..	..	..	14 297	17 341	..
		153	Grain mill and starch products; animal feeds	..	..	..	..	..	6 241	6 372	..
		154	Other food products	..	..	..	..	..	19 850	22 946	..
		155	Beverages	..	..	..	..	..	91 433	93 634	..
	16		Tobacco products	..	..	..	..	..	2 064	2 304	..
	17		Textiles	..	..	..	..	..	369 407	377 528	..
		171	Spinning, weaving and finishing of textiles	..	..	..	..	..	163 105	142 480	..
		172	Other textiles	..	..	..	..	..	99 331	113 588	..
		173	Knitted and crocheted fabrics and articles	..	..	..	..	..	106 971	121 460	..
	18		Wearing apparel and fur	..	..	..	..	..	336 212	493 161	..
		181	Wearing apparel, except fur apparel	..	..	..	..	..	332 762	489 808	..
		182	Dressing and dyeing of fur; articles of fur	..	..	..	..	..	3 450	3 353	..
	19		Leather and articles of leather; footwear	..	..	..	..	..	282 421	295 201	..
		191	Leather and articles of leather	..	..	..	..	..	6 795	6 972	..
		192	Footwear	..	..	..	..	..	275 626	288 229	..
	20		Wood & products of wood & cork, exc. furniture	..	..	..	..	..	178 548	191 819	..
		201	Sawmilling and planing of wood	..	..	..	..	..	13 693	20 757	..
		202	Wood products; articles of cork, straw, etc.	..	..	..	..	..	164 855	171 062	..
	21		Paper and paper products	..	..	..	..	..	192 421	187 006	..
	22		Printing, publishing and recorded media	..	..	..	..	..	13 626	10 358	..
		221	Publishing	..	..	..	..	..	3 023	4 568	..
		222	Printing and related activities	..	..	..	..	..	9 313	5 734	..
		223	Reproduction of recorded media	..	..	..	..	..	1 290	56	..
	23		Coke and petroleum products; nuclear fuel	..	..	..	..	..	101 563	108 019	..
		231	Coke oven products	..	..	..	..	..	..	..	..
		232	Refined petroleum products	..	..	..	..	..	101 563	108 019	..
		233	Processing of nuclear fuel	..	..	..	..	..	..	..	..
	24		Chemicals and chemical products	..	..	..	..	..	151 297	183 008	..
		241	Basic chemicals	..	..	..	..	..	107 144	123 584	..
		242	Other chemical products	..	..	..	..	..	32 999	44 340	..
		243	Man-made fibres	..	..	..	..	..	11 154	15 084	..
	25		Rubber and plastic products	..	..	..	..	..	59 463	69 174	..
		251	Rubber products	..	..	..	..	..	24 122	30 938	..
		252	Plastic products	..	..	..	..	..	35 341	38 236	..
	26		Other non-metallic mineral products	..	..	..	..	..	121 509	148 664	..
		261	Glass and glass products	..	..	..	..	..	29 637	37 938	..
		269	Non-metallic mineral products, nec	..	..	..	..	..	91 872	110 726	..
	27		Basic metals	..	..	..	..	..	56 137	62 379	..
		271	Basic iron and steel	..	..	..	..	..	29 139	27 804	..
		272	Basic precious and non-ferrous metals	..	..	..	..	..	8 749	9 731	..
		273	Casting of metals	..	..	..	..	..	18 249	24 844	..
	28		Fabricated metal products, exc. machinery	..	..	..	..	..	82 299	96 795	..
		281	Structural metal prod., tanks, reservoirs, etc.	..	..	..	..	..	12 273	15 020	..
		289	Other fabricated metal products	..	..	..	..	..	70 026	81 775	..
	29		Machinery and equipment, nec	..	..	..	..	..	138 244	146 941	..
		291	General purpose machinery	..	..	..	..	..	34 514	36 335	..
		292	Special purpose machinery	..	..	..	..	..	65 300	65 012	..
		293	Domestic appliances, nec	..	..	..	..	..	38 430	45 594	..
	30		Office, accounting & computing machinery	..	..	..	..	..	28	209	..
	31		Electrical machinery & apparatus, nec	..	..	..	..	..	145 779	196 838	..
		311	Electric motors, generators and transformers	..	..	..	..	..	11 256	11 400	..
		312	Electricity distribution & control apparatus	..	..	..	..	..	5 319	6 826	..
		313	Insulated wire and cable	..	..	..	..	..	54 438	67 392	..
		314	Accumulators, primary cells & primary batteries	..	..	..	..	..	7 870	8 614	..
		315	Electric lamps and lighting equipment	..	..	..	..	..	2 616	2 886	..
		319	Other electrical equipment, nec	..	..	..	..	..	64 280	99 720	..
	32		Radio, TV, communication equipment	..	..	..	..	..	202 784	225 105	..
		321	Electronic valves, tubes; other electronic components	..	..	..	..	..	105 861	100 922	..
		322	TV/radio transmitters; line teleph./telegr. apparatus	..	..	..	..	..	6 977	12 196	..
		323	TV/radio receivers; sound and video app.; assoc. gds.	..	..	..	..	..	89 946	111 987	..
	33		Medical, precision & optical instrum.; watch & clocks	..	..	..	..	..	17 463	16 882	..
		331	Medical appl.; instruments & appl. for measuring, etc.	..	..	..	..	..	8 632	13 597	..
		332	Optical instruments and photographic equipment	..	..	..	..	..	7 118	2 611	..
		333	Watches and clocks	..	..	..	..	..	1 713	674	..
	34		Motor vehicles, trailers and semi-trailers	..	..	..	..	..	556 463	588 710	..
		341	Motor vehicles	..	..	..	..	..	466 731	503 842	..
		342	Bodies for motor vehicles, trailers & semi-trailers	..	..	..	..	..	1 634	1 810	..
		343	Parts & accessories for motor vehicles & their engines	..	..	..	..	..	88 098	83 058	..
	35		Other transport equipment	..	..	..	..	..	84 438	67 184	..
		351	Building and repairing of ships and boats	..	..	..	..	..	65 810	47 242	..
		352	Railway and tramway locomotives and rolling stock	..	..	..	..	..	1 911	1 640	..
		353	Aircraft and spacecraft	..	..	..	..	..	9 622	9 452	..
		359	Transport equipment, nec	..	..	..	..	..	7 095	8 850	..
	36		Furniture; manufacturing, nec	..	..	..	..	..	52 623	65 924	..
		361	Furniture	..	..	..	..	..	31 138	36 121	..
		369	Manufacturing, nec	..	..	..	..	..	21 485	29 803	..
	37		Recycling	..	..	..	..	..	930	4 248	..
E	40_41		Electricity, gas and water supply	..	..	..	..	..	261	79	..
	40		Electricity, gas, steam and hot water supply	..	..	..	..	..	159	5	..
F	45		Construction	..	..	..	..	..	62 554	82 268	..

Table PO.10 **IMPORTATIONS - IMPORTS**

Milions de Esc (Prix courants)

1991	1992	1993	1994	1995	1996	1997	1998		CITI révision 3		
..	..	..	..	..	6 675	16 525	..	Activites extractives	C	10_14	
..	..	..	..	..	..	..	..	Extraction de charbon, de lignite et de tourbe		10	
..	..	..	..	..	..	..	..	Pétrole brut et gaz naturel ; activités annexes		11	
..	..	..	..	..	..	..	..	Extraction de Minerais d'Uranium et de Thorium		12	
..	..	..	..	..	2 490	..	..	Extraction de minerais métalliques		13	
..	..	..	..	..	4 168	12 473	..	Autres activités extractives		14	
..	..	..	..	..	2 669 299	2 401 385	..	Total activites manufacturieres	D	15_37	
..	..	..	..	..	278 297	327 095	..	Produits alimentaires et boissons		15	
..	..	..	..	..	102 100	154 359	..	Viande, poisson, fruits, légumes, huiles et graisses			151
..	..	..	..	..	19 723	19 344	..	Produits laitiers			152
..	..	..	..	..	36 850	39 512	..	Travail des grains; prod. amyl. ; alim. pour animaux			153
..	..	..	..	..	79 230	77 230	..	Autres produits alimentaires			154
..	..	..	..	..	40 394	36 650	..	Boissons			155
..	..	..	..	..	7 595	18 598	..	Produits à base de tabac		16	
..	..	..	..	..	197 521	195 696	..	Textiles		17	
..	..	..	..	..	101 524	96 580	..	Filature, tissage et achèvement des textiles			171
..	..	..	..	..	49 772	51 000	..	Autres articles textiles			172
..	..	..	..	..	46 225	48 116	..	Etoffes et articles de bonneterie			173
..	..	..	..	..	102 865	107 500	..	Articles d'habillement et fourrure		18	
..	..	..	..	..	101 207	105 563	..	Articles d'habillement autres qu'en fourrure			181
..	..	..	..	..	1 658	1 937	..	Prép. et teinture des fourrures ; articles en fourrure			182
..	..	..	..	..	95 975	110 459	..	Cuir et articles en cuir ; chaussures		19	
..	..	..	..	..	20 056	23 918	..	Cuir et articles en cuir sauf chaussures			191
..	..	..	..	..	75 919	86 541	..	Chaussures			192
..	..	..	..	..	39 220	51 008	..	Bois, articles en bois, sauf meubles		20	
..	..	..	..	..	8 208	17 797	..	Sciage et rabotage du bois			201
..	..	..	..	..	31 012	33 211	..	Articles en bois, liège, vannerie, sparterie			202
..	..	..	..	..	81 612	72 300	..	Papier, articles en papier et carton		21	
..	..	..	..	..	41 654	43 066	..	Edition, imprimerie et reproduction		22	
..	..	..	..	..	18 862	20 943	..	Edition			221
..	..	..	..	..	21 626	21 377	..	Imprimerie et activités annexes			222
..	..	..	..	..	1 166	746	..	Reproduction de supports enregistrés			223
..	..	..	..	..	321 634	..	..	Coke. et prod. pétroliers ; comb. nucléaires		23	
..	..	..	..	..	..	..	..	Cokéfaction			231
..	..	..	..	..	321 634	..	..	Produits pétroliers raffinés			232
..	..	..	..	..	..	..	..	Traitement de combustibles nucléaires			233
..	..	..	..	..	221 118	227 446	..	Produits chimiques		24	
..	..	..	..	..	94 554	90 556	..	Produits chimiques de base			241
..	..	..	..	..	118 535	123 222	..	Autres produits chimiques			242
..	..	..	..	..	8 029	13 668	..	Fibres synthétiques ou artificielles			243
..	..	..	..	..	61 926	70 792	..	Art. en caoutchouc et matières plastiques		25	
..	..	..	..	..	14 095	16 248	..	Articles en caoutchouc			251
..	..	..	..	..	47 831	54 544	..	Articles en matières plastiques			252
..	..	..	..	..	56 176	77 851	..	Autres produits minéraux non métalliques		26	
..	..	..	..	..	16 211	18 532	..	Verre et articles en verre			261
..	..	..	..	..	39 965	59 319	..	Produits minéraux non métalliques, nca			269
..	..	..	..	..	68 512	86 885	..	Produits métallurgiques de base		27	
..	..	..	..	..	47 028	59 347	..	Sidérurgie et première transformation de l'acier			271
..	..	..	..	..	17 969	21 222	..	Première transf. métaux précieux et non ferreux			272
..	..	..	..	..	3 515	6 316	..	Fonderie			273
..	..	..	..	..	64 400	82 923	..	Ouvrages en métaux, sauf machines		28	
..	..	..	..	..	12 524	17 959	..	Constr. et menuis. métal.; réservoirs; générateurs			281
..	..	..	..	..	51 876	64 964	..	Autres ouvrages en métaux			289
..	..	..	..	..	89 974	94 583	..	Machines et matériels, nca		29	
..	..	..	..	..	33 147	29 542	..	Machines d'usage général			291
..	..	..	..	..	34 201	32 217	..	Machines d'usage spécifique			292
..	..	..	..	..	22 626	32 824	..	Appareils domestiques, nca			293
..	..	..	..	..	3 869	1 769	..	Mach. de bureau, comptables, information		30	
..	..	..	..	..	89 459	117 572	..	Machines et appareils électriques, nca		31	
..	..	..	..	..	4 839	4 418	..	Moteurs, génératrices, transformateurs électriques			311
..	..	..	..	..	5 735	9 007	..	Matériel électrique de distribution et de commande			312
..	..	..	..	..	36 899	45 793	..	Fils et câbles électriques isolés			313
..	..	..	..	..	6 353	5 552	..	Accumulateurs et piles électriques			314
..	..	..	..	..	3 254	3 788	..	Lampes électriques et appareils d'éclairage			315
..	..	..	..	..	32 379	49 014	..	Autres matériels électriques, nca.			319
..	..	..	..	..	192 709	230 906	..	Equipements et appareils de radio, TV, etc.		32	
..	..	..	..	..	87 411	94 578	..	Tubes et valves électroniques;autres composants			321
..	..	..	..	..	11 095	23 991	..	Emetteurs radio/TV; app. téléphonie et télégraphie			322
..	..	..	..	..	94 203	112 337	..	Récepteurs radio/TV, son et image; art. associés			323
..	..	..	..	..	17 069	15 525	..	Instr. médicaux, de précision ; optique ; horlog.		33	
..	..	..	..	..	9 383	10 528	..	Instruments médicaux et de précision, etc.			331
..	..	..	..	..	7 330	4 685	..	Instruments d'optique ; matériel photographique			332
..	..	..	..	..	356	312	..	Horlogerie			333
..	..	..	..	..	550 932	372 134	..	Véhicules automobiles; remorques, etc.		34	
..	..	..	..	..	415 579	275 844	..	Véhicules automobiles			341
..	..	..	..	..	2 913	4 791	..	Carross. automobiles, remorques, semi-remorques			342
..	..	..	..	..	132 440	91 499	..	Pièces et accessoires pour automobiles et moteurs			343
..	..	..	..	..	23 497	30 980	..	Autres matériels de transport		35	
..	..	..	..	..	12 535	12 227	..	Construction et réparation de navires			351
..	..	..	..	..	2 503	9 136	..	Matériel ferroviaire roulant			352
..	..	..	..	..	4 802	5 718	..	Construction aéronautique et spatiale			353
..	..	..	..	..	3 657	3 899	..	Autres équipements de transport			359
..	..	..	..	..	61 361	64 191	..	Meubles ; activités de fabrication, nca		36	
..	..	..	..	..	27 942	29 534	..	Meubles			361
..	..	..	..	..	33 419	34 657	..	Activités de fabrication, nca			369
..	..	..	..	..	1 924	2 106	..	Récupération		37	
..	..	..	..	..	48 586	62 694	..	Electricité, gaz et eau	E	40_41	
..	..	..	..	..	47 875	61 906	..	Prod. et dist. d'électricité gaz, vapeur d'eau chaude		40	
..	..	..	..	..	45 186	75 843	..	Construction	F	45	

Table SP.1 **PRODUCTION AT FACTOR COSTS - PRODUCTION AU COUT DES FACTEURS**

Billions of Ptas (Current Prices)

ISIC revision 3				1991	1992	1993	1994	1995	1996	1997	1998
C	10_14		Mining and quarrying	..	..	574	576	652	632	621	600
	10		Mining of coal and lignite; extraction of peat	..	..	288	285	290	294	246	225
	11		Crude petroleum & natural gas; service activities	..	..	..	..	..	..	..	..
	12		Mining of uranium and thorium ores	..	..	..	..	..	..	..	..
	13		Mining of metal ores	..	..	24	22	37	36	38	30
	14		Other mining and quarrying	..	..	208	220	279	279	317	321
D	15_37		Total manufacturing	..	..	30 568	34 872	39 708	41 835	46 004	52 934
	15		Food products and beverages	..	..	7 010	7 685	8 294	8 640	9 304	9 499
		151	Production of meat, fish, fruit, vegetables, oils & fats	..	..	2 671	2 971	3 306	3 433	3 826	3 805
		152	Dairy products	..	..	754	855	915	921	953	1 025
		153	Grain mill and starch products; animal feeds	..	..	975	1 018	1 087	1 098	1 252	1 166
		154	Other food products	..	..	1 322	1 461	1 518	1 663	1 678	1 737
		155	Beverages	..	..	1 287	1 380	1 468	1 524	1 594	1 766
	16		Tobacco products	..	..	261	292	283	272	324	380
	17		Textiles	..	..	910	1 044	1 161	1 214	1 383	1 369
		171	Spinning, weaving and finishing of textiles	..	..	514	589	650	678	780	716
		172	Other textiles	..	..	222	288	329	334	384	417
		173	Knitted and crocheted fabrics and articles	..	..	174	167	182	203	219	237
	18		Wearing apparel and fur	..	..	812	873	863	895	1 050	1 087
		181	Wearing apparel, except fur apparel	..	..	774	824	824	853	997	1 041
		182	Dressing and dyeing of fur; articles of fur	..	..	38	49	40	42	53	46
	19		Leather and articles of leather; footwear	..	..	604	728	816	874	1 002	1 019
		191	Leather and articles of leather	..	..	208	254	226	257	256	243
		192	Footwear	..	..	396	475	589	617	746	777
	20		Wood & products of wood & cork, exc. furniture	..	..	672	740	889	920	1 018	1 151
		201	Sawmilling and planing of wood	..	..	91	97	121	121	148	155
		202	Wood products; articles of cork, straw, etc.	..	..	581	643	768	799	870	995
	21		Paper and paper products	..	..	808	965	1 266	1 187	1 229	1 297
	22		Printing, publishing and recorded media	..	..	1 353	1 498	1 660	1 752	1 922	2 098
		221	Publishing	..	..	698	805	843	870	985	1 035
		222	Printing and related activities	..	..	640	683	804	867	921	1 043
		223	Reproduction of recorded media	..	..	16	10	13	16	16	20
	23		Coke and petroleum products; nuclear fuel	..	..	1 125	1 157	1 255	1 605	1 793	1 516
		231	Coke oven products	..	..	..	..	..	..	..	..
		232	Refined petroleum products	..	..	1 098	1 130	1 227	1 561	1 754	1 473
		233	Processing of nuclear fuel	..	..	..	..	..	..	..	35
	24		Chemicals and chemical products	..	..	2 959	3 470	3 883	3 948	4 231	4 457
		241	Basic chemicals	..	..	971	1 323	1 496	1 504	1 585	1 590
		242	Other chemical products	..	..	1 903	2 031	2 249	2 318	2 499	2 708
		243	Man-made fibres	..	..	85	116	138	126	147	159
	25		Rubber and plastic products	..	..	1 038	1 262	1 534	1 610	1 732	1 958
		251	Rubber products	..	..	302	359	433	470	493	586
		252	Plastic products	..	..	736	904	1 101	1 140	1 239	1 371
	26		Other non-metallic mineral products	..	..	1 828	2 004	2 221	2 232	2 502	2 879
		261	Glass and glass products	..	..	244	293	342	339	374	394
		269	Non-metallic mineral products, nec	..	..	1 583	1 711	1 879	1 892	2 128	2 485
	27		Basic metals	..	..	1 485	1 764	2 312	2 093	2 420	2 534
		271	Basic iron and steel	..	..	912	1 085	1 481	1 255	1 433	1 509
		272	Basic precious and non-ferrous metals	..	..	362	445	548	563	659	678
		273	Casting of metals	..	..	212	234	283	276	328	347
	28		Fabricated metal products, exc. machinery	..	..	1 849	2 099	2 583	2 809	3 096	3 426
		281	Structural metal prod., tanks, reservoirs, etc.	..	..	655	770	947	1 016	1 072	1 234
		289	Other fabricated metal products	..	..	1 194	1 329	1 636	1 793	2 024	2 191
	29		Machinery and equipment, nec	..	..	1 410	1 717	1 963	2 221	2 363	2 747
		291	General purpose machinery	..	..	605	705	858	986	1 069	1 239
		292	Special purpose machinery	..	..	539	687	765	876	928	1 089
		293	Domestic appliances, nec	..	..	265	325	340	359	366	419
	30		Office, accounting & computing machinery	..	..	251	306	358	391	367	391
	31		Electrical machinery & apparatus, nec	..	..	891	1 014	1 195	1 338	1 440	1 674
		311	Electric motors, generators and transformers	..	..	139	154	182	241	258	302
		312	Electricity distribution & control apparatus	..	..	145	160	203	205	224	243
		313	Insulated wire and cable	..	..	113	149	179	158	175	187
		314	Accumulators, primary cells & primary batteries	..	..	54	57	56	60	64	71
		315	Electric lamps and lighting equipment	..	..	109	113	126	143	153	176
		319	Other electrical equipment, nec	..	..	332	382	449	530	567	695
	32		Radio, TV, communication equipment	..	..	499	552	630	717	752	807
		321	Electronic valves, tubes; other electronic components	..	..	111	150	170	169	187	194
		322	TV/radio transmitters; line teleph./telegr. apparatus	..	..	256	250	278	340	315	342
		323	TV/radio receivers; sound and video app.; assoc. gds.	..	..	132	151	181	208	251	271
	33		Medical, precision & optical instrum.; watch & clocks	..	..	245	257	283	315	344	399
		331	Medical appl.; instruments & appl. for measuring, etc.	..	..	206	215	243	274	301	338
		332	Optical instruments and photographic equipment	..	..	33	39	35	38	41	43
		333	Watches and clocks	..	..	6	3	6	3	2	18
	34		Motor vehicles, trailers and semi-trailers	..	..	2 914	3 750	4 397	4 763	5 456	6 089
		341	Motor vehicles	..	..	2 155	2 790	3 221	3 485	3 896	4 429
		342	Bodies for motor vehicles, trailers & semi-trailers	..	..	92	127	174	155	198	239
		343	Parts & accessories for motor vehicles & their engines	..	..	668	834	1 001	1 122	1 362	1 421
	35		Other transport equipment	..	..	592	572	640	709	817	940
		351	Building and repairing of ships and boats	..	..	267	251	289	351	403	446
		352	Railway and tramway locomotives and rolling stock	..	..	126	108	133	121	139	152
		353	Aircraft and spacecraft	..	..	143	146	139	151	182	208
		359	Transport equipment, nec	..	..	56	66	79	86	94	135
	36		Furniture; manufacturing, nec	..	..	1 024	1 077	1 174	1 296	1 422	1 678
		361	Furniture	..	..	739	777	850	926	1 027	1 254
		369	Manufacturing, nec	..	..	285	301	325	370	395	424
	37		Recycling	..	..	30	46	47	32	36	27
E	40_41		Electricity, gas and water supply	..	..	2 730	2 774	3 010	3 119	3 079	2 912
	40		Electricity, gas, steam and hot water supply	..	..	2 521	2 536	2 747	2 830	2 750	2 561
F	45		Construction	..	..	..	..	..	..	..	..

Table SP.2 **VALEUR AJOUTÉE AU COUT DES FACTEURS - VALUE ADDED AT FACTOR COSTS**

Milliards de Ptas (Prix courants)

1991	1992	1993	1994	1995	1996	1997	1998			CITI révision 3	
..	..	309	310	332	320	275	268	Activites extractives	C	10_14	
..	..	197	192	191	192	140	130	Extraction de charbon, de lignite et de tourbe		10	
..	..							Pétrole brut et gaz naturel ; activités annexes		11	
..	..							Extraction de Minerais d'Uranium et de Thorium		12	
..	..	8	8	12	10	13	8	Extraction de minerais métalliques		13	
..	..	86	92	114	111	115	123	Autres activités extractives		14	
..	..	9 777	10 731	11 760	12 278	13 394	16 446	Total activites manufacturieres	D	15_37	
..	..	1 751	1 834	1 859	1 944	2 031	2 136	Produits alimentaires et boissons		15	
..	..	487	520	522	555	573	649	Viande, poisson, fruits, légumes, huiles et graisses			151
..	..	171	198	189	195	205	202	Produits laitiers			152
..	..	123	134	147	151	164	140	Travail des grains; prod. amyl. ; alim. pour animaux			153
..	..	514	539	535	583	598	595	Autres produits alimentaires			154
..	..	455	443	466	461	492	550	Boissons			155
..	..	76	99	87	73	89	102	Produits à base de tabac		16	
..	..	313	350	367	379	417	434	Textiles		17	
..	..	178	191	204	213	231	236	Filature, tissage et achèvement des textiles			171
..	..	77	98	103	102	116	120	Autres articles textiles			172
..	..	58	60	60	64	69	78	Etoffes et articles de bonneterie			173
..	..	294	285	289	302	326	374	Articles d'habillement et fourrure		18	
..	..	282	272	279	292	316	363	Articles d'habillement autres qu'en fourrure			181
..	..	12	13	10	10	10	10	Prép. et teinture des fourrures ; articles en fourrure			182
..	..	158	173	179	204	210	225	Cuir et articles en cuir ; chaussures		19	
..	..	59	60	58	66	62	58	Cuir et articles en cuir sauf chaussures			191
..	..	99	113	121	138	147	167	Chaussures			192
..	..	229	239	275	290	311	342	Bois, articles en bois, sauf meubles		20	
..	..	28	28	35	35	39	39	Sciage et rabotage du bois			201
..	..	201	211	241	255	271	303	Articles en bois, liège, vannerie, sparterie			202
..	..	246	293	375	361	388	408	Papier, articles en papier et carton		21	
..	..	570	604	629	690	765	815	Edition, imprimerie et reproduction		22	
..	..	263	304	292	319	382	386	Edition			221
..	..	301	297	333	365	378	422	Imprimerie et activités annexes			222
..	..	5	4	4	5	6	7	Reproduction de supports enregistrés			223
..	..	187	150	183	191	264	260	Coke. et prod. pétroliers ; comb. nucléaires		23	
..	..							Cokéfaction			231
..	..	177	144	171	178	250	247	Produits pétroliers raffinés			232
..	..						11	Traitement de combustibles nucléaires			233
..	..	957	1 097	1 210	1 199	1 322	1 378	Produits chimiques		24	
..	..	251	356	426	401	448	458	Produits chimiques de base			241
..	..	677	695	739	761	828	868	Autres produits chimiques			242
..	..	29	46	44	37	46	52	Fibres synthétiques ou artificielles			243
..	..	413	480	532	588	614	688	Art. en caoutchouc et matières plastiques		25	
..	..	139	162	182	203	213	238	Articles en caoutchouc			251
..	..	274	317	350	384	401	450	Articles en matières plastiques			252
..	..	708	818	902	877	983	1 113	Autres produits minéraux non métalliques		26	
..	..	114	135	151	145	156	164	Verre et articles en verre			261
..	..	594	683	751	732	827	950	Produits minéraux non métalliques, nca			269
..	..	383	460	657	535	624	642	Produits métallurgiques de base		27	
..	..	224	280	421	320	360	372	Sidérurgie et première transformation de l'acier			271
..	..	72	92	131	110	146	143	Première transf. métaux précieux et non ferreux			272
..	..	88	88	105	105	118	127	Fonderie			273
..	..	741	822	954	1 052	1 152	1 286	Ouvrages en métaux, sauf machines		28	
..	..	253	286	309	352	367	426	Constr. et menuis. métal.; réservoirs; générateurs			281
..	..	488	536	645	699	785	860	Autres ouvrages en métaux			289
..	..	595	674	725	809	856	964	Machines et matériels, nca		29	
..	..	269	300	341	382	400	448	Machines d'usage général			291
..	..	229	264	287	322	343	388	Machines d'usage spécifique			292
..	..	97	109	97	105	113	128	Appareils domestiques, nca			293
..	..	104	115	114	135	108	122	Mach. de bureau, comptables, information		30	
..	..	350	385	426	478	505	557	Machines et appareils électriques, nca		31	
..	..	51	57	60	78	88	85	Moteurs, génératrices, transformateurs électriques			311
..	..	60	69	78	87	91	98	Matériel électrique de distribution et de commande			312
..	..	35	44	53	45	45	47	Fils et câbles électriques isolés			313
..	..	22	20	20	21	21	25	Accumulateurs et piles électriques			314
..	..	39	38	43	47	48	58	Lampes électriques et appareils d'éclairage			315
..	..	142	157	173	200	211	244	Autres matériels électriques, nca.			319
..	..	200	219	202	225	223	219	Equipements et appareils de radio, TV, etc.		32	
..	..	52	68	65	68	77	77	Tubes et valves électroniques;autres composants			321
..	..	118	114	102	116	102	94	Emetteurs radio/TV; app. téléphonie et télégraphie			322
..	..	30	36	34	41	44	47	Récepteurs radio/TV, son et image; art. associés			323
..	..	110	112	118	128	143	147	Instr. médicaux, de précision ; optique ; horlog.		33	
..	..	92	92	101	110	124	124	Instruments médicaux et de précision, etc.			331
..	..	16	19	15	16	18	18	Instruments d'optique ; matériel photographique			332
..	..	2	1	2	1	1	5	Horlogerie			333
..	..	744	887	1 005	1 101	1 259	1 321	Véhicules automobiles; remorques, etc.		34	
..	..	470	558	628	691	770	819	Véhicules automobiles			341
..	..	30	34	41	43	51	60	Carross. automobiles, remorques, semi-remorques			342
..	..	244	294	336	367	437	442	Pièces et accessoires pour automobiles et moteurs			343
..	..	244	234	245	252	291	299	Autres matériels de transport		35	
..	..	101	103	107	106	114	111	Construction et réparation de navires			351
..	..	46	35	45	48	60	54	Matériel ferroviaire roulant			352
..	..	80	78	73	79	95	103	Construction aéronautique et spatiale			353
..	..	17	18	20	20	23	31	Autres équipements de transport			359
..	..	398	390	418	459	505	565	Meubles ; activités de fabrication, nca		36	
..	..	288	284	304	335	366	423	Meubles			361
..	..	110	106	114	124	139	143	Activités de fabrication, nca			369
..	..	9	10	10	7	9	6	Récupération		37	
..	..	1 671	1 659	1 725	1 858	1 802	1 774	Electricité, gaz et eau	E	40_41	
..	..	1 563	1 536	1 598	1 715	1 642	1 599	Prod. et dist. d'électricité gaz, vapeur d'eau chaude		40	
..	..	..	..	..	..	..	..	Construction	F	45	

SPAIN

Table SP.3 **INVESTMENT - INVESTISSEMENT**

Billions of Ptas (Current Prices)

ISIC revision 3				1991	1992	1993	1994	1995	1996	1997	1998
C	10_14		Mining and quarrying	..	..	57	62	72	73	72	58
	10		Mining of coal and lignite; extraction of peat	..	..	36	38	33	38	33	22
	11		Crude petroleum & natural gas; service activities	..	..	..	..	..	..	..	..
	12		Mining of uranium and thorium ores	..	..	..	..	..	..	..	..
	13		Mining of metal ores	..	..	2	0	2	3	0	2
	14		Other mining and quarrying	..	..	17	19	35	31	38	34
D	15_37		Total manufacturing	..	..	1 322	1 250	1 502	1 787	1 855	2 740
	15		Food products and beverages	..	..	249	212	255	310	302	381
		151	Production of meat, fish, fruit, vegetables, oils & fats	..	..	81	83	93	106	108	134
		152	Dairy products	..	..	29	14	25	35	38	36
		153	Grain mill and starch products; animal feeds	..	..	26	18	22	24	28	36
		154	Other food products	..	..	64	58	62	82	74	92
		155	Beverages	..	..	50	38	54	63	55	84
	16		Tobacco products	..	..	10	7	7	6	7	2
	17		Textiles	..	..	33	55	49	53	71	75
		171	Spinning, weaving and finishing of textiles	..	..	17	28	27	31	41	43
		172	Other textiles	..	..	10	20	15	14	17	19
		173	Knitted and crocheted fabrics and articles	..	..	6	7	7	8	13	13
	18		Wearing apparel and fur	..	..	11	20	16	7	23	30
		181	Wearing apparel, except fur apparel	..	..	10	20	16	7	21	28
		182	Dressing and dyeing of fur; articles of fur	..	..	1	1	0	0	2	2
	19		Leather and articles of leather; footwear	..	..	9	11	12	15	19	24
		191	Leather and articles of leather	..	..	2	4	4	5	6	5
		192	Footwear	..	..	7	7	8	11	13	19
	20		Wood & products of wood & cork, exc. furniture	..	..	20	24	33	38	56	60
		201	Sawmilling and planing of wood	..	..	2	2	4	5	5	8
		202	Wood products; articles of cork, straw, etc.	..	..	18	21	29	33	51	53
	21		Paper and paper products	..	..	47	50	62	80	74	103
	22		Printing, publishing and recorded media	..	..	57	44	74	87	88	114
		221	Publishing	..	..	19	19	22	22	21	22
		222	Printing and related activities	..	..	38	24	50	64	67	91
		223	Reproduction of recorded media	..	..	0	1	1	0	0	1
	23		Coke and petroleum products; nuclear fuel	..	..	42	34	32	53	37	45
		231	Coke oven products	..	..	..	..	..	..	..	..
		232	Refined petroleum products	..	..	41	33	31	52	36	44
		233	Processing of nuclear fuel	..	..	..	..	..	..	..	1
	24		Chemicals and chemical products	..	..	131	131	163	206	244	234
		241	Basic chemicals	..	..	51	51	79	125	135	134
		242	Other chemical products	..	..	78	78	78	76	86	87
		243	Man-made fibres	..	..	2	2	6	5	23	13
	25		Rubber and plastic products	..	..	57	65	78	84	100	122
		251	Rubber products	..	..	16	15	21	20	27	23
		252	Plastic products	..	..	41	51	57	64	72	98
	26		Other non-metallic mineral products	..	..	94	86	128	137	135	175
		261	Glass and glass products	..	..	14	10	26	23	26	18
		269	Non-metallic mineral products, nec	..	..	79	76	102	114	109	157
	27		Basic metals	..	..	62	52	83	128	95	102
		271	Basic iron and steel	..	..	41	25	43	63	66	69
		272	Basic precious and non-ferrous metals	..	..	13	15	10	46	12	10
		273	Casting of metals	..	..	9	12	30	18	17	22
	28		Fabricated metal products, exc. machinery	..	..	92	84	96	128	139	180
		281	Structural metal prod., tanks, reservoirs, etc.	..	..	24	18	18	24	34	40
		289	Other fabricated metal products	..	..	68	65	78	104	105	140
	29		Machinery and equipment, nec	..	..	44	56	57	71	83	116
		291	General purpose machinery	..	..	21	19	28	32	36	48
		292	Special purpose machinery	..	..	13	24	20	28	30	45
		293	Domestic appliances, nec	..	..	10	13	9	11	16	23
	30		Office, accounting & computing machinery	..	..	31	14	11	15	15	15
	31		Electrical machinery & apparatus, nec	..	..	30	30	43	45	53	59
		311	Electric motors, generators and transformers	..	..	2	2	4	5	5	8
		312	Electricity distribution & control apparatus	..	..	7	4	8	7	8	6
		313	Insulated wire and cable	..	..	3	4	5	6	7	8
		314	Accumulators, primary cells & primary batteries	..	..	2	3	2	2	3	4
		315	Electric lamps and lighting equipment	..	..	4	5	6	5	7	7
		319	Other electrical equipment, nec	..	..	10	12	18	19	23	26
	32		Radio, TV, communication equipment	..	..	18	17	21	33	38	29
		321	Electronic valves, tubes; other electronic components	..	..	7	7	10	21	23	19
		322	TV/radio transmitters; line teleph./telegr. apparatus	..	..	5	5	7	6	8	7
		323	TV/radio receivers; sound and video app.; assoc. gds.	..	..	7	5	4	6	7	3
	33		Medical, precision & optical instrum.; watch & clocks	..	..	4	4	8	9	10	14
		331	Medical appl.; instruments & appl. for measuring, etc.	..	..	6	3	7	8	9	13
		332	Optical instruments and photographic equipment	..	..		0	1	1	1	1
		333	Watches and clocks	..	..	0	0	0	0	0	0
	34		Motor vehicles, trailers and semi-trailers	..	..	226	204	211	211	185	233
		341	Motor vehicles	..	..	194	173	160	152	124	149
		342	Bodies for motor vehicles, trailers & semi-trailers	..	..	3	2	3	5	2	8
		343	Parts & accessories for motor vehicles & their engines	..	..	29	29	49	54	58	76
	35		Other transport equipment	..	..	18	17	17	21	29	40
		351	Building and repairing of ships and boats	..	..	7	7	7	8	10	16
		352	Railway and tramway locomotives and rolling stock	..	..	6	3	..	3	4	5
		353	Aircraft and spacecraft	..	..	4	5	6	6	12	16
		359	Transport equipment, nec	..	..	2	2	5	3	4	3
	36		Furniture; manufacturing, nec	..	..	34	33	43	49	50	65
		361	Furniture	..	..	24	24	35	32	32	48
		369	Manufacturing, nec	..	..	10	8	8	17	18	17
	37		Recycling	..	..	1	1	2	1	1	2
E	40_41		Electricity, gas and water supply	..	..	395	456	447	432	468	462
	40		Electricity, gas, steam and hot water supply	..	..	362	425	407	400	412	407
F	45		Construction	..	..	..	..	..	..	..	..

Table SP.4 **EMPLOI - EMPLOYMENT**

Unités

1991	1992	1993	1994	1995	1996	1997	1998		CITI révision 3		
..	..	55 724	50 054	49 528	48 079	46 257	43 130	Activites extractives	C	10_14	
..	..	32 727	30 265	27 879	26 822	24 892	20 667	Extraction de charbon, de lignite et de tourbe		10	
..	..	..	..	..	..	..	..	Pétrole brut et gaz naturel ; activités annexes		11	
..	..	..	..	..	..	..	..	Extraction de Minerais d'Uranium et de Thorium		12	
..	..	3 216	2 120	2 296	2 077	1 867	1 701	Extraction de minerais métalliques		13	
..	..	18 830	16 825	18 534	18 692	18 984	20 243	Autres activités extractives		14	
..	..	2 222 134	2 197 577	2 205 099	2 273 344	2 316 532	2 516 327	Total activites manufacturieres	D	15_37	
..	..	369 467	368 759	363 618	360 903	361 162	364 059	Produits alimentaires et boissons		15	
..	..	124 195	120 738	119 964	118 879	123 314	129 681	Viande, poisson, fruits, légumes, huiles et graisses			151
..	..	27 880	29 377	28 503	26 303	26 207	26 402	Produits laitiers			152
..	..	20 728	20 442	19 765	19 869	20 858	20 172	Travail des grains; prod. amyl. ; alim. pour animaux			153
..	..	142 552	146 330	147 357	150 060	145 391	142 829	Autres produits alimentaires			154
..	..	54 113	51 872	48 030	45 791	45 392	44 975	Boissons			155
..	..	10 558	9 711	9 306	8 791	9 285	9 353	Produits à base de tabac		16	
..	..	96 566	100 447	96 261	101 665	100 455	102 372	Textiles		17	
..	..	51 096	51 830	50 021	52 650	49 985	51 294	Filature, tissage et achèvement des textiles			171
..	..	26 151	28 623	28 555	29 928	31 171	30 444	Autres articles textiles			172
..	..	19 320	19 994	17 685	19 087	19 299	20 633	Etoffes et articles de bonneterie			173
..	..	132 166	123 455	119 238	118 392	129 278	138 286	Articles d'habillement et fourrure		18	
..	..	128 074	119 487	115 845	115 025	125 937	135 289	Articles d'habillement autres qu'en fourrure			181
..	..	4 092	3 968	3 393	3 367	3 342	2 997	Prép. et teinture des fourrures ; articles en fourrure			182
..	..	57 920	61 172	61 515	70 667	69 497	74 662	Cuir et articles en cuir ; chaussures		19	
..	..	17 864	17 669	17 324	17 850	16 907	16 289	Cuir et articles en cuir sauf chaussures			191
..	..	40 055	43 503	44 191	52 817	52 590	58 373	Chaussures			192
..	..	88 339	86 145	89 911	94 019	92 111	97 286	Bois, articles en bois, sauf meubles		20	
..	..	10 818	10 887	12 101	12 057	12 440	12 336	Sciage et rabotage du bois			201
..	..	77 521	75 258	77 810	81 962	79 671	84 950	Articles en bois, liège, vannerie, sparterie			202
..	..	48 058	47 562	46 488	50 482	51 376	50 619	Papier, articles en papier et carton		21	
..	..	117 012	114 251	115 859	123 841	126 274	131 053	Edition, imprimerie et reproduction		22	
..	..	38 043	38 035	37 987	40 690	43 260	43 758	Edition			221
..	..	78 097	75 524	77 106	82 324	82 224	86 377	Imprimerie et activités annexes			222
..	..	872	693	766	827	790	918	Reproduction de supports enregistrés			223
..	..	8 306	7 741	7 464	8 486	9 231	8 791	Coke. et prod. pétroliers ; comb. nucléaires		23	
..	..	..	..	..	..	..	..	Cokéfaction			231
..	..	7 950	7 397	7 110	7 767	8 516	8 072	Produits pétroliers raffinés			232
..	..	..	..	..	..	..	594	Traitement de combustibles nucléaires			233
..	..	138 575	134 967	130 027	128 688	129 889	130 231	Produits chimiques		24	
..	..	35 939	35 281	32 347	33 338	33 336	32 992	Produits chimiques de base			241
..	..	96 819	93 733	91 727	89 557	90 613	91 222	Autres produits chimiques			242
..	..	5 817	5 954	5 953	5 793	5 940	6 016	Fibres synthétiques ou artificielles			243
..	..	88 288	93 922	95 614	102 053	100 807	105 911	Art. en caoutchouc et matières plastiques		25	
..	..	26 561	28 693	29 014	30 214	29 701	31 095	Articles en caoutchouc			251
..	..	61 727	65 229	66 600	71 839	71 106	74 816	Articles en matières plastiques			252
..	..	155 849	153 533	156 436	156 444	161 144	165 020	Autres produits minéraux non métalliques		26	
..	..	23 977	23 126	23 374	22 485	23 785	23 587	Verre et articles en verre			261
..	..	131 872	130 407	133 062	133 958	137 359	141 433	Produits minéraux non métalliques, nca			269
..	..	80 186	72 521	72 566	67 123	67 956	69 917	Produits métallurgiques de base		27	
..	..	45 332	38 574	38 026	33 978	34 232	35 469	Sidérurgie et première transformation de l'acier			271
..	..	14 449	14 463	13 447	13 607	13 664	14 065	Première transf. métaux précieux et non ferreux			272
..	..	20 406	19 484	21 093	19 538	20 060	20 383	Fonderie			273
..	..	215 096	221 441	232 804	250 127	260 077	280 985	Ouvrages en métaux, sauf machines		28	
..	..	89 093	93 412	96 921	105 880	104 946	117 946	Constr. et menuis. métal.; réservoirs; générateurs			281
..	..	126 003	128 029	135 882	144 247	155 131	163 039	Autres ouvrages en métaux			289
..	..	138 744	141 068	142 556	148 003	152 324	162 562	Machines et matériels, nca		29	
..	..	60 555	59 484	61 831	64 772	68 279	72 341	Machines d'usage général			291
..	..	58 478	62 054	61 793	65 252	65 962	70 268	Machines d'usage spécifique			292
..	..	19 711	19 530	18 932	17 980	18 083	19 953	Appareils domestiques, nca			293
..	..	7 921	8 320	9 711	8 965	7 782	7 784	Mach. de bureau, comptables, information		30	
..	..	76 714	75 802	74 516	78 633	80 118	87 622	Machines et appareils électriques, nca		31	
..	..	12 060	10 971	11 473	11 289	11 420	11 993	Moteurs, génératrices, transformateurs électriques			311
..	..	13 197	12 761	13 321	13 627	12 887	13 699	Matériel électrique de distribution et de commande			312
..	..	8 142	7 918	7 536	6 433	6 785	7 322	Fils et câbles électriques isolés			313
..	..	3 379	3 198	2 507	2 413	2 617	2 773	Accumulateurs et piles électriques			314
..	..	10 316	9 979	9 481	10 544	11 045	11 448	Lampes électriques et appareils d'éclairage			315
..	..	29 621	30 976	30 198	34 326	35 364	40 387	Autres matériels électriques, nca.			319
..	..	30 326	29 655	28 922	30 429	29 756	29 042	Equipements et appareils de radio, TV, etc.		32	
..	..	9 887	10 625	10 964	11 812	11 509	11 587	Tubes et valves électroniques;autres composants			321
..	..	14 664	13 202	11 992	12 473	11 625	10 724	Emetteurs radio/TV; app. téléphonie et télégraphie			322
..	..	5 775	5 828	5 966	6 144	6 621	6 731	Récepteurs radio/TV, son et image; art. associés			323
..	..	24 310	21 888	22 569	23 528	24 057	25 953	Instr. médicaux, de précision ; optique ; horlog.		33	
..	..	20 441	18 417	19 070	20 257	20 983	22 213	Instruments médicaux et de précision, etc.			331
..	..	3 285	3 128	2 940	2 908	2 868	2 788	Instruments d'optique ; matériel photographique			332
..	..	585	344	559	363	206	952	Horlogerie			333
..	..	141 518	136 146	138 313	141 163	149 123	154 053	Véhicules automobiles; remorques, etc.		34	
..	..	83 710	76 453	75 457	75 513	77 078	80 153	Véhicules automobiles			341
..	..	9 748	10 169	9 608	9 929	11 397	12 048	Carross. automobiles, remorques, semi-remorques			342
..	..	48 060	49 524	53 248	55 721	60 648	61 852	Pièces et accessoires pour automobiles et moteurs			343
..	..	51 969	50 400	51 724	52 528	53 935	52 257	Autres matériels de transport		35	
..	..	29 443	28 156	29 619	30 580	31 089	29 206	Construction et réparation de navires			351
..	..	7 794	7 657	7 664	7 421	8 005	7 526	Matériel ferroviaire roulant			352
..	..	10 949	10 739	10 627	10 644	10 770	11 214	Construction aéronautique et spatiale			353
..	..	3 783	3 849	3 815	3 884	4 072	4 312	Autres équipements de transport			359
..	..	142 194	136 634	137 935	146 971	149 369	159 960	Meubles ; activités de fabrication, nca		36	
..	..	110 295	105 990	107 133	114 857	117 019	126 996	Meubles			361
..	..	31 899	30 644	30 801	32 114	32 350	32 964	Activités de fabrication, nca			369
..	..	2 052	2 037	1 746	1 443	1 526	1 243	Récupération		37	
..	..	72 025	69 642	68 067	67 166	66 145	64 178	Electricité, gaz et eau	E	40_41	
..	..	56 301	53 560	51 219	50 021	48 197	45 240	Prod. et dist. d'électricité gaz, vapeur d'eau chaude		40	
..	..	..	..	..	..	..	..	Construction	F	45	

SPAIN

Table SP.5 COMPENSATION OF LABOUR - SALAIRES ET CHARGES SOCIALES

Billions of Ptas (Current Prices)

ISIC revision 3				1991	1992	1993	1994	1995	1996	1997	1998
C	10_14		Mining and quarrying	..	..	225	212	216	215	215	..
	10		Mining of coal and lignite; extraction of peat	..	..	150	148	139	..	..	..
	11		Crude petroleum & natural gas; service activities	..	..	5	5	5	..	..	..
	12		Mining of uranium and thorium ores	..	..	..	..	..	..	..	..
	13		Mining of metal ores	..	..	15	8	12	..	..	..
	14		Other mining and quarrying	..	..	54	51	59	..	..	..
D	15_37		Total manufacturing	..	..	7 119	7 188	7 404	8 519	8 833	9 364
	15		Food products and beverages	..	..	1 039	1 064	1 064	1 082	1 128	1 175
		151	Production of meat, fish, fruit, vegetables, oils & fats	..	..	305	304	316	321	341	366
		152	Dairy products	..	..	102	114	113	109	109	113
		153	Grain mill and starch products; animal feeds	..	..	68	69	67	69	76	78
		154	Other food products	..	..	334	348	347	365	377	389
		155	Beverages	..	..	230	229	221	218	226	230
	16		Tobacco products	..	..	56	53	50	49	51	55
	17		Textiles	..	..	234	245	247	266	272	278
		171	Spinning, weaving and finishing of textiles	..	..	136	136	139	150	148	150
		172	Other textiles	..	..	57	65	68	70	76	77
		173	Knitted and crocheted fabrics and articles	..	..	42	44	41	46	48	52
	18		Wearing apparel and fur	..	..	228	229	224	236	255	273
		181	Wearing apparel, except fur apparel	..	..	219	219	216	228	246	266
		182	Dressing and dyeing of fur; articles of fur	..	..	8	9	8	8	9	8
	19		Leather and articles of leather; footwear	..	..	110	116	125	141	145	157
		191	Leather and articles of leather	..	..	42	42	41	45	44	45
		192	Footwear	..	..	69	73	83	96	100	112
	20		Wood & products of wood & cork, exc. furniture	..	..	169	167	185	197	204	221
		201	Sawmilling and planing of wood	..	..	20	19	22	23	26	26
		202	Wood products; articles of cork, straw, etc.	..	..	149	148	163	174	179	195
	21		Paper and paper products	..	..	175	175	181	194	210	215
	22		Printing, publishing and recorded media	..	..	386	396	402	440	465	500
		221	Publishing	..	..	172	183	177	192	209	223
		222	Printing and related activities	..	..	211	210	222	245	253	273
		223	Reproduction of recorded media	..	..	3	2	2	3	3	3
	23		Coke and petroleum products; nuclear fuel	..	..	60	59	60	70	80	79
		231	Coke oven products	..	..	..	..	..	..	..	..
		232	Refined petroleum products	..	..	58	57	58	66	76	74
		233	Processing of nuclear fuel	..	..	..	..	..	..	..	5
	24		Chemicals and chemical products	..	..	700	704	705	717	718	755
		241	Basic chemicals	..	..	205	204	188	202	200	205
		242	Other chemical products	..	..	468	470	487	485	488	516
		243	Man-made fibres	..	..	27	29	30	29	31	34
	25		Rubber and plastic products	..	..	331	338	353	401	398	433
		251	Rubber products	..	..	135	124	131	156	148	158
		252	Plastic products	..	..	195	213	221	245	251	274
	26		Other non-metallic mineral products	..	..	482	479	509	522	557	601
		261	Glass and glass products	..	..	84	81	85	87	93	98
		269	Non-metallic mineral products, nec	..	..	398	398	424	435	464	503
	27		Basic metals	..	..	342	327	331	319	330	354
		271	Basic iron and steel	..	..	195	184	186	175	181	195
		272	Basic precious and non-ferrous metals	..	..	72	70	67	69	70	76
		273	Casting of metals	..	..	74	74	77	75	79	83
	28		Fabricated metal products, exc. machinery	..	..	583	615	677	772	800	887
		281	Structural metal prod., tanks, reservoirs, etc.	..	..	202	228	238	289	280	318
		289	Other fabricated metal products	..	..	381	388	438	483	520	569
	29		Machinery and equipment, nec	..	..	487	517	536	573	596	660
		291	General purpose machinery	..	..	216	226	241	267	279	307
		292	Special purpose machinery	..	..	195	214	221	236	246	272
		293	Domestic appliances, nec	..	..	75	77	74	70	71	81
	30		Office, accounting & computing machinery	..	..	81	79	81	66	60	63
	31		Electrical machinery & apparatus, nec	..	..	292	288	292	307	326	354
		311	Electric motors, generators and transformers	..	..	50	48	47	50	54	59
		312	Electricity distribution & control apparatus	..	..	49	49	52	55	56	59
		313	Insulated wire and cable	..	..	29	31	30	27	28	30
		314	Accumulators, primary cells & primary batteries	..	..	17	16	14	14	14	15
		315	Electric lamps and lighting equipment	..	..	31	30	29	33	34	38
		319	Other electrical equipment, nec	..	..	116	115	120	128	141	154
	32		Radio, TV, communication equipment	..	..	145	141	144	164	148	153
		321	Electronic valves, tubes; other electronic components	..	..	36	40	43	43	43	47
		322	TV/radio transmitters; line teleph./telegr. apparatus	..	..	85	79	78	97	77	77
		323	TV/radio receivers; sound and video app.; assoc. gds.	..	..	24	22	23	24	27	30
	33		Medical, precision & optical instrum.; watch & clocks	..	..	88	76	83	88	94	100
		331	Medical appl.; instruments & appl. for measuring, etc.	..	..	73	63	70	75	81	84
		332	Optical instruments and photographic equipment	..	..	14	13	11	12	12	12
		333	Watches and clocks	..	..	2	1	2	1	1	3
	34		Motor vehicles, trailers and semi-trailers	..	..	599	592	610	653	714	748
		341	Motor vehicles	..	..	384	367	363	388	414	437
		342	Bodies for motor vehicles, trailers & semi-trailers	..	..	27	29	29	31	34	40
		343	Parts & accessories for motor vehicles & their engines	..	..	187	196	218	234	266	271
	35		Other transport equipment	..	..	219	216	227	239	250	252
		351	Building and repairing of ships and boats	..	..	114	114	121	126	130	127
		352	Railway and tramway locomotives and rolling stock	..	..	37	36	38	39	43	43
		353	Aircraft and spacecraft	..	..	54	52	54	58	60	64
		359	Transport equipment, nec	..	..	14	15	14	15	16	17
	36		Furniture; manufacturing, nec	..	..	309	305	315	345	360	403
		361	Furniture	..	..	234	230	237	261	273	311
		369	Manufacturing, nec	..	..	75	75	78	84	87	91
	37		Recycling	..	..	6	6	6	5	5	4
E	40_41		Electricity, gas and water supply	..	..	430	433	434	458	450	..
	40		Electricity, gas, steam and hot water supply	..	..	364	360	354	..	..	..
F	45		Construction	..	..	..	..	..	..	..	..

Table SP.6 **SALAIRES ET TRAITEMENTS - TOTAL WAGES AND SALARIES**

Milliards de Ptas (Prix courants)

1991	1992	1993	1994	1995	1996	1997	1998		CITI révision 3		
..	..	160	150	153	154	155	..	Activites extractives	C	10_14	
..	..	..	..	..	..	..	..	Extraction de charbon, de lignite et de tourbe		10	
..	..	..	..	..	..	..	..	Pétrole brut et gaz naturel ; activités annexes		11	
..	..	..	..	..	..	..	..	Extraction de Minerais d'Uranium et de Thorium		12	
..	..	..	..	..	..	..	..	Extraction de minerais métalliques		13	
..	..	..	..	..	..	..	..	Autres activités extractives		14	
..	..	5 816	5 882	6 114	6 463	6 724	7 163	Total activites manufacturieres	D	15_37	
..	..	792	809	813	831	867	905	Produits alimentaires et boissons		15	
..	..	233	232	241	248	262	283	Viande, poisson, fruits, légumes, huiles et graisses			151
..	..	78	86	86	82	83	85	Produits laitiers			152
..	..	52	53	51	53	59	60	Travail des grains; prod. amyl. ; alim. pour animaux			153
..	..	253	263	265	280	289	300	Autres produits alimentaires			154
..	..	176	175	170	168	174	177	Boissons			155
..	..	40	37	37	36	37	40	Produits à base de tabac		16	
..	..	179	187	189	205	209	216	Textiles		17	
..	..	103	103	105	114	112	115	Filature, tissage et achèvement des textiles			171
..	..	43	49	52	55	59	60	Autres articles textiles			172
..	..	32	34	31	36	37	40	Etoffes et articles de bonneterie			173
..	..	175	175	175	186	199	215	Articles d'habillement et fourrure		18	
..	..	168	168	168	179	193	209	Articles d'habillement autres qu'en fourrure			181
..	..	6	7	7	7	7	6	Prép. et teinture des fourrures ; articles en fourrure			182
..	..	84	88	96	109	112	123	Cuir et articles en cuir ; chaussures		19	
..	..	32	33	32	35	34	35	Cuir et articles en cuir sauf chaussures			191
..	..	52	56	64	75	78	88	Chaussures			192
..	..	126	125	142	152	157	171	Bois, articles en bois, sauf meubles		20	
..	..	15	14	17	17	20	20	Sciage et rabotage du bois			201
..	..	112	111	125	134	138	151	Articles en bois, liège, vannerie, sparterie			202
..	..	134	134	140	150	162	166	Papier, articles en papier et carton		21	
..	..	302	310	317	352	370	401	Edition, imprimerie et reproduction		22	
..	..	137	145	141	155	166	180	Edition			221
..	..	163	163	174	195	201	218	Imprimerie et activités annexes			222
..	..	2	2	2	2	2	3	Reproduction de supports enregistrés			223
..	..	42	41	42	50	56	56	Coke. et prod. pétroliers ; comb. nucléaires		23	
..	..	..	..	..	..	..	..	Cokéfaction			231
..	..	41	40	41	46	52	52	Produits pétroliers raffinés			232
..	..	..	..	..	..	..	3	Traitement de combustibles nucléaires			233
..	..	533	536	538	549	550	579	Produits chimiques		24	
..	..	152	150	138	151	150	155	Produits chimiques de base			241
..	..	360	363	378	375	377	399	Autres produits chimiques			242
..	..	21	22	22	23	23	25	Fibres synthétiques ou artificielles			243
..	..	233	253	267	290	299	328	Art. en caoutchouc et matières plastiques		25	
..	..	84	89	95	101	104	114	Articles en caoutchouc			251
..	..	149	163	172	189	194	214	Articles en matières plastiques			252
..	..	365	362	388	399	426	462	Autres produits minéraux non métalliques		26	
..	..	64	60	64	65	69	73	Verre et articles en verre			261
..	..	301	302	324	334	357	389	Produits minéraux non métalliques, nca			269
..	..	249	237	245	239	248	265	Produits métallurgiques de base		27	
..	..	143	134	136	131	135	146	Sidérurgie et première transformation de l'acier			271
..	..	51	50	51	51	53	56	Première transf. métaux précieux et non ferreux			272
..	..	55	53	59	57	60	64	Fonderie			273
..	..	441	464	519	582	620	691	Ouvrages en métaux, sauf machines		28	
..	..	153	169	182	209	217	248	Constr. et menuis. métal.; réservoirs; générateurs			281
..	..	288	295	337	373	403	443	Autres ouvrages en métaux			289
..	..	370	393	410	442	458	509	Machines et matériels, nca		29	
..	..	166	172	185	207	214	237	Machines d'usage général			291
..	..	148	162	169	182	190	212	Machines d'usage spécifique			292
..	..	56	58	56	53	53	60	Appareils domestiques, nca			293
..	..	51	65	66	53	48	50	Mach. de bureau, comptables, information		30	
..	..	219	219	224	233	247	271	Machines et appareils électriques, nca		31	
..	..	38	37	38	39	42	46	Moteurs, génératrices, transformateurs électriques			311
..	..	38	37	40	42	43	46	Matériel électrique de distribution et de commande			312
..	..	23	24	23	20	21	23	Fils et câbles électriques isolés			313
..	..	13	12	11	10	10	11	Accumulateurs et piles électriques			314
..	..	24	22	22	25	26	29	Lampes électriques et appareils d'éclairage			315
..	..	85	87	91	97	105	116	Autres matériels électriques, nca.			319
..	..	114	109	111	130	114	119	Equipements et appareils de radio, TV, etc.		32	
..	..	28	31	33	34	33	37	Tubes et valves électroniques;autres composants			321
..	..	67	61	60	78	60	60	Emetteurs radio/TV; app. téléphonie et télégraphie			322
..	..	19	17	18	19	21	23	Récepteurs radio/TV, son et image; art. associés			323
..	..	68	58	64	68	73	77	Instr. médicaux, de précision ; optique ; horlog.		33	
..	..	56	48	54	58	63	65	Instruments médicaux et de précision, etc.			331
..	..	11	10	8	9	9	9	Instruments d'optique ; matériel photographique			332
..	..	1	1	1	1	1	3	Horlogerie			333
..	..	441	435	457	490	526	553	Véhicules automobiles; remorques, etc.		34	
..	..	280	267	269	290	300	319	Véhicules automobiles			341
..	..	21	22	22	23	26	31	Carross. automobiles, remorques, semi-remorques			342
..	..	141	146	165	177	200	203	Pièces et accessoires pour automobiles et moteurs			343
..	..	160	158	168	178	188	190	Autres matériels de transport		35	
..	..	81	81	88	93	98	95	Construction et réparation de navires			351
..	..	28	26	28	29	32	32	Matériel ferroviaire roulant			352
..	..	41	39	41	44	46	49	Construction aéronautique et spatiale			353
..	..	11	11	11	12	12	13	Autres équipements de transport			359
..	..	234	231	243	267	280	314	Meubles ; activités de fabrication, nca		36	
..	..	176	174	182	202	212	243	Meubles			361
..	..	57	57	61	65	68	72	Activités de fabrication, nca			369
..	..	4	4	4	4	4	3	Récupération		37	
..	..	300	301	306	316	318	..	Electricité, gaz et eau	E	40_41	
..	..	..	..	..	..	..	..	Prod. et dist. d'électricité gaz, vapeur d'eau chaude		40	
..	..	..	..	..	..	..	..	Construction	F	45	

SPAIN

Table SP.7 SUPPLEMENTS TO WAGES & SALARIES - CHARGES SOCIALES DES EMPLOYEURS

Millions of Ptas (Current Prices)

ISIC revision 3				1991	1992	1993	1994	1995	1996	1997	1998
C	10_14		Mining and quarrying	..	..	65	62	63	61	60	53
	10		Mining of coal and lignite; extraction of peat	..	..	46	46	44	43	42	33
	11		Crude petroleum & natural gas; service activities	..	..	..	..	..	..	..	..
	12		Mining of uranium and thorium ores	..	..	..	..	..	..	..	..
	13		Mining of metal ores	..	..	5	2	3	3	2	2
	14		Other mining and quarrying	..	..	13	13	14	15	15	17
D	15_37		Total manufacturing	..	..	1 763	1 756	1 750	1 853	1 917	2 201
	15		Food products and beverages	..	..	247	255	251	251	261	270
		151	Production of meat, fish, fruit, vegetables, oils & fats	..	..	72	72	75	73	78	83
		152	Dairy products	..	..	24	28	27	27	27	28
		153	Grain mill and starch products; animal feeds	..	..	16	16	15	16	17	18
		154	Other food products	..	..	81	85	82	85	87	89
		155	Beverages	..	..	54	54	51	50	52	53
	16		Tobacco products	..	..	16	16	14	14	14	15
	17		Textiles	..	..	56	58	58	61	63	62
		171	Spinning, weaving and finishing of textiles	..	..	33	33	33	35	35	35
		172	Other textiles	..	..	13	15	15	16	17	16
		173	Knitted and crocheted fabrics and articles	..	..	10	10	9	10	11	11
	18		Wearing apparel and fur	..	..	53	53	50	50	56	58
		181	Wearing apparel, except fur apparel	..	..	51	51	48	48	54	57
		182	Dressing and dyeing of fur; articles of fur	..	..	2	2	2	2	2	2
	19		Leather and articles of leather; footwear	..	..	27	27	29	32	32	34
		191	Leather and articles of leather	..	..	10	10	10	10	10	10
		192	Footwear	..	..	16	17	19	21	22	24
	20		Wood & products of wood & cork, exc. furniture	..	..	42	42	44	45	47	50
		201	Sawmilling and planing of wood	..	..	5	5	6	5	6	6
		202	Wood products; articles of cork, straw, etc.	..	..	37	37	38	40	41	44
	21		Paper and paper products	..	..	41	42	42	45	49	49
	22		Printing, publishing and recorded media	..	..	84	86	84	88	95	99
		221	Publishing	..	..	36	39	36	38	43	43
		222	Printing and related activities	..	..	48	47	48	50	52	55
		223	Reproduction of recorded media	..	..	1	0	0	1	1	1
	23		Coke and petroleum products; nuclear fuel	..	..	17	18	18	20	25	23
		231	Coke oven products	..	..	..	..	..	..	..	..
		232	Refined petroleum products	..	..	17	18	17	19	24	22
		233	Processing of nuclear fuel	..	..	..	..	..	..	..	1
	24		Chemicals and chemical products	..	..	167	168	167	168	168	176
		241	Basic chemicals	..	..	53	54	50	51	50	50
		242	Other chemical products	..	..	107	107	109	111	110	117
		243	Man-made fibres	..	..	6	7	7	7	8	9
	25		Rubber and plastic products	..	..	98	85	86	111	100	105
		251	Rubber products	..	..	51	35	37	56	43	44
		252	Plastic products	..	..	46	50	49	56	56	60
	26		Other non-metallic mineral products	..	..	117	117	121	123	131	139
		261	Glass and glass products	..	..	20	21	22	22	24	25
		269	Non-metallic mineral products, nec	..	..	97	96	99	100	107	115
	27		Basic metals	..	..	93	90	85	80	82	89
		271	Basic iron and steel	..	..	53	50	50	43	45	49
		272	Basic precious and non-ferrous metals	..	..	21	20	16	18	18	20
		273	Casting of metals	..	..	19	20	19	18	19	20
	28		Fabricated metal products, exc. machinery	..	..	142	151	158	190	180	196
		281	Structural metal prod., tanks, reservoirs, etc.	..	..	49	59	56	80	63	70
		289	Other fabricated metal products	..	..	92	93	102	110	117	126
	29		Machinery and equipment, nec	..	..	117	124	126	131	138	150
		291	General purpose machinery	..	..	51	53	56	60	64	70
		292	Special purpose machinery	..	..	48	51	51	54	56	60
		293	Domestic appliances, nec	..	..	19	19	18	17	18	20
	30		Office, accounting & computing machinery	..	..	31	14	14	13	12	13
	31		Electrical machinery & apparatus, nec	..	..	73	69	68	74	78	84
		311	Electric motors, generators and transformers	..	..	12	11	10	12	12	13
		312	Electricity distribution & control apparatus	..	..	11	11	12	13	13	13
		313	Insulated wire and cable	..	..	7	7	7	7	6	7
		314	Accumulators, primary cells & primary batteries	..	..	4	4	3	3	4	4
		315	Electric lamps and lighting equipment	..	..	8	7	7	8	8	8
		319	Other electrical equipment, nec	..	..	31	28	29	31	36	38
	32		Radio, TV, communication equipment	..	..	31	32	33	34	34	34
		321	Electronic valves, tubes; other electronic components	..	..	8	9	10	10	10	11
		322	TV/radio transmitters; line teleph./telegr. apparatus	..	..	18	18	18	19	18	17
		323	TV/radio receivers; sound and video app.; assoc. gds.	..	..	5	5	5	6	6	7
	33		Medical, precision & optical instrum.; watch & clocks	..	..	20	18	19	20	21	22
		331	Medical appl.; instruments & appl. for measuring, etc.	..	..	17	15	16	17	18	19
		332	Optical instruments and photographic equipment	..	..	3	3	3	3	3	3
		333	Watches and clocks	..	..	0	0	0	0	0	1
	34		Motor vehicles, trailers and semi-trailers	..	..	157	157	154	163	188	195
		341	Motor vehicles	..	..	105	101	94	98	114	118
		342	Bodies for motor vehicles, trailers & semi-trailers	..	..	7	7	7	7	8	9
		343	Parts & accessories for motor vehicles & their engines	..	..	46	50	53	57	66	68
	35		Other transport equipment	..	..	58	58	58	60	62	62
		351	Building and repairing of ships and boats	..	..	33	33	32	33	33	32
		352	Railway and tramway locomotives and rolling stock	..	..	9	9	10	10	11	11
		353	Aircraft and spacecraft	..	..	13	13	13	14	14	15
		359	Transport equipment, nec	..	..	3	4	3	4	4	4
	36		Furniture; manufacturing, nec	..	..	75	73	72	78	80	88
		361	Furniture	..	..	58	56	55	60	61	69
		369	Manufacturing, nec	..	..	17	18	17	19	19	19
	37		Recycling	..	..	1	1	1	1	1	1
E	40_41		Electricity, gas and water supply	..	..	131	131	127	142	131	132
	40		Electricity, gas, steam and hot water supply	..	..	111	110	104	116	104	105
F	45		Construction	..	..	..	..	..	..	..	..

Table SP.8 **HEURES OUVRÉES, SALARIÉS - HOURS WORKED BY EMPLOYEES**

Milliers

1991	1992	1993	1994	1995	1996	1997	1998			CITI révision 3
..	..	86 223	75 896	74 731	74 472	71 466	67 397	Activites extractives	C	**10_14**
..	..	46 857	42 516	38 114	37 830	35 378	29 507	Extraction de charbon, de lignite et de tourbe		**10**
..	..	..	..	..	..	..	..	Pétrole brut et gaz naturel ; activités annexes		**11**
..	..	..	..	..	..	..	..	Extraction de Minerais d'Uranium et de Thorium		**12**
..	..	5 170	3 580	3 923	3 916	3 414	3 172	Extraction de minerais métalliques		**13**
..	..	32 556	28 385	31 353	31 919	31 811	33 819	Autres activités extractives		**14**
..	..	3 696 926	3 633 857	3 637 130	3 745 930	3 811 755	4 137 161	Total activites manufacturieres	D	**15_37**
..	..	616 410	609 799	591 672	584 698	588 795	593 752	Produits alimentaires et boissons		**15**
..	..	213 237	203 519	201 042	198 671	206 901	217 436	Viande, poisson, fruits, légumes, huiles et graisses		151
..	..	49 497	51 761	49 690	45 854	45 691	45 991	Produits laitiers		152
..	..	36 367	35 761	34 617	34 371	35 916	35 132	Travail des grains; prod. amyl. ; alim. pour animaux		153
..	..	224 727	230 219	224 540	228 578	222 918	219 012	Autres produits alimentaires		154
..	..	92 583	88 538	81 784	77 224	77 370	76 182	Boissons		155
..	..	16 286	13 726	14 423	15 516	14 380	13 270	Produits à base de tabac		**16**
..	..	163 359	169 251	163 166	170 741	169 303	169 540	Textiles		**17**
..	..	87 432	88 867	85 995	89 946	85 967	87 003	Filature, tissage et achèvement des textiles		171
..	..	43 284	46 848	48 076	48 932	51 267	48 796	Autres articles textiles		172
..	..	32 643	33 537	29 095	31 863	32 069	33 740	Etoffes et articles de bonneterie		173
..	..	216 311	198 246	188 078	190 488	200 423	221 460	Articles d'habillement et fourrure		**18**
..	..	210 018	192 205	182 934	185 351	195 083	216 850	Articles d'habillement autres qu'en fourrure		181
..	..	6 293	6 041	5 144	5 137	5 339	4 610	Prép. et teinture des fourrures ; articles en fourrure		182
..	..	98 780	101 135	102 639	118 100	116 633	122 726	Cuir et articles en cuir ; chaussures		**19**
..	..	29 927	28 796	27 839	28 642	27 372	26 287	Cuir et articles en cuir sauf chaussures		191
..	..	68 854	72 340	74 800	89 458	89 261	96 439	Chaussures		192
..	..	139 890	131 869	137 922	143 454	142 131	150 037	Bois, articles en bois, sauf meubles		**20**
..	..	17 378	17 344	18 387	18 485	19 342	19 026	Sciage et rabotage du bois		201
..	..	122 512	114 525	119 534	124 969	122 789	131 011	Articles en bois, liège, vannerie, sparterie		202
..	..	84 231	83 031	81 012	86 784	88 696	86 455	Papier, articles en papier et carton		**21**
..	..	192 645	188 573	187 816	199 463	201 275	207 982	Edition, imprimerie et reproduction		**22**
..	..	62 602	64 847	60 943	65 703	69 853	70 884	Edition		221
..	..	128 531	122 488	125 536	132 297	130 052	135 492	Imprimerie et activités annexes		222
..	..	1 511	1 237	1 337	1 463	1 370	1 605	Reproduction de supports enregistrés		223
..	..	13 837	12 988	12 443	13 661	14 726	14 827	Coke. et prod. pétroliers ; comb. nucléaires		**23**
..	..	..	..	..	..	..	..	Cokéfaction		231
..	..	13 240	12 410	11 844	12 460	13 556	13 620	Produits pétroliers raffinés		232
..	..	..	..	..	..	..	983	Traitement de combustibles nucléaires		233
..	..	240 407	232 374	226 404	223 540	224 646	225 341	Produits chimiques		**24**
..	..	63 971	61 804	56 790	58 454	58 407	57 374	Produits chimiques de base		241
..	..	166 420	160 536	160 198	154 748	155 998	157 703	Autres produits chimiques		242
..	..	10 016	10 034	9 416	10 338	10 241	10 264	Fibres synthétiques ou artificielles		243
..	..	151 137	159 861	161 052	174 354	172 293	180 197	Art. en caoutchouc et matières plastiques		**25**
..	..	45 457	47 008	47 974	49 839	49 026	51 485	Articles en caoutchouc		251
..	..	105 680	112 853	113 078	124 514	123 267	128 712	Articles en matières plastiques		252
..	..	264 459	259 984	265 543	265 352	271 733	277 616	Autres produits minéraux non métalliques		**26**
..	..	40 181	38 962	40 063	38 410	40 520	40 376	Verre et articles en verre		261
..	..	224 278	221 022	225 480	226 942	231 213	237 240	Produits minéraux non métalliques, nca		269
..	..	133 450	120 436	120 427	113 745	115 723	120 813	Produits métallurgiques de base		**27**
..	..	74 941	62 793	61 764	57 597	57 704	61 822	Sidérurgie et première transformation de l'acier		271
..	..	25 177	24 951	22 986	23 296	23 185	24 099	Première transf. métaux précieux et non ferreux		272
..	..	33 332	32 692	35 677	32 852	34 834	34 893	Fonderie		273
..	..	348 898	355 003	375 228	400 444	413 932	446 423	Ouvrages en métaux, sauf machines		**28**
..	..	141 446	145 726	152 202	163 420	162 748	180 906	Constr. et menuis. métal.; réservoirs; générateurs		281
..	..	207 452	209 276	223 026	237 024	251 184	265 516	Autres ouvrages en métaux		289
..	..	232 628	234 677	240 509	248 351	251 941	270 990	Machines et matériels, nca		**29**
..	..	103 294	101 282	106 613	109 997	115 559	125 162	Machines d'usage général		291
..	..	96 448	100 451	101 613	107 823	106 520	112 988	Machines d'usage spécifique		292
..	..	32 887	32 944	32 283	30 530	29 862	32 840	Appareils domestiques, nca		293
..	..	13 762	14 444	16 946	15 670	13 931	13 650	Mach. de bureau, comptables, information		**30**
..	..	130 725	129 521	127 317	133 852	136 647	147 952	Machines et appareils électriques, nca		**31**
..	..	20 162	18 726	19 125	19 446	18 948	20 123	Moteurs, génératrices, transformateurs électriques		311
..	..	22 075	21 570	22 988	23 411	22 547	23 223	Matériel électrique de distribution et de commande		312
..	..	14 269	13 819	13 164	11 003	11 851	12 774	Fils et câbles électriques isolés		313
..	..	5 866	5 569	4 333	4 192	4 427	4 827	Accumulateurs et piles électriques		314
..	..	17 463	16 448	15 722	17 215	17 965	18 695	Lampes électriques et appareils d'éclairage		315
..	..	50 891	53 388	51 985	58 585	60 909	68 310	Autres matériels électriques, nca.		319
..	..	52 340	50 909	48 979	51 331	49 901	49 639	Equipements et appareils de radio, TV, etc.		**32**
..	..	16 706	18 360	18 430	19 627	18 915	19 446	Tubes et valves électroniques;autres composants		321
..	..	25 549	22 309	20 157	21 175	19 539	18 400	Emetteurs radio/TV; app. téléphonie et télégraphie		322
..	..	10 086	10 240	10 393	10 529	11 447	11 793	Récepteurs radio/TV, son et image; art. associés		323
..	..	40 027	35 634	36 995	38 484	39 223	42 599	Instr. médicaux, de précision ; optique ; horlog.		**33**
..	..	33 393	29 617	31 066	32 753	33 840	36 064	Instruments médicaux et de précision, etc.		331
..	..	5 790	5 455	5 113	5 115	5 044	4 916	Instruments d'optique ; matériel photographique		332
..	..	844	563	816	616	339	1 619	Horlogerie		333
..	..	226 055	228 790	234 615	241 095	256 522	263 675	Véhicules automobiles; remorques, etc.		**34**
..	..	129 613	126 867	125 035	126 652	131 247	136 800	Véhicules automobiles		341
..	..	16 156	16 540	15 926	16 781	18 988	20 454	Carross. automobiles, remorques, semi-remorques		342
..	..	80 286	85 384	93 655	97 662	106 287	106 422	Pièces et accessoires pour automobiles et moteurs		343
..	..	85 973	80 540	81 833	83 629	86 387	85 412	Autres matériels de transport		**35**
..	..	47 581	45 639	45 917	47 200	47 831	46 169	Construction et réparation de navires		351
..	..	13 085	10 521	11 745	12 642	13 722	13 263	Matériel ferroviaire roulant		352
..	..	18 987	18 465	17 819	17 394	17 851	18 637	Construction aéronautique et spatiale		353
..	..	6 321	5 916	6 352	6 393	6 983	7 343	Autres équipements de transport		359
..	..	231 905	219 702	219 243	230 806	239 959	254 968	Meubles ; activités de fabrication, nca		**36**
..	..	178 999	170 070	169 423	179 874	187 858	202 267	Meubles		361
..	..	52 906	49 632	49 820	50 932	52 101	52 700	Activités de fabrication, nca		369
..	..	3 410	3 364	2 866	2 375	2 556	1 999	Récupération		**37**
..	..	121 086	115 939	114 792	113 080	109 107	108 441	Electricité, gaz et eau	E	**40_41**
..	..	94 202	88 725	85 808	83 694	78 491	75 804	Prod. et dist. d'électricité gaz, vapeur d'eau chaude		**40**
..	..	..	..	..	..	..	..	Construction	F	**45**

Table SP.9 ESTABLISHMENTS - ÉTABLISSEMENTS

Units

ISIC revision 3				1991	1992	1993	1994	1995	1996	1997	1998
C	10_14		Mining and quarrying	..	..	1 873	1 655	1 809	1 833	1 871	1 996
	10		Mining of coal and lignite; extraction of peat	..	..	170	152	148	145	135	117
	11		Crude petroleum & natural gas; service activities	..	..	..	..	..	..	..	..
	12		Mining of uranium and thorium ores	..	..	..	..	..	..	..	..
	13		Mining of metal ores	..	..	45	41	37	36	37	28
	14		Other mining and quarrying	..	..	1 644	1 448	1 611	1 642	1 690	1 842
D	15_37		Total manufacturing	..	..	144 133	150 414	149 137	162 973	159 170	163 943
	15		Food products and beverages	..	..	25 006	27 320	27 507	28 184	26 898	26 287
		151	Production of meat, fish, fruit, vegetables, oils & fats	..	..	5 738	6 013	6 142	6 108	5 954	6 254
		152	Dairy products	..	..	1 040	1 047	1 031	921	1 030	911
		153	Grain mill and starch products; animal feeds	..	..	1 344	1 466	1 393	1 461	1 484	1 402
		154	Other food products	..	..	14 183	15 955	16 377	17 019	15 904	15 251
		155	Beverages	..	..	2 703	2 838	2 565	2 675	2 526	2 469
	16		Tobacco products	..	..	36	35	34	36	37	39
	17		Textiles	..	..	6 151	6 360	5 625	6 949	6 801	6 592
		171	Spinning, weaving and finishing of textiles	..	..	2 465	2 496	2 160	2 839	2 677	2 507
		172	Other textiles	..	..	2 527	2 606	2 359	2 832	2 845	2 755
		173	Knitted and crocheted fabrics and articles	..	..	1 158	1 259	1 106	1 278	1 278	1 329
	18		Wearing apparel and fur	..	..	10 167	10 612	10 638	10 567	11 014	11 561
		181	Wearing apparel, except fur apparel	..	..	9 626	10 057	10 205	10 118	10 549	11 138
		182	Dressing and dyeing of fur; articles of fur	..	..	541	555	433	449	465	424
	19		Leather and articles of leather; footwear	..	..	4 952	5 088	5 300	6 777	5 853	6 352
		191	Leather and articles of leather	..	..	1 494	1 551	1 573	1 697	1 487	1 432
		192	Footwear	..	..	3 458	3 537	3 727	5 081	4 367	4 920
	20		Wood & products of wood & cork, exc. furniture	..	..	11 825	12 062	12 316	13 417	12 141	12 316
		201	Sawmilling and planing of wood	..	..	1 410	1 432	1 610	1 529	1 429	1 482
		202	Wood products; articles of cork, straw, etc.	..	..	10 415	10 630	10 707	11 888	10 712	10 834
	21		Paper and paper products	..	..	1 607	1 931	1 475	1 933	1 952	1 854
	22		Printing, publishing and recorded media	..	..	11 569	11 615	12 044	13 479	13 253	13 563
		221	Publishing	..	..	2 421	2 228	2 606	2 989	3 136	3 101
		222	Printing and related activities	..	..	9 109	9 351	9 403	10 444	10 069	10 413
		223	Reproduction of recorded media	..	..	39	35	36	46	47	49
	23		Coke and petroleum products; nuclear fuel	..	..	27	25	20	40	24	24
		231	Coke oven products	..	..	..	..	..	..	..	..
		232	Refined petroleum products	..	..	22	21	16	33	18	18
		233	Processing of nuclear fuel	..	..	..	..	..	..	..	2
	24		Chemicals and chemical products	..	..	3 621	3 651	3 294	3 761	4 021	3 939
		241	Basic chemicals	..	..	946	947	768	842	1 062	1 108
		242	Other chemical products	..	..	2 639	2 671	2 489	2 873	2 915	2 789
		243	Man-made fibres	..	..	37	33	36	46	44	42
	25		Rubber and plastic products	..	..	3 972	4 426	4 279	4 760	4 601	4 396
		251	Rubber products	..	..	748	804	716	889	681	769
		252	Plastic products	..	..	3 224	3 622	3 563	3 870	3 920	3 628
	26		Other non-metallic mineral products	..	..	9 702	10 155	10 012	10 290	10 761	10 402
		261	Glass and glass products	..	..	1 026	1 103	1 163	1 058	1 071	1 055
		269	Non-metallic mineral products, nec	..	..	8 676	9 051	8 849	9 232	9 691	9 346
	27		Basic metals	..	..	1 446	1 558	1 591	1 536	1 497	1 386
		271	Basic iron and steel	..	..	552	453	444	480	454	451
		272	Basic precious and non-ferrous metals	..	..	243	308	258	313	310	311
		273	Casting of metals	..	..	650	797	889	743	733	624
	28		Fabricated metal products, exc. machinery	..	..	21 963	23 377	23 380	26 070	25 826	27 119
		281	Structural metal prod., tanks, reservoirs, etc.	..	..	11 590	12 158	12 287	13 963	12 890	14 229
		289	Other fabricated metal products	..	..	10 373	11 219	11 092	12 107	12 936	12 890
	29		Machinery and equipment, nec	..	..	8 239	8 672	8 651	9 267	9 360	9 389
		291	General purpose machinery	..	..	3 367	3 045	3 125	3 447	3 473	3 503
		292	Special purpose machinery	..	..	4 589	5 324	5 197	5 484	5 605	5 564
		293	Domestic appliances, nec	..	..	283	303	329	335	283	323
	30		Office, accounting & computing machinery	..	..	116	122	136	154	183	222
	31		Electrical machinery & apparatus, nec	..	..	3 252	3 086	3 153	3 751	3 242	3 651
		311	Electric motors, generators and transformers	..	..	421	436	469	471	442	399
		312	Electricity distribution & control apparatus	..	..	391	429	359	496	350	380
		313	Insulated wire and cable	..	..	155	139	157	149	164	170
		314	Accumulators, primary cells & primary batteries	..	..	17	16	14	12	15	19
		315	Electric lamps and lighting equipment	..	..	898	819	753	892	860	858
		319	Other electrical equipment, nec	..	..	1 371	1 247	1 401	1 731	1 411	1 826
	32		Radio, TV, communication equipment	..	..	741	831	662	750	752	738
		321	Electronic valves, tubes; other electronic components	..	..	444	492	426	521	520	509
		322	TV/radio transmitters; line teleph./telegr. apparatus	..	..	143	149	121	117	120	132
		323	TV/radio receivers; sound and video app.; assoc. gds.	..	..	154	191	116	111	112	98
	33		Medical, precision & optical instrum.; watch & clocks	..	..	1 922	1 754	1 766	1 894	1 893	2 025
		331	Medical appl.; instruments & appl. for measuring, etc.	..	..	1 736	1 628	1 527	1 738	1 800	1 911
		332	Optical instruments and photographic equipment	..	..	97	76	126	81	75	67
		333	Watches and clocks	..	..	88	50	113	75	18	47
	34		Motor vehicles, trailers and semi-trailers	..	..	1 446	1 666	1 628	1 708	1 718	1 683
		341	Motor vehicles	..	..	53	60	63	63	74	78
		342	Bodies for motor vehicles, trailers & semi-trailers	..	..	610	674	636	640	685	621
		343	Parts & accessories for motor vehicles & their engines	..	..	784	932	928	1 004	958	984
	35		Other transport equipment	..	..	1 120	1 214	1 075	1 287	1 250	1 161
		351	Building and repairing of ships and boats	..	..	882	963	879	1 102	1 056	967
		352	Railway and tramway locomotives and rolling stock	..	..	48	50	50	49	50	48
		353	Aircraft and spacecraft	..	..	53	59	53	52	51	51
		359	Transport equipment, nec	..	..	136	142	92	84	93	95
	36		Furniture; manufacturing, nec	..	..	15 064	14 633	14 415	16 234	15 962	15 776
		361	Furniture	..	..	11 843	11 365	11 133	12 645	12 600	12 331
		369	Manufacturing, nec	..	..	3 221	3 268	3 282	3 589	3 362	3 444
	37		Recycling	..	..	189	221	136	129	131	107
E	40_41		Electricity, gas and water supply	..	..	1 166	1 158	1 264	1 274	1 365	1 366
	40		Electricity, gas, steam and hot water supply	..	..	706	694	758	741	781	823
F	45		Construction	..	..	..	..	..	..	..	..

Table SP.10 **ENTREPRISES - ENTERPRISES**

Unités

1991	1992	1993	1994	1995	1996	1997	1998			CITI révision 3	
..	..	1 834	1 559	1 696	1 670	1 681	1 829	Activites extractives	C	10_14	
..	..	168	145	141	131	120	103	Extraction de charbon, de lignite et de tourbe		10	
..	..	..	..	..	..	..	..	Pétrole brut et gaz naturel ; activités annexes		11	
..	..	..	..	..	..	..	..	Extraction de Minerais d'Uranium et de Thorium		12	
..	..	44	38	35	34	35	25	Extraction de minerais métalliques		13	
..	..	1 608	1 365	1 508	1 497	1 518	1 694	Autres activités extractives		14	
..	..	142 913	147 603	146 051	159 589	155 743	160 289	Total activites manufacturieres	D	15_37	
..	..	24 754	26 567	26 760	27 180	26 133	25 563	Produits alimentaires et boissons		15	
..	..	5 687	5 910	6 055	5 930	5 802	6 124	Viande, poisson, fruits, légumes, huiles et graisses			151
..	..	1 010	1 011	997	890	986	868	Produits laitiers			152
..	..	1 299	1 386	1 320	1 393	1 397	1 323	Travail des grains; prod. amyl. ; alim. pour animaux			153
..	..	14 124	15 584	15 965	16 453	15 558	14 903	Autres produits alimentaires			154
..	..	2 664	2 765	2 466	2 585	2 427	2 394	Boissons			155
..	..	23	22	22	25	24	25	Produits à base de tabac		16	
..	..	6 114	6 224	5 523	6 836	6 720	6 489	Textiles		17	
..	..	2 451	2 440	2 104	2 783	2 645	2 474	Filature, tissage et achèvement des textiles			171
..	..	2 512	2 559	2 329	2 790	2 809	2 721	Autres articles textiles			172
..	..	1 152	1 226	1 092	1 264	1 270	1 299	Etoffes et articles de bonneterie			173
..	..	10 147	10 535	10 325	10 498	10 589	11 442	Articles d'habillement et fourrure		18	
..	..	9 610	9 994	9 907	10 052	10 126	11 026	Articles d'habillement autres qu'en fourrure			181
..	..	538	543	419	447	463	416	Prép. et teinture des fourrures ; articles en fourrure			182
..	..	4 950	5 071	5 263	6 748	5 831	6 328	Cuir et articles en cuir ; chaussures		19	
..	..	1 493	1 542	1 554	1 677	1 475	1 425	Cuir et articles en cuir sauf chaussures			191
..	..	3 457	3 529	3 709	5 071	4 356	4 902	Chaussures			192
..	..	11 777	11 996	12 271	13 338	12 056	12 283	Bois, articles en bois, sauf meubles		20	
..	..	1 401	1 424	1 595	1 517	1 416	1 475	Sciage et rabotage du bois			201
..	..	10 384	10 590	10 678	11 829	10 642	10 808	Articles en bois, liège, vannerie, sparterie			202
..	..	1 567	1 819	1 411	1 863	1 883	1 783	Papier, articles en papier et carton		21	
..	..	11 498	11 486	11 907	13 266	13 085	13 455	Edition, imprimerie et reproduction		22	
..	..	2 368	2 206	2 582	2 900	3 122	3 046	Edition			221
..	..	9 092	9 246	9 293	10 323	9 919	10 364	Imprimerie et activités annexes			222
..	..	39	35	36	46	47	49	Reproduction de supports enregistrés			223
..	..	21	18	14	33	16	16	Coke. et prod. pétroliers ; comb. nucléaires		23	
..	..	..	..	..	..	..	..	Cokéfaction			231
..	..	16	14	10	26	10	10	Produits pétroliers raffinés			232
..	..	..	..	..	..	..	2	Traitement de combustibles nucléaires			233
..	..	3 483	3 422	3 099	3 553	3 802	3 664	Produits chimiques		24	
..	..	861	818	655	725	930	924	Produits chimiques de base			241
..	..	2 589	2 576	2 413	2 785	2 830	2 702	Autres produits chimiques			242
..	..	35	31	34	45	43	41	Fibres synthétiques ou artificielles			243
..	..	3 911	4 358	4 202	4 694	4 534	4 322	Art. en caoutchouc et matières plastiques		25	
..	..	737	791	691	874	663	748	Articles en caoutchouc			251
..	..	3 176	3 567	3 513	3 820	3 871	3 575	Articles en matières plastiques			252
..	..	9 483	9 671	9 313	9 618	9 934	9 630	Autres produits minéraux non métalliques		26	
..	..	994	1 064	1 110	1 002	1 032	1 021	Verre et articles en verre			261
..	..	8 491	8 608	8 204	8 617	8 903	8 610	Produits minéraux non métalliques, nca			269
..	..	1 424	1 530	1 539	1 500	1 458	1 347	Produits métallurgiques de base		27	
..	..	544	443	426	464	437	435	Sidérurgie et première transformation de l'acier			271
..	..	233	297	247	300	295	298	Première transf. métaux précieux et non ferreux			272
..	..	648	791	867	735	725	614	Fonderie			273
..	..	21 870	23 220	23 182	25 815	25 624	26 932	Ouvrages en métaux, sauf machines		28	
..	..	11 567	12 122	12 204	13 878	12 817	14 163	Constr. et menuis. métal.; réservoirs; générateurs			281
..	..	10 305	11 135	10 981	11 939	12 808	12 770	Autres ouvrages en métaux			289
..	..	8 168	8 414	8 458	9 059	9 187	9 184	Machines et matériels, nca		29	
..	..	3 318	2 854	2 976	3 290	3 336	3 359	Machines d'usage général			291
..	..	4 571	5 268	5 165	5 445	5 580	5 513	Machines d'usage spécifique			292
..	..	278	292	317	323	272	313	Appareils domestiques, nca			293
..	..	106	121	136	151	179	209	Mach. de bureau, comptables, information		30	
..	..	3 220	3 029	3 097	3 674	3 153	3 557	Machines et appareils électriques, nca		31	
..	..	412	427	461	459	428	380	Moteurs, génératrices, transformateurs électriques			311
..	..	387	407	348	460	335	364	Matériel électrique de distribution et de commande			312
..	..	152	132	152	145	161	165	Fils et câbles électriques isolés			313
..	..	13	11	9	7	10	12	Accumulateurs et piles électriques			314
..	..	895	818	749	890	857	852	Lampes électriques et appareils d'éclairage			315
..	..	1 361	1 235	1 379	1 712	1 362	1 788	Autres matériels électriques, nca.			319
..	..	739	825	656	745	746	732	Equipements et appareils de radio, TV, etc.		32	
..	..	443	491	424	520	519	508	Tubes et valves électroniques;autres composants			321
..	..	143	146	118	114	115	127	Emetteurs radio/TV; app. téléphonie et télégraphie			322
..	..	154	190	116	111	112	98	Récepteurs radio/TV, son et image; art. associés			323
..	..	1 906	1 737	1 748	1 865	1 869	2 000	Instr. médicaux, de précision ; optique ; horlog.		33	
..	..	1 734	1 623	1 522	1 726	1 791	1 900	Instruments médicaux et de précision, etc.			331
..	..	83	64	113	66	60	54	Instruments d'optique ; matériel photographique			332
..	..	88	50	113	74	18	46	Horlogerie			333
..	..	1 409	1 608	1 572	1 646	1 651	1 608	Véhicules automobiles; remorques, etc.		34	
..	..	44	51	55	55	67	70	Véhicules automobiles			341
..	..	610	672	636	640	677	619	Carross. automobiles, remorques, semi-remorques			342
..	..	758	891	883	953	909	923	Pièces et accessoires pour automobiles et moteurs			343
..	..	1 105	1 188	1 048	1 260	1 220	1 133	Autres matériels de transport		35	
..	..	875	954	870	1 093	1 043	955	Construction et réparation de navires			351
..	..	43	41	41	39	40	39	Matériel ferroviaire roulant			352
..	..	50	52	46	45	44	43	Construction aéronautique et spatiale			353
..	..	136	141	91	83	93	95	Autres équipements de transport			359
..	..	15 051	14 527	14 372	16 096	15 921	15 725	Meubles ; activités de fabrication, nca		36	
..	..	11 830	11 302	11 104	12 536	12 573	12 305	Meubles			361
..	..	3 221	3 226	3 268	3 560	3 348	3 420	Activités de fabrication, nca			369
..	..	187	215	133	126	128	104	Récupération		37	
..	..	1 072	969	1 061	1 061	1 130	1 139	Electricité, gaz et eau	E	40_41	
..	..	640	592	646	621	649	691	Prod. et dist. d'électricité gaz, vapeur d'eau chaude		40	
..	..	..	..	..	..	..	..	Construction	F	45	

SWEDEN

Table SN.1 **PRODUCTION - PRODUCTION**

Millions of SKr (Current Prices)

ISIC revision 3				1991	1992	1993	1994	1995	1996	1997	1998
C	10_14		Mining and quarrying	8 973	8 651	8 524	9 022	10 329	..	..	..
	10		Mining of coal and lignite; extraction of peat	457	493	493	594	617	662	..	..
	11		Crude petroleum & natural gas; service activities	..	..	..	..	..	..	..	..
	13		Mining of metal ores	5 828	5 827	5 625	6 080	6 996	6 751	..	..
	14		Other mining and quarrying	2 687	2 331	2 406	2 348	2 716	2 367	..	..
D	15_37		Total manufacturing	672 030	632 671	668 857	791 543	939 359	..	..	..
	15		Food products and beverages	88 989	87 868	91 065	98 816	100 634	102 694	112 522	108 882
		151	Production of meat, fish, fruit, vegetables, oils & fats	34 859	34 682	36 411	38 080	37 935	36 513	42 876	41 242
		152	Dairy products	17 064	16 301	17 241	18 547	18 926	19 103	20 366	21 014
		153	Grain mill and starch products; animal feeds	7 197	7 122	7 296	7 295	7 078	7 622	5 842	6 601
		154	Other food products	20 721	20 376	20 900	23 891	25 483	25 685	31 552	30 072
		155	Beverages	9 149	9 388	9 216	11 004	11 212	11 067	11 886	12 381
	16		Tobacco products	2 169	2 433	2 306	2 479	2 497	..	..	2 428
	17		Textiles	6 938	6 369	6 406	7 005	7 632	7 289	8 735	9 087
		171	Spinning, weaving and finishing of textiles	2 250	1 993	2 027	1 967	2 118	1 872	2 144	1 982
		172	Other textiles	3 900	3 542	3 558	4 155	4 509	4 324	5 318	5 809
		173	Knitted and crocheted fabrics and articles	787	835	820	883	1 005	1 093	1 273	1 296
	18		Wearing apparel and fur	1 706	1 660	1 288	1 610	1 647	1 469	3 180	2 702
		181	Wearing apparel, except fur apparel	1 692	1 660	..	..	..	..	3 070	2 599
		182	Dressing and dyeing of fur; articles of fur	14	..	..	..	..	..	110	103
	19		Leather and articles of leather; footwear	892	836	907	992	1 080	1 008	1 507	1 482
		191	Leather and articles of leather	525	488	565	725	788	710	..	1 303
		192	Footwear	367	348	341	267	292	298	496	515
	20		Wood & products of wood & cork, exc. furniture	39 788	32 420	32 589	39 456	43 706	41 016	59 456	61 928
		201	Sawmilling and planing of wood	19 507	18 774	21 260	26 958	29 627	26 801	38 517	38 433
		202	Wood products; articles of cork, straw, etc.	20 281	13 646	11 329	12 498	14 079	14 215	20 939	23 494
	21		Paper and paper products	67 274	63 721	67 149	79 226	99 045	86 259	91 524	93 782
	22		Printing, publishing and recorded media	39 610	37 939	39 599	43 387	47 364	46 223	60 375	62 703
		221	Publishing	17 455	17 060	18 807	25 691	28 133	26 967	34 340	35 418
		222	Printing and related activities	22 155	20 879	20 792	17 697	19 231	19 127	25 773	26 955
		223	Reproduction of recorded media	..	..	..	..	..	128	262	331
	23		Coke and petroleum products; nuclear fuel	24 254	21 210	26 010	22 757	21 826	26 374	6 118	5 926
		231	Coke oven products	..	..	..	..	..	..	33	22
		232	Refined petroleum products	23 593	20 469	25 186	21 979	21 154	26 374	6 085	5 904
		233	Processing of nuclear fuel	..	..	..	..	..	..	..	..
	24		Chemicals and chemical products	45 286	44 701	51 394	59 283	67 673	65 778	83 324	90 701
		241	Basic chemicals	21 007	19 385	21 316	23 714	28 188	23 725	32 664	33 044
		242	Other chemical products	..	..	..	35 569	39 485	42 054	50 660	57 657
		243	Man-made fibres	..	..	..	..	..	..	380	..
	25		Rubber and plastic products	15 672	14 923	15 277	18 803	21 664	23 115	27 919	29 491
		251	Rubber products	3 560	3 867	3 883	5 260	6 171	6 092	7 725	8 447
		252	Plastic products	12 112	11 056	11 395	13 543	15 493	17 023	20 194	21 045
	26		Other non-metallic mineral products	16 825	14 973	13 632	14 819	16 408	16 234	7 831	7 744
		261	Glass and glass products	3 891	3 616	3 407	3 831	4 138	4 292	5 210	5 210
		269	Non-metallic mineral products, nec	12 935	11 358	10 225	10 987	12 270	..	..	16 097
	27		Basic metals	47 102	43 416	51 325	65 279	80 736	67 848	58 205	53 780
		271	Basic iron and steel	35 354	33 532	39 868	51 570	64 118	53 267	44 788	40 262
		272	Basic precious and non-ferrous metals	11 695	9 853	11 288	13 516	16 339	14 232	12 804	12 811
		273	Casting of metals	54	31	169	193	279	350	614	706
	28		Fabricated metal products, exc. machinery	37 568	32 452	32 388	39 850	49 512	52 689	78 068	86 057
		281	Structural metal prod., tanks, reservoirs, etc.	13 995	11 056	10 783	12 690	14 464	15 250	17 423	19 870
		289	Other fabricated metal products	23 572	21 396	21 605	27 160	35 048	37 438	60 646	66 187
	29		Machinery and equipment, nec	73 476	67 510	68 116	85 169	101 775	101 096	136 175	141 389
		291	General purpose machinery	36 180	34 192	33 652	41 869	50 764	47 442	65 969	66 955
		292	Special purpose machinery	31 062	27 613	28 364	36 265	43 629	47 115	61 131	63 516
		293	Domestic appliances, nec	6 235	5 705	6 100	7 034	7 382	6 538	9 076	10 918
	30		Office, accounting & computing machinery	7 512	6 134	4 649	4 002	3 595	3 612	5 353	6 225
	31		Electrical machinery & apparatus, nec	18 637	16 918	17 575	19 493	23 056	23 135	31 407	27 917
		311	Electric motors, generators and transformers	2 319	2 290	2 617	2 940	4 317	5 666	8 709	5 541
		312	Electricity distribution & control apparatus	5 114	4 737	4 503	5 079	6 046	5 659	8 138	6 461
		313	Insulated wire and cable	4 187	3 509	3 660	5 591	5 945	5 838	6 615	7 065
		314	Accumulators, primary cells & primary batteries	831	861	1 061	1 157	1 457	1 464	1 644	1 639
		315	Electric lamps and lighting equipment	2 361	2 221	2 186	2 166	2 312	2 173	2 952	3 313
		319	Other electrical equipment, nec	3 824	3 300	3 548	2 559	2 979	2 336	3 349	3 899
	32		Radio, TV, communication equipment	23 915	25 008	31 784	45 380	65 708	86 575	115 216	135 133
		321	Electronic valves, tubes; other electronic components	2 089	1 871	2 473	3 001	3 431	3 839	4 119	8 809
		322	TV/radio transmitters; line teleph./telegr. apparatus	17 580	22 024	28 222	40 480	59 657	77 529	105 533	120 572
		323	TV/radio receivers; sound and video app.; assoc. gds.	4 246	1 113	1 090	1 899	2 620	5 207	5 564	5 752
	33		Medical, precision & optical instrum.; watch & clocks	13 756	13 267	16 471	20 560	24 894	24 873	31 925	34 875
		331	Medical appl.; instruments & appl. for measuring, etc.	12 950	12 692	..	19 617	23 796	23 616	30 375	32 319
		332	Optical instruments and photographic equipment	806	575	836	943	1 098	1 257	1 336	2 319
		333	Watches and clocks	..	..	..	..	..	..	214	237
	34		Motor vehicles, trailers and semi-trailers	66 881	66 020	67 574	89 497	118 264	112 215	141 147	154 051
		341	Motor vehicles	47 685	48 152	49 356	65 303	87 729	80 248	117 590	118 469
		342	Bodies for motor vehicles, trailers & semi-trailers	5 797	6 084	5 674	7 553	9 614	9 883	3 063	3 269
		343	Parts & accessories for motor vehicles & their engines	13 399	11 784	12 544	16 641	20 921	22 084	20 494	32 314
	35		Other transport equipment	18 389	17 972	17 125	17 074	20 869	23 164	25 959	26 307
		351	Building and repairing of ships and boats	3 051	2 872	2 522	2 836	3 442	5 273	5 300	5 235
		352	Railway and tramway locomotives and rolling stock	4 431	4 221	4 280	4 484	4 717	4 691	5 425	4 126
		353	Aircraft and spacecraft	10 032	9 847	9 230	8 420	11 250	11 760	13 387	15 007
		359	Transport equipment, nec	875	1 032	1 093	1 334	1 459	1 440	1 848	1 939
	36		Furniture; manufacturing, nec	15 190	14 589	13 808	15 985	18 997	18 276	24 403	28 187
		361	Furniture	12 669	11 860	11 191	13 412	16 140	15 543	19 744	22 973
		369	Manufacturing, nec	2 521	2 729	2 617	2 572	2 857	..	..	..
	37		Recycling	203	329	420	622	775	840	2 327	2 570
E	40_41		Electricity, gas and water supply	..	..	..	..	..	..	..	..
	40		Electricity, gas, steam and hot water supply	..	..	..	..	..	..	..	..
	41		Collection, purification and distribution of water	..	..	..	..	..	..	..	..
F	45		Construction	..	..	..	..	..	..	..	..

Note: Before 1996, ISIC 222 includes 223; 332 includes 333. In 1992, ISIC 14 includes part of 13; 181 includes 182. From 1991 to 1996 ISIC 14 includes 11. In 1994, 1995 and 1996, ISIC 242 includes 243.

Table SN.2 **VALEUR AJOUTÉE - VALUE ADDED**

Millions de SKr (Prix courants)

1991	1992	1993	1994	1995	1996	1997	1998			CITI révision 3	
3 433	3 124	3 587	4 342	4 894	..	..	..	Activites extractives	C	10_14	
193	261	224	279	253	212	..	..	Extraction de charbon, de lignite et de tourbe		10	
..	..	..	..	..	..	..	..	Pétrole brut et gaz naturel ; activités annexes		11	
2 100	1 928	2 432	3 128	3 586	2 755	..	..	Extraction de minerais métalliques		13	
1 140	935	931	934	1 055	972	..	..	Autres activités extractives		14	
225 241	213 175	229 430	268 560	307 953	..	..	..	Total activites manufacturieres	D	15_37	
21 836	22 912	22 268	22 702	22 919	27 483	29 784	27 737	Produits alimentaires et boissons		15	
6 645	7 552	7 425	6 871	7 648	7 479	9 006	8 802	Viande, poisson, fruits, légumes, huiles et graisses			151
2 855	3 082	2 443	2 560	2 974	4 263	3 603	3 968	Produits laitiers			152
1 158	1 094	1 293	1 136	1 190	1 243	1 225	1 192	Travail des grains; prod. amyl. ; alim. pour animaux			153
8 110	8 070	8 281	8 858	7 940	9 150	11 807	11 348	Autres produits alimentaires			154
3 068	3 114	2 827	3 276	3 166	3 489	4 143	4 009	Boissons			155
1 209	1 392	1 290	1 443	1 416	..	..	1 581	Produits à base de tabac		16	
2 728	2 582	2 574	2 840	2 977	2 889	3 338	3 375	Textiles		17	
897	790	779	750	805	673	784	712	Filature, tissage et achèvement des textiles			171
1 523	1 451	1 512	1 743	1 768	1 807	2 133	2 305	Autres articles textiles			172
309	340	283	347	404	410	421	357	Etoffes et articles de bonneterie			173
762	666	548	578	616	607	1 000	1 023	Articles d'habillement et fourrure		18	
755	666	..	..	..	..	973	996	Articles d'habillement autres qu'en fourrure			181
7	..	..	..	..	..	27	27	Prép. et teinture des fourrures ; articles en fourrure			182
355	331	345	349	372	341	494	465	Cuir et articles en cuir ; chaussures		19	
207	193	215	247	269	235	..	421	Cuir et articles en cuir sauf chaussures			191
148	138	130	102	103	106	168	165	Chaussures			192
10 927	8 627	9 570	12 359	12 384	9 591	16 584	16 124	Bois, articles en bois, sauf meubles		20	
4 647	3 969	5 442	7 974	7 553	5 025	9 644	8 340	Sciage et rabotage du bois			201
6 280	4 658	4 128	4 386	4 831	4 566	6 939	7 784	Articles en bois, liège, vannerie, sparterie			202
19 530	17 776	21 034	26 238	38 414	29 435	31 094	31 705	Papier, articles en papier et carton		21	
14 870	14 857	14 918	16 264	17 469	16 348	21 854	22 752	Edition, imprimerie et reproduction		22	
5 204	5 458	5 653	9 149	10 097	8 767	11 663	12 509	Edition			221
9 666	9 399	9 264	7 116	7 372	7 526	10 111	10 149	Imprimerie et activités annexes			222
..	..	..	..	..	55	79	93	Reproduction de supports enregistrés			223
5 844	2 489	3 633	3 110	2 907	3 678	2 128	2 304	Coke. et prod. pétroliers ; comb. nucléaires		23	
..	..	..	..	..	..	7	7	Cokéfaction			231
5 502	2 187	3 225	2 523	2 297	3 678	2 121	2 297	Produits pétroliers raffinés			232
..	..	..	..	..	..	..	..	Traitement de combustibles nucléaires			233
18 678	20 063	23 539	25 932	27 719	24 254	30 070	32 955	Produits chimiques		24	
6 585	6 280	7 209	8 410	10 564	8 136	10 427	10 546	Produits chimiques de base			241
..	..	..	17 522	17 155	16 118	19 642	22 410	Autres produits chimiques			242
..	..	..	..	..	..	99	..	Fibres synthétiques ou artificielles			243
6 425	6 169	6 085	7 577	8 096	8 193	10 452	11 163	Art. en caoutchouc et matières plastiques		25	
1 565	1 578	1 543	2 166	2 321	2 088	3 081	3 467	Articles en caoutchouc			251
4 860	4 591	4 543	5 412	5 775	6 105	7 371	7 697	Articles en matières plastiques			252
7 719	6 733	6 139	6 226	6 982	6 946	3 121	3 234	Autres produits minéraux non métalliques		26	
1 646	1 468	1 500	1 572	1 769	1 658	2 035	2 120	Verre et articles en verre			261
6 073	5 265	4 639	4 654	5 213	..	..	5 772	Produits minéraux non métalliques, nca			269
11 157	10 571	13 019	17 427	21 708	16 442	16 238	15 006	Produits métallurgiques de base		27	
8 721	8 319	10 536	14 525	18 432	13 435	12 999	11 859	Sidérurgie et première transformation de l'acier			271
2 408	2 231	2 396	2 814	3 147	2 838	2 968	2 841	Première transf. métaux précieux et non ferreux			272
27	21	87	88	130	169	271	306	Fonderie			273
16 192	14 039	13 597	16 738	20 647	21 365	32 399	35 857	Ouvrages en métaux, sauf machines		28	
5 364	4 161	3 827	4 538	5 218	5 337	6 257	6 752	Constr. et menuis. métal.; réservoirs; générateurs			281
10 828	9 879	9 770	12 201	15 429	16 028	26 142	29 105	Autres ouvrages en métaux			289
28 905	30 697	27 781	35 242	39 594	37 997	49 965	50 574	Machines et matériels, nca		29	
13 990	16 552	14 034	18 091	20 407	18 122	24 279	23 769	Machines d'usage général			291
12 754	12 045	11 747	14 622	16 752	17 696	22 723	22 764	Machines d'usage spécifique			292
2 161	2 100	2 000	2 529	2 435	2 178	2 963	4 041	Appareils domestiques, nca			293
3 056	2 326	1 887	1 797	1 590	1 548	2 028	2 128	Mach. de bureau, comptables, information		30	
7 357	7 419	7 623	7 619	8 863	8 609	11 549	9 776	Machines et appareils électriques, nca		31	
1 177	1 157	1 213	1 310	1 702	2 092	3 440	1 841	Moteurs, génératrices, transformateurs électriques			311
2 041	2 234	2 320	2 126	2 525	2 016	2 805	2 491	Matériel électrique de distribution et de commande			312
1 443	1 311	1 325	1 748	1 765	1 947	2 042	2 278	Fils et câbles électriques isolés			313
329	404	465	459	636	533	690	643	Accumulateurs et piles électriques			314
889	829	804	859	924	966	1 168	1 208	Lampes électriques et appareils d'éclairage			315
1 477	1 484	1 496	1 117	1 310	1 055	1 404	1 317	Autres matériels électriques, nca.			319
8 248	8 929	11 142	13 001	16 742	20 285	47 533	28 283	Equipements et appareils de radio, TV, etc.		32	
1 252	974	1 307	1 278	2 001	2 236	1 588	2 994	Tubes et valves électroniques;autres composants			321
5 983	7 582	9 477	11 201	14 152	16 789	44 430	23 559	Emetteurs radio/TV; app. téléphonie et télégraphie			322
1 014	373	359	523	590	1 260	1 515	1 730	Récepteurs radio/TV, son et image; art. associés			323
7 039	6 727	8 105	9 680	10 753	11 808	13 496	13 851	Instr. médicaux, de précision ; optique ; horlog.		33	
6 559	6 363	..	9 161	10 159	11 167	12 848	12 750	Instruments médicaux et de précision, etc.			331
479	364	450	519	594	642	574	1 024	Instruments d'optique ; matériel photographique			332
..	..	..	..	..	..	73	78	Horlogerie			333
19 018	14 797	22 082	28 768	31 444	29 612	37 430	51 606	Véhicules automobiles; remorques, etc.		34	
12 057	8 255	15 308	20 520	21 206	19 174	29 074	40 359	Véhicules automobiles			341
2 524	2 403	2 337	2 426	3 268	3 051	1 032	1 080	Carross. automobiles, remorques, semi-remorques			342
4 436	4 139	4 436	5 822	6 970	7 387	7 323	10 168	Pièces et accessoires pour automobiles et moteurs			343
7 741	7 510	7 052	6 784	7 251	7 607	5 232	9 567	Autres matériels de transport		35	
1 181	1 081	1 113	1 130	1 054	1 440	1 599	1 712	Construction et réparation de navires			351
1 906	1 732	1 710	1 897	1 764	1 900	1 318	1 242	Matériel ferroviaire roulant			352
4 265	4 269	3 809	3 249	3 916	3 762	1 722	5 951	Construction aéronautique et spatiale			353
389	428	420	507	517	506	592	661	Autres équipements de transport			359
5 555	5 413	5 031	5 641	6 782	6 586	8 593	9 071	Meubles ; activités de fabrication, nca		36	
4 367	4 223	3 895	4 602	5 442	5 265	6 851	7 164	Meubles			361
1 188	1 190	1 137	1 039	1 341	..	..	..	Activités de fabrication, nca			369
90	152	168	243	306	295	471	469	Récupération		37	
..	..	..	..	..	..	..	..	Electricité, gaz et eau	E	40_41	
..	..	..	..	..	..	..	..	Prod. et dist. d'électricité gaz, vapeur d'eau chaude		40	
..	..	..	..	..	..	..	..	Captage, épuration et distribution d'eau		41	
..	..	..	..	..	..	..	..	Construction	F	45	

Note: Before 1996, ISIC 222 includes 223; 332 includes 333. In 1992, ISIC 14 includes part of 13; 181 includes 182. From 1991 to 1996 ISIC 14 includes 11. In 1994, 1995 and 1996, ISIC 242 includes 243.

SWEDEN

Table SN.3 INVESTMENT - INVESTISSEMENT

Millions of SKr (Current Prices)

ISIC revision 3				1991	1992	1993	1994	1995	1996	1997	1998
C	10_14		Mining and quarrying	..	..	..	..	..	..	..	..
	10		Mining of coal and lignite; extraction of peat	..	..	..	..	..	..	..	..
	11		Crude petroleum & natural gas; service activities	..	..	..	..	..	..	..	..
	13		Mining of metal ores	..	..	..	..	..	..	..	..
	14		Other mining and quarrying	..	..	..	..	..	..	..	..
D	15_37		Total manufacturing	..	..	..	..	..	..	..	..
	15		Food products and beverages	..	..	..	..	..	..	..	5 205
		151	Production of meat, fish, fruit, vegetables, oils & fats	..	..	..	..	..	..	1 724	1 860
		152	Dairy products	..	..	..	..	..	..	627	737
		153	Grain mill and starch products; animal feeds	..	..	..	..	..	..	282	235
		154	Other food products	..	..	..	..	..	..	1 581	1 432
		155	Beverages	..	..	..	..	..	..	656	985
	16		Tobacco products	..	..	..	..	..	..	..	42
	17		Textiles	..	..	..	..	..	..	437	549
		171	Spinning, weaving and finishing of textiles	..	..	..	..	..	..	69	152
		172	Other textiles	..	..	..	..	..	..	299	324
		173	Knitted and crocheted fabrics and articles	..	..	..	..	..	..	68	74
	18		Wearing apparel and fur	..	..	..	..	..	..	88	85
		181	Wearing apparel, except fur apparel	..	..	..	..	..	..	85	82
		182	Dressing and dyeing of fur; articles of fur	..	..	..	..	..	..	2	3
	19		Leather and articles of leather; footwear	..	..	..	..	..	..	50	68
		191	Leather and articles of leather	..	..	..	..	..	..	..	64
		192	Footwear	..	..	..	..	..	..	17	26
	20		Wood & products of wood & cork, exc. furniture	..	..	..	..	..	..	4 218	4 021
		201	Sawmilling and planing of wood	..	..	..	..	..	..	3 029	2 832
		202	Wood products; articles of cork, straw, etc.	..	..	..	..	..	..	1 189	1 188
	21		Paper and paper products	..	..	..	..	..	..	10 284	8 477
	22		Printing, publishing and recorded media	..	..	..	..	..	..	3 005	3 204
		221	Publishing	..	..	..	..	..	..	1 094	1 229
		222	Printing and related activities	..	..	..	..	..	..	1 889	1 963
		223	Reproduction of recorded media	..	..	..	..	..	..	22	12
	23		Coke and petroleum products; nuclear fuel	..	..	..	..	..	..	587	894
		231	Coke oven products	..	..	..	..	..	..	10	2
		232	Refined petroleum products	..	..	..	..	..	..	577	892
		233	Processing of nuclear fuel	..	..	..	..	..	..	..	..
	24		Chemicals and chemical products	..	..	..	..	..	..	8 014	7 707
		241	Basic chemicals	..	..	..	..	..	..	3 167	2 808
		242	Other chemical products	..	..	..	..	..	..	4 847	4 898
		243	Man-made fibres	..	..	..	..	..	..	10	..
	25		Rubber and plastic products	..	..	..	..	..	..	2 316	2 217
		251	Rubber products	..	..	..	..	..	..	441	505
		252	Plastic products	..	..	..	..	..	..	1 874	1 712
	26		Other non-metallic mineral products	..	..	..	..	..	..	262	398
		261	Glass and glass products	..	..	..	..	..	..	159	302
		269	Non-metallic mineral products, nec	..	..	..	..	..	..	..	1 010
	27		Basic metals	..	..	..	..	..	..	3 635	3 738
		271	Basic iron and steel	..	..	..	..	..	..	3 050	3 373
		272	Basic precious and non-ferrous metals	..	..	..	..	..	..	546	301
		273	Casting of metals	..	..	..	..	..	..	37	64
	28		Fabricated metal products, exc. machinery	..	..	..	..	..	..	5 787	6 214
		281	Structural metal prod., tanks, reservoirs, etc.	..	..	..	..	..	..	835	923
		289	Other fabricated metal products	..	..	..	..	..	..	4 952	5 291
	29		Machinery and equipment, nec	..	..	..	..	..	..	5 413	5 254
		291	General purpose machinery	..	..	..	..	..	..	2 632	2 476
		292	Special purpose machinery	..	..	..	..	..	..	2 377	2 331
		293	Domestic appliances, nec	..	..	..	..	..	..	405	447
	30		Office, accounting & computing machinery	..	..	..	..	..	..	229	273
	31		Electrical machinery & apparatus, nec	..	..	..	..	..	..	1 389	1 234
		311	Electric motors, generators and transformers	..	..	..	..	..	..	467	179
		312	Electricity distribution & control apparatus	..	..	..	..	..	..	264	414
		313	Insulated wire and cable	..	..	..	..	..	..	319	265
		314	Accumulators, primary cells & primary batteries	..	..	..	..	..	..	79	65
		315	Electric lamps and lighting equipment	..	..	..	..	..	..	131	133
		319	Other electrical equipment, nec	..	..	..	..	..	..	128	177
	32		Radio, TV, communication equipment	..	..	..	..	..	..	3 081	3 233
		321	Electronic valves, tubes; other electronic components	..	..	..	..	..	..	208	679
		322	TV/radio transmitters; line teleph./telegr. apparatus	..	..	..	..	..	..	2 574	2 204
		323	TV/radio receivers; sound and video app.; assoc. gds.	..	..	..	..	..	..	299	349
	33		Medical, precision & optical instrum.; watch & clocks	..	..	..	..	..	..	1 238	1 266
		331	Medical appl.; instruments & appl. for measuring, etc.	..	..	..	..	..	..	1 146	1 177
		332	Optical instruments and photographic equipment	..	..	..	..	..	..	78	78
		333	Watches and clocks	..	..	..	..	..	..	15	10
	34		Motor vehicles, trailers and semi-trailers	..	..	..	..	..	..	9 223	8 747
		341	Motor vehicles	..	..	..	..	..	..	7 805	6 793
		342	Bodies for motor vehicles, trailers & semi-trailers	..	..	..	..	..	..	119	122
		343	Parts & accessories for motor vehicles & their engines	..	..	..	..	..	..	1 300	1 832
	35		Other transport equipment	..	..	..	..	..	..	1 478	1 212
		351	Building and repairing of ships and boats	..	..	..	..	..	..	276	231
		352	Railway and tramway locomotives and rolling stock	..	..	..	..	..	..	110	251
		353	Aircraft and spacecraft	..	..	..	..	..	..	1 018	644
		359	Transport equipment, nec	..	..	..	..	..	..	75	86
	36		Furniture; manufacturing, nec	..	..	..	..	..	..	1 243	1 231
		361	Furniture	..	..	..	..	..	..	984	1 004
		369	Manufacturing, nec	..	..	..	..	..	..	..	..
	37		Recycling	..	..	..	..	..	..	134	103
E	40_41		Electricity, gas and water supply	..	..	..	..	..	..	..	..
	40		Electricity, gas, steam and hot water supply	..	..	..	..	..	..	..	..
	41		Collection, purification and distribution of water	..	..	..	..	..	..	..	..
F	45		Construction	..	..	..	..	..	..	..	..

Table SN.4 DÉPENSES DE MACHINES ET D'ÉQUIPEMENT - INVESTMENT IN MACHINERY & EQUIPMENT

Millions de SKr (Prix courants)

1991	1992	1993	1994	1995	1996	1997	1998			CITI révision 3	
..	..	..	..	..	..	..	..	Activites extractives	C	10_14	
..	..	..	..	..	..	..	..	Extraction de charbon, de lignite et de tourbe		10	
..	..	..	..	..	..	..	..	Pétrole brut et gaz naturel ; activités annexes		11	
..	..	..	..	..	..	..	..	Extraction de minerais métalliques		13	
..	..	..	..	..	..	..	..	Autres activités extractives		14	
..	..	..	..	..	..	..	..	Total activites manufacturieres	D	15_37	
..	..	..	..	..	..	..	4 103	Produits alimentaires et boissons		15	
..	..	..	..	..	..	1 170	1 444	Viande, poisson, fruits, légumes, huiles et graisses			151
..	..	..	..	..	..	560	667	Produits laitiers			152
..	..	..	..	..	..	201	190	Travail des grains; prod. amyl. ; alim. pour animaux			153
..	..	..	..	..	..	1 246	1 159	Autres produits alimentaires			154
..	..	..	..	..	..	500	681	Boissons			155
..	..	..	..	..	..	..	39	Produits à base de tabac		16	
..	..	..	..	..	..	343	379	Textiles		17	
..	..	..	..	..	..	56	95	Filature, tissage et achèvement des textiles			171
..	..	..	..	..	..	234	235	Autres articles textiles			172
..	..	..	..	..	..	54	49	Etoffes et articles de bonneterie			173
..	..	..	..	..	..	61	58	Articles d'habillement et fourrure		18	
..	..	..	..	..	..	60	57	Articles d'habillement autres qu'en fourrure			181
..	..	..	..	..	..	1	2	Prép. et teinture des fourrures ; articles en fourrure			182
..	..	..	..	..	..	32	40	Cuir et articles en cuir ; chaussures		19	
..	..	..	..	..	..	..	28	Cuir et articles en cuir sauf chaussures			191
..	..	..	..	..	..	12	18	Chaussures			192
..	..	..	..	..	..	3 041	2 519	Bois, articles en bois, sauf meubles		20	
..	..	..	..	..	..	2 193	1 691	Sciage et rabotage du bois			201
..	..	..	..	..	..	847	829	Articles en bois, liège, vannerie, sparterie			202
..	..	..	..	..	..	9 328	7 121	Papier, articles en papier et carton		21	
..	..	..	..	..	..	2 323	2 679	Edition, imprimerie et reproduction		22	
..	..	..	..	..	..	847	1 037	Edition			221
..	..	..	..	..	..	1 469	1 631	Imprimerie et activités annexes			222
..	..	..	..	..	..	7	11	Reproduction de supports enregistrés			223
..	..	..	..	..	..	547	818	Coke. et prod. pétroliers ; comb. nucléaires		23	
..	..	..	..	..	..	6	2	Cokéfaction			231
..	..	..	..	..	..	541	816	Produits pétroliers raffinés			232
..	..	..	..	..	..	..	..	Traitement de combustibles nucléaires			233
..	..	..	..	..	..	5 972	5 818	Produits chimiques		24	
..	..	..	..	..	..	2 717	2 453	Produits chimiques de base			241
..	..	..	..	..	..	3 254	3 365	Autres produits chimiques			242
..	..	..	..	..	..	10	..	Fibres synthétiques ou artificielles			243
..	..	..	..	..	..	1 787	1 726	Art. en caoutchouc et matières plastiques		25	
..	..	..	..	..	..	382	418	Articles en caoutchouc			251
..	..	..	..	..	..	1 405	1 308	Articles en matières plastiques			252
..	..	..	..	..	..	200	279	Autres produits minéraux non métalliques		26	
..	..	..	..	..	..	132	195	Verre et articles en verre			261
..	..	..	..	..	..	..	801	Produits minéraux non métalliques, nca			269
..	..	..	..	..	..	3 113	3 251	Produits métallurgiques de base		27	
..	..	..	..	..	..	2 691	2 938	Sidérurgie et première transformation de l'acier			271
..	..	..	..	..	..	402	272	Première transf. métaux précieux et non ferreux			272
..	..	..	..	..	..	21	42	Fonderie			273
..	..	..	..	..	..	4 314	4 601	Ouvrages en métaux, sauf machines		28	
..	..	..	..	..	..	555	677	Constr. et menuis. métal.; réservoirs; générateurs			281
..	..	..	..	..	..	3 759	3 922	Autres ouvrages en métaux			289
..	..	..	..	..	..	4 014	4 394	Machines et matériels, nca		29	
..	..	..	..	..	..	1 967	2 110	Machines d'usage général			291
..	..	..	..	..	..	1 716	1 854	Machines d'usage spécifique			292
..	..	..	..	..	..	330	429	Appareils domestiques, nca			293
..	..	..	..	..	..	193	214	Mach. de bureau, comptables, information		30	
..	..	..	..	..	..	1 181	1 038	Machines et appareils électriques, nca		31	
..	..	..	..	..	..	419	123	Moteurs, génératrices, transformateurs électriques			311
..	..	..	..	..	..	222	387	Matériel électrique de distribution et de commande			312
..	..	..	..	..	..	274	238	Fils et câbles électriques isolés			313
..	..	..	..	..	..	73	62	Accumulateurs et piles électriques			314
..	..	..	..	..	..	95	95	Lampes électriques et appareils d'éclairage			315
..	..	..	..	..	..	99	133	Autres matériels électriques, nca.			319
..	..	..	..	..	..	2 580	2 923	Equipements et appareils de radio, TV, etc.		32	
..	..	..	..	..	..	175	611	Tubes et valves électroniques;autres composants			321
..	..	..	..	..	..	2 166	2 042	Emetteurs radio/TV; app. téléphonie et télégraphie			322
..	..	..	..	..	..	238	270	Récepteurs radio/TV, son et image; art. associés			323
..	..	..	..	..	..	1 048	1 051	Instr. médicaux, de précision ; optique ; horlog.		33	
..	..	..	..	..	..	984	980	Instruments médicaux et de précision, etc.			331
..	..	..	..	..	..	58	61	Instruments d'optique ; matériel photographique			332
..	..	..	..	..	..	5	10	Horlogerie			333
..	..	..	..	..	..	8 032	7 789	Véhicules automobiles; remorques, etc.		34	
..	..	..	..	..	..	6 903	6 111	Véhicules automobiles			341
..	..	..	..	..	..	68	87	Carross. automobiles, remorques, semi-remorques			342
..	..	..	..	..	..	1 062	1 590	Pièces et accessoires pour automobiles et moteurs			343
..	..	..	..	..	..	972	961	Autres matériels de transport		35	
..	..	..	..	..	..	211	147	Construction et réparation de navires			351
..	..	..	..	..	..	106	201	Matériel ferroviaire roulant			352
..	..	..	..	..	..	604	553	Construction aéronautique et spatiale			353
..	..	..	..	..	..	51	59	Autres équipements de transport			359
..	..	..	..	..	..	835	912	Meubles ; activités de fabrication, nca		36	
..	..	..	..	..	..	631	717	Meubles			361
..	..	..	..	..	..	..	..	Activités de fabrication, nca			369
..	..	..	..	..	..	71	84	Récupération		37	
..	..	..	..	..	..	..	..	Electricité, gaz et eau	E	40_41	
..	..	..	..	..	..	..	..	Prod. et dist. d'électricité gaz, vapeur d'eau chaude		40	
..	..	..	..	..	..	..	..	Captage, épuration et distribution d'eau		41	
..	..	..	..	..	..	..	..	Construction	F	45	

Table SN.5 **EMPLOYMENT, EMPLOYEES - EMPLOI, SALARIÉS**

Units

ISIC revision 3				1991	1992	1993	1994	1995	1996	1997	1998
C	10_14		Mining and quarrying	9 094	8 110	7 163	7 054	7 127	..	..	..
	10		Mining of coal and lignite; extraction of peat	431	442	428	520	500	449	..	..
	11		Crude petroleum & natural gas; service activities	..	..	..	..	..	..	..	..
	13		Mining of metal ores	6 167	5 541	4 779	4 803	4 798	4 794	..	..
	14		Other mining and quarrying	2 496	2 127	1 956	1 731	1 829	1 900	..	..
D	15_37		Total manufacturing	703 045	632 068	579 246	593 413	628 168	..	..	..
	15		Food products and beverages	66 405	61 874	59 795	58 314	58 254	58 341	..	67 523
		151	Production of meat, fish, fruit, vegetables, oils & fats	24 264	22 593	22 725	21 653	22 049	20 706	23 027	23 463
		152	Dairy products	9 930	8 908	8 501	8 608	8 678	8 859	9 960	10 061
		153	Grain mill and starch products; animal feeds	2 224	2 116	2 081	1 792	1 792	1 889	2 037	2 293
		154	Other food products	25 128	23 568	22 140	22 115	21 225	21 742	26 048	26 167
		155	Beverages	4 859	4 689	4 348	4 146	4 510	4 342	6 499	6 658
	16		Tobacco products	1 077	988	886	792	778	..	..	1 119
	17		Textiles	11 442	9 873	8 690	8 506	8 495	8 143	9 180	10 336
		171	Spinning, weaving and finishing of textiles	3 530	3 022	2 701	2 361	2 444	2 244	2 384	2 018
		172	Other textiles	6 310	5 354	4 807	5 031	4 895	4 731	5 527	7 189
		173	Knitted and crocheted fabrics and articles	1 602	1 497	1 182	1 114	1 156	1 168	1 269	1 129
	18		Wearing apparel and fur	4 219	3 265	2 187	2 140	2 070	2 019	5 471	5 115
		181	Wearing apparel, except fur apparel	4 176	3 265	..	..	..	..	5 368	5 005
		182	Dressing and dyeing of fur; articles of fur	43	..	..	..	..	..	103	110
	19		Leather and articles of leather; footwear	1 456	1 303	1 164	1 127	1 157	1 161	1 551	1 418
		191	Leather and articles of leather	828	717	668	723	771	770	..	1 334
		192	Footwear	628	586	496	404	386	391	586	525
	20		Wood & products of wood & cork, exc. furniture	37 209	31 819	27 420	28 416	29 601	29 175	38 062	39 635
		201	Sawmilling and planing of wood	14 868	14 261	13 566	14 473	15 052	14 751	17 379	17 863
		202	Wood products; articles of cork, straw, etc.	22 341	17 558	13 854	13 943	14 549	14 424	20 683	21 772
	21		Paper and paper products	49 843	45 681	42 889	42 612	43 811	43 278	45 891	45 049
	22		Printing, publishing and recorded media	51 628	44 877	42 979	45 287	45 424	45 759	54 847	54 793
		221	Publishing	15 801	12 952	13 445	25 013	26 435	26 685	28 768	28 650
		222	Printing and related activities	35 827	31 925	29 534	20 274	18 989	18 949	25 847	25 897
		223	Reproduction of recorded media	..	..	..	..	..	125	232	246
	23		Coke and petroleum products; nuclear fuel	2 405	2 307	2 400	2 814	2 845	2 943	2 797	2 973
		231	Coke oven products	..	..	..	..	..	..	9	14
		232	Refined petroleum products	1 861	1 755	1 804	1 777	1 757	2 943	2 788	2 959
		233	Processing of nuclear fuel	..	..	..	..	..	..	..	..
	24		Chemicals and chemical products	32 986	30 943	28 968	31 384	33 053	35 573	38 570	40 013
		241	Basic chemicals	13 614	12 383	11 581	10 820	11 514	10 998	14 026	13 973
		242	Other chemical products	..	..	..	20 564	21 539	24 575	24 544	26 040
		243	Man-made fibres	..	..	..	..	..	..	187	..
	25		Rubber and plastic products	21 830	19 934	17 952	19 635	20 730	22 091	24 397	25 180
		251	Rubber products	6 373	5 755	5 028	5 722	6 068	6 144	7 207	7 567
		252	Plastic products	15 457	14 179	12 924	13 913	14 662	15 947	17 190	17 613
	26		Other non-metallic mineral products	22 419	19 228	16 180	15 611	16 340	16 745	7 912	7 692
		261	Glass and glass products	5 882	5 138	4 292	4 219	4 537	4 568	5 271	5 250
		269	Non-metallic mineral products, nec	16 537	14 090	11 888	11 392	11 803	..	..	13 110
	27		Basic metals	37 745	35 167	32 792	31 818	32 953	32 065	28 512	28 229
		271	Basic iron and steel	30 459	28 472	26 459	25 621	26 544	25 853	22 486	22 090
		272	Basic precious and non-ferrous metals	7 189	6 617	6 016	5 942	6 073	5 817	5 421	5 434
		273	Casting of metals	97	78	317	255	336	395	605	705
	28		Fabricated metal products, exc. machinery	54 925	46 861	42 543	45 542	51 465	54 448	94 982	96 584
		281	Structural metal prod., tanks, reservoirs, etc.	17 791	14 370	12 755	13 179	14 142	14 351	16 772	17 791
		289	Other fabricated metal products	37 134	32 491	29 788	32 363	37 323	40 097	78 210	78 793
	29		Machinery and equipment, nec	102 752	92 249	80 582	83 917	89 480	89 156	104 302	107 579
		291	General purpose machinery	49 651	48 221	39 956	43 665	46 257	41 966	50 247	51 294
		292	Special purpose machinery	43 608	36 310	33 422	32 924	35 963	40 620	46 111	45 739
		293	Domestic appliances, nec	9 493	7 718	7 204	7 328	7 260	6 570	7 944	10 546
	30		Office, accounting & computing machinery	5 972	5 318	4 874	4 314	3 624	3 304	4 092	3 764
	31		Electrical machinery & apparatus, nec	24 807	22 958	19 619	18 569	21 042	21 702	26 351	22 638
		311	Electric motors, generators and transformers	3 764	3 268	3 075	3 344	4 756	5 903	7 629	4 583
		312	Electricity distribution & control apparatus	7 364	6 595	5 592	5 166	5 575	5 377	6 565	5 533
		313	Insulated wire and cable	4 397	3 722	3 247	3 313	3 710	3 860	4 120	4 396
		314	Accumulators, primary cells & primary batteries	1 452	1 428	1 270	1 293	1 392	1 412	1 504	1 457
		315	Electric lamps and lighting equipment	3 706	3 678	2 900	2 763	2 834	2 725	3 177	3 237
		319	Other electrical equipment, nec	4 124	4 267	3 535	2 690	2 775	2 425	3 356	3 432
	32		Radio, TV, communication equipment	31 751	28 812	28 959	30 555	36 072	39 431	40 804	45 740
		321	Electronic valves, tubes; other electronic components	4 622	3 781	3 963	3 836	4 850	4 905	4 336	7 504
		322	TV/radio transmitters; line teleph./telegr. apparatus	21 024	23 148	23 883	25 102	29 284	30 863	33 405	34 447
		323	TV/radio receivers; sound and video app.; assoc. gds.	6 105	1 883	1 113	1 617	1 938	3 663	3 063	3 789
	33		Medical, precision & optical instrum.; watch & clocks	20 672	17 797	19 557	20 851	22 527	22 967	24 605	25 185
		331	Medical appl.; instruments & appl. for measuring, etc.	19 080	16 615	..	19 533	21 144	21 526	23 109	22 873
		332	Optical instruments and photographic equipment	1 592	1 182	1 192	1 318	1 383	1 441	1 307	2 143
		333	Watches and clocks	..	..	..	..	..	..	189	169
	34		Motor vehicles, trailers and semi-trailers	72 867	67 441	59 331	61 402	66 778	68 077	68 765	70 545
		341	Motor vehicles	45 945	43 826	37 556	38 590	40 988	40 950	48 523	45 433
		342	Bodies for motor vehicles, trailers & semi-trailers	9 169	8 519	7 303	7 545	8 470	8 136	2 675	2 947
		343	Parts & accessories for motor vehicles & their engines	17 753	15 096	14 472	15 267	17 320	18 991	17 567	22 165
	35		Other transport equipment	27 032	23 544	21 792	20 945	21 394	21 989	20 971	21 249
		351	Building and repairing of ships and boats	4 758	4 339	3 523	3 326	3 600	3 679	4 547	4 861
		352	Railway and tramway locomotives and rolling stock	6 326	5 193	4 890	4 724	4 812	4 950	3 501	3 341
		353	Aircraft and spacecraft	14 480	12 558	12 138	11 649	11 635	12 154	11 523	11 569
		359	Transport equipment, nec	1 468	1 454	1 241	1 246	1 347	1 206	1 400	1 478
	36		Furniture; manufacturing, nec	21 421	19 503	17 439	18 446	19 834	18 267	29 220	30 684
		361	Furniture	17 444	15 693	13 899	15 280	16 480	15 178	23 225	24 688
		369	Manufacturing, nec	3 977	3 810	3 540	3 166	3 354	..	..	..
	37		Recycling	182	326	248	416	441	485	697	736
E	40_41		Electricity, gas and water supply	..	..	..	..	..	..	..	..
	40		Electricity, gas, steam and hot water supply	..	..	..	..	..	..	..	..
	41		Collection, purification and distribution of water	..	..	..	..	..	..	..	..
F	45		Construction	..	..	..	..	..	..	..	..

Note: Before 1996, ISIC 222 includes 223; 332 includes 333. In 1992, ISIC 14 includes part of 13;
181 includes 182. From 1991 to 1996 ISIC 14 includes 11. In 1994, 1995 and 1996, ISIC 242 includes 243.

Table SN.6 **SALAIRES ET CHARGES SOCIALES, SALARIÉS - COMPENSATION OF LABOUR, EMPLOYEES**

Millions de SKr (Prix courants)

1991	1992	1993	1994	1995	1996	1997	1998		CITI révision 3		
..	..	..	..	..	..	..	..	Activites extractives	C	**10_14**	
..	..	..	..	..	..	..	..	Extraction de charbon, de lignite et de tourbe		**10**	
..	..	..	..	..	..	..	..	Pétrole brut et gaz naturel ; activités annexes		**11**	
..	..	..	..	..	..	..	..	Extraction de minerais métalliques		**13**	
..	..	..	..	..	..	..	..	Autres activités extractives		**14**	
..	..	..	..	..	..	..	..	Total activites manufacturieres	D	**15_37**	
..	..	..	..	..	..	..	19 585	Produits alimentaires et boissons		**15**	
..	..	..	..	..	..	6 512	6 660	Viande, poisson, fruits, légumes, huiles et graisses			151
..	..	..	..	..	..	2 734	3 150	Produits laitiers			152
..	..	..	..	..	..	629	677	Travail des grains; prod. amyl. ; alim. pour animaux			153
..	..	..	..	..	..	7 412	7 287	Autres produits alimentaires			154
..	..	..	..	..	..	2 113	2 240	Boissons			155
..	..	..	..	..	..	..	429	Produits à base de tabac		**16**	
..	..	..	..	..	..	2 320	2 613	Textiles		**17**	
..	..	..	..	..	..	580	524	Filature, tissage et achèvement des textiles			171
..	..	..	..	..	..	1 450	1 822	Autres articles textiles			172
..	..	..	..	..	..	289	267	Etoffes et articles de bonneterie			173
..	..	..	..	..	..	1 087	1 097	Articles d'habillement et fourrure		**18**	
..	..	..	..	..	..	1 069	1 080	Articles d'habillement autres qu'en fourrure			181
..	..	..	..	..	..	18	17	Prép. et teinture des fourrures ; articles en fourrure			182
..	..	..	..	..	..	358	348	Cuir et articles en cuir ; chaussures		**19**	
..	..	..	..	..	..	..	315	Cuir et articles en cuir sauf chaussures			191
..	..	..	..	..	..	128	127	Chaussures			192
..	..	..	..	..	..	10 327	11 215	Bois, articles en bois, sauf meubles		**20**	
..	..	..	..	..	..	5 015	5 286	Sciage et rabotage du bois			201
..	..	..	..	..	..	5 313	5 929	Articles en bois, liège, vannerie, sparterie			202
..	..	..	..	..	..	16 207	16 087	Papier, articles en papier et carton		**21**	
..	..	..	..	..	..	16 868	17 243	Edition, imprimerie et reproduction		**22**	
..	..	..	..	..	..	9 228	9 551	Edition			221
..	..	..	..	..	..	7 581	7 623	Imprimerie et activités annexes			222
..	..	..	..	..	..	59	69	Reproduction de supports enregistrés			223
..	..	..	..	..	..	1 095	1 217	Coke. et prod. pétroliers ; comb. nucléaires		**23**	
..	..	..	..	..	..	2	3	Cokéfaction			231
..	..	..	..	..	..	1 092	1 213	Produits pétroliers raffinés			232
..	..	..	..	..	..	..	..	Traitement de combustibles nucléaires			233
..	..	..	..	..	..	15 423	16 690	Produits chimiques		**24**	
..	..	..	..	..	..	5 393	5 705	Produits chimiques de base			241
..	..	..	..	..	..	10 030	10 986	Autres produits chimiques			242
..	..	..	..	..	..	61	..	Fibres synthétiques ou artificielles			243
..	..	..	..	..	..	6 951	7 549	Art. en caoutchouc et matières plastiques		**25**	
..	..	..	..	..	..	2 041	2 373	Articles en caoutchouc			251
..	..	..	..	..	..	4 910	5 177	Articles en matières plastiques			252
..	..	..	..	..	..	2 308	2 328	Autres produits minéraux non métalliques		**26**	
..	..	..	..	..	..	1 541	1 585	Verre et articles en verre			261
..	..	..	..	..	..	..	3 933	Produits minéraux non métalliques, nca			269
..	..	..	..	..	..	9 776	9 924	Produits métallurgiques de base		**27**	
..	..	..	..	..	..	7 784	7 810	Sidérurgie et première transformation de l'acier			271
..	..	..	..	..	..	1 826	1 916	Première transf. métaux précieux et non ferreux			272
..	..	..	..	..	..	166	199	Fonderie			273
..	..	..	..	..	..	25 314	27 666	Ouvrages en métaux, sauf machines		**28**	
..	..	..	..	..	..	4 713	5 219	Constr. et menuis. métal.; réservoirs; générateurs			281
..	..	..	..	..	..	20 602	22 449	Autres ouvrages en métaux			289
..	..	..	..	..	..	34 635	36 398	Machines et matériels, nca		**29**	
..	..	..	..	..	..	17 081	17 586	Machines d'usage général			291
..	..	..	..	..	..	14 910	15 345	Machines d'usage spécifique			292
..	..	..	..	..	..	2 645	3 467	Appareils domestiques, nca			293
..	..	..	..	..	..	1 381	1 429	Mach. de bureau, comptables, information		**30**	
..	..	..	..	..	..	8 727	7 262	Machines et appareils électriques, nca		**31**	
..	..	..	..	..	..	2 653	1 521	Moteurs, génératrices, transformateurs électriques			311
..	..	..	..	..	..	2 308	1 763	Matériel électrique de distribution et de commande			312
..	..	..	..	..	..	1 405	1 508	Fils et câbles électriques isolés			313
..	..	..	..	..	..	478	504	Accumulateurs et piles électriques			314
..	..	..	..	..	..	852	918	Lampes électriques et appareils d'éclairage			315
..	..	..	..	..	..	1 031	1 048	Autres matériels électriques, nca.			319
..	..	..	..	..	..	16 091	19 296	Equipements et appareils de radio, TV, etc.		**32**	
..	..	..	..	..	..	1 205	2 569	Tubes et valves électroniques;autres composants			321
..	..	..	..	..	..	13 886	15 608	Emetteurs radio/TV; app. téléphonie et télégraphie			322
..	..	..	..	..	..	1 000	1 119	Récepteurs radio/TV, son et image; art. associés			323
..	..	..	..	..	..	8 761	9 636	Instr. médicaux, de précision ; optique ; horlog.		**33**	
..	..	..	..	..	..	8 287	8 700	Instruments médicaux et de précision, etc.			331
..	..	..	..	..	..	420	887	Instruments d'optique ; matériel photographique			332
..	..	..	..	..	..	54	50	Horlogerie			333
..	..	..	..	..	..	23 950	22 521	Véhicules automobiles; remorques, etc.		**34**	
..	..	..	..	..	..	18 031	14 267	Véhicules automobiles			341
..	..	..	..	..	..	742	819	Carross. automobiles, remorques, semi-remorques			342
..	..	..	..	..	..	5 177	7 435	Pièces et accessoires pour automobiles et moteurs			343
..	..	..	..	..	..	7 206	7 467	Autres matériels de transport		**35**	
..	..	..	..	..	..	1 274	1 456	Construction et réparation de navires			351
..	..	..	..	..	..	1 207	1 243	Matériel ferroviaire roulant			352
..	..	..	..	..	..	4 329	4 318	Construction aéronautique et spatiale			353
..	..	..	..	..	..	395	451	Autres équipements de transport			359
..	..	..	..	..	..	6 970	7 981	Meubles ; activités de fabrication, nca		**36**	
..	..	..	..	..	..	5 525	6 453	Meubles			361
..	..	..	..	..	..	..	..	Activités de fabrication, nca			369
..	..	..	..	..	..	195	210	Récupération		**37**	
..	..	..	..	..	..	..	..	Electricité, gaz et eau	E	**40_41**	
..	..	..	..	..	..	..	..	Prod. et dist. d'électricité gaz, vapeur d'eau chaude		**40**	
..	..	..	..	..	..	..	..	Captage, épuration et distribution d'eau		**41**	
..	..	..	..	..	..	..	..	Construction	F	**45**	

SWEDEN

Table SN.7 WAGES AND SALARIES, EMPLOYEES - SALAIRES ET TRAITEMENTS, SALARIÉS

Millions of SKr (Current Prices)

ISIC revision 3				1991	1992	1993	1994	1995	1996	1997	1998
C	10_14		Mining and quarrying	1 683	1 552	1 451	1 495	1 617	..	..	..
	10		Mining of coal and lignite; extraction of peat	70	77	76	102	93	90	..	..
	11		Crude petroleum & natural gas; service activities	..	..	..	..	..	..	..	..
	13		Mining of metal ores	1 168	1 090	1 016	1 059	1 148	1 156	..	..
	14		Other mining and quarrying	444	385	359	335	376	401	..	..
D	15_37		Total manufacturing	114 490	112 207	107 851	115 665	130 100	..	..	..
	15		Food products and beverages	9 675	9 933	9 974	10 276	10 849	11 688	..	13 725
		151	Production of meat, fish, fruit, vegetables, oils & fats	3 364	3 499	3 689	3 752	4 024	4 073	4 610	4 708
		152	Dairy products	1 369	1 398	1 326	1 437	1 553	1 701	1 831	2 170
		153	Grain mill and starch products; animal feeds	366	371	384	343	378	413	439	476
		154	Other food products	3 733	3 766	3 707	3 902	3 981	4 339	5 195	5 144
		155	Beverages	843	898	868	842	914	963	1 457	1 546
	16		Tobacco products	168	181	197	175	181	..	..	320
	17		Textiles	1 521	1 438	1 349	1 367	1 462	1 489	1 627	1 837
		171	Spinning, weaving and finishing of textiles	497	465	430	381	419	394	407	371
		172	Other textiles	846	790	761	824	859	900	1 016	1 277
		173	Knitted and crocheted fabrics and articles	179	183	158	161	184	195	205	189
	18		Wearing apparel and fur	479	420	301	296	320	331	764	782
		181	Wearing apparel, except fur apparel	474	420	..	..	..	..	752	770
		182	Dressing and dyeing of fur; articles of fur	4	..	..	..	..	..	12	12
	19		Leather and articles of leather; footwear	187	176	166	167	192	196	255	248
		191	Leather and articles of leather	108	99	97	110	130	131	..	224
		192	Footwear	79	78	69	57	62	64	92	91
	20		Wood & products of wood & cork, exc. furniture	5 669	5 079	4 618	4 992	5 522	5 799	7 293	7 943
		201	Sawmilling and planing of wood	2 277	2 306	2 322	2 605	2 849	2 971	3 514	3 720
		202	Wood products; articles of cork, straw, etc.	3 393	2 773	2 296	2 387	2 673	2 828	3 779	4 223
	21		Paper and paper products	8 773	8 696	8 430	8 828	9 515	10 397	11 218	11 155
	22		Printing, publishing and recorded media	8 738	8 331	8 361	8 745	9 379	9 853	11 831	12 122
		221	Publishing	2 928	2 805	2 956	4 819	5 397	5 594	6 442	6 668
		222	Printing and related activities	5 810	5 526	5 404	3 925	3 981	4 231	5 347	5 403
		223	Reproduction of recorded media	..	..	..	..	..	29	42	51
	23		Coke and petroleum products; nuclear fuel	515	515	568	687	746	809	748	827
		231	Coke oven products	..	..	..	..	..	..	2	2
		232	Refined petroleum products	403	404	439	425	448	809	746	825
		233	Processing of nuclear fuel	..	..	..	..	..	..	..	..
	24		Chemicals and chemical products	6 165	6 201	6 185	7 100	7 783	9 008	10 444	11 278
		241	Basic chemicals	2 543	2 465	2 453	2 424	2 705	2 779	3 656	3 870
		242	Other chemical products	..	..	..	4 676	5 079	6 229	6 788	7 408
		243	Man-made fibres	..	..	..	..	..	..	44	..
	25		Rubber and plastic products	3 320	3 245	3 066	3 540	3 958	4 525	4 867	5 335
		251	Rubber products	961	898	843	1 015	1 137	1 219	1 396	1 683
		252	Plastic products	2 359	2 347	2 223	2 525	2 821	3 305	3 470	3 652
	26		Other non-metallic mineral products	3 562	3 357	2 942	2 972	3 240	3 438	1 607	1 644
		261	Glass and glass products	939	869	769	807	884	955	1 071	1 122
		269	Non-metallic mineral products, nec	2 623	2 487	2 173	2 165	2 356	..	..	2 752
	27		Basic metals	6 246	6 348	6 181	6 551	7 182	7 410	6 779	6 914
		271	Basic iron and steel	5 080	5 225	5 053	5 332	5 836	6 031	5 383	5 419
		272	Basic precious and non-ferrous metals	1 150	1 110	1 076	1 174	1 282	1 300	1 277	1 355
		273	Casting of metals	15	14	52	45	63	79	119	140
	28		Fabricated metal products, exc. machinery	8 559	7 749	7 261	8 240	9 935	11 203	17 762	19 467
		281	Structural metal prod., tanks, reservoirs, etc.	2 888	2 436	2 226	2 426	2 725	2 919	3 335	3 748
		289	Other fabricated metal products	5 671	5 313	5 036	5 814	7 210	8 284	14 427	15 718
	29		Machinery and equipment, nec	16 962	17 113	15 269	16 644	19 054	20 185	23 988	25 192
		291	General purpose machinery	8 278	9 209	7 717	8 724	9 930	9 700	11 885	12 157
		292	Special purpose machinery	7 412	6 737	6 416	6 677	7 815	9 259	10 296	10 617
		293	Domestic appliances, nec	1 273	1 167	1 135	1 243	1 309	1 226	1 806	2 418
	30		Office, accounting & computing machinery	1 118	1 122	1 055	977	821	766	956	988
	31		Electrical machinery & apparatus, nec	4 071	4 023	3 753	3 588	4 356	4 735	5 967	5 080
		311	Electric motors, generators and transformers	643	572	590	669	991	1 220	1 786	1 055
		312	Electricity distribution & control apparatus	1 234	1 224	1 080	1 005	1 213	1 314	1 553	1 236
		313	Insulated wire and cable	714	652	647	675	767	849	968	1 043
		314	Accumulators, primary cells & primary batteries	249	250	239	234	291	305	341	358
		315	Electric lamps and lighting equipment	520	525	463	472	515	523	592	649
		319	Other electrical equipment, nec	711	800	734	533	579	524	727	739
	32		Radio, TV, communication equipment	5 478	5 269	5 643	6 362	7 784	9 142	10 441	12 417
		321	Electronic valves, tubes; other electronic components	708	707	728	716	995	1 147	851	1 762
		322	TV/radio transmitters; line teleph./telegr. apparatus	3 807	4 311	4 710	5 341	6 379	7 256	8 891	9 865
		323	TV/radio receivers; sound and video app.; assoc. gds.	963	251	205	305	410	739	699	790
	33		Medical, precision & optical instrum.; watch & clocks	3 773	3 598	4 057	4 655	5 201	5 596	6 067	6 668
		331	Medical appl.; instruments & appl. for measuring, etc.	3 513	3 412	..	4 387	4 907	5 253	5 740	6 030
		332	Optical instruments and photographic equipment	260	186	218	268	294	343	288	603
		333	Watches and clocks	..	..	..	..	..	..	38	35
	34		Motor vehicles, trailers and semi-trailers	11 824	11 971	11 260	12 027	14 323	15 693	16 284	15 338
		341	Motor vehicles	7 741	8 087	7 526	7 817	9 165	9 959	12 132	9 593
		342	Bodies for motor vehicles, trailers & semi-trailers	1 402	1 420	1 298	1 422	1 748	1 865	526	592
		343	Parts & accessories for motor vehicles & their engines	2 682	2 464	2 436	2 788	3 410	3 870	3 627	5 153
	35		Other transport equipment	4 677	4 450	4 372	4 369	4 678	5 035	4 933	5 191
		351	Building and repairing of ships and boats	798	794	663	649	749	801	904	1 026
		352	Railway and tramway locomotives and rolling stock	1 068	915	900	932	1 026	1 131	844	862
		353	Aircraft and spacecraft	2 598	2 519	2 588	2 552	2 641	2 851	2 906	2 986
		359	Transport equipment, nec	212	223	221	236	262	251	279	316
	36		Furniture; manufacturing, nec	2 973	2 931	2 792	3 024	3 524	3 479	4 934	5 656
		361	Furniture	2 398	2 346	2 203	2 485	2 909	2 872	3 916	4 574
		369	Manufacturing, nec	575	585	588	539	615	..	..	..
	37		Recycling	35	62	52	84	95	109	137	149
E	40_41		Electricity, gas and water supply	..	..	..	..	..	..	..	..
	40		Electricity, gas, steam and hot water supply	..	..	..	..	..	..	..	..
	41		Collection, purification and distribution of water	..	..	..	..	..	..	..	..
F	45		Construction	..	..	..	..	..	..	..	..

Note: Before 1996, ISIC 222 includes 223; 332 includes 333. In 1992, ISIC 14 includes part of 13; 181 includes 182. From 1991 to 1996 ISIC 14 includes 11. In 1994, 1995 and 1996, ISIC 242 includes 243.

Table SN.8 **CHARGES SOCIALES DES EMPLOYEURS - SUPPLEMENTS TO WAGES & SALARIES**

Millions de SKr (Prix courants)

1991	1992	1993	1994	1995	1996	1997	1998			CITI révision 3
797	696	586	620	682	..	..	..	Activites extractives	C	10_14
30	29	26	35	37	34	..	..	Extraction de charbon, de lignite et de tourbe		10
..	..	..	..	..	..	..	..	Pétrole brut et gaz naturel ; activités annexes		11
554	488	410	444	486	497	..	..	Extraction de minerais métalliques		13
213	179	149	141	159	177	..	..	Autres activités extractives		14
54 351	50 740	45 079	47 714	55 531	..	..	..	Total activites manufacturieres	D	15_37
4 507	4 389	4 158	4 226	4 590	4 905	..	5 860	Produits alimentaires et boissons		15
1 563	1 537	1 482	1 501	1 648	1 705	1 901	1 952	Viande, poisson, fruits, légumes, huiles et graisses		151
655	598	580	633	688	722	903	980	Produits laitiers		152
167	164	158	141	157	175	190	200	Travail des grains; prod. amyl. ; alim. pour animaux		153
1 749	1 698	1 554	1 609	1 717	1 799	2 217	2 143	Autres produits alimentaires		154
373	391	384	343	380	424	655	694	Boissons		155
83	78	75	66	70	..	..	109	Produits à base de tabac		16
714	625	550	557	606	631	693	776	Textiles		17
238	200	166	145	165	157	173	153	Filature, tissage et achèvement des textiles		171
396	347	322	348	366	395	435	545	Autres articles textiles		172
81	78	61	64	74	79	85	78	Etoffes et articles de bonneterie		173
217	181	119	122	133	134	322	315	Articles d'habillement et fourrure		18
215	181	..	..	..	..	317	310	Articles d'habillement autres qu'en fourrure		181
2	..	..	..	..	..	5	5	Prép. et teinture des fourrures ; articles en fourrure		182
84	74	65	64	76	81	103	100	Cuir et articles en cuir ; chaussures		19
48	42	38	43	52	56	..	91	Cuir et articles en cuir sauf chaussures		191
36	32	27	22	24	25	36	36	Chaussures		192
2 538	2 184	1 824	1 944	2 241	2 340	3 034	3 272	Bois, articles en bois, sauf meubles		20
1 034	1 002	924	1 025	1 177	1 204	1 500	1 566	Sciage et rabotage du bois		201
1 504	1 182	899	919	1 064	1 136	1 533	1 706	Articles en bois, liège, vannerie, sparterie		202
4 172	3 917	3 606	3 703	4 116	4 443	4 988	4 932	Papier, articles en papier et carton		21
3 976	3 599	3 255	3 392	3 827	4 191	5 037	5 121	Edition, imprimerie et reproduction		22
1 359	1 271	1 170	1 871	2 223	2 449	2 786	2 883	Edition		221
2 617	2 328	2 084	1 521	1 604	1 730	2 234	2 221	Imprimerie et activités annexes		222
..	..	..	..	..	12	17	18	Reproduction de supports enregistrés		223
244	232	216	279	317	351	347	389	Coke. et prod. pétroliers ; comb. nucléaires		23
..	..	..	..	..	..	1	1	Cokéfaction		231
190	179	166	169	192	351	346	388	Produits pétroliers raffinés		232
..	..	..	..	..	..	..	..	Traitement de combustibles nucléaires		233
3 108	2 927	2 674	3 144	3 596	4 352	4 979	5 413	Produits chimiques		24
1 242	1 113	1 034	1 001	1 176	1 202	1 737	1 835	Produits chimiques de base		241
..	..	..	2 144	2 420	3 150	3 243	3 578	Autres produits chimiques		242
..	..	..	..	..	..	18	..	Fibres synthétiques ou artificielles		243
1 579	1 481	1 308	1 498	1 701	1 931	2 084	2 214	Art. en caoutchouc et matières plastiques		25
452	449	396	466	520	561	645	690	Articles en caoutchouc		251
1 127	1 032	912	1 032	1 181	1 370	1 439	1 524	Articles en matières plastiques		252
1 720	1 523	1 248	1 198	1 367	1 471	700	683	Autres produits minéraux non métalliques		26
427	369	296	292	344	396	470	463	Verre et articles en verre		261
1 292	1 155	952	906	1 023	..	..	1 182	Produits minéraux non métalliques, nca		269
2 996	2 832	2 503	2 676	3 043	3 196	2 997	3 010	Produits métallurgiques de base		27
2 443	2 320	2 054	2 199	2 500	2 628	2 401	2 390	Sidérurgie et première transformation de l'acier		271
547	506	429	459	516	535	549	561	Première transf. métaux précieux et non ferreux		272
7	6	20	18	27	33	47	59	Fonderie		273
4 020	3 475	2 975	3 280	4 119	4 687	7 552	8 199	Ouvrages en métaux, sauf machines		28
1 275	1 032	872	951	1 124	1 182	1 377	1 470	Constr. et menuis. métal.; réservoirs; générateurs		281
2 745	2 443	2 103	2 329	2 995	3 505	6 174	6 729	Autres ouvrages en métaux		289
8 157	7 733	6 473	6 959	8 165	8 650	10 647	11 206	Machines et matériels, nca		29
4 054	4 119	3 223	3 624	4 255	4 101	5 195	5 428	Machines d'usage général		291
3 479	3 063	2 772	2 862	3 368	4 045	4 613	4 729	Machines d'usage spécifique		292
623	551	478	473	542	504	839	1 049	Appareils domestiques, nca		293
573	536	420	397	337	323	425	440	Mach. de bureau, comptables, information		30
1 923	1 823	1 602	1 495	1 880	2 009	2 760	2 182	Machines et appareils électriques, nca		31
296	240	227	257	399	526	867	466	Moteurs, générateurs, transformateurs électriques		311
646	636	550	490	585	558	756	527	Matériel électrique de distribution et de commande		312
315	282	248	254	314	362	437	465	Fils et câbles électriques isolés		313
108	103	97	85	132	128	137	146	Accumulateurs et piles électriques		314
227	220	192	196	215	227	260	270	Lampes électriques et appareils d'éclairage		315
330	342	288	212	235	208	304	309	Autres matériels électriques, nca.		319
2 421	2 283	2 309	2 601	3 318	4 203	5 650	6 879	Equipements et appareils de radio, TV, etc.		32
350	309	300	299	422	487	354	807	Tubes et valves électroniques;autres composants		321
1 612	1 867	1 930	2 189	2 729	3 404	4 994	5 743	Emetteurs radio/TV; app. téléphonie et télégraphie		322
460	107	78	113	166	312	301	329	Récepteurs radio/TV, son et image; art. associés		323
1 830	1 650	1 757	1 942	2 204	2 372	2 694	2 968	Instr. médicaux, de précision ; optique ; horlog.		33
1 698	1 566	..	1 824	2 074	2 220	2 547	2 670	Instruments médicaux et de précision, etc.		331
132	83	101	117	130	152	132	283	Instruments d'optique ; matériel photographique		332
..	..	..	..	..	..	15	15	Horlogerie		333
5 876	5 848	4 917	5 020	6 366	6 879	7 666	7 183	Véhicules automobiles; remorques, etc.		34
4 080	4 153	3 396	3 344	4 256	4 474	5 899	4 674	Véhicules automobiles		341
622	621	506	541	694	781	217	227	Carross. automobiles, remorques, semi-remorques		342
1 173	1 074	1 014	1 135	1 416	1 624	1 549	2 282	Pièces et accessoires pour automobiles et moteurs		343
2 240	2 073	1 912	1 938	1 990	2 157	2 272	2 276	Autres matériels de transport		35
376	327	261	250	301	324	370	430	Construction et réparation de navires		351
491	408	383	388	411	422	363	380	Matériel ferroviaire roulant		352
1 279	1 238	1 176	1 207	1 178	1 310	1 423	1 332	Construction aéronautique et spatiale		353
94	100	93	93	101	101	116	134	Autres équipements de transport		359
1 357	1 249	1 094	1 181	1 427	1 432	2 036	2 325	Meubles ; activités de fabrication, nca		36
1 085	995	852	957	1 165	1 159	1 608	1 879	Meubles		361
272	254	242	224	262	..	..	..	Activités de fabrication, nca		369
16	27	21	33	40	47	57	61	Récupération		37
..	..	..	..	..	..	..	..	Electricité, gaz et eau	E	40_41
..	..	..	..	..	..	..	..	Prod. et dist. d'électricité gaz, vapeur d'eau chaude		40
..	..	..	..	..	..	..	..	Captage, épuration et distribution d'eau		41
..	..	..	..	..	..	..	..	Construction	F	45

Note: Before 1996, ISIC 222 includes 223; 332 includes 333. In 1992, ISIC 14 includes part of 13;
181 includes 182. From 1991 to 1996 ISIC 14 includes 11. In 1994, 1995 and 1996, ISIC 242 includes 243.

SWEDEN

Table SN.9 ESTABLISHMENTS - ÉTABLISSEMENTS

Units

ISIC revision 3				1991	1992	1993	1994	1995	1996	1997	1998
C	10_14		Mining and quarrying	188	175	166	158	160	..	..	..
	10		Mining of coal and lignite; extraction of peat	14	16	14	21	22	18	..	..
	11		Crude petroleum & natural gas; service activities	..	..	..	..	..	..	..	..
	13		Mining of metal ores	33	29	24	23	23	21	..	..
	14		Other mining and quarrying	141	130	128	114	115	110	..	..
D	15_37		Total manufacturing	9 284	8 754	8 153	8 210	8 510	..	..	..
	15		Food products and beverages	869	856	841	829	845	842	..	..
		151	Production of meat, fish, fruit, vegetables, oils & fats	284	285	291	283	305	284	..	..
		152	Dairy products	96	84	77	75	76	76	..	..
		153	Grain mill and starch products; animal feeds	68	64	60	55	54	57	..	..
		154	Other food products	387	388	380	383	377	387	..	..
		155	Beverages	34	35	33	33	33	33	..	..
	16		Tobacco products	9	8	8	8	8	..	..	..
	17		Textiles	241	224	202	188	187	175	..	..
		171	Spinning, weaving and finishing of textiles	50	44	43	39	38	38	..	..
		172	Other textiles	147	138	121	114	112	103	..	..
		173	Knitted and crocheted fabrics and articles	44	42	38	35	37	34	..	..
	18		Wearing apparel and fur	127	100	71	62	60	60	..	..
		181	Wearing apparel, except fur apparel	124	100	..	..	..	..	..	..
		182	Dressing and dyeing of fur; articles of fur	3	..	..	..	..	..	..	..
	19		Leather and articles of leather; footwear	44	38	39	32	34	35	..	..
		191	Leather and articles of leather	22	19	19	18	20	21	..	..
		192	Footwear	22	19	20	14	14	14	..	..
	20		Wood & products of wood & cork, exc. furniture	797	773	673	688	709	714	..	..
		201	Sawmilling and planing of wood	386	381	357	370	373	362	..	..
		202	Wood products; articles of cork, straw, etc.	411	392	316	318	336	352	..	..
	21		Paper and paper products	267	266	252	248	256	249	..	..
	22		Printing, publishing and recorded media	934	931	892	920	913	948	..	..
		221	Publishing	250	287	282	317	331	329	..	..
		222	Printing and related activities	684	644	610	603	582	613	..	..
		223	Reproduction of recorded media	..	..	..	..	..	6	..	..
	23		Coke and petroleum products; nuclear fuel	23	20	21	19	17	16	..	..
		231	Coke oven products	..	..	..	..	..	..	..	..
		232	Refined petroleum products	22	19	18	17	16	16	..	..
		233	Processing of nuclear fuel	..	..	..	..	..	..	..	..
	24		Chemicals and chemical products	320	313	308	322	315	317	..	..
		241	Basic chemicals	146	142	136	140	140	140	..	..
		242	Other chemical products	..	..	..	182	175	177	..	..
		243	Man-made fibres	..	..	..	..	..	..	..	..
	25		Rubber and plastic products	386	372	346	378	376	407	..	..
		251	Rubber products	72	65	64	76	77	79	..	..
		252	Plastic products	314	307	282	302	299	328	..	..
	26		Other non-metallic mineral products	419	410	390	376	356	357	..	..
		261	Glass and glass products	58	57	49	45	44	48	..	..
		269	Non-metallic mineral products, nec	361	353	341	331	312	..	..	..
	27		Basic metals	204	194	166	173	170	159	..	..
		271	Basic iron and steel	154	144	114	126	117	105	..	..
		272	Basic precious and non-ferrous metals	45	44	42	39	44	43	..	..
		273	Casting of metals	5	6	10	8	9	11	..	..
	28		Fabricated metal products, exc. machinery	1 510	1 316	1 228	1 252	1 415	1 540	..	..
		281	Structural metal prod., tanks, reservoirs, etc.	538	460	420	402	423	435	..	..
		289	Other fabricated metal products	972	856	808	850	992	1 105	..	..
	29		Machinery and equipment, nec	1 388	1 277	1 152	1 168	1 201	1 222	..	..
		291	General purpose machinery	686	661	612	633	652	612	..	..
		292	Special purpose machinery	639	562	488	486	502	576	..	..
		293	Domestic appliances, nec	63	54	52	49	47	34	..	..
	30		Office, accounting & computing machinery	61	59	54	51	48	52	..	..
	31		Electrical machinery & apparatus, nec	364	371	315	319	332	336	..	..
		311	Electric motors, generators and transformers	44	43	38	35	45	61	..	..
		312	Electricity distribution & control apparatus	108	102	88	85	88	91	..	..
		313	Insulated wire and cable	36	35	34	33	36	41	..	..
		314	Accumulators, primary cells & primary batteries	9	8	10	11	12	10	..	..
		315	Electric lamps and lighting equipment	69	71	60	64	66	64	..	..
		319	Other electrical equipment, nec	98	112	85	91	85	69	..	..
	32		Radio, TV, communication equipment	177	144	144	143	160	173	..	..
		321	Electronic valves, tubes; other electronic components	55	50	49	41	49	55	..	..
		322	TV/radio transmitters; line teleph./telegr. apparatus	94	70	67	70	77	82	..	..
		323	TV/radio receivers; sound and video app.; assoc. gds.	28	24	28	32	34	36	..	..
	33		Medical, precision & optical instrum.; watch & clocks	195	185	232	237	266	267	..	..
		331	Medical appl.; instruments & appl. for measuring, etc.	167	159	..	212	237	240	..	..
		332	Optical instruments and photographic equipment	28	26	21	25	29	27	..	..
		333	Watches and clocks	..	..	..	..	..	..	..	..
	34		Motor vehicles, trailers and semi-trailers	293	267	256	247	258	269	..	..
		341	Motor vehicles	44	41	36	34	37	38	..	..
		342	Bodies for motor vehicles, trailers & semi-trailers	72	76	68	60	59	69	..	..
		343	Parts & accessories for motor vehicles & their engines	177	150	152	153	162	162	..	..
	35		Other transport equipment	166	164	143	140	154	159	..	..
		351	Building and repairing of ships and boats	63	55	48	44	55	50	..	..
		352	Railway and tramway locomotives and rolling stock	52	48	44	42	44	47	..	..
		353	Aircraft and spacecraft	25	30	25	25	24	28	..	..
		359	Transport equipment, nec	26	31	26	29	31	34	..	..
	36		Furniture; manufacturing, nec	481	456	409	397	412	389	..	..
		361	Furniture	385	367	333	331	345	325	..	..
		369	Manufacturing, nec	96	89	76	66	67	..	..	..
	37		Recycling	9	10	11	13	18	21	..	..
E	40_41		Electricity, gas and water supply	..	..	..	..	..	..	..	..
	40		Electricity, gas, steam and hot water supply	..	..	..	..	..	..	..	..
	41		Collection, purification and distribution of water	..	..	..	..	..	..	..	..
F	45		Construction	..	..	..	..	..	..	..	..

Note: Before 1996, ISIC 222 includes 223; 332 includes 333. In 1992, ISIC 14 includes part of 13; 181 includes 182. From 1991 to 1996 ISIC 14 includes 11. In 1994, 1995 and 1996, ISIC 242 includes 243.

Table SN.10 **ENTREPRISES - ENTERPRISES**

Unités

1991	1992	1993	1994	1995	1996	1997	1998			CITI révision 3	
..	..	..	..	..	..	..	..	Activites extractives	C	10_14	
..	..	..	..	..	..	..	..	Extraction de charbon, de lignite et de tourbe		10	
..	..	..	..	..	..	..	..	Pétrole brut et gaz naturel ; activités annexes		11	
..	..	..	..	..	..	..	..	Extraction de minerais métalliques		13	
..	..	..	..	..	..	..	..	Autres activités extractives		14	
..	..	..	..	..	..	..	..	Total activites manufacturieres	D	15_37	
..	..	..	..	..	..	..	2 888	Produits alimentaires et boissons		15	
..	..	..	..	..	..	779	813	Viande, poisson, fruits, légumes, huiles et graisses			151
..	..	..	..	..	..	85	97	Produits laitiers			152
..	..	..	..	..	..	230	235	Travail des grains; prod. amyl. ; alim. pour animaux			153
..	..	..	..	..	..	1 710	1 669	Autres produits alimentaires			154
..	..	..	..	..	..	79	75	Boissons			155
..	..	..	..	..	..	..	1	Produits à base de tabac		16	
..	..	..	..	..	..	1 502	1 594	Textiles		17	
..	..	..	..	..	..	346	365	Filature, tissage et achèvement des textiles			171
..	..	..	..	..	..	1 013	1 092	Autres articles textiles			172
..	..	..	..	..	..	143	137	Etoffes et articles de bonneterie			173
..	..	..	..	..	..	1 589	1 638	Articles d'habillement et fourrure		18	
..	..	..	..	..	..	1 466	1 514	Articles d'habillement autres qu'en fourrure			181
..	..	..	..	..	..	123	124	Prép. et teinture des fourrures ; articles en fourrure			182
..	..	..	..	..	..	385	408	Cuir et articles en cuir ; chaussures		19	
..	..	..	..	..	..	..	485	Cuir et articles en cuir sauf chaussures			191
..	..	..	..	..	..	150	147	Chaussures			192
..	..	..	..	..	..	5 344	5 508	Bois, articles en bois, sauf meubles		20	
..	..	..	..	..	..	1 380	1 406	Sciage et rabotage du bois			201
..	..	..	..	..	..	3 964	4 102	Articles en bois, liège, vannerie, sparterie			202
..	..	..	..	..	..	383	396	Papier, articles en papier et carton		21	
..	..	..	..	..	..	8 008	8 189	Edition, imprimerie et reproduction		22	
..	..	..	..	..	..	3 778	3 953	Edition			221
..	..	..	..	..	..	4 093	4 066	Imprimerie et activités annexes			222
..	..	..	..	..	..	137	170	Reproduction de supports enregistrés			223
..	..	..	..	..	..	41	43	Coke. et prod. pétroliers ; comb. nucléaires		23	
..	..	..	..	..	..	8	10	Cokéfaction			231
..	..	..	..	..	..	33	33	Produits pétroliers raffinés			232
..	..	..	..	..	..	..	..	Traitement de combustibles nucléaires			233
..	..	..	..	..	..	769	796	Produits chimiques		24	
..	..	..	..	..	..	179	177	Produits chimiques de base			241
..	..	..	..	..	..	590	619	Autres produits chimiques			242
..	..	..	..	..	..	8	..	Fibres synthétiques ou artificielles			243
..	..	..	..	..	..	1 447	1 475	Art. en caoutchouc et matières plastiques		25	
..	..	..	..	..	..	240	255	Articles en caoutchouc			251
..	..	..	..	..	..	1 207	1 220	Articles en matières plastiques			252
..	..	..	..	..	..	693	747	Autres produits minéraux non métalliques		26	
..	..	..	..	..	..	248	261	Verre et articles en verre			261
..	..	..	..	..	..	..	1 203	Produits minéraux non métalliques, nca			269
..	..	..	..	..	..	330	337	Produits métallurgiques de base		27	
..	..	..	..	..	..	151	154	Sidérurgie et première transformation de l'acier			271
..	..	..	..	..	..	88	88	Première transf. métaux précieux et non ferreux			272
..	..	..	..	..	..	91	95	Fonderie			273
..	..	..	..	..	..	9 317	9 451	Ouvrages en métaux, sauf machines		28	
..	..	..	..	..	..	1 197	1 185	Constr. et menuis. métal.; réservoirs; générateurs			281
..	..	..	..	..	..	8 120	8 266	Autres ouvrages en métaux			289
..	..	..	..	..	..	4 772	4 843	Machines et matériels, nca		29	
..	..	..	..	..	..	2 033	2 027	Machines d'usage général			291
..	..	..	..	..	..	2 644	2 724	Machines d'usage spécifique			292
..	..	..	..	..	..	95	92	Appareils domestiques, nca			293
..	..	..	..	..	..	428	428	Mach. de bureau, comptables, information		30	
..	..	..	..	..	..	1 242	1 265	Machines et appareils électriques, nca		31	
..	..	..	..	..	..	244	253	Moteurs, génératrices, transformateurs électriques			311
..	..	..	..	..	..	275	282	Matériel électrique de distribution et de commande			312
..	..	..	..	..	..	79	75	Fils et câbles électriques isolés			313
..	..	..	..	..	..	30	34	Accumulateurs et piles électriques			314
..	..	..	..	..	..	231	238	Lampes électriques et appareils d'éclairage			315
..	..	..	..	..	..	383	383	Autres matériels électriques, nca.			319
..	..	..	..	..	..	556	592	Equipements et appareils de radio, TV, etc.		32	
..	..	..	..	..	..	282	294	Tubes et valves électroniques;autres composants			321
..	..	..	..	..	..	122	136	Emetteurs radio/TV; app. téléphonie et télégraphie			322
..	..	..	..	..	..	152	162	Récepteurs radio/TV, son et image; art. associés			323
..	..	..	..	..	..	1 948	1 950	Instr. médicaux, de précision ; optique ; horlog.		33	
..	..	..	..	..	..	1 785	1 782	Instruments médicaux et de précision, etc.			331
..	..	..	..	..	..	132	140	Instruments d'optique ; matériel photographique			332
..	..	..	..	..	..	31	28	Horlogerie			333
..	..	..	..	..	..	644	662	Véhicules automobiles; remorques, etc.		34	
..	..	..	..	..	..	104	101	Véhicules automobiles			341
..	..	..	..	..	..	195	197	Carross. automobiles, remorques, semi-remorques			342
..	..	..	..	..	..	345	364	Pièces et accessoires pour automobiles et moteurs			343
..	..	..	..	..	..	1 230	1 268	Autres matériels de transport		35	
..	..	..	..	..	..	988	1 011	Construction et réparation de navires			351
..	..	..	..	..	..	21	22	Matériel ferroviaire roulant			352
..	..	..	..	..	..	112	114	Construction aéronautique et spatiale			353
..	..	..	..	..	..	109	121	Autres équipements de transport			359
..	..	..	..	..	..	4 581	4 815	Meubles ; activités de fabrication, nca		36	
..	..	..	..	..	..	2 765	2 836	Meubles			361
..	..	..	..	..	..	..	..	Activités de fabrication, nca			369
..	..	..	..	..	..	117	133	Récupération		37	
..	..	..	..	..	..	..	..	Electricité, gaz et eau	E	40_41	
..	..	..	..	..	..	..	..	Prod. et dist. d'électricité gaz, vapeur d'eau chaude		40	
..	..	..	..	..	..	..	..	Captage, épuration et distribution d'eau		41	
..	..	..	..	..	..	..	..	Construction	F	45	

Table SW.1 **PRODUCTION AT FACTOR COSTS - PRODUCTION AU COUT DES FACTEURS**

Millions of SF (Current Prices)

ISIC revision 3				1991	1992	1993	1994	1995	1996	1997	1998
C	10_14		Mining and quarrying	..	..	..	..	..	..	..	..
	10		Mining of coal and lignite; extraction of peat	..	..	..	..	..	..	..	..
	11		Crude petroleum & natural gas; service activities	..	..	..	..	..	..	..	..
	13		Mining of metal ores	..	..	..	..	..	..	..	..
	14		Other mining and quarrying	..	..	..	..	..	..	1 350	1 896
D	15_37		Total manufacturing	..	..	..	..	..	..	..	..
	15		Food products and beverages	..	..	..	..	..	..	21 843	21 982
		151	Production of meat, fish, fruit, vegetables, oils & fats	..	..	..	..	..	..	..	..
		152	Dairy products	..	..	..	..	..	..	..	..
		153	Grain mill and starch products; animal feeds	..	..	..	..	..	..	..	..
		154	Other food products	..	..	..	..	..	..	..	..
		155	Beverages	..	..	..	..	..	..	..	..
	16		Tobacco products	..	..	..	..	..	..	..	..
	17		Textiles	..	..	..	..	..	..	3 296	3 502
		171	Spinning, weaving and finishing of textiles	..	..	..	..	..	..	..	..
		172	Other textiles	..	..	..	..	..	..	..	..
		173	Knitted and crocheted fabrics and articles	..	..	..	..	..	..	..	..
	18		Wearing apparel and fur	..	..	..	..	..	..	1 517	1 777
		181	Wearing apparel, except fur apparel	..	..	..	..	..	..	..	..
		182	Dressing and dyeing of fur; articles of fur	..	..	..	..	..	..	..	..
	19		Leather and articles of leather; footwear	..	..	..	..	..	..	609	441
		191	Leather and articles of leather	..	..	..	..	..	..	..	..
		192	Footwear	..	..	..	..	..	..	..	..
	20		Wood & products of wood & cork, exc. furniture	..	..	..	..	..	..	6 300	6 501
		201	Sawmilling and planing of wood	..	..	..	..	..	..	..	..
		202	Wood products; articles of cork, straw, etc.	..	..	..	..	..	..	..	..
	21		Paper and paper products	..	..	..	..	..	..	4 340	4 594
	22		Printing, publishing and recorded media	..	..	..	..	..	..	10 990	10 371
		221	Publishing	..	..	..	..	..	..	..	..
		222	Printing and related activities	..	..	..	..	..	..	..	..
		223	Reproduction of recorded media	..	..	..	..	..	..	..	..
	23		Coke and petroleum products; nuclear fuel	..	..	..	..	..	..	..	..
		231	Coke oven products	..	..	..	..	..	..	..	..
		232	Refined petroleum products	..	..	..	..	..	..	..	..
		233	Processing of nuclear fuel	..	..	..	..	..	..	..	..
	24		Chemicals and chemical products	..	..	..	..	..	..	41 284	43 396
		241	Basic chemicals	..	..	..	..	..	..	..	..
		242	Other chemical products	..	..	..	..	..	..	..	..
		243	Man-made fibres	..	..	..	..	..	..	..	..
	25		Rubber and plastic products	..	..	..	..	..	..	5 269	5 667
		251	Rubber products	..	..	..	..	..	..	..	..
		252	Plastic products	..	..	..	..	..	..	..	..
	26		Other non-metallic mineral products	..	..	..	..	..	..	4 437	4 588
		261	Glass and glass products	..	..	..	..	..	..	..	..
		269	Non-metallic mineral products, nec	..	..	..	..	..	..	..	..
	27		Basic metals	..	..	..	..	..	..	5 114	4 712
		271	Basic iron and steel	..	..	..	..	..	..	..	..
		272	Basic precious and non-ferrous metals	..	..	..	..	..	..	..	..
		273	Casting of metals	..	..	..	..	..	..	..	..
	28		Fabricated metal products, exc. machinery	..	..	..	..	..	..	16 038	17 041
		281	Structural metal prod., tanks, reservoirs, etc.	..	..	..	..	..	..	..	..
		289	Other fabricated metal products	..	..	..	..	..	..	..	..
	29		Machinery and equipment, nec	..	..	..	..	..	..	30 621	33 064
		291	General purpose machinery	..	..	..	..	..	..	..	..
		292	Special purpose machinery	..	..	..	..	..	..	..	..
		293	Domestic appliances, nec	..	..	..	..	..	..	..	..
	30		Office, accounting & computing machinery	..	..	..	..	..	..	..	..
	31		Electrical machinery & apparatus, nec	..	..	..	..	..	..	13 812	13 650
		311	Electric motors, generators and transformers	..	..	..	..	..	..	..	..
		312	Electricity distribution & control apparatus	..	..	..	..	..	..	..	..
		313	Insulated wire and cable	..	..	..	..	..	..	..	..
		314	Accumulators, primary cells & primary batteries	..	..	..	..	..	..	..	..
		315	Electric lamps and lighting equipment	..	..	..	..	..	..	..	..
		319	Other electrical equipment, nec	..	..	..	..	..	..	..	..
	32		Radio, TV, communication equipment	..	..	..	..	..	..	4 484	5 059
		321	Electronic valves, tubes; other electronic components	..	..	..	..	..	..	..	..
		322	TV/radio transmitters; line teleph./telegr. apparatus	..	..	..	..	..	..	..	..
		323	TV/radio receivers; sound and video app.; assoc. gds.	..	..	..	..	..	..	..	..
	33		Medical, precision & optical instrum.; watch & clocks	..	..	..	..	..	..	16 754	20 049
		331	Medical appl.; instruments & appl. for measuring, etc.	..	..	..	..	..	..	..	..
		332	Optical instruments and photographic equipment	..	..	..	..	..	..	..	..
		333	Watches and clocks	..	..	..	..	..	..	..	..
	34		Motor vehicles, trailers and semi-trailers	..	..	..	..	..	..	1 107	1 189
		341	Motor vehicles	..	..	..	..	..	..	..	..
		342	Bodies for motor vehicles, trailers & semi-trailers	..	..	..	..	..	..	..	..
		343	Parts & accessories for motor vehicles & their engines	..	..	..	..	..	..	..	..
	35		Other transport equipment	..	..	..	..	..	..	1 703	2 121
		351	Building and repairing of ships and boats	..	..	..	..	..	..	..	..
		352	Railway and tramway locomotives and rolling stock	..	..	..	..	..	..	..	..
		353	Aircraft and spacecraft	..	..	..	..	..	..	..	..
		359	Transport equipment, nec	..	..	..	..	..	..	..	..
	36		Furniture; manufacturing, nec	..	..	..	..	..	..	5 685	5 758
		361	Furniture	..	..	..	..	..	..	..	..
		369	Manufacturing, nec	..	..	..	..	..	..	..	..
	37		Recycling	..	..	..	..	..	..	568	540
E	40_41		Electricity, gas and water supply	..	..	..	..	..	..	..	..
	40		Electricity, gas, steam and hot water supply	..	..	..	..	..	..	21 363	21 169
	41		Collection, purification and distribution of water	..	..	..	..	..	..	..	..
F	45		Construction	..	..	..	..	..	..	41 288	46 195

Table SW.2 **VALEUR AJOUTÉE AU COUT DES FACTEURS - VALUE ADDED AT FACTOR COSTS**

Millions de SF (Prix courants)

1991	1992	1993	1994	1995	1996	1997	1998		CITI révision 3		
..	..	..	..	..	..	..	..	Activites extractives	C	10_14	
..	..	..	..	..	..	..	..	Extraction de charbon, de lignite et de tourbe		10	
..	..	..	..	..	..	..	..	Pétrole brut et gaz naturel ; activités annexes		11	
..	..	..	..	..	..	..	..	Extraction de minerais métalliques		13	
..	..	..	..	..	..	713	947	Autres activités extractives		14	
..	..	..	..	..	..	..	..	Total activites manufacturieres	D	15_37	
..	..	..	..	..	..	5 438	5 667	Produits alimentaires et boissons		15	
..	..	..	..	..	..	..	..	Viande, poisson, fruits, légumes, huiles et graisses			151
..	..	..	..	..	..	..	..	Produits laitiers			152
..	..	..	..	..	..	..	..	Travail des grains; prod. amyl. ; alim. pour animaux			153
..	..	..	..	..	..	..	..	Autres produits alimentaires			154
..	..	..	..	..	..	..	..	Boissons			155
..	..	..	..	..	..	..	..	Produits à base de tabac		16	
..	..	..	..	..	..	1 367	1 464	Textiles		17	
..	..	..	..	..	..	..	..	Filature, tissage et achèvement des textiles			171
..	..	..	..	..	..	..	..	Autres articles textiles			172
..	..	..	..	..	..	..	..	Etoffes et articles de bonneterie			173
..	..	..	..	..	..	571	541	Articles d'habillement et fourrure		18	
..	..	..	..	..	..	..	..	Articles d'habillement autres qu'en fourrure			181
..	..	..	..	..	..	..	..	Prép. et teinture des fourrures ; articles en fourrure			182
..	..	..	..	..	..	220	202	Cuir et articles en cuir ; chaussures		19	
..	..	..	..	..	..	..	..	Cuir et articles en cuir sauf chaussures			191
..	..	..	..	..	..	..	..	Chaussures			192
..	..	..	..	..	..	3 017	2 931	Bois, articles en bois, sauf meubles		20	
..	..	..	..	..	..	..	..	Sciage et rabotage du bois			201
..	..	..	..	..	..	..	..	Articles en bois, liège, vannerie, sparterie			202
..	..	..	..	..	..	1 601	1 738	Papier, articles en papier et carton		21	
..	..	..	..	..	..	5 501	5 379	Edition, imprimerie et reproduction		22	
..	..	..	..	..	..	..	..	Edition			221
..	..	..	..	..	..	..	..	Imprimerie et activités annexes			222
..	..	..	..	..	..	..	..	Reproduction de supports enregistrés			223
..	..	..	..	..	..	..	..	Coke. et prod. pétroliers ; comb. nucléaires		23	
..	..	..	..	..	..	..	..	Cokéfaction			231
..	..	..	..	..	..	..	..	Produits pétroliers raffinés			232
..	..	..	..	..	..	..	..	Traitement de combustibles nucléaires			233
..	..	..	..	..	..	12 296	13 026	Produits chimiques		24	
..	..	..	..	..	..	..	..	Produits chimiques de base			241
..	..	..	..	..	..	..	..	Autres produits chimiques			242
..	..	..	..	..	..	..	..	Fibres synthétiques ou artificielles			243
..	..	..	..	..	..	2 178	2 294	Art. en caoutchouc et matières plastiques		25	
..	..	..	..	..	..	..	..	Articles en caoutchouc			251
..	..	..	..	..	..	..	..	Articles en matières plastiques			252
..	..	..	..	..	..	1 709	1 690	Autres produits minéraux non métalliques		26	
..	..	..	..	..	..	..	..	Verre et articles en verre			261
..	..	..	..	..	..	..	..	Produits minéraux non métalliques, nca			269
..	..	..	..	..	..	1 524	1 493	Produits métallurgiques de base		27	
..	..	..	..	..	..	..	..	Sidérurgie et première transformation de l'acier			271
..	..	..	..	..	..	..	..	Première transf. métaux précieux et non ferreux			272
..	..	..	..	..	..	..	..	Fonderie			273
..	..	..	..	..	..	7 954	8 143	Ouvrages en métaux, sauf machines		28	
..	..	..	..	..	..	..	..	Constr. et menuis. métal.; réservoirs; générateurs			281
..	..	..	..	..	..	..	..	Autres ouvrages en métaux			289
..	..	..	..	..	..	11 660	11 824	Machines et matériels, nca		29	
..	..	..	..	..	..	..	..	Machines d'usage général			291
..	..	..	..	..	..	..	..	Machines d'usage spécifique			292
..	..	..	..	..	..	..	..	Appareils domestiques, nca			293
..	..	..	..	..	..	..	..	Mach. de bureau, comptables, information		30	
..	..	..	..	..	..	4 356	4 488	Machines et appareils électriques, nca		31	
..	..	..	..	..	..	..	..	Moteurs, génératrices, transformateurs électriques			311
..	..	..	..	..	..	..	..	Matériel électrique de distribution et de commande			312
..	..	..	..	..	..	..	..	Fils et câbles électriques isolés			313
..	..	..	..	..	..	..	..	Accumulateurs et piles électriques			314
..	..	..	..	..	..	..	..	Lampes électriques et appareils d'éclairage			315
..	..	..	..	..	..	..	..	Autres matériels électriques, nca.			319
..	..	..	..	..	..	1 664	1 764	Equipements et appareils de radio, TV, etc.		32	
..	..	..	..	..	..	..	..	Tubes et valves électroniques;autres composants			321
..	..	..	..	..	..	..	..	Emetteurs radio/TV; app. téléphonie et télégraphie			322
..	..	..	..	..	..	..	..	Récepteurs radio/TV, son et image; art. associés			323
..	..	..	..	..	..	7 147	7 798	Instr. médicaux, de précision ; optique ; horlog.		33	
..	..	..	..	..	..	..	..	Instruments médicaux et de précision, etc.			331
..	..	..	..	..	..	..	..	Instruments d'optique ; matériel photographique			332
..	..	..	..	..	..	..	..	Horlogerie			333
..	..	..	..	..	..	405	402	Véhicules automobiles; remorques, etc.		34	
..	..	..	..	..	..	..	..	Véhicules automobiles			341
..	..	..	..	..	..	..	..	Carross. automobiles, remorques, semi-remorques			342
..	..	..	..	..	..	..	..	Pièces et accessoires pour automobiles et moteurs			343
..	..	..	..	..	..	677	734	Autres matériels de transport		35	
..	..	..	..	..	..	..	..	Construction et réparation de navires			351
..	..	..	..	..	..	..	..	Matériel ferroviaire roulant			352
..	..	..	..	..	..	..	..	Construction aéronautique et spatiale			353
..	..	..	..	..	..	..	..	Autres équipements de transport			359
..	..	..	..	..	..	2 430	2 325	Meubles ; activités de fabrication, nca		36	
..	..	..	..	..	..	..	..	Meubles			361
..	..	..	..	..	..	..	..	Activités de fabrication, nca			369
..	..	..	..	..	..	191	187	Récupération		37	
..	..	..	..	..	..	..	..	Electricité, gaz et eau	E	40_41	
..	..	..	..	..	..	9 914	9 355	Prod. et dist. d'électricité gaz, vapeur d'eau chaude		40	
..	..	..	..	..	..	..	..	Captage, épuration et distribution d'eau		41	
..	..	..	..	..	..	21 667	23 302	Construction	F	45	

Table SW.3 **EMPLOYMENT - EMPLOI**

Units

ISIC revision 3				1991	1992	1993	1994	1995	1996	1997	1998
C	10_14		Mining and quarrying	..	..	..	..	..	..	..	..
	10		Mining of coal and lignite; extraction of peat	..	..	..	..	..	..	..	..
	11		Crude petroleum & natural gas; service activities	..	..	..	..	..	..	..	..
	13		Mining of metal ores	..	..	..	..	..	..	..	..
	14		Other mining and quarrying	..	..	..	..	..	..	5 470	5 596
D	15_37		Total manufacturing	..	..	..	..	..	..	..	..
	15		Food products and beverages	..	..	..	..	..	..	53 744	54 194
		151	Production of meat, fish, fruit, vegetables, oils & fats	..	..	..	..	..	..	..	..
		152	Dairy products	..	..	..	..	..	..	..	..
		153	Grain mill and starch products; animal feeds	..	..	..	..	..	..	..	..
		154	Other food products	..	..	..	..	..	..	..	..
		155	Beverages	..	..	..	..	..	..	..	..
	16		Tobacco products	..	..	..	..	..	..	..	..
	17		Textiles	..	..	..	..	..	..	16 982	17 204
		171	Spinning, weaving and finishing of textiles	..	..	..	..	..	..	..	..
		172	Other textiles	..	..	..	..	..	..	..	..
		173	Knitted and crocheted fabrics and articles	..	..	..	..	..	..	..	..
	18		Wearing apparel and fur	..	..	..	..	..	..	8 934	7 805
		181	Wearing apparel, except fur apparel	..	..	..	..	..	..	..	..
		182	Dressing and dyeing of fur; articles of fur	..	..	..	..	..	..	..	..
	19		Leather and articles of leather; footwear	..	..	..	..	..	..	3 639	3 202
		191	Leather and articles of leather	..	..	..	..	..	..	..	..
		192	Footwear	..	..	..	..	..	..	..	..
	20		Wood & products of wood & cork, exc. furniture	..	..	..	..	..	..	38 859	39 363
		201	Sawmilling and planing of wood	..	..	..	..	..	..	..	..
		202	Wood products; articles of cork, straw, etc.	..	..	..	..	..	..	..	..
	21		Paper and paper products	..	..	..	..	..	..	14 789	15 103
	22		Printing, publishing and recorded media	..	..	..	..	..	..	53 467	50 076
		221	Publishing	..	..	..	..	..	..	..	..
		222	Printing and related activities	..	..	..	..	..	..	..	..
		223	Reproduction of recorded media	..	..	..	..	..	..	..	..
	23		Coke and petroleum products; nuclear fuel	..	..	..	..	..	..	..	..
		231	Coke oven products	..	..	..	..	..	..	..	..
		232	Refined petroleum products	..	..	..	..	..	..	..	..
		233	Processing of nuclear fuel	..	..	..	..	..	..	..	..
	24		Chemicals and chemical products	..	..	..	..	..	..	64 279	65 962
		241	Basic chemicals	..	..	..	..	..	..	..	..
		242	Other chemical products	..	..	..	..	..	..	..	..
		243	Man-made fibres	..	..	..	..	..	..	..	..
	25		Rubber and plastic products	..	..	..	..	..	..	21 368	21 636
		251	Rubber products	..	..	..	..	..	..	..	..
		252	Plastic products	..	..	..	..	..	..	..	..
	26		Other non-metallic mineral products	..	..	..	..	..	..	18 112	17 133
		261	Glass and glass products	..	..	..	..	..	..	..	..
		269	Non-metallic mineral products, nec	..	..	..	..	..	..	..	..
	27		Basic metals	..	..	..	..	..	..	14 374	14 528
		271	Basic iron and steel	..	..	..	..	..	..	..	..
		272	Basic precious and non-ferrous metals	..	..	..	..	..	..	..	..
		273	Casting of metals	..	..	..	..	..	..	..	..
	28		Fabricated metal products, exc. machinery	..	..	..	..	..	..	82 165	83 009
		281	Structural metal prod., tanks, reservoirs, etc.	..	..	..	..	..	..	..	..
		289	Other fabricated metal products	..	..	..	..	..	..	..	..
	29		Machinery and equipment, nec	..	..	..	..	..	..	106 003	106 510
		291	General purpose machinery	..	..	..	..	..	..	..	..
		292	Special purpose machinery	..	..	..	..	..	..	..	..
		293	Domestic appliances, nec	..	..	..	..	..	..	..	..
	30		Office, accounting & computing machinery	..	..	..	..	..	..	..	..
	31		Electrical machinery & apparatus, nec	..	..	..	..	..	..	43 382	44 199
		311	Electric motors, generators and transformers	..	..	..	..	..	..	..	..
		312	Electricity distribution & control apparatus	..	..	..	..	..	..	..	..
		313	Insulated wire and cable	..	..	..	..	..	..	..	..
		314	Accumulators, primary cells & primary batteries	..	..	..	..	..	..	..	..
		315	Electric lamps and lighting equipment	..	..	..	..	..	..	..	..
		319	Other electrical equipment, nec	..	..	..	..	..	..	..	..
	32		Radio, TV, communication equipment	..	..	..	..	..	..	15 205	16 202
		321	Electronic valves, tubes; other electronic components	..	..	..	..	..	..	..	..
		322	TV/radio transmitters; line teleph./telegr. apparatus	..	..	..	..	..	..	..	..
		323	TV/radio receivers; sound and video app.; assoc. gds.	..	..	..	..	..	..	..	..
	33		Medical, precision & optical instrum.; watch & clocks	..	..	..	..	..	..	61 206	63 559
		331	Medical appl.; instruments & appl. for measuring, etc.	..	..	..	..	..	..	..	..
		332	Optical instruments and photographic equipment	..	..	..	..	..	..	..	..
		333	Watches and clocks	..	..	..	..	..	..	..	..
	34		Motor vehicles, trailers and semi-trailers	..	..	..	..	..	..	4 568	4 493
		341	Motor vehicles	..	..	..	..	..	..	..	..
		342	Bodies for motor vehicles, trailers & semi-trailers	..	..	..	..	..	..	..	..
		343	Parts & accessories for motor vehicles & their engines	..	..	..	..	..	..	..	..
	35		Other transport equipment	..	..	..	..	..	..	7 526	6 984
		351	Building and repairing of ships and boats	..	..	..	..	..	..	..	..
		352	Railway and tramway locomotives and rolling stock	..	..	..	..	..	..	..	..
		353	Aircraft and spacecraft	..	..	..	..	..	..	..	..
		359	Transport equipment, nec	..	..	..	..	..	..	..	..
	36		Furniture; manufacturing, nec	..	..	..	..	..	..	26 780	26 340
		361	Furniture	..	..	..	..	..	..	..	..
		369	Manufacturing, nec	..	..	..	..	..	..	..	..
	37		Recycling	..	..	..	..	..	..	1 902	1 871
E	40_41		Electricity, gas and water supply	..	..	..	..	..	..	..	..
	40		Electricity, gas, steam and hot water supply	..	..	..	..	..	..	20 891	20 765
	41		Collection, purification and distribution of water	..	..	..	..	..	..	..	..
F	45		Construction	..	..	..	..	..	..	289 447	299 150

Table SW.4 **SALAIRES ET CHARGES SOCIALES - COMPENSATION OF LABOUR**

Millions de SF (Prix courants)

1991	1992	1993	1994	1995	1996	1997	1998			CITI révision 3	
..	..	..	..	..	..	..	..	Activites extractives	C	10_14	
..	..	..	..	..	..	..	..	Extraction de charbon, de lignite et de tourbe		10	
..	..	..	..	..	..	..	..	Pétrole brut et gaz naturel ; activités annexes		11	
..	..	..	..	..	..	..	..	Extraction de minerais métalliques		13	
..	..	..	..	..	..	459	481	Autres activités extractives		14	
..	..	..	..	..	..	..	..	Total activites manufacturieres	D	15_37	
..	..	..	..	..	..	3 737	3 728	Produits alimentaires et boissons		15	
..	..	..	..	..	..	..	..	Viande, poisson, fruits, légumes, huiles et graisses			151
..	..	..	..	..	..	..	..	Produits laitiers			152
..	..	..	..	..	..	..	..	Travail des grains; prod. amyl. ; alim. pour animaux			153
..	..	..	..	..	..	..	..	Autres produits alimentaires			154
..	..	..	..	..	..	..	..	Boissons			155
..	..	..	..	..	..	..	..	Produits à base de tabac		16	
..	..	..	..	..	..	1 065	1 151	Textiles		17	
..	..	..	..	..	..	..	..	Filature, tissage et achèvement des textiles			171
..	..	..	..	..	..	..	..	Autres articles textiles			172
..	..	..	..	..	..	..	..	Etoffes et articles de bonneterie			173
..	..	..	..	..	..	480	476	Articles d'habillement et fourrure		18	
..	..	..	..	..	..	..	..	Articles d'habillement autres qu'en fourrure			181
..	..	..	..	..	..	..	..	Prép. et teinture des fourrures ; articles en fourrure			182
..	..	..	..	..	..	190	190	Cuir et articles en cuir ; chaussures		19	
..	..	..	..	..	..	..	..	Cuir et articles en cuir sauf chaussures			191
..	..	..	..	..	..	..	..	Chaussures			192
..	..	..	..	..	..	2 305	2 189	Bois, articles en bois, sauf meubles		20	
..	..	..	..	..	..	..	..	Sciage et rabotage du bois			201
..	..	..	..	..	..	..	..	Articles en bois, liège, vannerie, sparterie			202
..	..	..	..	..	..	1 118	1 201	Papier, articles en papier et carton		21	
..	..	..	..	..	..	4 364	4 137	Edition, imprimerie et reproduction		22	
..	..	..	..	..	..	..	..	Edition			221
..	..	..	..	..	..	..	..	Imprimerie et activités annexes			222
..	..	..	..	..	..	..	..	Reproduction de supports enregistrés			223
..	..	..	..	..	..	..	..	Coke. et prod. pétroliers ; comb. nucléaires		23	
..	..	..	..	..	..	..	..	Cokéfaction			231
..	..	..	..	..	..	..	..	Produits pétroliers raffinés			232
..	..	..	..	..	..	..	..	Traitement de combustibles nucléaires			233
..	..	..	..	..	..	6 482	6 929	Produits chimiques		24	
..	..	..	..	..	..	..	..	Produits chimiques de base			241
..	..	..	..	..	..	..	..	Autres produits chimiques			242
..	..	..	..	..	..	..	..	Fibres synthétiques ou artificielles			243
..	..	..	..	..	..	1 636	1 601	Art. en caoutchouc et matières plastiques		25	
..	..	..	..	..	..	..	..	Articles en caoutchouc			251
..	..	..	..	..	..	..	..	Articles en matières plastiques			252
..	..	..	..	..	..	1 316	1 256	Autres produits minéraux non métalliques		26	
..	..	..	..	..	..	..	..	Verre et articles en verre			261
..	..	..	..	..	..	..	..	Produits minéraux non métalliques, nca			269
..	..	..	..	..	..	1 109	1 124	Produits métallurgiques de base		27	
..	..	..	..	..	..	..	..	Sidérurgie et première transformation de l'acier			271
..	..	..	..	..	..	..	..	Première transf. métaux précieux et non ferreux			272
..	..	..	..	..	..	..	..	Fonderie			273
..	..	..	..	..	..	5 890	6 095	Ouvrages en métaux, sauf machines		28	
..	..	..	..	..	..	..	..	Constr. et menuis. métal.; réservoirs; générateurs			281
..	..	..	..	..	..	..	..	Autres ouvrages en métaux			289
..	..	..	..	..	..	9 002	9 099	Machines et matériels, nca		29	
..	..	..	..	..	..	..	..	Machines d'usage général			291
..	..	..	..	..	..	..	..	Machines d'usage spécifique			292
..	..	..	..	..	..	..	..	Appareils domestiques, nca			293
..	..	..	..	..	..	..	..	Mach. de bureau, comptables, information		30	
..	..	..	..	..	..	3 472	3 561	Machines et appareils électriques, nca		31	
..	..	..	..	..	..	..	..	Moteurs, génératrices, transformateurs électriques			311
..	..	..	..	..	..	..	..	Matériel électrique de distribution et de commande			312
..	..	..	..	..	..	..	..	Fils et câbles électriques isolés			313
..	..	..	..	..	..	..	..	Accumulateurs et piles électriques			314
..	..	..	..	..	..	..	..	Lampes électriques et appareils d'éclairage			315
..	..	..	..	..	..	..	..	Autres matériels électriques, nca.			319
..	..	..	..	..	..	1 274	1 333	Equipements et appareils de radio, TV, etc.		32	
..	..	..	..	..	..	..	..	Tubes et valves électroniques;autres composants			321
..	..	..	..	..	..	..	..	Emetteurs radio/TV; app. téléphonie et télégraphie			322
..	..	..	..	..	..	..	..	Récepteurs radio/TV, son et image; art. associés			323
..	..	..	..	..	..	4 893	5 027	Instr. médicaux, de précision ; optique ; horlog.		33	
..	..	..	..	..	..	..	..	Instruments médicaux et de précision, etc.			331
..	..	..	..	..	..	..	..	Instruments d'optique ; matériel photographique			332
..	..	..	..	..	..	..	..	Horlogerie			333
..	..	..	..	..	..	322	322	Véhicules automobiles; remorques, etc.		34	
..	..	..	..	..	..	..	..	Véhicules automobiles			341
..	..	..	..	..	..	..	..	Carross. automobiles, remorques, semi-remorques			342
..	..	..	..	..	..	..	..	Pièces et accessoires pour automobiles et moteurs			343
..	..	..	..	..	..	594	628	Autres matériels de transport		35	
..	..	..	..	..	..	..	..	Construction et réparation de navires			351
..	..	..	..	..	..	..	..	Matériel ferroviaire roulant			352
..	..	..	..	..	..	..	..	Construction aéronautique et spatiale			353
..	..	..	..	..	..	..	..	Autres équipements de transport			359
..	..	..	..	..	..	1 845	1 747	Meubles ; activités de fabrication, nca		36	
..	..	..	..	..	..	..	..	Meubles			361
..	..	..	..	..	..	..	..	Activités de fabrication, nca			369
..	..	..	..	..	..	126	125	Récupération		37	
..	..	..	..	..	..	..	..	Electricité, gaz et eau	E	40_41	
..	..	..	..	..	..	2 048	2 057	Prod. et dist. d'électricité gaz, vapeur d'eau chaude		40	
..	..	..	..	..	..	..	..	Captage, épuration et distribution d'eau		41	
..	..	..	..	..	..	17 927	18 970	Construction	F	45	

SWITZERLAND

Table SW.5 **TOTAL WAGES AND SALARIES - SALAIRES ET TRAITEMENTS**

Millions of SF (Current Prices)

ISIC revision 3				1991	1992	1993	1994	1995	1996	1997	1998
C	10_14		Mining and quarrying	..	..	..	..	..	..	..	..
	10		Mining of coal and lignite; extraction of peat	..	..	..	..	..	..	..	..
	11		Crude petroleum & natural gas; service activities	..	..	..	..	..	..	..	..
	13		Mining of metal ores	..	..	..	..	..	..	..	..
	14		Other mining and quarrying	..	..	..	..	..	..	381	399
D	15_37		Total manufacturing	..	..	..	..	..	..	..	..
	15		Food products and beverages	..	..	..	..	..	..	3 189	3 193
		151	Production of meat, fish, fruit, vegetables, oils & fats	..	..	..	..	..	..	..	..
		152	Dairy products	..	..	..	..	..	..	..	..
		153	Grain mill and starch products; animal feeds	..	..	..	..	..	..	..	..
		154	Other food products	..	..	..	..	..	..	..	..
		155	Beverages	..	..	..	..	..	..	..	..
	16		Tobacco products	..	..	..	..	..	..	..	..
	17		Textiles	..	..	..	..	..	..	918	984
		171	Spinning, weaving and finishing of textiles	..	..	..	..	..	..	..	..
		172	Other textiles	..	..	..	..	..	..	..	..
		173	Knitted and crocheted fabrics and articles	..	..	..	..	..	..	..	..
	18		Wearing apparel and fur	..	..	..	..	..	..	403	404
		181	Wearing apparel, except fur apparel	..	..	..	..	..	..	..	..
		182	Dressing and dyeing of fur; articles of fur	..	..	..	..	..	..	..	..
	19		Leather and articles of leather; footwear	..	..	..	..	..	..	167	171
		191	Leather and articles of leather	..	..	..	..	..	..	..	..
		192	Footwear	..	..	..	..	..	..	..	..
	20		Wood & products of wood & cork, exc. furniture	..	..	..	..	..	..	1 984	1 869
		201	Sawmilling and planing of wood	..	..	..	..	..	..	..	..
		202	Wood products; articles of cork, straw, etc.	..	..	..	..	..	..	..	..
	21		Paper and paper products	..	..	..	..	..	..	963	1 037
	22		Printing, publishing and recorded media	..	..	..	..	..	..	3 758	3 561
		221	Publishing	..	..	..	..	..	..	..	..
		222	Printing and related activities	..	..	..	..	..	..	..	..
		223	Reproduction of recorded media	..	..	..	..	..	..	..	..
	23		Coke and petroleum products; nuclear fuel	..	..	..	..	..	..	..	..
		231	Coke oven products	..	..	..	..	..	..	..	..
		232	Refined petroleum products	..	..	..	..	..	..	..	..
		233	Processing of nuclear fuel	..	..	..	..	..	..	..	..
	24		Chemicals and chemical products	..	..	..	..	..	..	5 464	5 828
		241	Basic chemicals	..	..	..	..	..	..	..	..
		242	Other chemical products	..	..	..	..	..	..	..	..
		243	Man-made fibres	..	..	..	..	..	..	..	..
	25		Rubber and plastic products	..	..	..	..	..	..	1 406	1 373
		251	Rubber products	..	..	..	..	..	..	..	..
		252	Plastic products	..	..	..	..	..	..	..	..
	26		Other non-metallic mineral products	..	..	..	..	..	..	1 125	1 068
		261	Glass and glass products	..	..	..	..	..	..	..	..
		269	Non-metallic mineral products, nec	..	..	..	..	..	..	..	..
	27		Basic metals	..	..	..	..	..	..	949	963
		271	Basic iron and steel	..	..	..	..	..	..	..	..
		272	Basic precious and non-ferrous metals	..	..	..	..	..	..	..	..
		273	Casting of metals	..	..	..	..	..	..	..	..
	28		Fabricated metal products, exc. machinery	..	..	..	..	..	..	5 024	5 194
		281	Structural metal prod., tanks, reservoirs, etc.	..	..	..	..	..	..	..	..
		289	Other fabricated metal products	..	..	..	..	..	..	..	..
	29		Machinery and equipment, nec	..	..	..	..	..	..	7 758	7 811
		291	General purpose machinery	..	..	..	..	..	..	..	..
		292	Special purpose machinery	..	..	..	..	..	..	..	..
		293	Domestic appliances, nec	..	..	..	..	..	..	..	..
	30		Office, accounting & computing machinery	..	..	..	..	..	..	..	..
	31		Electrical machinery & apparatus, nec	..	..	..	..	..	..	3 014	3 077
		311	Electric motors, generators and transformers	..	..	..	..	..	..	..	..
		312	Electricity distribution & control apparatus	..	..	..	..	..	..	..	..
		313	Insulated wire and cable	..	..	..	..	..	..	..	..
		314	Accumulators, primary cells & primary batteries	..	..	..	..	..	..	..	..
		315	Electric lamps and lighting equipment	..	..	..	..	..	..	..	..
		319	Other electrical equipment, nec	..	..	..	..	..	..	..	..
	32		Radio, TV, communication equipment	..	..	..	..	..	..	1 091	1 136
		321	Electronic valves, tubes; other electronic components	..	..	..	..	..	..	..	..
		322	TV/radio transmitters; line teleph./telegr. apparatus	..	..	..	..	..	..	..	..
		323	TV/radio receivers; sound and video app.; assoc. gds.	..	..	..	..	..	..	..	..
	33		Medical, precision & optical instrum.; watch & clocks	..	..	..	..	..	..	4 138	4 192
		331	Medical appl.; instruments & appl. for measuring, etc.	..	..	..	..	..	..	..	..
		332	Optical instruments and photographic equipment	..	..	..	..	..	..	..	..
		333	Watches and clocks	..	..	..	..	..	..	..	..
	34		Motor vehicles, trailers and semi-trailers	..	..	..	..	..	..	277	279
		341	Motor vehicles	..	..	..	..	..	..	..	..
		342	Bodies for motor vehicles, trailers & semi-trailers	..	..	..	..	..	..	..	..
		343	Parts & accessories for motor vehicles & their engines	..	..	..	..	..	..	..	..
	35		Other transport equipment	..	..	..	..	..	..	517	543
		351	Building and repairing of ships and boats	..	..	..	..	..	..	..	..
		352	Railway and tramway locomotives and rolling stock	..	..	..	..	..	..	..	..
		353	Aircraft and spacecraft	..	..	..	..	..	..	..	..
		359	Transport equipment, nec	..	..	..	..	..	..	..	..
	36		Furniture; manufacturing, nec	..	..	..	..	..	..	1 554	1 480
		361	Furniture	..	..	..	..	..	..	..	..
		369	Manufacturing, nec	..	..	..	..	..	..	..	..
	37		Recycling	..	..	..	..	..	..	107	104
E	40_41		Electricity, gas and water supply	..	..	..	..	..	..	..	..
	40		Electricity, gas, steam and hot water supply	..	..	..	..	..	..	1 715	1 726
	41		Collection, purification and distribution of water	..	..	..	..	..	..	..	..
F	45		Construction	..	..	..	..	..	..	14 997	15 920

SUISSE

Table SW.6 CHARGES SOCIALES DES EMPLOYEURS - SUPPLEMENTS TO WAGES & SALARIES

Millions de SF (Prix courants)

1991	1992	1993	1994	1995	1996	1997	1998		CITI révision 3		
..	..	..	..	..	..	..	..	Activites extractives	C	10_14	
..	..	..	..	..	..	..	..	Extraction de charbon, de lignite et de tourbe		10	
..	..	..	..	..	..	..	..	Pétrole brut et gaz naturel ; activités annexes		11	
..	..	..	..	..	..	..	..	Extraction de minerais métalliques		13	
..	..	..	..	..	..	78	82	Autres activités extractives		14	
..	..	..	..	..	..	..	..	Total activites manufacturieres	D	15_37	
..	..	..	..	..	..	548	536	Produits alimentaires et boissons		15	
..	..	..	..	..	..	..	..	Viande, poisson, fruits, légumes, huiles et graisses			151
..	..	..	..	..	..	..	..	Produits laitiers			152
..	..	..	..	..	..	..	..	Travail des grains; prod. amyl. ; alim. pour animaux			153
..	..	..	..	..	..	..	..	Autres produits alimentaires			154
..	..	..	..	..	..	..	..	Boissons			155
..	..	..	..	..	..	..	..	Produits à base de tabac		16	
..	..	..	..	..	..	147	168	Textiles		17	
..	..	..	..	..	..	..	..	Filature, tissage et achèvement des textiles			171
..	..	..	..	..	..	..	..	Autres articles textiles			172
..	..	..	..	..	..	..	..	Etoffes et articles de bonneterie			173
..	..	..	..	..	..	77	72	Articles d'habillement et fourrure		18	
..	..	..	..	..	..	..	..	Articles d'habillement autres qu'en fourrure			181
..	..	..	..	..	..	..	..	Prép. et teinture des fourrures ; articles en fourrure			182
..	..	..	..	..	..	23	19	Cuir et articles en cuir ; chaussures		19	
..	..	..	..	..	..	..	..	Cuir et articles en cuir sauf chaussures			191
..	..	..	..	..	..	..	..	Chaussures			192
..	..	..	..	..	..	321	320	Bois, articles en bois, sauf meubles		20	
..	..	..	..	..	..	..	..	Sciage et rabotage du bois			201
..	..	..	..	..	..	..	..	Articles en bois, liège, vannerie, sparterie			202
..	..	..	..	..	..	154	163	Papier, articles en papier et carton		21	
..	..	..	..	..	..	606	576	Edition, imprimerie et reproduction		22	
..	..	..	..	..	..	..	..	Edition			221
..	..	..	..	..	..	..	..	Imprimerie et activités annexes			222
..	..	..	..	..	..	..	..	Reproduction de supports enregistrés			223
..	..	..	..	..	..	..	..	Coke. et prod. pétroliers ; comb. nucléaires		23	
..	..	..	..	..	..	..	..	Cokéfaction			231
..	..	..	..	..	..	..	..	Produits pétroliers raffinés			232
..	..	..	..	..	..	..	..	Traitement de combustibles nucléaires			233
..	..	..	..	..	..	1 017	1 101	Produits chimiques		24	
..	..	..	..	..	..	..	..	Produits chimiques de base			241
..	..	..	..	..	..	..	..	Autres produits chimiques			242
..	..	..	..	..	..	..	..	Fibres synthétiques ou artificielles			243
..	..	..	..	..	..	230	229	Art. en caoutchouc et matières plastiques		25	
..	..	..	..	..	..	..	..	Articles en caoutchouc			251
..	..	..	..	..	..	..	..	Articles en matières plastiques			252
..	..	..	..	..	..	191	188	Autres produits minéraux non métalliques		26	
..	..	..	..	..	..	..	..	Verre et articles en verre			261
..	..	..	..	..	..	..	..	Produits minéraux non métalliques, nca			269
..	..	..	..	..	..	159	160	Produits métallurgiques de base		27	
..	..	..	..	..	..	..	..	Sidérurgie et première transformation de l'acier			271
..	..	..	..	..	..	..	..	Première transf. métaux précieux et non ferreux			272
..	..	..	..	..	..	..	..	Fonderie			273
..	..	..	..	..	..	866	901	Ouvrages en métaux, sauf machines		28	
..	..	..	..	..	..	..	..	Constr. et menuis. métal.; réservoirs; générateurs			281
..	..	..	..	..	..	..	..	Autres ouvrages en métaux			289
..	..	..	..	..	..	1 244	1 287	Machines et matériels, nca		29	
..	..	..	..	..	..	..	..	Machines d'usage général			291
..	..	..	..	..	..	..	..	Machines d'usage spécifique			292
..	..	..	..	..	..	..	..	Appareils domestiques, nca			293
..	..	..	..	..	..	..	..	Mach. de bureau, comptables, information		30	
..	..	..	..	..	..	457	484	Machines et appareils électriques, nca		31	
..	..	..	..	..	..	..	..	Moteurs, génératrices, transformateurs électriques			311
..	..	..	..	..	..	..	..	Matériel électrique de distribution et de commande			312
..	..	..	..	..	..	..	..	Fils et câbles électriques isolés			313
..	..	..	..	..	..	..	..	Accumulateurs et piles électriques			314
..	..	..	..	..	..	..	..	Lampes électriques et appareils d'éclairage			315
..	..	..	..	..	..	..	..	Autres matériels électriques, nca.			319
..	..	..	..	..	..	183	197	Equipements et appareils de radio, TV, etc.		32	
..	..	..	..	..	..	..	..	Tubes et valves électroniques;autres composants			321
..	..	..	..	..	..	..	..	Emetteurs radio/TV; app. téléphonie et télégraphie			322
..	..	..	..	..	..	..	..	Récepteurs radio/TV, son et image; art. associés			323
..	..	..	..	..	..	755	835	Instr. médicaux, de précision ; optique ; horlog.		33	
..	..	..	..	..	..	..	..	Instruments médicaux et de précision, etc.			331
..	..	..	..	..	..	..	..	Instruments d'optique ; matériel photographique			332
..	..	..	..	..	..	..	..	Horlogerie			333
..	..	..	..	..	..	45	43	Véhicules automobiles; remorques, etc.		34	
..	..	..	..	..	..	..	..	Véhicules automobiles			341
..	..	..	..	..	..	..	..	Carross. automobiles, remorques, semi-remorques			342
..	..	..	..	..	..	..	..	Pièces et accessoires pour automobiles et moteurs			343
..	..	..	..	..	..	77	85	Autres matériels de transport		35	
..	..	..	..	..	..	..	..	Construction et réparation de navires			351
..	..	..	..	..	..	..	..	Matériel ferroviaire roulant			352
..	..	..	..	..	..	..	..	Construction aéronautique et spatiale			353
..	..	..	..	..	..	..	..	Autres équipements de transport			359
..	..	..	..	..	..	292	267	Meubles ; activités de fabrication, nca		36	
..	..	..	..	..	..	..	..	Meubles			361
..	..	..	..	..	..	..	..	Activités de fabrication, nca			369
..	..	..	..	..	..	19	21	Récupération		37	
..	..	..	..	..	..	..	..	Electricité, gaz et eau	E	40_41	
..	..	..	..	..	..	333	331	Prod. et dist. d'électricité gaz, vapeur d'eau chaude		40	
..	..	..	..	..	..	..	..	Captage, épuration et distribution d'eau		41	
..	..	..	..	..	..	2 930	3 050	Construction	F	45	

Table UK.1 **PRODUCTION - PRODUCTION**

Millions of £ (Current Prices)

ISIC revision 3				1991	1992	1993	1994	1995	1996	1997	1998
C	10_14		Mining and quarrying	..	..	6 633	5 196	5 719	6 013	25 136	24 866
	10		Mining of coal and lignite; extraction of peat	..	..	..	2 049	2 531	2 306	1 996	1 720
	11		Crude petroleum & natural gas; service activities	..	..	13 842	15 942	17 829	20 995	19 535	19 282
	13		Mining of metal ores	..	..	..	..	10	11	22	8
	14		Other mining and quarrying	..	..	..	3 147	3 179	3 696	3 583	3 857
D	15_37		Total manufacturing	..	..	357 071	385 952	418 599	442 124	427 779	423 528
	15		Food products and beverages	..	..	58 571	60 746	63 690	66 087	62 706	60 195
		151	Production of meat, fish, fruit, vegetables, oils & fats	..	..	17 216	17 930	18 946	19 986	18 712	16 971
		152	Dairy products	..	..	7 632	7 714	8 644	7 750	7 065	7 086
		153	Grain mill and starch products; animal feeds	..	..	8 428	8 229	8 142	9 000	7 594	7 323
		154	Other food products	..	..	14 557	15 852	15 680	17 044	16 884	16 671
		155	Beverages	..	..	10 737	11 021	12 278	12 306	12 450	12 144
	16		Tobacco products	..	..	7 591	7 542	8 682	9 407	9 038	8 340
	17		Textiles	..	..	9 339	9 914	10 530	11 133	10 510	9 712
		171	Spinning, weaving and finishing of textiles	..	..	3 742	4 181	3 803	3 935	3 771	3 427
		172	Other textiles	..	..	3 609	3 734	4 439	4 725	4 553	4 378
		173	Knitted and crocheted fabrics and articles	..	..	1 988	1 999	2 288	2 472	2 185	1 907
	18		Wearing apparel and fur	..	..	5 388	5 730	6 286	6 160	7 356	6 449
		181	Wearing apparel, except fur apparel	..	..	5 342	5 704	6 253	6 125	7 321	6 422
		182	Dressing and dyeing of fur; articles of fur	..	..	47	26	33	36	34	27
	19		Leather and articles of leather; footwear	..	..	2 541	2 712	2 294	2 353	1 885	1 675
		191	Leather and articles of leather	..	..	976	970	878	821	760	721
		192	Footwear	..	..	1 565	1 741	1 416	1 532	1 124	955
	20		Wood & products of wood & cork, exc. furniture	..	..	4 581	5 490	5 759	5 602	5 552	5 263
		201	Sawmilling and planing of wood	..	..	1 257	1 694	1 385	924	874	806
		202	Wood products; articles of cork, straw, etc.	..	..	3 324	3 796	4 374	4 678	4 677	4 457
	21		Paper and paper products	..	..	10 458	11 698	13 420	13 282	11 672	11 411
	22		Printing, publishing and recorded media	..	..	21 600	23 718	25 505	27 378	28 956	29 849
		221	Publishing	..	..	11 124	11 952	13 245	14 206	14 721	15 884
		222	Printing and related activities	..	..	9 867	11 165	11 471	12 248	13 238	13 076
		223	Reproduction of recorded media	..	..	609	601	790	924	997	888
	23		Coke and petroleum products; nuclear fuel	..	..	22 254	21 851	22 335	25 941	24 923	24 900
		231	Coke oven products	..	..	117	141	..	74	61	36
		232	Refined petroleum products	..	..	20 729	20 328	20 383	24 208	23 295	23 099
		233	Processing of nuclear fuel	..	..	1 408	1 381	..	1 659	1 567	1 765
	24		Chemicals and chemical products	..	..	37 910	40 412	44 325	45 260	40 855	38 765
		241	Basic chemicals	..	..	14 408	15 169	17 970	17 574	15 038	14 132
		242	Other chemical products	..	..	22 394	24 268	25 188	26 582	24 974	23 828
		243	Man-made fibres	..	..	1 108	975	1 167	1 105	844	805
	25		Rubber and plastic products	..	..	14 919	16 611	17 968	19 147	19 014	18 905
		251	Rubber products	..	..	3 185	3 422	3 825	4 297	3 401	3 686
		252	Plastic products	..	..	11 734	13 189	14 143	14 851	15 612	15 219
	26		Other non-metallic mineral products	..	..	9 160	10 649	11 762	11 257	11 313	10 771
		261	Glass and glass products	..	..	2 245	2 527	2 999	2 810	2 675	2 477
		269	Non-metallic mineral products, nec	..	..	6 915	8 122	8 763	8 447	8 638	8 294
	27		Basic metals	..	..	15 173	16 866	19 572	19 170	17 253	16 880
		271	Basic iron and steel	..	..	8 485	9 719	10 938	10 214	10 180	9 975
		272	Basic precious and non-ferrous metals	..	..	4 975	5 118	6 556	6 614	4 961	4 973
		273	Casting of metals	..	..	1 713	2 030	2 078	2 342	2 112	1 932
	28		Fabricated metal products, exc. machinery	..	..	18 828	21 165	23 555	24 498	24 309	25 855
		281	Structural metal prod., tanks, reservoirs, etc.	..	..	5 855	6 265	6 105	6 517	7 570	8 166
		289	Other fabricated metal products	..	..	12 973	14 900	17 450	17 981	16 739	17 690
	29		Machinery and equipment, nec	..	..	26 284	28 712	33 612	34 643	33 157	32 406
		291	General purpose machinery	..	..	13 719	15 234	17 551	17 912	16 512	16 777
		292	Special purpose machinery	..	..	10 273	11 001	13 506	14 010	13 912	13 103
		293	Domestic appliances, nec	..	..	2 291	2 477	2 555	2 721	2 732	2 526
	30		Office, accounting & computing machinery	..	..	12 765	12 154	12 733	14 270	13 127	13 799
	31		Electrical machinery & apparatus, nec	..	..	11 315	12 145	12 583	13 636	13 834	13 433
		311	Electric motors, generators and transformers	..	..	1 919	1 946	2 173	2 294	2 665	2 218
		312	Electricity distribution & control apparatus	..	..	2 731	3 056	3 432	3 560	3 525	3 674
		313	Insulated wire and cable	..	..	1 675	1 900	2 070	2 007	1 992	1 579
		314	Accumulators, primary cells & primary batteries	..	..	574	633	577	691	556	542
		315	Electric lamps and lighting equipment	..	..	1 341	1 422	1 392	1 511	1 550	1 580
		319	Other electrical equipment, nec	..	..	3 075	3 190	2 939	3 574	3 547	3 840
	32		Radio, TV, communication equipment	..	..	9 633	12 428	15 047	16 050	16 463	16 266
		321	Electronic valves, tubes; other electronic components	..	..	3 282	4 214	5 095	4 905	5 696	4 708
		322	TV/radio transmitters; line teleph./telegr. apparatus	..	..	3 390	4 712	5 609	6 591	6 489	7 773
		323	TV/radio receivers; sound and video app.; assoc. gds.	..	..	2 961	3 502	4 343	4 554	4 278	3 786
	33		Medical, precision & optical instrum.; watch & clocks	..	..	8 624	8 698	9 611	10 819	10 303	10 256
		331	Medical appl.; instruments & appl. for measuring, etc.	..	..	7 705	7 608	8 430	9 449	8 914	8 905
		332	Optical instruments and photographic equipment	..	..	817	936	1 039	1 222	1 264	1 228
		333	Watches and clocks	..	..	103	155	142	148	125	123
	34		Motor vehicles, trailers and semi-trailers	..	..	26 577	30 760	33 217	36 857	33 924	34 983
		341	Motor vehicles	..	..	19 786	22 656	23 783	27 243	23 357	23 815
		342	Bodies for motor vehicles, trailers & semi-trailers	..	..	1 638	2 092	2 367	2 358	2 600	2 430
		343	Parts & accessories for motor vehicles & their engines	..	..	5 153	6 012	7 066	7 255	7 966	8 738
	35		Other transport equipment	..	..	13 839	13 536	12 688	15 378	17 351	19 023
		351	Building and repairing of ships and boats	..	..	3 144	2 503	2 257	2 610	2 879	3 067
		352	Railway and tramway locomotives and rolling stock	..	..	1 189	1 221	779	1 143	1 165	922
		353	Aircraft and spacecraft	..	..	9 143	9 488	9 221	11 189	12 885	14 617
		359	Transport equipment, nec	..	..	364	324	431	436	422	418
	36		Furniture; manufacturing, nec	..	..	9 722	11 537	12 095	12 402	12 989	13 218
		361	Furniture	..	..	6 393	7 214	7 353	7 994	8 652	8 819
		369	Manufacturing, nec	..	..	3 329	4 323	4 742	4 408	4 337	4 399
	37		Recycling	..	..	..	877	1 331	1 394	1 291	1 173
E	40_41		Electricity, gas and water supply	..	..	42 043	42 205	45 371	44 708	43 496	45 062
	40		Electricity, gas, steam and hot water supply	..	..	38 724	38 771	41 729	41 148	39 319	40 772
	41		Collection, purification and distribution of water	..	..	3 319	3 435	3 642	3 560	4 177	4 290
F	45		Construction	..	..	78 944	70 582	80 560	82 218	97 796	103 500

Table UK.2 **VALEUR AJOUTÉE AU COUT DES FACTEURS - VALUE ADDED AT FACTOR COSTS**

Millions de £ (Prix courants)

1991	1992	1993	1994	1995	1996	1997	1998		CITI révision 3		
..	..	2 319	1 812	2 254	2 485	16 381	14 020	Activites extractives	C	10_14	
..	..	1 228	753	888	906	871	691	Extraction de charbon, de lignite et de tourbe		10	
..	..	9 522	11 292	12 427	15 839	13 789	11 762	Pétrole brut et gaz naturel ; activités annexes		11	
..	..	..	..	4	4	10	4	Extraction de minerais métalliques		13	
..	..	1 092	1 059	1 362	1 575	1 710	1 563	Autres activités extractives		14	
..	..	115 177	125 741	135 386	138 753	146 210	147 332	Total activites manufacturieres	D	15_37	
..	..	15 450	16 313	16 137	17 616	18 452	17 674	Produits alimentaires et boissons		15	
..	..	4 248	4 418	4 438	4 811	4 922	4 724	Viande, poisson, fruits, légumes, huiles et graisses			151
..	..	1 457	1 576	1 456	1 548	1 479	1 441	Produits laitiers			152
..	..	1 731	1 830	1 658	1 747	1 928	1 706	Travail des grains; prod. amyl. ; alim. pour animaux			153
..	..	5 206	5 704	5 347	6 025	6 579	6 390	Autres produits alimentaires			154
..	..	2 808	2 785	3 238	3 486	3 544	3 413	Boissons			155
..	..	1 290	1 296	1 503	1 661	1 318	1 377	Produits à base de tabac		16	
..	..	3 602	3 698	3 937	4 136	4 126	3 883	Textiles		17	
..	..	1 386	1 486	1 375	1 405	1 426	1 355	Filature, tissage et achèvement des textiles			171
..	..	1 377	1 374	1 612	1 729	1 731	1 693	Autres articles textiles			172
..	..	839	837	951	1 002	969	835	Etoffes et articles de bonneterie			173
..	..	2 219	2 466	2 437	2 553	3 411	2 675	Articles d'habillement et fourrure		18	
..	..	2 201	2 453	2 421	2 545	3 401	2 667	Articles d'habillement autres qu'en fourrure			181
..	..	18	13	16	8	10	7	Prép. et teinture des fourrures ; articles en fourrure			182
..	..	970	976	789	872	771	659	Cuir et articles en cuir ; chaussures		19	
..	..	346	297	271	263	250	258	Cuir et articles en cuir sauf chaussures			191
..	..	625	679	518	609	521	401	Chaussures			192
..	..	1 436	1 598	1 814	1 747	2 157	2 196	Bois, articles en bois, sauf meubles		20	
..	..	259	341	285	252	309	336	Sciage et rabotage du bois			201
..	..	1 177	1 258	1 529	1 495	1 848	1 860	Articles en bois, liège, vannerie, sparterie			202
..	..	3 580	4 044	4 503	4 222	4 153	3 912	Papier, articles en papier et carton		21	
..	..	9 969	10 758	11 733	12 181	13 266	14 185	Edition, imprimerie et reproduction		22	
..	..	5 128	5 710	5 967	5 901	6 329	7 208	Edition			221
..	..	4 712	4 899	5 378	5 844	6 442	6 508	Imprimerie et activités annexes			222
..	..	129	150	389	437	495	468	Reproduction de supports enregistrés			223
..	..	2 347	2 270	2 912	2 479	1 945	2 252	Coke. et prod. pétroliers ; comb. nucléaires		23	
..	..	34	40	..	23	18	13	Cokéfaction			231
..	..	1 565	1 600	1 753	1 573	1 214	1 359	Produits pétroliers raffinés			232
..	..	748	630	..	883	714	880	Traitement de combustibles nucléaires			233
..	..	12 615	13 511	15 412	15 267	14 313	14 031	Produits chimiques		24	
..	..	3 618	4 175	5 589	5 334	4 169	4 333	Produits chimiques de base			241
..	..	8 624	9 051	9 387	9 494	9 840	9 393	Autres produits chimiques			242
..	..	373	285	435	438	304	304	Fibres synthétiques ou artificielles			243
..	..	5 820	6 382	6 554	6 987	7 692	7 857	Art. en caoutchouc et matières plastiques		25	
..	..	1 371	1 457	1 502	1 569	1 524	1 653	Articles en caoutchouc			251
..	..	4 449	4 925	5 052	5 417	6 168	6 205	Articles en matières plastiques			252
..	..	3 665	4 533	4 977	4 864	4 977	4 754	Autres produits minéraux non métalliques		26	
..	..	889	1 025	1 156	1 263	1 268	1 077	Verre et articles en verre			261
..	..	2 775	3 508	3 820	3 601	3 709	3 676	Produits minéraux non métalliques, nca			269
..	..	4 284	4 883	5 757	5 003	4 847	4 636	Produits métallurgiques de base		27	
..	..	2 380	2 719	3 341	2 622	2 682	2 397	Sidérurgie et première transformation de l'acier			271
..	..	1 125	1 287	1 532	1 410	1 169	1 345	Première transf. métaux précieux et non ferreux			272
..	..	779	876	884	970	996	894	Fonderie			273
..	..	7 935	8 729	10 133	10 010	11 162	12 320	Ouvrages en métaux, sauf machines		28	
..	..	2 104	2 223	2 306	2 365	3 189	3 466	Constr. et menuis. métal.; réservoirs; générateurs			281
..	..	5 832	6 506	7 827	7 644	7 973	8 854	Autres ouvrages en métaux			289
..	..	9 575	10 553	12 015	12 265	13 161	13 050	Machines et matériels, nca		29	
..	..	5 257	5 849	6 575	6 520	7 128	7 213	Machines d'usage général			291
..	..	3 544	3 823	4 652	4 882	5 098	4 870	Machines d'usage spécifique			292
..	..	775	882	787	863	934	966	Appareils domestiques, nca			293
..	..	2 571	2 861	2 906	2 929	2 706	3 210	Mach. de bureau, comptables, information		30	
..	..	4 220	4 715	4 818	4 953	5 595	5 292	Machines et appareils électriques, nca		31	
..	..	692	754	844	931	1 059	808	Moteurs, génératrices, transformateurs électriques			311
..	..	1 128	1 205	1 322	1 360	1 596	1 738	Matériel électrique de distribution et de commande			312
..	..	629	771	723	629	734	547	Fils et câbles électriques isolés			313
..	..	219	250	223	233	202	175	Accumulateurs et piles électriques			314
..	..	490	491	473	488	685	593	Lampes électriques et appareils d'éclairage			315
..	..	1 062	1 244	1 234	1 312	1 319	1 430	Autres matériels électriques, nca.			319
..	..	3 400	4 244	5 003	5 028	5 703	5 968	Equipements et appareils de radio, TV, etc.		32	
..	..	1 391	1 733	2 047	1 690	2 478	1 901	Tubes et valves électroniques;autres composants			321
..	..	1 288	1 667	1 876	2 344	2 104	3 129	Emetteurs radio/TV; app. téléphonie et télégraphie			322
..	..	720	844	1 079	994	1 121	939	Récepteurs radio/TV, son et image; art. associés			323
..	..	3 594	3 925	3 991	4 288	4 707	4 718	Instr. médicaux, de précision ; optique ; horlog.		33	
..	..	3 185	3 469	3 428	3 720	4 042	4 125	Instruments médicaux et de précision, etc.			331
..	..	367	401	505	506	622	552	Instruments d'optique ; matériel photographique			332
..	..	42	55	59	62	42	41	Horlogerie			333
..	..	7 218	7 225	8 343	9 469	9 671	9 273	Véhicules automobiles; remorques, etc.		34	
..	..	4 908	4 322	4 953	6 220	5 848	5 179	Véhicules automobiles			341
..	..	495	680	786	709	801	782	Carross. automobiles, remorques, semi-remorques			342
..	..	1 815	2 223	2 604	2 540	3 023	3 312	Pièces et accessoires pour automobiles et moteurs			343
..	..	5 764	6 393	5 083	5 168	6 344	7 479	Autres matériels de transport		35	
..	..	1 413	1 475	1 149	1 174	1 137	1 332	Construction et réparation de navires			351
..	..	319	338	239	382	376	..	Matériel ferroviaire roulant			352
..	..	3 919	4 471	3 523	3 457	4 685	..	Construction aéronautique et spatiale			353
..	..	113	109	171	156	145	127	Autres équipements de transport			359
..	..	3 655	4 180	4 367	4 733	5 381	5 649	Meubles ; activités de fabrication, nca		36	
..	..	2 405	2 634	2 659	3 084	3 548	3 779	Meubles			361
..	..	1 249	1 546	1 708	1 649	1 833	1 870	Activités de fabrication, nca			369
..	..	..	186	264	321	352	286	Récupération		37	
..	..	15 944	15 651	14 961	15 355	15 451	14 688	Electricité, gaz et eau	E	40_41	
..	..	13 698	13 312	12 450	12 951	12 763	11 807	Prod. et dist. d'électricité gaz, vapeur d'eau chaude		40	
..	..	2 246	2 339	2 511	2 404	2 688	2 881	Captage, épuration et distribution d'eau		41	
..	..	19 604	19 907	24 153	25 717	33 524	34 149	Construction	F	45	

UNITED KINGDOM

Table UK.3 **INVESTMENT - INVESTISSEMENT**

Millions of £ (Current Prices)

ISIC revision 3				1991	1992	1993	1994	1995	1996	1997	1998
C	**10_14**		Mining and quarrying	..	..	347	272	298	474	4 799	6 327
	10		Mining of coal and lignite; extraction of peat	..	..	219	73	92	157	141	91
	11		Crude petroleum & natural gas; service activities	..	..	4 664	3 751	4 444	4 440	4 397	5 914
	13		Mining of metal ores	..	..	..	..	..	1	1	
	14		Other mining and quarrying	..	..	127	199	..	316	261	321
D	**15_37**		Total manufacturing	..	..	13 048	14 011	17 406	17 837	20 589	20 533
	15		Food products and beverages	..	..	2 235	2 206	2 349	2 278	2 559	2 608
		151	Production of meat, fish, fruit, vegetables, oils & fats	..	..	555	565	655	606	696	665
		152	Dairy products	..	..	216	180	201	203	246	237
		153	Grain mill and starch products; animal feeds	..	..	261	239	258	228	289	289
		154	Other food products	..	..	687	713	653	604	775	689
		155	Beverages	..	..	517	509	582	637	553	728
	16		Tobacco products	..	..	86	85	95	68	107	146
	17		Textiles	..	..	319	364	352	445	344	422
		171	Spinning, weaving and finishing of textiles	..	..	118	177	149	140	135	153
		172	Other textiles	..	..	113	104	126	161	145	203
		173	Knitted and crocheted fabrics and articles	..	..	88	82	77	144	65	66
	18		Wearing apparel and fur	..	..	79	107	115	130	162	132
		181	Wearing apparel, except fur apparel	..	..	78	106	115	130	162	131
		182	Dressing and dyeing of fur; articles of fur	..	..	1	1	..	..	..	1
	19		Leather and articles of leather; footwear	..	..	53	66	61	49	49	42
		191	Leather and articles of leather	..	..	20	21	13	23	20	20
		192	Footwear	..	..	33	44	48	26	29	22
	20		Wood & products of wood & cork, exc. furniture	..	..	132	131	158	170	237	254
		201	Sawmilling and planing of wood	..	..	33	35	41	24	40	26
		202	Wood products; articles of cork, straw, etc.	..	..	99	96	117	146	197	228
	21		Paper and paper products	..	..	577	677	803	633	728	626
	22		Printing, publishing and recorded media	..	..	1 067	1 084	1 318	1 151	1 359	1 334
		221	Publishing	..	..	396	362	487	327	399	419
		222	Printing and related activities	..	..	646	694	763	775	908	860
		223	Reproduction of recorded media	..	..	26	28	68	49	52	55
	23		Coke and petroleum products; nuclear fuel	..	..	722	580	752	717	704	728
		231	Coke oven products	..	..	2	3	..	..	..	..
		232	Refined petroleum products	..	..	..	317	361	321	230	208
		233	Processing of nuclear fuel	..	..	..	260	391	..	..	..
	24		Chemicals and chemical products	..	..	1 882	1 889	2 286	2 758	2 917	3 303
		241	Basic chemicals	..	..	732	670	848	1 269	1 279	1 548
		242	Other chemical products	..	..	1 087	1 157	1 346	1 368	1 549	1 627
		243	Man-made fibres	..	..	64	62	92	121	90	127
	25		Rubber and plastic products	..	..	768	765	953	984	1 072	1 098
		251	Rubber products	..	..	130	114	132	181	150	162
		252	Plastic products	..	..	638	651	820	803	922	936
	26		Other non-metallic mineral products	..	..	413	423	608	672	571	631
		261	Glass and glass products	..	..	142	135	186	190	160	174
		269	Non-metallic mineral products, nec	..	..	271	288	422	482	411	456
	27		Basic metals	..	..	364	382	525	626	680	688
		271	Basic iron and steel	..	..	183	206	287	301	375	396
		272	Basic precious and non-ferrous metals	..	..	113	92	121	143	182	171
		273	Casting of metals	..	..	67	84	116	182	123	121
	28		Fabricated metal products, exc. machinery	..	..	599	747	845	946	1 065	1 352
		281	Structural metal prod., tanks, reservoirs, etc.	..	..	117	111	159	..	204	323
		289	Other fabricated metal products	..	..	482	636	686	..	861	1 029
	29		Machinery and equipment, nec	..	..	714	873	1 080	1 178	1 109	990
		291	General purpose machinery	..	..	361	446	570	607	540	566
		292	Special purpose machinery	..	..	261	330	407	454	461	331
		293	Domestic appliances, nec	..	..	92	97	103	116	108	..
	30		Office, accounting & computing machinery	..	..	270	335	364	291	680	679
	31		Electrical machinery & apparatus, nec	..	..	400	346	406	459	693	782
		311	Electric motors, generators and transformers	..	..	82	58	49	68	101	102
		312	Electricity distribution & control apparatus	..	..	111	76	106	122	147	310
		313	Insulated wire and cable	..	..	66	58	55	74	106	88
		314	Accumulators, primary cells & primary batteries	..	..	22	24	31	29	37	49
		315	Electric lamps and lighting equipment	..	..	38	36	43	48	70	58
		319	Other electrical equipment, nec	..	..	80	95	123	117	231	176
	32		Radio, TV, communication equipment	..	..	547	748	1 260	1 070	1 358	1 139
		321	Electronic valves, tubes; other electronic components	..	..	283	400	815	711	695	602
		322	TV/radio transmitters; line teleph./telegr. apparatus	..	..	147	195	267	204	428	333
		323	TV/radio receivers; sound and video app.; assoc. gds.	..	..	117	152	177	155	234	204
	33		Medical, precision & optical instrum.; watch & clocks	..	..	276	364	340	375	401	410
		331	Medical appl.; instruments & appl. for measuring, etc.	..	..	..	328	264	290	262	312
		332	Optical instruments and photographic equipment	..	..	33	32	67	83	134	96
		333	Watches and clocks	..	..	..	4	10	2	5	2
	34		Motor vehicles, trailers and semi-trailers	..	..	988	1 267	1 935	1 944	2 606	2 047
		341	Motor vehicles	..	..	711	975	1 570	1 520	2 036	1 455
		342	Bodies for motor vehicles, trailers & semi-trailers	..	..	34	39	26	45	51	61
		343	Parts & accessories for motor vehicles & their engines	..	..	243	253	338	380	519	531
	35		Other transport equipment	..	..	313	245	368	470	631	702
		351	Building and repairing of ships and boats	..	..	37	27	33	148	76	49
		352	Railway and tramway locomotives and rolling stock	..	..	15	..	65	19	26	27
		353	Aircraft and spacecraft	..	..	247	..	249	287	517	610
		359	Transport equipment, nec	..	..	15	8	21	16	13	17
	36		Furniture; manufacturing, nec	..	..	243	304	369	385	480	458
		361	Furniture	..	..	139	171	171	262	302	273
		369	Manufacturing, nec	..	..	104	133	198	..	178	186
	37		Recycling	..	..	..	24	65	37	77	63
E	**40_41**		Electricity, gas and water supply	..	..	5 303	5 277	4 795	4 175	5 102	5 583
	40		Electricity, gas, steam and hot water supply	..	..	3 788	3 959	3 520	2 623	3 269	3 786
	41		Collection, purification and distribution of water	..	..	1 514	1 317	1 276	1 552	1 833	1 797
F	**45**		Construction	..	..	1 118	1 250	1 299	1 187	1 875	1 783

Table UK.4 DÉPENSES DE MACHINES ET D'ÉQUIPEMENT - INVESTMENT IN MACHINERY & EQUIPMENT

Millions de £ (Prix courants)

1991	1992	1993	1994	1995	1996	1997	1998			CITI révision 3	
..	..	..	..	..	434	5 110	5 656	Activites extractives	C	10_14	
..	..	..	..	..	153	145	109	Extraction de charbon, de lignite et de tourbe		10	
..	..	..	..	..	..	4 698	5 291	Pétrole brut et gaz naturel ; activités annexes		11	
..	..	..	..	..	1	..	..	Extraction de minerais métalliques		13	
..	..	..	150	..	280	266	256	Autres activités extractives		14	
..	..	10 426	11 241	13 409	16 915	19 731	19 142	Total activites manufacturieres	D	15_37	
..	..	1 615	1 606	1 776	2 100	2 435	2 321	Produits alimentaires et boissons		15	
..	..	362	388	460	536	615	578	Viande, poisson, fruits, légumes, huiles et graisses			151
..	..	177	148	160	199	228	202	Produits laitiers			152
..	..	212	195	194	228	286	248	Travail des grains; prod. amyl. ; alim. pour animaux			153
..	..	491	526	532	557	716	615	Autres produits alimentaires			154
..	..	373	349	430	581	589	678	Boissons			155
..	..	53	44	74	70	105	144	Produits à base de tabac		16	
..	..	266	313	306	489	365	432	Textiles		17	
..	..	99	146	128	170	133	168	Filature, tissage et achèvement des textiles			171
..	..	95	96	97	167	155	194	Autres articles textiles			172
..	..	72	71	82	152	77	70	Etoffes et articles de bonneterie			173
..	..	72	96	100	149	167	129	Articles d'habillement et fourrure		18	
..	..	71	95	99	149	167	128	Articles d'habillement autres qu'en fourrure			181
..	..	1	1	..	..	..	1	Prép. et teinture des fourrures ; articles en fourrure			182
..	..	46	54	51	49	55	50	Cuir et articles en cuir ; chaussures		19	
..	..	18	18	12	23	21	22	Cuir et articles en cuir sauf chaussures			191
..	..	27	36	39	26	34	29	Chaussures			192
..	..	82	74	80	175	225	253	Bois, articles en bois, sauf meubles		20	
..	..	10	11	12	23	45	34	Sciage et rabotage du bois			201
..	..	72	63	69	152	180	218	Articles en bois, liège, vannerie, sparterie			202
..	..	473	582	716	575	750	756	Papier, articles en papier et carton		21	
..	..	798	856	848	1 203	1 441	1 482	Edition, imprimerie et reproduction		22	
..	..	257	272	297	330	342	460	Edition			221
..	..	518	557	499	828	1 042	968	Imprimerie et activités annexes			222
..	..	23	27	52	46	56	54	Reproduction de supports enregistrés			223
..	..	620	518	475	614	659	680	Coke. et prod. pétroliers ; comb. nucléaires		23	
..	..	..	3	..	..	..	..	Cokéfaction			231
..	..	317	291	293	299	215	197	Produits pétroliers raffinés			232
..	..	..	223	182	..	..	..	Traitement de combustibles nucléaires			233
..	..	1 458	1 473	1 868	2 540	2 786	2 980	Produits chimiques		24	
..	..	628	607	758	1 197	1 261	1 369	Produits chimiques de base			241
..	..	770	818	1 019	1 222	1 440	1 499	Autres produits chimiques			242
..	..	60	48	91	120	85	112	Fibres synthétiques ou artificielles			243
..	..	637	622	716	964	1 008	1 059	Art. en caoutchouc et matières plastiques		25	
..	..	111	102	120	166	153	157	Articles en caoutchouc			251
..	..	526	520	595	798	855	902	Articles en matières plastiques			252
..	..	534	547	473	663	583	619	Autres produits minéraux non métalliques		26	
..	..	134	127	160	..	159	170	Verre et articles en verre			261
..	..	399	421	313	477	424	449	Produits minéraux non métalliques, nca			269
..	..	..	..	483	639	664	627	Produits métallurgiques de base		27	
..	..	..	..	268	301	366	362	Sidérurgie et première transformation de l'acier			271
..	..	104	84	115	152	182	159	Première transf. métaux précieux et non ferreux			272
..	..	53	66	100	186	116	106	Fonderie			273
..	..	450	518	565	938	1 032	1 102	Ouvrages en métaux, sauf machines		28	
..	..	77	72	88	150	188	206	Constr. et menuis. métal.; réservoirs; générateurs			281
..	..	373	446	476	788	843	896	Autres ouvrages en métaux			289
..	..	538	644	832	1 089	1 091	1 104	Machines et matériels, nca		29	
..	..	253	308	464	554	514	582	Machines d'usage général			291
..	..	207	255	287	422	472	431	Machines d'usage spécifique			292
..	..	78	81	80	112	105	..	Appareils domestiques, nca			293
..	..	212	272	308	256	614	617	Mach. de bureau, comptables, information		30	
..	..	307	296	342	458	651	492	Machines et appareils électriques, nca		31	
..	..	44	44	38	66	72	81	Moteurs, génératrices, transformateurs électriques			311
..	..	85	70	98	121	131	141	Matériel électrique de distribution et de commande			312
..	..	61	52	56	71	94	73	Fils et câbles électriques isolés			313
..	..	21	21	28	29	31	29	Accumulateurs et piles électriques			314
..	..	30	32	32	47	78	45	Lampes électriques et appareils d'éclairage			315
..	..	65	77	90	124	246	122	Autres matériels électriques, nca.			319
..	..	470	653	872	920	1 125	939	Equipements et appareils de radio, TV, etc.		32	
..	..	241	347	522	580	574	495	Tubes et valves électroniques;autres composants			321
..	..	131	177	235	217	385	284	Emetteurs radio/TV; app. téléphonie et télégraphie			322
..	..	99	129	115	123	167	161	Récepteurs radio/TV, son et image; art. associés			323
..	..	208	274	237	341	394	378	Instr. médicaux, de précision ; optique ; horlog.		33	
..	..	183	240	187	273	271	291	Instruments médicaux et de précision, etc.			331
..	..	25	30	48	65	118	86	Instruments d'optique ; matériel photographique			332
..	..	..	3	3	3	5	1	Horlogerie			333
..	..	995	1 199	1 740	1 782	2 518	1 998	Véhicules automobiles; remorques, etc.		34	
..	..	762	955	1 446	1 370	2 014	1 431	Véhicules automobiles			341
..	..	19	22	15	36	37	68	Carross. automobiles, remorques, semi-remorques			342
..	..	214	222	279	376	467	498	Pièces et accessoires pour automobiles et moteurs			343
..	..	244	196	253	438	501	593	Autres matériels de transport		35	
..	..	20	13	33	136	38	57	Construction et réparation de navires			351
..	..	2	..	15	17	24	25	Matériel ferroviaire roulant			352
..	..	207	202	189	271	426	496	Construction aéronautique et spatiale			353
..	..	14	8	15	13	12	15	Autres équipements de transport			359
..	..	191	231	245	420	482	418	Meubles ; activités de fabrication, nca		36	
..	..	108	130	134	277	306	236	Meubles			361
..	..	84	102	111	143	177	181	Activités de fabrication, nca			369
..	..	..	23	49	41	80	63	Récupération		37	
..	..	..	..	..	3 191	3 666	3 724	Electricité, gaz et eau	E	40_41	
..	..	..	..	..	2 585	2 924	3 019	Prod. et dist. d'électricité gaz, vapeur d'eau chaude		40	
..	..	461	389	390	606	742	705	Captage, épuration et distribution d'eau		41	
..	..	667	636	627	1 610	1 922	2 050	Construction	F	45	

Table UK.5 EMPLOYMENT, EMPLOYEES - EMPLOI, SALARIÉS

Thousands

ISIC revision 3				1991	1992	1993	1994	1995	1996	1997	1998
C	10_14		Mining and quarrying	..	..	..	..	..	..	..	..
	10		Mining of coal and lignite; extraction of peat	..	..	48	14	17	19	16	16
	11		Crude petroleum & natural gas; service activities	..	..	..	..	..	..	..	30
	13		Mining of metal ores	..	..	..	..	..	..	..	..
	14		Other mining and quarrying	..	..	31	31	..	36	31	29
D	15_37		Total manufacturing	..	..	4 275	4 348	4 415	4 235	4 183	4 083
	15		Food products and beverages	..	..	557	551	522	486	480	457
		151	Production of meat, fish, fruit, vegetables, oils & fats	..	..	188	194	190	174	170	162
		152	Dairy products	..	..	48	47	43	41	39	34
		153	Grain mill and starch products; animal feeds	..	..	36	34	34	34	33	30
		154	Other food products	..	..	228	220	199	186	184	178
		155	Beverages	..	..	58	57	56	50	54	52
	16		Tobacco products	..	..	9	9	9	10	8	7
	17		Textiles	..	..	203	195	196	194	190	166
		171	Spinning, weaving and finishing of textiles	..	..	73	72	65	64	60	52
		172	Other textiles	..	..	75	71	77	78	83	74
		173	Knitted and crocheted fabrics and articles	..	..	55	52	54	52	48	39
	18		Wearing apparel and fur	..	..	186	173	183	160	162	132
		181	Wearing apparel, except fur apparel	..	..	185	173	182	159	162	131
		182	Dressing and dyeing of fur; articles of fur	..	..	1	1	1	1	1	..
	19		Leather and articles of leather; footwear	..	..	58	59	54	39	35	30
		191	Leather and articles of leather	..	..	18	17	16	13	13	12
		192	Footwear	..	..	41	42	39	27	23	18
	20		Wood & products of wood & cork, exc. furniture	..	..	73	83	87	86	85	83
		201	Sawmilling and planing of wood	..	..	15	17	16	13	13	12
		202	Wood products; articles of cork, straw, etc.	..	..	58	66	70	73	72	71
	21		Paper and paper products	..	..	127	128	126	122	117	113
	22		Printing, publishing and recorded media	..	..	318	342	353	343	330	361
		221	Publishing	..	..	139	148	151	142	147	157
		222	Printing and related activities	..	..	175	190	197	194	175	197
		223	Reproduction of recorded media	..	..	4	4	6	7	8	8
	23		Coke and petroleum products; nuclear fuel	..	..	30	28	27	28	26	28
		231	Coke oven products	..	..	1	1	..	1	1	..
		232	Refined petroleum products	..	..	13	13	13	14	13	16
		233	Processing of nuclear fuel	..	..	16	14	..	14	12	..
	24		Chemicals and chemical products	..	..	274	271	281	267	268	258
		241	Basic chemicals	..	..	84	80	84	78	82	73
		242	Other chemical products	..	..	182	183	191	182	181	178
		243	Man-made fibres	..	..	8	8	7	7	5	7
	25		Rubber and plastic products	..	..	235	239	247	249	260	238
		251	Rubber products	..	..	52	50	51	53	52	47
		252	Plastic products	..	..	183	189	196	197	208	191
	26		Other non-metallic mineral products	..	..	151	159	166	156	151	142
		261	Glass and glass products	..	..	38	41	45	39	38	38
		269	Non-metallic mineral products, nec	..	..	113	119	122	117	113	104
	27		Basic metals	..	..	148	142	139	139	138	128
		271	Basic iron and steel	..	..	73	71	67	68	72	66
		272	Basic precious and non-ferrous metals	..	..	38	34	34	34	29	27
		273	Casting of metals	..	..	38	37	38	38	37	35
	28		Fabricated metal products, exc. machinery	..	..	375	397	441	409	381	413
		281	Structural metal prod., tanks, reservoirs, etc.	..	..	96	93	96	94	95	100
		289	Other fabricated metal products	..	..	279	304	345	315	286	313
	29		Machinery and equipment, nec	..	..	393	402	421	398	398	396
		291	General purpose machinery	..	..	213	220	229	216	212	211
		292	Special purpose machinery	..	..	141	142	156	146	148	151
		293	Domestic appliances, nec	..	..	39	40	37	36	38	33
	30		Office, accounting & computing machinery	..	..	80	74	63	59	61	59
	31		Electrical machinery & apparatus, nec	..	..	188	199	195	194	194	185
		311	Electric motors, generators and transformers	..	..	36	37	37	35	37	34
		312	Electricity distribution & control apparatus	..	..	50	53	53	51	50	51
		313	Insulated wire and cable	..	..	22	24	26	26	25	22
		314	Accumulators, primary cells & primary batteries	..	..	7	8	8	8	7	7
		315	Electric lamps and lighting equipment	..	..	23	24	23	23	25	22
		319	Other electrical equipment, nec	..	..	50	54	49	52	51	49
	32		Radio, TV, communication equipment	..	..	119	135	138	138	141	126
		321	Electronic valves, tubes; other electronic components	..	..	46	52	56	54	64	52
		322	TV/radio transmitters; line teleph./telegr. apparatus	..	..	39	47	44	48	39	38
		323	TV/radio receivers; sound and video app.; assoc. gds.	..	..	34	36	39	36	38	36
	33		Medical, precision & optical instrum.; watch & clocks	..	..	145	136	139	141	137	151
		331	Medical appl.; instruments & appl. for measuring, etc.	..	..	128	120	122	123	118	130
		332	Optical instruments and photographic equipment	..	..	15	14	15	17	17	19
		333	Watches and clocks	..	..	2	3	3	2	2	2
	34		Motor vehicles, trailers and semi-trailers	..	..	222	229	240	245	249	229
		341	Motor vehicles	..	..	115	114	116	123	120	97
		342	Bodies for motor vehicles, trailers & semi-trailers	..	..	25	29	30	27	28	30
		343	Parts & accessories for motor vehicles & their engines	..	..	82	87	94	95	101	102
	35		Other transport equipment	..	..	197	179	167	161	162	168
		351	Building and repairing of ships and boats	..	..	48	42	40	39	41	40
		352	Railway and tramway locomotives and rolling stock	..	..	17	15	12	13	12	18
		353	Aircraft and spacecraft	..	..	126	115	109	103	104	104
		359	Transport equipment, nec	..	..	6	6	7	6	6	7
	36		Furniture; manufacturing, nec	..	..	189	215	212	204	200	203
		361	Furniture	..	..	123	133	129	133	132	128
		369	Manufacturing, nec	..	..	66	83	83	71	69	75
	37		Recycling	..	..	..	4	8	8	9	12
E	40_41		Electricity, gas and water supply	..	..	209	197	174	143	136	212
	40		Electricity, gas, steam and hot water supply	..	..	174	162	140	113	103	101
	41		Collection, purification and distribution of water	..	..	35	35	34	31	33	..
F	45		Construction	..	..	955	861	989	1 007	864	1 071

Table UK.6 **EMPLOI, FEMMES - EMPLOYMENT, FEMALES**

Milliers

1991	1992	1993	1994	1995	1996	1997	1998			CITI révision 3	
..	..	..	..	..	..	..	..	Activites extractives	C	**10_14**	
..	..	2	..	..	..	..	..	Extraction de charbon, de lignite et de tourbe		**10**	
..	..	..	..	..	..	..	..	Pétrole brut et gaz naturel ; activités annexes		**11**	
..	..	..	..	..	..	..	..	Extraction de minerais métalliques		**13**	
..	..	3	..	4	..	..	..	Autres activités extractives		**14**	
..	..	1 312	..	1 227	..	..	..	Total activites manufacturieres	D	**15_37**	
..	..	..	..	195	..	..	164	Produits alimentaires et boissons		**15**	
..	..	75	..	72	..	..	57	Viande, poisson, fruits, légumes, huiles et graisses			151
..	..	14	..	16	..	..	11	Produits laitiers			152
..	..	..	..	13	..	..	9	Travail des grains; prod. amyl. ; alim. pour animaux			153
..	..	105	..	73	..	..	72	Autres produits alimentaires			154
..	..	15	..	21	..	..	15	Boissons			155
..	..	3	..	2	..	..	2	Produits à base de tabac		**16**	
..	..	91	..	88	..	..	68	Textiles		**17**	
..	..	24	..	28	..	..	17	Filature, tissage et achèvement des textiles			171
..	..	35	..	33	..	..	31	Autres articles textiles			172
..	..	33	..	27	..	..	19	Etoffes et articles de bonneterie			173
..	..	142	..	137	..	..	95	Articles d'habillement et fourrure		**18**	
..	..	141	..	137	..	..	95	Articles d'habillement autres qu'en fourrure			181
..	..		..	1	..	..		Prép. et teinture des fourrures ; articles en fourrure			182
..	..	33	..	24	..	..	13	Cuir et articles en cuir ; chaussures		**19**	
..	..	13	..	7	..	..	5	Cuir et articles en cuir sauf chaussures			191
..	..	21	..	17	..	..	8	Chaussures			192
..	..	14	..	14	..	..	13	Bois, articles en bois, sauf meubles		**20**	
..	..	2	..	3	..	..	2	Sciage et rabotage du bois			201
..	..	12	..	11	..	..	11	Articles en bois, liège, vannerie, sparterie			202
..	..	38	..	36	..	..	..	Papier, articles en papier et carton		**21**	
..	..	132	..	128	..	..	153	Edition, imprimerie et reproduction		**22**	
..	..	75	..	57	..	..	75	Edition			221
..	..	54	..	68	..	..	75	Imprimerie et activités annexes			222
..	..	2	..	2	..	..	3	Reproduction de supports enregistrés			223
..	..	..	..	5	..	..	5	Coke. et prod. pétroliers ; comb. nucléaires		**23**	
..	..	..	..	..	..	..	..	Cokéfaction			231
..	..	3	..	3	..	..	3	Produits pétroliers raffinés			232
..	..	3	..	3	..	..	..	Traitement de combustibles nucléaires			233
..	..	88	..	85	..	..	83	Produits chimiques		**24**	
..	..	17	..	25	..	..	17	Produits chimiques de base			241
..	..	70	..	58	..	..	65	Autres produits chimiques			242
..	..	2	..	2	..	..	1	Fibres synthétiques ou artificielles			243
..	..	61	..	57	..	..	58	Art. en caoutchouc et matières plastiques		**25**	
..	..	9	..	12	..	..	9	Articles en caoutchouc			251
..	..	52	..	45	..	..	49	Articles en matières plastiques			252
..	..	40	..	33	..	..	31	Autres produits minéraux non métalliques		**26**	
..	..	8	..	9	..	..	8	Verre et articles en verre			261
..	..	32	..	24	..	..	23	Produits minéraux non métalliques, nca			269
..	..	..	..	14	..	..	19	Produits métallurgiques de base		**27**	
..	..	..	..	7	..	..	8	Sidérurgie et première transformation de l'acier			271
..	..	6	..	4	..	..	5	Première transf. métaux précieux et non ferreux			272
..	..	4	..	4	..	..	6	Fonderie			273
..	..	74	..	71	..	..	70	Ouvrages en métaux, sauf machines		**28**	
..	..	13	..	15	..	..	14	Constr. et menuis. métal.; réservoirs; générateurs			281
..	..	61	..	56	..	..	56	Autres ouvrages en métaux			289
..	..	75	..	79	..	..	76	Machines et matériels, nca		**29**	
..	..	38	..	42	..	..	39	Machines d'usage général			291
..	..	24	..	29	..	..	28	Machines d'usage spécifique			292
..	..	13	..	7	..	..	9	Appareils domestiques, nca			293
..	..	23	..	23	..	..	18	Mach. de bureau, comptables, information		**30**	
..	..	60	..	57	..	..	63	Machines et appareils électriques, nca		**31**	
..	..	9	..	9	..	..	10	Moteurs, génératrices, transformateurs électriques			311
..	..	17	..	14	..	..	18	Matériel électrique de distribution et de commande			312
..	..	7	..	7	..	..	9	Fils et câbles électriques isolés			313
..	..	2	..	3	..	..	2	Accumulateurs et piles électriques			314
..	..	10	..	7	..	..	9	Lampes électriques et appareils d'éclairage			315
..	..	17	..	18	..	..	17	Autres matériels électriques, nca.			319
..	..	44	..	48	..	..	37	Equipements et appareils de radio, TV, etc.		**32**	
..	..	19	..	24	..	..	16	Tubes et valves électroniques;autres composants			321
..	..	11	..	13	..	..	11	Emetteurs radio/TV; app. téléphonie et télégraphie			322
..	..	13	..	11	..	..	11	Récepteurs radio/TV, son et image; art. associés			323
..	..	48	..	45	..	..	48	Instr. médicaux, de précision ; optique ; horlog.		**33**	
..	..	40	..	40	..	..	40	Instruments médicaux et de précision, etc.			331
..	..	7	..	5	..	..	7	Instruments d'optique ; matériel photographique			332
..	..	1	..	1	..	..	1	Horlogerie			333
..	..	34	..	36	..	..	29	Véhicules automobiles; remorques, etc.		**34**	
..	..	9	..	18	..	..	8	Véhicules automobiles			341
..	..	2	..	2	..	..	4	Carross. automobiles, remorques, semi-remorques			342
..	..	22	..	14	..	..	17	Pièces et accessoires pour automobiles et moteurs			343
..	..	21	..	18	..	..	17	Autres matériels de transport		**35**	
..	..	4	..	4	..	..	4	Construction et réparation de navires			351
..	..	1	..	1	..	..	1	Matériel ferroviaire roulant			352
..	..	15	..	11	..	..	11	Construction aéronautique et spatiale			353
..	..	1	..	1	..	..	1	Autres équipements de transport			359
..	..	58	..	32	..	..	61	Meubles ; activités de fabrication, nca		**36**	
..	..	29	..	12	..	..	35	Meubles			361
..	..	30	..	20	..	..	27	Activités de fabrication, nca			369
..	..	..	..	2	..	..	2	Récupération		**37**	
..	..	..	..	..	..	..	..	Electricité, gaz et eau	E	**40_41**	
..	..	..	..	..	..	..	..	Prod. et dist. d'électricité gaz, vapeur d'eau chaude		**40**	
..	..	..	..	8	..	..	..	Captage, épuration et distribution d'eau		**41**	
..	..	..	..	158	..	..	..	Construction	F	**45**	

UNITED KINGDOM

Table UK.7 **COMPENSATION OF LABOUR, EMPLOYEES - SALAIRES ET CHARGES SOCIALES, SALARIÉS**

Millions of £ (Current Prices)

ISIC revision 3				1991	1992	1993	1994	1995	1996	1997	1998
C	10_14		Mining and quarrying	..	..	1 988	965	1 023	1 175	2 577	2 599
	10		Mining of coal and lignite; extraction of peat	..	..	1 434	388	462	511	484	461
	11		Crude petroleum & natural gas; service activities	..	..	1 181	1 316	1 239	1 106	1 446	1 456
	13		Mining of metal ores	..	..	..	..	..	6	7	2
	14		Other mining and quarrying	..	..	554	578	..	657	641	679
D	15_37		Total manufacturing	..	..	71 056	74 804	76 712	79 181	82 892	86 680
	15		Food products and beverages	..	..	7 910	8 198	8 056	8 169	8 883	8 965
		151	Production of meat, fish, fruit, vegetables, oils & fats	..	..	2 456	2 567	2 560	2 584	2 642	2 704
		152	Dairy products	..	..	790	777	701	743	765	784
		153	Grain mill and starch products; animal feeds	..	..	724	718	703	739	799	760
		154	Other food products	..	..	2 850	2 929	2 839	2 914	3 391	3 416
		155	Beverages	..	..	1 090	1 208	1 254	1 189	1 286	1 301
	16		Tobacco products	..	..	266	278	281	328	359	358
	17		Textiles	..	..	2 434	2 475	2 518	2 716	2 876	2 816
		171	Spinning, weaving and finishing of textiles	..	..	933	979	898	966	951	956
		172	Other textiles	..	..	926	918	988	1 098	1 306	1 265
		173	Knitted and crocheted fabrics and articles	..	..	574	579	633	652	620	594
	18		Wearing apparel and fur	..	..	1 573	1 603	1 623	1 590	1 731	1 668
		181	Wearing apparel, except fur apparel	..	..	1 563	1 593	1 615	1 584	1 724	1 664
		182	Dressing and dyeing of fur; articles of fur	..	..	10	10	8	6	7	4
	19		Leather and articles of leather; footwear	..	..	693	665	621	476	478	462
		191	Leather and articles of leather	..	..	263	204	201	170	170	178
		192	Footwear	..	..	430	461	420	306	308	284
	20		Wood & products of wood & cork, exc. furniture	..	..	998	1 185	1 200	1 212	1 267	1 360
		201	Sawmilling and planing of wood	..	..	206	256	192	169	174	177
		202	Wood products; articles of cork, straw, etc.	..	..	792	929	1 008	1 043	1 093	1 182
	21		Paper and paper products	..	..	2 255	2 354	2 382	2 445	2 421	2 447
	22		Printing, publishing and recorded media	..	..	5 904	6 747	6 800	7 393	7 341	7 840
		221	Publishing	..	..	2 677	2 994	3 075	3 295	3 338	3 631
		222	Printing and related activities	..	..	3 140	3 677	3 594	3 912	3 792	4 046
		223	Reproduction of recorded media	..	..	87	76	131	187	211	164
	23		Coke and petroleum products; nuclear fuel	..	..	792	800	799	841	787	914
		231	Coke oven products	..	..	20	24	..	12	12	10
		232	Refined petroleum products	..	..	375	374	409	413	370	416
		233	Processing of nuclear fuel	..	..	397	402	..	415	405	489
	24		Chemicals and chemical products	..	..	6 103	6 215	6 655	6 626	6 852	7 038
		241	Basic chemicals	..	..	2 071	2 027	2 033	2 102	2 230	2 329
		242	Other chemical products	..	..	3 856	3 970	4 429	4 361	4 484	4 548
		243	Man-made fibres	..	..	176	219	193	163	138	161
	25		Rubber and plastic products	..	..	3 678	3 898	4 083	4 326	4 631	4 825
		251	Rubber products	..	..	906	930	957	1 023	1 020	1 109
		252	Plastic products	..	..	2 772	2 967	3 126	3 303	3 611	3 716
	26		Other non-metallic mineral products	..	..	2 394	2 584	2 772	2 694	2 798	2 714
		261	Glass and glass products	..	..	632	663	709	697	708	694
		269	Non-metallic mineral products, nec	..	..	1 762	1 921	2 062	1 998	2 090	2 020
	27		Basic metals	..	..	2 688	2 728	2 855	2 946	3 017	3 142
		271	Basic iron and steel	..	..	1 421	1 463	1 527	1 572	1 693	1 761
		272	Basic precious and non-ferrous metals	..	..	682	635	666	682	616	676
		273	Casting of metals	..	..	586	630	662	693	709	706
	28		Fabricated metal products, exc. machinery	..	..	6 105	6 445	6 880	7 068	6 970	7 611
		281	Structural metal prod., tanks, reservoirs, etc.	..	..	1 692	1 683	1 646	1 744	1 881	2 146
		289	Other fabricated metal products	..	..	4 413	4 761	5 234	5 324	5 089	5 466
	29		Machinery and equipment, nec	..	..	6 983	7 435	7 946	7 928	8 345	8 612
		291	General purpose machinery	..	..	3 844	4 083	4 384	4 359	4 457	4 686
		292	Special purpose machinery	..	..	2 595	2 764	2 991	2 982	3 245	3 339
		293	Domestic appliances, nec	..	..	545	589	571	586	643	587
	30		Office, accounting & computing machinery	..	..	1 993	1 683	1 315	1 310	1 464	1 692
	31		Electrical machinery & apparatus, nec	..	..	2 989	3 167	3 193	3 365	3 511	3 657
		311	Electric motors, generators and transformers	..	..	551	600	594	623	718	653
		312	Electricity distribution & control apparatus	..	..	793	867	914	923	893	1 085
		313	Insulated wire and cable	..	..	369	392	402	418	429	381
		314	Accumulators, primary cells & primary batteries	..	..	130	136	139	143	136	145
		315	Electric lamps and lighting equipment	..	..	346	349	346	363	423	440
		319	Other electrical equipment, nec	..	..	800	823	798	894	913	954
	32		Radio, TV, communication equipment	..	..	1 982	2 447	2 568	2 767	2 975	3 187
		321	Electronic valves, tubes; other electronic components	..	..	719	857	971	1 029	1 324	1 143
		322	TV/radio transmitters; line teleph./telegr. apparatus	..	..	771	1 000	998	1 140	962	1 341
		323	TV/radio receivers; sound and video app.; assoc. gds.	..	..	492	590	599	597	689	703
	33		Medical, precision & optical instrum.; watch & clocks	..	..	2 529	2 444	2 528	2 720	2 816	3 024
		331	Medical appl.; instruments & appl. for measuring, etc.	..	..	2 270	2 172	2 224	2 407	2 488	2 629
		332	Optical instruments and photographic equipment	..	..	230	233	258	282	297	362
		333	Watches and clocks	..	..	29	39	45	31	31	33
	34		Motor vehicles, trailers and semi-trailers	..	..	4 294	4 697	5 048	5 457	5 702	6 298
		341	Motor vehicles	..	..	2 574	2 723	2 894	3 224	3 182	3 573
		342	Bodies for motor vehicles, trailers & semi-trailers	..	..	387	473	506	495	552	546
		343	Parts & accessories for motor vehicles & their engines	..	..	1 332	1 500	1 648	1 738	1 968	2 179
	35		Other transport equipment	..	..	3 932	3 621	3 514	3 591	4 174	4 552
		351	Building and repairing of ships and boats	..	..	939	784	762	800	908	959
		352	Railway and tramway locomotives and rolling stock	..	..	305	292	247	270	267	265
		353	Aircraft and spacecraft	..	..	2 609	2 460	2 410	2 432	2 901	3 218
		359	Transport equipment, nec	..	..	79	85	95	88	97	110
	36		Furniture; manufacturing, nec	..	..	2 561	3 077	2 958	3 085	3 350	3 352
		361	Furniture	..	..	1 749	1 933	1 888	2 062	2 265	2 302
		369	Manufacturing, nec	..	..	812	1 143	1 071	1 023	1 085	1 049
	37		Recycling	..	..	..	58	117	128	144	146
E	40_41		Electricity, gas and water supply	..	..	5 279	5 018	4 724	3 928	3 852	4 055
	40		Electricity, gas, steam and hot water supply	..	..	4 641	4 352	4 072	3 310	3 126	3 249
	41		Collection, purification and distribution of water	..	..	638	665	653	618	726	806
F	45		Construction	..	..	13 899	14 342	..	16 264	18 711	18 686

Table UK.8 **SALAIRES ET TRAITEMENTS, SALARIÉS - WAGES AND SALARIES, EMPLOYEES**

Millions de £ (Prix courants)

1991	1992	1993	1994	1995	1996	1997	1998		CITI révision 3		
..	..	1 777	844	895	1 018	2 280	2 274	Activites extractives	C	**10_14**	
..	..	1 294	339	401	443	418	403	Extraction de charbon, de lignite et de tourbe		**10**	
..	..	..	..	..	..	1 293	1 275	Pétrole brut et gaz naturel ; activités annexes		**11**	
..	..	..	..	..	6	6	2	Extraction de minerais métalliques		**13**	
..	..	483	505	..	570	563	594	Autres activités extractives		**14**	
..	..	62 726	65 900	67 518	69 783	73 385	76 449	Total activites manufacturieres	D	**15_37**	
..	..	7 030	7 265	7 119	7 243	7 916	7 984	Produits alimentaires et boissons		**15**	
..	..	2 213	2 322	2 300	2 325	2 396	2 446	Viande, poisson, fruits, légumes, huiles et graisses			151
..	..	703	684	617	663	690	701	Produits laitiers			152
..	..	637	626	616	644	701	665	Travail des grains; prod. amyl. ; alim. pour animaux			153
..	..	2 526	2 581	2 485	2 586	3 016	3 058	Autres produits alimentaires			154
..	..	951	1 052	1 101	1 024	1 113	1 113	Boissons			155
..	..	232	244	248	286	297	289	Produits à base de tabac		**16**	
..	..	2 161	2 193	2 245	2 421	2 566	2 497	Textiles		**17**	
..	..	826	860	799	861	846	840	Filature, tissage et achèvement des textiles			171
..	..	823	817	875	972	1 159	1 122	Autres articles textiles			172
..	..	513	517	572	587	560	535	Etoffes et articles de bonneterie			173
..	..	1 423	1 445	1 475	1 451	1 580	1 513	Articles d'habillement et fourrure		**18**	
..	..	1 415	1 437	1 468	1 446	1 574	1 509	Articles d'habillement autres qu'en fourrure			181
..	..	8	9	7	6	6	4	Prép. et teinture des fourrures ; articles en fourrure			182
..	..	622	593	569	426	429	410	Cuir et articles en cuir ; chaussures		**19**	
..	..	241	182	180	152	152	159	Cuir et articles en cuir sauf chaussures			191
..	..	381	411	389	274	277	252	Chaussures			192
..	..	889	1 042	1 095	1 080	1 136	1 224	Bois, articles en bois, sauf meubles		**20**	
..	..	179	227	175	150	155	159	Sciage et rabotage du bois			201
..	..	710	815	920	930	981	1 066	Articles en bois, liège, vannerie, sparterie			202
..	..	1 988	2 067	2 095	2 150	2 129	2 154	Papier, articles en papier et carton		**21**	
..	..	5 205	5 966	5 962	6 540	6 502	6 917	Edition, imprimerie et reproduction		**22**	
..	..	2 357	2 639	2 708	2 908	2 955	3 200	Edition			221
..	..	2 769	3 258	3 135	3 464	3 357	3 569	Imprimerie et activités annexes			222
..	..	79	69	119	168	190	148	Reproduction de supports enregistrés			223
..	..	670	675	670	714	669	809	Coke. et prod. pétroliers ; comb. nucléaires		**23**	
..	..	17	22	..	11	10	8	Cokéfaction			231
..	..	318	318	345	355	317	354	Produits pétroliers raffinés			232
..	..	335	336	..	349	341	447	Traitement de combustibles nucléaires			233
..	..	5 253	5 339	5 736	5 702	5 929	6 079	Produits chimiques		**24**	
..	..	1 755	1 722	1 716	1 770	1 893	1 991	Produits chimiques de base			241
..	..	3 340	3 432	3 860	3 794	3 915	3 944	Autres produits chimiques			242
..	..	157	186	160	139	121	143	Fibres synthétiques ou artificielles			243
..	..	3 257	3 445	3 613	3 837	4 096	4 241	Art. en caoutchouc et matières plastiques		**25**	
..	..	798	816	832	897	886	950	Articles en caoutchouc			251
..	..	2 459	2 629	2 781	2 940	3 210	3 291	Articles en matières plastiques			252
..	..	2 115	2 281	2 437	2 374	2 484	2 394	Autres produits minéraux non métalliques		**26**	
..	..	549	578	616	610	628	617	Verre et articles en verre			261
..	..	1 565	1 702	1 821	1 764	1 856	1 777	Produits minéraux non métalliques, nca			269
..	..	2 365	2 388	2 501	2 584	2 668	2 770	Produits métallurgiques de base		**27**	
..	..	1 252	1 281	1 333	1 376	1 496	1 554	Sidérurgie et première transformation de l'acier			271
..	..	596	553	584	598	544	590	Première transf. métaux précieux et non ferreux			272
..	..	517	554	584	611	629	626	Fonderie			273
..	..	5 451	5 743	6 146	6 277	6 220	6 748	Ouvrages en métaux, sauf machines		**28**	
..	..	1 518	1 512	1 482	1 548	1 680	1 896	Constr. et menuis. métal.; réservoirs; générateurs			281
..	..	3 933	4 232	4 665	4 729	4 540	4 852	Autres ouvrages en métaux			289
..	..	6 177	6 571	6 993	6 993	7 355	7 548	Machines et matériels, nca		**29**	
..	..	3 402	3 607	3 864	3 845	3 919	4 092	Machines d'usage général			291
..	..	2 292	2 439	2 625	2 628	2 864	2 935	Machines d'usage spécifique			292
..	..	484	524	504	521	572	521	Appareils domestiques, nca			293
..	..	1 732	1 460	1 160	1 156	1 296	1 491	Mach. de bureau, comptables, information		**30**	
..	..	2 661	2 803	2 820	2 980	3 122	3 249	Machines et appareils électriques, nca		**31**	
..	..	489	525	521	546	635	576	Moteurs, génératrices, transformateurs électriques			311
..	..	706	770	813	818	793	963	Matériel électrique de distribution et de commande			312
..	..	328	351	361	376	384	339	Fils et câbles électriques isolés			313
..	..	112	117	122	124	118	125	Accumulateurs et piles électriques			314
..	..	312	313	307	324	377	393	Lampes électriques et appareils d'éclairage			315
..	..	715	728	696	792	814	852	Autres matériels électriques, nca.			319
..	..	1 774	2 196	2 281	2 453	2 642	2 820	Equipements et appareils de radio, TV, etc.		**32**	
..	..	643	763	858	917	1 169	1 016	Tubes et valves électroniques;autres composants			321
..	..	689	899	883	996	850	1 176	Emetteurs radio/TV; app. téléphonie et télégraphie			322
..	..	442	534	541	540	623	628	Récepteurs radio/TV, son et image; art. associés			323
..	..	2 236	2 165	2 211	2 388	2 476	2 646	Instr. médicaux, de précision ; optique ; horlog.		**33**	
..	..	2 006	1 922	1 940	2 110	2 185	2 292	Instruments médicaux et de précision, etc.			331
..	..	205	208	230	250	264	325	Instruments d'optique ; matériel photographique			332
..	..	25	34	41	28	28	29	Horlogerie			333
..	..	3 725	4 066	4 337	4 719	5 065	5 549	Véhicules automobiles; remorques, etc.		**34**	
..	..	2 199	2 319	2 455	2 747	2 828	3 149	Véhicules automobiles			341
..	..	344	418	452	441	491	487	Carross. automobiles, remorques, semi-remorques			342
..	..	1 181	1 328	1 431	1 532	1 746	1 913	Pièces et accessoires pour automobiles et moteurs			343
..	..	3 487	3 175	3 081	3 148	3 671	3 991	Autres matériels de transport		**35**	
..	..	833	685	663	703	814	857	Construction et réparation de navires			351
..	..	272	262	219	245	240	235	Matériel ferroviaire roulant			352
..	..	2 313	2 152	2 114	2 121	2 530	2 799	Construction aéronautique et spatiale			353
..	..	70	76	85	79	87	99	Autres équipements de transport			359
..	..	2 273	2 731	2 625	2 748	3 010	2 996	Meubles ; activités de fabrication, nca		**36**	
..	..	1 559	1 723	1 676	1 842	2 036	2 056	Meubles			361
..	..	713	1 008	949	906	974	940	Activités de fabrication, nca			369
..	..	..	48	98	112	126	131	Récupération		**37**	
..	..	4 321	4 103	3 878	3 336	3 335	3 521	Electricité, gaz et eau	E	**40_41**	
..	..	3 782	3 543	3 329	2 806	2 704	2 828	Prod. et dist. d'électricité gaz, vapeur d'eau chaude		**40**	
..	..	539	561	549	529	630	693	Captage, épuration et distribution d'eau		**41**	
..	..	12 373	12 761	..	14 469	16 799	16 594	Construction	F	**45**	

UNITED KINGDOM

Table UK.9 **SUPPLEMENTS TO WAGES & SALARIES - CHARGES SOCIALES DES EMPLOYEURS**

Millions of £ (Current Prices)

ISIC revision 3				1991	1992	1993	1994	1995	1996	1997	1998
C	10_14		Mining and quarrying	..	..	211	121	128	156	296	325
	10		Mining of coal and lignite; extraction of peat	..	..	140	49	60	68	66	58
	11		Crude petroleum & natural gas; service activities	..	..	..	254	239	213	153	181
	13		Mining of metal ores	..	..	..	..	..	1	1	..
	14		Other mining and quarrying	..	..	71	73	..	87	77	85
D	15_37		Total manufacturing	..	..	8 330	8 904	9 196	9 399	9 510	10 231
	15		Food products and beverages	..	..	881	934	937	926	968	981
		151	Production of meat, fish, fruit, vegetables, oils & fats	..	..	243	245	260	258	246	257
		152	Dairy products	..	..	87	93	84	80	75	83
		153	Grain mill and starch products; animal feeds	..	..	87	92	87	94	99	95
		154	Other food products	..	..	324	348	354	328	375	358
		155	Beverages	..	..	139	156	153	165	173	188
	16		Tobacco products	..	..	33	34	33	41	62	69
	17		Textiles	..	..	273	282	273	296	310	319
		171	Spinning, weaving and finishing of textiles	..	..	108	119	99	105	105	116
		172	Other textiles	..	..	104	101	113	125	146	143
		173	Knitted and crocheted fabrics and articles	..	..	61	62	61	65	59	60
	18		Wearing apparel and fur	..	..	150	158	148	139	151	155
		181	Wearing apparel, except fur apparel	..	..	148	156	147	139	150	155
		182	Dressing and dyeing of fur; articles of fur	..	..	1	1	1	1	1	..
	19		Leather and articles of leather; footwear	..	..	71	72	53	51	49	52
		191	Leather and articles of leather	..	..	21	22	21	18	17	19
		192	Footwear	..	..	49	50	32	33	31	32
	20		Wood & products of wood & cork, exc. furniture	..	..	110	144	106	132	131	135
		201	Sawmilling and planing of wood	..	..	27	30	17	19	19	19
		202	Wood products; articles of cork, straw, etc.	..	..	82	114	88	113	112	117
	21		Paper and paper products	..	..	266	286	288	294	291	293
	22		Printing, publishing and recorded media	..	..	699	780	838	853	839	923
		221	Publishing	..	..	320	355	367	387	383	431
		222	Printing and related activities	..	..	370	418	460	448	435	477
		223	Reproduction of recorded media	..	..	8	7	11	19	21	16
	23		Coke and petroleum products; nuclear fuel	..	..	122	125	129	127	118	105
		231	Coke oven products	..	..	3	2	..	2	2	1
		232	Refined petroleum products	..	..	57	56	64	58	53	62
		233	Processing of nuclear fuel	..	..	63	66	..	67	64	42
	24		Chemicals and chemical products	..	..	851	876	919	923	923	959
		241	Basic chemicals	..	..	315	305	317	332	337	338
		242	Other chemical products	..	..	516	537	569	567	569	604
		243	Man-made fibres	..	..	19	33	33	24	17	17
	25		Rubber and plastic products	..	..	421	453	470	489	535	584
		251	Rubber products	..	..	109	115	126	126	134	158
		252	Plastic products	..	..	312	339	345	364	400	425
	26		Other non-metallic mineral products	..	..	279	303	334	320	314	320
		261	Glass and glass products	..	..	83	85	93	86	80	77
		269	Non-metallic mineral products, nec	..	..	197	218	241	234	234	243
	27		Basic metals	..	..	323	340	354	362	349	372
		271	Basic iron and steel	..	..	168	182	194	196	197	207
		272	Basic precious and non-ferrous metals	..	..	86	82	82	84	72	86
		273	Casting of metals	..	..	69	77	77	82	80	79
	28		Fabricated metal products, exc. machinery	..	..	654	701	733	791	750	864
		281	Structural metal prod., tanks, reservoirs, etc.	..	..	174	172	164	196	201	250
		289	Other fabricated metal products	..	..	480	530	570	595	550	614
	29		Machinery and equipment, nec	..	..	806	864	953	934	991	1 064
		291	General purpose machinery	..	..	442	476	520	515	538	594
		292	Special purpose machinery	..	..	303	324	366	355	382	404
		293	Domestic appliances, nec	..	..	61	64	67	65	71	65
	30		Office, accounting & computing machinery	..	..	261	223	155	155	168	201
	31		Electrical machinery & apparatus, nec	..	..	329	363	373	385	390	409
		311	Electric motors, generators and transformers	..	..	62	75	74	77	82	77
		312	Electricity distribution & control apparatus	..	..	87	97	100	105	99	122
		313	Insulated wire and cable	..	..	41	41	41	42	45	41
		314	Accumulators, primary cells & primary batteries	..	..	18	18	17	19	18	20
		315	Electric lamps and lighting equipment	..	..	35	36	39	39	46	47
		319	Other electrical equipment, nec	..	..	86	95	102	102	99	102
	32		Radio, TV, communication equipment	..	..	208	252	287	314	333	367
		321	Electronic valves, tubes; other electronic components	..	..	76	94	113	112	155	127
		322	TV/radio transmitters; line teleph./telegr. apparatus	..	..	82	102	116	145	112	165
		323	TV/radio receivers; sound and video app.; assoc. gds.	..	..	49	56	58	57	66	75
	33		Medical, precision & optical instrum.; watch & clocks	..	..	294	280	317	332	340	378
		331	Medical appl.; instruments & appl. for measuring, etc.	..	..	265	250	284	297	304	337
		332	Optical instruments and photographic equipment	..	..	25	25	28	31	33	37
		333	Watches and clocks	..	..	4	5	4	3	3	4
	34		Motor vehicles, trailers and semi-trailers	..	..	569	631	711	738	637	749
		341	Motor vehicles	..	..	375	404	439	478	353	424
		342	Bodies for motor vehicles, trailers & semi-trailers	..	..	43	55	54	54	61	59
		343	Parts & accessories for motor vehicles & their engines	..	..	151	172	217	206	223	266
	35		Other transport equipment	..	..	445	446	433	444	503	561
		351	Building and repairing of ships and boats	..	..	106	100	99	97	94	101
		352	Railway and tramway locomotives and rolling stock	..	..	33	30	28	25	27	30
		353	Aircraft and spacecraft	..	..	296	308	296	312	372	419
		359	Transport equipment, nec	..	..	9	9	11	10	10	11
	36		Furniture; manufacturing, nec	..	..	288	346	333	337	340	356
		361	Furniture	..	..	190	210	212	221	229	247
		369	Manufacturing, nec	..	..	98	136	121	117	111	109
	37		Recycling	..	..	..	10	20	16	18	15
E	40_41		Electricity, gas and water supply	..	..	958	914	846	592	517	534
	40		Electricity, gas, steam and hot water supply	..	..	859	810	743	504	421	421
	41		Collection, purification and distribution of water	..	..	99	105	103	89	96	113
F	45		Construction	..	..	1 526	1 582	..	1 795	1 911	2 092

Table UK.10 **ÉTABLISSEMENTS - ESTABLISHMENTS**

Unités

1991	1992	1993	1994	1995	1996	1997	1998			CITI révision 3	
..	..	1 192	1 124	1 097	1 098	1 665	1 584	Activites extractives	C	10_14	
..	..	193	225	211	185	202	187	Extraction de charbon, de lignite et de tourbe		10	
..	..	180	193	191	176	365	355	Pétrole brut et gaz naturel ; activités annexes		11	
..	..	..	..	13	15	48	38	Extraction de minerais métalliques		13	
..	..	999	899	873	898	1 050	1 004	Autres activités extractives		14	
..	..	160 991	171 291	171 518	164 808	169 663	169 356	Total activites manufacturieres	D	15_37	
..	..	10 120	10 561	8 807	8 265	8 320	8 228	Produits alimentaires et boissons		15	
..	..	3 145	3 381	2 946	2 630	2 559	2 405	Viande, poisson, fruits, légumes, huiles et graisses			151
..	..	806	812	849	720	692	666	Produits laitiers			152
..	..	789	902	812	746	762	730	Travail des grains; prod. amyl. ; alim. pour animaux			153
..	..	4 671	4 728	3 536	3 450	3 562	3 687	Autres produits alimentaires			154
..	..	709	738	664	719	745	740	Boissons			155
..	..	43	24	16	29	25	21	Produits à base de tabac		16	
..	..	7 907	6 541	6 321	6 347	6 330	6 205	Textiles		17	
..	..	2 820	2 920	1 571	1 558	1 580	1 525	Filature, tissage et achèvement des textiles			171
..	..	3 907	2 443	3 680	3 730	3 720	3 677	Autres articles textiles			172
..	..	1 180	1 178	1 070	1 059	1 030	1 003	Etoffes et articles de bonneterie			173
..	..	8 208	8 658	8 672	8 377	8 212	7 630	Articles d'habillement et fourrure		18	
..	..	8 148	8 590	8 611	8 324	8 173	7 591	Articles d'habillement autres qu'en fourrure			181
..	..	60	68	61	53	39	39	Prép. et teinture des fourrures ; articles en fourrure			182
..	..	2 349	1 834	1 561	1 450	1 304	1 208	Cuir et articles en cuir ; chaussures		19	
..	..	1 678	1 204	1 036	928	816	746	Cuir et articles en cuir sauf chaussures			191
..	..	671	630	525	522	488	462	Chaussures			192
..	..	8 506	7 862	8 256	8 434	8 561	8 578	Bois, articles en bois, sauf meubles		20	
..	..	2 164	2 107	1 492	1 347	1 268	1 170	Sciage et rabotage du bois			201
..	..	6 342	5 755	6 764	7 087	7 293	7 408	Articles en bois, liège, vannerie, sparterie			202
..	..	3 362	3 554	3 209	3 086	3 053	2 967	Papier, articles en papier et carton		21	
..	..	25 979	27 828	26 074	27 842	29 101	29 538	Edition, imprimerie et reproduction		22	
..	..	8 664	7 296	7 189	8 027	8 787	9 122	Edition			221
..	..	17 257	20 458	18 762	19 296	19 521	19 405	Imprimerie et activités annexes			222
..	..	58	74	123	519	793	1 011	Reproduction de supports enregistrés			223
..	..	213	238	469	295	264	255	Coke. et prod. pétroliers ; comb. nucléaires		23	
..	..	24	40	33	29	22	17	Cokéfaction			231
..	..	160	188	387	260	234	230	Produits pétroliers raffinés			232
..	..	29	10	49	6	8	8	Traitement de combustibles nucléaires			233
..	..	4 110	4 166	4 591	4 199	4 159	4 065	Produits chimiques		24	
..	..	1 758	1 703	1 531	1 454	1 418	1 360	Produits chimiques de base			241
..	..	2 304	2 401	2 996	2 686	2 690	2 656	Autres produits chimiques			242
..	..	48	62	64	59	51	49	Fibres synthétiques ou artificielles			243
..	..	5 552	6 183	6 901	7 060	7 263	7 255	Art. en caoutchouc et matières plastiques		25	
..	..	625	703	878	787	801	811	Articles en caoutchouc			251
..	..	4 927	5 480	6 023	6 273	6 462	6 444	Articles en matières plastiques			252
..	..	3 798	4 233	5 248	4 985	5 252	5 272	Autres produits minéraux non métalliques		26	
..	..	981	1 182	1 697	1 485	1 617	1 645	Verre et articles en verre			261
..	..	2 817	3 051	3 551	3 500	3 635	3 627	Produits minéraux non métalliques, nca			269
..	..	3 816	2 552	2 770	2 751	2 733	2 659	Produits métallurgiques de base		27	
..	..	1 565	651	875	935	1 015	1 001	Sidérurgie et première transformation de l'acier			271
..	..	1 542	1 169	1 077	970	893	828	Première transf. métaux précieux et non ferreux			272
..	..	709	732	818	846	825	830	Fonderie			273
..	..	28 643	30 356	29 021	28 665	29 195	29 653	Ouvrages en métaux, sauf machines		28	
..	..	4 027	3 983	4 727	3 802	4 044	4 184	Constr. et menuis. métal.; réservoirs; générateurs			281
..	..	24 616	26 373	24 294	24 863	25 151	25 469	Autres ouvrages en métaux			289
..	..	12 699	13 989	15 900	13 826	14 426	14 330	Machines et matériels, nca		29	
..	..	5 819	6 457	7 106	6 507	6 762	6 652	Machines d'usage général			291
..	..	6 289	6 829	7 992	6 759	7 051	7 084	Machines d'usage spécifique			292
..	..	591	703	802	560	613	594	Appareils domestiques, nca			293
..	..	1 775	1 832	2 327	1 394	1 615	1 561	Mach. de bureau, comptables, information		30	
..	..	6 030	6 657	5 728	5 525	5 851	5 877	Machines et appareils électriques, nca		31	
..	..	1 844	1 627	1 247	1 151	1 142	1 095	Moteurs, génératrices, transformateurs électriques			311
..	..	557	713	895	931	987	1 038	Matériel électrique de distribution et de commande			312
..	..	415	390	375	360	381	401	Fils et câbles électriques isolés			313
..	..	52	65	93	89	109	101	Accumulateurs et piles électriques			314
..	..	347	352	441	514	592	621	Lampes électriques et appareils d'éclairage			315
..	..	2 815	3 510	2 677	2 480	2 640	2 621	Autres matériels électriques, nca.			319
..	..	3 350	3 737	3 283	2 969	3 010	2 940	Equipements et appareils de radio, TV, etc.		32	
..	..	618	660	1 110	1 077	980	989	Tubes et valves électroniques;autres composants			321
..	..	1 253	1 364	1 096	866	927	880	Emetteurs radio/TV; app. téléphonie et télégraphie			322
..	..	1 479	1 713	1 077	1 026	1 103	1 071	Récepteurs radio/TV, son et image; art. associés			323
..	..	4 010	4 671	6 765	5 574	5 833	5 648	Instr. médicaux, de précision ; optique ; horlog.		33	
..	..	3 587	4 174	5 997	4 801	5 015	4 829	Instruments médicaux et de précision, etc.			331
..	..	292	350	604	634	683	690	Instruments d'optique ; matériel photographique			332
..	..	131	147	164	139	135	129	Horlogerie			333
..	..	1 607	1 898	4 109	2 500	2 726	2 750	Véhicules automobiles; remorques, etc.		34	
..	..	466	578	946	547	600	594	Véhicules automobiles			341
..	..	457	629	920	687	767	764	Carross. automobiles, remorques, semi-remorques			342
..	..	684	691	2 243	1 266	1 359	1 392	Pièces et accessoires pour automobiles et moteurs			343
..	..	2 414	2 833	3 172	2 672	2 896	2 786	Autres matériels de transport		35	
..	..	1 619	1 920	1 599	1 559	1 535	1 520	Construction et réparation de navires			351
..	..	85	96	114	100	91	93	Matériel ferroviaire roulant			352
..	..	541	616	1 212	766	971	875	Construction aéronautique et spatiale			353
..	..	169	201	247	247	299	298	Autres équipements de transport			359
..	..	16 500	20 953	17 815	17 803	18 609	18 929	Meubles ; activités de fabrication, nca		36	
..	..	7 002	8 338	7 174	7 317	7 536	7 667	Meubles			361
..	..	9 498	12 615	10 641	10 486	11 073	11 262	Activités de fabrication, nca			369
..	..	..	131	503	760	925	1 001	Récupération		37	
..	..	185	183	237	265	313	331	Electricité, gaz et eau	E	40_41	
..	..	127	127	162	179	215	236	Prod. et dist. d'électricité gaz, vapeur d'eau chaude		40	
..	..	58	56	75	86	98	95	Captage, épuration et distribution d'eau		41	
..	..	191 396	187 621	186 962	180 470	178 937	179 868	Construction	F	45	

UNITED KINGDOM

Table UK.11 ENTERPRISES - ENTREPRISES

Units

ISIC revision 3				1991	1992	1993	1994	1995	1996	1997	1998
C	10_14		Mining and quarrying	..	..	1 035	965	945	939	1 627	1 548
	10		Mining of coal and lignite; extraction of peat	..	..	176	211	192	166	202	187
	11		Crude petroleum & natural gas; service activities	..	..	..	..	..	..	364	353
	13		Mining of metal ores	..	..	..	..	13	14	48	38
	14		Other mining and quarrying	..	..	862	756	743	761	1 013	970
D	15_37		Total manufacturing	..	..	153 486	162 733	163 788	157 178	168 892	168 272
	15		Food products and beverages	..	..	9 541	9 962	8 294	7 767	8 194	8 073
		151	Production of meat, fish, fruit, vegetables, oils & fats	..	..	2 723	2 927	2 730	2 492	2 536	2 376
		152	Dairy products	..	..	754	764	799	689	677	650
		153	Grain mill and starch products; animal feeds	..	..	654	746	750	699	743	718
		154	Other food products	..	..	4 532	4 581	3 412	3 324	3 499	3 615
		155	Beverages	..	..	655	696	629	667	739	738
	16		Tobacco products	..	..	37	17	11	13	24	20
	17		Textiles	..	..	7 605	6 253	6 046	6 078	6 271	6 131
		171	Spinning, weaving and finishing of textiles	..	..	2 461	2 563	1 450	1 457	1 551	1 499
		172	Other textiles	..	..	3 506	2 176	3 604	3 654	3 706	3 659
		173	Knitted and crocheted fabrics and articles	..	..	1 048	1 043	1 032	1 021	1 014	988
	18		Wearing apparel and fur	..	..	8 094	8 539	8 560	8 276	8 182	7 598
		181	Wearing apparel, except fur apparel	..	..	7 332	..	8 226	8 223	8 143	7 559
		182	Dressing and dyeing of fur; articles of fur	..	..	58	68	61	53	39	39
	19		Leather and articles of leather; footwear	..	..	2 300	1 778	1 512	1 407	1 299	1 206
		191	Leather and articles of leather	..	..	1 515	1 082	996	911	815	745
		192	Footwear	..	..	648	598	497	500	484	461
	20		Wood & products of wood & cork, exc. furniture	..	..	8 419	7 762	8 177	8 357	8 551	8 570
		201	Sawmilling and planing of wood	..	..	2 135	2 071	1 467	1 326	1 263	1 165
		202	Wood products; articles of cork, straw, etc.	..	..	5 748	5 207	6 757	7 041	7 288	7 405
	21		Paper and paper products	..	..	3 144	3 324	2 983	2 857	2 948	2 870
	22		Printing, publishing and recorded media	..	..	25 243	26 967	25 223	26 913	29 072	29 512
		221	Publishing	..	..	8 312	6 908	6 782	7 567	8 775	9 113
		222	Printing and related activities	..	..	16 970	20 098	18 434	18 945	19 507	19 394
		223	Reproduction of recorded media	..	..	56	72	121	509	792	1 010
	23		Coke and petroleum products; nuclear fuel	..	..	198	220	452	280	258	248
		231	Coke oven products	..	..	24	38	33	29	22	17
		232	Refined petroleum products	..	..	148	175	373	248	230	225
		233	Processing of nuclear fuel	..	..	26	7	46	3	6	6
	24		Chemicals and chemical products	..	..	3 788	3 834	4 283	3 888	4 105	3 990
		241	Basic chemicals	..	..	1 650	1 604	1 429	1 356	1 385	1 323
		242	Other chemical products	..	..	1 987	2 056	2 821	2 541	2 672	2 640
		243	Man-made fibres	..	..	44	57	57	52	48	45
	25		Rubber and plastic products	..	..	5 228	5 828	6 586	6 755	7 203	7 203
		251	Rubber products	..	..	574	641	830	730	790	806
		252	Plastic products	..	..	4 676	5 212	5 782	6 044	6 413	6 403
	26		Other non-metallic mineral products	..	..	3 534	3 970	5 014	4 760	5 222	5 246
		261	Glass and glass products	..	..	947	1 149	1 669	1 457	1 615	1 643
		269	Non-metallic mineral products, nec	..	..	2 366	2 580	3 339	3 306	3 607	3 603
	27		Basic metals	..	..	3 612	2 339	2 586	2 575	2 689	2 601
		271	Basic iron and steel	..	..	1 383	535	831	873	985	967
		272	Basic precious and non-ferrous metals	..	..	1 485	1 112	1 026	920	880	813
		273	Casting of metals	..	..	661	680	775	806	824	827
	28		Fabricated metal products, exc. machinery	..	..	28 083	29 623	28 393	28 035	29 164	29 616
		281	Structural metal prod., tanks, reservoirs, etc.	..	..	3 606	3 544	4 432	3 740	4 041	4 180
		289	Other fabricated metal products	..	..	22 078	23 539	23 876	24 380	25 123	25 444
	29		Machinery and equipment, nec	..	..	12 153	13 379	15 350	13 333	14 386	14 279
		291	General purpose machinery	..	..	5 069	5 622	6 711	6 259	6 731	6 628
		292	Special purpose machinery	..	..	5 619	6 096	7 685	6 649	7 047	7 078
		293	Domestic appliances, nec	..	..	568	684	777	539	608	590
	30		Office, accounting & computing machinery	..	..	1 758	1 810	2 303	1 371	1 612	1 558
	31		Electrical machinery & apparatus, nec	..	..	5 808	6 425	5 497	5 265	5 819	5 846
		311	Electric motors, generators and transformers	..	..	1 805	1 592	1 205	1 105	1 139	1 090
		312	Electricity distribution & control apparatus	..	..	518	667	858	889	978	1 030
		313	Insulated wire and cable	..	..	402	376	355	337	373	396
		314	Accumulators, primary cells & primary batteries	..	..	46	60	89	83	109	101
		315	Electric lamps and lighting equipment	..	..	333	335	428	497	592	621
		319	Other electrical equipment, nec	..	..	2 778	3 472	2 639	2 432	2 628	2 611
	32		Radio, TV, communication equipment	..	..	3 270	3 643	3 168	2 864	3 003	2 927
		321	Electronic valves, tubes; other electronic components	..	..	596	631	1 063	1 034	977	986
		322	TV/radio transmitters; line teleph./telegr. apparatus	..	..	1 233	1 342	1 069	843	924	874
		323	TV/radio receivers; sound and video app.; assoc. gds.	..	..	1 462	1 697	1 057	1 005	1 102	1 071
	33		Medical, precision & optical instrum.; watch & clocks	..	..	3 823	4 477	6 599	5 387	5 806	5 622
		331	Medical appl.; instruments & appl. for measuring, etc.	..	..	3 130	3 665	5 623	4 644	4 989	4 810
		332	Optical instruments and photographic equipment	..	..	282	338	593	620	682	689
		333	Watches and clocks	..	..	129	144	161	137	135	129
	34		Motor vehicles, trailers and semi-trailers	..	..	1 509	1 795	4 007	2 398	2 697	2 717
		341	Motor vehicles	..	..	460	569	940	539	597	593
		342	Bodies for motor vehicles, trailers & semi-trailers	..	..	443	612	905	673	767	764
		343	Parts & accessories for motor vehicles & their engines	..	..	622	628	2 176	1 196	1 333	1 364
	35		Other transport equipment	..	..	2 297	2 704	3 060	2 547	2 874	2 769
		351	Building and repairing of ships and boats	..	..	1 595	1 886	1 574	1 530	1 535	1 520
		352	Railway and tramway locomotives and rolling stock	..	..	75	86	106	89	91	93
		353	Aircraft and spacecraft	..	..	475	549	1 148	695	949	858
		359	Transport equipment, nec	..	..	150	181	245	244	299	298
	36		Furniture; manufacturing, nec	..	..	16 342	20 726	17 607	17 593	18 589	18 910
		361	Furniture	..	..	6 893	8 189	7 054	7 200	7 517	7 649
		369	Manufacturing, nec	..	..	8 631	11 455	10 635	10 410	11 072	11 262
	37		Recycling	..	..	..	129	496	747	924	1 001
E	40_41		Electricity, gas and water supply	..	..	162	163	212	224	305	320
	40		Electricity, gas, steam and hot water supply	..	..	109	113	142	147	212	230
	41		Collection, purification and distribution of water	..	..	53	51	71	79	93	90
F	45		Construction	..	..	189 264	184 493	184 299	178 106	178 917	179 850

Table UK.12 **EXPORTATIONS - EXPORTS**

Millions de £ (Prix courants)

1991	1992	1993	1994	1995	1996	1997	1998			CITI révision 3	
6 649	6 414	8 123	9 170	9 853	11 279	9 976	7 263	Activites extractives	C	10_14	
81	52	60	56	57	71	73	53	Extraction de charbon, de lignite et de tourbe		10	
4 432	4 428	5 035	6 059	6 640	7 680	6 559	4 567	Pétrole brut et gaz naturel ; activités annexes		11	
7	11	15	15	30	19	27	16	Extraction de minerais métalliques		13	
2 130	1 924	3 013	3 039	3 126	3 508	3 317	2 623	Autres activités extractives		14	
93 172	96 818	105 114	120 325	139 282	151 763	156 917	152 266	Total activites manufacturieres	D	15_37	
5 826	6 553	7 287	8 263	9 127	9 132	8 997	8 402	Produits alimentaires et boissons		15	
1 349	1 516	1 758	2 249	2 653	2 283	2 055	1 903	Viande, poisson, fruits, légumes, huiles et graisses			151
434	521	605	664	794	715	723	715	Produits laitiers			152
467	545	553	624	658	776	768	779	Travail des grains; prod. amyl. ; alim. pour animaux			153
1 199	1 387	1 609	1 717	1 860	2 009	1 997	1 962	Autres produits alimentaires			154
2 374	2 582	2 758	3 010	3 162	3 349	3 452	3 045	Boissons			155
764	947	635	855	1 127	1 200	1 206	1 047	Produits à base de tabac		16	
2 442	2 604	2 703	3 175	3 448	3 656	3 607	3 295	Textiles		17	
1 140	1 216	1 272	1 500	1 603	1 639	1 617	1 497	Filature, tissage et achèvement des textiles			171
824	872	906	1 020	1 144	1 259	1 252	1 132	Autres articles textiles			172
478	515	524	655	701	758	738	666	Etoffes et articles de bonneterie			173
1 570	1 714	1 833	2 204	2 534	2 776	2 675	2 477	Articles d'habillement et fourrure		18	
1 529	1 672	1 789	2 154	2 472	2 692	2 600	2 433	Articles d'habillement autres qu'en fourrure			181
40	42	42	50	62	83	75	45	Prép. et teinture des fourrures ; articles en fourrure			182
611	649	746	879	953	1 087	1 082	959	Cuir et articles en cuir ; chaussures		19	
611	649	746	879	953	1 087	1 082	959	Cuir et articles en cuir sauf chaussures			191
315	341	392	471	516	602	605	543	Chaussures			192
136	152	147	202	224	263	284	289	Bois, articles en bois, sauf meubles		20	
18	20	16	30	30	34	33	39	Sciage et rabotage du bois			201
140	145	140	172	195	229	251	250	Articles en bois, liège, vannerie, sparterie			202
1 647	1 770	1 823	2 071	2 430	2 428	2 391	2 274	Papier, articles en papier et carton		21	
1 509	1 529	1 780	2 083	2 194	2 471	2 442	2 353	Edition, imprimerie et reproduction		22	
1 177	1 217	1 429	1 656	1 788	2 010	2 052	1 963	Edition			221
332	312	351	426	406	461	390	390	Imprimerie et activités annexes			222
..	..	..	..	..	..	..	..	Reproduction de supports enregistrés			223
2 975	2 923	3 406	3 066	2 974	3 570	4 074	2 560	Coke. et prod. pétroliers ; comb. nucléaires		23	
26	23	16	24	24	22	23	21	Cokéfaction			231
2 594	2 367	2 957	2 682	2 610	3 218	3 696	2 241	Produits pétroliers raffinés			232
355	533	433	360	340	330	354	297	Traitement de combustibles nucléaires			233
14 467	15 594	17 728	19 454	21 925	23 209	22 636	22 837	Produits chimiques		24	
1 514	1 522	1 702	1 981	2 308	2 252	2 091	2 083	Produits chimiques de base			241
7 746	8 585	9 905	11 043	12 593	13 704	13 476	13 738	Autres produits chimiques			242
660	660	620	662	766	808	760	754	Fibres synthétiques ou artificielles			243
2 549	2 769	2 885	3 318	3 889	4 186	4 111	4 199	Art. en caoutchouc et matières plastiques		25	
926	1 014	1 092	1 221	1 389	1 515	1 440	1 528	Articles en caoutchouc			251
1 623	1 755	1 793	2 097	2 500	2 671	2 671	2 671	Articles en matières plastiques			252
1 384	1 380	1 462	1 673	1 975	2 135	2 127	2 002	Autres produits minéraux non métalliques		26	
458	446	461	522	599	607	629	600	Verre et articles en verre			261
933	942	1 006	1 152	1 375	1 528	1 499	1 402	Produits minéraux non métalliques, nca			269
5 560	5 270	5 653	6 525	8 091	7 552	7 142	6 414	Produits métallurgiques de base		27	
3 225	3 183	3 449	3 926	4 736	4 413	3 943	3 571	Sidérurgie et première transformation de l'acier			271
2 335	2 087	2 204	2 600	3 355	3 139	3 199	2 843	Première transf. métaux précieux et non ferreux			272
..	..	..	..	..	..	..	..	Fonderie			273
2 286	2 320	2 339	2 763	3 231	3 551	3 662	3 764	Ouvrages en métaux, sauf machines		28	
529	518	597	688	791	871	1 005	901	Constr. et menuis. métal.; réservoirs; générateurs			281
1 730	1 783	1 731	2 075	2 440	2 680	2 657	2 863	Autres ouvrages en métaux			289
11 225	11 559	12 381	13 630	16 254	17 884	18 436	18 230	Machines et matériels, nca		29	
3 069	3 172	3 602	3 805	4 382	4 768	5 055	5 103	Machines d'usage général			291
5 269	5 416	5 633	6 208	7 463	8 193	8 491	8 374	Machines d'usage spécifique			292
505	548	600	701	818	899	858	797	Appareils domestiques, nca			293
6 593	6 641	8 069	9 588	11 799	12 265	13 229	12 728	Mach. de bureau, comptables, information		30	
3 445	3 635	4 075	4 884	5 657	6 495	6 709	6 691	Machines et appareils électriques, nca		31	
770	873	973	1 072	1 284	1 465	1 508	1 575	Moteurs, génératrices, transformateurs électriques			311
841	874	972	1 119	1 284	1 474	1 580	1 581	Matériel électrique de distribution et de commande			312
338	355	388	483	608	615	614	612	Fils et câbles électriques isolés			313
200	170	183	228	263	342	353	356	Accumulateurs et piles électriques			314
215	228	277	333	383	437	460	457	Lampes électriques et appareils d'éclairage			315
1 081	1 135	1 282	1 650	1 834	2 162	2 195	2 110	Autres matériels électriques, nca.			319
4 878	5 219	7 102	9 322	12 228	13 295	13 223	13 998	Equipements et appareils de radio, TV, etc.		32	
2 083	2 524	3 742	4 729	5 954	5 845	4 906	4 721	Tubes et valves électroniques;autres composants			321
897	929	1 243	2 122	3 340	4 059	5 040	6 360	Emetteurs radio/TV; app. téléphonie et télégraphie			322
1 898	1 767	2 117	2 470	2 933	3 390	3 277	2 917	Récepteurs radio/TV, son et image; art. associés			323
3 858	4 053	4 369	4 857	5 470	6 132	6 606	6 577	Instr. médicaux, de précision ; optique ; horlog.		33	
3 332	3 416	3 747	4 084	4 653	5 203	5 615	5 639	Instruments médicaux et de précision, etc.			331
439	509	463	553	575	664	698	733	Instruments d'optique ; matériel photographique			332
87	128	159	220	242	265	293	205	Horlogerie			333
9 685	10 075	9 450	11 088	13 380	16 172	16 633	16 384	Véhicules automobiles; remorques, etc.		34	
5 990	6 265	6 350	7 267	9 138	11 096	11 199	11 100	Véhicules automobiles			341
256	253	253	340	384	460	445	392	Carross. automobiles, remorques, semi-remorques			342
3 438	3 557	2 848	3 482	3 858	4 616	4 988	4 891	Pièces et accessoires pour automobiles et moteurs			343
7 696	7 191	6 817	7 243	6 942	8 517	11 606	11 020	Autres matériels de transport		35	
316	292	242	505	443	582	1 315	514	Construction et réparation de navires			351
104	89	133	494	104	128	95	175	Matériel ferroviaire roulant			352
7 152	6 675	6 294	6 074	6 207	7 600	9 982	10 150	Construction aéronautique et spatiale			353
637	581	465	169	189	208	213	182	Autres équipements de transport			359
2 067	2 270	2 424	3 181	3 431	3 787	4 040	3 766	Meubles ; activités de fabrication, nca		36	
538	592	551	665	794	922	931	946	Meubles			361
1 527	1 676	1 871	2 516	2 637	2 865	3 108	2 820	Activités de fabrication, nca			369
..	..	..	..	..	..	..	..	Récupération		37	
..	..	..	..	2	2	1	4	Electricité, gaz et eau	E	40_41	
..	..	..	..	2	2	1	4	Prod. et dist. d'électricité gaz, vapeur d'eau chaude		40	
..	..	..	..	..	..	..	..	Captage, épuration et distribution d'eau		41	
..	..	..	..	..	..	..	..	Construction	F	45	

UNITED KINGDOM

Table UK.13 **IMPORTS - IMPORTATIONS**

Millions of £ (Current Prices)

ISIC revision 3				1991	1992	1993	1994	1995	1996	1997	1998
C	10_14		Mining and quarrying	7 751	7 420	8 343	7 246	7 419	8 684	8 384	6 492
	10		Mining of coal and lignite; extraction of peat	736	738	643	586	587	658	682	667
	11		Crude petroleum & natural gas; service activities	4 335	4 131	4 400	3 475	3 345	4 150	3 743	2 213
	13		Mining of metal ores	566	551	508	485	595	626	641	583
	14		Other mining and quarrying	2 013	1 921	2 705	2 627	2 849	3 179	3 167	2 953
D	15_37		Total manufacturing	103 371	110 368	117 944	132 882	152 114	166 119	169 901	173 926
	15		Food products and beverages	9 496	10 513	10 642	11 415	12 681	13 898	13 313	13 253
		151	Production of meat, fish, fruit, vegetables, oils & fats	4 676	5 029	4 975	5 151	5 993	6 645	5 993	5 775
		152	Dairy products	847	1 097	997	1 096	1 097	1 195	1 114	1 134
		153	Grain mill and starch products; animal feeds	568	650	694	689	756	848	882	781
		154	Other food products	1 863	2 094	2 272	2 566	2 764	2 871	2 834	2 880
		155	Beverages	1 540	1 641	1 704	1 914	2 071	2 338	2 491	2 683
	16		Tobacco products	161	161	166	186	222	240	209	165
	17		Textiles	4 082	4 352	4 354	4 915	5 286	5 709	5 841	5 723
		171	Spinning, weaving and finishing of textiles	2 151	2 224	2 199	2 575	2 738	2 810	2 659	2 457
		172	Other textiles	1 190	1 278	1 248	1 412	1 563	1 733	1 776	1 863
		173	Knitted and crocheted fabrics and articles	741	849	908	928	985	1 167	1 406	1 403
	18		Wearing apparel and fur	3 451	3 683	3 780	3 998	4 386	5 163	5 554	5 958
		181	Wearing apparel, except fur apparel	3 434	3 662	3 764	3 974	4 359	5 115	5 505	5 926
		182	Dressing and dyeing of fur; articles of fur	19	23	18	24	27	47	49	32
	19		Leather and articles of leather; footwear	1 630	1 638	1 745	2 048	2 184	2 558	2 694	2 645
		191	Leather and articles of leather	1 630	1 638	1 745	2 048	2 184	2 558	2 694	2 645
		192	Footwear	1 170	1 152	1 206	1 435	1 522	1 792	1 939	1 897
	20		Wood & products of wood & cork, exc. furniture	1 855	1 862	1 953	2 383	2 203	2 279	2 393	2 267
		201	Sawmilling and planing of wood	1 005	986	1 046	1 328	1 115	1 155	1 251	1 120
		202	Wood products; articles of cork, straw, etc.	861	884	913	1 055	1 088	1 124	1 143	1 147
	21		Paper and paper products	4 468	4 416	4 201	4 844	6 281	5 696	5 217	5 149
	22		Printing, publishing and recorded media	1 067	1 119	1 136	1 194	1 270	1 409	1 473	1 519
		221	Publishing	742	790	888	859	921	1 047	1 082	1 126
		222	Printing and related activities	325	329	248	336	350	362	391	393
		223	Reproduction of recorded media	..	..	..	..	..	..	..	..
	23		Coke and petroleum products; nuclear fuel	2 432	1 982	2 000	1 996	1 753	2 127	1 879	1 646
		231	Coke oven products	19	26	41	31	40	58	55	47
		232	Refined petroleum products	2 009	1 630	1 649	1 622	1 463	1 724	1 584	1 379
		233	Processing of nuclear fuel	404	326	310	343	250	345	240	220
	24		Chemicals and chemical products	11 316	12 135	13 357	14 971	18 371	18 992	18 494	18 428
		241	Basic chemicals	2 203	2 160	2 266	2 677	3 416	3 174	2 978	2 686
		242	Other chemical products	4 809	5 461	6 228	6 951	8 071	8 916	9 117	9 496
		243	Man-made fibres	546	597	595	671	767	788	805	723
	25		Rubber and plastic products	3 079	3 381	3 369	3 961	4 545	4 699	4 638	4 823
		251	Rubber products	915	1 050	1 110	1 316	1 528	1 650	1 528	1 578
		252	Plastic products	2 164	2 331	2 259	2 645	3 016	3 048	3 110	3 245
	26		Other non-metallic mineral products	1 464	1 490	1 414	1 648	1 771	1 891	1 913	1 976
		261	Glass and glass products	653	676	669	746	826	844	779	782
		269	Non-metallic mineral products, nec	818	820	750	902	946	1 047	1 133	1 194
	27		Basic metals	5 824	5 702	5 951	6 587	8 485	8 558	8 101	8 122
		271	Basic iron and steel	2 649	2 538	2 486	2 978	3 753	3 760	3 434	3 354
		272	Basic precious and non-ferrous metals	3 172	3 161	3 464	3 609	4 732	4 797	4 667	4 769
		273	Casting of metals	..	..	..	..	..	..	..	..
	28		Fabricated metal products, exc. machinery	2 626	2 660	2 590	2 973	3 425	3 677	3 756	3 945
		281	Structural metal prod., tanks, reservoirs, etc.	548	470	423	454	550	593	639	666
		289	Other fabricated metal products	2 060	2 178	2 158	2 519	2 875	3 084	3 117	3 278
	29		Machinery and equipment, nec	10 238	10 878	10 802	12 208	14 216	14 950	15 551	15 715
		291	General purpose machinery	2 501	2 820	3 155	3 432	3 786	4 115	4 163	4 331
		292	Special purpose machinery	4 520	4 646	4 406	4 975	6 080	6 184	6 534	6 286
		293	Domestic appliances, nec	1 055	1 075	1 069	1 180	1 287	1 284	1 335	1 430
	30		Office, accounting & computing machinery	7 590	8 358	10 022	10 707	12 400	12 847	13 977	15 666
	31		Electrical machinery & apparatus, nec	3 592	4 064	4 493	5 460	6 497	6 957	7 185	7 154
		311	Electric motors, generators and transformers	701	840	844	979	1 201	1 465	1 524	1 639
		312	Electricity distribution & control apparatus	974	1 096	1 174	1 294	1 517	1 613	1 651	1 668
		313	Insulated wire and cable	381	406	486	561	729	781	724	670
		314	Accumulators, primary cells & primary batteries	222	248	352	418	492	440	514	537
		315	Electric lamps and lighting equipment	327	383	424	500	563	588	583	656
		319	Other electrical equipment, nec	988	1 090	1 213	1 708	1 994	2 071	2 189	1 984
	32		Radio, TV, communication equipment	5 997	6 503	7 836	9 264	11 999	15 153	14 232	13 979
		321	Electronic valves, tubes; other electronic components	2 781	3 068	3 901	4 376	5 801	7 737	6 493	5 659
		322	TV/radio transmitters; line teleph./telegr. apparatus	1 038	1 212	1 498	2 090	3 046	4 028	4 503	4 965
		323	TV/radio receivers; sound and video app.; assoc. gds.	2 179	2 222	2 436	2 798	3 152	3 388	3 236	3 355
	33		Medical, precision & optical instrum.; watch & clocks	3 743	3 923	4 296	4 584	5 175	5 933	5 928	6 135
		331	Medical appl.; instruments & appl. for measuring, etc.	2 769	2 877	3 094	3 256	3 711	4 288	4 254	4 531
		332	Optical instruments and photographic equipment	656	689	757	857	963	1 087	1 106	1 079
		333	Watches and clocks	318	358	446	472	501	558	569	525
	34		Motor vehicles, trailers and semi-trailers	10 772	12 796	14 699	16 940	19 167	21 705	22 658	23 672
		341	Motor vehicles	6 545	7 934	9 504	11 112	12 254	13 628	15 879	16 869
		342	Bodies for motor vehicles, trailers & semi-trailers	139	154	168	233	230	246	252	313
		343	Parts & accessories for motor vehicles & their engines	4 087	4 709	5 026	5 594	6 683	7 831	6 527	6 490
	35		Other transport equipment	5 137	4 918	4 900	5 976	5 037	6 222	8 718	9 708
		351	Building and repairing of ships and boats	306	237	64	61	313	147	170	232
		352	Railway and tramway locomotives and rolling stock	91	148	243	730	223	86	115	246
		353	Aircraft and spacecraft	4 407	4 200	4 224	4 764	4 047	5 464	7 818	8 565
		359	Transport equipment, nec	565	532	520	420	454	525	615	665
	36		Furniture; manufacturing, nec	3 351	3 836	4 236	4 626	4 758	5 458	6 176	6 279
		361	Furniture	989	1 036	942	1 081	1 177	1 375	1 467	1 663
		369	Manufacturing, nec	2 361	2 799	3 292	3 545	3 582	4 083	4 709	4 616
	37		Recycling	..	..	..	..	..	..	..	..
E	40_41		Electricity, gas and water supply	372	371	427	388	416	391	406	374
	40		Electricity, gas, steam and hot water supply	372	371	427	388	416	391	406	374
	41		Collection, purification and distribution of water	..	..	..	..	..	..	..	..
F	45		Construction	..	..	..	..	..	..	..	..

PART C

PARTIE C

ISIC REVISION 3, SERVICES

CITI REVISION 3, SERVICES

Structural business statistics and foreign trade statistics
Statistiques structurelles sur les entreprises et statistiques du commerce extérieur

List of countries
Liste des pays

Note :

More variables and sectors are available on the electronic version of the publication.

D'autres variables et d'autres secteurs sont disponibles dans la version électronique de cette publication.

AUSTRALIA

Table AS.S1 **PRODUCTION - PRODUCTION**

Millions of A$ (Current Prices)

ISIC revision 3				1990-91	1991-92	1992-93	1993-94	1994-95	1995-96	1996-97	1997-98
G	50_52		Wholesale & retail trade; repairs; etc.	..	..	231 268	253 222	271 010	293 767	313 951	..
	50		Sale, maint.,repair vehic/cycle, person., h/hold gd.	..	..	..	..	..	..	..	..
		501	Sale of motor cycle	..	..	..	..	..	..	..	..
		502	Maintenance and repair of motor vehicles	..	..	..	..	..	..	..	..
		503	Sale of motor vehicle parts and accessories	..	..	..	..	..	..	..	..
		504	Sale, maint., repair motor cycle, parts & acces.	..	..	..	..	..	..	..	..
		505	Retail sale of automative fuel	..	..	..	..	..	..	..	..
	51		W/sale. trade & commission exc motor veh. & cycle	..	..	..	..	..	..	..	..
		511	Wholesale on a fee or contract basis	..	..	..	..	..	..	..	..
		512	W/sale agric. raw mat.,live animal,food, bev., tobac.	..	..	..	..	..	..	..	..
		513	Wholesale of household goods	..	..	..	..	..	..	..	..
		514	W/sale. of non-agric. intermed. prod., waste, scrap	..	..	..	..	..	..	..	..
		515	Wholesale of machinery, equipment & supplies	..	..	..	..	..	..	..	..
		519	Other wholesale	..	..	..	..	..	..	..	..
	52		Retail trade exc. motor veh. & cyc.; repair h/hold gds.	..	..	..	..	..	..	..	..
		521	Non-specialised retail trade in stores	..	..	..	..	..	..	..	..
		522	Retail sale food, bev., tobac. in specialised stores	..	..	38 564	40 058	47 640	50 412	57 719	..
		523	Other retail trade of new goods specialised stores	..	..	..	..	..	..	..	..
		524	Retail sale of second-hand goods in stores	..	..	..	..	..	..	..	..
		525	Retail trade not in stores	..	..	..	..	..	..	..	..
		526	Repair of personal and household goods	..	..	..	..	..	..	..	..
H	55		Hotels and restaurants	..	..	23 312	22 076	25 084	26 338	27 406	..
		551	Hotels; camping ; other short-stay accommodation	..	..	..	..	..	..	..	..
		552	Restaurants, bars and canteens	..	..	..	..	..	..	..	..
I	60_64		Transport, storage and communications	..	..	54 088	57 023	61 806	67 025	71 467	..
	60		Land transport, transport via pipelines	..	..	14 338	15 979	17 124	18 966	20 049	..
		601	Transport via railways	..	..	3 615	3 428	3 983	3 854	5 190	..
		602	Other land transport	..	..	10 723	12 551	13 141	15 112	14 859	..
		603	Transport via pipelines	..	..	..	..	..	..	..	..
	61		Water transport	..	..	..	..	..	..	..	..
		611	Sea and coastal water transport	..	..	..	..	..	..	..	..
		612	Inland water transport	..	..	..	..	..	..	..	..
	62		Air transport	..	..	..	..	..	..	..	..
		621	Scheduled air transport	..	..	..	..	..	..	..	..
		622	Non-scheduled air transport	..	..	..	..	..	..	..	..
	63		Support. & aux. transport act., travel agencies activ.	..	..	9 703	8 701	9 912	10 194	10 794	..
	64		Post and telecommunications	..	..	15 937	17 463	19 883	21 709	23 723	..
		641	Post and courier activities	..	..	..	..	..	..	..	..
		642	Telecommunications	..	..	..	..	..	..	..	..
J	65_67		Financial intermediation	..	..	16 730	17 678	19 817	25 885	24 597	..
	65		Financial intermediation, exc. insur., pension fund	..	..	7 869	9 069	8 991	8 426	9 710	..
		651	Monetary intermediation	..	..	..	..	..	..	..	..
		659	Other financial intermediation	..	..	..	..	..	..	..	..
	66		Insurance pension fund, exc. compul. social sec.	..	..	2 067	1 931	2 536	2 264	2 663	..
	67		Activities auxiliary to financial intermediation	..	..	6 795	6 677	8 290	15 194	12 223	..
		671	Act. auxil to fin. intermed., exc. insur. & pens.fund	..	..	..	..	..	..	..	..
		672	Act. auxiliary to insurance and pension funding	..	..	..	..	..	..	..	..
K	70_74		Real estate, renting and business activities	..	..	43 578	36 733	48 542	65 195	71 301	..
	70		Real estate activities	..	..	..	..	..	..	..	..
		701	Real estate activ. with own or leased property	..	..	..	..	..	..	..	..
		702	Real estate activities on a fee or contract basis	..	..	..	..	..	..	..	..
	71		Renting of mach. without operator; rent. of pers. gds.	..	..	..	..	..	..	..	..
		711	Renting of transport equipment	..	..	..	..	..	..	..	..
		712	Renting of other machinery and equipment	..	..	..	..	..	..	..	..
		713	Renting of personal and household goods nec	..	..	..	..	..	..	..	..
	72		Computer and related activities	..	..	..	..	..	..	..	..
		721	Hardware consultancy	..	..	..	..	..	..	..	..
		722	Software consultancy and supply	..	..	..	..	..	..	..	..
		723	Data processing	..	..	..	..	..	..	..	..
		724	Data base activities	..	..	..	..	..	..	..	..
		725	Mainten.; repair of office, account. & comp. mach.	..	..	..	..	..	..	..	..
		729	Other computer related activities	..	..	..	..	..	..	..	..
	73		Research and development	..	..	..	..	..	..	..	..
		731	R & D experim. on natural sciences & engineering	..	..	..	..	..	..	..	..
		732	R & D experim. on social sciences & humanities	..	..	..	..	..	..	..	..
	74		Other business activities	..	..	..	..	..	..	..	..
		741	Legal, acc., book-keep., audit act.; tax consult., etc	..	..	..	..	..	..	..	..
		742	Architectural, engineering, & other techn. activities	..	..	..	..	..	..	..	..
		743	Advertising	..	..	..	..	..	..	..	..
		749	Business activities nec	..	..	..	..	..	..	..	..
M	80		Education	..	..	..	..	..	..	..	..
		801	Primary education	..	..	..	..	..	..	..	..
		802	Secondary education	..	..	..	..	..	..	..	..
		803	Higher education	..	..	..	..	..	..	..	..
		809	Adult and other education	..	..	..	..	..	..	..	..
N	85		Health and social work	..	..	..	..	..	..	..	..
		851	Human health activities	..	..	..	..	..	..	..	..
		852	Veterinary activities	..	..	..	..	..	..	..	..
		853	Social work activities	..	..	..	..	..	..	..	..
O	90_93		Other commun., soc. & pers. services activities	..	..	17 234	19 520	22 104	23 842	26 161	..
	90		Sewage & refuse disposal, sanitation, etc.	..	..	..	..	..	..	..	..
	91		Activities of member organisations nec	..	..	3 260	4 231	4 163	4 434	4 175	..
		912	Activities of trade unions	..	..	..	..	..	..	..	..
	92		Recreational, cultural and sporting activities	..	..	9 751	9 964	12 371	13 998	16 163	..
		921	Motion picture, radio, TV, other entertainment act.	..	..	..	..	..	..	..	..
		922	News agency activities	..	..	..	..	..	..	..	..
		923	Library, archives, museums, other cultural activ.	..	..	..	..	..	..	..	..
		924	Sporting and other recreational activities	..	..	5 150	4 385	5 711	7 787	9 403	..
	93		Other service activities	..	..	4 222	5 324	5 570	5 409	5 823	..

Note:Data cover the fiscal year ending 30 June.

Data for ISIC 50-52 include businesses mainly engaged in selling take-away food, which should be included in ISIC 55. ISIC 75 is not covered by the inquiry.

Data for ISIC 70-74 include certain activities which are primary to ISIC 93.

Table AS.S2 **VALEUR AJOUTÉE - VALUE ADDED**

Millions de A$ (Prix courants)

1990-91	1991-92	1992-93	1993-94	1994-95	1995-96	1996-97	1997-98		CITI révision 3		
..	..	..	..	..	..	32 588	36 288	Com. de gros, de detail ; réparation etc.	G	50_52	
..	..	..	..	..	..	..	..	Com., entr., répar. d'auto. et moto ; carburants		50	
..	..	..	..	..	..	..	..	Commerce de véhicules automobiles			501
..	..	..	..	..	..	..	..	Entretien et réparation de véhicules automobiles			502
..	..	..	..	..	..	..	..	Com. de pièces et access. de véhic. automobiles			503
..	..	..	..	..	..	..	..	Com., entret. et répar. de motocycles et access.			504
..	..	..	..	..	..	..	..	Com. de détail de carburants pour automobiles			505
..	..	..	..	..	..	..	..	Com. de gros et act. inter., sauf auto. et moto		51	
..	..	..	..	..	..	..	..	Activités d'intermédiaries de commerce de gros			511
..	..	..	..	..	..	..	..	Com. gros: prod. agr. bruts, anim. viv., prod. alim.			512
..	..	..	..	..	..	..	..	Com. gros: biens de consom., sauf alimentaires			513
..	..	..	..	..	..	..	..	Com. gros: prod. inter. non agr., déchets, débris			514
..	..	..	..	..	..	..	..	Com. gros: machines, équipments et fournitures			515
..	..	..	..	..	..	..	..	Autres commerces de gros			519
..	..	..	..	..	..	..	..	Com. détail, sauf auto, moto; répar. art. pers., dom.		52	
..	..	..	..	..	..	..	..	Commerce de détail non spécialisé en magasin			521
..	..	..	..	..	..	5 327	6 104	Com. détail: prod. aliment. en mag. spécialisés			522
..	..	..	..	..	..	..	..	Autre com. détail: prod. neufs en mag. spécialisés			523
..	..	..	..	..	..	..	..	Commerces de détail de biens d'occasion			524
..	..	..	..	..	..	..	..	Commerces de détail autres qu'en magasin			525
..	..	..	..	..	..	..	..	Réparation d'articles personnels et domestiques			526
..	..	..	..	..	..	7 650	7 929	Hotels et restaurants	H	55	
..	..	..	..	..	..	..	..	Hôtels; camping; autres équip. courts séjours			551
..	..	..	..	..	..	..	..	Restaurants, cafés et cantines			552
..	..	..	..	..	..	26 405	28 268	Transport, entreposage, communications	I	60_64	
..	..	..	..	..	..	8 034	8 855	Transports terrestres; transports par conduites		60	
..	..	..	..	..	..	3 425	3 545	Transports par chemin de fer			601
..	..	..	..	..	..	4 609	5 310	Autres transports terrestres			602
..	..	..	..	..	..	..	..	Transports par conduites			603
..	..	..	..	..	..	..	..	Transports par eau		61	
..	..	..	..	..	..	..	..	Transports maritimes et côtiers			611
..	..	..	..	..	..	..	..	Transports par voies navigables intérieures			612
..	..	..	..	..	..	..	..	Transports aériens		62	
..	..	..	..	..	..	..	..	Transports aériens réguliers			621
..	..	..	..	..	..	..	..	Transports aériens non réguliers			622
..	..	..	..	..	..	4 140	3 794	Act. annex. et aux. des transp.; act. des ag. de voy.		63	
..	..	..	..	..	..	10 350	11 205	Postes et télécommunications		64	
..	..	..	..	..	..	..	..	Activités de poste et de courrier			641
..	..	..	..	..	..	..	..	Télécommunications			642
..	..	..	..	..	..	..	..	Intermédiation financière	J	65_67	
..	..	..	..	..	..	..	..	Interm. financ., sauf assur. et caisses de retraite		65	
..	..	..	..	..	..	..	..	Intermédiation monétaire			651
..	..	..	..	..	..	..	..	Autres intermédiations financières			659
..	..	..	..	..	..	..	..	Act. d'assur., caisses de retr., sauf séc. soc. oblig.		66	
..	..	..	..	..	..	3 132	3 449	Act. auxiliaires de l'intermédiation financière		67	
..	..	..	..	..	..	..	..	Act. aux. de l'interm. fin., sauf assur., et retraite			671
..	..	..	..	..	..	..	..	Act. aux. des assurances et caisses de retraite			672
..	..	..	..	..	..	19 931	19 953	Immobilier, locations, services aux entreprises	K	70_74	
..	..	..	..	..	..	..	..	Activités immobilières		70	
..	..	..	..	..	..	..	..	Act. immobilières sur biens propres ou loués			701
..	..	..	..	..	..	..	..	Activités immobilières pour comptes de tiers			702
..	..	..	..	..	..	..	..	Locat. mach., équip. sans opér., biens pers. et dom.		71	
..	..	..	..	..	..	..	..	Location de matériel de transport			711
..	..	..	..	..	..	..	..	Location de machines ét equipements nca			712
..	..	..	..	..	..	..	..	Loc. d'articles personnels et domestiques nca			713
..	..	..	..	..	..	..	..	Activités informatiques et activités rattachées		72	
..	..	..	..	..	..	..	..	Conseil en matériel informatique			721
..	..	..	..	..	..	..	..	Production de logiciels			722
..	..	..	..	..	..	..	..	Traitement de données informatiques			723
..	..	..	..	..	..	..	..	Activités de banques de données			724
..	..	..	..	..	..	..	..	Entr., répar. mach. de bureau et compt. et inform.			725
..	..	..	..	..	..	..	..	Autres activités rattachées à l'informatique			729
..	..	..	..	..	..	..	..	Recherche-développement		73	
..	..	..	..	..	..	..	..	Rech. et dév. expér. en sc. phys., natur. et ingén.			731
..	..	..	..	..	..	..	..	Rech. et dév. expér. en sciences soc. et humaines			732
..	..	..	..	..	..	..	..	Autres activités de services aux entreprises		74	
..	..	..	..	..	..	..	..	Act. jur., compt., audit.; cons. fisc.; ét. march., soud.			741
..	..	..	..	..	..	..	..	Act. d'achitect., d'ingén.; autres act. techniques			742
..	..	..	..	..	..	..	..	Publicité			743
..	..	..	..	..	..	..	..	Activités de services aux entreprises nca.			749
..	..	..	..	..	..	..	..	Education	M	80	
..	..	..	..	..	..	..	..	Activités d'enseignement primaire			801
..	..	..	..	..	..	..	..	Activités d'enseignement secondaire			802
..	..	..	..	..	..	..	..	Activités d'enseignement supérieur			803
..	..	..	..	..	..	..	..	Act. de formation perman.; autres act. d'enseign.			809
..	..	..	..	..	..	..	..	Santé et action sociale	N	85	
..	..	..	..	..	..	..	..	Activités de santé humaine			851
..	..	..	..	..	..	..	..	Activités vétérinaires			852
..	..	..	..	..	..	..	..	Activités d'action sociale			853
..	..	..	..	..	..	7 121	8 383	Autres activités de serv. coll., soc. et personnels	O	90_93	
..	..	..	..	..	..	..	..	Assain. et enlèv. des ordures; voierie, act. simil.		90	
..	..	..	..	..	..	979	1 134	Activités associatives diverses		91	
..	..	..	..	..	..	..	..	Activités de syndicats de salariés			912
..	..	..	..	..	..	4 063	4 946	Activités récréatives, culturelles et sportives		92	
..	..	..	..	..	..	..	..	Act. cinémat., de radio, télé.; autres act. de spect.			921
..	..	..	..	..	..	..	..	Activités d'agences de presse			922
..	..	..	..	..	..	..	..	Act. des biblio., arch., musées, jardins bot. et zool.			923
..	..	..	..	..	..	2 215	2 014	Activités sportives et autres activités récréatives			924
..	..	..	..	..	..	2 079	2 303	Autres activités de services		93	

Note:Data cover the fiscal year ending 30 June.

Data for ISIC 50-52 include businesses mainly engaged in selling take-away food, which should be included in ISIC 55. ISIC 75 is not covered by the inquiry.

Data for ISIC 70-74 include certain activities which are primary to ISIC 93.

Table AS.S3 **EMPLOYMENT - EMPLOI**

Units

ISIC revision 3				1990-91	1991-92	1992-93	1993-94	1994-95	1995-96	1996-97	1997-98
G	**50_52**		Wholesale & retail trade; repairs; etc.	..	..	1111 000	1141 000	1210 000	1240 000	1236 000	..
	50		Sale, maint.,repair vehic/cycle, person., h/hold gd.	..	..	..	..	..	..	..	..
		501	Sale of motor cycle	..	..	..	..	..	..	..	..
		502	Maintenance and repair of motor vehicles	..	..	..	..	..	..	..	..
		503	Sale of motor vehicle parts and accessories	..	..	..	..	..	..	..	..
		504	Sale, maint., repair motor cycle, parts & acces.	..	..	..	..	..	..	..	..
		505	Retail sale of automative fuel	..	..	..	..	..	..	..	..
	51		W/sale. trade & commission exc motor veh. & cycle	..	..	..	..	..	..	..	..
		511	Wholesale on a fee or contract basis	..	..	..	..	..	..	..	..
		512	W/sale agric. raw mat.,live animal,food, bev., tobac.	..	..	..	..	..	..	..	..
		513	Wholesale of household goods	..	..	..	..	..	..	..	..
		514	W/sale. of non-agric. intermed. prod., waste, scrap	..	..	..	..	..	..	..	..
		515	Wholesale of machinery, equipment & supplies	..	..	..	..	..	..	..	..
		519	Other wholesale	..	..	..	..	..	..	..	..
	52		Retail trade exc. motor veh. & cyc.; repair h/hold gds.	..	..	..	..	..	..	..	..
		521	Non-specialised retail trade in stores	..	..	..	..	..	..	..	..
		522	Retail sale food, bev., tobac. in specialised stores	..	..	315 000	323 000	384 000	432 000	445 000	..
		523	Other retail trade of new goods specialised stores	..	..	..	..	..	..	..	..
		524	Retail sale of second-hand goods in stores	..	..	..	..	..	..	..	..
		525	Retail trade not in stores	..	..	..	..	..	..	..	..
		526	Repair of personal and household goods	..	..	..	..	..	..	..	..
H	**55**		Hotels and restaurants	..	..	355 000	385 000	393 000	403 000	431 000	..
		551	Hotels; camping ; other short-stay accommodation	..	..	..	..	..	..	..	..
		552	Restaurants, bars and canteens	..	..	..	..	..	..	..	..
I	**60_64**		Transport, storage and communications	..	..	416 000	409 000	433 000	445 000	449 000	..
	60		Land transport, transport via pipelines	..	..	184 000	174 000	179 000	181 000	175 000	..
		601	Transport via railways	..	..	64 000	54 000	52 000	49 000	46 000	..
		602	Other land transport	..	..	120 000	120 000	127 000	132 000	129 000	..
		603	Transport via pipelines	..	..	..	..	..	..	..	..
	61		Water transport	..	..	..	..	..	..	..	..
		611	Sea and coastal water transport	..	..	..	..	..	..	..	..
		612	Inland water transport	..	..	..	..	..	..	..	..
	62		Air transport	..	..	..	..	..	..	..	..
		621	Scheduled air transport	..	..	..	..	..	..	..	..
		622	Non-scheduled air transport	..	..	..	..	..	..	..	..
	63		Support. & aux. transport act., travel agencies activ.	..	..	57 000	56 000	63 000	60 000	82 000	..
	64		Post and telecommunications	..	..	116 000	117 000	127 000	138 000	126 000	..
		641	Post and courier activities	..	..	..	..	..	..	..	..
		642	Telecommunications	..	..	..	..	..	..	..	..
J	**65_67**		Financial intermediation	..	..	298 000	298 000	290 000	310 000	308 000	..
	65		Financial intermediation, exc. insur., pension fund	..	..	183 000	174 000	178 000	180 000	180 000	..
		651	Monetary intermediation	..	..	..	..	..	..	..	..
		659	Other financial intermediation	..	..	..	..	..	..	..	..
	66		Insurance pension fund, exc. compul. social sec.	..	..	53 000	57 000	55 000	51 000	57 000	..
	67		Activities auxiliary to financial intermediation	..	..	62 000	67 000	58 000	79 000	71 000	..
		671	Act. auxil to fin. intermed., exc. insur. & pens.fund	..	..	..	..	..	..	..	..
		672	Act. auxiliary to insurance and pension funding	..	..	..	..	..	..	..	..
K	**70_74**		Real estate, renting and business activities	..	..	577 000	556 000	637 000	774 000	824 000	..
	70		Real estate activities	..	..	..	..	..	..	..	..
		701	Real estate activ. with own or leased property	..	..	..	..	..	..	..	..
		702	Real estate activities on a fee or contract basis	..	..	..	..	..	..	..	..
	71		Renting of mach. without operator; rent. of pers. gds.	..	..	..	..	..	..	..	..
		711	Renting of transport equipment	..	..	..	..	..	..	..	..
		712	Renting of other machinery and equipment	..	..	..	..	..	..	..	..
		713	Renting of personal and household goods nec	..	..	..	..	..	..	..	..
	72		Computer and related activities	..	..	..	..	..	..	..	..
		721	Hardware consultancy	..	..	..	..	..	..	..	..
		722	Software consultancy and supply	..	..	..	..	..	..	..	..
		723	Data processing	..	..	..	..	..	..	..	..
		724	Data base activities	..	..	..	..	..	..	..	..
		725	Mainten.; repair of office, account. & comp. mach.	..	..	..	..	..	..	..	..
		729	Other computer related activities	..	..	..	..	..	..	..	..
	73		Research and development	..	..	..	..	..	..	..	..
		731	R & D experim. on natural sciences & engineering	..	..	..	..	..	..	..	..
		732	R & D experim. on social sciences & humanities	..	..	..	..	..	..	..	..
	74		Other business activities	..	..	..	..	..	..	..	..
		741	Legal, acc., book-keep., audit act.; tax consult., etc	..	..	..	..	..	..	..	..
		742	Architectural, engineering, & other techn. activities	..	..	..	..	..	..	..	..
		743	Advertising	..	..	..	..	..	..	..	..
		749	Business activities nec	..	..	..	..	..	..	..	..
M	**80**		Education	..	..	..	..	..	..	177 526	..
		801	Primary education	..	..	..	..	..	..	..	..
		802	Secondary education	..	..	..	..	..	..	..	..
		803	Higher education	..	..	..	..	..	..	..	..
		809	Adult and other education	..	..	..	..	..	..	..	..
N	**85**		Health and social work	..	..	..	..	..	..	522 122	..
		851	Human health activities	..	..	..	..	..	..	..	..
		852	Veterinary activities	..	..	..	..	..	..	..	..
		853	Social work activities	..	..	..	..	..	..	..	..
O	**90_93**		Other commun., soc. & pers. services activities	..	..	274 000	281 000	326 000	355 000	365 000	..
	90		Sewage & refuse disposal, sanitation, etc.	..	..	..	..	..	..	..	..
	91		Activities of member organisations nec	..	..	61 000	62 000	67 000	76 000	76 000	..
		912	Activities of trade unions	..	..	..	..	..	..	..	..
	92		Recreational, cultural and sporting activities	..	..	123 000	126 000	158 000	169 000	174 000	..
		921	Motion picture, radio, TV, other entertainment act.	..	..	..	..	..	..	..	..
		922	News agency activities	..	..	..	..	..	..	..	..
		923	Library, archives, museums, other cultural activ.	..	..	..	..	..	..	..	..
		924	Sporting and other recreational activities	..	..	88 000	80 000	103 000	118 000	123 000	..
	93		Other service activities	..	..	90 000	93 000	101 000	110 000	114 000	..

Note:Data cover the fiscal year ending 30 June.

Data for ISIC 50-52 include businesses mainly engaged in selling take-away food, which should be included in ISIC 55. ISIC 75 is not covered by the inquiry.

Data for ISIC 70-74 include certain activities which are primary to ISIC 93.

Table AS.S4 SALAIRES ET TRAITEMENTS, SALARIÉS - WAGES AND SALARIES, EMPLOYEES

Millions de A$ (Prix courants)

1990-91	1991-92	1992-93	1993-94	1994-95	1995-96	1996-97	1997-98	CITI révision 3			
..	..	23 573	26 568	28 724	30 623	..	..	Com. de gros, de detail ; réparation etc.	G	50_52	
..	..	..	..	..	..	..	..	Com., entr., répar. d'auto. et moto ; carburants		50	
..	..	..	..	..	..	..	..	Commerce de véhicules automobiles			501
..	..	..	..	..	..	..	..	Entretien et réparation de véhicules automobiles			502
..	..	..	..	..	..	..	..	Com. de pièces et access. de véhic. automobiles			503
..	..	..	..	..	..	..	..	Com., entret. et répar. de motocycles et access.			504
..	..	..	..	..	..	..	..	Com. de détail de carburants pour automobiles			505
..	..	..	..	..	..	..	..	Com. de gros et act. inter., sauf auto. et moto		51	
..	..	..	..	..	..	..	..	Activités d'intermédiaries de commerce de gros			511
..	..	..	..	..	..	..	..	Com. gros: prod. agr. bruts, anim. viv., prod. alim.			512
..	..	..	..	..	..	..	..	Com. gros: biens de consom., sauf alimentaires			513
..	..	..	..	..	..	..	..	Com. gros: prod. inter. non agr., déchets, débris			514
..	..	..	..	..	..	..	..	Com. gros: machines, équipments et fournitures			515
..	..	..	..	..	..	..	..	Autres commerces de gros			519
..	..	..	..	..	..	..	..	Com. détail, sauf auto, moto; répar. art. pers., dom.		52	
..	..	..	..	..	..	..	..	Commerce de détail non spécialisé en magasin			521
..	..	..	..	..	..	..	..	Com. détail: prod. aliment. en mag. spécialisés			522
..	..	..	..	..	..	..	..	Autre com. détail: prod. neufs en mag. spécialisés			523
..	..	..	..	..	..	..	..	Commerces de détail de biens d'occasion			524
..	..	..	..	..	..	..	..	Commerces de détail autres qu'en magasin			525
..	..	..	..	..	..	..	..	Réparation d'articles personnels et domestiques			526
..	..	4 832	5 518	5 919	6 301	..	..	Hotels et restaurants	H	55	
..	..	..	..	..	..	..	..	Hôtels; camping; autres équip. courts séjours			551
..	..	..	..	..	..	..	..	Restaurants, cafés et cantines			552
..	..	..	..	..	..	..	..	Transport, entreposage, communications	I	60_64	
..	..	14 053	14 582	15 679	15 774	..	..	Transports terrestres; transports par conduites		60	
..	..	..	..	..	..	..	..	Transports par chemin de fer			601
..	..	..	..	..	..	..	..	Autres transports terrestres			602
..	..	..	..	..	..	..	..	Transports par conduites			603
..	..	..	..	..	..	..	..	Transports par eau		61	
..	..	..	..	..	..	..	..	Transports maritimes et côtiers			611
..	..	..	..	..	..	..	..	Transports par voies navigables intérieures			612
..	..	..	..	..	..	..	..	Transports aériens		62	
..	..	..	..	..	..	..	..	Transports aériens réguliers			621
..	..	..	..	..	..	..	..	Transports aériens non réguliers			622
..	..	..	..	..	..	..	..	Act. annex. et aux. des transp.; act. des ag. de voy.		63	
..	..	..	..	..	..	..	..	Postes et télécommunications		64	
..	..	..	..	..	..	..	..	Activités de poste et de courrier			641
..	..	..	..	..	..	..	..	Télécommunications			642
..	..	10 218	10 958	11 464	11 970	..	..	Intermédiation financière	J	65_67	
..	..	..	..	..	..	..	..	Interm. financ., sauf assur. et caisses de retraite		65	
..	..	..	..	..	..	..	..	Intermédiation monétaire			651
..	..	..	..	..	..	..	..	Autres intermédiations financières			659
..	..	..	..	..	..	..	..	Act. d'assur., caisses de retr., sauf séc. soc. oblig.		66	
..	..	..	..	..	..	..	..	Act. auxiliaires de l'intermédiation financière		67	
..	..	..	..	..	..	..	..	Act. aux. de l'interm. fin., sauf assur., et retraite			671
..	..	..	..	..	..	..	..	Act. aux. des assurances et caisses de retraite			672
..	..	14 331	12 916	16 761	22 261	..	..	Immobilier, locations, services aux entreprises	K	70_74	
..	..	..	..	..	..	..	..	Activités immobilières		70	
..	..	..	..	..	..	..	..	Act. immobilières sur biens propres ou loués			701
..	..	..	..	..	..	..	..	Activités immobilières pour comptes de tiers			702
..	..	..	..	..	..	..	..	Locat. mach., équip. sans opér., biens pers. et dom.		71	
..	..	..	..	..	..	..	..	Location de matériel de transport			711
..	..	..	..	..	..	..	..	Location de machines ét equipements nca			712
..	..	..	..	..	..	..	..	Loc. d'articles personnels et domestiques nca			713
..	..	..	..	..	..	..	..	Activités informatiques et activités rattachées		72	
..	..	..	..	..	..	..	..	Conseil en matériel informatique			721
..	..	..	..	..	..	..	..	Production de logiciels			722
..	..	..	..	..	..	..	..	Traitement de données informatiques			723
..	..	..	..	..	..	..	..	Activités de banques de données			724
..	..	..	..	..	..	..	..	Entr., répar. mach. de bureau et compt. et inform.			725
..	..	..	..	..	..	..	..	Autres activités rattachées à l'informatique			729
..	..	..	..	..	..	..	..	Recherche-développement		73	
..	..	..	..	..	..	..	..	Rech. et dév. expér. en sc. phys., natur. et ingén.			731
..	..	..	..	..	..	..	..	Rech. et dév. expér. en sciences soc. et humaines			732
..	..	..	..	..	..	..	..	Autres activités de services aux entreprises		74	
..	..	..	..	..	..	..	..	Act. jur., compt., audit.; cons. fisc.; ét. march., soud.			741
..	..	..	..	..	..	..	..	Act. d'achitect., d'ingén.; autres act. techniques			742
..	..	..	..	..	..	..	..	Publicité			743
..	..	..	..	..	..	..	..	Activités de services aux entreprises nca.			749
..	..	..	..	..	..	..	..	Education	M	80	
..	..	..	..	..	..	..	..	Activités d'enseignement primaire			801
..	..	..	..	..	..	..	..	Activités d'enseignement secondaire			802
..	..	..	..	..	..	..	..	Activités d'enseignement supérieur			803
..	..	..	..	..	..	..	..	Act. de formation perman.; autres act. d'enseign.			809
..	..	..	..	..	..	..	..	Santé et action sociale	N	85	
..	..	..	..	..	..	..	..	Activités de santé humaine			851
..	..	..	..	..	..	..	..	Activités vétérinaires			852
..	..	..	..	..	..	..	..	Activités d'action sociale			853
..	..	..	..	..	..	..	..	Autres activités de serv. coll., soc. et personnels	O	90_93	
..	..	..	..	..	..	..	..	Assain. et enlèv. des ordures; voierie, act. simil.		90	
..	..	..	..	..	..	..	..	Activités associatives diverses		91	
..	..	..	..	..	..	..	..	Activités de syndicats de salariés			912
..	..	..	..	..	..	..	..	Activités récréatives, culturelles et sportives		92	
..	..	..	..	..	..	..	..	Act. cinémat., de radio, télé.; autres act. de spect.			921
..	..	..	..	..	..	..	..	Activités d'agences de presse			922
..	..	..	..	..	..	..	..	Act. des biblio., arch., musées, jardins bot. et zool.			923
..	..	..	..	..	..	..	..	Activités sportives et autres activités récréatives			924
..	..	..	..	..	..	..	..	Autres activités de services		93	

Note:Data cover the fiscal year ending 30 June.

Data for ISIC 50-52 include businesses mainly engaged in selling take-away food, which should be included in ISIC 55. ISIC 75 is not covered by the inquiry.

Data for ISIC 70-74 include certain activities which are primary to ISIC 93.

AUSTRIA

Table AU.S1 **PRODUCTION - PRODUCTION**

Millions of Sch (Current Prices)

ISIC revision 3				1991	1992	1993	1994	1995	1996	1997	1998
G	50_52		Wholesale & retail trade; repairs; etc.	..	..	..	..	478 247	..	1680 439	1729 499
	50		Sale, maint.,repair vehic/cycle, person., h/hold gd.	..	..	..	..	..	..	246 608	266 295
		501	Sale of motor cycle	..	..	..	..	..	..	156 198	169 542
		502	Maintenance and repair of motor vehicles	..	..	..	..	..	..	38 585	40 872
		503	Sale of motor vehicle parts and accessories	..	..	..	..	..	..	29 807	32 521
		504	Sale, maint., repair motor cycle, parts & acces.	..	..	..	..	..	..	3 865	4 592
		505	Retail sale of automative fuel	..	..	..	..	..	..	18 154	18 767
	51		W/sale. trade & commission exc motor veh. & cycle	..	..	..	..	..	..	988 065	991 917
		511	Wholesale on a fee or contract basis	..	..	..	..	..	..	9 733	10 242
		512	W/sale agric. raw mat.,live animal,food, bev., tobac.	..	..	..	..	..	..	252 165	244 789
		513	Wholesale of household goods	..	..	..	..	..	..	204 043	216 882
		514	W/sale. of non-agric. intermed. prod., waste, scrap	..	..	..	..	..	..	311 471	305 268
		515	Wholesale of machinery, equipment & supplies	..	..	..	..	..	..	175 062	186 933
		519	Other wholesale	..	..	..	..	..	..	35 591	27 802
	52		Retail trade exc. motor veh. & cyc.; repair h/hold gds.	..	..	..	..	..	..	445 766	471 287
		521	Non-specialised retail trade in stores	..	..	..	..	..	..	134 213	131 380
		522	Retail sale food, bev., tobac. in specialised stores	..	..	..	..	..	..	34 343	41 780
		523	Other retail trade of new goods specialised stores	..	..	..	..	..	..	257 827	279 235
		524	Retail sale of second-hand goods in stores	..	..	..	..	..	..	1 519	1 597
		525	Retail trade not in stores	..	..	..	..	..	..	16 251	15 867
		526	Repair of personal and household goods	..	..	..	..	..	..	1 612	1 428
H	55		Hotels and restaurants	..	..	..	..	111 403	..	114 679	121 482
		551	Hotels; camping ; other short-stay accommodation	..	..	..	..	..	..	64 805	66 788
		552	Restaurants, bars and canteens	..	..	..	..	..	..	49 874	54 694
I	60_64		Transport, storage and communications	..	..	..	..	..	..	324 827	346 400
	60		Land transport, transport via pipelines	..	..	..	..	204 040	..	97 069	105 748
		601	Transport via railways	..	..	..	..	..	..	..	..
		602	Other land transport	..	..	..	..	..	..	73 345	80 762
		603	Transport via pipelines	..	..	..	..	..	..	..	..
	61		Water transport	..	..	..	..	..	..	1 028	1 087
		611	Sea and coastal water transport	..	..	..	..	..	..	..	..
		612	Inland water transport	..	..	..	..	..	..	..	..
	62		Air transport	..	..	..	..	..	..	23 559	25 959
		621	Scheduled air transport	..	..	..	..	..	..	22 124	25 437
		622	Non-scheduled air transport	..	..	..	..	..	..	1 435	522
	63		Support. & aux. transport act., travel agencies activ.	..	..	..	..	..	..	122 306	126 291
	64		Post and telecommunications	..	..	..	..	..	..	80 865	87 315
		641	Post and courier activities	..	..	..	..	..	..	2 269	26 485
		642	Telecommunications	..	..	..	..	..	..	78 596	60 831
J	65_67		Financial intermediation	..	..	..	..	228 556	..	555 042	589 657
	65		Financial intermediation, exc. insur., pension fund	..	..	..	..	..	..	382 550	401 885
		651	Monetary intermediation	..	..	..	..	..	..	348 892	362 892
		659	Other financial intermediation	..	..	..	..	..	..	33 658	38 993
	66		Insurance pension fund, exc. compul. social sec.	..	..	..	..	..	..	166 815	181 357
	67		Activities auxiliary to financial intermediation	..	..	..	..	..	..	5 677	6 416
		671	Act. auxil to fin. intermed., exc. insur. & pens.fund	..	..	..	..	..	..	477	689
		672	Act. auxiliary to insurance and pension funding	..	..	..	..	..	..	5 200	5 727
K	70_74		Real estate, renting and business activities	..	..	..	..	211 941	..	301 311	330 696
	70		Real estate activities	..	..	..	..	..	..	77 473	83 581
		701	Real estate activ. with own or leased property	..	..	..	..	..	..	71 498	75 805
		702	Real estate activities on a fee or contract basis	..	..	..	..	..	..	5 975	7 776
	71		Renting of mach. without operator; rent. of pers. gds.	..	..	..	..	..	..	26 081	26 872
		711	Renting of transport equipment	..	..	..	..	..	..	15 019	15 253
		712	Renting of other machinery and equipment	..	..	..	..	..	..	7 009	8 181
		713	Renting of personal and household goods nec	..	..	..	..	..	..	4 053	3 437
	72		Computer and related activities	..	..	..	..	..	..	32 003	41 929
		721	Hardware consultancy	..	..	..	..	..	..	..	..
		722	Software consultancy and supply	..	..	..	..	..	..	11 611	12 367
		723	Data processing	..	..	..	..	..	..	18 500	27 580
		724	Data base activities	..	..	..	..	..	..	..	..
		725	Mainten.; repair of office, account. & comp. mach.	..	..	..	..	..	..	1 445	1 711
		729	Other computer related activities	..	..	..	..	..	..	..	..
	73		Research and development	..	..	..	..	..	..	837	942
		731	R & D experim. on natural sciences & engineering	..	..	..	..	..	..	750	938
		732	R & D experim. on social sciences & humanities	..	..	..	..	..	..	87	4
	74		Other business activities	..	..	..	..	..	..	164 917	177 373
		741	Legal, acc., book-keep., audit act.; tax consult., etc	..	..	..	..	..	..	53 664	58 858
		742	Architectural, engineering, & other techn. activities	..	..	..	..	..	..	45 117	49 416
		743	Advertising	..	..	..	..	..	..	27 287	27 098
		749	Business activities nec	..	..	..	..	..	..	38 849	42 000
M	80		Education	..	..	..	..	..	..	..	..
		801	Primary education	..	..	..	..	..	..	..	..
		802	Secondary education	..	..	..	..	..	..	..	..
		803	Higher education	..	..	..	..	..	..	..	..
		809	Adult and other education	..	..	..	..	..	..	..	..
N	85		Health and social work	..	..	..	..	..	..	..	..
		851	Human health activities	..	..	..	..	..	..	..	..
		852	Veterinary activities	..	..	..	..	..	..	..	..
		853	Social work activities	..	..	..	..	..	..	..	..
O	90_93		Other commun., soc. & pers. services activities	..	..	..	..	..	..	..	..
	90		Sewage & refuse disposal, sanitation, etc.	..	..	..	..	..	..	..	..
	91		Activities of member organisations nec	..	..	..	..	..	..	..	..
		912	Activities of trade unions	..	..	..	..	..	..	..	..
	92		Recreational, cultural and sporting activities	..	..	..	..	..	..	..	..
		921	Motion picture, radio, TV, other entertainment act.	..	..	..	..	..	..	..	..
		922	News agency activities	..	..	..	..	..	..	..	..
		923	Library, archives, museums, other cultural activ.	..	..	..	..	..	..	..	..
		924	Sporting and other recreational activities	..	..	..	..	..	..	..	..
	93		Other service activities	..	..	..	..	..	..	..	..

Table AU.S2 **VALEUR AJOUTÉE AU COUT DES FACTEURS - VALUE ADDED AT FACTOR COSTS**

Millions de Sch (Prix courants)

1991	1992	1993	1994	1995	1996	1997	1998		CITI révision 3		
..	..	..	..	..	..	261 100	273 596	Com. de gros, de detail ; réparation etc.	G	50_52	
..	..	..	..	254 460	..	37 008	38 647	Com., entr., répar. d'auto. et moto ; carburants		50	
..	..	..	..	..	..	18 128	18 437	Commerce de véhicules automobiles			501
..	..	..	..	..	..	9 745	11 120	Entretien et réparation de véhicules automobiles			502
..	..	..	..	..	..	6 104	5 606	Com. de pièces et access. de véhic. automobiles			503
..	..	..	..	..	..	522	529	Com., entret. et répar. de motocycles et access.			504
..	..	..	..	..	..	2 508	2 955	Com. de détail de carburants pour automobiles			505
..	..	..	..	..	..	134 077	137 810	Com. de gros et act. inter., sauf auto. et moto		51	
..	..	..	..	..	..	4 663	4 637	Activités d'intermédiaries de commerce de gros			511
..	..	..	..	..	..	27 301	27 829	Com. gros: prod. agr. bruts, anim. viv., prod. alim.			512
..	..	..	..	..	..	32 444	33 220	Com. gros: biens de consom., sauf alimentaires			513
..	..	..	..	..	..	34 151	33 314	Com. gros: prod. inter. non agr., déchets, débris			514
..	..	..	..	..	..	33 766	37 379	Com. gros: machines, équipements et fournitures			515
..	..	..	..	..	..	1 752	1 430	Autres commerces de gros			519
..	..	..	..	..	..	90 015	97 139	Com. détail, sauf auto, moto; répar. art. pers., dom.		52	
..	..	..	..	..	..	19 145	20 600	Commerce de détail non spécialisé en magasin			521
..	..	..	..	..	..	5 907	7 640	Com. détail: prod. aliment. en mag. spécialisés			522
..	..	..	..	..	..	60 113	64 629	Autre com. détail: prod. neufs en mag. spécialisés			523
..	..	..	..	..	..	475	516	Commerces de détail de biens d'occasion			524
..	..	..	..	..	..	3 547	3 082	Commerces de détail autres qu'en magasin			525
..	..	..	..	..	..	828	672	Réparation d'articles personnels et domestiques			526
..	..	..	..	52 552	..	54 735	57 235	Hotels et restaurants	H	55	
..	..	..	..	..	..	32 661	33 937	Hôtels; camping; autres équip. courts séjours			551
..	..	..	..	..	..	22 074	23 298	Restaurants, cafés et cantines			552
..	..	..	..	..	..	152 681	161 323	Transport, entreposage, communications	I	60_64	
..	..	..	..	139 130	..	74 414	78 813	Transports terrestres; transports par conduites		60	
..	..	..	..	..	..	..	..	Transports par chemin de fer			601
..	..	..	..	..	..	40 161	41 593	Autres transports terrestres			602
..	..	..	..	..	..	..	..	Transports par conduites			603
..	..	..	..	..	..	278	245	Transports par eau		61	
..	..	..	..	..	..	..	..	Transports maritimes et côtiers			611
..	..	..	..	..	..	..	..	Transports par voies navigables intérieures			612
..	..	..	..	..	..	6 200	7 778	Transports aériens		62	
..	..	..	..	..	..	5 579	7 461	Transports aériens réguliers			621
..	..	..	..	..	..	621	316	Transports aériens non réguliers			622
..	..	..	..	..	..	22 976	23 676	Act. annex. et aux. des transp.; act. des ag. de voy.		63	
..	..	..	..	..	..	48 813	50 812	Postes et télécommunications		64	
..	..	..	..	..	..	407	18 172	Activités de poste et de courrier			641
..	..	..	..	..	..	48 406	32 641	Télécommunications			642
..	..	..	..	142 120	..	142 484	145 710	Intermédiation financière	J	65_67	
..	..	..	..	..	..	110 087	113 440	Interm. financ., sauf assur. et caisses de retraite		65	
..	..	..	..	..	..	105 271	105 520	Intermédiation monétaire			651
..	..	..	..	..	..	4 816	7 920	Autres intermédiations financières			659
..	..	..	..	..	..	29 458	28 913	Act. d'assur., caisses de retr., sauf séc. soc. oblig.		66	
..	..	..	..	..	..	2 939	3 357	Act. auxiliaires de l'intermédiation financière		67	
..	..	..	..	..	..	320	339	Act. aux. de l'interm. fin., sauf assur., et retraite			671
..	..	..	..	..	..	2 619	3 018	Act. aux. des assurances et caisses de retraite			672
..	..	..	..	129 264	..	146 189	166 112	Immobilier, locations, services aux entreprises	K	70_74	
..	..	..	..	..	..	30 560	37 751	Activités immobilières		70	
..	..	..	..	..	..	27 286	32 957	Act. immobilières sur biens propres ou loués			701
..	..	..	..	..	..	3 274	4 794	Activités immobilières pour comptes de tiers			702
..	..	..	..	..	..	17 211	18 972	Locat. mach., équip. sans opér., biens pers. et dom.		71	
..	..	..	..	..	..	10 259	11 379	Location de matériel de transport			711
..	..	..	..	..	..	5 033	6 131	Location de machines ét equipements nca			712
..	..	..	..	..	..	1 918	1 461	Loc. d'articles personnels et domestiques nca			713
..	..	..	..	..	..	14 176	18 117	Activités informatiques et activités rattachées		72	
..	..	..	..	..	..	..	..	Conseil en matériel informatique			721
..	..	..	..	..	..	5 904	6 137	Production de logiciels			722
..	..	..	..	..	..	7 424	11 173	Traitement de données informatiques			723
..	..	..	..	..	..	..	..	Activités de banques de données			724
..	..	..	..	..	..	573	655	Entr., répar. mach. de bureau et compt. et inform.			725
..	..	..	..	..	..	..	..	Autres activités rattachées à l'informatique			729
..	..	..	..	..	..	790	813	Recherche-développement		73	
..	..	..	..	..	..	735	811	Rech. et dév. expér. en sc. phys., natur. et ingén.			731
..	..	..	..	..	..	54	2	Rech. et dév. expér. en sciences soc. et humaines			732
..	..	..	..	..	..	83 453	90 460	Autres activités de services aux entreprises		74	
..	..	..	..	..	..	32 544	34 715	Act. jur., compt., audit.; cons. fisc.; ét. march., soud.			741
..	..	..	..	..	..	19 927	23 215	Act. d'achitect., d'ingén.; autres act. techniques			742
..	..	..	..	..	..	7 372	7 370	Publicité			743
..	..	..	..	..	..	23 611	25 159	Activités de services aux entreprises nca.			749
..	..	..	..	..	..	..	..	Education	M	80	
..	..	..	..	..	..	..	..	Activités d'enseignement primaire			801
..	..	..	..	..	..	..	..	Activités d'enseignement secondaire			802
..	..	..	..	..	..	..	..	Activités d'enseignement supérieur			803
..	..	..	..	..	..	..	..	Act. de formation perman.; autres act. d'enseign.			809
..	..	..	..	..	..	..	..	Santé et action sociale	N	85	
..	..	..	..	..	..	..	..	Activités de santé humaine			851
..	..	..	..	..	..	..	..	Activités vétérinaires			852
..	..	..	..	..	..	..	..	Activités d'action sociale			853
..	..	..	..	..	..	..	..	Autres activités de serv. coll., soc. et personnels	O	90_93	
..	..	..	..	..	..	..	..	Assain. et enlèv. des ordures; voierie, act. simil.		90	
..	..	..	..	..	..	..	..	Activités associatives diverses		91	
..	..	..	..	..	..	..	..	Activités de syndicats de salariés			912
..	..	..	..	..	..	..	..	Activités récréatives, culturelles et sportives		92	
..	..	..	..	..	..	..	..	Act. cinémat., de radio, télé.; autres act. de spect.			921
..	..	..	..	..	..	..	..	Activités d'agences de presse			922
..	..	..	..	..	..	..	..	Act. des biblio., arch., musées, jardins bot. et zool.			923
..	..	..	..	..	..	..	..	Activités sportives et autres activités récréatives			924
..	..	..	..	..	..	..	..	Autres activités de services		93	

AUSTRIA

Table AU.S3 **INVESTMENT - INVESTISSEMENT**

Millions of Sch (Current Prices)

ISIC revision 3				1991	1992	1993	1994	1995	1996	1997	1998
G	50_52		Wholesale & retail trade; repairs; etc.	..	..	..	..	43 414	..	35 190	36 754
	50		Sale, maint.,repair vehic/cycle, person., h/hold gd.	..	..	..	..	..	..	4 754	5 502
		501	Sale of motor cycle	..	..	..	..	..	..	2 237	2 601
		502	Maintenance and repair of motor vehicles	..	..	..	..	..	..	1 151	1 448
		503	Sale of motor vehicle parts and accessories	..	..	..	..	..	..	862	1 071
		504	Sale, maint., repair motor cycle, parts & acces.	..	..	..	..	..	..	50	55
		505	Retail sale of automative fuel	..	..	..	..	..	..	454	327
	51		W/sale. trade & commission exc motor veh. & cycle	..	..	..	..	..	..	17 814	18 740
		511	Wholesale on a fee or contract basis	..	..	..	..	..	..	411	352
		512	W/sale agric. raw mat.,live animal,food, bev., tobac.	..	..	..	..	..	..	4 951	4 795
		513	Wholesale of household goods	..	..	..	..	..	..	3 386	3 075
		514	W/sale. of non-agric. intermed. prod., waste, scrap	..	..	..	..	..	..	4 943	5 716
		515	Wholesale of machinery, equipment & supplies	..	..	..	..	..	..	3 935	4 634
		519	Other wholesale	..	..	..	..	..	..	189	168
	52		Retail trade exc. motor veh. & cyc.; repair h/hold gds.	..	..	..	..	..	..	12 621	12 512
		521	Non-specialised retail trade in stores	..	..	..	..	..	..	3 666	3 134
		522	Retail sale food, bev., tobac. in specialised stores	..	..	..	..	..	..	597	983
		523	Other retail trade of new goods specialised stores	..	..	..	..	..	..	7 873	8 075
		524	Retail sale of second-hand goods in stores	..	..	..	..	..	..	54	45
		525	Retail trade not in stores	..	..	..	..	..	..	363	226
		526	Repair of personal and household goods	..	..	..	..	..	..	68	50
H	55		Hotels and restaurants	..	..	..	..	12 917	..	10 408	10 780
		551	Hotels; camping ; other short-stay accommodation	..	..	..	..	..	..	7 198	7 815
		552	Restaurants, bars and canteens	..	..	..	..	..	..	3 209	2 965
I	60_64		Transport, storage and communications	..	..	..	..	..	..	56 402	69 729
	60		Land transport, transport via pipelines	..	..	..	..	48 854	..	26 332	29 394
		601	Transport via railways	..	..	..	..	..	..	..	..
		602	Other land transport	..	..	..	..	..	..	14 333	15 585
		603	Transport via pipelines	..	..	..	..	..	..	..	..
	61		Water transport	..	..	..	..	..	..	111	57
		611	Sea and coastal water transport	..	..	..	..	..	..	..	..
		612	Inland water transport	..	..	..	..	..	..	..	..
	62		Air transport	..	..	..	..	..	..	6 769	8 871
		621	Scheduled air transport	..	..	..	..	..	..	6 665	8 845
		622	Non-scheduled air transport	..	..	..	..	..	..	104	27
	63		Support. & aux. transport act., travel agencies activ.	..	..	..	..	..	..	8 399	9 030
	64		Post and telecommunications	..	..	..	..	..	..	14 790	22 378
		641	Post and courier activities	..	..	..	..	..	..	45	1 111
		642	Telecommunications	..	..	..	..	..	..	14 746	21 266
J	65_67		Financial intermediation	..	..	..	..	20 846	..	23 501	23 807
	65		Financial intermediation, exc. insur., pension fund	..	..	..	..	..	..	19 168	18 096
		651	Monetary intermediation	..	..	..	..	..	..	13 061	13 776
		659	Other financial intermediation	..	..	..	..	..	..	6 106	4 320
	66		Insurance pension fund, exc. compul. social sec.	..	..	..	..	..	..	4 038	5 419
	67		Activities auxiliary to financial intermediation	..	..	..	..	..	..	295	293
		671	Act. auxil to fin. intermed., exc. insur. & pens.fund	..	..	..	..	..	..	52	56
		672	Act. auxiliary to insurance and pension funding	..	..	..	..	..	..	243	237
K	70_74		Real estate, renting and business activities	..	..	..	..	82 042	..	70 675	72 698
	70		Real estate activities	..	..	..	..	..	..	42 669	41 553
		701	Real estate activ. with own or leased property	..	..	..	..	..	..	41 879	39 561
		702	Real estate activities on a fee or contract basis	..	..	..	..	..	..	790	1 992
	71		Renting of mach. without operator; rent. of pers. gds.	..	..	..	..	..	..	16 052	17 358
		711	Renting of transport equipment	..	..	..	..	..	..	11 576	12 181
		712	Renting of other machinery and equipment	..	..	..	..	..	..	3 821	4 551
		713	Renting of personal and household goods nec	..	..	..	..	..	..	655	627
	72		Computer and related activities	..	..	..	..	..	..	2 154	3 710
		721	Hardware consultancy	..	..	..	..	..	..	..	..
		722	Software consultancy and supply	..	..	..	..	..	..	472	668
		723	Data processing	..	..	..	..	..	..	1 607	2 965
		724	Data base activities	..	..	..	..	..	..	..	..
		725	Mainten.; repair of office, account. & comp. mach.	..	..	..	..	..	..	59	46
		729	Other computer related activities	..	..	..	..	..	..	..	..
	73		Research and development	..	..	..	..	..	..	115	101
		731	R & D experim. on natural sciences & engineering	..	..	..	..	..	..	113	101
		732	R & D experim. on social sciences & humanities	..	..	..	..	..	..	2	0
	74		Other business activities	..	..	..	..	..	..	9 685	9 976
		741	Legal, acc., book-keep., audit act.; tax consult., etc	..	..	..	..	..	..	4 485	3 854
		742	Architectural, engineering, & other techn. activities	..	..	..	..	..	..	2 571	3 183
		743	Advertising	..	..	..	..	..	..	940	1 037
		749	Business activities nec	..	..	..	..	..	..	1 689	1 901
M	80		Education	..	..	..	..	..	..	..	..
		801	Primary education	..	..	..	..	..	..	..	..
		802	Secondary education	..	..	..	..	..	..	..	..
		803	Higher education	..	..	..	..	..	..	..	..
		809	Adult and other education	..	..	..	..	..	..	..	..
N	85		Health and social work	..	..	..	..	..	..	..	..
		851	Human health activities	..	..	..	..	..	..	..	..
		852	Veterinary activities	..	..	..	..	..	..	..	..
		853	Social work activities	..	..	..	..	..	..	..	..
O	90_93		Other commun., soc. & pers. services activities	..	..	..	..	..	..	..	..
	90		Sewage & refuse disposal, sanitation, etc.	..	..	..	..	..	..	..	..
	91		Activities of member organisations nec	..	..	..	..	..	..	..	..
		912	Activities of trade unions	..	..	..	..	..	..	..	..
	92		Recreational, cultural and sporting activities	..	..	..	..	..	..	..	..
		921	Motion picture, radio, TV, other entertainment act.	..	..	..	..	..	..	..	..
		922	News agency activities	..	..	..	..	..	..	..	..
		923	Library, archives, museums, other cultural activ.	..	..	..	..	..	..	..	..
		924	Sporting and other recreational activities	..	..	..	..	..	..	..	..
	93		Other service activities	..	..	..	..	..	..	..	..

Table AU.S4 **DÉPENSES DE MACHINES ET D'ÉQUIPEMENT - INVESTMENT IN MACHINERY & EQUIPMENT**

Millions de Sch (Prix courants)

1991	1992	1993	1994	1995	1996	1997	1998		CITI révision 3		
..	..	..	..	22 565	..	18 519	19 641	Com. de gros, de detail ; réparation etc.	G	50_52	
..	..	..	..	..	..	2 209	2 335	Com., entr., répar. d'auto. et moto ; carburants		50	
..	..	..	..	..	..	1 109	1 203	Commerce de véhicules automobiles			501
..	..	..	..	..	..	456	549	Entretien et réparation de véhicules automobiles			502
..	..	..	..	..	..	312	336	Com. de pièces et access. de véhic. automobiles			503
..	..	..	..	..	..	29	25	Com., entret. et répar. de motocycles et access.			504
..	..	..	..	..	..	304	223	Com. de détail de carburants pour automobiles			505
..	..	..	..	..	..	10 353	10 933	Com. de gros et act. inter., sauf auto. et moto		51	
..	..	..	..	..	..	277	254	Activités d'intermédiaries de commerce de gros			511
..	..	..	..	..	..	2 548	2 692	Com. gros: prod. agr. bruts, anim. viv., prod. alim.			512
..	..	..	..	..	..	2 344	1 858	Com. gros: biens de consom., sauf alimentaires			513
..	..	..	..	..	..	2 411	2 766	Com. gros: prod. inter. non agr., déchets, débris			514
..	..	..	..	..	..	2 714	3 303	Com. gros: machines, équipments et fournitures			515
..	..	..	..	..	..	59	60	Autres commerces de gros			519
..	..	..	..	..	..	5 957	6 373	Com. détail, sauf auto, moto; répar. art. pers., dom.		52	
..	..	..	..	..	..	1 608	1 636	Commerce de détail non spécialisé en magasin			521
..	..	..	..	..	..	418	691	Com. détail: prod. aliment. en mag. spécialisés			522
..	..	..	..	..	..	3 651	3 825	Autre com. détail: prod. neufs en mag. spécialisés			523
..	..	..	..	..	..	36	24	Commerces de détail de biens d'occasion			524
..	..	..	..	..	..	208	165	Commerces de détail autres qu'en magasin			525
..	..	..	..	..	..	37	32	Réparation d'articles personnels et domestiques			526
..	..	..	..	4 563	..	3 876	4 200	Hotels et restaurants	H	55	
..	..	..	..	..	..	2 252	2 596	Hôtels; camping; autres équip. courts séjours			551
..	..	..	..	..	..	1 623	1 604	Restaurants, cafés et cantines			552
..	..	..	..	..	..	33 887	36 863	Transport, entreposage, communications	I	60_64	
..	..	..	..	32 638	..	12 760	14 029	Transports terrestres; transports par conduites		60	
..	..	..	..	..	..	..	..	Transports par chemin de fer			601
..	..	..	..	..	..	9 883	11 286	Autres transports terrestres			602
..	..	..	..	..	..	..	..	Transports par conduites			603
..	..	..	..	..	..	109	54	Transports par eau		61	
..	..	..	..	..	..	..	..	Transports maritimes et côtiers			611
..	..	..	..	..	..	..	..	Transports par voies navigables intérieures			612
..	..	..	..	..	..	6 560	8 701	Transports aériens		62	
..	..	..	..	..	..	6 458	8 676	Transports aériens réguliers			621
..	..	..	..	..	..	102	25	Transports aériens non réguliers			622
..	..	..	..	..	..	1 920	1 824	Act. annex. et aux. des transp.; act. des ag. de voy.		63	
..	..	..	..	..	..	12 538	12 255	Postes et télécommunications		64	
..	..	..	..	..	..	27	541	Activités de poste et de courrier			641
..	..	..	..	..	..	12 511	11 714	Télécommunications			642
..	..	..	..	898	..	12 376	11 100	Intermédiation financière	J	65_67	
..	..	..	..	..	..	11 278	10 286	Interm. financ., sauf assur. et caisses de retraite		65	
..	..	..	..	..	..	5 449	6 785	Intermédiation monétaire			651
..	..	..	..	..	..	5 829	3 501	Autres intermédiations financières			659
..	..	..	..	..	..	937	645	Act. d'assur., caisses de retr., sauf séc. soc. oblig.		66	
..	..	..	..	..	..	161	169	Act. auxiliaires de l'intermédiation financière		67	
..	..	..	..	..	..	29	29	Act. aux. de l'interm. fin., sauf assur., et retraite			671
..	..	..	..	..	..	132	141	Act. aux. des assurances et caisses de retraite			672
..	..	..	..	20 013	..	23 681	26 194	Immobilier, locations, services aux entreprises	K	70_74	
..	..	..	..	..	..	1 967	2 748	Activités immobilières		70	
..	..	..	..	..	..	1 791	2 563	Act. immobilières sur biens propres ou loués			701
..	..	..	..	..	..	176	185	Activités immobilières pour comptes de tiers			702
..	..	..	..	..	..	14 987	16 233	Locat. mach., équip. sans opér., biens pers. et dom.		71	
..	..	..	..	..	..	11 113	11 683	Location de matériel de transport			711
..	..	..	..	..	..	3 409	4 074	Location de machines ét equipements nca			712
..	..	..	..	..	..	465	477	Loc. d'articles personnels et domestiques nca			713
..	..	..	..	..	..	1 420	2 400	Activités informatiques et activités rattachées		72	
..	..	..	..	..	..	..	..	Conseil en matériel informatique			721
..	..	..	..	..	..	308	367	Production de logiciels			722
..	..	..	..	..	..	1 056	1 973	Traitement de données informatiques			723
..	..	..	..	..	..	..	..	Activités de banques de données			724
..	..	..	..	..	..	43	35	Entr., répar. mach. de bureau et compt. et inform.			725
..	..	..	..	..	..	..	..	Autres activités rattachées à l'informatique			729
..	..	..	..	..	..	72	84	Recherche-développement		73	
..	..	..	..	..	..	71	84	Rech. et dév. expér. en sc. phys., natur. et ingén.			731
..	..	..	..	..	..	2	..	Rech. et dév. expér. en sciences soc. et humaines			732
..	..	..	..	..	..	5 235	4 729	Autres activités de services aux entreprises		74	
..	..	..	..	..	..	2 554	2 000	Act. jur., compt., audit.; cons. fisc.; ét. march., soud.			741
..	..	..	..	..	..	1 075	1 139	Act. d'achitect., d'ingén.; autres act. techniques			742
..	..	..	..	..	..	659	645	Publicité			743
..	..	..	..	..	..	948	945	Activités de services aux entreprises nca.			749
..	..	..	..	..	..	..	..	Education	M	80	
..	..	..	..	..	..	..	..	Activités d'enseignement primaire			801
..	..	..	..	..	..	..	..	Activités d'enseignement secondaire			802
..	..	..	..	..	..	..	..	Activités d'enseignement supérieur			803
..	..	..	..	..	..	..	..	Act. de formation perman.; autres act. d'enseign.			809
..	..	..	..	..	..	..	..	Santé et action sociale	N	85	
..	..	..	..	..	..	..	..	Activités de santé humaine			851
..	..	..	..	..	..	..	..	Activités vétérinaires			852
..	..	..	..	..	..	..	..	Activités d'action sociale			853
..	..	..	..	..	..	..	..	Autres activités de serv. coll., soc. et personnels	O	90_93	
..	..	..	..	..	..	..	..	Assain. et enlèv. des ordures; voierie, act. simil.		90	
..	..	..	..	..	..	..	..	Activités associatives diverses		91	
..	..	..	..	..	..	..	..	Activités de syndicats de salariés			912
..	..	..	..	..	..	..	..	Activités récréatives, culturelles et sportives		92	
..	..	..	..	..	..	..	..	Act. cinémat., de radio, télé.; autres act. de spect.			921
..	..	..	..	..	..	..	..	Activités d'agences de presse			922
..	..	..	..	..	..	..	..	Act. des biblio., arch., musées, jardins bot. et zool.			923
..	..	..	..	..	..	..	..	Activités sportives et autres activités récréatives			924
..	..	..	..	..	..	..	..	Autres activités de services		93	

Table AU.S5 **EMPLOYMENT, EMPLOYEES - EMPLOI, SALARIÉS**

Units

ISIC revision 3				1991	1992	1993	1994	1995	1996	1997	1998
G	50_52		Wholesale & retail trade; repairs; etc.	..	..	..	..	472 321	..	467 069	490 404
	50		Sale, maint.,repair vehic/cycle, person., h/hold gd.	..	..	..	..	..	..	67 424	71 162
		501	Sale of motor cycle	..	..	..	..	..	..	29 693	31 782
		502	Maintenance and repair of motor vehicles	..	..	..	..	..	..	21 755	22 983
		503	Sale of motor vehicle parts and accessories	..	..	..	..	..	..	8 668	8 818
		504	Sale, maint., repair motor cycle, parts & acces.	..	..	..	..	..	..	759	844
		505	Retail sale of automative fuel	..	..	..	..	..	..	6 549	6 735
	51		W/sale. trade & commission exc motor veh. & cycle	..	..	..	..	..	..	177 767	184 112
		511	Wholesale on a fee or contract basis	..	..	..	..	..	..	4 244	4 919
		512	W/sale agric. raw mat.,live animal,food, bev., tobac.	..	..	..	..	..	..	50 938	50 396
		513	Wholesale of household goods	..	..	..	..	..	..	41 817	41 916
		514	W/sale. of non-agric. intermed. prod., waste, scrap	..	..	..	..	..	..	40 696	41 449
		515	Wholesale of machinery, equipment & supplies	..	..	..	..	..	..	37 822	43 140
		519	Other wholesale	..	..	..	..	..	..	2 249	2 293
	52		Retail trade exc. motor veh. & cyc.; repair h/hold gds.	..	..	..	..	..	..	221 878	235 129
		521	Non-specialised retail trade in stores	..	..	..	..	..	..	61 465	62 575
		522	Retail sale food, bev., tobac. in specialised stores	..	..	..	..	..	..	10 914	13 675
		523	Other retail trade of new goods specialised stores	..	..	..	..	..	..	141 159	150 964
		524	Retail sale of second-hand goods in stores	..	..	..	..	..	..	848	823
		525	Retail trade not in stores	..	..	..	..	..	..	5 687	5 589
		526	Repair of personal and household goods	..	..	..	..	..	..	1 806	1 503
H	55		Hotels and restaurants	..	..	..	..	145 024	..	146 597	150 600
		551	Hotels; camping ; other short-stay accommodation	..	..	..	..	..	..	81 594	82 067
		552	Restaurants, bars and canteens	..	..	..	..	..	..	65 002	68 533
I	60_64		Transport, storage and communications	..	..	..	..	..	..	232 649	231 797
	60		Land transport, transport via pipelines	..	..	..	..	218 078	..	129 755	129 885
		601	Transport via railways	..	..	..	..	..	..	..	..
		602	Other land transport	..	..	..	..	..	..	71 872	73 389
		603	Transport via pipelines	..	..	..	..	..	..	..	..
	61		Water transport	..	..	..	..	..	..	184	241
		611	Sea and coastal water transport	..	..	..	..	..	..	..	..
		612	Inland water transport	..	..	..	..	..	..	..	..
	62		Air transport	..	..	..	..	..	..	6 761	7 343
		621	Scheduled air transport	..	..	..	..	..	..	6 410	7 179
		622	Non-scheduled air transport	..	..	..	..	..	..	351	164
	63		Support. & aux. transport act., travel agencies activ.	..	..	..	..	..	..	30 507	31 132
	64		Post and telecommunications	..	..	..	..	..	..	65 442	63 196
		641	Post and courier activities	..	..	..	..	..	..	976	39 860
		642	Telecommunications	..	..	..	..	..	..	64 466	23 336
J	65_67		Financial intermediation	..	..	..	..	115 555	..	112 462	112 258
	65		Financial intermediation, exc. insur., pension fund	..	..	..	..	..	..	76 484	76 326
		651	Monetary intermediation	..	..	..	..	..	..	73 336	73 339
		659	Other financial intermediation	..	..	..	..	..	..	3 148	2 986
	66		Insurance pension fund, exc. compul. social sec.	..	..	..	..	..	..	32 296	31 733
	67		Activities auxiliary to financial intermediation	..	..	..	..	..	..	3 682	4 200
		671	Act. auxil to fin. intermed., exc. insur. & pens.fund	..	..	..	..	..	..	469	528
		672	Act. auxiliary to insurance and pension funding	..	..	..	..	..	..	3 213	3 671
K	70_74		Real estate, renting and business activities	..	..	..	..	160 424	..	173 054	193 614
	70		Real estate activities	..	..	..	..	..	..	18 035	22 102
		701	Real estate activ. with own or leased property	..	..	..	..	..	..	13 314	14 071
		702	Real estate activities on a fee or contract basis	..	..	..	..	..	..	4 721	8 031
	71		Renting of mach. without operator; rent. of pers. gds.	..	..	..	..	..	..	4 669	4 824
		711	Renting of transport equipment	..	..	..	..	..	..	1 457	1 347
		712	Renting of other machinery and equipment	..	..	..	..	..	..	1 473	1 884
		713	Renting of personal and household goods nec	..	..	..	..	..	..	1 738	1 593
	72		Computer and related activities	..	..	..	..	..	..	15 109	18 824
		721	Hardware consultancy	..	..	..	..	..	..	..	..
		722	Software consultancy and supply	..	..	..	..	..	..	6 692	7 285
		723	Data processing	..	..	..	..	..	..	7 287	10 256
		724	Data base activities	..	..	..	..	..	..	..	..
		725	Mainten.; repair of office, account. & comp. mach.	..	..	..	..	..	..	955	1 136
		729	Other computer related activities	..	..	..	..	..	..	..	..
	73		Research and development	..	..	..	..	..	..	1 281	1 469
		731	R & D experim. on natural sciences & engineering	..	..	..	..	..	..	1 224	1 462
		732	R & D experim. on social sciences & humanities	..	..	..	..	..	..	57	7
	74		Other business activities	..	..	..	..	..	..	133 960	146 395
		741	Legal, acc., book-keep., audit act.; tax consult., etc	..	..	..	..	..	..	39 190	41 521
		742	Architectural, engineering, & other techn. activities	..	..	..	..	..	..	25 156	28 430
		743	Advertising	..	..	..	..	..	..	7 382	8 895
		749	Business activities nec	..	..	..	..	..	..	62 232	67 548
M	80		Education	..	..	..	..	..	..	..	..
		801	Primary education	..	..	..	..	..	..	..	..
		802	Secondary education	..	..	..	..	..	..	..	..
		803	Higher education	..	..	..	..	..	..	..	..
		809	Adult and other education	..	..	..	..	..	..	..	..
N	85		Health and social work	..	..	..	..	..	..	..	..
		851	Human health activities	..	..	..	..	..	..	..	..
		852	Veterinary activities	..	..	..	..	..	..	..	..
		853	Social work activities	..	..	..	..	..	..	..	..
O	90_93		Other commun., soc. & pers. services activities	..	..	..	..	..	..	..	..
	90		Sewage & refuse disposal, sanitation, etc.	..	..	..	..	..	..	..	..
	91		Activities of member organisations nec	..	..	..	..	..	..	..	..
		912	Activities of trade unions	..	..	..	..	..	..	..	..
	92		Recreational, cultural and sporting activities	..	..	..	..	..	..	..	..
		921	Motion picture, radio, TV, other entertainment act.	..	..	..	..	..	..	..	..
		922	News agency activities	..	..	..	..	..	..	..	..
		923	Library, archives, museums, other cultural activ.	..	..	..	..	..	..	..	..
		924	Sporting and other recreational activities	..	..	..	..	..	..	..	..
	93		Other service activities	..	..	..	..	..	..	..	..

Table AU.S6 **EMPLOI, FEMMES - EMPLOYMENT, FEMALES**

Unités

1991	1992	1993	1994	1995	1996	1997	1998		CITI révision 3		
..	..	..	..	238 488	..	244 682	259 899	Com. de gros, de detail ; réparation etc.	G	**50_52**	
..	..	..	..	..	..	14 298	15 344	Com., entr., répar. d'auto. et moto ; carburants		**50**	
..	..	..	..	..	..	5 243	5 681	Commerce de véhicules automobiles			501
..	..	..	..	..	..	4 057	4 425	Entretien et réparation de véhicules automobiles			502
..	..	..	..	..	..	2 097	1 858	Com. de pièces et access. de véhic. automobiles			503
..	..	..	..	..	..	252	242	Com., entret. et répar. de motocycles et access.			504
..	..	..	..	..	..	2 648	3 138	Com. de détail de carburants pour automobiles			505
..	..	..	..	..	..	71 274	72 855	Com. de gros et act. inter., sauf auto. et moto		**51**	
..	..	..	..	..	..	2 280	2 621	Activités d'intermédiaries de commerce de gros			511
..	..	..	..	..	..	22 360	22 348	Com. gros: prod. agr. bruts, anim. viv., prod. alim.			512
..	..	..	..	..	..	21 037	20 505	Com. gros: biens de consom., sauf alimentaires			513
..	..	..	..	..	..	12 418	12 898	Com. gros: prod. inter. non agr., déchets, débris			514
..	..	..	..	..	..	11 942	13 324	Com. gros: machines, équipments et fournitures			515
..	..	..	..	..	..	1 237	1 160	Autres commerces de gros			519
..	..	..	..	..	..	159 110	171 700	Com. détail, sauf auto, moto; répar. art. pers., dom.		**52**	
..	..	..	..	..	..	50 059	51 697	Commerce de détail non spécialisé en magasin			521
..	..	..	..	..	..	7 519	9 901	Com. détail: prod. aliment. en mag. spécialisés			522
..	..	..	..	..	..	96 899	105 672	Autre com. détail: prod. neufs en mag. spécialisés			523
..	..	..	..	..	..	419	406	Commerces de détail de biens d'occasion			524
..	..	..	..	..	..	3 593	3 394	Commerces de détail autres qu'en magasin			525
..	..	..	..	..	..	620	631	Réparation d'articles personnels et domestiques			526
..	..	..	..	87 195	..	88 880	93 264	Hotels et restaurants	H	**55**	
..	..	..	..	..	..	50 401	51 110	Hôtels; camping; autres équip. courts séjours			551
..	..	..	..	..	..	38 479	42 153	Restaurants, cafés et cantines			552
..	..	..	..	..	..	45 979	47 566	Transport, entreposage, communications	I	**60_64**	
..	..	..	..	40 594	..	14 025	14 583	Transports terrestres; transports par conduites		**60**	
..	..	..	..	..	..	..	..	Transports par chemin de fer			601
..	..	..	..	..	..	10 381	11 052	Autres transports terrestres			602
..	..	..	..	..	..	..	..	Transports par conduites			603
..	..	..	..	..	..	27	35	Transports par eau		**61**	
..	..	..	..	..	..	..	..	Transports maritimes et côtiers			611
..	..	..	..	..	..	..	..	Transports par voies navigables intérieures			612
..	..	..	..	..	..	3 152	3 782	Transports aériens		**62**	
..	..	..	..	..	..	3 097	3 707	Transports aériens réguliers			621
..	..	..	..	..	..	55	75	Transports aériens non réguliers			622
..	..	..	..	..	..	11 664	12 021	Act. annex. et aux. des transp.; act. des ag. de voy.		**63**	
..	..	..	..	..	..	17 110	17 145	Postes et télécommunications		**64**	
..	..	..	..	..	..	312	11 485	Activités de poste et de courrier			641
..	..	..	..	..	..	16 798	5 660	Télécommunications			642
..	..	..	..	55 293	..	53 705	53 502	Intermédiation financière	J	**65_67**	
..	..	..	..	..	..	39 914	39 810	Interm. financ., sauf assur. et caisses de retraite		**65**	
..	..	..	..	..	..	38 206	38 109	Intermédiation monétaire			651
..	..	..	..	..	..	1 708	1 701	Autres intermédiations financières			659
..	..	..	..	..	..	11 592	11 201	Act. d'assur., caisses de retr., sauf séc. soc. oblig.		**66**	
..	..	..	..	..	..	2 199	2 492	Act. auxiliaires de l'intermédiation financière		**67**	
..	..	..	..	..	..	237	219	Act. aux. de l'interm. fin., sauf assur., et retraite			671
..	..	..	..	..	..	1 962	2 273	Act. aux. des assurances et caisses de retraite			672
..	..	..	..	84 534	..	92 730	102 542	Immobilier, locations, services aux entreprises	K	**70_74**	
..	..	..	..	..	..	10 752	13 959	Activités immobilières		**70**	
..	..	..	..	..	..	7 581	8 310	Act. immobilières sur biens propres ou loués			701
..	..	..	..	..	..	3 171	5 648	Activités immobilières pour comptes de tiers			702
..	..	..	..	..	..	1 926	1 913	Locat. mach., équip. sans opér., biens pers. et dom.		**71**	
..	..	..	..	..	..	672	606	Location de matériel de transport			711
..	..	..	..	..	..	551	596	Location de machines ét equipements nca			712
..	..	..	..	..	..	702	712	Loc. d'articles personnels et domestiques nca			713
..	..	..	..	..	..	4 399	5 781	Activités informatiques et activités rattachées		**72**	
..	..	..	..	..	..	..	..	Conseil en matériel informatique			721
..	..	..	..	..	..	1 633	2 172	Production de logiciels			722
..	..	..	..	..	..	2 537	3 272	Traitement de données informatiques			723
..	..	..	..	..	..	..	..	Activités de banques de données			724
..	..	..	..	..	..	181	268	Entr., répar. mach. de bureau et compt. et inform.			725
..	..	..	..	..	..	..	..	Autres activités rattachées à l'informatique			729
..	..	..	..	..	..	384	384	Recherche-développement		**73**	
..	..	..	..	..	..	356	384	Rech. et dév. expér. en sc. phys., natur. et ingén.			731
..	..	..	..	..	..	28	..	Rech. et dév. expér. en sciences soc. et humaines			732
..	..	..	..	..	..	75 268	80 505	Autres activités de services aux entreprises		**74**	
..	..	..	..	..	..	28 950	30 084	Act. jur., compt., audit.; cons. fisc.; ét. march., soud.			741
..	..	..	..	..	..	8 608	10 091	Act. d'achitect., d'ingén.; autres act. techniques			742
..	..	..	..	..	..	4 351	5 320	Publicité			743
..	..	..	..	..	..	33 360	35 011	Activités de services aux entreprises nca.			749
..	..	..	..	..	..	..	..	Education	M	**80**	
..	..	..	..	..	..	..	..	Activités d'enseignement primaire			801
..	..	..	..	..	..	..	..	Activités d'enseignement secondaire			802
..	..	..	..	..	..	..	..	Activités d'enseignement supérieur			803
..	..	..	..	..	..	..	..	Act. de formation perman.; autres act. d'enseign.			809
..	..	..	..	..	..	..	..	Santé et action sociale	N	**85**	
..	..	..	..	..	..	..	..	Activités de santé humaine			851
..	..	..	..	..	..	..	..	Activités vétérinaires			852
..	..	..	..	..	..	..	..	Activités d'action sociale			853
..	..	..	..	..	..	..	..	Autres activités de serv. coll., soc. et personnels	O	**90_93**	
..	..	..	..	..	..	..	..	Assain. et enlèv. des ordures; voierie, act. simil.		**90**	
..	..	..	..	..	..	..	..	Activités associatives diverses		**91**	
..	..	..	..	..	..	..	..	Activités de syndicats de salariés			912
..	..	..	..	..	..	..	..	Activités récréatives, culturelles et sportives		**92**	
..	..	..	..	..	..	..	..	Act. cinémat., de radio, télé.; autres act. de spect.			921
..	..	..	..	..	..	..	..	Activités d'agences de presse			922
..	..	..	..	..	..	..	..	Act. des biblio., arch., musées, jardins bot. et zool.			923
..	..	..	..	..	..	..	..	Activités sportives et autres activités récréatives			924
..	..	..	..	..	..	..	..	Autres activités de services		**93**	

AUSTRIA

Table AU.S7 COMPENSATION OF LABOUR, EMPLOYEES - SALAIRES ET CHARGES SOCIALES, SALARIÉS

Millions of Sch (Current Prices)

ISIC revision 3				1991	1992	1993	1994	1995	1996	1997	1998
G	50_52		Wholesale & retail trade; repairs; etc.	..	..	..	..	174 391	..	177 422	185 944
	50		Sale, maint.,repair vehic/cycle, person., h/hold gd.	..	..	..	..	..	..	25 376	26 947
		501	Sale of motor cycle	..	..	..	..	..	..	12 600	13 446
		502	Maintenance and repair of motor vehicles	..	..	..	..	..	..	7 199	7 668
		503	Sale of motor vehicle parts and accessories	..	..	..	..	..	..	3 502	3 732
		504	Sale, maint., repair motor cycle, parts & acces.	..	..	..	..	..	..	283	321
		505	Retail sale of automative fuel	..	..	..	..	..	..	1 790	1 780
	51		W/sale. trade & commission exc motor veh. & cycle	..	..	..	..	..	..	88 126	92 012
		511	Wholesale on a fee or contract basis	..	..	..	..	..	..	2 315	2 445
		512	W/sale agric. raw mat.,live animal,food, bev., tobac.	..	..	..	..	..	..	20 101	19 768
		513	Wholesale of household goods	..	..	..	..	..	..	21 965	22 461
		514	W/sale. of non-agric. intermed. prod., waste, scrap	..	..	..	..	..	..	20 474	21 353
		515	Wholesale of machinery, equipment & supplies	..	..	..	..	..	..	22 171	24 951
		519	Other wholesale	..	..	..	..	..	..	1 100	1 035
	52		Retail trade exc. motor veh. & cyc.; repair h/hold gds.	..	..	..	..	..	..	63 920	66 984
		521	Non-specialised retail trade in stores	..	..	..	..	..	..	15 851	15 769
		522	Retail sale food, bev., tobac. in specialised stores	..	..	..	..	..	..	2 928	3 627
		523	Other retail trade of new goods specialised stores	..	..	..	..	..	..	42 229	44 861
		524	Retail sale of second-hand goods in stores	..	..	..	..	..	..	233	219
		525	Retail trade not in stores	..	..	..	..	..	..	2 120	2 097
		526	Repair of personal and household goods	..	..	..	..	..	..	559	411
H	55		Hotels and restaurants	..	..	..	..	35 027	..	35 775	36 404
		551	Hotels; camping ; other short-stay accommodation	..	..	..	..	..	..	21 283	20 612
		552	Restaurants, bars and canteens	..	..	..	..	..	..	14 492	15 793
I	60_64		Transport, storage and communications	..	..	..	..	..	..	100 351	103 266
	60		Land transport, transport via pipelines	..	..	..	..	102 184	..	53 482	55 114
		601	Transport via railways	..	..	..	..	..	..	..	..
		602	Other land transport	..	..	..	..	..	..	26 248	27 240
		603	Transport via pipelines	..	..	..	..	..	..	..	..
	61		Water transport	..	..	..	..	..	..	130	191
		611	Sea and coastal water transport	..	..	..	..	..	..	..	..
		612	Inland water transport	..	..	..	..	..	..	..	..
	62		Air transport	..	..	..	..	..	..	4 683	5 018
		621	Scheduled air transport	..	..	..	..	..	..	4 371	4 846
		622	Non-scheduled air transport	..	..	..	..	..	..	311	171
	63		Support. & aux. transport act., travel agencies activ.	..	..	..	..	..	..	15 098	15 974
	64		Post and telecommunications	..	..	..	..	..	..	26 959	26 969
		641	Post and courier activities	..	..	..	..	..	..	316	16 037
		642	Telecommunications	..	..	..	..	..	..	26 643	10 931
J	65_67		Financial intermediation	..	..	..	..	76 936	..	79 623	81 442
	65		Financial intermediation, exc. insur., pension fund	..	..	..	..	..	..	56 940	57 953
		651	Monetary intermediation	..	..	..	..	..	..	53 938	55 183
		659	Other financial intermediation	..	..	..	..	..	..	3 002	2 770
	66		Insurance pension fund, exc. compul. social sec.	..	..	..	..	..	..	20 929	21 544
	67		Activities auxiliary to financial intermediation	..	..	..	..	..	..	1 754	1 946
		671	Act. auxil to fin. intermed., exc. insur. & pens.fund	..	..	..	..	..	..	196	266
		672	Act. auxiliary to insurance and pension funding	..	..	..	..	..	..	1 557	1 680
K	70_74		Real estate, renting and business activities	..	..	..	..	64 035	..	73 333	83 040
	70		Real estate activities	..	..	..	..	..	..	8 292	9 582
		701	Real estate activ. with own or leased property	..	..	..	..	..	..	6 313	6 646
		702	Real estate activities on a fee or contract basis	..	..	..	..	..	..	1 979	2 935
	71		Renting of mach. without operator; rent. of pers. gds.	..	..	..	..	..	..	1 950	1 825
		711	Renting of transport equipment	..	..	..	..	..	..	766	727
		712	Renting of other machinery and equipment	..	..	..	..	..	..	706	682
		713	Renting of personal and household goods nec	..	..	..	..	..	..	479	416
	72		Computer and related activities	..	..	..	..	..	..	9 188	11 388
		721	Hardware consultancy	..	..	..	..	..	..	..	..
		722	Software consultancy and supply	..	..	..	..	..	..	4 186	4 320
		723	Data processing	..	..	..	..	..	..	4 404	6 410
		724	Data base activities	..	..	..	..	..	..	..	..
		725	Mainten.; repair of office, account. & comp. mach.	..	..	..	..	..	..	501	575
		729	Other computer related activities	..	..	..	..	..	..	..	..
	73		Research and development	..	..	..	..	..	..	844	895
		731	R & D experim. on natural sciences & engineering	..	..	..	..	..	..	814	895
		732	R & D experim. on social sciences & humanities	..	..	..	..	..	..	30	1
	74		Other business activities	..	..	..	..	..	..	53 059	59 350
		741	Legal, acc., book-keep., audit act.; tax consult., etc	..	..	..	..	..	..	18 127	19 746
		742	Architectural, engineering, & other techn. activities	..	..	..	..	..	..	13 301	15 610
		743	Advertising	..	..	..	..	..	..	3 677	4 341
		749	Business activities nec	..	..	..	..	..	..	17 954	19 652
M	80		Education	..	..	..	..	..	..	..	..
		801	Primary education	..	..	..	..	..	..	..	..
		802	Secondary education	..	..	..	..	..	..	..	..
		803	Higher education	..	..	..	..	..	..	..	..
		809	Adult and other education	..	..	..	..	..	..	..	..
N	85		Health and social work	..	..	..	..	..	..	..	..
		851	Human health activities	..	..	..	..	..	..	..	..
		852	Veterinary activities	..	..	..	..	..	..	..	..
		853	Social work activities	..	..	..	..	..	..	..	..
O	90_93		Other commun., soc. & pers. services activities	..	..	..	..	..	..	..	..
	90		Sewage & refuse disposal, sanitation, etc.	..	..	..	..	..	..	..	..
	91		Activities of member organisations nec	..	..	..	..	..	..	..	..
		912	Activities of trade unions	..	..	..	..	..	..	..	..
	92		Recreational, cultural and sporting activities	..	..	..	..	..	..	..	..
		921	Motion picture, radio, TV, other entertainment act.	..	..	..	..	..	..	..	..
		922	News agency activities	..	..	..	..	..	..	..	..
		923	Library, archives, museums, other cultural activ.	..	..	..	..	..	..	..	..
		924	Sporting and other recreational activities	..	..	..	..	..	..	..	..
	93		Other service activities	..	..	..	..	..	..	..	..

Table AU.S8 **SALAIRES ET TRAITEMENTS, SALARIÉS - WAGES AND SALARIES, EMPLOYEES**

Millions de Sch (Prix courants)

1991	1992	1993	1994	1995	1996	1997	1998			CITI révision 3	
..	..	..	..	135 668	..	137 857	144 997	Com. de gros, de detail ; réparation etc.	G	50_52	
..	..	..	..	..	..	19 659	20 942	Com., entr., répar. d'auto. et moto ; carburants		50	
..	..	..	..	..	..	9 747	10 428	Commerce de véhicules automobiles			501
..	..	..	..	..	..	5 564	5 966	Entretien et réparation de véhicules automobiles			502
..	..	..	..	..	..	2 724	2 905	Com. de pièces et access. de véhic. automobiles			503
..	..	..	..	..	..	220	252	Com., entret. et répar. de motocycles et access.			504
..	..	..	..	..	..	1 404	1 391	Com. de détail de carburants pour automobiles			505
..	..	..	..	..	..	68 395	71 755	Com. de gros et act. inter., sauf auto. et moto		51	
..	..	..	..	..	..	1 817	1 956	Activités d'intermédiaries de commerce de gros			511
..	..	..	..	..	..	15 454	15 190	Com. gros: prod. agr. bruts, anim. viv., prod. alim.			512
..	..	..	..	..	..	17 015	17 513	Com. gros: biens de consom., sauf alimentaires			513
..	..	..	..	..	..	15 770	16 610	Com. gros: prod. inter. non agr., déchets, débris			514
..	..	..	..	..	..	17 473	19 673	Com. gros: machines, équipments et fournitures			515
..	..	..	..	..	..	866	813	Autres commerces de gros			519
..	..	..	..	..	..	49 802	52 301	Com. détail, sauf auto, moto; répar. art. pers., dom.		52	
..	..	..	..	..	..	12 337	12 290	Commerce de détail non spécialisé en magasin			521
..	..	..	..	..	..	2 281	2 825	Com. détail: prod. aliment. en mag. spécialisés			522
..	..	..	..	..	..	32 938	35 071	Autre com. détail: prod. neufs en mag. spécialisés			523
..	..	..	..	..	..	184	172	Commerces de détail de biens d'occasion			524
..	..	..	..	..	..	1 625	1 618	Commerces de détail autres qu'en magasin			525
..	..	..	..	..	..	437	324	Réparation d'articles personnels et domestiques			526
..	..	..	..	26 699	..	27 481	28 021	Hotels et restaurants	H	55	
..	..	..	..	..	..	16 271	15 796	Hôtels; camping; autres équip. courts séjours			551
..	..	..	..	..	..	11 210	12 225	Restaurants, cafés et cantines			552
..	..	..	..	..	..	77 863	79 700	Transport, entreposage, communications	I	60_64	
..	..	..	..	73 153	..	41 270	42 375	Transports terrestres; transports par conduites		60	
..	..	..	..	..	..	..	..	Transports par chemin de fer			601
..	..	..	..	..	..	19 897	20 675	Autres transports terrestres			602
..	..	..	..	..	..	..	..	Transports par conduites			603
..	..	..	..	..	..	94	142	Transports par eau		61	
..	..	..	..	..	..	..	..	Transports maritimes et côtiers			611
..	..	..	..	..	..	..	..	Transports par voies navigables intérieures			612
..	..	..	..	..	..	3 627	3 831	Transports aériens		62	
..	..	..	..	..	..	3 376	3 696	Transports aériens réguliers			621
..	..	..	..	..	..	251	135	Transports aériens non réguliers			622
..	..	..	..	..	..	11 678	12 264	Act. annex. et aux. des transp.; act. des ag. de voy.		63	
..	..	..	..	..	..	21 194	21 088	Postes et télécommunications		64	
..	..	..	..	..	..	247	12 528	Activités de poste et de courrier			641
..	..	..	..	..	..	20 947	8 560	Télécommunications			642
..	..	..	..	55 120	..	56 447	57 652	Intermédiation financière	J	65_67	
..	..	..	..	..	..	39 474	39 895	Interm. financ., sauf assur. et caisses de retraite		65	
..	..	..	..	..	..	37 293	37 844	Intermédiation monétaire			651
..	..	..	..	..	..	2 181	2 051	Autres intermédiations financières			659
..	..	..	..	..	..	15 584	16 221	Act. d'assur., caisses de retr., sauf séc. soc. oblig.		66	
..	..	..	..	..	..	1 389	1 537	Act. auxiliaires de l'intermédiation financière		67	
..	..	..	..	..	..	154	210	Act. aux. de l'interm. fin., sauf assur., et retraite			671
..	..	..	..	..	..	1 235	1 326	Act. aux. des assurances et caisses de retraite			672
..	..	..	..	50 160	..	57 500	65 381	Immobilier, locations, services aux entreprises	K	70_74	
..	..	..	..	..	..	6 457	7 431	Activités immobilières		70	
..	..	..	..	..	..	4 905	5 135	Act. immobilières sur biens propres ou loués			701
..	..	..	..	..	..	1 552	2 296	Activités immobilières pour comptes de tiers			702
..	..	..	..	..	..	1 514	1 422	Locat. mach., équip. sans opér., biens pers. et dom.		71	
..	..	..	..	..	..	595	566	Location de matériel de transport			711
..	..	..	..	..	..	551	530	Location de machines ét equipements nca			712
..	..	..	..	..	..	368	326	Loc. d'articles personnels et domestiques nca			713
..	..	..	..	..	..	7 275	8 979	Activités informatiques et activités rattachées		72	
..	..	..	..	..	..	..	..	Conseil en matériel informatique			721
..	..	..	..	..	..	3 343	3 411	Production de logiciels			722
..	..	..	..	..	..	3 463	5 054	Traitement de données informatiques			723
..	..	..	..	..	..	..	..	Activités de banques de données			724
..	..	..	..	..	..	394	449	Entr., répar. mach. de bureau et compt. et inform.			725
..	..	..	..	..	..	..	..	Autres activités rattachées à l'informatique			729
..	..	..	..	..	..	658	688	Recherche-développement		73	
..	..	..	..	..	..	635	687	Rech. et dév. expér. en sc. phys., natur. et ingén.			731
..	..	..	..	..	..	23	..	Rech. et dév. expér. en sciences soc. et humaines			732
..	..	..	..	..	..	41 595	46 863	Autres activités de services aux entreprises		74	
..	..	..	..	..	..	14 275	15 765	Act. jur., compt., audit.; cons. fisc.; ét. march., soud.			741
..	..	..	..	..	..	10 582	12 449	Act. d'achitect., d'ingén.; autres act. techniques			742
..	..	..	..	..	..	2 897	3 425	Publicité			743
..	..	..	..	..	..	13 841	15 225	Activités de services aux entreprises nca.			749
..	..	..	..	..	..	..	..	Education	M	80	
..	..	..	..	..	..	..	..	Activités d'enseignement primaire			801
..	..	..	..	..	..	..	..	Activités d'enseignement secondaire			802
..	..	..	..	..	..	..	..	Activités d'enseignement supérieur			803
..	..	..	..	..	..	..	..	Act. de formation perman.; autres act. d'enseign.			809
..	..	..	..	..	..	..	..	Santé et action sociale	N	85	
..	..	..	..	..	..	..	..	Activités de santé humaine			851
..	..	..	..	..	..	..	..	Activités vétérinaires			852
..	..	..	..	..	..	..	..	Activités d'action sociale			853
..	..	..	..	..	..	..	..	Autres activités de serv. coll., soc. et personnels	O	90_93	
..	..	..	..	..	..	..	..	Assain. et enlèv. des ordures; voierie, act. simil.		90	
..	..	..	..	..	..	..	..	Activités associatives diverses		91	
..	..	..	..	..	..	..	..	Activités de syndicats de salariés			912
..	..	..	..	..	..	..	..	Activités récréatives, culturelles et sportives		92	
..	..	..	..	..	..	..	..	Act. cinémat., de radio, télé.; autres act. de spect.			921
..	..	..	..	..	..	..	..	Activités d'agences de presse			922
..	..	..	..	..	..	..	..	Act. des biblio., arch., musées, jardins bot. et zool.			923
..	..	..	..	..	..	..	..	Activités sportives et autres activités récréatives			924
..	..	..	..	..	..	..	..	Autres activités de services		93	

AUSTRIA

Table AU.S9 SUPPLEMENTS TO WAGES & SALARIES - CHARGES SOCIALES DES EMPLOYEURS

Millions of Sch (Current Prices)

ISIC revision 3				1991	1992	1993	1994	1995	1996	1997	1998
G	50_52		Wholesale & retail trade; repairs; etc.	..	..	..	..	38 723	..	39 565	40 947
	50		Sale, maint.,repair vehic/cycle, person., h/hold gd.	..	..	..	..	..	..	5 716	6 006
		501	Sale of motor cycle	..	..	..	..	..	..	2 853	3 018
		502	Maintenance and repair of motor vehicles	..	..	..	..	..	..	1 635	1 702
		503	Sale of motor vehicle parts and accessories	..	..	..	..	..	..	779	828
		504	Sale, maint., repair motor cycle, parts & acces.	..	..	..	..	..	..	63	69
		505	Retail sale of automative fuel	..	..	..	..	..	..	387	389
	51		W/sale. trade & commission exc motor veh. & cycle	..	..	..	..	..	..	19 731	20 258
		511	Wholesale on a fee or contract basis	..	..	..	..	..	..	498	489
		512	W/sale agric. raw mat.,live animal,food, bev., tobac.	..	..	..	..	..	..	4 647	4 578
		513	Wholesale of household goods	..	..	..	..	..	..	4 950	4 949
		514	W/sale. of non-agric. intermed. prod., waste, scrap	..	..	..	..	..	..	4 704	4 743
		515	Wholesale of machinery, equipment & supplies	..	..	..	..	..	..	4 698	5 277
		519	Other wholesale	..	..	..	..	..	..	234	222
	52		Retail trade exc. motor veh. & cyc.; repair h/hold gds.	..	..	..	..	..	..	14 118	14 683
		521	Non-specialised retail trade in stores	..	..	..	..	..	..	3 514	3 479
		522	Retail sale food, bev., tobac. in specialised stores	..	..	..	..	..	..	647	802
		523	Other retail trade of new goods specialised stores	..	..	..	..	..	..	9 291	9 790
		524	Retail sale of second-hand goods in stores	..	..	..	..	..	..	49	47
		525	Retail trade not in stores	..	..	..	..	..	..	495	479
		526	Repair of personal and household goods	..	..	..	..	..	..	122	87
H	55		Hotels and restaurants	..	..	..	..	8 329	..	8 294	8 383
		551	Hotels; camping ; other short-stay accommodation	..	..	..	..	..	..	5 012	4 816
		552	Restaurants, bars and canteens	..	..	..	..	..	..	3 282	3 568
I	60_64		Transport, storage and communications	..	..	..	..	..	..	22 488	23 566
	60		Land transport, transport via pipelines	..	..	..	..	29 031	..	12 212	12 739
		601	Transport via railways	..	..	..	..	..	..	..	..
		602	Other land transport	..	..	..	..	..	..	6 351	6 565
		603	Transport via pipelines	..	..	..	..	..	..	..	..
	61		Water transport	..	..	..	..	..	..	36	49
		611	Sea and coastal water transport	..	..	..	..	..	..	..	..
		612	Inland water transport	..	..	..	..	..	..	..	..
	62		Air transport	..	..	..	..	..	..	1 056	1 187
		621	Scheduled air transport	..	..	..	..	..	..	995	1 150
		622	Non-scheduled air transport	..	..	..	..	..	..	61	37
	63		Support. & aux. transport act., travel agencies activ.	..	..	..	..	..	..	3 420	3 710
	64		Post and telecommunications	..	..	..	..	..	..	5 765	5 881
		641	Post and courier activities	..	..	..	..	..	..	69	3 509
		642	Telecommunications	..	..	..	..	..	..	5 696	2 371
J	65_67		Financial intermediation	..	..	..	..	21 815	..	23 176	23 790
	65		Financial intermediation, exc. insur., pension fund	..	..	..	..	..	..	17 466	18 058
		651	Monetary intermediation	..	..	..	..	..	..	16 645	17 339
		659	Other financial intermediation	..	..	..	..	..	..	821	719
	66		Insurance pension fund, exc. compul. social sec.	..	..	..	..	..	..	5 346	5 323
	67		Activities auxiliary to financial intermediation	..	..	..	..	..	..	364	409
		671	Act. auxil to fin. intermed., exc. insur. & pens.fund	..	..	..	..	..	..	42	55
		672	Act. auxiliary to insurance and pension funding	..	..	..	..	..	..	323	354
K	70_74		Real estate, renting and business activities	..	..	..	..	13 875	..	15 833	17 658
	70		Real estate activities	..	..	..	..	..	..	1 835	2 151
		701	Real estate activ. with own or leased property	..	..	..	..	..	..	1 408	1 512
		702	Real estate activities on a fee or contract basis	..	..	..	..	..	..	427	639
	71		Renting of mach. without operator; rent. of pers. gds.	..	..	..	..	..	..	436	403
		711	Renting of transport equipment	..	..	..	..	..	..	171	161
		712	Renting of other machinery and equipment	..	..	..	..	..	..	154	152
		713	Renting of personal and household goods nec	..	..	..	..	..	..	111	90
	72		Computer and related activities	..	..	..	..	..	..	1 913	2 410
		721	Hardware consultancy	..	..	..	..	..	..	..	..
		722	Software consultancy and supply	..	..	..	..	..	..	844	909
		723	Data processing	..	..	..	..	..	..	941	1 357
		724	Data base activities	..	..	..	..	..	..	..	..
		725	Mainten.; repair of office, account. & comp. mach.	..	..	..	..	..	..	107	126
		729	Other computer related activities	..	..	..	..	..	..	..	..
	73		Research and development	..	..	..	..	..	..	185	208
		731	R & D experim. on natural sciences & engineering	..	..	..	..	..	..	179	207
		732	R & D experim. on social sciences & humanities	..	..	..	..	..	..	6	..
	74		Other business activities	..	..	..	..	..	..	11 464	12 487
		741	Legal, acc., book-keep., audit act.; tax consult., etc	..	..	..	..	..	..	3 852	3 982
		742	Architectural, engineering, & other techn. activities	..	..	..	..	..	..	2 719	3 162
		743	Advertising	..	..	..	..	..	..	780	916
		749	Business activities nec	..	..	..	..	..	..	4 113	4 427
M	80		Education	..	..	..	..	..	..	..	..
		801	Primary education	..	..	..	..	..	..	..	..
		802	Secondary education	..	..	..	..	..	..	..	..
		803	Higher education	..	..	..	..	..	..	..	..
		809	Adult and other education	..	..	..	..	..	..	..	..
N	85		Health and social work	..	..	..	..	..	..	..	..
		851	Human health activities	..	..	..	..	..	..	..	..
		852	Veterinary activities	..	..	..	..	..	..	..	..
		853	Social work activities	..	..	..	..	..	..	..	..
O	90_93		Other commun., soc. & pers. services activities	..	..	..	..	..	..	..	..
	90		Sewage & refuse disposal, sanitation, etc.	..	..	..	..	..	..	..	..
	91		Activities of member organisations nec	..	..	..	..	..	..	..	..
		912	Activities of trade unions	..	..	..	..	..	..	..	..
	92		Recreational, cultural and sporting activities	..	..	..	..	..	..	..	..
		921	Motion picture, radio, TV, other entertainment act.	..	..	..	..	..	..	..	..
		922	News agency activities	..	..	..	..	..	..	..	..
		923	Library, archives, museums, other cultural activ.	..	..	..	..	..	..	..	..
		924	Sporting and other recreational activities	..	..	..	..	..	..	..	..
	93		Other service activities	..	..	..	..	..	..	..	..

Table AU.S10 **ENTREPRISES - ENTERPRISES**

Unités

1991	1992	1993	1994	1995	1996	1997	1998		CITI révision 3		
..	..	..	..	64 531	..	60 783	64 039	Com. de gros, de detail ; réparation etc.	G	50_52	
..	..	..	..	..	..	7 711	8 244	Com., entr., répar. d'auto. et moto ; carburants		50	
..	..	..	..	..	..	1 819	2 094	Commerce de véhicules automobiles			501
..	..	..	..	..	..	3 048	3 272	Entretien et réparation de véhicules automobiles			502
..	..	..	..	..	..	751	784	Com. de pièces et access. de véhic. automobiles			503
..	..	..	..	..	..	212	229	Com., entret. et répar. de motocycles et access.			504
..	..	..	..	..	..	1 881	1 865	Com. de détail de carburants pour automobiles			505
..	..	..	..	..	..	18 416	19 409	Com. de gros et act. inter., sauf auto. et moto		51	
..	..	..	..	..	..	4 234	4 251	Activités d'intermédiaries de commerce de gros			511
..	..	..	..	..	..	2 591	2 819	Com. gros: prod. agr. bruts, anim. viv., prod. alim.			512
..	..	..	..	..	..	4 115	4 132	Com. gros: biens de consom., sauf alimentaires			513
..	..	..	..	..	..	2 877	3 093	Com. gros: prod. inter. non agr., déchets, débris			514
..	..	..	..	..	..	4 092	4 642	Com. gros: machines, équipments et fournitures			515
..	..	..	..	..	..	507	472	Autres commerces de gros			519
..	..	..	..	..	..	34 656	36 386	Com. détail, sauf auto, moto; répar. art. pers., dom.		52	
..	..	..	..	..	..	4 291	4 530	Commerce de détail non spécialisé en magasin			521
..	..	..	..	..	..	4 916	4 981	Com. détail: prod. aliment. en mag. spécialisés			522
..	..	..	..	..	..	21 874	23 350	Autre com. détail: prod. neufs en mag. spécialisés			523
..	..	..	..	..	..	952	983	Commerces de détail de biens d'occasion			524
..	..	..	..	..	..	1 631	1 355	Commerces de détail autres qu'en magasin			525
..	..	..	..	..	..	992	1 188	Réparation d'articles personnels et domestiques			526
..	..	..	..	38 768	..	36 605	39 245	Hotels et restaurants	H	55	
..	..	..	..	..	..	16 360	16 892	Hôtels; camping; autres équip. courts séjours			551
..	..	..	..	..	..	20 245	22 352	Restaurants, cafés et cantines			552
..	..	..	..	..	..	11 561	11 614	Transport, entreposage, communications	I	60_64	
..	..	..	..	11 271	..	9 058	8 948	Transports terrestres; transports par conduites		60	
..	..	..	..	..	..	11	18	Transports par chemin de fer			601
..	..	..	..	..	..	9 046	8 928	Autres transports terrestres			602
..	..	..	..	..	..	2	2	Transports par conduites			603
..	..	..	..	..	..	75	74	Transports par eau		61	
..	..	..	..	..	..	17	14	Transports maritimes et côtiers			611
..	..	..	..	..	..	58	60	Transports par voies navigables intérieures			612
..	..	..	..	..	..	81	66	Transports aériens		62	
..	..	..	..	..	..	4	4	Transports aériens réguliers			621
..	..	..	..	..	..	77	62	Transports aériens non réguliers			622
..	..	..	..	..	..	2 009	2 094	Act. annex. et aux. des transp.; act. des ag. de voy.		63	
..	..	..	..	..	..	338	433	Postes et télécommunications		64	
..	..	..	..	..	..	205	274	Activités de poste et de courrier			641
..	..	..	..	..	..	133	159	Télécommunications			642
..	..	..	..	2 709	..	3 422	4 026	Intermédiation financière	J	65_67	
..	..	..	..	..	..	1 003	1 017	Interm. financ., sauf assur. et caisses de retraite		65	
..	..	..	..	..	..	888	898	Intermédiation monétaire			651
..	..	..	..	..	..	115	119	Autres intermédiations financières			659
..	..	..	..	..	..	146	147	Act. d'assur., caisses de retr., sauf séc. soc. oblig.		66	
..	..	..	..	..	..	2 272	2 863	Act. auxiliaires de l'intermédiation financière		67	
..	..	..	..	..	..	138	179	Act. aux. de l'interm. fin., sauf assur., et retraite			671
..	..	..	..	..	..	2 134	2 683	Act. aux. des assurances et caisses de retraite			672
..	..	..	..	30 900	..	32 868	38 610	Immobilier, locations, services aux entreprises	K	70_74	
..	..	..	..	..	..	3 282	3 898	Activités immobilières		70	
..	..	..	..	..	..	1 648	2 001	Act. immobilières sur biens propres ou loués			701
..	..	..	..	..	..	1 634	1 897	Activités immobilières pour comptes de tiers			702
..	..	..	..	..	..	1 329	1 400	Locat. mach., équip. sans opér., biens pers. et dom.		71	
..	..	..	..	..	..	263	236	Location de matériel de transport			711
..	..	..	..	..	..	332	397	Location de machines ét equipements nca			712
..	..	..	..	..	..	735	766	Loc. d'articles personnels et domestiques nca			713
..	..	..	..	..	..	4 642	6 195	Activités informatiques et activités rattachées		72	
..	..	..	..	..	..	3	..	Conseil en matériel informatique			721
..	..	..	..	..	..	1 717	1 781	Production de logiciels			722
..	..	..	..	..	..	2 812	4 207	Traitement de données informatiques			723
..	..	..	..	..	..	14	1	Activités de banques de données			724
..	..	..	..	..	..	85	186	Entr., répar. mach. de bureau et compt. et inform.			725
..	..	..	..	..	..	11	19	Autres activités rattachées à l'informatique			729
..	..	..	..	..	..	54	57	Recherche-développement		73	
..	..	..	..	..	..	19	43	Rech. et dév. expér. en sc. phys., natur. et ingén.			731
..	..	..	..	..	..	34	14	Rech. et dév. expér. en sciences soc. et humaines			732
..	..	..	..	..	..	23 561	27 060	Autres activités de services aux entreprises		74	
..	..	..	..	..	..	8 797	9 381	Act. jur., compt., audit.; cons. fisc.; ét. march., soud.			741
..	..	..	..	..	..	7 389	8 727	Act. d'achitect., d'ingén.; autres act. techniques			742
..	..	..	..	..	..	2 819	3 666	Publicité			743
..	..	..	..	..	..	4 556	5 285	Activités de services aux entreprises nca.			749
..	..	..	..	..	..	..	..	Education	M	80	
..	..	..	..	..	..	..	..	Activités d'enseignement primaire			801
..	..	..	..	..	..	..	..	Activités d'enseignement secondaire			802
..	..	..	..	..	..	..	..	Activités d'enseignement supérieur			803
..	..	..	..	..	..	..	..	Act. de formation perman.; autres act. d'enseign.			809
..	..	..	..	..	..	..	..	Santé et action sociale	N	85	
..	..	..	..	..	..	..	..	Activités de santé humaine			851
..	..	..	..	..	..	..	..	Activités vétérinaires			852
..	..	..	..	..	..	..	..	Activités d'action sociale			853
..	..	..	..	..	..	..	..	Autres activités de serv. coll., soc. et personnels	O	90_93	
..	..	..	..	..	..	..	..	Assain. et enlèv. des ordures; voierie, act. simil.		90	
..	..	..	..	..	..	..	..	Activités associatives diverses		91	
..	..	..	..	..	..	..	..	Activités de syndicats de salariés			912
..	..	..	..	..	..	..	..	Activités récréatives, culturelles et sportives		92	
..	..	..	..	..	..	..	..	Act. cinémat., de radio, télé.; autres act. de spect.			921
..	..	..	..	..	..	..	..	Activités d'agences de presse			922
..	..	..	..	..	..	..	..	Act. des biblio., arch., musées, jardins bot. et zool.			923
..	..	..	..	..	..	..	..	Activités sportives et autres activités récréatives			924
..	..	..	..	..	..	..	..	Autres activités de services		93	

Table CZ.S1 **PRODUCTION - PRODUCTION**

Millions of CK (Current Prices)

ISIC revision 3				1991	1992	1993	1994	1995	1996	1997	1998
G	**50_52**		Wholesale & retail trade; repairs; etc.	..	..	..	..	349 226	434 076	381 153	402 663
	50		Sale, maint.,repair vehic/cycle, person., h/hold gd.	..	..	..	..	..	..	46 338	53 363
		501	Sale of motor cycle	..	..	..	..	..	..	14 296	16 018
		502	Maintenance and repair of motor vehicles	..	..	..	..	..	..	18 003	20 919
		503	Sale of motor vehicle parts and accessories	..	..	..	..	..	..	3 798	5 248
		504	Sale, maint., repair motor cycle, parts & acces.	..	..	..	..	..	..	291	211
		505	Retail sale of automative fuel	..	..	..	..	..	..	9 950	10 967
	51		W/sale. trade & commission exc motor veh. & cycle	..	..	..	..	..	..	212 864	233 544
		511	Wholesale on a fee or contract basis	..	..	..	..	..	..	28 155	32 703
		512	W/sale agric. raw mat.,live animal,food, bev., tobac.	..	..	..	..	..	..	21 869	32 199
		513	Wholesale of household goods	..	..	..	..	..	..	36 583	42 476
		514	W/sale. of non-agric. intermed. prod., waste, scrap	..	..	..	..	..	..	44 041	37 329
		515	Wholesale of machinery, equipment & supplies	..	..	..	..	..	..	30 489	20 654
		519	Other wholesale	..	..	..	..	..	..	51 727	68 183
	52		Retail trade exc. motor veh. & cyc.; repair h/hold gds.	..	..	..	..	..	..	121 951	115 756
		521	Non-specialised retail trade in stores	..	..	..	..	..	..	64 574	48 653
		522	Retail sale food, bev., tobac. in specialised stores	..	..	..	..	..	..	4 525	5 950
		523	Other retail trade of new goods specialised stores	..	..	..	..	..	..	42 543	46 522
		524	Retail sale of second-hand goods in stores	..	..	..	..	..	..	962	1 633
		525	Retail trade not in stores	..	..	..	..	..	..	6 469	10 118
		526	Repair of personal and household goods	..	..	..	..	..	..	2 878	2 880
H	**55**		Hotels and restaurants	..	..	..	..	47 562	63 382	57 394	61 893
		551	Hotels; camping ; other short-stay accommodation	..	..	..	..	..	..	17 960	17 441
		552	Restaurants, bars and canteens	..	..	..	..	..	..	39 434	44 452
I	**60_64**		Transport, storage and communications	..	..	..	..	..	..	246 444	289 251
	60		Land transport, transport via pipelines	..	..	..	..	181 433	224 424	107 083	123 302
		601	Transport via railways	..	..	..	..	..	..	30 687	32 003
		602	Other land transport	..	..	..	..	..	..	69 386	79 169
		603	Transport via pipelines	..	..	..	..	..	..	7 010	12 130
	61		Water transport	..	..	..	..	..	..	2 147	1 973
		611	Sea and coastal water transport	..	..	..	..	..	..	317	..
		612	Inland water transport	..	..	..	..	..	..	1 830	..
	62		Air transport	..	..	..	..	..	..	10 773	13 215
		621	Scheduled air transport	..	..	..	..	..	..	10 249	11 392
		622	Non-scheduled air transport	..	..	..	..	..	..	524	1 823
	63		Support. & aux. transport act., travel agencies activ.	..	..	..	..	..	..	56 370	68 376
	64		Post and telecommunications	..	..	..	..	..	..	70 070	82 386
		641	Post and courier activities	..	..	..	..	..	..	12 270	13 168
		642	Telecommunications	..	..	..	..	..	..	57 800	69 218
J	**65_67**		Financial intermediation	..	..	..	..	132 747	113 378	98 091	126 341
	65		Financial intermediation, exc. insur., pension fund	..	..	..	..	107 897	85 654	79 812	90 879
		651	Monetary intermediation	..	..	..	..	96 900	69 114	72 275	83 188
		659	Other financial intermediation	..	..	..	..	10 997	16 540	7 537	7 691
	66		Insurance pension fund, exc. compul. social sec.	..	..	..	..	19 044	22 039	12 166	28 344
	67		Activities auxiliary to financial intermediation	..	..	..	..	5 806	5 685	6 113	7 118
		671	Act. auxil to fin. intermed., exc. insur. & pens.fund	..	..	..	..	..	..	4 314	4 993
		672	Act. auxiliary to insurance and pension funding	..	..	..	..	..	..	1 799	2 125
K	**70_74**		Real estate, renting and business activities	..	..	..	..	202 374	236 956	273 932	295 953
	70		Real estate activities	..	..	..	..	..	..	43 442	55 890
		701	Real estate activ. with own or leased property	..	..	..	..	..	..	30 751	41 155
		702	Real estate activities on a fee or contract basis	..	..	..	..	..	..	12 691	14 734
	71		Renting of mach. without operator; rent. of pers. gds.	..	..	..	..	..	..	6 921	9 164
		711	Renting of transport equipment	..	..	..	..	..	..	2 834	2 682
		712	Renting of other machinery and equipment	..	..	..	..	..	..	3 035	3 821
		713	Renting of personal and household goods nec	..	..	..	..	..	..	1 052	2 661
	72		Computer and related activities	..	..	..	..	..	..	32 724	35 109
		721	Hardware consultancy	..	..	..	..	..	..	1 776	3 269
		722	Software consultancy and supply	..	..	..	..	..	..	19 986	22 771
		723	Data processing	..	..	..	..	..	..	4 459	4 638
		724	Data base activities	..	..	..	..	..	..	99	104
		725	Mainten.; repair of office, account. & comp. mach.	..	..	..	..	..	..	2 517	3 718
		729	Other computer related activities	..	..	..	..	..	..	3 886	609
	73		Research and development	..	..	..	..	..	..	4 608	4 621
		731	R & D experim. on natural sciences & engineering	..	..	..	..	..	..	4 604	4 620
		732	R & D experim. on social sciences & humanities	..	..	..	..	..	..	4	1
	74		Other business activities	..	..	..	..	..	..	186 238	191 170
		741	Legal, acc., book-keep., audit act.; tax consult., etc	..	..	..	..	..	..	37 226	46 130
		742	Architectural, engineering, & other techn. activities	..	..	..	..	..	..	61 156	64 432
		743	Advertising	..	..	..	..	..	..	31 515	34 910
		749	Business activities nec	..	..	..	..	..	..	56 341	45 698
M	**80**		Education	..	..	..	..	..	..	4 190	4 736
		801	Primary education	..	..	..	..	..	..	50	160
		802	Secondary education	..	..	..	..	..	..	787	1 079
		803	Higher education	..	..	..	..	..	..	33	64
		809	Adult and other education	..	..	..	..	..	..	3 320	3 433
N	**85**		Health and social work	..	..	..	..	..	..	31 642	33 258
		851	Human health activities	..	..	..	..	..	..	30 684	31 445
		852	Veterinary activities	..	..	..	..	..	..	825	1 372
		853	Social work activities	..	..	..	..	..	..	132	441
O	**90_93**		Other commun., soc. & pers. services activities	..	..	..	..	..	..	50 583	58 839
	90		Sewage & refuse disposal, sanitation, etc.	..	..	..	..	..	..	13 581	13 739
	91		Activities of member organisations nec	..	..	..	..	..	..	..	..
		912	Activities of trade unions	..	..	..	..	..	..	..	..
	92		Recreational, cultural and sporting activities	..	..	..	..	..	..	30 892	37 232
		921	Motion picture, radio, TV, other entertainment act.	..	..	..	..	..	..	10 432	15 184
		922	News agency activities	..	..	..	..	..	..	876	663
		923	Library, archives, museums, other cultural activ.	..	..	..	..	..	..	372	445
		924	Sporting and other recreational activities	..	..	..	..	..	..	19 213	20 940
	93		Other service activities	..	..	..	..	..	..	6 110	7 869

Table CZ.S2 **VALEUR AJOUTÉE - VALUE ADDED**

Millions de CK (Prix courants)

1991	1992	1993	1994	1995	1996	1997	1998		**CITI révision 3**		
..	..	..	..	157 036	189 372	146 417	168 655	Com. de gros, de detail ; réparation etc.	G	**50_52**	
..	..	..	..	..	..	14 626	17 752	Com., entr., répar. d'auto. et moto ; carburants		**50**	
..	..	..	..	..	..	5 316	4 734	Commerce de véhicules automobiles			501
..	..	..	..	..	..	2 201	5 247	Entretien et réparation de véhicules automobiles			502
..	..	..	..	..	..	1 769	2 632	Com. de pièces et access. de véhic. automobiles			503
..	..	..	..	..	..	158	..	Com., entret. et répar. de motocycles et access.			504
..	..	..	..	..	..	5 182	5 224	Com. de détail de carburants pour automobiles			505
..	..	..	..	..	..	86 470	98 485	Com. de gros et act. inter., sauf auto. et moto		**51**	
..	..	..	..	..	..	11 240	8 419	Activités d'intermédiaries de commerce de gros			511
..	..	..	..	..	..	8 572	13 280	Com. gros: prod. agr. bruts, anim. viv., prod. alim.			512
..	..	..	..	..	..	13 494	17 338	Com. gros: biens de consom., sauf alimentaires			513
..	..	..	..	..	..	20 237	17 347	Com. gros: prod. inter. non agr., déchets, débris			514
..	..	..	..	..	..	13 769	10 467	Com. gros: machines, équipments et fournitures			515
..	..	..	..	..	..	19 158	31 634	Autres commerces de gros			519
..	..	..	..	..	..	45 321	52 418	Com. détail, sauf auto, moto; répar. art. pers., dom.		**52**	
..	..	..	..	..	..	25 364	25 157	Commerce de détail non spécialisé en magasin			521
..	..	..	..	..	..	1 549	2 921	Com. détail: prod. aliment. en mag. spécialisés			522
..	..	..	..	..	..	15 072	20 028	Autre com. détail: prod. neufs en mag. spécialisés			523
..	..	..	..	..	..	345	664	Commerces de détail de biens d'occasion			524
..	..	..	..	..	..	2 342	2 764	Commerces de détail autres qu'en magasin			525
..	..	..	..	..	..	649	884	Réparation d'articles personnels et domestiques			526
..	..	..	..	15 133	23 892	15 251	15 124	Hotels et restaurants	H	**55**	
..	..	..	..	..	..	7 640	7 315	Hôtels; camping; autres équip. courts séjours			551
..	..	..	..	..	..	7 611	7 809	Restaurants, cafés et cantines			552
..	..	..	..	..	..	88 696	109 068	Transport, entreposage, communications	I	**60_64**	
..	..	..	..	64 645	79 723	35 436	44 095	Transports terrestres; transports par conduites		**60**	
..	..	..	..	..	..	13 259	15 847	Transports par chemin de fer			601
..	..	..	..	..	..	16 871	17 657	Autres transports terrestres			602
..	..	..	..	..	..	5 306	10 591	Transports par conduites			603
..	..	..	..	..	..	445	454	Transports par eau		**61**	
..	..	..	..	..	..	84	..	Transports maritimes et côtiers			611
..	..	..	..	..	..	361	..	Transports par voies navigables intérieures			612
..	..	..	..	..	..	3 081	4 301	Transports aériens		**62**	
..	..	..	..	..	..	3 073	3 969	Transports aériens réguliers			621
..	..	..	..	..	..	8	332	Transports aériens non réguliers			622
..	..	..	..	..	..	9 373	11 431	Act. annex. et aux. des transp.; act. des ag. de voy.		**63**	
..	..	..	..	..	..	40 361	48 786	Postes et télécommunications		**64**	
..	..	..	..	..	..	8 629	9 253	Activités de poste et de courrier			641
..	..	..	..	..	..	31 732	39 533	Télécommunications			642
..	..	..	..	95 649	73 187	59 953	81 256	Intermédiation financière	J	**65_67**	
..	..	..	..	77 593	53 292	51 342	57 572	Interm. financ., sauf assur. et caisses de retraite		**65**	
..	..	..	..	75 292	46 523	51 205	57 367	Intermédiation monétaire			651
..	..	..	..	2 301	6 769	137	205	Autres intermédiations financières			659
..	..	..	..	15 918	17 875	6 987	22 793	Act. d'assur., caisses de retr., sauf séc. soc. oblig.		**66**	
..	..	..	..	2 138	2 020	1 624	891	Act. auxiliaires de l'intermédiation financière		**67**	
..	..	..	..	..	..	945	216	Act. aux. de l'interm. fin., sauf assur., et retraite			671
..	..	..	..	..	..	679	675	Act. aux. des assurances et caisses de retraite			672
..	..	..	..	84 247	102 629	84 532	109 697	Immobilier, locations, services aux entreprises	K	**70_74**	
..	..	..	..	..	..	12 550	21 081	Activités immobilières		**70**	
..	..	..	..	..	..	10 011	15 991	Act. immobilières sur biens propres ou loués			701
..	..	..	..	..	..	2 539	5 090	Activités immobilières pour comptes de tiers			702
..	..	..	..	..	..	2 983	5 034	Locat. mach., équip. sans opér., biens pers. et dom.		**71**	
..	..	..	..	..	..	1 036	1 013	Location de matériel de transport			711
..	..	..	..	..	..	1 454	2 028	Location de machines ét equipements nca			712
..	..	..	..	..	..	493	1 993	Loc. d'articles personnels et domestiques nca			713
..	..	..	..	..	..	12 899	14 338	Activités informatiques et activités rattachées		**72**	
..	..	..	..	..	..	752	1 090	Conseil en matériel informatique			721
..	..	..	..	..	..	8 494	9 442	Production de logiciels			722
..	..	..	..	..	..	2 225	2 424	Traitement de données informatiques			723
..	..	..	..	..	..	40	37	Activités de banques de données			724
..	..	..	..	..	..	819	1 124	Entr., répar. mach. de bureau et compt. et inform.			725
..	..	..	..	..	..	569	220	Autres activités rattachées à l'informatique			729
..	..	..	..	..	..	1 952	2 092	Recherche-développement		**73**	
..	..	..	..	..	..	1 950	2 092	Rech. et dév. expér. en sc. phys., natur. et ingén.			731
..	..	..	..	..	..	2	0	Rech. et dév. expér. en sciences soc. et humaines			732
..	..	..	..	..	..	54 146	67 152	Autres activités de services aux entreprises		**74**	
..	..	..	..	..	..	13 880	19 330	Act. jur., compt., audit.; cons. fisc.; ét. march., soud.			741
..	..	..	..	..	..	18 045	23 139	Act. d'achitect., d'ingén.; autres act. techniques			742
..	..	..	..	..	..	4 716	4 576	Publicité			743
..	..	..	..	..	..	17 505	20 107	Activités de services aux entreprises nca.			749
..	..	..	..	..	..	1 327	1 326	Education	M	**80**	
..	..	..	..	..	..	..	32	Activités d'enseignement primaire			801
..	..	..	..	..	..	145	248	Activités d'enseignement secondaire			802
..	..	..	..	..	..	..	..	Activités d'enseignement supérieur			803
..	..	..	..	..	..	1 203	1 048	Act. de formation perman.; autres act. d'enseign.			809
..	..	..	..	..	..	14 873	16 405	Santé et action sociale	N	**85**	
..	..	..	..	..	..	14 540	15 825	Activités de santé humaine			851
..	..	..	..	..	..	276	386	Activités vétérinaires			852
..	..	..	..	..	..	58	193	Activités d'action sociale			853
..	..	..	..	..	..	20 196	23 221	Autres activités de serv. coll., soc. et personnels	O	**90_93**	
..	..	..	..	..	..	5 613	5 828	Assain. et enlèv. des ordures; voierie, act. simil.		**90**	
..	..	..	..	..	..	..	..	Activités associatives diverses		**91**	
..	..	..	..	..	..	..	..	Activités de syndicats de salariés			912
..	..	..	..	..	..	12 403	14 432	Activités récréatives, culturelles et sportives		**92**	
..	..	..	..	..	..	2 837	3 696	Act. cinémat., de radio, télé.; autres act. de spect.			921
..	..	..	..	..	..	313	189	Activités d'agences de presse			922
..	..	..	..	..	..	97	71	Act. des biblio., arch., musées, jardins bot. et zool.			923
..	..	..	..	..	..	9 157	10 476	Activités sportives et autres activités récréatives			924
..	..	..	..	..	..	2 181	2 961	Autres activités de services		**93**	

Table CZ.S3 **INVESTMENT - INVESTISSEMENT**

Millions of CK (Current Prices)

ISIC revision 3				1991	1992	1993	1994	1995	1996	1997	1998
G	50_52		Wholesale & retail trade; repairs; etc.	..	..	..	..	32 486	53 017	61 290	53 226
	50		Sale, maint.,repair vehic/cycle, person., h/hold gd.	..	..	..	..	..	..	15 947	11 803
		501	Sale of motor cycle	..	..	..	..	..	..	3 404	4 182
		502	Maintenance and repair of motor vehicles	..	..	..	..	..	..	2 911	1 645
		503	Sale of motor vehicle parts and accessories	..	..	..	..	..	..	877	898
		504	Sale, maint., repair motor cycle, parts & acces.	..	..	..	..	..	..	116	51
		505	Retail sale of automative fuel	..	..	..	..	..	..	8 639	5 027
	51		W/sale. trade & commission exc motor veh. & cycle	..	..	..	..	..	..	23 629	21 917
		511	Wholesale on a fee or contract basis	..	..	..	..	..	..	2 201	1 707
		512	W/sale agric. raw mat.,live animal,food, bev., tobac.	..	..	..	..	..	..	2 233	3 070
		513	Wholesale of household goods	..	..	..	..	..	..	4 349	2 069
		514	W/sale. of non-agric. intermed. prod., waste, scrap	..	..	..	..	..	..	4 426	4 089
		515	Wholesale of machinery, equipment & supplies	..	..	..	..	..	..	4 140	2 581
		519	Other wholesale	..	..	..	..	..	..	6 280	8 401
	52		Retail trade exc. motor veh. & cyc.; repair h/hold gds.	..	..	..	..	..	..	21 714	19 506
		521	Non-specialised retail trade in stores	..	..	..	..	..	..	14 103	9 846
		522	Retail sale food, bev., tobac. in specialised stores	..	..	..	..	..	..	574	427
		523	Other retail trade of new goods specialised stores	..	..	..	..	..	..	6 478	8 570
		524	Retail sale of second-hand goods in stores	..	..	..	..	..	..	20	22
		525	Retail trade not in stores	..	..	..	..	..	..	447	614
		526	Repair of personal and household goods	..	..	..	..	..	..	92	27
H	55		Hotels and restaurants	..	..	..	..	3 139	5 261	6 556	4 665
		551	Hotels; camping ; other short-stay accommodation	..	..	..	..	..	..	4 633	2 121
		552	Restaurants, bars and canteens	..	..	..	..	..	..	1 923	2 544
I	60_64		Transport, storage and communications	..	..	..	..	..	..	83 313	73 192
	60		Land transport, transport via pipelines	..	..	..	..	40 011	66 159	25 241	27 324
		601	Transport via railways	..	..	..	..	..	..	11 855	14 105
		602	Other land transport	..	..	..	..	..	..	10 096	10 856
		603	Transport via pipelines	..	..	..	..	..	..	3 290	2 363
	61		Water transport	..	..	..	..	..	..	2 024	294
		611	Sea and coastal water transport	..	..	..	..	..	..	1 930	..
		612	Inland water transport	..	..	..	..	..	..	94	..
	62		Air transport	..	..	..	..	..	..	1 404	713
		621	Scheduled air transport	..	..	..	..	..	..	1 343	629
		622	Non-scheduled air transport	..	..	..	..	..	..	61	84
	63		Support. & aux. transport act., travel agencies activ.	..	..	..	..	..	..	4 752	4 036
	64		Post and telecommunications	..	..	..	..	..	..	49 892	40 825
		641	Post and courier activities	..	..	..	..	..	..	873	1 057
		642	Telecommunications	..	..	..	..	..	..	49 019	39 768
J	65_67		Financial intermediation	..	..	..	..	55 008	63 439	54 712	68 892
	65		Financial intermediation, exc. insur., pension fund	..	..	..	..	51 294	61 032	51 985	66 474
		651	Monetary intermediation	..	..	..	..	18 019	13 618	11 782	8 020
		659	Other financial intermediation	..	..	..	..	33 275	47 414	40 203	58 454
	66		Insurance pension fund, exc. compul. social sec.	..	..	..	..	2 837	2 031	2 256	2 150
	67		Activities auxiliary to financial intermediation	..	..	..	..	877	376	471	268
		671	Act. auxil to fin. intermed., exc. insur. & pens.fund	..	..	..	..	..	..	375	212
		672	Act. auxiliary to insurance and pension funding	..	..	..	..	..	..	96	56
K	70_74		Real estate, renting and business activities	..	..	..	..	22 118	38 182	40 508	43 332
	70		Real estate activities	..	..	..	..	..	..	20 523	24 216
		701	Real estate activ. with own or leased property	..	..	..	..	..	..	16 958	17 779
		702	Real estate activities on a fee or contract basis	..	..	..	..	..	..	3 566	6 437
	71		Renting of mach. without operator; rent. of pers. gds.	..	..	..	..	..	..	3 344	6 441
		711	Renting of transport equipment	..	..	..	..	..	..	964	1 572
		712	Renting of other machinery and equipment	..	..	..	..	..	..	1 899	2 320
		713	Renting of personal and household goods nec	..	..	..	..	..	..	481	2 549
	72		Computer and related activities	..	..	..	..	..	..	1 892	1 679
		721	Hardware consultancy	..	..	..	..	..	..	89	93
		722	Software consultancy and supply	..	..	..	..	..	..	1 307	905
		723	Data processing	..	..	..	..	..	..	339	550
		724	Data base activities	..	..	..	..	..	..	3	4
		725	Mainten.; repair of office, account. & comp. mach.	..	..	..	..	..	..	95	102
		729	Other computer related activities	..	..	..	..	..	..	58	24
	73		Research and development	..	..	..	..	..	..	526	492
		731	R & D experim. on natural sciences & engineering	..	..	..	..	..	..	526	492
		732	R & D experim. on social sciences & humanities	..	..	..	..	..	..	..	..
	74		Other business activities	..	..	..	..	..	..	14 222	10 504
		741	Legal, acc., book-keep., audit act.; tax consult., etc	..	..	..	..	..	..	3 517	4 350
		742	Architectural, engineering, & other techn. activities	..	..	..	..	..	..	4 605	3 008
		743	Advertising	..	..	..	..	..	..	1 090	759
		749	Business activities nec	..	..	..	..	..	..	5 011	2 386
M	80		Education	..	..	..	..	..	..	419	260
		801	Primary education	..	..	..	..	..	..	57	10
		802	Secondary education	..	..	..	..	..	..	78	62
		803	Higher education	..	..	..	..	..	..	13	5
		809	Adult and other education	..	..	..	..	..	..	271	183
N	85		Health and social work	..	..	..	..	..	..	2 118	2 108
		851	Human health activities	..	..	..	..	..	..	2 098	2 045
		852	Veterinary activities	..	..	..	..	..	..	13	47
		853	Social work activities	..	..	..	..	..	..	7	16
O	90_93		Other commun., soc. & pers. services activities	..	..	..	..	..	..	5 002	6 400
	90		Sewage & refuse disposal, sanitation, etc.	..	..	..	..	..	..	2 496	3 487
	91		Activities of member organisations nec	..	..	..	..	..	..	..	..
		912	Activities of trade unions	..	..	..	..	..	..	..	..
	92		Recreational, cultural and sporting activities	..	..	..	..	..	..	2 165	2 257
		921	Motion picture, radio, TV, other entertainment act.	..	..	..	..	..	..	716	720
		922	News agency activities	..	..	..	..	..	..	103	26
		923	Library, archives, museums, other cultural activ.	..	..	..	..	..	..	30	16
		924	Sporting and other recreational activities	..	..	..	..	..	..	1 316	1 495
	93		Other service activities	..	..	..	..	..	..	341	657

Table CZ.S4 **EMPLOI, SALARIÉS - EMPLOYMENT, EMPLOYEES**

Unités

1991	1992	1993	1994	1995	1996	1997	1998		CITI révision 3		
..	..	..	..	..	..	519 000	493 000	Com. de gros, de detail ; réparation etc.	**G**	**50_52**	
..	..	..	..	..	..	55 000	55 000	Com., entr., répar. d'auto. et moto ; carburants		**50**	
..	..	..	..	..	..	15 000	15 000	Commerce de véhicules automobiles			501
..	..	..	..	..	..	26 000	23 000	Entretien et réparation de véhicules automobiles			502
..	..	..	..	..	..	5 000	7 000	Com. de pièces et access. de véhic. automobiles			503
..	..	..	..	..	..	..	..	Com., entret. et répar. de motocycles et access.			504
..	..	..	..	..	..	9 000	10 000	Com. de détail de carburants pour automobiles			505
..	..	..	..	..	..	185 000	176 000	Com. de gros et act. inter., sauf auto. et moto		**51**	
..	..	..	..	..	..	23 000	17 000	Activités d'intermédiaries de commerce de gros			511
..	..	..	..	..	..	27 000	35 000	Com. gros: prod. agr. bruts, anim. viv., prod. alim.			512
..	..	..	..	..	..	34 000	34 000	Com. gros: biens de consom., sauf alimentaires			513
..	..	..	..	..	..	31 000	31 000	Com. gros: prod. inter. non agr., déchets, débris			514
..	..	..	..	..	..	23 000	15 000	Com. gros: machines, équipments et fournitures			515
..	..	..	..	..	..	47 000	44 000	Autres commerces de gros			519
..	..	..	..	..	..	279 000	262 000	Com. détail, sauf auto, moto; répar. art. pers., dom.		**52**	
..	..	..	..	..	..	172 000	132 000	Commerce de détail non spécialisé en magasin			521
..	..	..	..	..	..	11 000	20 000	Com. détail: prod. aliment. en mag. spécialisés			522
..	..	..	..	..	..	82 000	96 000	Autre com. détail: prod. neufs en mag. spécialisés			523
..	..	..	..	..	..	1 000	2 000	Commerces de détail de biens d'occasion			524
..	..	..	..	..	..	9 000	7 000	Commerces de détail autres qu'en magasin			525
..	..	..	..	..	..	4 000	5 000	Réparation d'articles personnels et domestiques			526
..	..	..	..	..	..	113 000	119 000	Hotels et restaurants	**H**	**55**	
..	..	..	..	..	..	34 000	30 000	Hôtels; camping; autres équip. courts séjours			551
..	..	..	..	..	..	79 000	89 000	Restaurants, cafés et cantines			552
..	..	..	..	..	..	313 000	299 000	Transport, entreposage, communications	**I**	**60_64**	
..	..	..	..	..	..	204 000	196 000	Transports terrestres; transports par conduites		**60**	
..	..	..	..	..	..	100 000	95 000	Transports par chemin de fer			601
..	..	..	..	..	..	102 000	99 000	Autres transports terrestres			602
..	..	..	..	..	..	2 000	2 000	Transports par conduites			603
..	..	..	..	..	..	2 000	2 000	Transports par eau		**61**	
..	..	..	..	..	..	..	..	Transports maritimes et côtiers			611
..	..	..	..	..	..	2 000	..	Transports par voies navigables intérieures			612
..	..	..	..	..	..	5 000	5 000	Transports aériens		**62**	
..	..	..	..	..	..	5 000	5 000	Transports aériens réguliers			621
..	..	..	..	..	..	..	..	Transports aériens non réguliers			622
..	..	..	..	..	..	22 000	22 000	Act. annex. et aux. des transp.; act. des ag. de voy.		**63**	
..	..	..	..	..	..	80 000	74 000	Postes et télécommunications		**64**	
..	..	..	..	..	..	48 000	46 000	Activités de poste et de courrier			641
..	..	..	..	..	..	32 000	28 000	Télécommunications			642
..	..	..	..	80 000	83 000	82 000	82 000	Intermédiation financière	**J**	**65_67**	
..	..	..	..	63 000	65 000	63 000	61 000	Interm. financ., sauf assur. et caisses de retraite		**65**	
..	..	..	..	60 000	61 000	58 000	55 000	Intermédiation monétaire			651
..	..	..	..	3 000	4 000	5 000	6 000	Autres intermédiations financières			659
..	..	..	..	14 000	15 000	16 000	17 000	Act. d'assur., caisses de retr., sauf séc. soc. oblig.		**66**	
..	..	..	..	3 000	3 000	3 000	4 000	Act. auxiliaires de l'intermédiation financière		**67**	
..	..	..	..	..	..	2 000	3 000	Act. aux. de l'interm. fin., sauf assur., et retraite			671
..	..	..	..	..	..	1 000	1 000	Act. aux. des assurances et caisses de retraite			672
..	..	..	..	..	..	254 000	252 000	Immobilier, locations, services aux entreprises	**K**	**70_74**	
..	..	..	..	..	..	37 000	38 000	Activités immobilières		**70**	
..	..	..	..	..	..	26 000	23 000	Act. immobilières sur biens propres ou loués			701
..	..	..	..	..	..	11 000	14 000	Activités immobilières pour comptes de tiers			702
..	..	..	..	..	..	5 000	4 000	Locat. mach., équip. sans opér., biens pers. et dom.		**71**	
..	..	..	..	..	..	2 000	1 000	Location de matériel de transport			711
..	..	..	..	..	..	2 000	2 000	Location de machines ét equipements nca			712
..	..	..	..	..	..	1 000	1 000	Loc. d'articles personnels et domestiques nca			713
..	..	..	..	..	..	25 000	26 000	Activités informatiques et activités rattachées		**72**	
..	..	..	..	..	..	1 000	2 000	Conseil en matériel informatique			721
..	..	..	..	..	..	15 000	15 000	Production de logiciels			722
..	..	..	..	..	..	6 000	5 000	Traitement de données informatiques			723
..	..	..	..	..	..	..	..	Activités de banques de données			724
..	..	..	..	..	..	3 000	3 000	Entr., répar. mach. de bureau et compt. et inform.			725
..	..	..	..	..	..	1 000	..	Autres activités rattachées à l'informatique			729
..	..	..	..	..	..	7 000	7 000	Recherche-développement		**73**	
..	..	..	..	..	..	7 000	7 000	Rech. et dév. expér. en sc. phys., natur. et ingén.			731
..	..	..	..	..	..	..	..	Rech. et dév. expér. en sciences soc. et humaines			732
..	..	..	..	..	..	179 000	178 000	Autres activités de services aux entreprises		**74**	
..	..	..	..	..	..	29 000	30 000	Act. jur., compt., audit.; cons. fisc.; ét. march., soud.			741
..	..	..	..	..	..	45 000	47 000	Act. d'achitect., d'ingén.; autres act. techniques			742
..	..	..	..	..	..	10 000	9 000	Publicité			743
..	..	..	..	..	..	96 000	92 000	Activités de services aux entreprises nca.			749
..	..	..	..	..	..	12 000	13 000	Education	**M**	**80**	
..	..	..	..	..	..	1 000	1 000	Activités d'enseignement primaire			801
..	..	..	..	..	..	6 000	7 000	Activités d'enseignement secondaire			802
..	..	..	..	..	..		1 000	Activités d'enseignement supérieur			803
..	..	..	..	..	..	5 000	5 000	Act. de formation perman.; autres act. d'enseign.			809
..	..	..	..	..	..	58 000	56 000	Santé et action sociale	**N**	**85**	
..	..	..	..	..	..	57 000	55 000	Activités de santé humaine			851
..	..	..	..	..	..	..	1 000	Activités vétérinaires			852
..	..	..	..	..	..	..	1 000	Activités d'action sociale			853
..	..	..	..	..	..	48 000	52 000	Autres activités de serv. coll., soc. et personnels	**O**	**90_93**	
..	..	..	..	..	..	16 000	20 000	Assain. et enlèv. des ordures; voierie, act. simil.		**90**	
..	..	..	..	..	..	..	..	Activités associatives diverses		**91**	
..	..	..	..	..	..	..	..	Activités de syndicats de salariés			912
..	..	..	..	..	..	17 000	16 000	Activités récréatives, culturelles et sportives		**92**	
..	..	..	..	..	..	6 000	5 000	Act. cinémat., de radio, télé.; autres act. de spect.			921
..	..	..	..	..	..	1 000	1 000	Activités d'agences de presse			922
..	..	..	..	..	..	..	..	Act. des biblio., arch., musées, jardins bot. et zool.			923
..	..	..	..	..	..	10 000	10 000	Activités sportives et autres activités récréatives			924
..	..	..	..	..	..	15 000	16 000	Autres activités de services		**93**	

Table CZ.S5 **WAGES AND SALARIES, EMPLOYEES - SALAIRES ET TRAITEMENTS, SALARIÉS**

Millions of CK (Current Prices)

ISIC revision 3				1991	1992	1993	1994	1995	1996	1997	1998
G	50_52		Wholesale & retail trade; repairs; etc.	..	..	..	..	..	..	62 204	65 276
	50		Sale, maint.,repair vehic/cycle, person., h/hold gd.	..	..	..	..	..	..	6 763	7 708
		501	Sale of motor cycle	..	..	..	..	..	..	2 293	2 565
		502	Maintenance and repair of motor vehicles	..	..	..	..	..	..	2 583	2 771
		503	Sale of motor vehicle parts and accessories	..	..	..	..	..	..	740	1 007
		504	Sale, maint., repair motor cycle, parts & acces.	..	..	..	..	..	..	44	25
		505	Retail sale of automative fuel	..	..	..	..	..	..	1 103	1 340
	51		W/sale. trade & commission exc motor veh. & cycle	..	..	..	..	..	..	31 169	32 711
		511	Wholesale on a fee or contract basis	..	..	..	..	..	..	2 796	3 149
		512	W/sale agric. raw mat.,live animal,food, bev., tobac.	..	..	..	..	..	..	3 683	5 003
		513	Wholesale of household goods	..	..	..	..	..	..	5 511	6 283
		514	W/sale. of non-agric. intermed. prod., waste, scrap	..	..	..	..	..	..	5 797	5 440
		515	Wholesale of machinery, equipment & supplies	..	..	..	..	..	..	5 352	3 232
		519	Other wholesale	..	..	..	..	..	..	8 030	9 604
	52		Retail trade exc. motor veh. & cyc.; repair h/hold gds.	..	..	..	..	..	..	24 272	24 857
		521	Non-specialised retail trade in stores	..	..	..	..	..	..	13 495	12 075
		522	Retail sale food, bev., tobac. in specialised stores	..	..	..	..	..	..	889	1 380
		523	Other retail trade of new goods specialised stores	..	..	..	..	..	..	8 248	9 784
		524	Retail sale of second-hand goods in stores	..	..	..	..	..	..	110	159
		525	Retail trade not in stores	..	..	..	..	..	..	1 033	1 001
		526	Repair of personal and household goods	..	..	..	..	..	..	497	458
H	55		Hotels and restaurants	..	..	..	..	..	..	8 423	8 985
		551	Hotels; camping ; other short-stay accommodation	..	..	..	..	..	..	3 430	3 290
		552	Restaurants, bars and canteens	..	..	..	..	..	..	4 993	5 695
I	60_64		Transport, storage and communications	..	..	..	..	..	..	41 721	44 366
	60		Land transport, transport via pipelines	..	..	..	..	..	..	25 900	27 519
		601	Transport via railways	..	..	..	..	..	..	13 238	14 097
		602	Other land transport	..	..	..	..	..	..	12 237	12 973
		603	Transport via pipelines	..	..	..	..	..	..	425	449
	61		Water transport	..	..	..	..	..	..	286	241
		611	Sea and coastal water transport	..	..	..	..	..	..	35	..
		612	Inland water transport	..	..	..	..	..	..	251	..
	62		Air transport	..	..	..	..	..	..	889	1 004
		621	Scheduled air transport	..	..	..	..	..	..	839	946
		622	Non-scheduled air transport	..	..	..	..	..	..	50	58
	63		Support. & aux. transport act., travel agencies activ.	..	..	..	..	..	..	3 405	3 981
	64		Post and telecommunications	..	..	..	..	..	..	11 242	11 621
		641	Post and courier activities	..	..	..	..	..	..	4 919	5 311
		642	Telecommunications	..	..	..	..	..	..	6 323	6 310
J	65_67		Financial intermediation	..	..	..	..	13 686	16 515	18 678	20 622
	65		Financial intermediation, exc. insur., pension fund	..	..	..	..	11 023	13 144	14 517	15 799
		651	Monetary intermediation	..	..	..	..	10 231	12 151	13 070	14 184
		659	Other financial intermediation	..	..	..	..	792	993	1 447	1 615
	66		Insurance pension fund, exc. compul. social sec.	..	..	..	..	1 987	2 678	3 291	3 846
	67		Activities auxiliary to financial intermediation	..	..	..	..	676	693	870	977
		671	Act. auxil to fin. intermed., exc. insur. & pens.fund	..	..	..	..	..	..	670	723
		672	Act. auxiliary to insurance and pension funding	..	..	..	..	..	..	200	254
K	70_74		Real estate, renting and business activities	..	..	..	..	..	..	35 796	38 944
	70		Real estate activities	..	..	..	..	..	..	4 512	5 179
		701	Real estate activ. with own or leased property	..	..	..	..	..	..	2 934	3 375
		702	Real estate activities on a fee or contract basis	..	..	..	..	..	..	1 578	1 803
	71		Renting of mach. without operator; rent. of pers. gds.	..	..	..	..	..	..	674	652
		711	Renting of transport equipment	..	..	..	..	..	..	195	154
		712	Renting of other machinery and equipment	..	..	..	..	..	..	349	352
		713	Renting of personal and household goods nec	..	..	..	..	..	..	130	146
	72		Computer and related activities	..	..	..	..	..	..	5 509	6 119
		721	Hardware consultancy	..	..	..	..	..	..	255	437
		722	Software consultancy and supply	..	..	..	..	..	..	3 478	4 050
		723	Data processing	..	..	..	..	..	..	935	995
		724	Data base activities	..	..	..	..	..	..	22	18
		725	Mainten.; repair of office, account. & comp. mach.	..	..	..	..	..	..	462	518
		729	Other computer related activities	..	..	..	..	..	..	358	101
	73		Research and development	..	..	..	..	..	..	1 195	1 254
		731	R & D experim. on natural sciences & engineering	..	..	..	..	..	..	1 194	1 254
		732	R & D experim. on social sciences & humanities	..	..	..	..	..	..	1	..
	74		Other business activities	..	..	..	..	..	..	23 906	25 740
		741	Legal, acc., book-keep., audit act.; tax consult., etc	..	..	..	..	..	..	4 823	6 214
		742	Architectural, engineering, & other techn. activities	..	..	..	..	..	..	8 043	8 765
		743	Advertising	..	..	..	..	..	..	1 645	1 935
		749	Business activities nec	..	..	..	..	..	..	9 394	8 827
M	80		Education	..	..	..	..	..	..	1 249	1 596
		801	Primary education	..	..	..	..	..	..	56	109
		802	Secondary education	..	..	..	..	..	..	712	871
		803	Higher education	..	..	..	..	..	..	26	73
		809	Adult and other education	..	..	..	..	..	..	455	544
N	85		Health and social work	..	..	..	..	..	..	5 419	5 691
		851	Human health activities	..	..	..	..	..	..	5 361	5 583
		852	Veterinary activities	..	..	..	..	..	..	43	64
		853	Social work activities	..	..	..	..	..	..	15	44
O	90_93		Other commun., soc. & pers. services activities	..	..	..	..	..	..	5 397	6 114
	90		Sewage & refuse disposal, sanitation, etc.	..	..	..	..	..	..	1 892	2 495
	91		Activities of member organisations nec	..	..	..	..	..	..	..	..
		912	Activities of trade unions	..	..	..	..	..	..	..	..
	92		Recreational, cultural and sporting activities	..	..	..	..	..	..	2 427	2 485
		921	Motion picture, radio, TV, other entertainment act.	..	..	..	..	..	..	869	990
		922	News agency activities	..	..	..	..	..	..	156	117
		923	Library, archives, museums, other cultural activ.	..	..	..	..	..	..	31	14
		924	Sporting and other recreational activities	..	..	..	..	..	..	1 371	1 365
	93		Other service activities	..	..	..	..	..	..	1 078	1 134

Table CZ.S6 ENTREPRISES - ENTERPRISES

Unités

1991	1992	1993	1994	1995	1996	1997	1998		CITI révision 3		
..	..	..	..	..	..	204 962	223 404	Com. de gros, de detail ; réparation etc.	G	**50_52**	
..	..	..	..	..	..	20 085	24 559	Com., entr., répar. d'auto. et moto ; carburants		**50**	
..	..	..	..	..	..	1 098	3 306	Commerce de véhicules automobiles			501
..	..	..	..	..	..	17 168	16 453	Entretien et réparation de véhicules automobiles			502
..	..	..	..	..	..	789	2 786	Com. de pièces et access. de véhic. automobiles			503
..	..	..	..	..	..	69	448	Com., entret. et répar. de motocycles et access.			504
..	..	..	..	..	..	961	1 566	Com. de détail de carburants pour automobiles			505
..	..	..	..	..	..	61 484	66 556	Com. de gros et act. inter., sauf auto. et moto		**51**	
..	..	..	..	..	..	33 351	33 698	Activités d'intermédiaries de commerce de gros			511
..	..	..	..	..	..	1 753	6 608	Com. gros: prod. agr. bruts, anim. viv., prod. alim.			512
..	..	..	..	..	..	3 591	4 855	Com. gros: biens de consom., sauf alimentaires			513
..	..	..	..	..	..	2 887	3 725	Com. gros: prod. inter. non agr., déchets, débris			514
..	..	..	..	..	..	2 894	1 685	Com. gros: machines, équipments et fournitures			515
..	..	..	..	..	..	17 008	15 985	Autres commerces de gros			519
..	..	..	..	..	..	123 393	132 289	Com. détail, sauf auto, moto; répar. art. pers., dom.		**52**	
..	..	..	..	..	..	85 126	58 128	Commerce de détail non spécialisé en magasin			521
..	..	..	..	..	..	2 508	10 241	Com. détail: prod. aliment. en mag. spécialisés			522
..	..	..	..	..	..	22 124	43 520	Autre com. détail: prod. neufs en mag. spécialisés			523
..	..	..	..	..	..	777	3 499	Commerces de détail de biens d'occasion			524
..	..	..	..	..	..	9 904	12 756	Commerces de détail autres qu'en magasin			525
..	..	..	..	..	..	2 954	4 145	Réparation d'articles personnels et domestiques			526
..	..	..	..	..	..	33 509	40 887	Hotels et restaurants	H	**55**	
..	..	..	..	..	..	7 613	7 900	Hôtels; camping; autres équip. courts séjours			551
..	..	..	..	..	..	25 896	32 987	Restaurants, cafés et cantines			552
..	..	..	..	..	..	34 340	39 137	Transport, entreposage, communications	I	**60_64**	
..	..	..	..	..	..	28 067	30 672	Transports terrestres; transports par conduites		**60**	
..	..	..	..	..	..	53	54	Transports par chemin de fer			601
..	..	..	..	..	..	28 002	30 607	Autres transports terrestres			602
..	..	..	..	..	..	12	11	Transports par conduites			603
..	..	..	..	..	..	85	101	Transports par eau		**61**	
..	..	..	..	..	..	4	..	Transports maritimes et côtiers			611
..	..	..	..	..	..	81	..	Transports par voies navigables intérieures			612
..	..	..	..	..	..	66	35	Transports aériens		**62**	
..	..	..	..	..	..	25	19	Transports aériens réguliers			621
..	..	..	..	..	..	41	16	Transports aériens non réguliers			622
..	..	..	..	..	..	5 740	7 915	Act. annex. et aux. des transp.; act. des ag. de voy.		**63**	
..	..	..	..	..	..	382	414	Postes et télécommunications		**64**	
..	..	..	..	..	..	173	178	Activités de poste et de courrier			641
..	..	..	..	..	..	209	236	Télécommunications			642
..	..	..	..	4 966	4 705	6 239	9 174	Intermédiation financière	J	**65_67**	
..	..	..	..	673	775	695	691	Interm. financ., sauf assur. et caisses de retraite		**65**	
..	..	..	..	58	54	55	52	Intermédiation monétaire			651
..	..	..	..	615	721	640	639	Autres intermédiations financières			659
..	..	..	..	78	77	74	73	Act. d'assur., caisses de retr., sauf séc. soc. oblig.		**66**	
..	..	..	..	4 215	3 853	5 470	8 410	Act. auxiliaires de l'intermédiation financière		**67**	
..	..	..	..	..	..	470	490	Act. aux. de l'interm. fin., sauf assur., et retraite			671
..	..	..	..	..	..	5 000	7 920	Act. aux. des assurances et caisses de retraite			672
..	..	..	..	..	..	162 510	180 408	Immobilier, locations, services aux entreprises	K	**70_74**	
..	..	..	..	..	..	16 099	15 712	Activités immobilières		**70**	
..	..	..	..	..	..	10 130	9 605	Act. immobilières sur biens propres ou loués			701
..	..	..	..	..	..	5 969	6 107	Activités immobilières pour comptes de tiers			702
..	..	..	..	..	..	3 321	2 817	Locat. mach., équip. sans opér., biens pers. et dom.		**71**	
..	..	..	..	..	..	1 367	1 262	Location de matériel de transport			711
..	..	..	..	..	..	643	767	Location de machines ét equipements nca			712
..	..	..	..	..	..	1 311	788	Loc. d'articles personnels et domestiques nca			713
..	..	..	..	..	..	18 999	16 694	Activités informatiques et activités rattachées		**72**	
..	..	..	..	..	..	1 013	1 290	Conseil en matériel informatique			721
..	..	..	..	..	..	10 920	7 957	Production de logiciels			722
..	..	..	..	..	..	4 502	3 594	Traitement de données informatiques			723
..	..	..	..	..	..	48	82	Activités de banques de données			724
..	..	..	..	..	..	1 971	3 416	Entr., répar. mach. de bureau et compt. et inform.			725
..	..	..	..	..	..	545	355	Autres activités rattachées à l'informatique			729
..	..	..	..	..	..	207	267	Recherche-développement		**73**	
..	..	..	..	..	..	200	250	Rech. et dév. expér. en sc. phys., natur. et ingén.			731
..	..	..	..	..	..	7	17	Rech. et dév. expér. en sciences soc. et humaines			732
..	..	..	..	..	..	123 884	144 918	Autres activités de services aux entreprises		**74**	
..	..	..	..	..	..	44 340	57 687	Act. jur., compt., audit.; cons. fisc.; ét. march., soud.			741
..	..	..	..	..	..	22 901	34 627	Act. d'achitect., d'ingén.; autres act. techniques			742
..	..	..	..	..	..	7 882	7 226	Publicité			743
..	..	..	..	..	..	48 761	45 378	Activités de services aux entreprises nca.			749
..	..	..	..	..	..	7 426	8 281	Education	M	**80**	
..	..	..	..	..	..	157	184	Activités d'enseignement primaire			801
..	..	..	..	..	..	394	447	Activités d'enseignement secondaire			802
..	..	..	..	..	..	18	22	Activités d'enseignement supérieur			803
..	..	..	..	..	..	6 857	7 628	Act. de formation perman.; autres act. d'enseign.			809
..	..	..	..	..	..	21 866	21 374	Santé et action sociale	N	**85**	
..	..	..	..	..	..	19 886	18 978	Activités de santé humaine			851
..	..	..	..	..	..	1 516	1 796	Activités vétérinaires			852
..	..	..	..	..	..	464	600	Activités d'action sociale			853
..	..	..	..	..	..	35 537	38 408	Autres activités de serv. coll., soc. et personnels	O	**90_93**	
..	..	..	..	..	..	1 215	1 966	Assain. et enlèv. des ordures; voirie, act. simil.		**90**	
..	..	..	..	..	..	..	..	Activités associatives diverses		**91**	
..	..	..	..	..	..	..	..	Activités de syndicats de salariés			912
..	..	..	..	..	..	14 315	10 522	Activités récréatives, culturelles et sportives		**92**	
..	..	..	..	..	..	8 523	5 599	Act. cinémat., de radio, télé.; autres act. de spect.			921
..	..	..	..	..	..	234	135	Activités d'agences de presse			922
..	..	..	..	..	..	300	194	Act. des biblio., arch., musées, jardins bot. et zool.			923
..	..	..	..	..	..	5 258	4 594	Activités sportives et autres activités récréatives			924
..	..	..	..	..	..	20 007	25 920	Autres activités de services		**93**	

DENMARK

Table DN.S1 EMPLOYMENT, EMPLOYEES - EMPLOI, SALARIÉS

Units

ISIC revision 3				1991	1992	1993	1994	1995	1996	1997	1998
G	50_52		Wholesale & retail trade; repairs; etc.	..	..	344 521	356 982	373 750	383 754	393 253	403 499
	50		Sale, maint.,repair vehic/cycle, person., h/hold gd.	..	..	..	..	..	58 609	60 141	60 878
		501	Sale of motor cycle	..	..	..	..	..	26 597	27 211	27 374
		502	Maintenance and repair of motor vehicles	..	..	..	..	..	13 790	14 057	14 050
		503	Sale of motor vehicle parts and accessories	..	..	..	..	..	5 061	5 233	5 554
		504	Sale, maint., repair motor cycle, parts & acces.	..	..	..	..	..	287	326	359
		505	Retail sale of automative fuel	..	..	..	..	..	12 874	13 314	13 541
	51		W/sale. trade & commission exc motor veh. & cycle	..	..	..	..	..	156 812	162 317	167 171
		511	Wholesale on a fee or contract basis	..	..	..	..	..	4 599	4 996	5 287
		512	W/sale agric. raw mat.,live animal,food, bev., tobac.	..	..	..	..	..	..	..	31 499
		513	Wholesale of household goods	..	..	..	..	..	32 011	32 389	33 120
		514	W/sale. of non-agric. intermed. prod., waste, scrap	..	..	..	..	..	31 795	33 227	33 876
		515	Wholesale of machinery, equipment & supplies	..	..	..	..	..	51 478	53 717	56 485
		519	Other wholesale	..	..	..	..	..	6 799	7 133	6 904
	52		Retail trade exc. motor veh. & cyc.; repair h/hold gds.	..	..	..	..	..	168 333	170 795	175 450
		521	Non-specialised retail trade in stores	..	..	..	..	..	76 023	76 840	79 694
		522	Retail sale food, bev., tobac. in specialised stores	..	..	..	..	..	12 594	12 278	11 915
		523	Other retail trade of new goods specialised stores	..	..	..	..	..	..	..	79 893
		524	Retail sale of second-hand goods in stores	..	..	..	..	..	386	446	484
		525	Retail trade not in stores	..	..	..	..	..	1 168	1 174	1 231
		526	Repair of personal and household goods	..	..	..	..	..	2 213	2 255	2 233
H	55		Hotels and restaurants	..	..	66 950	71 442	75 871	78 119	81 577	82 931
		551	Hotels; camping ; other short-stay accommodation	..	..	..	..	..	..	..	22 947
		552	Restaurants, bars and canteens	..	..	..	..	..	..	..	59 984
I	60_64		Transport, storage and communications	..	..	..	..	..	180 123	180 444	184 606
	60		Land transport, transport via pipelines	..	..	171 797	173 069	173 867	68 373	68 055	69 142
		601	Transport via railways	..	..	..	..	..	11 789	10 649	10 338
		602	Other land transport	..	..	..	..	..	56 584	57 406	58 804
		603	Transport via pipelines	..	..	..	..	..	..	..	..
	61		Water transport	..	..	..	..	..	16 883	16 098	15 627
		611	Sea and coastal water transport	..	..	..	..	..	16 798	16 006	15 507
		612	Inland water transport	..	..	..	..	..	85	92	120
	62		Air transport	..	..	..	..	..	10 592	10 950	11 274
		621	Scheduled air transport	..	..	..	..	..	7 984	8 075	8 467
		622	Non-scheduled air transport	..	..	..	..	..	2 608	2 875	2 807
	63		Support. & aux. transport act., travel agencies activ.	..	..	..	..	..	29 549	30 314	31 490
	64		Post and telecommunications	..	..	..	..	..	54 726	55 027	57 073
		641	Post and courier activities	..	..	..	..	..	37 057	36 407	37 544
		642	Telecommunications	..	..	..	..	..	17 669	18 620	19 529
J	65_67		Financial intermediation	..	..	84 843	82 631	82 317	80 024	78 687	80 465
	65		Financial intermediation, exc. insur., pension fund	..	..	..	..	..	57 139	56 593	57 735
		651	Monetary intermediation	..	..	..	..	..	47 671	46 664	46 998
		659	Other financial intermediation	..	..	..	..	..	..	..	10 737
	66		Insurance pension fund, exc. compul. social sec.	..	..	..	..	..	18 606	18 472	18 956
	67		Activities auxiliary to financial intermediation	..	..	..	..	..	4 279	3 622	3 774
		671	Act. auxil to fin. intermed., exc. insur. & pens.fund	..	..	..	..	..	2 781	2 033	2 110
		672	Act. auxiliary to insurance and pension funding	..	..	..	..	..	1 498	1 589	1 664
K	70_74		Real estate, renting and business activities	..	..	189 439	207 304	212 999	223 127	236 497	255 940
	70		Real estate activities	..	..	..	..	..	33 286	34 234	34 669
		701	Real estate activ. with own or leased property	..	..	..	..	..	..	..	22 604
		702	Real estate activities on a fee or contract basis	..	..	..	..	..	11 404	11 812	12 065
	71		Renting of mach. without operator; rent. of pers. gds.	..	..	..	..	..	6 426	5 910	6 725
		711	Renting of transport equipment	..	..	..	..	..	..	..	1 430
		712	Renting of other machinery and equipment	..	..	..	..	..	2 369	2 274	2 742
		713	Renting of personal and household goods nec	..	..	..	..	..	2 663	2 227	2 553
	72		Computer and related activities	..	..	..	..	..	21 736	25 591	28 838
		721	Hardware consultancy	..	..	..	..	..	636	919	1 003
		722	Software consultancy and supply	..	..	..	..	..	12 351	14 600	19 772
		723	Data processing	..	..	..	..	..	6 638	7 778	5 700
		724	Data base activities	..	..	..	..	..	409	480	550
		725	Mainten.; repair of office, account. & comp. mach.	..	..	..	..	..	1 043	1 016	719
		729	Other computer related activities	..	..	..	..	..	..	..	1 094
	73		Research and development	..	..	..	..	..	9 137	9 510	10 263
		731	R & D experim. on natural sciences & engineering	..	..	..	..	..	7 853	8 154	8 823
		732	R & D experim. on social sciences & humanities	..	..	..	..	..	1 284	1 356	1 440
	74		Other business activities	..	..	..	..	..	152 542	161 252	175 445
		741	Legal, acc., book-keep., audit act.; tax consult., etc	..	..	..	..	..	38 383	39 540	41 227
		742	Architectural, engineering, & other techn. activities	..	..	..	..	..	..	..	37 007
		743	Advertising	..	..	..	..	..	..	..	14 226
		749	Business activities nec	..	..	..	..	..	..	..	..
M	80		Education	..	..	..	..	..	226 269	230 841	233 926
		801	Primary education	..	..	..	..	..	102 629	105 698	107 367
		802	Secondary education	..	..	..	..	..	44 364	45 831	46 377
		803	Higher education	..	..	..	..	..	35 200	36 721	37 017
		809	Adult and other education	..	..	..	..	..	..	..	43 165
N	85		Health and social work	..	..	..	..	..	444 660	469 078	483 608
		851	Human health activities	..	..	..	..	..	144 409	144 430	146 654
		852	Veterinary activities	..	..	..	..	..	1 453	1 463	1 507
		853	Social work activities	..	..	..	..	..	298 798	323 185	335 447
O	90_93		Other commun., soc. & pers. services activities	..	..	..	..	..	136 976	152 391	153 202
	90		Sewage & refuse disposal, sanitation, etc.	..	..	..	..	..	25 184	25 539	24 860
	91		Activities of member organisations nec	..	..	..	..	..	56 621	57 161	57 283
		912	Activities of trade unions	..	..	..	..	..	19 669	18 780	18 578
	92		Recreational, cultural and sporting activities	..	..	..	..	..	53 728	54 859	56 484
		921	Motion picture, radio, TV, other entertainment act.	..	..	..	..	..	..	..	20 399
		922	News agency activities	..	..	..	..	..	505	538	580
		923	Library, archives, museums, other cultural activ.	..	..	..	..	..	15 760	15 852	15 729
		924	Sporting and other recreational activities	..	..	..	..	..	..	..	19 776
	93		Other service activities	..	..	..	..	..	1 443	14 832	14 575

Table DN.S2 **EMPLOI, FEMMES - EMPLOYMENT, FEMALES**

Unités

1991	1992	1993	1994	1995	1996	1997	1998		CITI révision 3		
..	..	161 720	163 018	168 507	..	..	178 990	Com. de gros, de detail ; réparation etc.	G	**50_52**	
..	..	..	..	..	..	..	14 497	Com., entr., répar. d'auto. et moto ; carburants		**50**	
..	..	..	..	..	..	..	3 993	Commerce de véhicules automobiles			501
..	..	..	..	..	..	..	2 667	Entretien et réparation de véhicules automobiles			502
..	..	..	..	..	..	..	1 085	Com. de pièces et access. de véhic. automobiles			503
..	..	..	..	..	..	..	79	Com., entret. et répar. de motocycles et access.			504
..	..	..	..	..	..	..	6 673	Com. de détail de carburants pour automobiles			505
..	..	..	..	..	..	..	53 956	Com. de gros et act. inter., sauf auto. et moto		**51**	
..	..	..	..	..	..	..	2 353	Activités d'intermédiaries de commerce de gros			511
..	..	..	..	..	..	..	9 435	Com. gros: prod. agr. bruts, anim. viv., prod. alim.			512
..	..	..	..	..	..	..	16 389	Com. gros: biens de consom., sauf alimentaires			513
..	..	..	..	..	..	..	9 303	Com. gros: prod. inter. non agr., déchets, débris			514
..	..	..	..	..	..	..	14 217	Com. gros: machines, équipments et fournitures			515
..	..	..	..	..	..	..	2 259	Autres commerces de gros			519
..	..	..	..	..	..	..	110 537	Com. détail, sauf auto, moto; répar. art. pers., dom.		**52**	
..	..	..	..	..	..	..	46 550	Commerce de détail non spécialisé en magasin			521
..	..	..	..	..	..	..	9 001	Com. détail: prod. aliment. en mag. spécialisés			522
..	..	..	..	..	..	..	52 754	Autre com. détail: prod. neufs en mag. spécialisés			523
..	..	..	..	..	..	..	532	Commerces de détail de biens d'occasion			524
..	..	..	..	..	..	..	1 082	Commerces de détail autres qu'en magasin			525
..	..	..	..	..	..	..	618	Réparation d'articles personnels et domestiques			526
..	..	46 028	47 999	49 955	50 824	52 379	52 865	Hotels et restaurants	H	**55**	
..	..	..	..	..	..	..	14 322	Hôtels; camping; autres équip. courts séjours			551
..	..	..	..	..	..	..	38 543	Restaurants, cafés et cantines			552
..	..	..	..	..	..	..	52 419	Transport, entreposage, communications	I	**60_64**	
..	..	49 066	47 528	47 885	11 054	10 771	11 058	Transports terrestres; transports par conduites		**60**	
..	..	..	..	..	..	..	2 397	Transports par chemin de fer			601
..	..	..	..	..	..	..	8 661	Autres transports terrestres			602
..	..	..	..	..	..	..	..	Transports par conduites			603
..	..	..	..	..	..	..	3 248	Transports par eau		**61**	
..	..	..	..	..	..	..	3 207	Transports maritimes et côtiers			611
..	..	..	..	..	..	..	41	Transports par voies navigables intérieures			612
..	..	..	..	..	..	..	4 137	Transports aériens		**62**	
..	..	..	..	..	..	..	3 116	Transports aériens réguliers			621
..	..	..	..	..	..	..	1 021	Transports aériens non réguliers			622
..	..	..	..	..	..	..	11 151	Act. annex. et aux. des transp.; act. des ag. de voy.		**63**	
..	..	..	..	..	..	..	22 825	Postes et télécommunications		**64**	
..	..	..	..	..	..	..	14 885	Activités de poste et de courrier			641
..	..	..	..	..	..	..	7 940	Télécommunications			642
..	..	44 927	42 349	42 036	40 698	..	40 838	Intermédiation financière	J	**65_67**	
..	..	..	..	..	..	..	30 221	Interm. financ., sauf assur. et caisses de retraite		**65**	
..	..	..	..	..	..	..	25 353	Intermédiation monétaire			651
..	..	..	..	..	..	..	4 868	Autres intermédiations financières			659
..	..	..	..	..	..	..	8 968	Act. d'assur., caisses de retr., sauf séc. soc. oblig.		**66**	
..	..	..	..	..	..	..	1 649	Act. auxiliaires de l'intermédiation financière		**67**	
..	..	..	..	..	..	..	849	Act. aux. de l'interm. fin., sauf assur., et retraite			671
..	..	..	..	..	..	..	800	Act. aux. des assurances et caisses de retraite			672
..	..	98 864	104 765	109 280	..	..	126 686	Immobilier, locations, services aux entreprises	K	**70_74**	
..	..	..	..	..	..	..	13 956	Activités immobilières		**70**	
..	..	..	..	..	..	..	8 172	Act. immobilières sur biens propres ou loués			701
..	..	..	..	..	..	..	5 784	Activités immobilières pour comptes de tiers			702
..	..	..	..	..	..	..	2 365	Locat. mach., équip. sans opér., biens pers. et dom.		**71**	
..	..	..	..	..	..	..	532	Location de matériel de transport			711
..	..	..	..	..	..	..	598	Location de machines ét equipements nca			712
..	..	..	..	..	..	..	1 235	Loc. d'articles personnels et domestiques nca			713
..	..	..	..	..	..	..	8 217	Activités informatiques et activités rattachées		**72**	
..	..	..	..	..	..	..	200	Conseil en matériel informatique			721
..	..	..	..	..	..	..	5 587	Production de logiciels			722
..	..	..	..	..	..	..	1 737	Traitement de données informatiques			723
..	..	..	..	..	..	..	251	Activités de banques de données			724
..	..	..	..	..	..	..	109	Entr., répar. mach. de bureau et compt. et inform.			725
..	..	..	..	..	..	..	333	Autres activités rattachées à l'informatique			729
..	..	..	..	..	..	..	4 856	Recherche-développement		**73**	
..	..	..	..	..	..	..	4 107	Rech. et dév. expér. en sc. phys., natur. et ingén.			731
..	..	..	..	..	..	..	749	Rech. et dév. expér. en sciences soc. et humaines			732
..	..	..	..	..	..	..	97 292	Autres activités de services aux entreprises		**74**	
..	..	..	..	..	..	..	23 484	Act. jur., compt., audit.; cons. fisc.; ét. march., soud.			741
..	..	..	..	..	..	..	13 512	Act. d'achitect., d'ingén.; autres act. techniques			742
..	..	..	..	..	..	..	6 893	Publicité			743
..	..	..	..	..	..	..	..	Activités de services aux entreprises nca.			749
..	..	..	..	..	..	..	134 337	Education	M	**80**	
..	..	..	..	..	..	..	68 476	Activités d'enseignement primaire			801
..	..	..	..	..	..	..	24 366	Activités d'enseignement secondaire			802
..	..	..	..	..	..	..	16 090	Activités d'enseignement supérieur			803
..	..	..	..	..	..	..	25 405	Act. de formation perman.; autres act. d'enseign.			809
..	..	..	..	..	..	..	405 583	Santé et action sociale	N	**85**	
..	..	..	..	..	..	..	122 533	Activités de santé humaine			851
..	..	..	..	..	..	..	1 367	Activités vétérinaires			852
..	..	..	..	..	..	..	281 683	Activités d'action sociale			853
..	..	..	..	..	..	..	77 881	Autres activités de serv. coll., soc. et personnels	O	**90_93**	
..	..	..	..	..	..	..	3 077	Assain. et enlèv. des ordures; voierie, act. simil.		**90**	
..	..	..	..	..	..	..	28 721	Activités associatives diverses		**91**	
..	..	..	..	..	..	..	9 687	Activités de syndicats de salariés			912
..	..	..	..	..	..	..	28 966	Activités récréatives, culturelles et sportives		**92**	
..	..	..	..	..	..	..	9 109	Act. cinémat., de radio, télé.; autres act. de spect.			921
..	..	..	..	..	..	..	297	Activités d'agences de presse			922
..	..	..	..	..	..	..	10 444	Act. des biblio., arch., musées, jardins bot. et zool.			923
..	..	..	..	..	..	..	9 116	Activités sportives et autres activités récréatives			924
..	..	..	..	..	..	..	17 117	Autres activités de services		**93**	

DENMARK

Table DN.S3 WAGES AND SALARIES, EMPLOYEES - SALAIRES ET TRAITEMENTS, SALARIÉS

Millions of DKr (Current Prices)

ISIC revision 3				1991	1992	1993	1994	1995	1996	1997	1998
G	50_52		Wholesale & retail trade; repairs; etc.	..	..	57 254	60 165	64 279	68 033	72 681	77 672
	50		Sale, maint.,repair vehic/cycle, person., h/hold gd.	..	..	..	..	..	9 345	10 045	10 633
		501	Sale of motor cycle	..	..	..	..	..	5 180	5 564	5 839
		502	Maintenance and repair of motor vehicles	..	..	..	..	..	2 172	2 341	2 449
		503	Sale of motor vehicle parts and accessories	..	..	..	..	..	1 016	1 098	1 218
		504	Sale, maint., repair motor cycle, parts & acces.	..	..	..	..	..	57	65	76
		505	Retail sale of automative fuel	..	..	..	..	..	920	977	1 051
	51		W/sale. trade & commission exc motor veh. & cycle	..	..	..	..	..	37 392	40 131	43 255
		511	Wholesale on a fee or contract basis	..	..	..	..	..	1 060	1 195	1 323
		512	W/sale agric. raw mat.,live animal,food, bev., tobac.	..	..	..	..	..	..	..	7 276
		513	Wholesale of household goods	..	..	..	..	..	7 385	7 793	8 272
		514	W/sale. of non-agric. intermed. prod., waste, scrap	..	..	..	..	..	7 427	7 959	8 415
		515	Wholesale of machinery, equipment & supplies	..	..	..	..	..	13 713	14 831	16 272
		519	Other wholesale	..	..	..	..	..	1 522	1 645	1 697
	52		Retail trade exc. motor veh. & cyc.; repair h/hold gds.	..	..	..	..	..	21 296	22 505	23 784
		521	Non-specialised retail trade in stores	..	..	..	..	..	8 574	8 989	9 493
		522	Retail sale food, bev., tobac. in specialised stores	..	..	..	..	..	1 213	1 249	1 231
		523	Other retail trade of new goods specialised stores	..	..	..	..	..	..	..	12 361
		524	Retail sale of second-hand goods in stores	..	..	..	..	..	10 907	11 599	66
		525	Retail trade not in stores	..	..	..	..	..	172	207	218
		526	Repair of personal and household goods	..	..	..	..	..	381	400	415
H	55		Hotels and restaurants	..	..	6 677	7 020	7 416	7 838	8 183	8 532
		551	Hotels; camping ; other short-stay accommodation	..	..	..	..	..	..	..	2 640
		552	Restaurants, bars and canteens	..	..	..	..	..	..	..	5 892
I	60_64		Transport, storage and communications	..	..	..	..	..	..	..	41 664
	60		Land transport, transport via pipelines	..	..	33 825	35 123	36 025	13 866	14 074	14 464
		601	Transport via railways	..	..	..	..	..	2 817	2 629	2 626
		602	Other land transport	..	..	..	..	..	11 048	11 445	11 839
		603	Transport via pipelines	..	..	..	..	..	..	..	..
	61		Water transport	..	..	..	..	..	4 200	4 167	4 089
		611	Sea and coastal water transport	..	..	..	..	..	4 179	4 143	4 061
		612	Inland water transport	..	..	..	..	..	21	24	28
	62		Air transport	..	..	..	..	..	3 436	3 684	3 946
		621	Scheduled air transport	..	..	..	..	..	2 614	2 732	2 903
		622	Non-scheduled air transport	..	..	..	..	..	822	953	1 043
	63		Support. & aux. transport act., travel agencies activ.	..	..	..	..	..	6 454	6 841	7 240
	64		Post and telecommunications	..	..	..	..	..	10 394	11 244	11 925
		641	Post and courier activities	..	..	..	..	..	5 844	6 171	6 296
		642	Telecommunications	..	..	..	..	..	4 550	5 073	5 629
J	65_67		Financial intermediation	..	..	21 595	21 337	21 746	22 184	..	23 478
	65		Financial intermediation, exc. insur., pension fund	..	..	..	..	..	15 180	15 444	16 213
		651	Monetary intermediation	..	..	..	..	..	12 543	12 674	13 142
		659	Other financial intermediation	..	..	..	..	..	..	..	3 070
	66		Insurance pension fund, exc. compul. social sec.	..	..	..	..	..	5 678	5 813	6 032
	67		Activities auxiliary to financial intermediation	..	..	..	..	..	1 326	1 177	1 233
		671	Act. auxil to fin. intermed., exc. insur. & pens.fund	..	..	..	..	..	947	759	781
		672	Act. auxiliary to insurance and pension funding	..	..	..	..	..	378	418	452
K	70_74		Real estate, renting and business activities	..	..	36 149	39 355	40 990	..	..	..
	70		Real estate activities	..	..	..	..	..	5 115	5 533	5 812
		701	Real estate activ. with own or leased property	..	..	..	..	..	..	..	3 746
		702	Real estate activities on a fee or contract basis	..	..	..	..	..	1 744	1 933	2 066
	71		Renting of mach. without operator; rent. of pers. gds.	..	..	..	..	..	1 021	1 024	1 154
		711	Renting of transport equipment	..	..	..	..	..	..	..	288
		712	Renting of other machinery and equipment	..	..	..	..	..	479	514	610
		713	Renting of personal and household goods nec	..	..	..	..	..	268	233	256
	72		Computer and related activities	..	..	..	..	..	6 678	8 246	9 775
		721	Hardware consultancy	..	..	..	..	..	204	306	363
		722	Software consultancy and supply	..	..	..	..	..	3 692	4 634	6 614
		723	Data processing	..	..	..	..	..	2 193	2 656	2 064
		724	Data base activities	..	..	..	..	..	87	101	128
		725	Mainten.; repair of office, account. & comp. mach.	..	..	..	..	..	286	297	221
		729	Other computer related activities	..	..	..	..	..	..	..	385
	73		Research and development	..	..	..	..	..	2 148	2 337	2 592
		731	R & D experim. on natural sciences & engineering	..	..	..	..	..	1 887	2 054	2 291
		732	R & D experim. on social sciences & humanities	..	..	..	..	..	262	283	301
	74		Other business activities	..	..	..	..	..	28 931	31 739	34 540
		741	Legal, acc., book-keep., audit act.; tax consult., etc	..	..	..	..	..	9 086	9 695	10 683
		742	Architectural, engineering, & other techn. activities	..	..	..	..	..	..	..	11 131
		743	Advertising	..	..	..	..	..	..	..	2 607
		749	Business activities nec	..	..	..	..	..	..	..	..
M	80		Education	..	..	..	..	..	41 448	43 476	45 522
		801	Primary education	..	..	..	..	..	20 314	21 233	22 166
		802	Secondary education	..	..	..	..	..	9 487	10 034	10 435
		803	Higher education	..	..	..	..	..	5 931	6 227	6 643
		809	Adult and other education	..	..	..	..	..	..	..	6 278
N	85		Health and social work	..	..	..	..	..	75 322	81 316	86 837
		851	Human health activities	..	..	..	..	..	28 967	30 307	32 003
		852	Veterinary activities	..	..	..	..	..	238	248	264
		853	Social work activities	..	..	..	..	..	46 117	50 761	54 570
O	90_93		Other commun., soc. & pers. services activities	..	..	..	..	..	..	..	26 646
	90		Sewage & refuse disposal, sanitation, etc.	..	..	..	..	..	4 966	5 198	5 245
	91		Activities of member organisations nec	..	..	..	..	..	9 046	9 261	9 771
		912	Activities of trade unions	..	..	..	..	..	3 717	3 685	3 884
	92		Recreational, cultural and sporting activities	..	..	..	..	..	8 522	8 982	9 515
		921	Motion picture, radio, TV, other entertainment act.	..	..	..	..	..	..	..	4 439
		922	News agency activities	..	..	..	..	..	126	137	145
		923	Library, archives, museums, other cultural activ.	..	..	..	..	..	2 480	2 563	2 641
		924	Sporting and other recreational activities	..	..	..	..	..	..	..	2 290
	93		Other service activities	..	..	..	..	..	265	2 115	2 115

Table DN.S4 ÉTABLISSEMENTS - ESTABLISHMENTS

Unités

1991	1992	1993	1994	1995	1996	1997	1998	CITI révision 3			
..	..	63 891	62 483	62 248	62 081	61 070	58 635	Com. de gros, de detail ; réparation etc.	G	50_52	
..	..	..	..	..	10 329	10 320	10 097	Com., entr., répar. d'auto. et moto ; carburants		50	
..	..	..	..	..	2 748	2 709	2 655	Commerce de véhicules automobiles			501
..	..	..	..	..	5 320	5 313	5 159	Entretien et réparation de véhicules automobiles			502
..	..	..	..	..	730	738	745	Com. de pièces et access. de véhic. automobiles			503
..	..	..	..	..	113	124	122	Com., entret. et répar. de motocycles et access.			504
..	..	..	..	..	1 418	1 436	1 416	Com. de détail de carburants pour automobiles			505
..	..	..	..	..	19 721	19 299	18 586	Com. de gros et act. inter., sauf auto. et moto		51	
..	..	..	..	..	1 538	1 523	1 473	Activités d'intermédiaries de commerce de gros			511
..	..	..	..	..	..	..	3 506	Com. gros: prod. agr. bruts, anim. viv., prod. alim.			512
..	..	..	..	..	4 904	4 754	4 542	Com. gros: biens de consom., sauf alimentaires			513
..	..	..	..	..	3 355	3 280	3 176	Com. gros: prod. inter. non agr., déchets, débris			514
..	..	..	..	..	4 994	4 926	4 847	Com. gros: machines, équipments et fournitures			515
..	..	..	..	..	1 161	1 141	1 042	Autres commerces de gros			519
..	..	..	..	..	32 031	31 451	29 952	Com. détail, sauf auto, moto; répar. art. pers., dom.		52	
..	..	..	..	..	5 193	5 004	4 717	Commerce de détail non spécialisé en magasin			521
..	..	..	..	..	4 251	3 950	3 619	Com. détail: prod. aliment. en mag. spécialisés			522
..	..	..	..	..	..	..	18 488	Autre com. détail: prod. neufs en mag. spécialisés			523
..	..	..	..	..	914	924	877	Commerces de détail de biens d'occasion			524
..	..	..	..	..	805	892	776	Commerces de détail autres qu'en magasin			525
..	..	..	..	..	1 553	1 522	1 475	Réparation d'articles personnels et domestiques			526
..	..	11 535	11 473	11 755	11 921	12 027	11 338	Hotels et restaurants	H	55	
..	..	..	..	..	..	..	1 648	Hôtels; camping; autres équip. courts séjours			551
..	..	..	..	..	..	..	9 690	Restaurants, cafés et cantines			552
..	..	..	..	..	15 742	..	15 106	Transport, entreposage, communications	I	60_64	
..	..	16 065	15 674	15 629	10 998	10 759	10 443	Transports terrestres; transports par conduites		60	
..	..	..	..	..	338	323	294	Transports par chemin de fer			601
..	..	..	..	..	10 660	10 436	10 149	Autres transports terrestres			602
..	..	..	..	..	..	..	1	Transports par conduites			603
..	..	..	..	..	600	576	589	Transports par eau		61	
..	..	..	..	..	572	545	558	Transports maritimes et côtiers			611
..	..	..	..	..	28	31	31	Transports par voies navigables intérieures			612
..	..	..	..	..	150	143	127	Transports aériens		62	
..	..	..	..	..	79	72	67	Transports aériens réguliers			621
..	..	..	..	..	71	..	60	Transports aériens non réguliers			622
..	..	..	..	..	2 295	2 302	2 340	Act. annex. et aux. des transp.; act. des ag. de voy.		63	
..	..	..	..	..	1 699	1 691	1 607	Postes et télécommunications		64	
..	..	..	..	..	1 434	1 360	1 280	Activités de poste et de courrier			641
..	..	..	..	..	265	331	327	Télécommunications			642
..	..	4 920	4 873	4 698	4 594	..	..	Intermédiation financière	J	65_67	
..	..	..	..	..	3 395	3 410	3 392	Interm. financ., sauf assur. et caisses de retraite		65	
..	..	..	..	..	2 245	2 231	2 198	Intermédiation monétaire			651
..	..	..	..	..	..	1 179	1 194	Autres intermédiations financières			659
..	..	..	..	..	721	700	631	Act. d'assur., caisses de retr., sauf séc. soc. oblig.		66	
..	..	..	..	..	478	446	433	Act. auxiliaires de l'intermédiation financière		67	
..	..	..	..	..	193	155	147	Act. aux. de l'interm. fin., sauf assur., et retraite			671
..	..	..	..	..	285	291	286	Act. aux. des assurances et caisses de retraite			672
..	..	44 352	44 859	45 690	46 517	..	45 370	Immobilier, locations, services aux entreprises	K	70_74	
..	..	..	..	..	14 316	14 353	14 170	Activités immobilières		70	
..	..	..	..	..	..	..	10 299	Act. immobilières sur biens propres ou loués			701
..	..	..	..	..	3 830	3 914	3 871	Activités immobilières pour comptes de tiers			702
..	..	..	..	..	1 691	1 629	1 611	Locat. mach., équip. sans opér., biens pers. et dom.		71	
..	..	..	..	..	..	..	311	Location de matériel de transport			711
..	..	..	..	..	764	781	782	Location de machines ét equipements nca			712
..	..	..	..	..	584	527	518	Loc. d'articles personnels et domestiques nca			713
..	..	..	..	..	3 505	3 848	3 753	Activités informatiques et activités rattachées		72	
..	..	..	..	..	171	208	197	Conseil en matériel informatique			721
..	..	..	..	..	2 735	2 928	2 909	Production de logiciels			722
..	..	..	..	..	329	381	333	Traitement de données informatiques			723
..	..	..	..	..	35	46	49	Activités de banques de données			724
..	..	..	..	..	140	146	133	Entr., répar. mach. de bureau et compt. et inform.			725
..	..	..	..	..	..	139	132	Autres activités rattachées à l'informatique			729
..	..	..	..	..	292	245	276	Recherche-développement		73	
..	..	..	..	..	227	191	220	Rech. et dév. expér. en sc. phys., natur. et ingén.			731
..	..	..	..	..	65	54	56	Rech. et dév. expér. en sciences soc. et humaines			732
..	..	..	..	..	26 713	26 475	25 560	Autres activités de services aux entreprises		74	
..	..	..	..	..	8 459	8 637	8 404	Act. jur., compt., audit.; cons. fisc.; ét. march., soud.			741
..	..	..	..	..	..	..	5 509	Act. d'achitect., d'ingén.; autres act. techniques			742
..	..	..	..	..	..	2 506	2 396	Publicité			743
..	..	..	..	..	..	..	..	Activités de services aux entreprises nca.			749
..	..	..	..	..	6 802	6 844	6 869	Education	M	80	
..	..	..	..	..	2 930	2 921	2 939	Activités d'enseignement primaire			801
..	..	..	..	..	577	570	618	Activités d'enseignement secondaire			802
..	..	..	..	..	247	233	216	Activités d'enseignement supérieur			803
..	..	..	..	..	..	3 120	3 096	Act. de formation perman.; autres act. d'enseign.			809
..	..	..	..	..	20 469	20 795	21 001	Santé et action sociale	N	85	
..	..	..	..	..	8 129	8 262	8 215	Activités de santé humaine			851
..	..	..	..	..	534	523	527	Activités vétérinaires			852
..	..	..	..	..	11 806	12 010	12 259	Activités d'action sociale			853
..	..	..	..	..	15 542	..	21 616	Autres activités de serv. coll., soc. et personnels	O	90_93	
..	..	..	..	..	867	888	875	Assain. et enlèv. des ordures; voierie, act. simil.		90	
..	..	..	..	..	8 075	8 001	6 463	Activités associatives diverses		91	
..	..	..	..	..	1 502	1 459	1 437	Activités de syndicats de salariés			912
..	..	..	..	..	6 322	6 457	6 360	Activités récréatives, culturelles et sportives		92	
..	..	..	..	..	..	2 058	2 026	Act. cinémat., de radio, télé.; autres act. de spect.			921
..	..	..	..	..	181	205	201	Activités d'agences de presse			922
..	..	..	..	..	828	830	830	Act. des biblio., arch., musées, jardins bot. et zool.			923
..	..	..	..	..	..	..	3 303	Activités sportives et autres activités récréatives			924
..	..	..	..	..	278	8 312	7 918	Autres activités de services		93	

Table FN.S1 **PRODUCTION - PRODUCTION**

Millions of Mk (Current Prices)

ISIC revision 3				1991	1992	1993	1994	1995	1996	1997	1998
G	50_52		Wholesale & retail trade; repairs; etc.	..	..	..	..	74 958	83 073	88 685	97 841
	50		Sale, maint.,repair vehic/cycle, person., h/hold gd.	..	..	..	..	10 071	10 926	11 722	13 165
		501	Sale of motor cycle	..	..	..	..	3 886	4 287	4 616	5 334
		502	Maintenance and repair of motor vehicles	..	..	..	..	2 796	3 000	3 231	3 457
		503	Sale of motor vehicle parts and accessories	..	..	..	..	1 970	2 211	2 333	2 532
		504	Sale, maint., repair motor cycle, parts & acces.	..	..	..	..	76	98	120	144
		505	Retail sale of automative fuel	..	..	..	..	1 343	1 330	1 423	1 698
	51		W/sale. trade & commission exc motor veh. & cycle	..	..	..	..	36 828	42 089	46 149	50 586
		511	Wholesale on a fee or contract basis	..	..	..	..	2 363	2 466	2 968	3 079
		512	W/sale agric. raw mat.,live animal,food, bev., tobac.	..	..	..	..	3 652	4 406	4 825	5 420
		513	Wholesale of household goods	..	..	..	..	6 716	8 005	8 627	9 061
		514	W/sale. of non-agric. intermed. prod., waste, scrap	..	..	..	..	8 703	9 354	9 873	10 893
		515	Wholesale of machinery, equipment & supplies	..	..	..	..	11 770	13 647	14 653	15 517
		519	Other wholesale	..	..	..	..	3 624	4 210	5 204	6 616
	52		Retail trade exc. motor veh. & cyc.; repair h/hold gds.	..	..	..	..	28 059	30 058	30 814	34 090
		521	Non-specialised retail trade in stores	..	..	..	..	14 796	15 041	15 225	16 549
		522	Retail sale food, bev., tobac. in specialised stores	..	..	..	..	1 111	1 216	1 248	1 400
		523	Other retail trade of new goods specialised stores	..	..	..	..	11 005	12 453	12 916	14 628
		524	Retail sale of second-hand goods in stores	..	..	..	..	91	119	119	143
		525	Retail trade not in stores	..	..	..	..	674	835	878	907
		526	Repair of personal and household goods	..	..	..	..	383	394	427	462
H	55		Hotels and restaurants	..	..	..	..	10 714	11 936	12 516	13 556
		551	Hotels; camping ; other short-stay accommodation	..	..	..	..	4 117	4 396	4 589	5 025
		552	Restaurants, bars and canteens	..	..	..	..	6 597	7 540	7 926	8 531
I	60_64		Transport, storage and communications	..	..	..	..	66 981	79 154	85 621	106 164
	60		Land transport, transport via pipelines	..	..	..	..	17 919	25 974	26 040	31 298
		601	Transport via railways	..	..	..	..	..	..	..	..
		602	Other land transport	..	..	..	..	..	..	..	27 196
		603	Transport via pipelines	..	..	..	..	..	..	..	..
	61		Water transport	..	..	..	..	8 925	9 455	10 094	11 747
		611	Sea and coastal water transport	..	..	..	..	8 824	9 355	9 983	11 639
		612	Inland water transport	..	..	..	..	101	100	110	108
	62		Air transport	..	..	..	..	5 451	6 533	7 116	8 223
		621	Scheduled air transport	..	..	..	..	5 360	6 428	7 001	8 054
		622	Non-scheduled air transport	..	..	..	..	92	104	114	169
	63		Support. & aux. transport act., travel agencies activ.	..	..	..	..	18 320	18 293	20 769	27 058
	64		Post and telecommunications	..	..	..	..	16 365	18 899	21 603	27 838
		641	Post and courier activities	..	..	..	..	6 014	6 245	6 563	6 892
		642	Telecommunications	..	..	..	..	10 351	12 654	15 040	20 946
J	65_67		Financial intermediation	..	..	..	..	..	..	..	..
	65		Financial intermediation, exc. insur., pension fund	..	..	..	..	..	..	..	..
		651	Monetary intermediation	..	..	..	..	..	..	..	..
		659	Other financial intermediation	..	..	..	..	..	..	..	..
	66		Insurance pension fund, exc. compul. social sec.	..	..	..	..	..	..	..	..
	67		Activities auxiliary to financial intermediation	..	..	..	..	..	..	..	..
		671	Act. auxil to fin. intermed., exc. insur. & pens.fund	..	..	..	..	..	..	..	..
		672	Act. auxiliary to insurance and pension funding	..	..	..	..	..	..	..	..
K	70_74		Real estate, renting and business activities	..	..	..	..	52 479	58 918	62 888	82 253
	70		Real estate activities	..	..	..	..	11 247	13 349	13 237	16 469
		701	Real estate activ. with own or leased property	..	..	..	..	7 165	8 880	8 322	10 380
		702	Real estate activities on a fee or contract basis	..	..	..	..	4 082	4 469	4 915	6 089
	71		Renting of mach. without operator; rent. of pers. gds.	..	..	..	..	1 584	2 071	2 194	2 742
		711	Renting of transport equipment	..	..	..	..	643	730	863	1 055
		712	Renting of other machinery and equipment	..	..	..	..	660	915	984	1 219
		713	Renting of personal and household goods nec	..	..	..	..	227	221	172	196
	72		Computer and related activities	..	..	..	..	8 732	9 871	10 782	15 031
		721	Hardware consultancy	..	..	..	..	41	107	117	277
		722	Software consultancy and supply	..	..	..	..	5 192	5 977	6 847	9 761
		723	Data processing	..	..	..	..	2 699	3 028	3 162	3 928
		724	Data base activities	..	..	..	..	378	470	491	668
		725	Mainten.; repair of office, account. & comp. mach.	..	..	..	..	397	257	161	391
		729	Other computer related activities	..	..	..	..	26	31	5	5
	73		Research and development	..	..	..	..	372	464	634	714
		731	R & D experim. on natural sciences & engineering	..	..	..	..	346	449	618	697
		732	R & D experim. on social sciences & humanities	..	..	..	..	26	14	16	17
	74		Other business activities	..	..	..	..	30 544	33 164	36 040	47 298
		741	Legal, acc., book-keep., audit act.; tax consult., etc	..	..	..	..	9 817	10 451	12 087	17 565
		742	Architectural, engineering, & other techn. activities	..	..	..	..	9 943	10 558	10 369	12 192
		743	Advertising	..	..	..	..	4 260	4 760	5 224	7 408
		749	Business activities nec	..	..	..	..	6 523	7 396	8 359	10 134
M	80		Education	..	..	..	..	..	..	..	1 862
		801	Primary education	..	..	..	..	..	..	..	0
		802	Secondary education	..	..	..	..	..	..	..	216
		803	Higher education	..	..	..	..	..	..	..	151
		809	Adult and other education	..	..	..	..	..	..	..	1 495
N	85		Health and social work	..	..	..	..	4 907	5 302	5 683	7 144
		851	Human health activities	..	..	..	..	4 197	4 554	4 854	5 955
		852	Veterinary activities	..	..	..	..	265	275	282	340
		853	Social work activities	..	..	..	..	445	473	547	849
O	90_93		Other commun., soc. & pers. services activities	..	..	..	..	..	..	..	24 426
	90		Sewage & refuse disposal, sanitation, etc.	..	..	..	..	1 602	1 988	2 010	2 367
	91		Activities of member organisations nec	..	..	..	..	..	..	..	43
		912	Activities of trade unions	..	..	..	..	..	..	..	..
	92		Recreational, cultural and sporting activities	..	..	..	..	8 501	13 399	15 198	18 490
		921	Motion picture, radio, TV, other entertainment act.	..	..	..	..	1 552	1 513	1 689	3 280
		922	News agency activities	..	..	..	..	226	243	261	299
		923	Library, archives, museums, other cultural activ.	..	..	..	..	37	35	37	34
		924	Sporting and other recreational activities	..	..	..	..	5 399	8 167	9 401	10 675
	93		Other service activities	..	..	..	..	2 149	2 774	2 882	3 524

Table FN.S2 **VALEUR AJOUTÉE AU COUT DES FACTEURS - VALUE ADDED AT FACTOR COSTS**

Millions de Mk (Prix courants)

1991	1992	1993	1994	1995	1996	1997	1998		CITI révision 3		
..	..	..	..	47 015	50 873	54 012	57 497	Com. de gros, de detail ; réparation etc.	G	50_52	
..	..	..	..	6 665	7 042	7 735	8 707	Com., entr., répar. d'auto. et moto ; carburants		50	
..	..	..	..	2 449	2 630	2 916	3 376	Commerce de véhicules automobiles			501
..	..	..	..	1 942	2 081	2 272	2 419	Entretien et réparation de véhicules automobiles			502
..	..	..	..	1 379	1 437	1 581	1 711	Com. de pièces et access. de véhic. automobiles			503
..	..	..	..	42	53	63	80	Com., entret. et répar. de motocycles et access.			504
..	..	..	..	853	840	902	1 122	Com. de détail de carburants pour automobiles			505
..	..	..	..	22 732	25 249	26 813	27 587	Com. de gros et act. inter., sauf auto. et moto		51	
..	..	..	..	1 458	1 500	1 589	1 574	Activités d'intermédiaries de commerce de gros			511
..	..	..	..	2 014	2 414	2 496	2 543	Com. gros: prod. agr. bruts, anim. viv., prod. alim.			512
..	..	..	..	4 042	4 502	4 783	4 800	Com. gros: biens de consom., sauf alimentaires			513
..	..	..	..	5 787	5 990	6 128	6 279	Com. gros: prod. inter. non agr., déchets, débris			514
..	..	..	..	7 657	8 721	9 425	9 976	Com. gros: machines, équipments et fournitures			515
..	..	..	..	1 774	2 122	2 392	2 415	Autres commerces de gros			519
..	..	..	..	17 618	18 582	19 464	21 203	Com. détail, sauf auto, moto; répar. art. pers., dom.		52	
..	..	..	..	9 405	9 351	9 875	10 472	Commerce de détail non spécialisé en magasin			521
..	..	..	..	729	787	804	839	Com. détail: prod. aliment. en mag. spécialisés			522
..	..	..	..	6 897	7 795	8 111	9 196	Autre com. détail: prod. neufs en mag. spécialisés			523
..	..	..	..	42	57	57	68	Commerces de détail de biens d'occasion			524
..	..	..	..	281	329	332	312	Commerces de détail autres qu'en magasin			525
..	..	..	..	264	262	286	315	Réparation d'articles personnels et domestiques			526
..	..	..	..	6 379	6 847	7 311	7 860	Hotels et restaurants	H	55	
..	..	..	..	2 166	2 218	2 310	2 498	Hôtels; camping; autres équip. courts séjours			551
..	..	..	..	4 213	4 629	5 000	5 362	Restaurants, cafés et cantines			552
..	..	..	..	31 166	38 494	39 947	44 190	Transport, entreposage, communications	I	60_64	
..	..	..	..	11 008	16 215	15 500	16 905	Transports terrestres; transports par conduites		60	
..	..	..	..	..	..	..	..	Transports par chemin de fer			601
..	..	..	..	..	..	..	14 210	Autres transports terrestres			602
..	..	..	..	..	..	..	..	Transports par conduites			603
..	..	..	..	3 059	3 359	3 457	3 746	Transports par eau		61	
..	..	..	..	2 999	3 305	3 399	3 692	Transports maritimes et côtiers			611
..	..	..	..	60	54	58	54	Transports par voies navigables intérieures			612
..	..	..	..	2 525	2 726	3 230	3 174	Transports aériens		62	
..	..	..	..	2 494	2 681	3 199	3 120	Transports aériens réguliers			621
..	..	..	..	31	46	31	53	Transports aériens non réguliers			622
..	..	..	..	4 195	4 827	5 277	6 383	Act. annex. et aux. des transp.; act. des ag. de voy.		63	
..	..	..	..	10 378	11 366	12 484	13 982	Postes et télécommunications		64	
..	..	..	..	4 182	4 170	4 207	4 234	Activités de poste et de courrier			641
..	..	..	..	6 196	7 196	8 277	9 748	Télécommunications			642
..	..	..	..	..	..	..	..	Intermédiation financière	J	65_67	
..	..	..	..	..	..	..	..	Interm. financ., sauf assur. et caisses de retraite		65	
..	..	..	..	..	..	..	..	Intermédiation monétaire			651
..	..	..	..	..	..	..	..	Autres intermédiations financières			659
..	..	..	..	..	..	..	..	Act. d'assur., caisses de retr., sauf séc. soc. oblig.		66	
..	..	..	..	..	..	..	..	Act. auxiliaires de l'intermédiation financière		67	
..	..	..	..	..	..	..	..	Act. aux. de l'interm. fin., sauf assur., et retraite			671
..	..	..	..	..	..	..	..	Act. aux. des assurances et caisses de retraite			672
..	..	..	..	26 665	30 522	31 853	42 306	Immobilier, locations, services aux entreprises	K	70_74	
..	..	..	..	4 891	6 180	4 884	6 876	Activités immobilières		70	
..	..	..	..	2 597	3 548	2 194	3 388	Act. immobilières sur biens propres ou loués			701
..	..	..	..	2 294	2 632	2 690	3 489	Activités immobilières pour comptes de tiers			702
..	..	..	..	806	1 024	1 106	1 424	Locat. mach., équip. sans opér., biens pers. et dom.		71	
..	..	..	..	336	375	416	482	Location de matériel de transport			711
..	..	..	..	374	490	504	689	Location de machines ét equipements nca			712
..	..	..	..	77	90	72	88	Loc. d'articles personnels et domestiques nca			713
..	..	..	..	4 555	5 083	5 690	7 557	Activités informatiques et activités rattachées		72	
..	..	..	..	21	59	57	151	Conseil en matériel informatique			721
..	..	..	..	2 790	3 083	3 559	4 750	Production de logiciels			722
..	..	..	..	1 415	1 629	1 843	2 275	Traitement de données informatiques			723
..	..	..	..	158	185	162	259	Activités de banques de données			724
..	..	..	..	158	110	68	120	Entr., répar. mach. de bureau et compt. et inform.			725
..	..	..	..	13	16	1	1	Autres activités rattachées à l'informatique			729
..	..	..	..	240	267	303	281	Recherche-développement		73	
..	..	..	..	224	257	291	269	Rech. et dév. expér. en sc. phys., natur. et ingén.			731
..	..	..	..	16	10	12	12	Rech. et dév. expér. en sciences soc. et humaines			732
..	..	..	..	16 174	17 967	19 871	26 168	Autres activités de services aux entreprises		74	
..	..	..	..	5 755	6 365	7 288	11 567	Act. jur., compt., audit.; cons. fisc.; ét. march., soud.			741
..	..	..	..	4 801	5 605	5 762	6 516	Act. d'achitect., d'ingén.; autres act. techniques			742
..	..	..	..	1 280	1 341	1 573	1 879	Publicité			743
..	..	..	..	4 338	4 657	5 248	6 206	Activités de services aux entreprises nca.			749
..	..	..	..	..	..	..	1 016	Education	M	80	
..	..	..	..	..	..	..		Activités d'enseignement primaire			801
..	..	..	..	..	..	..	111	Activités d'enseignement secondaire			802
..	..	..	..	..	..	..	79	Activités d'enseignement supérieur			803
..	..	..	..	..	..	..	827	Act. de formation perman.; autres act. d'enseign.			809
..	..	..	..	3 132	3 379	3 512	4 329	Santé et action sociale	N	85	
..	..	..	..	2 689	2 908	3 010	3 611	Activités de santé humaine			851
..	..	..	..	139	142	140	152	Activités vétérinaires			852
..	..	..	..	304	329	362	566	Activités d'action sociale			853
..	..	..	..	..	..	..	10 991	Autres activités de serv. coll., soc. et personnels	O	90_93	
..	..	..	..	884	1 188	1 076	1 183	Assain. et enlèv. des ordures; voierie, act. simil.		90	
..	..	..	..	..	..	..	21	Activités associatives diverses		91	
..	..	..	..	..	..	..	..	Activités de syndicats de salariés			912
..	..	..	..	3 296	7 083	7 381	8 019	Activités récréatives, culturelles et sportives		92	
..	..	..	..	587	689	765	955	Act. cinémat., de radio, télé.; autres act. de spect.			921
..	..	..	..	102	118	119	134	Activités d'agences de presse			922
..	..	..	..	17	14	16	8	Act. des biblio., arch., musées, jardins bot. et zool.			923
..	..	..	..	2 174	4 400	4 660	5 069	Activités sportives et autres activités récréatives			924
..	..	..	..	1 195	1 511	1 543	1 768	Autres activités de services		93	

FINLAND

Table FN.S3 **INVESTMENT - INVESTISSEMENT**

Millions of Mk (Current Prices)

ISIC revision 3				1991	1992	1993	1994	1995	1996	1997	1998
G	50_52		Wholesale & retail trade; repairs; etc.	..	..	..	..	5 552	7 983	8 099	8 152
	50		Sale, maint.,repair vehic/cycle, person., h/hold gd.	..	..	..	..	988	1 524	1 561	1 699
		501	Sale of motor cycle	..	..	..	..	502	825	771	792
		502	Maintenance and repair of motor vehicles	..	..	..	..	333	446	487	612
		503	Sale of motor vehicle parts and accessories	..	..	..	..	72	158	187	169
		504	Sale, maint., repair motor cycle, parts & acces.	..	..	..	..	4	5	7	7
		505	Retail sale of automative fuel	..	..	..	..	78	90	109	120
	51		W/sale. trade & commission exc motor veh. & cycle	..	..	..	..	2 397	4 379	4 299	3 731
		511	Wholesale on a fee or contract basis	..	..	..	..	92	137	155	186
		512	W/sale agric. raw mat.,live animal,food, bev., tobac.	..	..	..	..	172	774	358	320
		513	Wholesale of household goods	..	..	..	..	328	327	433	455
		514	W/sale. of non-agric. intermed. prod., waste, scrap	..	..	..	..	792	1 238	1 797	1 051
		515	Wholesale of machinery, equipment & supplies	..	..	..	..	683	1 017	1 157	1 179
		519	Other wholesale	..	..	..	..	331	886	399	541
	52		Retail trade exc. motor veh. & cyc.; repair h/hold gds.	..	..	..	..	2 167	2 080	2 239	2 722
		521	Non-specialised retail trade in stores	..	..	..	..	1 634	1 400	1 434	1 854
		522	Retail sale food, bev., tobac. in specialised stores	..	..	..	..	70	55	64	86
		523	Other retail trade of new goods specialised stores	..	..	..	..	426	578	688	723
		524	Retail sale of second-hand goods in stores	..	..	..	..	5	7	8	9
		525	Retail trade not in stores	..	..	..	..	13	16	18	16
		526	Repair of personal and household goods	..	..	..	..	19	24	27	33
H	55		Hotels and restaurants	..	..	..	..	691	851	894	1 107
		551	Hotels; camping ; other short-stay accommodation	..	..	..	..	279	358	376	581
		552	Restaurants, bars and canteens	..	..	..	..	411	493	518	526
I	60_64		Transport, storage and communications	..	..	..	..	..	..	..	12 978
	60		Land transport, transport via pipelines	..	..	..	..	..	..	..	5 009
		601	Transport via railways	..	..	..	..	..	..	..	..
		602	Other land transport	..	..	..	..	..	..	..	4 398
		603	Transport via pipelines	..	..	..	..	..	..	..	..
	61		Water transport	..	..	..	..	751	936	1 304	1 129
		611	Sea and coastal water transport	..	..	..	..	736	919	1 298	1 118
		612	Inland water transport	..	..	..	..	16	17	7	11
	62		Air transport	..	..	..	..	226	1 049	851	1 262
		621	Scheduled air transport	..	..	..	..	195	983	807	1 225
		622	Non-scheduled air transport	..	..	..	..	31	66	44	37
	63		Support. & aux. transport act., travel agencies activ.	..	..	..	..	707	908	770	1 195
	64		Post and telecommunications	..	..	..	..	3 305	3 811	4 092	4 383
		641	Post and courier activities	..	..	..	..	210	172	296	336
		642	Telecommunications	..	..	..	..	3 094	3 638	3 796	4 047
J	65_67		Financial intermediation	..	..	..	..	4 279	3 132	3 922	4 863
	65		Financial intermediation, exc. insur., pension fund	..	..	..	..	4 060	2 774	3 706	4 614
		651	Monetary intermediation	..	..	..	..	758	425	428	410
		659	Other financial intermediation	..	..	..	..	3 303	2 348	3 278	4 204
	66		Insurance pension fund, exc. compul. social sec.	..	..	..	..	180	307	146	103
	67		Activities auxiliary to financial intermediation	..	..	..	..	39	52	70	147
		671	Act. auxil to fin. intermed., exc. insur. & pens.fund	..	..	..	..	36	48	66	140
		672	Act. auxiliary to insurance and pension funding	..	..	..	..	3	4	4	7
K	70_74		Real estate, renting and business activities	..	..	..	..	7 833	6 321	7 162	14 126
	70		Real estate activities	..	..	..	..	5 118	3 698	3 662	10 314
		701	Real estate activ. with own or leased property	..	..	..	..	4 841	3 079	3 337	9 709
		702	Real estate activities on a fee or contract basis	..	..	..	..	276	620	325	606
	71		Renting of mach. without operator; rent. of pers. gds.	..	..	..	..	651	762	727	1 300
		711	Renting of transport equipment	..	..	..	..	381	482	375	799
		712	Renting of other machinery and equipment	..	..	..	..	111	130	182	258
		713	Renting of personal and household goods nec	..	..	..	..	26	32	26	27
	72		Computer and related activities	..	..	..	..	507	444	527	589
		721	Hardware consultancy	..	..	..	..	3	8	7	7
		722	Software consultancy and supply	..	..	..	..	280	230	233	302
		723	Data processing	..	..	..	..	194	186	246	240
		724	Data base activities	..	..	..	..	17	15	36	33
		725	Mainten.; repair of office, account. & comp. mach.	..	..	..	..	12	5	5	8
		729	Other computer related activities	..	..	..	..	0	1	0	0
	73		Research and development	..	..	..	..	28	34	46	63
		731	R & D experim. on natural sciences & engineering	..	..	..	..	27	34	46	62
		732	R & D experim. on social sciences & humanities	..	..	..	..	1	1	0	1
	74		Other business activities	..	..	..	..	1 530	1 383	2 200	1 860
		741	Legal, acc., book-keep., audit act.; tax consult., etc	..	..	..	..	761	512	1 168	704
		742	Architectural, engineering, & other techn. activities	..	..	..	..	304	355	380	438
		743	Advertising	..	..	..	..	118	114	128	171
		749	Business activities nec	..	..	..	..	347	401	524	547
M	80		Education	..	..	..	..	80	90	87	118
		801	Primary education	..	..	..	..	0	0	0	0
		802	Secondary education	..	..	..	..	18	26	8	14
		803	Higher education	..	..	..	..	..	1	5	3
		809	Adult and other education	..	..	..	..	63	63	74	101
N	85		Health and social work	..	..	..	..	197	222	249	335
		851	Human health activities	..	..	..	..	149	175	195	247
		852	Veterinary activities	..	..	..	..	15	12	9	8
		853	Social work activities	..	..	..	..	33	35	45	79
O	90_93		Other commun., soc. & pers. services activities	..	..	..	..	690	1 116	1 512	1 466
	90		Sewage & refuse disposal, sanitation, etc.	..	..	..	..	266	272	298	381
	91		Activities of member organisations nec	..	..	..	..	1	1	1	2
		912	Activities of trade unions	..	..	..	..	..	..	..	..
	92		Recreational, cultural and sporting activities	..	..	..	..	423	843	1 214	942
		921	Motion picture, radio, TV, other entertainment act.	..	..	..	..	99	101	103	159
		922	News agency activities	..	..	..	..	11	11	18	14
		923	Library, archives, museums, other cultural activ.	..	..	..	..	2	2	19	3
		924	Sporting and other recreational activities	..	..	..	..	177	355	427	393
	93		Other service activities	..	..	..	..	90	119	100	142

Table FN.S4 **DÉPENSES DE MACHINES ET D'ÉQUIPEMENT - INVESTMENT IN MACHINERY & EQUIPMENT**

Millions de Mk (Prix courants)

1991	1992	1993	1994	1995	1996	1997	1998		CITI révision 3		
..	..	..	..	3 880	4 837	5 212	5 958	Com. de gros, de detail ; réparation etc.	G	50_52	
..	..	..	..	820	1 001	1 173	1 330	Com., entr., répar. d'auto. et moto ; carburants		50	
..	..	..	..	438	535	643	654	Commerce de véhicules automobiles			501
..	..	..	..	272	309	358	470	Entretien et réparation de véhicules automobiles			502
..	..	..	..	50	83	91	106	Com. de pièces et access. de véhic. automobiles			503
..	..	..	..	2	4	6	5	Com., entret. et répar. de motocycles et access.			504
..	..	..	..	58	70	76	96	Com. de détail de carburants pour automobiles			505
..	..	..	..	1 838	2 401	2 655	2 732	Com. de gros et act. inter., sauf auto. et moto		51	
..	..	..	..	85	127	106	164	Activités d'intermédiaries de commerce de gros			511
..	..	..	..	148	328	276	249	Com. gros: prod. agr. bruts, anim. viv., prod. alim.			512
..	..	..	..	233	286	294	333	Com. gros: biens de consom., sauf alimentaires			513
..	..	..	..	534	736	904	676	Com. gros: prod. inter. non agr., déchets, débris			514
..	..	..	..	615	756	873	980	Com. gros: machines, équipments et fournitures			515
..	..	..	..	221	168	200	330	Autres commerces de gros			519
..	..	..	..	1 222	1 435	1 384	1 895	Com. détail, sauf auto, moto; répar. art. pers., dom.		52	
..	..	..	..	780	888	803	1 170	Commerce de détail non spécialisé en magasin			521
..	..	..	..	67	52	61	80	Com. détail: prod. aliment. en mag. spécialisés			522
..	..	..	..	346	452	470	589	Autre com. détail: prod. neufs en mag. spécialisés			523
..	..	..	..	4	6	6	9	Commerces de détail de biens d'occasion			524
..	..	..	..	7	14	17	16	Commerces de détail autres qu'en magasin			525
..	..	..	..	18	23	26	32	Réparation d'articles personnels et domestiques			526
..	..	..	..	472	515	576	860	Hotels et restaurants	H	55	
..	..	..	..	160	149	216	414	Hôtels; camping; autres équip. courts séjours			551
..	..	..	..	312	366	360	446	Restaurants, cafés et cantines			552
..	..	..	..	7 100	9 150	9 351	11 152	Transport, entreposage, communications	I	60_64	
..	..	..	..	2 703	3 235	3 214	4 413	Transports terrestres; transports par conduites		60	
..	..	..	..	..	..	..	..	Transports par chemin de fer			601
..	..	..	..	..	..	..	3 840	Autres transports terrestres			602
..	..	..	..	..	..	..	..	Transports par conduites			603
..	..	..	..	708	921	1 284	1 087	Transports par eau		61	
..	..	..	..	696	905	1 278	1 077	Transports maritimes et côtiers			611
..	..	..	..	13	16	6	10	Transports par voies navigables intérieures			612
..	..	..	..	216	982	653	1 003	Transports aériens		62	
..	..	..	..	185	917	618	967	Transports aériens réguliers			621
..	..	..	..	31	66	36	36	Transports aériens non réguliers			622
..	..	..	..	372	455	477	628	Act. annex. et aux. des transp.; act. des ag. de voy.		63	
..	..	..	..	3 101	3 557	3 723	4 020	Postes et télécommunications		64	
..	..	..	..	144	129	192	266	Activités de poste et de courrier			641
..	..	..	..	2 957	3 428	3 531	3 754	Télécommunications			642
..	..	..	..	4 029	2 779	3 486	4 653	Intermédiation financière	J	65_67	
..	..	..	..	3 902	2 532	3 293	4 445	Interm. financ., sauf assur. et caisses de retraite		65	
..	..	..	..	634	280	270	308	Intermédiation monétaire			651
..	..	..	..	3 268	2 251	3 023	4 137	Autres intermédiations financières			659
..	..	..	..	94	203	132	87	Act. d'assur., caisses de retr., sauf séc. soc. oblig.		66	
..	..	..	..	33	45	61	122	Act. auxiliaires de l'intermédiation financière		67	
..	..	..	..	30	41	57	115	Act. aux. de l'interm. fin., sauf assur., et retraite			671
..	..	..	..	3	4	4	7	Act. aux. des assurances et caisses de retraite			672
..	..	..	..	2 829	2 739	3 200	3 868	Immobilier, locations, services aux entreprises	K	70_74	
..	..	..	..	384	411	383	495	Activités immobilières		70	
..	..	..	..	212	187	176	231	Act. immobilières sur biens propres ou loués			701
..	..	..	..	172	224	207	264	Activités immobilières pour comptes de tiers			702
..	..	..	..	642	747	669	1 274	Locat. mach., équip. sans opér., biens pers. et dom.		71	
..	..	..	..	380	477	345	788	Location de matériel de transport			711
..	..	..	..	108	121	161	244	Location de machines ét equipements nca			712
..	..	..	..	24	31	24	26	Loc. d'articles personnels et domestiques nca			713
..	..	..	..	502	414	509	580	Activités informatiques et activités rattachées		72	
..	..	..	..	2	7	6	6	Conseil en matériel informatique			721
..	..	..	..	276	202	220	296	Production de logiciels			722
..	..	..	..	194	185	242	238	Traitement de données informatiques			723
..	..	..	..	17	15	36	32	Activités de banques de données			724
..	..	..	..	12	5	4	8	Entr., répar. mach. de bureau et compt. et inform.			725
..	..	..	..	0	1	0	0	Autres activités rattachées à l'informatique			729
..	..	..	..	27	30	41	62	Recherche-développement		73	
..	..	..	..	27	29	40	61	Rech. et dév. expér. en sc. phys., natur. et ingén.			731
..	..	..	..	1	1	0	1	Rech. et dév. expér. en sciences soc. et humaines			732
..	..	..	..	1 274	1 137	1 599	1 458	Autres activités de services aux entreprises		74	
..	..	..	..	607	402	781	478	Act. jur., compt., audit.; cons. fisc.; ét. march., soud.			741
..	..	..	..	277	312	329	385	Act. d'achitect., d'ingén.; autres act. techniques			742
..	..	..	..	96	100	104	138	Publicité			743
..	..	..	..	293	322	385	456	Activités de services aux entreprises nca.			749
..	..	..	..	47	65	68	99	Education	M	80	
..	..	..	..	0	0	0	0	Activités d'enseignement primaire			801
..	..	..	..	3	7	5	10	Activités d'enseignement secondaire			802
..	..	..	..		1	4	3	Activités d'enseignement supérieur			803
..	..	..	..	44	57	59	86	Act. de formation perman.; autres act. d'enseign.			809
..	..	..	..	175	189	199	276	Santé et action sociale	N	85	
..	..	..	..	143	158	169	223	Activités de santé humaine			851
..	..	..	..	15	12	6	8	Activités vétérinaires			852
..	..	..	..	17	20	24	45	Activités d'action sociale			853
..	..	..	..	513	863	1 142	1 143	Autres activités de serv. coll., soc. et personnels	O	90_93	
..	..	..	..	198	190	191	278	Assain. et enlèv. des ordures; voierie, act. simil.		90	
..	..	..	..	1	1	1	0	Activités associatives diverses		91	
..	..	..	..				..	Activités de syndicats de salariés			912
..	..	..	..	314	673	950	747	Activités récréatives, culturelles et sportives		92	
..	..	..	..	85	89	96	151	Act. cinémat., de radio, télé.; autres act. de spect.			921
..	..	..	..	11	10	14	14	Activités d'agences de presse			922
..	..	..	..	1	1	3	2	Act. des biblio., arch., musées, jardins bot. et zool.			923
..	..	..	..	103	254	262	267	Activités sportives et autres activités récréatives			924
..	..	..	..	69	103	84	117	Autres activités de services		93	

FINLAND

Table FN.S5 **EMPLOYMENT - EMPLOI**

Units

ISIC revision 3				1991	1992	1993	1994	1995	1996	1997	1998
G	50_52		Wholesale & retail trade; repairs; etc.	..	..	..	..	110 228	212 966	211 294	227 622
	50		Sale, maint.,repair vehic/cycle, person., h/hold gd.	..	..	..	..	18 948	30 677	31 236	34 102
		501	Sale of motor cycle	..	..	..	..	2 979	8 405	8 472	8 999
		502	Maintenance and repair of motor vehicles	..	..	..	..	7 618	10 421	10 372	11 987
		503	Sale of motor vehicle parts and accessories	..	..	..	..	3 497	6 183	6 256	6 463
		504	Sale, maint., repair motor cycle, parts & acces.	..	..	..	..	289	296	314	372
		505	Retail sale of automative fuel	..	..	..	..	4 566	5 371	5 822	6 281
	51		W/sale. trade & commission exc motor veh. & cycle	..	..	..	..	37 714	76 180	79 615	82 974
		511	Wholesale on a fee or contract basis	..	..	..	..	3 947	5 064	5 101	5 419
		512	W/sale agric. raw mat.,live animal,food, bev., tobac.	..	..	..	..	3 410	8 479	8 393	9 319
		513	Wholesale of household goods	..	..	..	..	8 740	14 676	14 863	14 929
		514	W/sale. of non-agric. intermed. prod., waste, scrap	..	..	..	..	6 542	16 136	16 422	16 929
		515	Wholesale of machinery, equipment & supplies	..	..	..	..	14 653	26 239	27 141	27 823
		519	Other wholesale	..	..	..	..	423	5 586	7 695	8 555
	52		Retail trade exc. motor veh. & cyc.; repair h/hold gds.	..	..	..	..	53 566	106 109	100 443	110 546
		521	Non-specialised retail trade in stores	..	..	..	..	17 715	55 322	49 559	55 107
		522	Retail sale food, bev., tobac. in specialised stores	..	..	..	..	1 417	3 436	3 335	3 747
		523	Other retail trade of new goods specialised stores	..	..	..	..	31 770	43 498	43 391	47 223
		524	Retail sale of second-hand goods in stores	..	..	..	..	492	600	629	721
		525	Retail trade not in stores	..	..	..	..	742	1 412	1 517	1 746
		526	Repair of personal and household goods	..	..	..	..	1 431	1 842	2 012	2 002
H	55		Hotels and restaurants	..	..	..	..	27 314	45 105	45 189	49 016
		551	Hotels; camping ; other short-stay accommodation	..	..	..	..	5 620	11 734	11 774	13 369
		552	Restaurants, bars and canteens	..	..	..	..	21 694	33 371	33 415	35 647
I	60_64		Transport, storage and communications	..	..	..	..	132 284	143 987	144 877	155 974
	60		Land transport, transport via pipelines	..	..	..	..	55 783	60 942	62 961	70 102
		601	Transport via railways	..	..	..	..	..	..	..	..
		602	Other land transport	..	..	..	..	..	..	..	59 804
		603	Transport via pipelines	..	..	..	..	..	..	..	..
	61		Water transport	..	..	..	..	8 662	8 513	8 361	8 162
		611	Sea and coastal water transport	..	..	..	..	8 396	8 247	8 103	7 913
		612	Inland water transport	..	..	..	..	266	265	258	249
	62		Air transport	..	..	..	..	7 169	8 900	8 885	9 291
		621	Scheduled air transport	..	..	..	..	7 064	8 788	8 757	9 143
		622	Non-scheduled air transport	..	..	..	..	104	112	129	147
	63		Support. & aux. transport act., travel agencies activ.	..	..	..	..	17 231	19 353	20 100	22 293
	64		Post and telecommunications	..	..	..	..	43 439	46 280	44 570	46 127
		641	Post and courier activities	..	..	..	..	28 286	29 998	27 678	28 065
		642	Telecommunications	..	..	..	..	15 153	16 281	16 893	18 062
J	65_67		Financial intermediation	..	..	..	..	43 483	44 362	43 721	37 033
	65		Financial intermediation, exc. insur., pension fund	..	..	..	..	32 898	31 276	30 304	25 964
		651	Monetary intermediation	..	..	..	..	30 013	27 577	26 986	23 322
		659	Other financial intermediation	..	..	..	..	2 886	3 699	3 317	2 642
	66		Insurance pension fund, exc. compul. social sec.	..	..	..	..	9 044	10 942	11 241	8 628
	67		Activities auxiliary to financial intermediation	..	..	..	..	1 541	2 144	2 176	2 441
		671	Act. auxil to fin. intermed., exc. insur. & pens.fund	..	..	..	..	1 238	1 775	1 798	2 068
		672	Act. auxiliary to insurance and pension funding	..	..	..	..	302	370	378	374
K	70_74		Real estate, renting and business activities	..	..	..	..	116 894	132 208	131 094	143 140
	70		Real estate activities	..	..	..	..	21 359	22 869	22 329	18 985
		701	Real estate activ. with own or leased property	..	..	..	..	8 487	8 842	7 899	4 154
		702	Real estate activities on a fee or contract basis	..	..	..	..	12 872	14 027	14 430	14 831
	71		Renting of mach. without operator; rent. of pers. gds.	..	..	..	..	2 132	2 497	2 566	2 625
		711	Renting of transport equipment	..	..	..	..	436	567	623	531
		712	Renting of other machinery and equipment	..	..	..	..	1 227	1 429	1 480	1 584
		713	Renting of personal and household goods nec	..	..	..	..	401	406	366	429
	72		Computer and related activities	..	..	..	..	15 924	17 068	18 139	22 255
		721	Hardware consultancy	..	..	..	..	164	317	292	316
		722	Software consultancy and supply	..	..	..	..	10 316	11 092	11 616	14 249
		723	Data processing	..	..	..	..	4 468	4 796	5 464	6 600
		724	Data base activities	..	..	..	..	257	341	389	617
		725	Mainten.; repair of office, account. & comp. mach.	..	..	..	..	652	447	366	462
		729	Other computer related activities	..	..	..	..	67	75	12	11
	73		Research and development	..	..	..	..	1 072	1 118	1 410	1 307
		731	R & D experim. on natural sciences & engineering	..	..	..	..	978	1 001	1 303	1 245
		732	R & D experim. on social sciences & humanities	..	..	..	..	94	117	107	62
	74		Other business activities	..	..	..	..	76 406	88 658	86 651	97 968
		741	Legal, acc., book-keep., audit act.; tax consult., etc	..	..	..	..	20 547	21 757	20 986	23 013
		742	Architectural, engineering, & other techn. activities	..	..	..	..	20 738	22 790	23 565	25 382
		743	Advertising	..	..	..	..	5 048	5 685	5 465	6 476
		749	Business activities nec	..	..	..	..	30 073	38 426	36 635	43 097
M	80		Education	..	..	..	..	4 513	5 355	5 799	4 837
		801	Primary education	..	..	..	..	217	270	244	
		802	Secondary education	..	..	..	..	1 309	1 323	1 477	586
		803	Higher education	..	..	..	..	150	501	804	640
		809	Adult and other education	..	..	..	..	2 837	3 261	3 275	3 610
N	85		Health and social work	..	..	..	..	15 471	16 394	16 695	19 586
		851	Human health activities	..	..	..	..	13 054	13 459	13 575	15 304
		852	Veterinary activities	..	..	..	..	715	764	776	813
		853	Social work activities	..	..	..	..	1 701	2 171	2 344	3 470
O	90_93		Other commun., soc. & pers. services activities	..	..	..	..	29 963	31 379	32 958	34 249
	90		Sewage & refuse disposal, sanitation, etc.	..	..	..	..	3 064	2 663	3 483	4 072
	91		Activities of member organisations nec	..	..	..	..	2 443	2 391	2 389	105
		912	Activities of trade unions	..	..	..	..				..
	92		Recreational, cultural and sporting activities	..	..	..	..	14 625	15 352	16 075	16 426
		921	Motion picture, radio, TV, other entertainment act.	..	..	..	..	3 483	3 026	4 345	4 377
		922	News agency activities	..	..	..	..	406	427	428	413
		923	Library, archives, museums, other cultural activ.	..	..	..	..	112	101	116	95
		924	Sporting and other recreational activities	..	..	..	..	4 308	4 494	4 826	5 232
	93		Other service activities	..	..	..	..	9 830	10 972	11 011	13 646

Table FN.S6 **SALAIRES ET CHARGES SOCIALES, SALARIÉS - COMPENSATION OF LABOUR, EMPLOYEES**

Millions de Mk (Prix courants)

1991	1992	1993	1994	1995	1996	1997	1998		CITI révision 3		
..	..	..	..	27 826	30 321	31 874	35 124	Com. de gros, de detail ; réparation etc.	G	50_52	
..	..	..	..	3 926	4 247	4 555	5 045	Com., entr., répar. d'auto. et moto ; carburants		50	
..	..	..	..	1 336	1 451	1 602	1 749	Commerce de véhicules automobiles			501
..	..	..	..	1 210	1 266	1 341	1 466	Entretien et réparation de véhicules automobiles			502
..	..	..	..	799	916	959	1 010	Com. de pièces et access. de véhic. automobiles			503
..	..	..	..	27	34	35	39	Com., entret. et répar. de motocycles et access.			504
..	..	..	..	554	580	619	782	Com. de détail de carburants pour automobiles			505
..	..	..	..	12 630	14 228	15 422	16 744	Com. de gros et act. inter., sauf auto. et moto		51	
..	..	..	..	608	739	784	839	Activités d'intermédiaries de commerce de gros			511
..	..	..	..	1 163	1 336	1 418	1 666	Com. gros: prod. agr. bruts, anim. viv., prod. alim.			512
..	..	..	..	2 267	2 591	2 696	2 838	Com. gros: biens de consom., sauf alimentaires			513
..	..	..	..	2 874	3 022	3 132	3 443	Com. gros: prod. inter. non agr., déchets, débris			514
..	..	..	..	4 733	5 484	5 950	6 258	Com. gros: machines, équipments et fournitures			515
..	..	..	..	986	1 057	1 442	1 701	Autres commerces de gros			519
..	..	..	..	11 270	11 846	11 897	13 335	Com. détail, sauf auto, moto; répar. art. pers., dom.		52	
..	..	..	..	6 433	6 401	6 309	6 928	Commerce de détail non spécialisé en magasin			521
..	..	..	..	433	471	488	528	Com. détail: prod. aliment. en mag. spécialisés			522
..	..	..	..	4 114	4 655	4 756	5 494	Autre com. détail: prod. neufs en mag. spécialisés			523
..	..	..	..	17	24	26	28	Commerces de détail de biens d'occasion			524
..	..	..	..	140	164	177	194	Commerces de détail autres qu'en magasin			525
..	..	..	..	132	130	142	161	Réparation d'articles personnels et domestiques			526
..	..	..	..	4 602	4 900	5 205	5 643	Hotels et restaurants	H	55	
..	..	..	..	1 570	1 639	1 588	1 743	Hôtels; camping; autres équip. courts séjours			551
..	..	..	..	3 032	3 261	3 616	3 900	Restaurants, cafés et cantines			552
..	..	..	..	17 870	22 737	22 858	25 345	Transport, entreposage, communications	I	60_64	
..	..	..	..	5 236	9 063	8 450	9 505	Transports terrestres; transports par conduites		60	
..	..	..	..	..	..	..	..	Transports par chemin de fer			601
..	..	..	..	..	..	..	7 378	Autres transports terrestres			602
..	..	..	..	..	..	..	..	Transports par conduites			603
..	..	..	..	1 841	1 875	1 868	1 945	Transports par eau		61	
..	..	..	..	1 806	1 841	1 833	1 910	Transports maritimes et côtiers			611
..	..	..	..	35	34	35	35	Transports par voies navigables intérieures			612
..	..	..	..	1 563	1 959	2 167	2 321	Transports aériens		62	
..	..	..	..	1 546	1 935	2 148	2 291	Transports aériens réguliers			621
..	..	..	..	17	23	19	30	Transports aériens non réguliers			622
..	..	..	..	2 912	3 342	3 574	4 202	Act. annex. et aux. des transp.; act. des ag. de voy.		63	
..	..	..	..	6 317	6 498	6 799	7 371	Postes et télécommunications		64	
..	..	..	..	3 621	3 473	3 553	3 725	Activités de poste et de courrier			641
..	..	..	..	2 696	3 025	3 245	3 646	Télécommunications			642
..	..	..	..	..	..	..	..	Intermédiation financière	J	65_67	
..	..	..	..	..	..	..	..	Interm. financ., sauf assur. et caisses de retraite		65	
..	..	..	..	..	..	..	..	Intermédiation monétaire			651
..	..	..	..	..	..	..	..	Autres intermédiations financières			659
..	..	..	..	..	..	..	..	Act. d'assur., caisses de retr., sauf séc. soc. oblig.		66	
..	..	..	..	..	..	..	..	Act. auxiliaires de l'intermédiation financière		67	
..	..	..	..	..	..	..	..	Act. aux. de l'interm. fin., sauf assur., et retraite			671
..	..	..	..	..	..	..	..	Act. aux. des assurances et caisses de retraite			672
..	..	..	..	17 050	18 416	19 743	23 501	Immobilier, locations, services aux entreprises	K	70_74	
..	..	..	..	2 220	2 377	2 323	2 821	Activités immobilières		70	
..	..	..	..	478	565	522	680	Act. immobilières sur biens propres ou loués			701
..	..	..	..	1 742	1 812	1 801	2 141	Activités immobilières pour comptes de tiers			702
..	..	..	..	265	394	336	394	Locat. mach., équip. sans opér., biens pers. et dom.		71	
..	..	..	..	62	79	77	91	Location de matériel de transport			711
..	..	..	..	155	217	205	242	Location de machines ét equipements nca			712
..	..	..	..	42	54	38	43	Loc. d'articles personnels et domestiques nca			713
..	..	..	..	3 249	3 604	4 067	5 217	Activités informatiques et activités rattachées		72	
..	..	..	..	10	44	46	48	Conseil en matériel informatique			721
..	..	..	..	2 004	2 263	2 608	3 422	Production de logiciels			722
..	..	..	..	1 039	1 162	1 303	1 540	Traitement de données informatiques			723
..	..	..	..	52	54	72	125	Activités de banques de données			724
..	..	..	..	135	69	38	81	Entr., répar. mach. de bureau et compt. et inform.			725
..	..	..	..	7	11	1	0	Autres activités rattachées à l'informatique			729
..	..	..	..	187	208	255	261	Recherche-développement		73	
..	..	..	..	174	200	246	251	Rech. et dév. expér. en sc. phys., natur. et ingén.			731
..	..	..	..	13	9	10	10	Rech. et dév. expér. en sciences soc. et humaines			732
..	..	..	..	11 129	11 833	12 761	14 807	Autres activités de services aux entreprises		74	
..	..	..	..	3 360	3 527	3 625	4 171	Act. jur., compt., audit.; cons. fisc.; ét. march., soud.			741
..	..	..	..	3 685	3 999	4 167	4 702	Act. d'achitect., d'ingén.; autres act. techniques			742
..	..	..	..	823	843	986	1 192	Publicité			743
..	..	..	..	3 262	3 464	3 983	4 742	Activités de services aux entreprises nca.			749
..	..	..	..	..	..	..	752	Education	M	80	
..	..	..	..	..	..	..	0	Activités d'enseignement primaire			801
..	..	..	..	..	..	..	122	Activités d'enseignement secondaire			802
..	..	..	..	..	..	..	69	Activités d'enseignement supérieur			803
..	..	..	..	..	..	..	561	Act. de formation perman.; autres act. d'enseign.			809
..	..	..	..	1 462	1 674	1 752	2 135	Santé et action sociale	N	85	
..	..	..	..	1 228	1 406	1 460	1 704	Activités de santé humaine			851
..	..	..	..	42	47	45	50	Activités vétérinaires			852
..	..	..	..	192	222	247	380	Activités d'action sociale			853
..	..	..	..	..	..	..	4 423	Autres activités de serv. coll., soc. et personnels	O	90_93	
..	..	..	..	475	684	553	623	Assain. et enlèv. des ordures; voierie, act. simil.		90	
..	..	..	..	..	..	..	21	Activités associatives diverses		91	
..	..	..	..	..	..	..	..	Activités de syndicats de salariés			912
..	..	..	..	2 302	2 526	2 651	3 006	Activités récréatives, culturelles et sportives		92	
..	..	..	..	377	461	493	608	Act. cinémat., de radio, télé.; autres act. de spect.			921
..	..	..	..	85	97	90	102	Activités d'agences de presse			922
..	..	..	..	13	14	15	12	Act. des biblio., arch., musées, jardins bot. et zool.			923
..	..	..	..	530	614	618	710	Activités sportives et autres activités récréatives			924
..	..	..	..	530	674	698	773	Autres activités de services		93	

FINLAND

Table FN.S7 WAGES AND SALARIES, EMPLOYEES - SALAIRES ET TRAITEMENTS, SALARIÉS

Millions of Mk (Current Prices)

ISIC revision 3				1991	1992	1993	1994	1995	1996	1997	1998
G	50_52		Wholesale & retail trade; repairs; etc.	..	..	..	..	25 046	23 849	25 046	27 414
	50		Sale, maint.,repair vehic/cycle, person., h/hold gd.	..	..	..	..	2 975	3 319	3 542	3 906
		501	Sale of motor cycle	..	..	..	..	985	1 127	1 232	1 338
		502	Maintenance and repair of motor vehicles	..	..	..	..	920	995	1 050	1 133
		503	Sale of motor vehicle parts and accessories	..	..	..	..	621	713	745	786
		504	Sale, maint., repair motor cycle, parts & acces.	..	..	..	..	20	27	28	30
		505	Retail sale of automative fuel	..	..	..	..	430	456	487	619
	51		W/sale. trade & commission exc motor veh. & cycle	..	..	..	..	9 481	11 229	12 157	13 047
		511	Wholesale on a fee or contract basis	..	..	..	..	455	580	620	654
		512	W/sale agric. raw mat.,live animal,food, bev., tobac.	..	..	..	..	892	1 023	1 086	1 304
		513	Wholesale of household goods	..	..	..	..	1 726	2 043	2 110	2 214
		514	W/sale. of non-agric. intermed. prod., waste, scrap	..	..	..	..	2 110	2 372	2 452	2 671
		515	Wholesale of machinery, equipment & supplies	..	..	..	..	3 555	4 298	4 667	4 856
		519	Other wholesale	..	..	..	..	744	913	1 222	1 348
	52		Retail trade exc. motor veh. & cyc.; repair h/hold gds.	..	..	..	..	8 628	9 301	9 347	10 460
		521	Non-specialised retail trade in stores	..	..	..	..	4 904	4 998	4 921	5 421
		522	Retail sale food, bev., tobac. in specialised stores	..	..	..	..	311	367	374	412
		523	Other retail trade of new goods specialised stores	..	..	..	..	3 202	3 683	3 777	4 329
		524	Retail sale of second-hand goods in stores	..	..	..	..	8	18	20	22
		525	Retail trade not in stores	..	..	..	..	104	130	142	152
		526	Repair of personal and household goods	..	..	..	..	100	104	113	123
H	55		Hotels and restaurants	..	..	..	..	3 595	3 875	4 103	4 467
		551	Hotels; camping ; other short-stay accommodation	..	..	..	..	1 212	1 283	1 251	1 378
		552	Restaurants, bars and canteens	..	..	..	..	2 383	2 592	2 853	3 089
I	60_64		Transport, storage and communications	..	..	..	..	17 771	17 842	17 771	19 799
	60		Land transport, transport via pipelines	..	..	..	..	4 046	7 001	6 513	7 303
		601	Transport via railways	..	..	..	..	..	..	..	..
		602	Other land transport	..	..	..	..	..	..	..	5 694
		603	Transport via pipelines	..	..	..	..	..	..	..	..
	61		Water transport	..	..	..	..	1 424	1 468	1 476	1 547
		611	Sea and coastal water transport	..	..	..	..	1 397	1 442	1 449	1 520
		612	Inland water transport	..	..	..	..	27	26	28	27
	62		Air transport	..	..	..	..	1 203	1 523	1 645	1 785
		621	Scheduled air transport	..	..	..	..	1 190	1 504	1 630	1 762
		622	Non-scheduled air transport	..	..	..	..	13	19	15	23
	63		Support. & aux. transport act., travel agencies activ.	..	..	..	..	2 213	2 640	2 764	3 256
	64		Post and telecommunications	..	..	..	..	4 895	5 210	5 373	5 908
		641	Post and courier activities	..	..	..	..	2 787	2 845	2 862	3 021
		642	Telecommunications	..	..	..	..	2 108	2 365	2 510	2 886
J	65_67		Financial intermediation	..	..	..	..	..	..	..	..
	65		Financial intermediation, exc. insur., pension fund	..	..	..	..	..	..	..	..
		651	Monetary intermediation	..	..	..	..	..	..	..	..
		659	Other financial intermediation	..	..	..	..	..	..	..	..
	66		Insurance pension fund, exc. compul. social sec.	..	..	..	..	..	..	..	..
	67		Activities auxiliary to financial intermediation	..	..	..	..	..	..	..	..
		671	Act. auxil to fin. intermed., exc. insur. & pens.fund	..	..	..	..	..	..	..	..
		672	Act. auxiliary to insurance and pension funding	..	..	..	..	..	..	..	..
K	70_74		Real estate, renting and business activities	..	..	..	..	15 556	14 490	15 556	18 507
	70		Real estate activities	..	..	..	..	1 709	1 857	1 845	2 197
		701	Real estate activ. with own or leased property	..	..	..	..	368	436	421	527
		702	Real estate activities on a fee or contract basis	..	..	..	..	1 341	1 422	1 424	1 670
	71		Renting of mach. without operator; rent. of pers. gds.	..	..	..	..	204	304	255	306
		711	Renting of transport equipment	..	..	..	..	48	61	59	70
		712	Renting of other machinery and equipment	..	..	..	..	119	167	155	187
		713	Renting of personal and household goods nec	..	..	..	..	32	42	29	33
	72		Computer and related activities	..	..	..	..	2 490	2 839	3 213	4 149
		721	Hardware consultancy	..	..	..	..	8	35	37	38
		722	Software consultancy and supply	..	..	..	..	1 525	1 777	2 050	2 712
		723	Data processing	..	..	..	..	807	921	1 039	1 235
		724	Data base activities	..	..	..	..	40	43	56	100
		725	Mainten.; repair of office, account. & comp. mach.	..	..	..	..	104	54	30	64
		729	Other computer related activities	..	..	..	..	6	8	0	0
	73		Research and development	..	..	..	..	144	162	204	205
		731	R & D experim. on natural sciences & engineering	..	..	..	..	134	156	196	198
		732	R & D experim. on social sciences & humanities	..	..	..	..	10	7	8	8
	74		Other business activities	..	..	..	..	8 542	9 328	10 039	11 650
		741	Legal, acc., book-keep., audit act.; tax consult., etc	..	..	..	..	2 518	2 753	2 812	3 210
		742	Architectural, engineering, & other techn. activities	..	..	..	..	2 854	3 153	3 278	3 740
		743	Advertising	..	..	..	..	634	662	767	936
		749	Business activities nec	..	..	..	..	2 535	2 760	3 181	3 765
M	80		Education	..	..	..	..	..	..	..	586
		801	Primary education	..	..	..	..	..	..	..	0
		802	Secondary education	..	..	..	..	..	..	..	95
		803	Higher education	..	..	..	..	..	..	..	54
		809	Adult and other education	..	..	..	..	..	..	..	437
N	85		Health and social work	..	..	..	..	1 126	1 328	1 392	1 664
		851	Human health activities	..	..	..	..	945	1 115	1 164	1 330
		852	Veterinary activities	..	..	..	..	32	37	35	39
		853	Social work activities	..	..	..	..	148	175	193	296
O	90_93		Other commun., soc. & pers. services activities	..	..	..	..	..	..	..	3 504
	90		Sewage & refuse disposal, sanitation, etc.	..	..	..	..	360	531	430	479
	91		Activities of member organisations nec	..	..	..	..	..	..	..	16
		912	Activities of trade unions	..	..	..	..	..	..	..	..
	92		Recreational, cultural and sporting activities	..	..	..	..	1 864	2 002	2 115	2 406
		921	Motion picture, radio, TV, other entertainment act.	..	..	..	..	290	352	378	475
		922	News agency activities	..	..	..	..	66	74	69	80
		923	Library, archives, museums, other cultural activ.	..	..	..	..	10	10	11	10
		924	Sporting and other recreational activities	..	..	..	..	435	458	470	556
	93		Other service activities	..	..	..	..	406	523	547	602

Table FN.S8 CHARGES SOCIALES DES EMPLOYEURS - SUPPLEMENTS TO WAGES & SALARIES

Millions de Mk (Prix courants)

1991	1992	1993	1994	1995	1996	1997	1998	CITI révision 3			
..	..	..	..	6 760	6 551	6 806	7 710	Com. de gros, de detail ; réparation etc.	G	50_52	
..	..	..	..	952	936	1 003	1 138	Com., entr., répar. d'auto. et moto ; carburants		50	
..	..	..	..	352	327	363	411	Commerce de véhicules automobiles			501
..	..	..	..	290	273	282	332	Entretien et réparation de véhicules automobiles			502
..	..	..	..	179	207	217	224	Com. de pièces et access. de véhic. automobiles			503
..	..	..	..	7	6	7	9	Com., entret. et répar. de motocycles et access.			504
..	..	..	..	124	123	134	163	Com. de détail de carburants pour automobiles			505
..	..	..	..	3 154	3 097	3 305	3 697	Com. de gros et act. inter., sauf auto. et moto		51	
..	..	..	..	154	160	172	185	Activités d'intermédiaries de commerce de gros			511
..	..	..	..	271	294	303	361	Com. gros: prod. agr. bruts, anim. viv., prod. alim.			512
..	..	..	..	541	568	577	624	Com. gros: biens de consom., sauf alimentaires			513
..	..	..	..	762	647	660	772	Com. gros: prod. inter. non agr., déchets, débris			514
..	..	..	..	1 184	1 167	1 278	1 401	Com. gros: machines, équipments et fournitures			515
..	..	..	..	242	260	314	353	Autres commerces de gros			519
..	..	..	..	2 654	2 517	2 499	2 875	Com. détail, sauf auto, moto; répar. art. pers., dom.		52	
..	..	..	..	1 541	1 384	1 350	1 507	Commerce de détail non spécialisé en magasin			521
..	..	..	..	123	103	108	117	Com. détail: prod. aliment. en mag. spécialisés			522
..	..	..	..	914	967	973	1 165	Autre com. détail: prod. neufs en mag. spécialisés			523
..	..	..	..	9	5	6	6	Commerces de détail de biens d'occasion			524
..	..	..	..	36	33	33	42	Commerces de détail autres qu'en magasin			525
..	..	..	..	32	26	28	38	Réparation d'articles personnels et domestiques			526
..	..	..	..	1 007	1 017	1 099	1 176	Hotels et restaurants	H	55	
..	..	..	..	358	347	341	365	Hôtels; camping; autres équip. courts séjours			551
..	..	..	..	649	670	757	811	Restaurants, cafés et cantines			552
..	..	..	..	4 111	4 622	4 996	5 546	Transport, entreposage, communications	I	60_64	
..	..	..	..	1 190	1 763	1 865	2 202	Transports terrestres; transports par conduites		60	
..	..	..	..	..	..	..	..	Transports par chemin de fer			601
..	..	..	..	..	..	..	1 684	Autres transports terrestres			602
..	..	..	..	..	..	..	..	Transports par conduites			603
..	..	..	..	418	407	391	398	Transports par eau		61	
..	..	..	..	409	399	384	391	Transports maritimes et côtiers			611
..	..	..	..	8	8	7	8	Transports par voies navigables intérieures			612
..	..	..	..	360	436	522	536	Transports aériens		62	
..	..	..	..	356	431	518	529	Transports aériens réguliers			621
..	..	..	..	4	4	4	7	Transports aériens non réguliers			622
..	..	..	..	699	726	788	946	Act. annex. et aux. des transp.; act. des ag. de voy.		63	
..	..	..	..	1 444	1 291	1 429	1 463	Postes et télécommunications		64	
..	..	..	..	834	630	694	704	Activités de poste et de courrier			641
..	..	..	..	610	661	735	759	Télécommunications			642
..	..	..	..	..	..	..	..	Intermédiation financière	J	65_67	
..	..	..	..	..	..	..	..	Interm. financ., sauf assur. et caisses de retraite		65	
..	..	..	..	..	..	..	..	Intermédiation monétaire			651
..	..	..	..	..	..	..	..	Autres intermédiations financières			659
..	..	..	..	..	..	..	..	Act. d'assur., caisses de retr., sauf séc. soc. oblig.		66	
..	..	..	..	..	..	..	..	Act. auxiliaires de l'intermédiation financière		67	
..	..	..	..	..	..	..	..	Act. aux. de l'interm. fin., sauf assur., et retraite			671
..	..	..	..	..	..	..	..	Act. aux. des assurances et caisses de retraite			672
..	..	..	..	3 969	3 905	4 169	4 994	Immobilier, locations, services aux entreprises	K	70_74	
..	..	..	..	511	522	484	624	Activités immobilières		70	
..	..	..	..	110	130	106	154	Act. immobilières sur biens propres ou loués			701
..	..	..	..	401	392	377	471	Activités immobilières pour comptes de tiers			702
..	..	..	..	61	90	82	88	Locat. mach., équip. sans opér., biens pers. et dom.		71	
..	..	..	..	14	18	19	20	Location de matériel de transport			711
..	..	..	..	36	51	50	54	Location de machines ét equipements nca			712
..	..	..	..	10	12	9	10	Loc. d'articles personnels et domestiques nca			713
..	..	..	..	760	766	858	1 068	Activités informatiques et activités rattachées		72	
..	..	..	..	2	9	9	11	Conseil en matériel informatique			721
..	..	..	..	480	494	576	709	Production de logiciels			722
..	..	..	..	233	234	250	305	Traitement de données informatiques			723
..	..	..	..	12	11	16	25	Activités de banques de données			724
..	..	..	..	32	15	8	17	Entr., répar. mach. de bureau et compt. et inform.			725
..	..	..	..	1	3	0	0	Autres activités rattachées à l'informatique			729
..	..	..	..	43	45	54	56	Recherche-développement		73	
..	..	..	..	40	43	52	54	Rech. et dév. expér. en sc. phys., natur. et ingén.			731
..	..	..	..	3	2	2	2	Rech. et dév. expér. en sciences soc. et humaines			732
..	..	..	..	2 593	2 482	2 692	3 157	Autres activités de services aux entreprises		74	
..	..	..	..	844	772	807	961	Act. jur., compt., audit.; cons. fisc.; ét. march., soud.			741
..	..	..	..	834	831	876	963	Act. d'achitect., d'ingén.; autres act. techniques			742
..	..	..	..	189	178	212	256	Publicité			743
..	..	..	..	727	701	797	978	Activités de services aux entreprises nca.			749
..	..	..	..	..	..	..	166	Education	M	80	
..	..	..	..	..	..	..	0	Activités d'enseignement primaire			801
..	..	..	..	..	..	..	27	Activités d'enseignement secondaire			802
..	..	..	..	..	..	..	15	Activités d'enseignement supérieur			803
..	..	..	..	..	..	..	123	Act. de formation perman.; autres act. d'enseign.			809
..	..	..	..	337	353	379	470	Santé et action sociale	N	85	
..	..	..	..	283	297	315	375	Activités de santé humaine			851
..	..	..	..	10	10	10	11	Activités vétérinaires			852
..	..	..	..	44	46	54	85	Activités d'action sociale			853
..	..	..	..	..	..	..	919	Autres activités de serv. coll., soc. et personnels	O	90_93	
..	..	..	..	116	149	125	144	Assain. et enlèv. des ordures; voierie, act. simil.		90	
..	..	..	..	..	..	..	5	Activités associatives diverses		91	
..	..	..	..	..	..	..	..	Activités de syndicats de salariés			912
..	..	..	..	438	523	520	599	Activités récréatives, culturelles et sportives		92	
..	..	..	..	87	108	116	133	Act. cinémat., de radio, télé.; autres act. de spect.			921
..	..	..	..	20	22	21	22	Activités d'agences de presse			922
..	..	..	..	3	3	3	3	Act. des biblio., arch., musées, jardins bot. et zool.			923
..	..	..	..	95	147	148	153	Activités sportives et autres activités récréatives			924
..	..	..	..	124	148	152	171	Autres activités de services		93	

FINLAND

Table FN.S9 **ESTABLISHMENTS - ÉTABLISSEMENTS**

Units

ISIC revision 3				1991	1992	1993	1994	1995	1996	1997	1998
G	**50_52**		Wholesale & retail trade; repairs; etc.	..	..	..	..	54 486	57 597	59 178	..
	50		Sale, maint.,repair vehic/cycle, person., h/hold gd.	..	..	..	..	9 516	10 119	10 425	..
		501	Sale of motor cycle	..	..	..	..	1 733	1 701	1 847	..
		502	Maintenance and repair of motor vehicles	..	..	..	..	4 577	5 131	5 304	..
		503	Sale of motor vehicle parts and accessories	..	..	..	..	1 567	1 660	1 652	..
		504	Sale, maint., repair motor cycle, parts & acces.	..	..	..	..	159	168	172	..
		505	Retail sale of automative fuel	..	..	..	..	1 480	1 459	1 450	..
	51		W/sale. trade & commission exc motor veh. & cycle	..	..	..	..	16 668	17 928	18 745	..
		511	Wholesale on a fee or contract basis	..	..	..	..	3 531	3 992	4 228	..
		512	W/sale agric. raw mat.,live animal,food, bev., tobac.	..	..	..	..	1 555	1 796	1 793	..
		513	Wholesale of household goods	..	..	..	..	3 674	3 911	4 105	..
		514	W/sale. of non-agric. intermed. prod., waste, scrap	..	..	..	..	3 011	3 146	3 320	..
		515	Wholesale of machinery, equipment & supplies	..	..	..	..	4 465	4 567	4 727	..
		519	Other wholesale	..	..	..	..	432	516	572	..
	52		Retail trade exc. motor veh. & cyc.; repair h/hold gds.	..	..	..	..	28 302	29 550	30 008	..
		521	Non-specialised retail trade in stores	..	..	..	..	7 912	7 657	7 493	..
		522	Retail sale food, bev., tobac. in specialised stores	..	..	..	..	1 651	1 778	1 850	..
		523	Other retail trade of new goods specialised stores	..	..	..	..	15 958	16 787	17 274	..
		524	Retail sale of second-hand goods in stores	..	..	..	..	629	749	782	..
		525	Retail trade not in stores	..	..	..	..	739	1 010	986	..
		526	Repair of personal and household goods	..	..	..	..	1 413	1 569	1 623	..
H	**55**		Hotels and restaurants	..	..	..	..	11 360	12 220	12 730	..
		551	Hotels; camping ; other short-stay accommodation	..	..	..	..	1 426	1 581	1 671	..
		552	Restaurants, bars and canteens	..	..	..	..	9 934	10 639	11 059	..
I	**60_64**		Transport, storage and communications	..	..	..	..	25 066	26 449	26 558	..
	60		Land transport, transport via pipelines	..	..	..	..	19 626	20 998	21 070	..
		601	Transport via railways	..	..	..	..	..	..	..	..
		602	Other land transport	..	..	..	..	..	..	..	..
		603	Transport via pipelines	..	..	..	..	..	..	..	..
	61		Water transport	..	..	..	..	347	357	357	..
		611	Sea and coastal water transport	..	..	..	..	230	243	240	..
		612	Inland water transport	..	..	..	..	117	114	117	..
	62		Air transport	..	..	..	..	123	119	134	..
		621	Scheduled air transport	..	..	..	..	63	59	70	..
		622	Non-scheduled air transport	..	..	..	..	60	60	64	..
	63		Support. & aux. transport act., travel agencies activ.	..	..	..	..	1 908	2 014	2 183	..
	64		Post and telecommunications	..	..	..	..	3 062	2 961	2 814	..
		641	Post and courier activities	..	..	..	..	2 520	2 399	2 246	..
		642	Telecommunications	..	..	..	..	542	562	568	..
J	**65_67**		Financial intermediation	..	..	..	..	4 409	4 785	4 778	..
	65		Financial intermediation, exc. insur., pension fund	..	..	..	..	2 793	2 940	2 708	..
		651	Monetary intermediation	..	..	..	..	2 060	1 991	1 676	..
		659	Other financial intermediation	..	..	..	..	733	949	1 032	..
	66		Insurance pension fund, exc. compul. social sec.	..	..	..	..	690	781	790	..
	67		Activities auxiliary to financial intermediation	..	..	..	..	926	1 064	1 280	..
		671	Act. auxil to fin. intermed., exc. insur. & pens.fund	..	..	..	..	860	974	1 152	..
		672	Act. auxiliary to insurance and pension funding	..	..	..	..	66	90	128	..
K	**70_74**		Real estate, renting and business activities	..	..	..	..	33 132	36 216	38 924	..
	70		Real estate activities	..	..	..	..	6 931	7 308	7 750	..
		701	Real estate activ. with own or leased property	..	..	..	..	2 917	2 968	3 111	..
		702	Real estate activities on a fee or contract basis	..	..	..	..	4 014	4 340	4 639	..
	71		Renting of mach. without operator; rent. of pers. gds.	..	..	..	..	1 099	1 164	1 205	..
		711	Renting of transport equipment	..	..	..	..	309	338	359	..
		712	Renting of other machinery and equipment	..	..	..	..	542	594	591	..
		713	Renting of personal and household goods nec	..	..	..	..	218	197	217	..
	72		Computer and related activities	..	..	..	..	2 845	3 118	3 395	..
		721	Hardware consultancy	..	..	..	..	118	143	164	..
		722	Software consultancy and supply	..	..	..	..	2 164	2 365	2 614	..
		723	Data processing	..	..	..	..	249	268	280	..
		724	Data base activities	..	..	..	..	36	66	86	..
		725	Mainten.; repair of office, account. & comp. mach.	..	..	..	..	268	261	241	..
		729	Other computer related activities	..	..	..	..	10	15	10	..
	73		Research and development	..	..	..	..	154	185	204	..
		731	R & D experim. on natural sciences & engineering	..	..	..	..	134	162	177	..
		732	R & D experim. on social sciences & humanities	..	..	..	..	20	23	27	..
	74		Other business activities	..	..	..	..	22 103	24 441	26 370	..
		741	Legal, acc., book-keep., audit act.; tax consult., etc	..	..	..	..	8 591	9 550	10 264	..
		742	Architectural, engineering, & other techn. activities	..	..	..	..	5 704	6 293	6 698	..
		743	Advertising	..	..	..	..	2 078	2 160	2 314	..
		749	Business activities nec	..	..	..	..	5 730	6 438	7 094	..
M	**80**		Education	..	..	..	..	1 146	1 296	1 357	..
		801	Primary education	..	..	..	..	12	9	9	..
		802	Secondary education	..	..	..	..	38	48	47	..
		803	Higher education	..	..	..	..		10	20	..
		809	Adult and other education	..	..	..	..	1 096	1 229	1 281	..
N	**85**		Health and social work	..	..	..	..	9 491	10 100	10 790	..
		851	Human health activities	..	..	..	..	8 259	8 690	9 199	..
		852	Veterinary activities	..	..	..	..	684	742	752	..
		853	Social work activities	..	..	..	..	548	668	839	..
O	**90_93**		Other commun., soc. & pers. services activities	..	..	..	..	13 857	15 610	16 518	..
	90		Sewage & refuse disposal, sanitation, etc.	..	..	..	..	789	879	892	..
	91		Activities of member organisations nec	..	..	..	..	381	447	444	..
		912	Activities of trade unions	..	..	..	..	..	..	..	..
	92		Recreational, cultural and sporting activities	..	..	..	..	3 437	3 815	4 188	..
		921	Motion picture, radio, TV, other entertainment act.	..	..	..	..	1 560	1 722	1 926	..
		922	News agency activities	..	..	..	..	78	78	76	..
		923	Library, archives, museums, other cultural activ.	..	..	..	..	36	41	46	..
		924	Sporting and other recreational activities	..	..	..	..	1 547	1 797	1 955	..
	93		Other service activities	..	..	..	..	9 250	10 469	10 994	..

Table FN.S10 ENTREPRISES - ENTERPRISES

Unités

1991	1992	1993	1994	1995	1996	1997	1998	CITI révision 3		
..	..	..	..	45 839	48 921	50 409	52 270	Com. de gros, de detail ; réparation etc.	G	**50_52**
..	..	..	..	7 986	8 574	8 873	9 453	Com., entr., répar. d'auto. et moto ; carburants		**50**
..	..	..	..	1 220	1 322	1 381	1 497	Commerce de véhicules automobiles		501
..	..	..	..	4 110	4 541	4 779	5 017	Entretien et réparation de véhicules automobiles		502
..	..	..	..	1 313	1 397	1 399	1 445	Com. de pièces et access. de véhic. automobiles		503
..	..	..	..	156	164	166	183	Com., entret. et répar. de motocycles et access.		504
..	..	..	..	1 187	1 150	1 148	1 311	Com. de détail de carburants pour automobiles		505
..	..	..	..	14 800	16 028	16 895	17 358	Com. de gros et act. inter., sauf auto. et moto		**51**
..	..	..	..	3 499	3 933	4 162	4 478	Activités d'intermédiaries de commerce de gros		511
..	..	..	..	1 283	1 405	1 495	1 548	Com. gros: prod. agr. bruts, anim. viv., prod. alim.		512
..	..	..	..	3 435	3 679	3 861	3 871	Com. gros: biens de consom., sauf alimentaires		513
..	..	..	..	2 456	2 617	2 787	2 855	Com. gros: prod. inter. non agr., déchets, débris		514
..	..	..	..	3 739	3 951	4 073	4 073	Com. gros: machines, équipments et fournitures		515
..	..	..	..	388	443	517	533	Autres commerces de gros		519
..	..	..	..	23 053	24 319	24 641	25 459	Com. détail, sauf auto, moto; répar. art. pers., dom.		**52**
..	..	..	..	4 924	4 854	4 684	4 757	Commerce de détail non spécialisé en magasin		521
..	..	..	..	1 329	1 414	1 435	1 443	Com. détail: prod. aliment. en mag. spécialisés		522
..	..	..	..	14 086	14 857	15 260	15 913	Autre com. détail: prod. neufs en mag. spécialisés		523
..	..	..	..	620	690	746	804	Commerces de détail de biens d'occasion		524
..	..	..	..	721	977	939	943	Commerces de détail autres qu'en magasin		525
..	..	..	..	1 373	1 527	1 577	1 599	Réparation d'articles personnels et domestiques		526
..	..	..	..	8 930	9 752	10 204	10 714	Hotels et restaurants	H	**55**
..	..	..	..	1 159	1 282	1 359	1 395	Hôtels; camping; autres équip. courts séjours		551
..	..	..	..	7 771	8 470	8 845	9 319	Restaurants, cafés et cantines		552
..	..	..	..	..	..	..	23 791	Transport, entreposage, communications	I	**60_64**
..	..	..	..	..	..	..	21 256	Transports terrestres; transports par conduites		**60**
..	..	..	..	..	..	..	4	Transports par chemin de fer		601
..	..	..	..	..	..	..	21 251	Autres transports terrestres		602
..	..	..	..	..	..	..	1	Transports par conduites		603
..	..	..	..	290	304	305	306	Transports par eau		**61**
..	..	..	..	182	198	195	202	Transports maritimes et côtiers		611
..	..	..	..	108	106	110	104	Transports par voies navigables intérieures		612
..	..	..	..	78	78	83	65	Transports aériens		**62**
..	..	..	..	20	21	23	4	Transports aériens réguliers		621
..	..	..	..	58	57	60	61	Transports aériens non réguliers		622
..	..	..	..	1 287	1 375	1 524	1 686	Act. annex. et aux. des transp.; act. des ag. de voy.		**63**
..	..	..	..	351	390	438	478	Postes et télécommunications		**64**
..	..	..	..	206	228	244	268	Activités de poste et de courrier		641
..	..	..	..	145	162	194	210	Télécommunications		642
..	..	..	..	2 115	2 544	2 822	2 893	Intermédiation financière	J	**65_67**
..	..	..	..	1 027	1 242	1 315	1 395	Interm. financ., sauf assur. et caisses de retraite		**65**
..	..	..	..	360	354	357	271	Intermédiation monétaire		651
..	..	..	..	667	888	958	1 124	Autres intermédiations financières		659
..	..	..	..	181	265	267	100	Act. d'assur., caisses de retr., sauf séc. soc. oblig.		**66**
..	..	..	..	907	1 037	1 240	1 398	Act. auxiliaires de l'intermédiation financière		**67**
..	..	..	..	845	951	1 120	1 246	Act. aux. de l'interm. fin., sauf assur., et retraite		671
..	..	..	..	62	86	120	152	Act. aux. des assurances et caisses de retraite		672
..	..	..	..	31 422	34 539	37 230	40 118	Immobilier, locations, services aux entreprises	K	**70_74**
..	..	..	..	6 652	7 045	7 452	8 154	Activités immobilières		**70**
..	..	..	..	2 823	2 901	3 055	3 565	Act. immobilières sur biens propres ou loués		701
..	..	..	..	3 829	4 144	4 397	4 589	Activités immobilières pour comptes de tiers		702
..	..	..	..	901	976	1 026	1 084	Locat. mach., équip. sans opér., biens pers. et dom.		**71**
..	..	..	..	278	305	324	343	Location de matériel de transport		711
..	..	..	..	427	472	479	501	Location de machines ét equipements nca		712
..	..	..	..	167	170	192	212	Loc. d'articles personnels et domestiques nca		713
..	..	..	..	2 579	2 880	3 158	3 459	Activités informatiques et activités rattachées		**72**
..	..	..	..	118	136	158	220	Conseil en matériel informatique		721
..	..	..	..	2 018	2 235	2 466	2 655	Production de logiciels		722
..	..	..	..	190	202	216	213	Traitement de données informatiques		723
..	..	..	..	35	66	83	118	Activités de banques de données		724
..	..	..	..	208	226	225	239	Entr., répar. mach. de bureau et compt. et inform.		725
..	..	..	..	10	15	10	14	Autres activités rattachées à l'informatique		729
..	..	..	..	143	225	244	214	Recherche-développement		**73**
..	..	..	..	123	192	209	191	Rech. et dév. expér. en sc. phys., natur. et ingén.		731
..	..	..	..	20	33	35	23	Rech. et dév. expér. en sciences soc. et humaines		732
..	..	..	..	21 147	23 413	25 350	27 207	Autres activités de services aux entreprises		**74**
..	..	..	..	8 272	9 199	9 952	10 645	Act. jur., compt., audit.; cons. fisc.; ét. march., soud.		741
..	..	..	..	5 385	5 912	6 299	6 686	Act. d'achitect., d'ingén.; autres act. techniques		742
..	..	..	..	2 042	2 121	2 277	2 456	Publicité		743
..	..	..	..	5 448	6 181	6 822	7 420	Activités de services aux entreprises nca.		749
..	..	..	..	1 091	1 221	1 272	1 446	Education	M	**80**
..	..	..	..	11	8	8	3	Activités d'enseignement primaire		801
..	..	..	..	27	32	33	22	Activités d'enseignement secondaire		802
..	..	..	..		3	5	3	Activités d'enseignement supérieur		803
..	..	..	..	1 053	1 178	1 226	1 418	Act. de formation perman.; autres act. d'enseign.		809
..	..	..	..	9 092	9 659	10 348	11 276	Santé et action sociale	N	**85**
..	..	..	..	7 875	8 269	8 783	9 438	Activités de santé humaine		851
..	..	..	..	678	739	748	760	Activités vétérinaires		852
..	..	..	..	539	651	817	1 078	Activités d'action sociale		853
..	..	..	..	13 110	14 949	15 883	16 651	Autres activités de serv. coll., soc. et personnels	O	**90_93**
..	..	..	..	746	829	840	893	Assain. et enlèv. des ordures; voierie, act. simil.		**90**
..	..	..	..	307	385	386	51	Activités associatives diverses		**91**
..	..	..	..	..	..	..	..	Activités de syndicats de salariés		912
..	..	..	..	3 006	3 448	3 436	4 219	Activités récréatives, culturelles et sportives		**92**
..	..	..	..	1 395	1 598	1 797	2 008	Act. cinémat., de radio, télé.; autres act. de spect.		921
..	..	..	..	69	68	67	67	Activités d'agences de presse		922
..	..	..	..	30	36	41	39	Act. des biblio., arch., musées, jardins bot. et zool.		923
..	..	..	..	1 380	1 611	1 785	1 938	Activités sportives et autres activités récréatives		924
..	..	..	..	9 051	10 287	10 821	11 488	Autres activités de services		**93**

Table FR.S1 **PRODUCTION - PRODUCTION**

Millions of FF (Current Prices)

ISIC revision 3				1991	1992	1993	1994	1995	1996	1997	1998
G	50_52		Wholesale & retail trade; repairs; etc.	..	..	1111 269	1101 105	1193 650	1282 954	1330 122	1165 189
	50		Sale, maint.,repair vehic/cycle, person., h/hold gd.	..	..	141 772	143 765	148 164	176 016	164 383	171 818
		501	Sale of motor cycle	..	..	81 991	80 663	84 418	100 538	91 561	100 547
		502	Maintenance and repair of motor vehicles	..	..	30 061	32 661	31 661	33 857	34 060	31 746
		503	Sale of motor vehicle parts and accessories	..	..	20 700	21 216	22 294	27 267	26 606	27 072
		504	Sale, maint., repair motor cycle, parts & acces.	..	..	3 464	3 177	3 602	5 444	5 365	4 761
		505	Retail sale of automative fuel	..	..	5 556	6 049	6 189	8 911	6 791	7 692
	51		W/sale. trade & commission exc motor veh. & cycle	..	..	527 444	519 269	578 209	615 008	645 642	477 636
		511	Wholesale on a fee or contract basis	..	..	39 876	21 965	56 326	56 827	81 356	67 146
		512	W/sale agric. raw mat.,live animal,food, bev., tobac.	..	..	145 810	150 142	147 649	163 906	161 050	97 862
		513	Wholesale of household goods	..	..	104 470	106 576	116 274	127 479	129 552	108 461
		514	W/sale. of non-agric. intermed. prod., waste, scrap	..	..	106 225	109 405	114 183	116 789	118 112	93 288
		515	Wholesale of machinery, equipment & supplies	..	..	107 992	106 674	117 039	120 811	127 744	94 206
		519	Other wholesale	..	..	1 590	1 480	1 521	3 174	1 682	1 669
	52		Retail trade exc. motor veh. & cyc.; repair h/hold gds.	..	..	442 053	438 071	467 277	491 931	520 097	515 735
		521	Non-specialised retail trade in stores	..	..	145 554	143 593	159 354	166 081	175 177	186 108
		522	Retail sale food, bev., tobac. in specialised stores	..	..	29 661	28 828	28 518	30 092	32 013	27 660
		523	Other retail trade of new goods specialised stores	..	..	222 299	215 815	227 027	243 782	257 711	248 633
		524	Retail sale of second-hand goods in stores	..	..	3 774	4 277	4 618	5 415	5 853	5 331
		525	Retail trade not in stores	..	..	33 735	38 610	40 215	39 796	42 561	40 620
		526	Repair of personal and household goods	..	..	7 031	6 948	7 546	6 766	6 782	7 384
H	55		Hotels and restaurants	..	..	227 005	230 133	229 180	237 061	255 844	300 790
		551	Hotels; camping ; other short-stay accommodation	..	..	67 833	71 997	67 428	69 302	72 927	98 265
		552	Restaurants, bars and canteens	..	..	159 172	158 136	161 752	167 759	182 918	202 525
I	60_64		Transport, storage and communications	..	..	..	..	..	..	..	..
	60		Land transport, transport via pipelines	..	..	..	..	..	..	..	..
		601	Transport via railways	..	..	..	..	..	..	..	..
		602	Other land transport	..	..	..	..	..	..	..	..
		603	Transport via pipelines	..	..	..	..	..	..	..	..
	61		Water transport	..	..	..	..	..	..	..	..
		611	Sea and coastal water transport	..	..	..	..	..	..	..	..
		612	Inland water transport	..	..	..	..	..	..	..	..
	62		Air transport	..	..	..	..	..	..	..	..
		621	Scheduled air transport	..	..	..	..	..	..	..	..
		622	Non-scheduled air transport	..	..	..	..	..	..	..	..
	63		Support. & aux. transport act., travel agencies activ.	..	..	42 933	45 278	43 223	53 532	..	..
	64		Post and telecommunications	..	..	15 444	14 378	17 359	26 442	38 574	66 656
		641	Post and courier activities	..	..	..	1 430	1 530	2 060	2 415	2 517
		642	Telecommunications	..	..	15 444	12 947	15 829	24 383	36 159	64 139
J	65_67		Financial intermediation	..	..	..	..	..	..	..	..
	65		Financial intermediation, exc. insur., pension fund	..	..	..	..	..	..	..	..
		651	Monetary intermediation	..	..	..	..	..	..	..	..
		659	Other financial intermediation	..	..	..	..	..	..	..	..
	66		Insurance pension fund, exc. compul. social sec.	..	..	..	..	..	..	..	..
	67		Activities auxiliary to financial intermediation	..	..	..	..	..	..	..	..
		671	Act. auxil to fin. intermed., exc. insur. & pens.fund	..	..	..	..	..	..	..	..
		672	Act. auxiliary to insurance and pension funding	..	..	..	..	..	..	..	..
K	70_74		Real estate, renting and business activities	..	..	1048 591	1095 175	1123 355	1367 676	1433 803	1544 843
	70		Real estate activities	..	..	242 597	246 354	234 551	275 824	280 885	300 282
		701	Real estate activ. with own or leased property	..	..	187 215	177 471	179 254	217 882	218 706	238 575
		702	Real estate activities on a fee or contract basis	..	..	55 383	68 883	55 297	57 942	62 179	61 708
	71		Renting of mach. without operator; rent. of pers. gds.	..	..	65 983	69 848	69 875	85 357	83 986	97 231
		711	Renting of transport equipment	..	..	25 190	26 756	27 389	37 521	34 920	46 237
		712	Renting of other machinery and equipment	..	..	19 086	18 806	19 995	23 585	25 125	26 526
		713	Renting of personal and household goods nec	..	..	9 211	9 841	10 055	11 448	11 755	13 160
	72		Computer and related activities	..	..	114 597	117 475	125 893	129 168	144 293	174 079
		721	Hardware consultancy	..	..	32 904	35 245	36 925	32 938	42 963	51 008
		722	Software consultancy and supply	..	..	32 903	35 013	37 430	45 995	52 699	70 152
		723	Data processing	..	..	39 758	36 110	37 527	37 467	35 205	37 101
		724	Data base activities	..	..	1 659	2 759	2 441	2 563	3 239	3 873
		725	Mainten.; repair of office, account. & comp. mach.	..	..	7 374	8 348	11 570	10 204	10 187	11 946
		729	Other computer related activities	..	..	..	..	..	..	..	..
	73		Research and development	..	..	..	..	..	..	..	..
		731	R & D experim. on natural sciences & engineering	..	..	..	..	..	..	..	..
		732	R & D experim. on social sciences & humanities	..	..	..	..	..	..	..	..
	74		Other business activities	..	..	625 414	661 498	693 037	877 327	924 639	973 250
		741	Legal, acc., book-keep., audit act.; tax consult., etc	..	..	212 948	235 368	240 731	366 438	372 372	384 110
		742	Architectural, engineering, & other techn. activities	..	..	136 008	136 714	143 488	163 387	181 585	174 991
		743	Advertising	..	..	89 907	93 215	93 887	103 004	111 465	120 180
		749	Business activities nec	..	..	186 551	196 201	214 931	244 499	259 218	293 969
M	80		Education	..	..	..	..	..	..	..	..
		801	Primary education	..	..	..	..	..	..	..	..
		802	Secondary education	..	..	..	..	..	..	..	..
		803	Higher education	..	..	..	..	..	..	..	..
		809	Adult and other education	..	..	..	..	..	..	..	..
N	85		Health and social work	..	..	..	..	..	..	..	..
		851	Human health activities	..	..	..	..	..	..	..	..
		852	Veterinary activities	..	..	..	..	..	..	..	..
		853	Social work activities	..	..	..	..	..	..	..	..
O	90_93		Other commun., soc. & pers. services activities	..	..	129 428	138 242	148 227	172 305	186 440	201 533
	90		Sewage & refuse disposal, sanitation, etc.	..	..	20 139	22 884	25 828	26 770	29 180	34 241
	91		Activities of member organisations nec	..	..	..	..	..	..	..	..
		912	Activities of trade unions	..	..	..	..	..	..	..	..
	92		Recreational, cultural and sporting activities	..	..	64 963	71 326	76 675	98 034	106 425	115 563
		921	Motion picture, radio, TV, other entertainment act.	..	..	24 684	26 980	30 767	41 777	44 821	48 690
		922	News agency activities	..	..	3 185	3 627	3 518	3 749	3 888	4 483
		923	Library, archives, museums, other cultural activ.	..	..	..	..	..	..	..	..
		924	Sporting and other recreational activities	..	..	..	..	..	..	..	..
	93		Other service activities	..	..	44 325	44 033	45 724	47 501	50 834	51 729

Table FR.S2 **VALEUR AJOUTÉE - VALUE ADDED**

Millions de FF (Prix courants)

1991	1992	1993	1994	1995	1996	1997	1998			CITI révision 3	
..	..	641 712	626 574	675 116	740 550	778 324	721 369	Com. de gros, de detail ; réparation etc.	G	50_52	
..	..	85 437	85 068	87 279	94 547	100 594	103 703	Com., entr., répar. d'auto. et moto ; carburants		50	
..	..	48 016	47 391	48 831	53 009	52 684	58 129	Commerce de véhicules automobiles			501
..	..	18 246	18 537	18 250	17 952	22 724	20 256	Entretien et réparation de véhicules automobiles			502
..	..	13 689	13 669	14 490	16 345	16 963	17 395	Com. de pièces et access. de véhic. automobiles			503
..	..	2 023	1 941	2 207	2 591	3 282	2 984	Com., entret. et répar. de motocycles et access.			504
..	..	3 464	3 530	3 501	4 650	4 941	4 939	Com. de détail de carburants pour automobiles			505
..	..	278 560	270 528	302 479	351 732	342 358	286 734	Com. de gros et act. inter., sauf auto. et moto		51	
..	..	20 155	10 318	28 894	33 491	42 570	40 250	Activités d'intermédiaries de commerce de gros			511
..	..	66 621	65 186	66 108	82 329	74 199	51 024	Com. gros: prod. agr. bruts, anim. viv., prod. alim.			512
..	..	55 592	55 234	59 795	70 702	67 322	63 464	Com. gros: biens de consom., sauf alimentaires			513
..	..	59 938	63 284	63 080	71 112	66 852	56 978	Com. gros: prod. inter. non agr., déchets, débris			514
..	..	62 500	61 794	68 624	75 313	74 292	62 670	Com. gros: machines, équipments et fournitures			515
..	..	469	687	650	2 456	774	1 277	Autres commerces de gros			519
..	..	277 715	270 979	285 357	294 272	335 373	330 933	Com. détail, sauf auto, moto; répar. art. pers., dom.		52	
..	..	91 142	88 122	95 610	99 724	107 904	115 054	Commerce de détail non spécialisé en magasin			521
..	..	19 970	19 164	19 126	18 778	23 529	21 028	Com. détail: prod. aliment. en mag. spécialisés			522
..	..	145 398	140 105	146 627	151 355	174 146	167 779	Autre com. détail: prod. neufs en mag. spécialisés			523
..	..	1 862	2 161	2 374	2 955	4 081	3 191	Commerces de détail de biens d'occasion			524
..	..	15 209	17 386	17 366	16 916	20 249	18 761	Commerces de détail autres qu'en magasin			525
..	..	4 134	4 041	4 255	4 544	5 464	5 120	Réparation d'articles personnels et domestiques			526
..	..	107 088	109 123	108 293	113 495	119 726	143 926	Hotels et restaurants	H	55	
..	..	32 584	35 246	32 973	34 686	35 169	49 260	Hôtels; camping; autres équip. courts séjours			551
..	..	74 504	73 877	75 320	78 809	84 558	94 666	Restaurants, cafés et cantines			552
..	..	..	..	..	..	..	..	Transport, entreposage, communications	I	60_64	
..	..	..	..	..	..	..	..	Transports terrestres; transports par conduites		60	
..	..	..	..	..	..	..	..	Transports par chemin de fer			601
..	..	..	..	..	..	..	..	Autres transports terrestres			602
..	..	..	..	..	..	..	..	Transports par conduites			603
..	..	..	..	..	..	..	..	Transports par eau		61	
..	..	..	..	..	..	..	..	Transports maritimes et côtiers			611
..	..	..	..	..	..	..	..	Transports par voies navigables intérieures			612
..	..	..	..	..	..	..	..	Transports aériens		62	
..	..	..	..	..	..	..	..	Transports aériens réguliers			621
..	..	..	..	..	..	..	..	Transports aériens non réguliers			622
..	..	..	..	..	..	..	..	Act. annex. et aux. des transp.; act. des ag. de voy.		63	
..	..	7 117	4 772	4 685	6 286	6 093	11 545	Postes et télécommunications		64	
..	..	..	776	844	1 104	1 213	1 149	Activités de poste et de courrier			641
..	..	7 117	3 996	3 840	5 183	4 880	10 395	Télécommunications			642
..	..	..	..	..	..	..	..	Intermédiation financière	J	65_67	
..	..	..	..	..	..	..	..	Interm. financ., sauf assur. et caisses de retraite		65	
..	..	..	..	..	..	..	..	Intermédiation monétaire			651
..	..	..	..	..	..	..	..	Autres intermédiations financières			659
..	..	..	..	..	..	..	..	Act. d'assur., caisses de retr., sauf séc. soc. oblig.		66	
..	..	..	..	..	..	..	..	Act. auxiliaires de l'intermédiation financière		67	
..	..	..	..	..	..	..	..	Act. aux. de l'interm. fin., sauf assur., et retraite			671
..	..	..	..	..	..	..	..	Act. aux. des assurances et caisses de retraite			672
..	..	525 761	563 933	571 877	699 782	729 503	758 076	Immobilier, locations, services aux entreprises	K	70_74	
..	..	98 556	108 898	102 022	130 465	130 541	140 059	Activités immobilières		70	
..	..	74 415	79 132	76 096	101 636	98 674	107 680	Act. immobilières sur biens propres ou loués			701
..	..	24 141	29 766	25 926	28 829	31 867	32 379	Activités immobilières pour comptes de tiers			702
..	..	41 708	42 062	42 050	52 073	50 989	57 178	Locat. mach., équip. sans opér., biens pers. et dom.		71	
..	..	16 190	15 978	16 161	22 539	19 617	27 591	Location de matériel de transport			711
..	..	10 721	10 371	11 374	14 143	16 053	14 633	Location de machines ét equipements nca			712
..	..	5 515	5 696	5 904	6 741	6 880	6 893	Loc. d'articles personnels et domestiques nca			713
..	..	59 877	63 253	65 893	68 703	77 384	88 081	Activités informatiques et activités rattachées		72	
..	..	17 748	20 212	21 019	18 493	24 002	26 490	Conseil en matériel informatique			721
..	..	18 078	20 017	21 162	26 653	31 291	38 929	Production de logiciels			722
..	..	19 999	18 310	18 318	18 144	16 670	16 615	Traitement de données informatiques			723
..	..	733	936	1 088	1 054	1 315	1 351	Activités de banques de données			724
..	..	3 319	3 778	4 307	4 358	4 108	4 697	Entr., répar. mach. de bureau et compt. et inform.			725
..	..	..	..	..	..	..	..	Autres activités rattachées à l'informatique			729
..	..	..	..	..	..	..	..	Recherche-développement		73	
..	..	..	..	..	..	..	..	Rech. et dév. expér. en sc. phys., natur. et ingén.			731
..	..	..	..	..	..	..	..	Rech. et dév. expér. en sciences soc. et humaines			732
..	..	325 621	349 721	361 912	448 543	470 588	472 759	Autres activités de services aux entreprises		74	
..	..	120 834	128 675	125 307	181 407	183 728	171 731	Act. jur., compt., audit.; cons. fisc.; ét. march., soud.			741
..	..	62 940	65 039	65 489	79 293	83 627	78 766	Act. d'achitect., d'ingén.; autres act. techniques			742
..	..	25 566	27 225	27 791	31 880	34 535	33 111	Publicité			743
..	..	116 282	128 782	143 326	155 963	168 699	189 151	Activités de services aux entreprises nca.			749
..	..	..	..	..	..	..	..	Education	M	80	
..	..	..	..	..	..	..	..	Activités d'enseignement primaire			801
..	..	..	..	..	..	..	..	Activités d'enseignement secondaire			802
..	..	..	..	..	..	..	..	Activités d'enseignement supérieur			803
..	..	..	..	..	..	..	..	Act. de formation perman.; autres act. d'enseign.			809
..	..	..	..	..	..	..	..	Santé et action sociale	N	85	
..	..	..	..	..	..	..	..	Activités de santé humaine			851
..	..	..	..	..	..	..	..	Activités vétérinaires			852
..	..	..	..	..	..	..	..	Activités d'action sociale			853
..	..	74 742	78 408	85 873	88 956	93 981	96 484	Autres activités de serv. coll., soc. et personnels	O	90_93	
..	..	10 101	11 092	12 161	12 520	13 623	14 595	Assain. et enlèv. des ordures; voierie, act. simil.		90	
..	..	..	..	..	..	..	..	Activités associatives diverses		91	
..	..	..	..	..	..	..	..	Activités de syndicats de salariés			912
..	..	36 949	40 328	45 414	46 707	49 372	49 494	Activités récréatives, culturelles et sportives		92	
..	..	13 883	15 512	17 473	19 891	21 498	20 576	Act. cinémat., de radio, télé.; autres act. de spect.			921
..	..	1 999	2 217	2 265	2 313	2 375	2 656	Activités d'agences de presse			922
..	..	..	..	..	..	..	..	Act. des biblio., arch., musées, jardins bot. et zool.			923
..	..	..	..	..	..	..	..	Activités sportives et autres activités récréatives			924
..	..	27 691	26 989	28 297	29 729	30 986	32 395	Autres activités de services		93	

Table FR.S3 **INVESTMENT - INVESTISSEMENT**

Millions of FF (Current Prices)

ISIC revision 3				1991	1992	1993	1994	1995	1996	1997	1998
G	50_52		Wholesale & retail trade; repairs; etc.	..	..	61 403	69 850	63 409	60 758	78 552	67 296
	50		Sale, maint.,repair vehic/cycle, person., h/hold gd.	..	..	7 128	7 868	8 307	8 249	8 324	7 981
		501	Sale of motor cycle	..	..	4 640	5 031	5 563	5 030	5 099	5 181
		502	Maintenance and repair of motor vehicles	..	..	1 374	1 567	1 337	1 339	1 485	1 122
		503	Sale of motor vehicle parts and accessories	..	..	872	967	1 109	1 518	1 281	1 201
		504	Sale, maint., repair motor cycle, parts & acces.	..	..	88	140	127	164	184	236
		505	Retail sale of automative fuel	..	..	154	164	171	199	275	241
	51		W/sale. trade & commission exc motor veh. & cycle	..	..	22 708	21 881	23 347	23 797	26 152	20 432
		511	Wholesale on a fee or contract basis	..	..	1 282	432	1 370	1 635	1 718	1 521
		512	W/sale agric. raw mat.,live animal,food, bev., tobac.	..	..	7 944	7 983	7 120	7 149	8 178	4 513
		513	Wholesale of household goods	..	..	3 239	3 360	3 892	3 440	4 272	3 949
		514	W/sale. of non-agric. intermed. prod., waste, scrap	..	..	4 700	4 648	5 245	5 793	6 463	5 515
		515	Wholesale of machinery, equipment & supplies	..	..	4 561	4 462	4 653	4 835	4 571	4 197
		519	Other wholesale	..	..	37	97	56	86	59	44
	52		Retail trade exc. motor veh. & cyc.; repair h/hold gds.	..	..	31 567	40 100	31 755	28 712	44 075	38 883
		521	Non-specialised retail trade in stores	..	..	15 209	24 658	15 565	14 295	27 380	21 985
		522	Retail sale food, bev., tobac. in specialised stores	..	..	1 463	1 515	1 757	963	1 270	1 314
		523	Other retail trade of new goods specialised stores	..	..	13 321	12 020	11 772	11 779	13 145	13 979
		524	Retail sale of second-hand goods in stores	..	..	161	210	172	223	364	223
		525	Retail trade not in stores	..	..	1 151	1 465	2 168	1 247	1 624	1 193
		526	Repair of personal and household goods	..	..	262	233	323	206	292	190
H	55		Hotels and restaurants	..	..	18 716	14 980	14 674	12 993	..	13 970
		551	Hotels; camping ; other short-stay accommodation	..	..	8 448	6 684	6 754	7 232	..	6 751
		552	Restaurants, bars and canteens	..	..	10 269	8 296	7 920	5 761	..	7 220
I	60_64		Transport, storage and communications	..	..	..	..	..	..	..	..
	60		Land transport, transport via pipelines	..	..	..	..	..	..	..	..
		601	Transport via railways	..	..	..	..	..	..	..	..
		602	Other land transport	..	..	..	..	..	..	..	..
		603	Transport via pipelines	..	..	..	..	..	..	..	..
	61		Water transport	..	..	..	..	..	..	..	..
		611	Sea and coastal water transport	..	..	..	..	..	..	..	..
		612	Inland water transport	..	..	..	..	..	..	..	..
	62		Air transport	..	..	..	..	..	..	..	..
		621	Scheduled air transport	..	..	..	..	..	..	..	..
		622	Non-scheduled air transport	..	..	..	..	..	..	..	..
	63		Support. & aux. transport act., travel agencies activ.	..	..	459	421	464	534	..	..
	64		Post and telecommunications	..	..	2 699	3 060	5 063	5 670	9 744	18 241
		641	Post and courier activities	..	..	..	54	65	52	64	59
		642	Telecommunications	..	..	2 699	3 007	4 998	5 619	9 680	18 182
J	65_67		Financial intermediation	..	..	..	..	..	..	..	..
	65		Financial intermediation, exc. insur., pension fund	..	..	..	..	..	..	..	..
		651	Monetary intermediation	..	..	..	..	..	..	..	..
		659	Other financial intermediation	..	..	..	..	..	..	..	..
	66		Insurance pension fund, exc. compul. social sec.	..	..	..	..	..	..	..	..
	67		Activities auxiliary to financial intermediation	..	..	..	..	..	..	..	..
		671	Act. auxil to fin. intermed., exc. insur. & pens.fund	..	..	..	..	..	..	..	..
		672	Act. auxiliary to insurance and pension funding	..	..	..	..	..	..	..	..
K	70_74		Real estate, renting and business activities	..	..	149 770	156 536	145 144	157 401	..	188 795
	70		Real estate activities	..	..	79 385	92 353	79 370	68 546	..	70 984
		701	Real estate activ. with own or leased property	..	..	74 708	83 149	72 005	62 338	..	67 832
		702	Real estate activities on a fee or contract basis	..	..	4 677	9 205	7 365	6 207	..	3 152
	71		Renting of mach. without operator; rent. of pers. gds.	..	..	37 736	41 802	40 802	59 200	58 171	74 061
		711	Renting of transport equipment	..	..	21 227	24 272	25 313	33 195	33 628	48 977
		712	Renting of other machinery and equipment	..	..	5 966	6 145	7 096	10 069	13 368	15 015
		713	Renting of personal and household goods nec	..	..	2 066	2 018	2 101	2 876	2 263	2 753
	72		Computer and related activities	..	..	3 435	3 591	4 115	4 105	4 720	5 766
		721	Hardware consultancy	..	..	731	834	695	675	1 079	1 180
		722	Software consultancy and supply	..	..	701	760	871	1 305	1 242	1 731
		723	Data processing	..	..	1 770	1 689	2 172	1 762	2 030	2 439
		724	Data base activities	..	..	50	109	68	58	93	95
		725	Mainten.; repair of office, account. & comp. mach.	..	..	182	200	310	305	276	321
		729	Other computer related activities	..	..	..	..	..	..	..	..
	73		Research and development	..	..	..	..	..	..	..	..
		731	R & D experim. on natural sciences & engineering	..	..	..	..	..	..	..	..
		732	R & D experim. on social sciences & humanities	..	..	..	..	..	..	..	..
	74		Other business activities	..	..	29 214	18 791	20 857	25 550	28 144	37 984
		741	Legal, acc., book-keep., audit act.; tax consult., etc	..	..	13 935	8 323	10 842	13 425	16 059	18 737
		742	Architectural, engineering, & other techn. activities	..	..	4 591	3 477	3 704	4 147	4 632	6 653
		743	Advertising	..	..	1 379	1 496	1 429	1 462	1 419	2 148
		749	Business activities nec	..	..	9 309	5 495	4 883	6 518	6 035	10 447
M	80		Education	..	..	..	..	..	..	..	..
		801	Primary education	..	..	..	..	..	..	..	..
		802	Secondary education	..	..	..	..	..	..	..	..
		803	Higher education	..	..	..	..	..	..	..	..
		809	Adult and other education	..	..	..	..	..	..	..	..
N	85		Health and social work	..	..	..	..	..	..	..	..
		851	Human health activities	..	..	..	..	..	..	..	..
		852	Veterinary activities	..	..	..	..	..	..	..	..
		853	Social work activities	..	..	..	..	..	..	..	..
O	90_93		Other commun., soc. & pers. services activities	..	..	7 696	7 791	8 089	8 013	..	12 801
	90		Sewage & refuse disposal, sanitation, etc.	..	..	1 903	2 007	2 570	2 592	4 229	4 121
	91		Activities of member organisations nec	..	..	..	..	..	..	..	..
		912	Activities of trade unions	..	..	..	..	..	..	..	..
	92		Recreational, cultural and sporting activities	..	..	3 200	3 421	3 363	3 499	..	5 197
		921	Motion picture, radio, TV, other entertainment act.	..	..	1 242	1 144	1 332	1 605	..	2 812
		922	News agency activities	..	..	113	112	105	109	..	119
		923	Library, archives, museums, other cultural activ.	..	..	..	..	..	..	..	..
		924	Sporting and other recreational activities	..	..	..	..	..	..	..	..
	93		Other service activities	..	..	2 593	2 363	2 156	1 923	..	3 484

Table FR.S4 **EMPLOI, SALARIÉS - EMPLOYMENT, EMPLOYEES**

Unités

1991	1992	1993	1994	1995	1996	1997	1998			CITI révision 3	
..	..	2 448 391	2 397 919	2 500 758	2 681 584	2 800 571	2 538 243	Com. de gros, de detail ; réparation etc.	G	50_52	
..	..	358 769	364 347	365 001	398 341	404 800	396 101	Com., entr., répar. d'auto. et moto ; carburants		50	
..	..	201 395	197 182	198 953	211 552	205 655	206 811	Commerce de véhicules automobiles			501
..	..	78 740	84 090	80 046	85 148	95 973	87 310	Entretien et réparation de véhicules automobiles			502
..	..	51 469	53 739	54 301	63 533	63 140	63 523	Com. de pièces et access. de véhic. automobiles			503
..	..	7 919	7 820	9 119	10 762	11 879	11 891	Com., entret. et répar. de motocycles et access.			504
..	..	19 246	21 516	22 582	27 346	28 153	26 566	Com. de détail de carburants pour automobiles			505
..	..	857 663	808 557	863 373	932 358	958 114	754 436	Com. de gros et act. inter., sauf auto. et moto		51	
..	..	51 919	15 390	52 035	68 895	70 455	65 433	Activités d'intermédiaries de commerce de gros			511
..	..	227 691	217 892	213 540	225 603	227 902	149 312	Com. gros: prod. agr. bruts, anim. viv., prod. alim.			512
..	..	163 504	163 124	169 567	192 351	191 081	170 888	Com. gros: biens de consom., sauf alimentaires			513
..	..	174 873	174 711	182 712	184 079	196 097	157 871	Com. gros: prod. inter. non agr., déchets, débris			514
..	..	190 069	187 132	194 554	206 529	217 951	175 434	Com. gros: machines, équipments et fournitures			515
..	..	1 776	2 356	2 694	6 062	2 861	2 893	Autres commerces de gros			519
..	..	1 231 959	1 225 015	1 272 384	1 350 885	1 437 657	1 387 706	Com. détail, sauf auto, moto; répar. art. pers., dom.		52	
..	..	463 794	462 927	503 132	519 551	530 994	540 081	Commerce de détail non spécialisé en magasin			521
..	..	73 630	69 935	71 010	76 788	89 382	81 172	Com. détail: prod. aliment. en mag. spécialisés			522
..	..	595 076	586 077	604 603	648 269	688 477	666 409	Autre com. détail: prod. neufs en mag. spécialisés			523
..	..	5 819	6 583	7 651	10 768	12 949	9 692	Commerces de détail de biens d'occasion			524
..	..	78 065	83 018	69 556	76 219	92 180	70 796	Commerces de détail autres qu'en magasin			525
..	..	15 575	16 475	16 432	19 290	23 675	19 556	Réparation d'articles personnels et domestiques			526
..	..	518 101	533 107	543 189	508 633	557 988	575 508	Hotels et restaurants	H	55	
..	..	157 348	165 510	153 482	139 614	148 901	176 238	Hôtels; camping; autres équip. courts séjours			551
..	..	360 752	367 597	389 706	369 019	409 088	399 270	Restaurants, cafés et cantines			552
..	..	..	..	..	..	..	..	Transport, entreposage, communications	I	60_64	
..	..	..	..	..	..	..	..	Transports terrestres; transports par conduites		60	
..	..	..	..	..	..	..	..	Transports par chemin de fer			601
..	..	..	..	..	..	..	..	Autres transports terrestres			602
..	..	..	..	..	..	..	..	Transports par conduites			603
..	..	..	..	..	..	..	..	Transports par eau		61	
..	..	..	..	..	..	..	..	Transports maritimes et côtiers			611
..	..	..	..	..	..	..	..	Transports par voies navigables intérieures			612
..	..	..	..	..	..	..	..	Transports aériens		62	
..	..	..	..	..	..	..	..	Transports aériens réguliers			621
..	..	..	..	..	..	..	..	Transports aériens non réguliers			622
..	..	29 037	30 196	28 714	31 883	..	..	Act. annex. et aux. des transp.; act. des ag. de voy.		63	
..	..	10 154	10 400	11 123	15 669	20 335	30 801	Postes et télécommunications		64	
..	..	..	3 481	4 147	5 152	5 762	5 522	Activités de poste et de courrier			641
..	..	10 154	6 919	6 976	10 517	14 573	25 279	Télécommunications			642
..	..	..	..	..	..	..	..	Intermédiation financière	J	65_67	
..	..	..	..	..	..	..	..	Interm. financ., sauf assur. et caisses de retraite		65	
..	..	..	..	..	..	..	..	Intermédiation monétaire			651
..	..	..	..	..	..	..	..	Autres intermédiations financières			659
..	..	..	..	..	..	..	..	Act. d'assur., caisses de retr., sauf séc. soc. oblig.		66	
..	..	..	..	..	..	..	..	Act. auxiliaires de l'intermédiation financière		67	
..	..	..	..	..	..	..	..	Act. aux. de l'interm. fin., sauf assur., et retraite			671
..	..	..	..	..	..	..	..	Act. aux. des assurances et caisses de retraite			672
..	..	1 663 440	1 743 547	1 816 324	1 835 191	1 960 422	2 064 605	Immobilier, locations, services aux entreprises	K	70_74	
..	..	183 576	180 201	185 392	178 516	188 979	191 931	Activités immobilières		70	
..	..	109 018	108 679	109 518	102 624	107 854	104 765	Act. immobilières sur biens propres ou loués			701
..	..	74 558	71 523	75 874	75 893	81 125	87 166	Activités immobilières pour comptes de tiers			702
..	..	47 488	50 254	50 085	52 967	54 759	60 462	Locat. mach., équip. sans opér., biens pers. et dom.		71	
..	..	11 702	13 295	13 630	15 399	15 439	17 827	Location de matériel de transport			711
..	..	16 274	16 024	15 948	16 000	16 269	17 726	Location de machines ét equipements nca			712
..	..	18 128	19 246	19 015	19 960	21 676	22 809	Loc. d'articles personnels et domestiques nca			713
..	..	164 613	167 740	172 331	171 859	189 112	223 625	Activités informatiques et activités rattachées		72	
..	..	48 705	53 325	54 005	46 074	57 158	69 463	Conseil en matériel informatique			721
..	..	48 760	50 148	53 734	62 131	73 158	91 319	Production de logiciels			722
..	..	54 992	50 142	48 721	47 471	41 591	43 633	Traitement de données informatiques			723
..	..	1 891	2 267	2 589	2 547	2 979	4 119	Activités de banques de données			724
..	..	10 265	11 858	13 283	13 637	14 225	15 091	Entr., répar. mach. de bureau et compt. et inform.			725
..	..	..	..	..	..	..	..	Autres activités rattachées à l'informatique			729
..	..	..	..	..	..	..	..	Recherche-développement		73	
..	..	..	..	..	..	..	..	Rech. et dév. expér. en sc. phys., natur. et ingén.			731
..	..	..	..	..	..	..	..	Rech. et dév. expér. en sciences soc. et humaines			732
..	..	1 267 764	1 345 351	1 408 516	1 431 848	1 527 571	1 588 587	Autres activités de services aux entreprises		74	
..	..	307 530	316 985	320 159	369 578	369 073	319 430	Act. jur., compt., audit.; cons. fisc.; ét. march., soud.			741
..	..	173 262	179 727	185 204	197 105	199 901	195 740	Act. d'achitect., d'ingén.; autres act. techniques			742
..	..	82 083	82 201	79 165	81 896	89 169	94 356	Publicité			743
..	..	704 889	766 438	823 989	783 269	869 427	979 061	Activités de services aux entreprises nca.			749
..	..	..	..	..	..	..	..	Education	M	80	
..	..	..	..	..	..	..	..	Activités d'enseignement primaire			801
..	..	..	..	..	..	..	..	Activités d'enseignement secondaire			802
..	..	..	..	..	..	..	..	Activités d'enseignement supérieur			803
..	..	..	..	..	..	..	..	Act. de formation perman.; autres act. d'enseign.			809
..	..	..	..	..	..	..	..	Santé et action sociale	N	85	
..	..	..	..	..	..	..	..	Activités de santé humaine			851
..	..	..	..	..	..	..	..	Activités vétérinaires			852
..	..	..	..	..	..	..	..	Activités d'action sociale			853
..	..	243 938	243 312	259 250	241 440	268 973	255 944	Autres activités de serv. coll., soc. et personnels	O	90_93	
..	..	34 291	37 222	39 661	39 452	41 215	46 369	Assain. et enlèv. des ordures; voierie, act. simil.		90	
..	..	..	..	..	..	..	..	Activités associatives diverses		91	
..	..	..	..	..	..	..	..	Activités de syndicats de salariés			912
..	..	60 909	61 173	70 169	57 413	64 401	63 966	Activités récréatives, culturelles et sportives		92	
..	..	28 901	30 111	34 115	26 116	31 170	29 273	Act. cinémat., de radio, télé.; autres act. de spect.			921
..	..	4 293	4 488	4 417	4 608	4 470	5 185	Activités d'agences de presse			922
..	..	..	..	..	..	..	..	Act. des biblio., arch., musées, jardins bot. et zool.			923
..	..	..	..	..	..	..	..	Activités sportives et autres activités récréatives			924
..	..	148 738	144 918	149 421	144 575	163 358	145 609	Autres activités de services		93	

FRANCE

Table FR.S5 COMPENSATION OF LABOUR, EMPLOYEES - SALAIRES ET CHARGES SOCIALES, SALARIÉS

Millions of FF (Current Prices)

ISIC revision 3				1991	1992	1993	1994	1995	1996	1997	1998
G	50_52		Wholesale & retail trade; repairs; etc.	..	..	310 840	301 726	324 857	453 403	481 371	440 012
	50		Sale, maint.,repair vehic/cycle, person., h/hold gd.	..	..	40 691	40 445	41 619	65 839	67 999	68 265
		501	Sale of motor cycle	..	..	28 544	27 763	28 449	37 936	37 591	38 983
		502	Maintenance and repair of motor vehicles	..	..	3 188	3 355	3 431	11 369	13 008	12 160
		503	Sale of motor vehicle parts and accessories	..	..	7 598	7 897	8 084	11 593	11 961	11 956
		504	Sale, maint., repair motor cycle, parts & acces.	..	..	497	464	519	1 550	1 835	1 739
		505	Retail sale of automative fuel	..	..	864	965	1 136	3 392	3 604	3 427
	51		W/sale. trade & commission exc motor veh. & cycle	..	..	154 555	148 131	160 811	207 315	216 073	179 604
		511	Wholesale on a fee or contract basis	..	..	8 936	3 213	9 309	16 490	17 422	17 031
		512	W/sale agric. raw mat.,live animal,food, bev., tobac.	..	..	36 833	36 023	35 773	44 569	45 425	30 549
		513	Wholesale of household goods	..	..	28 236	28 924	30 463	41 728	42 174	38 768
		514	W/sale. of non-agric. intermed. prod., waste, scrap	..	..	32 120	32 544	34 750	40 694	44 262	37 509
		515	Wholesale of machinery, equipment & supplies	..	..	39 899	38 927	41 592	51 702	54 818	47 668
		519	Other wholesale	..	..	198	297	361	1 108	601	644
	52		Retail trade exc. motor veh. & cyc.; repair h/hold gds.	..	..	115 594	113 151	122 428	180 249	197 299	192 144
		521	Non-specialised retail trade in stores	..	..	62 789	61 156	66 344	70 308	73 295	76 483
		522	Retail sale food, bev., tobac. in specialised stores	..	..	2 848	2 727	3 187	8 436	10 383	9 053
		523	Other retail trade of new goods specialised stores	..	..	41 557	40 068	43 608	88 079	97 961	92 700
		524	Retail sale of second-hand goods in stores	..	..	78	81	126	1 162	1 422	1 157
		525	Retail trade not in stores	..	..	6 901	7 741	7 667	9 477	11 126	9 902
		526	Repair of personal and household goods	..	..	1 421	1 378	1 496	2 787	3 113	2 848
H	55		Hotels and restaurants	..	..	..	..	..	69 050	80 719	86 864
		551	Hotels; camping ; other short-stay accommodation	..	..	..	..	..	20 320	22 809	28 225
		552	Restaurants, bars and canteens	..	..	..	..	..	48 731	57 910	58 639
I	60_64		Transport, storage and communications	..	..	..	..	..	..	..	..
	60		Land transport, transport via pipelines	..	..	..	..	..	..	..	..
		601	Transport via railways	..	..	..	..	..	..	..	..
		602	Other land transport	..	..	..	..	..	..	..	..
		603	Transport via pipelines	..	..	..	..	..	..	..	..
	61		Water transport	..	..	..	..	..	..	..	..
		611	Sea and coastal water transport	..	..	..	..	..	..	..	..
		612	Inland water transport	..	..	..	..	..	..	..	..
	62		Air transport	..	..	..	..	..	..	..	..
		621	Scheduled air transport	..	..	..	..	..	..	..	..
		622	Non-scheduled air transport	..	..	..	..	..	..	..	..
	63		Support. & aux. transport act., travel agencies activ.	..	..	..	..	..	..	..	..
	64		Post and telecommunications	..	..	..	..	..	4 532	5 863	8 899
		641	Post and courier activities	..	..	..	..	..	842	955	936
		642	Telecommunications	..	..	..	..	..	3 690	4 908	7 963
J	65_67		Financial intermediation	..	..	..	..	..	..	..	..
	65		Financial intermediation, exc. insur., pension fund	..	..	..	..	..	..	..	..
		651	Monetary intermediation	..	..	..	..	..	..	..	..
		659	Other financial intermediation	..	..	..	..	..	..	..	..
	66		Insurance pension fund, exc. compul. social sec.	..	..	..	..	..	..	..	..
	67		Activities auxiliary to financial intermediation	..	..	..	..	..	..	..	..
		671	Act. auxil to fin. intermed., exc. insur. & pens.fund	..	..	..	..	..	..	..	..
		672	Act. auxiliary to insurance and pension funding	..	..	..	..	..	..	..	..
K	70_74		Real estate, renting and business activities	..	..	..	..	..	416 678	447 922	472 930
	70		Real estate activities	..	..	..	..	..	38 441	41 795	43 466
		701	Real estate activ. with own or leased property	..	..	..	..	..	21 291	22 623	22 780
		702	Real estate activities on a fee or contract basis	..	..	..	..	..	17 150	19 172	20 686
	71		Renting of mach. without operator; rent. of pers. gds.	..	..	..	..	..	10 565	10 822	12 270
		711	Renting of transport equipment	..	..	..	..	..	3 285	3 333	3 903
		712	Renting of other machinery and equipment	..	..	..	..	..	3 472	3 559	3 854
		713	Renting of personal and household goods nec	..	..	..	..	..	3 324	3 497	3 860
	72		Computer and related activities	..	..	..	..	..	54 118	59 303	70 257
		721	Hardware consultancy	..	..	..	..	..	15 214	18 717	22 346
		722	Software consultancy and supply	..	..	..	..	..	20 621	23 813	30 480
		723	Data processing	..	..	..	..	..	13 970	12 448	12 380
		724	Data base activities	..	..	..	..	..	657	820	1 111
		725	Mainten.; repair of office, account. & comp. mach.	..	..	..	..	..	3 656	3 505	3 940
		729	Other computer related activities	..	..	..	..	..	..	..	..
	73		Research and development	..	..	..	..	..	..	..	..
		731	R & D experim. on natural sciences & engineering	..	..	..	..	..	..	..	..
		732	R & D experim. on social sciences & humanities	..	..	..	..	..	..	..	..
	74		Other business activities	..	..	..	..	..	313 555	336 003	346 936
		741	Legal, acc., book-keep., audit act.; tax consult., etc	..	..	..	..	..	113 231	118 326	101 484
		742	Architectural, engineering, & other techn. activities	..	..	..	..	..	54 217	58 504	57 277
		743	Advertising	..	..	..	..	..	21 717	23 027	24 671
		749	Business activities nec	..	..	..	..	..	124 390	136 145	163 504
M	80		Education	..	..	..	..	..	..	..	..
		801	Primary education	..	..	..	..	..	..	..	..
		802	Secondary education	..	..	..	..	..	..	..	..
		803	Higher education	..	..	..	..	..	..	..	..
		809	Adult and other education	..	..	..	..	..	..	..	..
N	85		Health and social work	..	..	..	..	..	..	..	..
		851	Human health activities	..	..	..	..	..	..	..	..
		852	Veterinary activities	..	..	..	..	..	..	..	..
		853	Social work activities	..	..	..	..	..	..	..	..
O	90_93		Other commun., soc. & pers. services activities	..	..	..	..	..	45 460	51 717	52 496
	90		Sewage & refuse disposal, sanitation, etc.	..	..	..	..	..	7 894	8 469	9 733
	91		Activities of member organisations nec	..	..	..	..	..	..	..	..
		912	Activities of trade unions	..	..	..	..	..	..	..	..
	92		Recreational, cultural and sporting activities	..	..	..	..	..	21 218	22 854	24 258
		921	Motion picture, radio, TV, other entertainment act.	..	..	..	..	..	9 185	10 084	10 655
		922	News agency activities	..	..	..	..	..	1 950	1 999	2 230
		923	Library, archives, museums, other cultural activ.	..	..	..	..	..	..	..	..
		924	Sporting and other recreational activities	..	..	..	..	..	..	..	..
	93		Other service activities	..	..	..	..	..	16 349	20 395	18 505

Table FR.S6 **SALAIRES ET TRAITEMENTS, SALARIÉS - WAGES AND SALARIES, EMPLOYEES**

Millions de FF (Prix courants)

1991	1992	1993	1994	1995	1996	1997	1998	CITI révision 3			
..	..	220 823	213 807	231 096	325 714	345 416	316 211	Com. de gros, de detail ; réparation etc.	G	50_52	
..	..	28 668	28 372	29 311	46 892	48 270	48 406	Com., entr., répar. d'auto. et moto ; carburants		50	
..	..	20 086	19 460	20 026	26 850	26 502	27 352	Commerce de véhicules automobiles			501
..	..	2 239	2 346	2 406	8 177	9 260	8 772	Entretien et réparation de véhicules automobiles			502
..	..	5 375	5 544	5 678	8 252	8 521	8 495	Com. de pièces et access. de véhic. automobiles			503
..	..	355	330	371	1 122	1 330	1 261	Com., entret. et répar. de motocycles et access.			504
..	..	614	693	830	2 491	2 657	2 526	Com. de détail de carburants pour automobiles			505
..	..	109 135	104 207	113 374	146 694	153 117	127 242	Com. de gros et act. inter., sauf auto. et moto		51	
..	..	6 329	2 264	6 526	11 628	12 260	11 992	Activités d'intermédiaries de commerce de gros			511
..	..	25 923	25 334	25 199	31 501	32 197	21 680	Com. gros: prod. agr. bruts, anim. viv., prod. alim.			512
..	..	20 130	20 534	21 673	29 850	30 209	27 673	Com. gros: biens de consom., sauf alimentaires			513
..	..	22 557	22 783	24 445	28 645	31 219	26 381	Com. gros: prod. inter. non agr., déchets, débris			514
..	..	28 190	27 337	29 264	36 490	38 765	33 820	Com. gros: machines, équipments et fournitures			515
..	..	134	210	251	798	427	464	Autres commerces de gros			519
..	..	83 021	81 228	88 411	132 128	144 029	140 563	Com. détail, sauf auto, moto; répar. art. pers., dom.		52	
..	..	45 185	43 954	48 082	51 750	54 279	56 056	Commerce de détail non spécialisé en magasin			521
..	..	2 013	1 919	2 248	6 146	7 576	6 703	Com. détail: prod. aliment. en mag. spécialisés			522
..	..	29 765	28 726	31 385	64 404	70 609	67 626	Autre com. détail: prod. neufs en mag. spécialisés			523
..	..	56	58	91	871	1 054	868	Commerces de détail de biens d'occasion			524
..	..	4 998	5 601	5 550	6 946	8 244	7 251	Commerces de détail autres qu'en magasin			525
..	..	1 002	971	1 055	2 011	2 267	2 059	Réparation d'articles personnels et domestiques			526
..	..	45 582	46 695	47 540	51 314	59 529	64 553	Hotels et restaurants	H	55	
..	..	14 600	15 236	14 516	15 091	16 842	20 918	Hôtels; camping; autres équip. courts séjours			551
..	..	30 983	31 459	33 024	36 223	42 687	43 635	Restaurants, cafés et cantines			552
..	..	..	..	..	..	..	..	Transport, entreposage, communications	I	60_64	
..	..	..	..	..	..	..	..	Transports terrestres; transports par conduites		60	
..	..	..	..	..	..	..	..	Transports par chemin de fer			601
..	..	..	..	..	..	..	..	Autres transports terrestres			602
..	..	..	..	..	..	..	..	Transports par conduites			603
..	..	..	..	..	..	..	..	Transports par eau		61	
..	..	..	..	..	..	..	..	Transports maritimes et côtiers			611
..	..	..	..	..	..	..	..	Transports par voies navigables intérieures			612
..	..	..	..	..	..	..	..	Transports aériens		62	
..	..	..	..	..	..	..	..	Transports aériens réguliers			621
..	..	..	..	..	..	..	..	Transports aériens non réguliers			622
..	..	3 810	3 975	3 808	4 815	..	..	Act. annex. et aux. des transp.; act. des ag. de voy.		63	
..	..	2 129	1 791	1 911	3 136	4 060	6 119	Postes et télécommunications		64	
..	..	..	352	423	614	696	674	Activités de poste et de courrier			641
..	..	2 129	1 439	1 488	2 522	3 364	5 445	Télécommunications			642
..	..	..	..	..	..	..	..	Intermédiation financière	J	65_67	
..	..	..	..	..	..	..	..	Interm. financ., sauf assur. et caisses de retraite		65	
..	..	..	..	..	..	..	..	Intermédiation monétaire			651
..	..	..	..	..	..	..	..	Autres intermédiations financières			659
..	..	..	..	..	..	..	..	Act. d'assur., caisses de retr., sauf séc. soc. oblig.		66	
..	..	..	..	..	..	..	..	Act. auxiliaires de l'intermédiation financière		67	
..	..	..	..	..	..	..	..	Act. aux. de l'interm. fin., sauf assur., et retraite			671
..	..	..	..	..	..	..	..	Act. aux. des assurances et caisses de retraite			672
..	..	228 430	241 657	256 231	298 798	320 798	338 090	Immobilier, locations, services aux entreprises	K	70_74	
..	..	24 352	24 687	25 433	27 451	29 769	30 996	Activités immobilières		70	
..	..	13 802	14 066	14 194	15 040	15 975	16 064	Act. immobilières sur biens propres ou loués			701
..	..	10 550	10 621	11 240	12 411	13 795	14 932	Activités immobilières pour comptes de tiers			702
..	..	6 036	6 492	6 491	7 447	7 645	8 602	Locat. mach., équip. sans opér., biens pers. et dom.		71	
..	..	1 585	1 829	1 880	2 300	2 325	2 717	Location de matériel de transport			711
..	..	2 241	2 264	2 265	2 433	2 500	2 679	Location de machines ét equipements nca			712
..	..	1 955	2 033	2 043	2 374	2 519	2 753	Loc. d'articles personnels et domestiques nca			713
..	..	32 680	33 797	34 994	37 991	41 482	49 058	Activités informatiques et activités rattachées		72	
..	..	10 031	11 308	11 518	10 682	13 104	15 634	Conseil en matériel informatique			721
..	..	10 293	10 599	11 325	14 553	16 690	21 234	Production de logiciels			722
..	..	10 265	9 375	9 344	9 718	8 643	8 620	Traitement de données informatiques			723
..	..	301	389	420	466	580	787	Activités de banques de données			724
..	..	1 790	2 126	2 387	2 572	2 466	2 783	Entr., répar. mach. de bureau et compt. et inform.			725
..	..	..	..	..	..	..	..	Autres activités rattachées à l'informatique			729
..	..	..	..	..	..	..	..	Recherche-développement		73	
..	..	..	..	..	..	..	..	Rech. et dév. expér. en sc. phys., natur. et ingén.			731
..	..	..	..	..	..	..	..	Rech. et dév. expér. en sciences soc. et humaines			732
..	..	165 363	176 681	189 313	225 910	241 902	249 434	Autres activités de services aux entreprises		74	
..	..	55 083	57 618	59 270	80 255	83 569	72 238	Act. jur., compt., audit.; cons. fisc.; ét. march., soud.			741
..	..	30 468	31 725	33 363	38 363	41 243	40 458	Act. d'achitect., d'ingén.; autres act. techniques			742
..	..	12 870	13 127	13 488	15 530	16 469	17 487	Publicité			743
..	..	66 942	74 211	83 193	91 762	100 621	119 251	Activités de services aux entreprises nca.			749
..	..	..	..	..	..	..	..	Education	M	80	
..	..	..	..	..	..	..	..	Activités d'enseignement primaire			801
..	..	..	..	..	..	..	..	Activités d'enseignement secondaire			802
..	..	..	..	..	..	..	..	Activités d'enseignement supérieur			803
..	..	..	..	..	..	..	..	Act. de formation perman.; autres act. d'enseign.			809
..	..	..	..	..	..	..	..	Santé et action sociale	N	85	
..	..	..	..	..	..	..	..	Activités de santé humaine			851
..	..	..	..	..	..	..	..	Activités vétérinaires			852
..	..	..	..	..	..	..	..	Activités d'action sociale			853
..	..	25 851	26 695	29 123	32 879	37 124	37 820	Autres activités de serv. coll., soc. et personnels	O	90_93	
..	..	4 138	4 517	5 011	5 447	5 872	6 693	Assain. et enlèv. des ordures; voierie, act. simil.		90	
..	..	..	..	..	..	..	..	Activités associatives diverses		91	
..	..	..	..	..	..	..	..	Activités de syndicats de salariés			912
..	..	10 579	11 195	12 586	15 096	16 265	17 196	Activités récréatives, culturelles et sportives		92	
..	..	4 299	4 666	5 445	6 536	7 185	7 526	Act. cinémat., de radio, télé.; autres act. de spect.			921
..	..	875	942	1 004	1 436	1 474	1 643	Activités d'agences de presse			922
..	..	..	..	..	..	..	..	Act. des biblio., arch., musées, jardins bot. et zool.			923
..	..	..	..	..	..	..	..	Activités sportives et autres activités récréatives			924
..	..	11 135	10 983	11 526	12 336	14 988	13 932	Autres activités de services		93	

FRANCE

Table FR.S7 **ENTERPRISES - ENTREPRISES**

Units

ISIC revision 3				1991	1992	1993	1994	1995	1996	1997	1998
G	50_52		Wholesale & retail trade; repairs; etc.	..	..	531 445	502 571	528 533	521 130	543 677	512 634
	50		Sale, maint.,repair vehic/cycle, person., h/hold gd.	..	..	73 228	73 329	72 363	71 987	72 346	70 322
		501	Sale of motor cycle	..	..	26 472	25 447	25 622	24 022	24 398	22 850
		502	Maintenance and repair of motor vehicles	..	..	31 313	32 913	31 976	30 409	30 314	30 852
		503	Sale of motor vehicle parts and accessories	..	..	5 617	5 372	5 370	6 781	6 827	6 369
		504	Sale, maint., repair motor cycle, parts & acces.	..	..	3 681	3 555	3 984	3 951	4 121	4 020
		505	Retail sale of automative fuel	..	..	6 145	6 042	5 411	6 824	6 686	6 231
	51		W/sale. trade & commission exc motor veh. & cycle	..	..	107 434	85 370	112 316	119 874	130 144	121 715
		511	Wholesale on a fee or contract basis	..	..	21 914	1 142	22 427	21 043	29 125	27 853
		512	W/sale agric. raw mat.,live animal,food, bev., tobac.	..	..	24 737	23 872	23 977	24 672	25 112	22 092
		513	Wholesale of household goods	..	..	21 821	22 538	24 434	28 609	27 690	27 912
		514	W/sale. of non-agric. intermed. prod., waste, scrap	..	..	12 918	12 433	13 510	13 977	15 235	12 826
		515	Wholesale of machinery, equipment & supplies	..	..	19 609	18 974	20 532	22 228	24 626	22 781
		519	Other wholesale	..	..	598	744	1 681	3 531	2 330	2 899
	52		Retail trade exc. motor veh. & cyc.; repair h/hold gds.	..	..	350 783	343 872	343 854	329 269	341 187	320 597
		521	Non-specialised retail trade in stores	..	..	30 777	27 545	28 892	26 990	27 108	23 860
		522	Retail sale food, bev., tobac. in specialised stores	..	..	51 511	50 139	47 536	45 469	47 068	44 905
		523	Other retail trade of new goods specialised stores	..	..	192 652	187 445	181 269	174 105	182 515	171 188
		524	Retail sale of second-hand goods in stores	..	..	11 786	11 765	12 127	12 532	12 529	12 069
		525	Retail trade not in stores	..	..	49 882	53 977	59 041	55 215	56 686	53 609
		526	Repair of personal and household goods	..	..	14 175	13 001	14 989	14 958	15 281	14 966
H	55		Hotels and restaurants	..	..	164 485	161 562	161 967	158 630	174 529	183 650
		551	Hotels; camping ; other short-stay accommodation	..	..	31 521	30 949	29 630	28 561	30 236	36 959
		552	Restaurants, bars and canteens	..	..	132 964	130 613	132 337	130 069	144 293	146 691
I	60_64		Transport, storage and communications	..	..	..	..	..	..	..	..
	60		Land transport, transport via pipelines	..	..	..	..	..	..	..	..
		601	Transport via railways	..	..	..	..	..	..	..	..
		602	Other land transport	..	..	..	..	..	..	..	..
		603	Transport via pipelines	..	..	..	..	..	..	..	..
	61		Water transport	..	..	..	..	..	..	..	..
		611	Sea and coastal water transport	..	..	..	..	..	..	..	..
		612	Inland water transport	..	..	..	..	..	..	..	..
	62		Air transport	..	..	..	..	..	..	..	..
		621	Scheduled air transport	..	..	..	..	..	..	..	..
		622	Non-scheduled air transport	..	..	..	..	..	..	..	..
	63		Support. & aux. transport act., travel agencies activ.	..	..	2 524	2 765	2 279	3 035	..	..
	64		Post and telecommunications	..	..	297	1 574	1 433	1 446	1 838	2 162
		641	Post and courier activities	..	..	..	1 190	1 060	945	1 096	1 152
		642	Telecommunications	..	..	297	384	373	501	742	1 010
J	65_67		Financial intermediation	..	..	..	..	..	..	..	..
	65		Financial intermediation, exc. insur., pension fund	..	..	..	..	..	..	..	..
		651	Monetary intermediation	..	..	..	..	..	..	..	..
		659	Other financial intermediation	..	..	..	..	..	..	..	..
	66		Insurance pension fund, exc. compul. social sec.	..	..	..	..	..	..	..	..
	67		Activities auxiliary to financial intermediation	..	..	..	..	..	..	..	..
		671	Act. auxil to fin. intermed., exc. insur. & pens.fund	..	..	..	..	..	..	..	..
		672	Act. auxiliary to insurance and pension funding	..	..	..	..	..	..	..	..
K	70_74		Real estate, renting and business activities	..	..	278 618	284 015	294 146	313 019	310 806	368 265
	70		Real estate activities	..	..	54 995	54 163	55 041	55 226	59 792	69 760
		701	Real estate activ. with own or leased property	..	..	22 935	20 832	21 406	23 118	25 456	34 489
		702	Real estate activities on a fee or contract basis	..	..	32 060	33 331	33 635	32 108	34 336	35 271
	71		Renting of mach. without operator; rent. of pers. gds.	..	..	12 276	12 704	12 950	13 474	13 442	18 065
		711	Renting of transport equipment	..	..	2 160	1 918	1 799	2 121	1 907	5 949
		712	Renting of other machinery and equipment	..	..	6 602	6 858	7 015	7 346	7 440	4 921
		713	Renting of personal and household goods nec	..	..	3 321	3 769	3 978	3 724	3 908	6 940
	72		Computer and related activities	..	..	21 154	21 083	21 854	20 244	21 074	19 617
		721	Hardware consultancy	..	..	8 231	8 517	8 366	5 855	6 743	6 271
		722	Software consultancy and supply	..	..	6 070	6 427	7 183	8 679	8 827	7 459
		723	Data processing	..	..	4 971	4 145	4 220	3 796	3 424	3 265
		724	Data base activities	..	..	811	760	724	575	762	942
		725	Mainten.; repair of office, account. & comp. mach.	..	..	1 071	1 234	1 361	1 339	1 318	1 680
		729	Other computer related activities	..	..	..	..	..	..	..	..
	73		Research and development	..	..	..	..	..	..	..	..
		731	R & D experim. on natural sciences & engineering	..	..	..	..	..	..	..	..
		732	R & D experim. on social sciences & humanities	..	..	..	..	..	..	..	..
	74		Other business activities	..	..	190 193	196 065	204 301	224 075	216 498	260 823
		741	Legal, acc., book-keep., audit act.; tax consult., etc	..	..	73 265	78 160	82 073	90 361	95 620	120 597
		742	Architectural, engineering, & other techn. activities	..	..	47 675	46 869	47 986	51 382	51 381	57 921
		743	Advertising	..	..	13 729	12 868	14 829	15 950	15 223	13 834
		749	Business activities nec	..	..	55 524	58 168	59 413	66 382	54 274	68 471
M	80		Education	..	..	..	..	..	..	..	..
		801	Primary education	..	..	..	..	..	..	..	..
		802	Secondary education	..	..	..	..	..	..	..	..
		803	Higher education	..	..	..	..	..	..	..	..
		809	Adult and other education	..	..	..	..	..	..	..	..
N	85		Health and social work	..	..	..	..	..	..	..	..
		851	Human health activities	..	..	..	..	..	..	..	..
		852	Veterinary activities	..	..	..	..	..	..	..	..
		853	Social work activities	..	..	..	..	..	..	..	..
O	90_93		Other commun., soc. & pers. services activities	..	..	87 510	89 194	93 101	92 451	96 682	96 016
	90		Sewage & refuse disposal, sanitation, etc.	..	..	1 125	1 237	1 181	1 279	1 382	1 437
	91		Activities of member organisations nec	..	..	..	..	..	..	..	..
		912	Activities of trade unions	..	..	..	..	..	..	..	..
	92		Recreational, cultural and sporting activities	..	..	5 781	6 370	6 775	7 093	7 356	8 199
		921	Motion picture, radio, TV, other entertainment act.	..	..	4 160	4 443	4 933	5 242	5 390	6 010
		922	News agency activities	..	..	601	939	821	743	731	929
		923	Library, archives, museums, other cultural activ.	..	..	..	..	..	..	..	..
		924	Sporting and other recreational activities	..	..	..	..	..	..	..	..
	93		Other service activities	..	..	80 604	81 587	85 145	84 079	87 944	86 380

Table FR.S8 **EXPORTATIONS - EXPORTS**

Millions de FF (Prix courants)

1991	1992	1993	1994	1995	1996	1997	1998	CITI révision 3			
..	..	250 829	233 356	280 174	288 852	365 258	329 365	Com. de gros, de detail ; réparation etc.	G	50_52	
..	..	12 295	11 124	12 421	14 259	25 063	15 604	Com., entr., répar. d'auto. et moto ; carburants		50	
..	..	8 405	7 792	8 006	7 672	15 678	8 155	Commerce de véhicules automobiles			501
..	..	93	170	26	48	1 289	223	Entretien et réparation de véhicules automobiles			502
..	..	3 606	3 014	4 231	6 019	6 632	6 843	Com. de pièces et access. de véhic. automobiles			503
..	..	190	147	155	488	962	341	Com., entret. et répar. de motocycles et access.			504
..	..	..	0	3	32	503	41	Com. de détail de carburants pour automobiles			505
..	..	238 254	221 977	267 754	258 201	320 773	296 005	Com. de gros et act. inter., sauf auto. et moto		51	
..	..	14 360	1 876	12 574	20 639	21 440	30 947	Activités d'intermédiaries de commerce de gros			511
..	..	124 174	116 157	133 237	119 270	149 029	122 860	Com. gros: prod. agr. bruts, anim. viv., prod. alim.			512
..	..	34 857	40 905	42 764	44 924	54 421	58 088	Com. gros: biens de consom., sauf alimentaires			513
..	..	32 611	30 433	36 855	34 363	41 417	39 028	Com. gros: prod. inter. non agr., déchets, débris			514
..	..	23 165	25 696	33 643	31 352	46 508	34 997	Com. gros: machines, équipments et fournitures			515
..	..	4 161	2 724	3 533	2 781	2 378	3 650	Autres commerces de gros			519
..	..	280	255	..	16 392	19 421	17 756	Com. détail, sauf auto, moto; répar. art. pers., dom.		52	
..	..	..	..	..	845	1 688	1 360	Commerce de détail non spécialisé en magasin			521
..	..	49	..	..	554	2 093	922	Com. détail: prod. aliment. en mag. spécialisés			522
..	..	180	191	..	10 055	9 618	9 183	Autre com. détail: prod. neufs en mag. spécialisés			523
..	..	..	..	..	2 051	1 793	2 101	Commerces de détail de biens d'occasion			524
..	..	50	25	..	2 838	4 038	4 156	Commerces de détail autres qu'en magasin			525
..	..	0	39	..	49	192	34	Réparation d'articles personnels et domestiques			526
..	..	..	..	..	..	..	..	Hotels et restaurants	H	55	
..	..	..	..	..	..	..	..	Hôtels; camping; autres équip. courts séjours			551
..	..	..	..	..	..	..	..	Restaurants, cafés et cantines			552
..	..	..	..	..	..	..	..	Transport, entreposage, communications	I	60_64	
..	..	..	..	..	..	..	..	Transports terrestres; transports par conduites		60	
..	..	..	..	..	..	..	..	Transports par chemin de fer			601
..	..	..	..	..	..	..	..	Autres transports terrestres			602
..	..	..	..	..	..	..	..	Transports par conduites			603
..	..	..	..	..	..	..	..	Transports par eau		61	
..	..	..	..	..	..	..	..	Transports maritimes et côtiers			611
..	..	..	..	..	..	..	..	Transports par voies navigables intérieures			612
..	..	..	..	..	..	..	..	Transports aériens		62	
..	..	..	..	..	..	..	..	Transports aériens réguliers			621
..	..	..	..	..	..	..	..	Transports aériens non réguliers			622
..	..	..	..	..	..	..	..	Act. annex. et aux. des transp.; act. des ag. de voy.		63	
..	..	1 230	1 581	2 187	1 920	1 447	4 977	Postes et télécommunications		64	
..	..	..	105	97	..	..	130	Activités de poste et de courrier			641
..	..	1 230	1 476	2 091	1 920	1 447	4 848	Télécommunications			642
..	..	..	..	..	..	..	..	Intermédiation financière	J	65_67	
..	..	..	..	..	..	..	..	Interm. financ., sauf assur. et caisses de retraite		65	
..	..	..	..	..	..	..	..	Intermédiation monétaire			651
..	..	..	..	..	..	..	..	Autres intermédiations financières			659
..	..	..	..	..	..	..	..	Act. d'assur., caisses de retr., sauf séc. soc. oblig.		66	
..	..	..	..	..	..	..	..	Act. auxiliaires de l'intermédiation financière		67	
..	..	..	..	..	..	..	..	Act. aux. de l'interm. fin., sauf assur., et retraite			671
..	..	..	..	..	..	..	..	Act. aux. des assurances et caisses de retraite			672
..	..	53 015	54 821	54 382	73 914	145 705	143 508	Immobilier, locations, services aux entreprises	K	70_74	
..	..	..	..	..	..	..	..	Activités immobilières		70	
..	..	..	..	..	..	..	..	Act. immobilières sur biens propres ou loués			701
..	..	..	..	..	..	..	..	Activités immobilières pour comptes de tiers			702
..	..	..	..	..	718	1 018	5 422	Locat. mach., équip. sans opér., biens pers. et dom.		71	
..	..	..	..	..	718	1 018	3 283	Location de matériel de transport			711
..	..	..	..	..	..	..	1 749	Location de machines ét equipements nca			712
..	..	..	..	..	..	..	241	Loc. d'articles personnels et domestiques nca			713
..	..	6 408	5 945	7 544	9 373	10 866	18 877	Activités informatiques et activités rattachées		72	
..	..	1 776	1 459	1 718	1 896	2 691	4 088	Conseil en matériel informatique			721
..	..	3 044	3 297	4 007	5 207	6 575	10 152	Production de logiciels			722
..	..	1 004	937	1 453	1 834	1 164	3 330	Traitement de données informatiques			723
..	..	114	133	136	163	203	236	Activités de banques de données			724
..	..	470	119	230	273	233	1 072	Entr., répar. mach. de bureau et compt. et inform.			725
..	..	..	..	..	..	..	..	Autres activités rattachées à l'informatique			729
..	..	..	..	..	..	..	..	Recherche-développement		73	
..	..	..	..	..	..	..	..	Rech. et dév. expér. en sc. phys., natur. et ingén.			731
..	..	..	..	..	..	..	..	Rech. et dév. expér. en sciences soc. et humaines			732
..	..	46 607	48 876	46 838	63 824	133 822	119 209	Autres activités de services aux entreprises		74	
..	..	13 346	16 210	15 643	23 440	75 357	58 077	Act. jur., compt., audit.; cons. fisc.; ét. march., soud.			741
..	..	24 091	21 462	21 927	29 280	45 554	42 805	Act. d'achitect., d'ingén.; autres act. techniques			742
..	..	3 058	3 309	3 214	3 744	5 156	6 797	Publicité			743
..	..	6 113	7 896	6 053	7 360	7 755	11 530	Activités de services aux entreprises nca.			749
..	..	..	..	..	..	..	..	Education	M	80	
..	..	..	..	..	..	..	..	Activités d'enseignement primaire			801
..	..	..	..	..	..	..	..	Activités d'enseignement secondaire			802
..	..	..	..	..	..	..	..	Activités d'enseignement supérieur			803
..	..	..	..	..	..	..	..	Act. de formation perman.; autres act. d'enseign.			809
..	..	..	..	..	..	..	..	Santé et action sociale	N	85	
..	..	..	..	..	..	..	..	Activités de santé humaine			851
..	..	..	..	..	..	..	..	Activités vétérinaires			852
..	..	..	..	..	..	..	..	Activités d'action sociale			853
..	..	2 995	3 319	3 664	4 111	6 019	8 241	Autres activités de serv. coll., soc. et personnels	O	90_93	
..	..	156	155	208	256	169	426	Assain. et enlèv. des ordures; voierie, act. simil.		90	
..	..	..	..	..	..	..	..	Activités associatives diverses		91	
..	..	..	..	..	..	..	..	Activités de syndicats de salariés			912
..	..	2 839	3 164	3 456	3 855	5 850	7 815	Activités récréatives, culturelles et sportives		92	
..	..	1 885	1 740	1 904	2 766	3 265	4 543	Act. cinémat., de radio, télé.; autres act. de spect.			921
..	..	..	..	..	326	961	1 091	Activités d'agences de presse			922
..	..	..	..	..	..	..	..	Act. des biblio., arch., musées, jardins bot. et zool.			923
..	..	..	..	..	..	..	..	Activités sportives et autres activités récréatives			924
..	..	..	..	..	..	..	..	Autres activités de services		93	

Table FR.S9 **IMPORTS - IMPORTATIONS**

Millions of FF (Current Prices)

ISIC revision 3				1991	1992	1993	1994	1995	1996	1997	1998
G	50_52		Wholesale & retail trade; repairs; etc.	..	..	..	..	..	..	..	..
	50		Sale, maint.,repair vehic/cycle, person., h/hold gd.	..	..	54 510	63 226	62 814	66 598	60 806	77 036
		501	Sale of motor cycle	..	..	46 620	55 514	55 032	57 606	50 411	65 023
		502	Maintenance and repair of motor vehicles	..	..	44	56	87	102	73	254
		503	Sale of motor vehicle parts and accessories	..	..	5 605	5 620	5 709	5 679	7 509	7 923
		504	Sale, maint., repair motor cycle, parts & acces.	..	..	2 240	2 034	1 977	3 209	2 814	3 828
		505	Retail sale of automative fuel	..	..	..	2	10	1	..	8
	51		W/sale. trade & commission exc motor veh. & cycle	..	..	313 677	327 169	389 967	274 449	404 304	274 113
		511	Wholesale on a fee or contract basis	..	..	16 469	14 028	21 726	21 659	27 180	16 712
		512	W/sale agric. raw mat.,live animal,food, bev., tobac.	..	..	56 488	57 250	66 491	41 006	60 930	34 528
		513	Wholesale of household goods	..	..	81 186	83 917	95 677	71 380	105 935	89 880
		514	W/sale. of non-agric. intermed. prod., waste, scrap	..	..	75 787	83 933	97 838	60 195	87 631	63 691
		515	Wholesale of machinery, equipment & supplies	..	..	68 459	69 384	85 185	62 134	98 685	51 579
		519	Other wholesale	..	..	2 044	2 321	2 525	1 933	2 477	2 701
	52		Retail trade exc. motor veh. & cyc.; repair h/hold gds.	..	..	411	349	..	1 350	992	..
		521	Non-specialised retail trade in stores	..	..	0	..	..	1 017	206	..
		522	Retail sale food, bev., tobac. in specialised stores	..	..	59	..	..	18	29	1
		523	Other retail trade of new goods specialised stores	..	..	348	210	..	305	730	253
		524	Retail sale of second-hand goods in stores	..	..	..	..	..	8	0	..
		525	Retail trade not in stores	..	..	5	45	..	3	23	15
		526	Repair of personal and household goods	..	..	..	93	..	..	4	..
H	55		Hotels and restaurants	..	..	..	..	..	..	..	..
		551	Hotels; camping ; other short-stay accommodation	..	..	..	..	..	..	..	..
		552	Restaurants, bars and canteens	..	..	..	..	..	..	..	..
I	60_64		Transport, storage and communications	..	..	..	..	..	..	..	..
	60		Land transport, transport via pipelines	..	..	..	..	..	..	..	..
		601	Transport via railways	..	..	..	..	..	..	..	..
		602	Other land transport	..	..	..	..	..	..	..	..
		603	Transport via pipelines	..	..	..	..	..	..	..	..
	61		Water transport	..	..	..	..	..	..	..	..
		611	Sea and coastal water transport	..	..	..	..	..	..	..	..
		612	Inland water transport	..	..	..	..	..	..	..	..
	62		Air transport	..	..	..	..	..	..	..	..
		621	Scheduled air transport	..	..	..	..	..	..	..	..
		622	Non-scheduled air transport	..	..	..	..	..	..	..	..
	63		Support. & aux. transport act., travel agencies activ.	..	..	..	..	..	..	..	..
	64		Post and telecommunications	..	..	..	..	..	170	790	..
		641	Post and courier activities	..	..	..	..	..	..	..	..
		642	Telecommunications	..	..	..	..	..	170	790	..
J	65_67		Financial intermediation	..	..	..	..	..	..	..	..
	65		Financial intermediation, exc. insur., pension fund	..	..	..	..	..	..	..	..
		651	Monetary intermediation	..	..	..	..	..	..	..	..
		659	Other financial intermediation	..	..	..	..	..	..	..	..
	66		Insurance pension fund, exc. compul. social sec.	..	..	..	..	..	..	..	..
	67		Activities auxiliary to financial intermediation	..	..	..	..	..	..	..	..
		671	Act. auxil to fin. intermed., exc. insur. & pens.fund	..	..	..	..	..	..	..	..
		672	Act. auxiliary to insurance and pension funding	..	..	..	..	..	..	..	..
K	70_74		Real estate, renting and business activities	..	..	..	..	..	10 487	9 770	..
	70		Real estate activities	..	..	..	..	..	..	..	..
		701	Real estate activ. with own or leased property	..	..	..	..	..	..	..	..
		702	Real estate activities on a fee or contract basis	..	..	..	..	..	..	..	..
	71		Renting of mach. without operator; rent. of pers. gds.	..	..	..	..	..	59	69	..
		711	Renting of transport equipment	..	..	..	..	..	59	69	..
		712	Renting of other machinery and equipment	..	..	..	..	..	..	..	..
		713	Renting of personal and household goods nec	..	..	..	..	..	..	..	..
	72		Computer and related activities	..	..	..	..	..	2 672	2 661	..
		721	Hardware consultancy	..	..	..	..	..	602	587	..
		722	Software consultancy and supply	..	..	..	..	..	979	988	..
		723	Data processing	..	..	..	..	..	371	488	..
		724	Data base activities	..	..	..	..	..	97	25	..
		725	Mainten.; repair of office, account. & comp. mach.	..	..	..	..	..	622	574	..
		729	Other computer related activities	..	..	..	..	..	..	..	..
	73		Research and development	..	..	..	..	..	..	..	..
		731	R & D experim. on natural sciences & engineering	..	..	..	..	..	..	..	..
		732	R & D experim. on social sciences & humanities	..	..	..	..	..	..	..	..
	74		Other business activities	..	..	..	..	..	7 756	7 039	..
		741	Legal, acc., book-keep., audit act.; tax consult., etc	..	..	..	..	..	2 689	2 864	..
		742	Architectural, engineering, & other techn. activities	..	..	..	..	..	3 130	2 132	..
		743	Advertising	..	..	..	..	..	789	771	..
		749	Business activities nec	..	..	..	..	..	1 148	1 273	..
M	80		Education	..	..	..	..	..	..	..	..
		801	Primary education	..	..	..	..	..	..	..	..
		802	Secondary education	..	..	..	..	..	..	..	..
		803	Higher education	..	..	..	..	..	..	..	..
		809	Adult and other education	..	..	..	..	..	..	..	..
N	85		Health and social work	..	..	..	..	..	..	..	..
		851	Human health activities	..	..	..	..	..	..	..	..
		852	Veterinary activities	..	..	..	..	..	..	..	..
		853	Social work activities	..	..	..	..	..	..	..	..
O	90_93		Other commun., soc. & pers. services activities	..	..	..	..	..	3 366	3 757	..
	90		Sewage & refuse disposal, sanitation, etc.	..	..	..	..	..	46	59	..
	91		Activities of member organisations nec	..	..	..	..	..	..	..	..
		912	Activities of trade unions	..	..	..	..	..	..	..	..
	92		Recreational, cultural and sporting activities	..	..	..	..	..	3 320	..	4 971
		921	Motion picture, radio, TV, other entertainment act.	..	..	..	..	..	1 366	..	2 609
		922	News agency activities	..	..	..	..	..	20	..	9
		923	Library, archives, museums, other cultural activ.	..	..	..	..	..	..	..	..
		924	Sporting and other recreational activities	..	..	..	..	..	..	..	..
	93		Other service activities	..	..	..	..	..	..	..	..

Table HU.S1 **PRODUCTION - PRODUCTION**

Millions de Ft (Prix courants)

1991	1992	1993	1994	1995	1996	1997	1998		CITI révision 3		
..	470 160	565 547	692 312	889 214	1 086 069	1 347 295	1 719 543	Com. de gros, de detail ; réparation etc.	G	50_52	
..	..	47 259	72 260	113 206	150 054	173 968	233 949	Com., entr., répar. d'auto. et moto ; carburants		50	
..	..	12 657	21 367	56 060	86 050	99 648	132 671	Commerce de véhicules automobiles			501
..	..	18 260	25 088	24 637	25 947	27 940	38 587	Entretien et réparation de véhicules automobiles			502
..	..	5 859	8 375	11 302	14 961	19 734	31 068	Com. de pièces et access. de véhic. automobiles			503
..	..	10 483	17 430	21 207	23 096	26 646	1 163	Com., entret. et répar. de motocycles et access.			504
..	..	..	..	..	..	..	30 460	Com. de détail de carburants pour automobiles			505
..	..	338 782	403 561	531 548	664 005	839 113	998 654	Com. de gros et act. inter., sauf auto. et moto		51	
..	..	..	..	..	..	..	94 776	Activités d'intermédiaries de commerce de gros			511
..	..	..	..	..	..	..	268 885	Com. gros: prod. agr. bruts, anim. viv., prod. alim.			512
..	..	..	..	..	..	..	169 552	Com. gros: biens de consom., sauf alimentaires			513
..	..	..	..	..	..	..	126 730	Com. gros: prod. inter. non agr., déchets, débris			514
..	..	..	..	..	..	..	60 478	Com. gros: machines, équipments et fournitures			515
..	..	109 923	134 339	164 709	193 376	226 980	278 233	Autres commerces de gros			519
..	..	179 506	216 491	244 460	272 010	334 214	486 940	Com. détail, sauf auto, moto; répar. art. pers., dom.		52	
..	..	70 596	83 601	100 748	112 739	120 411	161 746	Commerce de détail non spécialisé en magasin			521
..	..	..	..	..	..	..	21 895	Com. détail: prod. aliment. en mag. spécialisés			522
..	..	..	..	..	..	..	262 219	Autre com. détail: prod. neufs en mag. spécialisés			523
..	..	..	..	..	..	..	6 777	Commerces de détail de biens d'occasion			524
..	..	..	..	..	..	..	20 505	Commerces de détail autres qu'en magasin			525
..	..	..	..	..	..	..	13 798	Réparation d'articles personnels et domestiques			526
..	50 042	57 171	69 361	83 206	98 275	126 618	149 330	Hotels et restaurants	H	55	
..	..	30 509	37 568	46 438	53 961	66 107	87 061	Hôtels; camping; autres équip. courts séjours			551
..	..	26 662	31 793	36 768	44 314	60 511	62 269	Restaurants, cafés et cantines			552
..	..	335 289	448 664	588 671	760 778	985 165	1 160 714	Transport, entreposage, communications	I	60_64	
..	170 916	146 710	206 224	263 131	319 862	389 552	492 699	Transports terrestres; transports par conduites		60	
..	..	33 911	71 789	98 955	116 374	139 054	164 467	Transports par chemin de fer			601
..	..	112 787	134 416	164 151	203 447	250 478	328 084	Autres transports terrestres			602
..	..		19	25	41	20	148	Transports par conduites			603
..	..	4 937	6 288	7 491	7 754	7 953	9 435	Transports par eau		61	
..	..		1 981	2 207	1 845	647	211	Transports maritimes et côtiers			611
..	..	4 234	4 307	5 284	5 909	7 306	9 224	Transports par voies navigables intérieures			612
..	..	26 725	32 097	41 916	52 341	63 499	75 878	Transports aériens		62	
..	..	26 482	31 774	41 529	51 932	62 510	72 204	Transports aériens réguliers			621
..	..	243	323	387	409	989	24 494	Transports aériens non réguliers			622
..	..	37 840	44 418	56 297	76 545	107 165	101 414	Act. annex. et aux. des transp.; act. des ag. de voy.		63	
..	..	119 077	159 637	219 836	304 276	416 996	481 288	Postes et télécommunications		64	
..	..	35 903	44 495	52 133	58 375	70 337	78 197	Activités de poste et de courrier			641
..	..	77 968	109 083	160 795	236 247	332 464	403 091	Télécommunications			642
..	140 875	..	..	..	..	..	25 116	Intermédiation financière	J	65_67	
..	..	..	..	..	..	..	13 690	Interm. financ., sauf assur. et caisses de retraite		65	
..	..	..	..	..	..	..	21	Intermédiation monétaire			651
..	..	..	..	..	..	..	13 669	Autres intermédiations financières			659
..	..	..	..	..	..	..	43	Act. d'assur., caisses de retr., sauf séc. soc. oblig.		66	
..	..	..	..	..	..	..	11 383	Act. auxiliaires de l'intermédiation financière		67	
..	..	..	..	..	..	..	6 237	Act. aux. de l'interm. fin., sauf assur., et retraite			671
..	..	..	..	..	..	..	5 146	Act. aux. des assurances et caisses de retraite			672
..	208 986	260 085	334 004	422 425	546 616	755 950	744 001	Immobilier, locations, services aux entreprises	K	70_74	
..	..	26 877	36 099	58 631	92 748	138 729	190 246	Activités immobilières		70	
..	..	..	..	..	..	..	143 445	Act. immobilières sur biens propres ou loués			701
..	..	..	..	..	..	..	46 801	Activités immobilières pour comptes de tiers			702
..	..	25 632	34 133	44 248	33 591	34 995	44 727	Locat. mach., équip. sans opér., biens pers. et dom.		71	
..	..	8 236	14 481	21 260	11 980	12 403	20 258	Location de matériel de transport			711
..	..	12 625	14 147	16 966	15 449	16 514	19 275	Location de machines ét equipements nca			712
..	..	..	..	..	..	..	5 194	Loc. d'articles personnels et domestiques nca			713
..	..	31 255	39 206	47 527	64 217	91 027	134 995	Activités informatiques et activités rattachées		72	
..	..	1 997	1 478	1 709	2 005	3 880	6 496	Conseil en matériel informatique			721
..	..	12 265	18 316	22 120	32 138	46 256	69 823	Production de logiciels			722
..	..	6 699	7 299	7 040	9 382	12 788	16 318	Traitement de données informatiques			723
..	..	2 571	2 895	3 553	4 641	5 083	4 212	Activités de banques de données			724
..	..	4 573	5 013	6 805	7 858	11 776	9 750	Entr., répar. mach. de bureau et compt. et inform.			725
..	..	3 150	4 205	6 300	8 193	11 244	28 396	Autres activités rattachées à l'informatique			729
..	..	13 991	14 073	16 594	20 774	25 329	26 782	Recherche-développement		73	
..	..	13 569	13 520	15 734	19 538	23 761	25 592	Rech. et dév. expér. en sc. phys., natur. et ingén.			731
..	..	422	553	860	1 236	1 568	1 190	Rech. et dév. expér. en sciences soc. et humaines			732
..	..	162 330	210 493	255 425	335 286	465 870	553 755	Autres activités de services aux entreprises		74	
..	..	42 587	61 094	77 400	100 568	179 138	199 110	Act. jur., compt., audit.; cons. fisc.; ét. march., soud.			741
..	..	64 369	71 453	83 999	101 745	127 150	148 315	Act. d'achitect., d'ingén.; autres act. techniques			742
..	..	10 362	15 222	18 605	24 182	41 246	52 855	Publicité			743
..	..	45 012	62 724	75 421	108 791	118 336	..	Activités de services aux entreprises nca.			749
..	..	9 010	11 693	14 108	15 946	21 122	25 573	Education	M	80	
..	..	42	59	82	118	183	453	Activités d'enseignement primaire			801
..	..	215	404	627	649	1 372	1 595	Activités d'enseignement secondaire			802
..	..	..	318	459	619	710	2 202	Activités d'enseignement supérieur			803
..	..	..	..	..	..	..	21 323	Act. de formation perman.; autres act. d'enseign.			809
..	..	6 685	13 435	20 865	31 478	37 564	54 403	Santé et action sociale	N	85	
..	..	6 183	12 654	19 880	30 160	35 757	52 079	Activités de santé humaine			851
..	..	..	532	779	934	1 057	1 277	Activités vétérinaires			852
..	..	..	249	206	384	750	1 047	Activités d'action sociale			853
..	..	..	..	110 421	132 596	196 128	230 450	Autres activités de serv. coll., soc. et personnels	O	90_93	
..	..	19 015	26 855	32 055	34 998	48 545	65 465	Assain. et enlèv. des ordures; voierie, act. simil.		90	
..	..	..	..	237	..	..	2 797	Activités associatives diverses		91	
..	..	..	..	11	..	..	242	Activités de syndicats de salariés			912
..	..	42 744	49 250	61 066	77 606	124 824	133 192	Activités récréatives, culturelles et sportives		92	
..	..	18 286	20 029	23 613	32 920	69 462	65 408	Act. cinémat., de radio, télé.; autres act. de spect.			921
..	..	..	..	..	..	..	3 194	Activités d'agences de presse			922
..	..	..	..	..	..	..	1 262	Act. des biblio., arch., musées, jardins bot. et zool.			923
..	..	..	..	..	..	..	63 328	Activités sportives et autres activités récréatives			924
..	..	11 466	12 569	17 063	19 992	22 759	28 996	Autres activités de services		93	

Table HU.S2 **VALUE ADDED - VALEUR AJOUTÉE**

Millions of Ft (Current Prices)

ISIC revision 3				1991	1992	1993	1994	1995	1996	1997	1998
G	50_52		Wholesale & retail trade; repairs; etc.	..	184 754	229 309	277 228	351 304	418 692	594 329	750 919
	50		Sale, maint.,repair vehic/cycle, person., h/hold gd.	..	..	16 041	26 893	42 744	65 491	81 210	106 605
		501	Sale of motor cycle	..	..	4 538	8 002	24 429	42 772	48 711	62 135
		502	Maintenance and repair of motor vehicles	..	..	7 074	9 191	8 683	8 559	10 957	15 467
		503	Sale of motor vehicle parts and accessories	..	..	2 008	3 267	5 100	6 713	10 597	16 603
		504	Sale, maint., repair motor cycle, parts & acces.	..	..	2 421	6 433	4 532	7 447	10 945	526
		505	Retail sale of automative fuel	..	..	..	..	..	..	..	11 874
	51		W/sale. trade & commission exc motor veh. & cycle	..	..	122 501	144 943	198 398	235 951	356 202	422 442
		511	Wholesale on a fee or contract basis	..	..	..	..	..	..	..	40 389
		512	W/sale agric. raw mat.,live animal,food, bev., tobac.	..	..	..	..	..	..	..	92 026
		513	Wholesale of household goods	..	..	..	..	..	..	..	72 431
		514	W/sale. of non-agric. intermed. prod., waste, scrap	..	..	..	..	..	..	..	48 952
		515	Wholesale of machinery, equipment & supplies	..	..	..	..	..	..	..	31 713
		519	Other wholesale	..	..	..	..	..	..	..	136 931
	52		Retail trade exc. motor veh. & cyc.; repair h/hold gds.	..	..	90 767	105 392	110 162	117 250	156 917	221 872
		521	Non-specialised retail trade in stores	..	..	36 271	43 247	47 486	53 109	61 046	78 582
		522	Retail sale food, bev., tobac. in specialised stores	..	..	8 408	9 768	9 429	9 801	11 757	7 280
		523	Other retail trade of new goods specialised stores	..	..	..	..	..	..	..	118 407
		524	Retail sale of second-hand goods in stores	..	..	..	..	..	..	..	4 239
		525	Retail trade not in stores	..	..	..	..	..	..	..	7 131
		526	Repair of personal and household goods	..	..	..	..	..	..	..	6 233
H	55		Hotels and restaurants	..	24 685	27 712	32 239	39 810	46 781	66 660	76 753
		551	Hotels; camping ; other short-stay accommodation	..	..	17 372	20 898	27 355	30 740	37 268	51 621
		552	Restaurants, bars and canteens	..	..	10 340	11 341	12 455	16 041	29 392	25 132
I	60_64		Transport, storage and communications	..	..	170 152	220 782	303 235	390 682	522 880	641 896
	60		Land transport, transport via pipelines	..	88 528	70 511	99 817	133 462	155 472	193 468	246 630
		601	Transport via railways	..	..	20 002	38 566	60 327	67 497	79 228	94 786
		602	Other land transport	..	..	50 498	61 238	73 119	87 945	114 229	151 775
		603	Transport via pipelines	..	..		13	16	30	11	69
	61		Water transport	..	..	2 085	1 780	2 688	1 708	2 579	4 175
		611	Sea and coastal water transport	..	..			171			78
		612	Inland water transport	..	..	2 057	1 837	2 517	2 527	3 371	4 097
	62		Air transport	..	..	3 979	4 659	6 825	7 737	10 135	13 060
		621	Scheduled air transport	..	..	3 974	4 562	6 756	7 681	9 607	11 725
		622	Non-scheduled air transport	..	..	5	97	69	56	528	..
	63		Support. & aux. transport act., travel agencies activ.	..	..	14 736	17 776	23 484	31 470	48 364	46 125
	64		Post and telecommunications	..	..	78 841	96 750	136 776	194 295	268 334	331 906
		641	Post and courier activities	..	..	28 009	34 625	39 475	43 346	53 303	59 515
		642	Telecommunications	..	..	47 301	57 863	92 454	144 560	206 767	272 391
J	65_67		Financial intermediation	..	71 809	..	..	..	..	..	..
	65		Financial intermediation, exc. insur., pension fund	..	..	..	..	..	..	..	9 478
		651	Monetary intermediation	..	..	..	..	..	..	..	18
		659	Other financial intermediation	..	..	..	..	..	..	..	9 460
	66		Insurance pension fund, exc. compul. social sec.	..	..	..	..	..	..	..	
	67		Activities auxiliary to financial intermediation	..	..	..	..	..	..	..	6 682
		671	Act. auxil to fin. intermed., exc. insur. & pens.fund	..	..	..	..	..	..	..	4 778
		672	Act. auxiliary to insurance and pension funding	..	..	..	..	..	..	..	1 904
K	70_74		Real estate, renting and business activities	..	105 353	137 497	173 921	212 173	268 252	406 897	503 225
	70		Real estate activities	..	..	13 339	17 171	30 271	44 294	72 474	103 563
		701	Real estate activ. with own or leased property	..	..	..	..	..	..	..	81 325
		702	Real estate activities on a fee or contract basis	..	..	..	..	..	..	..	22 238
	71		Renting of mach. without operator; rent. of pers. gds.	..	..	19 080	24 534	31 324	21 238	22 463	28 758
		711	Renting of transport equipment	..	..	6 380	11 504	17 638	7 920	8 402	14 275
		712	Renting of other machinery and equipment	..	..	9 792	10 009	10 549	10 380	11 206	11 922
		713	Renting of personal and household goods nec	..	..	1 064	526	520	447	440	2 561
	72		Computer and related activities	..	..	15 230	20 129	22 477	31 287	47 005	70 746
		721	Hardware consultancy	..	..	780	444	489	791	2 158	3 415
		722	Software consultancy and supply	..	..	6 482	10 476	10 803	16 139	25 191	37 393
		723	Data processing	..	..	3 173	3 767	3 768	4 959	7 269	9 817
		724	Data base activities	..	..	1 294	1 349	1 835	2 416	2 581	2 192
		725	Mainten.; repair of office, account. & comp. mach.	..	..	2 024	2 230	2 880	3 449	5 587	5 151
		729	Other computer related activities	..	..	..	..	..	..	..	12 778
	73		Research and development	..	..	6 663	6 644	7 684	9 395	13 110	14 190
		731	R & D experim. on natural sciences & engineering	..	..	6 426	6 384	7 275	8 901	12 422	13 617
		732	R & D experim. on social sciences & humanities	..	..	237	260	409	494	688	573
	74		Other business activities	..	..	83 185	105 443	120 417	162 038	251 845	285 968
		741	Legal, acc., book-keep., audit act.; tax consult., etc	..	..	21 665	31 863	37 579	49 668	96 124	97 243
		742	Architectural, engineering, & other techn. activities	..	..	34 717	35 082	39 936	49 656	68 392	79 880
		743	Advertising	..	..	..	..	..	..	..	22 723
		749	Business activities nec	..	..	23 134	32 634	37 018	55 295	63 957	..
M	80		Education	..	..	4 599	6 009	6 766	7 634	11 075	13 468
		801	Primary education	..	..	11	6	27	18	64	253
		802	Secondary education	..	..	11	77	191	184	411	673
		803	Higher education	..	..	119	178	238	351	285	1 111
		809	Adult and other education	..	..	..	..	..	..	..	11 431
N	85		Health and social work	..	..	3 147	6 478	9 843	15 964	21 635	30 059
		851	Human health activities	..	..	2 923	6 196	9 584	15 654	20 991	29 171
		852	Veterinary activities	..	..	189	192	178	242	462	641
		853	Social work activities	..	..	35	90	81	68	182	247
O	90_93		Other commun., soc. & pers. services activities	..	..	27 572	33 889	40 266	46 273	66 422	90 298
	90		Sewage & refuse disposal, sanitation, etc.	..	..	9 674	14 873	17 947	20 041	28 590	38 327
	91		Activities of member organisations nec	..	..	..	..	1	..	..	1 248
		912	Activities of trade unions	..	..	..	..	4	..	..	44
	92		Recreational, cultural and sporting activities	..	..	12 433	13 202	15 171	17 960	26 908	36 493
		921	Motion picture, radio, TV, other entertainment act.	..	..	5 055	4 237	3 519	6 153	10 214	16 754
		922	News agency activities	..	..	37	60	73	184	361	1 347
		923	Library, archives, museums, other cultural activ.	..	..	40	126	113	96	224	672
		924	Sporting and other recreational activities	..	..	207		404	136	1 318	17 720
	93		Other service activities	..	..	5 465	5 814	7 147	8 272	10 924	14 230

Table HU.S3 **EMPLOI, SALARIÉS - EMPLOYMENT, EMPLOYEES**

Unités

1991	1992	1993	1994	1995	1996	1997	1998			CITI révision 3	
..	251 510	227 887	189 049	346 416	367 471	348 276	..	Com. de gros, de detail ; réparation etc.	G	**50_52**	
..	..	..	..	36 195	39 533	41 065	..	Com., entr., répar. d'auto. et moto ; carburants		**50**	
..	..	..	..	11 968	15 736	17 050	..	Commerce de véhicules automobiles			501
..	..	..	..	14 568	13 138	12 445	..	Entretien et réparation de véhicules automobiles			502
..	..	..	..	4 989	5 734	6 145	..	Com. de pièces et access. de véhic. automobiles			503
..	..	..	..	4 670	4 925	5 425	..	Com., entret. et répar. de motocycles et access.			504
..	..	..	..	..	..	..	..	Com. de détail de carburants pour automobiles			505
..	..	..	..	146 231	147 198	150 257	..	Com. de gros et act. inter., sauf auto. et moto		**51**	
..	..	..	..	12 001	13 160	12 317	..	Activités d'intermédiaries de commerce de gros			511
..	..	..	..	..	..	..	..	Com. gros: prod. agr. bruts, anim. viv., prod. alim.			512
..	..	..	..	9 694	10 218	11 572	..	Com. gros: biens de consom., sauf alimentaires			513
..	..	..	..	29 570	30 284	33 258	..	Com. gros: prod. inter. non agr., déchets, débris			514
..	..	..	..	7 222	7 399	5 978	..	Com. gros: machines, équipments et fournitures			515
..	..	..	..	35 220	34 534	33 873	..	Autres commerces de gros			519
..	..	..	..	163 990	180 740	156 954	..	Com. détail, sauf auto, moto; répar. art. pers., dom.		**52**	
..	..	..	..	75 849	81 481	70 860	..	Commerce de détail non spécialisé en magasin			521
..	..	..	..	18 423	22 482	17 772	..	Com. détail: prod. aliment. en mag. spécialisés			522
..	..	..	..	15 538	15 165	15 618	..	Autre com. détail: prod. neufs en mag. spécialisés			523
..	..	..	..	41 681	48 964	42 646	..	Commerces de détail de biens d'occasion			524
..	..	..	..	2 418	2 780	1 937	..	Commerces de détail autres qu'en magasin			525
..	..	..	..	5 199	5 382	4 060	..	Réparation d'articles personnels et domestiques			526
..	57 407	57 692	52 921	55 223	54 519	57 989	..	Hotels et restaurants	H	**55**	
..	..	..	..	..	..	..	..	Hôtels; camping; autres équip. courts séjours			551
..	..	..	..	..	..	..	..	Restaurants, cafés et cantines			552
..	258 226	266 000	248 213	223 986	217 485	205 189	..	Transport, entreposage, communications	I	**60_64**	
..	258 226	266 000	248 213	147 624	143 495	132 952	..	Transports terrestres; transports par conduites		**60**	
..	..	..	..	72 326	69 154	61 652	..	Transports par chemin de fer			601
..	..	..	..	75 273	74 304	71 279	..	Autres transports terrestres			602
..	..	..	..	25	37	21	..	Transports par conduites			603
..	..	..	..	2 776	2 657	2 548	..	Transports par eau		**61**	
..	..	..	..	49	46	50	..	Transports maritimes et côtiers			611
..	..	..	..	2 727	2 611	2 498	..	Transports par voies navigables intérieures			612
..	..	..	..	3 765	3 616	3 719	..	Transports aériens		**62**	
..	..	..	..	3 712	3 572	3 676	..	Transports aériens réguliers			621
..	..	..	..	53	44	43	..	Transports aériens non réguliers			622
..	..	..	..	22 538	24 413	25 967	..	Act. annex. et aux. des transp.; act. des ag. de voy.		**63**	
..	..	..	..	76 362	73 990	72 237	..	Postes et télécommunications		**64**	
..	..	..	..	51 351	49 821	48 331	..	Activités de poste et de courrier			641
..	..	..	..	23 134	22 273	22 146	..	Télécommunications			642
..	53 296	61 192	63 829	..	..	..	..	Intermédiation financière	J	**65_67**	
..	..	..	..	..	..	..	..	Interm. financ., sauf assur. et caisses de retraite		**65**	
..	..	..	..	..	..	..	..	Intermédiation monétaire			651
..	..	..	..	..	..	..	..	Autres intermédiations financières			659
..	..	..	..	..	..	..	..	Act. d'assur., caisses de retr., sauf séc. soc. oblig.		**66**	
..	..	..	..	..	..	..	..	Act. auxiliaires de l'intermédiation financière		**67**	
..	..	..	..	..	..	..	..	Act. aux. de l'interm. fin., sauf assur., et retraite			671
..	..	..	..	..	..	..	..	Act. aux. des assurances et caisses de retraite			672
..	91 467	90 095	79 580	126 370	157 611	146 728	..	Immobilier, locations, services aux entreprises	K	**70_74**	
..	..	..	..	19 184	21 490	25 783	..	Activités immobilières		**70**	
..	..	..	..	9 776	10 189	11 270	..	Act. immobilières sur biens propres ou loués			701
..	..	..	..	9 408	11 301	14 513	..	Activités immobilières pour comptes de tiers			702
..	..	..	..	5 191	5 193	5 156	..	Locat. mach., équip. sans opér., biens pers. et dom.		**71**	
..	..	..	..	1 166	975	995	..	Location de matériel de transport			711
..	..	..	..	2 115	2 370	2 351	..	Location de machines ét equipements nca			712
..	..	..	..	..	..	..	..	Loc. d'articles personnels et domestiques nca			713
..	..	..	..	16 045	17 406	18 182	..	Activités informatiques et activités rattachées		**72**	
..	..	..	..	676	657	784	..	Conseil en matériel informatique			721
..	..	..	..	7 386	7 693	8 817	..	Production de logiciels			722
..	..	..	..	3 321	4 127	3 560	..	Traitement de données informatiques			723
..	..	..	..	845	923	954	..	Activités de banques de données			724
..	..	..	..	2 271	2 351	2 426	..	Entr., répar. mach. de bureau et compt. et inform.			725
..	..	..	..	1 546	1 655	1 641	..	Autres activités rattachées à l'informatique			729
..	..	..	..	6 188	6 081	5 865	..	Recherche-développement		**73**	
..	..	..	..	5 891	5 738	5 504	..	Rech. et dév. expér. en sc. phys., natur. et ingén.			731
..	..	..	..	297	343	361	..	Rech. et dév. expér. en sciences soc. et humaines			732
..	..	..	..	107 186	136 121	120 945	..	Autres activités de services aux entreprises		**74**	
..	..	..	..	24 571	37 201	31 294	..	Act. jur., compt., audit.; cons. fisc.; ét. march., soud.			741
..	..	..	..	31 285	31 054	29 565	..	Act. d'achitect., d'ingén.; autres act. techniques			742
..	..	..	..	3 846	4 246	4 343	..	Publicité			743
..	..	..	..	47 484	63 620	55 743	..	Activités de services aux entreprises nca.			749
..	..	..	..	6 123	6 237	6 467	..	Education	M	**80**	
..	..	..	..	87	202	167	..	Activités d'enseignement primaire			801
..	..	..	..	767	644	1 092	..	Activités d'enseignement secondaire			802
..	..	..	..	142	181	183	..	Activités d'enseignement supérieur			803
..	..	..	..	..	..	..	..	Act. de formation perman.; autres act. d'enseign.			809
..	..	..	..	12 707	17 890	16 403	..	Santé et action sociale	N	**85**	
..	..	..	..	11 949	16 716	14 649	..	Activités de santé humaine			851
..	..	..	..	301	349	362	..	Activités vétérinaires			852
..	..	..	..	457	825	1 392	..	Activités d'action sociale			853
..	..	..	..	29 944	29 176	27 822	..	Autres activités de serv. coll., soc. et personnels	O	**90_93**	
..	..	..	..	15 662	14 934	14 862	..	Assain. et enlèv. des ordures; voierie, act. simil.		**90**	
..	..	..	..	112	..	..	..	Activités associatives diverses		**91**	
..	..	..	..	5	..	..	..	Activités de syndicats de salariés			912
..	..	..	..	14 893	15 811	20 728	..	Activités récréatives, culturelles et sportives		**92**	
..	..	..	..	7 921	8 377	13 074	..	Act. cinémat., de radio, télé.; autres act. de spect.			921
..	..	..	..	98	150	186	..	Activités d'agences de presse			922
..	..	..	..	139	130	161	..	Act. des biblio., arch., musées, jardins bot. et zool.			923
..	..	..	..	..	..	..	..	Activités sportives et autres activités récréatives			924
..	..	..	..	..	..	..	..	Autres activités de services		**93**	

HUNGARY

Table HU.S4 WAGES AND SALARIES, EMPLOYEES - SALAIRES ET TRAITEMENTS, SALARIÉS

Millions of Ft (Current Prices)

ISIC revision 3				1991	1992	1993	1994	1995	1996	1997	1998
G	50_52		Wholesale & retail trade; repairs; etc.	..	..	..	..	130 341	155 191	178 782	..
	50		Sale, maint.,repair vehic/cycle, person., h/hold gd.	..	..	..	..	12 279	15 875	18 975	..
		501	Sale of motor cycle	..	..	..	..	4 688	7 645	9 011	..
		502	Maintenance and repair of motor vehicles	..	..	..	..	4 115	4 086	4 508	..
		503	Sale of motor vehicle parts and accessories	..	..	..	..	1 367	1 657	2 164	..
		504	Sale, maint., repair motor cycle, parts & acces.	..	..	..	..	2 109	2 487	3 292	..
		505	Retail sale of automative fuel	..	..	..	..	..	..	..	..
	51		W/sale. trade & commission exc motor veh. & cycle	..	..	..	..	67 714	81 710	96 399	..
		511	Wholesale on a fee or contract basis	..	..	..	..	4 047	5 403	5 676	..
		512	W/sale agric. raw mat.,live animal,food, bev., tobac.	..	..	..	..	..	..	..	..
		513	Wholesale of household goods	..	..	..	..	..	..	..	..
		514	W/sale. of non-agric. intermed. prod., waste, scrap	..	..	..	..	13 182	15 701	20 192	..
		515	Wholesale of machinery, equipment & supplies	..	..	..	..	4 820	6 171	6 577	..
		519	Other wholesale	..	..	..	..	..	..	..	..
	52		Retail trade exc. motor veh. & cyc.; repair h/hold gds.	..	..	..	..	50 348	57 606	63 408	..
		521	Non-specialised retail trade in stores	..	..	..	..	22 837	25 503	29 333	..
		522	Retail sale food, bev., tobac. in specialised stores	..	..	..	..	5 225	5 938	6 677	..
		523	Other retail trade of new goods specialised stores	..	..	..	..	..	..	..	..
		524	Retail sale of second-hand goods in stores	..	..	..	..	..	..	..	..
		525	Retail trade not in stores	..	..	..	..	932	1 145	1 027	..
		526	Repair of personal and household goods	..	..	..	..	1 213	1 465	1 442	..
H	55		Hotels and restaurants	..	..	..	..	16 989	19 068	24 323	..
		551	Hotels; camping ; other short-stay accommodation	..	..	..	..	..	..	..	..
		552	Restaurants, bars and canteens	..	..	..	..	..	..	..	..
I	60_64		Transport, storage and communications	..	..	..	..	121 025	145 377	169 646	..
	60		Land transport, transport via pipelines	..	..	..	..	66 730	78 436	86 498	..
		601	Transport via railways	..	..	..	..	33 612	39 440	41 776	..
		602	Other land transport	..	..	..	..	33 118	38 996	44 722	..
		603	Transport via pipelines	..	..	..	..				..
	61		Water transport	..	..	..	..	1 229	1 474	1 670	..
		611	Sea and coastal water transport	..	..	..	..	74	80	118	..
		612	Inland water transport	..	..	..	..	1 155	1 394	1 552	..
	62		Air transport	..	..	..	..	3 157	3 994	4 958	..
		621	Scheduled air transport	..	..	..	..	3 144	3 981	4 923	..
		622	Non-scheduled air transport	..	..	..	..	13	13	35	..
	63		Support. & aux. transport act., travel agencies activ.	..	..	..	..	10 076	12 976	16 587	..
	64		Post and telecommunications	..	..	..	..	39 833	48 497	59 933	..
		641	Post and courier activities	..	..	..	..	20 591	23 736	27 881	..
		642	Telecommunications	..	..	..	..	17 810	22 992	30 029	..
J	65_67		Financial intermediation	..	..	..	..	..	..	..	..
	65		Financial intermediation, exc. insur., pension fund	..	..	..	..	..	..	..	..
		651	Monetary intermediation	..	..	..	..	..	..	..	..
		659	Other financial intermediation	..	..	..	..	..	..	..	..
	66		Insurance pension fund, exc. compul. social sec.	..	..	..	..	..	..	..	..
	67		Activities auxiliary to financial intermediation	..	..	..	..	..	..	..	..
		671	Act. auxil to fin. intermed., exc. insur. & pens.fund	..	..	..	..	..	..	..	..
		672	Act. auxiliary to insurance and pension funding	..	..	..	..	..	..	..	..
K	70_74		Real estate, renting and business activities	..	..	..	..	63 454	80 322	101 564	..
	70		Real estate activities	..	..	..	..	9 263	13 082	16 654	..
		701	Real estate activ. with own or leased property	..	..	..	..	..	..	..	..
		702	Real estate activities on a fee or contract basis	..	..	..	..	4 106	6 355	8 164	..
	71		Renting of mach. without operator; rent. of pers. gds.	..	..	..	..	2 127	2 439	2 693	..
		711	Renting of transport equipment	..	..	..	..	..	..	..	..
		712	Renting of other machinery and equipment	..	..	..	..	..	..	..	..
		713	Renting of personal and household goods nec	..	..	..	..	..	..	..	..
	72		Computer and related activities	..	..	..	..	6 673	8 762	12 016	..
		721	Hardware consultancy	..	..	..	..	229	195	308	..
		722	Software consultancy and supply	..	..	..	..	2 653	3 600	5 792	..
		723	Data processing	..	..	..	..	1 577	2 314	2 573	..
		724	Data base activities	..	..	..	..	546	677	829	..
		725	Mainten.; repair of office, account. & comp. mach.	..	..	..	..	968	1 055	1 657	..
		729	Other computer related activities	..	..	..	..	..	..	..	..
	73		Research and development	..	..	..	..	3 165	3 638	4 136	..
		731	R & D experim. on natural sciences & engineering	..	..	..	..	3 000	3 441	3 935	..
		732	R & D experim. on social sciences & humanities	..	..	..	..	165	197	201	..
	74		Other business activities	..	..	..	..	42 226	52 401	66 065	..
		741	Legal, acc., book-keep., audit act.; tax consult., etc	..	..	..	..	11 907	15 271	21 206	..
		742	Architectural, engineering, & other techn. activities	..	..	..	..	..	..	..	..
		743	Advertising	..	..	..	..	..	..	..	..
		749	Business activities nec	..	..	..	..	..	..	..	..
M	80		Education	..	..	..	..	2 779	2 835	3 165	..
		801	Primary education	..	..	..	..	17	26	29	..
		802	Secondary education	..	..	..	..	345	352	609	..
		803	Higher education	..	..	..	..	105	138	126	..
		809	Adult and other education	..	..	..	..	..	..	..	..
N	85		Health and social work	..	..	..	..	3 569	5 136	6 168	..
		851	Human health activities	..	..	..	..	3 455	4 912	5 796	..
		852	Veterinary activities	..	..	..	..	49	53	71	..
		853	Social work activities	..	..	..	..	65	171	301	..
O	90_93		Other commun., soc. & pers. services activities	..	..	..	..	17 456	19 306	29 111	..
	90		Sewage & refuse disposal, sanitation, etc.	..	..	..	..	6 885	7 839	9 500	..
	91		Activities of member organisations nec	..	..	..	..	99			..
		912	Activities of trade unions	..	..	..	..	4			..
	92		Recreational, cultural and sporting activities	..	..	..	..	6 483	7 004	14 553	..
		921	Motion picture, radio, TV, other entertainment act.	..	..	..	..	..	..	..	..
		922	News agency activities	..	..	..	..	..	..	..	..
		923	Library, archives, museums, other cultural activ.	..	..	..	..	..	..	..	..
		924	Sporting and other recreational activities	..	..	..	..	246	221	378	..
	93		Other service activities	..	..	..	..	3 989	4 463	5 058	..

Table HU.S5 **ÉTABLISSEMENTS - ESTABLISHMENTS**

Unités

1991	1992	1993	1994	1995	1996	1997	1998		CITI révision 3		
..	39 818	48 957	58 530	64 441	72 167	72 478	77 445	Com. de gros, de detail ; réparation etc.	G	50_52	
..	..	4 638	6 079	6 974	7 870	8 376	9 193	Com., entr., répar. d'auto. et moto ; carburants		50	
..	..	1 424	2 068	2 451	2 729	2 955	2 916	Commerce de véhicules automobiles			501
..	..	1 677	2 012	2 226	2 482	2 594	2 998	Entretien et réparation de véhicules automobiles			502
..	..	1 088	1 327	1 510	1 751	1 896	2 345	Com. de pièces et access. de véhic. automobiles			503
..	..	449	672	787	908	931	146	Com., entret. et répar. de motocycles et access.			504
..	..	..	..	..	..	..	788	Com. de détail de carburants pour automobiles			505
..	..	25 614	28 798	31 220	33 855	34 639	32 658	Com. de gros et act. inter., sauf auto. et moto		51	
..	..	..	..	..	..	..	6 471	Activités d'intermédiaires de commerce de gros			511
..	..	..	..	..	..	..	4 697	Com. gros: prod. agr. bruts, anim. viv., prod. alim.			512
..	..	..	..	..	..	..	5 114	Com. gros: biens de consom., sauf alimentaires			513
..	..	..	..	..	..	..	2 680	Com. gros: prod. inter. non agr., déchets, débris			514
..	..	..	..	..	..	..	1 380	Com. gros: machines, équipments et fournitures			515
..	..	8 128	8 904	9 594	10 305	10 425	12 316	Autres commerces de gros			519
..	..	18 705	23 653	26 247	30 442	29 463	35 594	Com. détail, sauf auto, moto; répar. art. pers., dom.		52	
..	..	6 923	7 825	..	9 181	9 019	9 288	Commerce de détail non spécialisé en magasin			521
..	..	1 585	1 853	2 107	2 439	2 555	2 509	Com. détail: prod. aliment. en mag. spécialisés			522
..	..	..	..	..	..	..	18 435	Autre com. détail: prod. neufs en mag. spécialisés			523
..	..	..	..	..	..	..	785	Commerces de détail de biens d'occasion			524
..	..	..	..	..	..	..	2 996	Commerces de détail autres qu'en magasin			525
..	..	..	..	..	..	..	1 581	Réparation d'articles personnels et domestiques			526
..	4 984	5 942	6 825	7 530	8 381	8 820	9 874	Hotels et restaurants	H	55	
..	..	1 011	1 296	1 517	1 595	1 708	1 980	Hôtels; camping; autres équip. courts séjours			551
..	..	4 931	5 529	6 013	6 786	7 112	7 894	Restaurants, cafés et cantines			552
..	..	5 265	6 418	7 076	8 288	8 401	9 441	Transport, entreposage, communications	I	60_64	
..	2 180	2 774	3 590	4 018	4 783	4 697	5 763	Transports terrestres; transports par conduites		60	
..	..	8	8	9	7	9	17	Transports par chemin de fer			601
..	..	2 764	3 579	4 003	4 771	4 685	5 742	Autres transports terrestres			602
..	..		3	6	5	3	4	Transports par conduites			603
..	..	34	43	49	52	57	69	Transports par eau		61	
..	..		4	3	3	4	7	Transports maritimes et côtiers			611
..	..	33	39	46	49	53	62	Transports par voies navigables intérieures			612
..	..	38	41	49	56	54	59	Transports aériens		62	
..	..	29	33	40	44	44	6	Transports aériens réguliers			621
..	..	9	8	9	12	10	..	Transports aériens non réguliers			622
..	..	2 091	2 327	2 485	2 828	2 989	2 931	Act. annex. et aux. des transp.; act. des ag. de voy.		63	
..	..	328	417	475	569	604	619	Postes et télécommunications		64	
..	..	87	125	166	220	232	207	Activités de poste et de courrier			641
..	..	214	249	259	273	274	412	Télécommunications			642
..	2 221	..	..	..	..	..	516	Intermédiation financière	J	65_67	
..	..	..	..	..	..	..	191	Interm. financ., sauf assur. et caisses de retraite		65	
..	..	..	..	..	..	..	1	Intermédiation monétaire			651
..	..	..	..	..	..	..	190	Autres intermédiations financières			659
..	..	..	..	..	..	..	3	Act. d'assur., caisses de retr., sauf séc. soc. oblig.		66	
..	..	..	..	..	..	..	322	Act. auxiliaires de l'intermédiation financière		67	
..	..	..	..	..	..	..	96	Act. aux. de l'interm. fin., sauf assur., et retraite			671
..	..	..	..	..	..	..	226	Act. aux. des assurances et caisses de retraite			672
..	23 320	28 374	37 158	41 767	51 933	51 162	60 916	Immobilier, locations, services aux entreprises	K	70_74	
..	..	1 686	2 820	3 901	5 590	6 743	8 955	Activités immobilières		70	
..	..	839	1 635	2 308	3 181	3 435	6 209	Act. immobilières sur biens propres ou loués			701
..	..	847	1 185	1 593	2 409	3 308	2 746	Activités immobilières pour comptes de tiers			702
..	..	910	1 136	1 224	1 313	1 280	1 464	Locat. mach., équip. sans opér., biens pers. et dom.		71	
..	..	220	323	339	351	360	450	Location de matériel de transport			711
..	..	..	..	..	..	..	552	Location de machines ét equipements nca			712
..	..	50	43	38	37	37	462	Loc. d'articles personnels et domestiques nca			713
..	..	3 590	4 279	4 834	6 075	6 955	8 268	Activités informatiques et activités rattachées		72	
..	..	204	229	268	358	455	586	Conseil en matériel informatique			721
..	..	1 793	2 191	2 532	3 216	3 747	4 269	Production de logiciels			722
..	..	582	652	724	861	924	1 055	Traitement de données informatiques			723
..	..	182	210	225	263	262	214	Activités de banques de données			724
..	..	452	552	601	749	822	969	Entr., répar. mach. de bureau et compt. et inform.			725
..	..	..	..	..	..	..	1 175	Autres activités rattachées à l'informatique			729
..	..	1 155	1 130	1 185	1 275	1 372	1 334	Recherche-développement		73	
..	..	1 075	1 022	1 032	1 075	1 130	1 051	Rech. et dév. expér. en sc. phys., natur. et ingén.			731
..	..	80	108	153	200	242	283	Rech. et dév. expér. en sciences soc. et humaines			732
..	..	21 033	27 793	30 623	37 680	34 812	40 895	Autres activités de services aux entreprises		74	
..	..	6 719	8 572	9 954	12 502	14 054	17 689	Act. jur., compt., audit.; cons. fisc.; ét. march., soud.			741
..	..	6 512	7 160	7 518	8 334	8 688	9 630	Act. d'achitect., d'ingén.; autres act. techniques			742
..	..	..	..	..	..	..	2 323	Publicité			743
..	..	..	..	..	..	..	..	Activités de services aux entreprises nca.			749
..	..	1 411	1 679	1 932	2 590	3 105	3 628	Education	M	80	
..	..	28	49	52	91	92	104	Activités d'enseignement primaire			801
..	..	40	65	89	110	142	193	Activités d'enseignement secondaire			802
..	..	29	42	55	113	147	216	Activités d'enseignement supérieur			803
..	..	..	..	..	..	..	3 115	Act. de formation perman.; autres act. d'enseign.			809
..	..	1 580	3 595	5 024	6 713	6 504	7 906	Santé et action sociale	N	85	
..	..	1 438	3 407	4 809	6 413	6 200	7 520	Activités de santé humaine			851
..	..	90	137	152	220	213	265	Activités vétérinaires			852
..	..	52	51	63	80	91	121	Activités d'action sociale			853
..	..	5 276	6 531	7 774	9 406	9 825	11 562	Autres activités de serv. coll., soc. et personnels	O	90_93	
..	..	445	483	496	565	602	663	Assain. et enlèv. des ordures; voierie, act. simil.		90	
..	..	..	..	43	..	..	126	Activités associatives diverses		91	
..	..	..	..	3	..	..	4	Activités de syndicats de salariés			912
..	..	3 514	4 355	5 459	6 747	7 445	8 573	Activités récréatives, culturelles et sportives		92	
..	..	2 185	2 838	3 470	4 287	4 660	5 397	Act. cinémat., de radio, télé.; autres act. de spect.			921
..	..	51	78	141	202	253	427	Activités d'agences de presse			922
..	..	37	48	55	69	74	131	Act. des biblio., arch., musées, jardins bot. et zool.			923
..	..	325	391	460	619	717	2 618	Activités sportives et autres activités récréatives			924
..	..	1 317	1 693	1 776	2 094	1 778	2 200	Autres activités de services		93	

ITALY

Table IT.S1 PRODUCTION - PRODUCTION

Billions of L (Current Prices)

ISIC revision 3				1991	1992	1993	1994	1995	1996	1997	1998
G	**50_52**		Wholesale & retail trade; repairs; etc.	..	..	..	..	..	1066 170	730 968	..
	50		Sale, maint.,repair vehic/cycle, person., h/hold gd.	..	..	..	..	..	186 120	129 803	..
		501	Sale of motor cycle	..	..	..	..	..	..	77 833	..
		502	Maintenance and repair of motor vehicles	..	..	..	..	..	..	20 707	..
		503	Sale of motor vehicle parts and accessories	..	..	..	..	..	..	9 108	..
		504	Sale, maint., repair motor cycle, parts & acces.	..	..	..	..	..	..	2 879	..
		505	Retail sale of automative fuel	..	..	..	..	..	..	19 276	..
	51		W/sale. trade & commission exc motor veh. & cycle	..	..	..	..	..	557 640	390 939	..
		511	Wholesale on a fee or contract basis	..	..	..	..	..	..	28 842	..
		512	W/sale agric. raw mat.,live animal,food, bev., tobac.	..	..	..	..	..	..	81 902	..
		513	Wholesale of household goods	..	..	..	..	..	..	96 401	..
		514	W/sale. of non-agric. intermed. prod., waste, scrap	..	..	..	..	..	..	135 581	..
		515	Wholesale of machinery, equipment & supplies	..	..	..	..	..	..	30 446	..
		519	Other wholesale	..	..	..	..	..	..	17 768	..
	52		Retail trade exc. motor veh. & cyc.; repair h/hold gds.	..	..	..	..	..	322 410	210 226	..
		521	Non-specialised retail trade in stores	..	..	..	..	..	..	95 653	..
		522	Retail sale food, bev., tobac. in specialised stores	..	..	..	..	..	..	13 762	..
		523	Other retail trade of new goods specialised stores	..	..	..	..	..	..	89 598	..
		524	Retail sale of second-hand goods in stores	..	..	..	..	..	..	339	..
		525	Retail trade not in stores	..	..	..	..	..	..	8 836	..
		526	Repair of personal and household goods	..	..	..	..	..	..	2 037	..
H	**55**		Hotels and restaurants	..	..	..	..	..	65 029	68 518	..
		551	Hotels; camping ; other short-stay accommodation	..	..	..	..	..	20 198	21 714	..
		552	Restaurants, bars and canteens	..	..	..	..	..	44 831	46 804	..
I	**60_64**		Transport, storage and communications	..	..	..	..	..	..	195 465	..
	60		Land transport, transport via pipelines	..	..	..	..	..	67 057	66 374	..
		601	Transport via railways	..	..	..	..	..	..	7 574	..
		602	Other land transport	..	..	..	..	..	..	58 599	..
		603	Transport via pipelines	..	..	..	..	..	..	201	..
	61		Water transport	..	..	..	..	..	7 470	6 087	..
		611	Sea and coastal water transport	..	..	..	..	..	..	5 661	..
		612	Inland water transport	..	..	..	..	..	..	426	..
	62		Air transport	..	..	..	..	..	9 447	13 590	..
		621	Scheduled air transport	..	..	..	..	..	..	12 293	..
		622	Non-scheduled air transport	..	..	..	..	..	..	1 297	..
	63		Support. & aux. transport act., travel agencies activ.	..	..	..	..	..	53 647	54 990	..
	64		Post and telecommunications	..	..	..	..	..	51 354	54 423	..
		641	Post and courier activities	..	..	..	..	..	12 703	12 420	..
		642	Telecommunications	..	..	..	..	..	38 650	42 003	..
J	**65_67**		Financial intermediation	..	..	..	..	..	..	..	..
	65		Financial intermediation, exc. insur., pension fund	..	..	..	..	..	..	..	..
		651	Monetary intermediation	..	..	..	..	..	..	..	..
		659	Other financial intermediation	..	..	..	..	..	..	..	..
	66		Insurance pension fund, exc. compul. social sec.	..	..	..	..	..	..	..	..
	67		Activities auxiliary to financial intermediation	..	..	..	..	..	..	..	..
		671	Act. auxil to fin. intermed., exc. insur. & pens.fund	..	..	..	..	..	..	..	..
		672	Act. auxiliary to insurance and pension funding	..	..	..	..	..	..	..	..
K	**70_74**		Real estate, renting and business activities	..	..	..	..	..	193 119	202 961	..
	70		Real estate activities	..	..	..	..	..	28 608	33 996	..
		701	Real estate activ. with own or leased property	..	..	..	..	..	24 790	29 900	..
		702	Real estate activities on a fee or contract basis	..	..	..	..	..	3 818	4 097	..
	71		Renting of mach. without operator; rent. of pers. gds.	..	..	..	..	..	3 117	1 845	..
		711	Renting of transport equipment	..	..	..	..	..	..	664	..
		712	Renting of other machinery and equipment	..	..	..	..	..	..	832	..
		713	Renting of personal and household goods nec	..	..	..	..	..	..	349	..
	72		Computer and related activities	..	..	..	..	..	31 590	31 510	..
		721	Hardware consultancy	..	..	..	..	..	699	1 933	..
		722	Software consultancy and supply	..	..	..	..	..	15 443	16 325	..
		723	Data processing	..	..	..	..	..	7 923	7 908	..
		724	Data base activities	..	..	..	..	..	454	857	..
		725	Mainten.; repair of office, account. & comp. mach.	..	..	..	..	..	4 569	1 652	..
		729	Other computer related activities	..	..	..	..	..	2 503	2 835	..
	73		Research and development	..	..	..	..	..	3 084	3 381	..
		731	R & D experim. on natural sciences & engineering	..	..	..	..	..	2 825	3 046	..
		732	R & D experim. on social sciences & humanities	..	..	..	..	..	259	335	..
	74		Other business activities	..	..	..	..	..	126 720	132 227	..
		741	Legal, acc., book-keep., audit act.; tax consult., etc	..	..	..	..	..	41 767	43 161	..
		742	Architectural, engineering, & other techn. activities	..	..	..	..	..	29 134	30 310	..
		743	Advertising	..	..	..	..	..	7 979	8 751	..
		749	Business activities nec	..	..	..	..	..	47 840	50 006	..
M	**80**		Education	..	..	..	..	..	2 902	3 037	..
		801	Primary education	..	..	..	..	..	235	291	..
		802	Secondary education	..	..	..	..	..	960	1 009	..
		803	Higher education	..	..	..	..	..	230	315	..
		809	Adult and other education	..	..	..	..	..	1 478	1 423	..
N	**85**		Health and social work	..	..	..	..	..	42 763	43 239	..
		851	Human health activities	..	..	..	..	..	39 104	39 350	..
		852	Veterinary activities	..	..	..	..	..	584	442	..
		853	Social work activities	..	..	..	..	..	3 075	3 447	..
O	**90_93**		Other commun., soc. & pers. services activities	..	..	..	..	..	..	54 262	..
	90		Sewage & refuse disposal, sanitation, etc.	..	..	..	..	..	10 787	11 039	..
	91		Activities of member organisations nec	..	..	..	..	..	..	..	..
		912	Activities of trade unions	..	..	..	..	..	..	..	..
	92		Recreational, cultural and sporting activities	..	..	..	..	..	26 485	29 021	..
		921	Motion picture, radio, TV, other entertainment act.	..	..	..	..	..	..	22 649	..
		922	News agency activities	..	..	..	..	..	..	1 378	..
		923	Library, archives, museums, other cultural activ.	..	..	..	..	..	..	111	..
		924	Sporting and other recreational activities	..	..	..	..	..	..	4 882	..
	93		Other service activities	..	..	..	..	..	11 891	14 202	..

Table IT.S2 **INVESTISSEMENT - INVESTMENT**

Milliards de L (Prix courants)

1991	1992	1993	1994	1995	1996	1997	1998		CITI révision 3		
..	..	..	..	..	24 840	23 374	..	Com. de gros, de detail ; réparation etc.	G	**50_52**	
..	..	..	..	..	2 744	3 016	..	Com., entr., répar. d'auto. et moto ; carburants		**50**	
..	..	..	..	..	..	737	..	Commerce de véhicules automobiles			501
..	..	..	..	..	..	1 656	..	Entretien et réparation de véhicules automobiles			502
..	..	..	..	..	..	229	..	Com. de pièces et access. de véhic. automobiles			503
..	..	..	..	..	..	77	..	Com., entret. et répar. de motocycles et access.			504
..	..	..	..	..	..	316	..	Com. de détail de carburants pour automobiles			505
..	..	..	..	..	12 646	10 773	..	Com. de gros et act. inter., sauf auto. et moto		**51**	
..	..	..	..	..	..	2 091	..	Activités d'intermédiaries de commerce de gros			511
..	..	..	..	..	..	1 921	..	Com. gros: prod. agr. bruts, anim. viv., prod. alim.			512
..	..	..	..	..	..	2 545	..	Com. gros: biens de consom., sauf alimentaires			513
..	..	..	..	..	..	2 790	..	Com. gros: prod. inter. non agr., déchets, débris			514
..	..	..	..	..	..	974	..	Com. gros: machines, équipments et fournitures			515
..	..	..	..	..	..	453	..	Autres commerces de gros			519
..	..	..	..	..	9 450	9 585	..	Com. détail, sauf auto, moto; répar. art. pers., dom.		**52**	
..	..	..	..	..	..	3 350	..	Commerce de détail non spécialisé en magasin			521
..	..	..	..	..	..	972	..	Com. détail: prod. aliment. en mag. spécialisés			522
..	..	..	..	..	..	4 152	..	Autre com. détail: prod. neufs en mag. spécialisés			523
..	..	..	..	..	..	13	..	Commerces de détail de biens d'occasion			524
..	..	..	..	..	..	919	..	Commerces de détail autres qu'en magasin			525
..	..	..	..	..	..	179	..	Réparation d'articles personnels et domestiques			526
..	..	..	..	..	4 772	5 774	..	Hotels et restaurants	H	**55**	
..	..	..	..	..	2 039	2 624	..	Hôtels; camping; autres équip. courts séjours			551
..	..	..	..	..	2 733	3 150	..	Restaurants, cafés et cantines			552
..	..	..	..	..	21 918	23 483	..	Transport, entreposage, communications	I	**60_64**	
..	..	..	..	..	8 065	9 682	..	Transports terrestres; transports par conduites		**60**	
..	..	..	..	..	..	4 142	..	Transports par chemin de fer			601
..	..	..	..	..	..	5 488	..	Autres transports terrestres			602
..	..	..	..	..	..	53	..	Transports par conduites			603
..	..	..	..	..	2 121	1 006	..	Transports par eau		**61**	
..	..	..	..	..	..	928	..	Transports maritimes et côtiers			611
..	..	..	..	..	..	78	..	Transports par voies navigables intérieures			612
..	..	..	..	..	288	417	..	Transports aériens		**62**	
..	..	..	..	..	..	393	..	Transports aériens réguliers			621
..	..	..	..	..	..	24	..	Transports aériens non réguliers			622
..	..	..	..	..	2 464	2 264	..	Act. annex. et aux. des transp.; act. des ag. de voy.		**63**	
..	..	..	..	..	8 979	10 114	..	Postes et télécommunications		**64**	
..	..	..	..	..	209	251	..	Activités de poste et de courrier			641
..	..	..	..	..	8 771	9 863	..	Télécommunications			642
..	..	..	..	..	..	..	..	Intermédiation financière	J	**65_67**	
..	..	..	..	..	..	..	..	Interm. financ., sauf assur. et caisses de retraite		**65**	
..	..	..	..	..	..	..	..	Intermédiation monétaire			651
..	..	..	..	..	..	..	..	Autres intermédiations financières			659
..	..	..	..	..	..	..	..	Act. d'assur., caisses de retr., sauf séc. soc. oblig.		**66**	
..	..	..	..	..	..	..	..	Act. auxiliaires de l'intermédiation financière		**67**	
..	..	..	..	..	..	..	..	Act. aux. de l'interm. fin., sauf assur., et retraite			671
..	..	..	..	..	..	..	..	Act. aux. des assurances et caisses de retraite			672
..	..	..	..	..	14 494	11 896	..	Immobilier, locations, services aux entreprises	K	**70_74**	
..	..	..	..	..	5 536	3 791	..	Activités immobilières		**70**	
..	..	..	..	..	5 268	3 527	..	Act. immobilières sur biens propres ou loués			701
..	..	..	..	..	268	264	..	Activités immobilières pour comptes de tiers			702
..	..	..	..	..	1 453	412	..	Locat. mach., équip. sans opér., biens pers. et dom.		**71**	
..	..	..	..	..	..	211	..	Location de matériel de transport			711
..	..	..	..	..	..	143	..	Location de machines ét equipements nca			712
..	..	..	..	..	..	58	..	Loc. d'articles personnels et domestiques nca			713
..	..	..	..	..	1 413	1 633	..	Activités informatiques et activités rattachées		**72**	
..	..	..	..	..	16	118	..	Conseil en matériel informatique			721
..	..	..	..	..	513	662	..	Production de logiciels			722
..	..	..	..	..	636	603	..	Traitement de données informatiques			723
..	..	..	..	..	43	14	..	Activités de banques de données			724
..	..	..	..	..	126	65	..	Entr., répar. mach. de bureau et compt. et inform.			725
..	..	..	..	..	79	171	..	Autres activités rattachées à l'informatique			729
..	..	..	..	..	243	272	..	Recherche-développement		**73**	
..	..	..	..	..	237	250	..	Rech. et dév. expér. en sc. phys., natur. et ingén.			731
..	..	..	..	..	6	22	..	Rech. et dév. expér. en sciences soc. et humaines			732
..	..	..	..	..	5 849	5 789	..	Autres activités de services aux entreprises		**74**	
..	..	..	..	..	2 017	2 012	..	Act. jur., compt., audit.; cons. fisc.; ét. march., soud.			741
..	..	..	..	..	1 355	1 518	..	Act. d'achitect., d'ingén.; autres act. techniques			742
..	..	..	..	..	263	248	..	Publicité			743
..	..	..	..	..	2 213	2 011	..	Activités de services aux entreprises nca.			749
..	..	..	..	..	425	188	..	Education	M	**80**	
..	..	..	..	..	10	10	..	Activités d'enseignement primaire			801
..	..	..	..	..	161	36	..	Activités d'enseignement secondaire			802
..	..	..	..	..	5	32	..	Activités d'enseignement supérieur			803
..	..	..	..	..	250	110	..	Act. de formation perman.; autres act. d'enseign.			809
..	..	..	..	..	2 223	2 208	..	Santé et action sociale	N	**85**	
..	..	..	..	..	2 075	1 994	..	Activités de santé humaine			851
..	..	..	..	..	41	31	..	Activités vétérinaires			852
..	..	..	..	..	107	183	..	Activités d'action sociale			853
..	..	..	..	..	3 990	4 260	..	Autres activités de serv. coll., soc. et personnels	O	**90_93**	
..	..	..	..	..	929	1 102	..	Assain. et enlèv. des ordures; voierie, act. simil.		**90**	
..	..	..	..	..	..	..	..	Activités associatives diverses		**91**	
..	..	..	..	..	..	..	..	Activités de syndicats de salariés			912
..	..	..	..	..	1 744	1 765	..	Activités récréatives, culturelles et sportives		**92**	
..	..	..	..	..	..	1 297	..	Act. cinémat., de radio, télé.; autres act. de spect.			921
..	..	..	..	..	..	43	..	Activités d'agences de presse			922
..	..	..	..	..	..	33	..	Act. des biblio., arch., musées, jardins bot. et zool.			923
..	..	..	..	..	..	392	..	Activités sportives et autres activités récréatives			924
..	..	..	..	..	1 318	1 393	..	Autres activités de services		**93**	

ITALY

Table IT.S3 **EMPLOYMENT - EMPLOI**

Units

ISIC revision 3				1991	1992	1993	1994	1995	1996	1997	1998
G	**50_52**		Wholesale & retail trade; repairs; etc.	..	..	..	..	..	2935 266	2955 299	..
	50		Sale, maint.,repair vehic/cycle, person., h/hold gd.	..	..	..	..	..	434 748	431 162	..
		501	Sale of motor cycle	..	..	..	..	..	..	82 026	..
		502	Maintenance and repair of motor vehicles	..	..	..	..	..	..	230 344	..
		503	Sale of motor vehicle parts and accessories	..	..	..	..	..	..	41 478	..
		504	Sale, maint., repair motor cycle, parts & acces.	..	..	..	..	..	..	18 462	..
		505	Retail sale of automative fuel	..	..	..	..	..	..	58 852	..
	51		W/sale. trade & commission exc motor veh. & cycle	..	..	..	..	..	990 739	978 884	..
		511	Wholesale on a fee or contract basis	..	..	..	..	..	..	288 430	..
		512	W/sale agric. raw mat.,live animal,food, bev., tobac.	..	..	..	..	..	..	187 872	..
		513	Wholesale of household goods	..	..	..	..	..	..	213 252	..
		514	W/sale. of non-agric. intermed. prod., waste, scrap	..	..	..	..	..	..	156 183	..
		515	Wholesale of machinery, equipment & supplies	..	..	..	..	..	..	85 461	..
		519	Other wholesale	..	..	..	..	..	..	47 686	..
	52		Retail trade exc. motor veh. & cyc.; repair h/hold gds.	..	..	..	..	..	1509 779	1545 253	..
		521	Non-specialised retail trade in stores	..	..	..	..	..	..	380 050	..
		522	Retail sale food, bev., tobac. in specialised stores	..	..	..	..	..	..	194 733	..
		523	Other retail trade of new goods specialised stores	..	..	..	..	..	..	791 121	..
		524	Retail sale of second-hand goods in stores	..	..	..	..	..	..	4 524	..
		525	Retail trade not in stores	..	..	..	..	..	..	135 295	..
		526	Repair of personal and household goods	..	..	..	..	..	..	39 530	..
H	**55**		Hotels and restaurants	..	..	..	..	..	736 503	718 501	..
		551	Hotels; camping ; other short-stay accommodation	..	..	..	..	..	219 963	209 443	..
		552	Restaurants, bars and canteens	..	..	..	..	..	516 540	509 058	..
I	**60_64**		Transport, storage and communications	..	..	..	..	..	1086 739	1087 907	..
	60		Land transport, transport via pipelines	..	..	..	..	..	561 708	568 434	..
		601	Transport via railways	..	..	..	..	..	..	148 512	..
		602	Other land transport	..	..	..	..	..	..	419 426	..
		603	Transport via pipelines	..	..	..	..	..	..	496	..
	61		Water transport	..	..	..	..	..	22 281	22 556	..
		611	Sea and coastal water transport	..	..	..	..	..	..	16 335	..
		612	Inland water transport	..	..	..	..	..	..	6 221	..
	62		Air transport	..	..	..	..	..	18 892	20 627	..
		621	Scheduled air transport	..	..	..	..	..	..	18 942	..
		622	Non-scheduled air transport	..	..	..	..	..	..	1 685	..
	63		Support. & aux. transport act., travel agencies activ.	..	..	..	..	..	194 271	196 172	..
	64		Post and telecommunications	..	..	..	..	..	289 587	280 118	..
		641	Post and courier activities	..	..	..	..	..	193 289	185 381	..
		642	Telecommunications	..	..	..	..	..	96 298	94 737	..
J	**65_67**		Financial intermediation	..	..	..	..	..	..	..	..
	65		Financial intermediation, exc. insur., pension fund	..	..	..	..	..	..	..	..
		651	Monetary intermediation	..	..	..	..	..	..	..	..
		659	Other financial intermediation	..	..	..	..	..	..	..	..
	66		Insurance pension fund, exc. compul. social sec.	..	..	..	..	..	..	..	..
	67		Activities auxiliary to financial intermediation	..	..	..	..	..	..	..	..
		671	Act. auxil to fin. intermed., exc. insur. & pens.fund	..	..	..	..	..	..	..	..
		672	Act. auxiliary to insurance and pension funding	..	..	..	..	..	..	..	..
K	**70_74**		Real estate, renting and business activities	..	..	..	..	..	1521 573	1574 974	..
	70		Real estate activities	..	..	..	..	..	165 122	174 002	..
		701	Real estate activ. with own or leased property	..	..	..	..	..	123 509	133 152	..
		702	Real estate activities on a fee or contract basis	..	..	..	..	..	41 613	40 850	..
	71		Renting of mach. without operator; rent. of pers. gds.	..	..	..	..	..	15 145	14 974	..
		711	Renting of transport equipment	..	..	..	..	..	..	4 854	..
		712	Renting of other machinery and equipment	..	..	..	..	..	..	5 273	..
		713	Renting of personal and household goods nec	..	..	..	..	..	..	4 847	..
	72		Computer and related activities	..	..	..	..	..	200 298	202 380	..
		721	Hardware consultancy	..	..	..	..	..	4 830	6 547	..
		722	Software consultancy and supply	..	..	..	..	..	85 909	90 168	..
		723	Data processing	..	..	..	..	..	80 648	78 322	..
		724	Data base activities	..	..	..	..	..	1 179	2 184	..
		725	Mainten.; repair of office, account. & comp. mach.	..	..	..	..	..	17 523	11 891	..
		729	Other computer related activities	..	..	..	..	..	10 209	13 268	..
	73		Research and development	..	..	..	..	..	18 955	19 116	..
		731	R & D experim. on natural sciences & engineering	..	..	..	..	..	16 811	16 993	..
		732	R & D experim. on social sciences & humanities	..	..	..	..	..	2 144	2 123	..
	74		Other business activities	..	..	..	..	..	1122 053	1164 502	..
		741	Legal, acc., book-keep., audit act.; tax consult., etc	..	..	..	..	..	373 084	383 731	..
		742	Architectural, engineering, & other techn. activities	..	..	..	..	..	258 178	263 505	..
		743	Advertising	..	..	..	..	..	30 198	30 565	..
		749	Business activities nec	..	..	..	..	..	460 593	486 701	..
M	**80**		Education	..	..	..	..	..	44 163	41 624	..
		801	Primary education	..	..	..	..	..	7 958	8 407	..
		802	Secondary education	..	..	..	..	..	13 419	12 606	..
		803	Higher education	..	..	..	..	..	2 087	2 202	..
		809	Adult and other education	..	..	..	..	..	20 699	18 409	..
N	**85**		Health and social work	..	..	..	..	..	441 563	448 361	..
		851	Human health activities	..	..	..	..	..	363 223	366 462	..
		852	Veterinary activities	..	..	..	..	..	11 855	9 418	..
		853	Social work activities	..	..	..	..	..	66 485	72 481	..
O	**90_93**		Other commun., soc. & pers. services activities	..	..	..	..	..	489 374	489 682	..
	90		Sewage & refuse disposal, sanitation, etc.	..	..	..	..	..	62 685	60 484	..
	91		Activities of member organisations nec	..	..	..	..	..	..	..	..
		912	Activities of trade unions	..	..	..	..	..	..	..	..
	92		Recreational, cultural and sporting activities	..	..	..	..	..	139 651	130 784	..
		921	Motion picture, radio, TV, other entertainment act.	..	..	..	..	..	..	88 998	..
		922	News agency activities	..	..	..	..	..	..	5 089	..
		923	Library, archives, museums, other cultural activ.	..	..	..	..	..	..	1 185	..
		924	Sporting and other recreational activities	..	..	..	..	..	..	35 512	..
	93		Other service activities	..	..	..	..	..	287 038	298 414	..

Table IT.S4 **SALAIRES ET CHARGES SOCIALES - COMPENSATION OF LABOUR**

Milliards de L (Prix courants)

1991	1992	1993	1994	1995	1996	1997	1998	CITI révision 3			
..	..	..	..	..	61 241	63 459	..	Com. de gros, de detail ; réparation etc.	G	**50_52**	
..	..	..	..	..	8 722	9 376	..	Com., entr., répar. d'auto. et moto ; carburants		**50**	
..	..	..	..	..	..	3 504	..	Commerce de véhicules automobiles			501
..	..	..	..	..	..	3 428	..	Entretien et réparation de véhicules automobiles			502
..	..	..	..	..	..	1 324	..	Com. de pièces et access. de véhic. automobiles			503
..	..	..	..	..	..	270	..	Com., entret. et répar. de motocycles et access.			504
..	..	..	..	..	..	851	..	Com. de détail de carburants pour automobiles			505
..	..	..	..	..	29 502	28 561	..	Com. de gros et act. inter., sauf auto. et moto		**51**	
..	..	..	..	..	..	1 440	..	Activités d'intermédiaries de commerce de gros			511
..	..	..	..	..	..	5 793	..	Com. gros: prod. agr. bruts, anim. viv., prod. alim.			512
..	..	..	..	..	..	8 536	..	Com. gros: biens de consom., sauf alimentaires			513
..	..	..	..	..	..	6 487	..	Com. gros: prod. inter. non agr., déchets, débris			514
..	..	..	..	..	..	4 405	..	Com. gros: machines, équipments et fournitures			515
..	..	..	..	..	..	1 900	..	Autres commerces de gros			519
..	..	..	..	..	23 017	25 521	..	Com. détail, sauf auto, moto; répar. art. pers., dom.		**52**	
..	..	..	..	..	..	11 664	..	Commerce de détail non spécialisé en magasin			521
..	..	..	..	..	..	1 058	..	Com. détail: prod. aliment. en mag. spécialisés			522
..	..	..	..	..	..	11 911	..	Autre com. détail: prod. neufs en mag. spécialisés			523
..	..	..	..	..	..	38	..	Commerces de détail de biens d'occasion			524
..	..	..	..	..	..	595	..	Commerces de détail autres qu'en magasin			525
..	..	..	..	..	..	255	..	Réparation d'articles personnels et domestiques			526
..	..	..	..	..	14 344	14 577	..	Hotels et restaurants	H	**55**	
..	..	..	..	..	6 164	6 299	..	Hôtels; camping; autres équip. courts séjours			551
..	..	..	..	..	8 180	8 278	..	Restaurants, cafés et cantines			552
..	..	..	..	..	58 964	59 945	..	Transport, entreposage, communications	I	**60_64**	
..	..	..	..	..	27 639	27 640	..	Transports terrestres; transports par conduites		**60**	
..	..	..	..	..	..	11 527	..	Transports par chemin de fer			601
..	..	..	..	..	..	16 063	..	Autres transports terrestres			602
..	..	..	..	..	..	50	..	Transports par conduites			603
..	..	..	..	..	1 479	1 601	..	Transports par eau		**61**	
..	..	..	..	..	..	1 210	..	Transports maritimes et côtiers			611
..	..	..	..	..	..	391	..	Transports par voies navigables intérieures			612
..	..	..	..	..	2 058	2 274	..	Transports aériens		**62**	
..	..	..	..	..	..	2 130	..	Transports aériens réguliers			621
..	..	..	..	..	..	144	..	Transports aériens non réguliers			622
..	..	..	..	..	10 266	9 858	..	Act. annex. et aux. des transp.; act. des ag. de voy.		**63**	
..	..	..	..	..	17 521	18 572	..	Postes et télécommunications		**64**	
..	..	..	..	..	10 335	10 925	..	Activités de poste et de courrier			641
..	..	..	..	..	7 185	7 647	..	Télécommunications			642
..	..	..	..	..	..	..	..	Intermédiation financière	J	**65_67**	
..	..	..	..	..	..	..	..	Interm. financ., sauf assur. et caisses de retraite		**65**	
..	..	..	..	..	..	..	..	Intermédiation monétaire			651
..	..	..	..	..	..	..	..	Autres intermédiations financières			659
..	..	..	..	..	..	..	..	Act. d'assur., caisses de retr., sauf séc. soc. oblig.		**66**	
..	..	..	..	..	..	..	..	Act. auxiliaires de l'intermédiation financière		**67**	
..	..	..	..	..	..	..	..	Act. aux. de l'interm. fin., sauf assur., et retraite			671
..	..	..	..	..	..	..	..	Act. aux. des assurances et caisses de retraite			672
..	..	..	..	..	37 383	37 522	..	Immobilier, locations, services aux entreprises	K	**70_74**	
..	..	..	..	..	1 826	1 449	..	Activités immobilières		**70**	
..	..	..	..	..	1 287	1 018	..	Act. immobilières sur biens propres ou loués			701
..	..	..	..	..	539	432	..	Activités immobilières pour comptes de tiers			702
..	..	..	..	..	356	219	..	Locat. mach., équip. sans opér., biens pers. et dom.		**71**	
..	..	..	..	..	..	92	..	Location de matériel de transport			711
..	..	..	..	..	..	93	..	Location de machines et equipements nca			712
..	..	..	..	..	..	35	..	Loc. d'articles personnels et domestiques nca			713
..	..	..	..	..	8 652	8 931	..	Activités informatiques et activités rattachées		**72**	
..	..	..	..	..	182	400	..	Conseil en matériel informatique			721
..	..	..	..	..	4 527	4 748	..	Production de logiciels			722
..	..	..	..	..	2 690	2 757	..	Traitement de données informatiques			723
..	..	..	..	..	63	160	..	Activités de banques de données			724
..	..	..	..	..	858	428	..	Entr., répar. mach. de bureau et compt. et inform.			725
..	..	..	..	..	331	438	..	Autres activités rattachées à l'informatique			729
..	..	..	..	..	963	1 029	..	Recherche-développement		**73**	
..	..	..	..	..	937	1 008	..	Rech. et dév. expér. en sc. phys., natur. et ingén.			731
..	..	..	..	..	26	21	..	Rech. et dév. expér. en sciences soc. et humaines			732
..	..	..	..	..	25 586	25 893	..	Autres activités de services aux entreprises		**74**	
..	..	..	..	..	7 621	8 385	..	Act. jur., compt., audit.; cons. fisc.; ét. march., soud.			741
..	..	..	..	..	4 050	4 076	..	Act. d'achitect., d'ingén.; autres act. techniques			742
..	..	..	..	..	1 219	1 294	..	Publicité			743
..	..	..	..	..	12 697	12 139	..	Activités de services aux entreprises nca.			749
..	..	..	..	..	993	897	..	Education	M	**80**	
..	..	..	..	..	167	179	..	Activités d'enseignement primaire			801
..	..	..	..	..	462	421	..	Activités d'enseignement secondaire			802
..	..	..	..	..	49	54	..	Activités d'enseignement supérieur			803
..	..	..	..	..	315	243	..	Act. de formation perman.; autres act. d'enseign.			809
..	..	..	..	..	8 823	9 111	..	Santé et action sociale	N	**85**	
..	..	..	..	..	6 755	7 020	..	Activités de santé humaine			851
..	..	..	..	..	45	26	..	Activités vétérinaires			852
..	..	..	..	..	2 023	2 065	..	Activités d'action sociale			853
..	..	..	..	..	11 160	11 168	..	Autres activités de serv. coll., soc. et personnels	O	**90_93**	
..	..	..	..	..	3 572	3 567	..	Assain. et enlèv. des ordures; voierie, act. simil.		**90**	
..	..	..	..	..	..	..	..	Activités associatives diverses		**91**	
..	..	..	..	..	..	..	..	Activités de syndicats de salariés			912
..	..	..	..	..	4 852	4 582	..	Activités récréatives, culturelles et sportives		**92**	
..	..	..	..	..	..	3 587	..	Act. cinémat., de radio, télé.; autres act. de spect.			921
..	..	..	..	..	..	169	..	Activités d'agences de presse			922
..	..	..	..	..	..	34	..	Act. des biblio., arch., musées, jardins bot. et zool.			923
..	..	..	..	..	..	792	..	Activités sportives et autres activités récréatives			924
..	..	..	..	..	2 736	3 019	..	Autres activités de services		**93**	

Table IT.S5 ENTERPRISES - ENTREPRISES

Units

ISIC revision 3				1991	1992	1993	1994	1995	1996	1997	1998
G	**50_52**		Wholesale & retail trade; repairs; etc.	..	..	..	..	..	1238 242	1253 706	..
	50		Sale, maint.,repair vehic/cycle, person., h/hold gd.	..	..	..	..	..	164 389	158 034	..
		501	Sale of motor cycle	..	..	..	..	..	..	12 674	..
		502	Maintenance and repair of motor vehicles	..	..	..	..	..	..	100 613	..
		503	Sale of motor vehicle parts and accessories	..	..	..	..	..	..	10 848	..
		504	Sale, maint., repair motor cycle, parts & acces.	..	..	..	..	..	..	9 219	..
		505	Retail sale of automative fuel	..	..	..	..	..	..	24 680	..
	51		W/sale. trade & commission exc motor veh. & cycle	..	..	..	..	..	365 134	383 152	..
		511	Wholesale on a fee or contract basis	..	..	..	..	..	..	230 754	..
		512	W/sale agric. raw mat.,live animal,food, bev., tobac.	..	..	..	..	..	..	46 653	..
		513	Wholesale of household goods	..	..	..	..	..	..	47 691	..
		514	W/sale. of non-agric. intermed. prod., waste, scrap	..	..	..	..	..	..	30 659	..
		515	Wholesale of machinery, equipment & supplies	..	..	..	..	..	..	15 345	..
		519	Other wholesale	..	..	..	..	..	..	12 050	..
	52		Retail trade exc. motor veh. & cyc.; repair h/hold gds.	..	..	..	..	..	708 719	712 520	..
		521	Non-specialised retail trade in stores	..	..	..	..	..	..	82 322	..
		522	Retail sale food, bev., tobac. in specialised stores	..	..	..	..	..	..	120 733	..
		523	Other retail trade of new goods specialised stores	..	..	..	..	..	..	375 492	..
		524	Retail sale of second-hand goods in stores	..	..	..	..	..	..	3 053	..
		525	Retail trade not in stores	..	..	..	..	..	..	103 872	..
		526	Repair of personal and household goods	..	..	..	..	..	..	27 048	..
H	**55**		Hotels and restaurants	..	..	..	..	..	211 729	219 526	..
		551	Hotels; camping ; other short-stay accommodation	..	..	..	..	..	36 694	38 892	..
		552	Restaurants, bars and canteens	..	..	..	..	..	175 035	180 634	..
I	**60_64**		Transport, storage and communications	..	..	..	..	..	..	155 950	..
	60		Land transport, transport via pipelines	..	..	..	..	..	139 097	132 413	..
		601	Transport via railways	..	..	..	..	..	..	137	..
		602	Other land transport	..	..	..	..	..	..	132 271	..
		603	Transport via pipelines	..	..	..	..	..	..	5	..
	61		Water transport	..	..	..	..	..	902	1 102	..
		611	Sea and coastal water transport	..	..	..	..	..	..	444	..
		612	Inland water transport	..	..	..	..	..	..	658	..
	62		Air transport	..	..	..	..	..	84	153	..
		621	Scheduled air transport	..	..	..	..	..	..	59	..
		622	Non-scheduled air transport	..	..	..	..	..	..	94	..
	63		Support. & aux. transport act., travel agencies activ.	..	..	..	..	..	19 928	20 553	..
	64		Post and telecommunications	..	..	..	..	..	1 557	1 729	..
		641	Post and courier activities	..	..	..	..	..	1 316	1 411	..
		642	Telecommunications	..	..	..	..	..	241	318	..
J	**65_67**		Financial intermediation	..	..	..	..	..	..	..	..
	65		Financial intermediation, exc. insur., pension fund	..	..	..	..	..	..	..	..
		651	Monetary intermediation	..	..	..	..	..	..	..	..
		659	Other financial intermediation	..	..	..	..	..	..	..	..
	66		Insurance pension fund, exc. compul. social sec.	..	..	..	..	..	..	..	..
	67		Activities auxiliary to financial intermediation	..	..	..	..	..	..	..	..
		671	Act. auxil to fin. intermed., exc. insur. & pens.fund	..	..	..	..	..	..	..	..
		672	Act. auxiliary to insurance and pension funding	..	..	..	..	..	..	..	..
K	**70_74**		Real estate, renting and business activities	..	..	..	..	..	666 474	706 975	..
	70		Real estate activities	..	..	..	..	..	104 895	120 506	..
		701	Real estate activ. with own or leased property	..	..	..	..	..	79 393	93 502	..
		702	Real estate activities on a fee or contract basis	..	..	..	..	..	25 502	27 004	..
	71		Renting of mach. without operator; rent. of pers. gds.	..	..	..	..	..	7 967	8 924	..
		711	Renting of transport equipment	..	..	..	..	..	..	2 649	..
		712	Renting of other machinery and equipment	..	..	..	..	..	..	2 735	..
		713	Renting of personal and household goods nec	..	..	..	..	..	..	3 540	..
	72		Computer and related activities	..	..	..	..	..	47 268	50 507	..
		721	Hardware consultancy	..	..	..	..	..	1 054	1 067	..
		722	Software consultancy and supply	..	..	..	..	..	16 604	18 218	..
		723	Data processing	..	..	..	..	..	20 726	21 579	..
		724	Data base activities	..	..	..	..	..	306	338	..
		725	Mainten.; repair of office, account. & comp. mach.	..	..	..	..	..	4 106	3 628	..
		729	Other computer related activities	..	..	..	..	..	4 472	5 677	..
	73		Research and development	..	..	..	..	..	6 214	6 312	..
		731	R & D experim. on natural sciences & engineering	..	..	..	..	..	4 570	4 696	..
		732	R & D experim. on social sciences & humanities	..	..	..	..	..	1 644	1 616	..
	74		Other business activities	..	..	..	..	..	500 130	520 726	..
		741	Legal, acc., book-keep., audit act.; tax consult., etc	..	..	..	..	..	193 881	198 926	..
		742	Architectural, engineering, & other techn. activities	..	..	..	..	..	183 542	188 155	..
		743	Advertising	..	..	..	..	..	10 858	11 381	..
		749	Business activities nec	..	..	..	..	..	111 849	122 264	..
M	**80**		Education	..	..	..	..	..	11 892	12 249	..
		801	Primary education	..	..	..	..	..	1 535	1 885	..
		802	Secondary education	..	..	..	..	..	1 333	1 395	..
		803	Higher education	..	..	..	..	..	810	888	..
		809	Adult and other education	..	..	..	..	..	8 214	8 081	..
N	**85**		Health and social work	..	..	..	..	..	205 477	203 755	..
		851	Human health activities	..	..	..	..	..	193 091	192 145	..
		852	Veterinary activities	..	..	..	..	..	8 899	8 107	..
		853	Social work activities	..	..	..	..	..	3 487	3 503	..
O	**90_93**		Other commun., soc. & pers. services activities	..	..	..	..	..	217 984	220 900	..
	90		Sewage & refuse disposal, sanitation, etc.	..	..	..	..	..	3 702	3 734	..
	91		Activities of member organisations nec	..	..	..	..	..	..	..	..
		912	Activities of trade unions	..	..	..	..	..	..	..	..
	92		Recreational, cultural and sporting activities	..	..	..	..	..	51 915	53 025	..
		921	Motion picture, radio, TV, other entertainment act.	..	..	..	..	..	..	32 326	..
		922	News agency activities	..	..	..	..	..	..	3 178	..
		923	Library, archives, museums, other cultural activ.	..	..	..	..	..	..	386	..
		924	Sporting and other recreational activities	..	..	..	..	..	..	17 135	..
	93		Other service activities	..	..	..	..	..	162 367	164 141	..

Table KR.S1 PRODUCTION - PRODUCTION

Millions de Won (Prix courants)

1991	1992	1993	1994	1995	1996	1997	1998		CITI révision 3		
..	..	..	..	..	..	..	..	Com. de gros, de detail ; réparation etc.	G	**50_52**	
..	..	..	..	38 924 892	38 890 497	38 818 267	..	Com., entr., répar. d'auto. et moto ; carburants		**50**	
..	..	..	..	19 442 595	17 222 511	16 124 205	..	Commerce de véhicules automobiles			501
..	..	..	..	3 958 316	4 308 524	4 776 909	..	Entretien et réparation de véhicules automobiles			502
..	..	..	..	4 140 505	3 643 857	3 711 500	..	Com. de pièces et access. de véhic. automobiles			503
..	..	..	..	574 268	526 984	533 569	..	Com., entret. et répar. de motocycles et access.			504
..	..	..	..	10 809 208	13 188 621	13 672 084	..	Com. de détail de carburants pour automobiles			505
..	..	..	..	103 897 777	120 886 952	126 005 491	..	Com. de gros et act. inter., sauf auto. et moto		**51**	
..	..	..	..	..	..	..	..	Activités d'intermédiaries de commerce de gros			511
..	..	..	..	30 904 576	29 933 111	32 269 211	31 742 711	Com. gros: prod. agr. bruts, anim. viv., prod. alim.			512
..	..	..	..	24 062 683	27 591 124	27 916 274	27 228 785	Com. gros: biens de consom., sauf alimentaires			513
..	..	..	..	32 385 119	43 242 606	44 221 314	42 320 081	Com. gros: prod. inter. non agr., déchets, débris			514
..	..	..	..	13 986 027	17 997 461	19 471 353	17 962 228	Com. gros: machines, équipments et fournitures			515
..	..	..	..	2 559 372	2 122 650	2 127 339	..	Autres commerces de gros			519
..	..	..	..	84 596 214	89 427 603	94 694 366	92 945 584	Com. détail, sauf auto, moto; répar. art. pers., dom.		**52**	
..	..	..	..	18 049 650	19 878 012	24 884 379	25 338 347	Commerce de détail non spécialisé en magasin			521
..	..	..	..	17 725 844	15 229 572	14 975 085	15 105 980	Com. détail: prod. aliment. en mag. spécialisés			522
..	..	..	..	47 548 990	49 383 995	49 372 672	44 797 337	Autre com. détail: prod. neufs en mag. spécialisés			523
..	..	..	..	66 476	114 549	125 728	..	Commerces de détail de biens d'occasion			524
..	..	..	..	..	2 904 714	3 203 511	3 003 535	Commerces de détail autres qu'en magasin			525
..	..	..	..	1 205 254	1 916 761	2 132 991	1 696 850	Réparation d'articles personnels et domestiques			526
..	..	..	27 908 893	..	33 210 552	35 313 633	31 902 410	Hotels et restaurants	H	**55**	
..	..	..	3 801 384	..	5 061 143	5 083 689	4 645 975	Hôtels; camping; autres équip. courts séjours			551
..	..	..	24 107 509	..	28 149 409	30 229 944	27 256 435	Restaurants, cafés et cantines			552
..	..	..	..	..	..	..	..	Transport, entreposage, communications	I	**60_64**	
..	..	..	15 045 877	17 156 975	18 583 773	19 935 077	19 837 423	Transports terrestres; transports par conduites		**60**	
..	..	..	1 677 737	1 800 036	1 991 664	2 165 583	2 110 351	Transports par chemin de fer			601
..	..	..	13 356 708	15 340 518	16 565 709	17 729 030	17 686 501	Autres transports terrestres			602
..	..	..	11 432	16 421	26 400	40 464	40 571	Transports par conduites			603
..	..	..	6 950 095	8 780 642	9 920 881	13 058 106	17 011 982	Transports par eau		**61**	
..	..	..	6 877 965	8 704 961	9 834 622	12 967 890	16 912 610	Transports maritimes et côtiers			611
..	..	..	72 130	75 681	86 259	90 216	99 372	Transports par voies navigables intérieures			612
..	..	..	3 803 643	4 385 599	4 657 335	5 495 889	5 774 548	Transports aériens		**62**	
..	..	..	3 758 507	4 332 154	4 586 198	5 413 255	5 701 640	Transports aériens réguliers			621
..	..	..	45 136	53 445	71 137	82 634	72 908	Transports aériens non réguliers			622
..	..	..	3 380 736	3 956 125	4 559 942	5 036 308	4 698 887	Act. annex. et aux. des transp.; act. des ag. de voy.		**63**	
..	..	..	..	..	..	14 812 076	17 702 815	Postes et télécommunications		**64**	
..	..	..	..	..	..	978 930	992 839	Activités de poste et de courrier			641
..	..	..	..	..	..	13 833 146	16 709 976	Télécommunications			642
..	..	..	..	..	..	..	..	Intermédiation financière	J	**65_67**	
..	..	..	..	..	..	..	..	Interm. financ., sauf assur. et caisses de retraite		**65**	
..	..	..	..	..	..	..	..	Intermédiation monétaire			651
..	..	..	..	..	..	..	..	Autres intermédiations financières			659
..	..	..	..	..	..	..	..	Act. d'assur., caisses de retr., sauf séc. soc. oblig.		**66**	
..	..	..	..	..	..	..	..	Act. auxiliaires de l'intermédiation financière		**67**	
..	..	..	..	..	..	..	..	Act. aux. de l'interm. fin., sauf assur., et retraite			671
..	..	..	..	..	..	..	..	Act. aux. des assurances et caisses de retraite			672
..	..	..	..	..	..	..	..	Immobilier, locations, services aux entreprises	K	**70_74**	
..	..	..	8 027 592	..	11 742 827	15 214 806	14 101 161	Activités immobilières		**70**	
..	..	..	5 689 881	..	5 920 914	9 133 740	8 685 349	Act. immobilières sur biens propres ou loués			701
..	..	..	2 337 711	..	5 821 913	6 081 066	5 415 812	Activités immobilières pour comptes de tiers			702
..	..	..	6 759 437	..	9 720 855	11 550 277	10 197 055	Locat. mach., équip. sans opér., biens pers. et dom.		**71**	
..	..	..	108 975	..	158 249	165 132	146 302	Location de matériel de transport			711
..	..	..	5 685 410	..	8 645 383	10 456 105	9 143 936	Location de machines ét equipements nca			712
..	..	..	965 052	..	917 223	929 040	906 817	Loc. d'articles personnels et domestiques nca			713
..	..	..	3 031 570	..	3 511 315	3 912 575	4 206 983	Activités informatiques et activités rattachées		**72**	
..	..	..	1 355 815	..	1 563 773	1 631 309	1 749 315	Conseil en matériel informatique			721
..	..	..	1 430 514	..	1 623 889	1 857 059	2 035 257	Production de logiciels			722
..	..	..	111 450	..	105 256	135 583	130 260	Traitement de données informatiques			723
..	..	..	44 642	..	88 118	111 577	103 316	Activités de banques de données			724
..	..	..	62 264	..	117 444	160 124	171 348	Entr., répar. mach. de bureau et compt. et inform.			725
..	..	..	26 885	..	12 835	16 923	17 487	Autres activités rattachées à l'informatique			729
..	..	..	..	..	3 376 958	..	..	Recherche-développement		**73**	
..	..	..	..	..	3 081 420	..	..	Rech. et dév. expér. en sc. phys., natur. et ingén.			731
..	..	..	..	..	295 538	..	..	Rech. et dév. expér. en sciences soc. et humaines			732
..	..	..	8 798 327	..	16 016 777	17 375 805	15 759 050	Autres activités de services aux entreprises		**74**	
..	..	..	1 631 364	..	2 996 554	3 077 385	3 219 560	Act. jur., compt., audit.; cons. fisc.; ét. march., soud.			741
..	..	..	3 163 439	..	6 878 169	7 836 350	6 965 334	Act. d'achitect., d'ingén.; autres act. techniques			742
..	..	..	1 611 403	..	2 042 219	2 112 475	1 743 599	Publicité			743
..	..	..	2 392 121	..	4 099 835	4 349 595	3 830 557	Activités de services aux entreprises nca.			749
..	..	..	..	..	25 154 585	..	..	Education	M	**80**	
..	..	..	..	..	5 581 441	..	..	Activités d'enseignement primaire			801
..	..	..	..	..	6 763 361	..	..	Activités d'enseignement secondaire			802
..	..	..	..	..	7 823 841	..	..	Activités d'enseignement supérieur			803
..	..	..	..	..	4 985 942	..	..	Act. de formation perman.; autres act. d'enseign.			809
..	..	..	..	..	..	..	..	Santé et action sociale	N	**85**	
..	..	..	..	..	..	..	..	Activités de santé humaine			851
..	..	..	..	..	..	..	..	Activités vétérinaires			852
..	..	..	..	..	..	..	..	Activités d'action sociale			853
..	..	..	..	..	..	..	..	Autres activités de serv. coll., soc. et personnels	O	**90_93**	
..	..	..	..	1 114 889	1 003 183	1 153 826	1 188 280	Assain. et enlèv. des ordures; voierie, act. simil.		**90**	
..	..	..	..	..	2 111 904	..	..	Activités associatives diverses		**91**	
..	..	..	..	..	138 975	..	..	Activités de syndicats de salariés			912
..	..	..	..	11 745 329	13 602 486	15 654 531	15 015 715	Activités récréatives, culturelles et sportives		**92**	
..	..	..	..	4 730 235	5 564 867	5 943 578	6 011 540	Act. cinémat., de radio, télé.; autres act. de spect.			921
..	..	..	..	55 003	84 547	85 229	75 214	Activités d'agences de presse			922
..	..	..	..	341 504	415 284	470 384	448 825	Act. des biblio., arch., musées, jardins bot. et zool.			923
..	..	..	..	6 618 587	7 537 788	9 155 340	8 480 136	Activités sportives et autres activités récréatives			924
..	..	..	..	4 530 538	4 887 540	5 097 601	5 013 202	Autres activités de services		**93**	

KOREA

Table KR.S2 EMPLOYMENT - EMPLOI

Units

ISIC revision 3				1991	1992	1993	1994	1995	1996	1997	1998
G	**50_52**		Wholesale & retail trade; repairs; etc.	..	..	..	..	..	..	..	..
	50		Sale, maint.,repair vehic/cycle, person., h/hold gd.	..	..	..	..	290 298	280 403	266 333	48 818
		501	Sale of motor cycle	..	..	..	..	55 407	49 082	47 476	10 457
		502	Maintenance and repair of motor vehicles	..	..	..	..	115 723	122 071	114 553	18 355
		503	Sale of motor vehicle parts and accessories	..	..	..	..	38 136	36 336	34 900	9 688
		504	Sale, maint., repair motor cycle, parts & acces.	..	..	..	..	11 417	9 303	9 503	1 637
		505	Retail sale of automative fuel	..	..	..	..	69 615	63 611	59 901	8 681
	51		W/sale. trade & commission exc motor veh. & cycle	..	..	..	..	675 331	623 555	667 877	602 762
		511	Wholesale on a fee or contract basis	..	..	..	..	..	..	..	..
		512	W/sale agric. raw mat.,live animal,food, bev., tobac.	..	..	..	..	196 808	166 091	180 851	167 917
		513	Wholesale of household goods	..	..	..	..	214 722	198 462	215 439	188 273
		514	W/sale. of non-agric. intermed. prod., waste, scrap	..	..	..	..	160 286	149 141	150 678	133 067
		515	Wholesale of machinery, equipment & supplies	..	..	..	..	91 996	101 062	112 877	108 303
		519	Other wholesale	..	..	..	..	11 519	8 799	8 032	5 202
	52		Retail trade exc. motor veh. & cyc.; repair h/hold gds.	..	..	..	..	1 540 520	1 593 682	1 655 283	1 460 359
		521	Non-specialised retail trade in stores	..	..	..	..	155 203	164 181	202 268	123 182
		522	Retail sale food, bev., tobac. in specialised stores	..	..	..	..	463 642	430 691	436 740	435 206
		523	Other retail trade of new goods specialised stores	..	..	..	..	859 201	826 143	820 950	719 462
		524	Retail sale of second-hand goods in stores	..	..	..	..	2 860	3 332	4 046	5 555
		525	Retail trade not in stores	..	..	..	..	..	108 653	127 294	111 411
		526	Repair of personal and household goods	..	..	..	..	59 614	60 682	63 985	65 543
H	**55**		Hotels and restaurants	..	..	..	1 233 002	..	1 386 425	1 403 877	1 333 579
		551	Hotels; camping ; other short-stay accommodation	..	..	..	135 215	..	131 581	129 111	115 659
		552	Restaurants, bars and canteens	..	..	..	1 097 787	..	1 254 844	1 274 766	1 217 920
I	**60_64**		Transport, storage and communications	..	..	..	..	..	..	..	..
	60		Land transport, transport via pipelines	..	..	..	..	..	..	..	..
		601	Transport via railways	..	..	..	..	..	..	..	..
		602	Other land transport	..	..	..	..	..	..	..	..
		603	Transport via pipelines	..	..	..	..	..	..	..	..
	61		Water transport	..	..	..	..	..	..	..	..
		611	Sea and coastal water transport	..	..	..	..	..	..	..	..
		612	Inland water transport	..	..	..	..	..	..	..	..
	62		Air transport	..	..	..	..	..	..	..	..
		621	Scheduled air transport	..	..	..	..	..	..	..	..
		622	Non-scheduled air transport	..	..	..	..	..	..	..	..
	63		Support. & aux. transport act., travel agencies activ.	..	..	..	..	..	..	..	..
	64		Post and telecommunications	..	..	..	..	..	..	..	..
		641	Post and courier activities	..	..	..	..	..	..	..	..
		642	Telecommunications	..	..	..	..	..	..	..	..
J	**65_67**		Financial intermediation	..	..	..	..	..	..	..	..
	65		Financial intermediation, exc. insur., pension fund	..	..	..	..	..	..	..	..
		651	Monetary intermediation	..	..	..	..	..	..	..	..
		659	Other financial intermediation	..	..	..	..	..	..	..	..
	66		Insurance pension fund, exc. compul. social sec.	..	..	..	..	..	..	..	..
	67		Activities auxiliary to financial intermediation	..	..	..	..	..	..	..	..
		671	Act. auxil to fin. intermed., exc. insur. & pens.fund	..	..	..	..	..	..	..	..
		672	Act. auxiliary to insurance and pension funding	..	..	..	..	..	..	..	..
K	**70_74**		Real estate, renting and business activities	..	..	..	598 550	..	778 031	798 877	798 986
	70		Real estate activities	..	..	..	155 311	..	239 574	255 242	246 168
		701	Real estate activ. with own or leased property	..	..	..	26 417	..	27 936	35 667	33 800
		702	Real estate activities on a fee or contract basis	..	..	..	128 894	..	211 638	219 575	212 368
	71		Renting of mach. without operator; rent. of pers. gds.	..	..	..	68 094	..	66 047	65 946	65 424
		711	Renting of transport equipment	..	..	..	2 712	..	3 597	3 471	3 839
		712	Renting of other machinery and equipment	..	..	..	11 646	..	12 391	11 871	10 611
		713	Renting of personal and household goods nec	..	..	..	53 736	..	50 059	50 604	50 974
	72		Computer and related activities	..	..	..	35 690	..	45 233	51 103	58 235
		721	Hardware consultancy	..	..	..	11 999	..	13 726	11 927	12 796
		722	Software consultancy and supply	..	..	..	18 196	..	23 637	29 868	35 354
		723	Data processing	..	..	..	2 508	..	2 772	2 418	2 406
		724	Data base activities	..	..	..	891	..	1 789	2 312	2 460
		725	Mainten.; repair of office, account. & comp. mach.	..	..	..	1 561	..	2 988	4 127	4 742
		729	Other computer related activities	..	..	..	535	..	321	451	477
	73		Research and development	..	..	..	..	..	45 789	..	..
		731	R & D experim. on natural sciences & engineering	..	..	..	..	..	39 685	..	..
		732	R & D experim. on social sciences & humanities	..	..	..	..	..	6 104	..	..
	74		Other business activities	..	..	..	339 455	..	381 388	426 586	429 159
		741	Legal, acc., book-keep., audit act.; tax consult., etc	..	..	..	61 260	..	75 725	78 072	88 122
		742	Architectural, engineering, & other techn. activities	..	..	..	93 390	..	120 286	148 386	124 447
		743	Advertising	..	..	..	21 972	..	25 552	25 889	25 772
		749	Business activities nec	..	..	..	162 833	..	159 825	174 239	190 818
M	**80**		Education	..	..	..	..	..	818 038	..	..
		801	Primary education	..	..	..	..	..	225 816	..	..
		802	Secondary education	..	..	..	..	..	228 140	..	..
		803	Higher education	..	..	..	..	..	111 854	..	..
		809	Adult and other education	..	..	..	..	..	252 228	..	..
N	**85**		Health and social work	..	..	..	..	..	..	..	..
		851	Human health activities	..	..	..	..	..	..	..	..
		852	Veterinary activities	..	..	..	..	..	..	..	..
		853	Social work activities	..	..	..	..	..	..	..	..
O	**90_93**		Other commun., soc. & pers. services activities	..	..	..	..	..	..	..	..
	90		Sewage & refuse disposal, sanitation, etc.	..	..	..	..	25 573	22 605	23 544	21 813
	91		Activities of member organisations nec	..	..	..	..	..	83 616	..	..
		912	Activities of trade unions	..	..	..	..	..	18 400	..	..
	92		Recreational, cultural and sporting activities	..	..	..	..	239 736	262 929	262 500	249 985
		921	Motion picture, radio, TV, other entertainment act.	..	..	..	..	104 779	122 328	120 901	113 973
		922	News agency activities	..	..	..	..	733	958	1 000	816
		923	Library, archives, museums, other cultural activ.	..	..	..	..	18 215	17 422	19 146	18 453
		924	Sporting and other recreational activities	..	..	..	..	116 009	122 221	121 453	116 743
	93		Other service activities	..	..	..	..	252 790	258 673	263 725	259 491

Table KR.S3 ÉTABLISSEMENTS - ESTABLISHMENTS

Unités

1991	1992	1993	1994	1995	1996	1997	1998		CITI révision 3		
..	..	..	..	970 960	981 047	980 662	878 513	Com. de gros, de detail ; réparation etc.	G	50_52	
..	..	..	..	65 461	67 547	69 224	11 436	Com., entr., répar. d'auto. et moto ; carburants		50	
..	..	..	..	4 205	4 400	4 422	787	Commerce de véhicules automobiles			501
..	..	..	..	33 714	35 418	36 636	6 187	Entretien et réparation de véhicules automobiles			502
..	..	..	..	12 728	12 322	12 349	2 660	Com. de pièces et access. de véhic. automobiles			503
..	..	..	..	6 092	5 942	5 933	946	Com., entret. et répar. de motocycles et access.			504
..	..	..	..	8 722	9 465	9 884	856	Com. de détail de carburants pour automobiles			505
..	..	..	..	144 243	144 622	147 746	143 991	Com. de gros et act. inter., sauf auto. et moto		51	
..	..	..	..	..	..	..	..	Activités d'intermédiaries de commerce de gros			511
..	..	..	..	37 049	38 496	40 426	40 892	Com. gros: prod. agr. bruts, anim. viv., prod. alim.			512
..	..	..	..	47 708	47 365	46 357	41 918	Com. gros: biens de consom., sauf alimentaires			513
..	..	..	..	36 810	36 439	36 849	37 477	Com. gros: prod. inter. non agr., déchets, débris			514
..	..	..	..	21 346	21 462	23 319	23 207	Com. gros: machines, équipments et fournitures			515
..	..	..	..	1 330	860	795	497	Autres commerces de gros			519
..	..	..	..	761 256	768 878	763 692	723 086	Com. détail, sauf auto, moto; répar. art. pers., dom.		52	
..	..	..	..	40 756	16 659	20 119	16 778	Commerce de détail non spécialisé en magasin			521
..	..	..	..	257 177	281 041	273 326	268 972	Com. détail: prod. aliment. en mag. spécialisés			522
..	..	..	..	429 707	425 819	422 360	386 157	Autre com. détail: prod. neufs en mag. spécialisés			523
..	..	..	..	1 698	1 903	2 210	2 945	Commerces de détail de biens d'occasion			524
..	..	..	..	..	11 593	12 497	11 889	Commerces de détail autres qu'en magasin			525
..	..	..	..	31 918	31 863	33 180	36 345	Réparation d'articles personnels et domestiques			526
..	..	..	461 073	..	554 725	583 869	577 312	Hotels et restaurants	H	55	
..	..	..	31 081	..	33 551	34 883	35 247	Hôtels; camping; autres équip. courts séjours			551
..	..	..	429 992	..	521 174	548 986	542 065	Restaurants, cafés et cantines			552
..	..	..	..	..	..	..	..	Transport, entreposage, communications	I	60_64	
..	..	..	167 388	173 226	180 197	184 930	189 032	Transports terrestres; transports par conduites		60	
..	..	..	3	4	4	4	5	Transports par chemin de fer			601
..	..	..	167 383	173 220	180 191	184 924	189 026	Autres transports terrestres			602
..	..	..	2	2	2	2	1	Transports par conduites			603
..	..	..	417	394	398	390	374	Transports par eau		61	
..	..	..	323	294	298	300	280	Transports maritimes et côtiers			611
..	..	..	94	100	100	90	94	Transports par voies navigables intérieures			612
..	..	..	8	7	9	9	9	Transports aériens		62	
..	..	..	2	2	2	2	2	Transports aériens réguliers			621
..	..	..	6	5	7	7	7	Transports aériens non réguliers			622
..	..	..	3 807	3 913	4 162	4 581	4 534	Act. annex. et aux. des transp.; act. des ag. de voy.		63	
..	..	..	..	..	..	247	347	Postes et télécommunications		64	
..	..	..	..	..	..	1	1	Activités de poste et de courrier			641
..	..	..	..	..	..	246	346	Télécommunications			642
..	..	..	..	..	..	..	..	Intermédiation financière	J	65_67	
..	..	..	..	..	..	..	..	Interm. financ., sauf assur. et caisses de retraite		65	
..	..	..	..	..	..	..	..	Intermédiation monétaire			651
..	..	..	..	..	..	..	..	Autres intermédiations financières			659
..	..	..	..	..	..	..	..	Act. d'assur., caisses de retr., sauf séc. soc. oblig.		66	
..	..	..	..	..	..	..	..	Act. auxiliaires de l'intermédiation financière		67	
..	..	..	..	..	..	..	..	Act. aux. de l'interm. fin., sauf assur., et retraite			671
..	..	..	..	..	..	..	..	Act. aux. des assurances et caisses de retraite			672
..	..	..	..	..	142 771	147 383	148 010	Immobilier, locations, services aux entreprises	K	70_74	
..	..	..	46 550	..	50 148	52 734	51 872	Activités immobilières		70	
..	..	..	2 396	..	3 323	3 515	3 610	Act. immobilières sur biens propres ou loués			701
..	..	..	44 154	..	46 825	49 219	48 262	Activités immobilières pour comptes de tiers			702
..	..	..	41 775	..	41 110	39 680	39 298	Locat. mach., équip. sans opér., biens pers. et dom.		71	
..	..	..	595	..	786	833	902	Location de matériel de transport			711
..	..	..	1 708	..	2 300	2 474	2 534	Location de machines ét equipements nca			712
..	..	..	39 472	..	38 024	36 373	35 862	Loc. d'articles personnels et domestiques nca			713
..	..	..	1 957	..	2 958	3 835	4 340	Activités informatiques et activités rattachées		72	
..	..	..	100	..	126	133	136	Conseil en matériel informatique			721
..	..	..	1 220	..	1 871	2 413	2 749	Production de logiciels			722
..	..	..	136	..	165	189	166	Traitement de données informatiques			723
..	..	..	91	..	205	294	300	Activités de banques de données			724
..	..	..	288	..	553	735	917	Entr., répar. mach. de bureau et compt. et inform.			725
..	..	..	122	..	38	71	72	Autres activités rattachées à l'informatique			729
..	..	..	..	..	938	..	..	Recherche-développement		73	
..	..	..	..	..	568	..	..	Rech. et dév. expér. en sc. phys., natur. et ingén.			731
..	..	..	..	..	370	..	..	Rech. et dév. expér. en sciences soc. et humaines			732
..	..	..	39 292	..	47 617	51 134	52 500	Autres activités de services aux entreprises		74	
..	..	..	9 940	..	12 002	12 624	13 252	Act. jur., compt., audit.; cons. fisc.; ét. march., soud.			741
..	..	..	7 166	..	9 259	10 500	10 229	Act. d'achitect., d'ingén.; autres act. techniques			742
..	..	..	2 405	..	3 448	4 033	4 722	Publicité			743
..	..	..	19 781	..	22 908	23 977	24 297	Activités de services aux entreprises nca.			749
..	..	..	..	..	102 284	..	..	Education	M	80	
..	..	..	..	..	17 284	..	..	Activités d'enseignement primaire			801
..	..	..	..	..	4 620	..	..	Activités d'enseignement secondaire			802
..	..	..	..	..	738	..	..	Activités d'enseignement supérieur			803
..	..	..	..	..	79 642	..	..	Act. de formation perman.; autres act. d'enseign.			809
..	..	..	..	..	..	..	..	Santé et action sociale	N	85	
..	..	..	..	..	..	..	..	Activités de santé humaine			851
..	..	..	..	..	..	..	..	Activités vétérinaires			852
..	..	..	..	..	..	..	..	Activités d'action sociale			853
..	..	..	..	..	..	..	..	Autres activités de serv. coll., soc. et personnels	O	90_93	
..	..	..	..	1 523	1 449	1 610	1 740	Assain. et enlèv. des ordures; voierie, act. simil.		90	
..	..	..	..	..	12 295	..	..	Activités associatives diverses		91	
..	..	..	..	..	1 553	..	..	Activités de syndicats de salariés			912
..	..	..	..	85 802	89 724	94 473	97 637	Activités récréatives, culturelles et sportives		92	
..	..	..	..	38 473	40 594	42 578	44 580	Act. cinémat., de radio, télé.; autres act. de spect.			921
..	..	..	..	15	47	50	40	Activités d'agences de presse			922
..	..	..	..	5 796	5 412	5 749	5 740	Act. des biblio., arch., musées, jardins bot. et zool.			923
..	..	..	..	41 518	43 671	46 096	47 277	Activités sportives et autres activités récréatives			924
..	..	..	..	148 975	152 547	157 207	159 811	Autres activités de services		93	

LUXEMBOURG

Table LX.S1 **PRODUCTION - PRODUCTION**

Millions of LF (Current Prices)

ISIC revision 3				1991	1992	1993	1994	1995	1996	1997	1998
G	50_52		Wholesale & retail trade; repairs; etc.	76 147	76 551	79 417	84 132	87 838	92 509	95 277	..
	50		Sale, maint.,repair vehic/cycle, person., h/hold gd.	12 489	11 832	11 632	12 700	12 894	13 752	14 372	..
		501	Sale of motor cycle	8 371	7 430	6 938	7 304	7 545	8 283	8 360	..
		502	Maintenance and repair of motor vehicles	975	913	818	829	842	874	829	..
		503	Sale of motor vehicle parts and accessories	988	987	1 034	1 102	1 192	1 212	1 277	..
		504	Sale, maint., repair motor cycle, parts & acces.	103	107	126	121	141	128	131	..
		505	Retail sale of automative fuel	2 051	2 395	2 715	3 345	3 174	3 254	3 775	..
	51		W/sale. trade & commission exc motor veh. & cycle	37 272	37 856	39 659	42 445	45 456	47 760	48 289	..
		511	Wholesale on a fee or contract basis	..	..	..	..	..	..	..	..
		512	W/sale agric. raw mat.,live animal,food, bev., tobac.	..	..	..	..	..	..	..	..
		513	Wholesale of household goods	..	..	..	..	..	..	..	..
		514	W/sale. of non-agric. intermed. prod., waste, scrap	..	..	..	..	..	..	..	..
		515	Wholesale of machinery, equipment & supplies	..	..	..	..	..	..	..	..
		519	Other wholesale	..	..	..	..	..	..	..	..
	52		Retail trade exc. motor veh. & cyc.; repair h/hold gds.	26 387	26 862	28 126	28 986	29 488	30 998	32 616	..
		521	Non-specialised retail trade in stores	5 817	6 244	6 637	7 009	7 224	7 685	8 027	..
		522	Retail sale food, bev., tobac. in specialised stores	2 961	3 077	3 177	3 284	3 217	3 166	3 369	..
		523	Other retail trade of new goods specialised stores	16 797	16 651	17 378	17 654	18 012	19 083	20 131	..
		524	Retail sale of second-hand goods in stores	587	648	630	709	703	722	759	..
		525	Retail trade not in stores	..	..	..	..	..	..	..	..
		526	Repair of personal and household goods	225	242	304	330	333	341	330	..
H	55		Hotels and restaurants	..	..	..	..	..	..	..	..
		551	Hotels; camping ; other short-stay accommodation	..	..	..	..	..	..	..	..
		552	Restaurants, bars and canteens	..	..	..	..	..	..	..	..
I	60_64		Transport, storage and communications	..	..	..	..	..	..	..	..
	60		Land transport, transport via pipelines	..	..	..	..	..	..	..	..
		601	Transport via railways	..	..	..	..	..	..	..	..
		602	Other land transport	..	..	..	..	..	..	..	..
		603	Transport via pipelines	..	..	..	..	..	..	..	..
	61		Water transport	..	..	..	..	..	..	..	..
		611	Sea and coastal water transport	..	..	..	..	..	..	..	..
		612	Inland water transport	..	..	..	..	..	..	..	..
	62		Air transport	..	..	..	..	..	..	..	..
		621	Scheduled air transport	..	..	..	..	..	..	..	..
		622	Non-scheduled air transport	..	..	..	..	..	..	..	..
	63		Support. & aux. transport act., travel agencies activ.	..	..	..	..	..	..	..	..
	64		Post and telecommunications	..	..	..	..	..	..	..	..
		641	Post and courier activities	..	..	..	..	..	..	..	..
		642	Telecommunications	..	..	..	..	..	..	..	..
J	65_67		Financial intermediation	..	..	..	..	..	..	..	..
	65		Financial intermediation, exc. insur., pension fund	..	..	..	..	..	..	..	..
		651	Monetary intermediation	..	..	..	..	..	..	..	..
		659	Other financial intermediation	..	..	..	..	..	..	..	..
	66		Insurance pension fund, exc. compul. social sec.	..	..	..	..	..	..	..	..
	67		Activities auxiliary to financial intermediation	..	..	..	..	..	..	..	..
		671	Act. auxil to fin. intermed., exc. insur. & pens.fund	..	..	..	..	..	..	..	..
		672	Act. auxiliary to insurance and pension funding	..	..	..	..	..	..	..	..
K	70_74		Real estate, renting and business activities	..	..	..	..	..	..	..	..
	70		Real estate activities	..	..	..	..	..	..	..	..
		701	Real estate activ. with own or leased property	..	..	..	..	..	..	..	..
		702	Real estate activities on a fee or contract basis	..	..	..	..	..	..	..	..
	71		Renting of mach. without operator; rent. of pers. gds.	5 517	6 431	7 685	7 241	7 639	8 763	8 837	..
		711	Renting of transport equipment	..	..	..	..	..	..	..	..
		712	Renting of other machinery and equipment	..	..	..	..	..	..	..	..
		713	Renting of personal and household goods nec	..	..	..	..	..	..	..	..
	72		Computer and related activities	3 789	4 302	5 455	5 860	6 452	8 070	8 466	..
		721	Hardware consultancy	..	..	..	..	..	..	..	..
		722	Software consultancy and supply	..	..	..	..	..	..	..	..
		723	Data processing	..	..	..	..	..	..	..	..
		724	Data base activities	..	..	..	..	..	..	..	..
		725	Mainten.; repair of office, account. & comp. mach.	..	..	..	..	..	..	..	..
		729	Other computer related activities	..	..	..	..	..	..	..	..
	73		Research and development	..	..	..	..	..	..	..	..
		731	R & D experim. on natural sciences & engineering	..	..	..	..	..	..	..	..
		732	R & D experim. on social sciences & humanities	..	..	..	..	..	..	..	..
	74		Other business activities	..	..	..	..	..	..	..	..
		741	Legal, acc., book-keep., audit act.; tax consult., etc	10 811	11 607	13 022	14 960	17 728	21 161	24 276	..
		742	Architectural, engineering, & other techn. activities	..	..	..	..	..	..	..	..
		743	Advertising	1 009	1 139	1 228	1 248	1 426	1 569	1 634	..
		749	Business activities nec	..	..	..	..	..	..	..	..
M	80		Education	..	..	..	..	..	..	..	..
		801	Primary education	..	..	..	..	..	..	..	..
		802	Secondary education	..	..	..	..	..	..	..	..
		803	Higher education	..	..	..	..	..	..	..	..
		809	Adult and other education	..	..	..	..	..	..	..	..
N	85		Health and social work	..	..	..	..	..	..	..	..
		851	Human health activities	..	..	..	..	..	..	..	..
		852	Veterinary activities	..	..	..	..	..	..	..	..
		853	Social work activities	..	..	..	..	..	..	..	..
O	90_93		Other commun., soc. & pers. services activities	..	..	..	..	..	..	..	..
	90		Sewage & refuse disposal, sanitation, etc.	..	..	..	..	..	..	..	..
	91		Activities of member organisations nec	..	..	..	..	..	..	..	..
		912	Activities of trade unions	..	..	..	..	..	..	..	..
	92		Recreational, cultural and sporting activities	..	..	..	..	..	..	..	..
		921	Motion picture, radio, TV, other entertainment act.	..	..	..	..	..	..	..	..
		922	News agency activities	..	..	..	..	..	..	..	..
		923	Library, archives, museums, other cultural activ.	..	..	..	..	..	..	..	..
		924	Sporting and other recreational activities	..	..	..	..	..	..	..	..
	93		Other service activities	2 417	2 538	2 708	2 779	2 933	3 054	3 284	..

Note: Data for 1997 are provisional.

Table LX.S2 **VALEUR AJOUTÉE - VALUE ADDED**

Millions de LF (Prix courants)

1991	1992	1993	1994	1995	1996	1997	1998		CITI révision 3		
50 883	51 063	52 638	55 391	57 487	59 632	61 363	..	Com. de gros, de detail ; réparation etc.	G	50_52	
9 226	8 768	8 604	9 406	9 515	10 091	10 518	..	Com., entr., répar. d'auto. et moto ; carburants		50	
6 267	5 576	5 199	5 500	5 665	6 142	6 198	..	Commerce de véhicules automobiles			501
699	688	643	661	675	702	663	..	Entretien et réparation de véhicules automobiles			502
738	741	774	831	894	909	953	..	Com. de pièces et access. de véhic. automobiles			503
66	65	73	66	75	69	71	..	Com., entret. et répar. de motocycles et access.			504
1 457	1 697	1 915	2 349	2 206	2 269	2 633	..	Com. de détail de carburants pour automobiles			505
24 311	24 543	25 700	27 223	29 180	29 919	30 194	..	Com. de gros et act. inter., sauf auto. et moto		51	
..	..	..	..	..	..	..	..	Activités d'intermédiaries de commerce de gros			511
..	..	..	..	..	..	..	..	Com. gros: prod. agr. bruts, anim. viv., prod. alim.			512
..	..	..	..	..	..	..	..	Com. gros: biens de consom., sauf alimentaires			513
..	..	..	..	..	..	..	..	Com. gros: prod. inter. non agr., déchets, débris			514
..	..	..	..	..	..	..	..	Com. gros: machines, équipments et fournitures			515
..	..	..	..	..	..	..	..	Autres commerces de gros			519
17 346	17 753	18 334	18 762	18 792	19 622	20 651	..	Com. détail, sauf auto, moto; répar. art. pers., dom.		52	
3 835	4 098	4 273	4 483	4 537	4 791	5 005	..	Commerce de détail non spécialisé en magasin			521
2 018	2 130	2 198	2 286	2 199	2 152	2 291	..	Com. détail: prod. aliment. en mag. spécialisés			522
10 966	10 941	11 235	11 294	11 364	11 976	12 639	..	Autre com. détail: prod. neufs en mag. spécialisés			523
368	413	397	449	439	443	466	..	Commerces de détail de biens d'occasion			524
..	..	..	..	..	..	..	..	Commerces de détail autres qu'en magasin			525
159	170	230	249	253	260	250	..	Réparation d'articles personnels et domestiques			526
..	..	..	..	..	..	..	..	Hotels et restaurants	H	55	
..	..	..	..	..	..	..	..	Hôtels; camping; autres équip. courts séjours			551
..	..	..	..	..	..	..	..	Restaurants, cafés et cantines			552
..	..	..	..	..	..	..	..	Transport, entreposage, communications	I	60_64	
..	..	..	..	..	..	..	..	Transports terrestres; transports par conduites		60	
..	..	..	..	..	..	..	..	Transports par chemin de fer			601
..	..	..	..	..	..	..	..	Autres transports terrestres			602
..	..	..	..	..	..	..	..	Transports par conduites			603
..	..	..	..	..	..	..	..	Transports par eau		61	
..	..	..	..	..	..	..	..	Transports maritimes et côtiers			611
..	..	..	..	..	..	..	..	Transports par voies navigables intérieures			612
..	..	..	..	..	..	..	..	Transports aériens		62	
..	..	..	..	..	..	..	..	Transports aériens réguliers			621
..	..	..	..	..	..	..	..	Transports aériens non réguliers			622
..	..	..	..	..	..	..	..	Act. annex. et aux. des transp.; act. des ag. de voy.		63	
..	..	..	..	..	..	..	..	Postes et télécommunications		64	
..	..	..	..	..	..	..	..	Activités de poste et de courrier			641
..	..	..	..	..	..	..	..	Télécommunications			642
..	..	..	..	..	..	..	..	Intermédiation financière	J	65_67	
..	..	..	..	..	..	..	..	Interm. financ., sauf assur. et caisses de retraite		65	
..	..	..	..	..	..	..	..	Intermédiation monétaire			651
..	..	..	..	..	..	..	..	Autres intermédiations financières			659
..	..	..	..	..	..	..	..	Act. d'assur., caisses de retr., sauf séc. soc. oblig.		66	
..	..	..	..	..	..	..	..	Act. auxiliaires de l'intermédiation financière		67	
..	..	..	..	..	..	..	..	Act. aux. de l'interm. fin., sauf assur., et retraite			671
..	..	..	..	..	..	..	..	Act. aux. des assurances et caisses de retraite			672
..	..	..	..	..	..	..	..	Immobilier, locations, services aux entreprises	K	70_74	
..	..	..	..	..	..	..	..	Activités immobilières		70	
..	..	..	..	..	..	..	..	Act. immobilières sur biens propres ou loués			701
..	..	..	..	..	..	..	..	Activités immobilières pour comptes de tiers			702
3 621	4 353	5 333	5 061	5 387	6 053	6 131	..	Locat. mach., équip. sans opér., biens pers. et dom.		71	
..	..	..	..	..	..	..	..	Location de matériel de transport			711
..	..	..	..	..	..	..	..	Location de machines ét equipements nca			712
..	..	..	..	..	..	..	..	Loc. d'articles personnels et domestiques nca			713
2 119	2 433	2 968	3 204	3 439	4 321	4 519	..	Activités informatiques et activités rattachées		72	
..	..	..	..	..	..	..	..	Conseil en matériel informatique			721
..	..	..	..	..	..	..	..	Production de logiciels			722
..	..	..	..	..	..	..	..	Traitement de données informatiques			723
..	..	..	..	..	..	..	..	Activités de banques de données			724
..	..	..	..	..	..	..	..	Entr., répar. mach. de bureau et compt. et inform.			725
..	..	..	..	..	..	..	..	Autres activités rattachées à l'informatique			729
..	..	..	..	..	..	..	..	Recherche-développement		73	
..	..	..	..	..	..	..	..	Rech. et dév. expér. en sc. phys., natur. et ingén.			731
..	..	..	..	..	..	..	..	Rech. et dév. expér. en sciences soc. et humaines			732
..	..	..	..	..	..	..	..	Autres activités de services aux entreprises		74	
7 564	8 040	8 879	10 037	11 552	13 879	15 920	..	Act. jur., compt., audit.; cons. fisc.; ét. march., soud.			741
..	..	..	..	..	..	..	..	Act. d'achitect., d'ingén.; autres act. techniques			742
713	792	837	828	936	1 033	1 077	..	Publicité			743
..	..	..	..	..	..	..	..	Activités de services aux entreprises nca.			749
..	..	..	..	..	..	..	..	Education	M	80	
..	..	..	..	..	..	..	..	Activités d'enseignement primaire			801
..	..	..	..	..	..	..	..	Activités d'enseignement secondaire			802
..	..	..	..	..	..	..	..	Activités d'enseignement supérieur			803
..	..	..	..	..	..	..	..	Act. de formation perman.; autres act. d'enseign.			809
..	..	..	..	..	..	..	..	Santé et action sociale	N	85	
..	..	..	..	..	..	..	..	Activités de santé humaine			851
..	..	..	..	..	..	..	..	Activités vétérinaires			852
..	..	..	..	..	..	..	..	Activités d'action sociale			853
..	..	..	..	..	..	..	..	Autres activités de serv. coll., soc. et personnels	O	90_93	
..	..	..	..	..	..	..	..	Assain. et enlèv. des ordures; voierie, act. simil.		90	
..	..	..	..	..	..	..	..	Activités associatives diverses		91	
..	..	..	..	..	..	..	..	Activités de syndicats de salariés			912
..	..	..	..	..	..	..	..	Activités récréatives, culturelles et sportives		92	
..	..	..	..	..	..	..	..	Act. cinémat., de radio, télé.; autres act. de spect.			921
..	..	..	..	..	..	..	..	Activités d'agences de presse			922
..	..	..	..	..	..	..	..	Act. des biblio., arch., musées, jardins bot. et zool.			923
..	..	..	..	..	..	..	..	Activités sportives et autres activités récréatives			924
1 598	1 698	1 803	1 873	1 969	2 056	2 217	..	Autres activités de services		93	

Note: Data for 1997 are provisional.

Table LX.S3 EMPLOYMENT, EMPLOYEES - EMPLOI, SALARIÉS

Units

ISIC revision 3				1991	1992	1993	1994	1995	1996	1997	1998
G	50_52		Wholesale & retail trade; repairs; etc.	26 826	27 650	27 675	28 066	28 508	29 310	30 031	..
	50		Sale, maint.,repair vehic/cycle, person., h/hold gd.	4 530	4 673	4 682	4 831	4 851	4 891	5 075	..
		501	Sale of motor cycle	2 896	2 964	2 935	2 953	2 904	2 910	2 974	..
		502	Maintenance and repair of motor vehicles	383	379	337	370	388	380	407	..
		503	Sale of motor vehicle parts and accessories	476	459	464	490	483	493	554	..
		504	Sale, maint., repair motor cycle, parts & acces.	30	33	39	42	39	39	39	..
		505	Retail sale of automative fuel	745	838	909	975	1 037	1 070	1 101	..
	51		W/sale. trade & commission exc motor veh. & cycle	9 532	9 851	9 829	9 715	10 084	10 310	10 472	..
		511	Wholesale on a fee or contract basis	..	..	..	..	..	..	..	..
		512	W/sale agric. raw mat.,live animal,food, bev., tobac.	..	..	..	..	..	..	..	..
		513	Wholesale of household goods	..	..	..	..	..	..	..	..
		514	W/sale. of non-agric. intermed. prod., waste, scrap	..	..	..	..	..	..	..	..
		515	Wholesale of machinery, equipment & supplies	..	..	..	..	..	..	..	..
		519	Other wholesale	..	..	..	..	..	..	..	..
	52		Retail trade exc. motor veh. & cyc.; repair h/hold gds.	12 764	13 125	13 164	13 520	13 573	14 109	14 485	..
		521	Non-specialised retail trade in stores	3 349	3 495	3 486	3 639	3 685	4 045	4 181	..
		522	Retail sale food, bev., tobac. in specialised stores	1 596	1 573	1 609	1 663	1 629	1 643	1 584	..
		523	Other retail trade of new goods specialised stores	7 508	7 707	7 703	7 842	7 880	8 023	8 303	..
		524	Retail sale of second-hand goods in stores	204	242	244	250	254	269	290	..
		525	Retail trade not in stores	..	..	..	..	..	..	..	..
		526	Repair of personal and household goods	107	108	121	126	125	130	127	..
H	55		Hotels and restaurants	..	..	..	..	..	..	..	..
		551	Hotels; camping ; other short-stay accommodation	..	..	..	..	..	..	..	..
		552	Restaurants, bars and canteens	..	..	..	..	..	..	..	..
I	60_64		Transport, storage and communications	..	..	..	..	..	..	..	..
	60		Land transport, transport via pipelines	..	..	..	..	..	..	..	..
		601	Transport via railways	..	..	..	..	..	..	..	..
		602	Other land transport	..	..	..	..	..	..	..	..
		603	Transport via pipelines	..	..	..	..	..	..	..	..
	61		Water transport	..	..	..	..	..	..	..	..
		611	Sea and coastal water transport	..	..	..	..	..	..	..	..
		612	Inland water transport	..	..	..	..	..	..	..	..
	62		Air transport	..	..	..	..	..	..	..	..
		621	Scheduled air transport	..	..	..	..	..	..	..	..
		622	Non-scheduled air transport	..	..	..	..	..	..	..	..
	63		Support. & aux. transport act., travel agencies activ.	..	..	..	..	..	..	..	..
	64		Post and telecommunications	..	..	..	..	..	..	..	..
		641	Post and courier activities	..	..	..	..	..	..	..	..
		642	Telecommunications	..	..	..	..	..	..	..	..
J	65_67		Financial intermediation	..	..	..	..	..	..	..	..
	65		Financial intermediation, exc. insur., pension fund	..	..	..	..	..	..	..	..
		651	Monetary intermediation	..	..	..	..	..	..	..	..
		659	Other financial intermediation	..	..	..	..	..	..	..	..
	66		Insurance pension fund, exc. compul. social sec.	..	..	..	..	..	..	..	..
	67		Activities auxiliary to financial intermediation	..	..	..	..	..	..	..	..
		671	Act. auxil to fin. intermed., exc. insur. & pens.fund	..	..	..	..	..	..	..	..
		672	Act. auxiliary to insurance and pension funding	..	..	..	..	..	..	..	..
K	70_74		Real estate, renting and business activities	..	..	..	..	..	..	..	..
	70		Real estate activities	..	..	..	..	..	..	..	..
		701	Real estate activ. with own or leased property	..	..	..	..	..	..	..	..
		702	Real estate activities on a fee or contract basis	..	..	..	..	..	..	..	..
	71		Renting of mach. without operator; rent. of pers. gds.	357	395	376	395	411	454	457	..
		711	Renting of transport equipment	..	..	..	..	..	..	..	..
		712	Renting of other machinery and equipment	..	..	..	..	..	..	..	..
		713	Renting of personal and household goods nec	..	..	..	..	..	..	..	..
	72		Computer and related activities	914	1 096	1 319	1 415	1 694	2 154	2 484	..
		721	Hardware consultancy	..	..	..	..	..	..	..	..
		722	Software consultancy and supply	..	..	..	..	..	..	..	..
		723	Data processing	..	..	..	..	..	..	..	..
		724	Data base activities	..	..	..	..	..	..	..	..
		725	Mainten.; repair of office, account. & comp. mach.	..	..	..	..	..	..	..	..
		729	Other computer related activities	..	..	..	..	..	..	..	..
	73		Research and development	..	..	..	..	..	..	..	..
		731	R & D experim. on natural sciences & engineering	..	..	..	..	..	..	..	..
		732	R & D experim. on social sciences & humanities	..	..	..	..	..	..	..	..
	74		Other business activities	..	..	..	..	..	..	..	..
		741	Legal, acc., book-keep., audit act.; tax consult., etc	2 912	3 100	3 216	3 489	3 743	4 063	4 828	..
		742	Architectural, engineering, & other techn. activities	..	..	..	..	..	..	..	..
		743	Advertising	504	536	412	414	429	447	528	..
		749	Business activities nec	..	..	..	..	..	..	..	..
M	80		Education	..	..	..	..	..	..	..	..
		801	Primary education	..	..	..	..	..	..	..	..
		802	Secondary education	..	..	..	..	..	..	..	..
		803	Higher education	..	..	..	..	..	..	..	..
		809	Adult and other education	..	..	..	..	..	..	..	..
N	85		Health and social work	..	..	..	..	..	..	..	..
		851	Human health activities	..	..	..	..	..	..	..	..
		852	Veterinary activities	..	..	..	..	..	..	..	..
		853	Social work activities	..	..	..	..	..	..	..	..
O	90_93		Other commun., soc. & pers. services activities	..	..	..	..	..	..	..	..
	90		Sewage & refuse disposal, sanitation, etc.	..	..	..	..	..	..	..	..
	91		Activities of member organisations nec	..	..	..	..	..	..	..	..
		912	Activities of trade unions	..	..	..	..	..	..	..	..
	92		Recreational, cultural and sporting activities	..	..	..	..	..	..	..	..
		921	Motion picture, radio, TV, other entertainment act.	..	..	..	..	..	..	..	..
		922	News agency activities	..	..	..	..	..	..	..	..
		923	Library, archives, museums, other cultural activ.	..	..	..	..	..	..	..	..
		924	Sporting and other recreational activities	..	..	..	..	..	..	..	..
	93		Other service activities	1 698	1 752	1 779	1 814	1 816	1 826	1 836	..

Note: Data for 1997 are provisional.

Table LX.S4 **SALAIRES ET CHARGES SOCIALES, SALARIÉS - COMPENSATION OF LABOUR, EMPLOYEES**

Millions de LF (Prix courants)

1991	1992	1993	1994	1995	1996	1997	1998			CITI révision 3	
22 848	24 794	25 818	26 829	28 160	29 425	31 445	..	Com. de gros, de detail ; réparation etc.	G	50_52	
3 948	4 212	4 311	4 596	4 736	4 897	5 418	..	Com., entr., répar. d'auto. et moto ; carburants		50	
2 670	2 821	2 823	2 935	2 971	3 059	3 339	..	Commerce de véhicules automobiles			501
287	296	272	307	336	342	380	..	Entretien et réparation de véhicules automobiles			502
476	479	500	555	563	581	705	..	Com. de pièces et access. de véhic. automobiles			503
23	26	32	34	34	35	35	..	Com., entret. et répar. de motocycles et access.			504
491	590	684	766	832	879	958	..	Com. de détail de carburants pour automobiles			505
10 231	11 072	11 545	11 790	12 692	13 182	13 933	..	Com. de gros et act. inter., sauf auto. et moto		51	
..	..	..	..	..	..	..	..	Activités d'intermédiaries de commerce de gros			511
..	..	..	..	..	..	..	..	Com. gros: prod. agr. bruts, anim. viv., prod. alim.			512
..	..	..	..	..	..	..	..	Com. gros: biens de consom., sauf alimentaires			513
..	..	..	..	..	..	..	..	Com. gros: prod. inter. non agr., déchets, débris			514
..	..	..	..	..	..	..	..	Com. gros: machines, équipments et fournitures			515
..	..	..	..	..	..	..	..	Autres commerces de gros			519
8 669	9 510	9 962	10 443	10 732	11 346	12 095	..	Com. détail, sauf auto, moto; répar. art. pers., dom.		52	
2 153	2 440	2 539	2 684	2 758	3 098	3 298	..	Commerce de détail non spécialisé en magasin			521
1 040	1 084	1 142	1 214	1 209	1 256	1 278	..	Com. détail: prod. aliment. en mag. spécialisés			522
5 282	5 759	6 033	6 289	6 490	6 689	7 193	..	Autre com. détail: prod. neufs en mag. spécialisés			523
122	151	160	163	182	206	225	..	Commerces de détail de biens d'occasion			524
..	..	..	..	..	..	..	..	Commerces de détail autres qu'en magasin			525
71	76	88	94	92	98	101	..	Réparation d'articles personnels et domestiques			526
..	..	..	..	..	..	..	..	Hotels et restaurants	H	55	
..	..	..	..	..	..	..	..	Hôtels; camping; autres équip. courts séjours			551
..	..	..	..	..	..	..	..	Restaurants, cafés et cantines			552
..	..	..	..	..	..	..	..	Transport, entreposage, communications	I	60_64	
..	..	..	..	..	..	..	..	Transports terrestres; transports par conduites		60	
..	..	..	..	..	..	..	..	Transports par chemin de fer			601
..	..	..	..	..	..	..	..	Autres transports terrestres			602
..	..	..	..	..	..	..	..	Transports par conduites			603
..	..	..	..	..	..	..	..	Transports par eau		61	
..	..	..	..	..	..	..	..	Transports maritimes et côtiers			611
..	..	..	..	..	..	..	..	Transports par voies navigables intérieures			612
..	..	..	..	..	..	..	..	Transports aériens		62	
..	..	..	..	..	..	..	..	Transports aériens réguliers			621
..	..	..	..	..	..	..	..	Transports aériens non réguliers			622
..	..	..	..	..	..	..	..	Act. annex. et aux. des transp.; act. des ag. de voy.		63	
..	..	..	..	..	..	..	..	Postes et télécommunications		64	
..	..	..	..	..	..	..	..	Activités de poste et de courrier			641
..	..	..	..	..	..	..	..	Télécommunications			642
..	..	..	..	..	..	..	..	Intermédiation financière	J	65_67	
..	..	..	..	..	..	..	..	Interm. financ., sauf assur. et caisses de retraite		65	
..	..	..	..	..	..	..	..	Intermédiation monétaire			651
..	..	..	..	..	..	..	..	Autres intermédiations financières			659
..	..	..	..	..	..	..	..	Act. d'assur., caisses de retr., sauf séc. soc. oblig.		66	
..	..	..	..	..	..	..	..	Act. auxiliaires de l'intermédiation financière		67	
..	..	..	..	..	..	..	..	Act. aux. de l'interm. fin., sauf assur., et retraite			671
..	..	..	..	..	..	..	..	Act. aux. des assurances et caisses de retraite			672
..	..	..	..	..	..	..	..	Immobilier, locations, services aux entreprises	K	70_74	
..	..	..	..	..	..	..	..	Activités immobilières		70	
..	..	..	..	..	..	..	..	Act. immobilières sur biens propres ou loués			701
..	..	..	..	..	..	..	..	Activités immobilières pour comptes de tiers			702
464	517	525	559	591	677	706	..	Locat. mach., équip. sans opér., biens pers. et dom.		71	
..	..	..	..	..	..	..	..	Location de matériel de transport			711
..	..	..	..	..	..	..	..	Location de machines ét equipements nca			712
..	..	..	..	..	..	..	..	Loc. d'articles personnels et domestiques nca			713
1 276	1 657	2 092	2 314	2 863	3 684	4 435	..	Activités informatiques et activités rattachées		72	
..	..	..	..	..	..	..	..	Conseil en matériel informatique			721
..	..	..	..	..	..	..	..	Production de logiciels			722
..	..	..	..	..	..	..	..	Traitement de données informatiques			723
..	..	..	..	..	..	..	..	Activités de banques de données			724
..	..	..	..	..	..	..	..	Entr., répar. mach. de bureau et compt. et inform.			725
..	..	..	..	..	..	..	..	Autres activités rattachées à l'informatique			729
..	..	..	..	..	..	..	..	Recherche-développement		73	
..	..	..	..	..	..	..	..	Rech. et dév. expér. en sc. phys., natur. et ingén.			731
..	..	..	..	..	..	..	..	Rech. et dév. expér. en sciences soc. et humaines			732
..	..	..	..	..	..	..	..	Autres activités de services aux entreprises		74	
3 455	3 950	4 246	4 882	5 277	5 905	7 382	..	Act. jur., compt., audit.; cons. fisc.; ét. march., soud.			741
..	..	..	..	..	..	..	..	Act. d'achitect., d'ingén.; autres act. techniques			742
448	510	482	505	542	578	700	..	Publicité			743
..	..	..	..	..	..	..	..	Activités de services aux entreprises nca.			749
..	..	..	..	..	..	..	..	Education	M	80	
..	..	..	..	..	..	..	..	Activités d'enseignement primaire			801
..	..	..	..	..	..	..	..	Activités d'enseignement secondaire			802
..	..	..	..	..	..	..	..	Activités d'enseignement supérieur			803
..	..	..	..	..	..	..	..	Act. de formation perman.; autres act. d'enseign.			809
..	..	..	..	..	..	..	..	Santé et action sociale	N	85	
..	..	..	..	..	..	..	..	Activités de santé humaine			851
..	..	..	..	..	..	..	..	Activités vétérinaires			852
..	..	..	..	..	..	..	..	Activités d'action sociale			853
..	..	..	..	..	..	..	..	Autres activités de serv. coll., soc. et personnels	O	90_93	
..	..	..	..	..	..	..	..	Assain. et enlèv. des ordures; voierie, act. simil.		90	
..	..	..	..	..	..	..	..	Activités associatives diverses		91	
..	..	..	..	..	..	..	..	Activités de syndicats de salariés			912
..	..	..	..	..	..	..	..	Activités récréatives, culturelles et sportives		92	
..	..	..	..	..	..	..	..	Act. cinémat., de radio, télé.; autres act. de spect.			921
..	..	..	..	..	..	..	..	Activités d'agences de presse			922
..	..	..	..	..	..	..	..	Act. des biblio., arch., musées, jardins bot. et zool.			923
..	..	..	..	..	..	..	..	Activités sportives et autres activités récréatives			924
827	918	991	1 034	1 066	1 108	1 179	..	Autres activités de services		93	

Note: Data for 1997 are provisional

LUXEMBOURG

Table LX.S5 **ENTERPRISES - ENTREPRISES**

Units

ISIC revision 3				1991	1992	1993	1994	1995	1996	1997	1998
G	50_52		Wholesale & retail trade; repairs; etc.	5 655	5 722	5 926	6 116	6 208	6 449	6 487	..
	50		Sale, maint.,repair vehic/cycle, person., h/hold gd.	779	771	794	808	801	808	821	..
		501	Sale of motor cycle	299	306	333	346	348	357	372	..
		502	Maintenance and repair of motor vehicles	107	100	103	103	103	119	121	..
		503	Sale of motor vehicle parts and accessories	80	82	89	89	92	109	112	..
		504	Sale, maint., repair motor cycle, parts & acces.	11	10	9	11	10	10	12	..
		505	Retail sale of automative fuel	282	273	260	259	248	213	204	..
	51		W/sale. trade & commission exc motor veh. & cycle	1 793	1 873	2 064	2 242	2 406	2 757	2 808	..
		511	Wholesale on a fee or contract basis	..	..	..	..	..	..	..	..
		512	W/sale agric. raw mat.,live animal,food, bev., tobac.	..	..	..	..	..	..	..	..
		513	Wholesale of household goods	..	..	..	..	..	..	..	..
		514	W/sale. of non-agric. intermed. prod., waste, scrap	..	..	..	..	..	..	..	..
		515	Wholesale of machinery, equipment & supplies	..	..	..	..	..	..	..	..
		519	Other wholesale	..	..	..	..	..	..	..	..
	52		Retail trade exc. motor veh. & cyc.; repair h/hold gds.	3 083	3 078	3 068	3 066	3 001	2 884	2 858	..
		521	Non-specialised retail trade in stores	347	328	306	294	277	263	255	..
		522	Retail sale food, bev., tobac. in specialised stores	398	388	389	382	369	340	324	..
		523	Other retail trade of new goods specialised stores	2 025	2 058	2 075	2 089	2 059	1 999	2 001	..
		524	Retail sale of second-hand goods in stores	239	227	221	225	218	212	207	..
		525	Retail trade not in stores	..	..	..	..	..	..	..	..
		526	Repair of personal and household goods	74	77	77	76	78	70	71	..
H	55		Hotels and restaurants	..	..	..	..	..	..	..	..
		551	Hotels; camping ; other short-stay accommodation	..	..	..	..	..	..	..	..
		552	Restaurants, bars and canteens	..	..	..	..	..	..	..	..
I	60_64		Transport, storage and communications	..	..	..	..	..	..	..	..
	60		Land transport, transport via pipelines	..	..	..	..	..	..	..	..
		601	Transport via railways	..	..	..	..	..	..	..	..
		602	Other land transport	..	..	..	..	..	..	..	..
		603	Transport via pipelines	..	..	..	..	..	..	..	..
	61		Water transport	..	..	..	..	..	..	..	..
		611	Sea and coastal water transport	..	..	..	..	..	..	..	..
		612	Inland water transport	..	..	..	..	..	..	..	..
	62		Air transport	..	..	..	..	..	..	..	..
		621	Scheduled air transport	..	..	..	..	..	..	..	..
		622	Non-scheduled air transport	..	..	..	..	..	..	..	..
	63		Support. & aux. transport act., travel agencies activ.	..	..	..	..	..	..	..	..
	64		Post and telecommunications	..	..	..	..	..	..	..	..
		641	Post and courier activities	..	..	..	..	..	..	..	..
		642	Telecommunications	..	..	..	..	..	..	..	..
J	65_67		Financial intermediation	..	..	..	..	..	..	..	..
	65		Financial intermediation, exc. insur., pension fund	..	..	..	..	..	..	..	..
		651	Monetary intermediation	..	..	..	..	..	..	..	..
		659	Other financial intermediation	..	..	..	..	..	..	..	..
	66		Insurance pension fund, exc. compul. social sec.	..	..	..	..	..	..	..	..
	67		Activities auxiliary to financial intermediation	..	..	..	..	..	..	..	..
		671	Act. auxil to fin. intermed., exc. insur. & pens.fund	..	..	..	..	..	..	..	..
		672	Act. auxiliary to insurance and pension funding	..	..	..	..	..	..	..	..
K	70_74		Real estate, renting and business activities	..	..	..	..	..	..	..	..
	70		Real estate activities	..	..	..	..	..	..	..	..
		701	Real estate activ. with own or leased property	..	..	..	..	..	..	..	..
		702	Real estate activities on a fee or contract basis	..	..	..	..	..	..	..	..
	71		Renting of mach. without operator; rent. of pers. gds.	126	138	155	160	165	183	193	..
		711	Renting of transport equipment	..	..	..	..	..	..	..	..
		712	Renting of other machinery and equipment	..	..	..	..	..	..	..	..
		713	Renting of personal and household goods nec	..	..	..	..	..	..	..	..
	72		Computer and related activities	202	228	269	316	331	398	436	..
		721	Hardware consultancy	..	..	..	..	..	..	..	..
		722	Software consultancy and supply	..	..	..	..	..	..	..	..
		723	Data processing	..	..	..	..	..	..	..	..
		724	Data base activities	..	..	..	..	..	..	..	..
		725	Mainten.; repair of office, account. & comp. mach.	..	..	..	..	..	..	..	..
		729	Other computer related activities	..	..	..	..	..	..	..	..
	73		Research and development	..	..	..	..	..	..	..	..
		731	R & D experim. on natural sciences & engineering	..	..	..	..	..	..	..	..
		732	R & D experim. on social sciences & humanities	..	..	..	..	..	..	..	..
	74		Other business activities	..	..	..	..	..	..	..	..
		741	Legal, acc., book-keep., audit act.; tax consult., etc	874	960	1 063	1 207	1 318	1 601	1 668	..
		742	Architectural, engineering, & other techn. activities	..	..	..	..	..	..	..	..
		743	Advertising	150	157	164	165	181	186	199	..
		749	Business activities nec	..	..	..	..	..	..	..	..
M	80		Education	..	..	..	..	..	..	..	..
		801	Primary education	..	..	..	..	..	..	..	..
		802	Secondary education	..	..	..	..	..	..	..	..
		803	Higher education	..	..	..	..	..	..	..	..
		809	Adult and other education	..	..	..	..	..	..	..	..
N	85		Health and social work	..	..	..	..	..	..	..	..
		851	Human health activities	..	..	..	..	..	..	..	..
		852	Veterinary activities	..	..	..	..	..	..	..	..
		853	Social work activities	..	..	..	..	..	..	..	..
O	90_93		Other commun., soc. & pers. services activities	..	..	..	..	..	..	..	..
	90		Sewage & refuse disposal, sanitation, etc.	..	..	..	..	..	..	..	..
	91		Activities of member organisations nec	..	..	..	..	..	..	..	..
		912	Activities of trade unions	..	..	..	..	..	..	..	..
	92		Recreational, cultural and sporting activities	..	..	..	..	..	..	..	..
		921	Motion picture, radio, TV, other entertainment act.	..	..	..	..	..	..	..	..
		922	News agency activities	..	..	..	..	..	..	..	..
		923	Library, archives, museums, other cultural activ.	..	..	..	..	..	..	..	..
		924	Sporting and other recreational activities	..	..	..	..	..	..	..	..
	93		Other service activities	549	552	565	581	590	615	614	..

Note: Data for 1997 are provisional.

Table NZ.S1 **EMPLOI, SALARIÉS - EMPLOYMENT, EMPLOYEES**

Unités

1990-91	1991-92	1992-93	1993-94	1994-95	1995-96	1996-97	1997-98		CITI révision 3		
176 710	169 320	168 940	182 970	193 510	198 290	202 638	203 317	Com. de gros, de detail ; réparation etc.	G	50_52	
..	..	..	33 590	35 920	38 050	41 500	41 145	Com., entr., répar. d'auto. et moto ; carburants		50	
..	..	..	10 090	10 740	11 280	11 812	12 021	Commerce de véhicules automobiles			501
..	..	..	9 530	10 250	11 300	11 666	11 830	Entretien et réparation de véhicules automobiles			502
..	..	..	5 900	6 430	6 510	8 770	7 871	Com. de pièces et access. de véhic. automobiles			503
..	..	..	630	620	680	772	792	Com., entret. et répar. de motocycles et access.			504
..	..	..	7 440	7 880	8 300	8 481	8 633	Com. de détail de carburants pour automobiles			505
..	..	..	69 340	73 980	75 280	73 101	73 316	Com. de gros et act. inter., sauf auto. et moto		51	
..	..	..	..	..	..	..	..	Activités d'intermédiaries de commerce de gros			511
..	..	..	16 900	17 570	17 930	17 239	17 314	Com. gros: prod. agr. bruts, anim. viv., prod. alim.			512
..	..	..	13 510	13 690	14 000	12 746	13 121	Com. gros: biens de consom., sauf alimentaires			513
..	..	..	15 420	16 350	16 120	14 345	13 785	Com. gros: prod. inter. non agr., déchets, débris			514
..	..	..	18 520	20 990	22 150	22 679	23 016	Com. gros: machines, équipments et fournitures			515
..	..	..	5 000	5 390	5 080	6 093	6 082	Autres commerces de gros			519
..	..	..	80 040	83 610	84 950	93 535	95 584	Com. détail, sauf auto, moto; répar. art. pers., dom.		52	
..	..	..	29 200	32 160	33 930	10 994	12 047	Commerce de détail non spécialisé en magasin			521
..	..	..	12 180	11 820	11 470	41 863	42 927	Com. détail: prod. aliment. en mag. spécialisés			522
..	..	..	36 110	37 120	36 970	33 306	33 065	Autre com. détail: prod. neufs en mag. spécialisés			523
..	..	..	780	720	740	654	685	Commerces de détail de biens d'occasion			524
..	..	..	..	..	..	5 106	5 404	Commerces de détail autres qu'en magasin			525
..	..	..	1 780	1 790	1 830	1 613	1 457	Réparation d'articles personnels et domestiques			526
36 390	36 960	37 420	46 200	49 580	51 600	55 238	56 256	Hotels et restaurants	H	55	
..	..	..	14 950	15 560	15 920	16 576	16 229	Hôtels; camping; autres équip. courts séjours			551
..	..	..	31 250	34 020	35 680	38 662	40 027	Restaurants, cafés et cantines			552
..	..	..	65 340	67 160	69 960	78 356	78 362	Transport, entreposage, communications	I	60_64	
78 240	70 110	67 570	20 600	22 150	23 080	27 228	27 688	Transports terrestres; transports par conduites		60	
..	..	..	1 960	1 930	2 110	6 131	6 316	Transports par chemin de fer			601
..	..	..	18 650	20 210	20 960	21 088	21 365	Autres transports terrestres			602
..	..	..	3	..	3	10	7	Transports par conduites			603
..	..	..	2 130	2 040	1 950	990	1 119	Transports par eau		61	
..	..	..	1 910	1 800	1 720	739	769	Transports maritimes et côtiers			611
..	..	..	210	250	240	252	350	Transports par voies navigables intérieures			612
..	..	..	7 600	8 120	8 460	11 172	10 510	Transports aériens		62	
..	..	..	7 320	7 790	8 180	10 886	10 226	Transports aériens réguliers			621
..	..	..	290	320	270	286	284	Transports aériens non réguliers			622
..	..	..	14 730	15 660	16 310	16 223	16 814	Act. annex. et aux. des transp.; act. des ag. de voy.		63	
..	..	..	20 270	19 190	20 160	22 743	22 232	Postes et télécommunications		64	
..	..	..	8 870	9 170	10 060	10 586	10 946	Activités de poste et de courrier			641
..	..	..	11 400	10 020	10 110	12 158	11 286	Télécommunications			642
49 370	47 200	42 410	43 200	44 060	45 150	45 555	45 084	Intermédiation financière	J	65_67	
..	..	..	27 840	28 150	28 990	28 750	28 911	Interm. financ., sauf assur. et caisses de retraite		65	
..	..	..	24 570	24 790	25 090	25 571	26 856	Intermédiation monétaire			651
..	..	..	3 280	3 360	3 900	3 180	2 055	Autres intermédiations financières			659
..	..	..	8 810	9 110	9 330	9 863	8 593	Act. d'assur., caisses de retr., sauf séc. soc. oblig.		66	
..	..	..	6 550	6 800	6 820	6 942	7 580	Act. auxiliaires de l'intermédiation financière		67	
..	..	..	3 540	3 820	3 700	3 796	4 471	Act. aux. de l'interm. fin., sauf assur., et retraite			671
..	..	..	3 010	2 980	3 130	3 147	3 110	Act. aux. des assurances et caisses de retraite			672
74 750	73 430	70 220	86 840	92 010	96 800	102 532	109 270	Immobilier, locations, services aux entreprises	K	70_74	
..	..	..	6 980	6 520	7 170	10 510	9 619	Activités immobilières		70	
..	..	..	980	1 290	1 380	4 890	4 180	Act. immobilières sur biens propres ou loués			701
..	..	..	6 000	5 220	5 790	5 621	5 440	Activités immobilières pour comptes de tiers			702
..	..	..	4 290	4 740	5 120	5 364	5 724	Locat. mach., équip. sans opér., biens pers. et dom.		71	
..	..	..	1 480	1 650	1 860	2 011	2 229	Location de matériel de transport			711
..	..	..	1 900	2 130	2 220	2 270	2 410	Location de machines ét equipements nca			712
..	..	..	910	960	1 040	1 083	1 085	Loc. d'articles personnels et domestiques nca			713
..	..	..	5 780	5 670	6 620	7 084	6 707	Activités informatiques et activités rattachées		72	
..	..	..	2 810	3 010	3 960	4 685	5 232	Conseil en matériel informatique			721
..	..	..	..	..	..	..	..	Production de logiciels			722
..	..	..	1 710	1 360	1 550	2 018	1 048	Traitement de données informatiques			723
..	..	..	1 050	1 100	900	93	124	Activités de banques de données			724
..	..	..	210	190	200	288	303	Entr., répar. mach. de bureau et compt. et inform.			725
..	..	..					..	Autres activités rattachées à l'informatique			729
..	..	..	5 980	6 110	6 130	5 483	5 756	Recherche-développement		73	
..	..	..	..	..	..	..	..	Rech. et dév. expér. en sc. phys., natur. et ingén.			731
..	..	..	..	..	..	..	..	Rech. et dév. expér. en sciences soc. et humaines			732
..	..	..	63 800	68 990	71 780	62 416	69 508	Autres activités de services aux entreprises		74	
..	..	..	25 400	26 650	27 590	28 961	31 984	Act. jur., compt., audit.; cons. fisc.; ét. march., soud.			741
..	..	..	9 530	10 840	11 680	11 795	12 010	Act. d'achitect., d'ingén.; autres act. techniques			742
..	..	..	5 440	6 260	5 960	6 966	6 903	Publicité			743
..	..	..	23 430	25 230	26 550	26 269	30 453	Activités de services aux entreprises nca.			749
..	..	..	87 100	88 050	95 810	97 091	98 762	Education	M	80	
..	..	..	35 060	35 640	40 220	40 282	41 067	Activités d'enseignement primaire			801
..	..	..	21 490	21 150	22 970	21 872	21 997	Activités d'enseignement secondaire			802
..	..	..	23 910	23 720	24 110	26 243	26 747	Activités d'enseignement supérieur			803
..	..	..	6 650	7 550	8 520	8 695	8 952	Act. de formation perman.; autres act. d'enseign.			809
..	..	..	80 020	77 830	78 080	96 205	100 113	Santé et action sociale	N	85	
..	..	..	66 750	63 920	63 390	67 082	68 455	Activités de santé humaine			851
..	..	..	1 460	1 590	1 650	1 675	1 817	Activités vétérinaires			852
..	..	..	11 820	12 320	13 040	27 448	29 841	Activités d'action sociale			853
..	..	..	27 380	29 320	31 340	43 817	46 287	Autres activités de serv. coll., soc. et personnels	O	90_93	
..	..	..	2 050	2 010	2 210	1 425	1 575	Assain. et enlèv. des ordures; voierie, act. simil.		90	
..	..	..	..	..	..	9 075	9 252	Activités associatives diverses		91	
..	..	..	..	..	..	771	705	Activités de syndicats de salariés			912
..	..	..	25 330	27 310	29 140	1 823	1 934	Activités récréatives, culturelles et sportives		92	
..	..	..	6 370	7 210	7 360	7 949	7 448	Act. cinémat., de radio, télé.; autres act. de spect.			921
..	..	..	..	..	..	..	..	Activités d'agences de presse			922
..	..	..	3 790	4 030	5 160	1 652	3 697	Act. des biblio., arch., musées, jardins bot. et zool.			923
..	..	..	7 250	7 880	8 280	12 829	12 919	Activités sportives et autres activités récréatives			924
..	..	..	7 930	8 200	8 340	9 698	10 110	Autres activités de services		93	

Note: Break in series between 1995/96 and 1996/97.

NEW ZEALAND

Table NZ.S2 **EMPLOYMENT, FEMALES - EMPLOI, FEMMES**

Units

ISIC revision 3				1990-91	1991-92	1992-93	1993-94	1994-95	1995-96	1996-97	1997-98
G	50_52		Wholesale & retail trade; repairs; etc.	75 840	72 900	73 400	77 850	82 180	83 940	87 049	87 802
	50		Sale, maint.,repair vehic/cycle, person., h/hold gd.	..	..	..	6 660	6 960	7 530	8 488	8 607
		501	Sale of motor cycle	..	..	..	1 950	2 030	2 220	2 384	2 369
		502	Maintenance and repair of motor vehicles	..	..	..	1 400	1 460	1 590	1 780	1 924
		503	Sale of motor vehicle parts and accessories	..	..	..	1 020	1 100	1 100	1 474	1 328
		504	Sale, maint., repair motor cycle, parts & acces.	..	..	..	100	110	120	149	140
		505	Retail sale of automative fuel	..	..	..	2 170	2 250	2 500	2 701	2 848
	51		W/sale. trade & commission exc motor veh. & cycle	..	..	..	23 130	24 500	24 910	24 836	25 085
		511	Wholesale on a fee or contract basis	..	..	..	..	..	..	..	..
		512	W/sale agric. raw mat.,live animal,food, bev., tobac.	..	..	..	5 710	5 880	5 940	5 943	5 988
		513	Wholesale of household goods	..	..	..	6 590	6 720	6 950	6 480	6 780
		514	W/sale. of non-agric. intermed. prod., waste, scrap	..	..	..	3 650	3 860	3 790	3 426	3 297
		515	Wholesale of machinery, equipment & supplies	..	..	..	5 110	5 770	6 130	6 435	6 450
		519	Other wholesale	..	..	..	2 070	2 270	2 100	2 554	2 571
	52		Retail trade exc. motor veh. & cyc.; repair h/hold gds.	..	..	..	48 070	50 720	51 500	57 214	58 211
		521	Non-specialised retail trade in stores	..	..	..	18 340	20 040	20 930	7 483	8 040
		522	Retail sale food, bev., tobac. in specialised stores	..	..	..	6 350	6 430	6 240	24 236	25 077
		523	Other retail trade of new goods specialised stores	..	..	..	22 540	23 510	23 560	21 538	20 914
		524	Retail sale of second-hand goods in stores	..	..	..	420	350	380	314	333
		525	Retail trade not in stores	..	..	..	..	..	..	3 246	3 464
		526	Repair of personal and household goods	..	..	..	430	400	410	399	383
H	55		Hotels and restaurants	21 620	21 870	22 460	28 320	30 540	31 830	33 830	34 168
		551	Hotels; camping ; other short-stay accommodation	..	..	..	9 480	9 970	10 230	10 637	10 356
		552	Restaurants, bars and canteens	..	..	..	18 840	20 570	21 600	23 193	23 812
I	60_64		Transport, storage and communications	..	..	..	20 960	21 840	23 470	25 306	25 528
	60		Land transport, transport via pipelines	23 950	22 450	20 650	3 240	3 600	3 710	4 231	4 298
		601	Transport via railways	..	..	..	290	290	350	741	746
		602	Other land transport	..	..	..	2 960	3 300	3 360	3 490	3 551
		603	Transport via pipelines	..	..	..	..	..	..	1	1
	61		Water transport	..	..	..	240	280	310	265	314
		611	Sea and coastal water transport	..	..	..	160	190	210	167	165
		612	Inland water transport	..	..	..	80	95	100	99	149
	62		Air transport	..	..	..	3 050	3 310	3 520	3 845	3 804
		621	Scheduled air transport	..	..	..	2 980	3 230	3 460	3 787	3 742
		622	Non-scheduled air transport	..	..	..	70	75	55	58	62
	63		Support. & aux. transport act., travel agencies activ.	..	..	..	5 480	6 000	6 600	6 452	6 752
	64		Post and telecommunications	..	..	..	8 940	8 650	9 340	10 514	10 361
		641	Post and courier activities	..	..	..	4 760	4 960	5 430	5 782	5 988
		642	Telecommunications	..	..	..	4 180	3 690	3 900	4 732	4 373
J	65_67		Financial intermediation	29 950	28 550	25 750	25 660	26 060	26 800	27 384	27 211
	65		Financial intermediation, exc. insur., pension fund	..	..	..	17 310	17 460	18 000	18 213	18 387
		651	Monetary intermediation	..	..	..	15 820	15 870	16 110	16 659	17 406
		659	Other financial intermediation	..	..	..	1 480	1 580	1 900	1 554	981
	66		Insurance pension fund, exc. compul. social sec.	..	..	..	4 790	4 860	5 020	5 404	4 767
	67		Activities auxiliary to financial intermediation	..	..	..	3 570	3 760	3 770	3 768	4 058
		671	Act. auxil to fin. intermed., exc. insur. & pens.fund	..	..	..	1 930	2 090	1 990	1 984	2 363
		672	Act. auxiliary to insurance and pension funding	..	..	..	1 640	1 680	1 790	1 784	1 695
K	70_74		Real estate, renting and business activities	39 980	38 240	37 190	45 010	47 660	50 010	53 028	57 651
	70		Real estate activities	..	..	..	3 570	3 390	3 590	5 282	5 121
		701	Real estate activ. with own or leased property	..	..	..	210	260	300	1 944	1 857
		702	Real estate activities on a fee or contract basis	..	..	..	3 360	3 140	3 290	3 338	3 264
	71		Renting of mach. without operator; rent. of pers. gds.	..	..	..	1 630	1 770	1 970	2 009	2 221
		711	Renting of transport equipment	..	..	..	650	700	820	851	946
		712	Renting of other machinery and equipment	..	..	..	440	480	530	513	630
		713	Renting of personal and household goods nec	..	..	..	540	600	630	646	646
	72		Computer and related activities	..	..	..	2 220	2 120	2 350	2 468	2 308
		721	Hardware consultancy	..	..	..	910	960	1 260	1 512	1 734
		722	Software consultancy and supply	..	..	..	..	..	..	..	..
		723	Data processing	..	..	..	750	560	640	847	452
		724	Data base activities	..	..	..	520	540	420	44	60
		725	Mainten.; repair of office, account. & comp. mach.	..	..	..	45	45	45	66	63
		729	Other computer related activities	..	..	..	..	..	..	..	..
	73		Research and development	..	..	..	2 130	2 230	2 360	1 920	2 207
		731	R & D experim. on natural sciences & engineering	..	..	..	..	..	..	..	..
		732	R & D experim. on social sciences & humanities	..	..	..	..	..	..	..	..
	74		Other business activities	..	..	..	35 440	38 150	39 730	38 106	42 295
		741	Legal, acc., book-keep., audit act.; tax consult., etc	..	..	..	17 400	18 100	18 780	19 704	21 292
		742	Architectural, engineering, & other techn. activities	..	..	..	2 640	3 050	3 260	3 254	3 485
		743	Advertising	..	..	..	2 820	3 070	2 880	3 507	3 419
		749	Business activities nec	..	..	..	12 560	13 930	14 820	14 873	17 569
M	80		Education	..	..	..	55 900	57 270	63 620	64 374	65 714
		801	Primary education	..	..	..	27 310	27 980	32 110	32 274	32 847
		802	Secondary education	..	..	..	12 060	11 950	13 400	12 610	12 700
		803	Higher education	..	..	..	12 400	12 510	12 740	13 920	14 485
		809	Adult and other education	..	..	..	4 130	4 830	5 380	5 571	5 684
N	85		Health and social work	..	..	..	65 760	64 480	65 090	79 281	82 242
		851	Human health activities	..	..	..	53 850	52 010	51 880	54 477	55 337
		852	Veterinary activities	..	..	..	1 120	1 220	1 300	1 276	1 395
		853	Social work activities	..	..	..	10 790	11 250	11 910	23 529	25 510
O	90_93		Other commun., soc. & pers. services activities	..	..	..	15 260	16 520	17 390	24 183	25 341
	90		Sewage & refuse disposal, sanitation, etc.	..	..	..	320	340	420	283	342
	91		Activities of member organisations nec	..	..	..	..	..	..	5 768	5 919
		912	Activities of trade unions	..	..	..	..	..	..	381	371
	92		Recreational, cultural and sporting activities	..	..	..	14 930	16 180	16 970	944	1 017
		921	Motion picture, radio, TV, other entertainment act.	..	..	..	2 760	3 300	3 400	3 725	3 624
		922	News agency activities	..	..	..	..	..	..	..	..
		923	Library, archives, museums, other cultural activ.	..	..	..	2 400	2 520	2 910	967	1 579
		924	Sporting and other recreational activities	..	..	..	3 710	4 020	4 280	6 148	6 319
	93		Other service activities	..	..	..	6 070	6 330	6 380	6 735	6 999

Note: Break in series between 1995/96 and 1996/97.

Table NZ.S3 SALAIRES ET TRAITEMENTS, SALARIÉS - WAGES AND SALARIES, EMPLOYEES

Millions de NZ$ (Prix courants)

1990-91	1991-92	1992-93	1993-94	1994-95	1995-96	1996-97	1997-98		CITI révision 3		
..	..	..	4 151	4 457	4 813	5 195	5 313	Com. de gros, de detail ; réparation etc.	G	50_52	
..	..	..	732	786	847	879	903	Com., entr., répar. d'auto. et moto ; carburants		50	
..	..	..	278	304	330	346	330	Commerce de véhicules automobiles			501
..	..	..	166	183	203	208	220	Entretien et réparation de véhicules automobiles			502
..	..	..	146	156	163	169	183	Com. de pièces et access. de véhic. automobiles			503
..	..	..	12	13	17	20	22	Com., entret. et répar. de motocycles et access.			504
..	..	..	130	130	134	135	149	Com. de détail de carburants pour automobiles			505
..	..	..	2 070	2 244	2 417	2 592	2 574	Com. de gros et act. inter., sauf auto. et moto		51	
..	..	..	..	..	..	..	..	Activités d'intermédiaries de commerce de gros			511
..	..	..	464	487	520	554	562	Com. gros: prod. agr. bruts, anim. viv., prod. alim.			512
..	..	..	420	460	467	482	478	Com. gros: biens de consom., sauf alimentaires			513
..	..	..	453	489	536	597	557	Com. gros: prod. inter. non agr., déchets, débris			514
..	..	..	618	689	774	823	818	Com. gros: machines, équipments et fournitures			515
..	..	..	115	119	120	136	159	Autres commerces de gros			519
..	..	..	1 349	1 427	1 549	1 724	1 835	Com. détail, sauf auto, moto; répar. art. pers., dom.		52	
..	..	..	578	639	691	802	843	Commerce de détail non spécialisé en magasin			521
..	..	..	140	128	125	130	132	Com. détail: prod. aliment. en mag. spécialisés			522
..	..	..	597	623	696	758	823	Autre com. détail: prod. neufs en mag. spécialisés			523
..	..	..	5	4	3	5	5	Commerces de détail de biens d'occasion			524
..	..	..	..	..	..	..	..	Commerces de détail autres qu'en magasin			525
..	..	..	29	32	33	29	33	Réparation d'articles personnels et domestiques			526
..	..	..	721	794	884	912	944	Hotels et restaurants	H	55	
..	..	..	279	290	318	317	313	Hôtels; camping; autres équip. courts séjours			551
..	..	..	442	504	566	595	631	Restaurants, cafés et cantines			552
..	..	..	2 065	2 168	2 358	2 545	2 527	Transport, entreposage, communications	I	60_64	
..	..	..	504	559	598	619	610	Transports terrestres; transports par conduites		60	
..	..	..	64	72	83	86	90	Transports par chemin de fer			601
..	..	..	440	487	515	532	520	Autres transports terrestres			602
..	..	..	..	..	..	..	..	Transports par conduites			603
..	..	..	6	7	7	7	6	Transports par eau		61	
..	..	..	..	..	..	..	..	Transports maritimes et côtiers			611
..	..	..	6	7	7	7	6	Transports par voies navigables intérieures			612
..	..	..	329	361	386	395	405	Transports aériens		62	
..	..	..	327	358	384	392	401	Transports aériens réguliers			621
..	..	..	2	3	2	2	4	Transports aériens non réguliers			622
..	..	..	540	582	624	677	638	Act. annex. et aux. des transp.; act. des ag. de voy.		63	
..	..	..	685	660	742	848	868	Postes et télécommunications		64	
..	..	..	248	257	292	323	345	Activités de poste et de courrier			641
..	..	..	437	402	450	525	523	Télécommunications			642
..	..	..	1 516	1 668	1 799	1 808	1 744	Intermédiation financière	J	65_67	
..	..	..	1 006	1 097	1 205	1 188	1 110	Interm. financ., sauf assur. et caisses de retraite		65	
..	..	..	865	930	967	983	1 017	Intermédiation monétaire			651
..	..	..	141	168	239	205	93	Autres intermédiations financières			659
..	..	..	317	352	382	393	394	Act. d'assur., caisses de retr., sauf séc. soc. oblig.		66	
..	..	..	193	218	212	227	240	Act. auxiliaires de l'intermédiation financière		67	
..	..	..	110	126	122	142	168	Act. aux. de l'interm. fin., sauf assur., et retraite			671
..	..	..	83	93	90	85	72	Act. aux. des assurances et caisses de retraite			672
..	..	..	2 375	2 618	2 939	3 221	3 522	Immobilier, locations, services aux entreprises	K	70_74	
..	..	..	140	155	164	157	148	Activités immobilières		70	
..	..	..	35	49	57	55	48	Act. immobilières sur biens propres ou loués			701
..	..	..	105	105	107	102	100	Activités immobilières pour comptes de tiers			702
..	..	..	107	124	134	147	141	Locat. mach., équip. sans opér., biens pers. et dom.		71	
..	..	..	41	49	55	63	55	Location de matériel de transport			711
..	..	..	54	63	66	71	71	Location de machines ét equipements nca			712
..	..	..	11	12	13	13	15	Loc. d'articles personnels et domestiques nca			713
..	..	..	207	229	286	328	407	Activités informatiques et activités rattachées		72	
..	..	..	102	127	180	213	293	Conseil en matériel informatique			721
..	..	..	..	..	..	..	..	Production de logiciels			722
..	..	..	58	54	61	80	88	Traitement de données informatiques			723
..	..	..	43	43	36	15	10	Activités de banques de données			724
..	..	..	5	5	9	20	15	Entr., répar. mach. de bureau et compt. et inform.			725
..	..	..	..	..	..	..	..	Autres activités rattachées à l'informatique			729
..	..	..	215	225	212	212	222	Recherche-développement		73	
..	..	..	215	225	212	212	222	Rech. et dév. expér. en sc. phys., natur. et ingén.			731
..	..	..	..	..	..	..	..	Rech. et dév. expér. en sciences soc. et humaines			732
..	..	..	1 707	1 886	2 144	2 377	2 604	Autres activités de services aux entreprises		74	
..	..	..	759	808	910	1 042	1 124	Act. jur., compt., audit.; cons. fisc.; ét. march., soud.			741
..	..	..	290	335	405	450	453	Act. d'achitect., d'ingén.; autres act. techniques			742
..	..	..	144	153	162	180	185	Publicité			743
..	..	..	515	590	667	704	841	Activités de services aux entreprises nca.			749
..	..	..	2 765	2 409	3 142	3 350	3 580	Education	M	80	
..	..	..	940	965	1 078	1 139	1 286	Activités d'enseignement primaire			801
..	..	..	746	757	802	894	912	Activités d'enseignement secondaire			802
..	..	..	902	936	997	1 042	1 109	Activités d'enseignement supérieur			803
..	..	..	178	225	264	275	272	Act. de formation perman.; autres act. d'enseign.			809
..	..	..	2 425	2 482	2 642	2 867	2 933	Santé et action sociale	N	85	
..	..	..	1 943	1 972	2 067	2 253	2 249	Activités de santé humaine			851
..	..	..	44	49	55	51	54	Activités vétérinaires			852
..	..	..	438	461	521	563	630	Activités d'action sociale			853
..	..	..	818	902	1 021	1 042	1 091	Autres activités de serv. coll., soc. et personnels	O	90_93	
..	..	..	50	55	75	81	88	Assain. et enlèv. des ordures; voierie, act. simil.		90	
..	..	..	169	177	177	157	171	Activités associatives diverses		91	
..	..	..	24	22	14	11	15	Activités de syndicats de salariés			912
..	..	..	492	555	646	673	689	Activités récréatives, culturelles et sportives		92	
..	..	..	238	258	258	260	250	Act. cinémat., de radio, télé.; autres act. de spect.			921
..	..	..	..	..	..	..	..	Activités d'agences de presse			922
..	..	..	108	120	146	153	179	Act. des biblio., arch., musées, jardins bot. et zool.			923
..	..	..	146	178	242	260	260	Activités sportives et autres activités récréatives			924
..	..	..	107	115	122	130	142	Autres activités de services		93	

Note: Break in series between 1995/96 and 1996/97.

NEW ZEALAND

Table NZ.S4 HOURS WORKED BY EMPLOYEES - HEURES OUVRÉES, OUVRIERS

Thousands

ISIC revision 3				1990-91	1991-92	1992-93	1993-94	1994-95	1995-96	1996-97	1997-98
G	50_52		Wholesale & retail trade; repairs; etc.	..	..	..	310 557	326 562	341 694	356 940	357 750
	50		Sale, maint.,repair vehic/cycle, person., h/hold gd.	..	..	..	58 882	61 426	64 387	64 679	65 401
		501	Sale of motor cycle	..	..	..	20 605	21 491	22 499	22 309	21 178
		502	Maintenance and repair of motor vehicles	..	..	..	14 085	14 709	15 826	15 820	16 246
		503	Sale of motor vehicle parts and accessories	..	..	..	10 962	11 671	11 980	12 282	12 573
		504	Sale, maint., repair motor cycle, parts & acces.	..	..	..	982	1 134	1 384	1 502	1 664
		505	Retail sale of automative fuel	..	..	..	12 248	12 421	12 699	12 766	13 741
	51		W/sale. trade & commission exc motor veh. & cycle	..	..	..	126 404	134 396	139 355	142 947	138 329
		511	Wholesale on a fee or contract basis	..	..	..	..	..	..	..	..
		512	W/sale agric. raw mat.,live animal,food, bev., tobac.	..	..	..	30 392	31 696	32 772	33 460	35 975
		513	Wholesale of household goods	..	..	..	26 915	28 972	27 690	26 778	25 220
		514	W/sale. of non-agric. intermed. prod., waste, scrap	..	..	..	28 689	30 171	31 989	35 010	32 266
		515	Wholesale of machinery, equipment & supplies	..	..	..	33 175	36 283	39 840	40 133	39 014
		519	Other wholesale	..	..	..	7 234	7 275	7 064	7 567	8 854
	52		Retail trade exc. motor veh. & cyc.; repair h/hold gds.	..	..	..	125 271	130 741	137 952	149 314	154 020
		521	Non-specialised retail trade in stores	..	..	..	57 258	62 864	66 270	73 702	75 809
		522	Retail sale food, bev., tobac. in specialised stores	..	..	..	13 456	12 265	11 720	11 869	11 681
		523	Other retail trade of new goods specialised stores	..	..	..	51 880	52 785	57 217	61 208	63 989
		524	Retail sale of second-hand goods in stores	..	..	..	457	357	285	428	330
		525	Retail trade not in stores	..	..	..	..	..	..	..	..
		526	Repair of personal and household goods	..	..	..	2 220	2 469	2 461	2 107	2 210
H	55		Hotels and restaurants	..	..	..	66 301	72 029	78 172	78 611	79 736
		551	Hotels; camping ; other short-stay accommodation	..	..	..	24 500	24 725	26 062	25 427	24 635
		552	Restaurants, bars and canteens	..	..	..	41 801	47 304	52 111	53 183	55 102
I	60_64		Transport, storage and communications	..	..	..	124 006	128 644	136 461	141 612	138 067
	60		Land transport, transport via pipelines	..	..	..	38 111	41 771	43 279	43 517	41 988
		601	Transport via railways	..	..	..	3 555	3 895	4 117	4 134	4 174
		602	Other land transport	..	..	..	34 556	37 875	39 161	39 379	37 813
		603	Transport via pipelines	..	..	..	..	..	..	..	..
	61		Water transport	..	..	..	450	505	534	470	374
		611	Sea and coastal water transport	..	..	..	..	..	..	..	..
		612	Inland water transport	..	..	..	450	505	534	470	374
	62		Air transport	..	..	..	16 614	18 091	18 412	17 994	17 269
		621	Scheduled air transport	..	..	..	16 460	17 854	18 256	17 814	16 984
		622	Non-scheduled air transport	..	..	..	155	237	156	180	285
	63		Support. & aux. transport act., travel agencies activ.	..	..	..	29 996	31 608	33 310	35 132	34 129
	64		Post and telecommunications	..	..	..	38 835	36 669	40 925	44 499	44 307
		641	Post and courier activities	..	..	..	17 111	17 732	20 198	21 315	22 580
		642	Telecommunications	..	..	..	21 724	18 936	20 727	23 184	21 728
J	65_67		Financial intermediation	..	..	..	81 329	84 718	86 907	83 676	77 963
	65		Financial intermediation, exc. insur., pension fund	..	..	..	55 372	57 459	59 639	56 883	51 926
		651	Monetary intermediation	..	..	..	49 720	51 515	50 566	49 700	48 755
		659	Other financial intermediation	..	..	..	5 652	5 944	9 073	7 183	3 171
	66		Insurance pension fund, exc. compul. social sec.	..	..	..	16 693	17 489	18 104	17 442	16 734
	67		Activities auxiliary to financial intermediation	..	..	..	9 264	9 770	9 164	9 351	9 303
		671	Act. auxil to fin. intermed., exc. insur. & pens.fund	..	..	..	5 270	5 517	5 061	5 514	6 239
		672	Act. auxiliary to insurance and pension funding	..	..	..	3 994	4 253	4 103	3 837	3 063
K	70_74		Real estate, renting and business activities	..	..	..	136 671	147 816	158 460	166 838	178 676
	70		Real estate activities	..	..	..	7 974	8 418	8 621	7 761	7 361
		701	Real estate activ. with own or leased property	..	..	..	1 719	2 347	2 675	2 320	2 070
		702	Real estate activities on a fee or contract basis	..	..	..	6 255	6 071	5 946	5 441	5 291
	71		Renting of mach. without operator; rent. of pers. gds.	..	..	..	7 678	8 726	9 062	9 548	8 921
		711	Renting of transport equipment	..	..	..	2 767	3 101	3 403	3 631	3 064
		712	Renting of other machinery and equipment	..	..	..	3 789	4 360	4 325	4 647	4 463
		713	Renting of personal and household goods nec	..	..	..	1 122	1 265	1 334	1 270	1 394
	72		Computer and related activities	..	..	..	9 171	9 975	11 752	13 518	16 083
		721	Hardware consultancy	..	..	..	4 468	5 440	7 159	8 222	10 725
		722	Software consultancy and supply	..	..	..	..	..	..	..	..
		723	Data processing	..	..	..	2 454	2 251	2 583	3 656	4 135
		724	Data base activities	..	..	..	2 038	2 023	1 478	657	490
		725	Mainten.; repair of office, account. & comp. mach.	..	..	..	210	262	532	983	723
		729	Other computer related activities	..	..	..	..	..	..	..	..
	73		Research and development	..	..	..	10 929	11 216	10 235	9 855	9 909
		731	R & D experim. on natural sciences & engineering	..	..	..	10 929	11 216	10 235	9 855	9 909
		732	R & D experim. on social sciences & humanities	..	..	..	..	..	..	..	..
	74		Other business activities	..	..	..	100 920	109 481	118 790	126 155	136 402
		741	Legal, acc., book-keep., audit act.; tax consult., etc	..	..	..	41 332	42 590	45 228	49 875	52 480
		742	Architectural, engineering, & other techn. activities	..	..	..	15 352	17 283	19 945	21 705	21 052
		743	Advertising	..	..	..	6 711	7 034	7 522	8 002	7 509
		749	Business activities nec	..	..	..	37 524	42 574	46 095	46 573	55 360
M	80		Education	..	..	..	166 643	170 123	178 025	181 024	184 277
		801	Primary education	..	..	..	64 499	65 830	69 092	69 848	73 125
		802	Secondary education	..	..	..	42 835	43 022	44 320	44 839	45 121
		803	Higher education	..	..	..	46 549	46 512	48 079	47 997	49 383
		809	Adult and other education	..	..	..	12 761	14 759	16 534	18 341	16 648
N	85		Health and social work	..	..	..	155 155	153 394	157 627	164 161	164 315
		851	Human health activities	..	..	..	116 125	112 607	112 514	117 006	113 244
		852	Veterinary activities	..	..	..	2 750	3 089	3 400	3 163	3 234
		853	Social work activities	..	..	..	36 280	37 699	41 714	43 992	47 837
O	90_93		Other commun., soc. & pers. services activities	..	..	..	53 824	58 774	65 314	64 926	66 530
	90		Sewage & refuse disposal, sanitation, etc.	..	..	..	3 451	3 769	4 914	5 091	5 352
	91		Activities of member organisations nec	..	..	..	10 249	10 774	10 488	9 120	9 786
		912	Activities of trade unions	..	..	..	1 281	1 155	678	543	755
	92		Recreational, cultural and sporting activities	..	..	..	30 190	33 845	39 206	39 757	39 841
		921	Motion picture, radio, TV, other entertainment act.	..	..	..	12 731	13 424	13 259	12 961	12 151
		922	News agency activities	..	..	..	..	..	..	..	..
		923	Library, archives, museums, other cultural activ.	..	..	..	6 763	7 501	8 974	9 011	10 475
		924	Sporting and other recreational activities	..	..	..	10 696	12 920	16 973	17 785	17 215
	93		Other service activities	..	..	..	9 934	10 386	10 707	10 958	11 551

Table NZ.S5 **ÉTABLISSEMENTS - ESTABLISHMENTS**

Unités

1990-91	1991-92	1992-93	1993-94	1994-95	1995-96	1996-97	1997-98		CITI révision 3		
..	42 362	42 425	49 771	53 747	52 902	53 468	55 092	Com. de gros, de detail ; réparation etc.	G	50_52	
..	..	..	10 159	10 727	10 963	11 417	11 912	Com., entr., répar. d'auto. et moto ; carburants		50	
..	..	..	1 617	1 696	1 818	1 951	2 128	Commerce de véhicules automobiles			501
..	..	..	5 207	5 501	5 618	5 832	6 099	Entretien et réparation de véhicules automobiles			502
..	..	..	1 634	1 758	1 752	1 816	1 829	Com. de pièces et access. de véhic. automobiles			503
..	..	..	163	189	207	221	246	Com., entret. et répar. de motocycles et access.			504
..	..	..	1 538	1 583	1 568	1 597	1 610	Com. de détail de carburants pour automobiles			505
..	..	..	14 554	16 274	16 611	16 915	17 454	Com. de gros et act. inter., sauf auto. et moto		51	
..	..	..	..	..	..	..	..	Activités d'intermédiaries de commerce de gros			511
..	..	..	2 922	3 231	3 275	3 413	3 486	Com. gros: prod. agr. bruts, anim. viv., prod. alim.			512
..	..	..	2 613	2 930	3 035	3 063	3 156	Com. gros: biens de consom., sauf alimentaires			513
..	..	..	2 297	2 449	2 530	2 571	2 626	Com. gros: prod. inter. non agr., déchets, débris			514
..	..	..	3 679	4 200	4 329	4 534	4 674	Com. gros: machines, équipments et fournitures			515
..	..	..	3 043	3 464	3 442	3 334	3 512	Autres commerces de gros			519
..	..	..	25 058	26 746	25 328	25 136	25 726	Com. détail, sauf auto, moto; répar. art. pers., dom.		52	
..	..	..	622	650	645	652	649	Commerce de détail non spécialisé en magasin			521
..	..	..	6 061	6 251	5 798	5 851	5 860	Com. détail: prod. aliment. en mag. spécialisés			522
..	..	..	16 219	17 481	16 561	16 343	16 890	Autre com. détail: prod. neufs en mag. spécialisés			523
..	..	..	898	999	988	974	977	Commerces de détail de biens d'occasion			524
..	..	..	..	..	..	..	..	Commerces de détail autres qu'en magasin			525
..	..	..	1 258	1 365	1 336	1 316	1 350	Réparation d'articles personnels et domestiques			526
..	9 998	10 193	10 567	11 564	11 635	12 170	12 622	Hotels et restaurants	H	55	
..	..	..	2 538	2 777	3 008	3 209	3 396	Hôtels; camping; autres équip. courts séjours			551
..	..	..	8 029	8 787	8 627	8 961	9 226	Restaurants, cafés et cantines			552
..	..	..	12 946	14 353	14 862	15 344	16 037	Transport, entreposage, communications	I	60_64	
..	7 748	7 564	7 221	7 877	8 065	8 134	8 305	Transports terrestres; transports par conduites		60	
..	..	..	205	206	184	184	184	Transports par chemin de fer			601
..	..	..	7 015	7 670	7 880	7 947	8 117	Autres transports terrestres			602
..	..	..	1	1	1	3	4	Transports par conduites			603
..	..	..	108	102	102	175	212	Transports par eau		61	
..	..	..	71	67	70	139	166	Transports maritimes et côtiers			611
..	..	..	37	35	32	36	46	Transports par voies navigables intérieures			612
..	..	..	475	532	534	528	537	Transports aériens		62	
..	..	..	250	257	243	238	222	Transports aériens réguliers			621
..	..	..	225	275	291	290	315	Transports aériens non réguliers			622
..	..	..	2 453	2 726	2 722	2 764	2 913	Act. annex. et aux. des transp.; act. des ag. de voy.		63	
..	..	..	2 689	3 116	3 439	3 743	4 070	Postes et télécommunications		64	
..	..	..	2 177	2 555	2 841	3 063	3 298	Activités de poste et de courrier			641
..	..	..	512	561	598	680	772	Télécommunications			642
..	7 810	7 500	13 470	13 670	13 232	13 451	13 435	Intermédiation financière	J	65_67	
..	..	..	7 840	8 455	8 242	8 337	8 232	Interm. financ., sauf assur. et caisses de retraite		65	
..	..	..	2 015	1 989	1 809	1 728	1 672	Intermédiation monétaire			651
..	..	..	5 825	6 466	6 433	6 609	6 560	Autres intermédiations financières			659
..	..	..	2 880	2 486	2 305	2 278	2 150	Act. d'assur., caisses de retr., sauf séc. soc. oblig.		66	
..	..	..	2 750	2 729	2 685	2 836	3 053	Act. auxiliaires de l'intermédiation financière		67	
..	..	..	1 110	1 175	1 208	1 353	1 483	Act. aux. de l'interm. fin., sauf assur., et retraite			671
..	..	..	1 640	1 554	1 477	1 483	1 570	Act. aux. des assurances et caisses de retraite			672
..	24 846	25 816	30 939	37 153	40 906	44 209	49 786	Immobilier, locations, services aux entreprises	K	70_74	
..	..	..	6 964	9 316	10 140	11 145	12 161	Activités immobilières		70	
..	..	..	1 667	2 530	2 780	3 745	4 013	Act. immobilières sur biens propres ou loués			701
..	..	..	5 297	6 786	7 360	7 400	8 148	Activités immobilières pour comptes de tiers			702
..	..	..	2 137	2 462	2 597	2 755	3 046	Locat. mach., équip. sans opér., biens pers. et dom.		71	
..	..	..	719	874	932	965	1 096	Location de matériel de transport			711
..	..	..	923	1 067	1 157	1 268	1 403	Location de machines ét equipements nca			712
..	..	..	495	521	508	522	547	Loc. d'articles personnels et domestiques nca			713
..	..	..	2 166	2 698	3 405	3 873	4 773	Activités informatiques et activités rattachées		72	
..	..	..	1 848	2 351	3 040	3 494	4 333	Conseil en matériel informatique			721
..	..	..	..	..	..	..	..	Production de logiciels			722
..	..	..	205	196	194	178	176	Traitement de données informatiques			723
..	..	..	24	41	41	44	52	Activités de banques de données			724
..	..	..	89	110	130	157	212	Entr., répar. mach. de bureau et compt. et inform.			725
..	..	..	..	..	..	..	..	Autres activités rattachées à l'informatique			729
..	..	..	253	234	222	209	226	Recherche-développement		73	
..	..	..	..	..	..	..	..	Rech. et dév. expér. en sc. phys., natur. et ingén.			731
..	..	..	..	..	..	..	..	Rech. et dév. expér. en sciences soc. et humaines			732
..	..	..	19 419	22 443	24 542	26 227	29 580	Autres activités de services aux entreprises		74	
..	..	..	7 291	8 577	9 618	10 666	12 641	Act. jur., compt., audit.; cons. fisc.; ét. march., soud.			741
..	..	..	3 614	4 200	4 686	4 959	5 564	Act. d'achitect., d'ingén.; autres act. techniques			742
..	..	..	1 936	2 231	2 314	2 471	2 614	Publicité			743
..	..	..	6 578	7 435	7 924	8 131	8 761	Activités de services aux entreprises nca.			749
..	..	..	6 604	6 882	7 039	7 122	7 287	Education	M	80	
..	..	..	4 465	4 477	4 561	4 503	4 519	Activités d'enseignement primaire			801
..	..	..	379	374	381	395	407	Activités d'enseignement secondaire			802
..	..	..	104	112	119	115	130	Activités d'enseignement supérieur			803
..	..	..	1 656	1 919	1 978	2 109	2 231	Act. de formation perman.; autres act. d'enseign.			809
..	..	..	8 386	9 052	9 506	10 177	10 970	Santé et action sociale	N	85	
..	..	..	7 147	7 789	8 241	8 897	9 635	Activités de santé humaine			851
..	..	..	458	482	491	506	547	Activités vétérinaires			852
..	..	..	781	781	774	774	788	Activités d'action sociale			853
..	..	..	9 031	9 876	10 231	10 766	11 388	Autres activités de serv. coll., soc. et personnels	O	90_93	
..	..	..	691	727	721	779	826	Assain. et enlèv. des ordures; voierie, act. simil.		90	
..	..	..	..	..	..	..	..	Activités associatives diverses		91	
..	..	..	..	..	..	..	..	Activités de syndicats de salariés			912
..	..	..	5 098	5 616	5 990	6 155	6 569	Activités récréatives, culturelles et sportives		92	
..	..	..	1 925	2 234	2 518	2 692	2 922	Act. cinémat., de radio, télé.; autres act. de spect.			921
..	..	..	..	..	..	..	..	Activités d'agences de presse			922
..	..	..	438	463	546	574	611	Act. des biblio., arch., musées, jardins bot. et zool.			923
..	..	..	2 735	2 919	2 926	2 889	3 036	Activités sportives et autres activités récréatives			924
..	..	..	3 242	3 533	3 520	3 832	3 993	Autres activités de services		93	

Table NZ.S6 **ENTERPRISES - ENTREPRISES**

Units

ISIC revision 3				1990-91	1991-92	1992-93	1993-94	1994-95	1995-96	1996-97	1997-98
G	50_52		Wholesale & retail trade; repairs; etc.	43 447	45 346	45 414	42 464	46 259	45 405	45 401	46 873
	50		Sale, maint.,repair vehic/cycle, person., h/hold gd.	..	..	..	9 229	9 761	9 907	10 269	10 708
		501	Sale of motor cycle	..	..	..	1 498	1 577	1 689	1 791	1 949
		502	Maintenance and repair of motor vehicles	..	..	..	5 114	5 394	5 462	5 605	5 854
		503	Sale of motor vehicle parts and accessories	..	..	..	1 227	1 328	1 322	1 315	1 330
		504	Sale, maint., repair motor cycle, parts & acces.	..	..	..	161	187	206	218	240
		505	Retail sale of automative fuel	..	..	..	1 330	1 376	1 327	1 340	1 335
	51		W/sale. trade & commission exc motor veh. & cycle	..	..	..	11 871	13 506	13 836	13 771	14 272
		511	Wholesale on a fee or contract basis	..	..	..	..	..	..	..	..
		512	W/sale agric. raw mat.,live animal,food, bev., tobac.	..	..	..	2 321	2 602	2 653	2 689	2 774
		513	Wholesale of household goods	..	..	..	2 273	2 592	2 708	2 638	2 745
		514	W/sale. of non-agric. intermed. prod., waste, scrap	..	..	..	1 558	1 686	1 765	1 684	1 743
		515	Wholesale of machinery, equipment & supplies	..	..	..	2 928	3 385	3 485	3 624	3 718
		519	Other wholesale	..	..	..	2 847	3 292	3 285	3 136	3 292
	52		Retail trade exc. motor veh. & cyc.; repair h/hold gds.	..	..	..	21 528	23 166	21 834	24 246	24 798
		521	Non-specialised retail trade in stores	..	..	..	262	260	252	25	17
		522	Retail sale food, bev., tobac. in specialised stores	..	..	..	5 714	5 941	5 509	8 549	8 551
		523	Other retail trade of new goods specialised stores	..	..	..	13 616	14 814	13 952	9 491	9 430
		524	Retail sale of second-hand goods in stores	..	..	..	852	951	938	921	919
		525	Retail trade not in stores	..	..	..	..	..	..	4 048	4 624
		526	Repair of personal and household goods	..	..	..	1 154	1 259	1 234	1 212	1 257
H	55		Hotels and restaurants	6 764	7 014	7 204	9 900	10 909	10 906	11 234	11 656
		551	Hotels; camping ; other short-stay accommodation	..	..	..	2 388	2 619	2 842	2 887	3 067
		552	Restaurants, bars and canteens	..	..	..	7 550	8 324	8 114	8 347	8 589
I	60_64		Transport, storage and communications	..	..	..	10 818	12 218	12 834	13 195	13 827
	60		Land transport, transport via pipelines	10 457	11 384	11 330	6 703	7 338	7 540	7 596	7 745
		601	Transport via railways	..	..	..	3	4	1	1	1
		602	Other land transport	..	..	..	6 701	7 334	7 539	7 592	7 740
		603	Transport via pipelines	..	..	..	1	1	1	3	4
	61		Water transport	..	..	..	96	87	85	148	188
		611	Sea and coastal water transport	..	..	..	63	57	58	118	147
		612	Inland water transport	..	..	..	33	31	28	30	41
	62		Air transport	..	..	..	356	405	403	397	415
		621	Scheduled air transport	..	..	..	132	134	119	112	106
		622	Non-scheduled air transport	..	..	..	225	274	287	285	309
	63		Support. & aux. transport act., travel agencies activ.	..	..	..	1 762	2 041	2 143	2 038	2 152
	64		Post and telecommunications	..	..	..	1 948	2 389	2 702	3 016	3 327
		641	Post and courier activities	..	..	..	1 803	2 191	2 470	2 710	2 945
		642	Telecommunications	..	..	..	145	198	232	306	382
J	65_67		Financial intermediation	6 006	7 509	7 137	10 652	10 916	10 643	10 890	11 054
	65		Financial intermediation, exc. insur., pension fund	..	..	..	5 906	6 562	6 510	6 621	6 597
		651	Monetary intermediation	..	..	..	161	144	127	122	138
		659	Other financial intermediation	..	..	..	5 746	6 419	6 384	6 499	6 459
	66		Insurance pension fund, exc. compul. social sec.	..	..	..	2 337	1 968	1 782	1 770	1 718
	67		Activities auxiliary to financial intermediation	..	..	..	2 450	2 423	2 388	2 499	2 739
		671	Act. auxil to fin. intermed., exc. insur. & pens.fund	..	..	..	943	994	1 026	1 149	1 291
		672	Act. auxiliary to insurance and pension funding	..	..	..	1 509	1 431	1 364	1 350	1 448
K	70_74		Real estate, renting and business activities	21 331	23 900	24 907	28 842	34 946	38 688	67 642	76 916
	70		Real estate activities	..	..	..	6 515	8 868	9 672	36 239	41 041
		701	Real estate activ. with own or leased property	..	..	..	1 662	2 524	2 773	29 330	33 369
		702	Real estate activities on a fee or contract basis	..	..	..	4 854	6 344	6 899	6 909	7 672
	71		Renting of mach. without operator; rent. of pers. gds.	..	..	..	1 881	2 178	2 300	2 420	2 703
		711	Renting of transport equipment	..	..	..	618	757	807	833	962
		712	Renting of other machinery and equipment	..	..	..	824	952	1 040	1 124	1 261
		713	Renting of personal and household goods nec	..	..	..	439	469	453	463	480
	72		Computer and related activities	..	..	..	2 083	2 612	3 320	3 767	4 660
		721	Hardware consultancy	..	..	..	1 814	2 315	3 002	3 445	4 274
		722	Software consultancy and supply	..	..	..	..	..	..	..	..
		723	Data processing	..	..	..	177	169	168	148	147
		724	Data base activities	..	..	..	14	30	32	34	44
		725	Mainten.; repair of office, account. & comp. mach.	..	..	..	82	103	125	140	195
		729	Other computer related activities	..	..	..	..	..	..	..	..
	73		Research and development	..	..	..	126	118	110	103	122
		731	R & D experim. on natural sciences & engineering	..	..	..	..	..	..	..	..
		732	R & D experim. on social sciences & humanities	..	..	..	..	..	..	..	..
	74		Other business activities	..	..	..	18 269	21 219	23 344	20 131	22 817
		741	Legal, acc., book-keep., audit act.; tax consult., etc	..	..	..	6 898	8 142	9 215	10 145	12 090
		742	Architectural, engineering, & other techn. activities	..	..	..	3 343	3 907	4 399	4 602	5 188
		743	Advertising	..	..	..	1 897	2 187	2 271	2 405	2 548
		749	Business activities nec	..	..	..	6 156	7 008	7 484	7 519	8 135
M	80		Education	..	..	..	5 340	5 430	5 571	5 631	5 751
		801	Primary education	..	..	..	3 370	3 287	3 364	3 351	3 380
		802	Secondary education	..	..	..	375	369	376	355	351
		803	Higher education	..	..	..	69	78	73	71	76
		809	Adult and other education	..	..	..	1 535	1 705	1 769	1 854	1 944
N	85		Health and social work	..	..	..	7 438	8 074	8 504	10 160	11 060
		851	Human health activities	..	..	..	6 430	7 052	7 490	7 917	8 601
		852	Veterinary activities	..	..	..	358	373	386	406	445
		853	Social work activities	..	..	..	666	665	647	1 837	2 014
O	90_93		Other commun., soc. & pers. services activities	..	..	..	7 923	8 796	9 091	17 158	18 125
	90		Sewage & refuse disposal, sanitation, etc.	..	..	..	547	591	582	554	604
	91		Activities of member organisations nec	..	..	..	..	..	..	3 215	3 226
		912	Activities of trade unions	..	..	..	..	..	..	110	99
	92		Recreational, cultural and sporting activities	..	..	..	4 515	5 044	5 353	5 411	5 813
		921	Motion picture, radio, TV, other entertainment act.	..	..	..	1 772	2 086	2 380	2 533	2 771
		922	News agency activities	..	..	..	..	..	..	..	..
		923	Library, archives, museums, other cultural activ.	..	..	..	210	232	248	170	182
		924	Sporting and other recreational activities	..	..	..	2 533	2 726	2 725	2 708	2 860
	93		Other service activities	..	..	..	3 006	3 294	3 289	4 974	5 374

Note: Break in series between 1995/96 and 1996/97.

Table PL.S1 PRODUCTION - PRODUCTION

Millions de Zl (Prix courants)

1991	1992	1993	1994	1995	1996	1997	1998		CITI révision 3		
..	..	49 804	68 888	94 891	125 577	154 890	179 100	Com. de gros, de detail ; réparation etc.	G	**50_52**	
..	..	..	..	..	..	..	..	Com., entr., répar. d'auto. et moto ; carburants		**50**	
..	..	..	..	..	..	..	..	Commerce de véhicules automobiles			501
..	..	..	..	..	..	..	..	Entretien et réparation de véhicules automobiles			502
..	..	..	..	..	..	..	..	Com. de pièces et access. de véhic. automobiles			503
..	..	..	..	..	..	..	..	Com., entret. et répar. de motocycles et access.			504
..	..	..	..	..	..	..	..	Com. de détail de carburants pour automobiles			505
..	..	..	..	..	..	..	..	Com. de gros et act. inter., sauf auto. et moto		**51**	
..	..	..	..	..	..	..	..	Activités d'intermédiaries de commerce de gros			511
..	..	..	..	..	..	..	..	Com. gros: prod. agr. bruts, anim. viv., prod. alim.			512
..	..	..	..	..	..	..	..	Com. gros: biens de consom., sauf alimentaires			513
..	..	..	..	..	..	..	..	Com. gros: prod. inter. non agr., déchets, débris			514
..	..	..	..	..	..	..	..	Com. gros: machines, équipments et fournitures			515
..	..	..	..	..	..	..	..	Autres commerces de gros			519
..	..	..	..	..	..	..	..	Com. détail, sauf auto, moto; répar. art. pers., dom.		**52**	
..	..	..	..	..	..	..	..	Commerce de détail non spécialisé en magasin			521
..	..	..	..	..	..	..	..	Com. détail: prod. aliment. en mag. spécialisés			522
..	..	..	..	..	..	..	..	Autre com. détail: prod. neufs en mag. spécialisés			523
..	..	..	..	..	..	..	..	Commerces de détail de biens d'occasion			524
..	..	..	..	..	..	..	..	Commerces de détail autres qu'en magasin			525
..	..	..	..	..	..	..	..	Réparation d'articles personnels et domestiques			526
..	..	2 187	3 888	5 174	7 409	9 407	11 331	Hotels et restaurants	H	**55**	
..	..	..	..	..	..	..	..	Hôtels; camping; autres équip. courts séjours			551
..	..	..	..	..	..	..	..	Restaurants, cafés et cantines			552
..	..	22 548	29 492	38 505	47 615	59 774	71 617	Transport, entreposage, communications	I	**60_64**	
..	..	..	..	..	..	..	..	Transports terrestres; transports par conduites		**60**	
..	..	..	..	..	..	..	..	Transports par chemin de fer			601
..	..	..	..	..	..	..	..	Autres transports terrestres			602
..	..	..	..	..	..	..	..	Transports par conduites			603
..	..	..	..	..	..	..	..	Transports par eau		**61**	
..	..	..	..	..	..	..	..	Transports maritimes et côtiers			611
..	..	..	..	..	..	..	..	Transports par voies navigables intérieures			612
..	..	..	..	..	..	..	..	Transports aériens		**62**	
..	..	..	..	..	..	..	..	Transports aériens réguliers			621
..	..	..	..	..	..	..	..	Transports aériens non réguliers			622
..	..	..	..	..	..	..	..	Act. annex. et aux. des transp.; act. des ag. de voy.		**63**	
..	..	..	..	..	..	..	..	Postes et télécommunications		**64**	
..	..	..	..	..	..	..	..	Activités de poste et de courrier			641
..	..	..	..	..	..	..	..	Télécommunications			642
..	..	3 328	4 524	5 774	10 013	16 965	20 691	Intermédiation financière	J	**65_67**	
..	..	..	..	..	..	..	..	Interm. financ., sauf assur. et caisses de retraite		**65**	
..	..	..	..	..	..	..	..	Intermédiation monétaire			651
..	..	..	..	..	..	..	..	Autres intermédiations financières			659
..	..	..	..	..	..	..	..	Act. d'assur., caisses de retr., sauf séc. soc. oblig.		**66**	
..	..	..	..	..	..	..	..	Act. auxiliaires de l'intermédiation financière		**67**	
..	..	..	..	..	..	..	..	Act. aux. de l'interm. fin., sauf assur., et retraite			671
..	..	..	..	..	..	..	..	Act. aux. des assurances et caisses de retraite			672
..	..	19 695	28 350	39 860	51 905	72 706	99 182	Immobilier, locations, services aux entreprises	K	**70_74**	
..	..	..	..	..	..	..	..	Activités immobilières		**70**	
..	..	..	..	..	..	..	..	Act. immobilières sur biens propres ou loués			701
..	..	..	..	..	..	..	..	Activités immobilières pour comptes de tiers			702
..	..	..	..	..	..	..	..	Locat. mach., équip. sans opér., biens pers. et dom.		**71**	
..	..	..	..	..	..	..	..	Location de matériel de transport			711
..	..	..	..	..	..	..	..	Location de machines ét equipements nca			712
..	..	..	..	..	..	..	..	Loc. d'articles personnels et domestiques nca			713
..	..	..	..	..	..	..	..	Activités informatiques et activités rattachées		**72**	
..	..	..	..	..	..	..	..	Conseil en matériel informatique			721
..	..	..	..	..	..	..	..	Production de logiciels			722
..	..	..	..	..	..	..	..	Traitement de données informatiques			723
..	..	..	..	..	..	..	..	Activités de banques de données			724
..	..	..	..	..	..	..	..	Entr., répar. mach. de bureau et compt. et inform.			725
..	..	..	..	..	..	..	..	Autres activités rattachées à l'informatique			729
..	..	..	..	..	..	..	..	Recherche-développement		**73**	
..	..	..	..	..	..	..	..	Rech. et dév. expér. en sc. phys., natur. et ingén.			731
..	..	..	..	..	..	..	..	Rech. et dév. expér. en sciences soc. et humaines			732
..	..	..	..	..	..	..	..	Autres activités de services aux entreprises		**74**	
..	..	..	..	..	..	..	..	Act. jur., compt., audit.; cons. fisc.; ét. march., soud.			741
..	..	..	..	..	..	..	..	Act. d'achitect., d'ingén.; autres act. techniques			742
..	..	..	..	..	..	..	..	Publicité			743
..	..	..	..	..	..	..	..	Activités de services aux entreprises nca.			749
..	..	6 439	9 642	13 835	17 857	22 241	25 779	Education	M	**80**	
..	..	..	..	..	..	..	..	Activités d'enseignement primaire			801
..	..	..	..	..	..	..	..	Activités d'enseignement secondaire			802
..	..	..	..	..	..	..	..	Activités d'enseignement supérieur			803
..	..	..	..	..	..	..	..	Act. de formation perman.; autres act. d'enseign.			809
..	..	8 371	11 533	15 570	21 371	24 174	28 395	Santé et action sociale	N	**85**	
..	..	..	..	..	..	..	..	Activités de santé humaine			851
..	..	..	..	..	..	..	..	Activités vétérinaires			852
..	..	..	..	..	..	..	..	Activités d'action sociale			853
..	..	13 765	14 842	19 086	25 279	29 223	35 751	Autres activités de serv. coll., soc. et personnels	O	**90_93**	
..	..	..	..	..	..	..	..	Assain. et enlèv. des ordures; voierie, act. simil.		**90**	
..	..	..	..	..	..	..	..	Activités associatives diverses		**91**	
..	..	..	..	..	..	..	..	Activités de syndicats de salariés			912
..	..	..	..	..	..	..	..	Activités récréatives, culturelles et sportives		**92**	
..	..	..	..	..	..	..	..	Act. cinémat., de radio, télé.; autres act. de spect.			921
..	..	..	..	..	..	..	..	Activités d'agences de presse			922
..	..	..	..	..	..	..	..	Act. des biblio., arch., musées, jardins bot. et zool.			923
..	..	..	..	..	..	..	..	Activités sportives et autres activités récréatives			924
..	..	..	..	..	..	..	..	Autres activités de services		**93**	

POLAND

Table PL.S2 VALUE ADDED - VALEUR AJOUTÉE

Millions of Zl (Current Prices)

ISIC revision 3				1991	1992	1993	1994	1995	1996	1997	1998
G	50_52		Wholesale & retail trade; repairs; etc.	..	15 112	23 330	39 307	53 576	70 477	86 767	100 339
	50		Sale, maint.,repair vehic/cycle, person., h/hold gd.	..	..	..	..	..	..	..	..
		501	Sale of motor cycle	..	..	..	..	..	..	..	..
		502	Maintenance and repair of motor vehicles	..	..	..	..	..	..	..	..
		503	Sale of motor vehicle parts and accessories	..	..	..	..	..	..	..	..
		504	Sale, maint., repair motor cycle, parts & acces.	..	..	..	..	..	..	..	..
		505	Retail sale of automative fuel	..	..	..	..	..	..	..	..
	51		W/sale. trade & commission exc motor veh. & cycle	..	..	..	..	..	..	..	..
		511	Wholesale on a fee or contract basis	..	..	..	..	..	..	..	..
		512	W/sale agric. raw mat.,live animal,food, bev., tobac.	..	..	..	..	..	..	..	..
		513	Wholesale of household goods	..	..	..	..	..	..	..	..
		514	W/sale. of non-agric. intermed. prod., waste, scrap	..	..	..	..	..	..	..	..
		515	Wholesale of machinery, equipment & supplies	..	..	..	..	..	..	..	..
		519	Other wholesale	..	..	..	..	..	..	..	..
	52		Retail trade exc. motor veh. & cyc.; repair h/hold gds.	..	..	..	..	..	..	..	..
		521	Non-specialised retail trade in stores	..	..	..	..	..	..	..	..
		522	Retail sale food, bev., tobac. in specialised stores	..	..	..	..	..	..	..	..
		523	Other retail trade of new goods specialised stores	..	..	..	..	..	..	..	..
		524	Retail sale of second-hand goods in stores	..	..	..	..	..	..	..	..
		525	Retail trade not in stores	..	..	..	..	..	..	..	..
		526	Repair of personal and household goods	..	..	..	..	..	..	..	..
H	55		Hotels and restaurants	..	832	816	1 830	2 406	3 427	4 366	5 504
		551	Hotels; camping ; other short-stay accommodation	..	..	..	..	..	..	..	..
		552	Restaurants, bars and canteens	..	..	..	..	..	..	..	..
I	60_64		Transport, storage and communications	..	7 101	9 496	14 412	17 589	21 812	26 679	30 949
	60		Land transport, transport via pipelines	..	..	..	..	..	..	..	..
		601	Transport via railways	..	..	..	..	..	..	..	..
		602	Other land transport	..	..	..	..	..	..	..	..
		603	Transport via pipelines	..	..	..	..	..	..	..	..
	61		Water transport	..	..	..	..	..	..	..	..
		611	Sea and coastal water transport	..	..	..	..	..	..	..	..
		612	Inland water transport	..	..	..	..	..	..	..	..
	62		Air transport	..	..	..	..	..	..	..	..
		621	Scheduled air transport	..	..	..	..	..	..	..	..
		622	Non-scheduled air transport	..	..	..	..	..	..	..	..
	63		Support. & aux. transport act., travel agencies activ.	..	..	..	..	..	..	..	..
	64		Post and telecommunications	..	..	..	..	..	..	..	..
		641	Post and courier activities	..	..	..	..	..	..	..	..
		642	Telecommunications	..	..	..	..	..	..	..	..
J	65_67		Financial intermediation	..	598	902	2 125	2 625	3 512	5 708	7 717
	65		Financial intermediation, exc. insur., pension fund	..	..	..	..	..	..	..	..
		651	Monetary intermediation	..	..	..	..	..	..	..	..
		659	Other financial intermediation	..	..	..	..	..	..	..	..
	66		Insurance pension fund, exc. compul. social sec.	..	..	..	..	..	..	..	..
	67		Activities auxiliary to financial intermediation	..	..	..	..	..	..	..	..
		671	Act. auxil to fin. intermed., exc. insur. & pens.fund	..	..	..	..	..	..	..	..
		672	Act. auxiliary to insurance and pension funding	..	..	..	..	..	..	..	..
K	70_74		Real estate, renting and business activities	..	7 416	9 072	15 831	21 978	29 190	40 037	56 655
	70		Real estate activities	..	..	..	..	..	..	..	..
		701	Real estate activ. with own or leased property	..	..	..	..	..	..	..	..
		702	Real estate activities on a fee or contract basis	..	..	..	..	..	..	..	..
	71		Renting of mach. without operator; rent. of pers. gds.	..	..	..	..	..	..	..	..
		711	Renting of transport equipment	..	..	..	..	..	..	..	..
		712	Renting of other machinery and equipment	..	..	..	..	..	..	..	..
		713	Renting of personal and household goods nec	..	..	..	..	..	..	..	..
	72		Computer and related activities	..	..	..	..	..	..	..	..
		721	Hardware consultancy	..	..	..	..	..	..	..	..
		722	Software consultancy and supply	..	..	..	..	..	..	..	..
		723	Data processing	..	..	..	..	..	..	..	..
		724	Data base activities	..	..	..	..	..	..	..	..
		725	Mainten.; repair of office, account. & comp. mach.	..	..	..	..	..	..	..	..
		729	Other computer related activities	..	..	..	..	..	..	..	..
	73		Research and development	..	..	..	..	..	..	..	..
		731	R & D experim. on natural sciences & engineering	..	..	..	..	..	..	..	..
		732	R & D experim. on social sciences & humanities	..	..	..	..	..	..	..	..
	74		Other business activities	..	..	..	..	..	..	..	..
		741	Legal, acc., book-keep., audit act.; tax consult., etc	..	..	..	..	..	..	..	..
		742	Architectural, engineering, & other techn. activities	..	..	..	..	..	..	..	..
		743	Advertising	..	..	..	..	..	..	..	..
		749	Business activities nec	..	..	..	..	..	..	..	..
M	80		Education	..	..	5 000	7 655	10 914	13 861	17 678	20 556
		801	Primary education	..	..	..	..	..	..	..	..
		802	Secondary education	..	..	..	..	..	..	..	..
		803	Higher education	..	..	..	..	..	..	..	..
		809	Adult and other education	..	..	..	..	..	..	..	..
N	85		Health and social work	..	..	6 073	7 970	10 973	13 741	17 177	19 226
		851	Human health activities	..	..	..	..	..	..	..	..
		852	Veterinary activities	..	..	..	..	..	..	..	..
		853	Social work activities	..	..	..	..	..	..	..	..
O	90_93		Other commun., soc. & pers. services activities	..	..	9 105	8 010	10 437	13 548	15 576	19 239
	90		Sewage & refuse disposal, sanitation, etc.	..	..	..	..	..	..	..	..
	91		Activities of member organisations nec	..	..	..	..	..	..	..	..
		912	Activities of trade unions	..	..	..	..	..	..	..	..
	92		Recreational, cultural and sporting activities	..	..	..	..	..	..	..	..
		921	Motion picture, radio, TV, other entertainment act.	..	..	..	..	..	..	..	..
		922	News agency activities	..	..	..	..	..	..	..	..
		923	Library, archives, museums, other cultural activ.	..	..	..	..	..	..	..	..
		924	Sporting and other recreational activities	..	..	..	..	..	..	..	..
	93		Other service activities	..	..	..	..	..	..	..	..

TablePL.S3 INVESTISSEMENT - INVESTMENT

Milions de Zl (Prix courants)

1991	1992	1993	1994	1995	1996	1997	1998		CITI révision 3		
..	1 163	1 692	2 115	3 836	4 910	7 913	10 653	Com. de gros, de detail ; réparation etc.	G	50_52	
..	..	..	..	..	..	..	..	Com., entr., répar. d'auto. et moto ; carburants		50	
..	..	..	..	..	..	..	..	Commerce de véhicules automobiles			501
..	..	..	..	..	..	..	..	Entretien et réparation de véhicules automobiles			502
..	..	..	..	..	..	..	..	Com. de pièces et access. de véhic. automobiles			503
..	..	..	..	..	..	..	..	Com., entret. et répar. de motocycles et access.			504
..	..	..	..	..	..	..	..	Com. de détail de carburants pour automobiles			505
..	..	..	..	..	..	..	..	Com. de gros et act. inter., sauf auto. et moto		51	
..	..	..	..	..	..	..	..	Activités d'intermédiaries de commerce de gros			511
..	..	..	..	..	..	..	..	Com. gros: prod. agr. bruts, anim. viv., prod. alim.			512
..	..	..	..	..	..	..	..	Com. gros: biens de consom., sauf alimentaires			513
..	..	..	..	..	..	..	..	Com. gros: prod. inter. non agr., déchets, débris			514
..	..	..	..	..	..	..	..	Com. gros: machines, équipments et fournitures			515
..	..	..	..	..	..	..	..	Autres commerces de gros			519
..	..	..	..	..	..	..	..	Com. détail, sauf auto, moto; répar. art. pers., dom.		52	
..	..	..	..	..	..	..	..	Commerce de détail non spécialisé en magasin			521
..	..	..	..	..	..	..	..	Com. détail: prod. aliment. en mag. spécialisés			522
..	..	..	..	..	..	..	..	Autre com. détail: prod. neufs en mag. spécialisés			523
..	..	..	..	..	..	..	..	Commerces de détail de biens d'occasion			524
..	..	..	..	..	..	..	..	Commerces de détail autres qu'en magasin			525
..	..	..	..	..	..	..	..	Réparation d'articles personnels et domestiques			526
..	154	256	355	533	433	603	907	Hotels et restaurants	H	55	
..	..	..	..	..	..	..	..	Hôtels; camping; autres équip. courts séjours			551
..	..	..	..	..	..	..	..	Restaurants, cafés et cantines			552
..	1 747	2 512	3 268	5 077	7 849	11 405	13 200	Transport, entreposage, communications	I	60_64	
..	..	..	..	..	..	..	..	Transports terrestres; transports par conduites		60	
..	..	..	..	..	..	..	..	Transports par chemin de fer			601
..	..	..	..	..	..	..	..	Autres transports terrestres			602
..	..	..	..	..	..	..	..	Transports par conduites			603
..	..	..	..	..	..	..	..	Transports par eau		61	
..	..	..	..	..	..	..	..	Transports maritimes et côtiers			611
..	..	..	..	..	..	..	..	Transports par voies navigables intérieures			612
..	..	..	..	..	..	..	..	Transports aériens		62	
..	..	..	..	..	..	..	..	Transports aériens réguliers			621
..	..	..	..	..	..	..	..	Transports aériens non réguliers			622
..	..	..	..	..	..	..	..	Act. annex. et aux. des transp.; act. des ag. de voy.		63	
..	..	..	..	..	..	..	..	Postes et télécommunications		64	
..	..	..	..	..	..	..	..	Activités de poste et de courrier			641
..	..	..	..	..	..	..	..	Télécommunications			642
..	722	976	1 609	2 301	3 067	5 449	8 261	Intermédiation financière	J	65_67	
..	..	..	..	..	..	..	..	Interm. financ., sauf assur. et caisses de retraite		65	
..	..	..	..	..	..	..	..	Intermédiation monétaire			651
..	..	..	..	..	..	..	..	Autres intermédiations financières			659
..	..	..	..	..	..	..	..	Act. d'assur., caisses de retr., sauf séc. soc. oblig.		66	
..	..	..	..	..	..	..	..	Act. auxiliaires de l'intermédiation financière		67	
..	..	..	..	..	..	..	..	Act. aux. de l'interm. fin., sauf assur., et retraite			671
..	..	..	..	..	..	..	..	Act. aux. des assurances et caisses de retraite			672
..	4 862	4 660	5 044	6 065	7 243	10 391	14 750	Immobilier, locations, services aux entreprises	K	70_74	
..	..	..	..	..	..	..	..	Activités immobilières		70	
..	..	..	..	..	..	..	..	Act. immobilières sur biens propres ou loués			701
..	..	..	..	..	..	..	..	Activités immobilières pour comptes de tiers			702
..	..	..	..	..	..	..	..	Locat. mach., équip. sans opér., biens pers. et dom.		71	
..	..	..	..	..	..	..	..	Location de matériel de transport			711
..	..	..	..	..	..	..	..	Location de machines ét equipements nca			712
..	..	..	..	..	..	..	..	Loc. d'articles personnels et domestiques nca			713
..	..	..	..	..	..	..	..	Activités informatiques et activités rattachées		72	
..	..	..	..	..	..	..	..	Conseil en matériel informatique			721
..	..	..	..	..	..	..	..	Production de logiciels			722
..	..	..	..	..	..	..	..	Traitement de données informatiques			723
..	..	..	..	..	..	..	..	Activités de banques de données			724
..	..	..	..	..	..	..	..	Entr., répar. mach. de bureau et compt. et inform.			725
..	..	..	..	..	..	..	..	Autres activités rattachées à l'informatique			729
..	..	..	..	..	..	..	..	Recherche-développement		73	
..	..	..	..	..	..	..	..	Rech. et dév. expér. en sc. phys., natur. et ingén.			731
..	..	..	..	..	..	..	..	Rech. et dév. expér. en sciences soc. et humaines			732
..	..	..	..	..	..	..	..	Autres activités de services aux entreprises		74	
..	..	..	..	..	..	..	..	Act. jur., compt., audit.; cons. fisc.; ét. march., soud.			741
..	..	..	..	..	..	..	..	Act. d'achitect., d'ingén.; autres act. techniques			742
..	..	..	..	..	..	..	..	Publicité			743
..	..	..	..	..	..	..	..	Activités de services aux entreprises nca.			749
..	..	674	756	1 066	1 731	2 483	2 944	Education	M	80	
..	..	..	..	..	..	..	..	Activités d'enseignement primaire			801
..	..	..	..	..	..	..	..	Activités d'enseignement secondaire			802
..	..	..	..	..	..	..	..	Activités d'enseignement supérieur			803
..	..	..	..	..	..	..	..	Act. de formation perman.; autres act. d'enseign.			809
..	..	685	864	1 302	1 893	2 482	2 825	Santé et action sociale	N	85	
..	..	..	..	..	..	..	..	Activités de santé humaine			851
..	..	..	..	..	..	..	..	Activités vétérinaires			852
..	..	..	..	..	..	..	..	Activités d'action sociale			853
..	..	1 498	2 015	2 440	3 440	4 932	5 787	Autres activités de serv. coll., soc. et personnels	O	90_93	
..	..	..	..	..	..	..	..	Assain. et enlèv. des ordures; voierie, act. simil.		90	
..	..	..	..	..	..	..	..	Activités associatives diverses		91	
..	..	..	..	..	..	..	..	Activités de syndicats de salariés			912
..	..	..	..	..	..	..	..	Activités récréatives, culturelles et sportives		92	
..	..	..	..	..	..	..	..	Act. cinémat., de radio, télé.; autres act. de spect.			921
..	..	..	..	..	..	..	..	Activités d'agences de presse			922
..	..	..	..	..	..	..	..	Act. des biblio., arch., musées, jardins bot. et zool.			923
..	..	..	..	..	..	..	..	Activités sportives et autres activités récréatives			924
..	..	..	..	..	..	..	..	Autres activités de services		93	

POLAND

Table PL .S4 INVESTMENT IN MACHINERY & EQUIPMENT - DÉPENSES DE MACHINES ET D'ÉQUIPEMENT

Millioms of Zl (Current Prices)

ISIC revision 3				1991	1992	1993	1994	1995	1996	1997	1998
G	50_52		Wholesale & retail trade; repairs; etc.	..	..	550	605	1 446	1 775	2 692	3 412
	50		Sale, maint.,repair vehic/cycle, person., h/hold gd.	..	..	..	..	..	..	..	..
		501	Sale of motor cycle	..	..	..	..	..	..	..	..
		502	Maintenance and repair of motor vehicles	..	..	..	..	..	..	..	..
		503	Sale of motor vehicle parts and accessories	..	..	..	..	..	..	..	..
		504	Sale, maint., repair motor cycle, parts & acces.	..	..	..	..	..	..	..	..
		505	Retail sale of automative fuel	..	..	..	..	..	..	..	..
	51		W/sale. trade & commission exc motor veh. & cycle	..	..	..	..	..	..	..	..
		511	Wholesale on a fee or contract basis	..	..	..	..	..	..	..	..
		512	W/sale agric. raw mat.,live animal,food, bev., tobac.	..	..	..	..	..	..	..	..
		513	Wholesale of household goods	..	..	..	..	..	..	..	..
		514	W/sale. of non-agric. intermed. prod., waste, scrap	..	..	..	..	..	..	..	..
		515	Wholesale of machinery, equipment & supplies	..	..	..	..	..	..	..	..
		519	Other wholesale	..	..	..	..	..	..	..	..
	52		Retail trade exc. motor veh. & cyc.; repair h/hold gds.	..	..	..	..	..	..	..	..
		521	Non-specialised retail trade in stores	..	..	..	..	..	..	..	..
		522	Retail sale food, bev., tobac. in specialised stores	..	..	..	..	..	..	..	..
		523	Other retail trade of new goods specialised stores	..	..	..	..	..	..	..	..
		524	Retail sale of second-hand goods in stores	..	..	..	..	..	..	..	..
		525	Retail trade not in stores	..	..	..	..	..	..	..	..
		526	Repair of personal and household goods	..	..	..	..	..	..	..	..
H	55		Hotels and restaurants	..	..	66	74	141	155	183	266
		551	Hotels; camping ; other short-stay accommodation	..	..	..	..	..	..	..	..
		552	Restaurants, bars and canteens	..	..	..	..	..	..	..	..
I	60_64		Transport, storage and communications	..	..	971	1 212	1 903	2 495	4 015	5 735
	60		Land transport, transport via pipelines	..	..	..	..	..	..	..	..
		601	Transport via railways	..	..	..	..	..	..	..	..
		602	Other land transport	..	..	..	..	..	..	..	..
		603	Transport via pipelines	..	..	..	..	..	..	..	..
	61		Water transport	..	..	..	..	..	..	..	..
		611	Sea and coastal water transport	..	..	..	..	..	..	..	..
		612	Inland water transport	..	..	..	..	..	..	..	..
	62		Air transport	..	..	..	..	..	..	..	..
		621	Scheduled air transport	..	..	..	..	..	..	..	..
		622	Non-scheduled air transport	..	..	..	..	..	..	..	..
	63		Support. & aux. transport act., travel agencies activ.	..	..	..	..	..	..	..	..
	64		Post and telecommunications	..	..	..	..	..	..	..	..
		641	Post and courier activities	..	..	..	..	..	..	..	..
		642	Telecommunications	..	..	..	..	..	..	..	..
J	65_67		Financial intermediation	..	..	310	619	975	1 414	2 340	3 750
	65		Financial intermediation, exc. insur., pension fund	..	..	..	..	..	..	..	..
		651	Monetary intermediation	..	..	..	..	..	..	..	..
		659	Other financial intermediation	..	..	..	..	..	..	..	..
	66		Insurance pension fund, exc. compul. social sec.	..	..	..	..	..	..	..	..
	67		Activities auxiliary to financial intermediation	..	..	..	..	..	..	..	..
		671	Act. auxil to fin. intermed., exc. insur. & pens.fund	..	..	..	..	..	..	..	..
		672	Act. auxiliary to insurance and pension funding	..	..	..	..	..	..	..	..
K	70_74		Real estate, renting and business activities	..	..	432	378	712	904	1 275	1 872
	70		Real estate activities	..	..	..	..	..	..	..	..
		701	Real estate activ. with own or leased property	..	..	..	..	..	..	..	..
		702	Real estate activities on a fee or contract basis	..	..	..	..	..	..	..	..
	71		Renting of mach. without operator; rent. of pers. gds.	..	..	..	..	..	..	..	..
		711	Renting of transport equipment	..	..	..	..	..	..	..	..
		712	Renting of other machinery and equipment	..	..	..	..	..	..	..	..
		713	Renting of personal and household goods nec	..	..	..	..	..	..	..	..
	72		Computer and related activities	..	..	..	..	..	..	..	..
		721	Hardware consultancy	..	..	..	..	..	..	..	..
		722	Software consultancy and supply	..	..	..	..	..	..	..	..
		723	Data processing	..	..	..	..	..	..	..	..
		724	Data base activities	..	..	..	..	..	..	..	..
		725	Mainten.; repair of office, account. & comp. mach.	..	..	..	..	..	..	..	..
		729	Other computer related activities	..	..	..	..	..	..	..	..
	73		Research and development	..	..	..	..	..	..	..	..
		731	R & D experim. on natural sciences & engineering	..	..	..	..	..	..	..	..
		732	R & D experim. on social sciences & humanities	..	..	..	..	..	..	..	..
	74		Other business activities	..	..	..	..	..	..	..	..
		741	Legal, acc., book-keep., audit act.; tax consult., etc	..	..	..	..	..	..	..	..
		742	Architectural, engineering, & other techn. activities	..	..	..	..	..	..	..	..
		743	Advertising	..	..	..	..	..	..	..	..
		749	Business activities nec	..	..	..	..	..	..	..	..
M	80		Education	..	..	85	110	194	344	394	503
		801	Primary education	..	..	..	..	..	..	..	..
		802	Secondary education	..	..	..	..	..	..	..	..
		803	Higher education	..	..	..	..	..	..	..	..
		809	Adult and other education	..	..	..	..	..	..	..	..
N	85		Health and social work	..	..	231	290	531	749	1 028	1 149
		851	Human health activities	..	..	..	..	..	..	..	..
		852	Veterinary activities	..	..	..	..	..	..	..	..
		853	Social work activities	..	..	..	..	..	..	..	..
O	90_93		Other commun., soc. & pers. services activities	..	..	288	324	423	508	647	779
	90		Sewage & refuse disposal, sanitation, etc.	..	..	..	..	..	..	..	..
	91		Activities of member organisations nec	..	..	..	..	..	..	..	..
		912	Activities of trade unions	..	..	..	..	..	..	..	..
	92		Recreational, cultural and sporting activities	..	..	..	..	..	..	..	..
		921	Motion picture, radio, TV, other entertainment act.	..	..	..	..	..	..	..	..
		922	News agency activities	..	..	..	..	..	..	..	..
		923	Library, archives, museums, other cultural activ.	..	..	..	..	..	..	..	..
		924	Sporting and other recreational activities	..	..	..	..	..	..	..	..
	93		Other service activities	..	..	..	..	..	..	..	..

POLOGNE

Table PL .S5 EMPLOI, SALARIÉS - EMPLOYMENT, EMPLOYEES

Milliers

1991	1992	1993	1994	1995	1996	1997	1998			CITI révision 3	
..	1 042	999	996	1 079	1 120	1 245	1 321	Com. de gros, de detail ; réparation etc.	G	50_52	
..	..	74	90	90	114	123	141	Com., entr., répar. d'auto. et moto ; carburants		50	
..	..	..	..	..	..	..	..	Commerce de véhicules automobiles			501
..	..	..	..	..	..	..	..	Entretien et réparation de véhicules automobiles			502
..	..	..	..	..	..	..	..	Com. de pièces et access. de véhic. automobiles			503
..	..	..	..	..	..	..	..	Com., entret. et répar. de motocycles et access.			504
..	..	..	..	..	..	..	..	Com. de détail de carburants pour automobiles			505
..	..	380	392	477	480	525	585	Com. de gros et act. inter., sauf auto. et moto		51	
..	..	..	..	..	..	..	..	Activités d'intermédiaries de commerce de gros			511
..	..	..	..	..	..	..	..	Com. gros: prod. agr. bruts, anim. viv., prod. alim.			512
..	..	..	..	..	..	..	..	Com. gros: biens de consom., sauf alimentaires			513
..	..	..	..	..	..	..	..	Com. gros: prod. inter. non agr., déchets, débris			514
..	..	..	..	..	..	..	..	Com. gros: machines, équipments et fournitures			515
..	..	..	..	..	..	..	..	Autres commerces de gros			519
..	..	544	514	509	526	596	596	Com. détail, sauf auto, moto; répar. art. pers., dom.		52	
..	..	..	..	..	..	..	..	Commerce de détail non spécialisé en magasin			521
..	..	..	..	..	..	..	..	Com. détail: prod. aliment. en mag. spécialisés			522
..	..	..	..	..	..	..	..	Autre com. détail: prod. neufs en mag. spécialisés			523
..	..	..	..	..	..	..	..	Commerces de détail de biens d'occasion			524
..	..	..	..	..	..	..	..	Commerces de détail autres qu'en magasin			525
..	..	..	..	..	..	..	..	Réparation d'articles personnels et domestiques			526
..	102	105	123	128	134	139	149	Hotels et restaurants	H	55	
..	..	..	..	..	..	..	..	Hôtels; camping; autres équip. courts séjours			551
..	..	..	..	..	..	..	..	Restaurants, cafés et cantines			552
..	..	747	728	723	722	731	725	Transport, entreposage, communications	I	60_64	
..	795	479	473	465	457	459	447	Transports terrestres; transports par conduites		60	
..	..	..	..	..	..	..	..	Transports par chemin de fer			601
..	..	..	..	..	..	..	..	Autres transports terrestres			602
..	..	..	..	..	..	..	..	Transports par conduites			603
..	..	15	13	12	11	10	8	Transports par eau		61	
..	..	..	..	..	..	..	..	Transports maritimes et côtiers			611
..	..	..	..	..	..	..	..	Transports par voies navigables intérieures			612
..	..	5	5	..	5	5	5	Transports aériens		62	
..	..	..	..	..	..	..	..	Transports aériens réguliers			621
..	..	..	..	..	..	..	..	Transports aériens non réguliers			622
..	..	77	64	66	71	76	81	Act. annex. et aux. des transp.; act. des ag. de voy.		63	
..	..	171	173	176	179	181	185	Postes et télécommunications		64	
..	..	..	..	..	..	..	..	Activités de poste et de courrier			641
..	..	..	..	..	..	..	..	Télécommunications			642
..	186	210	225	237	249	260	271	Intermédiation financière	J	65_67	
..	..	149	160	165	173	182	186	Interm. financ., sauf assur. et caisses de retraite		65	
..	..	..	..	..	..	..	..	Intermédiation monétaire			651
..	..	..	..	..	..	..	..	Autres intermédiations financières			659
..	..	57	61	..	67	69	74	Act. d'assur., caisses de retr., sauf séc. soc. oblig.		66	
..	..	3	5	7	8	9	11	Act. auxiliaires de l'intermédiation financière		67	
..	..	..	..	..	..	..	..	Act. aux. de l'interm. fin., sauf assur., et retraite			671
..	..	..	..	..	..	..	..	Act. aux. des assurances et caisses de retraite			672
..	427	381	364	414	425	487	553	Immobilier, locations, services aux entreprises	K	70_74	
..	..	162	135	138	146	151	159	Activités immobilières		70	
..	..	..	..	..	..	..	..	Act. immobilières sur biens propres ou loués			701
..	..	..	..	..	..	..	..	Activités immobilières pour comptes de tiers			702
..	..	3	4	5	4	6	7	Locat. mach., équip. sans opér., biens pers. et dom.		71	
..	..	..	..	..	..	..	..	Location de matériel de transport			711
..	..	..	..	..	..	..	..	Location de machines ét equipements nca			712
..	..	..	..	..	..	..	..	Loc. d'articles personnels et domestiques nca			713
..	..	14	16	18	19	26	28	Activités informatiques et activités rattachées		72	
..	..	..	..	..	..	..	..	Conseil en matériel informatique			721
..	..	..	..	..	..	..	..	Production de logiciels			722
..	..	..	..	..	..	..	..	Traitement de données informatiques			723
..	..	..	..	..	..	..	..	Activités de banques de données			724
..	..	..	..	..	..	..	..	Entr., répar. mach. de bureau et compt. et inform.			725
..	..	..	..	..	..	..	..	Autres activités rattachées à l'informatique			729
..	..	36	58	56	57	52	51	Recherche-développement		73	
..	..	..	..	..	..	..	..	Rech. et dév. expér. en sc. phys., natur. et ingén.			731
..	..	..	..	..	..	..	..	Rech. et dév. expér. en sciences soc. et humaines			732
..	..	166	155	195	199	252	307	Autres activités de services aux entreprises		74	
..	..	..	..	..	..	..	..	Act. jur., compt., audit.; cons. fisc.; ét. march., soud.			741
..	..	..	..	..	..	..	..	Act. d'achitect., d'ingén.; autres act. techniques			742
..	..	..	..	..	..	..	..	Publicité			743
..	..	..	..	..	..	..	..	Activités de services aux entreprises nca.			749
..	..	805	836	842	845	848	897	Education	M	80	
..	..	..	..	..	..	..	..	Activités d'enseignement primaire			801
..	..	..	..	..	..	..	..	Activités d'enseignement secondaire			802
..	..	..	..	..	..	..	..	Activités d'enseignement supérieur			803
..	..	..	..	..	..	..	..	Act. de formation perman.; autres act. d'enseign.			809
..	..	945	955	962	964	964	969	Santé et action sociale	N	85	
..	..	..	..	..	..	..	..	Activités de santé humaine			851
..	..	..	..	..	..	..	..	Activités vétérinaires			852
..	..	..	..	..	..	..	..	Activités d'action sociale			853
..	..	351	309	243	251	370	253	Autres activités de serv. coll., soc. et personnels	O	90_93	
..	..	42	46	45	47	47	50	Assain. et enlèv. des ordures; voierie, act. simil.		90	
..	..	49	33	32	32	37	17	Activités associatives diverses		91	
..	..	..	..	..	..	..	..	Activités de syndicats de salariés			912
..	..	129	122	128	130	135	135	Activités récréatives, culturelles et sportives		92	
..	..	..	..	..	..	..	..	Act. cinémat., de radio, télé.; autres act. de spect.			921
..	..	..	..	..	..	..	..	Activités d'agences de presse			922
..	..	..	..	..	..	..	..	Act. des biblio., arch., musées, jardins bot. et zool.			923
..	..	..	..	..	..	..	..	Activités sportives et autres activités récréatives			924
..	..	..	35	38	42	54	51	Autres activités de services		93	

POLAND

Table .S6 EMPLOYMENT, FEMALES - EMPLOI, FEMMES

Thousands

ISIC revision 3				1991	1992	1993	1994	1995	1996	1997	1998
G	50_52		Wholesale & retail trade; repairs; etc.	..	..	922	969	989	969	1 077	1 124
	50		Sale, maint.,repair vehic/cycle, person., h/hold gd.	..	..	..	..	30	31	36	43
		501	Sale of motor cycle	..	..	..	..	..	..	..	..
		502	Maintenance and repair of motor vehicles	..	..	..	..	..	..	..	..
		503	Sale of motor vehicle parts and accessories	..	..	..	..	..	..	..	..
		504	Sale, maint., repair motor cycle, parts & acces.	..	..	..	..	..	..	..	..
		505	Retail sale of automative fuel	..	..	..	..	..	..	..	..
	51		W/sale. trade & commission exc motor veh. & cycle	..	..	..	..	289	284	296	313
		511	Wholesale on a fee or contract basis	..	..	..	..	..	..	..	..
		512	W/sale agric. raw mat.,live animal,food, bev., tobac.	..	..	..	..	..	..	..	..
		513	Wholesale of household goods	..	..	..	..	..	..	..	..
		514	W/sale. of non-agric. intermed. prod., waste, scrap	..	..	..	..	..	..	..	..
		515	Wholesale of machinery, equipment & supplies	..	..	..	..	..	..	..	..
		519	Other wholesale	..	..	..	..	..	..	..	..
	52		Retail trade exc. motor veh. & cyc.; repair h/hold gds.	..	..	..	..	671	654	745	768
		521	Non-specialised retail trade in stores	..	..	..	..	..	..	..	..
		522	Retail sale food, bev., tobac. in specialised stores	..	..	..	..	..	..	..	..
		523	Other retail trade of new goods specialised stores	..	..	..	..	..	..	..	..
		524	Retail sale of second-hand goods in stores	..	..	..	..	..	..	..	..
		525	Retail trade not in stores	..	..	..	..	..	..	..	..
		526	Repair of personal and household goods	..	..	..	..	..	..	..	..
H	55		Hotels and restaurants	..	..	97	113	124	126	132	149
		551	Hotels; camping ; other short-stay accommodation	..	..	..	..	..	..	..	..
		552	Restaurants, bars and canteens	..	..	..	..	..	..	..	..
I	60_64		Transport, storage and communications	..	..	235	252	253	248	253	259
	60		Land transport, transport via pipelines	..	..	..	..	44	43	46	75
		601	Transport via railways	..	..	..	..	..	..	..	..
		602	Other land transport	..	..	..	..	..	..	..	..
		603	Transport via pipelines	..	..	..	..	..	..	..	..
	61		Water transport	..	..	..	..	2	2	2	2
		611	Sea and coastal water transport	..	..	..	..	..	..	..	..
		612	Inland water transport	..	..	..	..	..	..	..	..
	62		Air transport	..	..	..	..	2	2	2	2
		621	Scheduled air transport	..	..	..	..	..	..	..	..
		622	Non-scheduled air transport	..	..	..	..	..	..	..	..
	63		Support. & aux. transport act., travel agencies activ.	..	..	..	..	99	98	101	78
	64		Post and telecommunications	..	..	..	..	107	103	103	103
		641	Post and courier activities	..	..	..	..	..	..	..	..
		642	Telecommunications	..	..	..	..	..	..	..	..
J	65_67		Financial intermediation	..	..	171	189	198	209	222	241
	65		Financial intermediation, exc. insur., pension fund	..	..	..	..	132	139	143	147
		651	Monetary intermediation	..	..	..	..	..	..	..	..
		659	Other financial intermediation	..	..	..	..	..	..	..	..
	66		Insurance pension fund, exc. compul. social sec.	..	..	..	..	56	56	57	59
	67		Activities auxiliary to financial intermediation	..	..	..	..	10	15	22	35
		671	Act. auxil to fin. intermed., exc. insur. & pens.fund	..	..	..	..	..	..	..	..
		672	Act. auxiliary to insurance and pension funding	..	..	..	..	..	..	..	..
K	70_74		Real estate, renting and business activities	..	..	247	242	254	269	315	350
	70		Real estate activities	..	..	..	..	76	83	85	93
		701	Real estate activ. with own or leased property	..	..	..	..	..	..	..	..
		702	Real estate activities on a fee or contract basis	..	..	..	..	..	..	..	..
	71		Renting of mach. without operator; rent. of pers. gds.	..	..	..	..	5	5	6	5
		711	Renting of transport equipment	..	..	..	..	..	..	..	..
		712	Renting of other machinery and equipment	..	..	..	..	..	..	..	..
		713	Renting of personal and household goods nec	..	..	..	..	..	..	..	..
	72		Computer and related activities	..	..	..	..	9	10	13	14
		721	Hardware consultancy	..	..	..	..	..	..	..	..
		722	Software consultancy and supply	..	..	..	..	..	..	..	..
		723	Data processing	..	..	..	..	..	..	..	..
		724	Data base activities	..	..	..	..	..	..	..	..
		725	Mainten.; repair of office, account. & comp. mach.	..	..	..	..	..	..	..	..
		729	Other computer related activities	..	..	..	..	..	..	..	..
	73		Research and development	..	..	..	..	33	33	33	32
		731	R & D experim. on natural sciences & engineering	..	..	..	..	..	..	..	..
		732	R & D experim. on social sciences & humanities	..	..	..	..	..	..	..	..
	74		Other business activities	..	..	..	..	131	138	179	207
		741	Legal, acc., book-keep., audit act.; tax consult., etc	..	..	..	..	..	..	..	..
		742	Architectural, engineering, & other techn. activities	..	..	..	..	..	..	..	..
		743	Advertising	..	..	..	..	..	..	..	..
		749	Business activities nec	..	..	..	..	..	..	..	..
M	80		Education	..	..	632	652	654	668	665	670
		801	Primary education	..	..	..	..	..	..	..	..
		802	Secondary education	..	..	..	..	..	..	..	..
		803	Higher education	..	..	..	..	..	..	..	..
		809	Adult and other education	..	..	..	..	..	..	..	..
N	85		Health and social work	..	..	798	815	824	828	842	840
		851	Human health activities	..	..	..	..	..	..	..	..
		852	Veterinary activities	..	..	..	..	..	..	..	..
		853	Social work activities	..	..	..	..	..	..	..	..
O	90_93		Other commun., soc. & pers. services activities	..	..	170	170	169	185	185	178
	90		Sewage & refuse disposal, sanitation, etc.	..	..	..	..	11	13	13	14
	91		Activities of member organisations nec	..	..	..	..	34	38	26	14
		912	Activities of trade unions	..	..	..	..	..	..	..	..
	92		Recreational, cultural and sporting activities	..	..	..	..	78	83	84	86
		921	Motion picture, radio, TV, other entertainment act.	..	..	..	..	..	..	..	..
		922	News agency activities	..	..	..	..	..	..	..	..
		923	Library, archives, museums, other cultural activ.	..	..	..	..	..	..	..	..
		924	Sporting and other recreational activities	..	..	..	..	..	..	..	..
	93		Other service activities	..	..	..	..	46	52	62	65

Table .S7 SALAIRES ET CHARGES SOCIALES, SALARIÉS - COMPENSATION OF LABOUR, EMPLOYEES

Millions de Zl (Prix courants)

1991	1992	1993	1994	1995	1996	1997	1998		CITI révision 3		
..	..	6 509	9 679	12 869	16 994	22 097	26 921	Com. de gros, de detail ; réparation etc.	G	50_52	
..	..	..	..	..	..	..	..	Com., entr., répar. d'auto. et moto ; carburants		50	
..	..	..	..	..	..	..	..	Commerce de véhicules automobiles			501
..	..	..	..	..	..	..	..	Entretien et réparation de véhicules automobiles			502
..	..	..	..	..	..	..	..	Com. de pièces et access. de véhic. automobiles			503
..	..	..	..	..	..	..	..	Com., entret. et répar. de motocycles et access.			504
..	..	..	..	..	..	..	..	Com. de détail de carburants pour automobiles			505
..	..	..	..	..	..	..	..	Com. de gros et act. inter., sauf auto. et moto		51	
..	..	..	..	..	..	..	..	Activités d'intermédiaries de commerce de gros			511
..	..	..	..	..	..	..	..	Com. gros: prod. agr. bruts, anim. viv., prod. alim.			512
..	..	..	..	..	..	..	..	Com. gros: biens de consom., sauf alimentaires			513
..	..	..	..	..	..	..	..	Com. gros: prod. inter. non agr., déchets, débris			514
..	..	..	..	..	..	..	..	Com. gros: machines, équipments et fournitures			515
..	..	..	..	..	..	..	..	Autres commerces de gros			519
..	..	..	..	..	..	..	..	Com. détail, sauf auto, moto; répar. art. pers., dom.		52	
..	..	..	..	..	..	..	..	Commerce de détail non spécialisé en magasin			521
..	..	..	..	..	..	..	..	Com. détail: prod. aliment. en mag. spécialisés			522
..	..	..	..	..	..	..	..	Autre com. détail: prod. neufs en mag. spécialisés			523
..	..	..	..	..	..	..	..	Commerces de détail de biens d'occasion			524
..	..	..	..	..	..	..	..	Commerces de détail autres qu'en magasin			525
..	..	..	..	..	..	..	..	Réparation d'articles personnels et domestiques			526
..	..	536	1 050	1 260	1 780	2 226	2 830	Hotels et restaurants	H	55	
..	..	..	..	..	..	..	..	Hôtels; camping; autres équip. courts séjours			551
..	..	..	..	..	..	..	..	Restaurants, cafés et cantines			552
..	..	5 551	7 201	9 350	11 813	15 396	18 103	Transport, entreposage, communications	I	60_64	
..	..	..	..	..	..	..	..	Transports terrestres; transports par conduites		60	
..	..	..	..	..	..	..	..	Transports par chemin de fer			601
..	..	..	..	..	..	..	..	Autres transports terrestres			602
..	..	..	..	..	..	..	..	Transports par conduites			603
..	..	..	..	..	..	..	..	Transports par eau		61	
..	..	..	..	..	..	..	..	Transports maritimes et côtiers			611
..	..	..	..	..	..	..	..	Transports par voies navigables intérieures			612
..	..	..	..	..	..	..	..	Transports aériens		62	
..	..	..	..	..	..	..	..	Transports aériens réguliers			621
..	..	..	..	..	..	..	..	Transports aériens non réguliers			622
..	..	..	..	..	..	..	..	Act. annex. et aux. des transp.; act. des ag. de voy.		63	
..	..	..	..	..	..	..	..	Postes et télécommunications		64	
..	..	..	..	..	..	..	..	Activités de poste et de courrier			641
..	..	..	..	..	..	..	..	Télécommunications			642
..	..	1 600	2 584	3 825	5 256	7 583	9 055	Intermédiation financière	J	65_67	
..	..	..	..	..	..	..	..	Interm. financ., sauf assur. et caisses de retraite		65	
..	..	..	..	..	..	..	..	Intermédiation monétaire			651
..	..	..	..	..	..	..	..	Autres intermédiations financières			659
..	..	..	..	..	..	..	..	Act. d'assur., caisses de retr., sauf séc. soc. oblig.		66	
..	..	..	..	..	..	..	..	Act. auxiliaires de l'intermédiation financière		67	
..	..	..	..	..	..	..	..	Act. aux. de l'interm. fin., sauf assur., et retraite			671
..	..	..	..	..	..	..	..	Act. aux. des assurances et caisses de retraite			672
..	..	4 033	4 243	6 847	8 958	12 457	15 909	Immobilier, locations, services aux entreprises	K	70_74	
..	..	..	..	..	..	..	..	Activités immobilières		70	
..	..	..	..	..	..	..	..	Act. immobilières sur biens propres ou loués			701
..	..	..	..	..	..	..	..	Activités immobilières pour comptes de tiers			702
..	..	..	..	..	..	..	..	Locat. mach., équip. sans opér., biens pers. et dom.		71	
..	..	..	..	..	..	..	..	Location de matériel de transport			711
..	..	..	..	..	..	..	..	Location de machines ét equipements nca			712
..	..	..	..	..	..	..	..	Loc. d'articles personnels et domestiques nca			713
..	..	..	..	..	..	..	..	Activités informatiques et activités rattachées		72	
..	..	..	..	..	..	..	..	Conseil en matériel informatique			721
..	..	..	..	..	..	..	..	Production de logiciels			722
..	..	..	..	..	..	..	..	Traitement de données informatiques			723
..	..	..	..	..	..	..	..	Activités de banques de données			724
..	..	..	..	..	..	..	..	Entr., répar. mach. de bureau et compt. et inform.			725
..	..	..	..	..	..	..	..	Autres activités rattachées à l'informatique			729
..	..	..	..	..	..	..	..	Recherche-développement		73	
..	..	..	..	..	..	..	..	Rech. et dév. expér. en sc. phys., natur. et ingén.			731
..	..	..	..	..	..	..	..	Rech. et dév. expér. en sciences soc. et humaines			732
..	..	..	..	..	..	..	..	Autres activités de services aux entreprises		74	
..	..	..	..	..	..	..	..	Act. jur., compt., audit.; cons. fisc.; ét. march., soud.			741
..	..	..	..	..	..	..	..	Act. d'achitect., d'ingén.; autres act. techniques			742
..	..	..	..	..	..	..	..	Publicité			743
..	..	..	..	..	..	..	..	Activités de services aux entreprises nca.			749
..	..	4 738	7 336	10 240	13 306	17 063	19 603	Education	M	80	
..	..	..	..	..	..	..	..	Activités d'enseignement primaire			801
..	..	..	..	..	..	..	..	Activités d'enseignement secondaire			802
..	..	..	..	..	..	..	..	Activités d'enseignement supérieur			803
..	..	..	..	..	..	..	..	Act. de formation perman.; autres act. d'enseign.			809
..	..	4 837	7 133	9 787	12 450	15 685	17 324	Santé et action sociale	N	85	
..	..	..	..	..	..	..	..	Activités de santé humaine			851
..	..	..	..	..	..	..	..	Activités vétérinaires			852
..	..	..	..	..	..	..	..	Activités d'action sociale			853
..	..	2 264	3 707	5 423	7 115	7 709	9 400	Autres activités de serv. coll., soc. et personnels	O	90_93	
..	..	..	..	..	..	..	..	Assain. et enlèv. des ordures; voierie, act. simil.		90	
..	..	..	..	..	..	..	..	Activités associatives diverses		91	
..	..	..	..	..	..	..	..	Activités de syndicats de salariés			912
..	..	..	..	..	..	..	..	Activités récréatives, culturelles et sportives		92	
..	..	..	..	..	..	..	..	Act. cinémat., de radio, télé.; autres act. de spect.			921
..	..	..	..	..	..	..	..	Activités d'agences de presse			922
..	..	..	..	..	..	..	..	Act. des biblio., arch., musées, jardins bot. et zool.			923
..	..	..	..	..	..	..	..	Activités sportives et autres activités récréatives			924
..	..	..	..	..	..	..	..	Autres activités de services		93	

POLAND

Table .S8 WAGES AND SALARIES, EMPLOYEES - SALAIRES ET TRAITEMENTS, SALARIÉS

Millions of Zl (Current Prices)

ISIC revision 3				1991	1992	1993	1994	1995	1996	1997	1998
G	50_52		Wholesale & retail trade; repairs; etc.	..	2 510	3 872	5 224	7 452	9 664	13 301	16 488
	50		Sale, maint.,repair vehic/cycle, person., h/hold gd.	..	..	293	458	606	974	1 300	1 788
		501	Sale of motor cycle	..	..	..	..	..	..	..	..
		502	Maintenance and repair of motor vehicles	..	..	..	..	..	..	..	..
		503	Sale of motor vehicle parts and accessories	..	..	..	..	..	..	..	..
		504	Sale, maint., repair motor cycle, parts & acces.	..	..	..	..	..	..	..	..
		505	Retail sale of automative fuel	..	..	..	..	..	..	..	..
	51		W/sale. trade & commission exc motor veh. & cycle	..	..	1 818	2 636	4 103	5 096	7 037	8 966
		511	Wholesale on a fee or contract basis	..	..	..	..	..	..	..	..
		512	W/sale agric. raw mat.,live animal,food, bev., tobac.	..	..	..	..	..	..	..	..
		513	Wholesale of household goods	..	..	..	..	..	..	..	..
		514	W/sale. of non-agric. intermed. prod., waste, scrap	..	..	..	..	..	..	..	..
		515	Wholesale of machinery, equipment & supplies	..	..	..	..	..	..	..	..
		519	Other wholesale	..	..	..	..	..	..	..	..
	52		Retail trade exc. motor veh. & cyc.; repair h/hold gds.	..	..	1 761	2 129	2 743	3 595	4 965	5 735
		521	Non-specialised retail trade in stores	..	..	..	..	..	..	..	..
		522	Retail sale food, bev., tobac. in specialised stores	..	..	..	..	..	..	..	..
		523	Other retail trade of new goods specialised stores	..	..	..	..	..	..	..	..
		524	Retail sale of second-hand goods in stores	..	..	..	..	..	..	..	..
		525	Retail trade not in stores	..	..	..	..	..	..	..	..
		526	Repair of personal and household goods	..	..	..	..	..	..	..	..
H	55		Hotels and restaurants	..	225	1 761	2 129	2 743	3 595	4 965	16 050
		551	Hotels; camping ; other short-stay accommodation	..	..	..	..	..	..	..	..
		552	Restaurants, bars and canteens	..	..	..	..	..	..	..	..
I	60_64		Transport, storage and communications	..	..	3 701	4 875	6 308	7 941	9 947	11 706
	60		Land transport, transport via pipelines	..	2 776	2 267	3 032	3 800	4 660	5 617	6 398
		601	Transport via railways	..	..	..	..	..	..	..	..
		602	Other land transport	..	..	..	..	..	..	..	..
		603	Transport via pipelines	..	..	..	..	..	..	..	..
	61		Water transport	..	..	74	88	110	139	143	136
		611	Sea and coastal water transport	..	..	..	..	..	..	..	..
		612	Inland water transport	..	..	..	..	..	..	..	..
	62		Air transport	..	..	43	58	..	113	163	192
		621	Scheduled air transport	..	..	..	..	..	..	..	..
		622	Non-scheduled air transport	..	..	..	..	..	..	..	..
	63		Support. & aux. transport act., travel agencies activ.	..	..	431	503	681	884	1 181	1 464
	64		Post and telecommunications	..	..	885	1 195	1 633	2 145	2 843	3 516
		641	Post and courier activities	..	..	..	..	..	..	..	..
		642	Telecommunications	..	..	..	..	..	..	..	..
J	65_67		Financial intermediation	..	966	1 470	2 072	2 850	3 904	5 118	6 387
	65		Financial intermediation, exc. insur., pension fund	..	..	1 095	1 546	2 100	2 872	3 758	4 594
		651	Monetary intermediation	..	..	..	..	..	..	..	..
		659	Other financial intermediation	..	..	..	..	..	..	..	..
	66		Insurance pension fund, exc. compul. social sec.	..	..	352	483	..	905	1 166	1 499
	67		Activities auxiliary to financial intermediation	..	..	24	44	81	127	194	295
		671	Act. auxil to fin. intermed., exc. insur. & pens.fund	..	..	..	..	..	..	..	..
		672	Act. auxiliary to insurance and pension funding	..	..	..	..	..	..	..	..
K	70_74		Real estate, renting and business activities	..	1 535	1 942	2 480	3 437	4 840	6 823	8 894
	70		Real estate activities	..	..	768	801	1 081	1 462	1 901	2 384
		701	Real estate activ. with own or leased property	..	..	..	..	..	..	..	..
		702	Real estate activities on a fee or contract basis	..	..	..	..	..	..	..	..
	71		Renting of mach. without operator; rent. of pers. gds.	..	..	14	23	35	49	91	120
		711	Renting of transport equipment	..	..	..	..	..	..	..	..
		712	Renting of other machinery and equipment	..	..	..	..	..	..	..	..
		713	Renting of personal and household goods nec	..	..	..	..	..	..	..	..
	72		Computer and related activities	..	..	90	136	207	305	472	686
		721	Hardware consultancy	..	..	..	..	..	..	..	..
		722	Software consultancy and supply	..	..	..	..	..	..	..	..
		723	Data processing	..	..	..	..	..	..	..	..
		724	Data base activities	..	..	..	..	..	..	..	..
		725	Mainten.; repair of office, account. & comp. mach.	..	..	..	..	..	..	..	..
		729	Other computer related activities	..	..	..	..	..	..	..	..
	73		Research and development	..	..	200	456	589	756	897	1 028
		731	R & D experim. on natural sciences & engineering	..	..	..	..	..	..	..	..
		732	R & D experim. on social sciences & humanities	..	..	..	..	..	..	..	..
	74		Other business activities	..	..	871	1 064	1 728	2 268	3 463	4 677
		741	Legal, acc., book-keep., audit act.; tax consult., etc	..	..	..	..	..	..	..	..
		742	Architectural, engineering, & other techn. activities	..	..	..	..	..	..	..	..
		743	Advertising	..	..	..	..	..	..	..	..
		749	Business activities nec	..	..	..	..	..	..	..	..
M	80		Education	..	..	3 384	4 594	6 248	8 019	9 946	11 682
		801	Primary education	..	..	..	..	..	..	..	..
		802	Secondary education	..	..	..	..	..	..	..	..
		803	Higher education	..	..	..	..	..	..	..	..
		809	Adult and other education	..	..	..	..	..	..	..	..
N	85		Health and social work	..	..	3 771	4 928	6 617	8 274	10 106	11 658
		851	Human health activities	..	..	..	..	..	..	..	..
		852	Veterinary activities	..	..	..	..	..	..	..	..
		853	Social work activities	..	..	..	..	..	..	..	..
O	90_93		Other commun., soc. & pers. services activities	..	..	1 119	1 388	1 876	2 395	3 199	3 472
	90		Sewage & refuse disposal, sanitation, etc.	..	..	189	260	323	426	536	676
	91		Activities of member organisations nec	..	..	211	195	249	318	422	270
		912	Activities of trade unions	..	..	..	..	..	..	..	..
	92		Recreational, cultural and sporting activities	..	..	611	793	1 102	1 381	1 835	2 085
		921	Motion picture, radio, TV, other entertainment act.	..	..	..	..	..	..	..	..
		922	News agency activities	..	..	..	..	..	..	..	..
		923	Library, archives, museums, other cultural activ.	..	..	..	..	..	..	..	..
		924	Sporting and other recreational activities	..	..	..	..	..	..	..	..
	93		Other service activities	..	..	108	141	202	270	406	441

Table .S9 HEURES OUVRÉES, OUVRIERS - HOURS WORKED BY EMPLOYEES

Units

1991	1992	1993	1994	1995	1996	1997	1998		CITI révision 3		
..	..	..	..	..	1 390 562	1 583 030	1 752 222	Com. de gros, de detail ; réparation etc.	G	50_52	
..	..	..	..	..	127 949	152 086	186 162	Com., entr., répar. d'auto. et moto ; carburants		50	
..	..	..	..	..	..	44 578	62 532	Commerce de véhicules automobiles			501
..	..	..	..	..	..	53 821	56 398	Entretien et réparation de véhicules automobiles			502
..	..	..	..	..	..	27 231	31 756	Com. de pièces et access. de véhic. automobiles			503
..	..	..	..	..	..	109	182	Com., entret. et répar. de motocycles et access.			504
..	..	..	..	..	..	26 347	35 296	Com. de détail de carburants pour automobiles			505
..	..	..	..	..	670 307	805 954	907 369	Com. de gros et act. inter., sauf auto. et moto		51	
..	..	..	..	..	..	27 088	34 521	Activités d'intermédiaries de commerce de gros			511
..	..	..	..	..	..	17 243	16 988	Com. gros: prod. agr. bruts, anim. viv., prod. alim.			512
..	..	..	..	..	..	132 431	146 053	Com. gros: biens de consom., sauf alimentaires			513
..	..	..	..	..	..	106 552	117 496	Com. gros: prod. inter. non agr., déchets, débris			514
..	..	..	..	..	..	131 502	145 455	Com. gros: machines, équipments et fournitures			515
..	..	..	..	..	..	..	..	Autres commerces de gros			519
..	..	..	..	..	592 305	624 990	658 691	Com. détail, sauf auto, moto; répar. art. pers., dom.		52	
..	..	..	..	..	..	403 635	402 506	Commerce de détail non spécialisé en magasin			521
..	..	..	..	..	..	64 424	70 405	Com. détail: prod. aliment. en mag. spécialisés			522
..	..	..	..	..	..	19 343	24 033	Autre com. détail: prod. neufs en mag. spécialisés			523
..	..	..	..	..	..	119 880	138 861	Commerces de détail de biens d'occasion			524
..	..	..	..	..	..	2 002	1 854	Commerces de détail autres qu'en magasin			525
..	..	..	..	..	..	11 261	16 899	Réparation d'articles personnels et domestiques			526
..	..	..	..	..	165 563	194 606	198 942	Hotels et restaurants	H	55	
..	..	..	..	..	27 247	51 488	56 470	Hôtels; camping; autres équip. courts séjours			551
..	..	..	..	..	53 183	49 586	44 647	Restaurants, cafés et cantines			552
..	..	..	..	..	1 242 448	1 251 786	1 224 433	Transport, entreposage, communications	I	60_64	
..	..	..	..	..	776 742	769 420	727 965	Transports terrestres; transports par conduites		60	
..	..	..	..	..	..	424 502	396 018	Transports par chemin de fer			601
..	..	..	..	..	..	343 481	330 500	Autres transports terrestres			602
..	..	..	..	..	..	1 438	1 447	Transports par conduites			603
..	..	..	..	..	22 305	18 261	14 398	Transports par eau		61	
..	..	..	..	..	..	15 453	12 561	Transports maritimes et côtiers			611
..	..	..	..	..	..	2 808	1 837	Transports par voies navigables intérieures			612
..	..	..	..	..	756	7 490	6 829	Transports aériens		62	
..	..	..	..	..	..	6 813	6 094	Transports aériens réguliers			621
..	..	..	..	..	..	677	735	Transports aériens non réguliers			622
..	..	..	..	..	119 569	132 053	138 699	Act. annex. et aux. des transp.; act. des ag. de voy.		63	
..	..	..	..	..	323 078	324 562	336 543	Postes et télécommunications		64	
..	..	..	..	..	..	178 465	186 269	Activités de poste et de courrier			641
..	..	..	..	..	..	146 097	150 273	Télécommunications			642
..	..	..	..	..	443 590	463 734	484 539	Intermédiation financière	J	65_67	
..	..	..	..	..	312 914	323 901	332 587	Interm. financ., sauf assur. et caisses de retraite		65	
..	..	..	..	..	290 747	295 393	300 885	Intermédiation monétaire			651
..	..	..	..	..	..	..	..	Autres intermédiations financières			659
..	..	..	..	..	118 722	122 193	131 471	Act. d'assur., caisses de retr., sauf séc. soc. oblig.		66	
..	..	..	..	..	11 955	17 640	20 481	Act. auxiliaires de l'intermédiation financière		67	
..	..	..	..	..	9 869	13 499	15 977	Act. aux. de l'interm. fin., sauf assur., et retraite			671
..	..	..	..	..	2 086	4 142	4 504	Act. aux. des assurances et caisses de retraite			672
..	..	..	..	..	688 121	801 692	889 363	Immobilier, locations, services aux entreprises	K	70_74	
..	..	..	..	..	259 293	271 006	269 190	Activités immobilières		70	
..	..	..	..	..	..	7 037	8 775	Act. immobilières sur biens propres ou loués			701
..	..	..	..	..	..	19 009	21 797	Activités immobilières pour comptes de tiers			702
..	..	..	..	..	6 830	7 836	9 342	Locat. mach., équip. sans opér., biens pers. et dom.		71	
..	..	..	..	..	..	776	1 413	Location de matériel de transport			711
..	..	..	..	..	..	3 237	3 519	Location de machines ét equipements nca			712
..	..	..	..	..	..	2 857	3 127	Loc. d'articles personnels et domestiques nca			713
..	..	..	..	..	31 668	38 529	48 073	Activités informatiques et activités rattachées		72	
..	..	..	..	..	..	2 316	2 628	Conseil en matériel informatique			721
..	..	..	..	..	..	21 967	28 499	Production de logiciels			722
..	..	..	..	..	..	7 542	7 764	Traitement de données informatiques			723
..	..	..	..	..	..	1 664	2 020	Activités de banques de données			724
..	..	..	..	..	..	2 951	3 564	Entr., répar. mach. de bureau et compt. et inform.			725
..	..	..	..	..	..	..	..	Autres activités rattachées à l'informatique			729
..	..	..	..	..	102 168	94 068	92 618	Recherche-développement		73	
..	..	..	..	..	97 181	89 073	87 666	Rech. et dév. expér. en sc. phys., natur. et ingén.			731
..	..	..	..	..	4 987	4 995	4 951	Rech. et dév. expér. en sciences soc. et humaines			732
..	..	..	..	..	288 163	390 254	470 141	Autres activités de services aux entreprises		74	
..	..	..	..	..	..	85 828	93 611	Act. jur., compt., audit.; cons. fisc.; ét. march., soud.			741
..	..	..	..	..	..	108 047	122 366	Act. d'achitect., d'ingén.; autres act. techniques			742
..	..	..	..	..	..	20 845	16 910	Publicité			743
..	..	..	..	..	..	..	..	Activités de services aux entreprises nca.			749
..	..	..	..	..	879 067	1 000 087	1 022 103	Education	M	80	
..	..	..	..	..	394 417	557 333	571 484	Activités d'enseignement primaire			801
..	..	..	..	..	258 956	219 690	233 739	Activités d'enseignement secondaire			802
..	..	..	..	..	204 927	204 062	196 776	Activités d'enseignement supérieur			803
..	..	..	..	..	..	..	..	Act. de formation perman.; autres act. d'enseign.			809
..	..	..	..	..	1 670 677	1 647 530	1 644 402	Santé et action sociale	N	85	
..	..	..	..	..	1 321 161	1 240 805	1 236 586	Activités de santé humaine			851
..	..	..	..	..	1 462	1 854	1 734	Activités vétérinaires			852
..	..	..	..	..	348 053	404 870	406 082	Activités d'action sociale			853
..	..	..	..	..	342 633	364 818	360 383	Autres activités de serv. coll., soc. et personnels	O	90_93	
..	..	..	..	..	77 216	86 747	86 218	Assain. et enlèv. des ordures; voierie, act. simil.		90	
..	..	..	..	..	..	..	..	Activités associatives diverses		91	
..	..	..	..	..	..	..	..	Activités de syndicats de salariés			912
..	..	..	..	..	208 033	218 104	218 696	Activités récréatives, culturelles et sportives		92	
..	..	..	..	..	..	7 661	8 702	Act. cinémat., de radio, télé.; autres act. de spect.			921
..	..	..	..	..	..	24 135	26 523	Activités d'agences de presse			922
..	..	..	..	..	..	71 178	71 222	Act. des biblio., arch., musées, jardins bot. et zool.			923
..	..	..	..	..	..	1 401	1 520	Activités sportives et autres activités récréatives			924
..	..	..	..	..	..	59 968	55 470	Autres activités de services		93	

POLAND

Table .S10 ENTERPRISES - ENTREPRISES

Units

ISIC revision 3				1991	1992	1993	1994	1995	1996	1997	1998
G	50_52		Wholesale & retail trade; repairs; etc.	..	..	..	..	..	..	964 524	1011 252
	50		Sale, maint.,repair vehic/cycle, person., h/hold gd.	..	..	..	..	..	..	101 301	107 672
		501	Sale of motor cycle	..	..	..	..	..	..	20 076	20 348
		502	Maintenance and repair of motor vehicles	..	..	..	..	..	..	60 660	64 969
		503	Sale of motor vehicle parts and accessories	..	..	..	..	..	..	16 757	18 172
		504	Sale, maint., repair motor cycle, parts & acces.	..	..	..	..	..	..	351	402
		505	Retail sale of automative fuel	..	..	..	..	..	..	3 457	3 781
	51		W/sale. trade & commission exc motor veh. & cycle	..	..	..	..	..	..	219 596	223 672
		511	Wholesale on a fee or contract basis	..	..	..	..	..	..	27 100	31 648
		512	W/sale agric. raw mat.,live animal,food, bev., tobac.	..	..	..	..	..	..	31 843	31 535
		513	Wholesale of household goods	..	..	..	..	..	..	26 401	26 634
		514	W/sale. of non-agric. intermed. prod., waste, scrap	..	..	..	..	..	..	22 732	24 677
		515	Wholesale of machinery, equipment & supplies	..	..	..	..	..	..	3 447	3 946
		519	Other wholesale	..	..	..	..	..	..	108 073	105 232
	52		Retail trade exc. motor veh. & cyc.; repair h/hold gds.	..	..	..	..	..	..	643 627	679 908
		521	Non-specialised retail trade in stores	..	..	..	..	..	..	170 437	177 706
		522	Retail sale food, bev., tobac. in specialised stores	..	..	..	..	..	..	50 597	52 784
		523	Other retail trade of new goods specialised stores	..	..	..	..	..	..	154 299	170 412
		524	Retail sale of second-hand goods in stores	..	..	..	..	..	..	15 736	15 736
		525	Retail trade not in stores	..	..	..	..	..	..	226 126	233 921
		526	Repair of personal and household goods	..	..	..	..	..	..	26 432	27 492
H	55		Hotels and restaurants	..	..	..	..	..	..	71 231	79 517
		551	Hotels; camping ; other short-stay accommodation	..	..	..	..	..	..	13 087	14 763
		552	Restaurants, bars and canteens	..	..	..	..	..	..	58 144	64 754
I	60_64		Transport, storage and communications	..	..	..	..	..	..	206 913	227 434
	60		Land transport, transport via pipelines	..	..	..	..	..	..	193 075	211 711
		601	Transport via railways	..	..	..	..	..	..	11	14
		602	Other land transport	..	..	..	..	..	..	81 416	86 167
		603	Transport via pipelines	..	..	..	..	..	..	4	4
	61		Water transport	..	..	..	..	..	..	190	326
		611	Sea and coastal water transport	..	..	..	..	..	..	63	63
		612	Inland water transport	..	..	..	..	..	..	127	263
	62		Air transport	..	..	..	..	..	..	33	31
		621	Scheduled air transport	..	..	..	..	..	..	6	6
		622	Non-scheduled air transport	..	..	..	..	..	..	27	25
	63		Support. & aux. transport act., travel agencies activ.	..	..	..	..	..	..	12 588	14 119
	64		Post and telecommunications	..	..	..	..	..	..	1 027	1 247
		641	Post and courier activities	..	..	..	..	..	..	256	303
		642	Telecommunications	..	..	..	..	..	..	771	944
J	65_67		Financial intermediation	..	..	..	..	..	..	44 251	61 114
	65		Financial intermediation, exc. insur., pension fund	..	..	..	..	..	..	3 740	4 419
		651	Monetary intermediation	..	..	..	..	..	..	1 697	1 751
		659	Other financial intermediation	..	..	..	..	..	..	2 043	2 668
	66		Insurance pension fund, exc. compul. social sec.	..	..	..	..	..	..	2 823	1 705
	67		Activities auxiliary to financial intermediation	..	..	..	..	..	..	37 688	54 990
		671	Act. auxil to fin. intermed., exc. insur. & pens.fund	..	..	..	..	..	..	3 112	3 821
		672	Act. auxiliary to insurance and pension funding	..	..	..	..	..	..	34 576	51 169
K	70_74		Real estate, renting and business activities	..	..	..	..	..	..	285 643	303 456
	70		Real estate activities	..	..	..	..	..	..	21 999	20 055
		701	Real estate activ. with own or leased property	..	..	..	..	..	..	4 627	5 603
		702	Real estate activities on a fee or contract basis	..	..	..	..	..	..	17 372	14 452
	71		Renting of mach. without operator; rent. of pers. gds.	..	..	..	..	..	..	10 481	10 392
		711	Renting of transport equipment	..	..	..	..	..	..	1 189	1 304
		712	Renting of other machinery and equipment	..	..	..	..	..	..	685	823
		713	Renting of personal and household goods nec	..	..	..	..	..	..	8 607	8 265
	72		Computer and related activities	..	..	..	..	..	..	20 364	23 872
		721	Hardware consultancy	..	..	..	..	..	..	1 702	2 033
		722	Software consultancy and supply	..	..	..	..	..	..	10 132	11 351
		723	Data processing	..	..	..	..	..	..	1 484	1 867
		724	Data base activities	..	..	..	..	..	..	776	1 018
		725	Mainten.; repair of office, account. & comp. mach.	..	..	..	..	..	..	5 075	6 020
		729	Other computer related activities	..	..	..	..	..	..	1 195	1 583
	73		Research and development	..	..	..	..	..	..	614	687
		731	R & D experim. on natural sciences & engineering	..	..	..	..	..	..	461	488
		732	R & D experim. on social sciences & humanities	..	..	..	..	..	..	153	199
	74		Other business activities	..	..	..	..	..	..	232 185	248 450
		741	Legal, acc., book-keep., audit act.; tax consult., etc	..	..	..	..	..	..	47 097	53 068
		742	Architectural, engineering, & other techn. activities	..	..	..	..	..	..	56 243	61 125
		743	Advertising	..	..	..	..	..	..	10 966	13 372
		749	Business activities nec	..	..	..	..	..	..	117 879	120 885
M	80		Education	..	..	..	..	..	..	22 587	28 512
		801	Primary education	..	..	..	..	..	..	259	284
		802	Secondary education	..	..	..	..	..	..	281	357
		803	Higher education	..	..	..	..	..	..	370	457
		809	Adult and other education	..	..	..	..	..	..	21 677	27 414
N	85		Health and social work	..	..	..	..	..	..	82 234	96 316
		851	Human health activities	..	..	..	..	..	..	74 592	88 107
		852	Veterinary activities	..	..	..	..	..	..	7 039	7 406
		853	Social work activities	..	..	..	..	..	..	603	803
O	90_93		Other commun., soc. & pers. services activities	..	..	..	..	..	..	83 023	93 890
	90		Sewage & refuse disposal, sanitation, etc.	..	..	..	..	..	..	3 579	4 027
	91		Activities of member organisations nec	..	..	..	..	..	..	120	153
		912	Activities of trade unions	..	..	..	..	..	..	5	3
	92		Recreational, cultural and sporting activities	..	..	..	..	..	..	33 344	37 777
		921	Motion picture, radio, TV, other entertainment act.	..	..	..	..	..	..	22 865	25 666
		922	News agency activities	..	..	..	..	..	..	387	596
		923	Library, archives, museums, other cultural activ.	..	..	..	..	..	..	1 166	1 275
		924	Sporting and other recreational activities	..	..	..	..	..	..	8 926	10 240
	93		Other service activities	..	..	..	..	..	..	45 980	51 933

Table PO.S1 **PRODUCTION - PRODUCTION**

Millions de Esc (Prix courants)

1991	1992	1993	1994	1995	1996	1997	1998		CITI révision 3		
..	..	..	..	..	12 873 193	13 865 740	16 372 935	Com. de gros, de detail ; réparation etc.	G	**50_52**	
..	..	..	..	..	2 314 083	2 546 805	3 026 759	Com., entr., répar. d'auto. et moto ; carburants		**50**	
..	..	..	..	..	939 932	1 035 153	1 545 438	Commerce de véhicules automobiles			501
..	..	..	..	..	368 752	481 084	337 178	Entretien et réparation de véhicules automobiles			502
..	..	..	..	..	297 228	319 624	327 798	Com. de pièces et access. de véhic. automobiles			503
..	..	..	..	..	138 980	127 470	127 696	Com., entret. et répar. de motocycles et access.			504
..	..	..	..	..	569 191	583 474	688 648	Com. de détail de carburants pour automobiles			505
..	..	..	..	..	6 212 366	6 714 915	7 988 247	Com. de gros et act. inter., sauf auto. et moto		**51**	
..	..	..	..	..	506 422	646 218	809 855	Activités d'intermédiaries de commerce de gros			511
..	..	..	..	..	1 740 420	1 819 186	2 082 756	Com. gros: prod. agr. bruts, anim. viv., prod. alim.			512
..	..	..	..	..	1 494 415	1 479 938	2 076 769	Com. gros: biens de consom., sauf alimentaires			513
..	..	..	..	..	1 313 173	1 430 720	1 468 781	Com. gros: prod. inter. non agr., déchets, débris			514
..	..	..	..	..	691 309	777 553	907 118	Com. gros: machines, équipments et fournitures			515
..	..	..	..	..	466 627	561 300	642 969	Autres commerces de gros			519
..	..	..	..	..	4 346 744	4 604 020	5 357 929	Com. détail, sauf auto, moto; répar. art. pers., dom.		**52**	
..	..	..	..	..	642 406	700 432	814 159	Commerce de détail non spécialisé en magasin			521
..	..	..	..	..	560 133	585 599	669 167	Com. détail: prod. aliment. en mag. spécialisés			522
..	..	..	..	..	2 993 883	3 177 737	3 739 569	Autre com. détail: prod. neufs en mag. spécialisés			523
..	..	..	..	..	12 546	12 444	10 718	Commerces de détail de biens d'occasion			524
..	..	..	..	..	85 462	78 676	82 362	Commerces de détail autres qu'en magasin			525
..	..	..	..	..	52 314	49 132	41 953	Réparation d'articles personnels et domestiques			526
..	..	..	..	..	1 143 494	1 167 444	1 180 466	Hotels et restaurants	H	**55**	
..	..	..	..	..	212 484	263 522	..	Hôtels; camping; autres équip. courts séjours			551
..	..	..	..	..	931 010	903 922	..	Restaurants, cafés et cantines			552
..	..	..	..	..	2 316 465	2 656 789	..	Transport, entreposage, communications	I	**60_64**	
..	..	..	..	..	636 101	838 300	..	Transports terrestres; transports par conduites		**60**	
..	..	..	..	..	..	..	..	Transports par chemin de fer			601
..	..	..	..	..	..	..	..	Autres transports terrestres			602
..	..	..	..	..	..	..	..	Transports par conduites			603
..	..	..	..	..	89 370	83 497	65 020	Transports par eau		**61**	
..	..	..	..	..	77 916	79 114	59 636	Transports maritimes et côtiers			611
..	..	..	..	..	11 454	4 383	5 384	Transports par voies navigables intérieures			612
..	..	..	..	..	195 078	222 243	256 229	Transports aériens		**62**	
..	..	..	..	..	192 494	219 233	..	Transports aériens réguliers			621
..	..	..	..	..	2 584	3 010	..	Transports aériens non réguliers			622
..	..	..	..	..	731 324	795 366	..	Act. annex. et aux. des transp.; act. des ag. de voy.		**63**	
..	..	..	..	..	664 592	717 383	867 386	Postes et télécommunications		**64**	
..	..	..	..	..	91 557	99 519	108 974	Activités de poste et de courrier			641
..	..	..	..	..	573 035	617 864	758 412	Télécommunications			642
..	..	..	..	..	..	..	..	Intermédiation financière	J	**65_67**	
..	..	..	..	..	..	..	..	Interm. financ., sauf assur. et caisses de retraite		**65**	
..	..	..	..	..	..	..	..	Intermédiation monétaire			651
..	..	..	..	..	..	..	..	Autres intermédiations financières			659
..	..	..	..	..	..	..	..	Act. d'assur., caisses de retr., sauf séc. soc. oblig.		**66**	
..	..	..	..	..	..	..	..	Act. auxiliaires de l'intermédiation financière		**67**	
..	..	..	..	..	..	..	..	Act. aux. de l'interm. fin., sauf assur., et retraite			671
..	..	..	..	..	..	..	..	Act. aux. des assurances et caisses de retraite			672
..	..	..	..	..	3 323 365	2 737 029	..	Immobilier, locations, services aux entreprises	K	**70_74**	
..	..	..	..	..	802 030	771 216	1 112 929	Activités immobilières		**70**	
..	..	..	..	..	350 800	431 384	..	Act. immobilières sur biens propres ou loués			701
..	..	..	..	..	451 230	339 832	..	Activités immobilières pour comptes de tiers			702
..	..	..	..	..	192 030	189 689	..	Locat. mach., équip. sans opér., biens pers. et dom.		**71**	
..	..	..	..	..	115 683	135 130	..	Location de matériel de transport			711
..	..	..	..	..	72 011	50 536	88 726	Location de machines ét equipements nca			712
..	..	..	..	..	4 336	4 023	8 135	Loc. d'articles personnels et domestiques nca			713
..	..	..	..	..	112 970	161 854	206 307	Activités informatiques et activités rattachées		**72**	
..	..	..	..	..	6 087	3 798	..	Conseil en matériel informatique			721
..	..	..	..	..	48 199	71 459	..	Production de logiciels			722
..	..	..	..	..	44 272	67 734	..	Traitement de données informatiques			723
..	..	..	..	..	475	2 705	..	Activités de banques de données			724
..	..	..	..	..	5 135	9 995	..	Entr., répar. mach. de bureau et compt. et inform.			725
..	..	..	..	..	8 802	6 163	..	Autres activités rattachées à l'informatique			729
..	..	..	..	..	538	1 274	1 429	Recherche-développement		**73**	
..	..	..	..	..	..	..	..	Rech. et dév. expér. en sc. phys., natur. et ingén.			731
..	..	..	..	..	..	..	..	Rech. et dév. expér. en sciences soc. et humaines			732
..	..	..	..	..	2 215 797	1 612 996	..	Autres activités de services aux entreprises		**74**	
..	..	..	..	..	1 050 773	456 369	..	Act. jur., compt., audit.; cons. fisc.; ét. march., soud.			741
..	..	..	..	..	291 087	315 364	..	Act. d'achitect., d'ingén.; autres act. techniques			742
..	..	..	..	..	475 160	388 961	517 715	Publicité			743
..	..	..	..	..	398 777	452 302	..	Activités de services aux entreprises nca.			749
..	..	..	..	..	116 779	129 023	..	Education	M	**80**	
..	..	..	..	..	21 734	22 515	..	Activités d'enseignement primaire			801
..	..	..	..	..	8 866	10 075	..	Activités d'enseignement secondaire			802
..	..	..	..	..	30 896	36 671	..	Activités d'enseignement supérieur			803
..	..	..	..	..	55 283	59 762	..	Act. de formation perman.; autres act. d'enseign.			809
..	..	..	..	..	219 020	263 523	..	Santé et action sociale	N	**85**	
..	..	..	..	..	211 013	254 725	..	Activités de santé humaine			851
..	..	..	..	..	2 876	2 726	..	Activités vétérinaires			852
..	..	..	..	..	5 131	6 072	..	Activités d'action sociale			853
..	..	..	..	..	362 872	315 996	..	Autres activités de serv. coll., soc. et personnels	O	**90_93**	
..	..	..	..	..	9 805	12 245	..	Assain. et enlèv. des ordures; voierie, act. simil.		**90**	
..	..	..	..	..	..	..	..	Activités associatives diverses		**91**	
..	..	..	..	..	..	..	..	Activités de syndicats de salariés			912
..	..	..	..	..	258 606	216 089	..	Activités récréatives, culturelles et sportives		**92**	
..	..	..	..	..	168 992	151 944	..	Act. cinémat., de radio, télé.; autres act. de spect.			921
..	..	..	..	..	8 682	5 411	..	Activités d'agences de presse			922
..	..	..	..	..	2 818	455	..	Act. des biblio., arch., musées, jardins bot. et zool.			923
..	..	..	..	..	78 114	58 279	..	Activités sportives et autres activités récréatives			924
..	..	..	..	..	94 461	87 662	..	Autres activités de services		**93**	

Table PO.S2 **VALUE ADDED - VALEUR AJOUTÉE**

Millions of Esc (Current Prices)

ISIC revision 3				1991	1992	1993	1994	1995	1996	1997	1998
G	50_52		Wholesale & retail trade; repairs; etc.	..	..	..	..	..	2 127 353	2 467 424	2 907 971
	50		Sale, maint.,repair vehic/cycle, person., h/hold gd.	..	..	..	..	..	358 597	402 477	439 545
		501	Sale of motor cycle	..	..	..	..	..	171 792	184 949	239 489
		502	Maintenance and repair of motor vehicles	..	..	..	..	..	92 217	104 824	92 637
		503	Sale of motor vehicle parts and accessories	..	..	..	..	..	49 534	65 146	53 582
		504	Sale, maint., repair motor cycle, parts & acces.	..	..	..	..	..	15 563	15 714	16 821
		505	Retail sale of automative fuel	..	..	..	..	..	29 490	31 844	37 016
	51		W/sale. trade & commission exc motor veh. & cycle	..	..	..	..	..	1 051 033	1 252 540	1 497 621
		511	Wholesale on a fee or contract basis	..	..	..	..	..	94 683	97 930	112 492
		512	W/sale agric. raw mat.,live animal,food, bev., tobac.	..	..	..	..	..	166 779	221 466	257 229
		513	Wholesale of household goods	..	..	..	..	..	268 669	304 462	389 814
		514	W/sale. of non-agric. intermed. prod., waste, scrap	..	..	..	..	..	289 795	383 767	436 701
		515	Wholesale of machinery, equipment & supplies	..	..	..	..	..	164 332	174 732	209 284
		519	Other wholesale	..	..	..	..	..	66 776	70 185	92 102
	52		Retail trade exc. motor veh. & cyc.; repair h/hold gds.	..	..	..	..	..	717 723	812 406	970 805
		521	Non-specialised retail trade in stores	..	..	..	..	..	157 237	179 591	200 355
		522	Retail sale food, bev., tobac. in specialised stores	..	..	..	..	..	61 623	62 780	75 225
		523	Other retail trade of new goods specialised stores	..	..	..	..	..	471 089	542 780	669 898
		524	Retail sale of second-hand goods in stores	..	..	..	..	..	1 522	2 107	1 676
		525	Retail trade not in stores	..	..	..	..	..	11 283	12 871	11 433
		526	Repair of personal and household goods	..	..	..	..	..	14 967	12 277	12 218
H	55		Hotels and restaurants	..	..	..	..	..	332 456	353 308	397 123
		551	Hotels; camping ; other short-stay accommodation	..	..	..	..	..	109 273	127 811	..
		552	Restaurants, bars and canteens	..	..	..	..	..	223 183	225 497	..
I	60_64		Transport, storage and communications	..	..	..	..	..	946 070	1 031 992	..
	60		Land transport, transport via pipelines	..	..	..	..	..	234 099	276 631	..
		601	Transport via railways	..	..	..	..	..	..	..	..
		602	Other land transport	..	..	..	..	..	..	..	..
		603	Transport via pipelines	..	..	..	..	..	..	..	..
	61		Water transport	..	..	..	..	..	19 210	14 112	14 681
		611	Sea and coastal water transport	..	..	..	..	..	16 117	13 187	13 211
		612	Inland water transport	..	..	..	..	..	3 092	925	1 470
	62		Air transport	..	..	..	..	..	74 097	82 543	97 638
		621	Scheduled air transport	..	..	..	..	..	73 539	81 386	..
		622	Non-scheduled air transport	..	..	..	..	..	558	1 157	..
	63		Support. & aux. transport act., travel agencies activ.	..	..	..	..	..	178 678	196 177	..
	64		Post and telecommunications	..	..	..	..	..	439 987	462 530	535 323
		641	Post and courier activities	..	..	..	..	..	71 044	77 796	82 349
		642	Telecommunications	..	..	..	..	..	368 943	384 734	452 974
J	65_67		Financial intermediation	..	..	..	..	..	..	..	..
	65		Financial intermediation, exc. insur., pension fund	..	..	..	..	..	..	..	..
		651	Monetary intermediation	..	..	..	..	..	..	..	..
		659	Other financial intermediation	..	..	..	..	..	..	..	..
	66		Insurance pension fund, exc. compul. social sec.	..	..	..	..	..	..	..	..
	67		Activities auxiliary to financial intermediation	..	..	..	..	..	..	..	..
		671	Act. auxil to fin. intermed., exc. insur. & pens.fund	..	..	..	..	..	..	..	..
		672	Act. auxiliary to insurance and pension funding	..	..	..	..	..	..	..	..
K	70_74		Real estate, renting and business activities	..	..	..	..	..	1 460 300	906 535	..
	70		Real estate activities	..	..	..	..	..	352 240	184 923	311 119
		701	Real estate activ. with own or leased property	..	..	..	..	..	89 114	139 385	..
		702	Real estate activities on a fee or contract basis	..	..	..	..	..	263 126	45 538	..
	71		Renting of mach. without operator; rent. of pers. gds.	..	..	..	..	..	119 276	123 051	..
		711	Renting of transport equipment	..	..	..	..	..	81 899	99 067	..
		712	Renting of other machinery and equipment	..	..	..	..	..	35 868	22 678	18 304
		713	Renting of personal and household goods nec	..	..	..	..	..	1 508	1 306	3 107
	72		Computer and related activities	..	..	..	..	..	40 783	58 015	72 026
		721	Hardware consultancy	..	..	..	..	..	2 679	1 272	..
		722	Software consultancy and supply	..	..	..	..	..	15 051	25 057	..
		723	Data processing	..	..	..	..	..	17 970	26 445	..
		724	Data base activities	..	..	..	..	..	79	782	..
		725	Mainten.; repair of office, account. & comp. mach.	..	..	..	..	..	1 884	3 406	..
		729	Other computer related activities	..	..	..	..	..	3 121	1 052	..
	73		Research and development	..	..	..	..	..	127	345	172
		731	R & D experim. on natural sciences & engineering	..	..	..	..	..	..	..	..
		732	R & D experim. on social sciences & humanities	..	..	..	..	..	..	..	..
	74		Other business activities	..	..	..	..	..	947 874	540 201	..
		741	Legal, acc., book-keep., audit act.; tax consult., etc	..	..	..	..	..	608 768	182 161	..
		742	Architectural, engineering, & other techn. activities	..	..	..	..	..	90 528	101 357	..
		743	Advertising	..	..	..	..	..	51 647	56 097	66 259
		749	Business activities nec	..	..	..	..	..	196 931	200 586	..
M	80		Education	..	..	..	..	..	56 729	65 299	..
		801	Primary education	..	..	..	..	..	13 854	13 990	..
		802	Secondary education	..	..	..	..	..	5 438	5 476	..
		803	Higher education	..	..	..	..	..	17 015	24 073	..
		809	Adult and other education	..	..	..	..	..	20 421	21 760	..
N	85		Health and social work	..	..	..	..	..	93 771	106 776	..
		851	Human health activities	..	..	..	..	..	90 794	103 866	..
		852	Veterinary activities	..	..	..	..	..	749	650	..
		853	Social work activities	..	..	..	..	..	2 228	2 260	..
O	90_93		Other commun., soc. & pers. services activities	..	..	..	..	..	93 248	92 766	..
	90		Sewage & refuse disposal, sanitation, etc.	..	..	..	..	..	4 285	4 928	..
	91		Activities of member organisations nec	..	..	..	..	..	..	..	..
		912	Activities of trade unions	..	..	..	..	..	..	..	..
	92		Recreational, cultural and sporting activities	..	..	..	..	..	46 528	48 377	..
		921	Motion picture, radio, TV, other entertainment act.	..	..	..	..	..	22 775	26 306	..
		922	News agency activities	..	..	..	..	..	2 264	1 573	..
		923	Library, archives, museums, other cultural activ.	..	..	..	..	..	1 479	145	..
		924	Sporting and other recreational activities	..	..	..	..	..	20 011	20 353	..
	93		Other service activities	..	..	..	..	..	42 435	39 461	..

Table PO.S3 **INVESTISSEMENT - INVESTMENT**

Millions de Esc (Prix courants)

1991	1992	1993	1994	1995	1996	1997	1998		CITI révision 3		
..	..	..	..	..	476 262	427 501	694 648	Com. de gros, de detail ; réparation etc.	G	50_52	
..	..	..	..	..	85 204	78 938	176 226	Com., entr., répar. d'auto. et moto ; carburants		50	
..	..	..	..	..	29 050	36 652	113 806	Commerce de véhicules automobiles			501
..	..	..	..	..	18 574	26 089	29 418	Entretien et réparation de véhicules automobiles			502
..	..	..	..	..	25 935	7 642	15 722	Com. de pièces et access. de véhic. automobiles			503
..	..	..	..	..	3 399	2 549	5 361	Com., entret. et répar. de motocycles et access.			504
..	..	..	..	..	8 245	6 006	11 919	Com. de détail de carburants pour automobiles			505
..	..	..	..	..	231 641	167 020	270 449	Com. de gros et act. inter., sauf auto. et moto		51	
..	..	..	..	..	15 606	14 429	18 428	Activités d'intermédiaries de commerce de gros			511
..	..	..	..	..	104 088	35 698	58 066	Com. gros: prod. agr. bruts, anim. viv., prod. alim.			512
..	..	..	..	..	27 265	30 581	53 419	Com. gros: biens de consom., sauf alimentaires			513
..	..	..	..	..	40 985	46 129	78 920	Com. gros: prod. inter. non agr., déchets, débris			514
..	..	..	..	..	21 797	27 924	43 540	Com. gros: machines, équipments et fournitures			515
..	..	..	..	..	21 900	12 260	18 075	Autres commerces de gros			519
..	..	..	..	..	159 416	181 543	247 973	Com. détail, sauf auto, moto; répar. art. pers., dom.		52	
..	..	..	..	..	38 442	26 239	80 462	Commerce de détail non spécialisé en magasin			521
..	..	..	..	..	19 463	10 996	20 056	Com. détail: prod. aliment. en mag. spécialisés			522
..	..	..	..	..	96 697	136 349	143 530	Autre com. détail: prod. neufs en mag. spécialisés			523
..	..	..	..	..	341	674	434	Commerces de détail de biens d'occasion			524
..	..	..	..	..	2 169	5 097	1 569	Commerces de détail autres qu'en magasin			525
..	..	..	..	..	2 304	2 188	1 920	Réparation d'articles personnels et domestiques			526
..	..	..	..	..	186 903	84 855	106 845	Hotels et restaurants	H	55	
..	..	..	..	..	37 074	32 957	..	Hôtels; camping; autres équip. courts séjours			551
..	..	..	..	..	149 829	51 898	..	Restaurants, cafés et cantines			552
..	..	..	..	..	382 130	398 644	..	Transport, entreposage, communications	I	60_64	
..	..	..	..	..	245 522	166 842	..	Transports terrestres; transports par conduites		60	
..	..	..	..	..	..	..	..	Transports par chemin de fer			601
..	..	..	..	..	..	..	..	Autres transports terrestres			602
..	..	..	..	..	..	..	..	Transports par conduites			603
..	..	..	..	..	16 846	15 112	15 549	Transports par eau		61	
..	..	..	..	..	13 954	11 142	14 194	Transports maritimes et côtiers			611
..	..	..	..	..	2 892	3 970	1 355	Transports par voies navigables intérieures			612
..	..	..	..	..		10 821	28 981	Transports aériens		62	
..	..	..	..	..		10 205	..	Transports aériens réguliers			621
..	..	..	..	..	598	616	..	Transports aériens non réguliers			622
..	..	..	..	..	61 980	135 434	..	Act. annex. et aux. des transp.; act. des ag. de voy.		63	
..	..	..	..	..	60 576	70 435	348 021	Postes et télécommunications		64	
..	..	..	..	..	5 933	6 704	21 463	Activités de poste et de courrier			641
..	..	..	..	..	54 643	63 731	326 558	Télécommunications			642
..	..	..	..	..	..	..	..	Intermédiation financière	J	65_67	
..	..	..	..	..	..	..	..	Interm. financ., sauf assur. et caisses de retraite		65	
..	..	..	..	..	..	..	..	Intermédiation monétaire			651
..	..	..	..	..	..	..	..	Autres intermédiations financières			659
..	..	..	..	..	..	..	..	Act. d'assur., caisses de retr., sauf séc. soc. oblig.		66	
..	..	..	..	..	..	..	..	Act. auxiliaires de l'intermédiation financière		67	
..	..	..	..	..	..	..	..	Act. aux. de l'interm. fin., sauf assur., et retraite			671
..	..	..	..	..	..	..	..	Act. aux. des assurances et caisses de retraite			672
..	..	..	..	..	770 996	275 529	..	Immobilier, locations, services aux entreprises	K	70_74	
..	..	..	..	..	382 693	43 018	322 012	Activités immobilières		70	
..	..	..	..	..	62 598	35 448	..	Act. immobilières sur biens propres ou loués			701
..	..	..	..	..	320 095	7 570	..	Activités immobilières pour comptes de tiers			702
..	..	..	..	..	78 000	63 613	..	Locat. mach., équip. sans opér., biens pers. et dom.		71	
..	..	..	..	..	42 979	50 127	..	Location de matériel de transport			711
..	..	..	..	..	34 352	12 233	9 086	Location de machines ét equipements nca			712
..	..	..	..	..	669	1 253	1 862	Loc. d'articles personnels et domestiques nca			713
..	..	..	..	..	26 866	10 471	12 524	Activités informatiques et activités rattachées		72	
..	..	..	..	..	393	112	..	Conseil en matériel informatique			721
..	..	..	..	..	2 555	6 228	..	Production de logiciels			722
..	..	..	..	..	22 643	3 262	..	Traitement de données informatiques			723
..	..	..	..	..	8	86	..	Activités de banques de données			724
..	..	..	..	..	361	295	..	Entr., répar. mach. de bureau et compt. et inform.			725
..	..	..	..	..	906	487	..	Autres activités rattachées à l'informatique			729
..	..	..	..	..	30	138	185	Recherche-développement		73	
..	..	..	..	..	..	..	..	Rech. et dév. expér. en sc. phys., natur. et ingén.			731
..	..	..	..	..	..	..	..	Rech. et dév. expér. en sciences soc. et humaines			732
..	..	..	..	..	283 407	158 289	..	Autres activités de services aux entreprises		74	
..	..	..	..	..	60 338	40 056	..	Act. jur., compt., audit.; cons. fisc.; ét. march., soud.			741
..	..	..	..	..	14 974	14 889	..	Act. d'achitect., d'ingén.; autres act. techniques			742
..	..	..	..	..	15 322	9 959	17 438	Publicité			743
..	..	..	..	..	192 773	93 386	..	Activités de services aux entreprises nca.			749
..	..	..	..	..	25 299	12 013	..	Education	M	80	
..	..	..	..	..	2 437	2 039	..	Activités d'enseignement primaire			801
..	..	..	..	..	1 394	1 441	..	Activités d'enseignement secondaire			802
..	..	..	..	..	15 627	5 828	..	Activités d'enseignement supérieur			803
..	..	..	..	..	5 841	2 705	..	Act. de formation perman.; autres act. d'enseign.			809
..	..	..	..	..	133 943	32 250	..	Santé et action sociale	N	85	
..	..	..	..	..	132 420	31 030	..	Activités de santé humaine			851
..	..	..	..	..	322	306	..	Activités vétérinaires			852
..	..	..	..	..	1 201	914	..	Activités d'action sociale			853
..	..	..	..	..	43 378	29 259	..	Autres activités de serv. coll., soc. et personnels	O	90_93	
..	..	..	..	..	2 669	6 627	..	Assain. et enlèv. des ordures; voierie, act. simil.		90	
..	..	..	..	..	..	..	..	Activités associatives diverses		91	
..	..	..	..	..	..	..	..	Activités de syndicats de salariés			912
..	..	..	..	..	22 068	17 218	..	Activités récréatives, culturelles et sportives		92	
..	..	..	..	..	9 841	16 518	..	Act. cinémat., de radio, télé.; autres act. de spect.			921
..	..	..	..	..	575	..	..	Activités d'agences de presse			922
..	..	..	..	..	289	66	..	Act. des biblio., arch., musées, jardins bot. et zool.			923
..	..	..	..	..	11 363	678	..	Activités sportives et autres activités récréatives			924
..	..	..	..	..	18 641	5 414	..	Autres activités de services		93	

Table PO.S4 **EMPLOYMENT, EMPLOYEES - EMPLOI, SALARIÉS**

Units

ISIC revision 3				1991	1992	1993	1994	1995	1996	1997	1998
G	50_52		Wholesale & retail trade; repairs; etc.	..	..	..	..	..	632 129	658 482	678 491
	50		Sale, maint.,repair vehic/cycle, person., h/hold gd.	..	..	..	..	..	119 928	124 754	119 616
		501	Sale of motor cycle	..	..	..	..	..	42 420	43 669	44 662
		502	Maintenance and repair of motor vehicles	..	..	..	..	..	41 795	44 697	36 616
		503	Sale of motor vehicle parts and accessories	..	..	..	..	..	16 208	17 639	16 856
		504	Sale, maint., repair motor cycle, parts & acces.	..	..	..	..	..	5 287	4 524	5 774
		505	Retail sale of automative fuel	..	..	..	..	..	14 218	14 225	15 708
	51		W/sale. trade & commission exc motor veh. & cycle	..	..	..	..	..	228 652	232 739	241 095
		511	Wholesale on a fee or contract basis	..	..	..	..	..	21 633	23 917	20 013
		512	W/sale agric. raw mat.,live animal,food, bev., tobac.	..	..	..	..	..	59 595	59 929	60 877
		513	Wholesale of household goods	..	..	..	..	..	62 355	61 629	64 419
		514	W/sale. of non-agric. intermed. prod., waste, scrap	..	..	..	..	..	36 624	39 153	39 754
		515	Wholesale of machinery, equipment & supplies	..	..	..	..	..	33 271	33 973	37 027
		519	Other wholesale	..	..	..	..	..	15 173	14 138	19 004
	52		Retail trade exc. motor veh. & cyc.; repair h/hold gds.	..	..	..	..	..	283 549	300 988	317 780
		521	Non-specialised retail trade in stores	..	..	..	..	..	57 435	68 309	65 317
		522	Retail sale food, bev., tobac. in specialised stores	..	..	..	..	..	31 054	28 684	30 599
		523	Other retail trade of new goods specialised stores	..	..	..	..	..	187 376	196 356	214 158
		524	Retail sale of second-hand goods in stores	..	..	..	..	..	674	881	765
		525	Retail trade not in stores	..	..	..	..	..	1 955	1 830	2 441
		526	Repair of personal and household goods	..	..	..	..	..	5 055	4 928	4 499
H	55		Hotels and restaurants	..	..	..	..	..	186 268	186 725	177 933
		551	Hotels; camping ; other short-stay accommodation	..	..	..	..	..	37 130	43 154	..
		552	Restaurants, bars and canteens	..	..	..	..	..	149 138	143 571	..
I	60_64		Transport, storage and communications	..	..	..	..	..	157 445	158 692	..
	60		Land transport, transport via pipelines	..	..	..	..	..	81 607	84 103	..
		601	Transport via railways	..	..	..	..	..	..	..	..
		602	Other land transport	..	..	..	..	..	..	..	..
		603	Transport via pipelines	..	..	..	..	..	..	..	..
	61		Water transport	..	..	..	..	..	2 294	2 068	2 016
		611	Sea and coastal water transport	..	..	..	..	..	1 255	1 131	1 388
		612	Inland water transport	..	..	..	..	..	1 039	937	628
	62		Air transport	..	..	..	..	..	9 777	9 718	10 390
		621	Scheduled air transport	..	..	..	..	..	9 674	9 614	..
		622	Non-scheduled air transport	..	..	..	..	..	103	104	..
	63		Support. & aux. transport act., travel agencies activ.	..	..	..	..	..	26 716	26 045	..
	64		Post and telecommunications	..	..	..	..	..	37 050	36 758	38 739
		641	Post and courier activities	..	..	..	..	..	16 053	16 271	16 969
		642	Telecommunications	..	..	..	..	..	20 997	20 487	21 770
J	65_67		Financial intermediation	..	..	..	..	..	..	..	..
	65		Financial intermediation, exc. insur., pension fund	..	..	..	..	..	..	..	..
		651	Monetary intermediation	..	..	..	..	..	..	..	..
		659	Other financial intermediation	..	..	..	..	..	..	..	..
	66		Insurance pension fund, exc. compul. social sec.	..	..	..	..	..	..	..	..
	67		Activities auxiliary to financial intermediation	..	..	..	..	..	..	..	..
		671	Act. auxil to fin. intermed., exc. insur. & pens.fund	..	..	..	..	..	..	..	..
		672	Act. auxiliary to insurance and pension funding	..	..	..	..	..	..	..	..
K	70_74		Real estate, renting and business activities	..	..	..	..	..	185 960	208 935	..
	70		Real estate activities	..	..	..	..	..	19 422	17 720	21 312
		701	Real estate activ. with own or leased property	..	..	..	..	..	10 434	10 468	..
		702	Real estate activities on a fee or contract basis	..	..	..	..	..	8 988	7 251	..
	71		Renting of mach. without operator; rent. of pers. gds.	..	..	..	..	..	7 521	7 022	..
		711	Renting of transport equipment	..	..	..	..	..	3 441	3 279	..
		712	Renting of other machinery and equipment	..	..	..	..	..	3 572	3 329	3 435
		713	Renting of personal and household goods nec	..	..	..	..	..	508	414	798
	72		Computer and related activities	..	..	..	..	..	8 106	10 606	11 108
		721	Hardware consultancy	..	..	..	..	..	316	394	..
		722	Software consultancy and supply	..	..	..	..	..	2 970	4 434	..
		723	Data processing	..	..	..	..	..	3 575	4 286	..
		724	Data base activities	..	..	..	..	..	36	137	..
		725	Mainten.; repair of office, account. & comp. mach.	..	..	..	..	..	650	896	..
		729	Other computer related activities	..	..	..	..	..	559	459	..
	73		Research and development	..	..	..	..	..	78	75	193
		731	R & D experim. on natural sciences & engineering	..	..	..	..	..	..	..	..
		732	R & D experim. on social sciences & humanities	..	..	..	..	..	..	..	..
	74		Other business activities	..	..	..	..	..	150 832	173 513	..
		741	Legal, acc., book-keep., audit act.; tax consult., etc	..	..	..	..	..	33 628	38 477	..
		742	Architectural, engineering, & other techn. activities	..	..	..	..	..	18 560	21 207	..
		743	Advertising	..	..	..	..	..	9 128	14 612	11 463
		749	Business activities nec	..	..	..	..	..	89 516	99 216	..
M	80		Education	..	..	..	..	..	25 790	29 394	..
		801	Primary education	..	..	..	..	..	8 762	9 194	..
		802	Secondary education	..	..	..	..	..	4 545	5 531	..
		803	Higher education	..	..	..	..	..	3 794	5 609	..
		809	Adult and other education	..	..	..	..	..	8 690	9 060	..
N	85		Health and social work	..	..	..	..	..	23 580	24 323	..
		851	Human health activities	..	..	..	..	..	20 920	21 477	..
		852	Veterinary activities	..	..	..	..	..	367	310	..
		853	Social work activities	..	..	..	..	..	2 293	2 536	..
O	90_93		Other commun., soc. & pers. services activities	..	..	..	..	..	41 184	38 650	..
	90		Sewage & refuse disposal, sanitation, etc.	..	..	..	..	..	1 448	1 352	..
	91		Activities of member organisations nec	..	..	..	..	..	..	..	..
		912	Activities of trade unions	..	..	..	..	..	..	..	..
	92		Recreational, cultural and sporting activities	..	..	..	..	..	16 518	16 009	..
		921	Motion picture, radio, TV, other entertainment act.	..	..	..	..	..	9 814	9 650	..
		922	News agency activities	..	..	..	..	..	416	393	..
		923	Library, archives, museums, other cultural activ.	..	..	..	..	..	498	79	..
		924	Sporting and other recreational activities	..	..	..	..	..	5 791	5 887	..
	93		Other service activities	..	..	..	..	..	23 217	21 289	..

Table PO.S5 **SALAIRES ET CHARGES SOCIALES, SALARIÉS - COMPENSATION OF LABOUR, EMPLOYEES**

Millions de Esc (Prix courants)

1991	1992	1993	1994	1995	1996	1997	1998		CITI révision 3		
..	..	..	..	..	1 225 634	1 342 970	1 460 383	Com. de gros, de detail ; réparation etc.	G	50_52	
..	..	..	..	..	229 276	255 525	256 376	Com., entr., répar. d'auto. et moto ; carburants		50	
..	..	..	..	..	101 638	110 967	122 723	Commerce de véhicules automobiles			501
..	..	..	..	..	65 634	74 347	62 616	Entretien et réparation de véhicules automobiles			502
..	..	..	..	..	31 213	38 121	34 889	Com. de pièces et access. de véhic. automobiles			503
..	..	..	..	..	8 829	8 505	8 829	Com., entret. et répar. de motocycles et access.			504
..	..	..	..	..	21 962	23 585	27 319	Com. de détail de carburants pour automobiles			505
..	..	..	..	..	549 187	590 322	655 515	Com. de gros et act. inter., sauf auto. et moto		51	
..	..	..	..	..	50 104	50 386	52 090	Activités d'intermédiaries de commerce de gros			511
..	..	..	..	..	114 879	119 428	129 021	Com. gros: prod. agr. bruts, anim. viv., prod. alim.			512
..	..	..	..	..	154 961	176 809	189 754	Com. gros: biens de consom., sauf alimentaires			513
..	..	..	..	..	96 581	102 674	109 193	Com. gros: prod. inter. non agr., déchets, débris			514
..	..	..	..	..	96 917	106 751	122 020	Com. gros: machines, équipments et fournitures			515
..	..	..	..	..	35 745	34 272	53 437	Autres commerces de gros			519
..	..	..	..	..	447 171	497 122	548 492	Com. détail, sauf auto, moto; répar. art. pers., dom.		52	
..	..	..	..	..	91 395	105 330	116 219	Commerce de détail non spécialisé en magasin			521
..	..	..	..	..	38 584	38 358	42 676	Com. détail: prod. aliment. en mag. spécialisés			522
..	..	..	..	..	302 738	339 343	375 866	Autre com. détail: prod. neufs en mag. spécialisés			523
..	..	..	..	..	1 192	1 536	1 326	Commerces de détail de biens d'occasion			524
..	..	..	..	..	4 535	4 253	5 173	Commerces de détail autres qu'en magasin			525
..	..	..	..	..	8 727	8 301	7 232	Réparation d'articles personnels et domestiques			526
..	..	..	..	..	240 321	257 724	253 871	Hotels et restaurants	H	55	
..	..	..	..	..	71 896	82 994	..	Hôtels; camping; autres équip. courts séjours			551
..	..	..	..	..	168 425	174 730	..	Restaurants, cafés et cantines			552
..	..	..	..	..	538 731	593 029	..	Transport, entreposage, communications	I	60_64	
..	..	..	..	..	212 537	237 325	..	Transports terrestres; transports par conduites		60	
..	..	..	..	..	..	..	..	Transports par chemin de fer			601
..	..	..	..	..	..	..	..	Autres transports terrestres			602
..	..	..	..	..	..	..	..	Transports par conduites			603
..	..	..	..	..	7 621	8 175	6 652	Transports par eau		61	
..	..	..	..	..	4 494	5 284	4 695	Transports maritimes et côtiers			611
..	..	..	..	..	3 127	2 891	1 957	Transports par voies navigables intérieures			612
..	..	..	..	..	59 558	65 395	76 145	Transports aériens		62	
..	..	..	..	..	59 175	64 961	..	Transports aériens réguliers			621
..	..	..	..	..	384	434	..	Transports aériens non réguliers			622
..	..	..	..	..	96 284	104 415	..	Act. annex. et aux. des transp.; act. des ag. de voy.		63	
..	..	..	..	..	162 731	177 719	206 490	Postes et télécommunications		64	
..	..	..	..	..	58 709	63 786	75 037	Activités de poste et de courrier			641
..	..	..	..	..	104 022	113 933	131 453	Télécommunications			642
..	..	..	..	..	..	..	..	Intermédiation financière	J	65_67	
..	..	..	..	..	..	..	..	Interm. financ., sauf assur. et caisses de retraite		65	
..	..	..	..	..	..	..	..	Intermédiation monétaire			651
..	..	..	..	..	..	..	..	Autres intermédiations financières			659
..	..	..	..	..	..	..	..	Act. d'assur., caisses de retr., sauf séc. soc. oblig.		66	
..	..	..	..	..	..	..	..	Act. auxiliaires de l'intermédiation financière		67	
..	..	..	..	..	..	..	..	Act. aux. de l'interm. fin., sauf assur., et retraite			671
..	..	..	..	..	..	..	..	Act. aux. des assurances et caisses de retraite			672
..	..	..	..	..	420 722	494 538	..	Immobilier, locations, services aux entreprises	K	70_74	
..	..	..	..	..	42 876	42 564	59 431	Activités immobilières		70	
..	..	..	..	..	26 352	27 126	..	Act. immobilières sur biens propres ou loués			701
..	..	..	..	..	16 524	15 438	..	Activités immobilières pour comptes de tiers			702
..	..	..	..	..	16 991	16 872	..	Locat. mach., équip. sans opér., biens pers. et dom.		71	
..	..	..	..	..	8 729	8 394	..	Location de matériel de transport			711
..	..	..	..	..	7 505	7 971	7 812	Location de machines ét equipements nca			712
..	..	..	..	..	756	507	1 631	Loc. d'articles personnels et domestiques nca			713
..	..	..	..	..	31 537	42 159	46 494	Activités informatiques et activités rattachées		72	
..	..	..	..	..	1 325	970	..	Conseil en matériel informatique			721
..	..	..	..	..	13 490	18 949	..	Production de logiciels			722
..	..	..	..	..	13 005	17 905	..	Traitement de données informatiques			723
..	..	..	..	..	84	377	..	Activités de banques de données			724
..	..	..	..	..	1 612	2 324	..	Entr., répar. mach. de bureau et compt. et inform.			725
..	..	..	..	..	2 020	1 634	..	Autres activités rattachées à l'informatique			729
..	..	..	..	..	128	139	872	Recherche-développement		73	
..	..	..	..	..	..	..	..	Rech. et dév. expér. en sc. phys., natur. et ingén.			731
..	..	..	..	..	..	..	..	Rech. et dév. expér. en sciences soc. et humaines			732
..	..	..	..	..	329 190	392 804	..	Autres activités de services aux entreprises		74	
..	..	..	..	..	100 025	126 704	..	Act. jur., compt., audit.; cons. fisc.; ét. march., soud.			741
..	..	..	..	..	57 503	71 106	..	Act. d'achitect., d'ingén.; autres act. techniques			742
..	..	..	..	..	31 643	34 842	39 389	Publicité			743
..	..	..	..	..	140 020	160 153	..	Activités de services aux entreprises nca.			749
..	..	..	..	..	58 164	70 109	..	Education	M	80	
..	..	..	..	..	16 485	17 057	..	Activités d'enseignement primaire			801
..	..	..	..	..	11 570	14 452	..	Activités d'enseignement secondaire			802
..	..	..	..	..	11 738	18 427	..	Activités d'enseignement supérieur			803
..	..	..	..	..	18 370	20 173	..	Act. de formation perman.; autres act. d'enseign.			809
..	..	..	..	..	43 413	51 419	..	Santé et action sociale	N	85	
..	..	..	..	..	40 415	47 446	..	Activités de santé humaine			851
..	..	..	..	..	426	380	..	Activités vétérinaires			852
..	..	..	..	..	2 572	3 592	..	Activités d'action sociale			853
..	..	..	..	..	86 192	82 969	..	Autres activités de serv. coll., soc. et personnels	O	90_93	
..	..	..	..	..	2 721	3 085	..	Assain. et enlèv. des ordures; voierie, act. simil.		90	
..	..	..	..	..	..	..	..	Activités associatives diverses		91	
..	..	..	..	..	..	..	..	Activités de syndicats de salariés			912
..	..	..	..	..	51 833	50 634	..	Activités récréatives, culturelles et sportives		92	
..	..	..	..	..	36 699	37 604	..	Act. cinémat., de radio, télé.; autres act. de spect.			921
..	..	..	..	..	2 452	1 735	..	Activités d'agences de presse			922
..	..	..	..	..	1 035	113	..	Act. des biblio., arch., musées, jardins bot. et zool.			923
..	..	..	..	..	11 648	11 181	..	Activités sportives et autres activités récréatives			924
..	..	..	..	..	31 638	29 251	..	Autres activités de services		93	

Table PO.S6 **WAGES AND SALARIES, EMPLOYEES - SALAIRES ET TRAITEMENTS, SALARIÉS**

Millions of Esc (Current Prices)

ISIC revision 3				1991	1992	1993	1994	1995	1996	1997	1998
G	50_52		Wholesale & retail trade; repairs; etc.	..	..	..	..	..	748 098	822 916	..
	50		Sale, maint.,repair vehic/cycle, person., h/hold gd.	..	..	..	..	..	144 095	159 955	..
		501	Sale of motor cycle	..	..	..	..	..	69 428	75 039	..
		502	Maintenance and repair of motor vehicles	..	..	..	..	..	37 746	43 392	..
		503	Sale of motor vehicle parts and accessories	..	..	..	..	..	17 814	21 810	..
		504	Sale, maint., repair motor cycle, parts & acces.	..	..	..	..	..	5 091	4 618	..
		505	Retail sale of automative fuel	..	..	..	..	..	14 016	15 096	..
	51		W/sale. trade & commission exc motor veh. & cycle	..	..	..	..	..	334 560	361 879	..
		511	Wholesale on a fee or contract basis	..	..	..	..	..	27 902	28 699	..
		512	W/sale agric. raw mat.,live animal,food, bev., tobac.	..	..	..	..	..	73 582	75 941	..
		513	Wholesale of household goods	..	..	..	..	..	97 051	107 610	..
		514	W/sale. of non-agric. intermed. prod., waste, scrap	..	..	..	..	..	56 791	63 758	..
		515	Wholesale of machinery, equipment & supplies	..	..	..	..	..	58 853	66 599	..
		519	Other wholesale	..	..	..	..	..	20 381	19 272	..
	52		Retail trade exc. motor veh. & cyc.; repair h/hold gds.	..	..	..	..	..	269 444	301 083	..
		521	Non-specialised retail trade in stores	..	..	..	..	..	64 348	74 534	..
		522	Retail sale food, bev., tobac. in specialised stores	..	..	..	..	..	21 265	19 972	..
		523	Other retail trade of new goods specialised stores	..	..	..	..	..	176 529	199 533	..
		524	Retail sale of second-hand goods in stores	..	..	..	..	..	458	747	..
		525	Retail trade not in stores	..	..	..	..	..	2 270	2 149	..
		526	Repair of personal and household goods	..	..	..	..	..	4 574	4 148	..
H	55		Hotels and restaurants	..	..	..	..	..	153 846	167 050	..
		551	Hotels; camping ; other short-stay accommodation	..	..	..	..	..	51 914	60 017	..
		552	Restaurants, bars and canteens	..	..	..	..	..	101 932	107 033	..
I	60_64		Transport, storage and communications	..	..	..	..	..	384 168	416 660	..
	60		Land transport, transport via pipelines	..	..	..	..	..	151 534	169 799	..
		601	Transport via railways	..	..	..	..	..	..	..	..
		602	Other land transport	..	..	..	..	..	..	..	..
		603	Transport via pipelines	..	..	..	..	..	..	..	..
	61		Water transport	..	..	..	..	..	5 465	5 930	..
		611	Sea and coastal water transport	..	..	..	..	..	3 078	3 732	..
		612	Inland water transport	..	..	..	..	..	2 387	2 198	..
	62		Air transport	..	..	..	..	..	42 736	47 343	..
		621	Scheduled air transport	..	..	..	..	..	42 505	47 089	..
		622	Non-scheduled air transport	..	..	..	..	..	231	254	..
	63		Support. & aux. transport act., travel agencies activ.	..	..	..	..	..	64 956	70 971	..
	64		Post and telecommunications	..	..	..	..	..	119 477	122 617	..
		641	Post and courier activities	..	..	..	..	..	42 414	46 326	..
		642	Telecommunications	..	..	..	..	..	77 063	76 291	..
J	65_67		Financial intermediation	..	..	..	..	..	..	..	..
	65		Financial intermediation, exc. insur., pension fund	..	..	..	..	..	..	..	..
		651	Monetary intermediation	..	..	..	..	..	..	..	..
		659	Other financial intermediation	..	..	..	..	..	..	..	..
	66		Insurance pension fund, exc. compul. social sec.	..	..	..	..	..	..	..	..
	67		Activities auxiliary to financial intermediation	..	..	..	..	..	..	..	..
		671	Act. auxil to fin. intermed., exc. insur. & pens.fund	..	..	..	..	..	..	..	..
		672	Act. auxiliary to insurance and pension funding	..	..	..	..	..	..	..	..
K	70_74		Real estate, renting and business activities	..	..	..	..	..	260 021	318 602	..
	70		Real estate activities	..	..	..	..	..	23 260	23 180	..
		701	Real estate activ. with own or leased property	..	..	..	..	..	14 162	14 648	..
		702	Real estate activities on a fee or contract basis	..	..	..	..	..	9 098	8 532	..
	71		Renting of mach. without operator; rent. of pers. gds.	..	..	..	..	..	11 305	11 235	..
		711	Renting of transport equipment	..	..	..	..	..	6 084	5 920	..
		712	Renting of other machinery and equipment	..	..	..	..	..	4 887	5 037	..
		713	Renting of personal and household goods nec	..	..	..	..	..	334	278	..
	72		Computer and related activities	..	..	..	..	..	19 861	27 504	..
		721	Hardware consultancy	..	..	..	..	..	529	497	..
		722	Software consultancy and supply	..	..	..	..	..	8 718	12 878	..
		723	Data processing	..	..	..	..	..	8 227	11 590	..
		724	Data base activities	..	..	..	..	..	53	240	..
		725	Mainten.; repair of office, account. & comp. mach.	..	..	..	..	..	991	1 286	..
		729	Other computer related activities	..	..	..	..	..	1 343	1 013	..
	73		Research and development	..	..	..	..	..	64	98	..
		731	R & D experim. on natural sciences & engineering	..	..	..	..	..	..	..	..
		732	R & D experim. on social sciences & humanities	..	..	..	..	..	..	..	..
	74		Other business activities	..	..	..	..	..	205 531	256 585	..
		741	Legal, acc., book-keep., audit act.; tax consult., etc	..	..	..	..	..	57 515	78 761	..
		742	Architectural, engineering, & other techn. activities	..	..	..	..	..	33 044	43 914	..
		743	Advertising	..	..	..	..	..	19 204	20 303	..
		749	Business activities nec	..	..	..	..	..	95 768	113 607	..
M	80		Education	..	..	..	..	..	42 043	51 683	..
		801	Primary education	..	..	..	..	..	12 314	12 688	..
		802	Secondary education	..	..	..	..	..	9 172	11 490	..
		803	Higher education	..	..	..	..	..	8 266	14 299	..
		809	Adult and other education	..	..	..	..	..	12 291	13 206	..
N	85		Health and social work	..	..	..	..	..	26 358	29 503	..
		851	Human health activities	..	..	..	..	..	24 322	26 890	..
		852	Veterinary activities	..	..	..	..	..	228	167	..
		853	Social work activities	..	..	..	..	..	1 808	2 446	..
O	90_93		Other commun., soc. & pers. services activities	..	..	..	..	..	52 930	52 560	..
	90		Sewage & refuse disposal, sanitation, etc.	..	..	..	..	..	1 780	2 160	..
	91		Activities of member organisations nec	..	..	..	..	..	..	..	..
		912	Activities of trade unions	..	..	..	..	..	..	..	..
	92		Recreational, cultural and sporting activities	..	..	..	..	..	34 268	34 705	..
		921	Motion picture, radio, TV, other entertainment act.	..	..	..	..	..	23 620	25 650	..
		922	News agency activities	..	..	..	..	..	1 943	1 293	..
		923	Library, archives, museums, other cultural activ.	..	..	..	..	..	719	55	..
		924	Sporting and other recreational activities	..	..	..	..	..	7 986	7 707	..
	93		Other service activities	..	..	..	..	..	16 882	15 695	..

Table PO.S7 **CHARGES SOCIALES DES EMPLOYEURS - SUPPLEMENTS TO WAGES & SALARIES**

Millions de Esc (Prix courants)

1991	1992	1993	1994	1995	1996	1997	1998	CITI révision 3			
..	..	..	..	..	276 720	303 725	..	Com. de gros, de detail ; réparation etc.	G	50_52	
..	..	..	..	..	52 757	57 638	..	Com., entr., répar. d'auto. et moto ; carburants		50	
..	..	..	..	..	24 306	26 227	..	Commerce de véhicules automobiles			501
..	..	..	..	..	14 281	15 778	..	Entretien et réparation de véhicules automobiles			502
..	..	..	..	..	7 547	8 818	..	Com. de pièces et access. de véhic. automobiles			503
..	..	..	..	..	1 744	1 730	..	Com., entret. et répar. de motocycles et access.			504
..	..	..	..	..	4 879	5 085	..	Com. de détail de carburants pour automobiles			505
..	..	..	..	..	130 741	142 621	..	Com. de gros et act. inter., sauf auto. et moto		51	
..	..	..	..	..	10 851	10 960	..	Activités d'intermédiaries de commerce de gros			511
..	..	..	..	..	26 195	26 927	..	Com. gros: prod. agr. bruts, anim. viv., prod. alim.			512
..	..	..	..	..	36 206	44 756	..	Com. gros: biens de consom., sauf alimentaires			513
..	..	..	..	..	24 654	25 293	..	Com. gros: prod. inter. non agr., déchets, débris			514
..	..	..	..	..	24 651	26 728	..	Com. gros: machines, équipments et fournitures			515
..	..	..	..	..	8 184	7 957	..	Autres commerces de gros			519
..	..	..	..	..	93 222	103 466	..	Com. détail, sauf auto, moto; répar. art. pers., dom.		52	
..	..	..	..	..	18 539	20 917	..	Commerce de détail non spécialisé en magasin			521
..	..	..	..	..	7 754	7 516	..	Com. détail: prod. aliment. en mag. spécialisés			522
..	..	..	..	..	63 622	71 982	..	Autre com. détail: prod. neufs en mag. spécialisés			523
..	..	..	..	..	328	320	..	Commerces de détail de biens d'occasion			524
..	..	..	..	..	1 077	967	..	Commerces de détail autres qu'en magasin			525
..	..	..	..	..	1 902	1 764	..	Réparation d'articles personnels et domestiques			526
..	..	..	..	..	51 909	56 166	..	Hotels et restaurants	H	55	
..	..	..	..	..	17 000	19 291	..	Hôtels; camping; autres équip. courts séjours			551
..	..	..	..	..	34 909	36 875	..	Restaurants, cafés et cantines			552
..	..	..	..	..	131 302	151 539	..	Transport, entreposage, communications	I	60_64	
..	..	..	..	..	47 502	52 537	..	Transports terrestres; transports par conduites		60	
..	..	..	..	..	..	..	..	Transports par chemin de fer			601
..	..	..	..	..	..	..	..	Autres transports terrestres			602
..	..	..	..	..	..	..	..	Transports par conduites			603
..	..	..	..	..	1 765	1 859	..	Transports par eau		61	
..	..	..	..	..	1 106	1 257	..	Transports maritimes et côtiers			611
..	..	..	..	..	659	602	..	Transports par voies navigables intérieures			612
..	..	..	..	..	16 632	17 867	..	Transports aériens		62	
..	..	..	..	..	16 533	17 742	..	Transports aériens réguliers			621
..	..	..	..	..	99	125	..	Transports aériens non réguliers			622
..	..	..	..	..	23 436	25 231	..	Act. annex. et aux. des transp.; act. des ag. de voy.		63	
..	..	..	..	..	41 967	54 045	..	Postes et télécommunications		64	
..	..	..	..	..	16 133	17 333	..	Activités de poste et de courrier			641
..	..	..	..	..	25 834	36 712	..	Télécommunications			642
..	..	..	..	..	..	..	..	Intermédiation financière	J	65_67	
..	..	..	..	..	..	..	..	Interm. financ., sauf assur. et caisses de retraite		65	
..	..	..	..	..	..	..	..	Intermédiation monétaire			651
..	..	..	..	..	..	..	..	Autres intermédiations financières			659
..	..	..	..	..	..	..	..	Act. d'assur., caisses de retr., sauf séc. soc. oblig.		66	
..	..	..	..	..	..	..	..	Act. auxiliaires de l'intermédiation financière		67	
..	..	..	..	..	..	..	..	Act. aux. de l'interm. fin., sauf assur., et retraite			671
..	..	..	..	..	..	..	..	Act. aux. des assurances et caisses de retraite			672
..	..	..	..	..	97 678	114 092	..	Immobilier, locations, services aux entreprises	K	70_74	
..	..	..	..	..	9 584	9 809	..	Activités immobilières		70	
..	..	..	..	..	5 944	6 494	..	Act. immobilières sur biens propres ou loués			701
..	..	..	..	..	3 640	3 315	..	Activités immobilières pour comptes de tiers			702
..	..	..	..	..	3 745	3 797	..	Locat. mach., équip. sans opér., biens pers. et dom.		71	
..	..	..	..	..	2 075	1 848	..	Location de matériel de transport			711
..	..	..	..	..	1 472	1 849	..	Location de machines ét equipements nca			712
..	..	..	..	..	198	100	..	Loc. d'articles personnels et domestiques nca			713
..	..	..	..	..	6 910	9 546	..	Activités informatiques et activités rattachées		72	
..	..	..	..	..	221	213	..	Conseil en matériel informatique			721
..	..	..	..	..	2 975	3 963	..	Production de logiciels			722
..	..	..	..	..	2 885	4 176	..	Traitement de données informatiques			723
..	..	..	..	..	18	98	..	Activités de banques de données			724
..	..	..	..	..	392	756	..	Entr., répar. mach. de bureau et compt. et inform.			725
..	..	..	..	..	419	340	..	Autres activités rattachées à l'informatique			729
..	..	..	..	..	26	29	..	Recherche-développement		73	
..	..	..	..	..	..	..	..	Rech. et dév. expér. en sc. phys., natur. et ingén.			731
..	..	..	..	..	..	..	..	Rech. et dév. expér. en sciences soc. et humaines			732
..	..	..	..	..	77 413	90 911	..	Autres activités de services aux entreprises		74	
..	..	..	..	..	22 602	28 409	..	Act. jur., compt., audit.; cons. fisc.; ét. march., soud.			741
..	..	..	..	..	14 869	17 738	..	Act. d'achitect., d'ingén.; autres act. techniques			742
..	..	..	..	..	7 605	9 107	..	Publicité			743
..	..	..	..	..	32 337	35 657	..	Activités de services aux entreprises nca.			749
..	..	..	..	..	11 303	13 381	..	Education	M	80	
..	..	..	..	..	3 022	3 145	..	Activités d'enseignement primaire			801
..	..	..	..	..	2 126	2 666	..	Activités d'enseignement secondaire			802
..	..	..	..	..	2 162	2 894	..	Activités d'enseignement supérieur			803
..	..	..	..	..	3 993	4 676	..	Act. de formation perman.; autres act. d'enseign.			809
..	..	..	..	..	10 255	11 206	..	Santé et action sociale	N	85	
..	..	..	..	..	9 611	10 366	..	Activités de santé humaine			851
..	..	..	..	..	90	74	..	Activités vétérinaires			852
..	..	..	..	..	554	766	..	Activités d'action sociale			853
..	..	..	..	..	21 835	19 699	..	Autres activités de serv. coll., soc. et personnels	O	90_93	
..	..	..	..	..	762	673	..	Assain. et enlèv. des ordures; voierie, act. simil.		90	
..	..	..	..	..	..	..	..	Activités associatives diverses		91	
..	..	..	..	..	..	..	..	Activités de syndicats de salariés			912
..	..	..	..	..	14 500	12 963	..	Activités récréatives, culturelles et sportives		92	
..	..	..	..	..	11 080	9 884	..	Act. cinémat., de radio, télé.; autres act. de spect.			921
..	..	..	..	..	435	350	..	Activités d'agences de presse			922
..	..	..	..	..	235	27	..	Act. des biblio., arch., musées, jardins bot. et zool.			923
..	..	..	..	..	2 750	2 702	..	Activités sportives et autres activités récréatives			924
..	..	..	..	..	6 573	6 063	..	Autres activités de services		93	

PORTUGAL

Table PO.S8 **ENTERPRISES - ENTREPRISES**

Units

ISIC revision 3				1991	1992	1993	1994	1995	1996	1997	1998
G	50_52		Wholesale & retail trade; repairs; etc.	..	..	..	..	..	199 027	196 306	212 078
	50		Sale, maint.,repair vehic/cycle, person., h/hold gd.	..	..	..	..	..	28 164	31 795	26 953
		501	Sale of motor cycle	..	..	..	..	..	4 540	4 589	4 227
		502	Maintenance and repair of motor vehicles	..	..	..	..	..	14 935	18 868	14 908
		503	Sale of motor vehicle parts and accessories	..	..	..	..	..	3 447	3 326	2 873
		504	Sale, maint., repair motor cycle, parts & acces.	..	..	..	..	..	3 174	3 109	2 938
		505	Retail sale of automative fuel	..	..	..	..	..	2 068	1 902	2 008
	51		W/sale. trade & commission exc motor veh. & cycle	..	..	..	..	..	52 704	47 882	49 309
		511	Wholesale on a fee or contract basis	..	..	..	..	..	18 568	14 088	15 072
		512	W/sale agric. raw mat.,live animal,food, bev., tobac.	..	..	..	..	..	10 142	10 227	10 550
		513	Wholesale of household goods	..	..	..	..	..	9 909	9 359	9 662
		514	W/sale. of non-agric. intermed. prod., waste, scrap	..	..	..	..	..	6 455	6 124	5 929
		515	Wholesale of machinery, equipment & supplies	..	..	..	..	..	4 261	4 445	4 401
		519	Other wholesale	..	..	..	..	..	3 368	3 638	3 695
	52		Retail trade exc. motor veh. & cyc.; repair h/hold gds.	..	..	..	..	..	118 159	116 630	135 816
		521	Non-specialised retail trade in stores	..	..	..	..	..	13 674	13 573	20 602
		522	Retail sale food, bev., tobac. in specialised stores	..	..	..	..	..	22 486	22 021	30 730
		523	Other retail trade of new goods specialised stores	..	..	..	..	..	65 314	65 939	70 850
		524	Retail sale of second-hand goods in stores	..	..	..	..	..	603	591	503
		525	Retail trade not in stores	..	..	..	..	..	8 424	7 510	6 858
		526	Repair of personal and household goods	..	..	..	..	..	7 657	6 995	6 273
H	55		Hotels and restaurants	..	..	..	..	..	63 297	60 935	56 048
		551	Hotels; camping ; other short-stay accommodation	..	..	..	..	..	2 740	3 440	..
		552	Restaurants, bars and canteens	..	..	..	..	..	60 557	57 494	..
I	60_64		Transport, storage and communications	..	..	..	..	..	20 377	19 551	..
	60		Land transport, transport via pipelines	..	..	..	..	..	17 506	16 907	..
		601	Transport via railways	..	..	..	..	..	..	..	..
		602	Other land transport	..	..	..	..	..	..	..	..
		603	Transport via pipelines	..	..	..	..	..	..	..	..
	61		Water transport	..	..	..	..	..	98	97	..
		611	Sea and coastal water transport	..	..	..	..	..	62	75	67
		612	Inland water transport	..	..	..	..	..	36	22	28
	62		Air transport	..	..	..	..	..	25	25	29
		621	Scheduled air transport	..	..	..	..	..	9	11	..
		622	Non-scheduled air transport	..	..	..	..	..	16	14	..
	63		Support. & aux. transport act., travel agencies activ.	..	..	..	..	..	2 616	2 387	..
	64		Post and telecommunications	..	..	..	..	..	132	136	175
		641	Post and courier activities	..	..	..	..	..	19	21	28
		642	Telecommunications	..	..	..	..	..	113	115	147
J	65_67		Financial intermediation	..	..	..	..	..	..	..	..
	65		Financial intermediation, exc. insur., pension fund	..	..	..	..	..	..	..	..
		651	Monetary intermediation	..	..	..	..	..	..	..	..
		659	Other financial intermediation	..	..	..	..	..	..	..	..
	66		Insurance pension fund, exc. compul. social sec.	..	..	..	..	..	..	..	..
	67		Activities auxiliary to financial intermediation	..	..	..	..	..	..	..	..
		671	Act. auxil to fin. intermed., exc. insur. & pens.fund	..	..	..	..	..	..	..	..
		672	Act. auxiliary to insurance and pension funding	..	..	..	..	..	..	..	..
K	70_74		Real estate, renting and business activities	..	..	..	..	..	55 275	49 196	..
	70		Real estate activities	..	..	..	..	..	9 032	8 965	8 927
		701	Real estate activ. with own or leased property	..	..	..	..	..	4 926	4 941	..
		702	Real estate activities on a fee or contract basis	..	..	..	..	..	4 106	4 025	..
	71		Renting of mach. without operator; rent. of pers. gds.	..	..	..	..	..	2 662	2 520	..
		711	Renting of transport equipment	..	..	..	..	..	403	363	..
		712	Renting of other machinery and equipment	..	..	..	..	..	1 676	1 657	1 289
		713	Renting of personal and household goods nec	..	..	..	..	..	583	500	338
	72		Computer and related activities	..	..	..	..	..	2 316	2 154	2 161
		721	Hardware consultancy	..	..	..	..	..	183	167	..
		722	Software consultancy and supply	..	..	..	..	..	679	757	..
		723	Data processing	..	..	..	..	..	801	698	..
		724	Data base activities	..	..	..	..	..	22	29	..
		725	Mainten.; repair of office, account. & comp. mach.	..	..	..	..	..	399	318	..
		729	Other computer related activities	..	..	..	..	..	232	184	..
	73		Research and development	..	..	..	..	..	41	34	32
		731	R & D experim. on natural sciences & engineering	..	..	..	..	..	..	..	..
		732	R & D experim. on social sciences & humanities	..	..	..	..	..	..	..	..
	74		Other business activities	..	..	..	..	..	41 225	35 524	..
		741	Legal, acc., book-keep., audit act.; tax consult., etc	..	..	..	..	..	9 470	8 956	..
		742	Architectural, engineering, & other techn. activities	..	..	..	..	..	3 672	3 691	..
		743	Advertising	..	..	..	..	..	2 986	2 688	2 433
		749	Business activities nec	..	..	..	..	..	25 097	20 188	..
M	80		Education	..	..	..	..	..	2 474	2 360	..
		801	Primary education	..	..	..	..	..	795	845	..
		802	Secondary education	..	..	..	..	..	87	95	..
		803	Higher education	..	..	..	..	..	46	36	..
		809	Adult and other education	..	..	..	..	..	1 546	1 385	..
N	85		Health and social work	..	..	..	..	..	5 018	5 091	..
		851	Human health activities	..	..	..	..	..	4 415	4 512	..
		852	Veterinary activities	..	..	..	..	..	227	214	..
		853	Social work activities	..	..	..	..	..	376	364	..
O	90_93		Other commun., soc. & pers. services activities	..	..	..	..	..	26 040	24 394	..
	90		Sewage & refuse disposal, sanitation, etc.	..	..	..	..	..	137	86	..
	91		Activities of member organisations nec	..	..	..	..	..	..	..	..
		912	Activities of trade unions	..	..	..	..	..	..	..	..
	92		Recreational, cultural and sporting activities	..	..	..	..	..	3 536	3 308	..
		921	Motion picture, radio, TV, other entertainment act.	..	..	..	..	..	1 840	1 744	..
		922	News agency activities	..	..	..	..	..	44	29	..
		923	Library, archives, museums, other cultural activ.	..	..	..	..	..	39	34	..
		924	Sporting and other recreational activities	..	..	..	..	..	1 613	1 500	..
	93		Other service activities	..	..	..	..	..	22 367	21 000	..

Table PO.S9 **EXPORTATIONS - EXPORTS**

Millions de Esc (Prix courants)

1991	1992	1993	1994	1995	1996	1997	1998			CITI révision 3
..	..	..	..	..	690 666	599 685	..	Com. de gros, de detail ; réparation etc.	G	**50_52**
..	..	..	..	..	29 034	38 492	..	Com., entr., répar. d'auto. et moto ; carburants		**50**
..	..	..	..	..	9 554	10 823	..	Commerce de véhicules automobiles		501
..	..	..	..	..	13	10 187	..	Entretien et réparation de véhicules automobiles		502
..	..	..	..	..	5 005	7 305	..	Com. de pièces et access. de véhic. automobiles		503
..	..	..	..	..	6 932	9 725	..	Com., entret. et répar. de motocycles et access.		504
..	..	..	..	..	7 530	452	..	Com. de détail de carburants pour automobiles		505
..	..	..	..	..	639 489	535 841	..	Com. de gros et act. inter., sauf auto. et moto		**51**
..	..	..	..	..	67 729	78 453	..	Activités d'intermédiaries de commerce de gros		511
..	..	..	..	..	70 694	108 887	..	Com. gros: prod. agr. bruts, anim. viv., prod. alim.		512
..	..	..	..	..	266 058	152 796	..	Com. gros: biens de consom., sauf alimentaires		513
..	..	..	..	..	98 065	79 751	..	Com. gros: prod. inter. non agr., déchets, débris		514
..	..	..	..	..	46 749	31 618	..	Com. gros: machines, équipments et fournitures		515
..	..	..	..	..	90 194	84 336	..	Autres commerces de gros		519
..	..	..	..	..	22 143	25 352	..	Com. détail, sauf auto, moto; répar. art. pers., dom.		**52**
..	..	..	..	..	10 030	10 896	..	Commerce de détail non spécialisé en magasin		521
..	..	..	..	..	1 205	969	..	Com. détail: prod. aliment. en mag. spécialisés		522
..	..	..	..	..	9 826	12 042	..	Autre com. détail: prod. neufs en mag. spécialisés		523
..	..	..	..	..	4	101	..	Commerces de détail de biens d'occasion		524
..	..	..	..	..	1 000	1 285	..	Commerces de détail autres qu'en magasin		525
..	..	..	..	..	78	59	..	Réparation d'articles personnels et domestiques		526
..	..	..	..	..	16 716	23 413	..	Hotels et restaurants	H	**55**
..	..	..	..	..	15 033	20 953	..	Hôtels; camping; autres équip. courts séjours		551
..	..	..	..	..	1 683	2 460	..	Restaurants, cafés et cantines		552
..	..	..	..	..	324 324	385 696	..	Transport, entreposage, communications	I	**60_64**
..	..	..	..	..	53 958	48 814	..	Transports terrestres; transports par conduites		**60**
..	..	..	..	..	..	..	..	Transports par chemin de fer		601
..	..	..	..	..	..	..	..	Autres transports terrestres		602
..	..	..	..	..	..	..	..	Transports par conduites		603
..	..	..	..	..	9 999	11 473	..	Transports par eau		**61**
..	..	..	..	..	4 355	11 473	..	Transports maritimes et côtiers		611
..	..	..	..	..	5 644	..	..	Transports par voies navigables intérieures		612
..	..	..	..	..	118 495	135 833	..	Transports aériens		**62**
..	..	..	..	..	117 956	135 459	..	Transports aériens réguliers		621
..	..	..	..	..	539	374	..	Transports aériens non réguliers		622
..	..	..	..	..	133 856	141 376	..	Act. annex. et aux. des transp.; act. des ag. de voy.		**63**
..	..	..	..	..	8 016	48 200	..	Postes et télécommunications		**64**
..	..	..	..	..	5 785	3 202	..	Activités de poste et de courrier		641
..	..	..	..	..	2 231	44 998	..	Télécommunications		642
..	..	..	..	..	..	..	..	Intermédiation financière	J	**65_67**
..	..	..	..	..	..	..	..	Interm. financ., sauf assur. et caisses de retraite		**65**
..	..	..	..	..	..	..	..	Intermédiation monétaire		651
..	..	..	..	..	..	..	..	Autres intermédiations financières		659
..	..	..	..	..	..	..	..	Act. d'assur., caisses de retr., sauf séc. soc. oblig.		**66**
..	..	..	..	..	..	..	..	Act. auxiliaires de l'intermédiation financière		**67**
..	..	..	..	..	..	..	..	Act. aux. de l'interm. fin., sauf assur., et retraite		671
..	..	..	..	..	..	..	..	Act. aux. des assurances et caisses de retraite		672
..	..	..	..	..	159 150	121 633	..	Immobilier, locations, services aux entreprises	K	**70_74**
..	..	..	..	..	2 604	2 051	..	Activités immobilières		**70**
..	..	..	..	..	2 391	1 219	..	Act. immobilières sur biens propres ou loués		701
..	..	..	..	..	213	832	..	Activités immobilières pour comptes de tiers		702
..	..	..	..	..	7 087	8 148	..	Locat. mach., équip. sans opér., biens pers. et dom.		**71**
..	..	..	..	..	6 638	7 759	..	Location de matériel de transport		711
..	..	..	..	..	429	381	..	Location de machines ét equipements nca		712
..	..	..	..	..	20	8	..	Loc. d'articles personnels et domestiques nca		713
..	..	..	..	..	7 597	15 431	..	Activités informatiques et activités rattachées		**72**
..	..	..	..	..	191	91	..	Conseil en matériel informatique		721
..	..	..	..	..	2 561	10 393	..	Production de logiciels		722
..	..	..	..	..	4 139	3 182	..	Traitement de données informatiques		723
..	..	..	..	..	..	1 619	..	Activités de banques de données		724
..	..	..	..	..	36	16	..	Entr., répar. mach. de bureau et compt. et inform.		725
..	..	..	..	..	670	130	..	Autres activités rattachées à l'informatique		729
..	..	..	..	..	4	51	..	Recherche-développement		**73**
..	..	..	..	..	..	..	..	Rech. et dév. expér. en sc. phys., natur. et ingén.		731
..	..	..	..	..	..	..	..	Rech. et dév. expér. en sciences soc. et humaines		732
..	..	..	..	..	141 858	95 952	..	Autres activités de services aux entreprises		**74**
..	..	..	..	..	91 836	42 642	..	Act. jur., compt., audit.; cons. fisc.; ét. march., soud.		741
..	..	..	..	..	24 078	35 227	..	Act. d'achitect., d'ingén.; autres act. techniques		742
..	..	..	..	..	8 025	6 036	..	Publicité		743
..	..	..	..	..	17 919	12 047	..	Activités de services aux entreprises nca.		749
..	..	..	..	..	452	298	..	Education	M	**80**
..	..	..	..	..	..	..	..	Activités d'enseignement primaire		801
..	..	..	..	..	47	..	..	Activités d'enseignement secondaire		802
..	..	..	..	..	..	..	..	Activités d'enseignement supérieur		803
..	..	..	..	..	405	298	..	Act. de formation perman.; autres act. d'enseign.		809
..	..	..	..	..	25	..	..	Santé et action sociale	N	**85**
..	..	..	..	..	..	..	..	Activités de santé humaine		851
..	..	..	..	..	25	..	..	Activités vétérinaires		852
..	..	..	..	..	..	..	..	Activités d'action sociale		853
..	..	..	..	..	2 916	2 454	..	Autres activités de serv. coll., soc. et personnels	O	**90_93**
..	..	..	..	..	630	21	..	Assain. et enlèv. des ordures; voierie, act. simil.		**90**
..	..	..	..	..	..	..	..	Activités associatives diverses		**91**
..	..	..	..	..	..	..	..	Activités de syndicats de salariés		912
..	..	..	..	..	1 791	2 393	..	Activités récréatives, culturelles et sportives		**92**
..	..	..	..	..	754	1 906	..	Act. cinémat., de radio, télé.; autres act. de spect.		921
..	..	..	..	..	531	259	..	Activités d'agences de presse		922
..	..	..	..	..	..	..	..	Act. des biblio., arch., musées, jardins bot. et zool.		923
..	..	..	..	..	506	228	..	Activités sportives et autres activités récréatives		924
..	..	..	..	..	495	40	..	Autres activités de services		**93**

PORTUGAL

Table PO.S10 **IMPORTS - IMPORTATIONS**

Millions of Esc (Current Prices)

ISIC revision 3				1991	1992	1993	1994	1995	1996	1997	1998
G	50_52		Wholesale & retail trade; repairs; etc.	..	..	..	..	..	2 958 091	3 076 570	..
	50		Sale, maint.,repair vehic/cycle, person., h/hold gd.	..	..	..	..	..	480 336	740 824	..
		501	Sale of motor cycle	..	..	..	..	..	350 290	477 348	..
		502	Maintenance and repair of motor vehicles	..	..	..	..	..	1 714	11 337	..
		503	Sale of motor vehicle parts and accessories	..	..	..	..	..	97 354	225 919	..
		504	Sale, maint., repair motor cycle, parts & acces.	..	..	..	..	..	30 856	25 871	..
		505	Retail sale of automative fuel	..	..	..	..	..	122	349	..
	51		W/sale. trade & commission exc motor veh. & cycle	..	..	..	..	..	2 240 286	2 112 910	..
		511	Wholesale on a fee or contract basis	..	..	..	..	..	218 651	101 507	..
		512	W/sale agric. raw mat.,live animal,food, bev., tobac.	..	..	..	..	..	457 631	401 433	..
		513	Wholesale of household goods	..	..	..	..	..	590 835	624 107	..
		514	W/sale. of non-agric. intermed. prod., waste, scrap	..	..	..	..	..	417 547	383 799	..
		515	Wholesale of machinery, equipment & supplies	..	..	..	..	..	377 539	378 584	..
		519	Other wholesale	..	..	..	..	..	178 083	223 480	..
	52		Retail trade exc. motor veh. & cyc.; repair h/hold gds.	..	..	..	..	..	237 470	222 836	..
		521	Non-specialised retail trade in stores	..	..	..	..	..	28 264	48 583	..
		522	Retail sale food, bev., tobac. in specialised stores	..	..	..	..	..	13 295	6 949	..
		523	Other retail trade of new goods specialised stores	..	..	..	..	..	180 711	155 655	..
		524	Retail sale of second-hand goods in stores	..	..	..	..	..	304	644	..
		525	Retail trade not in stores	..	..	..	..	..	12 889	10 510	..
		526	Repair of personal and household goods	..	..	..	..	..	2 007	495	..
H	55		Hotels and restaurants	..	..	..	..	..	23 014	5 542	..
		551	Hotels; camping ; other short-stay accommodation	..	..	..	..	..	6 048	2 315	..
		552	Restaurants, bars and canteens	..	..	..	..	..	16 966	3 227	..
I	60_64		Transport, storage and communications	..	..	..	..	..	168 261	236 541	..
	60		Land transport, transport via pipelines	..	..	..	..	..	34 598	45 419	..
		601	Transport via railways	..	..	..	..	..	..	..	..
		602	Other land transport	..	..	..	..	..	..	..	..
		603	Transport via pipelines	..	..	..	..	..	..	..	..
	61		Water transport	..	..	..	..	..	14 343	12 305	..
		611	Sea and coastal water transport	..	..	..	..	..	12 426	11 414	..
		612	Inland water transport	..	..	..	..	..	1 917	891	..
	62		Air transport	..	..	..	..	..	41 642	56 709	..
		621	Scheduled air transport	..	..	..	..	..	40 968	56 422	..
		622	Non-scheduled air transport	..	..	..	..	..	674	287	..
	63		Support. & aux. transport act., travel agencies activ.	..	..	..	..	..	59 289	59 242	..
	64		Post and telecommunications	..	..	..	..	..	18 389	62 866	..
		641	Post and courier activities	..	..	..	..	..	4 006	4 211	..
		642	Telecommunications	..	..	..	..	..	14 383	58 655	..
J	65_67		Financial intermediation	..	..	..	..	..	..	..	..
	65		Financial intermediation, exc. insur., pension fund	..	..	..	..	..	..	..	..
		651	Monetary intermediation	..	..	..	..	..	..	..	..
		659	Other financial intermediation	..	..	..	..	..	..	..	..
	66		Insurance pension fund, exc. compul. social sec.	..	..	..	..	..	..	..	..
	67		Activities auxiliary to financial intermediation	..	..	..	..	..	..	..	..
		671	Act. auxil to fin. intermed., exc. insur. & pens.fund	..	..	..	..	..	..	..	..
		672	Act. auxiliary to insurance and pension funding	..	..	..	..	..	..	..	..
K	70_74		Real estate, renting and business activities	..	..	..	..	..	112 048	79 776	..
	70		Real estate activities	..	..	..	..	..	2 340	5 984	..
		701	Real estate activ. with own or leased property	..	..	..	..	..	1 896	5 374	..
		702	Real estate activities on a fee or contract basis	..	..	..	..	..	444	610	..
	71		Renting of mach. without operator; rent. of pers. gds.	..	..	..	..	..	4 781	3 752	..
		711	Renting of transport equipment	..	..	..	..	..	1 081	1 404	..
		712	Renting of other machinery and equipment	..	..	..	..	..	3 683	2 083	..
		713	Renting of personal and household goods nec	..	..	..	..	..	17	265	..
	72		Computer and related activities	..	..	..	..	..	18 336	12 137	..
		721	Hardware consultancy	..	..	..	..	..	52	9	..
		722	Software consultancy and supply	..	..	..	..	..	5 656	4 694	..
		723	Data processing	..	..	..	..	..	11 577	7 145	..
		724	Data base activities	..	..	..	..	..	59	94	..
		725	Mainten.; repair of office, account. & comp. mach.	..	..	..	..	..	246	154	..
		729	Other computer related activities	..	..	..	..	..	746	41	..
	73		Research and development	..	..	..	..	..	51	..	..
		731	R & D experim. on natural sciences & engineering	..	..	..	..	..	..	..	..
		732	R & D experim. on social sciences & humanities	..	..	..	..	..	..	..	..
	74		Other business activities	..	..	..	..	..	86 540	57 903	..
		741	Legal, acc., book-keep., audit act.; tax consult., etc	..	..	..	..	..	50 081	18 293	..
		742	Architectural, engineering, & other techn. activities	..	..	..	..	..	13 623	18 698	..
		743	Advertising	..	..	..	..	..	8 300	9 388	..
		749	Business activities nec	..	..	..	..	..	14 536	11 524	..
M	80		Education	..	..	..	..	..	791	3 172	..
		801	Primary education	..	..	..	..	..	1	..	..
		802	Secondary education	..	..	..	..	..	..	..	..
		803	Higher education	..	..	..	..	..	..	2	..
		809	Adult and other education	..	..	..	..	..	790	3 170	..
N	85		Health and social work	..	..	..	..	..	756	1 102	..
		851	Human health activities	..	..	..	..	..	661	1 077	..
		852	Veterinary activities	..	..	..	..	..	94	25	..
		853	Social work activities	..	..	..	..	..	1	..	..
O	90_93		Other commun., soc. & pers. services activities	..	..	..	..	..	39 898	17 242	..
	90		Sewage & refuse disposal, sanitation, etc.	..	..	..	..	..	547	535	..
	91		Activities of member organisations nec	..	..	..	..	..	..	..	..
		912	Activities of trade unions	..	..	..	..	..	..	..	..
	92		Recreational, cultural and sporting activities	..	..	..	..	..	36 949	16 406	..
		921	Motion picture, radio, TV, other entertainment act.	..	..	..	..	..	11 973	10 643	..
		922	News agency activities	..	..	..	..	..	4 415	2 213	..
		923	Library, archives, museums, other cultural activ.	..	..	..	..	..	44	4	..
		924	Sporting and other recreational activities	..	..	..	..	..	20 517	3 546	..
	93		Other service activities	..	..	..	..	..	2 402	301	..

Table SN.S1 **PRODUCTION - PRODUCTION**

Millions de SKr (Prix courants)

1991	1992	1993	1994	1995	1996	1997	1998		CITI révision 3		
..	..	..	..	..	..	323 054	350 539	Com. de gros, de detail ; réparation etc.	G	**50_52**	
..	..	..	..	..	..	39 154	43 568	Com., entr., répar. d'auto. et moto ; carburants		**50**	
..	..	..	..	..	..	18 697	21 323	Commerce de véhicules automobiles			501
..	..	..	..	..	..	9 371	10 499	Entretien et réparation de véhicules automobiles			502
..	..	..	..	..	..	4 884	5 427	Com. de pièces et access. de véhic. automobiles			503
..	..	..	..	..	..	648	909	Com., entret. et répar. de motocycles et access.			504
..	..	..	..	..	..	5 554	5 410	Com. de détail de carburants pour automobiles			505
..	..	..	..	..	..	188 728	207 471	Com. de gros et act. inter., sauf auto. et moto		**51**	
..	..	..	..	..	..	9 189	17 980	Activités d'intermédiaries de commerce de gros			511
..	..	..	..	..	..	..	..	Com. gros: prod. agr. bruts, anim. viv., prod. alim.			512
..	..	..	..	..	..	46 531	47 010	Com. gros: biens de consom., sauf alimentaires			513
..	..	..	..	..	..	48 818	51 684	Com. gros: prod. inter. non agr., déchets, débris			514
..	..	..	..	..	..	51 140	59 081	Com. gros: machines, équipments et fournitures			515
..	..	..	..	..	..	2 336	1 891	Autres commerces de gros			519
..	..	..	..	..	..	95 172	99 500	Com. détail, sauf auto, moto; répar. art. pers., dom.		**52**	
..	..	..	..	..	..	32 450	33 476	Commerce de détail non spécialisé en magasin			521
..	..	..	..	..	..	6 469	6 349	Com. détail: prod. aliment. en mag. spécialisés			522
..	..	..	..	..	..	..	..	Autre com. détail: prod. neufs en mag. spécialisés			523
..	..	..	..	..	..	572	336	Commerces de détail de biens d'occasion			524
..	..	..	..	..	..	4 208	4 560	Commerces de détail autres qu'en magasin			525
..	..	..	..	..	..	2 944	3 153	Réparation d'articles personnels et domestiques			526
..	..	..	..	..	..	51 420	55 474	Hotels et restaurants	H	**55**	
..	..	..	..	..	..	..	..	Hôtels; camping; autres équip. courts séjours			551
..	..	..	..	..	..	..	..	Restaurants, cafés et cantines			552
..	..	..	..	..	..	335 268	343 958	Transport, entreposage, communications	I	**60_64**	
..	..	..	..	..	..	93 820	98 798	Transports terrestres; transports par conduites		**60**	
..	..	..	..	..	..	11 892	11 919	Transports par chemin de fer			601
..	..	..	..	..	..	81 919	86 869	Autres transports terrestres			602
..	..	..	..	..	..	9	10	Transports par conduites			603
..	..	..	..	..	..	27 833	29 858	Transports par eau		**61**	
..	..	..	..	..	..	27 136	28 947	Transports maritimes et côtiers			611
..	..	..	..	..	..	697	911	Transports par voies navigables intérieures			612
..	..	..	..	..	..	22 517	23 275	Transports aériens		**62**	
..	..	..	..	..	..	..	..	Transports aériens réguliers			621
..	..	..	..	..	..	..	..	Transports aériens non réguliers			622
..	..	..	..	..	..	113 915	104 507	Act. annex. et aux. des transp.; act. des ag. de voy.		**63**	
..	..	..	..	..	..	77 183	87 520	Postes et télécommunications		**64**	
..	..	..	..	..	..	..	..	Activités de poste et de courrier			641
..	..	..	..	..	..	..	..	Télécommunications			642
..	..	..	..	..	..	..	..	Intermédiation financière	J	**65_67**	
..	..	..	..	..	..	..	..	Interm. financ., sauf assur. et caisses de retraite		**65**	
..	..	..	..	..	..	..	..	Intermédiation monétaire			651
..	..	..	..	..	..	..	..	Autres intermédiations financières			659
..	..	..	..	..	..	..	..	Act. d'assur., caisses de retr., sauf séc. soc. oblig.		**66**	
..	..	..	..	..	..	..	..	Act. auxiliaires de l'intermédiation financière		**67**	
..	..	..	..	..	..	..	..	Act. aux. de l'interm. fin., sauf assur., et retraite			671
..	..	..	..	..	..	..	..	Act. aux. des assurances et caisses de retraite			672
..	..	..	..	..	..	425 834	476 912	Immobilier, locations, services aux entreprises	K	**70_74**	
..	..	..	..	..	..	145 000	150 430	Activités immobilières		**70**	
..	..	..	..	..	..	..	..	Act. immobilières sur biens propres ou loués			701
..	..	..	..	..	..	..	..	Activités immobilières pour comptes de tiers			702
..	..	..	..	..	..	16 690	16 660	Locat. mach., équip. sans opér., biens pers. et dom.		**71**	
..	..	..	..	..	..	..	..	Location de matériel de transport			711
..	..	..	..	..	..	8 664	8 404	Location de machines ét equipements nca			712
..	..	..	..	..	..	1 821	2 001	Loc. d'articles personnels et domestiques nca			713
..	..	..	..	..	..	56 473	74 933	Activités informatiques et activités rattachées		**72**	
..	..	..	..	..	..	..	..	Conseil en matériel informatique			721
..	..	..	..	..	..	..	..	Production de logiciels			722
..	..	..	..	..	..	..	..	Traitement de données informatiques			723
..	..	..	..	..	..	..	..	Activités de banques de données			724
..	..	..	..	..	..	..	..	Entr., répar. mach. de bureau et compt. et inform.			725
..	..	..	..	..	..	..	..	Autres activités rattachées à l'informatique			729
..	..	..	..	..	..	11 910	16 529	Recherche-développement		**73**	
..	..	..	..	..	..	..	..	Rech. et dév. expér. en sc. phys., natur. et ingén.			731
..	..	..	..	..	..	..	..	Rech. et dév. expér. en sciences soc. et humaines			732
..	..	..	..	..	..	195 762	218 359	Autres activités de services aux entreprises		**74**	
..	..	..	..	..	..	..	..	Act. jur., compt., audit.; cons. fisc.; ét. march., soud.			741
..	..	..	..	..	..	..	..	Act. d'achitect., d'ingén.; autres act. techniques			742
..	..	..	..	..	..	..	..	Publicité			743
..	..	..	..	..	..	..	..	Activités de services aux entreprises nca.			749
..	..	..	..	..	..	..	..	Education	M	**80**	
..	..	..	..	..	..	416	676	Activités d'enseignement primaire			801
..	..	..	..	..	..	346	463	Activités d'enseignement secondaire			802
..	..	..	..	..	..	178	237	Activités d'enseignement supérieur			803
..	..	..	..	..	..	8 138	9 723	Act. de formation perman.; autres act. d'enseign.			809
..	..	..	..	..	..	..	..	Santé et action sociale	N	**85**	
..	..	..	..	..	..	18 629	21 035	Activités de santé humaine			851
..	..	..	..	..	..	694	707	Activités vétérinaires			852
..	..	..	..	..	..	7 418	9 450	Activités d'action sociale			853
..	..	..	..	..	..	..	..	Autres activités de serv. coll., soc. et personnels	O	**90_93**	
..	..	..	..	..	..	6 447	7 415	Assain. et enlèv. des ordures; voierie, act. simil.		**90**	
..	..	..	..	..	..	..	..	Activités associatives diverses		**91**	
..	..	..	..	..	..	82	111	Activités de syndicats de salariés			912
..	..	..	..	..	..	..	..	Activités récréatives, culturelles et sportives		**92**	
..	..	..	..	..	..	19 718	22 396	Act. cinémat., de radio, télé.; autres act. de spect.			921
..	..	..	..	..	..	1 100	1 049	Activités d'agences de presse			922
..	..	..	..	..	..	286	286	Act. des biblio., arch., musées, jardins bot. et zool.			923
..	..	..	..	..	..	27 651	30 442	Activités sportives et autres activités récréatives			924
..	..	..	..	..	..	8 652	9 286	Autres activités de services		**93**	

SWEDEN

Table SN.S2 **VALUE ADDED - VALEUR AJOUTÉE**

Millions of SKr (Current Prices)

ISIC revision 3				1991	1992	1993	1994	1995	1996	1997	1998
G	50_52		Wholesale & retail trade; repairs; etc.	..	..	..	..	..	..	167 548	179 975
	50		Sale, maint.,repair vehic/cycle, person., h/hold gd.	..	..	..	..	..	..	23 843	24 730
		501	Sale of motor cycle	..	..	..	..	..	..	11 280	11 892
		502	Maintenance and repair of motor vehicles	..	..	..	..	..	..	5 762	5 951
		503	Sale of motor vehicle parts and accessories	..	..	..	..	..	..	2 719	2 869
		504	Sale, maint., repair motor cycle, parts & acces.	..	..	..	..	..	..	373	422
		505	Retail sale of automative fuel	..	..	..	..	..	..	3 709	3 595
	51		W/sale. trade & commission exc motor veh. & cycle	..	..	..	..	..	..	88 790	97 446
		511	Wholesale on a fee or contract basis	..	..	..	..	..	..	4 344	4 453
		512	W/sale agric. raw mat.,live animal,food, bev., tobac.	..	..	..	..	..	..	..	..
		513	Wholesale of household goods	..	..	..	..	..	..	20 616	22 541
		514	W/sale. of non-agric. intermed. prod., waste, scrap	..	..	..	..	..	..	24 300	25 531
		515	Wholesale of machinery, equipment & supplies	..	..	..	..	..	..	27 830	32 206
		519	Other wholesale	..	..	..	..	..	..	606	662
	52		Retail trade exc. motor veh. & cyc.; repair h/hold gds.	..	..	..	..	..	..	54 915	57 799
		521	Non-specialised retail trade in stores	..	..	..	..	..	..	19 310	20 029
		522	Retail sale food, bev., tobac. in specialised stores	..	..	..	..	..	..	4 003	3 998
		523	Other retail trade of new goods specialised stores	..	..	..	..	..	..	..	..
		524	Retail sale of second-hand goods in stores	..	..	..	..	..	..	277	136
		525	Retail trade not in stores	..	..	..	..	..	..	1 527	1 843
		526	Repair of personal and household goods	..	..	..	..	..	..	1 317	1 424
H	55		Hotels and restaurants	..	..	..	..	..	..	19 543	21 517
		551	Hotels; camping ; other short-stay accommodation	..	..	..	..	..	..	..	..
		552	Restaurants, bars and canteens	..	..	..	..	..	..	..	..
I	60_64		Transport, storage and communications	..	..	..	..	..	..	110 219	118 876
	60		Land transport, transport via pipelines	..	..	..	..	..	..	33 794	36 622
		601	Transport via railways	..	..	..	..	..	..	4 634	4 753
		602	Other land transport	..	..	..	..	..	..	29 156	31 867
		603	Transport via pipelines	..	..	..	..	..	..	4	3
	61		Water transport	..	..	..	..	..	..	6 918	6 871
		611	Sea and coastal water transport	..	..	..	..	..	..	6 683	6 605
		612	Inland water transport	..	..	..	..	..	..	235	266
	62		Air transport	..	..	..	..	..	..	6 812	7 873
		621	Scheduled air transport	..	..	..	..	..	..	..	..
		622	Non-scheduled air transport	..	..	..	..	..	..	..	..
	63		Support. & aux. transport act., travel agencies activ.	..	..	..	..	..	..	20 245	20 830
	64		Post and telecommunications	..	..	..	..	..	..	42 450	46 678
		641	Post and courier activities	..	..	..	..	..	..	..	..
		642	Telecommunications	..	..	..	..	..	..	..	..
J	65_67		Financial intermediation	..	..	..	..	..	..	..	..
	65		Financial intermediation, exc. insur., pension fund	..	..	..	..	..	..	..	..
		651	Monetary intermediation	..	..	..	..	..	..	..	..
		659	Other financial intermediation	..	..	..	..	..	..	..	..
	66		Insurance pension fund, exc. compul. social sec.	..	..	..	..	..	..	..	..
	67		Activities auxiliary to financial intermediation	..	..	..	..	..	..	..	..
		671	Act. auxil to fin. intermed., exc. insur. & pens.fund	..	..	..	..	..	..	..	..
		672	Act. auxiliary to insurance and pension funding	..	..	..	..	..	..	..	..
K	70_74		Real estate, renting and business activities	..	..	..	..	..	..	197 404	219 265
	70		Real estate activities	..	..	..	..	..	..	75 122	78 453
		701	Real estate activ. with own or leased property	..	..	..	..	..	..	..	..
		702	Real estate activities on a fee or contract basis	..	..	..	..	..	..	..	..
	71		Renting of mach. without operator; rent. of pers. gds.	..	..	..	..	..	..	6 994	6 662
		711	Renting of transport equipment	..	..	..	..	..	..	..	..
		712	Renting of other machinery and equipment	..	..	..	..	..	..	4 046	3 782
		713	Renting of personal and household goods nec	..	..	..	..	..	..	681	693
	72		Computer and related activities	..	..	..	..	..	..	27 686	35 804
		721	Hardware consultancy	..	..	..	..	..	..	..	..
		722	Software consultancy and supply	..	..	..	..	..	..	..	..
		723	Data processing	..	..	..	..	..	..	..	..
		724	Data base activities	..	..	..	..	..	..	..	..
		725	Mainten.; repair of office, account. & comp. mach.	..	..	..	..	..	..	..	..
		729	Other computer related activities	..	..	..	..	..	..	..	..
	73		Research and development	..	..	..	..	..	..	4 506	5 765
		731	R & D experim. on natural sciences & engineering	..	..	..	..	..	..	..	..
		732	R & D experim. on social sciences & humanities	..	..	..	..	..	..	..	..
	74		Other business activities	..	..	..	..	..	..	83 096	92 581
		741	Legal, acc., book-keep., audit act.; tax consult., etc	..	..	..	..	..	..	..	..
		742	Architectural, engineering, & other techn. activities	..	..	..	..	..	..	..	..
		743	Advertising	..	..	..	..	..	..	..	..
		749	Business activities nec	..	..	..	..	..	..	..	..
M	80		Education	..	..	..	..	..	..	..	..
		801	Primary education	..	..	..	..	..	..	278	440
		802	Secondary education	..	..	..	..	..	..	191	263
		803	Higher education	..	..	..	..	..	..	111	138
		809	Adult and other education	..	..	..	..	..	..	4 030	4 729
N	85		Health and social work	..	..	..	..	..	..	..	..
		851	Human health activities	..	..	..	..	..	..	11 394	12 703
		852	Veterinary activities	..	..	..	..	..	..	354	376
		853	Social work activities	..	..	..	..	..	..	5 414	6 757
O	90_93		Other commun., soc. & pers. services activities	..	..	..	..	..	..	..	..
	90		Sewage & refuse disposal, sanitation, etc.	..	..	..	..	..	..	2 861	3 192
	91		Activities of member organisations nec	..	..	..	..	..	..	..	..
		912	Activities of trade unions	..	..	..	..	..	..	24	30
	92		Recreational, cultural and sporting activities	..	..	..	..	..	..	..	..
		921	Motion picture, radio, TV, other entertainment act.	..	..	..	..	..	..	6 389	7 281
		922	News agency activities	..	..	..	..	..	..	476	399
		923	Library, archives, museums, other cultural activ.	..	..	..	..	..	..	116	119
		924	Sporting and other recreational activities	..	..	..	..	..	..	6 387	6 985
	93		Other service activities	..	..	..	..	..	..	4 433	4 810

Table SN.S3 **INVESTISSEMENT - INVESTMENT**

Millions de SKr (Prix courants)

1991	1992	1993	1994	1995	1996	1997	1998		CITI révision 3		
..	..	..	..	..	..	28 466	26 693	Com. de gros, de detail ; réparation etc.	G	50_52	
..	..	..	..	..	..	5 744	5 428	Com., entr., répar. d'auto. et moto ; carburants		50	
..	..	..	..	..	..	3 329	3 207	Commerce de véhicules automobiles			501
..	..	..	..	..	..	1 126	1 131	Entretien et réparation de véhicules automobiles			502
..	..	..	..	..	..	464	318	Com. de pièces et access. de véhic. automobiles			503
..	..	..	..	..	..	54	53	Com., entret. et répar. de motocycles et access.			504
..	..	..	..	..	..	772	718	Com. de détail de carburants pour automobiles			505
..	..	..	..	..	..	16 031	14 827	Com. de gros et act. inter., sauf auto. et moto		51	
..	..	..	..	..	..	934	1 080	Activités d'intermédiaries de commerce de gros			511
..	..	..	..	..	..	..	..	Com. gros: prod. agr. bruts, anim. viv., prod. alim.			512
..	..	..	..	..	..	2 571	2 152	Com. gros: biens de consom., sauf alimentaires			513
..	..	..	..	..	..	5 231	4 297	Com. gros: prod. inter. non agr., déchets, débris			514
..	..	..	..	..	..	3 907	5 263	Com. gros: machines, équipments et fournitures			515
..	..	..	..	..	..	208	134	Autres commerces de gros			519
..	..	..	..	..	..	6 691	6 438	Com. détail, sauf auto, moto; répar. art. pers., dom.		52	
..	..	..	..	..	..	2 428	2 343	Commerce de détail non spécialisé en magasin			521
..	..	..	..	..	..	595	536	Com. détail: prod. aliment. en mag. spécialisés			522
..	..	..	..	..	..	..	..	Autre com. détail: prod. neufs en mag. spécialisés			523
..	..	..	..	..	..	53	31	Commerces de détail de biens d'occasion			524
..	..	..	..	..	..	209	251	Commerces de détail autres qu'en magasin			525
..	..	..	..	..	..	118	98	Réparation d'articles personnels et domestiques			526
..	..	..	..	..	..	5 891	4 113	Hotels et restaurants	H	55	
..	..	..	..	..	..	..	..	Hôtels; camping; autres équip. courts séjours			551
..	..	..	..	..	..	..	..	Restaurants, cafés et cantines			552
..	..	..	..	..	..	41 063	41 569	Transport, entreposage, communications	I	60_64	
..	..	..	..	..	..	14 077	12 598	Transports terrestres; transports par conduites		60	
..	..	..	..	..	..	501	485	Transports par chemin de fer			601
..	..	..	..	..	..	13 575	12 111	Autres transports terrestres			602
..	..	..	..	..	..	1	2	Transports par conduites			603
..	..	..	..	..	..	3 296	1 687	Transports par eau		61	
..	..	..	..	..	..	3 152	1 564	Transports maritimes et côtiers			611
..	..	..	..	..	..	144	123	Transports par voies navigables intérieures			612
..	..	..	..	..	..	1 690	2 215	Transports aériens		62	
..	..	..	..	..	..	..	..	Transports aériens réguliers			621
..	..	..	..	..	..	..	..	Transports aériens non réguliers			622
..	..	..	..	..	..	9 402	9 669	Act. annex. et aux. des transp.; act. des ag. de voy.		63	
..	..	..	..	..	..	12 597	15 401	Postes et télécommunications		64	
..	..	..	..	..	..	..	..	Activités de poste et de courrier			641
..	..	..	..	..	..	..	..	Télécommunications			642
..	..	..	..	..	..	..	..	Intermédiation financière	J	65_67	
..	..	..	..	..	..	..	..	Interm. financ., sauf assur. et caisses de retraite		65	
..	..	..	..	..	..	..	..	Intermédiation monétaire			651
..	..	..	..	..	..	..	..	Autres intermédiations financières			659
..	..	..	..	..	..	..	..	Act. d'assur., caisses de retr., sauf séc. soc. oblig.		66	
..	..	..	..	..	..	..	..	Act. auxiliaires de l'intermédiation financière		67	
..	..	..	..	..	..	..	..	Act. aux. de l'interm. fin., sauf assur., et retraite			671
..	..	..	..	..	..	..	..	Act. aux. des assurances et caisses de retraite			672
..	..	..	..	..	..	124 508	111 040	Immobilier, locations, services aux entreprises	K	70_74	
..	..	..	..	..	..	101 394	90 780	Activités immobilières		70	
..	..	..	..	..	..	..	..	Act. immobilières sur biens propres ou loués			701
..	..	..	..	..	..	..	..	Activités immobilières pour comptes de tiers			702
..	..	..	..	..	..	5 893	4 769	Locat. mach., équip. sans opér., biens pers. et dom.		71	
..	..	..	..	..	..	..	..	Location de matériel de transport			711
..	..	..	..	..	..	2 881	2 330	Location de machines ét equipements nca			712
..	..	..	..	..	..	365	283	Loc. d'articles personnels et domestiques nca			713
..	..	..	..	..	..	2 751	3 098	Activités informatiques et activités rattachées		72	
..	..	..	..	..	..	..	..	Conseil en matériel informatique			721
..	..	..	..	..	..	..	..	Production de logiciels			722
..	..	..	..	..	..	..	..	Traitement de données informatiques			723
..	..	..	..	..	..	..	..	Activités de banques de données			724
..	..	..	..	..	..	..	..	Entr., répar. mach. de bureau et compt. et inform.			725
..	..	..	..	..	..	..	..	Autres activités rattachées à l'informatique			729
..	..	..	..	..	..	1 109	1 249	Recherche-développement		73	
..	..	..	..	..	..	..	..	Rech. et dév. expér. en sc. phys., natur. et ingén.			731
..	..	..	..	..	..	..	..	Rech. et dév. expér. en sciences soc. et humaines			732
..	..	..	..	..	..	13 361	11 144	Autres activités de services aux entreprises		74	
..	..	..	..	..	..	..	..	Act. jur., compt., audit.; cons. fisc.; ét. march., soud.			741
..	..	..	..	..	..	..	..	Act. d'achitect., d'ingén.; autres act. techniques			742
..	..	..	..	..	..	..	..	Publicité			743
..	..	..	..	..	..	..	..	Activités de services aux entreprises nca.			749
..	..	..	..	..	..	..	..	Education	M	80	
..	..	..	..	..	..	13	26	Activités d'enseignement primaire			801
..	..	..	..	..	..	33	30	Activités d'enseignement secondaire			802
..	..	..	..	..	..	17	16	Activités d'enseignement supérieur			803
..	..	..	..	..	..	731	613	Act. de formation perman.; autres act. d'enseign.			809
..	..	..	..	..	..	..	..	Santé et action sociale	N	85	
..	..	..	..	..	..	1 335	1 144	Activités de santé humaine			851
..	..	..	..	..	..	54	40	Activités vétérinaires			852
..	..	..	..	..	..	261	317	Activités d'action sociale			853
..	..	..	..	..	..	..	..	Autres activités de serv. coll., soc. et personnels	O	90_93	
..	..	..	..	..	..	1 252	953	Assain. et enlèv. des ordures; voierie, act. simil.		90	
..	..	..	..	..	..	..	..	Activités associatives diverses		91	
..	..	..	..	..	..	10	3	Activités de syndicats de salariés			912
..	..	..	..	..	..	..	..	Activités récréatives, culturelles et sportives		92	
..	..	..	..	..	..	1 288	1 418	Act. cinémat., de radio, télé.; autres act. de spect.			921
..	..	..	..	..	..	59	45	Activités d'agences de presse			922
..	..	..	..	..	..	29	52	Act. des biblio., arch., musées, jardins bot. et zool.			923
..	..	..	..	..	..	1 502	1 190	Activités sportives et autres activités récréatives			924
..	..	..	..	..	..	564	471	Autres activités de services		93	

SWEDEN

Table SN.S4 INVESTMENT IN MACHINERY & EQUIPMENT - DÉPENSES DE MACHINES ET D'ÉQUIPEMENT

Millions of SKr (Current Prices)

ISIC revision 3				1991	1992	1993	1994	1995	1996	1997	1998
G	50_52		Wholesale & retail trade; repairs; etc.	..	..	..	..	..	..	18 557	19 907
	50		Sale, maint.,repair vehic/cycle, person., h/hold gd.	..	..	..	..	..	..	4 075	3 864
		501	Sale of motor cycle	..	..	..	..	..	..	2 556	2 467
		502	Maintenance and repair of motor vehicles	..	..	..	..	..	..	766	736
		503	Sale of motor vehicle parts and accessories	..	..	..	..	..	..	224	220
		504	Sale, maint., repair motor cycle, parts & acces.	..	..	..	..	..	..	29	35
		505	Retail sale of automative fuel	..	..	..	..	..	..	501	406
	51		W/sale. trade & commission exc motor veh. & cycle	..	..	..	..	..	..	10 058	11 273
		511	Wholesale on a fee or contract basis	..	..	..	..	..	..	436	949
		512	W/sale agric. raw mat.,live animal,food, bev., tobac.	..	..	..	..	..	..	..	..
		513	Wholesale of household goods	..	..	..	..	..	..	1 560	1 455
		514	W/sale. of non-agric. intermed. prod., waste, scrap	..	..	..	..	..	..	3 298	2 986
		515	Wholesale of machinery, equipment & supplies	..	..	..	..	..	..	2 910	4 464
		519	Other wholesale	..	..	..	..	..	..	182	114
	52		Retail trade exc. motor veh. & cyc.; repair h/hold gds.	..	..	..	..	..	..	4 424	4 770
		521	Non-specialised retail trade in stores	..	..	..	..	..	..	1 718	1 934
		522	Retail sale food, bev., tobac. in specialised stores	..	..	..	..	..	..	400	382
		523	Other retail trade of new goods specialised stores	..	..	..	..	..	..	..	..
		524	Retail sale of second-hand goods in stores	..	..	..	..	..	..	39	11
		525	Retail trade not in stores	..	..	..	..	..	..	148	200
		526	Repair of personal and household goods	..	..	..	..	..	..	83	77
H	55		Hotels and restaurants	..	..	..	..	..	..	2 386	2 595
		551	Hotels; camping ; other short-stay accommodation	..	..	..	..	..	..	..	..
		552	Restaurants, bars and canteens	..	..	..	..	..	..	..	..
I	60_64		Transport, storage and communications	..	..	..	..	..	..	28 415	28 826
	60		Land transport, transport via pipelines	..	..	..	..	..	..	10 674	10 397
		601	Transport via railways	..	..	..	..	..	..	289	212
		602	Other land transport	..	..	..	..	..	..	10 384	10 183
		603	Transport via pipelines	..	..	..	..	..	..	1	2
	61		Water transport	..	..	..	..	..	..	2 761	1 509
		611	Sea and coastal water transport	..	..	..	..	..	..	2 650	1 407
		612	Inland water transport	..	..	..	..	..	..	112	102
	62		Air transport	..	..	..	..	..	..	1 414	1 746
		621	Scheduled air transport	..	..	..	..	..	..	..	..
		622	Non-scheduled air transport	..	..	..	..	..	..	..	..
	63		Support. & aux. transport act., travel agencies activ.	..	..	..	..	..	..	2 710	2 276
	64		Post and telecommunications	..	..	..	..	..	..	10 856	12 897
		641	Post and courier activities	..	..	..	..	..	..	..	..
		642	Telecommunications	..	..	..	..	..	..	..	..
J	65_67		Financial intermediation	..	..	..	..	..	..	..	..
	65		Financial intermediation, exc. insur., pension fund	..	..	..	..	..	..	..	..
		651	Monetary intermediation	..	..	..	..	..	..	..	..
		659	Other financial intermediation	..	..	..	..	..	..	..	..
	66		Insurance pension fund, exc. compul. social sec.	..	..	..	..	..	..	..	..
	67		Activities auxiliary to financial intermediation	..	..	..	..	..	..	..	..
		671	Act. auxil to fin. intermed., exc. insur. & pens.fund	..	..	..	..	..	..	..	..
		672	Act. auxiliary to insurance and pension funding	..	..	..	..	..	..	..	..
K	70_74		Real estate, renting and business activities	..	..	..	..	..	..	17 977	18 415
	70		Real estate activities	..	..	..	..	..	..	2 769	3 135
		701	Real estate activ. with own or leased property	..	..	..	..	..	..	..	..
		702	Real estate activities on a fee or contract basis	..	..	..	..	..	..	..	..
	71		Renting of mach. without operator; rent. of pers. gds.	..	..	..	..	..	..	4 186	3 924
		711	Renting of transport equipment	..	..	..	..	..	..	..	..
		712	Renting of other machinery and equipment	..	..	..	..	..	..	2 347	1 787
		713	Renting of personal and household goods nec	..	..	..	..	..	..	336	258
	72		Computer and related activities	..	..	..	..	..	..	2 328	2 819
		721	Hardware consultancy	..	..	..	..	..	..	..	..
		722	Software consultancy and supply	..	..	..	..	..	..	..	..
		723	Data processing	..	..	..	..	..	..	..	..
		724	Data base activities	..	..	..	..	..	..	..	..
		725	Mainten.; repair of office, account. & comp. mach.	..	..	..	..	..	..	..	..
		729	Other computer related activities	..	..	..	..	..	..	..	..
	73		Research and development	..	..	..	..	..	..	928	1 068
		731	R & D experim. on natural sciences & engineering	..	..	..	..	..	..	..	..
		732	R & D experim. on social sciences & humanities	..	..	..	..	..	..	..	..
	74		Other business activities	..	..	..	..	..	..	7 765	7 470
		741	Legal, acc., book-keep., audit act.; tax consult., etc	..	..	..	..	..	..	..	..
		742	Architectural, engineering, & other techn. activities	..	..	..	..	..	..	..	..
		743	Advertising	..	..	..	..	..	..	..	..
		749	Business activities nec	..	..	..	..	..	..	..	..
M	80		Education	..	..	..	..	..	..	..	..
		801	Primary education	..	..	..	..	..	..	11	21
		802	Secondary education	..	..	..	..	..	..	27	20
		803	Higher education	..	..	..	..	..	..	17	16
		809	Adult and other education	..	..	..	..	..	..	596	477
N	85		Health and social work	..	..	..	..	..	..	..	..
		851	Human health activities	..	..	..	..	..	..	760	805
		852	Veterinary activities	..	..	..	..	..	..	36	26
		853	Social work activities	..	..	..	..	..	..	136	166
O	90_93		Other commun., soc. & pers. services activities	..	..	..	..	..	..	..	..
	90		Sewage & refuse disposal, sanitation, etc.	..	..	..	..	..	..	847	748
	91		Activities of member organisations nec	..	..	..	..	..	..	..	..
		912	Activities of trade unions	..	..	..	..	..	..	5	2
	92		Recreational, cultural and sporting activities	..	..	..	..	..	..	..	..
		921	Motion picture, radio, TV, other entertainment act.	..	..	..	..	..	..	927	1 162
		922	News agency activities	..	..	..	..	..	..	50	37
		923	Library, archives, museums, other cultural activ.	..	..	..	..	..	..	16	16
		924	Sporting and other recreational activities	..	..	..	..	..	..	781	842
	93		Other service activities	..	..	..	..	..	..	438	326

Table SN.S5 **EMPLOI, SALARIÉS - EMPLOYMENT, EMPLOYEES**

Unités

1991	1992	1993	1994	1995	1996	1997	1998		CITI révision 3		
..	..	..	..	..	..	447 886	466 028	Com. de gros, de detail ; réparation etc.	G	50_52	
..	..	..	..	..	..	61 574	65 277	Com., entr., répar. d'auto. et moto ; carburants		50	
..	..	..	..	..	..	25 975	27 863	Commerce de véhicules automobiles			501
..	..	..	..	..	..	16 339	16 964	Entretien et réparation de véhicules automobiles			502
..	..	..	..	..	..	6 437	7 121	Com. de pièces et access. de véhic. automobiles			503
..	..	..	..	..	..	925	1 022	Com., entret. et répar. de motocycles et access.			504
..	..	..	..	..	..	11 898	12 307	Com. de détail de carburants pour automobiles			505
..	..	..	..	..	..	188 572	196 402	Com. de gros et act. inter., sauf auto. et moto		51	
..	..	..	..	..	..	8 283	8 054	Activités d'intermédiaries de commerce de gros			511
..	..	..	..	..	..	..	..	Com. gros: prod. agr. bruts, anim. viv., prod. alim.			512
..	..	..	..	..	..	43 218	45 537	Com. gros: biens de consom., sauf alimentaires			513
..	..	..	..	..	..	49 149	50 900	Com. gros: prod. inter. non agr., déchets, débris			514
..	..	..	..	..	..	56 949	60 759	Com. gros: machines, équipments et fournitures			515
..	..	..	..	..	..	1 388	1 437	Autres commerces de gros			519
..	..	..	..	..	..	197 740	204 349	Com. détail, sauf auto, moto; répar. art. pers., dom.		52	
..	..	..	..	..	..	78 353	78 661	Commerce de détail non spécialisé en magasin			521
..	..	..	..	..	..	14 347	14 172	Com. détail: prod. aliment. en mag. spécialisés			522
..	..	..	..	..	..	..	..	Autre com. détail: prod. neufs en mag. spécialisés			523
..	..	..	..	..	..	685	385	Commerces de détail de biens d'occasion			524
..	..	..	..	..	..	5 476	5 482	Commerces de détail autres qu'en magasin			525
..	..	..	..	..	..	3 653	3 827	Réparation d'articles personnels et domestiques			526
..	..	..	..	..	..	82 285	87 459	Hotels et restaurants	H	55	
..	..	..	..	..	..	..	..	Hôtels; camping; autres équip. courts séjours			551
..	..	..	..	..	..	..	..	Restaurants, cafés et cantines			552
..	..	..	..	..	..	269 304	274 473	Transport, entreposage, communications	I	60_64	
..	..	..	..	..	..	105 147	111 427	Transports terrestres; transports par conduites		60	
..	..	..	..	..	..	12 906	12 105	Transports par chemin de fer			601
..	..	..	..	..	..	92 233	99 313	Autres transports terrestres			602
..	..	..	..	..	..	8	9	Transports par conduites			603
..	..	..	..	..	..	14 065	14 382	Transports par eau		61	
..	..	..	..	..	..	13 142	13 337	Transports maritimes et côtiers			611
..	..	..	..	..	..	923	1 045	Transports par voies navigables intérieures			612
..	..	..	..	..	..	12 720	12 849	Transports aériens		62	
..	..	..	..	..	..	..	..	Transports aériens réguliers			621
..	..	..	..	..	..	..	..	Transports aériens non réguliers			622
..	..	..	..	..	..	46 780	47 300	Act. annex. et aux. des transp.; act. des ag. de voy.		63	
..	..	..	..	..	..	90 592	88 515	Postes et télécommunications		64	
..	..	..	..	..	..	..	..	Activités de poste et de courrier			641
..	..	..	..	..	..	..	..	Télécommunications			642
..	..	..	..	..	..	..	..	Intermédiation financière	J	65_67	
..	..	..	..	..	..	..	..	Interm. financ., sauf assur. et caisses de retraite		65	
..	..	..	..	..	..	..	..	Intermédiation monétaire			651
..	..	..	..	..	..	..	..	Autres intermédiations financières			659
..	..	..	..	..	..	..	..	Act. d'assur., caisses de retr., sauf séc. soc. oblig.		66	
..	..	..	..	..	..	..	..	Act. auxiliaires de l'intermédiation financière		67	
..	..	..	..	..	..	..	..	Act. aux. de l'interm. fin., sauf assur., et retraite			671
..	..	..	..	..	..	..	..	Act. aux. des assurances et caisses de retraite			672
..	..	..	..	..	..	328 767	360 409	Immobilier, locations, services aux entreprises	K	70_74	
..	..	..	..	..	..	51 064	51 120	Activités immobilières		70	
..	..	..	..	..	..	..	..	Act. immobilières sur biens propres ou loués			701
..	..	..	..	..	..	..	..	Activités immobilières pour comptes de tiers			702
..	..	..	..	..	..	8 615	8 398	Locat. mach., équip. sans opér., biens pers. et dom.		71	
..	..	..	..	..	..	..	..	Location de matériel de transport			711
..	..	..	..	..	..	5 302	4 832	Location de machines ét equipements nca			712
..	..	..	..	..	..	1 563	1 682	Loc. d'articles personnels et domestiques nca			713
..	..	..	..	..	..	50 995	63 191	Activités informatiques et activités rattachées		72	
..	..	..	..	..	..	..	..	Conseil en matériel informatique			721
..	..	..	..	..	..	..	..	Production de logiciels			722
..	..	..	..	..	..	..	..	Traitement de données informatiques			723
..	..	..	..	..	..	..	..	Activités de banques de données			724
..	..	..	..	..	..	..	..	Entr., répar. mach. de bureau et compt. et inform.			725
..	..	..	..	..	..	..	..	Autres activités rattachées à l'informatique			729
..	..	..	..	..	..	11 131	12 796	Recherche-développement		73	
..	..	..	..	..	..	..	..	Rech. et dév. expér. en sc. phys., natur. et ingén.			731
..	..	..	..	..	..	..	..	Rech. et dév. expér. en sciences soc. et humaines			732
..	..	..	..	..	..	206 962	224 904	Autres activités de services aux entreprises		74	
..	..	..	..	..	..	..	..	Act. jur., compt., audit.; cons. fisc.; ét. march., soud.			741
..	..	..	..	..	..	..	..	Act. d'achitect., d'ingén.; autres act. techniques			742
..	..	..	..	..	..	..	..	Publicité			743
..	..	..	..	..	..	..	..	Activités de services aux entreprises nca.			749
..	..	..	..	..	..	..	..	Education	M	80	
..	..	..	..	..	..	1 101	1 667	Activités d'enseignement primaire			801
..	..	..	..	..	..	637	777	Activités d'enseignement secondaire			802
..	..	..	..	..	..	292	330	Activités d'enseignement supérieur			803
..	..	..	..	..	..	16 838	13 542	Act. de formation perman.; autres act. d'enseign.			809
..	..	..	..	..	..	..	..	Santé et action sociale	N	85	
..	..	..	..	..	..	32 852	36 235	Activités de santé humaine			851
..	..	..	..	..	..	961	970	Activités vétérinaires			852
..	..	..	..	..	..	24 834	29 167	Activités d'action sociale			853
..	..	..	..	..	..	..	..	Autres activités de serv. coll., soc. et personnels	O	90_93	
..	..	..	..	..	..	6 299	6 447	Assain. et enlèv. des ordures; voierie, act. simil.		90	
..	..	..	..	..	..	..	..	Activités associatives diverses		91	
..	..	..	..	..	..	64	64	Activités de syndicats de salariés			912
..	..	..	..	..	..	..	..	Activités récréatives, culturelles et sportives		92	
..	..	..	..	..	..	22 709	23 661	Act. cinémat., de radio, télé.; autres act. de spect.			921
..	..	..	..	..	..	1 047	763	Activités d'agences de presse			922
..	..	..	..	..	..	553	574	Act. des biblio., arch., musées, jardins bot. et zool.			923
..	..	..	..	..	..	9 733	9 941	Activités sportives et autres activités récréatives			924
..	..	..	..	..	..	11 993	12 463	Autres activités de services		93	

SWEDEN

Table SN.S6 COMPENSATION OF LABOUR, EMPLOYEES - SALAIRES ET CHARGES SOCIALES, SALARIÉS

Millions of SKr (Current Prices)

ISIC revision 3				1991	1992	1993	1994	1995	1996	1997	1998
G	50_52		Wholesale & retail trade; repairs; etc.	..	..	..	..	..	..	117 867	128 075
	50		Sale, maint.,repair vehic/cycle, person., h/hold gd.	..	..	..	..	..	..	15 867	17 037
		501	Sale of motor cycle	..	..	..	..	..	..	7 329	8 106
		502	Maintenance and repair of motor vehicles	..	..	..	..	..	..	3 888	3 989
		503	Sale of motor vehicle parts and accessories	..	..	..	..	..	..	1 839	2 085
		504	Sale, maint., repair motor cycle, parts & acces.	..	..	..	..	..	..	220	244
		505	Retail sale of automative fuel	..	..	..	..	..	..	2 591	2 613
	51		W/sale. trade & commission exc motor veh. & cycle	..	..	..	..	..	..	60 873	67 376
		511	Wholesale on a fee or contract basis	..	..	..	..	..	..	2 621	2 695
		512	W/sale agric. raw mat.,live animal,food, bev., tobac.	..	..	..	..	..	..	..	..
		513	Wholesale of household goods	..	..	..	..	..	..	13 825	15 420
		514	W/sale. of non-agric. intermed. prod., waste, scrap	..	..	..	..	..	..	15 511	16 546
		515	Wholesale of machinery, equipment & supplies	..	..	..	..	..	..	19 956	23 254
		519	Other wholesale	..	..	..	..	..	..	394	429
	52		Retail trade exc. motor veh. & cyc.; repair h/hold gds.	..	..	..	..	..	..	41 127	43 662
		521	Non-specialised retail trade in stores	..	..	..	..	..	..	15 407	16 185
		522	Retail sale food, bev., tobac. in specialised stores	..	..	..	..	..	..	2 679	2 602
		523	Other retail trade of new goods specialised stores	..	..	..	..	..	..	..	..
		524	Retail sale of second-hand goods in stores	..	..	..	..	..	..	132	46
		525	Retail trade not in stores	..	..	..	..	..	..	1 309	1 481
		526	Repair of personal and household goods	..	..	..	..	..	..	903	962
H	55		Hotels and restaurants	..	..	..	..	..	..	14 760	16 064
		551	Hotels; camping ; other short-stay accommodation	..	..	..	..	..	..	..	..
		552	Restaurants, bars and canteens	..	..	..	..	..	..	..	..
I	60_64		Transport, storage and communications	..	..	..	..	..	..	78 742	83 953
	60		Land transport, transport via pipelines	..	..	..	..	..	..	27 237	28 640
		601	Transport via railways	..	..	..	..	..	..	4 242	4 100
		602	Other land transport	..	..	..	..	..	..	22 993	24 538
		603	Transport via pipelines	..	..	..	..	..	..	2	2
	61		Water transport	..	..	..	..	..	..	4 277	4 631
		611	Sea and coastal water transport	..	..	..	..	..	..	4 056	4 371
		612	Inland water transport	..	..	..	..	..	..	221	260
	62		Air transport	..	..	..	..	..	..	5 198	6 141
		621	Scheduled air transport	..	..	..	..	..	..	..	..
		622	Non-scheduled air transport	..	..	..	..	..	..	..	..
	63		Support. & aux. transport act., travel agencies activ.	..	..	..	..	..	..	14 460	14 844
	64		Post and telecommunications	..	..	..	..	..	..	27 570	29 697
		641	Post and courier activities	..	..	..	..	..	..	..	..
		642	Telecommunications	..	..	..	..	..	..	..	..
J	65_67		Financial intermediation	..	..	..	..	..	..	..	..
	65		Financial intermediation, exc. insur., pension fund	..	..	..	..	..	..	..	..
		651	Monetary intermediation	..	..	..	..	..	..	..	..
		659	Other financial intermediation	..	..	..	..	..	..	..	..
	66		Insurance pension fund, exc. compul. social sec.	..	..	..	..	..	..	..	..
	67		Activities auxiliary to financial intermediation	..	..	..	..	..	..	..	..
		671	Act. auxil to fin. intermed., exc. insur. & pens.fund	..	..	..	..	..	..	..	..
		672	Act. auxiliary to insurance and pension funding	..	..	..	..	..	..	..	..
K	70_74		Real estate, renting and business activities	..	..	..	..	..	..	104 158	121 974
	70		Real estate activities	..	..	..	..	..	..	14 099	14 727
		701	Real estate activ. with own or leased property	..	..	..	..	..	..	..	..
		702	Real estate activities on a fee or contract basis	..	..	..	..	..	..	..	..
	71		Renting of mach. without operator; rent. of pers. gds.	..	..	..	..	..	..	2 336	2 299
		711	Renting of transport equipment	..	..	..	..	..	..	..	..
		712	Renting of other machinery and equipment	..	..	..	..	..	..	1 547	1 420
		713	Renting of personal and household goods nec	..	..	..	..	..	..	349	387
	72		Computer and related activities	..	..	..	..	..	..	21 281	28 304
		721	Hardware consultancy	..	..	..	..	..	..	..	..
		722	Software consultancy and supply	..	..	..	..	..	..	..	..
		723	Data processing	..	..	..	..	..	..	..	..
		724	Data base activities	..	..	..	..	..	..	..	..
		725	Mainten.; repair of office, account. & comp. mach.	..	..	..	..	..	..	..	..
		729	Other computer related activities	..	..	..	..	..	..	..	..
	73		Research and development	..	..	..	..	..	..	4 010	5 288
		731	R & D experim. on natural sciences & engineering	..	..	..	..	..	..	..	..
		732	R & D experim. on social sciences & humanities	..	..	..	..	..	..	..	..
	74		Other business activities	..	..	..	..	..	..	62 431	71 357
		741	Legal, acc., book-keep., audit act.; tax consult., etc	..	..	..	..	..	..	..	..
		742	Architectural, engineering, & other techn. activities	..	..	..	..	..	..	..	..
		743	Advertising	..	..	..	..	..	..	..	..
		749	Business activities nec	..	..	..	..	..	..	..	..
M	80		Education	..	..	..	..	..	..	..	..
		801	Primary education	..	..	..	..	..	..	263	406
		802	Secondary education	..	..	..	..	..	..	170	231
		803	Higher education	..	..	..	..	..	..	100	121
		809	Adult and other education	..	..	..	..	..	..	3 111	3 520
N	85		Health and social work	..	..	..	..	..	..	..	..
		851	Human health activities	..	..	..	..	..	..	8 979	10 233
		852	Veterinary activities	..	..	..	..	..	..	228	250
		853	Social work activities	..	..	..	..	..	..	4 976	6 314
O	90_93		Other commun., soc. & pers. services activities	..	..	..	..	..	..	..	..
	90		Sewage & refuse disposal, sanitation, etc.	..	..	..	..	..	..	1 833	1 955
	91		Activities of member organisations nec	..	..	..	..	..	..	..	..
		912	Activities of trade unions	..	..	..	..	..	..	27	32
	92		Recreational, cultural and sporting activities	..	..	..	..	..	..	..	..
		921	Motion picture, radio, TV, other entertainment act.	..	..	..	..	..	..	6 159	6 769
		922	News agency activities	..	..	..	..	..	..	385	276
		923	Library, archives, museums, other cultural activ.	..	..	..	..	..	..	118	129
		924	Sporting and other recreational activities	..	..	..	..	..	..	1 975	2 136
	93		Other service activities	..	..	..	..	..	..	2 494	2 664

Table SN.S7 **SALAIRES ET TRAITEMENTS, SALARIÉS - WAGES AND SALARIES, EMPLOYEES**

Millions de SKr (Prix courants)

1991	1992	1993	1994	1995	1996	1997	1998			CITI révision 3
..	..	..	..	..	..	82 703	90 532	Com. de gros, de detail ; réparation etc.	G	50_52
..	..	..	..	..	..	11 130	12 041	Com., entr., répar. d'auto. et moto ; carburants		50
..	..	..	..	..	..	5 087	5 636	Commerce de véhicules automobiles		501
..	..	..	..	..	..	2 744	2 850	Entretien et réparation de véhicules automobiles		502
..	..	..	..	..	..	1 297	1 473	Com. de pièces et access. de véhic. automobiles		503
..	..	..	..	..	..	159	176	Com., entret. et répar. de motocycles et access.		504
..	..	..	..	..	..	1 844	1 906	Com. de détail de carburants pour automobiles		505
..	..	..	..	..	..	42 410	47 165	Com. de gros et act. inter., sauf auto. et moto		51
..	..	..	..	..	..	1 849	1 902	Activités d'intermédiaries de commerce de gros		511
..	..	..	..	..	..	..	..	Com. gros: prod. agr. bruts, anim. viv., prod. alim.		512
..	..	..	..	..	..	9 640	10 804	Com. gros: biens de consom., sauf alimentaires		513
..	..	..	..	..	..	10 725	11 555	Com. gros: prod. inter. non agr., déchets, débris		514
..	..	..	..	..	..	13 956	16 328	Com. gros: machines, équipments et fournitures		515
..	..	..	..	..	..	278	307	Autres commerces de gros		519
..	..	..	..	..	..	29 162	31 326	Com. détail, sauf auto, moto; répar. art. pers., dom.		52
..	..	..	..	..	..	11 093	11 690	Commerce de détail non spécialisé en magasin		521
..	..	..	..	..	..	1 831	1 852	Com. détail: prod. aliment. en mag. spécialisés		522
..	..	..	..	..	..	..	..	Autre com. détail: prod. neufs en mag. spécialisés		523
..	..	..	..	..	..	94	34	Commerces de détail de biens d'occasion		524
..	..	..	..	..	..	903	1 028	Commerces de détail autres qu'en magasin		525
..	..	..	..	..	..	633	681	Réparation d'articles personnels et domestiques		526
..	..	..	..	..	..	10 586	11 608	Hotels et restaurants	H	55
..	..	..	..	..	..	..	..	Hôtels; camping; autres équip. courts séjours		551
..	..	..	..	..	..	..	..	Restaurants, cafés et cantines		552
..	..	..	..	..	..	53 732	58 884	Transport, entreposage, communications	I	60_64
..	..	..	..	..	..	19 147	20 546	Transports terrestres; transports par conduites		60
..	..	..	..	..	..	2 884	2 806	Transports par chemin de fer		601
..	..	..	..	..	..	16 261	17 739	Autres transports terrestres		602
..	..	..	..	..	..	2	1	Transports par conduites		603
..	..	..	..	..	..	2 980	3 237	Transports par eau		61
..	..	..	..	..	..	2 821	3 051	Transports maritimes et côtiers		611
..	..	..	..	..	..	159	186	Transports par voies navigables intérieures		612
..	..	..	..	..	..	4 130	4 832	Transports aériens		62
..	..	..	..	..	..	..	..	Transports aériens réguliers		621
..	..	..	..	..	..	..	..	Transports aériens non réguliers		622
..	..	..	..	..	..	9 853	10 377	Act. annex. et aux. des transp.; act. des ag. de voy.		63
..	..	..	..	..	..	17 621	19 892	Postes et télécommunications		64
..	..	..	..	..	..	..	..	Activités de poste et de courrier		641
..	..	..	..	..	..	..	..	Télécommunications		642
..	..	..	..	..	..	..	..	Intermédiation financière	J	65_67
..	..	..	..	..	..	..	..	Interm. financ., sauf assur. et caisses de retraite		65
..	..	..	..	..	..	..	..	Intermédiation monétaire		651
..	..	..	..	..	..	..	..	Autres intermédiations financières		659
..	..	..	..	..	..	..	..	Act. d'assur., caisses de retr., sauf séc. soc. oblig.		66
..	..	..	..	..	..	..	..	Act. auxiliaires de l'intermédiation financière		67
..	..	..	..	..	..	..	..	Act. aux. de l'interm. fin., sauf assur., et retraite		671
..	..	..	..	..	..	..	..	Act. aux. des assurances et caisses de retraite		672
..	..	..	..	..	..	71 266	84 589	Immobilier, locations, services aux entreprises	K	70_74
..	..	..	..	..	..	9 629	10 125	Activités immobilières		70
..	..	..	..	..	..	..	..	Act. immobilières sur biens propres ou loués		701
..	..	..	..	..	..	..	..	Activités immobilières pour comptes de tiers		702
..	..	..	..	..	..	1 656	1 649	Locat. mach., équip. sans opér., biens pers. et dom.		71
..	..	..	..	..	..	..	..	Location de matériel de transport		711
..	..	..	..	..	..	1 099	1 026	Location de machines ét equipements nca		712
..	..	..	..	..	..	245	271	Loc. d'articles personnels et domestiques nca		713
..	..	..	..	..	..	14 749	19 783	Activités informatiques et activités rattachées		72
..	..	..	..	..	..	..	..	Conseil en matériel informatique		721
..	..	..	..	..	..	..	..	Production de logiciels		722
..	..	..	..	..	..	..	..	Traitement de données informatiques		723
..	..	..	..	..	..	..	..	Activités de banques de données		724
..	..	..	..	..	..	..	..	Entr., répar. mach. de bureau et compt. et inform.		725
..	..	..	..	..	..	..	..	Autres activités rattachées à l'informatique		729
..	..	..	..	..	..	2 752	3 581	Recherche-développement		73
..	..	..	..	..	..	..	..	Rech. et dév. expér. en sc. phys., natur. et ingén.		731
..	..	..	..	..	..	..	..	Rech. et dév. expér. en sciences soc. et humaines		732
..	..	..	..	..	..	42 480	49 451	Autres activités de services aux entreprises		74
..	..	..	..	..	..	..	..	Act. jur., compt., audit.; cons. fisc.; ét. march., soud.		741
..	..	..	..	..	..	..	..	Act. d'achitect., d'ingén.; autres act. techniques		742
..	..	..	..	..	..	..	..	Publicité		743
..	..	..	..	..	..	..	..	Activités de services aux entreprises nca.		749
..	..	..	..	..	..	..	..	Education	M	80
..	..	..	..	..	..	..	..	Activités d'enseignement primaire		801
..	..	..	..	..	..	..	..	Activités d'enseignement secondaire		802
..	..	..	..	..	..	..	..	Activités d'enseignement supérieur		803
..	..	..	..	..	..	..	..	Act. de formation perman.; autres act. d'enseign.		809
..	..	..	..	..	..	..	..	Santé et action sociale	N	85
..	..	..	..	..	..	..	..	Activités de santé humaine		851
..	..	..	..	..	..	..	..	Activités vétérinaires		852
..	..	..	..	..	..	..	..	Activités d'action sociale		853
..	..	..	..	..	..	..	..	Autres activités de serv. coll., soc. et personnels	O	90_93
..	..	..	..	..	..	..	..	Assain. et enlèv. des ordures; voierie, act. simil.		90
..	..	..	..	..	..	..	..	Activités associatives diverses		91
..	..	..	..	..	..	..	..	Activités de syndicats de salariés		912
..	..	..	..	..	..	..	..	Activités récréatives, culturelles et sportives		92
..	..	..	..	..	..	..	..	Act. cinémat., de radio, télé.; autres act. de spect.		921
..	..	..	..	..	..	..	..	Activités d'agences de presse		922
..	..	..	..	..	..	..	..	Act. des biblio., arch., musées, jardins bot. et zool.		923
..	..	..	..	..	..	..	..	Activités sportives et autres activités récréatives		924
..	..	..	..	..	..	..	..	Autres activités de services		93

SWEDEN

Table SN.S8 SUPPLEMENTS TO WAGES & SALARIES - CHARGES SOCIALES DES EMPLOYEURS

Millions of SKr (Current Prices)

ISIC revision 3				1991	1992	1993	1994	1995	1996	1997	1998
G	50_52		Wholesale & retail trade; repairs; etc.	..	..	..	..	..	..	35 164	37 543
	50		Sale, maint.,repair vehic/cycle, person., h/hold gd.	..	..	..	..	..	..	4 737	4 996
		501	Sale of motor cycle	..	..	..	..	..	..	2 242	2 470
		502	Maintenance and repair of motor vehicles	..	..	..	..	..	..	1 145	1 139
		503	Sale of motor vehicle parts and accessories	..	..	..	..	..	..	542	612
		504	Sale, maint., repair motor cycle, parts & acces.	..	..	..	..	..	..	61	68
		505	Retail sale of automative fuel	..	..	..	..	..	..	747	707
	51		W/sale. trade & commission exc motor veh. & cycle	..	..	..	..	..	..	18 463	20 210
		511	Wholesale on a fee or contract basis	..	..	..	..	..	..	772	793
		512	W/sale agric. raw mat.,live animal,food, bev., tobac.	..	..	..	..	..	..	..	..
		513	Wholesale of household goods	..	..	..	..	..	..	4 185	4 616
		514	W/sale. of non-agric. intermed. prod., waste, scrap	..	..	..	..	..	..	4 786	4 991
		515	Wholesale of machinery, equipment & supplies	..	..	..	..	..	..	6 000	6 926
		519	Other wholesale	..	..	..	..	..	..	116	122
	52		Retail trade exc. motor veh. & cyc.; repair h/hold gds.	..	..	..	..	..	..	11 965	12 336
		521	Non-specialised retail trade in stores	..	..	..	..	..	..	4 314	4 495
		522	Retail sale food, bev., tobac. in specialised stores	..	..	..	..	..	..	848	749
		523	Other retail trade of new goods specialised stores	..	..	..	..	..	..	..	..
		524	Retail sale of second-hand goods in stores	..	..	..	..	..	..	38	13
		525	Retail trade not in stores	..	..	..	..	..	..	406	454
		526	Repair of personal and household goods	..	..	..	..	..	..	270	282
H	55		Hotels and restaurants	..	..	..	..	..	..	4 175	4 456
		551	Hotels; camping ; other short-stay accommodation	..	..	..	..	..	..	..	..
		552	Restaurants, bars and canteens	..	..	..	..	..	..	..	..
I	60_64		Transport, storage and communications	..	..	..	..	..	..	25 011	25 069
	60		Land transport, transport via pipelines	..	..	..	..	..	..	8 090	8 093
		601	Transport via railways	..	..	..	..	..	..	1 358	1 294
		602	Other land transport	..	..	..	..	..	..	6 732	6 799
		603	Transport via pipelines	..	..	..	..	..	..	1	1
	61		Water transport	..	..	..	..	..	..	1 297	1 394
		611	Sea and coastal water transport	..	..	..	..	..	..	1 236	1 320
		612	Inland water transport	..	..	..	..	..	..	62	74
	62		Air transport	..	..	..	..	..	..	1 068	1 309
		621	Scheduled air transport	..	..	..	..	..	..	..	..
		622	Non-scheduled air transport	..	..	..	..	..	..	..	..
	63		Support. & aux. transport act., travel agencies activ.	..	..	..	..	..	..	4 607	4 467
	64		Post and telecommunications	..	..	..	..	..	..	9 948	9 805
		641	Post and courier activities	..	..	..	..	..	..	..	..
		642	Telecommunications	..	..	..	..	..	..	..	..
J	65_67		Financial intermediation	..	..	..	..	..	..	..	..
	65		Financial intermediation, exc. insur., pension fund	..	..	..	..	..	..	..	..
		651	Monetary intermediation	..	..	..	..	..	..	..	..
		659	Other financial intermediation	..	..	..	..	..	..	..	..
	66		Insurance pension fund, exc. compul. social sec.	..	..	..	..	..	..	..	..
	67		Activities auxiliary to financial intermediation	..	..	..	..	..	..	..	..
		671	Act. auxil to fin. intermed., exc. insur. & pens.fund	..	..	..	..	..	..	..	..
		672	Act. auxiliary to insurance and pension funding	..	..	..	..	..	..	..	..
K	70_74		Real estate, renting and business activities	..	..	..	..	..	..	32 892	37 385
	70		Real estate activities	..	..	..	..	..	..	4 471	4 601
		701	Real estate activ. with own or leased property	..	..	..	..	..	..	..	..
		702	Real estate activities on a fee or contract basis	..	..	..	..	..	..	..	..
	71		Renting of mach. without operator; rent. of pers. gds.	..	..	..	..	..	..	680	650
		711	Renting of transport equipment	..	..	..	..	..	..	..	..
		712	Renting of other machinery and equipment	..	..	..	..	..	..	449	394
		713	Renting of personal and household goods nec	..	..	..	..	..	..	104	116
	72		Computer and related activities	..	..	..	..	..	..	6 532	8 520
		721	Hardware consultancy	..	..	..	..	..	..	..	..
		722	Software consultancy and supply	..	..	..	..	..	..	..	..
		723	Data processing	..	..	..	..	..	..	..	..
		724	Data base activities	..	..	..	..	..	..	..	..
		725	Mainten.; repair of office, account. & comp. mach.	..	..	..	..	..	..	..	..
		729	Other computer related activities	..	..	..	..	..	..	..	..
	73		Research and development	..	..	..	..	..	..	1 258	1 707
		731	R & D experim. on natural sciences & engineering	..	..	..	..	..	..	..	..
		732	R & D experim. on social sciences & humanities	..	..	..	..	..	..	..	..
	74		Other business activities	..	..	..	..	..	..	19 952	21 906
		741	Legal, acc., book-keep., audit act.; tax consult., etc	..	..	..	..	..	..	..	..
		742	Architectural, engineering, & other techn. activities	..	..	..	..	..	..	..	..
		743	Advertising	..	..	..	..	..	..	..	..
		749	Business activities nec	..	..	..	..	..	..	..	..
M	80		Education	..	..	..	..	..	..	..	..
		801	Primary education	..	..	..	..	..	..	..	..
		802	Secondary education	..	..	..	..	..	..	..	..
		803	Higher education	..	..	..	..	..	..	..	..
		809	Adult and other education	..	..	..	..	..	..	..	..
N	85		Health and social work	..	..	..	..	..	..	..	..
		851	Human health activities	..	..	..	..	..	..	..	..
		852	Veterinary activities	..	..	..	..	..	..	..	..
		853	Social work activities	..	..	..	..	..	..	..	..
O	90_93		Other commun., soc. & pers. services activities	..	..	..	..	..	..	..	..
	90		Sewage & refuse disposal, sanitation, etc.	..	..	..	..	..	..	..	..
	91		Activities of member organisations nec	..	..	..	..	..	..	..	..
		912	Activities of trade unions	..	..	..	..	..	..	..	..
	92		Recreational, cultural and sporting activities	..	..	..	..	..	..	..	..
		921	Motion picture, radio, TV, other entertainment act.	..	..	..	..	..	..	..	..
		922	News agency activities	..	..	..	..	..	..	..	..
		923	Library, archives, museums, other cultural activ.	..	..	..	..	..	..	..	..
		924	Sporting and other recreational activities	..	..	..	..	..	..	..	..
	93		Other service activities	..	..	..	..	..	..	..	..

Table SN.S9 **ENTREPRISES - ENTERPRISES**

Unités

1991	1992	1993	1994	1995	1996	1997	1998			CITI révision 3	
..	..	..	..	..	..	119 521	119 431	Com. de gros, de detail ; réparation etc.	G	50_52	
..	..	..	..	..	..	18 534	18 556	Com., entr., répar. d'auto. et moto ; carburants		50	
..	..	..	..	..	..	3 819	3 781	Commerce de véhicules automobiles			501
..	..	..	..	..	..	10 089	10 148	Entretien et réparation de véhicules automobiles			502
..	..	..	..	..	..	1 704	1 732	Com. de pièces et access. de véhic. automobiles			503
..	..	..	..	..	..	630	668	Com., entret. et répar. de motocycles et access.			504
..	..	..	..	..	..	2 292	2 227	Com. de détail de carburants pour automobiles			505
..	..	..	..	..	..	42 295	42 228	Com. de gros et act. inter., sauf auto. et moto		51	
..	..	..	..	..	..	3 373	3 444	Activités d'intermédiaries de commerce de gros			511
..	..	..	..	..	..	..	..	Com. gros: prod. agr. bruts, anim. viv., prod. alim.			512
..	..	..	..	..	..	12 130	12 202	Com. gros: biens de consom., sauf alimentaires			513
..	..	..	..	..	..	11 053	10 816	Com. gros: prod. inter. non agr., déchets, débris			514
..	..	..	..	..	..	10 546	10 410	Com. gros: machines, équipments et fournitures			515
..	..	..	..	..	..	468	482	Autres commerces de gros			519
..	..	..	..	..	..	58 692	58 647	Com. détail, sauf auto, moto; répar. art. pers., dom.		52	
..	..	..	..	..	..	7 207	7 036	Commerce de détail non spécialisé en magasin			521
..	..	..	..	..	..	7 555	7 336	Com. détail: prod. aliment. en mag. spécialisés			522
..	..	..	..	..	..	..	..	Autre com. détail: prod. neufs en mag. spécialisés			523
..	..	..	..	..	..	1 476	1 585	Commerces de détail de biens d'occasion			524
..	..	..	..	..	..	2 854	3 388	Commerces de détail autres qu'en magasin			525
..	..	..	..	..	..	3 549	3 610	Réparation d'articles personnels et domestiques			526
..	..	..	..	..	..	19 731	19 942	Hotels et restaurants	H	55	
..	..	..	..	..	..	..	..	Hôtels; camping; autres équip. courts séjours			551
..	..	..	..	..	..	..	..	Restaurants, cafés et cantines			552
..	..	..	..	..	..	31 963	31 596	Transport, entreposage, communications	I	60_64	
..	..	..	..	..	..	26 443	25 841	Transports terrestres; transports par conduites		60	
..	..	..	..	..	..	23	24	Transports par chemin de fer			601
..	..	..	..	..	..	26 412	25 808	Autres transports terrestres			602
..	..	..	..	..	..	8	9	Transports par conduites			603
..	..	..	..	..	..	722	747	Transports par eau		61	
..	..	..	..	..	..	409	418	Transports maritimes et côtiers			611
..	..	..	..	..	..	313	329	Transports par voies navigables intérieures			612
..	..	..	..	..	..	213	211	Transports aériens		62	
..	..	..	..	..	..	..	..	Transports aériens réguliers			621
..	..	..	..	..	..	..	..	Transports aériens non réguliers			622
..	..	..	..	..	..	4 036	4 240	Act. annex. et aux. des transp.; act. des ag. de voy.		63	
..	..	..	..	..	..	549	557	Postes et télécommunications		64	
..	..	..	..	..	..	..	..	Activités de poste et de courrier			641
..	..	..	..	..	..	..	..	Télécommunications			642
..	..	..	..	..	..	..	..	Intermédiation financière	J	65_67	
..	..	..	..	..	..	..	..	Interm. financ., sauf assur. et caisses de retraite		65	
..	..	..	..	..	..	..	..	Intermédiation monétaire			651
..	..	..	..	..	..	..	..	Autres intermédiations financières			659
..	..	..	..	..	..	..	..	Act. d'assur., caisses de retr., sauf séc. soc. oblig.		66	
..	..	..	..	..	..	..	..	Act. auxiliaires de l'intermédiation financière		67	
..	..	..	..	..	..	..	..	Act. aux. de l'interm. fin., sauf assur., et retraite			671
..	..	..	..	..	..	..	..	Act. aux. des assurances et caisses de retraite			672
..	..	..	..	..	..	136 312	144 715	Immobilier, locations, services aux entreprises	K	70_74	
..	..	..	..	..	..	28 141	29 250	Activités immobilières		70	
..	..	..	..	..	..	..	..	Act. immobilières sur biens propres ou loués			701
..	..	..	..	..	..	..	..	Activités immobilières pour comptes de tiers			702
..	..	..	..	..	..	4 780	4 734	Locat. mach., équip. sans opér., biens pers. et dom.		71	
..	..	..	..	..	..	..	..	Location de matériel de transport			711
..	..	..	..	..	..	2 705	2 642	Location de machines ét equipements nca			712
..	..	..	..	..	..	1 015	1 085	Loc. d'articles personnels et domestiques nca			713
..	..	..	..	..	..	15 262	17 444	Activités informatiques et activités rattachées		72	
..	..	..	..	..	..	..	..	Conseil en matériel informatique			721
..	..	..	..	..	..	..	..	Production de logiciels			722
..	..	..	..	..	..	..	..	Traitement de données informatiques			723
..	..	..	..	..	..	..	..	Activités de banques de données			724
..	..	..	..	..	..	..	..	Entr., répar. mach. de bureau et compt. et inform.			725
..	..	..	..	..	..	..	..	Autres activités rattachées à l'informatique			729
..	..	..	..	..	..	1 552	1 736	Recherche-développement		73	
..	..	..	..	..	..	..	..	Rech. et dév. expér. en sc. phys., natur. et ingén.			731
..	..	..	..	..	..	..	..	Rech. et dév. expér. en sciences soc. et humaines			732
..	..	..	..	..	..	86 577	91 551	Autres activités de services aux entreprises		74	
..	..	..	..	..	..	..	..	Act. jur., compt., audit.; cons. fisc.; ét. march., soud.			741
..	..	..	..	..	..	..	..	Act. d'achitect., d'ingén.; autres act. techniques			742
..	..	..	..	..	..	..	..	Publicité			743
..	..	..	..	..	..	..	..	Activités de services aux entreprises nca.			749
..	..	..	..	..	..	..	..	Education	M	80	
..	..	..	..	..	..	88	102	Activités d'enseignement primaire			801
..	..	..	..	..	..	66	56	Activités d'enseignement secondaire			802
..	..	..	..	..	..	14	14	Activités d'enseignement supérieur			803
..	..	..	..	..	..	5 194	5 892	Act. de formation perman.; autres act. d'enseign.			809
..	..	..	..	..	..	..	..	Santé et action sociale	N	85	
..	..	..	..	..	..	10 606	11 578	Activités de santé humaine			851
..	..	..	..	..	..	700	696	Activités vétérinaires			852
..	..	..	..	..	..	2 221	2 483	Activités d'action sociale			853
..	..	..	..	..	..	..	..	Autres activités de serv. coll., soc. et personnels	O	90_93	
..	..	..	..	..	..	751	745	Assain. et enlèv. des ordures; voierie, act. simil.		90	
..	..	..	..	..	..	..	..	Activités associatives diverses		91	
..	..	..	..	..	..	19	19	Activités de syndicats de salariés			912
..	..	..	..	..	..	..	..	Activités récréatives, culturelles et sportives		92	
..	..	..	..	..	..	9 340	11 190	Act. cinémat., de radio, télé.; autres act. de spect.			921
..	..	..	..	..	..	145	150	Activités d'agences de presse			922
..	..	..	..	..	..	300	310	Act. des biblio., arch., musées, jardins bot. et zool.			923
..	..	..	..	..	..	5 481	6 157	Activités sportives et autres activités récréatives			924
..	..	..	..	..	..	17 064	18 141	Autres activités de services		93	

SWITZERLAND

Table SW.S1 **PRODUCTION - PRODUCTION**

Millions of SF (Current Prices)

ISIC revision 3				1991	1992	1993	1994	1995	1996	1997	1998
G	**50_52**		Wholesale & retail trade; repairs; etc.	..	..	..	..	..	..	73 990	78 217
	50		Sale, maint.,repair vehic/cycle, person., h/hold gd.	..	..	..	..	..	..	8 257	8 564
		501	Sale of motor cycle	..	..	..	..	..	..	..	..
		502	Maintenance and repair of motor vehicles	..	..	..	..	..	..	..	..
		503	Sale of motor vehicle parts and accessories	..	..	..	..	..	..	..	..
		504	Sale, maint., repair motor cycle, parts & acces.	..	..	..	..	..	..	..	..
		505	Retail sale of automative fuel	..	..	..	..	..	..	..	..
	51		W/sale. trade & commission exc motor veh. & cycle	..	..	..	..	..	..	36 338	38 686
		511	Wholesale on a fee or contract basis	..	..	..	..	..	..	..	..
		512	W/sale agric. raw mat.,live animal,food, bev., tobac.	..	..	..	..	..	..	..	..
		513	Wholesale of household goods	..	..	..	..	..	..	..	..
		514	W/sale. of non-agric. intermed. prod., waste, scrap	..	..	..	..	..	..	..	..
		515	Wholesale of machinery, equipment & supplies	..	..	..	..	..	..	..	..
		519	Other wholesale	..	..	..	..	..	..	..	..
	52		Retail trade exc. motor veh. & cyc.; repair h/hold gds.	..	..	..	..	..	..	29 395	30 968
		521	Non-specialised retail trade in stores	..	..	..	..	..	..	..	..
		522	Retail sale food, bev., tobac. in specialised stores	..	..	..	..	..	..	..	..
		523	Other retail trade of new goods specialised stores	..	..	..	..	..	..	..	..
		524	Retail sale of second-hand goods in stores	..	..	..	..	..	..	..	..
		525	Retail trade not in stores	..	..	..	..	..	..	..	..
		526	Repair of personal and household goods	..	..	..	..	..	..	..	..
H	**55**		Hotels and restaurants	..	..	..	..	..	..	18 537	19 311
		551	Hotels; camping ; other short-stay accommodation	..	..	..	..	..	..	..	..
		552	Restaurants, bars and canteens	..	..	..	..	..	..	..	..
I	**60_64**		Transport, storage and communications	..	..	..	..	..	..	..	..
	60		Land transport, transport via pipelines	..	..	..	..	..	..	5 722	7 289
		601	Transport via railways	..	..	..	..	..	..	..	..
		602	Other land transport	..	..	..	..	..	..	..	..
		603	Transport via pipelines	..	..	..	..	..	..	..	..
	61		Water transport	..	..	..	..	..	..	365	337
		611	Sea and coastal water transport	..	..	..	..	..	..	..	..
		612	Inland water transport	..	..	..	..	..	..	..	..
	62		Air transport	..	..	..	..	..	..	..	..
		621	Scheduled air transport	..	..	..	..	..	..	..	..
		622	Non-scheduled air transport	..	..	..	..	..	..	..	..
	63		Support. & aux. transport act., travel agencies activ.	..	..	..	..	..	..	8 560	8 886
	64		Post and telecommunications	..	..	..	..	..	..	..	..
		641	Post and courier activities	..	..	..	..	..	..	..	..
		642	Telecommunications	..	..	..	..	..	..	..	..
J	**65_67**		Financial intermediation	..	..	..	..	..	..	..	..
	65		Financial intermediation, exc. insur., pension fund	..	..	..	..	..	..	..	..
		651	Monetary intermediation	..	..	..	..	..	..	..	..
		659	Other financial intermediation	..	..	..	..	..	..	..	..
	66		Insurance pension fund, exc. compul. social sec.	..	..	..	..	..	..	..	..
	67		Activities auxiliary to financial intermediation	..	..	..	..	..	..	..	..
		671	Act. auxil to fin. intermed., exc. insur. & pens.fund	..	..	..	..	..	..	..	..
		672	Act. auxiliary to insurance and pension funding	..	..	..	..	..	..	..	..
K	**70_74**		Real estate, renting and business activities	..	..	..	..	..	..	51 612	59 778
	70		Real estate activities	..	..	..	..	..	..	6 249	7 473
		701	Real estate activ. with own or leased property	..	..	..	..	..	..	..	..
		702	Real estate activities on a fee or contract basis	..	..	..	..	..	..	..	..
	71		Renting of mach. without operator; rent. of pers. gds.	..	..	..	..	..	..	..	..
		711	Renting of transport equipment	..	..	..	..	..	..	..	..
		712	Renting of other machinery and equipment	..	..	..	..	..	..	..	..
		713	Renting of personal and household goods nec	..	..	..	..	..	..	..	..
	72		Computer and related activities	..	..	..	..	..	..	7 122	8 478
		721	Hardware consultancy	..	..	..	..	..	..	..	..
		722	Software consultancy and supply	..	..	..	..	..	..	..	..
		723	Data processing	..	..	..	..	..	..	..	..
		724	Data base activities	..	..	..	..	..	..	..	..
		725	Mainten.; repair of office, account. & comp. mach.	..	..	..	..	..	..	..	..
		729	Other computer related activities	..	..	..	..	..	..	..	..
	73		Research and development	..	..	..	..	..	..	1 486	1 215
		731	R & D experim. on natural sciences & engineering	..	..	..	..	..	..	..	..
		732	R & D experim. on social sciences & humanities	..	..	..	..	..	..	..	..
	74		Other business activities	..	..	..	..	..	..	36 754	42 612
		741	Legal, acc., book-keep., audit act.; tax consult., etc	..	..	..	..	..	..	..	..
		742	Architectural, engineering, & other techn. activities	..	..	..	..	..	..	..	..
		743	Advertising	..	..	..	..	..	..	..	..
		749	Business activities nec	..	..	..	..	..	..	..	..
M	**80**		Education	..	..	..	..	..	..	1 855	1 538
		801	Primary education	..	..	..	..	..	..	..	..
		802	Secondary education	..	..	..	..	..	..	..	..
		803	Higher education	..	..	..	..	..	..	..	..
		809	Adult and other education	..	..	..	..	..	..	..	..
N	**85**		Health and social work	..	..	..	..	..	..	2 861	2 869
		851	Human health activities	..	..	..	..	..	..	..	..
		852	Veterinary activities	..	..	..	..	..	..	..	..
		853	Social work activities	..	..	..	..	..	..	..	..
O	**90_93**		Other commun., soc. & pers. services activities	..	..	..	..	..	..	9 574	11 375
	90		Sewage & refuse disposal, sanitation, etc.	..	..	..	..	..	..	2 141	2 171
	91		Activities of member organisations nec	..	..	..	..	..	..	..	..
		912	Activities of trade unions	..	..	..	..	..	..	..	..
	92		Recreational, cultural and sporting activities	..	..	..	..	..	..	5 135	6 728
		921	Motion picture, radio, TV, other entertainment act.	..	..	..	..	..	..	..	..
		922	News agency activities	..	..	..	..	..	..	..	..
		923	Library, archives, museums, other cultural activ.	..	..	..	..	..	..	..	..
		924	Sporting and other recreational activities	..	..	..	..	..	..	..	..
	93		Other service activities	..	..	..	..	..	..	2 297	2 476

Table SW.S2 VALEUR AJOUTÉE AU COUT DES FACTEURS - VALUE ADDED AT FACTOR COSTS

Millions de SF (Prix courants)

1991	1992	1993	1994	1995	1996	1997	1998			CITI révision 3	
..	..	..	..	..	..	49 304	52 494	Com. de gros, de detail ; réparation etc.	G	50_52	
..	..	..	..	..	..	5 710	5 904	Com., entr., répar. d'auto. et moto ; carburants		50	
..	..	..	..	..	..	..	..	Commerce de véhicules automobiles			501
..	..	..	..	..	..	..	..	Entretien et réparation de véhicules automobiles			502
..	..	..	..	..	..	..	..	Com. de pièces et access. de véhic. automobiles			503
..	..	..	..	..	..	..	..	Com., entret. et répar. de motocycles et access.			504
..	..	..	..	..	..	..	..	Com. de détail de carburants pour automobiles			505
..	..	..	..	..	..	23 728	25 573	Com. de gros et act. inter., sauf auto. et moto		51	
..	..	..	..	..	..	..	..	Activités d'intermédiaries de commerce de gros			511
..	..	..	..	..	..	..	..	Com. gros: prod. agr. bruts, anim. viv., prod. alim.			512
..	..	..	..	..	..	..	..	Com. gros: biens de consom., sauf alimentaires			513
..	..	..	..	..	..	..	..	Com. gros: prod. inter. non agr., déchets, débris			514
..	..	..	..	..	..	..	..	Com. gros: machines, équipments et fournitures			515
..	..	..	..	..	..	..	..	Autres commerces de gros			519
..	..	..	..	..	..	19 865	21 017	Com. détail, sauf auto, moto; répar. art. pers., dom.		52	
..	..	..	..	..	..	..	..	Commerce de détail non spécialisé en magasin			521
..	..	..	..	..	..	..	..	Com. détail: prod. aliment. en mag. spécialisés			522
..	..	..	..	..	..	..	..	Autre com. détail: prod. neufs en mag. spécialisés			523
..	..	..	..	..	..	..	..	Commerces de détail de biens d'occasion			524
..	..	..	..	..	..	..	..	Commerces de détail autres qu'en magasin			525
..	..	..	..	..	..	..	..	Réparation d'articles personnels et domestiques			526
..	..	..	..	..	..	9 749	10 235	Hotels et restaurants	H	55	
..	..	..	..	..	..	..	..	Hôtels; camping; autres équip. courts séjours			551
..	..	..	..	..	..	..	..	Restaurants, cafés et cantines			552
..	..	..	..	..	..	..	..	Transport, entreposage, communications	I	60_64	
..	..	..	..	..	..	3 037	3 046	Transports terrestres; transports par conduites		60	
..	..	..	..	..	..	..	..	Transports par chemin de fer			601
..	..	..	..	..	..	..	..	Autres transports terrestres			602
..	..	..	..	..	..	..	..	Transports par conduites			603
..	..	..	..	..	..	289	278	Transports par eau		61	
..	..	..	..	..	..	..	..	Transports maritimes et côtiers			611
..	..	..	..	..	..	..	..	Transports par voies navigables intérieures			612
..	..	..	..	..	..	..	..	Transports aériens		62	
..	..	..	..	..	..	..	..	Transports aériens réguliers			621
..	..	..	..	..	..	..	..	Transports aériens non réguliers			622
..	..	..	..	..	..	3 872	3 560	Act. annex. et aux. des transp.; act. des ag. de voy.		63	
..	..	..	..	..	..	..	..	Postes et télécommunications		64	
..	..	..	..	..	..	..	..	Activités de poste et de courrier			641
..	..	..	..	..	..	..	..	Télécommunications			642
..	..	..	..	..	..	..	..	Intermédiation financière	J	65_67	
..	..	..	..	..	..	..	..	Interm. financ., sauf assur. et caisses de retraite		65	
..	..	..	..	..	..	..	..	Intermédiation monétaire			651
..	..	..	..	..	..	..	..	Autres intermédiations financières			659
..	..	..	..	..	..	..	..	Act. d'assur., caisses de retr., sauf séc. soc. oblig.		66	
..	..	..	..	..	..	..	..	Act. auxiliaires de l'intermédiation financière		67	
..	..	..	..	..	..	..	..	Act. aux. de l'interm. fin., sauf assur., et retraite			671
..	..	..	..	..	..	..	..	Act. aux. des assurances et caisses de retraite			672
..	..	..	..	..	..	33 305	38 131	Immobilier, locations, services aux entreprises	K	70_74	
..	..	..	..	..	..	4 381	5 235	Activités immobilières		70	
..	..	..	..	..	..	..	..	Act. immobilières sur biens propres ou loués			701
..	..	..	..	..	..	..	..	Activités immobilières pour comptes de tiers			702
..	..	..	..	..	..	..	..	Locat. mach., équip. sans opér., biens pers. et dom.		71	
..	..	..	..	..	..	..	..	Location de matériel de transport			711
..	..	..	..	..	..	..	..	Location de machines ét equipements nca			712
..	..	..	..	..	..	..	..	Loc. d'articles personnels et domestiques nca			713
..	..	..	..	..	..	4 525	4 615	Activités informatiques et activités rattachées		72	
..	..	..	..	..	..	..	..	Conseil en matériel informatique			721
..	..	..	..	..	..	..	..	Production de logiciels			722
..	..	..	..	..	..	..	..	Traitement de données informatiques			723
..	..	..	..	..	..	..	..	Activités de banques de données			724
..	..	..	..	..	..	..	..	Entr., répar. mach. de bureau et compt. et inform.			725
..	..	..	..	..	..	..	..	Autres activités rattachées à l'informatique			729
..	..	..	..	..	..	737	516	Recherche-développement		73	
..	..	..	..	..	..	..	..	Rech. et dév. expér. en sc. phys., natur. et ingén.			731
..	..	..	..	..	..	..	..	Rech. et dév. expér. en sciences soc. et humaines			732
..	..	..	..	..	..	23 662	27 765	Autres activités de services aux entreprises		74	
..	..	..	..	..	..	..	..	Act. jur., compt., audit.; cons. fisc.; ét. march., soud.			741
..	..	..	..	..	..	..	..	Act. d'achitect., d'ingén.; autres act. techniques			742
..	..	..	..	..	..	..	..	Publicité			743
..	..	..	..	..	..	..	..	Activités de services aux entreprises nca.			749
..	..	..	..	..	..	1 668	1 286	Education	M	80	
..	..	..	..	..	..	..	..	Activités d'enseignement primaire			801
..	..	..	..	..	..	..	..	Activités d'enseignement secondaire			802
..	..	..	..	..	..	..	..	Activités d'enseignement supérieur			803
..	..	..	..	..	..	..	..	Act. de formation perman.; autres act. d'enseign.			809
..	..	..	..	..	..	2 208	2 252	Santé et action sociale	N	85	
..	..	..	..	..	..	..	..	Activités de santé humaine			851
..	..	..	..	..	..	..	..	Activités vétérinaires			852
..	..	..	..	..	..	..	..	Activités d'action sociale			853
..	..	..	..	..	..	5 545	5 839	Autres activités de serv. coll., soc. et personnels	O	90_93	
..	..	..	..	..	..	893	794	Assain. et enlèv. des ordures; voierie, act. simil.		90	
..	..	..	..	..	..	..	..	Activités associatives diverses		91	
..	..	..	..	..	..	..	..	Activités de syndicats de salariés			912
..	..	..	..	..	..	3 002	3 497	Activités récréatives, culturelles et sportives		92	
..	..	..	..	..	..	..	..	Act. cinémat., de radio, télé.; autres act. de spect.			921
..	..	..	..	..	..	..	..	Activités d'agences de presse			922
..	..	..	..	..	..	..	..	Act. des biblio., arch., musées, jardins bot. et zool.			923
..	..	..	..	..	..	..	..	Activités sportives et autres activités récréatives			924
..	..	..	..	..	..	1 650	1 548	Autres activités de services		93	

Table SW.S3 **EMPLOYMENT - EMPLOI**

Units

ISIC revision 3				1991	1992	1993	1994	1995	1996	1997	1998
G	50_52		Wholesale & retail trade; repairs; etc.	..	..	..	..	..	..	525 942	533 621
	50		Sale, maint.,repair vehic/cycle, person., h/hold gd.	..	..	..	..	..	..	72 646	74 261
		501	Sale of motor cycle	..	..	..	..	..	..	..	..
		502	Maintenance and repair of motor vehicles	..	..	..	..	..	..	..	..
		503	Sale of motor vehicle parts and accessories	..	..	..	..	..	..	..	..
		504	Sale, maint., repair motor cycle, parts & acces.	..	..	..	..	..	..	..	..
		505	Retail sale of automative fuel	..	..	..	..	..	..	..	..
	51		W/sale. trade & commission exc motor veh. & cycle	..	..	..	..	..	..	178 920	181 942
		511	Wholesale on a fee or contract basis	..	..	..	..	..	..	..	..
		512	W/sale agric. raw mat.,live animal,food, bev., tobac.	..	..	..	..	..	..	..	..
		513	Wholesale of household goods	..	..	..	..	..	..	..	..
		514	W/sale. of non-agric. intermed. prod., waste, scrap	..	..	..	..	..	..	..	..
		515	Wholesale of machinery, equipment & supplies	..	..	..	..	..	..	..	..
		519	Other wholesale	..	..	..	..	..	..	..	..
	52		Retail trade exc. motor veh. & cyc.; repair h/hold gds.	..	..	..	..	..	..	274 376	277 418
		521	Non-specialised retail trade in stores	..	..	..	..	..	..	..	..
		522	Retail sale food, bev., tobac. in specialised stores	..	..	..	..	..	..	..	..
		523	Other retail trade of new goods specialised stores	..	..	..	..	..	..	..	..
		524	Retail sale of second-hand goods in stores	..	..	..	..	..	..	..	..
		525	Retail trade not in stores	..	..	..	..	..	..	..	..
		526	Repair of personal and household goods	..	..	..	..	..	..	..	..
H	55		Hotels and restaurants	..	..	..	..	..	..	163 487	168 650
		551	Hotels; camping ; other short-stay accommodation	..	..	..	..	..	..	..	..
		552	Restaurants, bars and canteens	..	..	..	..	..	..	..	..
I	60_64		Transport, storage and communications	..	..	..	..	..	..	..	..
	60		Land transport, transport via pipelines	..	..	..	..	..	..	34 081	35 360
		601	Transport via railways	..	..	..	..	..	..	..	..
		602	Other land transport	..	..	..	..	..	..	..	..
		603	Transport via pipelines	..	..	..	..	..	..	..	..
	61		Water transport	..	..	..	..	..	..	2 642	2 549
		611	Sea and coastal water transport	..	..	..	..	..	..	..	..
		612	Inland water transport	..	..	..	..	..	..	..	..
	62		Air transport	..	..	..	..	..	..	..	..
		621	Scheduled air transport	..	..	..	..	..	..	..	..
		622	Non-scheduled air transport	..	..	..	..	..	..	..	..
	63		Support. & aux. transport act., travel agencies activ.	..	..	..	..	..	..	38 978	37 128
	64		Post and telecommunications	..	..	..	..	..	..	..	..
		641	Post and courier activities	..	..	..	..	..	..	..	..
		642	Telecommunications	..	..	..	..	..	..	..	..
J	65_67		Financial intermediation	..	..	..	..	..	..	..	..
	65		Financial intermediation, exc. insur., pension fund	..	..	..	..	..	..	..	..
		651	Monetary intermediation	..	..	..	..	..	..	..	..
		659	Other financial intermediation	..	..	..	..	..	..	..	..
	66		Insurance pension fund, exc. compul. social sec.	..	..	..	..	..	..	..	..
	67		Activities auxiliary to financial intermediation	..	..	..	..	..	..	..	..
		671	Act. auxil to fin. intermed., exc. insur. & pens.fund	..	..	..	..	..	..	..	..
		672	Act. auxiliary to insurance and pension funding	..	..	..	..	..	..	..	..
K	70_74		Real estate, renting and business activities	..	..	..	..	..	..	266 225	276 108
	70		Real estate activities	..	..	..	..	..	..	15 446	17 243
		701	Real estate activ. with own or leased property	..	..	..	..	..	..	..	..
		702	Real estate activities on a fee or contract basis	..	..	..	..	..	..	..	..
	71		Renting of mach. without operator; rent. of pers. gds.	..	..	..	..	..	..	..	..
		711	Renting of transport equipment	..	..	..	..	..	..	..	..
		712	Renting of other machinery and equipment	..	..	..	..	..	..	..	..
		713	Renting of personal and household goods nec	..	..	..	..	..	..	..	..
	72		Computer and related activities	..	..	..	..	..	..	29 510	32 324
		721	Hardware consultancy	..	..	..	..	..	..	..	..
		722	Software consultancy and supply	..	..	..	..	..	..	..	..
		723	Data processing	..	..	..	..	..	..	..	..
		724	Data base activities	..	..	..	..	..	..	..	..
		725	Mainten.; repair of office, account. & comp. mach.	..	..	..	..	..	..	..	..
		729	Other computer related activities	..	..	..	..	..	..	..	..
	73		Research and development	..	..	..	..	..	..	4 319	4 104
		731	R & D experim. on natural sciences & engineering	..	..	..	..	..	..	..	..
		732	R & D experim. on social sciences & humanities	..	..	..	..	..	..	..	..
	74		Other business activities	..	..	..	..	..	..	216 950	222 437
		741	Legal, acc., book-keep., audit act.; tax consult., etc	..	..	..	..	..	..	..	..
		742	Architectural, engineering, & other techn. activities	..	..	..	..	..	..	..	..
		743	Advertising	..	..	..	..	..	..	..	..
		749	Business activities nec	..	..	..	..	..	..	..	..
M	80		Education	..	..	..	..	..	..	12 734	12 915
		801	Primary education	..	..	..	..	..	..	..	..
		802	Secondary education	..	..	..	..	..	..	..	..
		803	Higher education	..	..	..	..	..	..	..	..
		809	Adult and other education	..	..	..	..	..	..	..	..
N	85		Health and social work	..	..	..	..	..	..	31 438	31 842
		851	Human health activities	..	..	..	..	..	..	..	..
		852	Veterinary activities	..	..	..	..	..	..	..	..
		853	Social work activities	..	..	..	..	..	..	..	..
O	90_93		Other commun., soc. & pers. services activities	..	..	..	..	..	..	61 234	62 218
	90		Sewage & refuse disposal, sanitation, etc.	..	..	..	..	..	..	5 161	5 327
	91		Activities of member organisations nec	..	..	..	..	..	..	..	..
		912	Activities of trade unions	..	..	..	..	..	..	..	..
	92		Recreational, cultural and sporting activities	..	..	..	..	..	..	23 714	24 425
		921	Motion picture, radio, TV, other entertainment act.	..	..	..	..	..	..	..	..
		922	News agency activities	..	..	..	..	..	..	..	..
		923	Library, archives, museums, other cultural activ.	..	..	..	..	..	..	..	..
		924	Sporting and other recreational activities	..	..	..	..	..	..	..	..
	93		Other service activities	..	..	..	..	..	..	32 359	32 466

Table SW.S4 **SALAIRES ET CHARGES SOCIALES - COMPENSATION OF LABOUR**

Millions de SF (Prix courants)

1991	1992	1993	1994	1995	1996	1997	1998		CITI révision 3		
..	..	..	..	..	..	35 337	36 655	Com. de gros, de detail ; réparation etc.	G	**50_52**	
..	..	..	..	..	..	4 250	4 344	Com., entr., répar. d'auto. et moto ; carburants		**50**	
..	..	..	..	..	..	..	..	Commerce de véhicules automobiles			501
..	..	..	..	..	..	..	..	Entretien et réparation de véhicules automobiles			502
..	..	..	..	..	..	..	..	Com. de pièces et access. de véhic. automobiles			503
..	..	..	..	..	..	..	..	Com., entret. et répar. de motocycles et access.			504
..	..	..	..	..	..	..	..	Com. de détail de carburants pour automobiles			505
..	..	..	..	..	..	15 885	16 651	Com. de gros et act. inter., sauf auto. et moto		**51**	
..	..	..	..	..	..	..	..	Activités d'intermédiaries de commerce de gros			511
..	..	..	..	..	..	..	..	Com. gros: prod. agr. bruts, anim. viv., prod. alim.			512
..	..	..	..	..	..	..	..	Com. gros: biens de consom., sauf alimentaires			513
..	..	..	..	..	..	..	..	Com. gros: prod. inter. non agr., déchets, débris			514
..	..	..	..	..	..	..	..	Com. gros: machines, équipments et fournitures			515
..	..	..	..	..	..	..	..	Autres commerces de gros			519
..	..	..	..	..	..	15 203	15 660	Com. détail, sauf auto, moto; répar. art. pers., dom.		**52**	
..	..	..	..	..	..	..	..	Commerce de détail non spécialisé en magasin			521
..	..	..	..	..	..	..	..	Com. détail: prod. aliment. en mag. spécialisés			522
..	..	..	..	..	..	..	..	Autre com. détail: prod. neufs en mag. spécialisés			523
..	..	..	..	..	..	..	..	Commerces de détail de biens d'occasion			524
..	..	..	..	..	..	..	..	Commerces de détail autres qu'en magasin			525
..	..	..	..	..	..	..	..	Réparation d'articles personnels et domestiques			526
..	..	..	..	..	..	7 054	7 382	Hotels et restaurants	H	**55**	
..	..	..	..	..	..	..	..	Hôtels; camping; autres équip. courts séjours			551
..	..	..	..	..	..	..	..	Restaurants, cafés et cantines			552
..	..	..	..	..	..	..	..	Transport, entreposage, communications	I	**60_64**	
..	..	..	..	..	..	2 322	1 991	Transports terrestres; transports par conduites		**60**	
..	..	..	..	..	..	..	..	Transports par chemin de fer			601
..	..	..	..	..	..	..	..	Autres transports terrestres			602
..	..	..	..	..	..	..	..	Transports par conduites			603
..	..	..	..	..	..	231	207	Transports par eau		**61**	
..	..	..	..	..	..	..	..	Transports maritimes et côtiers			611
..	..	..	..	..	..	..	..	Transports par voies navigables intérieures			612
..	..	..	..	..	..	..	..	Transports aériens		**62**	
..	..	..	..	..	..	..	..	Transports aériens réguliers			621
..	..	..	..	..	..	..	..	Transports aériens non réguliers			622
..	..	..	..	..	..	2 859	2 642	Act. annex. et aux. des transp.; act. des ag. de voy.		**63**	
..	..	..	..	..	..	..	..	Postes et télécommunications		**64**	
..	..	..	..	..	..	..	..	Activités de poste et de courrier			641
..	..	..	..	..	..	..	..	Télécommunications			642
..	..	..	..	..	..	..	..	Intermédiation financière	J	**65_67**	
..	..	..	..	..	..	..	..	Interm. financ., sauf assur. et caisses de retraite		**65**	
..	..	..	..	..	..	..	..	Intermédiation monétaire			651
..	..	..	..	..	..	..	..	Autres intermédiations financières			659
..	..	..	..	..	..	..	..	Act. d'assur., caisses de retr., sauf séc. soc. oblig.		**66**	
..	..	..	..	..	..	..	..	Act. auxiliaires de l'intermédiation financière		**67**	
..	..	..	..	..	..	..	..	Act. aux. de l'interm. fin., sauf assur., et retraite			671
..	..	..	..	..	..	..	..	Act. aux. des assurances et caisses de retraite			672
..	..	..	..	..	..	22 867	25 712	Immobilier, locations, services aux entreprises	K	**70_74**	
..	..	..	..	..	..	1 434	1 563	Activités immobilières		**70**	
..	..	..	..	..	..	..	..	Act. immobilières sur biens propres ou loués			701
..	..	..	..	..	..	..	..	Activités immobilières pour comptes de tiers			702
..	..	..	..	..	..	..	..	Locat. mach., équip. sans opér., biens pers. et dom.		**71**	
..	..	..	..	..	..	..	..	Location de matériel de transport			711
..	..	..	..	..	..	..	..	Location de machines ét equipements nca			712
..	..	..	..	..	..	..	..	Loc. d'articles personnels et domestiques nca			713
..	..	..	..	..	..	3 343	3 476	Activités informatiques et activités rattachées		**72**	
..	..	..	..	..	..	..	..	Conseil en matériel informatique			721
..	..	..	..	..	..	..	..	Production de logiciels			722
..	..	..	..	..	..	..	..	Traitement de données informatiques			723
..	..	..	..	..	..	..	..	Activités de banques de données			724
..	..	..	..	..	..	..	..	Entr., répar. mach. de bureau et compt. et inform.			725
..	..	..	..	..	..	..	..	Autres activités rattachées à l'informatique			729
..	..	..	..	..	..	467	481	Recherche-développement		**73**	
..	..	..	..	..	..	..	..	Rech. et dév. expér. en sc. phys., natur. et ingén.			731
..	..	..	..	..	..	..	..	Rech. et dév. expér. en sciences soc. et humaines			732
..	..	..	..	..	..	17 622	20 190	Autres activités de services aux entreprises		**74**	
..	..	..	..	..	..	..	..	Act. jur., compt., audit.; cons. fisc.; ét. march., soud.			741
..	..	..	..	..	..	..	..	Act. d'achitect., d'ingén.; autres act. techniques			742
..	..	..	..	..	..	..	..	Publicité			743
..	..	..	..	..	..	..	..	Activités de services aux entreprises nca.			749
..	..	..	..	..	..	1 411	813	Education	M	**80**	
..	..	..	..	..	..	..	..	Activités d'enseignement primaire			801
..	..	..	..	..	..	..	..	Activités d'enseignement secondaire			802
..	..	..	..	..	..	..	..	Activités d'enseignement supérieur			803
..	..	..	..	..	..	..	..	Act. de formation perman.; autres act. d'enseign.			809
..	..	..	..	..	..	1 930	1 925	Santé et action sociale	N	**85**	
..	..	..	..	..	..	..	..	Activités de santé humaine			851
..	..	..	..	..	..	..	..	Activités vétérinaires			852
..	..	..	..	..	..	..	..	Activités d'action sociale			853
..	..	..	..	..	..	3 416	3 621	Autres activités de serv. coll., soc. et personnels	O	**90_93**	
..	..	..	..	..	..	431	435	Assain. et enlèv. des ordures; voierie, act. simil.		**90**	
..	..	..	..	..	..	..	..	Activités associatives diverses		**91**	
..	..	..	..	..	..	..	..	Activités de syndicats de salariés			912
..	..	..	..	..	..	1 995	2 158	Activités récréatives, culturelles et sportives		**92**	
..	..	..	..	..	..	..	..	Act. cinémat., de radio, télé.; autres act. de spect.			921
..	..	..	..	..	..	..	..	Activités d'agences de presse			922
..	..	..	..	..	..	..	..	Act. des biblio., arch., musées, jardins bot. et zool.			923
..	..	..	..	..	..	..	..	Activités sportives et autres activités récréatives			924
..	..	..	..	..	..	990	1 027	Autres activités de services		**93**	

SWITZERLAND

Table SW.S5 TOTAL WAGES AND SALARIES - SALAIRES ET TRAITEMENTS

Millions of SF (Current Prices)

ISIC revision 3				1991	1992	1993	1994	1995	1996	1997	1998
G	50_52		Wholesale & retail trade; repairs; etc.	..	..	..	..	..	..	30 357	31 373
	50		Sale, maint.,repair vehic/cycle, person., h/hold gd.	..	..	..	..	..	..	3 656	3 709
		501	Sale of motor cycle	..	..	..	..	..	..	..	..
		502	Maintenance and repair of motor vehicles	..	..	..	..	..	..	..	..
		503	Sale of motor vehicle parts and accessories	..	..	..	..	..	..	..	..
		504	Sale, maint., repair motor cycle, parts & acces.	..	..	..	..	..	..	..	..
		505	Retail sale of automative fuel	..	..	..	..	..	..	..	..
	51		W/sale. trade & commission exc motor veh. & cycle	..	..	..	..	..	..	13 627	14 211
		511	Wholesale on a fee or contract basis	..	..	..	..	..	..	..	..
		512	W/sale agric. raw mat.,live animal,food, bev., tobac.	..	..	..	..	..	..	..	..
		513	Wholesale of household goods	..	..	..	..	..	..	..	..
		514	W/sale. of non-agric. intermed. prod., waste, scrap	..	..	..	..	..	..	..	..
		515	Wholesale of machinery, equipment & supplies	..	..	..	..	..	..	..	..
		519	Other wholesale	..	..	..	..	..	..	..	..
	52		Retail trade exc. motor veh. & cyc.; repair h/hold gds.	..	..	..	..	..	..	13 075	13 453
		521	Non-specialised retail trade in stores	..	..	..	..	..	..	..	..
		522	Retail sale food, bev., tobac. in specialised stores	..	..	..	..	..	..	..	..
		523	Other retail trade of new goods specialised stores	..	..	..	..	..	..	..	..
		524	Retail sale of second-hand goods in stores	..	..	..	..	..	..	..	..
		525	Retail trade not in stores	..	..	..	..	..	..	..	..
		526	Repair of personal and household goods	..	..	..	..	..	..	..	..
H	55		Hotels and restaurants	..	..	..	..	..	..	6 059	6 439
		551	Hotels; camping ; other short-stay accommodation	..	..	..	..	..	..	..	..
		552	Restaurants, bars and canteens	..	..	..	..	..	..	..	..
I	60_64		Transport, storage and communications	..	..	..	..	..	..	..	..
	60		Land transport, transport via pipelines	..	..	..	..	..	..	2 016	1 706
		601	Transport via railways	..	..	..	..	..	..	..	..
		602	Other land transport	..	..	..	..	..	..	..	..
		603	Transport via pipelines	..	..	..	..	..	..	..	..
	61		Water transport	..	..	..	..	..	..	192	173
		611	Sea and coastal water transport	..	..	..	..	..	..	..	..
		612	Inland water transport	..	..	..	..	..	..	..	..
	62		Air transport	..	..	..	..	..	..	..	..
		621	Scheduled air transport	..	..	..	..	..	..	..	..
		622	Non-scheduled air transport	..	..	..	..	..	..	..	..
	63		Support. & aux. transport act., travel agencies activ.	..	..	..	..	..	..	2 443	2 247
	64		Post and telecommunications	..	..	..	..	..	..	..	..
		641	Post and courier activities	..	..	..	..	..	..	..	..
		642	Telecommunications	..	..	..	..	..	..	..	..
J	65_67		Financial intermediation	..	..	..	..	..	..	..	..
	65		Financial intermediation, exc. insur., pension fund	..	..	..	..	..	..	..	..
		651	Monetary intermediation	..	..	..	..	..	..	..	..
		659	Other financial intermediation	..	..	..	..	..	..	..	..
	66		Insurance pension fund, exc. compul. social sec.	..	..	..	..	..	..	..	..
	67		Activities auxiliary to financial intermediation	..	..	..	..	..	..	..	..
		671	Act. auxil to fin. intermed., exc. insur. & pens.fund	..	..	..	..	..	..	..	..
		672	Act. auxiliary to insurance and pension funding	..	..	..	..	..	..	..	..
K	70_74		Real estate, renting and business activities	..	..	..	..	..	..	19 444	21 873
	70		Real estate activities	..	..	..	..	..	..	1 211	1 298
		701	Real estate activ. with own or leased property	..	..	..	..	..	..	..	..
		702	Real estate activities on a fee or contract basis	..	..	..	..	..	..	..	..
	71		Renting of mach. without operator; rent. of pers. gds.	..	..	..	..	..	..	..	..
		711	Renting of transport equipment	..	..	..	..	..	..	..	..
		712	Renting of other machinery and equipment	..	..	..	..	..	..	..	..
		713	Renting of personal and household goods nec	..	..	..	..	..	..	..	..
	72		Computer and related activities	..	..	..	..	..	..	2 840	2 918
		721	Hardware consultancy	..	..	..	..	..	..	..	..
		722	Software consultancy and supply	..	..	..	..	..	..	..	..
		723	Data processing	..	..	..	..	..	..	..	..
		724	Data base activities	..	..	..	..	..	..	..	..
		725	Mainten.; repair of office, account. & comp. mach.	..	..	..	..	..	..	..	..
		729	Other computer related activities	..	..	..	..	..	..	..	..
	73		Research and development	..	..	..	..	..	..	409	411
		731	R & D experim. on natural sciences & engineering	..	..	..	..	..	..	..	..
		732	R & D experim. on social sciences & humanities	..	..	..	..	..	..	..	..
	74		Other business activities	..	..	..	..	..	..	14 984	17 247
		741	Legal, acc., book-keep., audit act.; tax consult., etc	..	..	..	..	..	..	..	..
		742	Architectural, engineering, & other techn. activities	..	..	..	..	..	..	..	..
		743	Advertising	..	..	..	..	..	..	..	..
		749	Business activities nec	..	..	..	..	..	..	..	..
M	80		Education	..	..	..	..	..	..	1 238	702
		801	Primary education	..	..	..	..	..	..	..	..
		802	Secondary education	..	..	..	..	..	..	..	..
		803	Higher education	..	..	..	..	..	..	..	..
		809	Adult and other education	..	..	..	..	..	..	..	..
N	85		Health and social work	..	..	..	..	..	..	1 658	1 662
		851	Human health activities	..	..	..	..	..	..	..	..
		852	Veterinary activities	..	..	..	..	..	..	..	..
		853	Social work activities	..	..	..	..	..	..	..	..
O	90_93		Other commun., soc. & pers. services activities	..	..	..	..	..	..	2 922	3 111
	90		Sewage & refuse disposal, sanitation, etc.	..	..	..	..	..	..	360	369
	91		Activities of member organisations nec	..	..	..	..	..	..	..	..
		912	Activities of trade unions	..	..	..	..	..	..	..	..
	92		Recreational, cultural and sporting activities	..	..	..	..	..	..	1 709	1 842
		921	Motion picture, radio, TV, other entertainment act.	..	..	..	..	..	..	..	..
		922	News agency activities	..	..	..	..	..	..	..	..
		923	Library, archives, museums, other cultural activ.	..	..	..	..	..	..	..	..
		924	Sporting and other recreational activities	..	..	..	..	..	..	..	..
	93		Other service activities	..	..	..	..	..	..	852	901

Table TU.S1 **PRODUCTION - PRODUCTION**

Milliards de TL (Prix courants)

1991	1992	1993	1994	1995	1996	1997	1998			CITI révision 3	
..	..	289 512	535 893	1 181 513	2 520 723	5 498 976	..	Com. de gros, de detail ; réparation etc.	G	50_52	
..	..	39 432	79 240	162 319	355 712	893 720	..	Com., entr., répar. d'auto. et moto ; carburants		50	
..	..	11 934	18 356	40 379	83 580	224 913	..	Commerce de véhicules automobiles			501
..	..	2 507	7 073	12 176	22 161	48 221	..	Entretien et réparation de véhicules automobiles			502
..	..	11 955	25 217	49 454	118 510	249 071	..	Com. de pièces et access. de véhic. automobiles			503
..	..	695	1 101	2 324	5 832	12 096	..	Com., entret. et répar. de motocycles et access.			504
..	..	12 341	27 493	57 985	125 629	359 419	..	Com. de détail de carburants pour automobiles			505
..	..	128 145	244 110	600 845	1 226 693	2 469 808	..	Com. de gros et act. inter., sauf auto. et moto		51	
..	..	3 402	4 530	9 546	20 150	45 571	..	Activités d'intermédiaries de commerce de gros			511
..	..	42 106	50 235	162 712	290 801	582 707	..	Com. gros: prod. agr. bruts, anim. viv., prod. alim.			512
..	..	33 930	83 550	196 535	394 634	795 240	..	Com. gros: biens de consom., sauf alimentaires			513
..	..	43 529	92 195	206 133	457 880	936 570	..	Com. gros: prod. inter. non agr., déchets, débris			514
..	..	4 202	12 550	23 179	57 561	98 803	..	Com. gros: machines, équipments et fournitures			515
..	..	976	1 049	2 739	5 666	10 918	..	Autres commerces de gros			519
..	..	121 935	212 543	418 349	938 317	2 135 448	..	Com. détail, sauf auto, moto; répar. art. pers., dom.		52	
..	..	36 014	58 517	116 951	267 923	657 376	..	Commerce de détail non spécialisé en magasin			521
..	..	11 892	18 986	39 176	93 095	201 277	..	Com. détail: prod. aliment. en mag. spécialisés			522
..	..	72 239	132 393	257 352	566 304	1 250 571	..	Autre com. détail: prod. neufs en mag. spécialisés			523
..	..	118	132	316	788	2 065	..	Commerces de détail de biens d'occasion			524
..	..	..	..	..	..	1 568	..	Commerces de détail autres qu'en magasin			525
..	..	1 673	2 516	4 555	10 207	22 592	..	Réparation d'articles personnels et domestiques			526
..	..	61 106	115 304	247 982	527 308	1 372 648	..	Hotels et restaurants	H	55	
..	..	19 520	39 448	90 079	209 296	557 964	..	Hôtels; camping; autres équip. courts séjours			551
..	..	41 585	75 856	157 903	318 013	814 684	..	Restaurants, cafés et cantines			552
..	..	45 239	197 856	316 449	658 595	1 437 168	..	Transport, entreposage, communications	I	60_64	
..	..	14 501	43 442	76 249	161 739	357 291	..	Transports terrestres; transports par conduites		60	
..	..	..	..	..	..	..	..	Transports par chemin de fer			601
..	..	14 501	43 442	76 249	161 739	357 291	..	Autres transports terrestres			602
..	..	..	..	..	..	..	..	Transports par conduites			603
..	..	13 286	60 878	83 411	119 008	287 602	..	Transports par eau		61	
..	..	13 286	60 878	83 411	119 008	287 602	..	Transports maritimes et côtiers			611
..	..	..	..	..	..	..	..	Transports par voies navigables intérieures			612
..	..	15 678	38 252	83 665	170 807	359 442	..	Transports aériens		62	
..	..	14 892	35 556	76 927	143 506	262 549	..	Transports aériens réguliers			621
..	..	786	2 697	6 738	27 301	96 894	..	Transports aériens non réguliers			622
..	..	1 774	55 283	73 124	207 041	432 832	..	Act. annex. et aux. des transp.; act. des ag. de voy.		63	
..	..	..	..	..	..	..	..	Postes et télécommunications		64	
..	..	..	..	..	..	..	..	Activités de poste et de courrier			641
..	..	..	..	..	..	..	..	Télécommunications			642
..	..	202 461	584 955	1 248 860	2 556 399	4 809 052	..	Intermédiation financière	J	65_67	
..	..	172 250	531 668	718 827	1 841 195	4 294 943	..	Interm. financ., sauf assur. et caisses de retraite		65	
..	..	164 544	511 315	681 559	1 728 007	3 791 569	..	Intermédiation monétaire			651
..	..	7 706	20 353	37 268	113 188	503 374	..	Autres intermédiations financières			659
..	..	25 633	39 772	80 440	180 432	385 394	..	Act. d'assur., caisses de retr., sauf séc. soc. oblig.		66	
..	..	4 578	13 515	449 592	534 772	128 716	..	Act. auxiliaires de l'intermédiation financière		67	
..	..	3 055	7 816	438 309	518 920	99 074	..	Act. aux. de l'interm. fin., sauf assur., et retraite			671
..	..	1 524	5 699	11 284	15 852	29 642	..	Act. aux. des assurances et caisses de retraite			672
..	..	27 313	48 442	108 955	267 447	601 397	..	Immobilier, locations, services aux entreprises	K	70_74	
..	..	1 556	2 788	2 635	4 394	8 810	..	Activités immobilières		70	
..	..	72	148	357	758	1 792	..	Act. immobilières sur biens propres ou loués			701
..	..	1 484	2 640	2 278	3 635	7 018	..	Activités immobilières pour comptes de tiers			702
..	..	133	235	432	490	1 046	..	Locat. mach., équip. sans opér., biens pers. et dom.		71	
..	..	1	2	..	..	..	..	Location de matériel de transport			711
..	..	27	76	193	20	28	..	Location de machines ét equipements nca			712
..	..	106	157	235	465	1 009	..	Loc. d'articles personnels et domestiques nca			713
..	..	725	1 661	3 187	8 118	26 990	..	Activités informatiques et activités rattachées		72	
..	..	..	..	218	283	814	..	Conseil en matériel informatique			721
..	..	465	592	1 300	4 688	18 938	..	Production de logiciels			722
..	..	..	..	..	..	..	..	Traitement de données informatiques			723
..	..	..	..	..	..	..	..	Activités de banques de données			724
..	..	141	303	454	856	2 123	..	Entr., répar. mach. de bureau et compt. et inform.			725
..	..	1	367	..	..	..	..	Autres activités rattachées à l'informatique			729
..	..	38	12	..	..	162	..	Recherche-développement		73	
..	..	..	..	..	..	..	..	Rech. et dév. expér. en sc. phys., natur. et ingén.			731
..	..	35	..	..	..	136	..	Rech. et dév. expér. en sciences soc. et humaines			732
..	..	24 861	43 746	102 684	254 412	564 389	..	Autres activités de services aux entreprises		74	
..	..	6 579	13 301	28 981	66 228	141 648	..	Act. jur., compt., audit.; cons. fisc.; ét. march., soud.			741
..	..	5 503	9 207	24 796	63 408	133 063	..	Act. d'achitect., d'ingén.; autres act. techniques			742
..	..	6 271	10 512	25 574	88 135	203 258	..	Publicité			743
..	..	6 507	10 727	23 334	36 642	86 419	..	Activités de services aux entreprises nca.			749
..	..	5 070	7 993	15 200	33 526	106 791	..	Education	M	80	
..	..	118	212	228	560	7 113	..	Activités d'enseignement primaire			801
..	..	1 306	2 189	4 803	10 395	47 284	..	Activités d'enseignement secondaire			802
..	..	..	..	..	..	..	..	Activités d'enseignement supérieur			803
..	..	3 645	5 591	10 169	22 570	52 395	..	Act. de formation perman.; autres act. d'enseign.			809
..	..	6 084	12 430	23 161	47 203	282 586	..	Santé et action sociale	N	85	
..	..	5 852	11 843	22 333	45 609	278 663	..	Activités de santé humaine			851
..	..	84	126	301	596	1 763	..	Activités vétérinaires			852
..	..	148	460	527	999	2 160	..	Activités d'action sociale			853
..	..	11 896	18 722	33 256	58 331	408 828	..	Autres activités de serv. coll., soc. et personnels	O	90_93	
..	..	6	9	37	38	160	..	Assain. et enlèv. des ordures; voierie, act. simil.		90	
..	..	..	..	..	..	..	..	Activités associatives diverses		91	
..	..	..	..	..	..	..	..	Activités de syndicats de salariés			912
..	..	4 674	6 651	13 613	23 803	337 182	..	Activités récréatives, culturelles et sportives		92	
..	..	3 723	4 371	7 654	15 104	312 043	..	Act. cinémat., de radio, télé.; autres act. de spect.			921
..	..	11	108	218	373	435	..	Activités d'agences de presse			922
..	..	..	..	..	..	..	..	Act. des biblio., arch., musées, jardins bot. et zool.			923
..	..	940	2 172	5 741	8 326	24 704	..	Activités sportives et autres activités récréatives			924
..	..	7 216	12 062	19 606	34 491	71 486	..	Autres activités de services		93	

Table TU.S2 **VALUE ADDED - VALEUR AJOUTÉE**

Billions of TL (Current Prices)

ISIC revision 3				1991	1992	1993	1994	1995	1996	1997	1998
G	50_52		Wholesale & retail trade; repairs; etc.	..	..	242 535	441 037	987 937	2 128 735	4 659 300	..
	50		Sale, maint.,repair vehic/cycle, person., h/hold gd.	..	..	33 140	63 539	130 053	289 183	743 184	..
		501	Sale of motor cycle	..	..	10 374	15 765	34 989	71 951	199 964	..
		502	Maintenance and repair of motor vehicles	..	..	1 824	4 189	7 383	14 496	31 645	..
		503	Sale of motor vehicle parts and accessories	..	..	10 013	20 473	41 694	100 691	213 530	..
		504	Sale, maint., repair motor cycle, parts & acces.	..	..	577	901	1 900	4 990	10 327	..
		505	Retail sale of automative fuel	..	..	10 351	22 210	44 087	97 055	287 718	..
	51		W/sale. trade & commission exc motor veh. & cycle	..	..	107 047	202 522	508 954	1 045 896	2 107 367	..
		511	Wholesale on a fee or contract basis	..	..	2 474	3 532	7 639	15 467	34 124	..
		512	W/sale agric. raw mat.,live animal,food, bev., tobac.	..	..	35 719	43 099	145 755	259 497	518 592	..
		513	Wholesale of household goods	..	..	27 598	66 501	157 972	329 467	661 900	..
		514	W/sale. of non-agric. intermed. prod., waste, scrap	..	..	37 009	77 972	176 106	387 150	802 350	..
		515	Wholesale of machinery, equipment & supplies	..	..	3 388	10 536	19 167	49 291	80 568	..
		519	Other wholesale	..	..	859	883	2 315	5 025	9 834	..
	52		Retail trade exc. motor veh. & cyc.; repair h/hold gds.	..	..	102 348	174 976	348 931	793 657	1 808 749	..
		521	Non-specialised retail trade in stores	..	..	31 105	49 164	96 003	226 027	558 102	..
		522	Retail sale food, bev., tobac. in specialised stores	..	..	10 308	15 623	32 842	80 490	172 926	..
		523	Other retail trade of new goods specialised stores	..	..	60 129	108 287	216 899	479 950	1 060 580	..
		524	Retail sale of second-hand goods in stores	..	..	96	111	272	711	1 873	..
		525	Retail trade not in stores	..	..	..	..	..	..	1 042	..
		526	Repair of personal and household goods	..	..	711	1 792	2 915	6 480	14 227	..
H	55		Hotels and restaurants	..	..	28 027	58 116	126 356	272 991	724 370	..
		551	Hotels; camping ; other short-stay accommodation	..	..	10 959	22 764	51 047	118 654	311 649	..
		552	Restaurants, bars and canteens	..	..	17 068	35 353	75 309	154 337	412 720	..
I	60_64		Transport, storage and communications	..	..	24 200	71 556	116 685	218 737	476 953	..
	60		Land transport, transport via pipelines	..	..	6 911	14 924	21 985	46 858	99 808	..
		601	Transport via railways	..	..	..	..	..	..	..	..
		602	Other land transport	..	..	6 911	14 924	21 985	46 858	99 808	..
		603	Transport via pipelines	..	..	..	..	..	..	..	..
	61		Water transport	..	..	7 230	29 954	31 912	60 078	129 421	..
		611	Sea and coastal water transport	..	..	7 230	29 954	31 912	60 078	129 421	..
		612	Inland water transport	..	..	..	..	..	..	..	..
	62		Air transport	..	..	9 129	8 916	44 516	52 311	138 813	..
		621	Scheduled air transport	..	..	8 820	8 318	43 876	40 523	100 213	..
		622	Non-scheduled air transport	..	..	308	598	640	11 789	38 600	..
	63		Support. & aux. transport act., travel agencies activ.	..	..	930	17 762	18 272	59 489	108 910	..
	64		Post and telecommunications	..	..	..	..	..	..	..	..
		641	Post and courier activities	..	..	..	..	..	..	..	..
		642	Telecommunications	..	..	..	..	..	..	..	..
J	65_67		Financial intermediation	..	..	95 803	169 256	360 870	868 047	1 884 303	..
	65		Financial intermediation, exc. insur., pension fund	..	..	86 041	145 466	315 619	776 973	1 648 128	..
		651	Monetary intermediation	..	..	79 088	129 699	284 854	680 291	1 361 903	..
		659	Other financial intermediation	..	..	6 954	15 767	30 765	96 682	286 225	..
	66		Insurance pension fund, exc. compul. social sec.	..	..	7 272	14 927	29 589	66 319	143 732	..
	67		Activities auxiliary to financial intermediation	..	..	2 490	8 863	15 661	24 755	92 443	..
		671	Act. auxil to fin. intermed., exc. insur. & pens.fund	..	..	1 532	5 282	8 572	14 359	72 126	..
		672	Act. auxiliary to insurance and pension funding	..	..	957	3 580	7 089	10 396	20 317	..
K	70_74		Real estate, renting and business activities	..	..	16 039	30 073	65 001	160 616	349 939	..
	70		Real estate activities	..	..	942	1 504	1 637	2 826	5 027	..
		701	Real estate activ. with own or leased property	..	..	30	72	158	301	333	..
		702	Real estate activities on a fee or contract basis	..	..	913	1 432	1 479	2 525	4 693	..
	71		Renting of mach. without operator; rent. of pers. gds.	..	..	54	101	212	319	689	..
		711	Renting of transport equipment	..	..	0	1	..	..	5	..
		712	Renting of other machinery and equipment	..	..	4	34	101	10	15	..
		713	Renting of personal and household goods nec	..	..	49	65	109	305	668	..
	72		Computer and related activities	..	..	459	892	1 596	3 335	11 073	..
		721	Hardware consultancy	..	..	..	..	72	85	276	..
		722	Software consultancy and supply	..	..	288	359	602	1 495	6 644	..
		723	Data processing	..	..	..	..	..	..	..	..
		724	Data base activities	..	..	..	..	..	..	..	..
		725	Mainten.; repair of office, account. & comp. mach.	..	..	65	130	212	419	932	..
		729	Other computer related activities	..	..	1	144	..	..	..	..
	73		Research and development	..	..	19	11	..	..	123	..
		731	R & D experim. on natural sciences & engineering	..	..	..	..	..	..	..	..
		732	R & D experim. on social sciences & humanities	..	..	16	..	..	..	101	..
	74		Other business activities	..	..	14 565	27 566	61 541	154 106	333 028	..
		741	Legal, acc., book-keep., audit act.; tax consult., etc	..	..	5 077	10 562	22 039	52 432	107 963	..
		742	Architectural, engineering, & other techn. activities	..	..	3 174	4 694	12 222	29 446	64 892	..
		743	Advertising	..	..	3 526	7 000	14 444	54 604	115 390	..
		749	Business activities nec	..	..	2 789	5 311	12 836	17 623	44 782	..
M	80		Education	..	..	3 018	5 210	9 227	19 598	65 104	..
		801	Primary education	..	..	72	150	168	389	4 752	..
		802	Secondary education	..	..	865	1 719	3 823	6 702	33 078	..
		803	Higher education	..	..	..	..	..	..	..	..
		809	Adult and other education	..	..	2 081	3 341	5 237	12 507	27 275	..
N	85		Health and social work	..	..	4 222	8 407	16 434	30 945	207 355	..
		851	Human health activities	..	..	4 082	8 009	15 929	29 990	204 805	..
		852	Veterinary activities	..	..	51	74	174	331	1 029	..
		853	Social work activities	..	..	89	324	330	623	1 522	..
O	90_93		Other commun., soc. & pers. services activities	..	..	7 511	11 973	22 114	39 080	209 228	..
	90		Sewage & refuse disposal, sanitation, etc.	..	..	2	5	23	23	126	..
	91		Activities of member organisations nec	..	..	..	..	..	..	..	..
		912	Activities of trade unions	..	..	..	..	..	..	..	..
	92		Recreational, cultural and sporting activities	..	..	3 060	4 071	8 676	16 802	163 699	..
		921	Motion picture, radio, TV, other entertainment act.	..	..	2 391	2 708	4 345	10 437	144 379	..
		922	News agency activities	..	..	6	22	109	196	232	..
		923	Library, archives, museums, other cultural activ.	..	..	..	..	..	..	..	..
		924	Sporting and other recreational activities	..	..	663	1 340	4 222	6 169	19 088	..
	93		Other service activities	..	..	4 449	7 897	13 415	22 254	45 403	..

Table TU.S3 **INVESTISSEMENT - INVESTMENT**

Milliards de TL (Prix courants)

1991	1992	1993	1994	1995	1996	1997	1998		CITI révision 3		
..	..	16 199	14 944	35 644	74 192	110 011	..	Com. de gros, de detail ; réparation etc.	G	**50_52**	
..	..	1 631	2 821	5 797	13 781	22 606	..	Com., entr., répar. d'auto. et moto ; carburants		**50**	
..	..	484	1 347	1 694	6 547	10 861	..	Commerce de véhicules automobiles			501
..	..	22	2	27	2	329	..	Entretien et réparation de véhicules automobiles			502
..	..	453	357	1 714	3 261	4 064	..	Com. de pièces et access. de véhic. automobiles			503
..	..	3	1	8	48	21	..	Com., entret. et répar. de motocycles et access.			504
..	..	670	1 113	2 354	3 924	7 331	..	Com. de détail de carburants pour automobiles			505
..	..	8 782	9 802	22 277	32 359	46 841	..	Com. de gros et act. inter., sauf auto. et moto		**51**	
..	..	350	86	392	847	1 531	..	Activités d'intermédiaries de commerce de gros			511
..	..	3 117	1 691	3 481	7 779	10 233	..	Com. gros: prod. agr. bruts, anim. viv., prod. alim.			512
..	..	1 589	2 669	6 365	9 757	14 509	..	Com. gros: biens de consom., sauf alimentaires			513
..	..	3 406	5 094	10 492	11 655	17 328	..	Com. gros: prod. inter. non agr., déchets, débris			514
..	..	235	245	1 500	2 105	3 085	..	Com. gros: machines, équipments et fournitures			515
..	..	84	17	46	216	154	..	Autres commerces de gros			519
..	..	5 787	2 322	7 570	28 052	40 564	..	Com. détail, sauf auto, moto; répar. art. pers., dom.		**52**	
..	..	2 966	657	3 794	15 676	23 135	..	Commerce de détail non spécialisé en magasin			521
..	..	312	224	230	735	777	..	Com. détail: prod. aliment. en mag. spécialisés			522
..	..	2 505	1 437	3 540	11 628	16 405	..	Autre com. détail: prod. neufs en mag. spécialisés			523
..	..	0	1	0	1	2	..	Commerces de détail de biens d'occasion			524
..	..	..	..	..	..	194	..	Commerces de détail autres qu'en magasin			525
..	..	3	3	5	12	52	..	Réparation d'articles personnels et domestiques			526
..	..	1 454	2 307	4 729	8 154	17 077	..	Hotels et restaurants	H	**55**	
..	..	1 197	2 064	3 811	6 616	13 794	..	Hôtels; camping; autres équip. courts séjours			551
..	..	257	243	918	1 538	3 283	..	Restaurants, cafés et cantines			552
..	..	5 711	38 554	51 650	114 352	212 656	..	Transport, entreposage, communications	I	**60_64**	
..	..	2 053	20 412	14 825	27 507	52 134	..	Transports terrestres; transports par conduites		**60**	
..	..	..	..	..	..	..	..	Transports par chemin de fer			601
..	..	2 053	20 412	14 825	27 507	52 134	..	Autres transports terrestres			602
..	..	..	..	..	..	..	..	Transports par conduites			603
..	..	3 564	10 209	27 591	53 625	61 905	..	Transports par eau		**61**	
..	..	3 564	10 209	27 591	53 625	61 905	..	Transports maritimes et côtiers			611
..	..	..	..	..	..	..	..	Transports par voies navigables intérieures			612
..	..	36	907	2 218	26 967	90 193	..	Transports aériens		**62**	
..	..	36	560	1 879	22 165	46 515	..	Transports aériens réguliers			621
..	..	..	347	340	4 802	43 678	..	Transports aériens non réguliers			622
..	..	57	7 026	7 015	6 254	8 424	..	Act. annex. et aux. des transp.; act. des ag. de voy.		**63**	
..	..	..	..	..	..	..	..	Postes et télécommunications		**64**	
..	..	..	..	..	..	..	..	Activités de poste et de courrier			641
..	..	..	..	..	..	..	..	Télécommunications			642
..	..	27 417	15 648	86 733	173 515	335 888	..	Intermédiation financière	J	**65_67**	
..	..	17 511	13 760	82 920	164 084	319 479	..	Interm. financ., sauf assur. et caisses de retraite		**65**	
..	..	13 386	8 973	56 769	101 378	152 791	..	Intermédiation monétaire			651
..	..	4 126	4 787	26 151	62 706	166 689	..	Autres intermédiations financières			659
..	..	8 472	656	2 141	6 912	5 787	..	Act. d'assur., caisses de retr., sauf séc. soc. oblig.		**66**	
..	..	1 434	1 232	1 673	2 519	10 622	..	Act. auxiliaires de l'intermédiation financière		**67**	
..	..	1 373	1 171	1 079	2 177	7 109	..	Act. aux. de l'interm. fin., sauf assur., et retraite			671
..	..	61	61	594	342	3 513	..	Act. aux. des assurances et caisses de retraite			672
..	..	918	595	2 424	4 142	11 294	..	Immobilier, locations, services aux entreprises	K	**70_74**	
..	..	6	13	25	37	23	..	Activités immobilières		**70**	
..	..	0	..	0	..	0	..	Act. immobilières sur biens propres ou loués			701
..	..	6	13	25	37	22	..	Activités immobilières pour comptes de tiers			702
..	..	1	0	2	3	4	..	Locat. mach., équip. sans opér., biens pers. et dom.		**71**	
..	..	..	0	..	..	..	..	Location de matériel de transport			711
..	..	1	0	1	2	3	..	Location de machines ét equipements nca			712
..	..	0	..	0	2	0	..	Loc. d'articles personnels et domestiques nca			713
..	..	14	19	56	121	819	..	Activités informatiques et activités rattachées		**72**	
..	..	..	..	4	1	50	..	Conseil en matériel informatique			721
..	..	10	11	40	100	693	..	Production de logiciels			722
..	..	..	..	..	..	..	..	Traitement de données informatiques			723
..	..	..	..	..	..	..	..	Activités de banques de données			724
..	..	3	3	10	19	59	..	Entr., répar. mach. de bureau et compt. et inform.			725
..	..	0	..	..	..	..	..	Autres activités rattachées à l'informatique			729
..	..	1	0	..	..	4	..	Recherche-développement		**73**	
..	..	..	..	..	..	..	..	Rech. et dév. expér. en sc. phys., natur. et ingén.			731
..	..	1	..	..	..	4	..	Rech. et dév. expér. en sciences soc. et humaines			732
..	..	889	551	2 341	3 981	10 444	..	Autres activités de services aux entreprises		**74**	
..	..	151	179	567	1 481	2 270	..	Act. jur., compt., audit.; cons. fisc.; ét. march., soud.			741
..	..	235	121	781	978	4 055	..	Act. d'achitect., d'ingén.; autres act. techniques			742
..	..	186	108	312	658	1 136	..	Publicité			743
..	..	317	143	682	864	2 983	..	Activités de services aux entreprises nca.			749
..	..	214	471	1 437	892	6 117	..	Education	M	**80**	
..	..	2	5	7	13	1 095	..	Activités d'enseignement primaire			801
..	..	60	219	1 043	317	3 073	..	Activités d'enseignement secondaire			802
..	..	..	..	..	..	..	..	Activités d'enseignement supérieur			803
..	..	151	247	387	563	1 948	..	Act. de formation perman.; autres act. d'enseign.			809
..	..	211	400	1 307	1 317	57 638	..	Santé et action sociale	N	**85**	
..	..	207	394	1 294	1 245	57 524	..	Activités de santé humaine			851
..	..	1	1	3	4	11	..	Activités vétérinaires			852
..	..	3	5	10	69	103	..	Activités d'action sociale			853
..	..	148	107	964	735	22 216	..	Autres activités de serv. coll., soc. et personnels	O	**90_93**	
..	..	0	0	1	1	5	..	Assain. et enlèv. des ordures; voierie, act. simil.		**90**	
..	..	..	..	..	..	..	..	Activités associatives diverses		**91**	
..	..	..	..	..	..	..	..	Activités de syndicats de salariés			912
..	..	126	92	355	633	21 908	..	Activités récréatives, culturelles et sportives		**92**	
..	..	119	90	329	589	21 501	..	Act. cinémat., de radio, télé.; autres act. de spect.			921
..	..	0	1	1	1	1	..	Activités d'agences de presse			922
..	..	..	..	..	..	..	..	Act. des biblio., arch., musées, jardins bot. et zool.			923
..	..	7	1	26	43	406	..	Activités sportives et autres activités récréatives			924
..	..	22	15	607	102	303	..	Autres activités de services		**93**	

TURKEY

Table TU.S4 **INVESTMENT IN MACHINERY & EQUIPMENT - DÉPENSES DE MACHINES ET D'ÉQUIPEMENT**

Billions of TL (Current Prices)

ISIC revision 3				1991	1992	1993	1994	1995	1996	1997	1998
G	50_52		Wholesale & retail trade; repairs; etc.	..	..	2 339	3 209	6 672	9 317	18 760	..
	50		Sale, maint.,repair vehic/cycle, person., h/hold gd.	..	..	407	631	1 400	1 799	3 067	..
		501	Sale of motor cycle	..	..	117	154	316	1 239	1 040	..
		502	Maintenance and repair of motor vehicles	..	..	14	2	2	1	329	..
		503	Sale of motor vehicle parts and accessories	..	..	90	42	348	116	425	..
		504	Sale, maint., repair motor cycle, parts & acces.	..	..	0	0	0	3	3	..
		505	Retail sale of automative fuel	..	..	186	434	733	441	1 269	..
	51		W/sale. trade & commission exc motor veh. & cycle	..	..	1 214	1 994	3 455	3 632	10 229	..
		511	Wholesale on a fee or contract basis	..	..	24	8	39	166	215	..
		512	W/sale agric. raw mat.,live animal,food, bev., tobac.	..	..	304	390	592	890	1 317	..
		513	Wholesale of household goods	..	..	169	572	308	289	1 035	..
		514	W/sale. of non-agric. intermed. prod., waste, scrap	..	..	650	983	2 373	873	7 094	..
		515	Wholesale of machinery, equipment & supplies	..	..	62	37	137	1 238	551	..
		519	Other wholesale	..	..	4	4	6	176	19	..
	52		Retail trade exc. motor veh. & cyc.; repair h/hold gds.	..	..	718	584	1 818	3 886	5 464	..
		521	Non-specialised retail trade in stores	..	..	305	200	1 171	1 621	4 068	..
		522	Retail sale food, bev., tobac. in specialised stores	..	..	48	67	59	381	34	..
		523	Other retail trade of new goods specialised stores	..	..	363	316	586	1 880	1 306	..
		524	Retail sale of second-hand goods in stores	..	..	0	..	..	..	..	..
		525	Retail trade not in stores	..	..	..	..	..	..	39	..
		526	Repair of personal and household goods	..	..	3	1	1	4	16	..
H	55		Hotels and restaurants	..	..	268	554	1 386	2 320	4 861	..
		551	Hotels; camping ; other short-stay accommodation	..	..	203	476	959	1 757	3 629	..
		552	Restaurants, bars and canteens	..	..	65	78	427	562	1 233	..
I	60_64		Transport, storage and communications	..	..	5 423	38 179	49 647	112 140	204 023	..
	60		Land transport, transport via pipelines	..	..	1 873	20 257	14 315	26 470	47 828	..
		601	Transport via railways	..	..	..	..	..	..	..	..
		602	Other land transport	..	..	1 873	20 257	14 315	26 470	47 828	..
		603	Transport via pipelines	..	..	..	..	..	..	..	..
	61		Water transport	..	..	3 461	10 208	26 646	53 600	61 484	..
		611	Sea and coastal water transport	..	..	3 461	10 208	26 646	53 600	61 484	..
		612	Inland water transport	..	..	..	..	..	..	..	..
	62		Air transport	..	..	34	905	2 078	26 830	87 788	..
		621	Scheduled air transport	..	..	34	558	1 739	22 045	44 346	..
		622	Non-scheduled air transport	..	..	..	347	339	4 785	43 442	..
	63		Support. & aux. transport act., travel agencies activ.	..	..	55	6 809	6 608	5 241	6 924	..
	64		Post and telecommunications	..	..	..	..	..	..	..	..
		641	Post and courier activities	..	..	..	..	..	..	..	..
		642	Telecommunications	..	..	..	..	..	..	..	..
J	65_67		Financial intermediation	..	..	3 882	4 966	10 069	34 420	153 918	..
	65		Financial intermediation, exc. insur., pension fund	..	..	2 448	4 618	9 759	33 067	151 529	..
		651	Monetary intermediation	..	..	1 329	2 430	2 953	5 998	22 036	..
		659	Other financial intermediation	..	..	1 119	2 188	6 806	27 069	129 493	..
	66		Insurance pension fund, exc. compul. social sec.	..	..	926	129	74	714	441	..
	67		Activities auxiliary to financial intermediation	..	..	508	219	236	638	1 948	..
		671	Act. auxil to fin. intermed., exc. insur. & pens.fund	..	..	499	209	177	619	1 909	..
		672	Act. auxiliary to insurance and pension funding	..	..	9	9	59	19	39	..
K	70_74		Real estate, renting and business activities	..	..	287	222	607	1 368	2 259	..
	70		Real estate activities	..	..	0	9	3	4	3	..
		701	Real estate activ. with own or leased property	..	..	..	..	..	..	..	..
		702	Real estate activities on a fee or contract basis	..	..	0	9	3	4	3	..
	71		Renting of mach. without operator; rent. of pers. gds.	..	..	0	0	1	3	2	..
		711	Renting of transport equipment	..	..	..	..	..	..	..	..
		712	Renting of other machinery and equipment	..	..	..	0	1	2	2	..
		713	Renting of personal and household goods nec	..	..	0	..	0	1	..	..
	72		Computer and related activities	..	..	8	7	12	28	66	..
		721	Hardware consultancy	..	..	..	..	..	..	..	..
		722	Software consultancy and supply	..	..	8	5	11	27	64	..
		723	Data processing	..	..	..	..	..	..	..	..
		724	Data base activities	..	..	..	..	..	..	..	..
		725	Mainten.; repair of office, account. & comp. mach.	..	..	0	0	0	1	3	..
		729	Other computer related activities	..	..	0	..	..	..	..	..
	73		Research and development	..	..	0	0	..	..	0	..
		731	R & D experim. on natural sciences & engineering	..	..	..	..	..	..	..	..
		732	R & D experim. on social sciences & humanities	..	..	0	..	..	..	0	..
	74		Other business activities	..	..	278	206	591	1 333	2 187	..
		741	Legal, acc., book-keep., audit act.; tax consult., etc	..	..	41	61	92	645	287	..
		742	Architectural, engineering, & other techn. activities	..	..	53	26	186	329	755	..
		743	Advertising	..	..	97	72	64	131	227	..
		749	Business activities nec	..	..	87	48	249	228	919	..
M	80		Education	..	..	30	102	98	112	633	..
		801	Primary education	..	..	0	0	0	2	28	..
		802	Secondary education	..	..	4	44	33	41	383	..
		803	Higher education	..	..	..	..	..	..	..	..
		809	Adult and other education	..	..	26	57	65	69	223	..
N	85		Health and social work	..	..	83	152	439	509	10 807	..
		851	Human health activities	..	..	82	151	438	504	10 796	..
		852	Veterinary activities	..	..	1	0	1	3	8	..
		853	Social work activities	..	..	1	1	0	2	3	..
O	90_93		Other commun., soc. & pers. services activities	..	..	63	51	572	334	12 263	..
	90		Sewage & refuse disposal, sanitation, etc.	..	..	..	0	1	..	1	..
	91		Activities of member organisations nec	..	..	..	..	..	..	..	..
		912	Activities of trade unions	..	..	..	..	..	..	..	..
	92		Recreational, cultural and sporting activities	..	..	49	44	131	311	12 161	..
		921	Motion picture, radio, TV, other entertainment act.	..	..	46	44	124	279	11 797	..
		922	News agency activities	..	..	0	0	0	..	0	..
		923	Library, archives, museums, other cultural activ.	..	..	..	..	..	..	..	..
		924	Sporting and other recreational activities	..	..	3	0	6	32	363	..
	93		Other service activities	..	..	13	8	441	23	101	..

Table TU.S5 **EMPLOI, SALARIÉS - EMPLOYMENT, EMPLOYEES**

Unités

1991	1992	1993	1994	1995	1996	1997	1998		CITI révision 3		
..	..	583 595	679 694	765 359	877 053	990 800	..	Com. de gros, de detail ; réparation etc.	G	50_52	
..	..	87 779	116 978	127 500	155 185	176 589	..	Com., entr., répar. d'auto. et moto ; carburants		50	
..	..	13 945	14 746	19 109	21 994	28 239	..	Commerce de véhicules automobiles			501
..	..	6 881	16 579	18 578	22 723	25 726	..	Entretien et réparation de véhicules automobiles			502
..	..	28 797	32 639	35 240	42 440	49 194	..	Com. de pièces et access. de véhic. automobiles			503
..	..	1 369	585	501	596	638	..	Com., entret. et répar. de motocycles et access.			504
..	..	36 787	52 429	54 072	67 432	72 792	..	Com. de détail de carburants pour automobiles			505
..	..	224 159	236 587	274 662	311 556	339 550	..	Com. de gros et act. inter., sauf auto. et moto		51	
..	..	6 080	6 282	7 587	8 469	9 995	..	Activités d'intermédiaries de commerce de gros			511
..	..	66 002	41 379	63 137	77 940	82 391	..	Com. gros: prod. agr. bruts, anim. viv., prod. alim.			512
..	..	75 024	88 575	91 758	103 498	114 953	..	Com. gros: biens de consom., sauf alimentaires			513
..	..	67 751	85 122	94 592	101 143	111 136	..	Com. gros: prod. inter. non agr., déchets, débris			514
..	..	8 143	14 530	16 859	19 738	20 253	..	Com. gros: machines, équipments et fournitures			515
..	..	1 159	699	729	768	822	..	Autres commerces de gros			519
..	..	271 657	326 129	363 197	410 312	474 661	..	Com. détail, sauf auto, moto; répar. art. pers., dom.		52	
..	..	68 405	76 515	84 624	94 138	125 701	..	Commerce de détail non spécialisé en magasin			521
..	..	23 097	25 444	28 473	30 068	31 365	..	Com. détail: prod. aliment. en mag. spécialisés			522
..	..	177 748	221 155	246 989	282 705	313 124	..	Autre com. détail: prod. neufs en mag. spécialisés			523
..	..	141	100	203	212	221	..	Commerces de détail de biens d'occasion			524
..	..	..	..	..	..	546	..	Commerces de détail autres qu'en magasin			525
..	..	2 266	2 915	2 908	3 189	3 704	..	Réparation d'articles personnels et domestiques			526
..	..	231 861	252 479	262 880	277 565	406 350	..	Hotels et restaurants	H	55	
..	..	70 717	74 066	79 089	85 832	130 676	..	Hôtels; camping; autres équip. courts séjours			551
..	..	161 144	178 413	183 791	191 733	275 674	..	Restaurants, cafés et cantines			552
..	..	84 774	79 812	78 850	87 143	99 246	..	Transport, entreposage, communications	I	60_64	
..	..	25 365	23 255	23 207	28 516	33 394	..	Transports terrestres; transports par conduites		60	
..	..	..	..	..	..	..	..	Transports par chemin de fer			601
..	..	25 365	23 255	23 207	28 516	33 394	..	Autres transports terrestres			602
..	..	..	..	..	..	..	..	Transports par conduites			603
..	..	22 246	24 452	23 266	19 971	23 110	..	Transports par eau		61	
..	..	22 246	24 452	23 266	19 971	23 110	..	Transports maritimes et côtiers			611
..	..	..	..	..	..	..	..	Transports par voies navigables intérieures			612
..	..	11 798	11 576	12 967	14 235	15 004	..	Transports aériens		62	
..	..	11 449	11 122	12 288	12 840	12 849	..	Transports aériens réguliers			621
..	..	349	454	679	1 395	2 155	..	Transports aériens non réguliers			622
..	..	25 365	20 529	19 410	24 421	27 738	..	Act. annex. et aux. des transp.; act. des ag. de voy.		63	
..	..	..	..	..	..	..	..	Postes et télécommunications		64	
..	..	..	..	..	..	..	..	Activités de poste et de courrier			641
..	..	..	..	..	..	..	..	Télécommunications			642
..	..	173 460	172 421	170 779	177 514	192 470	..	Intermédiation financière	J	65_67	
..	..	159 418	156 148	154 864	156 824	165 727	..	Interm. financ., sauf assur. et caisses de retraite		65	
..	..	158 582	155 163	153 941	155 461	161 081	..	Intermédiation monétaire			651
..	..	836	985	923	1 363	4 646	..	Autres intermédiations financières			659
..	..	5 359	5 967	6 216	7 588	8 074	..	Act. d'assur., caisses de retr., sauf séc. soc. oblig.		66	
..	..	8 683	10 306	9 699	13 102	18 669	..	Act. auxiliaires de l'intermédiation financière		67	
..	..	3 909	5 532	4 925	6 851	7 989	..	Act. aux. de l'interm. fin., sauf assur., et retraite			671
..	..	4 774	4 774	4 774	6 251	10 680	..	Act. aux. des assurances et caisses de retraite			672
..	..	71 219	76 375	87 231	98 168	113 891	..	Immobilier, locations, services aux entreprises	K	70_74	
..	..	1 536	1 851	2 021	2 439	2 744	..	Activités immobilières		70	
..	..	230	236	245	261	283	..	Act. immobilières sur biens propres ou loués			701
..	..	1 306	1 615	1 776	2 178	2 461	..	Activités immobilières pour comptes de tiers			702
..	..	697	699	382	203	231	..	Locat. mach., équip. sans opér., biens pers. et dom.		71	
..	..	5	4	..	..	..	..	Location de matériel de transport			711
..	..	160	167	197	18	18	..	Location de machines ét equipements nca			712
..	..	532	528	181	182	210	..	Loc. d'articles personnels et domestiques nca			713
..	..	1 064	1 231	1 224	2 028	2 807	..	Activités informatiques et activités rattachées		72	
..	..	..	..	58	99	106	..	Conseil en matériel informatique			721
..	..	383	406	386	1 149	1 878	..	Production de logiciels			722
..	..	..	..	..	..	..	..	Traitement de données informatiques			723
..	..	..	..	..	..	..	..	Activités de banques de données			724
..	..	316	341	378	382	414	..	Entr., répar. mach. de bureau et compt. et inform.			725
..	..	12	40	..	..	..	..	Autres activités rattachées à l'informatique			729
..	..	107	19	..	..	25	..	Recherche-développement		73	
..	..	..	..	..	..	..	..	Rech. et dév. expér. en sc. phys., natur. et ingén.			731
..	..	99	..	..	..	19	..	Rech. et dév. expér. en sciences soc. et humaines			732
..	..	67 815	72 575	83 604	93 488	108 084	..	Autres activités de services aux entreprises		74	
..	..	32 763	36 176	41 146	46 468	50 946	..	Act. jur., compt., audit.; cons. fisc.; ét. march., soud.			741
..	..	13 383	14 660	18 231	20 749	25 325	..	Act. d'achitect., d'ingén.; autres act. techniques			742
..	..	4 492	5 204	5 036	6 577	7 594	..	Publicité			743
..	..	17 177	16 535	19 191	19 694	24 219	..	Activités de services aux entreprises nca.			749
..	..	31 592	30 627	30 799	39 514	53 586	..	Education	M	80	
..	..	562	951	749	890	3 733	..	Activités d'enseignement primaire			801
..	..	5 296	7 093	7 410	8 450	12 666	..	Activités d'enseignement secondaire			802
..	..	..	..	..	..	..	..	Activités d'enseignement supérieur			803
..	..	25 734	22 583	22 640	30 174	37 187	..	Act. de formation perman.; autres act. d'enseign.			809
..	..	23 701	28 828	30 962	37 973	69 143	..	Santé et action sociale	N	85	
..	..	22 313	26 879	28 772	35 521	66 832	..	Activités de santé humaine			851
..	..	149	135	378	485	755	..	Activités vétérinaires			852
..	..	1 239	1 814	1 812	1 967	1 556	..	Activités d'action sociale			853
..	..	42 627	48 579	51 486	57 058	93 232	..	Autres activités de serv. coll., soc. et personnels	O	90_93	
..	..	42	29	47	39	85	..	Assain. et enlèv. des ordures; voierie, act. simil.		90	
..	..	..	..	..	..	..	..	Activités associatives diverses		91	
..	..	..	..	..	..	..	..	Activités de syndicats de salariés			912
..	..	10 422	9 080	11 684	12 619	37 156	..	Activités récréatives, culturelles et sportives		92	
..	..	7 178	5 613	6 132	6 264	29 028	..	Act. cinémat., de radio, télé.; autres act. de spect.			921
..	..	34	63	49	49	85	..	Activités d'agences de presse			922
..	..	..	..	..	..	..	..	Act. des biblio., arch., musées, jardins bot. et zool.			923
..	..	3 210	3 404	5 503	6 306	8 043	..	Activités sportives et autres activités récréatives			924
..	..	32 163	39 470	39 755	44 400	55 991	..	Autres activités de services		93	

TURKEY

Table TU.S6 COMPENSATION OF LABOUR, EMPLOYEES - SALAIRES ET CHARGES SOCIALES, SALARIÉS

Billions of TL (Current Prices)

ISIC revision 3				1991	1992	1993	1994	1995	1996	1997	1998
G	50_52		Wholesale & retail trade; repairs; etc.	..	..	29 011	47 553	111 865	233 107	477 875	..
	50		Sale, maint.,repair vehic/cycle, person., h/hold gd.	..	..	3 987	8 620	16 791	38 554	82 575	..
		501	Sale of motor cycle	..	..	1 114	2 394	4 650	10 042	21 132	..
		502	Maintenance and repair of motor vehicles	..	..	230	662	1 260	3 397	7 004	..
		503	Sale of motor vehicle parts and accessories	..	..	1 195	2 067	3 940	8 692	21 128	..
		504	Sale, maint., repair motor cycle, parts & acces.	..	..	60	22	38	94	182	..
		505	Retail sale of automative fuel	..	..	1 387	3 475	6 904	16 329	33 129	..
	51		W/sale. trade & commission exc motor veh. & cycle	..	..	13 326	21 106	49 726	100 936	193 016	..
		511	Wholesale on a fee or contract basis	..	..	333	400	785	1 699	3 371	..
		512	W/sale agric. raw mat.,live animal,food, bev., tobac.	..	..	4 825	2 789	12 115	22 915	42 385	..
		513	Wholesale of household goods	..	..	3 617	8 988	17 106	36 901	69 186	..
		514	W/sale. of non-agric. intermed. prod., waste, scrap	..	..	3 900	7 266	15 921	32 470	62 005	..
		515	Wholesale of machinery, equipment & supplies	..	..	502	1 465	3 546	6 590	15 363	..
		519	Other wholesale	..	..	149	198	254	362	705	..
	52		Retail trade exc. motor veh. & cyc.; repair h/hold gds.	..	..	11 698	17 827	45 348	93 616	202 284	..
		521	Non-specialised retail trade in stores	..	..	4 270	3 878	14 670	34 019	71 880	..
		522	Retail sale food, bev., tobac. in specialised stores	..	..	806	1 190	2 265	5 062	9 805	..
		523	Other retail trade of new goods specialised stores	..	..	6 548	12 650	28 237	54 129	119 550	..
		524	Retail sale of second-hand goods in stores	..	..	5	5	17	39	73	..
		525	Retail trade not in stores	..	..	..	..	..	..	224	..
		526	Repair of personal and household goods	..	..	70	103	158	367	753	..
H	55		Hotels and restaurants	..	..	1 690	3 429	6 095	11 852	143 536	..
		551	Hotels; camping ; other short-stay accommodation	..	..	900	1 777	3 017	5 968	65 383	..
		552	Restaurants, bars and canteens	..	..	791	1 652	3 079	5 884	78 153	..
I	60_64		Transport, storage and communications	..	..	6 466	10 809	22 922	44 342	104 631	..
	60		Land transport, transport via pipelines	..	..	1 117	1 861	2 865	7 174	15 123	..
		601	Transport via railways	..	..	..	..	..	..	..	..
		602	Other land transport	..	..	1 117	1 861	2 865	7 174	15 123	..
		603	Transport via pipelines	..	..	..	..	..	..	..	..
	61		Water transport	..	..	2 232	2 960	7 009	13 350	27 033	..
		611	Sea and coastal water transport	..	..	2 232	2 960	7 009	13 350	27 033	..
		612	Inland water transport	..	..	..	..	..	..	..	..
	62		Air transport	..	..	2 870	4 097	10 015	15 374	43 432	..
		621	Scheduled air transport	..	..	2 792	3 958	9 496	13 639	37 181	..
		622	Non-scheduled air transport	..	..	77	139	518	1 735	6 251	..
	63		Support. & aux. transport act., travel agencies activ.	..	..	248	1 891	3 033	8 444	19 044	..
	64		Post and telecommunications	..	..	..	..	..	..	..	..
		641	Post and courier activities	..	..	..	..	..	..	..	..
		642	Telecommunications	..	..	..	..	..	..	..	..
J	65_67		Financial intermediation	..	..	29 211	51 973	95 016	196 131	456 524	..
	65		Financial intermediation, exc. insur., pension fund	..	..	27 463	45 128	87 406	178 484	380 359	..
		651	Monetary intermediation	..	..	27 252	44 583	86 509	176 143	366 613	..
		659	Other financial intermediation	..	..	212	545	897	2 341	13 746	..
	66		Insurance pension fund, exc. compul. social sec.	..	..	1 205	2 241	4 362	9 864	20 160	..
	67		Activities auxiliary to financial intermediation	..	..	542	4 604	3 247	7 783	56 005	..
		671	Act. auxil to fin. intermed., exc. insur. & pens.fund	..	..	326	989	2 071	5 174	49 023	..
		672	Act. auxiliary to insurance and pension funding	..	..	216	3 615	1 176	2 609	6 983	..
K	70_74		Real estate, renting and business activities	..	..	3 360	6 791	11 511	26 692	60 686	..
	70		Real estate activities	..	..	57	142	240	450	1 174	..
		701	Real estate activ. with own or leased property	..	..	20	40	59	130	414	..
		702	Real estate activities on a fee or contract basis	..	..	37	103	181	320	760	..
	71		Renting of mach. without operator; rent. of pers. gds.	..	..	19	31	29	35	68	..
		711	Renting of transport equipment	..	..	0	0	..	..	..	..
		712	Renting of other machinery and equipment	..	..	4	8	14	3	4	..
		713	Renting of personal and household goods nec	..	..	15	23	15	32	63	..
	72		Computer and related activities	..	..	198	369	782	1 701	3 483	..
		721	Hardware consultancy	..	..	..	..	4	37	52	..
		722	Software consultancy and supply	..	..	83	119	259	718	2 001	..
		723	Data processing	..	..	..	..	..	..	..	..
		724	Data base activities	..	..	..	..	..	..	..	..
		725	Mainten.; repair of office, account. & comp. mach.	..	..	16	32	67	126	291	..
		729	Other computer related activities	..	..	0	12	..	..	..	..
	73		Research and development	..	..	7	2	..	..	20	..
		731	R & D experim. on natural sciences & engineering	..	..	..	..	..	..	..	..
		732	R & D experim. on social sciences & humanities	..	..	6	..	..	..	14	..
	74		Other business activities	..	..	3 080	6 245	10 459	24 505	55 942	..
		741	Legal, acc., book-keep., audit act.; tax consult., etc	..	..	1 115	2 157	4 427	9 714	22 022	..
		742	Architectural, engineering, & other techn. activities	..	..	1 090	1 739	3 516	7 695	17 061	..
		743	Advertising	..	..	349	568	1 027	2 898	6 164	..
		749	Business activities nec	..	..	526	1 782	1 489	4 197	10 695	..
M	80		Education	..	..	1 452	2 120	3 567	9 833	31 433	..
		801	Primary education	..	..	34	67	60	163	2 898	..
		802	Secondary education	..	..	572	867	1 538	4 109	14 595	..
		803	Higher education	..	..	..	..	..	..	..	..
		809	Adult and other education	..	..	845	1 185	1 970	5 561	13 941	..
N	85		Health and social work	..	..	874	1 809	3 601	8 214	82 996	..
		851	Human health activities	..	..	832	1 705	3 390	7 669	81 999	..
		852	Veterinary activities	..	..	5	7	32	75	223	..
		853	Social work activities	..	..	37	98	178	470	774	..
O	90_93		Other commun., soc. & pers. services activities	..	..	1 509	2 526	4 907	9 449	57 485	..
	90		Sewage & refuse disposal, sanitation, etc.	..	..	1	1	4	5	34	..
	91		Activities of member organisations nec	..	..	..	..	..	..	..	..
		912	Activities of trade unions	..	..	..	..	..	..	..	..
	92		Recreational, cultural and sporting activities	..	..	806	967	1 967	3 099	41 768	..
		921	Motion picture, radio, TV, other entertainment act.	..	..	674	688	1 242	1 862	39 034	..
		922	News agency activities	..	..	4	7	7	23	86	..
		923	Library, archives, museums, other cultural activ.	..	..	..	..	..	..	..	..
		924	Sporting and other recreational activities	..	..	128	272	719	1 214	2 648	..
	93		Other service activities	..	..	702	1 558	2 936	6 345	15 683	..

Table TU.S7 SALAIRES ET TRAITEMENTS, SALARIÉS - WAGES AND SALARIES, EMPLOYEES

Milliards de TL (Prix courants)

1991	1992	1993	1994	1995	1996	1997	1998		CITI révision 3		
..	..	24 205	38 543	91 831	185 194	385 304	..	Com. de gros, de detail ; réparation etc.	G	50_52	
..	..	3 333	6 636	13 845	31 123	66 969	..	Com., entr., répar. d'auto. et moto ; carburants		50	
..	..	917	1 423	3 789	7 953	16 967	..	Commerce de véhicules automobiles			501
..	..	209	566	1 079	2 847	5 846	..	Entretien et réparation de véhicules automobiles			502
..	..	1 003	1 730	3 275	7 157	17 330	..	Com. de pièces et access. de véhic. automobiles			503
..	..	55	19	33	79	150	..	Com., entret. et répar. de motocycles et access.			504
..	..	1 148	2 897	5 669	13 086	26 676	..	Com. de détail de carburants pour automobiles			505
..	..	11 208	17 293	40 699	78 690	154 616	..	Com. de gros et act. inter., sauf auto. et moto		51	
..	..	281	327	655	1 413	2 754	..	Activités d'intermédiaries de commerce de gros			511
..	..	4 082	2 057	9 931	18 162	33 362	..	Com. gros: prod. agr. bruts, anim. viv., prod. alim.			512
..	..	3 039	7 414	13 923	29 201	55 717	..	Com. gros: biens de consom., sauf alimentaires			513
..	..	3 251	6 096	13 081	24 277	49 733	..	Com. gros: prod. inter. non agr., déchets, débris			514
..	..	423	1 221	2 901	5 346	12 470	..	Com. gros: machines, équipments et fournitures			515
..	..	132	177	209	291	579	..	Autres commerces de gros			519
..	..	9 664	14 615	37 287	75 382	163 720	..	Com. détail, sauf auto, moto; répar. art. pers., dom.		52	
..	..	3 421	3 151	11 951	26 640	57 579	..	Commerce de détail non spécialisé en magasin			521
..	..	657	1 009	1 873	4 240	8 000	..	Com. détail: prod. aliment. en mag. spécialisés			522
..	..	5 522	10 363	23 313	44 162	97 275	..	Autre com. détail: prod. neufs en mag. spécialisés			523
..	..	4	4	14	32	61	..	Commerces de détail de biens d'occasion			524
..	..	..	..	..	..	175	..	Commerces de détail autres qu'en magasin			525
..	..	60	87	136	307	629	..	Réparation d'articles personnels et domestiques			526
..	..	7 320	13 829	27 542	52 991	115 592	..	Hotels et restaurants	H	55	
..	..	3 342	6 247	12 753	24 904	51 841	..	Hôtels; camping; autres équip. courts séjours			551
..	..	3 978	7 581	14 790	28 087	63 752	..	Restaurants, cafés et cantines			552
..	..	5 892	9 512	20 099	38 247	92 150	..	Transport, entreposage, communications	I	60_64	
..	..	951	1 592	2 451	6 112	12 698	..	Transports terrestres; transports par conduites		60	
..	..	..	..	..	..	..	..	Transports par chemin de fer			601
..	..	951	1 592	2 451	6 112	12 698	..	Autres transports terrestres			602
..	..	..	..	..	..	..	..	Transports par conduites			603
..	..	1 977	2 424	6 242	11 536	22 981	..	Transports par eau		61	
..	..	1 977	2 424	6 242	11 536	22 981	..	Transports maritimes et côtiers			611
..	..	..	..	..	..	..	..	Transports par voies navigables intérieures			612
..	..	2 751	3 863	8 821	13 372	40 210	..	Transports aériens		62	
..	..	2 680	3 742	8 383	11 900	34 848	..	Transports aériens réguliers			621
..	..	71	121	438	1 471	53 632	..	Transports aériens non réguliers			622
..	..	213	1 633	2 584	7 228	16 261	..	Act. annex. et aux. des transp.; act. des ag. de voy.		63	
..	..	..	..	..	..	..	..	Postes et télécommunications		64	
..	..	..	..	..	..	..	..	Activités de poste et de courrier			641
..	..	..	..	..	..	..	..	Télécommunications			642
..	..	24 755	42 057	85 377	174 256	382 648	..	Intermédiation financière	J	65_67	
..	..	23 217	38 840	78 694	158 284	343 138	..	Interm. financ., sauf assur. et caisses de retraite		65	
..	..	23 027	38 406	77 918	156 116	330 269	..	Intermédiation monétaire			651
..	..	191	435	776	2 168	12 868	..	Autres intermédiations financières			659
..	..	1 067	2 037	4 087	9 069	18 440	..	Act. d'assur., caisses de retr., sauf séc. soc. oblig.		66	
..	..	470	1 179	2 596	6 903	21 070	..	Act. auxiliaires de l'intermédiation financière		67	
..	..	285	728	1 688	4 698	15 544	..	Act. aux. de l'interm. fin., sauf assur., et retraite			671
..	..	185	452	908	2 205	5 526	..	Act. aux. des assurances et caisses de retraite			672
..	..	2 848	4 995	9 613	21 937	49 762	..	Immobilier, locations, services aux entreprises	K	70_74	
..	..	49	118	204	373	967	..	Activités immobilières		70	
..	..	16	31	52	106	338	..	Act. immobilières sur biens propres ou loués			701
..	..	33	87	152	267	629	..	Activités immobilières pour comptes de tiers			702
..	..	16	26	23	29	56	..	Locat. mach., équip. sans opér., biens pers. et dom.		71	
..	..	0	0	..	..	..	..	Location de matériel de transport			711
..	..	3	7	11	2	4	..	Location de machines ét equipements nca			712
..	..	13	19	12	27	51	..	Loc. d'articles personnels et domestiques nca			713
..	..	165	306	669	1 393	2 847	..	Activités informatiques et activités rattachées		72	
..	..	..	..	3	30	43	..	Conseil en matériel informatique			721
..	..	71	101	241	589	1 657	..	Production de logiciels			722
..	..	..	..	..	..	..	..	Traitement de données informatiques			723
..	..	..	..	..	..	..	..	Activités de banques de données			724
..	..	12	26	56	104	234	..	Entr., répar. mach. de bureau et compt. et inform.			725
..	..	0	10	..	..	..	..	Autres activités rattachées à l'informatique			729
..	..	6	2	..	..	17	..	Recherche-développement		73	
..	..	..	..	..	..	..	..	Rech. et dév. expér. en sc. phys., natur. et ingén.			731
..	..	5	..	..	..	11	..	Rech. et dév. expér. en sciences soc. et humaines			732
..	..	2 613	4 543	8 716	20 141	45 875	..	Autres activités de services aux entreprises		74	
..	..	956	1 831	3 731	8 114	18 258	..	Act. jur., compt., audit.; cons. fisc.; ét. march., soud.			741
..	..	937	1 402	2 909	6 191	13 924	..	Act. d'achitect., d'ingén.; autres act. techniques			742
..	..	282	486	856	2 404	5 036	..	Publicité			743
..	..	438	824	1 219	3 432	8 657	..	Activités de services aux entreprises nca.			749
..	..	1 155	1 769	2 891	7 994	25 665	..	Education	M	80	
..	..	29	58	50	134	2 376	..	Activités d'enseignement primaire			801
..	..	417	727	1 222	3 326	12 011	..	Activités d'enseignement secondaire			802
..	..	..	..	..	..	..	..	Activités d'enseignement supérieur			803
..	..	709	984	1 619	4 534	11 278	..	Act. de formation perman.; autres act. d'enseign.			809
..	..	734	1 521	2 986	6 749	68 826	..	Santé et action sociale	N	85	
..	..	698	1 432	2 811	6 298	68 004	..	Activités de santé humaine			851
..	..	5	6	27	63	183	..	Activités vétérinaires			852
..	..	32	84	148	388	639	..	Activités d'action sociale			853
..	..	1 342	2 187	4 016	7 816	49 380	..	Autres activités de serv. coll., soc. et personnels	O	90_93	
..	..	1	1	3	5	28	..	Assain. et enlèv. des ordures; voierie, act. simil.		90	
..	..	..	..	..	..	..	..	Activités associatives diverses		91	
..	..	..	..	..	..	..	..	Activités de syndicats de salariés			912
..	..	705	813	1 503	2 514	36 262	..	Activités récréatives, culturelles et sportives		92	
..	..	592	584	976	1 498	34 035	..	Act. cinémat., de radio, télé.; autres act. de spect.			921
..	..	3	6	5	18	71	..	Activités d'agences de presse			922
..	..	..	..	..	..	..	..	Act. des biblio., arch., musées, jardins bot. et zool.			923
..	..	110	223	522	999	2 156	..	Activités sportives et autres activités récréatives			924
..	..	636	1 373	2 509	5 297	13 091	..	Autres activités de services		93	

TURKEY

Table TU.S8 ESTABLISHMENTS - ÉTABLISSEMENTS

Units

ISIC revision 3				1991	1992	1993	1994	1995	1996	1997	1998
G	50_52		Wholesale & retail trade; repairs; etc.	..	..	618 760	630 951	650 311	683 353	702 494	..
	50		Sale, maint.,repair vehic/cycle, person., h/hold gd.	..	..	61 977	69 509	70 073	76 379	80 695	..
		501	Sale of motor cycle	..	..	3 583	3 364	3 194	3 289	3 646	..
		502	Maintenance and repair of motor vehicles	..	..	19 356	20 812	19 111	24 029	25 304	..
		503	Sale of motor vehicle parts and accessories	..	..	26 272	31 356	33 640	34 238	36 077	..
		504	Sale, maint., repair motor cycle, parts & acces.	..	..	3 467	3 864	4 066	4 327	4 557	..
		505	Retail sale of automative fuel	..	..	9 299	10 113	10 062	10 496	11 111	..
	51		W/sale. trade & commission exc motor veh. & cycle	..	..	75 645	72 137	74 108	77 063	79 096	..
		511	Wholesale on a fee or contract basis	..	..	4 915	5 714	6 031	6 395	6 445	..
		512	W/sale agric. raw mat.,live animal,food, bev., tobac.	..	..	19 025	16 619	16 788	17 396	18 113	..
		513	Wholesale of household goods	..	..	25 603	21 909	24 375	25 735	26 463	..
		514	W/sale. of non-agric. intermed. prod., waste, scrap	..	..	23 436	25 269	24 076	24 592	25 059	..
		515	Wholesale of machinery, equipment & supplies	..	..	1 877	1 868	2 157	2 236	2 275	..
		519	Other wholesale	..	..	789	758	681	709	741	..
	52		Retail trade exc. motor veh. & cyc.; repair h/hold gds.	..	..	481 138	489 305	506 130	529 911	542 703	..
		521	Non-specialised retail trade in stores	..	..	153 071	169 409	169 373	177 820	180 840	..
		522	Retail sale food, bev., tobac. in specialised stores	..	..	52 841	51 944	54 534	55 751	58 562	..
		523	Other retail trade of new goods specialised stores	..	..	261 269	253 467	267 394	280 072	286 584	..
		524	Retail sale of second-hand goods in stores	..	..	1 376	1 477	1 446	1 462	1 523	..
		525	Retail trade not in stores	..	..	..	..	..	..	56	..
		526	Repair of personal and household goods	..	..	12 581	13 008	13 383	14 806	15 138	..
H	55		Hotels and restaurants	..	..	136 906	140 723	144 845	149 060	152 776	..
		551	Hotels; camping ; other short-stay accommodation	..	..	7 913	8 703	8 621	9 019	10 525	..
		552	Restaurants, bars and canteens	..	..	128 993	132 020	136 224	140 041	142 251	..
I	60_64		Transport, storage and communications	..	..	3 977	4 386	4 425	4 886	5 631	..
	60		Land transport, transport via pipelines	..	..	2 076	1 197	1 071	1 270	1 464	..
		601	Transport via railways	..	..	..	..	..	..	..	..
		602	Other land transport	..	..	2 076	1 197	1 071	1 270	1 464	..
		603	Transport via pipelines	..	..	..	..	..	..	..	..
	61		Water transport	..	..	900	834	795	768	839	..
		611	Sea and coastal water transport	..	..	900	834	795	768	839	..
		612	Inland water transport	..	..	..	..	..	..	..	..
	62		Air transport	..	..	29	12	19	22	37	..
		621	Scheduled air transport	..	..	22	6	7	5	4	..
		622	Non-scheduled air transport	..	..	7	6	12	17	33	..
	63		Support. & aux. transport act., travel agencies activ.	..	..	972	2 343	2 540	2 826	3 291	..
	64		Post and telecommunications	..	..	..	..	..	..	..	..
		641	Post and courier activities	..	..	..	..	..	..	..	..
		642	Telecommunications	..	..	..	..	..	..	..	..
J	65_67		Financial intermediation	..	..	3 743	3 750	3 743	3 219	3 923	..
	65		Financial intermediation, exc. insur., pension fund	..	..	146	147	156	148	257	..
		651	Monetary intermediation	..	..	78	78	81	75	79	..
		659	Other financial intermediation	..	..	68	69	75	73	178	..
	66		Insurance pension fund, exc. compul. social sec.	..	..	57	56	58	62	63	..
	67		Activities auxiliary to financial intermediation	..	..	3 540	3 547	3 529	3 009	3 603	..
		671	Act. auxil to fin. intermed., exc. insur. & pens.fund	..	..	725	732	714	743	925	..
		672	Act. auxiliary to insurance and pension funding	..	..	2 815	2 815	2 815	2 266	2 678	..
K	70_74		Real estate, renting and business activities	..	..	48 655	50 960	54 497	58 322	61 014	..
	70		Real estate activities	..	..	4 172	4 834	4 779	5 029	5 489	..
		701	Real estate activ. with own or leased property	..	..	7	9	9	9	9	..
		702	Real estate activities on a fee or contract basis	..	..	4 165	4 825	4 770	5 020	5 480	..
	71		Renting of mach. without operator; rent. of pers. gds.	..	..	731	709	699	731	751	..
		711	Renting of transport equipment	..	..	3	3	2	..	..	..
		712	Renting of other machinery and equipment	..	..	21	22	23	5	5	..
		713	Renting of personal and household goods nec	..	..	707	684	674	725	745	..
	72		Computer and related activities	..	..	252	208	212	237	243	..
		721	Hardware consultancy	..	..	..	..	3	3	3	..
		722	Software consultancy and supply	..	..	54	44	46	61	62	..
		723	Data processing	..	..	..	..	..	..	..	..
		724	Data base activities	..	..	..	..	..	..	..	..
		725	Mainten.; repair of office, account. & comp. mach.	..	..	191	156	159	170	174	..
		729	Other computer related activities	..	..	3	4	..	..	..	..
	73		Research and development	..	..	15	4	..	..	4	..
		731	R & D experim. on natural sciences & engineering	..	..	..	..	..	..	..	..
		732	R & D experim. on social sciences & humanities	..	..	13	..	..	..	3	..
	74		Other business activities	..	..	43 485	45 205	48 805	52 323	54 527	..
		741	Legal, acc., book-keep., audit act.; tax consult., etc	..	..	24 171	26 064	28 097	30 445	31 560	..
		742	Architectural, engineering, & other techn. activities	..	..	5 732	5 691	6 354	6 852	7 263	..
		743	Advertising	..	..	1 714	1 766	1 916	2 034	2 209	..
		749	Business activities nec	..	..	11 868	11 684	12 438	12 992	13 495	..
M	80		Education	..	..	2 466	2 469	2 447	2 788	3 316	..
		801	Primary education	..	..	67	104	82	102	198	..
		802	Secondary education	..	..	71	81	85	87	134	..
		803	Higher education	..	..	..	..	..	..	..	..
		809	Adult and other education	..	..	2 328	2 284	2 280	2 599	2 984	..
N	85		Health and social work	..	..	19 949	21 437	21 989	23 531	24 476	..
		851	Human health activities	..	..	19 294	20 713	21 138	22 609	23 521	..
		852	Veterinary activities	..	..	396	401	493	548	567	..
		853	Social work activities	..	..	259	323	358	374	388	..
O	90_93		Other commun., soc. & pers. services activities	..	..	49 187	52 200	56 232	59 189	63 477	..
	90		Sewage & refuse disposal, sanitation, etc.	..	..	9	8	7	6	7	..
	91		Activities of member organisations nec	..	..	..	..	..	..	..	..
		912	Activities of trade unions	..	..	..	..	..	..	..	..
	92		Recreational, cultural and sporting activities	..	..	4 390	4 186	4 478	4 626	6 372	..
		921	Motion picture, radio, TV, other entertainment act.	..	..	1 547	1 324	1 458	1 482	2 988	..
		922	News agency activities	..	..	11	13	14	15	15	..
		923	Library, archives, museums, other cultural activ.	..	..	..	..	..	..	..	..
		924	Sporting and other recreational activities	..	..	2 832	2 849	3 006	3 129	3 369	..
	93		Other service activities	..	..	44 788	48 006	51 747	54 557	57 098	..

Table UK.S1 **PRODUCTION - PRODUCTION**

Millions de £ (Prix courants)

1991	1992	1993	1994	1995	1996	1997	1998		CITI révision 3		
..	..	484 619	532 815	628 171	663 645	674 988	676 063	Com. de gros, de detail ; réparation etc.	G	50_52	
..	..	80 748	89 360	100 742	108 805	119 065	123 130	Com., entr., répar. d'auto. et moto ; carburants		50	
..	..	52 879	59 220	69 177	76 920	84 395	87 035	Commerce de véhicules automobiles			501
..	..	8 501	9 520	8 930	8 263	8 008	8 976	Entretien et réparation de véhicules automobiles			502
..	..	7 631	8 547	9 437	10 066	10 741	10 962	Com. de pièces et access. de véhic. automobiles			503
..	..	899	1 007	1 054	1 019	1 248	1 274	Com., entret. et répar. de motocycles et access.			504
..	..	10 838	11 067	12 144	12 537	14 673	14 883	Com. de détail de carburants pour automobiles			505
..	..	255 342	286 805	342 207	352 889	348 434	331 935	Com. de gros et act. inter., sauf auto. et moto		51	
..	..	6 243	6 993	8 809	7 712	7 685	9 139	Activités d'intermédiaries de commerce de gros			511
..	..	51 485	60 093	67 309	69 071	72 150	67 668	Com. gros: prod. agr. bruts, anim. viv., prod. alim.			512
..	..	42 268	44 237	48 221	52 049	58 627	61 334	Com. gros: biens de consom., sauf alimentaires			513
..	..	81 337	92 554	113 560	139 600	118 413	100 689	Com. gros: prod. inter. non agr., déchets, débris			514
..	..	28 664	32 130	40 321	51 125	57 990	62 324	Com. gros: machines, équipments et fournitures			515
..	..	45 345	50 798	63 987	33 332	33 569	30 781	Autres commerces de gros			519
..	..	148 529	156 650	185 222	201 951	207 489	220 998	Com. détail, sauf auto, moto; répar. art. pers., dom.		52	
..	..	62 707	69 014	86 275	93 699	95 307	104 967	Commerce de détail non spécialisé en magasin			521
..	..	13 293	13 103	11 524	15 222	14 690	13 794	Com. détail: prod. aliment. en mag. spécialisés			522
..	..	60 848	63 138	74 384	81 375	85 151	88 900	Autre com. détail: prod. neufs en mag. spécialisés			523
..	..	1 064	1 258	1 165	1 579	1 695	2 063	Commerces de détail de biens d'occasion			524
..	..	10 533	9 981	11 586	9 619	10 003	10 450	Commerces de détail autres qu'en magasin			525
..	..	84	156	287	457	643	822	Réparation d'articles personnels et domestiques			526
..	..	36 404	38 349	39 057	41 666	45 844	47 946	Hotels et restaurants	H	55	
..	..	8 368	8 613	9 666	10 553	11 534	12 422	Hôtels; camping; autres équip. courts séjours			551
..	..	28 036	29 736	29 391	31 113	34 310	35 524	Restaurants, cafés et cantines			552
..	..	..	..	..	101 197	112 390	121 790	Transport, entreposage, communications	I	60_64	
..	..	..	..	..	26 855	30 347	30 749	Transports terrestres; transports par conduites		60	
..	..	..	..	..	2 710	5 453	5 504	Transports par chemin de fer			601
..	..	15 878	17 788	22 039	24 091	24 843	25 200	Autres transports terrestres			602
..	..	28	32	44	54	51	45	Transports par conduites			603
..	..	..	..	..	5 358	4 803	4 073	Transports par eau		61	
..	..	..	..	..	5 228	4 758	4 073	Transports maritimes et côtiers			611
..	..	..	..	..	130	45	..	Transports par voies navigables intérieures			612
..	..	..	..	..	13 288	14 471	15 379	Transports aériens		62	
..	..	..	..	..	..	..	..	Transports aériens réguliers			621
..	..	..	..	..	1 676	2 140	2 480	Transports aériens non réguliers			622
..	..	..	..	..	30 378	32 847	36 163	Act. annex. et aux. des transp.; act. des ag. de voy.		63	
..	..	19 453	21 273	22 784	25 448	29 967	35 426	Postes et télécommunications		64	
..	..	1 490	1 622	2 058	2 356	2 869	3 489	Activités de poste et de courrier			641
..	..	17 963	19 651	20 726	23 092	27 098	31 937	Télécommunications			642
..	..	..	..	..	..	..	..	Intermédiation financière	J	65_67	
..	..	..	..	..	..	..	..	Interm. financ., sauf assur. et caisses de retraite		65	
..	..	..	..	..	..	..	..	Intermédiation monétaire			651
..	..	..	..	..	..	..	..	Autres intermédiations financières			659
..	..	..	..	..	..	..	..	Act. d'assur., caisses de retr., sauf séc. soc. oblig.		66	
..	..	..	..	..	..	..	..	Act. auxiliaires de l'intermédiation financière		67	
..	..	..	..	..	..	..	..	Act. aux. de l'interm. fin., sauf assur., et retraite			671
..	..	..	..	..	..	..	..	Act. aux. des assurances et caisses de retraite			672
..	..	..	..	145 809	156 660	179 591	205 080	Immobilier, locations, services aux entreprises	K	70_74	
..	..	..	..	21 146	20 020	24 144	24 952	Activités immobilières		70	
..	..	..	..	15 656	15 541	18 637	19 287	Act. immobilières sur biens propres ou loués			701
..	..	..	..	5 490	4 479	5 507	5 665	Activités immobilières pour comptes de tiers			702
..	..	..	..	12 207	12 537	12 572	14 816	Locat. mach., équip. sans opér., biens pers. et dom.		71	
..	..	4 277	4 652	5 512	6 226	5 858	7 376	Location de matériel de transport			711
..	..	..	..	3 627	3 832	4 541	5 059	Location de machines ét equipements nca			712
..	..	..	..	1 927	1 579	1 503	1 682	Loc. d'articles personnels et domestiques nca			713
..	..	14 498	16 786	16 691	18 345	23 260	32 183	Activités informatiques et activités rattachées		72	
..	..	327	378	376	480	615	926	Conseil en matériel informatique			721
..	..	11 201	12 969	13 043	13 704	16 259	20 952	Production de logiciels			722
..	..	1 440	1 668	1 584	2 205	3 070	4 913	Traitement de données informatiques			723
..	..	..	..	..	..	..	..	Activités de banques de données			724
..	..	1 113	1 288	1 193	1 097	1 340	1 540	Entr., répar. mach. de bureau et compt. et inform.			725
..	..	417	483	495	859	1 976	3 852	Autres activités rattachées à l'informatique			729
..	..	2 710	4 019	4 513	4 621	4 730	5 141	Recherche-développement		73	
..	..	2 710	4 019	4 513	4 621	4 730	5 141	Rech. et dév. expér. en sc. phys., natur. et ingén.			731
..	..	..	..	..	..	..	..	Rech. et dév. expér. en sciences soc. et humaines			732
..	..	74 985	86 596	91 252	101 137	114 885	127 988	Autres activités de services aux entreprises		74	
..	..	22 338	25 453	25 016	28 328	32 440	39 064	Act. jur., compt., audit.; cons. fisc.; ét. march., soud.			741
..	..	15 479	15 820	16 536	19 677	24 393	25 826	Act. d'achitect., d'ingén.; autres act. techniques			742
..	..	9 550	11 544	12 940	13 716	15 418	15 988	Publicité			743
..	..	27 618	33 779	36 760	39 416	42 634	47 110	Activités de services aux entreprises nca.			749
..	..	4 277	5 236	15 467	13 651	17 052	15 873	Education	M	80	
..	..	..	..	..	..	..	..	Activités d'enseignement primaire			801
..	..	..	..	..	..	..	..	Activités d'enseignement secondaire			802
..	..	..	..	..	..	..	..	Activités d'enseignement supérieur			803
..	..	..	..	..	..	..	..	Act. de formation perman.; autres act. d'enseign.			809
..	..	..	..	10 992	11 155	15 141	12 967	Santé et action sociale	N	85	
..	..	..	..	..	..	..	..	Activités de santé humaine			851
..	..	..	..	..	..	..	..	Activités vétérinaires			852
..	..	..	..	..	..	..	..	Activités d'action sociale			853
..	..	..	..	..	..	..	..	Autres activités de serv. coll., soc. et personnels	O	90_93	
..	..	..	..	5 295	6 124	6 723	7 005	Assain. et enlèv. des ordures; voierie, act. simil.		90	
..	..	..	..	..	..	..	7 106	Activités associatives diverses		91	
..	..	..	..	..	..	..	631	Activités de syndicats de salariés			912
..	..	..	..	32 871	35 313	40 543	41 630	Activités récréatives, culturelles et sportives		92	
..	..	6 245	6 833	7 190	8 631	8 724	9 210	Act. cinémat., de radio, télé.; autres act. de spect.			921
..	..	513	619	634	759	840	1 084	Activités d'agences de presse			922
..	..	..	..	502	580	808	733	Act. des biblio., arch., musées, jardins bot. et zool.			923
..	..	13 275	16 734	17 190	17 661	19 948	20 114	Activités sportives et autres activités récréatives			924
..	..	7 887	9 642	11 311	11 074	11 487	11 001	Autres activités de services		93	

UNITED KINGDOM

Table UK.S2 **VALUE ADDED AT FACTOR COSTS - VALEUR AJOUTÉE AU COUT DES FACTEURS**

Millions of £ (Current Prices)

ISIC revision 3				1991	1992	1993	1994	1995	1996	1997	1998
G	**50_52**		Wholesale & retail trade; repairs; etc.	..	..	..	..	75 550	88 905	105 112	107 616
	50		Sale, maint.,repair vehic/cycle, person., h/hold gd.	..	..	..	..	13 787	14 816	17 804	17 464
		501	Sale of motor cycle	..	..	..	..	8 366	8 904	10 856	10 274
		502	Maintenance and repair of motor vehicles	..	..	..	..	2 364	2 588	2 898	3 317
		503	Sale of motor vehicle parts and accessories	..	..	..	..	1 740	2 034	2 449	2 373
		504	Sale, maint., repair motor cycle, parts & acces.	..	..	..	..	185	177	213	262
		505	Retail sale of automative fuel	..	..	..	..	1 132	1 113	1 388	1 238
	51		W/sale. trade & commission exc motor veh. & cycle	..	..	..	..	32 165	36 194	47 588	47 022
		511	Wholesale on a fee or contract basis	..	..	..	..	1 383	1 714	1 845	2 132
		512	W/sale agric. raw mat.,live animal,food, bev., tobac.	..	..	..	..	5 770	6 029	8 284	6 362
		513	Wholesale of household goods	..	..	..	..	7 148	8 049	11 163	10 762
		514	W/sale. of non-agric. intermed. prod., waste, scrap	..	..	..	..	7 013	7 546	9 058	9 680
		515	Wholesale of machinery, equipment & supplies	..	..	..	..	8 130	9 509	12 568	13 977
		519	Other wholesale	..	..	..	..	2 721	3 347	4 670	4 109
	52		Retail trade exc. motor veh. & cyc.; repair h/hold gds.	..	..	..	..	29 598	37 895	39 720	43 130
		521	Non-specialised retail trade in stores	..	..	..	..	12 588	14 756	17 004	19 124
		522	Retail sale food, bev., tobac. in specialised stores	..	..	..	..	2 120	2 487	2 589	2 710
		523	Other retail trade of new goods specialised stores	..	..	..	..	12 583	18 032	17 376	17 935
		524	Retail sale of second-hand goods in stores	..	..	..	..	234	295	437	491
		525	Retail trade not in stores	..	..	..	..	1 952	2 109	1 979	2 565
		526	Repair of personal and household goods	..	..	..	..	116	210	332	307
H	**55**		Hotels and restaurants	..	..	..	..	13 475	15 639	18 059	18 950
		551	Hotels; camping ; other short-stay accommodation	..	..	..	..	4 153	4 854	5 485	5 914
		552	Restaurants, bars and canteens	..	..	..	..	9 322	10 785	12 574	13 036
I	**60_64**		Transport, storage and communications	..	..	..	..	..	37 215	44 422	48 713
	60		Land transport, transport via pipelines	..	..	..	..	..	11 918	14 135	14 926
		601	Transport via railways	..	..	..	..	..	694	1 617	1 970
		602	Other land transport	..	..	..	..	10 285	11 188	12 482	12 935
		603	Transport via pipelines	..	..	..	..	22	36	36	21
	61		Water transport	..	..	..	..	..	1 886	1 649	1 291
		611	Sea and coastal water transport	..	..	..	..	..	1 886	1 649	1 291
		612	Inland water transport	..	..	..	..				
	62		Air transport	..	..	..	..	..	2 373	3 378	4 278
		621	Scheduled air transport	..	..	..	..	..	..	..	..
		622	Non-scheduled air transport	..	..	..	..	..	442	498	535
	63		Support. & aux. transport act., travel agencies activ.	..	..	..	..	..	8 126	8 995	10 052
	64		Post and telecommunications	..	..	..	..	11 858	12 912	16 265	18 166
		641	Post and courier activities	..	..	..	..	793	992	1 213	1 636
		642	Telecommunications	..	..	..	..	11 065	11 920	15 052	16 530
J	**65_67**		Financial intermediation	..	..	..	..	..	..	..	..
	65		Financial intermediation, exc. insur., pension fund	..	..	..	..	..	..	..	..
		651	Monetary intermediation	..	..	..	..	..	..	..	..
		659	Other financial intermediation	..	..	..	..	..	..	..	..
	66		Insurance pension fund, exc. compul. social sec.	..	..	..	..	..	..	..	..
	67		Activities auxiliary to financial intermediation	..	..	..	..	..	..	..	..
		671	Act. auxil to fin. intermed., exc. insur. & pens.fund	..	..	..	..	..	..	..	..
		672	Act. auxiliary to insurance and pension funding	..	..	..	..	..	..	..	..
K	**70_74**		Real estate, renting and business activities	..	..	..	..	..	..	..	..
	70		Real estate activities	..	..	..	..	..	..	..	..
		701	Real estate activ. with own or leased property	..	..	..	..	..	..	..	..
		702	Real estate activities on a fee or contract basis	..	..	..	..	..	..	..	..
	71		Renting of mach. without operator; rent. of pers. gds.	..	..	..	..	6 934	7 338	7 223	8 772
		711	Renting of transport equipment	..	..	..	..	3 174	3 736	3 626	4 302
		712	Renting of other machinery and equipment	..	..	..	..	1 906	2 096	2 555	2 961
		713	Renting of personal and household goods nec	..	..	..	..	1 023	844	738	871
	72		Computer and related activities	..	..	..	..	7 703	9 994	13 714	19 597
		721	Hardware consultancy	..	..	..	..	160	231	387	615
		722	Software consultancy and supply	..	..	..	..	6 030	7 717	9 819	13 191
		723	Data processing	..	..	..	..	692	1 161	1 867	2 948
		724	Data base activities	..	..	..	..	..	..	..	..
		725	Mainten.; repair of office, account. & comp. mach.	..	..	..	..	582	495	518	598
		729	Other computer related activities	..	..	..	..	239	390	1 123	2 245
	73		Research and development	..	..	..	..	1 730	2 342	2 391	2 317
		731	R & D experim. on natural sciences & engineering	..	..	..	..	1 730	2 342	2 391	2 317
		732	R & D experim. on social sciences & humanities	..	..	..	..	..	..	..	..
	74		Other business activities	..	..	..	..	47 440	54 891	62 958	70 281
		741	Legal, acc., book-keep., audit act.; tax consult., etc	..	..	..	..	15 581	18 614	21 166	26 078
		742	Architectural, engineering, & other techn. activities	..	..	..	..	8 009	9 828	13 264	13 796
		743	Advertising	..	..	..	..	2 602	2 974	3 438	3 538
		749	Business activities nec	..	..	..	..	21 248	23 475	25 090	26 869
M	**80**		Education	..	..	..	..	8 456	7 778	8 295	6 996
		801	Primary education	..	..	..	..	..	..	..	..
		802	Secondary education	..	..	..	..	..	..	..	..
		803	Higher education	..	..	..	..	..	..	..	..
		809	Adult and other education	..	..	..	..	..	..	..	..
N	**85**		Health and social work	..	..	..	..	7 504	8 086	10 275	8 198
		851	Human health activities	..	..	..	..	..	..	..	..
		852	Veterinary activities	..	..	..	..	..	..	..	..
		853	Social work activities	..	..	..	..	..	..	..	..
O	**90_93**		Other commun., soc. & pers. services activities	..	..	..	..	..	..	..	..
	90		Sewage & refuse disposal, sanitation, etc.	..	..	..	..	3 449	4 225	4 590	4 710
	91		Activities of member organisations nec	..	..	..	..	..	..	..	2 788
		912	Activities of trade unions	..	..	..	..	..	..	..	371
	92		Recreational, cultural and sporting activities	..	..	..	..	10 020	12 184	12 654	13 430
		921	Motion picture, radio, TV, other entertainment act.	..	..	..	..	3 542	4 692	4 408	4 389
		922	News agency activities	..	..	..	..	326	391	414	530
		923	Library, archives, museums, other cultural activ.	..	..	..	..	204	261	302	193
		924	Sporting and other recreational activities	..	..	..	..	3 586	4 122	5 011	4 589
	93		Other service activities	..	..	..	..	5 195	5 312	6 020	5 497

Table UK.S3 **INVESTISSEMENT - INVESTMENT**

Millions de £ (Prix courants)

1991	1992	1993	1994	1995	1996	1997	1998			CITI révision 3
..	..	8 234	8 317	9 310	9 073	11 003	12 524	Com. de gros, de detail ; réparation etc.	G	**50_52**
..	..	707	711	487	473	980	1 190	Com., entr., répar. d'auto. et moto ; carburants		**50**
..	..	470	447	291	306	536	570	Commerce de véhicules automobiles		501
..	..	67	64	56	58	228	340	Entretien et réparation de véhicules automobiles		502
..	..	68	65	79	77	157	137	Com. de pièces et access. de véhic. automobiles		503
..	..	8	8	9	9	18	16	Com., entret. et répar. de motocycles et access.		504
..	..	94	127	52	23	41	127	Com. de détail de carburants pour automobiles		505
..	..	2 938	3 174	3 092	2 977	3 748	4 213	Com. de gros et act. inter., sauf auto. et moto		**51**
..	..	94	102	105	98	185	191	Activités d'intermédiaries de commerce de gros		511
..	..	503	566	518	550	816	936	Com. gros: prod. agr. bruts, anim. viv., prod. alim.		512
..	..	709	720	663	413	673	785	Com. gros: biens de consom., sauf alimentaires		513
..	..	846	930	907	867	799	832	Com. gros: prod. inter. non agr., déchets, débris		514
..	..	476	518	540	757	890	1 154	Com. gros: machines, équipments et fournitures		515
..	..	310	338	359	292	385	315	Autres commerces de gros		519
..	..	4 589	4 432	5 731	5 623	6 275	7 121	Com. détail, sauf auto, moto; répar. art. pers., dom.		**52**
..	..	2 966	2 801	3 183	2 968	3 072	4 082	Commerce de détail non spécialisé en magasin		521
..	..	145	138	162	131	192	228	Com. détail: prod. aliment. en mag. spécialisés		522
..	..	1 348	1 410	2 233	2 377	2 788	2 535	Autre com. détail: prod. neufs en mag. spécialisés		523
..	..	12	6	15	10	26	19	Commerces de détail de biens d'occasion		524
..	..	117	77	131	125	178	211	Commerces de détail autres qu'en magasin		525
..	..	1	..	7	12	19	46	Réparation d'articles personnels et domestiques		526
..	..	1 588	1 788	2 271	2 546	3 412	3 858	Hotels et restaurants	H	**55**
..	..	567	542	659	731	1 203	1 292	Hôtels; camping; autres équip. courts séjours		551
..	..	1 021	1 246	1 612	1 815	2 209	2 566	Restaurants, cafés et cantines		552
..	..	..	..	..	12 085	..	15 642	Transport, entreposage, communications	I	**60_64**
..	..	..	..	..	1 672	2 477	2 702	Transports terrestres; transports par conduites		**60**
..	..	..	..	..	102	307	277	Transports par chemin de fer		601
..	..	1 122	1 428	1 697	1 570	2 170	2 425	Autres transports terrestres		602
..	..	..	..	..	..	..	..	Transports par conduites		603
..	..	..	..	..	424	221	175	Transports par eau		**61**
..	..	..	..	..	424	221	175	Transports maritimes et côtiers		611
..	..	..	..	..	35	7	..	Transports par voies navigables intérieures		612
..	..	..	..	..	1 557	..	2 142	Transports aériens		**62**
..	..	..	..	..	1 557	1 894	2 008	Transports aériens réguliers		621
..	..	..	..	..	..	..	134	Transports aériens non réguliers		622
..	..	..	..	..	2 280	2 708	3 087	Act. annex. et aux. des transp.; act. des ag. de voy.		**63**
..	..	3 501	4 282	4 662	6 152	7 741	7 536	Postes et télécommunications		**64**
..	..	67	82	105	90	81	105	Activités de poste et de courrier		641
..	..	3 434	4 200	4 557	6 062	7 660	7 431	Télécommunications		642
..	..	..	..	..	..	..	..	Intermédiation financière	J	**65_67**
..	..	..	..	..	..	..	..	Interm. financ., sauf assur. et caisses de retraite		**65**
..	..	..	..	..	..	..	..	Intermédiation monétaire		651
..	..	..	..	..	..	..	..	Autres intermédiations financières		659
..	..	..	..	..	..	..	..	Act. d'assur., caisses de retr., sauf séc. soc. oblig.		**66**
..	..	..	..	..	..	..	..	Act. auxiliaires de l'intermédiation financière		**67**
..	..	..	..	..	..	..	..	Act. aux. de l'interm. fin., sauf assur., et retraite		671
..	..	..	..	..	..	..	..	Act. aux. des assurances et caisses de retraite		672
..	..	..	..	13 333	10 567	6 952	10 623	Immobilier, locations, services aux entreprises	K	**70_74**
..	..	..	..	6 416	3 336		594	Activités immobilières		**70**
..	..	..	..	4 747	2 519		80	Act. immobilières sur biens propres ou loués		701
..	..	..	..	1 669	817	164	514	Activités immobilières pour comptes de tiers		702
..	..	..	..	3 098	3 188	3 026	4 009	Locat. mach., équip. sans opér., biens pers. et dom.		**71**
..	..	1 567	1 930	2 178	2 140	2 183	3 019	Location de matériel de transport		711
..	..	..	..	682	736	681	841	Location de machines ét equipements nca		712
..	..	..	..	238	312	162	149	Loc. d'articles personnels et domestiques nca		713
..	..	698	908	908	930	1 081	1 359	Activités informatiques et activités rattachées		**72**
..	..	12	16	16	16	11	31	Conseil en matériel informatique		721
..	..	484	629	630	549	651	824	Production de logiciels		722
..	..	173	225	219	246	210	259	Traitement de données informatiques		723
..	..	..	..	..	..	..	..	Activités de banques de données		724
..	..	8	11	11	32	33	34	Entr., répar. mach. de bureau et compt. et inform.		725
..	..	21	27	32	87	176	211	Autres activités rattachées à l'informatique		729
..	..	236	390	414	480	458	558	Recherche-développement		**73**
..	..	236	390	414	480	458	558	Rech. et dév. expér. en sc. phys., natur. et ingén.		731
..	..	..	..	..	..	..	..	Rech. et dév. expér. en sciences soc. et humaines		732
..	..	1 715	2 219	2 287	2 467	3 120	3 703	Autres activités de services aux entreprises		**74**
..	..	606	783	754	737	952	1 260	Act. jur., compt., audit.; cons. fisc.; ét. march., soud.		741
..	..	369	435	456	558	712	773	Act. d'achitect., d'ingén.; autres act. techniques		742
..	..	114	155	171	161	277	239	Publicité		743
..	..	626	846	906	1 011	1 179	1 431	Activités de services aux entreprises nca.		749
..	..	145	199	3 547	3 353	1 582	1 576	Education	M	**80**
..	..	..	..	..	..	..	..	Activités d'enseignement primaire		801
..	..	..	..	..	..	..	..	Activités d'enseignement secondaire		802
..	..	..	..	..	..	..	..	Activités d'enseignement supérieur		803
..	..	..	..	..	..	..	..	Act. de formation perman.; autres act. d'enseign.		809
..	..	..	..	1 110	844	1 313	1 040	Santé et action sociale	N	**85**
..	..	..	..	..	..	..	..	Activités de santé humaine		851
..	..	..	..	..	..	..	..	Activités vétérinaires		852
..	..	..	..	..	..	..	..	Activités d'action sociale		853
..	..	..	..	..	..	..	..	Autres activités de serv. coll., soc. et personnels	O	**90_93**
..	..	995	1 503	1 641	1 835	2 711	2 667	Assain. et enlèv. des ordures; voierie, act. simil.		**90**
..	..	..	..	..	..	..	281	Activités associatives diverses		**91**
..	..	..	..	..	..	..	6	Activités de syndicats de salariés		912
..	..	..	..	1 762	1 650	2 557	2 265	Activités récréatives, culturelles et sportives		**92**
..	..	220	272	292	442	594	507	Act. cinémat., de radio, télé.; autres act. de spect.		921
..	..	84	113	58	72	107	90	Activités d'agences de presse		922
..	..	..	..	112	103	170	217	Act. des biblio., arch., musées, jardins bot. et zool.		923
..	..	683	961	1 021	771	1 195	1 003	Activités sportives et autres activités récréatives		924
..	..	356	491	467	604	501	521	Autres activités de services		**93**

Note:ISIC 269 includes 271.

Table UK.S4 **EMPLOYMENT, EMPLOYEES - EMPLOI, SALARIÉS**

Thousands

ISIC revision 3				1991	1992	1993	1994	1995	1996	1997	1998
G	**50_52**		Wholesale & retail trade; repairs; etc.	..	..	3 610	3 680	3 780	3 819	3 965	3 982
	50		Sale, maint.,repair vehic/cycle, person., h/hold gd.	..	..	507	501	543	531	535	541
		501	Sale of motor cycle	..	..	328	327	374	317	323	226
		502	Maintenance and repair of motor vehicles	..	..	179	173	169	162	163	151
		503	Sale of motor vehicle parts and accessories	..	..	..	..	..	..	..	92
		504	Sale, maint., repair motor cycle, parts & acces.	..	..	..	..	..	..	..	15
		505	Retail sale of automative fuel	..	..	..	..	..	51	49	58
	51		W/sale. trade & commission exc motor veh. & cycle	..	..	885	923	992	1 023	1 079	1 083
		511	Wholesale on a fee or contract basis	..	..	30	36	45	44	47	47
		512	W/sale agric. raw mat.,live animal,food, bev., tobac.	..	..	208	212	213	211	208	208
		513	Wholesale of household goods	..	..	189	199	215	236	248	250
		514	W/sale. of non-agric. intermed. prod., waste, scrap	..	..	224	225	235	237	238	246
		515	Wholesale of machinery, equipment & supplies	..	..	180	187	197	217	242	245
		519	Other wholesale	..	..	54	65	86	79	95	87
	52		Retail trade exc. motor veh. & cyc.; repair h/hold gds.	..	..	2 218	2 256	2 244	2 265	2 352	2 357
		521	Non-specialised retail trade in stores	..	..	..	..	..	938	1 015	1 034
		522	Retail sale food, bev., tobac. in specialised stores	..	..	..	..	..	210	198	218
		523	Other retail trade of new goods specialised stores	..	..	..	..	..	985	1 008	965
		524	Retail sale of second-hand goods in stores	..	..	..	..	..	115	111	23
		525	Retail trade not in stores	..	..	..	..	..	..	..	95
		526	Repair of personal and household goods	..	..	..	..	..	18	19	23
H	**55**		Hotels and restaurants	..	..	1 146	1 160	1 151	1 223	1 274	1 351
		551	Hotels; camping ; other short-stay accommodation	..	..	276	281	272	283	280	325
		552	Restaurants, bars and canteens	..	..	869	879	879	940	994	1 026
I	**60_64**		Transport, storage and communications	..	..	1 283	1 278	1 289	1 310	1 330	1 357
	60		Land transport, transport via pipelines	..	..	478	471	465	461	454	474
		601	Transport via railways	..	..	130	115	93	60	34	48
		602	Other land transport	..	..	347	356	372	401	420	425
		603	Transport via pipelines	..	..	..	..	..	..	..	1
	61		Water transport	..	..	21	24	27	22	19	18
		611	Sea and coastal water transport	..	..	..	..	..	..	..	16
		612	Inland water transport	..	..	..	..	..	..	..	1
	62		Air transport	..	..	58	54	55	64	73	82
		621	Scheduled air transport	..	..	..	..	..	..	..	68
		622	Non-scheduled air transport	..	..	..	..	..	..	..	..
	63		Support. & aux. transport act., travel agencies activ.	..	..	306	308	313	333	343	334
	64		Post and telecommunications	..	..	420	420	429	430	440	449
		641	Post and courier activities	..	..	233	234	245	254	254	269
		642	Telecommunications	..	..	186	186	184	176	186	180
J	**65_67**		Financial intermediation	..	..	966	982	996	958	987	990
	65		Financial intermediation, exc. insur., pension fund	..	..	572	579	592	550	569	560
		651	Monetary intermediation	..	..	..	..	..	443	459	444
		659	Other financial intermediation	..	..	..	..	..	108	110	115
	66		Insurance pension fund, exc. compul. social sec.	..	..	219	221	212	217	230	240
	67		Activities auxiliary to financial intermediation	..	..	175	181	192	190	188	191
		671	Act. auxil to fin. intermed., exc. insur. & pens.fund	..	..	36	42	47	51	64	75
		672	Act. auxiliary to insurance and pension funding	..	..	139	139	145	139	123	116
K	**70_74**		Real estate, renting and business activities	..	..	2 467	2 573	2 926	3 033	3 246	3 361
	70		Real estate activities	..	..	244	257	268	262	282	297
		701	Real estate activ. with own or leased property	..	..	138	149	161	147	157	166
		702	Real estate activities on a fee or contract basis	..	..	106	108	107	115	124	131
	71		Renting of mach. without operator; rent. of pers. gds.	..	..	115	117	125	123	131	146
		711	Renting of transport equipment	..	..	..	..	..	..	..	42
		712	Renting of other machinery and equipment	..	..	..	..	..	..	..	69
		713	Renting of personal and household goods nec	..	..	..	..	..	..	..	35
	72		Computer and related activities	..	..	184	200	271	305	367	402
		721	Hardware consultancy	..	..	..	..	..	..	..	21
		722	Software consultancy and supply	..	..	..	..	..	..	..	240
		723	Data processing	..	..	..	..	..	..	..	39
		724	Data base activities	..	..	..	..	..	..	..	11
		725	Mainten.; repair of office, account. & comp. mach.	..	..	..	..	..	..	..	35
		729	Other computer related activities	..	..	..	..	..	..	..	56
	73		Research and development	..	..	94	91	86	97	89	98
		731	R & D experim. on natural sciences & engineering	..	..	..	..	..	..	..	93
		732	R & D experim. on social sciences & humanities	..	..	..	..	..	..	..	6
	74		Other business activities	..	..	1 829	1 908	2 176	2 246	2 377	2 419
		741	Legal, acc., book-keep., audit act.; tax consult., etc	..	..	520	527	611	631	639	676
		742	Architectural, engineering, & other techn. activities	..	..	396	350	348	319	309	305
		743	Advertising	..	..	57	58	75	85	84	88
		749	Business activities nec	..	..	391	403	420	1 212	1 345	1 350
M	**80**		Education	..	..	1 785	1 801	1 799	1 825	1 822	1 781
		801	Primary education	..	..	..	..	..	..	..	662
		802	Secondary education	..	..	..	..	..	..	..	530
		803	Higher education	..	..	..	..	..	..	..	408
		809	Adult and other education	..	..	..	..	..	..	..	181
N	**85**		Health and social work	..	..	2 385	2 396	2 426	2 455	2 456	2 428
		851	Human health activities	..	..	1 480	1 494	1 511	1 492	1 557	1 502
		852	Veterinary activities	..	..	..	..	..	..	..	24
		853	Social work activities	..	..	904	902	915	963	898	901
O	**90_93**		Other commun., soc. & pers. services activities	..	..	905	880	888	948	962	1 057
	90		Sewage & refuse disposal, sanitation, etc.	..	..	78	71	68	79	75	80
	91		Activities of member organisations nec	..	..	192	183	180	196	204	193
		912	Activities of trade unions	..	..	..	..	..	..	..	10
	92		Recreational, cultural and sporting activities	..	..	503	507	528	561	567	553
		921	Motion picture, radio, TV, other entertainment act.	..	..	130	133	154	163	167	150
		922	News agency activities	..	..	..	..	..	..	..	9
		923	Library, archives, museums, other cultural activ.	..	..	73	74	73	76	78	76
		924	Sporting and other recreational activities	..	..	299	299	302	322	323	318
	93		Other service activities	..	..	133	119	112	112	116	231

Table UK.S5 **EMPLOI, FEMMES - EMPLOYMENT, FEMALES**

Milliers

1991	1992	1993	1994	1995	1996	1997	1998		CITI révision 3		
..	..	1 881	1 928	1 982	1 958	2 018	2 025	Com. de gros, de detail ; réparation etc.	G	50_52	
..	..	126	124	138	119	107	114	Com., entr., répar. d'auto. et moto ; carburants		50	
..	..	93	93	107	76	67	49	Commerce de véhicules automobiles			501
..	..	33	31	31	29	27	26	Entretien et réparation de véhicules automobiles			502
..	..	..	..	..	..	..	19	Com. de pièces et access. de véhic. automobiles			503
..	..	..	..	..	..	..	3	Com., entret. et répar. de motocycles et access.			504
..	..	..	..	..	15	13	18	Com. de détail de carburants pour automobiles			505
..	..	293	300	314	298	331	325	Com. de gros et act. inter., sauf auto. et moto		51	
..	..	11	14	16	13	15	14	Activités d'intermédiaries de commerce de gros			511
..	..	71	71	70	64	64	64	Com. gros: prod. agr. bruts, anim. viv., prod. alim.			512
..	..	84	87	87	83	93	88	Com. gros: biens de consom., sauf alimentaires			513
..	..	55	54	58	58	58	65	Com. gros: prod. inter. non agr., déchets, débris			514
..	..	52	52	54	57	68	68	Com. gros: machines, équipments et fournitures			515
..	..	19	22	29	23	32	26	Autres commerces de gros			519
..	..	1 462	1 503	1 530	1 541	1 580	1 586	Com. détail, sauf auto, moto; répar. art. pers., dom.		52	
..	..	..	..	..	636	692	702	Commerce de détail non spécialisé en magasin			521
..	..	..	..	..	147	138	152	Com. détail: prod. aliment. en mag. spécialisés			522
..	..	..	..	..	671	672	646	Autre com. détail: prod. neufs en mag. spécialisés			523
..	..	..	..	..	76	70	16	Commerces de détail de biens d'occasion			524
..	..	..	..	..	..	..	59	Commerces de détail autres qu'en magasin			525
..	..	..	..	..	10	9	12	Réparation d'articles personnels et domestiques			526
..	..	731	743	752	800	814	802	Hotels et restaurants	H	55	
..	..	161	167	164	167	159	178	Hôtels; camping; autres équip. courts séjours			551
..	..	570	576	588	633	655	624	Restaurants, cafés et cantines			552
..	..	331	321	323	338	361	390	Transport, entreposage, communications	I	60_64	
..	..	90	79	72	74	73	79	Transports terrestres; transports par conduites		60	
..	..	41	27	13	10	5	7	Transports par chemin de fer			601
..	..	49	52	59	64	68	72	Autres transports terrestres			602
..	..	..	..	..	..	..	..	Transports par conduites			603
..	..	6	6	7	6	5	5	Transports par eau		61	
..	..	..	..	..	..	..	5	Transports maritimes et côtiers			611
..	..	..	..	..	..	..	..	Transports par voies navigables intérieures			612
..	..	24	23	24	28	33	37	Transports aériens		62	
..	..	..	..	..	..	..	31	Transports aériens réguliers			621
..	..	..	..	..	..	..	..	Transports aériens non réguliers			622
..	..	106	109	116	128	139	141	Act. annex. et aux. des transp.; act. des ag. de voy.		63	
..	..	105	103	104	103	111	128	Postes et télécommunications		64	
..	..	52	52	54	54	55	76	Activités de poste et de courrier			641
..	..	53	51	50	49	56	52	Télécommunications			642
..	..	538	546	547	534	545	497	Intermédiation financière	J	65_67	
..	..	341	346	357	339	348	293	Interm. financ., sauf assur. et caisses de retraite		65	
..	..	..	..	..	277	285	229	Intermédiation monétaire			651
..	..	..	..	..	62	62	64	Autres intermédiations financières			659
..	..	109	111	103	105	114	121	Act. d'assur., caisses de retr., sauf séc. soc. oblig.		66	
..	..	87	89	87	90	84	84	Act. auxiliaires de l'intermédiation financière		67	
..	..	14	17	19	21	27	31	Act. aux. de l'interm. fin., sauf assur., et retraite			671
..	..	73	73	68	69	57	53	Act. aux. des assurances et caisses de retraite			672
..	..	1 200	1 285	1 493	1 622	1 671	..	Immobilier, locations, services aux entreprises	K	70_74	
..	..	135	146	151	147	156	166	Activités immobilières		70	
..	..	75	83	90	83	89	95	Act. immobilières sur biens propres ou loués			701
..	..	60	62	61	64	67	71	Activités immobilières pour comptes de tiers			702
..	..	33	35	42	42	40	44	Locat. mach., équip. sans opér., biens pers. et dom.		71	
..	..	..	..	..	..	..	13	Location de matériel de transport			711
..	..	..	..	..	..	..	16	Location de machines ét equipements nca			712
..	..	..	..	..	..	..	15	Loc. d'articles personnels et domestiques nca			713
..	..	60	64	82	90	105	139	Activités informatiques et activités rattachées		72	
..	..	..	..	..	..	..	7	Conseil en matériel informatique			721
..	..	..	..	..	..	..	86	Production de logiciels			722
..	..	..	..	..	..	..	14	Traitement de données informatiques			723
..	..	..	..	..	..	..	4	Activités de banques de données			724
..	..	..	..	..	..	..	11	Entr., répar. mach. de bureau et compt. et inform.			725
..	..	..	..	..	..	..	17	Autres activités rattachées à l'informatique			729
..	..	36	33	33	36	35	41	Recherche-développement		73	
..	..	..	..	..	..	..	39	Rech. et dév. expér. en sc. phys., natur. et ingén.			731
..	..	..	..	..	..	..	3	Rech. et dév. expér. en sciences soc. et humaines			732
..	..	936	1 007	1 185	1 307	1 335	..	Autres activités de services aux entreprises		74	
..	..	314	324	374	403	392	392	Act. jur., compt., audit.; cons. fisc.; ét. march., soud.			741
..	..	107	100	127	148	139	139	Act. d'achitect., d'ingén.; autres act. techniques			742
..	..	27	29	41	49	49	50	Publicité			743
..	..	286	296	296	707	755	..	Activités de services aux entreprises nca.			749
..	..	1 257	1 273	1 278	1 295	1 301	1 300	Education	M	80	
..	..	..	..	..	..	..	585	Activités d'enseignement primaire			801
..	..	..	..	..	..	..	348	Activités d'enseignement secondaire			802
..	..	..	..	..	..	..	236	Activités d'enseignement supérieur			803
..	..	..	..	..	..	..	131	Act. de formation perman.; autres act. d'enseign.			809
..	..	1 957	1 956	1 957	1 965	1 958	2 428	Santé et action sociale	N	85	
..	..	1 209	1 213	1 213	1 189	1 239	1 502	Activités de santé humaine			851
..	..	..	..	..	..	..	19	Activités vétérinaires			852
..	..	748	743	744	776	720	728	Activités d'action sociale			853
..	..	495	478	486	512	513	572	Autres activités de serv. coll., soc. et personnels	O	90_93	
..	..	13	12	13	16	14	16	Assain. et enlèv. des ordures; voierie, act. simil.		90	
..	..	111	106	107	117	119	112	Activités associatives diverses		91	
..	..	..	..	..	..	..	6	Activités de syndicats de salariés			912
..	..	261	263	279	299	298	294	Activités récréatives, culturelles et sportives		92	
..	..	62	63	75	81	83	76	Act. cinémat., de radio, télé.; autres act. de spect.			921
..	..	..	..	..	..	..	4	Activités d'agences de presse			922
..	..	50	51	49	50	51	50	Act. des biblio., arch., musées, jardins bot. et zool.			923
..	..	149	150	155	168	164	165	Activités sportives et autres activités récréatives			924
..	..	110	96	87	80	82	150	Autres activités de services		93	

UNITED KINGDOM

Table UK.S6 **COMPENSATION OF LABOUR, EMPLOYEES - SALAIRES ET CHARGES SOCIALES, SALARIÉS**

Millions of £ (Current Prices)

ISIC revision 3				1991	1992	1993	1994	1995	1996	1997	1998
G	50_52		Wholesale & retail trade; repairs; etc.	..	..	31 151	33 284	39 674	42 109	49 214	54 147
	50		Sale, maint.,repair vehic/cycle, person., h/hold gd.	..	..	4 519	4 692	5 683	6 206	6 981	7 660
		501	Sale of motor cycle	..	..	3 141	3 287	3 068	3 381	3 766	4 048
		502	Maintenance and repair of motor vehicles	..	..	498	521	1 221	1 366	1 435	1 718
		503	Sale of motor vehicle parts and accessories	..	..	453	474	939	1 018	1 231	1 292
		504	Sale, maint., repair motor cycle, parts & acces.	..	..	54	56	85	69	93	101
		505	Retail sale of automative fuel	..	..	373	354	370	372	456	501
	51		W/sale. trade & commission exc motor veh. & cycle	..	..	12 907	13 743	16 218	16 800	21 055	23 351
		511	Wholesale on a fee or contract basis	..	..	524	554	731	692	792	1 056
		512	W/sale agric. raw mat.,live animal,food, bev., tobac.	..	..	2 257	2 483	2 729	2 624	3 405	3 553
		513	Wholesale of household goods	..	..	2 890	2 885	3 223	3 385	4 387	5 081
		514	W/sale. of non-agric. intermed. prod., waste, scrap	..	..	2 958	3 283	3 838	3 781	4 615	4 928
		515	Wholesale of machinery, equipment & supplies	..	..	3 028	3 197	3 974	4 618	5 874	7 050
		519	Other wholesale	..	..	1 250	1 341	1 723	1 700	1 982	1 683
	52		Retail trade exc. motor veh. & cyc.; repair h/hold gds.	..	..	13 725	14 849	17 773	19 103	21 178	23 136
		521	Non-specialised retail trade in stores	..	..	5 588	6 539	7 651	8 187	8 840	10 099
		522	Retail sale food, bev., tobac. in specialised stores	..	..	1 026	996	1 018	1 272	1 314	1 246
		523	Other retail trade of new goods specialised stores	..	..	6 243	6 450	8 032	8 491	9 759	10 388
		524	Retail sale of second-hand goods in stores	..	..	70	86	93	110	97	114
		525	Retail trade not in stores	..	..	784	735	912	894	1 039	1 074
		526	Repair of personal and household goods	..	..	14	43	68	149	128	215
H	55		Hotels and restaurants	..	..	6 513	7 296	7 742	8 214	9 199	10 146
		551	Hotels; camping ; other short-stay accommodation	..	..	1 753	1 979	2 296	2 434	2 569	2 804
		552	Restaurants, bars and canteens	..	..	4 760	5 317	5 446	5 780	6 630	7 342
I	60_64		Transport, storage and communications	..	..	..	..	..	20 538	23 646	24 795
	60		Land transport, transport via pipelines	..	..	..	..	..	7 222	8 839	9 004
		601	Transport via railways	..	..	..	..	..	614	1 341	1 403
		602	Other land transport	..	..	4 464	4 885	6 206	6 603	7 494	7 596
		603	Transport via pipelines	..	..	4	4	5	5	4	5
	61		Water transport	..	..	..	..	..	804	561	549
		611	Sea and coastal water transport	..	..	..	..	..	804	561	549
		612	Inland water transport	..	..	..	..	..	..	..	..
	62		Air transport	..	..	..	..	..	2 298	2 480	2 608
		621	Scheduled air transport	..	..	..	..	..	2 113	2 223	2 295
		622	Non-scheduled air transport	..	..	..	..	..	185	257	313
	63		Support. & aux. transport act., travel agencies activ.	..	..	..	..	..	3 911	4 784	5 385
	64		Post and telecommunications	..	..	4 433	4 661	5 419	6 303	6 982	7 249
		641	Post and courier activities	..	..	445	466	623	633	745	869
		642	Telecommunications	..	..	3 988	4 195	4 796	5 670	6 237	6 380
J	65_67		Financial intermediation	..	..	..	..	..	..	..	..
	65		Financial intermediation, exc. insur., pension fund	..	..	..	..	..	..	..	..
		651	Monetary intermediation	..	..	..	..	..	..	..	..
		659	Other financial intermediation	..	..	..	..	..	..	..	..
	66		Insurance pension fund, exc. compul. social sec.	..	..	..	..	..	..	..	..
	67		Activities auxiliary to financial intermediation	..	..	..	..	..	..	..	..
		671	Act. auxil to fin. intermed., exc. insur. & pens.fund	..	..	..	..	..	..	..	..
		672	Act. auxiliary to insurance and pension funding	..	..	..	..	..	..	..	..
K	70_74		Real estate, renting and business activities	..	..	..	..	34 976	38 961	47 417	57 825
	70		Real estate activities	..	..	..	..	2 828	2 902	3 843	4 226
		701	Real estate activ. with own or leased property	..	..	..	..	1 277	1 508	2 153	2 292
		702	Real estate activities on a fee or contract basis	..	..	..	..	1 551	1 394	1 690	1 934
	71		Renting of mach. without operator; rent. of pers. gds.	..	..	..	..	1 651	1 608	2 015	2 092
		711	Renting of transport equipment	..	..	411	431	443	544	650	634
		712	Renting of other machinery and equipment	..	..	..	..	696	708	1 012	1 074
		713	Renting of personal and household goods nec	..	..	..	..	512	356	353	384
	72		Computer and related activities	..	..	4 348	4 840	4 438	5 153	6 831	9 528
		721	Hardware consultancy	..	..	91	101	91	84	99	150
		722	Software consultancy and supply	..	..	3 242	3 609	3 233	3 714	4 756	6 574
		723	Data processing	..	..	509	567	586	793	1 109	1 491
		724	Data base activities	..	..	..	..	..	..	..	..
		725	Mainten.; repair of office, account. & comp. mach.	..	..	391	435	398	361	417	496
		729	Other computer related activities	..	..	115	128	130	201	450	817
	73		Research and development	..	..	1 037	1 479	1 740	1 815	1 947	2 256
		731	R & D experim. on natural sciences & engineering	..	..	1 037	1 479	1 740	1 815	1 947	2 256
		732	R & D experim. on social sciences & humanities	..	..	..	..	..	..	..	..
	74		Other business activities	..	..	21 445	23 956	24 195	27 446	32 724	39 674
		741	Legal, acc., book-keep., audit act.; tax consult., etc	..	..	6 791	7 443	6 979	8 239	9 743	11 735
		742	Architectural, engineering, & other techn. activities	..	..	4 548	4 509	4 561	5 377	6 714	7 723
		743	Advertising	..	..	1 066	1 239	1 332	1 373	1 847	2 104
		749	Business activities nec	..	..	9 040	10 765	11 323	12 457	14 420	18 112
M	80		Education	..	..	1 273	1 500	8 365	8 072	13 811	14 660
		801	Primary education	..	..	..	..	..	..	..	..
		802	Secondary education	..	..	..	..	..	..	..	..
		803	Higher education	..	..	..	..	..	..	..	..
		809	Adult and other education	..	..	..	..	..	..	..	..
N	85		Health and social work	..	..	..	..	4 070	4 625	6 614	7 244
		851	Human health activities	..	..	..	..	..	..	..	..
		852	Veterinary activities	..	..	..	..	..	..	..	..
		853	Social work activities	..	..	..	..	..	..	..	..
O	90_93		Other commun., soc. & pers. services activities	..	..	..	..	..	..	..	13 232
	90		Sewage & refuse disposal, sanitation, etc.	..	..	588	764	901	1 053	1 221	1 239
	91		Activities of member organisations nec	..	..	..	..	..	..	..	2 769
		912	Activities of trade unions	..	..	..	..	..	..	..	267
	92		Recreational, cultural and sporting activities	..	..	..	..	4 392	4 866	6 198	6 763
		921	Motion picture, radio, TV, other entertainment act.	..	..	932	982	936	1 119	1 291	1 419
		922	News agency activities	..	..	97	113	119	224	293	408
		923	Library, archives, museums, other cultural activ.	..	..	..	..	258	310	455	496
		924	Sporting and other recreational activities	..	..	1 492	1 808	1 738	1 913	2 335	2 698
	93		Other service activities	..	..	1 702	1 890	1 910	1 955	1 964	2 461

Table UK.S7 SALAIRES ET TRAITEMENTS, SALARIÉS - WAGES AND SALARIES, EMPLOYEES

Millions de £ (Prix courants)

1991	1992	1993	1994	1995	1996	1997	1998		CITI révision 3		
..	..	..	..	35 510	37 657	43 883	48 084	Com. de gros, de detail ; réparation etc.	G	50_52	
..	..	..	..	5 082	5 564	6 229	6 802	Com., entr., répar. d'auto. et moto ; carburants		50	
..	..	..	..	2 728	3 024	3 349	3 594	Commerce de véhicules automobiles			501
..	..	..	..	1 100	1 230	1 273	1 517	Entretien et réparation de véhicules automobiles			502
..	..	..	..	842	911	1 104	1 142	Com. de pièces et access. de véhic. automobiles			503
..	..	..	..	76	61	83	90	Com., entret. et répar. de motocycles et access.			504
..	..	..	..	336	338	420	459	Com. de détail de carburants pour automobiles			505
..	..	..	..	14 264	14 751	18 489	20 445	Com. de gros et act. inter., sauf auto. et moto		51	
..	..	..	..	613	607	658	940	Activités d'intermédiaries de commerce de gros			511
..	..	..	..	2 426	2 319	3 010	3 122	Com. gros: prod. agr. bruts, anim. viv., prod. alim.			512
..	..	..	..	2 843	2 985	3 856	4 449	Com. gros: biens de consom., sauf alimentaires			513
..	..	..	..	3 356	3 287	4 048	4 279	Com. gros: prod. inter. non agr., déchets, débris			514
..	..	..	..	3 504	4 050	5 160	6 178	Com. gros: machines, équipments et fournitures			515
..	..	..	..	1 522	1 503	1 757	1 477	Autres commerces de gros			519
..	..	..	..	16 164	17 342	19 165	20 837	Com. détail, sauf auto, moto; répar. art. pers., dom.		52	
..	..	..	..	7 064	7 494	8 087	9 165	Commerce de détail non spécialisé en magasin			521
..	..	..	..	930	1 167	1 206	1 137	Com. détail: prod. aliment. en mag. spécialisés			522
..	..	..	..	7 226	7 636	8 735	9 285	Autre com. détail: prod. neufs en mag. spécialisés			523
..	..	..	..	84	97	87	102	Commerces de détail de biens d'occasion			524
..	..	..	..	799	812	937	956	Commerces de détail autres qu'en magasin			525
..	..	..	..	61	135	114	193	Réparation d'articles personnels et domestiques			526
..	..	..	..	7 133	7 570	8 493	9 351	Hotels et restaurants	H	55	
..	..	..	..	2 104	2 233	2 364	2 566	Hôtels; camping; autres équip. courts séjours			551
..	..	..	..	5 029	5 337	6 129	6 785	Restaurants, cafés et cantines			552
..	..	..	..	..	18 222	20 758	21 979	Transport, entreposage, communications	I	60_64	
..	..	..	..	..	6 480	7 853	7 980	Transports terrestres; transports par conduites		60	
..	..	..	..	..	538	1 195	1 265	Transports par chemin de fer			601
..	..	..	..	5 565	5 937	6 654	6 710	Autres transports terrestres			602
..	..	..	..	5	5	4	5	Transports par conduites			603
..	..	..	..	..	721	511	493	Transports par eau		61	
..	..	..	..	..	721	511	493	Transports maritimes et côtiers			611
..	..	..	..	..	..	..	..	Transports par voies navigables intérieures			612
..	..	..	..	..	2 006	2 162	2 271	Transports aériens		62	
..	..	..	..	..	1 849	1 945	2 006	Transports aériens réguliers			621
..	..	..	..	..	157	217	265	Transports aériens non réguliers			622
..	..	..	..	..	3 481	4 262	4 764	Act. annex. et aux. des transp.; act. des ag. de voy.		63	
..	..	..	..	4 693	5 534	5 970	6 471	Postes et télécommunications		64	
..	..	..	..	555	575	676	782	Activités de poste et de courrier			641
..	..	..	..	4 138	4 959	5 294	5 689	Télécommunications			642
..	..	..	..	..	..	..	..	Intermédiation financière	J	65_67	
..	..	..	..	..	..	..	..	Interm. financ., sauf assur. et caisses de retraite		65	
..	..	..	..	..	..	..	..	Intermédiation monétaire			651
..	..	..	..	..	..	..	..	Autres intermédiations financières			659
..	..	..	..	..	..	..	..	Act. d'assur., caisses de retr., sauf séc. soc. oblig.		66	
..	..	..	..	..	..	..	..	Act. auxiliaires de l'intermédiation financière		67	
..	..	..	..	..	..	..	..	Act. aux. de l'interm. fin., sauf assur., et retraite			671
..	..	..	..	..	..	..	..	Act. aux. des assurances et caisses de retraite			672
..	..	..	..	31 176	34 826	42 217	51 283	Immobilier, locations, services aux entreprises	K	70_74	
..	..	..	..	2 523	2 570	3 382	3 650	Activités immobilières		70	
..	..	..	..	1 135	1 329	1 887	1 951	Act. immobilières sur biens propres ou loués			701
..	..	..	..	1 388	1 241	1 495	1 699	Activités immobilières pour comptes de tiers			702
..	..	..	..	1 488	1 445	1 807	1 861	Locat. mach., équip. sans opér., biens pers. et dom.		71	
..	..	..	..	397	485	579	564	Location de matériel de transport			711
..	..	..	..	619	634	902	948	Location de machines ét equipements nca			712
..	..	..	..	472	326	326	349	Loc. d'articles personnels et domestiques nca			713
..	..	..	..	3 871	4 508	6 004	8 285	Activités informatiques et activités rattachées		72	
..	..	..	..	81	73	89	130	Conseil en matériel informatique			721
..	..	..	..	2 817	3 239	4 181	5 692	Production de logiciels			722
..	..	..	..	506	698	960	1 311	Traitement de données informatiques			723
..	..	..	..	..	..	..	..	Activités de banques de données			724
..	..	..	..	354	322	375	431	Entr., répar. mach. de bureau et compt. et inform.			725
..	..	..	..	113	176	399	721	Autres activités rattachées à l'informatique			729
..	..	..	..	1 533	1 560	1 683	1 950	Recherche-développement		73	
..	..	..	..	1 533	1 560	1 683	1 950	Rech. et dév. expér. en sc. phys., natur. et ingén.			731
..	..	..	..	..	..	..	..	Rech. et dév. expér. en sciences soc. et humaines			732
..	..	..	..	21 652	24 710	29 289	35 492	Autres activités de services aux entreprises		74	
..	..	..	..	6 231	7 368	8 680	10 467	Act. jur., compt., audit.; cons. fisc.; ét. march., soud.			741
..	..	..	..	4 023	4 818	5 947	6 840	Act. d'achitect., d'ingén.; autres act. techniques			742
..	..	..	..	1 170	1 205	1 605	1 825	Publicité			743
..	..	..	..	10 228	11 319	13 057	16 360	Activités de services aux entreprises nca.			749
..	..	..	..	7 364	7 183	11 993	12 812	Education	M	80	
..	..	..	..	..	..	..	..	Activités d'enseignement primaire			801
..	..	..	..	..	..	..	..	Activités d'enseignement secondaire			802
..	..	..	..	..	..	..	..	Activités d'enseignement supérieur			803
..	..	..	..	..	..	..	..	Act. de formation perman.; autres act. d'enseign.			809
..	..	..	..	3 732	4 204	5 952	6 632	Santé et action sociale	N	85	
..	..	..	..	..	..	..	..	Activités de santé humaine			851
..	..	..	..	..	..	..	..	Activités vétérinaires			852
..	..	..	..	..	..	..	..	Activités d'action sociale			853
..	..	..	..	..	..	..	11 731	Autres activités de serv. coll., soc. et personnels	O	90_93	
..	..	..	..	789	918	1 072	1 087	Assain. et enlèv. des ordures; voierie, act. simil.		90	
..	..	..	..	..	..	..	2 387	Activités associatives diverses		91	
..	..	..	..	..	..	..	219	Activités de syndicats de salariés			912
..	..	..	..	3 937	4 348	5 532	6 036	Activités récréatives, culturelles et sportives		92	
..	..	..	..	847	1 007	1 154	1 241	Act. cinémat., de radio, télé.; autres act. de spect.			921
..	..	..	..	104	194	255	358	Activités d'agences de presse			922
..	..	..	..	232	275	394	431	Act. des biblio., arch., musées, jardins bot. et zool.			923
..	..	..	..	1 572	1 729	2 119	2 456	Activités sportives et autres activités récréatives			924
..	..	..	..	1 729	1 773	1 830	2 221	Autres activités de services		93	

UNITED KINGDOM

Table UK.S8 **SUPPLEMENTS TO WAGES & SALARIES - CHARGES SOCIALES DES EMPLOYEURS**

Millions of £ (Current Prices)

ISIC revision 3				1991	1992	1993	1994	1995	1996	1997	1998
G	**50_52**		Wholesale & retail trade; repairs; etc.	..	..	3 898	4 022	4 164	4 452	5 331	6 062
	50		Sale, maint.,repair vehic/cycle, person., h/hold gd.	..	..	533	556	601	642	752	858
		501	Sale of motor cycle	..	..	377	394	340	357	417	454
		502	Maintenance and repair of motor vehicles	..	..	60	62	121	136	162	201
		503	Sale of motor vehicle parts and accessories	..	..	54	56	97	107	127	150
		504	Sale, maint., repair motor cycle, parts & acces.	..	..	6	7	9	8	10	11
		505	Retail sale of automative fuel	..	..	36	37	34	34	36	42
	51		W/sale. trade & commission exc motor veh. & cycle	..	..	1 822	1 922	1 954	2 049	2 566	2 906
		511	Wholesale on a fee or contract basis	..	..	104	109	118	85	134	116
		512	W/sale agric. raw mat.,live animal,food, bev., tobac.	..	..	293	320	303	305	395	431
		513	Wholesale of household goods	..	..	397	394	380	400	531	632
		514	W/sale. of non-agric. intermed. prod., waste, scrap	..	..	437	477	482	494	567	649
		515	Wholesale of machinery, equipment & supplies	..	..	419	440	470	568	714	872
		519	Other wholesale	..	..	172	182	201	197	225	206
	52		Retail trade exc. motor veh. & cyc.; repair h/hold gds.	..	..	1 543	1 544	1 609	1 761	2 013	2 298
		521	Non-specialised retail trade in stores	..	..	557	600	587	693	753	934
		522	Retail sale food, bev., tobac. in specialised stores	..	..	89	102	88	105	108	109
		523	Other retail trade of new goods specialised stores	..	..	749	718	806	855	1 024	1 103
		524	Retail sale of second-hand goods in stores	..	..	9	10	9	13	10	12
		525	Retail trade not in stores	..	..	137	110	113	82	103	118
		526	Repair of personal and household goods	..	..	2	4	7	14	14	22
H	**55**		Hotels and restaurants	..	..	637	643	609	644	706	795
		551	Hotels; camping ; other short-stay accommodation	..	..	195	190	192	201	205	238
		552	Restaurants, bars and canteens	..	..	442	453	417	443	501	557
I	**60_64**		Transport, storage and communications	..	..	..	..	..	2 316	2 888	2 816
	60		Land transport, transport via pipelines	..	..	..	..	..	742	986	1 024
		601	Transport via railways	..	..	..	..	..	76	146	138
		602	Other land transport	..	..	501	549	641	666	840	886
		603	Transport via pipelines	..	..	..	..	..	..	..	..
	61		Water transport	..	..	..	..	..	83	50	56
		611	Sea and coastal water transport	..	..	..	..	..	83	50	56
		612	Inland water transport	..	..	..	..	..	2	1	..
	62		Air transport	..	..	..	..	..	292	318	337
		621	Scheduled air transport	..	..	..	..	..	264	278	289
		622	Non-scheduled air transport	..	..	..	..	..	28	40	48
	63		Support. & aux. transport act., travel agencies activ.	..	..	..	..	..	430	522	621
	64		Post and telecommunications	..	..	670	706	726	769	1 012	778
		641	Post and courier activities	..	..	53	56	68	58	69	87
		642	Telecommunications	..	..	617	650	658	711	943	691
J	**65_67**		Financial intermediation	..	..	..	..	..	..	..	..
	65		Financial intermediation, exc. insur., pension fund	..	..	..	..	..	..	..	..
		651	Monetary intermediation	..	..	..	..	..	..	..	..
		659	Other financial intermediation	..	..	..	..	..	..	..	..
	66		Insurance pension fund, exc. compul. social sec.	..	..	..	..	..	..	..	..
	67		Activities auxiliary to financial intermediation	..	..	..	..	..	..	..	..
		671	Act. auxil to fin. intermed., exc. insur. & pens.fund	..	..	..	..	..	..	..	..
		672	Act. auxiliary to insurance and pension funding	..	..	..	..	..	..	..	..
K	**70_74**		Real estate, renting and business activities	..	..	..	..	3 801	4 136	5 200	6 542
	70		Real estate activities	..	..	..	..	306	333	461	576
		701	Real estate activ. with own or leased property	..	..	..	..	142	180	266	341
		702	Real estate activities on a fee or contract basis	..	..	..	..	164	153	195	235
	71		Renting of mach. without operator; rent. of pers. gds.	..	..	..	..	178	167	213	235
		711	Renting of transport equipment	..	..	46	49	46	59	71	70
		712	Renting of other machinery and equipment	..	..	..	..	77	74	110	126
		713	Renting of personal and household goods nec	..	..	..	..	40	30	27	35
	72		Computer and related activities	..	..	620	689	567	645	827	1 243
		721	Hardware consultancy	..	..	11	12	10	11	10	20
		722	Software consultancy and supply	..	..	466	519	416	475	575	882
		723	Data processing	..	..	78	87	80	95	149	180
		724	Data base activities	..	..	..	..	..	..	..	..
		725	Mainten.; repair of office, account. & comp. mach.	..	..	48	53	44	39	42	65
		729	Other computer related activities	..	..	17	18	17	25	51	96
	73		Research and development	..	..	136	194	207	255	264	306
		731	R & D experim. on natural sciences & engineering	..	..	136	194	207	255	264	306
		732	R & D experim. on social sciences & humanities	..	..	..	..	..	..	..	..
	74		Other business activities	..	..	2 476	2 756	2 543	2 736	3 435	4 182
		741	Legal, acc., book-keep., audit act.; tax consult., etc	..	..	795	874	748	871	1 063	1 268
		742	Architectural, engineering, & other techn. activities	..	..	594	588	538	559	767	883
		743	Advertising	..	..	143	167	162	168	242	279
		749	Business activities nec	..	..	944	1 127	1 095	1 138	1 363	1 752
M	**80**		Education	..	..	151	178	1 001	889	1 818	1 848
		801	Primary education	..	..	..	..	..	..	..	..
		802	Secondary education	..	..	..	..	..	..	..	..
		803	Higher education	..	..	..	..	..	..	..	..
		809	Adult and other education	..	..	..	..	..	..	..	..
N	**85**		Health and social work	..	..	..	..	338	421	662	612
		851	Human health activities	..	..	..	..	..	..	..	..
		852	Veterinary activities	..	..	..	..	..	..	..	..
		853	Social work activities	..	..	..	..	..	..	..	..
O	**90_93**		Other commun., soc. & pers. services activities	..	..	..	..	..	..	..	..
	90		Sewage & refuse disposal, sanitation, etc.	..	..	82	106	112	135	149	152
	91		Activities of member organisations nec	..	..	..	..	..	..	..	382
		912	Activities of trade unions	..	..	..	..	..	..	..	48
	92		Recreational, cultural and sporting activities	..	..	..	..	455	518	666	727
		921	Motion picture, radio, TV, other entertainment act.	..	..	96	102	89	112	137	178
		922	News agency activities	..	..	14	16	15	30	38	50
		923	Library, archives, museums, other cultural activ.	..	..	..	..	26	35	61	65
		924	Sporting and other recreational activities	..	..	154	187	166	184	216	242
	93		Other service activities	..	..	174	197	181	182	134	240

Table UK.S9 **ÉTABLISSEMENTS - ESTABLISHMENTS**

Unités

1991	1992	1993	1994	1995	1996	1997	1998			CITI révision 3
..	..	..	..	..	520 565	518 930	512 045	Com. de gros, de detail ; réparation etc.	G	**50_52**
..	..	..	..	..	84 235	84 035	83 055	Com., entr., répar. d'auto. et moto ; carburants		**50**
..	..	..	..	..	40 645	38 235	35 090	Commerce de véhicules automobiles		501
..	..	..	..	..	21 660	23 355	25 230	Entretien et réparation de véhicules automobiles		502
..	..	..	..	..	10 475	11 345	12 010	Com. de pièces et access. de véhic. automobiles		503
..	..	..	..	..	1 240	1 360	1 560	Com., entret. et répar. de motocycles et access.		504
..	..	..	..	..	10 215	9 740	9 165	Com. de détail de carburants pour automobiles		505
..	..	..	..	..	132 555	135 775	133 960	Com. de gros et act. inter., sauf auto. et moto		**51**
..	..	..	..	..	13 235	14 860	15 440	Activités d'intermédiaries de commerce de gros		511
..	..	..	..	..	20 260	20 630	20 430	Com. gros: prod. agr. bruts, anim. viv., prod. alim.		512
..	..	..	..	..	30 235	30 800	30 065	Com. gros: biens de consom., sauf alimentaires		513
..	..	..	..	..	22 285	23 200	22 820	Com. gros: prod. inter. non agr., déchets, débris		514
..	..	..	..	..	14 775	17 260	18 585	Com. gros: machines, équipments et fournitures		515
..	..	..	..	..	31 765	29 025	26 620	Autres commerces de gros		519
..	..	..	..	..	303 775	299 120	295 030	Com. détail, sauf auto, moto; répar. art. pers., dom.		**52**
..	..	..	..	..	52 230	51 540	53 285	Commerce de détail non spécialisé en magasin		521
..	..	..	..	..	76 280	70 205	65 085	Com. détail: prod. aliment. en mag. spécialisés		522
..	..	..	..	..	158 820	157 705	155 635	Autre com. détail: prod. neufs en mag. spécialisés		523
..	..	..	..	..	6 210	6 670	6 810	Commerces de détail de biens d'occasion		524
..	..	..	..	..	6 760	8 960	10 255	Commerces de détail autres qu'en magasin		525
..	..	..	..	..	3 475	4 040	3 960	Réparation d'articles personnels et domestiques		526
..	..	..	..	..	134 595	143 070	145 620	Hotels et restaurants	H	**55**
..	..	..	..	..	17 310	16 565	16 525	Hôtels; camping; autres équip. courts séjours		551
..	..	..	..	..	117 285	126 505	129 095	Restaurants, cafés et cantines		552
..	..	..	..	..	92 835	93 825	94 075	Transport, entreposage, communications	I	**60_64**
..	..	..	..	..	50 125	50 225	49 730	Transports terrestres; transports par conduites		**60**
..	..	..	..	..	225	185	515	Transports par chemin de fer		601
..	..	..	..	..	49 870	49 970	49 190	Autres transports terrestres		602
..	..	..	..	..	30	70	25	Transports par conduites		603
..	..	..	..	..	1 300	1 450	1 320	Transports par eau		**61**
..	..	..	..	..	1 015	1 195	1 090	Transports maritimes et côtiers		611
..	..	..	..	..	..	..	..	Transports par voies navigables intérieures		612
..	..	..	..	..	1 405	1 365	1 285	Transports aériens		**62**
..	..	..	..	..	1 170	1 100	935	Transports aériens réguliers		621
..	..	..	..	..	235	265	350	Transports aériens non réguliers		622
..	..	..	..	..	26 410	25 970	25 845	Act. annex. et aux. des transp.; act. des ag. de voy.		**63**
..	..	..	..	..	13 595	14 815	15 895	Postes et télécommunications		**64**
..	..	..	..	..	6 445	8 035	9 080	Activités de poste et de courrier		641
..	..	..	..	..	7 150	6 780	6 815	Télécommunications		642
..	..	..	..	..	..	..	..	Intermédiation financière	J	**65_67**
..	..	..	..	..	..	..	..	Interm. financ., sauf assur. et caisses de retraite		**65**
..	..	..	..	..	..	..	..	Intermédiation monétaire		651
..	..	..	..	..	..	..	..	Autres intermédiations financières		659
..	..	..	..	..	..	..	..	Act. d'assur., caisses de retr., sauf séc. soc. oblig.		**66**
..	..	..	..	..	..	..	..	Act. auxiliaires de l'intermédiation financière		**67**
..	..	..	..	..	..	..	..	Act. aux. de l'interm. fin., sauf assur., et retraite		671
..	..	..	..	..	..	..	..	Act. aux. des assurances et caisses de retraite		672
..	..	..	..	..	371 820	406 965	437 830	Immobilier, locations, services aux entreprises	K	**70_74**
..	..	..	..	..	47 785	51 485	53 615	Activités immobilières		**70**
..	..	..	..	..	32 220	34 695	36 680	Act. immobilières sur biens propres ou loués		701
..	..	..	..	..	15 565	16 790	16 935	Activités immobilières pour comptes de tiers		702
..	..	..	..	..	13 770	16 845	17 665	Locat. mach., équip. sans opér., biens pers. et dom.		**71**
..	..	..	..	..	3 070	3 830	4 460	Location de matériel de transport		711
..	..	..	..	..	7 550	8 530	8 960	Location de machines ét equipements nca		712
..	..	..	..	..	3 150	4 485	4 245	Loc. d'articles personnels et domestiques nca		713
..	..	..	..	..	59 395	70 975	88 460	Activités informatiques et activités rattachées		**72**
..	..	..	..	..	2 350	3 335	4 735	Conseil en matériel informatique		721
..	..	..	..	..	50 030	51 515	54 515	Production de logiciels		722
..	..	..	..	..	1 365	1 270	1 315	Traitement de données informatiques		723
..	..	..	..	..	215	380	460	Activités de banques de données		724
..	..	..	..	..	1 170	1 450	1 515	Entr., répar. mach. de bureau et compt. et inform.		725
..	..	..	..	..	4 265	13 025	25 920	Autres activités rattachées à l'informatique		729
..	..	..	..	..	2 840	2 945	3 080	Recherche-développement		**73**
..	..	..	..	..	2 645	2 720	2 820	Rech. et dév. expér. en sc. phys., natur. et ingén.		731
..	..	..	..	..	195	225	260	Rech. et dév. expér. en sciences soc. et humaines		732
..	..	..	..	..	248 030	264 715	275 010	Autres activités de services aux entreprises		**74**
..	..	..	..	..	81 915	88 500	95 260	Act. jur., compt., audit.; cons. fisc.; ét. march., soud.		741
..	..	..	..	..	51 950	51 530	52 245	Act. d'achitect., d'ingén.; autres act. techniques		742
..	..	..	..	..	11 400	11 170	10 940	Publicité		743
..	..	..	..	..	102 765	113 515	116 565	Activités de services aux entreprises nca.		749
..	..	..	..	..	35 220	38 075	36 125	Education	M	**80**
..	..	..	..	..	21 750	22 585	19 650	Activités d'enseignement primaire		801
..	..	..	..	..	4 685	4 830	4 805	Activités d'enseignement secondaire		802
..	..	..	..	..	1 730	1 885	1 915	Activités d'enseignement supérieur		803
..	..	..	..	..	7 055	8 775	9 755	Act. de formation perman.; autres act. d'enseign.		809
..	..	..	..	..	31 585	34 035	34 185	Santé et action sociale	N	**85**
..	..	..	..	..	5 290	6 210	5 595	Activités de santé humaine		851
..	..	..	..	..	3 160	3 315	3 225	Activités vétérinaires		852
..	..	..	..	..	23 135	24 510	25 365	Activités d'action sociale		853
..	..	..	..	..	168 540	168 230	173 305	Autres activités de serv. coll., soc. et personnels	O	**90_93**
..	..	..	..	..	3 235	3 880	4 470	Assain. et enlèv. des ordures; voierie, act. simil.		**90**
..	..	..	..	..	14 675	16 005	16 750	Activités associatives diverses		**91**
..	..	..	..	..	520	555	550	Activités de syndicats de salariés		912
..	..	..	..	..	74 070	73 765	77 585	Activités récréatives, culturelles et sportives		**92**
..	..	..	..	..	37 060	36 775	37 415	Act. cinémat., de radio, télé.; autres act. de spect.		921
..	..	..	..	..	420	480	555	Activités d'agences de presse		922
..	..	..	..	..	5 195	5 765	6 125	Act. des biblio., arch., musées, jardins bot. et zool.		923
..	..	..	..	..	29 700	28 445	30 865	Activités sportives et autres activités récréatives		924
..	..	..	..	..	76 560	74 580	74 500	Autres activités de services		**93**

Table UK.S10 **ENTERPRISES - ENTREPRISES**

Units

ISIC revision 3				1991	1992	1993	1994	1995	1996	1997	1998
G	50_52		Wholesale & retail trade; repairs; etc.	..	..	397 203	392 766	398 275	396 035	410 572	408 206
	50		Sale, maint.,repair vehic/cycle, person., h/hold gd.	..	..	71 679	69 148	70 278	71 119	72 527	72 105
		501	Sale of motor cycle	..	..	39 957	38 693	39 473	36 383	34 664	31 196
		502	Maintenance and repair of motor vehicles	..	..	17 844	17 280	17 363	19 709	22 027	24 680
		503	Sale of motor vehicle parts and accessories	..	..	4 844	4 690	4 801	5 829	6 583	7 187
		504	Sale, maint., repair motor cycle, parts & acces.	..	..	1 249	1 209	1 276	1 231	1 552	1 782
		505	Retail sale of automative fuel	..	..	7 785	7 276	7 365	7 967	7 701	7 260
	51		W/sale. trade & commission exc motor veh. & cycle	..	..	123 683	127 055	118 611	113 854	119 712	120 101
		511	Wholesale on a fee or contract basis	..	..	12 208	12 412	11 928	13 283	14 929	15 734
		512	W/sale agric. raw mat.,live animal,food, bev., tobac.	..	..	18 130	18 964	17 449	17 342	18 143	18 167
		513	Wholesale of household goods	..	..	28 561	29 035	27 546	26 286	27 527	27 344
		514	W/sale. of non-agric. intermed. prod., waste, scrap	..	..	18 362	19 460	16 348	15 876	17 702	17 657
		515	Wholesale of machinery, equipment & supplies	..	..	12 592	12 791	12 285	12 158	14 313	15 124
		519	Other wholesale	..	..	33 830	34 393	33 055	28 909	27 098	26 075
	52		Retail trade exc. motor veh. & cyc.; repair h/hold gds.	..	..	201 841	196 563	209 386	211 062	218 333	216 000
		521	Non-specialised retail trade in stores	..	..	23 700	21 146	40 996	38 303	39 427	38 360
		522	Retail sale food, bev., tobac. in specialised stores	..	..	47 098	45 971	38 800	59 476	54 999	50 435
		523	Other retail trade of new goods specialised stores	..	..	93 675	89 085	93 775	99 545	105 342	106 230
		524	Retail sale of second-hand goods in stores	..	..	5 013	5 870	4 025	5 248	6 187	6 105
		525	Retail trade not in stores	..	..	31 683	33 169	30 217	6 446	8 773	10 373
		526	Repair of personal and household goods	..	..	672	1 322	1 575	2 045	3 605	4 497
H	55		Hotels and restaurants	..	..	110 471	108 459	112 398	110 583	110 245	111 815
		551	Hotels; camping ; other short-stay accommodation	..	..	14 665	15 272	15 628	15 257	14 844	14 809
		552	Restaurants, bars and canteens	..	..	95 806	93 187	96 770	95 326	95 401	97 006
I	60_64		Transport, storage and communications	..	..	..	..	..	..	74 726	76 061
	60		Land transport, transport via pipelines	..	..	..	..	..	47 935	47 396	47 130
		601	Transport via railways	..	..	..	..	..	107	116	121
		602	Other land transport	..	..	40 901	39 923	47 452	47 816	47 265	46 995
		603	Transport via pipelines	..	..	12	13	16	12	15	14
	61		Water transport	..	..	..	..	..	1 216	1 251	1 240
		611	Sea and coastal water transport	..	..	..	..	..	1 216	1 251	1 240
		612	Inland water transport	..	..	..	..	..	..	..	..
	62		Air transport	..	..	..	..	..	1 101	985	956
		621	Scheduled air transport	..	..	..	..	..	887	706	631
		622	Non-scheduled air transport	..	..	..	..	..	214	279	325
	63		Support. & aux. transport act., travel agencies activ.	..	..	..	..	..	17 376	17 310	16 913
	64		Post and telecommunications	..	..	..	..	..	..	7 784	9 822
		641	Post and courier activities	..	..	1 023	994	1 113	2 272	3 678	4 868
		642	Telecommunications	..	..	..	..	..	..	4 106	4 954
J	65_67		Financial intermediation	..	..	..	..	..	..	..	..
	65		Financial intermediation, exc. insur., pension fund	..	..	..	..	..	..	..	..
		651	Monetary intermediation	..	..	..	..	..	..	..	..
		659	Other financial intermediation	..	..	..	..	..	..	..	..
	66		Insurance pension fund, exc. compul. social sec.	..	..	..	..	..	..	..	..
	67		Activities auxiliary to financial intermediation	..	..	..	..	..	..	..	..
		671	Act. auxil to fin. intermed., exc. insur. & pens.fund	..	..	..	..	..	..	..	..
		672	Act. auxiliary to insurance and pension funding	..	..	..	..	..	..	..	..
K	70_74		Real estate, renting and business activities	..	..	..	..	341 376	..	403 014	433 500
	70		Real estate activities	..	..	..	..	56 516	56 914	59 027	59 576
		701	Real estate activ. with own or leased property	..	..	..	..	43 084	43 099	42 442	43 409
		702	Real estate activities on a fee or contract basis	..	..	..	..	13 432	13 815	16 585	16 167
	71		Renting of mach. without operator; rent. of pers. gds.	..	..	..	..	9 737	10 633	12 055	13 185
		711	Renting of transport equipment	..	..	1 865	1 828	2 117	2 415	3 162	3 641
		712	Renting of other machinery and equipment	..	..	..	..	5 053	5 859	6 275	6 915
		713	Renting of personal and household goods nec	..	..	..	..	1 864	1 702	1 930	1 994
	72		Computer and related activities	..	..	42 477	48 360	53 673	..	84 959	104 355
		721	Hardware consultancy	..	..	1 569	1 786	2 069	3 123	5 449	6 827
		722	Software consultancy and supply	..	..	37 672	42 890	47 406	51 195	54 607	58 157
		723	Data processing	..	..	537	611	675	1 486	2 893	2 714
		724	Data base activities	..	..	..	..	..	..	..	..
		725	Mainten.; repair of office, account. & comp. mach.	..	..	1 671	1 903	2 222	..	1 672	1 695
		729	Other computer related activities	..	..	1 028	1 170	1 301	8 180	20 338	34 962
	73		Research and development	..	..	2 186	2 330	2 424	2 469	2 627	2 716
		731	R & D experim. on natural sciences & engineering	..	..	2 186	2 330	2 424	2 469	2 627	2 716
		732	R & D experim. on social sciences & humanities	..	..	..	..	..	..	..	..
	74		Other business activities	..	..	197 767	198 016	219 026	228 450	244 346	253 668
		741	Legal, acc., book-keep., audit act.; tax consult., etc	..	..	64 871	65 367	73 659	79 381	88 164	94 883
		742	Architectural, engineering, & other techn. activities	..	..	49 809	43 941	49 624	52 598	53 773	54 300
		743	Advertising	..	..	9 419	10 290	11 058	10 274	11 189	11 345
		749	Business activities nec	..	..	73 668	78 418	84 685	86 197	91 220	93 140
M	80		Education	..	..	..	..	13 437	14 460	13 396	16 615
		801	Primary education	..	..	..	..	..	..	..	..
		802	Secondary education	..	..	..	..	..	..	..	..
		803	Higher education	..	..	..	..	..	..	..	..
		809	Adult and other education	..	..	..	..	..	..	..	..
N	85		Health and social work	..	..	..	..	34 791	31 592	33 559	38 165
		851	Human health activities	..	..	..	..	..	..	..	..
		852	Veterinary activities	..	..	..	..	..	..	..	..
		853	Social work activities	..	..	..	..	..	..	..	..
O	90_93		Other commun., soc. & pers. services activities	..	..	..	..	..	..	..	..
	90		Sewage & refuse disposal, sanitation, etc.	..	..	948	1 069	1 309	1 474	1 686	1 864
	91		Activities of member organisations nec	..	..	..	..	..	..	..	21 800
		912	Activities of trade unions	..	..	..	..	..	..	..	586
	92		Recreational, cultural and sporting activities	..	..	..	..	63 236	62 450	63 674	65 261
		921	Motion picture, radio, TV, other entertainment act.	..	..	30 148	31 824	35 456	36 910	37 468	38 288
		922	News agency activities	..	..	279	298	331	385	468	553
		923	Library, archives, museums, other cultural activ.	..	..	..	..	1 281	1 304	1 306	1 291
		924	Sporting and other recreational activities	..	..	18 160	18 882	25 249	22 230	22 129	22 414
	93		Other service activities	..	..	72 000	75 441	88 390	85 000	80 437	79 121

Table UK.S11 EXPORTATIONS - EXPORTS

Millions de £ (Prix courants)

1991	1992	1993	1994	1995	1996	1997	1998			CITI révision 3	
..	..	..	..	..	..	..	..	Com. de gros, de detail ; réparation etc.	G	50_52	
..	..	..	..	..	..	..	..	Com., entr., répar. d'auto. et moto ; carburants		50	
..	..	..	..	..	..	..	..	Commerce de véhicules automobiles			501
..	..	..	..	..	..	..	..	Entretien et réparation de véhicules automobiles			502
..	..	..	..	..	..	..	..	Com. de pièces et access. de véhic. automobiles			503
..	..	..	..	..	..	..	..	Com., entret. et répar. de motocycles et access.			504
..	..	..	..	..	..	..	..	Com. de détail de carburants pour automobiles			505
..	..	..	..	..	..	..	..	Com. de gros et act. inter., sauf auto. et moto		51	
..	..	..	..	..	..	..	..	Activités d'intermédiaries de commerce de gros			511
..	..	..	..	..	..	..	..	Com. gros: prod. agr. bruts, anim. viv., prod. alim.			512
..	..	..	..	..	..	..	..	Com. gros: biens de consom., sauf alimentaires			513
..	..	..	..	..	..	..	..	Com. gros: prod. inter. non agr., déchets, débris			514
..	..	..	..	..	..	..	..	Com. gros: machines, équipements et fournitures			515
..	..	..	..	..	..	..	..	Autres commerces de gros			519
..	..	..	..	..	..	..	..	Com. détail, sauf auto, moto; répar. art. pers., dom.		52	
..	..	..	..	..	..	..	..	Commerce de détail non spécialisé en magasin			521
..	..	..	..	..	..	..	..	Com. détail: prod. aliment. en mag. spécialisés			522
..	..	..	..	..	..	..	..	Autre com. détail: prod. neufs en mag. spécialisés			523
..	..	..	..	..	..	..	..	Commerces de détail de biens d'occasion			524
..	..	..	..	..	..	..	..	Commerces de détail autres qu'en magasin			525
..	..	..	..	..	..	..	..	Réparation d'articles personnels et domestiques			526
..	..	..	..	..	..	..	..	Hotels et restaurants	H	55	
..	..	..	..	..	..	..	..	Hôtels; camping; autres équip. courts séjours			551
..	..	..	..	..	..	..	..	Restaurants, cafés et cantines			552
..	..	..	..	..	..	..	..	Transport, entreposage, communications	I	60_64	
..	..	..	..	..	..	..	..	Transports terrestres; transports par conduites		60	
..	..	..	..	..	..	..	..	Transports par chemin de fer			601
..	..	..	..	..	..	..	..	Autres transports terrestres			602
..	..	..	..	..	..	..	..	Transports par conduites			603
..	..	..	..	..	..	..	..	Transports par eau		61	
..	..	..	..	..	..	..	..	Transports maritimes et côtiers			611
..	..	..	..	..	..	..	..	Transports par voies navigables intérieures			612
..	..	..	..	..	..	..	..	Transports aériens		62	
..	..	..	..	..	..	..	..	Transports aériens réguliers			621
..	..	..	..	..	..	..	..	Transports aériens non réguliers			622
..	..	..	..	..	..	..	..	Act. annex. et aux. des transp.; act. des ag. de voy.		63	
..	..	..	..	..	..	..	..	Postes et télécommunications		64	
..	..	..	..	..	..	..	..	Activités de poste et de courrier			641
..	..	..	..	..	..	..	..	Télécommunications			642
..	..	..	..	..	..	..	..	Intermédiation financière	J	65_67	
..	..	..	..	..	..	..	..	Interm. financ., sauf assur. et caisses de retraite		65	
..	..	..	..	..	..	..	..	Intermédiation monétaire			651
..	..	..	..	..	..	..	..	Autres intermédiations financières			659
..	..	..	..	..	..	..	..	Act. d'assur., caisses de retr., sauf séc. soc. oblig.		66	
..	..	..	..	..	..	..	..	Act. auxiliaires de l'intermédiation financière		67	
..	..	..	..	..	..	..	..	Act. aux. de l'interm. fin., sauf assur., et retraite			671
..	..	..	..	..	..	..	..	Act. aux. des assurances et caisses de retraite			672
..	..	..	..	..	..	..	..	Immobilier, locations, services aux entreprises	K	70_74	
..	..	..	..	..	..	..	..	Activités immobilières		70	
..	..	..	..	..	..	..	..	Act. immobilières sur biens propres ou loués			701
..	..	..	..	..	..	..	..	Activités immobilières pour comptes de tiers			702
..	..	..	..	..	..	..	..	Locat. mach., équip. sans opér., biens pers. et dom.		71	
..	..	..	..	..	..	..	..	Location de matériel de transport			711
..	..	..	..	..	..	..	..	Location de machines ét equipements nca			712
..	..	..	..	..	..	..	..	Loc. d'articles personnels et domestiques nca			713
340	349	436	555	662	460	406	359	Activités informatiques et activités rattachées		72	
..	..	..	..	..	..	..	..	Conseil en matériel informatique			721
340	349	436	555	662	460	406	359	Production de logiciels			722
..	..	..	..	..	..	..	..	Traitement de données informatiques			723
..	..	..	..	..	..	..	..	Activités de banques de données			724
..	..	..	..	..	..	..	..	Entr., répar. mach. de bureau et compt. et inform.			725
..	..	..	..	..	..	..	..	Autres activités rattachées à l'informatique			729
..	..	..	..	..	..	..	..	Recherche-développement		73	
..	..	..	..	..	..	..	..	Rech. et dév. expér. en sc. phys., natur. et ingén.			731
..	..	..	..	..	..	..	..	Rech. et dév. expér. en sciences soc. et humaines			732
36	68	38	47	48	54	43	44	Autres activités de services aux entreprises		74	
..	..	..	..	..	..	..	..	Act. jur., compt., audit.; cons. fisc.; ét. march., soud.			741
7	32	8	7	11	15	5	9	Act. d'achitect., d'ingén.; autres act. techniques			742
..	..	..	..	..	..	..	..	Publicité			743
29	36	30	39	37	39	38	34	Activités de services aux entreprises nca.			749
..	..	..	..	..	..	..	..	Education	M	80	
..	..	..	..	..	..	..	..	Activités d'enseignement primaire			801
..	..	..	..	..	..	..	..	Activités d'enseignement secondaire			802
..	..	..	..	..	..	..	..	Activités d'enseignement supérieur			803
..	..	..	..	..	..	..	..	Act. de formation perman.; autres act. d'enseign.			809
..	..	..	..	..	..	..	..	Santé et action sociale	N	85	
..	..	..	..	..	..	..	..	Activités de santé humaine			851
..	..	..	..	..	..	..	..	Activités vétérinaires			852
..	..	..	..	..	..	..	..	Activités d'action sociale			853
1 531	1 549	1 254	1 328	1 331	1 486	1 736	1 787	Autres activités de serv. coll., soc. et personnels	O	90_93	
..	..	..	..	..	..	..	..	Assain. et enlèv. des ordures; voierie, act. simil.		90	
..	..	..	..	..	..	..	..	Activités associatives diverses		91	
..	..	..	..	..	..	..	..	Activités de syndicats de salariés			912
34	33	33	35	27	291	309	326	Activités récréatives, culturelles et sportives		92	
34	33	33	35	27	291	309	326	Act. cinémat., de radio, télé.; autres act. de spect.			921
..	..	..	..	..	..	..	..	Activités d'agences de presse			922
..	..	..	..	..	..	..	..	Act. des biblio., arch., musées, jardins bot. et zool.			923
..	..	..	..	..	..	..	..	Activités sportives et autres activités récréatives			924
1 498	1 516	1 221	1 293	1 304	1 196	1 428	1 461	Autres activités de services		93	

Table UK.S12 **IMPORTS - IMPORTATIONS**

Millions of £ (Current Prices)

ISIC revision 3				1991	1992	1993	1994	1995	1996	1997	1998
G	50_52		Wholesale & retail trade; repairs; etc.	..	..	..	..	..	..	..	..
	50		Sale, maint.,repair vehic/cycle, person., h/hold gd.	..	..	..	..	..	..	..	..
		501	Sale of motor cycle	..	..	..	..	..	..	..	..
		502	Maintenance and repair of motor vehicles	..	..	..	..	..	..	..	..
		503	Sale of motor vehicle parts and accessories	..	..	..	..	..	..	..	..
		504	Sale, maint., repair motor cycle, parts & acces.	..	..	..	..	..	..	..	..
		505	Retail sale of automative fuel	..	..	..	..	..	..	..	..
	51		W/sale. trade & commission exc motor veh. & cycle	..	..	..	..	..	..	..	..
		511	Wholesale on a fee or contract basis	..	..	..	..	..	..	..	..
		512	W/sale agric. raw mat.,live animal,food, bev., tobac.	..	..	..	..	..	..	..	..
		513	Wholesale of household goods	..	..	..	..	..	..	..	..
		514	W/sale. of non-agric. intermed. prod., waste, scrap	..	..	..	..	..	..	..	..
		515	Wholesale of machinery, equipment & supplies	..	..	..	..	..	..	..	..
		519	Other wholesale	..	..	..	..	..	..	..	..
	52		Retail trade exc. motor veh. & cyc.; repair h/hold gds.	..	..	..	..	..	..	..	..
		521	Non-specialised retail trade in stores	..	..	..	..	..	..	..	..
		522	Retail sale food, bev., tobac. in specialised stores	..	..	..	..	..	..	..	..
		523	Other retail trade of new goods specialised stores	..	..	..	..	..	..	..	..
		524	Retail sale of second-hand goods in stores	..	..	..	..	..	..	..	..
		525	Retail trade not in stores	..	..	..	..	..	..	..	..
		526	Repair of personal and household goods	..	..	..	..	..	..	..	..
H	55		Hotels and restaurants	..	..	..	..	..	..	..	..
		551	Hotels; camping ; other short-stay accommodation	..	..	..	..	..	..	..	..
		552	Restaurants, bars and canteens	..	..	..	..	..	..	..	..
I	60_64		Transport, storage and communications	..	..	..	..	..	..	..	..
	60		Land transport, transport via pipelines	..	..	..	..	..	..	..	..
		601	Transport via railways	..	..	..	..	..	..	..	..
		602	Other land transport	..	..	..	..	..	..	..	..
		603	Transport via pipelines	..	..	..	..	..	..	..	..
	61		Water transport	..	..	..	..	..	..	..	..
		611	Sea and coastal water transport	..	..	..	..	..	..	..	..
		612	Inland water transport	..	..	..	..	..	..	..	..
	62		Air transport	..	..	..	..	..	..	..	..
		621	Scheduled air transport	..	..	..	..	..	..	..	..
		622	Non-scheduled air transport	..	..	..	..	..	..	..	..
	63		Support. & aux. transport act., travel agencies activ.	..	..	..	..	..	..	..	..
	64		Post and telecommunications	..	..	..	..	..	..	..	..
		641	Post and courier activities	..	..	..	..	..	..	..	..
		642	Telecommunications	..	..	..	..	..	..	..	..
J	65_67		Financial intermediation	..	..	..	..	..	..	..	..
	65		Financial intermediation, exc. insur., pension fund	..	..	..	..	..	..	..	..
		651	Monetary intermediation	..	..	..	..	..	..	..	..
		659	Other financial intermediation	..	..	..	..	..	..	..	..
	66		Insurance pension fund, exc. compul. social sec.	..	..	..	..	..	..	..	..
	67		Activities auxiliary to financial intermediation	..	..	..	..	..	..	..	..
		671	Act. auxil to fin. intermed., exc. insur. & pens.fund	..	..	..	..	..	..	..	..
		672	Act. auxiliary to insurance and pension funding	..	..	..	..	..	..	..	..
K	70_74		Real estate, renting and business activities	..	..	..	..	..	..	..	..
	70		Real estate activities	..	..	..	..	..	..	..	..
		701	Real estate activ. with own or leased property	..	..	..	..	..	..	..	..
		702	Real estate activities on a fee or contract basis	..	..	..	..	..	..	..	..
	71		Renting of mach. without operator; rent. of pers. gds.	..	..	..	..	..	..	..	..
		711	Renting of transport equipment	..	..	..	..	..	..	..	..
		712	Renting of other machinery and equipment	..	..	..	..	..	..	..	..
		713	Renting of personal and household goods nec	..	..	..	..	..	..	..	..
	72		Computer and related activities	414	540	595	721	754	545	550	601
		721	Hardware consultancy	..	..	..	..	..	..	..	..
		722	Software consultancy and supply	414	540	595	721	754	545	550	601
		723	Data processing	..	..	..	..	..	..	..	..
		724	Data base activities	..	..	..	..	..	..	..	..
		725	Mainten.; repair of office, account. & comp. mach.	..	..	..	..	..	..	..	..
		729	Other computer related activities	..	..	..	..	..	..	..	..
	73		Research and development	..	..	..	..	..	..	..	..
		731	R & D experim. on natural sciences & engineering	..	..	..	..	..	..	..	..
		732	R & D experim. on social sciences & humanities	..	..	..	..	..	..	..	..
	74		Other business activities	22	16	16	19	26	25	19	17
		741	Legal, acc., book-keep., audit act.; tax consult., etc	..	..	..	..	..	..	..	..
		742	Architectural, engineering, & other techn. activities	3	2	1	2	3	5	5	3
		743	Advertising	..	..	..	..	..	..	..	..
		749	Business activities nec	18	15	14	18	22	20	15	14
M	80		Education	..	..	..	..	..	..	..	..
		801	Primary education	..	..	..	..	..	..	..	..
		802	Secondary education	..	..	..	..	..	..	..	..
		803	Higher education	..	..	..	..	..	..	..	..
		809	Adult and other education	..	..	..	..	..	..	..	..
N	85		Health and social work	..	..	..	..	..	..	..	..
		851	Human health activities	..	..	..	..	..	..	..	..
		852	Veterinary activities	..	..	..	..	..	..	..	..
		853	Social work activities	..	..	..	..	..	..	..	..
O	90_93		Other commun., soc. & pers. services activities	1 236	1 398	1 099	1 305	1 043	984	1 274	1 860
	90		Sewage & refuse disposal, sanitation, etc.	..	..	..	..	..	..	..	..
	91		Activities of member organisations nec	..	..	..	..	..	..	..	..
		912	Activities of trade unions	..	..	..	..	..	..	..	..
	92		Recreational, cultural and sporting activities	10	8	6	7	5	188	252	353
		921	Motion picture, radio, TV, other entertainment act.	10	8	6	7	5	188	252	353
		922	News agency activities	..	..	..	..	..	..	..	..
		923	Library, archives, museums, other cultural activ.	..	..	..	..	..	..	..	..
		924	Sporting and other recreational activities	..	..	..	..	..	..	..	..
	93		Other service activities	1 225	1 390	1 093	1 297	1 038	796	1 021	1 507

PART D

PARTIE D

METHODOLOGICAL INFORMATION

INFORMATIONS METHODOLOGIQUES

Section I

Sources and Definitions
Sources et définitions

Table of contents - Table des matières

AUSTRALIA

The information is provided by the Australian Bureau of Statistics (ABS).

Industrial Classification

The Australia and New Zealand Standard Industrial Classification (ANZSIC) is used. There is a correspondence between ANZSIC and ISIC Revision 3 at division level. ANZSIC subdivisions, groups and classes are often combinations of ISIC sectors.

INDUSTRIAL STATISTICS

Source

Most data are derived from sample surveys: Annual Manufacturing Survey, Annual survey of Other Mining, Services to Mining and Water Supply industry. Both enterprises ("management units") and establishments are approached depending of the information requested. Surveys do not cover non-employing businesses. The period covered is the fiscal year starting on 1 July.

Concepts and Definitions of Variables

All variables, except employment, collected in the annual manufacturing survey and the annual mining and utilities collections have been designed to meet as nearly as practicable the requirements of the concepts of the 1993 version of the System of National Accounts (SNA 93).

Production includes the values of shipments of products or services produced by the unit, the receipts for industrial work done or industrial services rendered to others, the sales of goods shipped in the same condition as received (resale), the net change in stocks of finished products manufactured by the unit, in stocks of work in progress manufactured by the unit, and in stocks of goods purchased for resale in the same condition as received, the value of electricity sold, the sales of products manufactured by subcontractors where production is carried out on materials owned by the establishment and the value of capital assets manufactured for own use, advertising income.

Value added corresponds to industry gross product. It is defined as output minus the value of industrial materials and services for processing, the value of fuels and purchased electricity, the costs of industrial services received, the payments for contract and commission work and the payments for repair and maintenance work. Value Added is valued at basic prices.

Employment refers to the number of persons employed (including working proprietors and working partners of unincorporated businesses) at 30 June of each year. It includes all employees whether full time, part time, permanent, temporary or casual. It includes employees on paid leave at 30 June of each year. It excludes home-workers, contractors, consultants, other self-employed persons providing services, unpaid family workers and volunteers.

Wages and salaries, employees are defined as gross wages and salaries paid to employees for the year ended 30 June. They include wages, salaries, bonuses, commissions, overtime payments, penalty rates, leave payments and provisions for employee entitlements. They also include salaries and fees of directors and executives. They exclude payments to contractors, consultants and other self employed persons and the drawings of proprietors and partners of unincorporated businesses. They also exclude payroll-related taxes and reimbursements to employees for expenses incurred. Wages and salaries refer to gross earnings before taxation and other deductions.

Numbers of *establishments* are at 30 June of the year indicated. An establishment is the smallest accounting unit of a business, within a single Australian State or Territory, which controls its productive activities and maintains a specified range of detailed data.

Publications

i) *Manufacturing Industry,* Australia, (Preliminary) (ABS Catalogue No. 8201.0);
ii) *Manufacturing Industry,* Australia (ABS Catalogue No. 8221.0);
iii) *Manufacturing,* Australia (ABS Catalogue No. 8225.0);
iv) *Mining, Electricity and Gas Operations* (Preliminary) (ABS Catalogue No. 8401.0);
v) *Australian Mining Industry* (ABS Catalogue No. 8414.0)
vi) *Mining Operations,* Australia (ABS Catalogue No. 8415.0);
vii) *Electricity, Gas, Water and Sewerage Industries,* Australia (ABS Catalogue No. 8208.0);
viii) *Electricity, Gas, Water and Sewerage Operations,* Australia (ABS Catalogue No. 8226.0);

All publications are distributed by the ABS.

AUSTRIA

Data are supplied by Statistics Austria.

Industrial Classification

From 1995, the classification used is the Austrian version of NACE Revision 1, so called "ÖNACE 1995". Up to the 4-digit level there is a complete correspondence between ÖNACE 1995 and NACE Revision 1, the 4-digit level being further subdivided into sub-classes. The ÖNACE 1995 is in line with the ISIC Revision 3.

INDUSTRIAL STATISTICS

Source

As from 1995, data are not strictly comparable which the years before. Data for 1995 are the results of the Non-Agricultural Census. No survey was run for 1996 data. Data for 1997 and 1998 are the results of the annual Structural Censuses. Information on hours worked derives from the monthly survey "*Konjunkturstatistik - Produzierender Bereich*" (Statistics on short-term data - Manufacturing industry - NACE Sections C to F).

The statistical unit is the enterprise.

Concepts and Definitions of Variables

Production corresponds to turnover, including revenues out of capitalisation of services rendered for own account, excluding purchase of goods and services for resale in the same condition as received (without foreign freights). It includes net change in stocks of commercial articles, finished products and work in progress.

Value added at factor cost is equal to value added at basic prices, excluding taxes and duties (without taxes of products), including subsidies (without subsidies on products). The value added at basic prices is calculated as the production value less foreign freights, purchase of material for working and processing, expenditure for repairs and maintenance done by other enterprises, expenditure for working done by other enterprises, payments for agency workers, purchase of fuels, expenditure for rents, expenditure for operational leasing of goods, other operating costs, expenditure for low-valued assets, taxes of products, including net change in stocks of fuels, raw materials and consumables and subsidies on products.

Investment (capital formation) is investment expenditure, defined as the value of purchases of fixed assets, including own-account construction of assets. The value of sales of fixed assets is not deducted. Investments in land, in construction and alteration of buildings, in machinery and equipment, in tangible goods, in concessions, patents, licences, trade marks and in software are included. The valuation is at full cost including delivered price, cost of installation, fees and taxes (except turnover tax).

Investment in machinery and equipment is given in terms of investment expenditure defined as for total investment i.e. the value of the purchase of fixed assets plus own-account construction of assets. Included are transport equipment, industrial machinery and equipment, office machinery, equipment and furniture, professional instruments and equipment, alterations, improvement and repairs.

Employment data are given for employees and for females employed (on a fixed day: 31 December). *Employment, employees* include salaried employees, operatives, apprentices, homeworkers and part-time employees. The number of women employees is given under *Employment, females.*

Compensation of labour, employees covers wages and salaries and employers' contributions to social security schemes and pension funds.

Wages and salaries, employees include the gross remuneration of salaried employees (white- and blue-collar workers, apprentices, homeworkers and part-time employees). Data include regular and overtime payments, bonuses, cost-of-living allowances, vacation and sick leave pay, taxes and social insurance contributions payable by employees and deducted by employers, payments in kind and other allowances.

Employers' social costs cover legally required and voluntary contributions, such as social security contributions (health, old-age, accident, unemployment insurance), contributions to the equalisation fund for child allowance, contributions to the residential construction fund and lodging allowance, and non-obligatory payments (e.g. in case of accident, disease or distress; canteen services; mileage allowance).

Hours worked by employees relate to actual hours worked (i.e. all paid hours less paid holiday leave) by salaried employees, operatives and apprentices, according to the wage list. Included are overtime, waiting time, rest periods, sick leave and public holidays. Homeworkers and part-time workers are excluded.

The *enterprise* is generally the smallest legal unit that may consist of one or more establishments (local kind of activity unit).

Publications

All data to the 3-digit ISIC level are presented in the following publications:

i) *Industriestatistik*, 2. Teil, published by the Austrian Central Statistical Office in the framework of the publication series contribution for Austrian Statistics.

ii) *Gewerbestatistik*, 2. Teil, published by the Austrian *Central* Statistical Office in the framework of the publication series contribution for Austrian Statistics.

CZECH REPUBLIC

Data are supplied by the Czech Statistical Office (•eský Statistický Ú•ad).

Industrial Classification

The branch classification of economic activities (Odv•tvová Klasifikace Ekonomických •inností - OKE• or CZ-NACE -) was introduced in 1992 and updated in 1993. OKE• uses 5-digit codes; they are taken over from NACE Revision 1 down to the 4th digit, the 5th digit is used for national specifics. OKE• is identical to ISIC Revision 3 down to the 2nd digit.

INDUSTRY AND SERVICE STATISTICS

Source

Data are derived from an sample survey, the annual structural business survey. Random sampling is used, stratified by size of enterprise and 2-digit code of the classification. The survey covers all enterprises, irrespective of their size, which were active at least part of the year. For the branch of financial intermediation, four annual and three quarterly exhaustive surveys are conducted. The reference period is the calendar year. The statistical unit is the enterprise.

Concepts and Definitions of Variables

Production (gross output) is the sum of market and non-market outputs produced by domestic producers over a given period. It is equal to the sum of revenues from sales of own products, goods, works and services provided to third parts, capitalisation of products, goods, works and services provided for own use, value of fixed assets produced during the period by the unit for its own use and changes in stocks of own production and work in progress. Costs incurred on sold goods, subsidies and VAT are excluded. Other tax payable as a consequence of production and transport charges invoiced separately by the producer are included.

Value added is obtained by subtracting the value of inputs (the cost of materials, fuels, produced electricity, industrial and non-industrial services received and other supplies) from gross output.

Investment covers expenditures spent on the acquisition of tangible assets (by either purchase or own activity), together with the overall value of tangible assets acquired free of charge, by transfer according to relevant legislation, or by transfer from personal use to business. Major alterations, additions and improvements to existing assets which extend their normal economic life or raise their productivity are also included. Tangible assets are valued at purchasers' prices. Deductible VAT is excluded.

Employment, employees (average number) encompasses all categories of permanent, temporary and seasonal employees contracted for work in the enterprise. It excludes females on maternity leave and child-care leave, persons on parental leave, persons engaged by companies on the basis of work execution or working activity agreements and apprentices.

Wages and salaries, employees are the sum paid during the year, before deduction of taxes and pension contributions. Includes regular and overtime cash payments, bonuses, vacation etc., payable by employers.

Hours worked by operatives cover hours worked by manual workers in enterprises with 20 or more employees in the year s 1997 and 1998, 100 or more employees in the years 1995 and 1996.

The *enterprise* is the smallest combination of legal units, that is an organisational unit producing goods or services, which benefits from certain degree of autonomy in decision-making, especially for allocating of current resources. An enterprise carries out one or more activities at one or more locations.

Publications

- *i) Statistical Yearbook of the Czech Republic* - available every year in December
- *ii) Selected indicators of yearly survey 1997 and 1998 in the Czech Industry* (*Vybrané ukazatele ro•ního zpracování 1997 a 1998 v pr•myslu •eské republiky*)

Publications are distributed by the CSO.

DENMARK

The information was provided by Danmarks Statistik.

Industrial Classification

From the year 1993, and onwards, official statistical data for Denmark are collected and compiled in NACE revision 1.

INDUSTRY AND SERVICE STATISTICS

Source

The figures for number of establishments, employment and remuneration are derived from the register-based statistics of establishments and employment. All establishments with any employment in manufacturing are covered. The number of establishments includes establishments in activity at end-November. For remuneration, the reference period is the calendar year.

Concepts and Definitions of Variables

Employment, employees figures include all employees and all self-employed at end November in establishments included in the register (as mentioned above). Both full-time and part-time employed are included, and even jobs in these establishments, which are a person's sideline job, are included.

The figures for *wages and salaries, employees* include all outlays of these kind reported to the tax authorities in units operating during the year. This means that even outlays in establishments that are not in activity at end-November but were in activity during some part of the year are included.

The number of *establishments* is given as the population of local units. All establishments with any employment, in activity at end November, are covered.

Publications

- *i) Statistical News*, Series Industry and Energy;
- *ii)* Statistical Reports: *Industrial Accounts Statistics*;
- *iii) Statistical News*, Series General Economic Statistics and Internal Trade.

All are published by Danmarks Statistik.

FINLAND

Statistics Finland, Helsinki, has provided information for structural business statistics (industry and services) and foreign trade statistics.

Industrial Classification

The classification, derived from NACE Revision 1, is the Finnish Standard Industrial Classification 1995 (SIC-95). It aligns with the ISIC Revision 3.

I. INDUSTRY AND SERVICE STATISTICS

Source

Data are derived from three sources: 1) the Structural business statistics on industry (ISIC categories C to F) or the Structural business statistics on services (ISIC categories G to O); 2) Tax return files supplied by the National Board of Taxation; 3) the Business Register of Statistics Finland.

The structural surveys cover the fiscal year, between 1 April (year t) and 31 March (year t+1).

Industrial sectors:

The statistical unit is the establishment.

Since the year 1996, the size threshold of the survey is 20 employees. The number of units covered by the survey is about 2 500. The total number of units available in the administrative source is about 60 000. Estimates for non-response are obtained from tax files.

The method of data collection has been changed in 1995, therefore comparable data is available only for 1995-1998.

Service sectors:

The statistical unit is the enterprise.

The Structural business statistics on services covers all enterprises with more than 10, 20 or 50 employees depending on the activity. The number of units covered by the survey is about 4 700. The total number of units available in the administrative source is about 160 000. Estimates for non-response are obtained from tax files.

Concepts and Definitions of Variables

Production or output includes the values of shipments of goods and services produced by the unit, the receipts for industrial work done or industrial services rendered to others, the sales of goods shipped in the same condition as received (less the amount paid for these goods), the net change in stocks of finished goods, work in progress, and stocks of goods purchased for resale, the value of electricity sold, the sales of products manufactured by subcontractors, the value of capital assets manufactured for own use, and receipts for non-industrial services. It excludes income from sales of capital assets produced on own accounts. Direct government subsidies related to current production and import and export subsidies are included. Assistance in the form of interest subsidies is excluded. VAT, deductible VAT, commodity taxes invoiced to the client by the producer and non-commodity taxes paid by the producer are excluded. The valuation is at basic prices.

Value added at factor costs is the gross income from operating activities. It can be calculated from output by excluding the following items: value of industrial materials and services for processing, costs of industrial services received, and operating costs and payments for services of a non-industrial nature. It is valued in factor costs, excluding indirect taxes and including subsidies.

Investment (capital formation) is expenditure on the purchase of fixed assets plus own-account construction of assets. Fixed assets include investments in equipment goods, buildings, other construction and other fixed assets and exclude investments in land and water resources. They are investments with a productive life of at least one year, and include assets made by a unit's labour force for its own use; additions, alterations, improvements and repairs to existing assets; new assets (whether or not in use) and used assets. The valuation is at full cost incurred, including delivered price, cost of installation and fees and taxes.

Investment in machinery and equipment follows the same definition. Expenditure on transport equipment, industrial machinery and equipment, office machinery, equipment and furniture, professional instruments and equipment are included together with alterations and improvements to these items.

Employment relates to the total number of persons engaged and covers all persons who work for an employer and receive compensation in the form of wages, salaries, fees, gratuities, piecework pay or remuneration in kind. It includes part-time and seasonal workers, apprentices, homeworkers, working proprietors, working partners who draw regular earnings, unpaid family workers, directors of incorporated enterprises, persons who work outside the unit, belong to it and are paid by it, voluntary workers, persons on short term leave and persons on strike.

Employment, employees covers total number of salaried employees excluding working proprietors, working partners who draw regular earnings and unpaid family workers.

Hours worked by operatives are measured by the actual hours worked by manual workers for the output of the observation unit during the fiscal year, including overtime hours calculated in terms of actual hours spent at work, time spent at the place of work on work such as preparation of the workplace, repairs and maintenance, preparation and cleaning of tools and time spent at the place of work waiting or standing by for such reasons as lack of supply of work, breakdown of machinery or accidents. Are excluded hours paid but not actually worked such as annual leave, holidays and sick leave.

Compensation of labour, employees is defined as wages and salaries for employees and employers' social costs.

Wages and salaries, employees cover all paid regular and overtime wages and salaries, cash payments, bonuses, cost-of-living allowances, paid education, vacation and sick leave along with taxes and social insurance contributions payable by employees and deducted by employers, and payments in kind. It includes the total remuneration in cash or in kind, payable in return for work done during the year.

Employers' social costs are employers' legally required supplements to wages and salaries. They cover statutory social insurance contributions and voluntary social insurance contributions of a statutory nature.

II. FOREIGN TRADE STATISTICS

Source

Data are derived from Board of Customs statistics.

Concepts and Definitions of Variables

General trade is given, *exports* fob, *imports* cif. Both include state trading, commercial transactions in military equipment, foreign aid transactions, transactions in new ships and aircraft, electric energy and gas, and specifically exclude direct transit trade, temporary exports or imports, monetary transactions and minor transactions of less than 1 000 Mk.

Publications

i) *Preliminary Data on Finnish Industry*
ii) *Preliminary Data on Business Services*
iii) *Preliminary Data on Trade*
iv) *Preliminary Data on Telephone Communication*
v) *Statistics on the Structure of Industry and Construction*
vi) *Financial Statements Statistics*

All publications are distributed by Statistics Finland, Helsinki, in the series of *Official Statistics of Finland.*

FRANCE

Industrial statistics and foreign trade statistics have been supplied by SESSI (the statistical service of the Ministry of Industry).

Statistics on services have been supplied by the French Institut National de la Statistique et des Etudes Economiques (INSEE, Paris).

Industrial Classification

The French classification is the NAF-1993 (Nomenclature d'Activités Françaises), divided into five levels. NAF-1993 is fully compatible with NACE Revision 1 and with ISIC Revision 3.

I. INDUSTRIAL STATISTICS AND FOREIGN TRADE

Source

Statistics are derived from the annual business survey ('Enquête Annuelle d'Entreprise' - EAE) which covers all enterprises of at least 20 salaried employees or with an important turnover, whatever their size, excluding arms and farm-produce industries but covering energy, i.e. 22305 enterprises in 1998.

Concepts and Definitions of Variables

Production corresponds to the total value of goods and services, sold, stocked or immobilised, that have been produced during the year. It excludes resale of goods. Sold products include the margin realised by the enterprises. The immobilised production corresponds to the costs of work done by the enterprise on own account. Stocked production and immobilised production are estimated at producers' cost, excluding deductible VAT and subsidies.

Gross value added at factor costs is calculated as production, plus sold goods, less intermediate consumption (which corresponds to purchased goods, materials, fuels and other supplies, plus net change in stocks of goods, materials, fuels and other supplies, plus other external purchases and costs), plus subsidies and less taxes and assimilated payments.

Investment data is the sum of the value of acquisitions of either new or used assets and production of fixed capital. It includes fixed assets in progress but excludes intangible and financial assets and equipment financed by leasing. Its valuation is made at acquisition costs.

Employment, employees, which corresponds to the average number of employees over the twelve months of the year, includes all persons who receive wages and salaries from the enterprise, even if they work in another enterprise and whatever their status is (directors, homeworkers, ...). It does not include interim employed persons. Part time workers are counted in proportion to their part time.

Wages and salaries, employees correspond to the remuneration paid by the employer during the year to compensate for the work of his employees (excluding interim employed persons and workers on hire). They exclude employer's contributions to social security, which are given separately.

Employers' social costs are the social contributions paid by the employer during the year (sickness, maternity, old age, unemployment, ...). They have to be specified in statutory, collectively agreed or contractual arrangements.

Exports, collected by the EAE, are defined as exported goods and services declared by the enterprises of the sector. They include intra-Europe deliveries. They differ from exports registered by customs (which are FOB/CIF) in that they correspond to direct exports only (i.e. not through a third party). Furthermore, the sector to which exports are classified is defined by the enterprise's main activity (i.e. secondary activity is classified to the same sector as the main activity).

II. SERVICE STATISTICS

Source

Data are derived from an annual survey (Enquête Annuelle d'Entreprise, EAE).

In the EAE-trade 1993, sale and repair of motor vehicles and repair of personal and household goods were covered. In 1994, intermediates for wholesale trade and activities relating to bakery products were not surveyed. As from 1995, the EAE-trade surveyed the whole trade activities, retail trade and wholesale trade, including the commission trade, sale and repair of motor vehicles and the craftsmen activities such as pork-butcher's trade and bakery and pastry activities.

Up to 1997, the EAE-services were sent to all market-service enterprises (some variables have been newly defined in EAE-1997). As from 1998, the EAE-services questionnaires are regularly addressed to all enterprises with more than 30 salaried employees or with a turnover exceeding 30 Million Francs every year. Small enterprises under this threshold are surveyed every two years.

Concepts and Definitions of Variables

Sales activities:

Production corresponds to the total turnover excluding taxes, plus the stocked and immobilised production minus the costs of merchandises.

Value added is defined as value added at market prices. It is calculated as production less costs of materials and other external purchases and external costs.

Investment data is the sum of the value of acquisitions and production of fixed capital. It includes capital goods received from outside contribution. Only tangible assets are surveyed in the EAE.

Employment, employees is defined as the average of the quarterly numbers of salaried employees. Part time workers are counted in proportion to their part time.

Compensation, employees covers the wages and salaries paid to the employees by the employer during the year as well as employer's contributions to social security.

Exports and *imports* data are collected in a few sectors.

Services activities:

Production, or enlarged production, corresponds to the total turnover, as well as the value of the production stocked and immobilised. The stocked production corresponds to the production of services not yet sold (project and provision of services in progress). The immobilised production corresponds to the production of services done by the enterprise on own account and not for sale (films, TV programmes, software...).

Value added is defined as value added at market prices. It is calculated as production, less intermediate consumption (purchase of merchandises or materials, variation of stocks of merchandises and materials and other external purchases and external costs).

Investment data relates to the tangible assets only. It comprises investment in the form of acquisitions or creation of tangible elements during the business year and intended to be used on a permanent basis by the enterprise. It also includes tangible investment received by contributions (for example received from another enterprise in the process of a restructuring) during the business year and intended to be used on a permanent basis by the enterprise.

Employment, employees corresponds to the average number of salaried employees, at the end of each quarter, in full time equivalent in FTE: part time workers are counted in proportion to their part time.

Compensation of labour, employees corresponds to wages and salaries paid to the employees plus employer's contributions to social security. These two variables concern neither non-salaried employees, nor labour force made available to the unit by other units and interim employed persons.

Wages and salaries, employees covers the wages and salaries paid to employees by the employer during the year, including employees social contributions but excluding employer's contributions to social security.

Data for *exports* and *imports* (as from 1996 only) are collected in a few sectors.

Publications

i) *La situation dans l'industrie* (paper publication) ;

ii) *L'industrie française en 300 secteurs* (CD-ROM) ;

iii) Other thematic or synthetical publications.

All publications are distributed by the SESSI.

GERMANY

The information is supplied by the Federal Statistical Office (Statistisches Bundesamt) in Wiesbaden.

Data prior to 1990 refer to western Germany (Federal Republic of Germany before the unification of Germany, including West Berlin). Data from 1991 onwards refer to unified Germany, except for data for value added, investment, and investment in machinery and equipment, which continue to refer to the former western Germany.

Industrial Classification

The data are collected and compiled according to the national classification (Klassification der Wirtschaftszweige, Ausgabe 1993 - WZ93). WZ93 corresponds to NACE revision 1 and can be converted to ISIC revision 3.

INDUSTRIAL STATISTICS

Source

In general, the figures for mining and manufacturing are derived from the monthly inquiry carried out by the Office. The inquiry covers all local units belonging to enterprises with 20 or more persons, including production handicrafts and the industrial units of non-industrial activities.

Concepts and Definitions of Variables

Production relate to turnover from own production based on invoice values and include proceeds from goods sold as purchased, from the sale of waste, power and similar items. Proceeds from non-industrial activities (interest payments, rentals and dividends) are excluded. The value of shipments also excludes VAT or turnover taxes but includes excise duties, costs for freight packing, postage and other changes. Shipments between units of the same enterprise are not recorded.

Employment figures are annual averages of all persons engaged at month's end. Included are working proprietors, active business partners, unpaid family workers, provided they are working for at least one third of the usual working time. Part-time and temporary employees are covered as well as apprentices. Homeworkers are excluded.

Hours worked by operatives cover hours worked, including overtime, by manual workers which are defined as persons engaged who are subject to obligatory insurance in the workers' old age insurance funds. This includes industrial apprentices but not homeworkers. The following categories are also included: supervisors, casual and part-time workers.

Total wages and salaries exclude employer's contributions to social security. All payments for interim employed persons are included: contributions to company old age pensions institutions are excluded.

Publications

Fachserie 4, Reihe 4.1.1., *Beschäftigung, Umsatz und Energieversorgung der Unternehmen und Betriebe im Bergbau und im Verarbeitenden Gewerbe*, Statistisches Bundesamt, Wiesbaden.

GREECE

The information is supplied by the National Statistical Service of Greece (NSSG).

Industrial Classification

The estimates are based on national classification and are adjusted to ISIC as far as possible.

INDUSTRIAL STATISTICS

Source

The data result from annual surveys conducted by the National Statistical Service of Greece:
- the Annual Industrial Survey,
- the Annual Statistical Survey on Mines, Quarries and Salterns
- and the Annual Electricity and Gas Surveys.

For the industrial survey, manufacturing establishments with 10 or more persons engaged are covered on the basis of a complete enumeration of establishments with 20 or more persons engaged and a stratified random sample of establishments with 10 to 19 persons. Government establishments are not included in the industrial survey. Electricity and gas surveys cover the two enterprises engaged in production and distribution for public use. Steam and water supply are not covered. The base unit is the establishment except for electricity and gas, where the enterprise is used.

Concepts and Definitions of Variables

Production comprises the value of finished goods produced, receipts from contract work on account of others, receipts from repair work, receipts from the sale of goods sold in the same condition as purchased and other receipts (sales of self-produced steam, electric energy, processing waste, etc.; subsidies and reimbursements for re-export of products). The value is at ex-factory selling price, not including taxes.

Value added is calculated as the value of the output less the cost of materials and supplies, purchased fuels and electricity consumed, the purchase value of goods sold in the same condition as purchased, and payments for contract and repair work.

Investment (gross fixed capital formation) represents current expenditure on new and used fixed assets less sales and write-offs due to obsolescence. Own-account construction of fixed assets is included. The data for mining exclude salterns and the value of sales (insignificant) is generally not deducted.

Employment, employees covers salaried employees, defined as persons receiving a monthly salary on a contractual basis irrespective of the amount of the remuneration, as well as wage earners, defined as persons receiving a daily pay or paid on a piecework basis either in cash or in kind.

Days worked cover days worked by operatives, relating to wage earners as defined above.

Compensation of labour, employees covers wages, salaries, payments for overtime work and regular leave, bonuses, allowances, dismissal compensation and the value of receipts in kind.

Employers' social costs cover employers' contributions to social security schemes on behalf of employees and wage earners.

Publications

i) Annual Industrial Survey;
ii) Annual Statistical Survey on Mines, Quarries and Salterns;
iii) Statistical Yearbook;
iv) Monthly Statistical Bulletin.

All are published by the National Statistical Service of Greece, Athens.

HUNGARY

The information is supplied by the Hungarian Central Statistical Office.

Industrial Classification

Data are classified according to the Hungarian classification (TEÁOR 92), which is only fully compatible with ISIC Revision 3 at the two-digit level.

Source

From 1992 onwards, all enterprises where two of three criteria are met in two consecutive years -- over 100 employees, over 300 million HUF annual turnover, 150 million HUF of total assets -- are covered. Only investments above 10 million HUF are reported. The employment data result from the Annual Employment and Earnings Survey, which covers enterprises with 20 or more employees.

Concepts and Definitions of Variables

Employment, employees refers to the average staff number of paid employees. Only full-time workers are covered.

Wages and salaries, employees include bonuses and vacation pay. Benefits in kind and sick leave pay are excluded.

ICELAND

Data are provided by the National Economic Institute of Iceland (Þjóðhagsstofnun).

I. INDUSTRIAL STATISTICS

Industrial Classification

The national classification is based on ISIC Revision 1 from 1958; the data provided are converted to ISIC Revision 2 from 1968. NACE Revision 1 has been introduced and will be implemented in all statistics shortly.

Source

The National Economic Institute of Iceland annually draws a sample of financial statements out of tax records. The sample size covers around 50% of the total wage bill in various industries. The probability of an establishment being drawn into a sample increases with size. Establishments and kind-of-activity units are used as basic reference units.

For production and value added the stratified samples drawn are "grossed up" to the total size according to the proportion between the total wage bill and the sample wage bill. Employment figures as well as numbers of enterprises and establishments are based on administrative registers of total coverage. The reference period is the calendar year.

Concepts and Definitions of Variables

From 1992 onwards, *production* (gross output) is valued in basic prices, including all subsidies on products but excluding VAT; before 1992 gross output is valued in producers' prices, excluding all subsidies and VAT. The value of all products is taken, allowing for net changes of work in progress.

Value added uses the "national accounting" concept of value added. It is calculated as the value of gross output less the cost of materials, fuels and other supplies, of contract and commission work done by others, of repair and maintenance work done by others and of electricity purchased. It includes receipts for non-industrial services but excludes the cost of non-industrial services. Input estimates are made on a consumed basis. The valuation is the same as for production.

Employment is calculated as the number of work years, based on the number of insured weeks of work provided by a register-based tax-source. Included are working proprietors, active business partners, unpaid family workers, self-employed, salaried employees and wage earners. Full-time and part-time employees are covered, as well as clerical workers or supervisory personnel performing functions closely related to the production process. Employment data are converted to full-time equivalent (FTE).

Employment, employees includes all salaried employees, excluding working proprietors, active business partners and unpaid family workers.

Compensation of labour, employees is defined as the "national accounting" concept of "compensation of employees". It covers employees only, excluding the self-employed, and includes wages and salaries as well as employers' social contributions for all employees.

Publications

i) *Atvinnuvegaskyrslur (Industrial Statistics),* published annually;

ii) *Historical Statistics,* published every other year;

iii) *Compiling Icelandic National Accounts,* documentation of methods applied, output and expenditure approaches, Reykjavík, January 1996.

All the above are published by the National Economic Institute.

II. FOREIGN TRADE STATISTICS

Industrial Classification

Data are transmitted according to ISIC Revision 3.

Source

Data are obtained from the trade returns sent by exporters to the Customs Department.

Concepts and Definitions of Variables

Exports are derived from the export declarations compiled by the Custom authorities. The goods are classified according to the Harmonised System (HS-classification) and then grouped together into the principal activity classification used domestically. Exports include transactions concerning new ships and aircraft. They are given FOB/CIF.

IRELAND

Information was supplied by the Irish Central Statistics Office.

INDUSTRIAL STATISTICS

Source

Data are taken from the Annual Census of Industrial Local Units. This census relates to all units engaged in industrial activity which have on average three or more persons engaged during the year.

The extent to which separate returns are obtained in practice, however, depends on the availability of separate records in the business for the different local units.

Concepts and Definitions of Variables

Production (gross output) represents the net selling value of all goods manufactured in the year, whether sold or not, including work done and capital assets manufactured for own use. Operating subsidies related to the production or sales of the output are included in the value of gross output; excise duty and VAT are excluded. Prior to 1991, the value of capital work done on own account was included in gross output only in the case of two industries, namely electricity and turf production.

Value added (net output) is the difference between gross output and industrial input. Industrial input consists of the industrial materials, industrial services and fuel and power used in the production of the output. Valuation is exclusive of deductible VAT.

Investment is gross fixed capital formation (GFCF). Capital Assets (land, buildings, plant and equipment) are defined as goods with an expected useful life of more than one year intended for use by the local unit itself. Acquisitions include purchases from other local units and production by the local unit itself of capital goods for its own use. Major alterations, improvements and repairs that extend the useful life of an asset or increase its productivity are included. The value of work put in place during the year is included whether or not completed. Acquisitions are valued at total cost including installation charges and fees or duties but excluding deductible VAT and financial costs. Sales are valued at the price actually received excluding VAT. GFCF is the difference between the acquisitions and sales of capital assets.

Employment, employees corresponds to persons who are paid a fixed wage or salary. Persons at work or temporarily absent because of illness, holidays, strikes, etc. are included, as are part-time workers. Outside piece workers i.e. homeworkers are excluded. The numbers given for each year refer to a week in September.

Wages and salaries, employees are defined as the gross amount paid to all employees, excluding outside piece-workers, before deduction of income tax, employees' contributions to social security etc. Overtime pay, bonuses, commissions, holiday pay and sick pay are included. Payments to working proprietors and home workers are excluded.

The *establishments* include all separate local units of multi-location enterprises, even if separate details are not provided by the respondent.

ITALY

Information was provided by the Istituto Nazionale di Statistica (National Institute of Statistics), Rome.

Industrial Classification

The national NACE-related classification is not fully convertible to ISIC. However, national conversion tables are used to produce 3-digit ISIC data and an effort is made to produce reliable estimation at the 4-digit level.

INDUSTRY AND SERVICE STATISTICS

Source

The Institute annually conducts total surveys on gross product and capital investments in the industrial sector. Up to 1995, enterprises with 20 or more persons engaged were covered. No attempt was made to account for units outside the scope of the surveys. As from 1996, all enterprises are covered in the surveys. The inquiry covers the calendar year and takes the enterprise, classified by kind of activity, as its basic reference unit. Multi-unit enterprises with more than 250 persons engaged are required to furnish data separately for each unit. Up to 1993, figures on hours worked were taken from the Industrial Enterprise Survey. Starting from 1993, hours worked data are derived from a special survey on employment, wages and hours worked taken from a sample of 18 000 enterprises with 10 to 199 employees and all enterprises (around 3 700) with more than 200 employees.

Concepts and Definitions of Variables

Production, or gross output net of indirect taxes and subsidies, is valued in selling prices, excluding VAT. Electricity sold, fixed assets produced for own use and goods made by other enterprises are included, making allowance for changes in stocks of finished goods and of work in progress.

Value added, net of indirect subsidies and taxes, uses the "census" concept of value added. The valuation excludes all indirect taxes and subsidies. It is calculated as gross output plus receipts for non-industrial services, less the cost of materials, fuels and other supplies; contract and commission work done and repair and maintenance by others; electricity purchases; and non-industrial services. Input estimates are made on a received basis, making adjustment for stock changes in materials, fuels and other supplies.

Investment (gross fixed capital formation), value of sales not deducted, is defined as the value of purchases of fixed assets, including own-account construction. Fixed assets covered are those with a productive life of at least one year. Assets made by a unit's labour force for its own use; additions, alterations, improvements and repairs to existing assets; new assets (not yet in use, or previously used outside the country) and used assets are all included. The valuation is at full cost incurred, including installation expenses.

Investment in machinery and equipment is gross fixed capital formation in machinery and equipment, value of sales not deducted, is given for. Using the same definition, it covers transport equipment, industrial machinery, equipment and furniture, professional instruments and equipment. Alterations, improvements and repairs are also included.

Employment gives the number of persons engaged, based on the average number of employees through the year plus the total number of other persons engaged at four fixed dates. These dates are 31 March, 30 June, 30 September and 31 December. Working proprietors, active business partners, unpaid family workers, the self-employed, salaried employees and wage earners are all included.

Hours worked by operatives cover hours worked by manual workers, including overtime, rest periods and time on call. Annual leave, public holidays, sick leave and other paid leave are included. The following categories are excluded: employers, supervisors and apprentices.

Compensation of labour covers wages and salaries, and employers' social costs. It includes regular and overtime cash payments, bonuses and cost-of-living allowances, vacation and sick leave pay, payments in kind and allowances.

Employers' social costs are employers' contributions to social security schemes and to pension funds.

Establishments are kind-of-activity units.

Enterprises, available as from 1997, cover all Italian enterprises, including the enterprises with less than 20 employees.

Publications

Fatturato, prodotto lordo e investimenti delle imprese industriali: supplements to *Bollettino Mensile di Statistica*, ISTAT, Rome.

JAPAN

The information was supplied by the Japanese Ministry of International Trade and Industry (MITI) and the Ministry of Health and Welfare.

Industrial Classification

The data are classified according to the Japanese Standard Industrial Classification (JSIC Rev. 10). The JSIC Rev. 10 is consistent with ISIC Revision 3.

INDUSTRIAL STATISTICS

Source

Manufacturing figures are derived from MITI's Annual Census of Manufactures. This survey covers all establishments classified under Manufacturing except those belonging to the government and public service corporations.

Since 1981, however, surveys have been conducted on all establishments (including those with three employees or less) only in years where the last digit is 0, 3, 5 or 8. In all other years, the survey is limited to those establishments with four or more employees. Investment figures cover establishments with 30 or more persons engaged. No attempt is made to account for units not covered.

Concepts and Definitions of Variables

Production (gross output) represents the value of shipments of finished goods, adjusted for changes in stocks of finished goods and work in progress. Production of manufacturing is calculated as the value of shipments, plus the manufactured products' stocks at the end of the year minus the manufactured products' stocks at the beginning of the year, plus semi-manufactured and work-in-progress stocks at the end of the year minus semi-manufactured and work-in-progress stocks at the beginning of the year.

Value added is the value of products produced, less the value of excise taxes, less the value of raw materials, fuels and electricity consumed, and cost of contract and commission work done by others, less value depreciated.

Investment is gross fixed capital formation, defined as value of purchases of fixed assets, including own account construction, less value of sales. Fixed assets covered are those with a productive life of more than one year and include repairs to existing assets, new assets whether not yet in use, unused, or previously used abroad, and used assets. The valuation is at full cost incurred, including delivered price, cost of installation and fees and taxes. For own-produced assets, the cost of work put in progress covers labour costs, materials and supplies and includes an allocation for overheads. The amount realised is taken for sales of used assets.

Investment in machinery and equipment corresponds to the acquisition value of machinery and equipment (including attachments) out of tangible fixed assets.

Employment, employees comprises regular employees, as of 31st of December. Specifically included as regular employees are the following persons: temporary workers employed for a definite period of more than 30 days, temporary workers employed for 18 days or more in each month of November and December, directors who work full-time and are paid monthly and family workers who work full-time and are paid monthly.

Wages and salaries, employees relate to total cash payments, including basic wages, bonuses and other premiums and allowances paid to employees. Mining figures include payments to contract workers. Figures for manufacturing include employees' retirement and termination allowances; these payments are not included in data for operatives.

Publications

i) Census of Manufactures (annual), published by MITI.

KOREA

Data were provided by the National Statistical Office of Korea.

Industrial Classification

The Korean Standard Industrial Classification is the SIC (revised in 1992). It aligns with ISIC Revision 3.

I. INDUSTRIAL STATISTICS

Source

Data are drawn from the annual Mining and Manufacturing Survey. The survey is conducted annually. The statistical unit is the establishment. All establishments with five or more employees are covered: about 80000 were surveyed in 1998. The period covered is the calendar year.

Concepts and Definitions of Variables

Production uses the census concept, which only covers activities of an industrial nature. Valuation is at producers' prices, that is, those amounts receivable by the producer, less all commodity taxes, subsidies related to the production and separately invoiced transport charges. It comprises: the value of all products of the establishment; the net change between the beginning and the end of the reference period in the value of work in progress and finished and half-finished products; receipts for repairs and the value of industrial work done or industrial services rendered to others. The followings are not included: the net change between the beginning and the end of the reference period in the value of stock of goods to be shipped in the same condition as received; the value of goods shipped in the same condition as received less the amount paid for these goods and the value of fixed assets produced by the unit for its own use.

Value added uses the census concept, which is defined as the value of census output less the value of census input. Items in the later are the value of materials and supplies for production (including cost of all fuel and produced electricity, water) and the cost of industrial services received (mainly payments for contract and commission work and repair and maintenance work). Valuation is at producers' prices.

Investment refers to the value of fixed assets, purchased or constructed by the statistical unit's own labour force for its own use, during the reference year. The value of corresponding sales is not subtracted from it. The fixed assets covered are those (whether new or used) with a productive life of one year or more. Major additions, alterations and improvements to existing assets which extend their normal economic life or raise their productivity are also included. Transactions in fixed assets include: land; buildings, other construction and land improvements; transport equipment; machinery and other equipment. Assets acquired from others are valued at purchasers' prices, which cover all costs directly connected with the acquisition and installation of the items for use. Assets produced on own account are also valued in this manner. Assets produced by one establishment of a multi-establishment enterprise for use by another establishment of the same enterprise were valued by the receiving establishment as though purchased from outside the enterprise.

Investment in machinery and equipment is broken down by kind of equipment goods from whole investment for machinery.

Employment, employees includes employees, part-time workers and seasonal workers on the payroll, persons on short-term leave (sick leave, annual leave, vacation), persons on strike and persons carrying out repair and maintenance work in the unit on the behalf of other units. They exclude working proprietors, active business partners, unpaid family workers, directors of incorporated enterprises and members of shareholders' committees who are paid solely for their attendance at meetings, labour force made available to the concerned unit by other units and charged for (such as contract workers paid through contractors), persons on indefinite leave, military leave or pension and home-workers. The figures refer to the average number of employees during the reference year.

The share of salaried females is given under *Employment, females*.

Wages and salaries, employees include all money payments and payments in kind paid or supplied by the employer during the reference period to all employees counted on the payroll by way of remuneration for work done for the concerned unit. These payments include: direct wages and salaries and all payments in cash and for overtime; remuneration for time not worked, for example, bank holidays, paid vacation and sick leave; bonuses and gratuities, for example, holiday and performance bonuses; housing allowances and family allowances paid directly by the employer; payments in kind; taxes and social security contributions and similar contributions payable by the employee and deducted by the employer. They exclude social security, pension, retirement and other contributions payable by the employer.

Employers' social costs include all expenditures for employees' welfare and convenience.

An *establishment* is a unit that engages, under a single ownership or control, in one or predominately one, kind of activity at a single location; for example a workshop or factory.

II. SERVICE STATISTICS

Source

Data are drawn from annual surveys: Wholesale and Retail Trade Survey (50-55); Transportation Survey (60-65); Service Industry Survey (70-95). The statistical unit is the establishment. All establishments are surveyed in divisions and groups of the sections. A sample survey is conducted for classes. The period covered is the calendar year.

Concepts and Definitions of Variables

Production is gross output measured as the total value of all goods produced and services rendered to others by the establishment during the year. In practice the calculation is as follows: the value of shipments, plus the net variation in stocks of finished products and work-in-progress. It is valued at producers' prices, excluding any tax payable as a consequence of production and subsidies related to production and excluding separately invoiced transport charges.

Publications

i) Report on the Mining and Manufacturing Survey

ii) Annual report on the survey of wholesale & retail trade, service industries, transportation.

They are published each year by the National Statistical Office of the Republic of Korea.

LUXEMBOURG

Information was supplied by the Service central de la statistique et des études économiques (STATEC), Luxembourg.

Industrial Classification

As from 1993, the classification used is the NACE LUX which corresponds to NACE Revision 1 down to the 4th level and is further detailed at a 5th level for national needs. It is compatible with ISIC Revision 3. The survey covers all enterprises belonging to sections C to K of ISIC Revision 3.

INDUSTRY AND SERVICE STATISTICS

Source

Data are derived from annual sample surveys, administrative sources and annual accounts of large enterprises. Surveys are exhaustive for units with more than 50 salaried employees or with an annual turnover (excluding VAT) of more than 280 million Lux Francs. For the other units, statistics are obtained from sample surveys. The sample of units is taken from the Business Register of STATEC.

Concepts and Definitions of Variables

Definitions of variables are consistent with the Regulation (CE) No 2700/98 of the Commission of European Communities of 17/12/1998.

Production is defined as the turnover adjusted for changes in the value of stocks, less the acquisition of goods and services shipped in the same condition as received (resale), plus the immobilised production and other operating incomes (excluding subsidies). Are excluded value added taxes (VAT) invoiced to the client by the producer and deductible VAT on purchased of raw materials and intermediate products. The valuation is made at market prices.

Value added at factor costs relates to the gross value of operating activities taking into account subsidies and indirect taxes. It is calculated as turnover plus the immobilised output and other operating products, adjusted for changes in the value of stocks. Are not included the acquisition of goods and services as well as not deductible taxes on products related to turnover and taxes related to production. It excludes receipts and costs belonging to the business accounts under items 'financial receipts/charges' and 'exceptional receipts/charges'.

Employment, employees is the sum of the numbers of apprentices, wage earners and other employees. It excludes working proprietors and unpaid family workers as well as persons who work outside the unit, who belong to it and are paid by it. Homeworkers whose contribution is considered negligible are not covered. Employment data are not converted to full time equivalent (FTE).

Compensation of labour, employees is the sum of wages and salaries paid to salaried employees, and employers' social contributions.

Wages and salaries, employees relates to payments made to employees in cash or in kind in return for work done during the reference year and wages paid to manual workers, including overtime and night work cash payments. Are included gratuities, tips, severance payments, workplace and performance bonuses, ex gratia payments, thirteenth month pay, seniority, attendance, incentive and productivity bonuses, payments for day off (leave, holidays, sick leave), cost of vocational training for workers, travel and other similar expenses. Are also included taxes and social insurance contributions payable by employees but deducted by employers and value of free actions delivered to employees.

Employers' social costs are employers' social contributions to schemes for sickness, maternity leave, invalidity, old age, unemployment, work accidents and family allowances, costs related to medical care and drugs given free of charge and costs of maintaining schools and of sporting and recreational activities. The employers' contributions are to be specified in a statutory arrangement or in a collectively agreed arrangement.

Exports are valued FOB. They include government procurement, commercial transactions for military equipment and supplies, foreign aid transactions (except where second-hand goods are involved), transactions concerning new aircraft, transactions concerning electricity and gas, shipments of goods relating to transactions of no commercial interest.

Enterprises, for service sectors, is an approximation of the legal unit as defined in Regulation (EC, EURATOM) No 58/97 relating to structural business statistics.

Establishments, for industrial sectors, are kind of activity units, as defined in Regulation (EEC) No 696/93 relating to statistical units, which can supply information on production value, intermediate consumption, personnel costs, operating surplus, employment and gross fixed capital formation.

Publications

Annuaire statistique du Luxembourg, published by STATEC.

NETHERLANDS

Data were supplied by Statistics Netherlands (Centraal Bureau voor de Statistiek), Voorburg.

Industrial Classification

The classification used is the Standard Industrial Classification (Standaard Bedrijfs Indeling - SBI 1993). Besides a few derogations, it is fully compliant to NACE Rev. 1 up to the 4th digit.

INDUSTRY STATISTICS

Source

Data are based on various sources: surveys on Annual Production Statistics and on Investments in tangible fixed assets in industry, annual Census and VAT register. The surveys are sample surveys stratified by activity and size class. Enterprises with 0 employees are excluded. The Census covers all enterprises with more than 20 employees in Agriculture, Industry, Domestic Trade and Transport, and enterprises with more than 100 employees in Construction.

Concepts and Definitions of Variables

Production includes the value of shipments of goods and services produced by the unit, the receipts for industrial work done or industrial services rendered to others, the margin on sales of goods shipped in the same condition as received, the net change in stocks of finished goods and work in progress, the value of electricity sold, the sales of products manufactured by subcontractors, the value of capital assets manufactured for own use, and the receipts for non-industrial services. It excludes income from sales of capital assets produced on own accounts. Commodity taxes invoiced to the client by the producer are included. Data on production is only disseminated in market prices. However, since information on indirect taxes and products subsidies is available, factor costs values can be calculated.

Value added, valued at market prices, is calculated by excluding from output the following items: value of industrial materials and services for processing, costs of industrial services received, and operating costs and payments for services of a non-industrial nature. *Value added at factor costs* is the gross income from operating activities after adjusting for operating subsidies and indirect taxes. It can be calculated from turnover, plus other operating income, plus or minus the changes in stocks, minus the purchases of goods and services, minus other taxes on products which are linked to turnover but not deductible, minus the duties and taxes linked to production. Alternatively, it can be calculated from gross operating surplus by adding personnel costs. Income and expenditure classified as financial or extra-ordinary in company accounts is excluded from value added. Value added at factor costs is calculated "gross", as value adjustments (such depreciation) are not subtracted.

Investment refers to the value of new or used fixed assets purchased (including leasing and renting) or constructed by the unit's own labour force for its own use, during the reference year. Major additions, alterations and improvements to fixed assets carried out by third parties are also included. Transactions in fixed assets includes land, buildings, site preparation and other construction work, land which is not used for the establishment business, transport equipment, machinery and other equipment, as well as additions, alterations and improvements which prolong the service life or increase productive capacity of capital goods. Fixed assets are valued at acquisition costs, including transportation costs and cost of installation, taxes and insurance for goods in transit.

Investment in machinery and equipment is gross fixed capital formation in transport equipment (work trucks of the type used in factories, warehouses, etc., for short distance transport or for handling of goods), industrial machinery and equipment, office machinery, equipment and furniture, professional instruments and equipment, and alterations and improvements to the above.

Employment, employees includes all salaried employees who work for an enterprise and receive compensation. It includes part-time and seasonal workers, apprentices and persons on training contracts, homeworkers receiving compensation, directors of incorporated enterprises, persons on leave (short term leave as well as indefinite leave), persons on strike and persons performing military service. Data are not converted to full time equivalent (FTE).

Compensation of labour, employees corresponds to personnel costs, i.e. wages and salaries paid to employees plus employers' social costs.

Wages and salaries, employees refer to wages and salaries paid to employees in cash or in kind in return for work done during the reference year and wages paid to manual workers, including overtime and night work cash payments. Are included gratuities, tips, severance payments, workplace and performance bonuses, ex gratia payments, thirteenth month pay, seniority, attendance, incentive and productivity bonuses, payments for day off, such as leave, holidays, and sick leave. Are also included taxes and social insurance contributions payable by employees but deducted by employers.

Employers' social costs include both obligatory and voluntary contributions. They cover employers' social security contributions to schemes for sickness, maternity leave, invalidity, old age, unemployment, work accidents and family allowances, employers' contribution to life insurance for employees, costs related to medical care and drugs given free of charge and costs of maintaining schools and of sporting and recreational activities.

According to relevant EU regulations, the *number of enterprises* is defined as a count of the number of enterprises registered to the population concerned in the business register corrected for errors, in particular frame errors. Dormant units are excluded. It should include all units active during at least a part of the reference period.

Publications

i) *Summary of manufacturing industries*, annually
ii) *Production statistics of energy and water supply enterprises*
iii) *Production statistics of construction*
iv) *Construction figures*
v) *Monthly bulletin of distribution and service trade statistics*, Kerncijfers Detailhandel (Core data retail trade)

NEW ZEALAND

Statistics New Zealand, Wellington, supplied the information.

Industrial Classification

As from 1994, the classification used is the Australian and New Zealand Standard Industrial Classification (ANZSIC-96). There is a concordance up to the 4-digit level between ISIC revision 3 and ANZSIC-96, but not for all industries.

The ANZSIC-based series from 1987-1993 is likely to be of a lower quality than the series from 1994 onwards because ANZSIC codes were added to the data retrospectively.

For all variables, where a single ANZSIC category (ANZSIC96 or ANZSIC87) would normally translate to more than one ISIC revision 3 category, this concordance shows only the predominant ISIC revision 3 category. Predominance was determined by Statistics New Zealand's Classifications and Standards Section using knowledge of the two industrial classifications and the New Zealand economy to select the single ISIC revision 3 category that best represents a particular ANZSIC category (ANZSIC96 or ANZSIC87). Some accuracy is lost in this translation as not all ISIC revision 3 categories are represented.

For trade data, the Harmonised System Classification 1996 (HS) is also used. HS is concordant with ANZSIC-96; then ANZSIC-96 is concordant with ISIC revision 3.

I. INDUSTRY AND SERVICE STATISTICS

Source

Data are taken from the Annual Business Frame Update Survey (ABFUS). The survey is conducted in mid-February each year to update information held on Statistics New Zealand's Business Frame.

Concepts and Definitions of Variables

Employment, employees relates to the sum of the persons (manual and non-manual) who work for an employer and receive compensation in the form of wages, salaries, gratuities, piecework pay or remuneration in kind. It includes part-time workers and seasonal workers on the payroll, apprentices and persons on training contract, homeworkers receiving compensation, persons who work outside the unit, who belong to it and are paid by it, persons on short term leave, persons on strike, persons performing military service and persons employed on commission unless registered for GST (Goods and Services Tax). It excludes working proprietors and active business partners. Employment data is converted to full-time equivalent (FTE), i.e. they are the sum of full time employees, plus half part-time employees, at the end of February of the following year.

Employment, females is the number of full-time females employees, plus half part-time female employees.

The *hours worked by employees* correspond to the number of hours paid for rather than the number of hours worked.

Wages and salaries, employees correspond to the gross total payments to all employees. All kind of gratuities, tips, commissions, bonuses and other allowances, overtime and night work payments, paid leave, sick leave and expenses incurred by workers in the course of their work are included, as are taxes and social insurance contributions, payable by employees but deducted by employers. Costs of vocational training for workers are excluded.

Establishments correspond to geographic units.

Enterprises are single business entities operating in New Zealand. Enterprises in the private sector generally correspond to legal entities such as companies, trusts, estates, incorporated societies, partnerships, producer boards, sole proprietorships. Some enterprises are defined by reference to other characteristics, such as a set of common objectives or the existence of an organisational structure such as government departments, local authorities, churches, voluntary organisations, etc.

II. TRADE STATISTICS

Source

Data are taken from CusMod, an integrated technology system enabling importers and exporters to file electronically with customs. The period used is the calendar year.

For the purpose of overseas trade statistics, "New Zealand" includes the North, South, and Stewart Islands as well as the Antipodes, Auckland, Bounty, Campbell, Kermadec and Chatham Islands, Ross Dependency in Antarctica and the Snares.

Concepts and Definitions of Variables

Exports (including re-exports) are valued, in New Zealand dollars, free on board (fob), which is the value of goods at New Zealand ports before export. Values given in foreign currencies are converted by Statistics New Zealand into New Zealand dollars using weekly exchange rates when the statistics are compiled. Re-exports are exported goods which were earlier imported into New Zealand and which include less than 50% New Zealand content by value. The following are excluded from export data:

- goods consigned to New Zealand forces overseas or diplomatic representatives overseas;
- goods consigned for modification or repair;
- currency transactions in gold, silver, current coin;
- consignments valued under $1000;
- second hand clothing for foreign aid projects;
- replacements, short-shipped or short-packed goods;
- returnable containers and returnable samples;
- aircraft parts for use in New Zealand aircraft overseas and unserviceable parts removed from foreign aircraft and being returned overseas;
- temporary trade items and tourist effects leaving New Zealand.

Imports are valued in New Zealand dollars for both vfd (value for duty) and cif (cost including insurance and freight). These values are converted from the foreign currency when import documents are processed by the New Zealand Customs service. The exchange rates used are set by the Customs department on a fortnight basis. The following are excluded from import data:

- temporary imports into New Zealand such as tourist effects, returnable samples and containers, transit goods;
- fish landed by New Zealand vessels;
- goods (other than motor vehicles) for officials of overseas countries;

- passenger baggage imported permanently;
- consignments valued under $1000;
- currency transactions in gold, silver, current coin;
- goods on loan;
- replacements, short-shipped or short-packed goods recorded previously;
- goods imported for use by foreign armed forces.

Publications

i) *Business Activity Statistics*
ii) *Labour Market Statistics*
iii) *New Zealand Official Yearbook*
iv) *Key Statistics* (periodical)

NORWAY

Industrial statistics data are provided by Statistisk Sentralbyrå, Statistics Norway, Oslo.

Industrial Classification

The revised Norwegian Standard Industrial Classification (SIC 94) has been used from 1993. The classification has a six-level hierarchical structure. SIC 94 is identical with NACE Rev.1 down to the 4-digit level, the 5-digit level being a subdivision for national needs. It has a concordance with ISIC Revision 3.

INDUSTRIAL STATISTICS

Source

From 1994 the annual enquiry covers all establishments in mining, quarrying and manufacturing industries. Enterprises with individual proprietorship where the owners work individually ("one-man establishment"), as well as establishments with less than half a man-year are not included in this survey. In 1992 and 1993, the annual enquiry covered, in general, all establishments in mining, quarrying and manufacturing industries with 10 or more persons engaged. Previous annual enquiries covered, in general, all establishments with 5 or more persons engaged. Information on oil and gas extraction is not included. For small establishments (less than 10 persons), information is collected yearly from administrative registers. The reference unit is the establishment. The reference period is the calendar year.

Concepts and Definitions of Variables

Production (gross output) is valued in producers' prices, excluding all subsidies, and including all indirect taxes but VAT. Included is the value of all goods produced on own account, repair work for others, mounting or installation of own products and merchandise, contract work, *other* work, own-account investment work, own-account repair work, rental receipts and gross profit of goods sold in the same condition as purchased.

Value added is valued at market prices and equal gross value of production less cost of goods and services consumed. It excludes all subsidies and including all taxes but VAT. The cost of materials, fuels and other supplies, contract and commission work done by others, repair and maintenance work done by others and electricity purchased are deducted from gross output. Cost of non-industrial services and receipts for non-industrial services are included. Input estimates are made on a consumed basis.

Investment (gross fixed capital formation) is defined as the acquisition of fixed durable assets, new and used, with an expected productive life of more than one year, including own account investment work, less receipts from sales of fixed durable assets. Acquisition of machinery, transport equipment etc. is recorded at the time when the assets are received, whether or not put into operation during the year. Purchase of buildings, structures, land etc. is treated alike. As for construction work in progress, the value of work carried out during the year is recorded, whether or not the project is completed within the period. The cost of acquisition means actual expenses incurred during the year, including investment levy, irrespective of time of payment. The figures on gross fixed capital formation do not comprise all capital used in mining, quarrying and manufacturing.

Investment in machinery and equipment is also capital formation, and follows the definition for total investment. Included are: transport equipment, industrial machinery and equipment, *office* machinery, equipment and furniture, professional instruments and equipment, and alterations and improvements to these items.

Employment, employees gives the number of salaried employees and wage earners, calculated as the average number of persons working in the establishment throughout the year, whether full- or part-time. It comprises salaried managers and directors as well as transport workers, messengers, newsmen, watchmen and cleaning personnel but not homeworkers and persons on military leave. Persons on sick leave, vacation and strike are included.

Hours worked by employees cover hours worked by salaried employees, including overtime. Foremen and supervisors are included, also apprentices, casual workers and part-time workers.

Compensation of labour, employees covers wages and salaries in cash or kind, other benefits for the employees and social expenses levied by law. It does not include compensation of individual proprietors, partners or family workers without regular wages.

Wages and salaries, employees comprise all payments, whether in cash or in kind, made by the employer in connection with work done, to all persons included in the count of employees (also employed outworkers). It includes taxes, social insurance and pension contributions payable by the employee, but deducted by the employer. Bonuses, production awards, holiday allowances and wages paid during periods of sickness, military leave or other absences, and commission earned by salesmen and representatives, are also included.

As from 1994 data, *employers' social costs* include the other benefits for the employees and social expenses levied by law. Other benefits for the employees comprise employers' contributions to private pension, family allowance, health and casualty insurance, life insurance and similar schemes. Before 1994, data includes only social expenses levied by law.

An *establishment* is defined as a functional unit which at a single physical location is engaged mainly in activities within a specific activity group of the Norwegian Standard Industrial Classification.

Publications

i) Industristatistikk (annual) -- Mining, Quarrying and Manufacturing;

ii) Elektrisitetstatistikk (annual) -- Electricity and Gas Supply;

iii) Bygge- og anleggsstatistikk (annual) -- Construction.

Published by Statistics Norway.

POLAND

The Central Statistical Office (G•ówni Urzad Statystyczny), Warszawa, collects statistical data concerning economic activity of industrial enterprises.

Industrial classification

The Polish classification EKD (Europejska Klasyfikacja Dzia•alno•ci), introduced into the Polish statistics in 1993, has been used up to 1999. Since 2000, the PKD classification (Polska Klasyfikacja Dzia•alno•ci) is used. There is a direct correspondence with ISIC Revision 3 and NACE Revision 1 at the 1- and 2-digit levels.

INDUSTRY AND SERVICE STATISTICS

Source

A variety of surveys are conducted:

- the annual survey on economic activity;
- the annual survey on enterprises employing up to 9 persons;
- the annual survey on employment on a given day;
- and the survey on employment, earnings and work hours (which is carried out monthly, quarterly and annually.

The statistical unit used is the enterprise.

Concepts and Definitions of Variables

The value of sold *production* is defined as revenue from selling own products (goods and services) and value of manufactured goods which are not included in the sale, but are treated on an equal basis (for example value of goods and services used for representative and advertisement needs). There is no adjustment for works in progress and for changes in stock of finished products. Transactions within establishments of the same enterprise are not included. Data for 1992 and 1993 include turnover tax, VAT (introduced on 5 July 1993) and excise tax. For 1993, 1994 and 1995, data include VAT and excise tax. For 1995 and 1996 VAT and excise tax are not included and price supplements are accounted for.

Gross *value added* is calculated for all units of the national economy, irrespective how many persons they employ. Since 1994, gross value added is measured at basic prices (before it was done at producers' prices). This change was caused by the introduction of a new tax system (July 5, 1993), including VAT on goods and services, as well as excise tax. Basic prices exclude VAT and excise tax, but include subsidies on products. Gross value added is calculated as the value of output less the value of intermediate consumption. Different computation methods are applied for market and non-market units. Intermediate consumption is equal to the value of costs of used raw materials (including packaging), fuel, energy, coal, gas, external services (external processing, repair services, transport services, equipment renting, telecommunication, computer services, insurance services), expenditure on business trips and other expenditure (e.g. advertising costs, rents, costs of business ticket). The value of gross value added includes estimates of the "hidden economy" - concealed production (production which is not forbidden by law, but its scope, partly or entirely is concealed from public administration authorities).

Investment outlays are financial or tangible outlays of all units of the national economy (regardless of the number of persons employed), in order to create new fixed assets or improve (rebuild, enlarge, reconstruct, adapt or modernise) existing property, and outlays for primary equipment. Outlays for fixed assets include:

- buildings and structures, e.g. construction and assembly works, design and cost estimate documentation,
- machinery, technical equipment and tools (including instruments, movables and fittings),
- transport equipment,
- other, e.g. land amelioration, expenditures on purchase of land and second-hand fixed assets, livestock (since 1995), long-term planting and interests on investment credits and investment loans for the period of investment realisation (data given exclusively at current prices).

Other outlays include investments on primary equipment and costs of realisation of investment. These outlays do not include value of fixed assets.

Employment, employees refers to the number of salaried employees performing work providing earnings (in the main place of work), both full- and part-time workers. They include persons hired on the basis of employment contracts (labour contract, call-up, appointment or elections).

Wages and salaries, employees correspond to monetary payments as well as the value of benefits in kind or their equivalents due to employees for work performed. They include personal wages and salaries (excluding those of apprentices, home-workers and persons employed abroad), payments from profit for distribution and balance surplus in co-operatives, annual extra wages and salaries for employees of budgetary sphere entities and fees paid to selected groups of employees for performing work in accordance with a labour contract (journalists, producers of films and radio and television programmes).

Hours worked by employees are defined as normal and overtime working hours. They exclude paid hours not worked due to various reasons such as holidays, illness, occasional leaves.

An *enterprise* is a legal entity, i.e. legal persons, independent organisational entities without legal personality as well as natural persons conducting economic activities.

Publications

i) *Statistical Yearbook* (annual)
ii) *Labour Yearbook* (every two years)
iii) *Employment in national economy* (annual)

All publications are distributed by the Central Statistical Office, Warsaw.

PORTUGAL

Data are supplied by the National Institute of Statistics (Instituto Nacional de Estatística), Lisboa.

INDUSTRY AND SERVICE STATISTICS

Source

Data are taken from annual surveys in manufacturing and mining industries, electricity, gas, building and public works.

Data are collected via the Annual Enterprise Inquiry, an enterprise-based survey covering 65 000 enterprises in the mining, manufacturing and electricity, gas and water sectors.

Figures for production, value added, employment, investment and number of enterprises cover the entire industry population. Figures for employment-employees, investment in machinery and equipment, wages and salaries, supplements to wages and salaries and hours worked cover enterprises employing 100 persons or more.

Concepts and Definitions of Variables

Production is given in terms of gross value of production, defined as the total value of finished goods produced by the enterprise (intermediate goods sold as such are considered as finished products), plus the value of fixed assets produced for own use, industrial works done for other users, electricity sold, waste products sold to third persons, less the value of work in progress at the beginning of the year.

Value added is calculated as the gross value of output, plus the value of work done by the enterprise in the capacity of sub-contractor, less the value of materials, lubricants and energy consumed, hiring of machinery, and contract work done by others, the value of industrial work, repair and maintenance work done by others and non-industrial services received, and the value of duties paid for utilisation of manufacturing processes.

Investment corresponds to gross investment in all tangible goods during the reference period.

Employment, employees covers all salaried employees who work for an employer and who have a contract of employment and receive compensation in the form of wages, salaries, fees, gratuities, piecework pay or remuneration in kind. It includes the persons on short leave (vacation and sick, strike...), but does not include the persons on military service or on the retired list.

Compensation of labour, employees is defined as the total remuneration, in cash or in kind, payable by an employer to an employee in return for work done by the latter during the reference period plus supplements to salaries and wages, plus employer's contributions.

Wages and salaries, employees are defined as the remuneration, in cash or in kind, payable to employees counted on the payroll, in return for work done during the accounting period.

Enterprise means any production unit, whether it is subsidiary to an establishment classed in another sector of activity, is under direct private management or is a public sector body carrying out work under direct management.

Publications

i) *Industrial Statistics* (Volume I and Volume II);
ii) *Statistics on Buildings and Dwellings.*

SPAIN

The Spanish National Institute of Statistics (Instituto Nacional de Estadística, Ministerio de Economía y Hacienda, Madrid) supplied the information.

Industrial Classification

The industrial nomenclature used in the survey is the revised National Classification of Economic Activities – 1993 (Clasificación Nacional de Actividades Económicas - CNAE 1993). It is equivalent to NACE Revision 1 at the 4-digit level and has a correspondence with ISIC Revision 3.

INDUSTRIAL STATISTICS

Source

Statistics result from the annual Business Industrial Survey (Encuesta Industrial de Empresas), covering all industrial establishments employing one or more paid persons. Units with more than 20 persons engaged are surveyed exhaustively every year; units with less than 20 persons engaged are surveyed by sample. The total annual sample includes approximately 40 000 establishments.

Concepts and Definitions of Variables

Production follows the Spanish definition of gross output, this being the aggregate total of a) the output of goods and services for sale, including payments received for services and work performed by others and the value of electricity sold; b) goods resold in the same condition as received; c) payments for services rendered (rental of machines, technical assistance, etc.); d) products and work in progress; e) capital formation for own account (fixed assets produced for own use and the value of major repairs and improvements to own capital assets using own resources). The value of this gross output is calculated at producers' prices since operating subsidies are excluded from the initial amount and indirect taxes have been included.

Value added is calculated as the difference between the value of gross output at producers' prices and intermediate consumption at purchasers' prices. Intermediate consumption includes the cost of a) raw materials and consumables; b) energy consumed for power and heating, steam, electricity, gas, etc.; c) purchases of industrial and non-industrial services; d) goods purchased for resale in the same condition as received.

Investment (gross fixed capital formation) at current prices is calculated by subtracting the value of sales of own fixed assets from the value of purchases of new and used fixed assets, major repairs and improvements to fixed assets performed by others, construction of assets for own account and major repairs and improvements to fixed assets carried out using own materials and labour. The valuation of investment is made at acquisition costs. It includes transportation costs.

Employment, total is calculated as the number of persons engaged, including both paid and unpaid staff and office staff as well as operatives, with the exception of home-workers. Employment data are converted to full-time equivalent (FTE).

Hours worked by employees correspond to the total number of hours effectively worked by all salaried employees during the reference year.

Wages and salaries, total is gross salary and wages paid to all persons engaged during the reference year and payments in kind to the paid staff, excluding employers' social costs.

Employers' social costs are social contributions paid by the employer (social security, pensions, and other social costs, including employers' contributions to life insurance for employees). They have to be specified in a collectively agreed arrangement.

Publications

Encuesta Industrial de Empresas, Instituto Nacional de Estadística.

SWEDEN

The information is supplied by Statistics Sweden (Statistiska Centralbyran), Stockholm.

Industrial Classification

Data are classified according to the Swedish Standard for Industrial Classification 1992 (SE-SIC92) which is identical to ISIC revision 3 on levels 1-2 and to NACE revision 1 on levels 1-4.

INDUSTRY AND SERVICE STATISTICS

Source

Up to 1996, the data are taken from the Bureau's industrial statistics and are derived from its annual manufacturing inquiry. The statistical reference unit is the establishment. The survey covers manufacturing establishments (with five or more persons) in manufacturing enterprises employing ten or more persons and manufacturing establishments (with ten or more employees) in non-manufacturing enterprises. Non-respondents are estimated.

From the reference year 1997, data are produced according to the EU Regulation on Structural Business Statistics. The statistical reference unit is the enterprise. There is no cut-off any longer: all enterprises are included.

Concepts and Definitions of Variables

Production at factor costs is gross output, excluding all indirect taxes and including subsidies. The value of all products, net change of work in progress, services rendered, waste products and goods made by homeworkers are included.

Value added at factor costs is calculated as production less total costs except wages and salaries (including social costs). Input estimates are on a consumed basis.

Employment, employees data are taken from the average number of wage earners throughout the year, combined with the number of salaried employees at a certain date. They do not include working proprietors, active business partners and self-employed persons. All full- and part-time employees are included, as well as clerical workers or supervisory personnel performing functions closely related to the production process.

Wages and salaries, employees include regular and overtime cash payments made to salaried employees, bonuses and cost-of-living allowances, vacation and sick leave pay, as well as taxes and social insurance contributions, etc., payable by employees and deducted by employers and payments in kind. Social security costs are not included.

Publications

Manufacturing: Parts I and II in the series of *Official Statistics of Sweden* (SOS) published by Statistics Sweden.

Data are also partly obtainable from the statistical databases at Statistics Sweden.

SWITZERLAND

Statistics are prepared by the Swiss Federal Statistical Office (Office fédéral de la statistique - OFS - Bundesamt für Statistik), Neuchâtel.

Industrial Classification

The general classification of economic activities (NOmenclature Générale des Activités économiques - NOGA) came into effect in 1995. It replaces the 1985 classification (Nomenclature Générale des Activités Economiques - NGAE).

Up to the 4-digit level there is a complete correspondence between NOGA 1995 and NACE Revision 1, the 4-digit level being further subdivided into sub-classes for internal use. The NOGA 1995 is compatible with the ISIC Revision 3.

INDUSTRY AND SERVICE STATISTICS

Source

Data correspond to new information available in the framework of the business statistics for which no base allowing the calculation of a set of consistent data exists before 1997. They are not compatible with previous data given as National accounts based statistics.

Data are collected through the annual survey of production and value added ("Statistique de la production et de la valeur ajoutée"). It is based on a sample drawn from the business register (Registre des Etablissements et des Entreprises - REE). Data cover the calendar year.

The enterprise is the reference unit.

Concepts and Definitions of Variables

The value of *production* measures the exact output of the observed unit. It is defined as the turnover adjusted for changes in the value of stocks, less the acquisition of goods and services shipped in the same condition as received (resale), plus the immobilised production and other operating incomes. The valuation is made at market prices.

Value added at factor costs relates to the gross value of operating activities taking into account subsidies and indirect taxes. It is calculated as turnover plus the immobilised output and other operating products, adjusted for changes in the value of stocks, less the acquisition of goods and services and other taxes on products related to turnover but not deductible and taxes related to production.

Employment includes all persons engaged in an establishment and working at least 6 hours per week. It includes working proprietors, active business partners, unpaid family workers, apprentices, commercial travellers and persons who work outside the unit. Homeworkers, agents or experts working independently are excluded. Establishments where employees work less than 20 hours per week are not covered. Employment data are converted to full time equivalent (FTE).

Compensation of labour, employees is the sum of wages and salaries paid to salaried employees, and employers' social contributions.

Wages and salaries, employees relates to payments made in cash or in kind to salaried employees, including homeworkers, in return for work done during the reference year, however they are paid (on a time, piecework or job-work basis, regularly or not). Are included gratuities, tips, workplace and performance bonuses, ex gratia payments, thirteenth month pay, cost of living, housing, transport and family allowances, seniority, attendance, incentive and productivity bonuses, payments for day off, such as leave, holidays, and sick leave. Are also included taxes and social insurance contributions payable by employees but deducted by employers.

Employers' social costs are employers' social contributions to schemes for sickness, maternity leave, invalidity, old age, unemployment, work accidents and family allowances. The employers' contributions are to be specified in a statutory arrangement.

Publications

i) Statistiques de la production et de la valeur ajoutée/Produktions- und Wertschöpfungsstatistik
Publications are distributed by OFS.

TURKEY

Data are obtained from the State Institute of Statistics (Devlet Istatistik Enstitüsü), Ankara.

I. INDUSTRIAL STATISTICS

Industrial Classification

The Turkish classification of activity (Ulusal Faaliyet S•n•flamas•) is an adapted version of the ISIC revision 2 up to 4-digit level.

Source

Data are derived from a number of sources: the General Census of Industry and Business Establishments conducted in the years ending with a three, the Annual Manufacturing Industry Survey (for all manufacturing establishments in the public sector and establishments employing 10 or more persons in the private sector), the Small Sized Manufacturing Industry Statistics (for establishments with 1-9 persons engaged), the Annual Electricity Generation Survey, the Annual Electricity Distribution Survey, the Annual Municipality Water Statistics Survey, , the Annual Natural Gas Distribution Statistics Survey, the Construction and Installation Establishments Survey, the annual Mining and Quarrying Statistics Questionnaire, the annual Sand, Clay and Stone Questionnaire.

Concepts and Definitions of Variables

Production, or output, is the value of output calculated by subtracting the value of the beginning of the year stock (finished and semi-finished goods and goods purchased to be resold without any further processing) from the total of receipts from sales and services rendered to others, receipts from sales of transfers of electricity plus the end-of-year stock (finished and semi-finished goods and goods purchased to be resold without any further processing) and the production value of fixed assets produced by the establishments' staff for own use. Production is valued at producers' prices.

Value added, also valued at producers' prices, is obtained by subtracting the value of inputs from output. The value of inputs is calculated by subtracting the value of the end-of-year stock (raw materials, supplementary materials, packaging materials and fuel) from the total value of goods and services purchased or transferred, electricity purchased and the beginning-of-year stock (raw materials, supplementary materials, packaging materials and fuel).

Investment is gross additions to the fixed assets during the year. It is calculated by subtracting the sales value of fixed assets sold during the year from total expenditures made on new or used fixed assets purchased from the domestic market, fixed assets imported new or used, fixed assets produced by the establishments' own staff, parts of fixed assets installed during the year purchased on a bid basis, major repairs and expenditures made on fixed assets, studies and plans, drawings, machinery, equipment, motor vehicles and building, other construction, office equipment and furniture (used by the establishment and expected to have a productive life of more than one year and recorded in the capital accounts) and expenditures on land and land improvements.

Investment in machinery and equipment refers to gross additions to machinery and equipment during the year. It is calculated by subtracting the sales value of machinery and equipment sold during the year from the expenditures made on: new or used machinery and equipment purchased from the domestic market, machinery and equipment imported new or used, machinery and equipment produced by the establishments' own staff, part of machinery and equipment installed during the year purchased on a bid basis, major repairs and expenditures made on machinery and equipment, machinery and equipment which is used by the establishment and expected to have a productive life of more than one year.

Employment, total or the average number of persons engaged, is obtained by adding the number of owners, partners and unpaid family workers, who were active in November (in July in Mining and Quarrying; in August in Construction) in the establishments, to the average number of employees. Employees who were temporarily absent on this period because of illness, annual leave or strikes, etc. are included. Employment data are converted to full-time equivalent (FTE).

Employment, employees corresponds to the arythmetic average of the number of employees in February, May, August and November (average of the twelve months in Mining and Quarrying and in Construction). It covers the workers (operatives, and administrative and other employees) in the establishments whatever their title, who were paid wages and salaries and the like. Unpaid family workers, owners and partners are excluded.

The *hours worked by operatives* in one year are calculated by multiplying the average number of production workers by the hours worked in one shift and the number of days worked in a year.

Compensation of labour, employees includes all payments in cash or kind made to employees only during the reference year and employers' social costs.

W*ages and salaries, employees*, or annual payments to employees, include all payments in the form of wages and salaries and daily-wage before deduction of income tax, social security and pension fund premiums. It excludes social security and pension contributions and the like payable by the employer. It includes overtime payments, bonuses, indemnities and payments in kind.

Employers' social costs include payments for employees to social security, lay-off and severance pay, saving fund and housing allowances.

An *establishment* is a work place where single ownership or controlling status is established and generally where one type of economic activity occurs and where there are a register and accounting sources necessary to fill in the questionnaires.

II. SERVICE AND FOREIGN TRADE STATISTICS

Industrial Classification

The classification used is the ISIC revision 3.

Source

Data are derived from the annual manufacturing survey conducted by the State Institute of Statistics, Ankara, except for 1992 for which data are derived from the Census of Industry Establishments.

Concepts and Definitions of Variables

Production, or output, is the value of output calculated by subtracting the total value of sales of goods at the beginning of the year and goods purchased to be resold without any further processing from the total value of sales of goods and services, stock of trade goods at the end of the year and production value of fixed assets produced by the establishments' own staff.

Value added is obtained by subtracting the value of inputs from output. The value of inputs is calculated by subtracting the total value of stock of material and fuel at the end of the year and goods purchased to be resold without any further processing from the total value of the purchases of goods and services and stock of material and fuel at the beginning of the year.

Investment, defined as fixed capital investment, covers fixed assets used over one year and amortisation allocated to building, machinery and equipment in inventory stock, vehicle, office equipment and furniture in inventory stock and land. It also covers modifications, renewals, developments and major repair of those.

Investment in machinery and equipment refers to gross additions to machinery and equipment during the year. It is calculated by subtracting the sales value of machinery and equipment sold during the year from the expenditures made on those purchased as newly from the domestic market, those purchased as used from the domestic market, those imported as new or used, those made by establishment's own staff, and expenditures on slow assets.

Employment, employees corresponds to the arithmetic average of the number of employees in February, May, August and November. It covers the workers in the establishments whatever their title, who were paid wages and salaries and the like. Unpaid family workers, owners and partners are excluded. Employees who were temporarily absent on this day because of illness, annual leave or strikes, etc. are included.

W*ages and salaries, employees*, or annual payments to employees, includes all payments in the form of wages and salaries and daily-wage before deduction of income tax, social security and pension fund premiums. It excludes social security and pension contributions and the like payable by the employer. It also includes overtime payments, bonuses, indemnities and payments in kind.

Exports and *Import*s cover international movements of goods only across customs borders. Exports are fob (free on board) and imports cif (cost, insurance and freight).

An *establishment* is a work place where single ownership or controlling status is established and generally where one type of economic activity is carried out and which has the registers and accounting sources necessary to fill in the questionnaires.

Publications

i) *Annual Manufacturing Industry Statistics*, annual
ii) *Small-Sized Manufacturing Industry Statistics*, annual
iii) *Annual Electricity, Gas and Water Statistics*
iv) *1992-1997 Construction and Installation Establishments Statistics*
v) Mining and Quarrying Statistics

UNITED KINGDOM

I. INDUSTRIAL STATISTICS

The information is provided by the UK Office for National Statistics (formerly the Central Statistical Office).

Industrial Classification

Results of the annual censuses from 1973 to 1979 were all based on the Standard Industrial Classification Revised 1968.

Since 1980, censuses have been based on the Standard Industrial Classification Revised 1980. In order to provide a link between earlier results and those for 1980 and beyond, the 1979 results were reworked on the basis of the new classification.

In 1990, a new Standard Industrial Classification, SIC(92), based on NACE revision 1, was introduced. The 1993 Census was the first conducted on SIC(92).

Source

The data is collected in the Annual Sample Survey into Production and Construction (formerly known as Annual Census of Production and Annual Census of Construction).

Concepts and Definitions of Variables

Production, or gross output, is calculated by adjusting the value of total sales, work done and services rendered, by the net changes during the year of work in progress and goods on hand for sale and for 1996-1997 only, work of a capital nature by own staff. Total sales represents deliveries on sale of goods produced by establishments in the United Kingdom coming within the scope of the census and includes sales of goods made for establishments from materials given out by them to other organisations or to outworkers, and the sales of waste products and residues. The value of sales is the "net selling value", i.e. the amount charged to customers whether values "ex-works" or "delivered" less VAT, trade discounts, agents' commissions, etc., and allowances on returned goods. Where products attract excise duty, the value is inclusive of duty if goods are sold "duty-paid" and exclusive of duty if goods are sold in bond or exported. Sales of fixed assets are excluded.

Value added, or gross *value added at factor cost*, is calculated as the value of gross output less the cost of purchases of materials for use in production and packaging, fuels, goods for merchanting or factoring, the cost of industrial and non-industrial services received, rates and the cost of licensing of motor vehicles. The value of purchases is adjusted for changes during the year of stocks of materials, stores and fuel. The valuation is in factor values, i.e. excluding indirect taxes and including subsidies.

Investment, or gross fixed capital formation, is the value of purchases of fixed assets less the value of sales of such assets. The valuation is at full cost incurred, including the cost of installation and certain fees and taxes (namely legal fees, stamp duty, agents' commissions, etc.), and is inclusive of any amounts received or expected to be received in grants and/or allowances from government sources, statutory bodies or local authorities. In general, the value is that charged to capital account, together with any other amounts which rank as capital items for taxation purposes during the year. Where expenditure is spread over more than one census year, payments are included in the years in which they were made. Capital expenditure during the year in respect of production units where production had not started before the end of the year, and the value of capital goods produced for establishments' own use by establishments' own staff are included. The value of any assets acquired in taking over an existing business is excluded. The figures include non-deductible VAT but exclude deductible VAT. No allowance is made for depreciation, amortisation or obsolescence.

As from 1993, data for *investment in machinery and equipment* refers only to that manufactured or built for the establishment own use: it represents the value of all work of a capital nature carried out during the year by the staff of, and the use in, the business covered by the return.

Employment, employees gives the average number of full time and part time administrative, technical and clerical employees and operatives on payroll employed during the year of return. It does not include working proprietors, i.e. persons regarded as self-employed for national insurance purposes, members of their families who worked in the business without receiving a definite wage or salary for at least half the normal working hours and directors who worked in the business but did not receive a definite wage, salary or commission. It includes directors who received a definite wage, salary, fee or commission, managers, foremen, draughtsmen, editorial and advertising staff, travellers, all office employees and research and design employees except operatives.

Compensation of labour, employees represents amounts paid during the year to employees. They include wages and salaries and supplements. Employers' National Insurance contributions etc. are included, as is outworkers remuneration. Outworkers remuneration corresponds to amounts paid to outworkers, i.e. people who do work in their own homes generally on a piecework basis, whose names appear on the payroll. Amounts paid to outworkers by subcontractors are included in the cost of industrial services received. Estimates are not made for remuneration of outworkers for businesses not completing returns.

Wages and salaries, employees represent amounts paid during the year to employees. All overtime payments, bonuses, commissions, holiday pay and redundancy payments less any amounts reimbursed for this purpose from government sources are included. No deduction is made for income tax or employees' National Insurance contributions etc. Payments to working proprietors, payments in kind, travelling expenses, lodgings allowances etc. and employers' National Insurance contributions are excluded.

Employers' social costs include employers' National Insurance contributions, i.e. employers' National Insurance contributions under the Social Security Pensions Act 1975, commercial insurance premiums for policies providing pensions, superannuation or other retirement benefits, sickness benefit, personal accident benefits, disability benefits or death benefits for employees or their dependants.

Establishments are defined as the smallest unit which can provide the information normally required for an economic census.

Enterprises are defined as the smallest combination of legal units which have a certain degree of autonomy within an enterprise group. An enterprise group is simply a number of enterprises under common ownership or control.

II. FOREIGN TRADE STATISTICS

Industrial Classification

Customs and Excise compile statistics using the Standard International Trade Classification (SITC). They are then aggregated to the UK Standard Industrial Classification by the Office for National Statistics, which afterwards performs the conversion to the ISIC.

Source

Data are collected by Customs and Excise officials and subsequently passed on to the Office for National Statistics.

Concepts and Definitions of Variables

General trade is used. *Exports*, fob, are classified according to country of ultimate destination, *imports*, cif, according to country of consignment. State trading, commercial transactions in military equipment and supplies, foreign aid transactions, improvement and repair trade, transactions in new ships and aircraft, electric energy and gas, and private donations and gifts are included. Sea products landed direct from the high seas are also covered, though natural produce of the sea brought ashore by British registered vessels is excluded from imports. Direct transit trade and transactions in second-hand ships and aircraft are specifically excluded; temporary exports, monetary items, industrial gold and minor transactions are not.

Publications

i) *Census of Production Introductory Notes* (Business Monitor PA 1001)
ii) *Annual Abstract of Statistics*
iii) *Overseas Trade Statistics* (monthly)

The two first publications are distributed by the Office for National Statistics (London), the third one is published by HMSO.

ALLEMAGNE

Les informations émanent de l'Office Fédéral de Statistique (Statistisches Bundesamt) de Wiesbaden.

Les données relatives aux années antérieures à 1990 se réfèrent à l'Allemagne occidentale (République Fédérale d'Allemagne avant unification de l'Allemagne, y compris Berlin Ouest). Les données à partir de 1991 se réfèrent à l'Allemagne après unification, sauf les données concernant la valeur ajoutée, l'investissement et les dépenses en machines et équipement, qui continuent à ne couvrir que l'ancienne Allemagne occidentale.

Classification industrielle

Les données sont collectées et compilées selon la classification nationale (Klassification der Wirtschaftszweige, Ausgabe 1993 - WZ93). La classification WZ93 correspond à la NACE révision 1 et peut être convertie en CITI révision 3.

STATISTIQUES INDUSTRIELLES

Source

En règle générale, les chiffres concernant les secteurs extractifs et manufacturiers sont tirés de l'enquête mensuelle menée par l'Office de Statistique. Cette enquête s'adresse à toutes les unités locales des entreprises de 20 salariés ou plus, y compris les artisanats de production et les unités de ces entreprises qui sont engagées dans des activités non industrielles.

Concepts et définitions des variables

Les chiffres sur la *production* indiquent le chiffre d'affaires tiré de la production de l'unité tel qu'il ressort des factures, y compris les recettes au titre des marchandises réexpédiées en l'état, de la vente des déchets de fabrication, d'électricité et autres articles similaires. Les recettes au titre des activités non industrielles (paiements d'intérêt, loyers, dividendes, par exemple) ne sont pas prises en compte. La valeur des livraisons ne tient pas non plus compte des taxes sur la valeur ajoutée ou le chiffre d'affaires, mais elle inclut les droits d'accises, les coûts d'emballage, d'affranchissement et autres frais. Les livraisons entre unités de la même entreprise ne sont pas comptabilisées.

Les chiffres sur l'*emploi* indiquent le nombre annuel moyen de personnes occupées en fin de mois. Y sont inclus les propriétaires exploitants, les associés actifs, les travailleurs familiaux non rémunérés, à condition qu'ils travaillent pendant un nombre d'heures équivalent à un tiers au moins de l'horaire normal. Les salariés à temps partiel et temporaires sont pris en compte de même que les apprentis. Les travailleurs à domicile sont exclus.

Les *heures ouvrées, ouvriers* couvrent les heures travaillées, y compris les heures supplémentaires, par les travailleurs manuels qui sont définis comme les personnes occupées soumises à l'assurance obligatoire auprès des caisses de retraite des ouvriers. Cette définition englobe les apprentis, mais non les travailleurs à domicile. Elle comprend également les agents de maîtrise de même que les travailleurs occasionnels et à temps partiel.

Les données concernant les *salaires et traitements, total* ne comprennent pas les cotisations patronales de sécurité sociale. Toutes les sommes versées aux intérimaires sont incluses ; les cotisations aux caisses de retraite des sociétés sont exclues.

Publications

Fachserie 4, Reihe 4.1.1., *Beschäftigung, Umsatz und Energieversorgung der Unternehmen und Betriebe im Bergbau und im Verarbeitenden Gewerbe*, Statistisches Bundesamt, Wiesbaden.

AUSTRALIE

Les informations émanent du Bureau Australien de Statistiques (ABS).

Classification industrielle

La classification utilisée est l'«Australian and New Zealand Standard Industrial Classification» (ANZSIC), alignée sur la CITI révision 3 au niveau des divisions. Les subdivisions, groupes et classes de l'ANZSIC sont souvent des combinaisons de secteurs CITI.

STATISTIQUES INDUSTRIELLES

Source

La plupart des données sont tirées d'enquêtes par échantillon : «Annual Manufacturing Survey», «Annual survey of Other Mining, Services to mining and Water supply industry». L'unité de base est à la fois l'entreprise («management unit») et l'établissement, selon l'information requise. Les enquêtes ne couvrent pas les unités sans employés. La période couverte est l'exercice budgétaire qui débute le 1er juillet.

Concepts et définitions des variables

A l'exception des données d'emploi, toutes les variables collectées dans les enquêtes relatives aux industries manufacturières et extractives et à l'électricité, gaz et eau ont été étudiées de telle sorte qu'elles se rapprochent le plus possible des concepts de la version 1993 du Système des Comptes Nationaux (SCN 93).

La *production* inclut la valeur des livraisons de biens ou services produits par l'unité, les recettes au titre des travaux industriels effectués ou des services industriels rendus aux autres, les ventes de marchandises réexpédiées en l'état (reventes), la variation nette des stocks de produits finis fabriqués par l'unité, des stocks de travaux en cours fabriqués par l'unité et des stocks de marchandises achetées pour la revente en l'état, la valeur de l'électricité vendue, les ventes de produits manufacturés par les sous-traitants lorsqu'il s'agit de production à partir de matériaux appartenant à l'établissement, la valeur des actifs fixes produits par l'unité pour son propre compte et les recettes perçues au titre de la publicité.

La *valeur ajoutée* correspond au produit brut de la branche. Elle équivaut à la production moins la valeur des matériaux et autres fournitures industrielles nécessaires à la transformation, la valeur des combustibles et de l'électricité achetés, les coûts des services industriels reçus, les paiements pour les travaux exécutés sous contrat ou à la commande et les paiements pour les travaux de réparation et d'entretien. La valeur ajoutée est évaluée aux prix de base.

L'*emploi* indique le nombre de personnes occupées (y compris les propriétaires exploitants et les associés pour les entreprises non constituées en société) au 30 juin de chaque année. Sont compris tous les salariés qu'ils soient à plein temps ou à temps partiel, permanents, temporaires ou saisonniers. Les salariés en congé payé le 30 juin sont compris. En sont exclus les travailleurs à domicile, les sous-traitants, les consultants et les autres travailleurs indépendants fournissant des services ainsi que les travailleurs familiaux et les travailleurs bénévoles non rémunérés.

Les *salaires et traitements, salariés* couvrent les salaires et traitements versés à tous les salariés de l'établissement pour l'année se terminant le 30 juin. Ils comprennent les salaires, les traitements, les primes et commissions, les paiements pour heures supplémentaires, les amendes, les congés payés et les provisions pour charges sociales. Sont également compris les salaires et honoraires versés aux directeurs et cadres supérieurs. Les paiements versés aux sous-traitants, consultants et autres travailleurs indépendants, et aux propriétaires et associés d'entreprises individuelles sont exclus, ainsi que les taxes assises sur les salaires et les remboursements de frais aux salariés. Les salaires et traitements s'entendent bruts, avant impôt et autres prélèvements.

Le nombre d'*établissements* est comptabilisé au 30 juin de l'année considérée. L'établissement représente la plus petite unité comptable d'une entreprise au sein d'un état ou d'un territoire australien, ayant le contrôle de ses activités de production et disposant d'une liste déterminée de données détaillées.

Publications

i) *Manufacturing Industry,* Australia, (Preliminary) (ABS Catalogue No. 8201.0) ;
ii) *Manufacturing Industry,* Australia (ABS Catalogue No. 8221.0) ;
iii) *Manufacturing,* Australia (ABS Catalogue No. 8225.0) ;
iv) *Mining, Electricity and Gas Operations* (Preliminary) (ABS Catalogue No. 8401.0) ;
v) *Australian Mining Industry* (ABS Catalogue No. 8414.0) ;
vi) *Mining Operations,* Australia (ABS Catalogue No. 8415.0) ;
vii) *Electricity, Gas, Water and Sewerage Industries,* Australia (ABS Catalogue No. 8208.0) ;
viii) *Electricity, Gas, Water and Sewerage Operations,* Australia (ABS Catalogue No. 8226.0) ;

Toutes les publications sont distribuées par le Bureau Australien de Statistiques (ABS).

AUTRICHE

Les données émanent de l'office statistique d'Autriche (Statistics Austria).

Classification industrielle

A partir de 1995, la classification utilisée est la version autrichienne de la NACE révision 1, appelée "ÖNACE 1995". Elle est parfaitement comparable à la NACE révision 1 jusqu'au niveau à quatre chiffres, le quatrième niveau étant subdivisé en sous-classes. L'"ÖNACE 1995" est compatible avec la CITI révision 3.

STATISTIQUES INDUSTRIELLES

Source

A partir de 1995, les données ne sont pas strictement comparables à celles des années précédentes. Les données pour 1995 sont le résultat du recensement non agricole. Il n'y a pas eu d'enquête en 1996. Les données de 1997 et 1998 résultent des recensements structurels annuels.

Les statistiques relatives aux heures ouvrées proviennent de l'enquête mensuelle *«konjunkturstatistik - Produzierender Bereich» (Statistiques conjoncturelles - Industries manufacturières - NACE Sections C à F).*

L'unité statistique est l'entreprise.

Concepts et définitions des variables

La *production* correspond au chiffre d'affaires, y compris les revenus tirés de la capitalisation des services rendus pour son propre compte, non compris l'achat de biens et services destinés à la revente en l'état (sans frets étrangers). Elle comprend la variation nette des stocks d'articles commerciaux, des produits finis et des travaux en cours.

La *valeur ajoutée au coût des facteurs* est égale à la valeur ajoutée aux prix de base, non compris les impôts et redevances (sans les impôts sur les produits), y compris les subventions d'exploitation (sans les subventions sur les produits). La valeur ajoutée aux prix de base est calculée comme la valeur de la production moins les frets étrangers, les achats de matériaux pour les travaux et fabrications, les dépenses pour les réparations et l'entretien et pour les travaux effectués par d'autres entreprises, les dépenses pour les travailleurs intérimaires, l'achat de combustibles, les loyers, crédit-bail et autres coûts d'exploitation, les dépenses pour actifs faibles, les impôts sur les produits, y compris la variation nette des stocks de combustibles, de matières premières et matériaux consommés, les subventions sur les produits.

L'*investissement* (formation de capital) correspond aux dépenses d'équipement, qui sont définies comme étant la valeur des acquisitions d'actifs fixes, y compris ceux fabriqués par le personnel de l'établissement pour l'usage de ce dernier. La valeur des ventes d'actifs fixes n'est pas déduite. L'investissement comprend les terrains, les constructions et les modifications apportées aux constructions, les machines et équipements, les biens corporels, les concessions, les brevets, les licences, les marques de fabrique et les logiciels. L'investissement est évalué au prix de revient total, y compris le prix rendu, les frais d'installation, les droits et redevances (à l'exception de la taxe sur la valeur ajoutée).

Les *dépenses en machines et équipement* correspondent aux dépenses d'investissement, telles qu'elles sont définies ci-dessus, c'est-à-dire à la valeur des acquisitions d'actifs fixes, y compris ceux fabriqués par le personnel de l'établissement pour l'usage de ce dernier. Elles comprennent le matériel de transport, les machines et l'outillage industriels, les machines, les équipements et le mobilier de bureau, les instruments et équipements professionnels, les modifications, améliorations et réparations apportées à ces matériels.

Des informations sur l'emploi sont données pour le nombre de salariés et pour le nombre de femmes salariées (au 31 décembre). L'*emploi, salariés* comprend les salariés (employés et ouvriers), les apprentis, les travailleurs à domicile et les travailleurs à temps partiel. Le nombre de femmes salariées est donné sous le titre *Emploi, femmes*.

Les *salaires et charges sociales, salariés* couvrent à la fois les salaires et traitements, et les cotisations patronales aux régimes de Sécurité sociale et de retraite.

Les *salaires et traitements, salariés* comprennent la rémunération brute des salariés (travailleurs à col blanc et à col bleu, apprentis, travailleurs à domicile et travailleurs à temps partiel). Les données comprennent les paiements au titre des heures travaillées, y compris les heures supplémentaires, les primes, les indemnités de vie chère, les congés payés et congés-maladie, les impôts et contributions aux caisses d'assurance sociale, dus par les salariés mais déduits par l'employeur, les paiements en nature et autres indemnités.

Les *charges sociales des employeurs* englobent les dépenses exigées par la loi et les contributions volontaires, telles que cotisations de sécurité sociale (santé, vieillesse, accident, assurance chômage), les cotisations au fonds de péréquation des allocations familiales, les cotisations à la caisse de construction résidentielle et les indemnités de logement, ainsi que les paiements non obligatoires (par exemple en cas d'accident, de maladie ou de grande difficulté matérielle, les services de restauration, les indemnités kilométriques).

Les *heures ouvrées, salariés* correspondent aux heures effectivement travaillées (c'est-à-dire toutes les heures payées à l'exception des congés payés) par les salariés (employés et ouvriers), et par les apprentis, qui sont sur le registre des salaires. Sont inclus les heures supplémentaires, les temps d'attente, les périodes de repos, les congés maladie et les fêtes légales. Les travailleurs à domicile et les travailleurs à temps partiel sont exclus.

L'*entreprise* est généralement la plus petite unité légale. Elle peut comprendre un ou plusieurs établissements (unité d'activité économique locale).

Publications

Toutes les données sont fournies, avec une ventilation allant jusqu'aux positions à trois chiffres de la CITI, dans les publications suivantes :

i) *Industriestatistik*, 2. Teil ; publié par le Bureau central de statistique d'Autriche et sert, avec d'autres publications, à l'établissement des statistiques autrichiennes ;

ii) *Gewerbestatistik*, 2. Teil ; publié par le Bureau central de statistique d'Autriche et sert, avec d'autres publications, à l'établissement des statistiques autrichiennes

CORÉE

Les informations émanent de l'Office national statistique de Corée.

Classification industrielle

La classification coréenne type des activités industrielles est la SIC (révisée en 1992). Elle est compatible avec la CITI révision 3.

I. STATISTIQUES INDUSTRIELLES

Source

Les données sont tirées de l'enquête annuelle sur les industries extractives et manufacturières. L'enquête est menée chaque année. L'unité statistique est l'établissement. Tous les établissements comptant au moins 5 salariés sont couverts par l'enquête : environ 80000 en 1998. La période de référence est l'année civile.

Concepts et définitions des variables

La *production* correspond à la notion de production 'recensée', qui ne couvre que les activités de nature industrielle. L'évaluation est faite aux prix départ-usine, c'est-à-dire les montants reçus par le producteur à l'exclusion de tout impôt et de toute subvention liés à la production et à l'exclusion des frais de transport facturés séparément. Elle comprend la valeur de tous les produits de l'établissement, la variation nette entre le début et la fin de la période de référence de la valeur des stocks de produits finis et semi-finis et des travaux en cours, les recettes pour réparations, la valeur des travaux industriels faits et des services industriels rendus aux autres. Sont exclues la variation nette des stocks de marchandises achetées pour la revente en l'état, les ventes de marchandises réexpédiées en l'état (reventes) moins le montant payé pour ces biens et la valeur des actifs fixes produits par l'unité pour son propre compte.

La *valeur ajoutée* correspond à la notion de valeur ajoutée 'recensée', obtenue en soustrayant les coûts de production (ou entrants) de la valeur de la production recensée. Les coûts de production correspondent à la valeur des matériaux et fournitures pour la production (y compris le coût des combustibles et de l'électricité produite, et de l'eau) et au coût des services industriels reçus (principalement les paiements pour les travaux exécutés sous contrat ou à la commande et pour les travaux de réparation et d'entretien). L'évaluation est faite aux prix départ-usine.

L'*investissement* correspond à la valeur des actifs fixes, achetés ou construits par le personnel de l'unité statistique pour son usage propre au cours de l'année de référence. La valeur des ventes correspondantes n'est pas déduite. Par actifs fixes, il faut entendre les biens (neufs ou usagés) qui ont une durée de vie productive supérieure à un an. Sont comprises les adjonctions, transformations et améliorations d'actifs fixes existants qui allongent la durée d'utilisation normale ou augmentent leur capacité. Les terrains, bâtiments et autres constructions, y compris leur amélioration, le matériel de transport, les machines et autres équipements sont également inclus. Lors de l'achat d'actifs à d'autres établissements, l'évaluation est faite au prix d'acquisition, qui couvre tous les coûts directement liés à l'acquisition et l'installation des biens à utiliser. Les actifs construits par l'unité elle-même pour son propre usage sont évalués de la même manière. Les actifs produits par un établissement d'une entreprise à plusieurs établissements pour être utiliser par un autre établissement de la même entreprise sont évalués par l'établissement receveur comme s'ils étaient achetés à l'extérieur.

Les dépenses en machines et équipement sont ventilées par type de biens d'équipement à partir de l'investissement total consacré aux machines.

L'*emploi, salariés* inclut les salariés, les travailleurs à temps partiel, les travailleurs saisonniers qui figurent sur la liste des salariés de l'entreprise, les personnes en congés de courte durée (congé-maladie, congé annuel, vacances), les personnes en grève et les personnes effectuant des travaux de réparation et d'entretien dans l'unité pour le compte d'autres unités. Il exclut les propriétaires exploitants, les associés actifs, les travailleurs familiaux non rémunérés, les directeurs d'entreprises constituées en sociétés commerciales, les membres des conseils d'administration qui ne reçoivent que des jetons de présence pour les réunions, les personnes prêtées à l'unité observée par d'autres unités et payées pour ces travaux (tels que les travailleurs sous contrat payés par des sous-traitants), les personnes en congé de durée indéterminée, en congé militaire ou retraitées et les travailleurs à domicile. Les chiffres correspondent au nombre moyen de salariés pendant l'année de référence.

La part des femmes salariées est présentée sous *Emploi, femmes*.

Les *salaires et traitements, salariés* correspondent à toutes les sommes en espèces versées et avantages en nature fournis par l'employeur pendant la période de référence à tous les salariés figurant sur le registre du personnel de l'établissement en rémunération du travail effectué pour l'unité observée. Sont inclus les salaires et traitements directs et toutes les sommes en nature versées, ainsi que les paiements pour heures supplémentaires, les paiements des jours non travaillés comme les jours fériés, les congés payés et les congés-maladie, les gratifications et primes, comme les primes de vacances et de performance, les indemnités de logement et les allocations familiales payées directement par l'employeur, les avantages en nature, les impôts et contributions aux caisses d'assurance sociale et autres contributions dus par le salarié mais déduits par l'employeur. Ils ne comprennent pas les paiements au titre de la sécurité sociale, des pensions, retraites et autres contributions patronales.

Les *charges sociales des employeurs* sont les dépenses exposées pour la protection et le bien-être des salariés.

L'*établissement* est défini comme une unité physique avec un statut unique de propriété ou de contrôle, menant une seule activité, ou une activité de manière prédominante, sur un seul site, par exemple une usine ou un atelier.

II. STATISTIQUES DES SERVICES

Source

Les données sont tirées d'enquêtes annuelles : enquête sur le commerce de gros et de détail (50-55) ; enquête sur les transports (60-65) ; enquête sur l'industrie des services (70-95). L'unité statistique est l'établissement. Tous les établissements des divisions et groupes sont couverts par les enquêtes. Une enquête par échantillon est menée pour les classes. La période de référence est l'année civile.

Concepts et définitions des variables

La *production*, ou production brute, est mesurée par la valeur totale de tous les biens produits et de tous les services rendus par l'établissement au cours de l'année. Elle est calculée de la façon suivante : la valeur des livraisons, plus les variations nettes des stocks de produits finis et des travaux en cours pendant l'année. L'évaluation est faite aux prix départ-usine, à l'exclusion de tout impôt et de toute subvention liés à la production, et à l'exclusion des frais de transport facturés séparément.

Publications

i) Report on the Mining and Manufacturing Survey

ii) Annual report on the survey of wholesale & retail trade, service industries, transportation.

Ils sont publiés chaque année par l'Office national de statistique de la République de Corée.

DANEMARK

Les informations ci-après ont été fournies par Danmarks Statistik.

Classification industrielle

Depuis 1993, la collecte et la compilation des statistiques officielles danoises s'effectuent en NACE révision 1.

STATISTIQUES DE L'INDUSTRIE ET DES SERVICES

Source

Les données sur le nombre d'établissements, l'emploi et les rémunérations proviennent des statistiques des établissements et de l'emploi découlant des registres. Tous les établissements du secteur manufacturier sont couverts.

Concepts et définitions des variables

L'*emploi, salariés* comprend tous les salariés et tous les travailleurs indépendants employés fin novembre dans les établissements contenus dans le registre (tels que mentionnés ci-dessus). Sont inclus les salariés à temps plein et les salariés à temps partiel, y compris lorsque cet emploi constitue pour le salarié une occupation secondaire.

Les *salaires et traitements, salariés* comprennent toutes les dépenses des établissements en activité au cours de l'année, déclarées aux autorités fiscales. Il s'ensuit que sont comprises dans les chiffres les dépenses des établissements qui ne sont pas en activité fin novembre, mais l'ont été à un moment quelconque de l'année considérée.

Le nombre d'*établissements* est le nombre total d'unités locales. Tous les établissements en activité à la fin du mois de novembre, quel que soit le nombre de leurs salariés, sont comptabilisés.

Publications

i) *Statistical News*, Collection Industrie et Energie ;

ii) Rapports statistiques : *Industrial Accounts Statistics* ;

iii) *Statistical News*, Collection Statistiques Economiques Générales et Commerce Interne.

Toutes sont publiées par Danmarks Statistik.

ESPAGNE

Les informations émanent de l'Institut National de la Statistique d'Espagne (Instituto Nacional de Estadística, Ministerio de Economía y Hacienda, Madrid).

Classification industrielle

La classification industrielle utilisée dans l'enquête est la Classification nationale des activités économiques révisée - 1993 (Clasificación Nacional de Actividades Económicas - CNAE 1993). Elle est équivalente à la NACE révision 1 jusqu'au 4ème chiffre et est compatible avec la CITI révision 3.

STATISTIQUES INDUSTRIELLES

Source

Les statistiques proviennent de l'enquête industrielle annuelle des entreprises (Encuesta Industrial de Empresas) qui couvre tous les établissements employant au moins une personne.

Une enquête exhaustive est effectuée chaque année pour les établissements de plus de 20 salariés ; les établissements de moins de 20 salariés font l'objet d'une enquête par échantillon. L'échantillon total annuel comprend environ 40 000 établissements.

Concepts et définitions des variables

La *production* correspond à la production brute au sens de la définition espagnole. Elle est calculée comme la somme des éléments suivants : *a)* la valeur de la production de biens et services destinés à la vente, y compris les sommes perçues au titre de services et de travaux réalisés par des tiers et la valeur de l'électricité vendue ; *b)* les ventes de marchandises réexpédiées en l'état ; *c)* les recettes perçues au titre de services rendus (location de machines, assistance technique, etc.) ; *d)* la valeur des produits et travaux en cours ; *e)* la valeur de la formation de capital pour compte propre (biens d'équipement produits par l'entreprise pour son usage propre et grosses réparations et améliorations apportées par l'unité, avec ses propres moyens, à ces biens d'équipement). La production brute est évaluée au prix départ-usine, c'est-à-dire exclut les subventions d'exploitation et inclut les impôts indirects.

La *valeur ajoutée* est donnée par la différence entre la production brute aux prix départ-usine et la consommation intermédiaire aux prix d'acquisition. La consommation intermédiaire comprend le coût des matières premières et des matériaux consommés, la valeur de l'énergie consommée pour obtenir force motrice, chaleur, vapeur, électricité, gaz, etc. , le coût des services industriels et non industriels achetés et la valeur des marchandises achetées pour être revendues en l'état.

L'*investissement* (formation brute de capital fixe) aux prix courants est égal à la valeur des achats de biens d'équipement neufs et usagés, des grosses réparations et améliorations apportées aux biens d'équipement par des tiers, de la production de biens d'équipement pour compte propre et des grosses réparations et améliorations apportées à ces biens au moyen de ressources propres en matériel et en main-d'œuvre, diminuée de la valeur des ventes de biens de capital fixe propres. L'investissement est évalué aux coûts d'acquisition et inclut les coûts de transport.

L'*emploi, total* représente le nombre de personnes occupées. Il inclut les travailleurs rémunérés et non rémunérés qu'il s'agisse de personnel de bureau ou d'ouvriers, à l'exception des travailleurs à domicile. Les données d'emploi sont converties en équivalents plein-temps (EPT).

Les *heures ouvrées, salariés* correspondent au nombre total d'heures effectivement travaillées par tous les salariés au cours de l'année de référence.

Les *salaires et traitements, total* englobent l'ensemble des traitements et des salaires bruts versés à toutes les personnes occupées dans l'entreprise pendant l'année de référence. Ils comprennent également les prestations en nature versées au personnel rémunéré, à l'exclusion des charges sociales des employeurs.

Les *charges sociales des employeurs* représentent les cotisations sociales payées par les employeurs (sécurité sociale, retraite, autres coûts sociaux,). Elles comprennent les contributions patronales pour l'assurance-vie des salariés. Elles doivent être énoncées dans le cadre d'une convention collective.

Publications

Encuesta Industrial de Empresas

Cette publication est distribuée par l'Instituto Nacional de Estadística.

FINLANDE

Les informations relatives aux statistiques structurelles des entreprises (industrie et services) et aux statistiques du commerce extérieur émanent de l'Office statistique de Finlande à Helsinki.

Classification industrielle

La classification nationale, dérivée de la NACE révision 1, est la Classification industrielle type finlandaise de 1995 (SIC-95). Elle est alignée sur la CITI révision 3.

I. STATISTIQUES DE L'INDUSTRIE ET DES SERVICES

Source

Les données sont tirées de trois sources : 1) les statistiques structurelles d'entreprises dans l'industrie (catégories C à F de la CITI) ou les statistiques structurelles d'entreprises dans les services (catégories G à O de la CITI) ; 2) les registres fiscaux fournis par le bureau national des impôts ; 3) le registre d'entreprises de Statistics Finland.

L'enquête porte sur l'année fiscale entre le 1er avril de l'année t et le 31 mars de l'année t+1.

Secteurs industriels :

L'unité statistique de référence est l'établissement.

Depuis 1996, le seuil de collecte de l'enquête a été porté à 20 personnes. Le nombre d'unités couvertes par l'enquête est de 2 500 environ. Le nombre total d'unités disponibles dans la source administrative est d'environ 60 000. Les non-réponses sont estimées à partir des fichiers fiscaux. Du fait des changements dans les méthodes de collecte de données en 1995, seules les données relatives à la période 1995-1998 sont comparables entre elles.

Secteurs des services :

L'unité statistique de référence est l'entreprise.

Les statistiques structurelles d'entreprises dans les services couvrent la totalité des entreprises de plus de 10, 20 ou 50 salariés selon le secteur d'activité. Le nombre d'unités couvertes par l'enquête est de 4 700 environ. Le nombre total d'unités disponibles dans la source administrative est de 160 000 environ. Les non-réponses sont estimées à partir des fichiers fiscaux.

Concepts et définitions des variables

La *production* comprend la valeur des livraisons de biens ou services produits par l'unité, les recettes au titre des travaux industriels effectués ou des services industriels rendus aux autres, les ventes de marchandises réexpédiées en l'état (déduction faite des montants payés pour ces biens), la variation nette des stocks de produits finis fabriqués par l'unité, des stocks de travaux en cours fabriqués par l'unité et des stocks de marchandises achetées pour la revente en l'état, la valeur de l'électricité vendue, les ventes de produits manufacturés par les sous-traitants, la valeur des actifs fixes produits par l'unité pour son propre compte et les recettes perçues au titre de services non-industriels. Elle ne comprend pas les revenus provenant de ventes d'actifs fixes produits par l'unité pour son propre compte. Les subventions directes du gouvernement liées à la production et les subventions d'importations et d'exportations sont incluses. L'aide sous forme de bonifications d'intérêt est exclue. La taxe sur la valeur ajoutée (TVA) facturée au client par le producteur, la TVA déductible sur l'achat de matières premières, de produits intermédiaires et autres, les taxes affectées aux produits facturées au client par le producteur et les taxes non affectées aux produits payées par le producteur ne sont pas comprises. La production est évaluée au prix de base.

La *valeur ajoutée au coût des facteurs* correspond au revenu brut provenant d'activités d'exploitation. Elle peut se calculer à partir de la production en excluant la valeur des matériaux et autres fournitures industrielles nécessaires à la transformation, les coûts des services industriels reçus et les coûts d'exploitation et paiements pour les services de nature non industrielle. Elle est évaluée au coût des facteurs à l'exclusion de tous les impôts indirects, mais y compris toutes les subventions.

L'*investissement* (formation de capital) correspond aux acquisitions d'actifs fixes, y compris ceux fabriqués par le personnel de l'établissement pour l'usage de ce dernier. Par actifs fixes, il faut entendre les biens d'équipement, les bâtiments et autres constructions, et les autres actifs fixes ; sont exclus les investissements en terrains et en ressources en eau. Ils correspondent aux investissements dont la durée de vie productive est d'un an au moins, y compris ceux fabriqués par le personnel de l'établissement pour l'usage de ce dernier : les adjonctions, transformations, améliorations et réparations d'actifs existants, les actifs neufs (qu'ils soient ou non mis en service) et les actifs usagés. L'évaluation est faite au prix de revient total, y compris le prix rendu, les frais d'installation et les droits et redevances.

Les *dépenses en machines et équipement* sont obtenues sur la base de la même définition. Y sont incluses les dépenses en matériel de transport, en machines et matériels industriels, en machines, équipement et mobilier de bureau, en instruments et équipements professionnels ainsi que les modifications et améliorations apportées à ces matériels.

L'*emploi* comprend le nombre total de personnes occupées et englobe toutes les personnes qui travaillent pour un employeur et perçoivent une rétribution sous forme de salaire, honoraires, pourboires, paiements à la pièce ou rémunération en nature. Il comprend les travailleurs à temps partiel, les travailleurs saisonniers, les apprentis, les travailleurs à domicile, les propriétaires exploitants, les associés touchant des revenus réguliers, les travailleurs familiaux non rémunérés, les directeurs d'entreprises constituées en sociétés commerciales, les personnes travaillant à l'extérieur de l'unité tout en faisant partie de celle-ci et en étant rémunérées par elle, les travailleurs bénévoles ainsi que toutes les personnes en congé de courte durée ou en grève.

L'*emploi, salariés* couvre le nombre total de salariés, non compris les propriétaires exploitants, les associés touchant des revenus réguliers et les travailleurs familiaux non rémunérés.

Les *heures ouvrées, ouvriers* correspondent aux heures réellement effectuées par les ouvriers pour la production de l'unité d'observation au cours de l'année fiscale, y compris les heures supplémentaires calculées en terme d'heure de travail effectif, le temps passé sur le lieu de travail à des tâches telles que la préparation du lieu de travail, les réparations et l'entretien, la préparation et le nettoyage des outils et le temps passé sur le lieu de travail à attendre pour des raisons telles le manque d'offre de travail, des pannes de machines ou des accidents. En sont exclues les heures payées qui n'ont pas été effectivement travaillées : congés annuels, jours fériés, congés maladie.

Les *salaires et charges sociales, salariés* sont définis comme la somme des salaires et traitements des salariés et des charges sociales.

Les *salaires et traitements, salariés* comprennent le paiement des heures normales et supplémentaires, les paiements en espèces, les gratifications et indemnités de cherté de vie, la formation, les congés payés et les congés de maladie ainsi que les impôts et cotisations d'assurance sociale dus par les employés et retenus par les employeurs et les prestations en nature. Elle comprend la rémunération totale payable en espèces ou en nature pour le travail effectué durant l'année.

Les *charges sociales des employeurs* correspondent aux sommes que les employeurs sont légalement tenus de payer en sus des traitements et salaires. Ils couvrent les cotisations d'assurance sociale obligatoires et les cotisations d'assurance sociale volontaires de nature statutaire.

II. STATISTIQUES DU COMMERCE EXTÉRIEUR

Source

Les données sont tirées des statistiques du service des douanes.

Concepts et définitions des variables

Les données portent sur le commerce général, les *exportations* étant exprimées FAB et les *importations* CAF. Sont inclus dans les deux cas : le commerce d'État, les transactions commerciales d'équipements militaires, les transactions d'aide à l'extérieur, les transactions portant sur les navires et avions neufs, l'électricité et le gaz. Par contre, sont expressément exclus les échanges directs en transit, les exportations ou importations temporaires, les transactions monétaires et les opérations d'un montant inférieur à 1 000 Mk.

Publications

i) *Preliminary Data on Finnish Industry*
ii) *Preliminary Data on Business Services*
iii) *Preliminary Data on Trade*
iv) *Preliminary Data on Telephone Communication*
v) *Statistics on the Structure of Industry and Construction*
vi) *Financial Statements Statistics*

Toutes les publications sont distribuées par l'Office statistique de Finlande à Helsinki, dans la série *Official Statistics of Finland.*

FRANCE

Les statistiques industrielles et les statistiques du commerce extérieur sont fournies par le SESSI (Ministère de l'Economie, des Finances et de l'Industrie).

Les statistiques sur les services sont fournies par l'INSEE (Institut National de la Statistique et des Etudes Economiques, Paris).

Classification industrielle

La classification française est la Nomenclature des Activités Françaises (NAF-1993). Elle est parfaitement compatible avec la NACE révision 1 et avec la CITI révision 3.

I. STATISTIQUES INDUSTRIELLES ET STATISTIQUES DU COMMERCE EXTERIEUR

Source

Les statistiques proviennent de l'Enquête Annuelle d'Entreprise (EAE) qui couvre toutes les entreprises d'au moins 20 salariés et les donneurs d'ordre ayant un chiffre d'affaires important quelle que soit leur taille, à l'exception de l'armement et du secteur agro-alimentaire mais y compris l'énergie, soit 22305 entreprises en 1998.

Concepts et définitions des variables

La *production* correspond à la valeur totale des biens et services produits dans l'exercice, qu'ils soient vendus, stockés ou immobilisés. Elle exclut les reventes de biens. Les produits vendus incorporent la marge réalisée par l'entreprise. La production immobilisée correspond au coût des travaux faits par l'entreprise pour elle-même. La production stockée et la production immobilisée sont estimées au coût de production, non compris la TVA déductible et les subventions.

La *valeur ajoutée brute au coût des facteurs* se calcule comme suit : production, plus ventes de marchandises, moins consommation intermédiaire (c'est-à-dire les achats de marchandises, de matières premières et autres approvisionnements, plus la variation nette des stocks de marchandises, de matières premières et autres approvisionnements, plus les autres achats et charges externes), plus subventions d'exploitation, moins impôts, taxes et versements assimilés.

Les données d'*investissement* correspondent à la somme des dépenses consacrées par les entreprises à l'acquisition d'actifs neufs ou usagés ou à la création de moyens de production. Elles comprennent les immobilisations en cours, mais excluent les immobilisations incorporelles et financières et les équipements financés par crédit-bail. Elles sont évaluées aux coûts d'acquisition.

L'*emploi, salariés* est la moyenne annuelle sur les douze mois de l'année. Il comprend l'ensemble des personnes qui perçoivent un salaire de l'entreprise, même s'ils travaillent dans une autre entreprise et quel que soit leur statut (directeurs, travailleurs à domicile,...). Il ne comprend pas le personnel intérimaire. Les salariés qui ne sont pas à temps plein sont comptés au prorata de leur temps partiel.

Les *salaires et traitements, salariés* couvrent les rémunérations payées par l'employeur au cours de l'année en compensation du travail de ses propres employés (à l'exclusion du personnel loué et du personnel intérimaire). Ils n'incluent pas les charges sociales des employeurs, données séparément.

Les *charges sociales des employeurs* sont les contributions sociales que verse l'employeur au cours de l'exercice (maladie, maternité, vieillesse, chômage, ...). Elles doivent être énoncées dans le cadre de dispositions obligatoires, de conventions collectives ou d'accords contractuels.

Les *exportations*, collectées par l'enquête annuelle d'entreprises, sont les ventes à l'exportation de biens et services déclarées par les entreprises. Elles comprennent les livraisons intra-communautaires. Ces exportations diffèrent de celles qui sont recensées par les douanes (qui sont FAB/CAF) car elles correspondent aux seules exportations directes (c'est-à-dire qui ne passent pas par une société tiers) réalisées par les entreprises. De plus, les exportations d'un secteur comprennent les exportations des entreprises du secteur réalisées pour l'ensemble de leurs activités confondues.

II. STATISTIQUES DES SERVICES

Source

Les données proviennent de l'Enquête Annuelle d'Entreprise (EAE).

L'EAE-Commerce 1993 intègre le commerce et la réparation automobile ainsi que la réparation d'articles personnels et domestiques. En 1994, les intermédiaires du commerce de gros et les activités de boulangerie n'ont pas été couvertes par l'enquête. Depuis 1995, l'enquête concerne à nouveau la totalité du champ, le commerce de détail, le commerce de gros, y compris les intermédiaires du commerce de gros, le commerce et la réparation automobile et les activités artisanales relevant du commerce (charcuteries, boulangeries, pâtisseries).

Jusqu'en 1997, l'EAE-Services étaient envoyée à toutes les entreprises de services marchands (quelques variables ont été nouvellement définies dans l'EAE-1997). A partir de 1998, les questionnaires EAE-Services sont systématiquement adressés chaque année aux entreprises de plus de 30 salariés ou dont le chiffre d'affaires dépasse 30 millions de Francs. Les petites entreprises en dessous de ce seuil ne sont interrogées qu'une année sur deux.

Concepts et définitions des variables

Activités de commerce :

La *production* correspond à la valeur totale du chiffre d'affaires hors taxes, plus la valeur de la production stockée et de la production immobilisée, moins le coût des marchandises.

La *valeur ajoutée* est définie comme la valeur ajoutée aux prix du marché. Elle se calcule comme suit : production moins coût des matières premières et autres achats et charges externes.

Les données d'*investissement* correspondent aux investissements acquis ou créés hors apports, plus les investissements reçus par voie d'apports. Seuls les investissements corporels sont demandés par l'enquête.

L'*emploi, salariés* est égal à la moyenne annuelle des effectifs salariés trimestriels. Les travailleurs à temps partiel sont comptés en proportion de leur temps partiel.

Les *salaires et charges sociales, salariés* couvrent les rémunérations payées au cours de l'année par l'employeur à ses employés salariés ainsi que les contributions patronales aux caisses de sécurité sociale.

Les *exportations* et les *importations* ne sont collectées que dans certains secteurs.

Activités de services :

La *production*, ou production élargie, correspond à la valeur totale du chiffre d'affaires, plus la valeur de la production stockée et de la production immobilisée. La production stockée représente la production non encore vendue de services (études ou prestations de services en cours). La production immobilisée est la production de services faite en propre pour l'entreprise et non destinée à la vente (films, programmes T.V., logiciels…).

La *valeur ajoutée* est définie comme la valeur ajoutée aux prix du marché. Elle est la différence entre la production élargie et les consommations intermédiaires (c'est-à-dire les achats de marchandises ou de matières premières, la variation de stocks de marchandises ou de matières premières, les autres achats et charges externes).

Les dépenses d'*investissement* correspondent aux investissements corporels seulement. Elles se composent des investissements corporels totaux hors apports reçus qui représentent les entrées dans le patrimoine au cours de l'exercice, par acquisitions ou créations d'éléments corporels destinés à servir de façon durable à l'activité de l'entreprise, et des investissements corporels totaux reçus par apports au cours de l'exercice qui représentent les seules entrées dans le patrimoine par apports (reçus d'une autre entreprise par exemple dans le cadre d'une restructuration) et qui sont destinés à servir de façon durable à l'activité de l'entreprise.

L'emploi, salariés est la moyenne des nombres de salariés en équivalents plein-temps à la fin de chaque trimestre. Les travailleurs à temps partiel sont comptés en proportion de leur temps partiel.

Les *salaires et charges sociales, salariés* comprennent les rémunérations brutes des salariés y compris les contributions patronales aux caisses de sécurité sociale au cours de l'exercice. Ces deux variables ne concernent ni les non salariés, ni le personnel prêté par une autre entreprise et le personnel intérimaire.

Les *salaires et traitements, salariés* couvrent les traitements et salaires versées au cours de l'année par l'employeur à ses employés salariés, incluant les cotisations sociales des salariés mais excluant les contributions patronales aux caisses de sécurité sociale.

Les *exportations* et d'*importations* (depuis 1996) ne sont disponibles que pour certains secteurs.

Publications

i) *La situation dans l'industrie* (publication papier) ;
ii) *L'industrie française en 300 secteurs* (CD-ROM) ;
iii) Autres publications de synthèse ou thématiques.

Toutes les publications sont distribuées par le SESSI.

GRÈCE

Les informations proviennent de l'office national de statistique de Grèce (NSSG).

Classification industrielle

Les données sont établies en fonction de la classification nationale et ventilées selon la CITI, dans la mesure du possible.

STATISTIQUES INDUSTRIELLES

Source

Les données sont tirées d'enquêtes effectuées par l'Office National de Statistique de Grèce :

- l'Enquête Annuelle dans l'Industrie portant sur les établissements manufacturiers de 10 salariés et plus ; elle s'adresse à tous les établissements de 20 salariés ou plus et à un échantillon aléatoire stratifié des établissements employant de 10 à 19 personnes. Les établissements publics ne sont pas inclus dans l'échantillon.

- l'Enquête Statistique Annuelle dans les Industries Extractives («Mines, Quarries and Salterns»).

- les Enquêtes Annuelles dans l'Electricité et le Gaz qui couvrent les deux entreprises engagées dans la production et la distribution pour l'usage public. La distribution de vapeur et d'eau n'est pas incluse.

L'unité de référence est l'établissement sauf pour l'électricité et le gaz où l'unité est l'entreprise.

Concepts et définitions des variables

La *production* comprend la valeur de la production de produits finis, les recettes au titre des travaux effectués sous contrat pour le compte d'autres unités, des travaux de réparation et de la vente de marchandises réexpédiées en l'état, et d'autres opérations (ventes de vapeur, d'énergie électrique, de déchets de fabrication, etc., produits par l'unité ; subventions et remboursements au titre de la réexportation de marchandises). L'évaluation est faite au prix de vente départ usine, hors taxes.

La *valeur ajoutée* est donnée par la valeur de la production déduction faite du coût des matériaux et autres fournitures, des combustibles achetés et de l'électricité consommée, des marchandises réexpédiées en l'état et des travaux effectués sous contrat ainsi que des travaux de réparation.

L'*investissement* (formation brute de capital fixe) représente les dépenses courantes au titre d'actifs fixes, neufs ou usagés, déduction faite des ventes et des réductions de capital dues à l'obsolescence. La fabrication de biens d'équipement par le personnel de l'établissement pour l'usage de ce dernier est incluse. Les données relatives aux industries extractives ne comprennent pas l'extraction du sel et la valeur des ventes (insignifiante) n'est généralement pas déduite.

L'*emploi, salariés* couvre les employés, c'est-à-dire les personnes qui reçoivent une rémunération mensuelle sur une base contractuelle quel que soit le montant du salaire, et les ouvriers définis comme les personnes qui reçoivent un salaire journalier ou sont payées à la pièce, en espèces ou en nature.

Les *jours ouvrés* sont les jours de travail effectués par les ouvriers tels que définis ci-dessus.

Les *salaires et charges sociales, salariés* couvrent les salaires et traitements, le paiement des heures supplémentaires et des congés réguliers, les gratifications, les indemnités, l'indemnité de licenciement et la valeur des recettes en nature.

Les *charges sociales des employeurs* comprennent les contributions patronales aux caisses de sécurité sociale en faveur des employés et salariés.

Publications

i) *Enquête Annuelle dans l'Industrie* ;

ii) *Enquête Statistique Annuelle dans les Industries Extractives* («Mines, Quarries and Salterns») ;

iii) *Annuaire Statistique* ;

iv) *Bulletin Statistique Mensuel.*

Toutes sont publiées par l'Office National de Statistique de la Grèce, à Athènes.

HONGRIE

Les informations émanent de l'Office Central Statistique de Hongrie.

Classification industrielle

La classification hongroise est la TEÁOR 92, qui est compatible avec la CITI révision 3 jusqu'au 2ème chiffre.

STATISTIQUES INDUSTRIELLES

Source

A partir de 1992, toutes les entreprises qui répondent à deux des trois critères ci-dessous pendant deux années consécutives sont couvertes par l'enquête : plus de 100 salariés, plus de 300 millions de Forint de chiffre d'affaires annuel, 150 millions de Forint de capital total.

Seuls les investissements de plus de 10 millions de Forint sont déclarés.

Les données d'emploi résultent de l'enquête annuelle sur l'emploi et les salaires qui couvre les entreprises de 20 salariés et plus.

Concepts et définitions des variables

L'*emploi, salariés* se rapporte à la moyenne des effectifs de salariés rémunérés. Il ne couvre que l'emploi à plein temps.

Les *salaires et traitements, salariés* comprennent les primes et congés payés. Les prestations en nature et les congés de maladie sont exclus.

IRLANDE

Les informations ont été fournies par le Bureau central de statistiques de l'Irlande.

STATISTIQUES INDUSTRIELLES

Source

Les données sont tirées du «Census of Industrial Local Units» (recensement annuel des unités locales industrielles). Ce recensement porte sur toutes les unités menant une activité industrielle qui occupaient en moyenne trois personnes ou plus pendant l'année considérée.

La possibilité d'obtenir des déclarations distinctes pour chaque unité dépend toutefois de la disponibilité de relevés séparés dans l'entreprise pour les différentes unités locales.

Concepts et définitions des variables

La *production* (production brute) représente la valeur marchande nette de tous les biens fabriqués au cours de l'année, qu'ils soient ou non vendus, y compris la valeur du travail exécuté et des travaux d'équipement effectués pour compte propre. Les subventions d'exploitation relatives à la production ou à la vente de la production sont incluses dans la valeur de la production ; le droit d'accises et la TVA en sont exclus. Jusqu'en 1990, dans le cas de la production d'électricité et de la production de tourbe, la valeur des travaux d'équipement effectués pour compte propre était incluse dans la production brute.

La *valeur ajoutée* (production nette) correspond à la différence entre la production brute et la consommation intermédiaire de l'industrie. La consommation intermédiaire de l'industrie comprend les matériaux industriels, les services industriels ainsi que le combustible et l'énergie utilisés dans la production. L'évaluation ne tient pas compte de la TVA récupérable.

L'investissement correspond à la formation brute de capital fixe (FBCF). Par actifs fixes (terrains, bâtiments, installations et équipements), on entend les biens destinés à être utilisés par l'unité elle-même et dont la durée de vie utile prévue est supérieure à un an. Les acquisitions comprennent les achats auprès d'autres unités et la production de biens d'équipement par l'unité elle-même pour son propre usage. Les modifications, améliorations et réparations importantes qui prolongent la durée de vie utile d'un actif ou en accroissent la productivité sont incluses. La valeur du travail entrepris au cours de l'année, qu'il soit ou non achevé, est inclus. L'évaluation des acquisitions est faite au prix de revient total, y compris les frais d'installation et les droits ou redevances, mais à l'exclusion de la TVA récupérable et des frais financiers. L'évaluation des ventes est faite au prix réellement perçu, déduction faite de la TVA. La FBCF représente la différence entre les acquisitions et les ventes d'actifs fixes.

L'emploi, salariés couvre les personnes qui reçoivent un salaire ou un traitement fixe. Les personnes au travail ou celles qui sont temporairement absentes pour des raisons de maladie, de congé, de grève, etc. sont prises en compte, de même que les travailleurs à temps partiel. Les ouvriers à la pièce qui travaillent à domicile ne sont pas pris en compte. Les données se rapportent à une semaine de septembre.

Les *salaires et traitements, salariés* représentent le montant brut payé aux employés, à l'exception des ouvriers à la pièce qui travaillent à domicile, avant déduction des impôts sur le revenu, les cotisations de sécurité sociale dues par les salariés, etc. Sont inclus les paiements pour les heures supplémentaires, les gratifications, les commissions, les congés payés et les congés de maladie. Sont exclus les paiements aux propriétaires exploitants et aux travailleurs à domicile.

Par *établissement*, on entend toutes les unités locales situées sur des sites de production distincts des entreprises implantées sur plusieurs sites, même si le répondant ne fournit pas de détails particuliers à leur sujet.

ISLANDE

Les données émanent de l'institut économique national d'Islande (**Þ**jóðhagsstofnun).

STATISTIQUES INDUSTRIELLES

Classification industrielle

La classification nationale s'appuie sur la CITI révision 1 depuis 1958. Les données fournies sont converties en CITI révision 2 à partir de 1968. La NACE révision 1 a été adoptée et sera appliquée dans toutes les statistiques prochainement.

Source

L'institut économique national tire chaque année un échantillon d'états financiers à partir des registres de l'administration fiscale. L'échantillon couvre environ 50 pour cent de la masse salariale des diverses branches. La probabilité d'appartenir à l'échantillon augmente avec la taille de l'établissement. L'établissement et l'unité d'activité économique sont utilisés comme unités statistiques de base. La production et la valeur ajoutée de l'échantillon stratifié sont converties à l'échelle de l'ensemble du secteur industriel à l'aide de la part de l'échantillon dans la masse salariale totale. Les chiffres relatifs à l'emploi ainsi qu'aux entreprises et aux établissements sont fondés sur des registres administratifs qui couvrent la totalité du secteur industriel. La période de référence est l'année civile.

Concepts et définitions des variables

A partir de 1992, la *production* (production brute) est évaluée aux prix de base, y compris toutes les subventions sur les produits, mais à l'exclusion de la TVA ; avant 1992, elle était évaluée aux prix départ-usine, à l'exclusion de toutes les subventions et de la TVA. Elle comprend la valeur de tous les produits, compte tenu de la variation nette des travaux en cours.

La *valeur ajoutée*, aux prix de base, utilise la notion de valeur ajoutée de la comptabilité nationale. Elle est donnée par la valeur de la production brute aux prix de base, déduction faite du coût des matériaux, combustibles et autres fournitures, des travaux exécutés sous contrat ou à la commande par d'autres, des travaux de réparation et d'entretien faits par d'autres et de l'électricité achetée. Elle comprend les recettes perçues au titre de services non-industriels mais exclut le coût des services non-industriels. Les consommations intermédiaires sont estimées sur la base des quantités consommées.

L'*emploi* est mesuré par le nombre d'années de travail, en équivalents plein-temps, calculé d'après le nombre de semaines de travail assurées qui provient de sources fiscales se fondant sur les registres. Sont compris les propriétaires exploitants, les associés actifs, les travailleurs familiaux non rémunérés, les travailleurs indépendants et les salariés. Les données englobent les salariés à temps complet, les salariés à temps partiel et le personnel de bureau et d'encadrement dont les fonctions sont étroitement liées au processus de production.

L'*emploi, salariés* comprend tous les salariés et exclut les propriétaires exploitants, les associés actifs et les travailleurs familiaux non rémunérés.

Par *salaires et charges sociales, salariés* on entend la rémunération totale des salariés au sens de la comptabilité nationale. Elle couvre uniquement les salariés, les travailleurs indépendants étant exclus, et comprend les salaires et les cotisations sociales patronales versées aux employés.

II. STATISTIQUES DU COMMERCE EXTÉRIEUR

Classification industrielle

Les données sont transmises en CITI révision 3.

Source

Les données proviennent des déclarations faites par les exportateurs au Service des Douanes.

Concepts et définitions des variables

Les données sur les exportations sont tirées des déclarations d'exportations compilées par les services des douanes. Les biens sont classés selon le système harmonisé (SH), puis regroupés selon la classification des activités principales en usage au plan national. Les exportations comprennent les transactions portant sur les navires et sur les avions. Elles sont données FAB/CAF.

Publications

i) *Atvinnuvegaskyrslur (Statistiques industrielles)*, publiées annuellement ;
ii) *Historical Statistics (Statistiques rétrospectives)*, publiées tous les deux ans ;
iii) *Compiling Icelandic National Accounts* (L'établissement des comptes nationaux islandais), documentation sur les méthodes appliquées, sous l'optique de la production et des dépenses, Reykjavik, janvier 1996.

Ce sont toutes des publications du National Economic Institute.

ITALIE

Les données émanent de l'Istituto Nazionale di Statistica (Institut National de Statistique), Rome.

Classification industrielle

La classification nationale, inspirée de la NACE révision 1, n'est pas totalement convertible en CITI révision 3. Néanmoins, des tables nationales de conversion permettent d'établir des données au niveau 3 chiffres de la CITI révision 3 et de produire des estimations fiables au niveau 4 chiffres.

STATISTIQUES DE L'NDUSTRIE ET DES SERVICES

Source

L'Institut de statistique mène chaque année des enquêtes globales sur le produit brut et les investissements en biens d'équipement dans le secteur industriel. Jusqu'en 1995, ces enquêtes s'adressaient aux entreprises de 20 salariés ou plus ; aucune estimation n'était faite pour tenir compte des unités non couvertes. Depuis 1996, elles s'adressent à toutes les entreprises. Les enquêtes portent sur l'année civile et utilisent comme unité de référence l'entreprise, classées par type d'activité. Les entreprises de plus de 250 salariés qui se composent de plusieurs unités doivent fournir une réponse distincte pour chacune des unités. Jusqu'en 1993, les chiffres sur les heures ouvrées étaient tirés de l'enquête sur les entreprises industrielles. Depuis 1993, ils sont tirés d'une enquête spéciale sur l'emploi, les salaires et les heures ouvrées menée auprès d'un échantillon de 18 000 entreprises de 10 à 199 salariés et auprès de toutes les entreprises (environ 3 700) de 200 salariés et plus.

Concepts et définitions des variables

La *production*, ou production brute nette des impôts indirects et subventions, est évaluée aux prix de vente, hors TVA. L'électricité vendue, les actifs fixes produits par l'unité pour son propre usage et les biens fabriqués par d'autres entreprises sont inclus, compte tenu de la variation des stocks de produits finis et de la variation nette des travaux en cours.

La *valeur ajoutée*, nette des subventions et impôts indirects, est établie selon le concept de valeur ajoutée «recensée». Elle exclut tous impôts indirects et subventions. Elle correspond à la production brute majorée des recettes au titre des services non industriels, déduction faite du coût des matériaux, combustibles et autres fournitures, des travaux exécutés sous contrat ou à la commande par d'autres unités, des travaux de réparation et d'entretien réalisés par d'autres unités, de l'électricité achetée et des services non industriels. Les consommations intermédiaires sont estimées sur la base des quantités reçues, après ajustement pour tenir compte de la variation des stocks de matériaux, combustibles et autres fournitures.

L'*investissement* (formation brute de capital fixe), déduction non faite de la valeur des ventes, est égal à la valeur des acquisitions d'actifs fixes, y compris ceux produits par l'unité pour son compte. Les actifs fixes couverts sont ceux dont la durée de vie productive est au moins égale à un an. Sont pris en compte les actifs fabriqués par le personnel de l'établissement pour l'usage de ce dernier, les adjonctions, transformations, améliorations et réparations d'actifs existants, les actifs fixes neufs (non encore en service ou précédemment utilisés en dehors du pays) et les actifs usagés. L'évaluation est faite au prix de revient total, y compris les frais d'installation.

Les *dépenses en machines et équipement* correspondent à la formation brute de capital fixe en machines et équipement, valeur des ventes non déduite. Elles reposent sur la même définition que l'investissement total et englobent le matériel de transport, les machines, équipement et mobilier industriels, les instruments et équipements professionnels. Les modifications, améliorations et réparations sont également prises en compte.

L'*emploi* indique le nombre de personnes occupées, tel qu'il ressort du nombre moyen de salariés employés tout au long de l'année et du nombre total d'autres personnes employées à quatre dates précises, à savoir le 31 mars, le 30 juin, le 30 septembre et le 31 décembre. Les propriétaires exploitants, associés actifs, travailleurs familiaux non rémunérés, travailleurs indépendants, employés et salariés sont tous pris en compte.

Les *heures ouvrées, ouvriers* correspondent aux heures effectuées par les travailleurs manuels, y compris les heures supplémentaires, les heures de repos et les heures passées à attendre. Sont compris les congés annuels, les jours fériés, les congés de maladie et autres congés payés. Sont exclus : les employeurs, les agents de maîtrise et les apprentis.

Les *salaires et charges sociales* couvrent les traitements, les salaires et les charges sociales des employeurs. Ils comprennent le paiement en espèces des heures normales et supplémentaires, les gratifications et indemnités de cherté de vie, les congés payés et congés de maladie, les prestations en nature et les indemnités.

Les *charges sociales des employeurs* couvrent le coût total des contributions patronales aux caisses de sécurité sociale et aux caisses de retraite.

Les *établissements* sont définis en fonction du type d'activité de l'unité.

Les données sur les *entreprises*, disponibles à partir de 1997, couvrent toutes les entreprises italiennes, y compris celles qui emploient moins de 20 salariés.

Publications

Fatturato, prodotto lordo e investimenti delle imprese industriali

Ces suppléments au *Bollettino Mensile di Statistica* sont publiés par l'ISTAT, Rome.

JAPON

Les informations émanent du Ministère japonais du Commerce International et de l'Industrie (MITI) et du Ministère de la Santé et des Affaires Sociales.

Classification industrielle

Les données sont ventilées selon la classification japonaise type des activités industrielles (JSIC révision 10). La JSIC Rev. 10 est compatible avec la CITI Révision 3.

STATISTIQUES INDUSTRIELLES

Source

Les données relatives aux industries manufacturières proviennent du recensement annuel des industries manufacturières (Annual Census of Manufactures) du MITI qui couvre tous les établissements définis comme manufacturiers à l'exception de ceux appartenant à des entreprises publiques ou de service public.

Depuis 1981, tous les établissements (y compris ceux employant trois personnes ou moins) ne sont cependant plus interrogés que les années dont le chiffre se termine par 0, 3, 5 et 8. Les autres années, l'enquête se limite aux établissements employant quatre personnes ou plus.

Les données sur l'investissement concernent les établissements de 30 salariés ou plus. Aucune estimation n'est opérée pour tenir compte des unités non couvertes.

Concepts et définitions des variables

La *production* (production brute) indique la valeur des livraisons de produits finis ajustée pour tenir compte de la variation des stocks de produits finis et des travaux en cours. La production des entreprises manufacturières correspond à la valeur des livraisons, majorée de la valeur des stocks de produits manufacturés à la fin de l'année moins la valeur des stocks de produits manufacturés au début de l'année. S'y ajoute la variation nette des stocks de produits semi-manufacturés et des travaux en cours (stocks en fin d'année, moins stocks en début d'année).

La *valeur ajoutée* est donnée par la production déduction faite des impôts indirects, du coût des matières premières, des combustibles et de l'électricité consommée, du coût des travaux exécutés sous contrat ou à la commande par d'autres unités ainsi que des amortissements.

L'*investissement* correspond à la formation brute de capital fixe, définie comme la valeur des acquisitions d'actifs fixes, y compris ceux fabriqués par le personnel de l'établissement pour l'usage de ce dernier, déduction faite de la valeur des ventes. Les actifs fixes couverts sont ceux qui ont une durée de vie productive supérieure à un an et qui comprennent les réparations d'actifs existants, les actifs neufs non encore en service, inutilisés ou utilisés précédemment en dehors du pays, et les actifs usagés. L'évaluation est faite au prix de revient total, y compris le prix rendu, les frais d'installation et les droits et redevances. Dans le cas d'actifs produits par l'unité, le coût des travaux en cours couvre les coûts de main-d'œuvre, le coût des matières premières et des fournitures ainsi qu'une provision pour frais généraux. Pour l'évaluation des ventes d'actifs usagés, on se fonde sur le montant effectivement réalisé.

Les *dépenses en machines et équipement* correspondent à la part des acquisitions de machines et équipements (y compris les pièces) dans les actifs fixes corporels.

L'*emploi, salariés* comprend les salariés permanents au 31 décembre. Sont englobés dans les salariés permanents les travailleurs temporaires employés pendant une période de durée définie de plus de 30 jours, les travailleurs temporaires employés pendant au moins 18 jours au cours des mois de novembre et de décembre, les cadres de direction travaillant à plein temps et rémunérés au mois et les travailleurs familiaux travaillant à plein temps et rémunérés au mois.

Les *salaires et traitements, salariés* correspondent au total des paiements en espèces versés aux personnes salariées, et comprennent le salaire de base, les gratifications et autres primes, et les indemnités. Pour les industries minières et extractives, les chiffres comprennent les sommes versées aux travailleurs sous contrat. Pour les industries manufacturières, les chiffres couvrent les indemnités de départ à la retraite et de licenciement des employés, mais pas celles des ouvriers.

Publications

Census of Manufactures (annuel),
Cette publication est distribuée par le MITI.

LUXEMBOURG

Les informations émanent du Service central de la statistique et des études économiques (STATEC), Luxembourg.

Classification industrielle

A partir de 1993 la nomenclature utilisée est la NACE LUX qui correspond à la NACE Révision 1 jusqu'au 4ème niveau et est plus finement détaillée jusqu'à un 5ème niveau pour les besoins nationaux. Elle est compatible avec la CITI Révision 3. Les enquêtes couvrent les sections C à K de la CITI Révision 3.

STATISTIQUES DE L'INDUSTRIE ET DES SERVICES

Source

Les données proviennent d'enquêtes annuelles par échantillon, de sources administratives et des comptes annuels des grandes entreprises. Les enquêtes sont exhaustives pour les unités de plus de 50 salariés ou réalisant un chiffre d'affaires (non compris la TVA) de plus de 280 millions de Francs luxembourgeois. Pour les autres unités, les statistiques proviennent d'enquêtes par échantillon. L'échantillon des unités enquêtées est tiré du Répertoire des Entreprises du STATEC.

Concepts et définitions des variables

Les définitions des variables sont conformes au règlement (CE) No 2700/98 de la Commission des Communautés Européennes du 17/12/1998.

La *production* est définie comme le chiffre d'affaires corrigé de la variation des stocks, diminué des acquisitions de biens et services destinés à la revente, augmenté de la production immobilisée et des autres produits d'exploitation (à l'exclusion des subventions). Sont exclues la taxe sur la valeur ajoutée (TVA) facturée au client par le producteur et la TVA déductible sur l'achat de matières premières et de produits intermédiaires. La production est évaluée aux prix du marché.

La *valeur ajoutée aux coûts des facteurs* est le revenu brut des activités d'exploitation une fois pris en compte les subventions d'exploitation et les impôts indirects. Elle correspond au chiffre d'affaires, augmenté de la production immobilisée et des autres produits d'exploitation, corrigé de la variation des stocks, diminué des acquisitions de biens et services et des autres taxes sur les produits liés au chiffre d'affaires mais non déductibles et des impôts et taxes liés à la production. Elle exclut les recettes et les dépenses portées dans les comptes de l'entreprise aux postes financiers ou exceptionnels.

L'*emploi, salariés* correspond à la somme des nombres d'apprentis, de salariés et d'autres employés. Il exclut les propriétaires exploitants et les travailleurs familiaux non rémunérés, ainsi que les personnes qui travaillent à l'extérieur de l'unité tout en faisant partie de celle-ci et en étant rémunérées par elle. Les travailleurs à domicile, dont la contribution est considérée comme négligeable, sont exclus. Les données d'emploi ne sont pas converties en équivalents plein-temps.

Les *salaires et charges sociales, salariés* sont égaux à la somme des salaires et traitements versés aux salariés et des contributions sociales patronales.

Les *salaires et traitements, salariés* correspondent aux paiements versés aux salariés en espèces ou en nature en rémunération de leur travail au cours de la période de référence et des salaires versés aux travailleurs manuels, y compris les sommes en espèces pour les heures supplémentaires et le travail de nuit. Sont inclus les pourboires, les gratifications, les indemnités de licenciement, les primes liées au poste de travail et les primes de performance, les gratifications, le treizième mois, les primes d'ancienneté, d'assiduité, de stimulation et de productivité, les paiements pour jours non travaillés tels que les congés payés, vacances, et congés-maladie, les coûts de formation pour les travailleurs, les coûts occasionnés par les voyages, et autres coûts similaires. Les impôts et les contributions aux caisses d'assurance sociale, dus par les salariés mais déduits par l'employeur et la valeur des actions gratuites délivrées aux salariés sont également inclus.

Les *charges sociales des employeurs* sont les contributions patronales aux régimes de sécurité sociale, couvrant les risques de maladie, de maternité, d'invalidité, de vieillesse, de chômage, d'accident du travail ainsi que les prestations familiales, les coûts liés aux soins médicaux et aux médicaments donnés gratuitement, les coûts d'entretien des écoles, ceux relatifs aux sports et aux loisirs. Les contributions patronales doivent être énoncées dans le cadre d'une convention collective.

Les *exportations* sont évaluées FAB. Elles comprennent le commerce d'état, les transactions commerciales d'équipement et de fournitures militaires, les transactions d'aide à l'étranger (à l'exception des marchandises d'occasion), les transactions portant sur les avions neufs et sur l'électricité et le gaz et les livraisons de marchandises portant sur des opérations ne présentant aucun intérêt sur le plan commercial.

Les *établissements*, donnés pour les secteurs industriels, sont des unités d'activité économique, telles que définies par le Règlement (CEE) No. 696/93 relatif aux unités statistiques d'observation. Ils correspondent à une unité qui dispose d'un système d'information permettant de fournir la valeur de la production, des consommations intermédiaires, des frais de personnel, de l'excédent d'exploitation, ainsi que l'emploi et la formation brute de capital fixe.

Les *entreprises,* données pour les secteurs de services, sont une approximation du concept d'unité légale telle que définie par le Règlement (CEE, EURATOM) No. 58/97 relatif aux statistiques structurelles sur les entreprises.

Publications

Annuaire statistique du Luxembourg, publié par STATEC.

NORVÈGE

Les statistiques industrielles et les données de comptabilité nationale émanent du Bureau Central de statistiques de Norvège (Statistisk Sentralbyrå), à Oslo.

Classification industrielle

La classification industrielle type de Norvège, révisée, (SIC 94) a été adoptée à partir de 1993. La classification a une structure hiérarchique à 6 niveaux.

La SIC 94 est identique à la NACE révision 1 jusqu'au 4ème niveau, le 5ème chiffre étant une subdivision pour les besoins nationaux. Elle est compatible avec la CITI révision 3.

STATISTIQUES INDUSTRIELLES

Source

Depuis 1994, l'enquête annuelle s'adresse à tous les établissements des industries extractives et manufacturières. Les entreprises appartenant à un particulier qui y travaille seul (entreprise individuelle) ne sont pas couvertes, pas plus que les établissements où l'emploi est inférieur à une demi-année de travail. En 1992 et 1993, l'enquête annuelle touchait, en règle générale, tous les établissements des industries extractives et manufacturières de 10 salariés ou plus. Les enquêtes annuelles précédentes couvraient, en règle générale, tous les établissements de 5 salariés ou plus. Les informations sur l'extraction de pétrole et de gaz ne sont pas incluses.

Pour les petits établissements (de moins de 10 personnes), les informations proviennent chaque année des registres administratifs.

L'unité de référence est l'établissement. La période de référence est l'année civile.

Concepts et définitions des variables

La *production* correspond à la production brute évaluée aux prix départ-usine, à l'exclusion de toutes les subventions d'exploitation, mais y compris tous les impôts indirects, sauf la TVA. Elle inclut la valeur de tous les biens produits par l'établissement pour son propre compte, des travaux de réparation effectués pour d'autres unités, du montage ou de l'installation de ses propres biens et produits, des travaux exécutés sous contrat, des autres travaux, des travaux d'investissement et de réparation effectués par l'unité pour son propre compte, des revenus de locations, et des profits bruts tirés de la revente de biens en l'état.

La *valeur ajoutée* est évaluée aux prix du marché et est égale à la valeur brute de la production diminuée du coût des biens et services consommés. Elle exclut toutes les subventions, mais comprend tous les impôts indirects, sauf la TVA. Les coûts des matières premières, combustibles et autres fournitures, des travaux exécutés sous contrat ou à la commande par d'autres unités, des travaux de réparation et d'entretien exécutés par d'autres unités et de l'électricité achetée sont déduits de la valeur de la production brute. Les coûts des services non industriels sont inclus de même que les recettes à ce titre. Les données relatives aux consommations intermédiaires sont estimées sur la base des quantités consommées.

L'*investissement* (formation brute de capital fixe) est défini comme la valeur des acquisitions d'actifs fixes durables, neufs ou usagés, dont la durée de vie productive est d'un an au moins, y compris ceux fabriqués par le personnel de l'établissement pour l'usage de ce dernier, déduction faite de la valeur des ventes d'actifs durables. L'acquisition de machines, d'équipement de transport, etc. est enregistrée à la date de réception des actifs, qu'ils soient ou non mis en œuvre dans l'année. L'achat de bâtiments, autres ouvrages, terrains, etc. est traité de la même façon. Pour les travaux de construction en cours, la valeur des travaux effectués au cours de l'année est enregistrée même s'ils ne sont pas achevés dans l'année. Par coût d'acquisition on entend les dépenses réelles encourues pendant l'année, y compris les prélèvements sur l'investissement, quelle que soit la date de paiement. Les chiffres sur la formation de capital fixe ne comprennent pas tout le capital utilisé dans les industries extractives et manufacturières.

Les *dépenses en machines et équipement* correspondent, elles aussi, à la formation de capital et reposent sur la même définition que l'investissement total. Y sont inclus le matériel de transport, les machines et matériels industriels, les machines, équipement et mobilier de bureau, les instruments et équipements professionnels, ainsi que les modifications et améliorations apportées à ces matériels.

Un *établissement* se définit comme une unité fonctionnelle qui mène, en un seul endroit, une activité principale appartenant à un groupe d'activités spécifique de la classification industrielle type de Norvège.

L'*emploi, total* indique le nombre de personnes occupées et est calculé sur la base du nombre moyen de personnes travaillant dans l'établissement tout au long de l'année, à temps complet ou à temps partiel. Il comprend les propriétaires exploitants, les associés, les travailleurs familiaux non rémunérés, les salariés ainsi que les travailleurs indépendants mais ni les travailleurs à domicile ni les travailleurs appelés dans les forces armées. Sont incluses les personnes en congé de maladie, en congé annuel et en grève.

L'*emploi, salariés* indique le nombre de salariés et autres employés. Il est calculé sur la base du nombre moyen de personnes travaillant dans l'établissement tout au long de l'année, à temps complet ou à temps partiel. Il englobe les gérants et directeurs salariés aussi bien que les commissionnaires, messagers, livreurs de journaux, gardiens et les agents d'entretien mais ni les travailleurs à domicile ni les travailleurs appelés dans les forces armées. Sont incluses les personnes en congé de maladie, en congé annuel et en grève.

Les *heures ouvrées, salariés* correspondent aux heures effectuées par les salariés et comprennent les heures supplémentaires. Sont compris les contremaîtres et agents de maîtrise, les apprentis, les travailleurs occasionnels et les travailleurs à temps partiel.

Les *salaires et charges sociales, salariés* englobent les salaires et traitements versés en espèces ou en nature, les autres avantages versés aux salariés et les cotisations sociales obligatoires. Sont exclus les rémunérations des propriétaires d'entreprise individuelle, des associés et des travailleurs familiaux ordinairement non rémunérés.

Les *salaires et traitements, salariés* comprennent tous les paiements, qu'ils soient effectués en espèces ou en nature, versés par l'employeur en échange du travail effectué, à toutes personnes comptabilisées dans l'effectif (y compris les travailleurs à domicile). Ils comprennent les impôts et les cotisations d'assurance sociale et de retraite dus par le salarié, mais retenus par l'employeur. Sont également inclus les gratifications, les primes de production, les salaires payés au titre des congés payés, congés de maladie et congés pour service militaire et les commissions des vendeurs et des représentants.

Depuis 1994, les *charges sociales des employeurs* recouvrent les cotisations sociales obligatoires et les autres prestations versées aux salariés, à savoir les cotisations patronales aux régimes privés de pensions, les allocations familiales, les assurances maladie-accidents, les assurances sur la vie et les autres régimes similaires. Avant 1994, les données ne comprenaient que les cotisations sociales obligatoires.

Un *établissement* est défini comme une unité fonctionnelle qui est engagé, en un lieu unique, dans des activités qui appartiennent principalement à un groupe spécifique d'activités de la classification industrielle type de Norvège.

Publications

i) Industristatistikk (annuelle) -- Industries extractives et manufacturières ;

ii) Elektrisitetstatistikk (annuelle) -- Industries de l'électricité et du gaz ;

iii) Bygge- og anleggsstatistikk (annuelle) -- Construction.

Toutes sont publiées par le Bureau central de statistiques de Norvège.

NOUVELLE-ZÉLANDE

Les informations relatives aux statistiques industrielles émanent de Statistics New Zealand, Wellington.

Classification industrielle

La classification utilisée depuis 1994 est la classification industrielle type australienne et néo-zélandaise (Australian and New Zealand Standard Industrial Classification – ANZSIC-96). Il y a une correspondance jusqu'au 4ème chiffre entre la CITI révision 3 et l'ANZSIC-96, mais pas pour toutes les industries.

Les données relatives à la période 1987-1993 fondées sur l'ANZSIC sont probablement de moindre qualité que celles établies à compter de 1994 étant donné qu'elles ont été rétrospectivement reclassées selon l'ANZSIC.

Pour toutes les variables, quand une catégorie de l'ANZSIC (ANZSIC96 ou ANZSIC87) devrait normalement correspondre à plus d'une catégorie de la CITI révision 3, la concordance est uniquement établie avec la catégorie prédominante de la CITI révision 3. La règle de prédominance est établie par Statistics New Zealand («Classifications and Standards Section») qui retient, en fonction de sa connaissance des classifications industrielles et de l'économie néo-zélandaise, la catégorie de la CITI révision 3 qui représente le mieux telle ou telle catégorie de l'ANZSIC (ANZSIC96 ou ANZSIC87). Il en résulte une certaine imprécision puisque certaines catégories de la CITI révision 3 ne sont pas représentées.

Pour les statistiques sur le commerce extérieur, on utilise aussi le Système Harmonisé 1996 (HS) qui s'accorde avec l'ANZSIC-96, elle-même en concordance avec la CITI révision 3.

I. STATISTIQUES DE L'INDUSTRIE ET DES SERVICES

Source

Les données proviennent de l'enquête «Annual Business Frame Update Survey» (ABFUS). L'enquête est conduite chaque année à la mi-février pour mettre à jour les informations contenues dans le registre «Business Frame» de Statistics New Zealand.

Concepts et définitions des variables

L'*emploi, salariés* correspond à la somme des personnes (à col bleu et à col blanc) qui travaillent pour un employeur et perçoivent une rémunération en compensation sous forme de traitements, salaires, pourboires, paiement à la pièce ou rémunération en nature. Il comprend les travailleurs à temps partiel et les travailleurs saisonniers qui figurent sur la liste des employés de l'entreprise, les apprentis et les travailleurs sous contrat de formation, les travailleurs à domicile qui perçoivent une rémunération, les personnes qui travaillent à l'extérieur de l'unité tout en faisant partie de celle-ci et en étant rémunérées par elle, les personnes en congés de courte durée, les personnes en grève, les appelés du contingent et les personnes employées sous contrat qui ne sont pas enregistrées pour la taxe sur les biens et services (GST). Il exclut les propriétaires exploitants et les partenaires associés actifs. Les données d'emploi sont converties en équivalents plein-temps, c'est-à-dire qu'elles sont la somme des salariés occupés à plein temps, plus la moitié des salariés occupés à temps partiel à la fin du mois de février de l'année suivante.

L'*emploi, femmes* correspond au nombre de femmes employées à plein temps, plus la moitié du nombre de femmes employées à temps partiel.

Le nombre d'*heures ouvrées, salariés* correspond au nombre d'heures payées et non au nombre d'heures travaillées.

Les salaires et traitements, salariés correspondent au montant total brut payé à l'ensemble des salariés. Sont incluses tous les pourboires, commissions, primes, et autres indemnités, les paiements pour heures supplémentaires et travail de nuit, les congés payés, les congés-maladie et les dépenses incombant aux travailleurs au cours de leur travail, de même que les impôts et contributions aux caisses d'assurance sociale dus par les salariés mais déduits par l'employeur. Les coûts de formation pour les travailleurs sont exclus.

Par *établissements*, on entend des unités géographiques.

Par *entreprises*, on entend les entités juridiques individuelles en activité en Nouvelle-Zélande. Les entreprises du secteur privé correspondent généralement à des entités juridiques telles que sociétés, sociétés en nom collectif, fiducies, domaines agricoles, entreprises non constituées en société, centrales de production, entreprises individuelles, etc. Certaines entreprises sont définies par référence à d'autres caractéristiques, telles qu'un ensemble d'objectifs communs ou l'existence d'une structure organisationnelle. Dans cette catégorie sont compris les services de l'administration publique, les collectivités locales, les églises, les organisations caritatives, etc.

II. STATISTIQUES DU COMMERCE EXTERIEUR

Source

Les statistiques sur le commerce international proviennent d'un système intégré (CusMod) permettant aux exportateurs et aux exportateurs de répondre électroniquement aux Douanes. Les données sur le commerce international concernent l'année civile.

Pour les statistiques du commerce outre-mer, la "Nouvelle Zélande" inclut les îles du Nord, du Sud et Stewart ainsi que les Antipodes, Auckland, Bounty, Campbell, les îles Kermadec et Chatham, Ross Dependency dans l'Antarctique et les Snares.

Concepts et définitions des variables

Les *exportations* (y compris les réexportations) sont exprimées franco à bord (fab), c'est-à-dire à la valeur des biens dans les ports néo-zélandais avant exportation, et chiffrées en dollars de Nouvelle-Zélande. Les valeurs exprimées en monnaies étrangères sont converties par Statistics New Zealand en dollars néo-zélandais sur la base du taux de change de la semaine où les statistiques sont établies. Les réexportations sont des biens exportés après avoir été importés en Nouvelle-Zélande et qui comprennent moins de 50 pour cent de valeur ajoutée néo-zélandaise. Sont exclues des exportations les données concernant :

- les marchandises expédiées aux forces armées néo-zélandaises stationnées à l'étranger ou aux représentations diplomatiques néo-zélandaises à l'étranger ;
- les marchandises expédiées pour modification ou réparation ;
- les transactions monétaires en or, argent et monnaie courante ;
- les expéditions d'une valeur inférieure à 1 000 dollars ;
- les vêtements usagés destinés à l'aide humanitaire ;
- les marchandises de remplacement et celles non embarquées ;
- les emballages consignés et échantillons à rendre ;
- les pièces de rechange destinées aux avions néo-zélandais à l'étranger et les pièces non réparables démontées sur des avions étrangers et rendues au pays d'origine ;
- les articles en transit et les effets des touristes quittant la Nouvelle-Zélande.

Les *importations* sont exprimées en dollars néo-zélandais aussi bien en ce qui concerne la valeur en douane que la valeur caf (coût, assurance, fret). Une conversion est opérée à partir de la valeur en monnaie étrangère lorsque les documents d'importation sont traités par le service des douanes néo-zélandais. Le taux de change utilisé est fixé toutes les deux semaines par les douanes. Sont exclues des importations les données concernant :

- les marchandises importées temporairement en Nouvelle-Zélande telles que les effets des touristes, les échantillons et emballages à rendre, les marchandises en transit ;
- le poisson débarqué par les navires néo-zélandais ;
- les biens (autres que les véhicules à moteur) appartenant à des représentants officiels de pays étrangers ;
- les bagages de voyageurs importés de façon définitive ;
- les expéditions d'une valeur inférieure à 1 000 dollars ;
- les transactions monétaires en or, argent et monnaie courante ;
- les biens en prêt ;
- les marchandises de remplacement et celles non embarquées précédemment citées ;
- les biens importés destinés à l'usage de forces armées étrangères.

Publications

i) *Business Activity Statistics*
ii) *Labour Market Statistics*
iii) *New Zealand Official Yearbook*
iv) *Key Statistics* (périodique)

PAYS-BAS

Les données émanent du Bureau Central de Statistique des Pays-Bas (Centraal Bureau voor de Statistiek), Voorburg.

Classification industrielle

La classification utilisée est la Classification Industrielle Type (Standaard Bedrijfs Indeling - SBI 1993). Hormis quelques dérogations, elle est totalement compatible avec la NACE révision 1 jusqu'au 4ème chiffre.

STATISTIQUES SUR L'INDUSTRIE ET LES SERVICES

Source

Les données proviennent de plusieurs sources : l'enquête annuelle sur les statistiques de production, l'enquête sur les investissements en capital fixe dans l'industrie, le recensement annuel et le registre TVA. Les deux premières sources sont des enquêtes par échantillon stratifiées par activité et classe de taille. Les entreprises sans salarié ne sont pas couvertes. Le recensement annuel couvre toutes les entreprises de plus de 20 salariés dans l'agriculture, l'industrie, le commerce intérieur et les transports et les entreprises de plus de 100 salariés dans la construction.

Concepts et définitions des variables

La *production* inclut la valeur des livraisons de biens et services produits par l'unité, les recettes au titre des travaux industriels effectués ou des services industriels rendus aux autres, la marge sur les ventes de marchandises réexpédiées en l'état, la variation nette des stocks de produits finis et des travaux en cours, la valeur de l'électricité vendue, les ventes de produits manufacturés par des sous-traitants, la valeur des actifs fixes produits par l'unité pour son propre usage et les recettes perçues au titre de services non-industriels. Elle exclut les revenus provenant de ventes d'actifs fixes produits pour comptes propres. Les taxes affectées aux produits facturées au client par le producteur sont incluses. Les données relatives à la production ne sont disponibles qu'au prix du marché. Cependant, étant donné que l'information relative aux impôts indirects et aux subventions sur les produits existe, il est possible de calculer la production au coût des facteurs.

La *valeur ajoutée*, évaluée au prix du marché, est donnée par la valeur de la production brute déduction faite des éléments suivants : la valeur des matériaux industriels et des services nécessaires à la fabrication, le coût des services industriels reçus, les coûts d'exploitation et paiements pour les services de nature non industrielle. La *valeur ajoutée au coût des facteurs* correspond au revenu brut d'exploitation, après ajustement pour tenir compte des subventions d'exploitation et des impôts indirects. Il peut se calculer à partir du chiffre d'affaires, augmenté des autres revenus d'exploitation et des variations de stocks, diminué des achats de biens et services et des autres impôts sur les produits liés au chiffre d'affaires, mais non déductibles, ainsi que des droits et redevances applicables à la production. Un autre mode de calcul consiste à ajouter les dépenses de personnel à l'excédent brut d'exploitation. Les recettes et dépenses inscrites comme financières ou extraordinaires dans les comptes de l'entreprise sont exclues. La valeur ajoutée au coût des facteurs est une valeur «brute» puisque des ajustements (tels que l'amortissement) n'en sont pas défalqués.

L'*investissement* est défini comme la valeur des acquisitions d'actifs fixes neuf ou usagés (y compris les biens en location ou en leasing) ou de ceux fabriqués par le personnel de l'établissement pour l'usage de ce dernier, au cours de l'année de référence. Les adjonctions, transformations et améliorations d'actifs fixes effectuées par des tiers sont également incluses. Les transactions sur les actifs fixes couvrent les terrains, les bâtiments, la préparation des sites et autres travaux de construction, les terrains non utilisés pour les activités de l'établissement, les équipements de transport, les machines et autres équipements, ainsi que les adjonctions, transformations et améliorations qui allongent la durée d'utilisation ou augmentent la capacité des biens d'équipement. Les actifs fixes sont évalués aux coûts d'acquisition, y compris les coûts de transport et les frais d'installation, les taxes et l'assurance des marchandises en transit.

Les *dépenses en machines et équipement* correspondent à la formation brute de capital fixe en matériel de transport (du type de ceux qui sont utilisés dans les usines, entrepôts, etc., pour le transport sur courte distance ou la manutention des marchandises), machines et matériels industriels, machines, équipement et mobilier de bureau, instruments et équipements professionnels. Elles incluent les modifications et améliorations apportées à ces matériels.

L'emploi, salariés comprend tous les salariés qui travaillent pour une entreprise et perçoivent une rémunération. Il inclut les travailleurs à temps partiel et les travailleurs saisonniers, les apprentis et les personnes sous contrat de formation, les travailleurs à domicile qui perçoivent une rémunération, les directeurs d'entreprises constituées en sociétés commerciales, les personnes en congé (de courte durée ou de durée indéfinie), les personnes en grève et les appelés du contingent. Les données d'emploi ne sont pas converties en équivalents plein-temps (EPT).

Les *salaires et charges sociales, salariés* correspondent aux dépenses de personnel, c'est-à-dire aux salaires et traitements et aux charges sociales des employeurs.

Les *salaires et traitements, salariés* correspondent à la rémunération totale, en espèces ou en nature, payable aux salariés en échange du travail effectué pendant la période de référence et aux salaires versés aux ouvriers, y compris les paiements pour heures supplémentaires et travail de nuit. Sont inclus les pourboires, les indemnités de licenciement, les primes liées au poste de travail et les primes de performance, les gratifications et treizième mois, les primes d'ancienneté, d'assiduité, de stimulation et de productivité, les paiements des jours non travaillés tels que congés payés, vacances, congés-maladie. Sont également inclus les impôts et contributions aux caisses d'assurance sociale dus par les salariés mais déduits par l'employeur.

Les *charges sociales des employeurs* comprennent à la fois les dépenses exigées par la loi et les contributions volontaires. Elles couvrent les contributions patronales aux régimes de sécurité sociale pour les risques de maladie, maternité, invalidé, vieillesse, chômage, accidents du travail, les contributions patronales pour les prestations familiales, l'assurance-vie des salariés, les coûts relatifs aux soins médicaux et aux médicaments donnés gratuitement aux salariés, les coûts d'entretien des écoles et ceux relatifs aux sports et aux loisirs.

En accord avec les règlements de l'Union Européenne, le *nombre d'entreprises* correspond à celui des entreprises inscrites dans la catégorie concernée, sur le registre des entreprises, après correction des erreurs, en particulier les erreurs d'affectation. Les unités en sommeil ne sont pas prises en compte. Toutes les unités en activité pendant au moins une partie de la période de référence doivent être couvertes.

Publications

i) *Summary of manufacturing industries*, annuelle
ii) *Production statistics of energy and water supply enterprises*
iii) *Production statistics of construction*
iv) *Construction figures*
v) *Monthly bulletin of distribution and service trade statistics*, Kerncijfers Detailhandel (données de base du commerce de détail)

POLOGNE

Le Bureau Central de Statistiques (G•ówni Urzad Statystyczny), Warszawa, collecte les données statistiques relatives à l'activité économique des entreprises industrielles.

Classification industrielle

La classification polonaise EKD (Europejska Klasyfikacja Dzia•alno•ci), introduite dans les statistiques de Pologne en 1993, a été utilisée jusqu'en 1999. Depuis 2000, elle a été remplacée par la classification PKD (Polska Klasyfikacja Dzia•alno•ci). Il existe une correspondance directe avec la CITI révision 3 et avec la NACE révision 1, aux niveaux 1- et 2-chiffres.

STATISTIQUES DE L'INDUSTRIE ET DES SERVICES

Source

Plusieurs enquêtes sont menées :

- l'enquête annuelle sur l'activité économique ;
- l'enquête annuelle sur les entreprises employant jusqu'à 9 personnes ;
- l'enquête annuelle sur l'emploi pour un jour donné ;
- et l'enquête sur l'emploi, les revenus et les heures travaillées (conduite mensuellement, trimestriellement et annuellement).

L'unité statistique utilisée est l'entreprise.

Concepts et définitions des variables

La valeur de la *production* vendue est définie comme le revenu provenant de la vente des produits (biens et services) de l'entreprise et la valeur des biens fabriqués qui ne sont pas inclus dans les ventes mais sont traités sur une base égale (par exemple, la valeur des biens et services utilisés pour des besoins de représentation et de publicité). Aucun ajustement n'est effectué pour les travaux en cours et pour les variations des stocks de produits finis. Les transactions entre établissements d'une même entreprise ne sont pas incluses. Les données de 1992 et 1993 incluent la taxe sur le chiffre d'affaires, la TVA (introduite le 5 juillet 1993) et les droits d'accises. Pour 1993, 1994 et 1995 les données incluent la TVA et les droits d'accises. En 1995 et 1996, la TVA et les droits d'accises ne sont pas compris et les suppléments de prix sont comptés.

La variable *valeur ajoutée* brute est calculée pour toutes les unités de l'économie nationale, quel que soit le nombre de personnes qu'elles emploient. Depuis 1994, la valeur ajoutée brute est mesurée aux prix de base (antérieurement elle l'était aux prix départ-usine). Ce changement est dû à l'introduction d'un nouveau système de taxes (le 5 juillet 1993), incluant la TVA sur les biens et services ainsi que les droits d'accises. Les prix de base ne comprennent pas la TVA et les droits d'accises mais incluent les subventions sur les produits. La valeur ajoutée brute correspond à la valeur de la production diminuée de la valeur des consommations intermédiaires. Des méthodes de calcul différentes sont utilisées pour les unités marchandes et non marchandes. La consommation intermédiaire est égale à la valeur des matières premières utilisées (y compris les conditionnements), des combustibles, de l'énergie, du charbon, du gaz, des services fournis par d'autres unités (traitement externe, réparations, transports, location de matériel, télécommunications, services informatiques, assurances), des dépenses de voyages d'affaires et d'autres dépenses (frais de publicité, de location, d'achat de billets d'affaires, par exemple). Sont incluses dans la valeur ajoutée brute des estimations de «l'économie parallèle» - la production dissimulée (production non interdite par la loi, mais dont l'ampleur est cachée, en partie ou en totalité, à l'administration).

Les dépenses d'*investissement* comprennent les dépenses financières ou matérielles de toutes les unités de l'économie nationale, quel que soit le nombre de personnes qu'elles emploient, effectuées pour créer de nouveaux actifs ou pour améliorer (reconstruction, agrandissement, adaptation ou modernisation) les actifs existants, et les dépenses effectuées pour des équipements primaires. Les dépenses de capital fixe comprennent :

- bâtiments et autres ouvrages, c'est-à-dire travaux de construction et de montage, plans et devis ;
- machines, équipement technique et outils (y compris instruments, mobiliers et installations) ;
- matériel de transport ;
- autres, à savoir amélioration des terrains, dépenses d'achats de terrains et de capital fixe d'occasion, bétail (depuis 1995), plantations à long terme, intérêts sur les crédits et les emprunts pour la période de réalisation de l'investissement (données exclusivement à prix courants).

D'autres dépenses incluent les investissements sur l'équipement primaire et les coûts de réalisation de l'investissement mais n'incluent pas la valeur du capital fixe.

L'*emploi, salariés* se rapporte aux personnes salariées effectuant un travail fournissant des revenus (sur le lieu de travail principal), que ce soit à temps plein ou à temps partiel. Elles comprennent les personnes travaillant au titre d'un contrat (contrat de travail, appel sous les drapeaux, nomination ou élection).

Les *salaires et traitements, salariés* correspondent aux paiements en espèces et à la valeur des prestations en nature ou leur équivalent dus aux salariés en échange du travail effectué. Ils recouvrent les salaires et traitements individuels (à l'exclusion de ceux des apprentis, des travailleurs à domicile et des personnes employées à l'étranger), les participations aux bénéfices, les excédents de soldes dans les coopératives, les suppléments annuels de salaires pour les salariés d'entités ..., et autres types de rémunérations versées à des groupes de salariés particuliers effectuant un travail conformément à un contrat de travail (journalistes, producteurs de films et de programme TV ou radio).

Les *heures ouvrées, salariés* sont définies comme les heures de travail normales et supplémentaires. Elles ne comprennent pas les heures payées non travaillées pour diverses raisons comme les vacances, la maladie, les congés exceptionnels.

Une *entreprise* est une entité légale : elle peut être une personne légale, une entité organisationnelle indépendante sans caractère légal ou une personne menant une activité économique.

Publications

i) Statistical Yearbook (annuelle)

ii) Labour Yearbook (tous les deux ans)

iii) Employment in national economy (annuelle)

Toutes les publications sont distribuées par le Bureau Central de Statistiques de Varsovie.

PORTUGAL

Les données émanent de l'Institut National de Statistique (Instituto Nacional de Estatística), à Lisbonne.

STATISTIQUES DE L'INDUSTRIE ET DES SERVICES

Source

Les données sont tirées des enquêtes annuelles réalisées dans les industries manufacturières et extractives, l'électricité, le gaz, ainsi que dans le bâtiment et les travaux publics.

Les données sont tirées de l'Enquête Annuelle aux Entreprises, qui touche 65 000 entreprises dans l'industrie extractive, l'industrie manufacturière et le secteur électricité, gaz et eau. Les chiffres pour la production, la valeur ajoutée, l'emploi, les investissements et le nombre d'entreprises englobent la totalité du secteur industriel. Les chiffres pour l'emploi des salariés, les dépenses en machines et équipement, les salaires et traitements, les compléments des salaires et traitements et les heures travaillées n'englobent que les entreprises ayant 100 salariés ou plus.

Concepts et définitions des variables

La *production* est la valeur brute de production, définie comme la valeur totale des produits finis fabriqués par l'établissement/«entreprise» (les produits intermédiaires vendus en l'état sont considérés comme des produits finis), à laquelle s'ajoute celle des actifs fixes produits par l'unité, pour son propre usage, des travaux industriels exécutés pour des tiers, de l'électricité vendue, et des déchets de fabrication vendus à des tiers, déduction faite de la valeur des produits en cours de fabrication au début de l'année.

La *valeur ajoutée* est obtenue en ajoutant à la valeur brute de production les travaux de sous-traitance exécutés par l'entreprise, déduction faite de la valeur des matériaux, combustibles et énergie consommés, de la location de machines, des travaux de sous-traitance exécutés par des tiers, des travaux industriels ou d'entretien et de réparation exécutés par des tiers, ainsi que des services non industriels reçus, et des droits acquittés pour l'utilisation de procédés de fabrication.

L'*investissement* correspond à l'ensemble des investissements bruts effectués au cours de la période de référence dans tous les biens corporels.

L'*emploi, salarié* couvre le personnel salarié qui travaille pour le compte d'un employeur, disposant d'un contrat de travail et percevant une rémunération sous la forme d'un traitement, d'un salaire, d'émoluments, de gratifications, d'un salaire aux pièces ou d'une rémunération en nature. Il comprend les personnes qui sont absentes pour une courte durée (maladie, vacances, grève, etc.), mais non les personnes qui effectuent leur service militaire ou qui sont à la retraite.

Les *salaires et charges sociales, salariés* se définissent comme la rémunération totale, en espèces ou en nature, due par un employeur à un salarié en échange du travail effectué par celui-ci au cours de la période de référence à laquelle s'ajoutent les compléments des salaires et traitements et les cotisations patronales aux régimes de Sécurité sociale et de retraite.

Les *salaires et traitements, salariés* comprennent toutes les sommes en espèces versées et les avantages en nature fournis aux personnes comptées au nombre des salariés, en rémunération de leur travail au cours de l'exercice comptable.

Par *entreprise*, on entend toute entité productrice, qu'elle soit auxiliaire d'un établissement classé dans un autre secteur d'activité, qu'elle soit dirigée par des particuliers en régime d'administration directe ou encore qu'elle corresponde à une entité du secteur public exécutant des travaux en régime d'administration directe.

Publications

i) *Industrial Statistics* (volume I et volume II) ;
ii) *Statistics on Building and Dwellings.*

REPUBLIQUE TCHEQUE

Les données proviennent de l'Office Statistique Tchèque (•eský Statistický U•ad).

Classification industrielle

La classification industrielle des activités économiques (Odv•tvová klasifikace ekonomických •inností - OKEC or CZ-NACE -) a été introduite en 1992 et mise à jour en 1993. L'OKEC est divisée en 5 niveaux, les quatre premiers comparables à ceux de la NACE révision 1, le cinquième ajouté pour les besoins du pays. L'OKEC est identique à la CITI révision 3 jusqu'au niveau 2.

STATISTIQUES DE L'INDUSTRIE ET DES SERVICES

Source

Les données proviennent de l'enquête annuelle sur les structures de l'entreprise. Il s'agit d'une enquête par échantillon aléatoire, stratifiée par taille d'entreprise et par secteur industriel à 2-chiffres de l'OKEC. Toutes les entreprises, qui étaient actives au moins une partie de l'année, sont couvertes par l'enquête quelle que soit leur taille.

Pour la branche d'intermédiation financière, quatre enquêtes annuelles et trois enquêtes trimestrielles, exhaustives, sont menées. La période de référence est l'année civile.

L'unité statistique est l'entreprise.

Concepts et définitions des variables

La *production* (production brute) correspond à la valeur des biens et services marchands et non-marchands élaborés par des producteurs intérieurs sur une période donnée. Elle est égale à la somme des revenus tirés des ventes des produits, biens, travaux et services fournis à des tiers, de la capitalisation de produits, biens, travaux et services fournis pour son propre compte, de la valeur des actifs fixes produits par l'unité pour son propre compte et des variations de stocks de sa production et des travaux en cours. Les frais encourus sur les produits vendus, les subventions et la TVA sont exclus. Tout autre impôt lié à la production et les frais de transport facturés séparément par le producteur sont inclus.

La *valeur ajoutée* est obtenue en soustrayant la valeur des consommations intermédiaires (coûts des matières premières, combustibles, électricité produite, services industriels et non-industriels reçus et autres fournitures) de la valeur de la production brute.

L'*investissement* correspond à la somme des dépenses exposées pour l'acquisition d'actifs corporels (achat ou production propre), plus la valeur globale des actifs fixes acquis sans frais du fait soit d'un transfert légal, soit du transfert à l'entreprise d'un actif à usage personnel. Les adjonctions, transformations, améliorations et réparations allongeant la durée d'utilisation normale ou augmentant la capacité des biens d'équipement sont également incluses. Les actifs tangibles sont évalués au prix d'acquisition. La TVA déductible est exclue.

L'*emploi, salariés* (moyenne) inclut toutes les catégories de salariés permanents, temporaires et saisonniers ayant un contrat de travail dans l'entreprise. Les femmes en congé de maternité ou de soins pour enfants, les personnes en congé parental, les personnes engagées par les entreprises sur la base de l'exécution d'un travail ou de contrats de travail et les apprentis sont exclus.

Les *salaires et traitements, salariés* correspondent à la somme versée aux salariés pendant l'année, après déduction des impôts et contributions à la retraite. Ils comprennent les paiements en nature pour les heures travaillées, y compris les heures supplémentaires, les primes, congés-payés, etc ..., versés par les employeurs.

Les *heures ouvrées, ouvriers* couvrent les heures travaillées par les ouvriers dans les entreprises de 20 salariés ou plus en 1997 et 1998, de 100 salariés ou plus en 1995 et 1996.

L'*entreprise* est la plus petite combinaison d'unités légales, c'est à dire une unité organisationnelle qui produit des biens ou des services et qui bénéficie d'un certain degré d'autonomie dans la prise de décisions, en particulier pour allouer des ressources courantes. Une entreprise effectue une ou plusieurs activités en un ou plusieurs lieux.

Publications

i) *Annuaire statistique de la République tchèque* - disponible chaque année en décembre ;

ii) Choix d'indicateurs des enquêtes annuelles 1997 et 1998 en République tchèque (***Vybrané ukazatele ro•ního zpracování 1997 a 1998 v pr•myslu •eské republiky).***

Les publications sont distribuées par le CSO.

ROYAUME-UNI

I. STATISTIQUES INDUSTRIELLES

Les informations émanent du Bureau de statistiques nationales («Office for National Statistics» anciennement appelé «Central Statistical Office») du Royaume-Uni.

Classification industrielle

Les résultats des recensements annuels effectués entre 1973 et 1979 étaient ventilés selon la Classification inductrielle type, version révisée de 1968 (SIC-68). Depuis le recensement de 1980, les chiffres sont ventilés selon la version 1980 de cette classification. Afin de relier les données des premiers recensements à celles qui ont été obtenues en 1980 pour les années suivantes, les chiffres de 1979 ont été recalculés sur la base de la nouvelle classification.

Une nouvelle classification industrielle type (SIC-92) fondée sur la NACE révision 1 est applicable depuis 1990 et a été mis en œuvre pour la première fois pour le recensement de 1993.

Source

Les données proviennent de l'enquête annuelle par échantillon dans la production et la construction («Annual Sample Survey into Production and Construction», anciennement appelée «Annual Census of Production» et «Annual Census of Construction»).

Concepts et définitions des variables

La *production*, ou production brute, est donnée par la valeur des ventes totales, des travaux exécutés et des services rendus, corrigée de la variation nette, durant l'année considérée, des travaux en cours et des stocks de biens en vente ainsi que, pour 1996-1997 uniquement, des travaux d'investissement réalisés par le personnel de l'unité. Les ventes totales correspondent aux livraisons de biens produits par les établissements implantés au Royaume-Uni et couverts par le recensement, et comprennent les ventes de biens fabriqués à l'intention de ces établissements à partir de matières premières fournies par ces derniers à d'autres établissements ou à des travailleurs extérieurs, ainsi que les ventes de déchets et de résidus. La valeur des ventes correspond à la valeur nette des ventes, autrement dit la somme demandée aux clients -- prix départ-usine ou prix rendu -- déduction faite de la TVA, des remises, des commissions des représentants, etc., ainsi que des déductions sur biens rendus. Lorsque les produits sont soumis à un droit d'accises, celui-ci est inclus si les biens sont vendus "après paiement des droits" et exclu si les biens sont vendus sous douane ou s'ils sont exportés. Les ventes d'actifs fixes ne sont pas prises en compte.

La valeur ajoutée ou *valeur ajoutée brute au coût des facteurs*, correspond à la valeur de la production brute déduction faite du coût des matériaux utilisés dans la production et le conditionnement, des combustibles, des biens nécessaires pour la vente ou l'affacturage, du coût des services industriels et non industriels reçus, des impôts locaux et des taxes de mise en circulation des véhicules. La valeur des achats est corrigée des variations, durant l'année considérée, des stocks de matières premières, de fournitures et de combustibles. L'évaluation est faite au coût des facteurs à l'exclusion des impôts indirects mais y compris les subventions.

L'*investissement*, ou formation brute de capital fixe, représente la valeur des acquisitions d'actifs fixes déduction faite des ventes. L'évaluation est faite au prix de revient total, y compris les frais d'installation et certains droits et redevances (frais d'acte, droit de timbre, commissions de représentants, etc.) et comprend tout montant reçu ou attendu de l'état, d'organismes officiels ou des collectivités locales, sous forme de subventions et/ou déductions pour investissement. En règle générale, la valeur retenue correspond au montant porté au compte de capital auquel s'ajoute toute autre somme qui, aux fins de l'imposition durant l'année en question, est considérée comme du capital. Lorsque les dépenses sont étalées sur plusieurs années, les paiements sont comptabilisés l'année où ils ont été effectués. Sont incluses les dépenses d'équipement encourues durant l'année du recensement par des unités de production où la production n'a pas commencé avant la fin de l'année considérée, ainsi que la valeur des biens d'équipement produits par le personnel de l'établissement pour l'usage de ce dernier. Est exclue la valeur de tout actif acquis à l'occasion du rachat d'une entreprise existante. Les chiffres incluent le montant de la TVA non récupérable, mais ne comprennent pas celui de la TVA récupérable. Aucune provision n'est prévue pour amortissement ou obsolescence.

A partir de 1993, les données concernant les *dépenses en machines et équipements* se rapportent uniquement aux actifs fabriqués ou édifiés par l'établissement pour son propre usage ; elles sont égales à la valeur des travaux d'équipement au cours de l'année de référence effectués par le personnel de l'établissement à l'usage de ce dernier.

L'*emploi, total* représente le nombre de personnes occupées ; il indique le nombre moyen d'agents administratifs, de techniciens, d'employés de bureau et d'ouvriers travaillant à plein temps et à temps partiel inscrits sur le registre du personnel de l'établissement, ainsi que le nombre de propriétaires exploitants occupés durant l'année de référence. Sont inclus les travailleurs indépendants, les associés actifs, les travailleurs familiaux non rémunérés travaillant au moins la moitié de l'horaire normal, les salariés, à temps complet ou à temps partiel, mais sont exclus les travailleurs à domicile ainsi que les travailleurs occasionnels.

L'*emploi, salariés* indique le nombre moyen d'agents administratifs, de techniciens, d'employés de bureau et d'ouvriers travaillant à plein temps et à temps partiel inscrits sur le registre du personnel de l'établissement. Il ne comprend pas les propriétaires exploitants, c'est-à-dire les personnes considérées comme des travailleurs indépendants par l'assurance sociale, les membres de leurs familles travaillant dans l'entreprise sans véritablement recevoir ni salaire ni traitement bien défini pour au moins la moitié de l'horaire normal, ainsi que les directeurs travaillant dans l'entreprise qui n'ont pas reçu de salaire, traitement ou commission bien défini. Il comprend les dirigeants ayant perçu un salaire, un traitement, des honoraires ou une commission bien définis, les cadres supérieurs, les contremaîtres, les dessinateurs, les personnels des services d'édition ou de publicité, les représentants, tous les employés de bureau et des services de recherche et de conception, à l'exclusion des ouvriers.

Les *salaires et charges sociales, salariés* correspondent aux sommes versées durant l'année aux salariés. Ils comprennent les salaires et traitements et les suppléments. Les cotisations patronales au régime national d'assurance, etc. sont incluses de même que la rémunération des travailleurs extérieurs. Celle-ci correspond aux rémunérations versées aux personnes qui travaillent réellement à leur domicile, qui sont généralement rémunérées à la pièce et dont les noms figurent sur les états de paie. Les rémunérations versées à des travailleurs extérieurs par des sous-traitants sont incluses dans le coût des services industriels reçus. La rémunération des travailleurs extérieurs relative à des entreprises qui n'ont pas répondu ne fait pas l'objet d'une estimation.

Les *salaires et traitements, salariés* représentent les sommes versées pendant l'année aux salariés. Tous les paiements au titre d'heures supplémentaires, les gratifications, les commissions, les congés payés ainsi que les compensations pour perte d'emploi faites aux ouvriers en surnombre, après déduction des montants remboursés à cet effet de source gouvernementale, sont compris. Aucune déduction n'est effectuée au titre des impôts sur le revenu, des cotisations salariales au régime national d'assurance, etc. Sont exclus les sommes versées aux propriétaires exploitants, les prestations en nature, les frais de déplacement, les indemnités de logement, etc. et les cotisations patronales au régime national d'assurance.

Les *charges sociales des employeurs* comprennent les cotisations patronales au régime national d'assurance, prévues par la loi aux termes de la «Social Security Pensions Act» de 1975 ainsi que les primes versées, pour le compte des salariés ou des personnes à leur charge, aux caisses privées d'assurance à divers titres : pensions, retraite ou autres, assurance-maladie, assurance-accident, invalidité ou décès.

L'*établissement* est défini comme la plus petite unité qui puisse fournir l'information normalement requise dans le cadre d'un recensement.

L'entreprise est définie comme le plus petit ensemble d'entités juridiques ayant un certain degré d'autonomie au sein d'un groupe d'entreprises. Un groupe d'entreprises est constitué par l'ensemble des entreprises appartenant à une même entité ou contrôlées par une même entité.

II. STATISTIQUES DU COMMERCE EXTÉRIEUR

Classification industrielle

Les statistiques des douanes et de la régie sont établies selon la classification type du commerce international (CTCI). Elles sont ensuite regroupées selon la classification industrielle type du Royaume-Uni par le Bureau national de statistiques, qui les ventile alors selon la CITI.

Source

Les données sont recueillies par les fonctionnaires des douanes et de la régie puis transmises au Bureau national de statistiques («Office for National Statistics»).

Concepts et définitions des variables

Les données portent sur le commerce général, aussi bien pour les exportations que pour les importations. Les *exportations*, exprimées FAB, sont ventilées en fonction du pays de destination finale et les *importations*, exprimées CAF, selon le pays d'expédition. Sont inclus dans les deux cas : le commerce d'état, les transactions commerciales d'équipement et de fournitures militaires, les transactions relatives à l'aide extérieure, le commerce d'entretien et de réparation, les transactions portant sur les navires et avions neufs, l'électricité et le gaz, ainsi que les donations et dons privés. Les produits de la mer provenant directement de haute mer sont aussi pris en compte, mais les produits naturels de la mer amenés par des navires immatriculés au Royaume-Uni sont exclus des importations. Sont spécifiquement exclus : les échanges directs en transit et les transactions portant sur des navires et avions d'occasion, mais non les exportations temporaires, les articles monétaires, l'or industriel et les petites transactions.

Publications

i) *Census of Production Introductory Notes* (Business Monitor PA 1001)
ii) *Annual Abstract of Statistics*
iii) *Overseas Trade Statistics* (mensuel)

Les deux premières publications sont distribuées par le Bureau national de statistiques («Office for National Statistics - London»), la troisième par HMSO.

SUÈDE

Les données émanent du Bureau de statistique de Suède (Statistiska Centralbyran), à Stockholm.

Classification industrielle

Les données sont élaborées dans la classification industrielle type de Suède de 1992 (SE-SIC92) qui est identique à la CITI révision 3 aux niveaux 1 et 2 chiffres et à la NACE révision 1 aux niveaux 1 à 4 chiffres.

I. STATISTIQUES INDUSTRIELLES

Source

Jusqu'en 1996, les données proviennent du fichier de statistiques industrielles du Bureau et sont tirées de son enquête annuelle dans les industries manufacturières. L'unité statistique de référence est l'établissement. L'enquête s'adresse aux établissements manufacturiers (de 5 salariés ou plus) des entreprises manufacturières de dix salariés ou plus et aux établissements manufacturiers (de dix salariés ou plus) des entreprises non manufacturières. Des estimations sont effectuées pour tenir compte des réponses manquantes.

A partir de l'année de référence 1997, les données sont produites selon le Règlement Européen sur les Statistiques Structurelles d'Entreprises. L'unité statistique est l'entreprise. Il n'y a plus de limite à la collecte de données : toutes les entreprises sont incluses.

Concepts et définitions des variables

La *production au coût des facteurs* correspond à la production brute, à l'exclusion des impôts indirects mais y compris toutes les subventions. Elle inclut la valeur de tous les produits, la variation nette des travaux en cours, les services rendus, les déchets de fabrication et les biens fabriqués par les travailleurs à domicile.

La *valeur ajoutée au coût des facteurs* est égale à la production déduction faite des coûts totaux, hors salaires et traitements (charges sociales incluses). Les données relatives aux consommations sont estimées sur la base des quantités consommées.

Les données sur l'*emploi, salariés* correspondent au nombre moyen d'ouvriers occupés tout au long de l'année de référence auquel on ajoute le nombre d'employés occupés à certaines dates. Ne sont pas compris les propriétaires exploitants, les associés actifs et les travailleurs indépendants. Tous les salariés à temps complet et à temps partiel sont pris en compte, de même que le personnel de bureau et le personnel d'encadrement dont les fonctions sont étroitement liées au processus de production.

Les *salaires et traitements, salariés* comprennent les paiements versés en espèces aux salariés pour les heures normales et supplémentaires, les gratifications et indemnités de cherté de vie, les congés payés et congés de maladie, ainsi que les impôts et cotisations d'assurance sociale, etc., dus par les salariés et retenus par l'employeur ainsi que les prestations en nature. Les charges de sécurité sociale ne sont pas prises en compte.

Publications

i) Industries manufacturières : parties I et II de la série *Official Statistics of Sweden* (SOS)

Ces publications sont distribuées par le Bureau de statistique.

Une partie des séries est également disponible dans les banques de données du Bureau de statistique.

SUISSE

Les statistiques sont établies par l'Office fédéral de la statistique de Suisse (OFS - Bundesamt für Statistik), Neuchâtel.

Classification industrielle

La NOmenclature Générale des Activités économiques, NOGA, est entrée en vigueur en 1995. Elle remplace la Nomenclature Générale des Activités Economiques de 1985 (NGAE).

Elle est parfaitement comparable à la NACE révision 1 jusqu'au niveau à quatre chiffres, le quatrième niveau étant subdivisé en sous-classes pour une utilisation interne. La NOGA 1995 est compatible avec la CITI révision 3.

STATISTIQUES DE L'INDUSTRIE ET DES SERVICES

Source

Les données correspondent à de nouvelles informations disponibles dans le cadre des statistiques d'entreprises pour lesquelles aucune base permettant le calcul d'un ensemble cohérent de données n'existe avant 1997. Elles ne sont pas comparables aux données précédemment fournies dans le cadre des Comptes Nationaux.

Les informations proviennent de l'enquête annuelle 'Statistique de la production et de la valeur ajoutée' qui est une enquête par sondage. L'échantillon est tiré du Registre des Etablissements et des Entreprises (REE). Les données couvrent l'année civile.

L'unité de référence est l'entreprise.

Concepts et définitions des variables

La valeur de la *production* mesure le montant de ce qui a été effectivement produit par l'unité. Elle est définie comme le chiffre d'affaires corrigé de la variation des stocks, diminué des acquisitions de biens et services destinés à la revente, augmenté de la production immobilisée et des autres produits d'exploitation. La production est évaluée aux prix du marché.

La *valeur ajoutée au coût des facteurs* est le revenu brut des activités d'exploitation une fois pris en compte les subventions d'exploitation et les impôts indirects. Elle correspond au chiffre d'affaires, augmenté de la production immobilisée et des autres produits d'exploitation, corrigé de la variation des stocks, diminué des acquisitions de biens et services et des autres taxes sur les produits liés au chiffre d'affaires mais non déductibles et des impôts et taxes liés à la production.

L'*emploi* comprend la totalité des personnes engagées dans un établissement et travaillant au moins 6 heures par semaine. Elle comprend les propriétaires exploitants, les gérants, les travailleurs familiaux non rémunérés, les apprentis, les représentants et les autres personnes exerçant à l'extérieur de l'unité. Les travailleurs à domicile, les représentants ou les consultants travaillant en tant qu'indépendants sont exclus. Les établissements où l'on travaille moins de 20 heures par semaine ne sont pas couverts. Les données d'emploi sont converties en équivalents plein-temps (EPT).

Les *salaires et charges sociales, salariés* sont égaux à la somme des salaires et traitements versés aux salariés et des contributions sociales patronales.

Les *salaires et traitements, salariés* comprennent toutes les sommes en espèces versées et les avantages en nature fournis aux salariés, y compris les travailleurs à domicile, en rémunération de leur travail au cours de la période de référence, qu'ils soient rémunérés sur une base horaire, à la production ou à la pièce et qu'ils soient payés régulièrement ou non. Sont inclus les pourboires, les primes, les gratifications, le treizième mois, les allocations de logement, les indemnités de transport, les indemnités de vie chère, les allocations familiales, les primes d'ancienneté, d'assiduité, de stimulation et de productivité, les paiements pour jours non travaillés tels que les congés payés, vacances, et congés-maladie. Les impôts et les contributions aux caisses d'assurance sociale, dus par les salariés mais déduits par l'employeur sont également inclus.

Les *charges sociales des employeurs* sont les contributions patronales aux régimes de sécurité sociale, couvrant les risques de maladie, de maternité, d'invalidité, de vieillesse, de chômage, d'accident du travail ainsi que les prestations familiales. Les contributions patronales doivent être énoncées dans le cadre de dispositions obligatoires.

Publications

i) Statistiques de la production et de la valeur ajoutée/Produktions- und Wertschöpfungsstatistik

Les publications sont diffusées par l'Office Fédéral de la Statistique (OFS).

TURQUIE

Les données proviennent de l'Institut public des statistiques (Devlet Istatistik Enstitüsü) d'Ankara

I. STATISTIQUES INDUSTRIELLES

Classification industrielle

La classification turque (Ulusal Faaliyet S•n•flamas•) est une adaptation de la CITI révision 2 jusqu'au niveau 4 chiffres.

Source

Les données proviennent de plusieurs enquêtes : le Recensement des Etablissements de l'Industrie qui se fait chaque année se terminant par trois, l'enquête annuelle dans les industries manufacturières (pour tous les établissements du secteur public et pour les établissements du secteur privé employant 10 personnes ou plus), les statistiques des petites entreprises de secteur manufacturier (pour les établissements de 1 à 9 personnes), l'enquête annuelle sur la production d'électricité, l'enquête annuelle sur la distribution d'électricité, l'enquête annuelle sur les statistiques de l'eau, l'enquête annuelle sur la distribution de gaz naturel, l'enquête sur les établissements de construction et d'installation, le questionnaire statistique sur les industries extractives et le questionnaire annuel sur le sable, l'argile et la pierre.

Concepts et définitions des variables

La *production* est donnée par la valeur totale des ventes de biens et services rendus à d'autres unités et des recettes provenant des ventes et des transferts d'électricité, à laquelle on ajoute la valeur des stocks en fin d'année (de produits finis et semi-finis et des biens revendus en l'état) et celle des actifs fixes produits par l'unité pour son propre usage et dont on déduit la valeur des stocks en début d'année (de produits finis et semi-finis et des biens revendus en l'état). La production est évaluée au prix départ-usine.

La *valeur ajoutée*, également évaluée au prix départ-usine, est donnée par la production, déduction faite de la valeur des consommations intermédiaires. Cette dernière correspond à la valeur totale des produits et services achetés ou transférés, de l'électricité achetée et de la valeur des stocks en début d'année (de matières premières, fournitures, matériel d'emballage et combustibles) après déduction de la valeur de ces mêmes stocks en fin d'année.

L'*investissement* représente les acquisitions brutes d'actifs fixes au cours de l'année. Il est donné par le coût total des actifs fixes neufs ou usagés acquis sur le marché intérieur, des actifs fixes neufs ou usagés importés, des actifs fixes produits par le personnel de l'établissement, des actifs fixes installés pendant l'année et achetés aux enchères, des grosses réparations et des dépenses faites sur les actifs fixes, des études, plans et dessins, des machines, équipements, véhicules et constructions, du matériel et mobilier de bureau (à condition qu'ils soient utilisés dans l'établissement, qu'ils aient une espérance de vie productive supérieure à un an et qu'ils soient portés au compte de capital) et les dépenses d'acquisition et d'amélioration des terrains, déduction faite de la valeur des ventes d'actifs fixes au cours de l'année.

Les *dépenses en machines et équipement* représentent les acquisitions brutes de machines et d'équipement au cours de l'année. Elles correspondent au coût total des machines et équipement neufs ou usagés acquis sur le marché intérieur, des machines et équipement neufs ou usagés importés, des machines et équipement produits par le personnel de l'établissement, des machines et équipement installés pendant l'année et achetés aux enchères, des grosses réparations et dépenses faites sur les machines et équipement, des machines et équipement utilisés par l'établissement et dont l'espérance de vie productive est supérieure à un an, déduction faite de la valeur des ventes de machines et d'équipement au cours de l'année.

L'*emploi, total*, ou nombre moyen de personnes occupées, est donné par le nombre de propriétaires, d'associés et de travailleurs familiaux non rémunérés en activité dans l'établissement au mois de novembre (au mois de juillet dans les industries extractives ; au mois d'août dans la construction), auquel on ajoute le nombre moyen de salariés. Les salariés temporairement absents au moment de l'enquête pour raison de maladie, de congé annuel ou de grève, etc. sont inclus. Les données d'emploi sont converties en equivalents plein-temps (EPT).

L'*emploi, salariés* est égal à la moyenne arithmétique du nombre de salariés employés en février, mai, août et novembre (moyenne des douze mois dans les industries extractives et dans la construction). Il comprend les personnes travaillant dans l'établissement (travailleurs manuels, personnel administratif et autres salariés) quel que soit l'intitulé de leur emploi, qui ont perçu un salaire, un traitement ou une rémunération. Les travailleurs familiaux non rémunérés, les propriétaires et associés sont exclus.

Les *heures ouvrées, ouvriers* dans une année correspondent au produit du nombre moyen d'ouvriers par le nombre d'heures travaillées dans une équipe postée et le nombre de jours travaillés dans l'année.

Les salaires et charges sociales, salariés comprennent les rémunérations en espèces et prestations en nature versées aux seuls salariés pendant l'année de référence et les charges sociales des employeurs.

Les *salaires et traitements, salariés,* ou sommes versées aux salariés au cours de l'année, comprennent tous les paiements effectués au titre de salaires ou de gages journaliers, avant déduction de l'impôt sur le revenu et des cotisations d'assurance sociale et de retraite. Les cotisations patronales aux caisses de sécurité sociale, de retraite et autres sont exclues. En revanche sont inclus le paiement des heures supplémentaires, les gratifications, les indemnités et les prestations en nature.

Les charges sociales des employeurs comprennent les paiements effectués pour le compte des salariés à la caisse de sécurité sociale et à un régime d'épargne, les indemnités de licenciement et les allocations de logement.

Un *établissement* est un lieu de travail où il existe un statut unique de propriété ou de contrôle, où, en général, un seul type d'activité économique est réalisé et qui dispose des informations comptables et des registres nécessaires pour remplir les questionnaires.

II. STATISTIQUES DES SERVICES ET STATISTIQUES DU COMMERCE EXTÉRIEUR

Classification industrielle

La classification utilisée est la CITI révision 3.

Concepts et définitions de variables

La *production* est donnée par la valeur totale des ventes de biens et des services rendus aux autres, des stocks de biens marchands à la fin de l'année et de la valeur de fabrication des actifs fixes produits par le personnel de l'établissement, déduction faite de la valeur totale des stocks de biens au début de l'année (biens finis et semi-finis et biens revendus en l'état).

La *valeur ajoutée* est donnée par la production déduction faite de la valeur de la consommation intermédiaire. Cette dernière correspond à la valeur totale des biens et services achetés et des stocks de matériaux et de combustibles au début de l'année déduction faite de la valeur totale des stocks de matériaux et de combustibles à la fin de l'année et des biens achetés pour être revendus en l'état.

L'*investissement*, défini comme l'investissement en équipements fixes, recouvre les équipements utilisés pendant l'année et l'amortissement des bâtiments, des machines et des équipements en stock, des véhicules, des matériels et fournitures de bureau en stock et des terrains. Il recouvre aussi les modifications, rénovations, adjonctions et grosses réparations de ces biens.

Les *dépenses en machines et équipement* se rapportent aux adjonctions brutes de machines et équipements opérées au cours de l'année. Elles correspondent aux dépenses de machines et équipements neufs ou usagés achetés sur le marché intérieur, de machines et équipements neufs ou usagés importés, de machines et équipements fabriqués par le personnel de l'établissement et aux dépenses relatives à des actifs qui se déprécient lentement, déduction faite de la valeur des ventes de machines et équipements au cours de l'année.

L'*emploi, salariés* est la moyenne arithmétique du nombre de salariés employés en février, mai, août et novembre. Il comprend les personnes travaillant dans l'établissement, quel que soit l'intitulé de leur emploi, qui ont perçu un salaire, un traitement ou une rémunération similaire. Les travailleurs familiaux non rémunérés, les propriétaires et associés sont exclus. Les salariés temporairement absents le jour de l'enquête pour raison de maladie, de congé annuel ou de grève, etc. sont inclus.

Les *salaires et traitements, salariés,* ou sommes versées aux salariés au cours de l'année, comprennent tous les paiements effectués au titre de salaires ou de gages journaliers avant déduction de l'impôt sur le revenu et des cotisations d'assurance sociale et de retraite. Les cotisations patronales aux caisses de sécurité sociale, de retraite et autres sont exclues. Sont en revanche inclus le paiement des heures supplémentaires, les gratifications, les indemnités et les prestations en nature.

Les *exportations* et *importations* correspondent aux mouvements internationaux de biens par delà les frontières. Les exportations sont exprimées FAB (franco à bord) et les importations CAF (coût, assurance et fret).

Un *établissement* est un lieu de travail où il existe un statut unique de propriété ou de contrôle, où, en général, un seul type d'activité économique est réalisé et qui dispose des informations comptables et des registres nécessaires pour remplir les questionnaires.

Publications

i) *Annual Manufacturing Industry Statistics*, annuelle
ii) *Small-Sized Manufacturing Industry Statistics*, annuelle
iii) *Annual Electricity, Gas and Water Statistics*
iv) *1992-1997 Construction and Installation Establishments Statistics*
v) *Mining and Quarrying Statistics*

Section II

Exchange rates and purchasing powers parities for GDP
Taux de change et parités de pouvoir d'achat du PIB

Purchasing power parities for GDP
national currency per US dollar

Parités de pouvoir d'achat du PIB
monnaie nationale par dollar É-U

	1991	1992	1993	1994	1995	1996	1997	1998	
Australia	1.3666	1.3713	1.3530	1.3378	1.2871	1.2994	1.3202	1.3075	Australie
Austria	14.178	13.977	13.865	13.914	13.733	13.579	13.597	13.725	Autriche
Belgium	39.179	37.785	37.303	37.276	36.743	36.824	37.471	37.392	Belgique
Canada	1.2873	1.2817	1.2630	1.2508	1.1826	1.1853	1.1860	1.1642	Canada
Czech Republic	7.6474	8.3496	9.4610	10.3902	10.8113	11.6901	12.4078	13.4999	République Tchèque
Denmark	9.1818	9.1455	8.7859	8.7062	8.4167	8.3278	8.5447	8.5781	Danemark
Finland	6.2987	6.3534	6.0862	6.1446	5.8615	5.8858	6.0421	6.1437	Finlande
France	6.5130	6.4178	6.5728	6.6208	6.4602	6.5718	6.7221	6.7047	France
Germany	2.0942	2.0657	2.1029	2.0687	2.0156	2.0274	2.0077	2.0143	Allemagne
Greece	161.26	170.07	184.34	196.14	203.08	213.90	232.39	238.09	Grèce
Hungary	32.649	37.079	43.221	51.170	60.552	72.551	85.593	95.204	Hongrie
Iceland	85.725	83.132	82.927	84.028	75.872	76.755	80.133	83.427	Islande
Ireland	0.6668	0.6379	0.6548	0.6382	0.6347	0.6725	0.6752	0.7203	Irelande
Italy	1,463.3	1,458.7	1,533.8	1,532.9	1,550.3	1,583.0	1,643.5	1,620.3	Italie
Japan	193.06	188.16	184.31	180.59	169.94	165.62	165.05	163.52	Japon
Korea	590.73	585.60	591.45	618.16	615.20	629.24	641.36	665.39	Corée
Luxembourg	39.498	38.882	39.620	39.991	38.872	39.708	41.242	41.550	Luxembourg
Mexico	1.8854	2.0163	2.1218	2.2762	2.9573	3.7894	4.4821	5.1085	Mexique
Netherlands	2.1847	2.1355	2.1344	2.1235	2.0274	2.0447	2.0335	2.0032	Pays-Bas
New Zealand	1.5641	1.5119	1.5117	1.5014	1.4672	1.4779	1.4724	1.4667	Nouvelle Zélande
Norway	9.5977	8.9807	8.9309	9.1160	9.1442	9.1140	9.3184	9.5730	Norvège
Poland	0.4666	0.6040	0.7579	1.0003	1.1371	1.3647	1.6719	1.7496	Pologne
Portugal	109.99	115.56	116.96	118.07	119.07	122.39	122.22	128.63	Portugal
Spain	110.45	114.77	116.96	121.27	122.08	123.68	126.85	130.59	Espagne
Sweden	9.9492	9.7971	9.8332	9.8951	9.7281	9.6780	9.7884	9.8544	Suède
Switzerland	2.2320	2.1628	2.1316	2.0973	2.0109	2.0524	1.9864	1.9598	Suisse
Turkey	2,280.2	3,667.2	5,989.8	12,096.2	22,334.2	39,274.6	70,999.9	123,169.0	Turquie
United Kingdom	0.6353	0.6158	0.6373	0.6453	0.6539	0.6440	0.6506	0.6520	Royaume-Uni
United States	1.0000	1.0000	1.0000	1.0000	1.0000	1.0000	1.0000	1.0000	États-Unis

Exchange rates
national Currency per US dollar

Taux de Change
monnaie nationale par dollar É-U

Pays	1991	1992	1993	1994	1995	1996	1997	1998	
Australia	1.2840	1.3623	1.4730	1.3691	1.3496	1.2769	1.3480	1.5923	Australie
Austria	11.670	10.991	11.633	11.419	10.083	10.579	12.197	12.376	Autriche
Belgium	34.162	32.146	34.549	33.461	29.497	30.976	35.758	36.298	Belgique
Canada	1.1462	1.2089	1.2902	1.3659	1.3725	1.3638	1.3850	1.4835	Canada
Czech Republic	29.470	28.260	29.150	28.790	26.540	27.150	31.700	32.280	République Tchèque
Denmark	6.3928	6.0384	6.4822	6.3602	5.6038	5.7982	6.6041	6.6962	Danemark
Finland	4.0432	4.4858	5.7208	5.2229	4.3671	4.5922	5.1872	5.3453	Finlande
France	5.6409	5.2937	5.6623	5.5521	4.9908	5.1158	5.8373	5.8994	France
Germany	1.6588	1.562	1.6533	1.6229	1.433	1.5047	1.7342	1.7594	Allemagne
Greece	182.06	190.47	229.07	242.24	231.58	240.66	272.95	295.27	Grèce
Hungary	74.770	78.990	91.910	105.120	125.720	152.610	186.630	214.260	Hongrie
Iceland	59.096	57.619	67.642	69.990	64.766	66.691	70.969	71.174	Islande
Ireland	0.6216	0.5877	0.6833	0.6697	0.6245	0.6254	0.6604	0.703	Irelande
Italy	1,240.7	1232.03	1571.66	1612.66	1629.03	1543.04	1702.76	1736.4	Italie
Japan	134.50	126.67	111.18	102.23	94.07	108.82	121.00	130.89	Japon
Korea	733.23	780.01	802.44	804.27	771.40	804.42	950.51	1,400.48	Corée
Luxembourg	34.162	32.146	34.549	33.461	29.497	30.976	35.758	36.298	Luxembourg
Mexico	3.0225	3.095	3.1153	3.3886	6.4213	7.6007	7.9236	9.1533	Mexique
Netherlands	1.8697	1.7585	1.8572	1.8197	1.6053	1.6861	1.9509	1.9845	Pays-Bas
New Zealand	1.7292	1.86	1.8507	1.6871	1.524	1.4536	1.5126	1.8685	Nouvelle Zélande
Norway	6.4839	6.2145	7.0944	7.0565	6.3372	6.4574	7.0723	7.5453	Norvège
Poland	1.0578	1.3628	1.8136	2.2728	2.425	2.6956	3.2769	3.492	Pologne
Portugal	144.35	134.79	160.66	166.04	149.94	154.21	175.16	180.15	Portugal
Spain	103.93	102.40	127.24	133.96	124.69	126.69	146.41	149.38	Espagne
Sweden	6.0455	5.823	7.7854	7.7157	7.1336	6.7071	7.6346	7.9471	Suède
Switzerland	1.4337	1.4058	1.4774	1.3671	1.1821	1.2361	1.45	1.4497	Suisse
Turkey	4168.9	6860.5	10964.4	29778.1	45738.4	81281.0	151594.9	260472.6	Turquie
United Kingdom	0.5669	0.5697	0.666	0.6533	0.6336	0.6408	0.6105	0.6036	Royaume-Uni
United States	1.0000	1.0000	1.0000	1.0000	1.0000	1.0000	1.0000	1.0000	États-Unis

Exchange rates and purchasing power parities are given for inormation. If used to convert the data presented in this publication for purposes of international comparison, care should be taken to understand the limitations of the validity of such comparisons. these are due to the different survey and estimations techniques used in each country to compile national industrially disaggregated datasets

Les taux de change et les parités de pouvoir d'achat sont présentés pour information. Leur utilisation pour convertir les données contenues dans cette publication publication afin de permettre des comparaisons internationales devrait être accompagnée d'une compréhension des limites de telles comparaisons. Ces limites sont dues aux différentes techniques d'enquêts et d'estimations utilisées dans chaque pays pour compiler des ensembles de données nationales désagrégées par secteur.

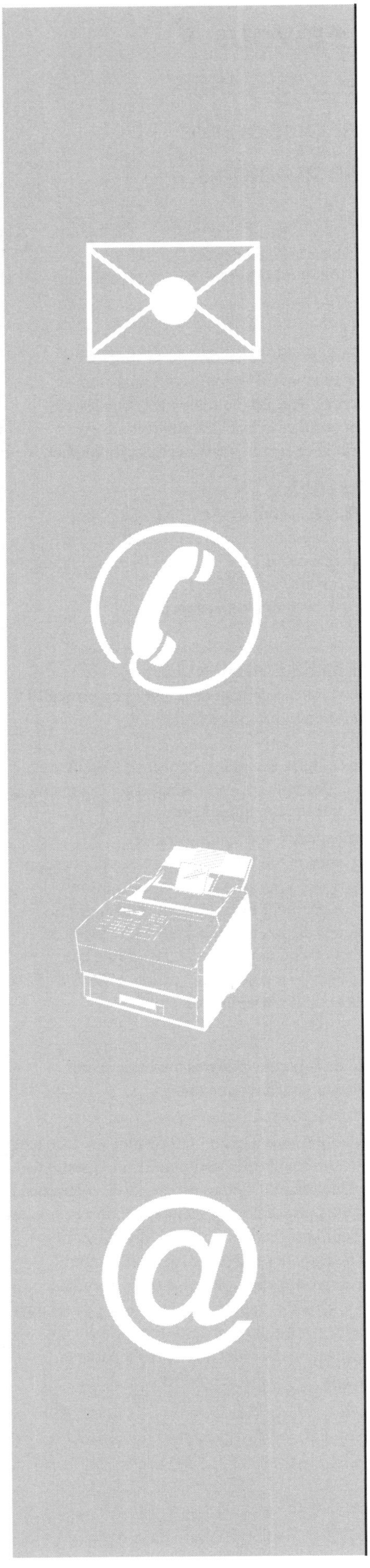

For more information about all OECD publications contact your nearest OECD Centre, or visit
www.oecd.org/bookshop

Pour plus d'informations sur les publications de l'OCDE, contactez votre Centre OCDE le plus proche ou visitez notre librairie en ligne : **www.oecd.org/bookshop**

Where to send your request:
Où passer commande :

In Central and Latin America / En Amérique centrale et en Amérique du Sud

OECD MEXICO CENTRE / CENTRE OCDE DE MEXICO
Edificio INFOTEC
Av. San Fernando No. 37 Col. Toriello Guerra
Tlalpan C.P. 14050, Mexico D.F.
Tel.: +525 281 38 10 Fax: + 525 606 13 07
E-mail: mexico.contact@oecd.org Internet: www.rtn.net.mx/ocde

In North America / En Amérique du Nord

OECD WASHINGTON CENTER / CENTRE OCDE DE WASHINGTON
2001 L Street N.W., Suite 650
Washington, DC 20036-4922
Tel.: +1 202 785-6323
Toll free / Numéro vert : +1 800 456-6323 Fax: +1 202 785-0350
E-mail: washington.contact@oecd.org Internet: www.oecdwash.org

In Japan / Au Japon

OECD TOKYO CENTRE / CENTRE OCDE DE TOKYO
Landic Akasaka Bldg.
2-3-4 Akasaka, Minato-ku
Tokyo 107-0052
Tel.: +81 3 3586 2016 Fax: +81 3 3584 7929
E-mail : center@oecdtokyo.org Internet: www.oecdtokyo.org

In the rest of the world / Dans le reste du monde

DVGmbH
Birkenmaarsstrasse 8
D-53340 Meckenheim
Germany
Tel.: +49 22 25 9 26 166/7/8 Fax: +49 22 25 9 26 169
E-mail: oecd@dvg.dsb.net

OECD Information Centre and Bookshop/
Centre d'information de l'OCDE et Librairie

OECD PARIS CENTRE / CENTRE OCDE DE PARIS
2 rue André-Pascal, 75775 Paris Cedex 16, France
Enquiries / Renseignements : Tel: +33 (0) 1 45 24 81 67
E-mail: sales@oecd.org

ONLINE BOOKSHOP / LIBRAIRIE EN LIGNE : **www.oecd.org/bookshop**
(secure payment with credit card / paiement sécurisé par carte de crédit)

OECD PUBLICATIONS, 2, rue André-Pascal, 75775 PARIS CEDEX 16
PRINTED IN FRANCE
(30 2001 03 3 P) ISBN 92-64-08599-8 – No. 51730 2001